CRC
Standard
Mathematical
Tables

26th Edition

Editor of Mathematics and Statistics

William H. Beyer, Ph.D.

Professor of Mathematics and Statistics
and Head of the Department of Mathematics
and Statistics, University of Akron, Akron, Ohio

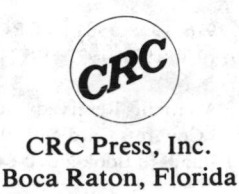

CRC Press, Inc.
Boca Raton, Florida

This book represents information obtained from authentic and highly regarded sources. Re-printed material is quoted with permission, and sources are indicated. A wide variety of references are listed. Every reasonable effort has been made to give reliable data and information, but the author and the publisher cannot assume responsibility for the validity of all materials or for the consequences of their use.

Direct all inquiries to CRC Press, Inc., 2000 Corporate Blvd., N.W., Boca Raton, Florida, 33431.

PREFACE

Consistent with CRC policy, this 26th Edition of the *CRC Standard Mathematical Tables* follows the tradition of its predecessors. The reorganization of the 25th Edition, with its new multisectional format, is found in this 26th Edition. The customary material contained in earlier editions, plus the new expository and tabular material included in the 25th Edition, is repeated. However, several desirable additions—expanded sections on Algebra of Sets, Vector Analysis, Finite Difference Interpolation Formulae, Normal Probability Function; new material on Complex Variables, Coefficients for Orthogonal Polynomials—are to be noted throughout this new edition. This additional material should prove to be beneficial to applied mathematicians and statisticians.

It is hoped that this 26th Edition will receive the same popular acceptance enjoyed by its predecessors. It is suggested that if the user desires more extensive and/or additional reference material than provided in this edition, he or she refer to other CRC publications such as the *CRC Handbook of Tables for Mathematics* (renamed the *CRC Handbook of Mathematical Sciences*), the *CRC Handbook of Tables for Probability and Statistics*, etc.

The Editor gratefully acknowledges the services rendered by Paul Gottehrer, Senior Editor, for the handling of the detail work so as to make the final production of this edition possible.

To the many users of current editions who sent in suggestions for alternations and additions, the Editorial Staff extends a special thanks. It is hoped that those interested will continue to send in suggestions and comments to assist in the continous improvement of the contents.

William H. Beyer
Editor, Mathematics and Statistics

TABLE OF CONTENTS

Greek letter	Greek name	English equivalent	Greek letter	Greek name	English equivalent
A α	Alpha	a	N ν	Nu	n
B β	Beta	b	Ξ ξ	Xi	x
Γ γ	Gamma	g	O o	Omicron	ŏ
Δ δ	Delta	d	Π π	Pi	p
E ε	Epsilon	ĕ	P ρ	Rho	r
Z ζ	Zeta	z	Σ σ ς	Sigma	s
H η	Eta	ē	T τ	Tau	t
Θ θ ϑ	Theta	th	Υ υ	Upsilon	u
I ι	Iota	i	Φ φ φ	Phi	ph
K κ	Kappa	k	X χ	Chi	ch
Λ λ	Lambda	l	Ψ ψ	Psi	ps
M μ	Mu	m	Ω ω	Omega	ō

THE NUMBER OF EACH DAY OF THE YEAR

Day of Mo.	Jan.	Feb.	Mar.	Apr.	May	Jun.	Jul.	Aug.	Sep.	Oct.	Nov.	Dec.	Day of Mo.
1	1	32	60	91	121	152	182	213	244	274	305	335	1
2	2	33	61	92	122	153	183	214	245	275	306	336	2
3	3	34	62	93	123	154	184	215	246	276	307	337	3
4	4	35	63	94	124	155	185	216	247	277	308	338	4
5	5	36	64	95	125	156	186	217	248	278	309	339	5
6	6	37	65	96	126	157	187	218	249	279	310	340	6
7	7	38	66	97	127	158	188	219	250	280	311	341	7
8	8	39	67	98	128	159	189	220	251	281	312	342	8
9	9	40	68	99	129	160	190	221	252	282	313	343	9
10	10	41	69	100	130	161	191	222	253	283	314	344	10
11	11	42	70	101	131	162	192	223	254	284	315	345	11
12	12	43	71	102	132	163	193	224	255	285	316	346	12
13	13	44	72	103	133	164	194	225	256	286	317	347	13
14	14	45	73	104	134	165	195	226	257	287	318	348	14
15	15	46	74	105	135	166	196	227	258	288	319	349	15
16	16	47	75	106	136	167	197	228	259	289	320	350	16
17	17	48	76	107	137	168	198	229	260	290	321	351	17
18	18	49	77	108	138	169	199	230	261	291	322	352	18
19	19	50	78	109	139	170	200	231	262	292	323	353	19
20	20	51	79	110	140	171	201	232	263	293	324	354	20
21	21	52	80	111	141	172	202	233	264	294	325	355	21
22	22	53	81	112	142	173	203	234	265	295	326	356	22
23	23	54	82	113	143	174	204	235	266	296	327	357	23
24	24	55	83	114	144	175	205	236	267	297	328	358	24
25	25	56	84	115	145	176	206	237	268	298	329	359	25
26	26	57	85	116	146	177	207	238	269	299	330	360	26
27	27	58	86	117	147	178	208	239	270	300	331	361	27
28	28	59	87	118	148	179	209	240	271	301	332	362	28
29	29	*	88	119	149	180	210	241	272	302	333	363	29
30	30		89	120	150	181	211	242	273	303	334	364	30
31	31		90		151		212	243		304		365	31

* In leap years, after February 28, add 1 to the tabulated number.

SI SYSTEM OF MEASUREMENT

SI, which is the abbreviation of the French words "Système Internationale d'Unites," is the accepted abbreviation for the International Metric System, which has seven base units, as shown below.

UNITS FOR A SYSTEM OF MEASURES AS USED INTERNATIONALLY

Quantity measured	Unit	Abbreviation
Length	meter	m
Mass	kilogram	kg
Time	second	s
Electric current	ampere	A
Temperature	degree Kelvin	K
Luminous intensity	candela	cd
Amount of substance	mole	mol

RECOMMENDED DECIMAL MULTIPLES AND SUBMULTIPLES

Multiples and submultiples	Prefixes	Symbols
10^{18}	exa	E
10^{15}	pecta	P
10^{12}	tera	T
10^{9}	giga	G
10^{6}	mega	M
10^{3}	kilo	k
10^{2}	hecto	h
10	deca	da
10^{-1}	deci	d
10^{-2}	centi	c
10^{-3}	milli	m
10^{-6}	micro	μ (greek mu)
10^{-9}	nano	n
10^{-12}	pico	p
10^{-15}	femto	f
10^{-18}	atto	a

Conversion Factors – Metric to English

To obtain	Multiply	By
Inches	Centimeters	0.3937007874
Feet	Meters	3.280839895
Yards	Meters	1.093613298
Miles	Kilometers	0.6213711922
Ounces	Grams	$3.527396195 \times 10^{-2}$
Pounds	Kilograms	2.204622622
Gallons (U.S. Liquid)	Liters	0.2641720524
Fluid ounces	Milliliters (cc)	$3.381402270 \times 10^{-2}$
Square inches	Square centimeters	0.1550003100
Square feet	Square meters	10.76391042
Square yards	Square meters	1.195990046
Cubic inches	Milliliters (cc)	$6.102374409 \times 10^{-2}$
Cubic feet	Cubic meters	35.31466672
Cubic yards	Cubic meters	1.307950619

Conversion Factors – English to Metric*

To obtain	Multiply	By
Microns	Mils	**25.4**
Centimeters	Inches	**2.54**
Meters	Feet	**0.3048**
Meters	Yards	**0.9144**
Kilometers	Miles	**1.609344**
Grams	Ounces	28.34952313
Kilograms	Pounds	**0.45359237**
Liters	Gallons (U.S. Liquid)	**3.785411784**
Milliliters (cc)	Fluid ounces	29.57352956
Square centimeters	Square inches	**6.4516**
Square meters	Square feet	**0.09290304**
Square meters	Square yards	**0.83612736**
Milliliters (cc)	Cubic inches	**16.387064**
Cubic meters	Cubic feet	$2.831684659 \times 10^{-2}$
Cubic meters	Cubic yards	0.764554858

Conversion Factors – General*

To obtain	Multiply	By
Atmospheres	Feet of water @ 4°C	2.950×10^{-2}
Atmospheres	Inches of mercury @ 0°C	3.342×10^{-2}
Atmospheres	Pounds per square inch	6.804×10^{-2}
BTU	Food-pounds	1.285×10^{-3}
BTU	Joules	9.480×10^{-4}
Cubic feet	Cords	**128**

*Boldface numbers are exact; others are given to ten significant figures where so indicated by the multiplier factor.

Conversion Factors – General (Continued)

To obtain	Multiply	By
Degree (angle)	Radians	57.2958
Ergs	Foot-pounds	1.356×10^7
Feet	Miles	**5280**
Feet of water @ 4°C	Atmospheres	33.90
Foot-pounds	Horsepower-hours	1.98×10^6
Foot-pounds	Kilowatt-hours	2.655×10^6
Food-pounds per min	Horsepower	3.3×10^4
Horsepower	Foot-pounds per sec	1.818×10^{-3}
Inches of mercury @ 0°C	Pounds per square inch	2.036
Joules	BTU	1054.8
Joules	Foot-pounds	1.35582
Kilowatts	BTU per min	1.758×10^{-2}
Kilowatts	Foot-pounds per min	2.26×10^{-5}
Kilowatts	Horsepower	0.745712
Knots	Miles per hour	0.86897624
Miles	Feet	1.894×10^{-4}
Nautical miles	Miles	0.86897624
Radians	Degrees	1.745×10^{-2}
Square feet	Acres	**43560**
Watts	BTU per min	17.5796

Temperature Factors

$$°F = 9/5\ (°C) + 32$$

Fahrenheit temperature = 1.8 (temperature in kelvins) −459.67

$$°C = 5/9\ [(°F) - 32]$$

Celsius temperature = temperature in kelvins −273.15

Fahrenheit temperature = 1.8 (Celsius temperature) +32

DECIMAL EQUIVALENTS OF FRACTIONS OF AN INCH

		1/64 =	0.015	625		11/32 22/64 =	0.343	75		43/64 =	0.671	875
	1/32	2/64 =	.031	25		23/64 =	.359	375	11/16 22/32	44/64 =	.687	5
		3/64 =	.046	875	3/8	12/32 24/64 =	.375			45/64 =	.703	125
1/16	2/32	4/64 =	.062	5		25/64 =	.390	625	23/32	46/64 =	.718	75
		5/64 =	.078	125		13/32 26/64 =	.406	25		47/64 =	.734	375
	3/32	6/64 =	.093	75		27/64 =	.421	875	3/4 24/32	48/64 =	.75	
		7/64 =	.109	375	7/16	14/32 28/64 =	.437	5		49/64 =	.765	625
1/8	4/32	8/64 =	.125			29/64 =	.453	125	25/32	50/64 =	.781	25
		9/64 =	.140	625		15/32 30/64 =	.468	75		51/64 =	.796	875
	5/32	10/64 =	.156	25		31/64 =	.484	375	13/16 26/32	52/64 =	.812	5
		11/64 =	.171	875	1/2	16/32 32/64 =	.50			53/64 =	.828	125
3/16	6/32	12/64 =	.187	5		33/64 =	.515	625	27/32	54/64 =	.843	75
		13/64 =	.203	125		17/32 34/64 =	.531	25		55/64 =	.859	375
	7/32	14/64 =	.218	75		35/64 =	.546	875	7/8 28/32	56/64 =	.875	
		15/64 =	.234	375	9/16	18/32 36/64 =	.562	5		57/64 =	.890	625
1/4	8/32	16/64 =	.25			37/64 =	.578	125	29/32	58/64 =	.906	25
		17/64 =	.265	625		19/32 38/64 =	.593	75		59/64 =	.921	875
	9/32	18/64 =	.281	25		39/64 =	.609	375	15/16 30/32	60/64 =	.937	5
		19/64 =	.296	875	5/8	20/32 40/64 =	.625			61/64 =	.953	125
5/16	10/32	20/64 =	.312	5		41/64 =	.640	625	31/32	62/64 =	.968	75
		21/64 =	.328	125		21/32 42/64 =	.656	25		63/64 =	.984	375

PHYSICAL CONSTANTS

Equatorial radius of the earth = 6378.388 km = 3963.34 miles (statute).
Polar radius of the earth, 6356.912 km = 3949.99 miles (statute).
1 degree of latitude at $40°$ = 69 miles.
1 international nautical mile = 1.15078 miles (statute) = 1852 m = 6076.115 ft.
Mean density of the earth = 5.522 g/cm^3 = 344.7 lb/ft^3.
Constant of gravitation, $(6.673 ± 0.003) × 10^{-8} cm^3 gm^{-1} s^{-2}$.
Acceleration due to gravity at sea level, latitude $45°$ = $980.665 cm/s^2$ = 32.1740 ft/sec^2.
Length of seconds pendulum at sea level, latitude $45°$ = 99.3574 cm = 39.1171 in.
1 knot (international) = 101.269 ft/min = 1.6878 ft/sec = 1.1508 miles (statute)/hr.
1 micron = 10^{-4} cm.
1 angstrom = 10^{-8} cm.
Mass of hydrogen atom = $(1.67339 ± 0.0031) × 10^{-24}$ g.
Density of mercury at $0°C$ = 13.5955 g/ml.
Density of water at $3.98°C$ = 1.000000 g/ml.
Density, maximum, of water, at $3.98°C$ = 0.999973 g/cm^3.
Density of dry air at $0°C$, 760 mm = 1.2929 g/liter.
Velocity of sound in dry air at $0°C$ = 331.36 m/s - 1087.1 ft/sec.
Velocity of light in vacuum = $(2.997925 ± 0.000002) × 10^{10}$ cm/s.
Heat of fusion of water $0°C$ = 79.71 cal/g.
Heat of vaporization of water $100°C$ = 539.55 cal/g.
Electrochemical equivalent of silver 0.001118 g/sec international amp.
Absolute wave length of red cadmium light in air at $15°C$, 760 mm pressure = 6438.4696 A.
Wave length of orange-red line of krypton 86 = 6057.802 A.

π CONSTANTS

$$\pi = 3.14159\ 26535\ 89793\ 23846\ 26433\ 83279\ 50288\ 41971\ 69399\ 37511$$
$$1/\pi = 0.31830\ 98861\ 83790\ 67153\ 77675\ 26745\ 02872\ 40689\ 19291\ 48091$$
$$\pi^2 = 9.8690\ 44010\ 89358\ 61883\ 44909\ 99876\ 15113\ 53136\ 99407\ 24079$$
$$\log_e\pi = 1.14472\ 98858\ 49400\ 17414\ 34273\ 51353\ 05871\ 16472\ 94812\ 91531$$
$$\log_{10}\pi = 0.49714\ 98726\ 94133\ 85435\ 12682\ 88290\ 89887\ 36516\ 78324\ 38044$$
$$\log_{10}\sqrt{2\pi} = 0.39908\ 99341\ 79057\ 52478\ 25035\ 91507\ 69595\ 02099\ 34102\ 92128$$

CONSTANTS INVOLVING e

$$e = 2.71828\ 18284\ 59045\ 23536\ 02874\ 71352\ 66249\ 77572\ 47093\ 69996$$
$$1/e = 0.36787\ 94411\ 71442\ 32159\ 55237\ 70161\ 46086\ 74458\ 11131\ 03177$$
$$e^2 = 7.38905\ 60989\ 30650\ 22723\ 04274\ 60575\ 00781\ 31803\ 15570\ 55185$$
$$M = \log_{10}e = 0.43429\ 44819\ 03251\ 82765\ 11289\ 18916\ 60508\ 22943\ 97005\ 80367$$
$$1/M = \log_e 10 = 2.30258\ 50929\ 94045\ 68401\ 79914\ 54684\ 36420\ 76011\ 01488\ 62877$$
$$\log_{10}M = 9.63778\ 43113\ 00536\ 78912\ 29674\ 98645\ -10$$

π^e AND e^π CONSTANTS

$$\pi^e = 22.45915\ 77183\ 61045\ 47342\ 71522$$
$$e^\pi = 23.14069\ 26327\ 79269\ 00572\ 90864$$
$$e^{-\pi} = 0.04321\ 39182\ 63772\ 24977\ 44177$$
$$e^{1/2\pi} = 4.81047\ 73809\ 65351\ 65547\ 30357$$
$$= e^{-1/2\pi} = 0.20787\ 95763\ 50761\ 90854\ 69556$$

NUMERICAL CONSTANTS

$$\sqrt{2} = 1.41421\ 35623\ 73095\ 04880\ 16887\ 24209\ 69807\ 85696\ 71875\ 37695$$
$$\sqrt[3]{2} = 1.25992\ 10498\ 94873\ 16476\ 72106\ 07278\ 22835\ 05702\ 51464\ 70151$$
$$\log_e 2 = 0.69314\ 71805\ 59945\ 30941\ 72321\ 21458\ 17656\ 80755\ 00134\ 36026$$
$$\log_{10}2 = 0.30102\ 99956\ 63981\ 19521\ 37388\ 94724\ 49302\ 67881\ 89881\ 46211$$
$$\sqrt{3} = 1.73205\ 08075\ 68877\ 29352\ 74463\ 41505\ 87236\ 69428\ 05253\ 81039$$
$$\sqrt[3]{3} = 1.44224\ 95703\ 07408\ 38232\ 16383\ 10780\ 10958\ 83918\ 69253\ 49935$$
$$\log_e 3 = 1.09861\ 22886\ 68109\ 69139\ 52452\ 36922\ 52570\ 46474\ 90557\ 82275$$
$$\log_{10}3 = 0.47712\ 12547\ 19662\ 43729\ 50279\ 03255\ 11530\ 92001\ 28864\ 19070$$

OTHER CONSTANTS

$$\text{Euler's Constant } \gamma = 0.57721\ 56649\ 01532\ 86061$$
$$\log_e \gamma = -0.54953\ 93129\ 81644\ 82234$$
$$\text{Golden Ratio } \phi = 1.61803\ 39887\ 49894\ 84820\ 45868\ 34365\ 63811\ 77203\ 09180$$

CONSTANTS AND CONVERSION FACTORS

NUMBERS CONTAINING π

	Number	Logarithm		Number	Logarithm
π	3.1415 927	0.4971 499	$2\pi^2$	19.7392 088	1.2953 297
2π	6.2831 853	0.7981 799	$\pi/180$	0.0174 533	8.2418 774 - 10
3π	9.4247 780	0.9742 711	$180/\pi$	57.2957 795	1.7581 226
4π	12.5663 706	1.0992 099	$4\pi^2$	39.4784 176	1.5963 597
8π	25.1327 412	1.4002 399	$1/\pi^2$	0.1013 212	9.0057 003 - 10
$\pi/2$	1.5707 963	0.1961 199	$1/(2\pi^2)$	0.0506 606	8.7046 703 - 10
$\pi/3$	1.0471 976	0.0200 286	$1/(4\pi^2)$	0.0253 303	8.4036 403 - 10
$\pi/4$	0.7853 982	9.8950 899 - 10	$\sqrt{\pi}$	1.7724 539	0.2485 749
$\pi/6$	0.5235 988	9.7189 986 - 10	$\sqrt{\dfrac{\pi}{2}}$	0.8862 269	9.9475 449 - 10
$\pi/8$	0.3926 991	9.5940 599 - 10	$\sqrt{\dfrac{\pi}{4}}$	0.4431 135	9.6465 149 - 10
$2\pi/3$	2.0943 951	0.3210 586	$\sqrt{\dfrac{\pi}{2}}$	1.2533 141	0.0980 599
$4\pi/3$	4.1887 902	0.6220 886	$\sqrt{\dfrac{2}{\pi}}$	0.7978 846	9.9019 401 - 10
$1/\pi$	0.3183 099	9.5028 501 - 10		31.0062 767	1.4914 496
$2/\pi$	0.6366 198	9.8038 801 - 10	π^3	1.4645 919	0.1657 166
$4/\pi$	1.2732 395	0.1049 101	$\sqrt[4]{\pi}$	0.6827 841	9.8342 834 - 10
$1/(2\pi)$	0.1591 549	9.2018 201 - 10	$1/\sqrt[3]{\pi}$	2.1450 294	0.3314 332
$1/(4\pi)$	0.0795 775	8.9007 901 - 10	$1/\sqrt[3]{\pi^2}$	0.5641 896	9.7514 251 - 10
$1/(6\pi)$	0.0530 516	8.7246 989 - 10	$1/\sqrt{\pi}$	0.3989 423	9.6009 101 - 10
$1/(8\pi)$	0.0397 887	8.5997 601 - 10	$1/\sqrt{2\pi}$	1.1283 792	0.0524 551
π^2	9.8696 044	0.9942 997	$2/\sqrt{\pi}$		

MULTIPLES OF $\dfrac{\pi}{2}$

n	$n\dfrac{\pi}{2}$	n	$n\dfrac{\pi}{2}$	n	$n\dfrac{\pi}{2}$	n	$n\dfrac{\pi}{2}$
1	1.57079 63268	26	40.84070 44967	51	80.11061 26665	76	119.38502 08364
2	3.14159 26536	27	42.41150 08235	52	81.68140 89933	77	120.95131 71632
3	4.71238 89804	28	43.98229 71503	53	83.25220 53201	78	122.52211 34900
4	6.28318 53072	29	45.55309 34771	54	84.82300 16469	79	124.09290 98168
5	7.85398 16340	30	47.12388 98038	55	86.39379 79737	80	125.66370 61436
6	9.42477 79608	31	48.69468 61306	56	87.96459 43005	81	127.23450 24704
7	10.99557 42876	32	50.26548 24574	57	89.53539 06273	82	128.80529 87972
8	12.56637 06144	33	51.83627 87842	58	91.10618 69541	83	130.37609 51240
9	14.13716 69412	34	53.40707 51110	59	92.67698 32809	84	131.94689 14508
10	15.70796 32679	35	54.97787 14378	60	94.24777 96077	85	133.51768 77776
11	17.27875 95947	36	56.54866 77646	61	95.81857 59345	86	135.08848 41044
12	18.84955 59215	37	58.11946 40914	62	97.38937 22613	87	136.65928 04312
13	20.42035 22483	38	59.69026 04182	63	98.96016 85881	88	138.23007 67580
14	21.99114 85751	39	61.26105 67450	64	100.53096 49149	89	139.80087 30847
15	23.56194 49019	40	62.83185 30718	65	102.10176 12417	90	141.37166 94115
16	25.13274 12287	41	64.40264 93986	66	103.67255 75685	91	142.94246 57383
17	26.70533 75555	42	65.97344 57254	67	105.24335 38953	92	144.51326 20651
18	28.27433 38823	43	67.54424 20522	68	106.81415 02221	93	146.08405 83919
19	29.84513 02091	44	69.11503 83790	69	108.38494 65488	94	147.65485 47187
20	31.41592 65359	45	70.68583 47058	70	109.95574 28765	95	149.22565 10455
21	32.98672 28627	46	72.25663 10326	71	111.52653 92024	96	150.79644 73723
22	34.55751 91895	47	73.82742 73594	72	113.09733 55292	97	152.36724 36991
23	36.12831 55163	48	75.39822 36862	73	114.66813 18560	98	153.93804 00259
24	37.69911 18431	49	76.96902 00129	74	116.23892 81828	99	155.50883 63527
25	39.26990 81699	50	78.53981 63397	75	117.80972 45096	100	157.07963 26795

FACTORS AND EXPANSIONS

$(a \pm b)^2 = a^2 \pm 2ab + b^2$.

$(a \pm b)^3 = a^3 \pm 3a^2 b + 3ab^2 \pm b^3$.

$(a \pm b)^4 = a^4 \pm 4a^3 b + 6a^2 b^2 \pm 4ab^3 + b^4$.

$a^2 - b^2 = (a - b)(a + b)$.

$a^2 + b^2 = (a + b\sqrt{-1})(a - b\sqrt{-1})$.

$a^3 + b^3 = (a - b)(a^2 + ab + b^2)$.

$a^3 + b^3 = (a + b)(a^2 - ab + b^2)$.

$a^4 + b^4 = (a^2 + ab\sqrt{2} + b^2)(a^2 - ab\sqrt{2} + b^2)$.

$a^n - b^n = (a - b)(a^{n-1} + a^{n-2}b + \ldots + b^{n-1})$.

$a^n - b^n = (a + b)(a^{n-1} - a^{n-2}b + \ldots - b^{n-1})$.

for even values of n.

$a^n + b^n = (a + b)(a^{n-1} - a^{n-2}b + \ldots + b^{n-1})$,

for odd values of n.

$a^4 + a^2 b^2 + b^4 = (a^2 + ab + b^2)(a^2 - ab + b^2)$.

$(a + b + c)^2 = a^2 + b^2 + c^2 + 2ab + 2ac + 2bc$.

$(a + b + c)^3 = a^3 + b^3 + c^3 + 3a^2(b + c) + 3b^2(a + c) + 3c^2(a + b) + 6abc$.

$(a + b + c + d + \ldots)^2 = a^2 + b^2 + c^2 + d^2 + \ldots + 2a(b + c + d + \ldots) + 2b(c + d + \ldots) + 2c(d + \ldots) + \ldots$

See also under Series.

POWERS AND ROOTS

$a^x \times a^y = a^{(x+y)}$.

$\dfrac{a^x}{a^y} = a^{(x-y)}$.

$(a^x)^y = a^{xy}$.

$\sqrt[x]{\sqrt[y]{a}} = \sqrt[xy]{a}$.

$a^0 = 1$ [if $a \neq 0$]

$a^{-x} = \dfrac{1}{a^x}$.

$a^{\frac{1}{x}} = \sqrt[x]{a}$.

$a^{\frac{x}{y}} = \sqrt[y]{a^x}$.

$(ab)^x = a^x b^x$.

$\left(\dfrac{a}{b}\right)^x = \dfrac{a^x}{b^x}$.

$\sqrt[x]{ab} = \sqrt[x]{a}\,\sqrt[x]{b}$.

$\sqrt[x]{\dfrac{a}{b}} = \dfrac{\sqrt[x]{a}}{\sqrt[x]{b}}$

PROPORTION

If $\dfrac{a}{b} = \dfrac{c}{d}$, then $\dfrac{a+b}{b} = \dfrac{c+d}{d}$,

$\dfrac{a-b}{b} = \dfrac{c-d}{d}$, $\dfrac{a-b}{a+b} = \dfrac{c-d}{c+d}$.

ARITHMETIC PROGRESSION*

An arithmetic progression is a sequence of numbers such that each number differs from the previous number by a constant amount, called the *common difference*.

If a_1 is the first term; a_n the nth term; d the common difference; n the number of terms; and s_n the sum of n terms—

$$a_n = a_1 + (n - 1)d, \quad s_n = \frac{n}{2}[a_1 + a_n].$$

$$s_n = \frac{n}{2}[2a_1 + (n - 1)d].$$

The arithmetic mean between a and b is given by $\dfrac{a + b}{2}$.

GEOMETRIC PROGRESSION*

A geometric progression is a sequence of numbers such that each number bears a constant ratio, called the *common ratio*, to the previous number.

If a_1 is the first term; a_n the nth term; r the common ratio; n the number of terms; and s_n the sum of n terms

$$a_n = a_1 r^{n-1}; \quad s_n = a_1 \frac{1 - r^n}{1 - r}$$

$$= a_1 \frac{r^n - 1}{r - 1}, \quad r \neq 1.$$

$$= \frac{a_1 - ra_n}{1 - r}$$

$$= \frac{ra_n - a_1}{r - 1}$$

If $|r| < 1$, then the sum of an infinite geometrical progression converges to the limiting value

$$\frac{a_1}{1 - r}, \quad \left[s_\infty = \lim_{n \to \infty} \frac{a_1(1 - r^n)}{1 - r} = \frac{a_1}{1 - r} \right]$$

The geometric mean between a and b is given by $\sqrt{ab}$.

*It is customary to represent a_n by l in a finite progression and refer to it as the last term.

HARMONIC PROGRESSION

A sequence of numbers whose reciprocals form an arithmetic progression is called an harmonic progression. Thus

$$\frac{1}{a_1}, \quad \frac{1}{a_1 + d}, \quad \frac{1}{a_1 + 2d}, \dots, \frac{1}{a_1 + (n - 1)d}, \dots,$$

where

$$\frac{1}{a_n} = \frac{1}{a_1 + (n - 1)d}$$

forms an harmonic progression. The harmonic mean between a and b is given by $\dfrac{2ab}{a + b}$.

If A, G, H respectively represent the arithmetic mean, geometric mean, and harmonic mean between a and b, then $G^2 = AH$.

QUADRATIC EQUATIONS

Any quadratic equation may be reduced to the form,—

$$ax^2 + bx + c = 0$$

Then

$$x = \frac{-b \pm \sqrt{b^2 - 4ac}}{2a}$$

If a, b, and c are real then:
If $b^2 - 4ac$ is positive, the roots are real and unequal;
If $b^2 - 4ac$ is zero, the roots are real and equal;
If $b^2 - 4ac$ is negative, the roots are imaginary and unequal.

CUBIC EQUATIONS

A cubic equation, $y^3 + py^2 + qy + r = 0$ may be reduced to the form,—

$$x^3 + ax + b = 0$$

by substituting for y the value, $x - \dfrac{p}{3}$. Here

$$a = \tfrac{1}{3}(3q - p^2) \text{ and } b = \tfrac{1}{27}(2p^3 - 9pq + 27r).$$

For solution let,—

$$A = \sqrt[3]{-\frac{b}{2} + \sqrt{\frac{b^2}{4} + \frac{a^3}{27}}}, \quad B = -\sqrt[3]{+\frac{b}{2} + \sqrt{\frac{b^2}{4} + \frac{a^3}{27}}},$$

then the values of x will be given by,

$$x = A + B, \quad -\frac{A + B}{2} + \frac{A - B}{2}\sqrt{-3}, \quad -\frac{A + B}{2} - \frac{A - B}{2}\sqrt{-3}.$$

If p, q, r are real, then:

If $\dfrac{b^2}{4} + \dfrac{a^3}{27} > 0$, there will be one real root and two conjugate complex roots;

If $\dfrac{b^2}{4} + \dfrac{a^3}{27} = 0$, there will be three real roots of which at least two are equal;

If $\dfrac{b^2}{4} + \dfrac{a^3}{27} < 0$, there will be three real and unequal roots.

Trignometric Solution of the Cubic Equation

The form $x^3 + ax + b = 0$ with $ab \neq 0$ can always be solved by transforming it to the trignometric identity

$$4 \cos^3 \theta - 3 \cos \theta - \cos (3\theta) \equiv 0.$$

Let $x = m \cos \theta$, then

$$x^3 + ax + b \equiv m^3 \cos^3 \theta + am \cos \theta + b \equiv 4 \cos^3 \theta - 3 \cos \theta - \cos (3\theta) \equiv 0.$$

Hence

$$\frac{4}{m^3} = -\frac{3}{am} = \frac{-\cos(3\theta)}{b},$$

from which follows that

$$m = 2\sqrt{-\frac{a}{3}}, \quad \cos (3\theta) = \frac{3b}{am}.$$

Any solution θ_1 which satisfies $\cos (3\theta) = \dfrac{3b}{am}$, will also have the solutions

$$\theta_1 + \frac{2\pi}{3} \quad \text{and} \quad \theta_1 + \frac{4\pi}{3}.$$

The roots of the cubic $x^3 + ax + b = 0$ are

$$2\sqrt{-\frac{a}{3}} \cos \theta_1, \quad 2\sqrt{-\frac{a}{3}} \cos\left(\theta_1 + \frac{2\pi}{3}\right), \quad 2\sqrt{-\frac{a}{3}} \cos\left(\theta_1 + \frac{4\pi}{3}\right).$$

Example where hyperbolic functions are necessary for solution with latter procedure

The roots of the equation $x^3 - x + 2 = 0$ may be found as follows:

Here

$$a = -1, \quad b = 2, \quad m = 2\sqrt{\tfrac{1}{3}} = 1.155$$

$$\cos (3\theta) = \frac{6}{-1.155} = -5.196$$

$$\cos (3\theta) = -\cos (3\theta - \pi) = -\cosh [i(3\theta - \pi)] = -5.196.$$

Using hyperbolic function tables for cosh $[i(3\theta - \pi)] = 5.196$, it is found that

$$i(3\theta - \pi) = 2.332.$$

Thus

$$3\theta - \pi = -i(2.332).$$

$$3\theta = \pi - i(2.332)$$

$$\theta_1 = \frac{\pi}{3} - i(0.777)$$

$$\theta_1 + \frac{2\pi}{3} = \pi - i(0.777)$$

$$\theta_1 + \frac{4\pi}{3} = \frac{5\pi}{3} - i(0.777)$$

$$\cos\theta_1 = \cos\left[\frac{\pi}{3} - i(0.777)\right]$$

$$= \left(\cos\frac{\pi}{3}\right)[\cos i(0.777)] + \left(\sin\frac{\pi}{3}\right)[\sin i(0.777)]$$

$$= \left(\cos\frac{\pi}{3}\right)(\cosh 0.777) + i\left(\sin\frac{\pi}{3}\right)(\sinh 0.777)$$

$$= (0.5)(1.317) + i(0.866)(0.858) = 0.659 + i(0.743).$$

Note that

$$\cos\mu = \cosh(i\mu) \quad \text{and} \quad \sin\mu = -i\sinh(i\mu).$$

Similarly

$$\cos\left(\theta_1 + \frac{2\pi}{3}\right) = \cos[\pi - i(0.777)]$$

$$= (\cos\pi)(\cosh 0.777) + i(\sin\pi)(\sinh 0.777)$$

$$= -1.317,$$

and

$$\cos\left(\theta_1 + \frac{4\pi}{3}\right) = \cos\left[\frac{5\pi}{3} - i(0.777)\right]$$

$$= \left(\cos\frac{5\pi}{3}\right)(\cosh 0.777) + i\left(\sin\frac{5\pi}{3}\right)(\sinh 0.777)$$

$$= (0.5)(1.317) - i(0.866)(0.858) = 0.659 - i(0.743).$$

The required roots are

$$1.155[0.659 + i(0.743)] = 0.760 + i(0.858)$$

$$(1.155)(-1.317) = -1.520$$

$$(1.155)[0.659 - i(0.743)] = 0.760 - i(0.858).$$

QUARTIC OR BIQUADRATIC EQUATION

A quartic equation,

$$x^4 + ax^3 + bx^2 + cx + d = 0,$$

has the *resolvent cubic equation*

$$y^3 - by^2 + (ac - 4d)y - a^2d + 4bd - c^2 = 0.$$

Let y be any root of this equation, and

$$R = \sqrt{\frac{a^2}{4} - b + y}.$$

If $R \neq 0$, then let

$$D = \sqrt{\frac{3a^2}{4} - R^2 - 2b + \frac{4ab - 8c - a^3}{4R}}$$

and

$$E = \sqrt{\frac{3a^2}{4} - R^2 - 2b - \frac{4ab - 8c - a^3}{4R}}$$

If $R = 0$, then let

$$D = \sqrt{\frac{3a^2}{4} - 2b + 2\sqrt{y^2 - 4d}}$$

and

$$E = \sqrt{\frac{3a^2}{4} - 2b - 2\sqrt{y^2 - 4d}}.$$

Then the four roots of the original equation are given by

$$x = -\frac{a}{4} + \frac{R}{2} \pm \frac{D}{2}$$

and

$$x = -\frac{a}{4} - \frac{R}{2} \pm \frac{E}{2}.$$

EQUATION $x^n = c$

Using DeMoivre's theorem:

$$(\cos \theta + i \sin \theta)^n = \cos n\theta + i \sin n\theta; \ i = \sqrt{-1},$$

the equation $x^n = c$ has n roots given by

$$x = \sqrt[n]{c} \left(\cos \frac{2m\pi}{n} + i \sin \frac{2m\pi}{n} \right) \text{if } c > 0,$$

or

$$x = \sqrt[n]{-c} \left(\cos \frac{(2m + 1)\pi}{n} + i \sin \frac{(2m + 1)\pi}{n} \right) \text{ if } c < 0,$$

where m takes the n values $0, 1, 2, \ldots (n - 1)$ giving n roots.

PARTIAL FRACTIONS

This section applies only to rational algebraic fractions with numerator of lower degree than the denominator. Improper fractions can be reduced to proper fractions by long division.

Every fraction may be expressed as the sum of component fractions whose denominators are factors of the denominator of the original fraction.

Let $N(x)$ = numerator, a polynomial of the form

$$n_0 + n_1 x + n_2 x^2 + \cdots + n_i x^i$$

I. *Non-repeated Linear Factors*

$$\frac{N(x)}{(x - a)G(x)} = \frac{A}{x - a} + \frac{F(x)}{G(x)}$$

$$A = \left[\frac{N(x)}{G(x)}\right]_{x=a}$$

$F(x)$ determined by methods discussed in the following sections.

Example:

$$\frac{x^2 + 3}{x(x - 2)(x^2 + 2x + 4)} = \frac{A}{x} + \frac{B}{x - 2} + \frac{F(x)}{x^2 + 2x + 4}$$

$$A = \left[\frac{x^2 + 3}{(x - 2)(x^2 + 2x + 4)}\right]_{x=0} = -\frac{3}{8}$$

$$B = \left[\frac{x^2 + 3}{x(x^2 + 2x + 4)}\right]_{x=2} = \frac{4 + 3}{2(4 + 4 + 4)} = \frac{7}{24}$$

II. *Repeated Linear Factors*

$$\frac{N(x)}{x^m G(x)} = \frac{A_0}{x^m} + \frac{A_1}{x^{m-1}} + \cdots + \frac{A_{m-1}}{x} + \frac{F(x)}{G(x)}$$

$$F(x) = f_0 + f_1 x + f_2 x^2 + \cdots, \quad G(x) = g_0 + g_1 x + g_2 x^2 + \cdots$$

$$A_0 = \frac{n_0}{g_0}, \quad A_1 = \frac{n_1 - A_0 g_1}{g_0}, \quad A_2 = \frac{n_2 - A_0 g_2 - A_1 g_1}{g_0}$$

General term:

$$A_k = \frac{1}{g_0}\left[n_k - \sum_{i=0}^{k-1} A_i g_{k-i}\right]$$

$$*m = 1 \begin{cases} f_0 = n_1 - A_0 g_1 \\ f_1 = n_2 - A_0 g_2 \\ f_j = n_{j+1} - A_0 g_{j+1} \end{cases}$$

$$m = 2 \begin{cases} f_0 = n_2 - A_0 g_2 - A_1 g_1 \\ f_1 = n_3 - A_0 g_3 - A_1 g_2 \\ f_j = n_{j+2} - [A_0 g_{j+2} + A_1 g_{j+1}] \end{cases}$$

$$m = 3 \begin{cases} f_0 = n_3 - A_0 g_3 - A_1 g_2 - A_2 g_1 \\ f_1 = n_3 - A_0 g_4 - A_1 g_3 - A_2 g_2 \\ f_j = n_{j+3} - [A_0 g_{j+3} + A_1 g_{j+2} + A_2 g_{j+1}] \end{cases}$$

$$\text{any } m: \quad f_j = n_{m+j} - \sum_{i=0}^{m-1} A_i g_{m+j-i}$$

Example:

$$\frac{x^2 + 1}{x^3(x^2 - 3x + 6)} = \frac{A_0}{x^3} + \frac{A_1}{x^2} + \frac{A_2}{x} + \frac{f_1 x + f_0}{x^2 - 3x + 6}$$

$$A_0 = \frac{1}{6}, \quad A_1 = \frac{0 - (\frac{1}{6})(-3)}{6} = \frac{1}{12},$$

$$A_2 = \frac{1 - (\frac{1}{6})(1) - (\frac{1}{12})(-3)}{6} = \frac{13}{72},$$

$$m = 3 \begin{cases} f_0 = 0 - \frac{1}{6}(0) + \frac{1}{12}(1) - \frac{13}{72}(-3) = \frac{11}{24} \\ f_1 = 0 - \frac{1}{6}(0) - \frac{1}{12}(0) - \frac{13}{72}(1) = -\frac{13}{72} \end{cases}$$

*Note: If $G(x)$ contains linear factors, $F(x)$ may be determined by previous section I.

III. *Repeated Linear Factors*

$$\frac{N(x)}{(x - a)^m G(x)} = \frac{A_0}{(x - a)^m} + \frac{A_1}{(x - a)^{m-1}} + \cdots + \frac{A_{m-1}}{(x - a)} + \frac{F(x)}{G(x)}$$

Change to form $\dfrac{N'(y)}{y^m G'(y)}$ by substitution of $x = y + a$. Resolve into partial fractions in terms of y as described in Section II. Then express in terms of x by substitution $y = x - a$.

Example:

$$\frac{x - 3}{(x - 2)^2(x^2 + x + 1)}$$

Let $x - 2 = y$, $x = y + 2$

$$\frac{(y + 2) - 3}{y^2[(y + 2)^2 + (y + 2) + 1]} = \frac{y - 1}{y^2(y^2 + 5y + 7)} = \frac{A_0}{y^2} + \frac{A_1}{y} + \frac{f_1 y + f_0}{y^2 + 5y + 7}$$

$$A_0 = -\frac{1}{7}, \quad A_1 = \frac{1 - (-\frac{1}{7})(5)}{7} = \frac{12}{49},$$

$$m = 2 \begin{cases} f_0 = 0 - (-\frac{1}{7})(1) - (\frac{12}{49})(5) = -\frac{53}{49} \\ f_1 = 0 - (-\frac{1}{7})(0) - (\frac{12}{49})(1) = -\frac{12}{49} \end{cases}$$

$$\therefore \quad \frac{y - 1}{y^2(y^2 + 5y + 7)} = \frac{-\frac{1}{7}}{y^2} + \frac{\frac{12}{49}}{y} + \frac{-\frac{12}{49} y - \frac{53}{49}}{y^2 + 5y + 7}$$

Let $y = x - 2$, then

$$\frac{x - 3}{(x - 2)^2(x^2 + x + 1)} = \frac{-\frac{1}{7}}{(x - 2)^2} + \frac{\frac{12}{35}}{(x - 2)} + \frac{-\frac{12}{49}(x - 2) - \frac{53}{49}}{x^2 + x + 1}$$

$$= -\frac{1}{7(x - 2)^2} + \frac{12}{35(x - 2)} + \frac{-12x - 29}{49(x^2 + x + 1)}$$

IV. *Repeated Linear Factors*

Alternative method of determining coefficients:

$$\frac{N(x)}{(x-a)^m G(x)} = \frac{A_0}{(x-a)^m} + \cdots + \frac{A_k}{(x-a)^{m-k}} + \cdots + \frac{A_{m-1}}{x-a} + \frac{F(x)}{G(x)}$$

$$A_k = \frac{1}{k!}\left\{D_x^k\left[\frac{N(x)}{G(x)}\right]\right\}_{x=a}$$

where D_x^k is the differentiating operator, and the derivative of zero order is defined as:

$$D_x^0 u = u.$$

V. *Factors of Higher Degree*

Factors of higher degree have the corresponding numerators indicated.

$$\frac{N(x)}{(x^2 + h_1 x + h_0)G(x)} = \frac{a_1 x + a_0}{x^2 + h_1 x + h_0} + \frac{F(x)}{G(x)}$$

$$\frac{N(x)}{(x^2 + h_1 x + h_0)^2 G(x)} = \frac{a_1 x + a_0}{(x^2 + h_1 x + h_0)^2} + \frac{b_1 x + b_0}{(x^2 + h_1 x + h_0)} + \frac{F(x)}{G(x)}$$

$$\frac{N(x)}{(x^3 + h_2 x^2 + h_1 x + h_0)G(x)} = \frac{a_2 x^2 + a_1 x + a_0}{x^3 + h_2 x^2 + h_1 x + h_0} + \frac{F(x)}{G(x)}$$

etc.

Problems of this type are determined first by solving for the coefficients due to linear factors as shown above, and then determining the remaining coefficients by the general methods given below.

VI. *General Methods for Evaluating Coefficients*

1.
$$\frac{N(x)}{D(x)} = \frac{N(x)}{G(x)H(x)L(x)} = \frac{A(x)}{G(x)} + \frac{B(x)}{H(x)} + \frac{C(x)}{L(x)} + \cdots$$

Multiply both sides of equation by $D(x)$ to clear fractions. Then collect terms, equate like powers of x, and solve the resulting simultaneous equations for the unknown coefficients.

2. Clear fractions as above. Then let x assume certain convenient values ($x = 1, 0, -1, \ldots$). Solve the resulting equations for the unknown coefficients.

3.
$$\frac{N(x)}{G(x)H(x)} = \frac{A(x)}{G(x)} + \frac{B(x)}{H(x)}$$

Then

$$\frac{N(x)}{G(x)H(x)} - \frac{A(x)}{G(x)} = \frac{B(x)}{H(x)}$$

If $A(x)$ can be determined, such as by Method I, then $B(x)$ can be found as above.

BASIC CONCEPTS IN ALGEBRA

Dr. W. E. Deskins

I. ALGEBRA OF SETS

1. Intuitively a set is a collection of objects called the elements of the set. Set and set membership are generally accepted as basic, undefined terms used to define and construct mathematical systems.

    The notation $a \in A$ indicates that a is an element of the set A. The notation $a \notin A$ means that a is not a member of A.

    A set is sometimes specified by listing its elements within a set of braces: $\{a\}$ is the set containing only the element a.

2. Set A is a subset of set B provided $a \in A$ implies $a \in B$. This is denoted by $A \subseteq B$. Every set has as a subset the empty or null set, denoted by ϕ, which has no elements.

3. Set A equals set B, written $A = B$, if and only if $A \subseteq B$ and $B \subseteq A$. A is a proper subset of B, sometimes indicated by $A \subset B$, if and only if $A \subseteq B$ and $A \neq B$; then B has at least one element which does not belong to A.

4. The Cartesian product of sets A and B, denoted by $A \times B$, is the set of all ordered pairs (a, b) where $a \in A$ and $b \in B$. A subset R of $A \times A$ is a binary relation on A, and this is an equivalence relation on A provided (i) $(a, a) \in R$ for every $a \in A$, (ii) $(a, b) \in R$ implies $(b, a) \in R$, and (iii) $(a, b) \in R$ and $(b, c) \in R$ imply $(a, c) \in R$. Ordinary equality of numbers, equality of sets, and congruence of plane figures are examples of equivalence relations.

5. A subset F of $A \times B$ is a *function* from A to B provided each element of A appears exactly once as the first element of a pair in F. A function F from A to B is *onto* provided each element of B appears at least once as the second element of a pair in F. It is *one-to-one* provided each element of B appears at most once as the second element of a pair in F. A function from $A \times A$ to A is a *binary operation* on A. Addition and multiplication of ordinary numbers are examples of binary operations.

6. If consideration is restricted to elements and subsets of a particular set I, then I is the universal set.

7. Common binary operations on subsets of I are: $A \cup B$, the union or join of sets A and B, is the set of all elements of I which belong to either A or B or both A and B.

    $A \cap B$, the intersection or meet of sets A and B, is the set of all elements of I which belong to both A and B.

    $A \setminus B$, the difference of sets A and B, is the set of elements of I which belong to A but not B.

    The difference $I \setminus A$ is denoted by A' and called the complement of A (relative to I). Except in dealing with the concept of complementation the use of a universal set is not essential to the above ideas.

8. Some theorems basic to the Algebra of Sets:

    Let A, B, and C be arbitrary subsets of a universal set I.

    (a) (Commutativity) $A \cup B = B \cup A$ and $A \cap B = B \cap A$.

    (b) (Associativity) $(A \cup B) \cup C = A \cup (B \cup C)$ and
         $(A \cap B) \cap C = A \cap (B \cap C)$.

(c) (Distributivity) $A \cap (B \cup C) = (A \cap B) \cup (A \cap C)$ and

$A \cup (B \cap C) = (A \cup B) \cap (A \cup C)$.

(d) (Idempotency) $A \cup A = A \cap A = A$.

(e) Properties of I and ϕ: $A \cap I = A \cup \phi = A$,

$A \cup I = I$, and

$A \cap \phi = \phi$.

(f) $(A \cap B) \cup (A \backslash B) = A$.

(g) $(A \backslash B) \cup B = A \cup B$.

(h) $A \subseteq A \cup B$.

(i) $A \cap B \subseteq A$.

(j) $A \cup B = A$ if and only if $B \subseteq A$.

(k) $A \cap B = A$ if and only if $A \subseteq B$.

(m) $A \backslash B = A \backslash (A \cap B)$.

(n) (DeMorgan's Theorem) $(A \backslash B) \cap (A \backslash C) = A \backslash (B \cup C)$ and

$(A \backslash B) \cup (A \backslash C) = A \backslash (B \cap C)$.

(o) $(A \cup B)' = A' \cap B'$ and $(A \cap B)' = A' \cup B'$.

(p) $A \cup A' = I$ and $A \cap A' = \phi$.

9. A mathematical system S is a set $S = \{E, O, A\}$ where E is a nonempty set of elements, O is a set of relations and operations on E, and A is a set of axioms, postulates, or assumptions concerning the elements of E and O.

10. The Algebra of Sets provides an example of a mathematical system called a Boolean Algebra (or Boolean Ring) which is defined as:

Set E of elements $a, b, c, \ldots$;

Set O of 2 binary operations $\oplus$ and $\otimes$; (Here $a \oplus b$ denotes the image of (a, b) under the binary operation.)

Set A of axioms for all a, b, c of E:

A_1.   The binary operations are commutative; i.e.,

$$a \oplus b = b \oplus a \quad \text{and} \quad a \otimes b = b \otimes a.$$

A_2.   Each binary operation is distributive over the other; i.e.,

$$a \oplus (b \otimes c) = (a \oplus b) \otimes (a \oplus c) \quad \text{and} \quad a \otimes (b \oplus c) = (a \otimes b) + (a \otimes c).$$

A_3.   There exist elements e and z in E such that for each $a \in E$, $a \oplus z = a$ and $a \otimes e = a$.

A_4.   For each $a \in E$ there exists an element $a' \in E$ such that $a \otimes a' = e$ and $a \oplus a' = z$.

In the algebra of subsets of a (universal) set I, ϕ plays the role of z, I that of e, $\cup$ that of $\oplus$, and $\cap$ that of $\otimes$.

11. A Boolean Algebra has the Principle of Duality: If the interchanges of $\left\{ \begin{array}{l} \oplus \text{ and } \otimes \\ e \text{ and } z \end{array} \right\}$ are made in a correct statement, then the result is also a correct statement.

12. In addition to the Algrebra of Sets which is a Boolean Algebra, other representations of Boolean Algebra that are interesting of themselves and valuable for their applications are:

(a) The Algebra of Symbolic Logic

(b) The Algebra of Switching Currents

Algebra of Sets	*Binary Operator*	*Symbolic Logic*	*Binary Operator*	*Switching Circuits*	*Binary Operator*
Union of 2 sets	$\cup$	Disjunction of 2 propositions	$\vee$	2 switches in $\parallel$	$+$
Intersection of 2 sets	$\cap$	Conjunction of 2 propositions	$\wedge$	2 switches in series	$\times$
Complement of a set A		Negation of a proposition T, F	$\sim$	on-off, or 1, 0	1 or $^-$

Both in symbolic logic and in switching circuits, we can consider "0" and "1" as the elements, in the former representing "False" and "True"; in the latter, "Off" and "On", satisfying the following "rules":

$$\left.\begin{array}{l} 0 + 0 = 0 \\ 1 + 1 = 1 \end{array}\right\} \text{ i.e. } a + a = a$$

$$\left.\begin{array}{l} 0 \times 0 = 0 \\ 1 \times 1 = 1 \end{array}\right\} \text{ i.e. } a \times a = a$$

$$0' = 1$$
$$1' = 0$$

$$\left.\begin{array}{l} 0 + 0' = 0 + 1 = 1 \\ 1 + 1' = 1 + 0 = 1 \end{array}\right\} \text{ i.e. } a + a' = 1$$

$$\left.\begin{array}{l} 0 \times 0' = 0 \times 1 = 0 \\ 1 \times 1' = 1 \times 0 = 0 \end{array}\right\} a \times a' = 0$$

In switching circuits:

Series Circuit

$$A \times B$$

Parallel Circuit

$$A + B$$

here, "0" represents an open circuit: ●⟋ ●—— and "1"
$$A = 0$$

a closed circuit: ●⟍●——
$$A = 1$$

In the Algebra of Symbolic Logic, we use *Truth Tables* to define the operations $\wedge$, $\vee$, $\sim$ as follows:

					Other Operators Used	
p	q	$p \wedge q$	$p \vee q$	$\sim p$	$p \rightarrow q$	$p \leftrightarrow q$
T	T	T	T	F	T	T
T	F	F	T	F	F	F
F	T	F	T	T	T	F
F	F	F	F	T	T	T

13. In order to re-emphasize the use of switching circuits and their relation to truth tables the following is included. Conventionally a "1" represents "True" and a "0" represents "False." The switching circuit symbols are $-$, $\cdot$, $+$, $\rightarrow$, $\equiv$ representing "Not," "And," "Or," "Implies," "Equivalent" respectively and their Truth Table Definitions are

p	q	$p \cdot q$	$p + q$	p	$p \rightarrow q$	$p \equiv q$
0	0	0	0	1	1	1
0	1	0	1	1	1	0
1	0	0	1	0	0	0
1	1	1	1	0	1	1

The comparison with the Algebra of Symbolic Logic being obvious. The "rules" for these circuits are as follows:

$$0 + 0 = 0$$
$$1 + 1 = 1$$
$$1 + 0 = 0 + 1 = 1$$
$$0 \cdot 0 = 0$$
$$1 \cdot 1 = 1$$
$$0 \cdot 1 = 1 \cdot 0 = 0$$
$$\bar{0} = 1$$
$$\bar{1} = 0$$

Mechanical switches or relays are represented by

$$\underline{\qquad\qquad} p \underline{\qquad\qquad} \quad \text{or} \quad \underline{\qquad\qquad} \bar{p} \underline{\qquad\qquad}$$

the former indicating that the circuit is closed, i.e. the switch is made, when $p = 1$ and the latter indicating the converse namely that the circuit is closed when $\bar{p} = 1$ or, what amounts to the same thing, when $p = 0$.

Electronic switches or gates are represented by more complex symbols—four in all, three of which are independent and can stand alone

"Or" $p + q$

"And" $p \cdot q$

"Not" $\bar{p}$

and one which represents the negation of an input or an output and is used with one of the above

"Not"   ◯

An example of its use on an input line is

or on an output is

and on both

The basic functions obtained from the two types of switching circuits are

All the above electronic circuits can be negated by simply adding a negating circle to the output as for example in

Alternative circuits however, which are direct analogues of their relay switching counterparts, are

The operation $+$ is sometimes referred to as the "Inclusive Or" and $\not\equiv$ as the "Exclusive Or", the former having the value "True" when both the inputs are "True"—see the truth table. Note that $p \not\equiv q$ is a shorthand for $\overline{p \equiv q}$.

14. Two sets A and B are equivalent (have same cardinal) if and only if there exists a one-to-one correspondence between the elements of the two sets. This is an equivalence relation on the collection of subsets of set I.

   A set is infinite if and only if it is equivalent with a proper subset of itself.

   A set is called countably (denumerably) infinite if it is equivalent with the set of all positive integers. The set of all rational numbers is countably infinite but the set of all real numbers is noncountably infinite. The cardinal of the set of all rational numbers is denoted by (aleph null); the cardinal of the set of reals is denoted by (aleph).

II. ABSTRACT ALGEBRAIC SYSTEMS

1. *Semigroup.*   A semigroup is a system $\{S, 0, A\}$; S is a nonempty set $\{a, b, c, \ldots\}$, θ consists of one binary operation on S, denoted by $*$, and A consists of the axiom
A_1.   Associativity: $a * (b * c) = (a * b) * c$ for all a, b, $c \in S$.
   Basic Theorem.   (*Generalized Associativity*).   If $a_1, a_2, \ldots, a_n$ are elements of S then all associations of the n elements yield the same "product". (For example,

$$a * ((b * c) * d) = a * (b * (c * d)) = (a * b) * (c * d), \text{etc.})$$

2. *Group.*   A group is a system $\{G, 0, A\}$; G is a nonempty set $\{a, b, c, \ldots\}$, θ consists of one binary operation denoted by $\circ$, and A consists of the axioms:
A_1.   Associativity: $a \circ (b \circ c) = (a \circ b) \circ c$ for all a, b, $c \in G$.
A_2.   Identity Element: G contains an element e having the property, $a \circ e = e \circ a = a$ for every $a \in G$.
A_3.   Inverse Element: For each $a \in G$ there is an element $a' \in G$ with the property, $a \circ a' = a' \circ a = e$.
   If the following additional axiom belongs to A,
A_4.   Commutativity: $a \circ b = b \circ a$ for all $a, b \in G$. Then the group is called Abelian (after Niels Henrik Abel).   Some basic theorems:
   (a) The element e (Axiom A_2) is unique. Then e is *the* identity element of G.
   (b) The element a' (Axiom A_3) is unique for each $a \in G$.   Then a' is *the* inverse of a in G.
   (c) The equation $a \circ x = b$ has a unique solution in G, viz., $x = a' \circ b$.
   (d) $(a')' = a$ and $(a \circ b)' = b' \circ a'$.
   (e) $a \circ b = a \circ c$ if and only if $b = c$.
      If a nonempty subset H of G satisfies the two conditions:
      H_1. $a \circ b \in H$ whenever $a, b \in H$.   (Closure)
      H_2. $a \in H$ if and only if $a' \in H$.
   then H is a *subgroup* of G.
      (Lagrange).   If G is a finite set then the number of elements in H divides the number of elements in G.

Example of group. Let G be the set of all one-to-one functions from a nonempty S onto itself. For any $f, g \in G$, define the function $f \circ g$ as the function which maps s onto $f(g(s))$, for each $s \in S$. Relative to this binary operation G is a group, the *symmetric group* of all permutations on S.

Each group is essentially a subgroup of the symmetric group of some set S.

3. *Ring.* A ring is a system $\{R, \theta, A\}$; R is a nonempty set $\{a, b, c, \ldots\}$, θ consists of two binary operations denoted by $+$ and $\times$, and A consists of the axioms:

A_0. Relative to addition (i.e., $+$) R is an Abelian group in which the identity element is denoted by z and the inverse of a is denoted by $-a$.

M_0. Relative to multiplication (i.e., $\times$) R is a semigroup.

D_1. Left distributive: $a \times (b + c) = (a \times b) + (a \times c)$, all $a, b, c \in R$.

D_2. Right distributive: $(b + c)a = (b \times a) + (c \times a)$, all $a, b, c \in R$.

EXAMPLE 1. The set of all integers (whole numbers) and ordinary addition and multiplication.

EXAMPLE 2. The set of all real functions continuous on the interval $0 \leq y \leq 1$, with addition and multiplication defined by $(f + g)(y) = f(y) + g(y)$, sum of real numbers, and $(f \times g)(y) = f(y) \times g(y)$, product of real numbers.

Special types of rings have been studied extensively.

3.1 *Integral Domain.* An integral domain is a ring R in which multiplication ($\times$) satisfies the additional assumptions:

M_1. Commutativity: $a \times b = b \times a$ for all a and b in R.

M_2. Multiplicative identity: R contains an element $e \neq z$ with the property $a \times e = e \times a = a$ for all a in R.

M_3. Cancellation: $a \times b = a \times c$ if and only if $b = c$.

An element u of integral domain R is a *unit* provided R contains v such that $u \times v = e$.

An element p of integral domain R is a *prime* (irreducible element) provided $p = a \times b$ implies that exactly one of the elements a or b is a unit.

The elements of integral domain R which differ from z and are neither units nor primes are *composites*.

In some integral domains (such as the ring of integers) each composite can be factored uniquely (up to unit factors) as the product of a finite set of primes. However in the integral domain of all entire functions this is not true.

3.2 *Field.* A field is an integral domain in which every element except z is a unit. In other words, the non-z elements form an Abelian group relative to multiplication ($\times$).

EXAMPLE 1. The rational field consisting of ordinary fractions, addition, and multiplication.

EXAMPLE 2. The set of all real numbers $a + b\sqrt{2}$, a and b rational. Then

$$(a + b\sqrt{2}) + (c + d\sqrt{2}) = (a + c) + (b + d)\sqrt{2} \text{ and}$$
$$(a + b\sqrt{2}) \times (c + d\sqrt{2}) = (ac + 2bd) + (ad + bc)\sqrt{2}.$$

Besides these well-known examples there exist finite fields (sometimes called Galois fields).

EXAMPLE 3. Let p be a prime integer. Denote by $GF(p)$ the p integers $0, 1, \ldots,$ $p - 1$. Define addition($\oplus$) of two of these elements a and b as the remainder of $a + b$ (ordinary addition) after division by p. (Thus $1 \oplus (p - 1) = 0$.)

Define $a \otimes b$, the product, to be the remainder of ab (ordinary multiplication) after division by p. (Thus, when $p = 3$, $2 \otimes 2 = 1$.) The resulting system $\{GF(p), \oplus, \otimes\}$ is the (modular) field of integers modulo p.

3.3 *Skew Field or Division Ring.* A skew field is a ring in which the non-z elements form a group relative to multiplication ($\times$).

The classical example of a skew field is the ring of real *quaternions*, first described by W. R. Hamilton. A quaternion is expressible in the form $ae + bi + cj + dk$ where $a, b, c,$ and d are real numbers and $e, i, j,$ and k are elements which commute with all real numbers and multiply as follows:

$$e \times e = e, \quad e \times i = i \times e = i, \quad e \times j = j \times e = j, \quad e \times k = k \times e = k;$$
$$i \times i = -e, \quad i \times j = k, \quad j \times i = -k, \quad i \times k = -j, \quad k \times i = j,$$
$$j \times j = -e, \quad j \times k = i, \quad k \times j = -i, \quad k \times k = -e.$$

These elements distribute over addition. e is generally identified with and written as the real number 1.

3.4 *Matric Ring.* The matric ring $M_n(R)$ over the ring R, where n is a positive integer, consists of all doubly-ordered sets of n^2 elements of R, written as an array

$$\begin{pmatrix} a_{1,1} & a_{1,2} \cdots a_{1,n} \\ a_{2,1} & a_{2,2} \cdots a_{2,n} \\ \vdots \\ a_{n,1} & \cdots \qquad a_{n,n} \end{pmatrix} = (a_{i,j})$$

with addition and multiplication defined as follows:

$$(a_{i,j}) + (b_{i,j}) = (a_{i,j} + b_{i,j})$$
$$(a_{i,j}) \times (b_{i,j}) = (c_{i,j})$$
where

$$c_{i,j} = \sum_{k=1}^{n} a_{i,k} b_{k,j}, \quad i = 1, \ldots, n \quad \text{and} \quad j = 1, \ldots, n.$$

If $n > 1$, then multiplication is noncommutative in general; i.e., $(a_{i,j}) \times (b_{i,j})$ can differ from $(b_{i,j}) \times (a_{i,j})$. Moreover, the product of two nonzero matrices can be the zero matrix (which consists of only the element z in all n^2 positions).

A similar useful method for forming a new ring from a known ring utilizes sequences.

3.5 *Power Series and Polynomial Ring.* Let R be a ring in which multiplication ($\times$) is commutative. The set $PS(R)$ of all sequences $(a_0, a_1, \ldots)$ with $a_i \in R$ is the power series ring of R, with addition and multiplication defined as

$$(a_0, a_1, \ldots) \oplus (b_0, b_1, \ldots) = (a_0 + b_0, \quad a_1 + b_1, \ldots) \text{ and}$$
$$(a_0, a_1, \ldots) \oplus (b_0, b_1, \ldots) = (c_0, c_1, \ldots)$$
where

$$c_0 = a_0 \times b_0, \quad c_1 = a_0 \times b_1 + a_1 \times b_0, \ldots, \text{ and,}$$
generally,

$$c_n = a_0 \times b_n + a_1 \times b_{n-1} + \cdots + a_n \times b_0.$$

The subset $P(R)$ of $PS(R)$ consisting of those sequences $(a_0, a_1, \ldots)$ in which at most only finitely many of the a_i differ from z, form a ring relative to the addition and multiplication just defined. This ring $\{P(R),\oplus,\otimes\}$, is the polynomial ring of R.

Some theorems for rings, fields, etc.

(a) In a ring R, if $a = b$ and $c = d$, then $a + c = b + d$ and $a \times c = b \times d$.

(b) In a ring R, $-(-a) = a$; $(-a) \times b = a \times (-b) = -(a \times b)$; and $(-a) \times (-b) = a \times b$, for all $a,b \in R$.

(c) In a ring R, $a \times z = z \times a = z$, for all $a \in R$.

(d) In a ring R the equation $a + x = b$ has a unique solution, viz., $x = -a + b$.

(f) In a field, skew field, or integral domain, $a \times b = z$ if and only if a and/or b equals z.

(g) A finite integral domain is a field.

(h) The polynomial ring of an integral domain is also an integral domain.

(i) The power series ring of an integral domain is also an integral domain.

(j) A ring is a field provided it is both an integral domain and a skew field.

(k) If R is a (skew) field, then the equation $a \times y = b$, $a \neq z$, has a unique solution $y = a' \times b$.

(l) The polynomial ring and the power series ring of a field are unique factorization domains.

4. *Vector Space.* A vector space $V(F)$ over a field F consists of a nonempty set V (the vectors), a binary operation ($\oplus$) on V, a function (called *scalar multiplication*) from the product set $F \times V$ onto V with the image of (a,ν) denoted by $a \circ \nu$, and the following axioms:

A_0. Relative to addition ($\oplus$) V is an Abelian group in which the identity element (vector) is denoted by z and the inverse of ν is denoted by $-\nu$.

M_1. $a \circ (b \circ \nu) = (ab) \circ \nu$ for all $a,b \in F$ and $\nu \in F$. (Here ab denotes the product of a and b in F.)

M_2 $1 \circ \nu = \nu$ for all $\nu \in V$ (Here 1 denotes the multiplicative indentity element of F.)

D_1. $a \circ (\mu \oplus \nu) = (a \circ \mu) \oplus (a \circ \nu)$ for all $a \in F$, $\mu, \nu \in V$.

D_2. $(a + b) \circ \nu = (a \circ \nu) \oplus (b \circ \nu)$ for all $a,b \in F$, $\nu \in V$. (Here $+$ denotes addition in the field F.)

The elements of F are referred to as *scalars.*

EXAMPLE 1.   The polynomial ring $P(F)$ of a field F is a vector space over F. In this example scalar multiplication is a special case of the multiplication defined for $P(F)$.

EXAMPLE 2.   Denote by $C_n(F)$ the set of all n-tuples, $(a_1, a_2, \ldots, a_n)$, n a positive integer, with all $a_i \in F$. Define

$$(a_1, \ldots, a_n) \oplus (b_1, \ldots, b_n) = (a_1 + b_1, \ldots, a_n + b_n) \quad \text{and}$$
$$c \circ (a_1, \ldots, a_n) = (c \times a_1, \ldots, c \times a_n),$$

where $+$ and $\times$ denote the addition and multiplication, respectively, of the field F. Relative to $\oplus$ and $\circ$, $C_n(F)$ is a vector space, called the *n-dimensional coordinate space* over F.

A vector space $V(F)$ is *n-dimensional* over F provided V contains n elements $\nu_1, \nu_2, \ldots, \nu_n$ such that each element $\nu \in V$ is uniquely expressible in the form

$$\nu = a_1 \circ \nu_1 \oplus a_2 \circ \nu_2 \oplus \cdots \oplus a_n \circ \nu_n$$

for some $a_1, a_2, \ldots, a_n \in F$.

Two vector spaces $V(F)$ and $W(F)$ over the field of scalars F are *isomorphic* provided there is a one-to-one correspondence between the elements of V and the elements of W which is preserved under the arithmetic of the two spaces.

Basic Theorem.   An n-dimensional vector space $V(F)$ is isomorphic with the coordinate space $C_n(F)$ (of Example 2, above).

MATRICES AND DETERMINANTS

Dr. R. E. Bargmann

1. GENERAL DEFINITIONS

1.1. A matrix is an array of numbers, consisting of m rows and n columns. It is usually denoted by a bold-face capital letter, e.g.,

$$\mathbf{A} \qquad \mathbf{\Sigma} \qquad \mathbf{M}$$

1.2. The (i, j) element of a matrix is the element occurring in row i and column j. It is usually denoted by a lower-case letter with subscripts, e.g.,

$$a_{ij} \qquad \sigma_{ij} \qquad m_{ij}$$

Exceptions to this convention will be stated where required.

1.3. A matrix is called rectangular if m (number of rows) $\neq n$ (number of columns).

1.4. A matrix is called square if $m = n$.

1.5a. In the transpose of a matrix $\mathbf{A}$, denoted by $\mathbf{A}'$, the element in the j'th row and i'th column of $\mathbf{A}$ is equal to the element in the i'th row and j'th column of $\mathbf{A}'$. Formally $(\mathbf{A}')_{ij} = (\mathbf{A})_{ji}$ where the symbol $(\mathbf{A}')_{ij}$ denotes the (i, j) element of $\mathbf{A}'$.

1.5b. The Hermitian conjugate of a matrix $\mathbf{A}$, denoted by $\mathbf{A}^H$ or $\mathbf{A}^\dagger$ is obtained by transposing $\mathbf{A}$ and replacing each element by its conjugate complex. Hence if

$$a_{kl} = u_{kl} + iv_{kl}$$

then

$$(\mathbf{A}^H)_{kl} = u_{lk} - iv_{lk}$$

where typical elements have been denoted by (k, l) to avoid confusion with $i = \sqrt{-1}$.

1.6a. A square matrix is called symmetric if $\mathbf{A} = \mathbf{A}'$.

1.6b. A square matrix is called Hermitian if $\mathbf{A} = \mathbf{A}^H$.

1.7. A matrix with m rows and 1 column is called a column vector and is usually denoted by bold faced, lower-case letters, e.g.,

$$\boldsymbol{\beta} \qquad \mathbf{x} \qquad \mathbf{a}$$

1.8. A matrix with one row and n columns is called a row vector and is usually denoted by a primed, bold faced, lower-case letter, e.g.,

$$\mathbf{a}' \qquad \mathbf{c}' \qquad \boldsymbol{\mu}'$$

1.9. A matrix with one row and one column is called a scalar, and is usually denoted by a lower-case letter, occasionally italicized.

1.10. The diagonal extending from upper left (NW) to lower right (SE) is called the principal diagonal of a square matrix.

1.11a. A matrix with all elements above the principal diagonal equal to zero is called a lower triangular matrix.

Example

$$\mathbf{T} = \begin{bmatrix} t_{11} & 0 & 0 \\ t_{21} & t_{22} & 0 \\ t_{31} & t_{32} & t_{33} \end{bmatrix} \text{ is lower triangular}$$

1.11b. The transpose of a lower triangular matrix is called an upper triangular matrix.

1.12. A square matrix with all off-diagonal elements equal to zero is called a diagonal matrix, denoted by the letter **D** with subscript indicating the typical element in the principal diagonal.

Example

$$\mathbf{D}_a = \begin{bmatrix} a_1 & 0 & 0 \\ 0 & a_2 & 0 \\ 0 & 0 & a_3 \end{bmatrix} \text{ is diagonal}$$

2. ADDITION, SUBTRACTION, AND MULTIPLICATION

2.1. Two matrices **A** and **B** can be added (subtracted) if the number of rows (columns) in **A** equals the number of rows (columns) in **B**.

$$\mathbf{A} \pm \mathbf{B} = \mathbf{C}$$

implies

$$a_{ij} \pm b_{ij} = c_{ij} \qquad i = 1, 2, \ldots m$$
$$j = 1, 2, \ldots n$$

2.2. Multiplication of a matrix or vector by a scalar implies multiplication of each element by the scalar. If

$$\mathbf{B} = \gamma \mathbf{A}$$

then

$$b_{ij} = \gamma a_{ij}$$

for all elements.

2.3a. Two matrices, **A** and **B**, can be multiplied if the number of columns in **A** equals the number of rows in **B**.

2.3b. Let **A** be of order $(m \times n)$ (have m rows and n columns) and **B** of order $(n \times p)$. Then the product of two matrices $\mathbf{C} = \mathbf{AB}$, is a matrix of order $(m \times p)$ with elements

$$c_{ij} = \sum_{k=1}^{n} a_{ik} b_{kj}$$

This states that c_{ij} is the scalar product of the i'th row vector of **A** and the j'th column vector of **B**.

Example

$$\begin{bmatrix} 3 & 4 & 2 \\ 2 & 3 & -1 \end{bmatrix} \begin{bmatrix} 1 & -2 & -4 \\ 0 & -1 & 2 \\ 6 & -3 & 9 \end{bmatrix} = \begin{bmatrix} 15 & -16 & 14 \\ -4 & -4 & -11 \end{bmatrix}$$

e.g.,

$$c_{23} = \begin{bmatrix} 2 & 3 & -1 \end{bmatrix} \begin{bmatrix} -4 \\ 2 \\ 9 \end{bmatrix}$$

$$= 2 \times (-4) + 3 \times 2 + (-1) \times 9 = -11$$

2.3c. In general, matrix multiplication is not commutative

$$\mathbf{AB} \neq \mathbf{BA}$$

2.3d. Matrix multiplication is associative

$$\mathbf{A(BC)} = \mathbf{(AB)C}$$

2.3e. The distributive law for multiplication and addition holds as in the case of scalars,

$$\mathbf{(A + B)C} = \mathbf{AC} + \mathbf{BC}$$
$$\mathbf{C(A + B)} = \mathbf{CA} + \mathbf{CB}$$

2.4. In some applications, the term-by-term product of two matrices **A** and **B** of identical order is defined as

$$\mathbf{C} = \mathbf{A} * \mathbf{B}$$

where

$$c_{ij} = a_{ij}b_{ij}$$

2.5. $\mathbf{(ABC)'} = \mathbf{C'B'A'}$

2.6. $\mathbf{(ABC)}^H = \mathbf{C}^H\mathbf{B}^H\mathbf{A}^H$

2.7. If both **A** and **B** are symmetric, then $\mathbf{(AB)'} = \mathbf{BA}$. Note that the product of two symmetric matrices is generally not symmetric.

3. RECOGNITION RULES AND SPECIAL FORMS

3.1. A column (row) vector with all elements equal to zero is called a null vector, and usually denoted by the symbol **0**.

3.2. A null matrix has all elements equal to zero.

3.3a. A diagonal matrix with all elements equal to one in the principal diagonal is called the identity matrix **I**.

3.3b. $\gamma\mathbf{I}$, i.e., a diagonal matrix with all diagonal elements equal to a constant γ, is called a scalar matrix.

3.4. A matrix which has only one element equal to one and all others equal to zero is called an elementary matrix $\mathbf{(EL)}_{ij}$.

Example

$$\mathbf{(EL)}_{23} = \begin{bmatrix} 0 & 0 & 0 & 0 & 0 \\ 0 & 0 & 1 & 0 & 0 \\ 0 & 0 & 0 & 0 & 0 \\ 0 & 0 & 0 & 0 & 0 \end{bmatrix}$$

The order of the matrix is usually implicit.

3.5a. The symbol **j** is reserved for a column vector with all elements equal to 1.

3.5b. The symbol **j'** is reserved for a row vector with all elements equal to 1.

3.6. An expression ending with a column vector is a column vector.

Example

$$\mathbf{ABx} = \mathbf{y}$$

(It is assumed that rule 2.3a is satisfied, else matrix multiplication would not be defined.)

3.7. An expression beginning with a row vector is a row vector.

Example

$$\mathbf{y'(A + BC)} = \mathbf{d'}$$

3.8. An expression beginning with a row vector and ending with a column vector, is a scalar.

Example

$$\mathbf{a'Bc} = \gamma$$

3.9a. If $\mathbf{Q}$ is a square matrix, the scalar $\mathbf{x'Qx}$ is called a quadratic form. If $\mathbf{Q}$ is non-symmetric, one can always find a symmetric matrix $\mathbf{Q^*}$ such that

$$\mathbf{x'Qx} = \mathbf{x'Q^*x}$$

where

$$(\mathbf{Q^*})_{ij} = \tfrac{1}{2}(q_{ij} + q_{ji})$$

3.9b. If $\mathbf{Q}$ is a square matrix the scalar $\mathbf{x}^H \mathbf{Qx}$ is called a Hermitian form.

3.10. A scalar $\mathbf{x'Qy}$ is called a bilinear form.

3.11. The scalar $\mathbf{x'x} = \Sigma x_i^2$, i.e., the sum of squares of all elements of $\mathbf{x}$.

3.12. The scalar $\mathbf{x'y} = \Sigma x_i y_i$, i.e., the sum of products of elements in $\mathbf{x}$ by those in $\mathbf{y}$. $\mathbf{x}$ and $\mathbf{y}$ have the same number of elements.

3.13. The scalar $\mathbf{x'D_w x} = \Sigma w_i x_i^2$ is called a weighted sum of squares.

3.14. The scalar $\mathbf{x'D_w y} = \Sigma w_i x_i y_i$ is called a weighted sum of products.

3.15a. The vector $\mathbf{Aj}$ is a column vector whose elements are the row sums of $\mathbf{A}$.

3.15b. The vector $\mathbf{j'A}$ is a row vector whose elements are the column sums of $\mathbf{A}$.

3.15c. The scalar $\mathbf{j'Aj}$ is the sum of all elements in $\mathbf{A}$. Schematically

A	Aj
j'A	**j'Aj**

3.16a. If $\mathbf{B} = \mathbf{D_w A}$; then $b_{ij} = w_i a_{ij}$.

3.16b. If $\mathbf{B} = \mathbf{AD_w}$; then $b_{ij} = a_{ij} w_j$.

3.17. Interchanging summation and matrix notation:

If

$$\mathbf{ABCD} = \mathbf{E}$$

then

$$e_{ij} = \sum_k \sum_l \sum_m a_{ik} b_{kl} c_{lm} d_{mj}$$

The second subscript of an element must coincide with the first of the next one. Reordering and transposing may be required.

Example

If

$$e_{ij} = \sum_k \sum_l \sum_m a_{kl} b_{ki} c_{jm} d_{ml}$$

$$= \sum_k \sum_l \sum_m b_{ki} a_{kl} d_{ml} c_{jm}$$

Then

$$\mathbf{E} = \mathbf{B'AD'C'}$$

3.18a. $\mathbf{A'A}$ is a symmetric matrix whose (i,j) element is the scalar product of the i'th column vector and the j'th column vector of $\mathbf{A'}$.

3.18b. $\mathbf{AA'}$ is a symmetric matrix whose (i,j) element is the scalar product of the i'th row vector and the j'th row vector of $\mathbf{A}$.

4. DETERMINANTS

4.1a. A determinant $|\mathbf{A}|$ or $\det(\mathbf{A})$ is a scalar function of a square matrix defined in such a way that

$$|\mathbf{A}|\,|\mathbf{B}| = |\mathbf{AB}|$$

and

$$\begin{vmatrix} a_{11} & a_{12} \\ a_{21} & a_{22} \end{vmatrix} = a_{11}a_{22} - a_{12}a_{21}$$

4.1b. $|\mathbf{A}| = |\mathbf{A}'|$

4.2.
$$\begin{vmatrix} a_{11} & a_{12} & a_{13} \\ a_{21} & a_{22} & a_{23} \\ a_{31} & a_{32} & a_{33} \end{vmatrix} = \begin{array}{l} a_{11}a_{22}a_{33} + a_{12}a_{23}a_{31} + a_{13}a_{21}a_{32} \\ - a_{13}a_{22}a_{31} - a_{11}a_{23}a_{32} - a_{12}a_{21}a_{33} \end{array}$$

4.3.
$$\begin{vmatrix} a_{11} & a_{12} & \cdots & a_{1n} \\ a_{21} & a_{22} & \cdots & a_{2n} \\ \cdot & & \cdots & \cdot \\ a_{n1} & a_{n2} & \cdots & a_{nm} \end{vmatrix} = \sum (-1)^{\delta} a_{1i_1} a_{2i_2} \cdots a_{ni_n}$$

where the sum is over all permutations

$$i_1 \neq i_2 \neq \cdots i_n$$

and δ denotes the number of exchanges necessary to bring the sequence $(i_1, i_2, \ldots i_n)$ back into the natural order $(1, 2, \ldots n)$.

4.4. If two rows (columns) in a matrix are exchanged, the determinant will change its sign.

4.5. A determinant does not change its value if a linear combination of other rows (columns) is added to any given row (column).

Example

$$\begin{vmatrix} a_{11} & a_{12} & a_{13} & a_{14} \\ b_{21} & b_{22} & b_{23} & b_{24} \\ a_{31} & a_{32} & a_{33} & a_{34} \\ a_{41} & a_{42} & a_{43} & a_{44} \end{vmatrix} = \begin{vmatrix} a_{11} & a_{12} & a_{13} & a_{14} \\ a_{21} & a_{22} & a_{23} & a_{24} \\ a_{31} & a_{32} & a_{33} & a_{34} \\ a_{41} & a_{42} & a_{43} & a_{44} \end{vmatrix}$$

where

$$b_{2i} = a_{2i} + \gamma_1 a_{1i} + \gamma_3 a_{3i} + \gamma_4 a_{4i}$$
$$i = 1, 2, 3, 4$$

$\gamma_1, \gamma_3, \gamma_4$ arbitrary.

4.6. If the i'th row (column) equals (a constant times) the j'th row (column) of a matrix, its determinant is equal to zero, $(i \neq j)$.

4.7. If, in a matrix $\mathbf{A}$, each element of a row (column) is multiplied by a constant γ, the determinant is multiplied by γ.

4.8. $|\gamma \mathbf{A}| = \gamma^n |\mathbf{A}|$ assuming that $\mathbf{A}$ is of order $(n \times n)$.

4.9. The cofactor of a square matrix $\mathbf{A}$, $\mathrm{cof}_{ij}(\mathbf{A})$ is the determinant of a matrix obtained by striking the i'th row and j'th column of $\mathbf{A}$ and choosing positive (negative) sign if $i + j$ is even (odd).

Example

$$\text{cof}_{23} \begin{bmatrix} 2 & 4 & 3 \\ 6 & 1 & 5 \\ -2 & 1 & 3 \end{bmatrix} = - \begin{vmatrix} 2 & 4 \\ -2 & 1 \end{vmatrix}$$

$$= -(2 + 8) = -10$$

4.10. (Laplace Development)

$$|\mathbf{A}| = a_{i1}\text{cof}_{i1}(\mathbf{A}) + a_{i2}\text{cof}_{i2}(\mathbf{A}) + \cdots + a_{in}\text{cof}_{in}(\mathbf{A})$$
$$= a_{1j}\text{cof}_{1j}(\mathbf{A}) + a_{2j}\text{cof}_{2j}(\mathbf{A}) + \cdots + a_{nj}\text{cof}_{nj}(\mathbf{A})$$

for any row i or any column j.

4.11. Numerical Evaluation of the determinant of a symmetric matrix.

Note: If $\mathbf{A}$ is non-symmetric, form $\mathbf{A'A}$ or $\mathbf{AA'}$ by rule 3.18, obtain its determinant, and take the square root.

("Forward Doolittle Scheme", "left side")

Let

$$p_{11} = a_{11}, \quad p_{12} = a_{12} = a_{21}, \dots p_{1n} = a_{1n}$$

$$\begin{array}{cccccc} p_{11} & p_{12} & p_{13} & \cdots & p_{1n} \\ \hline 1 & u_{12} & u_{13} & \cdots & u_{1n} \\ \hline & a_{22} & a_{23} & \cdots & a_{2n} \\ & p_{22} & p_{23} & \cdots & p_{2n} \\ \hline & 1 & u_{23} & \cdots & u_{2n} \\ \hline & & a_{33} & \cdots & a_{3n} \\ & & p_{33} & \cdots & p_{3n} \\ \hline & & 1 & \cdots & u_{3n} \\ & & \cdot & \cdots & \cdot \\ & & & & a_{nn} \\ & & & & p_{nn} \\ \hline & & & & 1 \end{array}$$

$$u_{1i} = p_{1i}/p_{11} \qquad\qquad i = 1, 2, \dots n$$
$$p_{2i} = a_{2i} - u_{12}p_{1i} \qquad\qquad i = 2, 3, \dots n$$
$$u_{2i} = p_{2i}/p_{22}$$
$$p_{3i} = a_{3i} - u_{13}p_{1i} - u_{23}p_{2i} \qquad\qquad i = 3, 4, \dots n$$
$$u_{3i} = p_{3i}/p_{33}$$
$$p_{ki} = a_{ki} - u_{1k}p_{1i} - u_{2k}p_{2i} - \cdots - u_{k-1,k}p_{k-1,i} \quad i = k, k+1, \dots n$$
$$k = 2, 3, \dots n$$
$$u_{ki} = p_{ki}/p_{kk}$$

If, at some stage, $p_{kk} = 0$, reordering of rows and columns may be required. If the matrix is positive-definite (see 8.16) (always true for $\mathbf{AA'}$ or $\mathbf{A'A}$, see rule 10.24), none of the

p_{kk} will be zero. The p_{ii} are called pivots. Then

$$|\mathbf{A}| = \prod_{i=1}^{n} p_{ii}$$

Further, if $\mathbf{A}$ is partitioned

$$\mathbf{A} = \begin{bmatrix} \mathbf{A}_{11} & \mathbf{A}_{12} \\ \mathbf{A}'_{12} & \mathbf{A}_{22} \end{bmatrix}$$

where $\mathbf{A}_{11}$ is of order $(k \times k)$, then

$$|\mathbf{A}_{11}| = \prod_{i=1}^{k} p_{ii}$$

(Numerical Examples: see 6.14.)

5.   SINGULARITY AND RANK

5.1. A matrix $\mathbf{A}$ is called singular if there exists a vector $\mathbf{x} \neq \mathbf{0}$ such that $\mathbf{A}\mathbf{x} = \mathbf{0}$ or $\mathbf{A}'\mathbf{x} = \mathbf{0}$. Note $\mathbf{x} \neq \mathbf{0}$ if a single element of $\mathbf{x}$ is unequal 0. If a matrix is not singular, it is called non-singular.

5.2. If a matrix $\mathbf{A}_1$ can be formed by selection of r rows and columns of $\mathbf{A}$ such that $\mathbf{A}_1\mathbf{x} \neq \mathbf{0}$ or $\mathbf{A}'_1\mathbf{x} \neq \mathbf{0}$ for every $\mathbf{x} \neq \mathbf{0}$, and if addition of an $(r + 1)st$ row and column would produce a singular matrix, r is called the rank of $\mathbf{A}$.

Example

$$\mathbf{A} = \begin{bmatrix} 2 & 4 & 6 \\ 1 & 3 & 7 \\ 3 & 7 & 13 \\ 1 & 1 & -1 \end{bmatrix}$$

Note that

$$[1, \quad 1, \quad -1] \begin{bmatrix} 2 & 4 & 6 \\ 1 & 3 & 7 \\ 3 & 7 & 13 \end{bmatrix} = [0 \quad 0 \quad 0]$$

and

$$[1, \quad -1, \quad -1] \begin{bmatrix} 2 & 4 & 6 \\ 1 & 3 & 7 \\ 1 & 1 & -1 \end{bmatrix} = [0 \quad 0 \quad 0]$$

but

$$\begin{bmatrix} 2 & 4 \\ 1 & 3 \end{bmatrix} \begin{bmatrix} x_1 \\ x_2 \end{bmatrix} \neq \begin{bmatrix} 0 \\ 0 \end{bmatrix}$$

or

$$[x_1 \quad x_2] \begin{bmatrix} 2 & 4 \\ 1 & 3 \end{bmatrix} \neq [0, \quad 0]$$

for any arbitrary

$$[x_1, \quad x_2] \neq [0, \quad 0].$$

Hence the matrix has rank 2.

5.3. If A has rank r and if A_1 is a non-singular submatrix consisting of r rows and columns of A, then A_1 is called a basis of A.

5.4a. The determinant of a square singular matrix is 0.

5.4b. The determinant of a non-singular matrix is $\neq 0$.

5.5. rank $(AB) \leq$ min [rank (A), rank (B)].

5.6. rank (AA') = rank $(A'A)$ = rank (A).

5.7. $|A'A| = |AA'| = |A|^2$ if A is square.

5.8. $|A'A| = |AA'| \geq 0$ for every A with real elements.

6. INVERSION

(Regular Case, non-singular matrices)

6.1. If A is square and non-singular ($|A| \neq 0$) there exists a unique matrix A^{-1} such that $AA^{-1} = A^{-1}A = I$.

6.2. $(ABC)^{-1} = C^{-1}B^{-1}A^{-1}$ (provided that all inverses exist).

6.3. $(A^{-1})' = (A')^{-1}$

6.4. $Ax = b$ is a system of linear equations. If A is square and non-singular, there exists a unique solution

$$x = A^{-1}b$$

6.5. $(\gamma A)^{-1} = (1/\gamma)A^{-1}$

6.6. $|A^{-1}| = 1/|A|$

6.7. $D_w^{-1} = D_{1/w}$ where D is a diagonal matrix.

6.8. If

$$A = B + uv'$$

then

$$A^{-1} = B^{-1} - \lambda yz'$$

where

$$y = B^{-1}u, \quad z' = v'B^{-1},$$

and

$$\lambda = 1/(1 + z'u)$$

Example 6.8.1

$$A = \begin{bmatrix} 4 & 2 & 4 & 5 \\ 3 & 9 & 12 & 15 \\ 2 & 4 & 11 & 10 \\ 1 & 2 & 4 & 10 \end{bmatrix}$$

This matrix can be written as

$$\begin{bmatrix} 3 & 0 & 0 & 0 \\ 0 & 3 & 0 & 0 \\ 0 & 0 & 3 & 0 \\ 0 & 0 & 0 & 5 \end{bmatrix} + \begin{bmatrix} 1 \\ 3 \\ 2 \\ 1 \end{bmatrix} [1 \quad 2 \quad 4 \quad 5] = B + uv'$$

$$\mathbf{B}^{-1} = \begin{bmatrix} 1/3 & 0 & 0 & 0 \\ 0 & 1/3 & 0 & 0 \\ 0 & 0 & 1/3 & 0 \\ 0 & 0 & 0 & 1/5 \end{bmatrix}$$

$$\mathbf{y} = \mathbf{B}^{-1}\mathbf{u} = \begin{bmatrix} 1/3 \\ 1 \\ 2/3 \\ 1/5 \end{bmatrix}$$

$$\mathbf{z}' = \mathbf{v}'\mathbf{B}^{-1} = [1/3 \quad 2/3 \quad 4/3 \quad 1]$$

$$\mathbf{z}'\mathbf{u} = 1/3 \times 1 + 2/3 \times 3 + 4/3 \times 2 + 1 \times 1 = 6$$

$$\lambda = 1/7$$

$$\mathbf{A}^{-1} = \begin{bmatrix} 1/3 & 0 & 0 & 0 \\ 0 & 1/3 & 0 & 0 \\ 0 & 0 & 1/3 & 0 \\ 0 & 0 & 0 & 1/5 \end{bmatrix} - (1/7) \begin{bmatrix} 1/3 \\ 1 \\ 2/3 \\ 1/5 \end{bmatrix} [1/3 \quad 2/3 \quad 4/3 \quad 1]$$

$$= (1/315) \begin{bmatrix} 100 & -10 & -20 & -15 \\ -15 & 75 & -60 & -45 \\ -10 & -20 & 65 & -30 \\ -3 & -6 & -12 & 54 \end{bmatrix}$$

(This rule is especially useful if all off-diagonal elements are equal, then $\mathbf{u} = k\mathbf{j}$ and $\mathbf{v}' = \mathbf{j}'$ and $\mathbf{B}$ is diagonal.)

6.9. Let $\mathbf{B}$ (elements b_{ij}) have a known inverse, $\mathbf{B}^{-1}$ (elements b^{ij}). Let $\mathbf{A} = \mathbf{B}$ except for one element $a_{rs} = b_{rs} + k$. Then the elements of $\mathbf{A}^{-1}$ are

$$a^{ij} = b^{ij} - \frac{kb^{ir}b^{sj}}{1 + kb^{sr}}.$$

6.10. (Partitioning)

Let

$$\mathbf{A} = \begin{matrix} (p) \\ (q) \end{matrix} \begin{bmatrix} \mathbf{B} & \mathbf{C} \\ \mathbf{D} & \mathbf{E} \end{bmatrix}$$

(p) (q) (letters in parentheses denote order of the submatrices)

Let $\mathbf{B}^{-1}$ and $\mathbf{E}^{-1}$ exist. Then

$$\mathbf{A}^{-1} = \begin{bmatrix} \mathbf{X} & \mathbf{Y} \\ \mathbf{Z} & \mathbf{U} \end{bmatrix}$$

where

$$\mathbf{X} = (\mathbf{B} - \mathbf{C}\mathbf{E}^{-1}\mathbf{D})^{-1}$$
$$\mathbf{U} = (\mathbf{E} - \mathbf{D}\mathbf{B}^{-1}\mathbf{C})^{-1}$$
$$\mathbf{Y} = -\mathbf{B}^{-1}\mathbf{C}\mathbf{U}$$
$$\mathbf{Z} = -\mathbf{E}^{-1}\mathbf{D}\mathbf{X}$$

6.11. (Partitioning of Determinants)

Let

$$|\mathbf{A}| = \begin{vmatrix} \mathbf{B} & \mathbf{C} \\ \mathbf{D} & \mathbf{E} \end{vmatrix} \qquad \text{(same structure as in 6.10)}$$

Then

$$|\mathbf{A}| = |\mathbf{E}| \, |(\mathbf{B} - \mathbf{C}\mathbf{E}^{-1}\mathbf{D})| = |\mathbf{B}| \, |(\mathbf{E} - \mathbf{D}\mathbf{B}^{-1}\mathbf{C})|$$

6.12. Let

$$\mathbf{A} = \mathbf{B} + \mathbf{U}\mathbf{V}$$

where $\mathbf{B}(n \times n)$ has an inverse

$\mathbf{U}$ is of order $(n \times k)$, with k usually very small

$\mathbf{V}$ is of order $(k \times n)$

(the special case for $k = 1$ is treated in 6.8).

Then
$$\mathbf{A}^{-1} = \mathbf{B}^{-1} - \mathbf{Y}\mathbf{\Lambda}\mathbf{Z}$$

where

$$\mathbf{Y} = \mathbf{B}^{-1}\mathbf{U}(n \times k)$$
$$\mathbf{Z} = \mathbf{V}\mathbf{B}^{-1}(k \times n)$$

and

$$\mathbf{\Lambda}(k \times k) = [\mathbf{I} + \mathbf{Z}\mathbf{U}]^{-1}$$

6.13. Let a_{ij} denote the elements of $\mathbf{A}$ and a^{ij} those of $\mathbf{A}^{-1}$. Then

$$a^{ij} = \mathrm{cof}_{ji}(\mathbf{A})/|\mathbf{A}|$$

where cof is the determinant defined in 4.9.

6.14. "Doolittle" Method of inverting symmetric matrices (see also 4.11). Let

$$p_{11} = a_{11}, \quad p_{12} = a_{12} = a_{21}, \ldots p_{1n} = a_{1n} = a_{n1}$$

Forward Solution

p_{11}	p_{12}	p_{13}	$\cdots$	p_{1n}	1				
1	u_{12}	u_{13}	$\cdots$	u_{1n}	u_{11}				
	a_{22}	a_{23}	$\cdots$	a_{2n}	0	1			
	p_{22}	p_{23}	$\cdots$	p_{2n}	p_{21}	p_{2II}			
	1	u_{23}	$\cdots$	u_{2n}	u_{21}	u_{2II}			
		a_{33}	$\cdots$	a_{3n}	0	0	1		
		p_{33}	$\cdots$	p_{3n}	p_{31}	p_{3II}	p_{3III}		
		1	$\cdots$	u_{3n}	u_{31}	u_{3II}	u_{3III}		
		$\cdot$	$\cdots$	$\cdot$	$\cdot$	$\cdots$	$\cdot$		
				a_{nn}	0	0	0	$\cdots$	1
				p_{nn}	p_{n1}	p_{nII}	p_{nIII}	$\cdots$	p_{nN}
				1	u_{n1}	u_{nII}	u_{nIII}	$\cdots$	u_{nN}

$$u_{1i} = p_{1i}/p_{11} \qquad\qquad\qquad i = 1, 2, \ldots n, \mathrm{I}$$

$$p_{2i} = a_{2i} - u_{12}p_{1i} \qquad\qquad i = 2, 3, \ldots n, \mathrm{I}, \mathrm{II}$$

$$u_{2i} = p_{2i}/p_{22}$$

$$p_{3i} = a_{3i} - u_{13}p_{1i} - u_{23}p_{2i} \qquad i = 3, 4, \ldots n, \mathrm{I}, \mathrm{II}, \mathrm{III}$$

$$u_{3i} = p_{3i}/p_{33}$$

$$p_{ki} = a_{ki} - u_{1k}p_{1i} - u_{2k}p_{2i} - \cdots - u_{k-1,k}p_{k-1,i} \quad i = k, k+1, \ldots n, \mathrm{I}, \mathrm{II}, \ldots K$$
$$k = 2, 3, \ldots n$$

$$u_{ki} = p_{ki}/p_{kk}$$

Backward Solution

(j refers to Arabic, J refers to Roman numerals)

The elements of $\mathbf{A}^{-1}$ are a^{ij}

$$a^{nj} = u_{nJ} \qquad\qquad\qquad\qquad\qquad\qquad j = 1, 2, \ldots n;$$
$$J = \mathrm{I}, \mathrm{II}, \ldots N$$

$$a^{n-1j} = u_{n-1,J} - u_{n-1,n}a^{nj} \qquad\qquad\qquad j = 1, 2, \ldots (n-1);$$
$$J = \mathrm{I}, \mathrm{II}, \ldots (N-1)$$

$$a^{n-2,j} = u_{n-2,J} - u_{n-2,n}a^{nj} - u_{n-2,n-1}a^{n-1,j} \qquad j = 1, 2, \ldots (n-2);$$
$$J = \mathrm{I}, \mathrm{II}, \ldots (N-2)$$

$$a^{n-k,j} = u_{n-k,J} - u_{n-k,n}a^{nj} - u_{n-k,n-1}a^{n-1,j} - \cdots - u_{n-k,n-k+1}a^{n-k+1,j} \qquad j = 1, 2, \ldots (n-k);$$
$$J = \mathrm{I}, \mathrm{II}, \ldots (N-k);$$
$$k = 1, 2, \ldots (n-1),$$

and $a^{ji} = a^{ij}$.

Numerical Example 6.14.1.

Invert the Matrix

$$\begin{bmatrix} 25 & 30 & -10 \\ 30 & 40 & -6 \\ -10 & -6 & 17 \end{bmatrix}$$

a_1	25	30	-10	1		
u_1	1	1.2	-0.4	0.04		
	a_2	40	-6	0	1	
	p_2	4	6	-1.2	1	
	u_2	1	1.5	-0.3	0.25	
		a_3	17	0	0	1
		p_3	4	2.2	-1.5	1
		u_3	1	0.55	-0.375	0.25

1.61	-1.125	0.55
-1.125	0.8125	-0.375
0.55	-0.375	0.25

Enter row a_1.

Elements in u_1 = Elements in a_1 divided by $a_{11}(=25)$.

Enter row a_2.

$$p_{22} = 40 - 1.2 \times 30 = 4$$
$$p_{23} = -6 - 1.2 \times (-10) = 6$$
$$p_{21} = 0 - 1.2 \times 1 = -1.2$$
$$p_{211} = 1$$

Elements in u_2 = Elements in p_2 divided by $p_{22}(=4)$.

Enter row a_3.

$$p_{33} = 17 - (-0.4) \times (-10) - 1.5 \times 6 = 4$$
$$p_{31} = 0 - (-0.4) \times 1 - 1.5 \times (-1.2) = 2.2$$
$$p_{311} = 0 - 1.5 \times 1 = -1.5$$
$$p_{3111} = 1$$

Elements in u_3 = Elements in p_3 divided by $p_{33}(=4)$.

Copy the right-hand side of the last (third) u – row as the last column below the double line.

$$a^{21} = -0.3 - 1.5 \times 0.55 = -1.125$$
$$a^{22} = 0.25 - 1.5 \times (-0.375) = 0.8125$$
$$a^{23} = 0 - 1.5 \times 0.25 = -0.375 \quad \text{(check against } a^{32}).$$

These are entered in the next to last (second) column below.

$$a^{11} = 0.04 - (-0.4) \times 0.55 - 1.2 \times (-1.125) = 1.61$$
$$a^{12} = 0 - (-0.4) \times (-0.375) - 1.2 \times 0.8125 = -1.125 \quad \text{(check against } a^{21})$$
$$a^{13} = 0 - (-0.4) \times (0.25) - 1.2 \times (-0.375) = 0.55 \quad \text{(check against } a^{31}).$$

6.15. A matrix is called orthogonal if $\mathbf{A}' = \mathbf{A}^{-1}$ (or $\mathbf{AA}' = \mathbf{I}$).

7. TRACES

7.1. If $\mathbf{A}$ is a square matrix then the trace of $\mathbf{A}$ is $tr\, \mathbf{A} = \sum_i a_{ii}$, i.e., the sum of the diagonal elements.

7.2. If $\mathbf{A}$ is of order $(m \times k)$ and $\mathbf{B}$ of order $(k \times m)$ then $tr(\mathbf{AB}) = tr(\mathbf{BA})$.

7.3. If $\mathbf{A}$ is of order $(m \times k)$, $\mathbf{B}$ of order $(k \times r)$ and $\mathbf{C}$ of order $(r \times m)$, then

$$tr(\mathbf{ABC}) = tr(\mathbf{BCA}) = tr(\mathbf{CAB}).$$

7.3a. If $\mathbf{b}$ is a column vector and $\mathbf{c}'$ a row vector, then

$$tr(\mathbf{Abc}') = tr(\mathbf{bc}'\mathbf{A}) = \mathbf{c}'\mathbf{Ab}$$

since the trace of a scalar is the scalar.

7.4. $tr(\mathbf{A} + \gamma \mathbf{B}) = tr\mathbf{A} + \gamma tr\mathbf{B}$; where γ is a scalar.

7.5. $tr(\mathbf{EL})_{ij}\mathbf{A} = tr\mathbf{A}(\mathbf{EL})_{ij} = a_{ji}$; where $(\mathbf{EL})_{ij}$ is an elementary matrix as defined in 3.4.

7.6. $tr(\mathbf{EL})_{ij}\mathbf{A}(\mathbf{EL})_{rs}\mathbf{B} = a_{jr}b_{si}$.

(These rules are useful in matrix differentiation)

7.7. The trace of the second order of a square matrix $\mathbf{A}$ is the sum of the determinants of all $\binom{n}{2}$ matrices of order (2×2) which can be formed by intersecting rows i and j with columns i and j.

$$tr_2 \mathbf{A} = \begin{vmatrix} a_{11} & a_{12} \\ a_{21} & a_{22} \end{vmatrix} + \begin{vmatrix} a_{11} & a_{13} \\ a_{31} & a_{33} \end{vmatrix}$$

$$+ \cdots + \begin{vmatrix} a_{11} & a_{1n} \\ a_{n1} & a_{nn} \end{vmatrix} + \begin{vmatrix} a_{22} & a_{23} \\ a_{32} & a_{33} \end{vmatrix}$$

$$+ \cdots + \begin{vmatrix} a_{22} & a_{2n} \\ a_{n2} & a_{nn} \end{vmatrix} + \cdots + \begin{vmatrix} a_{n-1,n-1} & a_{n-1,n} \\ a_{n,n-1} & a_{nn} \end{vmatrix}$$

7.8. The trace of the k'th order of a square matrix is the sum of the determinants of all $\binom{n}{k}$ matrices of order $(k \times k)$ which can be formed by intersecting any k rows of $\mathbf{A}$ with the same k columns.

$$tr_k \mathbf{A} = \sum \begin{vmatrix} a_{i_1 i_1} & a_{i_1 i_2} & \cdots & a_{i_1 i_k} \\ a_{i_2 i_1} & a_{i_2 i_2} & \cdots & a_{i_2 i_k} \\ \cdot & \cdot & \cdots & \cdot \\ a_{i_k i_1} & a_{i_k i_2} & \cdots & a_{i_k i_k} \end{vmatrix}$$

where the sum extends over all combinations of n elements taken k at a time in order

$$i_1 < i_2 < \cdots < i_k.$$

7.9. Rules 7.2 and 7.3 (cyclic exchange) are valid for trace of k'th order.
7.10. $tr_n \mathbf{A} = |\mathbf{A}|$ if $\mathbf{A}$ is of order $(n \times n)$.

8. CHARACTERISTIC ROOTS AND VECTORS

8.1. If $\mathbf{A}$ is a square matrix of order $(n \times n)$, then $|\mathbf{A} - \lambda \mathbf{I}| = 0$ is called the characteristic equation of the matrix $\mathbf{A}$. It is a polynomial of the n'th degree in λ.
8.2. The n roots of the characteristic equation (not necessarily distinct) are called the characteristic roots of $\mathbf{A}$

$$ch(\mathbf{A}) = \lambda_1, \lambda_2, \ldots \lambda_n$$

8.3. The characteristic equation of $\mathbf{A}$ can be obtained by the relation

$$\lambda^n - (tr\mathbf{A})\lambda^{n-1} + (tr_2\mathbf{A})\lambda^{n-2} - (tr_3\mathbf{A})\lambda^{n-3} \cdots - (-1)^n(tr_{n-1}\mathbf{A})\lambda + (-1)^n|\mathbf{A}| = 0$$

where tr_k is defined in 7.8.

Example 8.3.1

$$\mathbf{A} = \begin{bmatrix} 25 & 30 & -10 \\ 30 & 40 & -6 \\ -10 & -6 & 17 \end{bmatrix}$$

$tr\mathbf{A} = 25 + 40 + 17 = 82$
$tr_2\mathbf{A} = (25 \times 40 - 30 \times 30) + (25 \times 17 - 10 \times 10) + (40 \times 17 - 6 \times 6) = 1069$
$tr_3\mathbf{A} = |\mathbf{A}| = 25 \times 4 \times 4 = 400$
    (cf. 6.14 and procedure stated in 4.11)
Hence

$$\lambda^3 - 82\lambda^2 + 1069\lambda - 400 = 0$$

The solutions (by Newton iteration) are

$$\lambda_1 = 65.86108$$
$$\lambda_2 = 15.75339$$
$$\lambda_3 = 0.38553$$

These are the characteristic roots of A.

8.4. $ch(A + \gamma I) = \gamma + ch(A)$

8.5. $ch(AB) = ch(BA)$

(except that AB or BA may have additional roots equal to zero).

8.6. $ch(A^{-1}) = 1/ch(A)$

8.7. If $\lambda_1, \lambda_2, \ldots \lambda_n$ are the roots of A then

$$\sum_i \lambda_i = tr A$$

$$\sum_{i<j} \lambda_i \lambda_j = tr_2 A$$

$$\sum_{i<j<k} \lambda_i \lambda_j \lambda_k = tr_3 A$$

$$\prod_i \lambda_i = |A|$$

8.8. If x' denotes the radius vector (running coordinates $[x, y, z]$) and if a matrix Q is positive-definite, then

$$(x' - x_0')Q^{-1}(x - x_0) = 1$$

is the equation of an ellipsoid with center at $[x_0, y_0, z_0] = x_0'$ and semi-axes equal to the square roots of the characteristic roots of Q.

8.9. The characteristic roots of a triangular (or diagonal) matrix are the diagonal elements of the matrix.

8.10. If A is a real matrix with positive roots, then

$$ch_{min}(AA') \leq [ch_{min}(A)]^2 \leq [ch_{max}(A)]^2 \leq ch_{max}(AA')$$

where ch_{min} denotes the smallest and ch_{max} the largest root.

8.11. The ratio of two quadratic forms (B non-singular)

$$u = \frac{x'Ax}{x'Bx}$$

attains stationary values at the roots of $B^{-1}A$. In particular

$$u_{max} = ch_{max}(B^{-1}A) \quad \text{and} \quad u_{min} = ch_{min}(B^{-1}A)$$

8.12. The equation system

$$Ax = \lambda x$$

permits non-zero solutions only if λ is one of the characteristic roots of A. Such a solution x is called a characteristic vector.

8.13. If x is a solution to 8.12, so is γx for an arbitrary scalar γ.

8.14. A solution x which has unit length ($x'x = 1$) is called the eigenvector associated with the characteristic root λ of A. The vector is frequently denoted by e.

8.15. A real symmetric matrix has real roots.

8.16. A matrix A is called positive-definite (abbreviated p.d.) if the quadratic form $x'Ax > 0$ for every $x \neq 0$.

8.17. A matrix **A** is called positive-semidefinite (abbreviated p.s.d.) if the quadratic form $x'Ax > 0$ and/or $x'Ax = 0$ for some $x \neq 0$.

8.18. A positive-definite real symmetric matrix has only positive characteristic roots.

8.19. If a real symmetric matrix is positive-semidefinite, it has no negative roots. The number of non-zero roots equals the rank of the matrix.

8.20. If all roots of a real symmetric matrix are distinct, the associated eigenvectors are distinct.

8.21. The matrix of eigenvectors

$$\mathbf{E} = [e_1, e_2, \ldots e_n]$$

of a real symmetric matrix is (or can be chosen to be) orthogonal.

8.22. $\mathbf{AE} = \mathbf{ED}_\lambda$.

8.23. For a real symmetric matrix, $\mathbf{A} = \mathbf{ED}_\lambda\mathbf{E}'$ (decomposition into matrices of unit rank)

$$\mathbf{E'AE} = \mathbf{D}_\lambda$$

where $\mathbf{D}_\lambda$ denotes the diagonal matrix of characteristic roots ordered in the same way as the eigenvector columns in **E**.

8.24. If $f(\lambda)$ is a polynomial in λ, then

$$f(\mathbf{A}) = \mathbf{ED}_{f(\lambda)}\mathbf{E}^{-1}$$

where λ are the characteristic roots of **A** and **E** is the matrix of associated eigenvectors. If **A** is symmetric, $\mathbf{E}^{-1} = \mathbf{E}'$.

Example 8.24.1

Consider the matrix in 8.3 (and 6.14).

$$\mathbf{A} = \begin{bmatrix} 25 & 30 & -10 \\ 30 & 40 & -6 \\ -10 & -6 & 17 \end{bmatrix}$$

The characteristic roots were found in Example 8.3.1,

$$\lambda_1 = 65.86108 \qquad \lambda_2 = 15.75339 \qquad \lambda_3 = 0.38553.$$

To find some **x** such that $\mathbf{Ax} = \lambda_1 \mathbf{x}$, we arbitrarily set the first element of **x** equal to 1. Using only the first two rows of **A** we solve the equation system

$$25 + 30x_2 - 10x_3 = 65.86108$$
$$30 + 40x_2 - 6x_3 = 65.86108x_2$$

which yields $x_2 = 1.24294$ and $x_3 = -0.35729$. Substitution of these values into the third equation

$$-10 - 6x_2 + 17x_3 = 65.86108x_3$$

yields zero to five decimal places, indicating the accuracy of the first characteristic root. To reduce to unit length the characteristic vector

$$[1 \qquad 1.24294 \qquad -0.35729]$$

we divide each element by

$$\sqrt{1 + 1.24294^2 + 0.35729^2}$$

and thus obtain the first eigenvector

$$[0.61170 \qquad 0.76030 \qquad -0.21855]$$

This, written as a column vector, is e_1. Repeating the same process for the second and third eigenvectors we obtain

$$e_2 = \begin{bmatrix} -0.08659 \\ 0.33896 \\ 0.93681 \end{bmatrix} \qquad e_3 = \begin{bmatrix} 0.78634 \\ -0.55412 \\ 0.27318 \end{bmatrix}$$

The three vectors can be placed into the eigenvector matrix **E**, which is easily seen to be orthogonal.

9.   CONDITIONAL INVERSES

9.1. Any matrix **A** (singular or non-singular, rectangular or square) has some conditional or generalized inverse $A^{(-1)}$ defined by the relation

$$AA^{(-1)}A = A.$$

9.2. If (and only if) **A** is square and non-singular, $A^{(-1)}$ is unique and equals A^{-1}. Otherwise there will be infinitely many matrices $A^{(-1)}$ which satisfy the defining relation 9.1.

9.3a. If **A** is rectangular ($n \times m$) of rank m, with $m < n$, then $A^{(-1)}$ is of order ($m \times n$) and $A^{(-1)}A = I(m \times m)$. Then $A^{(-1)}$ is called an inverse from the left. $AA^{(-1)} \neq I$ in this case.

9.3b. If **A** is rectangular ($n \times m$) of rank n, with $m > n$, then $A^{(-1)}$ is of order ($m \times n$) and $AA^{(-1)} = I(n \times n)$. Then $A^{(-1)}$ is called an inverse to the right. In this case,

$$A^{(-1)}A \neq I.$$

9.3c. For a square, singular matrix, $AA^{(-1)} \neq I$ and $A^{(-1)}A \neq I$.

Example 9.3.1

$$A = \begin{bmatrix} 3 \\ 2 \\ 1 \end{bmatrix}$$

The row vector $[1/3 \quad 0 \quad 0]$ is an inverse from the left. The row vector

$$[x \quad y \quad (1 - 3x - 2y)]$$

is a conditional inverse of the above matrix **A** for any values of x and y. It is called the generalized inverse of **A**.

Example 9.3.2

$$A = \begin{bmatrix} 1 & 2 & 3 \\ 2 & 5 & 6 \\ 3 & 7 & 9 \end{bmatrix}$$

A conditional inverse is

$$A^{(-1)} = \begin{bmatrix} 5 & -2 & 0 \\ -2 & 1 & 0 \\ 0 & 0 & 0 \end{bmatrix}$$

Here it was obtained by inversion of the basis (the 2×2 matrix in the upper left-hand corner) and replacement of the other elements by zeros.

9.4. A square matrix $\mathbf{A}$ is called idempotent if $\mathbf{A}\mathbf{A} = \mathbf{A}^2 = \mathbf{A}$.

9.5. $\mathbf{A}\mathbf{A}^{(-1)}$ and $\mathbf{A}^{(-1)}\mathbf{A}$ are idempotent.

9.6. All characteristic roots of idempotent matrices are either zero or one.

9.7. A system of linear equations (m equations in n unknowns)

$$\mathbf{A}\mathbf{x} = \mathbf{b}$$

is called consistent if there exists some solution x which satisfies the equation system.

Example 9.7.1

The system

$$x + y = 2$$
$$2x + 2y = 4 \qquad \text{is consistent.}$$

Example 9.7.2

The system

$$x + y = 2$$
$$2x + 2y = 5 \qquad \text{is inconsistent,}$$

for no pair of values (x, y) will satisfy this system.

9.8. If, in a system of equations (rectangular or square)

$$\mathbf{A}\mathbf{x} = \mathbf{b}$$

$\mathbf{A}\mathbf{A}^{(-1)}\mathbf{b} = \mathbf{b}$ for some conditional inverse $\mathbf{A}^{(-1)}$, then $\mathbf{A}\mathbf{A}^{(-1)}\mathbf{b} = \mathbf{b}$ for every conditional inverse of $\mathbf{A}$, and $\mathbf{A}\mathbf{x} = \mathbf{b}$ is consistent. Conversely, if $\mathbf{A}\mathbf{A}^{(-1)}\mathbf{b} \neq \mathbf{b}$ for some conditional inverse $\mathbf{A}^{(-1)}$ then $\mathbf{A}\mathbf{A}^{(-1)}\mathbf{b} \neq \mathbf{b}$ for every conditional inverse of $\mathbf{A}$, and $\mathbf{A}\mathbf{x} = \mathbf{b}$ is inconsistent.

9.9. If $\mathbf{A}\mathbf{x} = \mathbf{b}$ is consistent, then $\mathbf{x} = \mathbf{A}^{(-1)}\mathbf{b}$ is a solution (generally a different one for each $\mathbf{A}^{(-1)}$).

9.10. Let $\mathbf{y}$ ($p \times 1$) be a set of linear functions of the solutions $\mathbf{x}$ ($n \times 1$) of a consistent system of equations $\mathbf{A}\mathbf{x} = \mathbf{b}$, given by the relation $\mathbf{y} = \mathbf{C}\mathbf{x}$. Then $\mathbf{y} = \mathbf{C}\mathbf{x}$ is called unique if the same values of $\mathbf{y}$ will result regardless which solution x is used.

Example 9.10.1

$$3x + 4y + 5z = 22$$
$$x + y + z = 6$$

is a consistent system. One solution would be

$$x = 3 \quad y = 2 \quad z = 1$$

Another solution is

$$x = 2 \quad y = 4 \quad z = 0$$

The linear function

$$\begin{bmatrix} 7 & 9 & 11 \end{bmatrix} \begin{bmatrix} x \\ y \\ z \end{bmatrix} = u$$

($7x + 9y + 11z = u$) will have the same value (50) regardless which of the two (or any other) solutions is substituted. Thus u is unique.

9.11. Let $\mathbf{Ax} = \mathbf{b}$ be a consistent system of equations. For $\mathbf{Cx} = \mathbf{y}$ to be a unique linear combination of the solution $\mathbf{x}$, it is necessary and sufficient that $\mathbf{CA}^{(-1)}\mathbf{A} = \mathbf{C}$. If this relation holds for some $\mathbf{A}^{(-1)}$ it will hold for every conditional inverse of $\mathbf{A}$. If it is violated for some $\mathbf{A}^{(-1)}$ it will be violated for every $\mathbf{A}^{(-1)}$, and y will be non-unique.

9.12. Let $\mathbf{A}$ be of rank r and select r rows and r columns which form a basis of $\mathbf{A}$. Then a conditional inverse of $\mathbf{A}$ can be obtained as follows: Invert the $(r \times r)$ matrix, place the inverse (without transposing) into the r rows corresponding to the column numbers and the r columns corresponding to the row numbers of the basis, and place zero into all remaining elements. Thus, if $\mathbf{A}$ is of order (5×4) and rank 3, and if rows 1, 2, 4 and columns 2, 3, 4 are selected as a basis, $\mathbf{A}^{(-1)}$, of order (4×5) will contain the inverse elements of the basis in rows 2, 3, 4 and columns 1, 2, 4, and zeros elsewhere. (See example 9.3.2)

9.13. If $\mathbf{A}$ is a square, singular matrix of order $(n \times n)$ and rank r, let $\mathbf{M}$ be a matrix of order $[n \times (n - r)]$ and $\mathbf{K}$ another matrix of order $[(n - r) \times n]$ chosen in such a way that $\mathbf{A} + \mathbf{MK}$ is non-singular. Then $(\mathbf{A} + \mathbf{MK})^{-1}$ is a conditional inverse of $\mathbf{A}$.

Example 9.13.1

$$\mathbf{A} = \begin{bmatrix} 3 & -1 & -1 & -1 \\ -1 & 3 & -1 & -1 \\ -1 & -1 & 3 & -1 \\ -1 & -1 & -1 & 3 \end{bmatrix}$$

is of order (4×4) and rank 3. Take $\mathbf{M} = \mathbf{j}$ (column vector of ones) and $\mathbf{K} = \mathbf{j}'$ (row vector of ones). Then $\mathbf{A} + \mathbf{MK} = \mathbf{A} + \mathbf{jj}' = 4\mathbf{I}$. Hence $(1/4)\,\mathbf{I}$ is a conditional inverse of $\mathbf{A}$.

9.14. The "Doolittle" method (see 6.14) can be employed to obtain a conditional inverse of a symmetric matrix. If, at any stage, the leading element of the p-row is zero, that cycle is disregarded.

Example 9.14.1

Invert, conditionally, the matrix

$$\mathbf{A} = \begin{bmatrix} 4 & 2 & -2 & 4 \\ 2 & 17 & 11 & 6 \\ -2 & 11 & 10 & 1 \\ 4 & 6 & 1 & 30 \end{bmatrix}$$

4	2	−2	4	1			
1	.5	−.5	1	.25			
	17	11	6	0	1		
	16	12	4	−.5	1		
	1	.75	.25	−.03125	.0625		
		10	1	0	0	1	
		0					
			30	0	0	0	1
			25	−.875	−.25	0	1
			1	−.035	−.01	0	.04

ALGEBRA

$$\begin{bmatrix} .29625 & -.0225 & 0 & -.035 \\ -.0225 & .065 & 0 & -.01 \\ 0 & 0 & 0 & 0 \\ -.035 & -.01 & 0 & .04 \end{bmatrix} = \mathbf{A}^{(-1)}$$

10. MATRIX DIFFERENTIATION

10.1a. If the elements of a matrix $\mathbf{Y}$ $(m \times n)$ are functions of a scalar, x, the expression

$$\partial \mathbf{Y}/\partial x$$

denotes a matrix of order $(m \times n)$ with elements $\partial y_{ij}/\partial x$.

10.1b. If the elements of a column (row) vector $\mathbf{y}$ $(\mathbf{y}')$ are functions of a scalar, x, the expression

$$\partial \mathbf{y}/\partial x \quad (\partial \mathbf{y}'/\partial x)$$

denotes a column (row) vector with elements $\partial y_i/\partial x$.

10.2a. If y is a scalar function of $m \times n$ variables, x_{ij}, arranged into a matrix $\mathbf{X}$, the expression

$$\partial y/\partial \mathbf{X}$$

denotes a matrix with elements $\partial y/\partial x_{ij}$.

(Note: Partial differentiation is performed with respect to the element in row i and column j of $\mathbf{X}$. If the same x-variable occurs in another place as, e.g., in a symmetric matrix, differentiation with respect to the distinct (repeated) variable is performed in two stages.)

Example 10.2.1

If $y = \mathbf{j}'\mathbf{X}\mathbf{j}$ (sum of all elements of a square matrix), $\partial y/\partial \mathbf{X}$ is a matrix of ones. If $\mathbf{X}$ is symmetric, one can introduce a new notation $x_{ij} = x_{ji} = z_{ij}$. Then

$$\begin{aligned} \partial y/\partial z_{ij} &= (\partial y/\partial x_{ij})(\partial x_{ij}/\partial z_{ij}) \\ &+ (\partial y/\partial x_{ji})(\partial x_{ji}/\partial z_{ij}) \\ &= 1 + 1 = 2 \quad (\text{if } i \neq j) \\ &= 1 \quad\quad\quad\ (\text{if } i = j) \end{aligned}$$

10.2b. If y is a scalar function of n variables, x_i, arranged into a column (row) vector $\mathbf{x}$ $(\mathbf{x}')$, the expression

$$\partial y/\partial \mathbf{x} \quad (\partial y/\partial \mathbf{x}')$$

denotes a column (row) vector with elements $\partial y/\partial x_i$.

10.3. If $\mathbf{y}$ is a column vector with m elements, each a function of n variables, x_i, arranged into a row vector $\mathbf{x}'$, the expression $\partial \mathbf{y}/\partial \mathbf{x}'$ denotes a matrix with m rows and n columns, with elements $\partial y_i/\partial x_j$.

10.4. $\partial \mathbf{Y}/\partial y_{ij} = (\mathbf{EL})_{ij}$ (see definition of $(\mathbf{EL})$ in 3.4).

10.5. $\partial \mathbf{U}\mathbf{V}/\partial x = (\partial \mathbf{U}/\partial x)\mathbf{V} + \mathbf{U}(\partial \mathbf{V}/\partial x)$.

10.6. $\partial \mathbf{A}\mathbf{Y}/\partial x = \mathbf{A}(\partial \mathbf{Y}/\partial x)$ (if elements of $\mathbf{A}$ are not functions of x).

10.7. $\partial \mathbf{Y}'/\partial y_{ij} = (\mathbf{EL})_{ji}$

10.8. $\partial \mathbf{A}'\mathbf{Y}\mathbf{A}/\partial x = \mathbf{A}'(\partial \mathbf{Y}/\partial x)\mathbf{A}$

10.9. $\partial \mathbf{Y}'\mathbf{A}\mathbf{Y}/\partial x = (\partial \mathbf{Y}'/\partial x)\mathbf{A}\mathbf{Y} + \mathbf{Y}'\mathbf{A}(\partial \mathbf{Y}/\partial x)$

10.10. $\partial \mathbf{a}'\mathbf{x}/\partial \mathbf{x} = \mathbf{a}$

10.11. $\partial \mathbf{x}'\mathbf{x}/\partial \mathbf{x} = 2\mathbf{x}$

10.12. $\partial x'Ax/\partial x = Ax + A'x$

10.13. (Chain Rule No. 1) $\partial y/\partial x' = (\partial y/\partial z')(\partial z/\partial x')$

10.14. $\partial Ax/\partial x' = A$

10.15. $\partial tr X/\partial X = I$

10.16. $\partial tr AX/\partial X = \partial tr XA/\partial X = A'$

10.17. $\partial tr AXB/\partial X = A'B'$

10.18. $\partial tr X'AX/\partial X = AX + A'X$

10.19. $\partial \log |X|/\partial X = (X')^{-1}$ (log to base e)

10.20. $\partial Y^{-1}/\partial x = -Y^{-1}(\partial Y/\partial x)Y^{-1}$

10.21. (Chain Rule No. 2)

$$\partial y/\partial x = tr(\partial y/\partial Z)(\partial Z'/\partial x)$$

where y and x are scalars. The scalar y is a function of $m \times n$ variables z_{ij}, and each of the z_{ij} is a function of x.

Example 10.21.1

Obtain $\log |R - FF'|/\partial F$, where R is symmetric.

By Chain Rule No. 2:

$\partial \log |R - FF'|/\partial f_{ij}$

$= tr[\partial \log |R - FF'|/\partial (R - FF')][\partial (R - FF')/\partial f_{ij}]$ (since R and FF' are symmetric)

$= tr(R - FF')^{-1}[\partial (R - FF')/\partial f_{ij}]$ (by 10.19)

$= tr(R - FF')^{-1}[-(\partial F/\partial f_{ij})F' - F(\partial F'/\partial f_{ij})]$ (by 10.5)

$= tr(R - FF')^{-1}[-(EL)_{ij}F' - F(EL)_{ji}]$ (by 10.4 and 10.7)

$= -tr(R - FF')^{-1}(EL)_{ij}F' - tr(R - FF')^{-1}F(EL)_{ji}$

$= -tr(F.I.)_{ij}F'(R - FF')^{-1} - tr(EL)_{ji}(R - FF')^{-1}F$ (by 7.3)

$= -[F'(R - FF')^{-1}]_{ji} - [(R - FF')^{-1}F]_{ij}$,

where $[\ ]_{ij}$ denotes the (i,j) element of the matrix in brackets (by 7.5),

$= -[(R - FF')^{-1}F]_{ij} - [(R - FF')^{-1}F]_{ij}$ (since $R - FF'$ is symmetric)

$= -2[(R - FF')^{-1}F]_{ij}$.

Hence, by definition 10.2a

$$\partial \log |R - FF'|/\partial F = -2(R - FF')^{-1}F.$$

10.22. $|\partial y/\partial x'| = J(y;x)$ is called the Jacobian or Functional Determinant used in variable transformation of multiple integrals. Formally, if y is a column vector with m elements, each a function of m variables x_i arranged into a row vector x',

$$dx_1 dx_2 \ldots dx_m = |\partial y/\partial x'|^{-1} dy_1 dy_2 \ldots dy_m.$$

10.23. For a scalar y (a function of m variables x_i) to attain a stationary value, it is necessary that

$$\partial y/\partial x = 0.$$

10.24. For a stationary value to be a minimum (maximum) it is necessary that

$$\partial (\partial y/\partial x)/\partial x'   (-\partial (\partial y/\partial x)/\partial x')$$

be a positive-definite matrix for the value of x satisfying 10.23.

Example 10.24.1

Find the values of β which minimize $u = x'x$ (the sum of squares of x_i) where $x = y - A\beta$ (with y and A known and fixed).

$$\partial u/\partial \beta' = (\partial u/\partial x')(\partial x/\partial \beta') \quad \text{(by Chain Rule No. 1)}$$
$$= -2x'A \quad \text{(by 10.11 and 10.14)}$$

Hence

$$\partial u/\partial \beta = -2A'x$$
$$= -2A'(y - A\beta).$$

Hence, for a stationary value, by 10.23, it is necessary that

$$A'A\hat{\beta} = A'y$$

where $\hat{\beta}$ denotes the values which make u stationary. Now,

$$\partial(\partial u/\partial \beta)/\partial \beta' = 2\partial(A'A\beta)/\partial \beta' = 2A'A.$$

If **A** has real elements, and if **A'A** is non-singular, then it is positive-definite (since, given an arbitrary real $x \neq 0$, $x'A'Ax = z'z$, with $z = Ax$; thus this is a sum of squares). Hence $\hat{\beta}$ minimizes u.

10.25. (Generalized Newton Iteration)

Let x_0' be an initial estimate (m elements) of the roots of the m equations

$$f(x') = 0$$

where the m elements of the column vector **f** are each functions of $x_1, x_2, \ldots x_m$. Then an improved root is

$$x_1 = x_0 - Q_0^{-1}f(x_0'),$$

where Q_0 is the matrix of derivatives $\partial f/\partial x'$ evaluated at $x = x_0$. The usual procedure consists of evaluating $f(x_0')$, then solving $Q_0 u = f(x_0')$ for **u**. Then $x_1 = x_0 - u$.

Example 10.25.1

Solve

$$f_1(x, y) = x^3 - x^2y + y^2 - 3.526 = 0$$
$$f_2(x, y) = x^3 + y^3 - 14.911 = 0$$

$$Q = \begin{bmatrix} 3x^2 - 2xy & 2y - x^2 \\ 3x^2 & 3y^2 \end{bmatrix}$$

Take $x_0 = 1, \quad y_0 = 2$

$$f_1(x_0, y_0) = -0.526$$
$$f_2(x_0, y_0) = -5.911$$

$$Q_0 = \begin{bmatrix} -1 & 3 \\ 3 & 12 \end{bmatrix}$$

$$-u + 3v = -0.526$$
$$3u + 12v = -5.911$$

yields $u = -0.55, \quad v = -0.36.$

Then,

$$x_1 = x_0 - u = 1.55$$
$$y_1 = y_0 - v = 2.36$$
$$f_1(x_1, y_1) = 0.0976$$
$$f_2(x_1, y_1) = 1.9572$$

$$\mathbf{Q}_1 = \begin{bmatrix} -0.1085 & 2.3175 \\ 7.2075 & 16.7088 \end{bmatrix}$$

$$-0.1085u + 2.3175v = 0.0976$$
$$7.2075u + 16.7088v = 1.9572$$

yields $u = 0.157$, $v = 0.049$.

Then

$$x_2 = x_1 - u = 1.393$$
$$y_2 = y_1 - v = 2.311$$
$$f_1(x_2, y_2) = 0.03337$$
$$f_2(x_2, y_2) = 0.13443$$

$$\mathbf{Q}_2 = \begin{bmatrix} -0.61710 & 2.68155 \\ 5.82135 & 16.02216 \end{bmatrix}$$

$$-0.61710u + 2.68155v = 0.03337$$
$$5.82135u + 16.02216v = 0.13443$$

yields $u = -0.0068$, $v = 0.0109$.

Then,

$$x_3 = x_2 - u = 1.3998$$

$$y_3 = y_2 - v = 2.3001$$

(The exact roots are $x = 1.4$ and $y = 2.3$).

11. STATISTICAL MATRIX FORMS

11.1. Let E denote the expectation operator, and let $\mathbf{y}$ be a set of p random variables. Then

$$E(\mathbf{y}) = \boldsymbol{\mu}$$

states that $E(y_i) = \mu_i$ $(i = 1, 2, \ldots p)$.

11.2. Let var denote variance. Then

$$\text{var}(\mathbf{y}) = \boldsymbol{\Sigma}$$

denotes a $p \times p$ symmetric matrix whose elements are $\text{cov}(y_i, y_j)$, and whose diagonal elements are $\text{var}(y_i)$, where cov denotes covariance.

11.3. $E(\mathbf{Ay} + \mathbf{b}) = \mathbf{A}E(\mathbf{y}) + \mathbf{b} = \mathbf{A}\boldsymbol{\mu} + \mathbf{b}$

11.4. $\text{var}(\mathbf{Ay} + \mathbf{b}) = \mathbf{A}\,\text{var}(\mathbf{y})\,\mathbf{A}' = \mathbf{A}\boldsymbol{\Sigma}\mathbf{A}'$

11.5. $\text{cov}(\mathbf{y}, \mathbf{z}')$ denotes a matrix with elements $\text{cov}(y_i, z_j)$.

$\text{cov}(\mathbf{z}, \mathbf{y}') = [\text{cov}(\mathbf{y}, \mathbf{z}')]'$.

11.6. $\text{cov}(\mathbf{Ay} + \mathbf{b}, \mathbf{z}'\mathbf{C} + \mathbf{d}') = \mathbf{A}\,\text{cov}(\mathbf{y}, \mathbf{z}')\,\mathbf{C}$

11.7. $\text{var}(\mathbf{y}) = E(\mathbf{yy}') - E(\mathbf{y})E(\mathbf{y}')$

11.8. $\text{cov}(\mathbf{y}, \mathbf{z}') = E(\mathbf{yz}') - E(\mathbf{y})E(\mathbf{z}')$

11.9. (Expected "sum of squares")

$$E(\mathbf{y}'\mathbf{Qy}) = tr[\mathbf{Q}\,\text{var}(\mathbf{y})] + E(\mathbf{y}')\mathbf{Q}E(\mathbf{y}).$$

11.10. If a matrix $\mathbf{Q}$ is symmetric and positive-definite, one can find a lower triangular matrix $\mathbf{T}$ (with positive diagonal terms, for uniqueness) such that $\mathbf{TT}' = \mathbf{Q}$. The matrices $\mathbf{T}$ and $\mathbf{T}^{-1}$ can be obtained from the Doolittle pattern (6.14) (Gauss elimination or square-root method) as follows: In each cycle, divide the p-row (left and right hand side) by $\sqrt{p_{ii}}$ (instead of p_{ii} for the u-row). Thus obtain rows designated as t-rows. The left-hand side (Arabic subscripts) is $\mathbf{T}'$, and the right-hand side (Roman subscripts) is $\mathbf{T}^{-1}$

11.11. If a coordinate system $\mathbf{x}$ is oblique, and if the cosines between reference vectors (scalar products of basis vectors of unit length) are stated in a symmetric matrix $\mathbf{Q}$, then $\mathbf{T}^{-1}\mathbf{x} = \mathbf{y}$ is an orthogonal system, where $\mathbf{T}$ is obtained from $\mathbf{Q}$ by 11.10.

11.12. The likelihood function of a sample of size n from a multivariate normal distribution (p responses), with common variance-covariance matrix $\Sigma(p \times p)$, and with means or main effects replaced by maximum-likelihood or least-squares estimates, can be written as

$$\log L = -\frac{np}{2} \log 2\pi - \frac{n}{2} \log |\Sigma| - \frac{n}{2} tr \Sigma^{-1}\mathbf{S}$$

where $\Sigma(p \times p)$ is the common variance-covariance matrix, and $\mathbf{S}$ its maximum-likelihood estimate (matrix of sums of squares and products due to error, divided by sample size n).

11.13. If Σ has a structure under a model or null hypothesis, and if elements of Σ are to be estimated, by maximum-likelihood, two cases can be distinguished: (11.14) Σ^{-1} has the same structure (intraclass correlation, mixed model, compound symmetry, factor analysis). (11.15) Σ^{-1} has a different structure (autocorrelation, Simplex structure).

11.14. If the structure of Σ and Σ^{-1} is identical, and if u and v are elements (or functions of elements) of Σ^{-1} then estimates of Σ can be obtained from the relations (usually requiring Newton iteration, see 10.25):

$$\partial \log L/\partial u = \frac{n}{2} tr \mathbf{A}(\Sigma - \mathbf{S})$$

where $\mathbf{A} = \partial \Sigma^{-1}/\partial u$, is frequently an elementary matrix (see 3.4 and, especially, the rules 7.5 and 7.6).

$$\partial^2 \log L/\partial u \partial v = \frac{n}{2} tr(\partial \mathbf{A}/\partial v)(\Sigma - \mathbf{S}) + \frac{n}{2} tr \mathbf{A}\Sigma^{-1}\mathbf{B}\Sigma^{-1}$$

where $\mathbf{B} = \partial \Sigma^{-1}/\partial v$. These rules are useful to obtain Newton iterations and asymptotic variance-covariance matrices of the estimates.

11.15. If the structures of Σ and Σ^{-1} are different, then an estimate of Σ can be obtained from the relations

$$\partial \log L/\partial x = -\frac{n}{2} tr \mathbf{A}(\Sigma^{-1} - \mathbf{Q}),$$

where

$$\mathbf{Q} = \Sigma^{-1}\mathbf{S}\Sigma^{-1}$$

and

$$\mathbf{A} = \partial \Sigma/\partial x \qquad \text{(see comments in 11.14)}.$$

$$\partial^2 \log L/\partial x \partial y = -\frac{n}{2} tr(\partial \mathbf{A}/\partial y)(\Sigma^{-1} - \mathbf{Q}) + \frac{n}{2} tr \mathbf{A}\Sigma^{-1}\mathbf{B}(\Sigma^{-1} - \mathbf{Q}) - \frac{n}{2} tr \mathbf{A}\mathbf{Q}\mathbf{B}\Sigma^{-1},$$

where

$$\mathbf{B} = \partial \Sigma/\partial y$$

x and y are elements (or functions of elements) of Σ. The comments of 11.14 apply, but the iterative procedure is considerably more complex.

Suggestions for further reading:

A. S. Householder, *The Theory of Matrices in Numerical Analysis*, Blaisdell, 1964.
S. R. Searle, *Matrix Algebra for the Biological Sciences*, Wiley, 1966.
P. H. Schonemann, Matrix differentiation of traces and determinants, *Psychometrika*, 1966.

POSITIVE POWERS OF TWO

n	2ⁿ
1	2
2	4
3	8
4	16
5	32
6	64
7	128
8	256
9	512
10	1024
11	2048
12	4096
13	8192
14	16384
15	32768
16	65536
17	13107 2
18	26214 4
19	52428 8
20	10485 76
21	20971 52
22	41943 04
23	83886 08
24	16777 216
25	33554 432
26	67108 864
27	13421 7728
28	26843 5456
29	53687 0912
30	10737 41824
31	21474 83648
32	42949 67296
33	85899 34592
34	17179 86918 4
35	34359 73836 8
36	68719 47673 0
37	13743 89534 72
38	27487 79069 44
39	54975 58138 88
40	10995 11627 776
41	21990 23255 552
42	43980 46511 104
43	87960 93022 208
44	17592 18604 4416
45	35184 37208 8832
46	70368 74417 7664
47	14073 74883 55328
48	28147 49767 10656
49	56294 99534 21312
50	11258 99906 84262 4

n	2ⁿ
51	22517 99813 68524 8
52	45035 99627 37049 6
53	90071 99254 74099 2
54	18014 39850 94819 84
55	36028 79701 89639 68
56	72057 59403 79279 36
57	14411 51880 75855 872
58	28823 03761 51711 744
59	57646 07523 03423 488
60	11529 21504 60684 6976
61	23058 43009 21369 3952
62	46116 86018 42738 7904
63	92233 72036 85477 5808
64	18446 74407 37095 51616
65	36893 48814 74191 03232
66	73786 97629 48382 06464
67	14757 39525 89676 41292 8
68	29514 79051 79352 82585 6
69	59029 58103 58705 65171 2
70	11805 91620 71741 13034 24
71	23611 83241 43482 26068 48
72	47223 66482 86964 52136 96
73	94447 32965 73929 04273 92
74	18889 46593 14785 80854 784
75	37778 93186 29571 61709 568
76	75557 86372 59143 23419 136
77	15111 57274 51828 64683 8272
78	30223 14549 03657 29367 6544
79	60446 29098 07314 58735 3088
80	12089 25819 61462 91747 06176
81	24178 51639 22925 83494 12352
82	48357 03278 45851 66988 24704
83	96714 06556 91703 33976 49408
84	19342 81311 38340 66795 29881 6
85	38685 62622 76681 33590 59763 2
86	77371 25245 53362 67181 19526 4
87	15474 25049 10672 53436 23905 28
88	30048 50098 21345 06872 47810 56
89	61897 00196 42690 13744 95621 12
90	12379 40039 28538 02748 99124 224
91	24758 80078 57076 05497 98248 448
92	49517 60157 14152 10995 96496 896
93	99035 20314 28304 21991 92993 792
94	19807 04062 85660 84398 38598 7584
95	39614 08125 71321 68796 77197 5168
96	79228 16251 42643 37593 54395 0336
97	15845 63250 28528 67518 70879 00672
98	31691 26500 57057 35037 41758 01344
99	63382 53001 14114 70074 83516 02688
100	12676 50600 22822 94014 96703 20537 6
101	25353 01200 45645 88029 93406 41075 2

COMBINATORIAL ANALYSIS

NEGATIVE POWERS OF TWO

n	2^{-n}
0	1.0
1	0.5
2	0.25
3	0.125
4	0.0625
5	0.03125
6	0.01562 5
7	0.00781 25
8	0.00390 625
9	0.00195 3125
10	0.00097 65625
11	0.00048 82812 5
12	0.00024 41406 25
13	0.00012 20703 125
14	0.00006 10351 5625
15	0.00003 05175 78125
16	0.00001 52587 89062 5
17	0.00000 76293 94531 25
18	0.00000 38146 97265 625
19	0.00000 19073 48632 8125
20	0.00000 09536 74316 40625
21	0.00000 04768 37158 20312 5
22	0.00000 02384 18579 10156 25
23	0.00000 01192 09289 55078 125
24	0.00000 00596 04644 77539 0625
25	0.00000 00298 02322 38769 53125
26	0.00000 00149 01161 19384 76562 5
27	0.00000 00074 50580 59692 38281 25
28	0.00000 00037 25290 29846 19140 625
29	0.00000 00018 62645 14923 09570 3125
30	0.00000 00009 31322 57461 54785 15625
31	0.00000 00004 65661 28730 77392 57812 5
32	0.00000 00002 32830 64365 38696 28906 25
33	0.00000 00001 16415 32182 69348 14453 125
34	0.00000 00000 58207 66091 34674 07226 5625
35	0.00000 00000 29103 83045 67337 03613 28125
36	0.00000 00000 14551 91522 83668 51806 64062 5
37	0.00000 00000 07275 95761 41834 25903 32031 25
38	0.00000 00000 03637 97880 70917 12951 66015 625
39	0.00000 00000 01818 98940 35458 56475 83007 8125
40	0.00000 00000 00909 49470 17729 28237 91503 90625
41	0.00000 00000 00454 74735 08864 64118 95751 95312 5
42	0.00000 00000 00227 37367 54432 32059 47875 97656 25
43	0.00000 00000 00113 68683 77216 16029 73937 98828 125
44	0.00000 00000 00056 84341 88608 08014 86968 99414 0625
45	0.00000 00000 00028 42170 94304 04007 43484 49707 03125
46	0.00000 00000 00014 21085 47152 02003 71742 24853 51562 5
47	0.00000 00000 00007 10542 73576 01001 85871 12426 75781 25
48	0.00000 00000 00003 55271 36788 00500 92935 56213 37890 625
49	0.00000 00000 00001 77635 68394 00250 46467 78106 68945 3125
50	0.00000 00000 00000 88817 84197 00125 23233 89053 34472 65625

SUMS OF POWERS OF INTEGERS, $\sum_{k=1}^{n} k^m$

$(m = 1, 2, 3, 4); 1 \leq n \leq 40$

n	Σk	Σk^2	Σk^3	Σk^4
1	1	1	1	1
2	3	5	9	17
3	6	14	36	98
4	10	30	100	354
5	15	55	225	979
6	21	91	441	2275
7	28	140	784	4676
8	36	204	1296	8772
9	45	285	2025	15333
10	55	385	3025	25333
11	66	506	4356	39974
12	78	650	6084	60710
13	91	819	8281	89271
14	105	1015	11025	127687
15	120	1240	14400	178312
16	136	1496	18496	243848
17	153	1785	23409	327369
18	171	2109	29241	432345
19	190	2470	36100	562666
20	210	2870	44100	722666
21	231	3311	53361	917147
22	253	3795	64009	1151403
23	276	4324	76176	1431244
24	300	4900	90000	1763020
25	325	5525	105625	2153645
26	351	6201	123201	2610621
27	378	6930	142884	3142062
28	406	7714	164836	3756718
29	435	8555	189225	4463999
30	465	9455	216225	5273999
31	496	10416	246016	6197520
32	528	11440	278784	7246096
33	561	12529	314721	8432017
34	595	13685	354025	9768353
35	630	14910	396900	11268978
36	666	16206	443556	12948594
37	703	17575	494209	14822755
38	741	19019	549081	16907891
39	780	20540	608400	19221332
40	820	22140	672400	21781332

SUMS OF POWERS OF THE FIRST n INTEGERS

$$\sum_{k=1}^{n} k = 1 + 2 + 3 + \cdots + n = \frac{n(n + 1)}{2}$$

$$\sum_{k=1}^{n} k^2 = 1^2 + 2^2 + 3^2 + \cdots + n^2 = \frac{n(n + 1)(2n + 1)}{6}$$

$$\sum_{k=1}^{n} k^3 = \frac{n^2(n + 1)^2}{4}$$

$$\sum_{k=1}^{n} k^4 = \frac{n}{30}(n + 1)(2n + 1)(3n^2 + 3n - 1).$$

$$\sum_{k=1}^{n} k^5 = \frac{n^2}{12}(n + 1)^2(2n^2 + 2n - 1).$$

$$\sum_{k=1}^{n} k^6 = \frac{n}{42}(n + 1)(2n + 1)(3n^4 + 6n^3 - 3n + 1).$$

$$\sum_{k=1}^{n} k^7 = \frac{n^2}{24}(n + 1)^2(3n^4 + 6n^3 - n^2 - 4n + 2).$$

$$\sum_{k=1}^{n} k^8 = \frac{n}{90}(n + 1)(2n + 1)(5n^6 + 15n^5 + 5n^4 - 15n^3 - n^2 + 9n - 3).$$

$$\sum_{k=1}^{n} k^9 = \frac{n^2}{20}(n + 1)^2(2n^6 + 6n^5 + n^4 - 8n^3 + n^2 + 6n - 3).$$

$$\sum_{k=1}^{n} k^{10} = \frac{n}{66}(n + 1)(2n + 1)(3n^8 + 12n^7 + 8n^6 - 18n^5$$
$$- 10n^4 + 24n^3 + 2n^2 - 15n + 5).$$

Note that

$$\sum_{k=1}^{n} k^p = 1^p + 2^p + 3^p + \cdots + n^p \text{ is a function of } n \text{ which can be conveniently generated}$$

by use of the *following proposition*

If
$$\sum_{k=1}^{n} k^p = a_1 n^{p+1} + a_2 n^p + a_3 n^{p-1} + \cdots + a_{p+1} n$$

then

$$\sum_{k=1}^{n} k^{p+1} = \frac{p + 1}{p + 2} a_1 n^{p+2} + \frac{p + 1}{p + 1} a_2 n^{p+1} + \frac{p + 1}{p} a_3 n^p$$
$$+ \cdots + \frac{p + 1}{2} a_{p+1} n^2 + \left[1 - (p + 1) \sum_{k=1}^{p+1} \frac{a_k}{(p + 3 - k)} \right] n$$

Example Since $\sum_{k=1}^{n} k = \frac{1}{2}n^2 + \frac{1}{2}n$, then

$$\sum_{k=1}^{n} k^2 = \frac{1}{3}n^3 + \frac{1}{2}n^2 + \frac{1}{6}n \text{ and from this result}$$

$$\sum_{k=1}^{n} k^3 = \frac{1}{4}n^4 + \frac{1}{2}n^3 + \frac{1}{4}n^2 \quad \text{etc.}$$

This proposition is extracted from a paper written by Michael A. Budin and Arnold J. Cantor entitled "Simplified Computation of Sums of Powers of Integers."

COMBINATORIAL ANALYSIS
SUMS OF RECIPROCAL POWERS*

n	$\zeta(n) = \sum\limits_{k=1}^{\infty} k^{-n}$			$\sum\limits_{k=1}^{\infty} (-1)^{k-1} k^{-n}$		
1	∞			0.69314	71805	59945	30942
2	1.64493	40668	48226	43637	0.82246	70334	24113	21824
3	1.20205	69031	59594	28540	0.90154	26773	69695	71405
4	1.08232	32337	11138	19152	0.94703	28294	97245	91758
5	1.03692	77551	43369	92633	0.97211	97704	46909	30594
6	1.01734	30619	84449	13971	0.98555	10912	97435	10410
7	1.00834	92773	81922	82684	0.99259	38199	22830	28267
8	1.00407	73561	97944	33938	0.99623	30018	52647	89923
9	1.00200	83928	26082	21442	0.99809	42975	41605	33077
10	1.00099	45751	27818	08534	0.99903	95075	98271	56564
11	1.00049	41886	04119	46456	0.99951	71434	98060	75414
12	1.00024	60865	53308	04830	0.99975	76851	43858	19085
13	1.00012	27133	47578	48915	0.99987	85427	63265	11549
14	1.00006	12481	35058	70483	0.99993	91703	45979	71817
15	1.00003	05882	36307	02049	0.99996	95512	13099	23808
16	1.00001	52822	59408	65187	0.99998	47642	14906	10644
17	1.00000	76371	97637	89976	0.99999	23782	92041	01198
18	1.00000	38172	93264	99984	0.99999	61878	69610	11348
19	1.00000	19082	12716	55394	0.99999	80935	08171	67511
20	1.00000	09539	62033	87280	0.99999	90466	11581	52212
21	1.00000	04769	32986	78781	0.99999	95232	58215	54282
22	1.00000	02384	50502	72773	0.99999	97616	13230	82255
23	1.00000	01192	19925	96531	0.99999	98808	01318	43950
24	1.00000	00596	08189	05126	0.99999	99403	98892	39463
25	1.00000	00298	03503	51465	0.99999	99701	98856	96283
26	1.00000	00149	01554	82837	0.99999	99850	99231	99657
27	1.00000	00074	50711	78984	0.99999	99925	49550	48496
28	1.00000	00037	25334	02479	0.99999	99962	74753	40011
29	1.00000	00018	62659	72351	0.99999	99981	37369	41811
30	1.00000	00009	31327	43242	0.99999	99990	68682	28145
31	1.00000	00004	65662	90650	0.99999	99995	34340	33145
32	1.00000	00002	32831	18337	0.99999	99997	67169	89595
33	1.00000	00001	16415	50173	0.99999	99998	83584	85805
34	1.00000	00000	58207	72088	0.99999	99999	41792	39905
35	1.00000	00000	29103	85044	0.99999	99999	70896	18953
36	1.00000	00000	14551	92189	0.99999	99999	85448	09143
37	1.00000	00000	07275	95984	0.99999	99999	92724	04461
38	1.00000	00000	03637	97955	0.99999	99999	96362	02193
39	1.00000	00000	01818	98965	0.99999	99999	98181	01084
40	1.00000	00000	00909	49478	0.99999	99999	99090	50538
41	1.00000	00000	00454	74738	0.99999	99999	99545	25268
42	1.00000	00000	00227	37368	0.99999	99999	99772	62633

For $n > 42$, $\sum\limits_{k=1}^{\infty} k^{-(n+1)} = \dfrac{1}{2}\left[1 + \sum\limits_{k=1}^{\infty} k^{-n}\right]$, $\sum\limits_{k=1}^{\infty} (-1)^{k-1} k^{-(n+1)} = \dfrac{1}{2}\left[1 + \sum\limits_{k=1}^{\infty} (-1)^{k-1} k^{-n}\right]$

*Note: By definition Riemann's Zeta Function is $\zeta(p) = \text{Zeta }(p) = 1 + \dfrac{1}{2^p} + \dfrac{1}{3^p} + \dfrac{1}{4^p} + \ldots$

SUMS OF RECIPROCAL POWERS

n	$\sum_{k=0}^{\infty} (2k+1)^{-n}$				$\sum_{k=0}^{\infty} (-1)^k (2k+1)^{-n}$			
1		∞			0.78539	81633	97448	310
2	1.23370	05501	36169	82735	0.91596	55941	77219	015
3	1.05179	97902	64644	99972	0.96894	61462	59369	380
4	1.01467	80316	04192	05455	0.98894	45517	41105	336
5	1.00452	37627	95139	61613	0.99615	78280	77088	064
6	1.00144	70766	40942	12191	0.99868	52222	18438	135
7	1.00047	15486	52376	55476	0.99955	45078	90539	909
8	1.00015	51790	25296	11930	0.99984	99902	46829	656
9	1.00005	13451	83843	77259	0.99994	96841	87220	090
10	1.00001	70413	63044	82549	0.99998	31640	26196	877
11	1.00000	56660	51090	10935	0.99999	43749	73823	699
12	1.00000	18858	48583	11958	0.99999	81223	50587	882
13	1.00000	06280	55421	80232	0.99999	93735	83771	841
14	1.00000	02092	40519	21150	0.99999	97910	87248	734
15	1.00000	00697	24703	12929	0.99999	99303	40842	624
16	1.00000	00232	37157	37916	0.99999	99767	75950	903
17	1.00000	00077	44839	45587	0.99999	99922	57782	104
18	1.00000	00025	81437	55666	0.99999	99974	19086	745
19	1.00000	00008	60444	11452	0.99999	99991	39660	745
20	1.00000	00002	86807	69746	0.99999	99997	13213	274
21	1.00000	00000	95601	16531	0.99999	99999	04403	029
22	1.00000	00000	31866	77514	0.99999	99999	68134	064
23	1.00000	00000	10622	20241	0.99999	99999	89377	965
24	1.00000	00000	03540	72294	0.99999	99999	96459	311
25	1.00000	00000	01180	23874	0.99999	99999	98819	768
26	1.00000	00000	00393	41247	0.99999	99999	99606	589
27	1.00000	00000	00131	13740	0.99999	99999	99868	863
28	1.00000	00000	00043	71245	0.99999	99999	99956	288
29	1.00000	00000	00014	57081	0.99999	99999	99985	429
30	1.00000	00000	00004	85694	0.99999	99999	99995	143
31	1.00000	00000	00001	61898	0.99999	99999	99998	381
32	1.00000	00000	00000	53966	0.99999	99999	99999	460
33	1.00000	00000	00000	17989	0.99999	99999	99999	820
34	1.00000	00000	00000	05996	0.99999	99999	99999	940
35	1.00000	00000	00000	01999	0.99999	99999	99999	980
36	1.00000	00000	00000	00666	0.99999	99999	99999	993
37	1.00000	00000	00000	00222	0.99999	99999	99999	998
38	1.00000	00000	00000	00074	0.99999	99999	99999	999
39	1.00000	00000	00000	00025				
40	1.00000	00000	00000	00008				
41	1.00000	00000	00000	00003				
42	1.00000	00000	00000	00001				

COMBINATORIAL ANALYSIS

Factorials, Exact Values

n	$n!$
0	1 (by definition)
1	1
2	2
3	6
4	24
5	120
6	720
7	5040
8	40,320
9	362,880
10	3,628,800
11	39,916,800
12	479,001,600
13	6,227,020,800
14	87,178,291,200
15	1,307,674,368,000
16	20,922,789,888,000
17	355,687,428,096,000
18	6,402,373,705,728,000
19	121,645,100,408,832,000
20	2,432,902,008,176,640,000
21	51,090,942,171,709,440,000
22	1,124,000,727,777,607,680,000
23	25,852,016,738,884,976,640,000
24	620,448,401,733,239,439,360,000
25	15,511,210,043,330,985,984,000,000
26	403,291,461,126,605,635,584,000,000
27	10,888,869,450,418,352,160,768,000,000
28	304,888,344,611,713,860,501,504,000,000
29	8,841,761,993,739,701,954,543,616,000,000
30	265,252,859,812,191,058,636,308,480,000,000
31	8.22284×10^{33}
32	2.63131×10^{35}
33	8.68332×10^{36}
34	2.95233×10^{38}
35	1.03331×10^{40}
36	3.71993×10^{41}
37	1.37638×10^{43}
38	5.23023×10^{44}
39	2.03979×10^{46}

$\underline{n} = n! = e^{-n} n^{n} \sqrt{2\pi n}.$ approximately, known as Stirling's formula

$\log_e n! = n \log_e n - n$, approximately.

FACTORIALS AND THEIR COMMON LOGARITHMS

This table presents values of $n! = n(n-1)(n-2) \ldots 2 \cdot 1$ and its logarithm for numbers from 1 to 100. The values of $n!$ are expressed exponentially to 5 significant figures.

n	$n!$	log $n!$	n	$n!$	log $n!$
			50	3.0414×10^{64}	64.48307
1	1,0000	0.00000	51	1.5511×10^{66}	66.19065
2	2,0000	0.30103	52	8.0658×10^{67}	67.90665
3	6,0000	0.77815	53	4.2749×10^{69}	69.63092
4	2.4000×10	1.38021	54	2.3084×10^{71}	71.36332
5	1.2000×10^{2}	2.07918	55	1.2696×10^{73}	73.10368
6	7.2000×10^{2}	2.85733	56	7.1100×10^{74}	74.85187
7	5.0400×10^{3}	3.70243	57	4.0527×10^{76}	76.60774
8	4.0320×10^{4}	4.60552	58	2.3506×10^{78}	78.37117
9	3.6288×10^{5}	5.55976	59	1.3868×10^{80}	80.14202
10	3.6288×10^{6}	6.55976	60	8.3210×10^{81}	81.92017
11	3.9917×10^{7}	7.60116	61	5.0758×10^{83}	83.70550
12	4.7900×10^{8}	8.68034	62	3.1470×10^{85}	85.49790
13	6.2270×10^{9}	9.79428	63	1.9826×10^{87}	87.29724
14	8.7178×10^{10}	10.94041	64	1.2689×10^{89}	89.10342
15	1.3077×10^{12}	12.11650	65	8.2477×10^{90}	90.91633
16	2.0923×10^{13}	13.32062	66	5.4434×10^{92}	92.73587
17	3.5569×10^{14}	14.55107	67	3.6471×10^{94}	94.56195
18	6.4024×10^{15}	15.80634	68	2.4800×10^{96}	96.39446
19	1.2165×10^{17}	17.08509	69	1.7112×10^{98}	98.23331
20	2.4329×10^{68}	18.38612	70	1.1979×10^{100}	100.07841
21	5.1091×10^{19}	19.70834	71	8.5048×10^{101}	101.92966
22	1.1240×10^{21}	21.05077	72	6.1234×10^{103}	103.78700
23	2.5852×10^{22}	22.41249	73	4.4701×10^{105}	105.65032
24	6.2045×10^{23}	23.79271	74	3.3079×10^{107}	107.51955
25	1.5511×10^{25}	25.19065	75	2.4809×10^{109}	109.39461
26	4.0329×10^{26}	26.60562	76	1.8855×10^{111}	111.27543
27	1.0889×10^{28}	28.03698	77	1.4518×10^{113}	113.16192
28	3.0489×10^{68}	29.48414	78	1.1324×10^{115}	115.05401
29	8.8418×10^{30}	30.94654	79	8.9462×10^{116}	116.95164
30	2.6525×10^{32}	32.42366	80	7.1569×10^{118}	118.85473
31	8.2228×10^{33}	33.91502	81	5.7971×10^{120}	120.76321
32	2.6313×10^{35}	35.42017	82	4.7536×10^{122}	122.67703
33	8.6833×10^{36}	36.93869	83	3.9455×10^{124}	124.59610
34	2.9523×10^{38}	38.47016	84	3.3142×10^{126}	126.52038
35	1.0333×10^{40}	40.01423	85	2.8171×10^{128}	128.44980
36	3.7199×10^{41}	41.57054	86	2.4227×10^{130}	130.38430
37	1.3764×10^{43}	43.13874	87	2.1078×10^{132}	132.32382
38	5.2302×10^{44}	44.71852	88	1.8548×10^{134}	134.26830
39	2.0398×10^{46}	46.30959	89	1.6508×10^{136}	136.21769
40	8.1592×10^{47}	47.91165	90	1.4857×10^{138}	138.17194
41	3.3453×10^{49}	49.52443	91	1.3520×10^{140}	140.13098
42	1.4050×10^{51}	51.14768	92	1.2438×10^{142}	142.09476
43	6.0415×10^{52}	52.78115	93	1.1568×10^{144}	144.06325
44	2.6583×10^{54}	54.42460	94	1.0874×10^{146}	146.03638
45	1.1962×10^{56}	56.07781	95	1.0330×10^{148}	148.01410
46	5.5026×10^{57}	57.74057	96	9.9168×10^{149}	149.99637
47	2.5862×10^{59}	59.41267	97	9.6193×10^{151}	151.98314
48	1.2414×10^{61}	61.09391	98	9.4269×10^{153}	153.97437
49	6.0828×10^{62}	62.78410	99	9.3326×10^{155}	155.97000
50	3.0414×10^{64}	64.48307	100	9.3326×10^{157}	157.97000

$$n! = \left(\frac{n}{e}\right)^{n} \sqrt{2n\pi} + h; n = 1,2,3, \ldots \left[0 < \frac{h}{n!} < \frac{1}{12n}\right], \lim_{n \to \infty} \frac{n! e^{n}}{n^{n+\frac{1}{2}}} = \sqrt{2\pi}, \lim_{n \to \infty} \frac{(n!)^{\frac{1}{n}}}{n} = \frac{1}{e}$$

RECIPROCALS OF FACTORIALS AND THEIR COMMON LOGARITHMS

This table presents the reciprocals of the factorials and their logarithms for numbers from 1 to 100.

n	$1/n!$	$\log(1/n!)$	n	$1/n!$	$\log(1/n!)$
1	1.	.00000	51	$.64470 \times 10^{-66}$	$\overline{67}.80935$
2	0.5	$\overline{1}.69897$	52	$.12398 \times 10^{-67}$	$\overline{68}.09335$
3	.16667	$\overline{1}.22185$	53	$.23392 \times 10^{-69}$	$\overline{70}.36908$
*4	$.41667 \times 10^{-1}$	$\overline{2}.61979$	54	$.43319 \times 10^{-71}$	$\overline{72}.63668$
5	$.83333 \times 10^{-2}$	$\overline{3}.92082$	55	$.78762 \times 10^{-73}$	$\overline{74}.89632$
6	$.13889 \times 10^{-2}$	$\overline{3}.14267$	56	$.14065 \times 10^{-74}$	$\overline{75}.14813$
7	$.19841 \times 10^{-3}$	$\overline{4}.29757$	57	$.24675 \times 10^{-76}$	$\overline{77}.39226$
8	$.24802 \times 10^{-4}$	$\overline{5}.39448$	58	$.42543 \times 10^{-78}$	$\overline{79}.62883$
9	$.27557 \times 10^{-5}$	$\overline{6}.44024$	59	$.72107 \times 10^{-80}$	$\overline{81}.85798$
10	$.27557 \times 10^{-6}$	$\overline{7}.44024$	60	$.12018 \times 10^{-81}$	$\overline{82}.07983$
11	$.25052 \times 10^{-7}$	$\overline{8}.39884$	61	$.19701 \times 10^{-83}$	$\overline{84}.29450$
12	$.20877 \times 10^{-8}$	$\overline{9}.31966$	62	$.31776 \times 10^{-85}$	$\overline{86}.50210$
13	$.16059 \times 10^{-9}$	$\overline{10}.20572$	63	$.50439 \times 10^{-87}$	$\overline{88}.70276$
14	$.11471 \times 10^{-10}$	$\overline{11}.05959$	64	$.78810 \times 10^{-89}$	$\overline{90}.89658$
15	$.76472 \times 10^{-12}$	$\overline{13}.88350$	65	$.12125 \times 10^{-90}$	$\overline{91}.08367$
16	$.47795 \times 10^{-13}$	$\overline{14}.67938$	66	$.18371 \times 10^{-92}$	$\overline{93}.26413$
17	$.28115 \times 10^{-14}$	$\overline{15}.44893$	67	$.27419 \times 10^{-94}$	$\overline{95}.43805$
18	$.15619 \times 10^{-15}$	$\overline{16}.19366$	68	$.40322 \times 10^{-96}$	$\overline{97}.60554$
19	$.82206 \times 10^{-17}$	$\overline{18}.91491$	69	$.58438 \times 10^{-98}$	$\overline{99}.76669$
20	$.41103 \times 10^{-18}$	$\overline{19}.61388$	70	$.83482 \times 10^{-100}$	$\overline{101}.92159$
21	$.19573 \times 10^{-19}$	$\overline{20}.29166$	71	$.11758 \times 10^{-101}$	$\overline{102}.07034$
22	$.88968 \times 10^{-21}$	$\overline{22}.94923$	72	$.16331 \times 10^{-103}$	$\overline{104}.21300$
23	$.38682 \times 10^{-22}$	$\overline{23}.58751$	73	$.22371 \times 10^{-105}$	$\overline{106}.34968$
24	$.16117 \times 10^{-23}$	$\overline{24}.20729$	74	$.30231 \times 10^{-107}$	$\overline{108}.48045$
25	$.64470 \times 10^{-25}$	$\overline{26}.80935$	75	$.40308 \times 10^{-109}$	$\overline{110}.60539$
26	$.24796 \times 10^{-26}$	$\overline{27}.39438$	76	$.53036 \times 10^{-111}$	$\overline{112}.72457$
27	$.91837 \times 10^{-28}$	$\overline{29}.96302$	77	$.68879 \times 10^{-113}$	$\overline{114}.83808$
28	$.32799 \times 10^{-29}$	$\overline{30}.51586$	78	$.88306 \times 10^{-115}$	$\overline{116}.94599$
29	$.11310 \times 10^{-30}$	$\overline{31}.05346$	79	$.11178 \times 10^{-116}$	$\overline{117}.04836$
30	$.37700 \times 10^{-32}$	$\overline{33}.57634$	80	$.13972 \times 10^{-118}$	$\overline{119}.14527$
31	$.12161 \times 10^{-33}$	$\overline{34}.08498$	81	$.17250 \times 10^{-120}$	$\overline{121}.23679$
32	$.38004 \times 10^{-35}$	$\overline{36}.57983$	82	$.21036 \times 10^{-122}$	$\overline{123}.32297$
33	$.11516 \times 10^{-36}$	$\overline{37}.06131$	83	$.25345 \times 10^{-124}$	$\overline{125}.40390$
34	$.33872 \times 10^{-38}$	$\overline{39}.52984$	84	$.30173 \times 10^{-126}$	$\overline{127}.47962$
35	$.96776 \times 10^{-40}$	$\overline{41}.98577$	85	$.35497 \times 10^{-128}$	$\overline{129}.55020$
36	$.26882 \times 10^{-41}$	$\overline{42}.42946$	86	$.41276 \times 10^{-130}$	$\overline{131}.61570$
37	$.72655 \times 10^{-43}$	$\overline{44}.86126$	87	$.47444 \times 10^{-132}$	$\overline{133}.67618$
38	$.19120 \times 10^{-44}$	$\overline{45}.28148$	88	$.53913 \times 10^{-134}$	$\overline{135}.73170$
39	$.49025 \times 10^{-46}$	$\overline{47}.69041$	89	$.60577 \times 10^{-136}$	$\overline{137}.78231$
40	$.12256 \times 10^{-47}$	$\overline{48}.08835$	90	$.67308 \times 10^{-138}$	$\overline{139}.82806$
41	$.29893 \times 10^{-49}$	$\overline{50}.47557$	91	$.73964 \times 10^{-140}$	$\overline{141}.86902$
42	$.71174 \times 10^{-51}$	$\overline{52}.85232$	92	$.80396 \times 10^{-142}$	$\overline{143}.90524$
43	$.16552 \times 10^{-52}$	$\overline{53}.21885$	93	$.86447 \times 10^{-144}$	$\overline{145}.93675$
44	$.37618 \times 10^{-54}$	$\overline{55}.57540$	94	$.91965 \times 10^{-146}$	$\overline{147}.96362$
45	$.83597 \times 10^{-56}$	$\overline{57}.92219$	95	$.96806 \times 10^{-148}$	$\overline{149}.98590$
46	$.18173 \times 10^{-57}$	$\overline{58}.25943$	96	$.10084 \times 10^{-149}$	$\overline{150}.00363$
47	$.38666 \times 10^{-59}$	$\overline{60}.58733$	97	$.10396 \times 10^{-151}$	$\overline{152}.01686$
48	$.80555 \times 10^{-61}$	$\overline{62}.90609$	98	$.10608 \times 10^{-153}$	$\overline{154}.02563$
49	$.16440 \times 10^{-62}$	$\overline{63}.21590$	99	$.10715 \times 10^{-155}$	$\overline{156}.03000$
50	$.32879 \times 10^{-64}$	$\overline{65}.51693$	100	$.10715 \times 10^{-157}$	$\overline{158}.03000$

* For example $\log \frac{1}{4!} = \overline{2}.61979 = .61979 - 2 = 8.61979 - 10.$

NUMBER OF PERMUTATIONS $P(n,m)$

This table contains the number of permutations of n distinct things taken m at a time, given by

$$P(n,m) = \frac{n!}{(n-m)!} = n(n-1) \cdots (n-m+1)$$

n \ m	0	1	2	3	4	5	6	7	8	9	10
0	1										
1	1	1									
2	1	2	2								
3	1	3	6	6							
4	1	4	12	24	24						
5	1	5	20	60	120	120					
6	1	6	30	120	360	720	720				
7	1	7	42	210	840	2520	5040	5040			
8	1	8	56	336	1680	6720	20160	40320	40320		
9	1	9	72	504	3024	15120	60480	1 81440	3 62880	3 62880	
10	1	10	90	720	5040	30240	1 51200	6 04800	18 14400	36 28800	36 28800
11	1	11	110	990	7920	55440	3 32640	16 63200	66 52800	199 58400	399 16800
12	1	12	132	1320	11880	95040	39 91680	39 91680	199 58400	199 58400	2395 00800
13	1	13	156	1716	17160	1 54440	12 35520	86 48640	518 91840	2594 59200	10378 36800
14	1	14	182	2184	24024	2 40240	21 62160	172 97280	1210 80960	7264 85760	36324 28800
15	1	15	210	2730	32760	3 60360	36 03600	324 32400	2594 59200	18162 14400	1 08972 86400

n \ m	11	12	13	14	15
8					
9					
10					
11	399 16800				
12	4790 01600	4790 01600			
13	31135 10400	62270 20800	62270 20800		
14	1 45297 15200	4 35891 45600	8 71782 91200	8 71782 91200	
15	5 44864 32000	21 79457 28000	65 38371 84000	130 76743 68000	130 76743 68000

NUMBER OF COMBINATIONS

This table contains the number of combinations of n distinct things taken m at a time, given by $\binom{n}{m} = C(n,m)$. For coefficients missing from the table, use the relation $\binom{n}{m} = \binom{n}{n-m}$.

n \ m	0	1	2	3	4	5	6	7	8
1	1	1							
2	1	2	1						
3	1	3	3	1					
4	1	4	6	4	1				
5	1	5	10	10	5	1			
6	1	6	15	20	15	6	1		
7	1	7	21	35	35	21	7	1	
8	1	8	28	56	70	56	28	8	1
9	1	9	36	84	126	126	84	36	9
10	1	10	45	120	210	252	210	120	45
11	1	11	55	165	330	462	462	330	165
12	1	12	66	220	495	792	924	792	495
13	1	13	78	286	715	1287	1716	1716	1287
14	1	14	91	364	1001	2002	3003	3432	3003
15	1	15	105	455	1365	3003	5005	6435	6435
16	1	16	120	560	1820	4368	8008	11440	12870
17	1	17	136	680	2380	6188	12376	19448	24310
18	1	18	153	816	3060	8568	18564	31824	43758
19	1	19	171	969	3876	11628	27132	50388	75582
20	1	20	190	1140	4845	15504	38760	77520	1 25970
21	1	21	210	1330	5985	20349	54264	1 16280	2 03490
22	1	22	231	1540	7315	26334	74613	1 70544	3 19770
23	1	23	253	1771	8855	33649	1 00947	2 45157	4 90314
24	1	24	276	2024	10626	42504	1 34596	3 46104	7 35471
25	1	25	300	2300	12650	53130	1 77100	4 80700	10 81575
26	1	26	325	2600	14950	65780	2 30230	6 57800	15 62275
27	1	27	351	2925	17550	80730	2 96010	8 88030	22 20075
28	1	28	378	3276	20475	98280	3 76740	11 84040	31 08105
29	1	29	406	3654	23751	1 18755	4 75020	15 60780	42 92145
30	1	30	435	4060	27405	1 42506	5 93775	20 35800	58 52925
31	1	31	465	4495	31465	1 69911	7 36281	26 29575	78 88725
32	1	32	496	4960	35960	2 01376	9 06192	33 65856	105 18300
33	1	33	528	5456	40920	2 37336	11 07568	42 72048	138 84156
34	1	34	561	5984	46376	2 78256	13 44904	53 79616	181 56204
35	1	35	595	6545	52360	3 24632	16 23160	67 24520	235 35820
36	1	36	630	7140	58905	3 76992	19 47792	83 47680	302 60340
37	1	37	666	7770	66045	4 35897	23 24784	102 95472	386 08020
38	1	38	703	8436	73815	5 01942	27 60681	126 20256	489 03492
39	1	39	741	9139	82251	5 75757	32 62623	153 80937	615 23748
40	1	40	780	9880	91390	6 58008	38 38380	186 43560	769 04685
41	1	41	820	10660	101270	7 49398	44 96388	224 81940	955 48245
42	1	42	861	11480	111930	8 50668	52 45786	269 78328	1180 30185
43	1	43	903	12341	123410	9 62598	60 96454	322 24114	1450 08513
44	1	44	946	13244	135751	10 86008	70 59052	383 20568	1772 32627
45	1	45	990	14190	148995	12 21759	81 45060	453 79620	2155 53195
46	1	46	1035	15180	163185	13 70754	93 66819	535 24680	2609 32815
47	1	47	1081	16215	178365	15 33939	107 37573	628 91499	3144 57495
48	1	48	1128	17296	194580	17 12304	122 71512	736 29072	3773 48994
49	1	49	1176	18424	211876	19 06884	139 83816	859 00584	4509 78066
50	1	50	1225	19600	230300	21 18760	158 90700	998 84400	5368 78650

NUMBER OF COMBINATIONS

$$\binom{n}{m} = C(n,m)$$

m n	9	10	11	12	13
9	1				
10	10	1			
11	55	11	1		
12	220	66	12	1	
13	715	286	78	13	1
14	2002	1001	364	91	14
15	5005	3003	1365	455	105
16	11440	8008	4368	1820	560
17	24310	19448	12376	6188	2380
18	48620	43758	31824	18564	8568
19	92378	92378	75582	50388	27132
20	1 67960	1 84756	1 67960	1 25970	77520
21	2 93930	3 52716	3 52716	2 93930	2 03490
22	4 97420	6 46646	7 05432	6 46646	4 97420
23	8 17190	11 44066	13 52078	13 52078	11 44066
24	13 07504	19 61256	24 96144	27 04156	24 96144
25	20 42975	32 68760	44 57400	52 00300	52 00300
26	31 24550	53 11735	77 26160	96 57700	104 00600
27	46 86825	84 36285	130 37895	173 83860	200 58300
28	69 06900	131 23110	214 74180	304 21755	374 42160
29	100 15005	200 30010	345 97290	518 95935	678 63915
30	143 07150	300 45015	546 27300	864 93225	1197 59850
31	201 60075	443 52165	846 72315	1411 20525	2062 53075
32	280 48800	645 12240	1290 24480	2257 92840	3473 73600
33	385 67100	925 61040	1935 36720	3548 17320	5731 66440
34	524 51256	1311 28140	2860 97760	5483 54040	9279 83760
35	706 07460	1835 79396	4172 25900	8344 51800	14763 37800
36	941 43280	2541 86856	6008 05296	12516 77700	23107 89600
37	1244 03620	3483 30136	8549 92152	18524 82996	35624 67300
38	1630 11640	4727 33756	12033 22288	27074 75148	54149 50296
39	2119 15132	6357 45396	16760 56044	39107 97436	81224 25444
40	2734 38880	8476 60528	23118 01440	55868 53480	1 20332 22880
41	3503 43565	11210 99408	31594 61968	78986 54920	1 76200 76360
42	4458 91810	14714 42973	42805 61376	1 10581 16888	2 55187 31280
43	5639 21995	19173 34783	57520 04349	1 53386 78264	3 65768 48168
44	7098 30508	24812 56778	76693 39132	2 10906 82613	5 19155 26432
45	8861 63135	31901 87286	1 01505 95910	2 87600 21745	7 30062 09045
46	11017 16330	40763 50421	1 33407 83196	3 89106 17655	10 17662 30790
47	13626 49145	51780 66751	1 74171 33617	5 22514 00851	14 06768 48445
48	16771 06640	65407 15896	2 25952 00368	6 96685 34468	19 29282 49296
49	20544 55634	82178 22536	2 91359 16264	9 22637 34836	26 25967 83764
50	25054 33700	1 02722 78170	3 73537 38800	12 13996 51100	35 48605 18600

NUMBER OF COMBINATIONS

$$\binom{n}{m} = C(n,m)$$

m \ n	14	15	16	17	18	19
14	1					
15	15	1				
16	120	16	1			
17	680	136	17	1		
18	3060	816	153	18	1	
19	11628	3876	969	171	19	1
20	38760	15504	4845	1140	190	20
21	1 16280	54264	20349	5985	1330	210
22	3 19770	1 70544	74613	26334	7315	1540
23	8 17190	4 90314	2 45157	1 00947	33649	8855
24	19 61256	13 07504	7 35471	3 46104	1 34596	42504
25	44 57400	32 68760	20 42975	10 81575	4 80700	1 77100
26	96 57700	77 26160	53 11735	31 24550	15 62275	6 57800
27	200 58300	173 83860	130 37895	84 36285	46 86825	22 20075
28	401 16600	374 42160	304 21755	214 74180	131 23110	69 06900
29	775 58760	775 58760	678 63915	518 95935	345 97290	200 30010
30	1454 22675	1551 17520	1454 22675	1197 59850	864 93225	546 27300
31	2651 82525	3005 40195	3005 40195	2651 82525	2062 53075	1411 20525
32	4714 35600	5657 22720	6010 80390	5657 22720	4714 35600	3473 73600
33	8188 09200	10371 58320	11668 03110	11668 03110	10371 58320	8188 09200
34	13919 75640	18559 67520	22039 61430	23336 06220	22039 61430	18559 67520
35	23199 59400	32479 43160	40599 28950	45375 67650	45375 67650	40599 28950
36	37962 97200	55679 02560	73078 72110	85974 96600	90751 35300	85974 96600
37	61070 86800	93641 99760	1 28757 74670	1 59053 68710	1 76726 31900	1 76726 31900
38	96695 54100	1 54712 86560	2 22399 74430	2 87811 43380	3 35780 00610	3 53452 63800
39	1 50845 04396	2 51408 40660	3 77112 60990	5 10211 17810	6 23591 43990	6 89232 64410
40	2 32069 29840	4 02253 45056	6 28521 01650	8 87323 78800	11 33802 61800	13 12824 08400
41	3 52401 52720	6 34322 74896	10 30774 46706	15 15844 80450	20 21126 40600	24 46626 70200
42	5 28602 29080	9 86724 27616	16 65097 21602	25 46619 27156	35 36971 21050	44 67753 10800
43	7 83789 60360	15 15326 56696	26 51821 49218	42 11716 48758	60 83590 48206	80 04724 31850
44	11 49558 08528	22 99116 17056	41 67148 05914	68 63537 97976	102 95306 96964	140 88314 80056
45	16 68713 34960	34 48674 25584	64 66264 22970	110 30686 03890	171 58844 94940	243 83621 77020
46	23 98775 44005	51 17387 60544	99 14938 48554	174 96950 26860	281 89530 98830	415 42466 71960
47	34 16437 74795	75 16163 04549	150 32326 09098	274 11888 75414	456 86481 25690	697 31997 70790
48	48 23206 23240	109 32600 79344	225 48489 13647	424 44214 84512	730 98370 01104	1154 18478 96480
49	67 52488 72536	157 55807 02584	334 81089 92991	649 92703 98159	1155 42584 85616	1885 16648 97584
50	93 78456 56300	225 08295 75120	492 36896 95575	984 73793 91150	1805 35288 83775	3040 59433 83200

NUMBER OF COMBINATIONS

$$\binom{n}{m} = C(n,m)$$

n \ m	20	21	22	23	24	25
20	1					
21	21	1				
22	231	22	1			
23	1771	253	23	1		
24	10626	2024	276	24	1	
25	53130	12650	2300	300	25	1
26	2 30230	65780	14950	2600	325	26
27	8 88030	2 96010	80730	17550	2925	351
28	31 08105	11 84040	3 76740	98280	20475	3276
29	100 15005	42 92145	15 60780	4 75620	1 18755	23751
30	300 45015	143 07150	58 52925	20 35800	5 93775	1 42506
31	846 72315	443 52165	201 60675	78 88725	26 29575	7 36281
32	2257 92840	1290 24480	645 12240	280 48800	105 18300	33 65856
33	5731 66440	3548 17320	1935 36720	925 61040	385 67100	138 84156
34	13919 75640	9279 83760	5483 54640	2860 97760	1311 28140	524 51256
35	32479 43160	23199 59400	14763 37500	8344 51800	4172 25900	1835 79396
36	73078 72110	55679 02560	37962 97200	23107 89600	12516 77700	6008 05296
37	1 59053 68710	1 28757 74670	93641 99760	61070 88800	35624 67300	18524 82996
38	3 35780 00610	2 87811 43380	2 22399 74430	1 54712 86560	96695 54100	54149 50296
39	6 89232 64410	6 23591 43990	5 10211 17810	3 77112 60990	2 51408 40660	1 50845 04396
40	13 78465 28820	13 12824 08400	11 33802 61500	8 87323 78800	6 28521 01650	4 02253 45056
41	26 91289 37220	26 91289 37220	24 46626 70200	20 21126 40600	15 15844 80450	10 30774 46706
42	51 37916 07420	53 82578 74440	51 37916 07420	44 67753 10800	35 36971 21050	25 46619 27156
43	96 05669 18220	105 20494 81860	105 20494 81560	96 05669 18220	80 04724 31850	60 83590 48206
44	176 10393 50070	201 26164 00080	210 40989 63720	201 26164 00080	176 10393 50070	140 88314 80056
45	316 98708 30126	377 36557 50150	411 67153 63500	411 67153 63800	377 36557 50150	316 98708 30126
46	560 82330 07146	694 35265 80276	789 03711 13550	823 34307 27600	789 03711 13950	694 35265 80276
47	976 24796 79106	1255 17595 87422	1483 38976 94226	1612 38018 41550	1612 38018 41550	1483 38976 94226
48	1673 56794 49896	2231 42392 66528	2738 56572 81648	3095 76995 35776	3224 76036 83100	3095 76995 35776
49	2827 75273 46376	3904 99187 16424	4969 98965 48176	5834 33568 17424	6320 53032 18876	6320 53032 18876
50	4712 92122 43960	6732 74460 62800	8874 98152 64600	10804 32533 65600	12154 86600 36300	12641 06064 37752

Properties of Binomial Coefficients

$$(1 + x)^n = \binom{n}{0} x^0 + \binom{n}{1} x^n + \binom{n}{2} x^2 + \ldots + \binom{n}{n} x^n$$

$$\binom{n}{m} + \binom{n}{m+1} = \binom{n+1}{m+1}, \binom{n}{m} = \binom{n}{n-m}$$

This leads to Pascal's triangle

$$\binom{n}{0} + \binom{n}{1} + \binom{n}{2} + \ldots + \binom{n}{n} = 2^n$$

$$\binom{n}{0} - \binom{n}{1} + \binom{n}{2} - \ldots (-1)^n \binom{n}{n} = 0$$

$$\binom{n}{n} + \binom{n+1}{n} + \binom{n+2}{n} + \ldots + \binom{n+m}{n} = \binom{n+m+1}{n+1}$$

$$\binom{n}{0} + \binom{n}{2} + \binom{n}{4} + \ldots = 2^{n-1}$$

$$\binom{n}{1} + \binom{n}{3} + \binom{n}{5} + \ldots = 2^{n-1}$$

$$\binom{n}{0}^2 + \binom{n}{1}^2 + \binom{n}{2}^2 + \ldots + \binom{n}{n}^2 = \binom{2n}{n}$$

$$\binom{m}{0}\binom{n}{p} + \binom{m}{1}\binom{n}{p-1} + \ldots + \binom{m}{p}\binom{n}{0} = \binom{m+n}{p}$$

$$(1)\binom{n}{1} + (2)\binom{n}{2} + (3)\binom{n}{3} + \ldots + (n)\binom{n}{n} = n2^{n-1}$$

$$(1)\binom{n}{1} - (2)\binom{n}{2} + (3)\binom{n}{3} - \ldots (-1)^{n+1} (n)\binom{n}{n} = 0$$

POSITIONAL NOTATION

In our ordinary system of writing numbers, the value of any digit depends on its position in the number. The value of a digit in any position is ten times the value of the same digit one position to the right, or one-tenth the value of the same digit one position to the left. Thus, for example,

$$173.246 = 1 \times 10^2 + 7 \times 10^1 + 3 + 2 \times \frac{1}{10} + 4 \times \frac{1}{10^2} + 6 \times \frac{1}{10^3}.$$

There is no reason that a number other than 10 cannot be used as the *base*, or *radix*, of the number system. In fact, bases of 2, 8, and 16 are commonly used in working with digital computers. When the base used is not clear from the context, it is usually indicated as a parenthesized subscript or merely as a subscript. Thus

$$743_{(8)} = 7 \times 8^2 + 4 \times 8 + 3 = 7 \times 64 + 4 \times 8 + 3 = 448 + 32 + 3 = 483_{(10)}$$

$$1011.101_{(2)} = 1 \times 2^3 + 0 \times 2^2 + 1 \times 2 + 1 + 1 \times \tfrac{1}{2} + 0 \times \tfrac{1}{4} + 1 \times \tfrac{1}{8} = 11.625_{(10)}$$

CHANGE OF BASE

In this section, it is assumed that all calculations will be performed in base 10, since this is the only base in which most people can easily compute. However, there is no logical reason that some other base could not be used for the computations.

To convert a number from another base into base 10:

Simply write down the digits of the number, with each one multiplied by its appropriate positional value. Then perform the indicated computations in base 10, and write down the answer.

For examples, see the two examples in the previous section.

To convert a number from base 10 into another base:

The part of the number to the left of the point and the part to the right must be operated on separately. For the integer part (the part to the left of the point):

a. Divide the number by the new base, getting an integer quotient and remainder.
b. Write down the remainder as the last digit of the number in the new base.
c. Using the quotient from the last division in place of the original number, repeat the above two steps until the quotient becomes zero.

For the fractional part (the part to the right of the point):

a. Multiply the number by the new base.
b. Write down the integral part of the product as the first digit of the fractional part in the new base.
c. Using the fractional part of the last product in place of the original number, repeat the above two steps until the product becomes an integer, or until the desired number of places have been computed.

Examples:

These examples show a convenient method of arranging the computations.

1. Convert $103.118_{(10)}$ to base 8.

```
8  |103|  7                                    .118
8   |12|  4                                       8
     1                    147.074324...        .944
                                                  8
```

The calculation of the fractional part could be carried out as far as desired. It is a non-terminating fraction which will eventually repeat itself.

```
7.552
   8
4.416
   8
3.328
   8
2.624
   8
4.992
```

$$103.118_{(10)} = 147.074324\ldots_{(8)}$$

The calculations may be further shortened by not writing down the multiplier and
divisor at each step of the algorithm, as shown in the next example.

2. Convert $275.824_{(10)}$ to base 5.

```
5  |275|  0                    .824
    |55|  0                   4.120
    |11|  1                   0.600
      2                       3.000
```

$$275.824_{(10)} = 2100.403_{(5)}$$

To convert from one base to another (neither of which is 10):

The easiest procedure is usually to convert first to base 10, and then to the desired
base. However, there are two exceptions to this:

1. If computational facility is possessed in either of the bases, it may be used instead of
   base 10, and the appropriate one of the above methods applied.
2. If the two bases are different powers of the same number, the conversion may be
   done digit-by-digit to the base which is the common root of both bases, and then
   digit-by-digit back to the other base.

Example: Convert $127.653_{(8)}$ to base 16. (For base 16, the letters A–F are
used for the digits $10_{(10)}-15_{(10)}$.)

The first step is to convert the number to base 2, simply by converting each digit to
its binary equivalent:

$$127.653_{(8)} = 001\ 010\ 111 \cdot 110\ 101\ 011_{(2)}$$

Now by simply regrouping the binary number into groups of four binary digits, starting
at the point, we convert to base 16:

$$127.653_{(8)} = 101\ 0111 \cdot 1101\ 0101\ 1_{(2)} = 57.D58_{(16)}$$

$10^{\pm n}$ IN OCTAL SCALE

10^n	n	10^{-n}	10^n	n	10^{-n}
1	0	1.000 000 000 000 000	112 402 762 000	10	0.000 000 000 006 676
12	1	0.063 146 314 631 463	1 351 035 564 000	11	0.000 000 000 000 500
144	2	0.005 075 341 217 270	16 432 451 210 000	12	0.000 000 000 000 043
1 750	3	0.000 406 111 564 571	221 441 634 520 000	13	0.000 000 000 000 003
23 420	4	0.000 032 155 613 531	2 657 142 036 440 000	14	0.000 000 000 000 000
303 240	5	0.000 002 476 132 611	34 327 724 461 500 000	15	0.000 000 000 000 000
3 641 100	6	0.000 000 206 157 364	434 157 115 760 200 000	16	0.000 000 000 000 000
46 113 200	7	0.000 000 015 327 745	5 432 127 413 542 400 000	17	0.000 000 000 000 000
575 360 400	8	0.000 000 001 257 144	67 405 553 164 731 000 000	18	0.000 000 000 000 000
7 346 545 000	9	0.000 000 000 104 560			

2^n IN DECIMAL SCALE

n	2^n	n	2^n	n	2^n
0.001	1.00069 33874 62581	0.01	1.00695 55500 56719	0.1	1.07177 34625 36293
0.002	1.00138 72557 11335	0.02	1.01395 94797 90029	0.2	1.14869 83549 97035
0.003	1.00208 16050 79633	0.03	1.02101 21257 07193	0.3	1.23114 44133 44916
0.004	1.00277 63359 01078	0.04	1.02811 38266 56067	0.4	1.31950 79107 72894.
0.005	1.00347 17845 09503	0.05	1.03526 49238 41377	0.5	1.4121 35623 73095
0.006	1.00416 75432 38973	0.06	1.04246 57608 41121	0.6	1.51571 65665 10398
0.007	1.00486 38204 23785	0.07	1.04971 66836 23067	0.7	1.62450 47927 12471
0.008	1.00556 05803 98468	0.08	1.05701 80405 61380	0.8	1.74110 11265 92248
0.009	1.00625 78234 97782	0.09	1.06437 01824 53360	0.9	1.86606 59830 73615

$n\log_{10} 2$, $n\log_2 10$ IN DECIMAL SCALE

n	$n\log_{10}2$	$n\log_2 10$	n	$n\log_{10}2$	$n\log_2 10$
1	0.30102 99957	3.32192 80949	6	1.80617 99740	19.93156 85693
2	0.60205 99913	6.64385 61898	7	2.10720 99696	23.25349 66642
3	0.90308 99870	9.96578 42847	8	2.40823 99653	26.57542 47591
4	1.20411 99827	13.28771 23795	9	2.70926 99610	29.89735 28540
5	1.50514 99783	16.60964 04744	10	3.01029 99566	33.21928 09480

ADDITION AND MULTIPLICATION TABLES

Binary Scale

Addition

$$0 + 0 = 0$$
$$0 + 1 = 1 + 0 = 1$$
$$1 + 1 = 10$$

Multiplication

$$0 \times 0 = 0$$
$$0 \times 1 = 1 \times 0 = 0$$
$$1 \times 1 = 1$$

Octal Scale

Addition

0	01	02	03	04	05	06	07
1	02	03	04	05	06	07	10
2	03	04	05	06	07	10	11
3	04	05	06	07	10	11	12
4	05	06	07	10	11	12	13
5	06	07	10	11	12	13	14
6	07	10	11	12	13	14	15
7	10	11	12	13	14	15	16

Multiplication

1	02	03	04	05	06	07
2	04	06	10	12	14	16
3	06	11	14	17	22	25
4	10	14	20	24	30	34
5	12	17	24	31	36	43
6	14	22	30	36	44	52
7	16	25	34	43	52	61

MATHEMATICAL CONSTANTS IN OCTAL SCALE

$\pi = (3.11037\ 552421)_{(8)}$

$\pi^{-1} = (0.24276\ 301556)_{(8)}$

$\sqrt{\pi} = (1.61337\ 611067)_{(8)}$

$\log_6 \pi = 1.11206\ 404435)_{(8)}$

$\log_2 \pi = (1.51544\ 163223)_{(8)}$

$\sqrt{10} = (3.12305\ 407267)_{(8)}$

$e = (2.55760\ 521305)_{(8)}$

$e^{-1} = (0.27426\ 530661)_{(8)}$

$\sqrt{e} = (1.51411\ 230704)_{(8)}$

$\log_{10} e = (0.33626\ 754251)_{(8)}$

$\log_2 e = (1.34252\ 166245)_{(8)}$

$\log_2 10 = (3.24464\ 741136)_{(8)}$

$\gamma = (0.44742\ 147707)_{(8)}$

$\log_6 \gamma = -(0.43127\ 233602)_{(8)}$

$\log_2 \gamma = -(0.62573\ 030645)_{(8)}$

$\sqrt{2} = (1.32404\ 746320)_{(8)}$

$\log_6 2 = (0.54271\ 027760)_{(8)}$

$\log_6 10 = (2.23273\ 067355)_{(8)}$

OCTAL-DECIMAL INTEGER CONVERSION TABLE

	0	1	2	3	4	5	6	7		0	1	2	3	4	5	6	7
0000	0000	0001	0002	0003	0004	0005	0006	0007	0400	0256	0257	0258	0259	0260	0261	0262	0263
0010	0008	0009	0010	0011	0012	0013	0014	0015	0410	0264	0265	0266	0267	0268	0269	0270	0271
0020	0016	0017	0018	0019	0020	0021	0022	0023	0420	0272	0273	0274	0275	0276	0277	0278	0279
0030	0024	0025	0026	0027	0028	0029	0030	0031	0430	0280	0281	0282	0283	0284	0285	0286	0287
0040	0032	0033	0034	0035	0036	0037	0038	0039	0440	0288	0289	0290	0291	0292	0293	0294	0295
0050	0040	0041	0042	0043	0044	0045	0046	0047	0450	0296	0297	0298	0299	0300	0301	0302	0303
0060	0048	0049	0050	0051	0052	0053	0054	0055	0460	0304	0305	0306	0307	0308	0309	0310	0311
0070	0056	0057	0058	0059	0060	0061	0062	0063	0470	0312	0313	0314	0315	0316	0317	0318	0319
0100	0064	0065	0066	0067	0068	0069	0070	0071	0500	0320	0321	0322	0323	0324	0325	0326	0327
0110	0072	0073	0074	0075	0076	0077	0078	0079	0510	0328	0329	0330	0331	0332	0333	0334	0335
0120	0080	0081	0082	0083	0084	0085	0086	0087	0520	0336	0337	0338	0339	0340	0341	0342	0343
0130	0088	0089	0090	0091	0092	0093	0094	0095	0530	0344	0345	0346	0347	0348	0349	0350	0351
0140	0096	0097	0098	0099	0100	0101	0102	0103	0540	0352	0353	0354	0355	0356	0357	0358	0359
0150	0104	0105	0106	0107	0108	0109	0110	0111	0550	0360	0361	0362	0363	0364	0365	0366	0367
0160	0112	0113	0114	0115	0116	0117	0118	0119	0560	0368	0369	0370	0371	0372	0373	0374	0375
0170	0120	0121	0122	0123	0124	0125	0126	0127	0570	0376	0377	0378	0379	0380	0381	0382	0383
0200	0128	0129	0130	0131	0132	0133	0134	0135	0600	0384	0385	0386	0387	0388	0389	0390	0391
0210	0136	0137	0138	0139	0140	0141	0142	0143	0610	0392	0393	0394	0395	0396	0397	0398	0399
0220	0144	0145	0146	0147	0148	0149	0150	0151	0620	0400	0401	0402	0403	0404	0405	0406	0407
0230	0152	0153	0154	0155	0156	0157	0158	0159	0630	0408	0409	0410	0411	0412	0413	0414	0415
0240	0160	0161	0162	0163	0164	0165	1066	0167	0640	0416	0417	0418	0419	0420	0421	0422	0423
0250	0168	0169	0170	0171	0172	0173	0174	0175	0650	0424	0425	0426	0427	0428	0429	0430	0431
0260	0176	0177	0178	0179	0180	0181	0182	0183	0660	0432	0433	0434	0435	0436	0437	0438	0439
0270	0184	0185	0186	0187	0188	0189	0190	0191	0670	0440	0441	0442	0443	0444	0445	0446	0447
0300	0192	0193	0194	0195	0196	0197	0198	0199	0700	0448	0449	0450	0451	0452	0453	0454	0455
0310	0200	0201	0202	0203	0204	0205	0206	0207	0710	0456	0457	0458	0459	0460	0461	0462	0463
0320	0208	0209	0210	0211	0212	0213	0214	0215	0720	0464	0465	0466	0467	0468	0469	0470	0471
0330	0216	0217	0218	0219	0220	0221	0222	0223	0730	0472	0473	0474	0475	0476	0477	0478	0479
0340	0224	0225	0226	0227	0228	0229	0230	0231	0740	0480	0481	0482	0483	0484	0485	0486	0487
0350	0232	0233	0234	0235	0236	0237	0238	0239	0750	0488	0489	0490	0491	0492	0493	0494	0495
0360	0240	0241	0242	0243	0244	0245	0246	0247	0760	0496	0497	0498	0499	0500	0501	0502	0503
0370	0248	0249	0250	0251	0252	0253	0254	0255	0770	0504	0505	0506	0507	0508	0509	0510	0511

```
0000      0000
  to        to
0777      0511
(Octal)  (Decimal)

Octal  Decimal
10000—4096
20000—8192
30000—12288
40000—16384
50000—20480
60000—24576
70000—28672
```

	0	1	2	3	4	5	6	7		0	1	2	3	4	5	6	7
1000	0512	0513	0514	0515	0516	0517	0518	0519	1400	0768	0769	0770	0771	0772	0773	0774	0775
1010	0520	0521	0522	0523	0524	0525	0526	0527	1410	0776	0777	0778	0779	0780	0781	0782	0783
1020	0528	0529	0530	0531	0532	0533	0534	0535	1420	0784	0785	0786	0787	0788	0789	0790	0791
1030	0536	0537	0538	0539	0540	0541	0542	0543	1430	0792	0793	0794	0795	0796	0797	0798	0799
1040	0544	0545	0546	0547	0548	0549	0550	0551	1440	0800	0801	0802	0803	0804	0805	0806	0807
1050	0552	0553	0554	0555	0556	0557	0558	0559	1450	0808	0809	0810	0811	0812	0813	0814	0815
1060	0560	0561	0562	0563	0564	0565	0566	0567	1460	0816	0817	0818	0819	0820	0821	0822	0823
1070	0568	0569	0570	0571	0572	0573	0574	0575	1470	0824	0825	0826	0827	0828	0829	0830	0831
1100	0576	0577	0578	0579	0580	0581	0582	0583	1500	0832	0833	0834	0835	0836	0837	0838	0839
1110	0584	0585	0586	0587	0588	0589	0590	0591	1510	0840	0841	0842	0843	0844	0845	0846	0847
1120	0592	0593	0594	0595	0596	0597	0598	0599	1520	0848	0849	0850	0851	0852	0853	0854	0855
1130	0600	0601	0602	0603	0604	0605	0606	0607	1530	0856	0857	0858	0859	0860	0861	0862	0863
1140	0608	0609	0610	0611	0612	0613	0614	0615	1540	0864	0865	0866	0867	0868	0869	0870	0871
1150	0616	0617	0618	0619	0620	0621	0622	0623	1550	0872	0873	0874	0875	0876	0877	0878	0879
1160	0624	0625	0626	0627	0628	0629	0630	0631	1560	0880	0881	0882	0883	0884	0885	0886	0887
1170	0632	0633	0634	0635	0636	0637	0638	0639	1570	0888	0889	0890	0891	0892	0893	0894	0895
1200	0640	0641	0642	0643	0644	0645	0646	0647	1600	0896	0897	0898	0899	0900	0901	0902	0903
1210	0648	0649	0650	0651	0652	0653	0654	0655	1610	0904	0905	0906	0907	0908	0909	0910	0911
1220	0656	0657	0658	0659	0660	0661	0662	0663	1620	0612	0913	0914	0915	0916	0917	0918	0919
1230	0664	0665	0666	0667	0668	0669	0670	0671	1630	0920	0921	0922	0923	0924	0925	0926	0927
1240	0672	0673	0674	0675	0676	0677	0678	0679	1640	0928	0929	0930	0931	0932	0933	0934	0935
1250	0680	0681	0682	0683	0684	0685	0686	0687	1650	0936	0937	0938	0939	0940	0941	0942	0943
1260	0688	0689	0690	0691	0692	0693	0694	0695	1660	0944	0945	0946	0947	0948	0949	0950	0951
1270	0696	0697	0698	0699	0700	0701	0702	0703	1670	0952	0953	0954	0955	0956	0957	0958	0959
1300	0704	0705	0706	0707	0708	0709	0710	0711	1700	0960	0961	0962	0963	0964	0965	0966	0967
1310	0712	0713	0714	0715	0716	0717	0718	0719	1710	0968	0969	0970	0971	0972	0973	0974	0975
1320	0720	0721	0722	0723	0724	0725	0726	0727	1720	0976	0977	0978	0979	0980	0981	0982	0983
1330	0728	0729	0730	0731	0732	0733	0734	0735	1730	0984	0985	0986	0987	0988	0989	0990	0991
1340	0736	0737	0738	0739	0740	0741	0742	0743	1740	0992	0993	0994	0995	0996	0997	0998	0999
1350	0744	0745	0746	0747	0748	0749	0750	0751	1750	1000	1001	1002	1003	1004	1005	1006	1007
1360	0752	0753	0754	0755	0756	0757	0758	0759	1760	1008	1009	1010	1011	1012	1013	1014	1015
1370	0760	0761	0762	0763	0764	0765	0766	0767	1770	1016	1017	1018	1019	1020	1021	1022	1023

```
1000      0512
  to        to
1777      1023
(Octal)  (Decimal)
```

OCTAL-DECIMAL INTEGER CONVERSION TABLE
(Continued)

	0	1	2	3	4	5	6	7		0	1	2	3	4	5	6	7
2000	1024	1025	1026	1027	1028	1029	1030	1031	2400	1280	1281	1282	1283	1284	1285	1286	1287
2010	1032	1033	1034	1035	1036	1037	1038	1039	2410	1288	1289	1290	1291	1292	1293	1294	1295
2020	1040	1041	1042	1043	1044	1045	1046	1047	2420	1296	1297	1298	1299	1300	1301	1302	1303
2030	1048	1049	1050	1051	1052	1053	1054	1045	2430	1304	1305	1306	1307	1308	1309	1310	1311
2040	1056	1057	1058	1059	1060	1061	1062	1063	2440	1312	1313	1314	1315	1316	1317	1318	1319
2050	1064	1065	1066	1067	1068	1069	1070	1071	2450	1320	1321	1322	1323	1324	1325	1326	1327
2060	1072	1073	1074	1075	1076	1077	1078	1079	2460	1328	1329	1330	1331	1332	1333	1334	1335
2070	1080	1081	1082	1083	1084	1085	1086	1087	2470	1336	1337	1338	1339	1340	1341	1342	1343
2100	1088	1089	1090	1091	1092	1093	1094	1095	2500	1344	1345	1346	1347	1348	1349	1350	1351
2110	1096	1097	1098	1099	1100	1101	1102	1103	2510	1352	1353	1354	1355	1356	1357	1358	1359
2120	1104	1105	1106	1107	1108	1109	1110	1111	2520	1360	1361	1362	1363	1364	1365	1366	1367
2130	1112	1113	1114	1115	1116	1117	1118	1119	2530	1368	1369	1370	1371	1372	1373	1374	1375
2140	1120	1121	1122	1123	1124	1125	1126	1127	2540	1376	1377	1378	1379	1380	1381	1382	1383
2150	1128	1129	1130	1131	1132	1133	1134	1135	2550	1384	1385	1386	1387	1388	1389	1390	1391
2160	1136	1137	1138	1139	1140	1141	1142	1143	2560	1392	1393	1394	1395	1396	1397	1398	1399
2170	1144	1145	1146	1147	1148	1149	1150	1151	2570	1400	1401	1402	1403	1404	1405	1406	1407
2200	1152	1153	1154	1155	1156	1157	1158	1159	2600	1408	1409	1410	1411	1412	1413	1414	1415
2210	1160	1161	1162	1163	1164	1165	1166	1167	2610	1416	1417	1418	1419	1420	1421	1422	1423
2220	1168	1169	1170	1171	1172	1173	1174	1175	2620	1424	1425	1426	1427	1428	1429	1430	1431
2230	1176	1177	1178	1179	1180	1181	1182	1183	2630	1432	1433	1434	1435	1436	1437	1438	1439
2240	1184	1185	1186	1187	1188	1189	1190	1191	2640	1440	1441	1442	1443	1444	1445	1446	1447
2250	1192	1193	1194	1195	1196	1197	1198	1199	2650	1448	1449	1450	1451	1452	1453	1454	1455
2260	1200	1201	1202	1203	1204	1205	1206	1207	2660	1456	1457	1458	1459	1460	1461	1462	1463
2270	1208	1209	1210	1211	1212	1213	1214	1215	2670	1464	1465	1466	1467	1468	1469	1470	1471
2300	1216	1217	1218	1219	1220	1221	1222	1223	2700	1472	1473	1474	1475	1476	1477	1478	1479
2310	1224	1225	1226	1227	1228	1229	1230	1231	2710	1480	1481	1482	1483	1484	1485	1486	1487
2320	1232	1233	1234	1235	1236	1237	1238	1239	2720	1488	1489	1490	1491	1492	1493	1494	1495
2330	1240	1241	1242	1243	1244	1245	1246	1247	2730	1496	1497	1498	1499	1500	1501	1502	1503
2340	1248	1249	1250	1251	1252	1253	1254	1255	2740	1504	1505	1506	1507	1508	1509	1510	1511
2350	1256	1257	1258	1259	1260	1261	1262	1263	2750	1512	1513	1514	1515	1516	1517	1518	1519
2360	1264	1265	1266	1267	1268	1269	1270	1271	2760	1520	1521	1522	1523	1524	1525	1526	1527
2370	1272	1273	1274	1275	1276	1277	1278	1279	2770	1528	1529	1530	1531	1532	1533	1534	1535

2000	1024
to	to
2772	1535
(Octal)	(Decimal)

Octal	Decimal
10000	4096
20000	8192
30000	12288
40000	16384
50000	20480
60000	24576
70000	28672

3000	1536
to	to
3777	2047
(Octal)	(Decimal)

	0	1	2	3	4	5	6	7		0	1	2	3	4	5	6	7
3000	1536	1537	1538	1539	1540	1541	1542	1543	3400	1792	1793	1794	1795	1796	1797	1798	1799
3010	1544	1545	1546	1547	1548	1549	1550	1551	3410	1800	1801	1802	1803	1804	1805	1806	1807
3020	1552	1553	1554	1555	1556	1557	1558	1559	3420	1808	1809	1810	1811	1812	1813	1814	1815
3030	1560	1561	1562	1563	1564	1565	1566	1567	3430	1816	1817	1818	1819	1820	1821	1822	1823
3040	1568	1569	1570	1571	1572	1573	1574	1575	3440	1824	1825	1826	1827	1828	1829	1830	1831
3050	1576	1577	1578	1579	1580	1581	1582	1583	3450	1832	1833	1834	1835	1836	1837	1838	1839
3060	1584	1585	1586	1587	1588	1589	1590	1591	3460	1840	1841	1842	1843	1844	1845	1846	1847
3070	1592	1593	1594	1595	1596	1597	1598	1599	3470	1848	1849	1850	1851	1852	1853	1954	1855
3100	1600	1601	1602	1603	1604	1605	1606	1607	3500	1856	1857	1858	1859	1860	1861	1862	1863
3110	1608	1609	1610	1611	1612	1613	1614	1615	3510	1864	1865	1866	1867	1868	1869	1870	1871
3120	1616	1617	1618	1619	1620	1621	1622	1623	3520	1872	1873	1874	1875	1876	1877	1878	1879
3130	1624	1625	1626	1627	1628	1629	1630	1631	3530	1880	1881	1882	1883	1884	1885	1886	1887
3140	1632	1633	1634	1635	1636	1637	1638	1639	3540	1888	1889	1890	1891	1892	1893	1894	1895
3150	1640	1641	1642	1643	1644	1645	1646	1647	3550	1896	1897	1898	1899	1900	1901	1902	1903
3160	1648	1649	1650	1651	1652	1653	1654	1655	3560	1904	1905	1906	1907	1908	1909	1910	1911
3170	1656	1657	1658	1659	1660	1661	1662	1663	3570	1912	1913	1914	1915	1916	1917	1918	1919
3200	1664	1665	1666	1667	1668	1669	1670	1671	3600	1920	1921	1922	1923	1924	1925	1926	1927
3210	1672	1673	1674	1675	1676	1677	1678	1679	3610	1928	1929	1930	1931	1932	1933	1934	1935
3220	1680	1681	1682	1683	1684	1685	1686	1687	3620	1936	1937	1938	1939	1940	1941	1942	1943
3230	1688	1689	1690	1691	1692	1693	1694	1695	3630	1944	1945	1946	1947	1948	1949	1950	1951
3240	1696	1697	1698	1699	1700	1701	1702	1703	3640	1952	1953	1954	1955	1956	1957	1958	1959
3250	1704	1705	1706	1707	1708	1709	1710	1711	3650	1960	1961	1962	1963	1964	1965	1966	1967
3260	1712	1713	1714	1715	1716	1717	1718	1719	3660	1968	1969	1970	1971	1972	1973	1974	1975
3270	1720	1721	1722	1723	1724	1725	1726	1727	3670	1976	1977	1978	1979	1980	1981	1982	1983
3300	1728	1729	1730	1731	1732	1733	1734	1735	3700	1984	1985	1986	1987	1988	1989	1990	1991
3310	1736	1737	1738	1739	1740	1741	1742	1743	3710	1992	1993	1994	1995	1996	1997	1998	1999
3320	1744	1745	1746	1747	1748	1749	1750	1751	3720	2000	2001	2002	2003	2004	2005	2006	2007
3330	1752	1753	1754	1755	1756	1757	1758	1759	3730	2008	2009	2010	2011	2012	2013	2014	2015
3340	1760	1761	1762	1763	1764	1765	1766	1767	3740	2016	2017	2018	2019	2020	2021	2022	2023
3350	1768	1769	1770	1771	1772	1773	1774	1775	3750	2024	2025	2026	2027	2028	2029	2030	2031
3360	1776	1777	1778	1779	1780	1781	1782	1783	3760	2032	2033	2034	2035	2036	2037	2038	2039
3370	1784	1785	1786	1787	1788	1789	1790	1791	3770	2040	2041	2042	2043	2044	2045	2046	2047

COMBINATORIAL ANALYSIS

OCTAL-DECIMAL INTEGER CONVERSION TABLE
(Continued)

	0	1	2	3	4	5	6	7		0	1	2	3	4	5	6	7
4000	2048	2049	2050	2051	2052	2053	2054	2055	4400	2304	2305	2306	2307	2308	2309	2310	2311
4010	2056	2057	2058	2059	2060	2061	2062	2063	4410	2312	2313	2314	2315	2316	2317	2318	2319
4020	2064	2065	2066	2067	2068	2069	2070	2071	4420	2320	2321	2322	2323	2324	2325	2326	2327
4030	2072	2073	2074	2075	2076	2077	2078	2079	4430	2328	2329	2330	2331	2332	2333	2334	2335
4040	2080	2081	2082	2083	2084	2085	2086	2087	4440	2336	2337	2338	2339	2340	2341	2342	2343
4050	2088	2089	2090	2091	2092	2093	2094	2095	4450	2344	2445	2346	2347	2348	2349	2350	2351
4060	2096	2097	2098	2099	2100	2101	2102	2103	4460	2352	2353	2354	2355	2356	2357	2358	2359
4070	2104	2105	2106	2107	2108	2109	2110	2111	4470	2360	2361	2362	2363	2364	2365	2366	2367
4100	2112	2113	2114	2115	2116	2117	2118	2119	4500	2368	2369	2370	2371	2372	2373	2374	2375
4110	2120	2121	2122	2123	2124	2125	2126	2127	4510	2376	2377	2378	2379	2380	2381	2382	2383
4120	2128	2129	2130	2131	2132	2133	2134	2135	4520	2384	2385	2386	2387	2388	2389	2390	2391
4130	2136	2137	2138	2139	2140	2141	2142	2143	4530	2392	2393	2394	2395	2396	2397	2398	2399
4140	2144	2145	2146	2147	2148	2149	2150	2151	4540	2400	2401	2402	2403	2404	2405	2406	2407
4150	2152	2153	2154	2155	2156	2157	2158	2159	4550	2408	2409	2410	2411	2412	2413	2114	2415
4160	2160	2161	2162	2163	2164	2165	2166	2167	4560	2416	2417	2418	2419	2420	2421	2422	2423
4170	2168	2169	2170	2171	2172	2173	2174	2175	4570	2424	2425	2426	2427	2428	2429	2430	2431
4200	2176	2177	2178	2179	2180	2181	2182	2183	4600	2432	2433	2434	2435	2436	2437	2438	2439
4210	2184	2185	2186	2187	2188	2189	2190	2191	4610	2440	2441	2442	2443	2444	2445	2446	2447
4220	2192	2193	2194	2195	2196	2197	2198	2199	4620	2448	2449	2450	2451	2452	2453	2454	2455
4230	2200	2201	2202	2203	2204	2205	2206	2207	4630	2456	2457	2458	2459	2460	2461	2462	2463
4240	2208	2209	2210	2211	2212	2213	2214	2215	4640	2464	2465	2466	2467	2468	2469	2470	2471
4250	2216	2217	2218	2219	2220	2221	2222	2223	4650	2472	2473	2474	2475	2476	2477	2478	2479
4260	2224	2225	2226	2227	2228	2229	2230	2231	4660	2480	2481	2482	2483	2484	2485	2486	2487
4270	2232	2233	2234	2235	2236	2237	2238	2239	4670	2488	2489	2490	2491	2492	2493	2494	2495
4300	2240	2241	2242	2243	2244	2245	2246	2247	4700	2496	2497	2498	2499	2500	2501	2502	2503
4310	2248	2249	2250	2251	2252	2253	2254	2255	4710	2504	2505	2506	2507	2508	2509	2510	2511
4320	2256	2257	2258	2259	2260	2261	2262	2263	4720	2512	2513	2514	2515	2416	2517	2518	2519
4330	2264	2265	2266	2267	2268	2269	2270	2271	4730	2520	2521	2522	2523	2524	2525	2526	2527
4340	2272	2273	2274	2275	2276	2277	2278	2279	4740	2528	2529	2530	2531	2532	2533	2534	2535
4350	2280	2281	2282	2283	2284	2285	2286	2287	4750	2536	2537	2538	2539	2540	2541	2542	2543
4360	2288	2289	2290	2291	2292	2293	2294	2295	4760	2544	2545	2546	2547	2548	2549	2550	2551
4370	2296	2297	2298	2299	2300	2301	2302	2303	4770	2552	2553	2554	2555	2556	2557	2558	2559

4000	2048
to	to
4777	2559
(Octal)	(Decimal)

Octal	Decimal
10000	4096
20000	8192
30000	12288
40000	16384
50000	20480
60000	24576
70000	28672

	0	1	2	3	4	5	6	7		0	1	2	3	4	5	6	7
5000	2560	2561	2562	2563	2564	2565	2566	2567	5400	2816	2817	2818	2819	2820	2821	2822	2823
5010	2568	2569	2570	2571	2572	2573	2574	2575	5410	2824	2825	2826	2827	2828	2829	2830	2831
5020	2576	2577	2578	2579	2580	2581	2582	2583	5420	2832	2833	2834	2835	2836	2837	2838	2839
5030	2584	2585	2586	2587	2588	2589	2490	2591	5430	2840	2841	2842	2843	2844	2845	2846	2847
5040	2592	2593	2594	2595	2596	2597	2598	2599	5440	2848	2849	2850	2851	2852	2853	2854	2855
5050	2600	2601	2602	2603	2604	2605	2606	2607	5450	2856	2857	2858	2859	2860	2861	2862	2863
5060	2608	2609	2610	2611	2612	2613	2614	2615	5460	2864	2865	2866	2867	2868	3869	2870	2871
5070	2616	2617	2618	2619	2620	2621	2622	2623	5470	2872	2873	2874	2875	2876	2877	2878	2879
5100	2624	2625	2626	2627	2628	2629	2630	2631	5500	2880	2881	2882	2883	2884	2885	2886	2887
5110	2632	2633	2634	2635	2636	2637	2638	2639	5510	2888	2889	2890	2891	2892	2893	2894	2895
5120	2640	2641	2642	2643	2644	2645	2646	2647	5520	2896	2897	2898	2899	2900	2901	2902	2903
5130	2648	2649	2650	2651	2652	2653	2654	2655	5530	2904	2905	2906	2907	2908	2909	2910	2911
5140	2656	2657	2658	2659	2660	2661	2662	2663	5540	2912	2913	2914	2915	2916	2917	2818	2919
5150	2664	2665	2666	2667	2668	2669	2670	2671	5550	2920	2921	2922	2923	2924	2925	2926	2927
5160	2672	2673	2674	2675	2676	2677	2678	2679	5560	2928	2929	2930	2931	2932	2933	2934	2935
5170	2680	2681	2682	2683	2684	2685	2686	2687	5570	2936	2937	2938	2939	2940	2941	2942	2943
5200	2688	2689	2690	2691	2692	2693	2694	2695	5600	2944	2945	2946	2947	2948	2949	2950	2951
5210	2696	2697	2698	2699	2700	2701	2702	2703	5610	2952	2953	2954	2955	2956	2957	2958	2959
5220	2704	2705	2706	2707	2708	2709	2710	2711	5620	2960	2961	2962	2963	2964	2965	2966	2967
5230	2712	2713	2714	2715	2716	2717	2718	2719	5630	2968	2969	2970	2971	2972	2973	2974	2975
5240	2720	2721	2722	2723	2724	2725	2726	2727	5640	2976	2977	2978	2979	2980	2981	2982	2983
5250	2728	2729	2730	2731	2732	2733	2734	2735	5650	2984	2985	2986	2987	2988	2989	2990	2991
5260	2736	2737	2738	2739	2740	2741	2742	2743	5660	2992	2993	2994	2995	2996	2997	2998	2999
5270	2744	2745	2746	2747	2748	2749	2750	2751	5670	3000	3001	3002	3003	3004	3005	3006	3007
5300	2752	2753	2754	2755	2756	2757	2758	2759	5700	3008	3009	3010	3011	3012	3013	3014	3015
5310	2760	2761	2762	2763	2764	2765	2766	2767	5710	3016	3017	3018	3019	3020	3021	3022	3023
5320	2768	2769	2770	2771	2772	2773	2774	2775	5720	3024	3025	3026	3027	3028	3029	3030	3031
5330	2776	2777	2778	2779	2780	2781	2782	2783	5730	3032	3033	3034	3035	2036	3037	3038	3039
5340	2784	2785	2786	2787	2788	2789	2790	2791	5740	3040	3041	3042	3043	3044	3045	3046	3047
5350	2792	2793	2794	2795	2796	2797	2798	2799	5750	3048	3049	3050	3051	3052	3053	3054	3055
5360	2800	2801	2802	2803	2804	2805	2806	2807	5760	3056	3057	3058	3059	3060	3061	3062	3063
5370	2808	2809	2810	2811	2812	2813	2814	2815	5770	3064	3065	3066	3067	3068	3069	3070	3071

5000	2560
to	to
5777	3071
(Octal)	(Decimal)

OCTAL-DECIMAL INTEGER CONVERSION TABLE
(Continued)

Upper section

Octal	0	1	2	3	4	5	6	7
6000	3072	3073	3074	3075	3076	3077	3078	3079
6010	3080	3081	3082	3083	3084	3085	3086	3087
6020	3088	3089	3090	3091	3092	3093	3094	3095
6030	3096	3097	3098	3099	3100	3101	3102	3103
6040	3104	3105	3106	3107	3108	3109	3110	3111
6050	3112	3113	3114	3115	3116	3117	3118	3119
6060	3120	3121	3122	3123	3124	3125	3126	3127
6070	3128	3129	3130	3131	2132	3133	3134	3135
6100	3136	3137	3138	3139	3140	3141	3142	3143
6110	3144	3145	3146	3147	3148	3149	3150	3151
6120	3152	3153	3154	3155	3156	3157	3158	3159
6130	3160	3161	3162	3163	3164	3165	3166	3167
6140	3168	3169	3170	3171	3172	3173	3174	3175
6150	3176	3177	3178	3179	3180	3181	3182	4183
6160	3184	3185	3186	3187	3188	3189	3190	3191
6170	3192	3193	3194	3195	3196	3197	3198	3199
6200	3200	3201	3202	3203	3204	3205	3206	3207
6210	3208	3209	3210	3211	3212	3213	3214	3215
6220	3216	3217	3218	3219	3220	3221	3222	3223
6230	3224	3225	3226	3227	3228	3229	3230	3231
6240	3232	3233	3234	3235	3236	3237	3238	3239
6250	3240	3241	3442	3243	3244	3245	3246	3247
6260	3248	3249	3250	3251	3252	3253	3254	3255
6270	3256	3257	3258	3259	3260	3261	3262	3263
6300	3264	3267	3266	3267	3268	3269	3270	3871
6310	3272	3273	3274	3275	3276	3277	3278	3279
6320	3280	3281	3282	3283	3284	3285	3286	3287
6330	3288	3289	3290	3291	3292	3293	3294	3295
6340	3296	3297	3298	3299	3300	3301	3302	3003
6350	3304	3305	3306	3307	3308	3309	3310	3311
6360	3312	3313	3314	3315	3316	3317	3318	3319
6370	3320	3321	3322	3323	3324	3325	3326	3327

Octal	0	1	2	3	4	5	6	7
6400	3328	3329	3330	3331	3332	3333	3334	3335
6410	3336	3337	3338	3339	3340	3341	3342	3343
6420	3344	3345	3346	3347	3348	3349	3350	3351
6430	3352	3353	3354	3355	3356	3357	3358	3359
6440	3360	3361	3362	3363	3364	3365	3366	3367
6450	3368	3369	3370	3371	3372	3373	3374	3375
6460	3376	3377	3378	3379	3380	3381	3382	3383
6470	3384	3385	3386	3387	3388	3389	3390	3391
6500	3392	3393	3394	3395	3396	3397	3398	3399
6510	3400	3401	3402	3403	3404	3405	3406	3407
6520	3408	3409	3410	3411	3412	3413	3414	3415
6530	3416	3417	3418	3419	3420	3421	3422	3423
6540	3424	3425	3426	3427	3428	3429	3430	3431
6550	3432	3433	3434	3435	3436	3437	3438	3439
6560	3440	3441	3442	3443	3444	3445	3446	3447
6570	3448	3449	3450	3451	3452	3453	3454	3455
6600	3456	3457	3458	3459	3460	3461	3462	3463
6610	3464	3465	3466	3467	3468	3469	3470	3471
6620	3472	3473	3474	3475	3476	3477	3478	3479
6630	3480	3481	3482	3483	3484	3485	3486	3487
6640	3488	3489	3490	3491	3492	3493	3494	3495
6650	3496	3497	3498	3499	3500	3501	3502	3503
6660	3504	3505	3506	3307	3508	3509	3510	3511
6670	3512	3513	3514	3515	3516	3517	3518	3519
6700	3520	3521	3522	3523	3524	3525	3526	3527
6710	3528	3529	3530	3531	3532	3533	3534	3535
6720	3536	3537	3538	3539	3540	3541	3542	3543
6730	3544	3545	3546	3547	3548	3549	3550	3551
6740	3552	3553	3554	3555	3556	3557	3558	3559
6750	3560	3561	3562	3563	3564	3565	3566	3567
6760	3568	3569	3570	3571	3572	3573	3574	3575
6770	3576	3577	3578	3579	3580	3581	3582	3583

Side note (upper section):

6000 3072
to to
6777 3583
(Octal) (Decimal)

Octal	Decimal
10000	4096
20000	8192
30000	12288
40000	16384
50000	20480
60000	24576
70000	28672

Lower section

7000 3584
to to
7777 4095
(Octal) (Decimal)

Octal	0	1	2	3	4	5	6	7
7000	3584	3585	3586	3587	3588	3589	3590	3591
7010	3592	3593	3594	3595	3596	3597	3598	3599
7020	3600	3601	3602	3603	3604	3605	3606	3607
7030	4608	3609	3610	3611	3612	3613	3614	3615
7040	3616	3617	3618	3619	3620	3621	3622	3623
7050	3624	3625	3626	3627	3628	3629	3630	3631
7060	2632	3633	3634	3635	3636	3637	3638	3639
7070	3640	3641	3642	3643	3644	3645	3646	3647
7100	3648	3649	3650	3651	3652	3653	3654	3655
7110	3656	3657	3658	3659	3660	3661	3662	3663
7120	3664	3665	3666	3667	3668	3669	3670	3671
7130	3672	3673	3674	3675	3676	3677	3678	3679
7140	3680	3681	3682	3683	3684	3685	3686	3687
7150	3688	3689	3690	3691	3692	3693	3694	3695
7160	3696	3697	3698	3699	3700	3701	3702	3903
7170	3704	3705	3706	3707	3708	3709	3710	3711
7200	3712	3713	3714	3715	3716	3717	3718	3719
7210	3720	3721	3722	3723	3724	3725	3726	3727
7220	3728	3729	3730	3731	3732	3733	3734	3735
7230	3736	3737	3738	3739	3740	3741	3742	3743
7240	3744	3745	3746	3747	3748	3749	3750	3751
7250	3752	3753	3754	3755	3756	3757	3758	3759
7260	3760	3761	3762	3763	3764	3765	3766	3767
7270	3768	3769	3770	3771	3772	3774	3774	3775
7300	3776	3777	3778	3779	3780	3781	3782	3783
7310	3784	3785	3786	3787	3788	3789	3790	3791
7320	3792	3893	3794	3795	3796	3797	3798	3799
7330	3800	3801	3802	3803	3804	3805	3806	3807
7340	3808	3809	3810	3811	3812	3813	3814	3815
7350	3816	3817	3818	3819	3820	3821	3822	3823
7360	3824	3825	3826	3827	3828	3829	3830	3831
7370	3832	3833	3834	3835	3836	3837	3838	3839

Octal	0	1	2	3	4	5	6	7
7400	3840	3841	3482	3843	3844	3845	3846	3847
7410	3848	3849	3450	3851	3852	3853	3854	3855
7420	3856	3857	3858	3859	3860	3861	3862	3863
7430	3864	3865	3866	3867	3868	3869	3870	3871
7440	3872	3873	3874	3875	3876	3877	3878	3879
7450	3880	3881	3882	3883	3884	3885	3886	3887
7460	3888	3889	3890	3891	3892	3893	3894	3895
7470	3896	3897	3898	3899	3900	3901	3902	3903
7500	3904	3905	3906	3907	3908	3909	3910	3911
7510	3912	3913	3914	3915	3916	3917	3918	3919
7520	3920	3921	3922	3923	3924	3925	3926	3927
7530	3928	3929	3930	3931	3932	3933	3934	3935
7540	3836	3937	3938	3939	3940	3941	3942	3943
7550	3944	3945	3946	3947	3948	3949	3950	3951
7560	3952	3953	3954	3955	3956	3957	3958	3959
7570	3960	3961	3962	3963	3964	3965	3966	3967
7600	3968	3969	4970	3971	3972	3973	3974	3975
7610	3976	3977	3978	3979	3980	3981	3982	3983
7620	3984	3985	3986	3987	3988	3989	3990	3991
7630	3992	3993	3994	3995	3996	3997	3998	3999
7640	4000	4001	4002	4003	4004	4005	4006	4007
7650	4008	4009	4010	4011	4012	4013	4014	4015
7660	4016	4017	4018	4019	4020	4021	4022	4023
7670	4024	4025	4026	4027	4028	4029	4030	4031
7700	4032	4033	4034	4035	4036	4037	4038	4039
7710	4040	4041	4042	4043	4044	4045	4046	4047
7720	4048	4049	4050	4051	4052	4053	4054	4055
7730	4056	4057	4058	4059	4060	4061	4062	4063
7740	4064	4065	4066	4067	4068	4069	4070	4071
7750	4072	4073	4074	4075	4076	4077	4078	4079
7760	4080	4081	4082	4083	4084	4085	4086	4087
7770	4088	4089	4090	4091	4092	4093	4094	4095

COMBINATORIAL ANALYSIS

OCTAL-DECIMAL FRACTION CONVERSION TABLE

This table covers the entries from $(.000)_8$ to $(.377)_8$. For entries from $(.400)_8$ to $(.777)_8$, cognizance should be made of the fact that $(.400)_8$ is $(.500)_{10}$. Hence if $(.637)_8$ is desired, find $(.237)_8$ in table, namely, $(.310456)_{10}$ and add $(.50000)_{10}$ for $(.400)_8$. Thus

$(.637)_8 = (.237)_8 + (.400)_8$
$= (.310456)_{10} + (.50000)_{10}$
$= (.810456)_{10}$.

OCTAL	DEC.	OCTAL	DEC.	OCTAL	DEC.	OCTAL	DEC.
.000	.000000	.100	.125000	.200	.250000	.300	.375000
.001	.001953	.101	.126953	.201	.251953	.301	.376953
.002	.003906	.102	.128906	.202	.253906	.302	.378906
.003	.005859	.103	.130859	.203	.255859	.303	.380859
.004	.007812	.104	.132812	.204	.257812	.304	.382812
.005	.009765	.105	.134765	.205	.259765	.305	.384765
.006	.011718	.106	.136718	.206	.261718	.306	.386718
.007	.013671	.107	.138671	.207	.263671	.307	.388671
.010	.015625	.110	.140625	.210	.265625	.310	.390625
.011	.017578	.111	.142578	.211	.267578	.311	.392578
.012	.019531	.112	.144531	.212	.269531	.312	.394531
.013	.021484	.113	.146484	.213	.271484	.313	.396484
.014	.023437	.114	.148437	.214	.273437	.314	.398437
.015	.025390	.115	.150390	.215	.275390	.315	.400490
.016	.027343	.116	.152343	.216	.277343	.316	.402343
.017	.029296	.117	.154296	.217	.279296	.317	.404296
.020	.031250	.120	.156250	.220	.281250	.320	.406250
.021	.033203	.121	.158203	.221	.283203	.321	.408203
.022	.035156	.122	.160156	.222	.285156	.322	.410156
.023	.037109	.123	.162109	.223	.287109	.323	.412109
.024	.039062	.124	.164062	.224	.289062	.324	.414062
.025	.041015	.125	.166015	.225	.291015	.325	.416015
.026	.042968	.126	.167968	.226	.292968	.326	.417968
.027	.044921	.127	.169921	.227	.294921	.327	.419921
.030	.046875	.130	.171875	.230	.294875	.330	.421875
.031	.048828	.131	.173828	.231	.298828	.331	.423828
.032	.050781	.132	.175781	.232	.300781	.332	.425781
.033	.052734	.133	.177734	.233	.302734	.333	.427734
.034	.054687	.134	.179687	.234	.304687	.334	.429687
.035	.056640	.135	.181640	.235	.306640	.335	.431640
.036	.058593	.136	.183593	.236	.308593	.336	.433593
.037	.060546	.137	.185546	.237	.310546	.337	.435546
.040	.062500	.140	.187500	.240	.312500	.340	.437500
.041	.064453	.141	.189453	.241	.314453	.341	.439453
.042	.066406	.142	.191406	.242	.316406	.342	.441406
.043	.068359	.143	.193359	.243	.318359	.343	.443359
.044	.070312	.144	.195312	.244	.320312	.344	.445312
.045	.072265	.145	.197265	.245	.322265	.345	.447265
.046	.074218	.146	.199218	.246	.324218	.346	.449218
.047	.076171	.147	.201171	.247	.326171	.347	.451171
.050	.078125	.150	.203125	.250	.328125	.350	.453125
.051	.080078	.151	.205078	.251	.330078	.351	.455078
.052	.082031	.152	.207031	.252	.332031	.352	.457031
.053	.083984	.153	.208984	.253	.333984	.353	.458984
.054	.085937	.154	.210937	.254	.335937	.354	.460937
.055	.087890	.155	.212890	.255	.337890	.355	.462890
.056	.089843	.156	.214843	.256	.339843	.356	.464843
.057	.091796	.157	.216796	.257	.341796	.357	.466796
.060	.093750	.160	.218750	.260	.343750	.360	.468750
.061	.095703	.161	.220703	.261	.345703	.361	.470703
.062	.097656	.162	.222656	.262	.347656	.362	.472656
.063	.099609	.163	.224609	.263	.349609	.363	.474609
.064	.101562	.164	.226562	.264	.351562	.364	.476562
.065	.103515	.165	.228515	.265	.353515	.365	.478515
.066	.105468	.166	.230468	.266	.355468	.366	.480468
.067	.107421	.167	.232421	.267	.357421	.367	.482421
.070	.109375	.170	.234375	.270	.359375	.370	.484375
.071	.111328	.171	.236328	.271	.361328	.371	.486328
.072	.113281	.172	.238281	.272	.363281	.372	.488281
.073	.115234	.173	.240234	.273	.365234	.373	.490234
.074	.117187	.174	.242187	.274	.367187	.374	.492187
.075	.119140	.175	.244140	.275	.369140	.375	.494140
.076	.121093	.176	.246093	.276	.371093	.376	.496093
.077	.123046	.177	.248046	.277	.373046	.377	.498046

OCTAL-DECIMAL FRACTION CONVERSION TABLE (Continued)

OCTAL	DEC.	OCTAL	DEC.	OCTAL	DEC.	OCTAL	DEC.
.000000	.000000	.000100	.000244	.000200	.000488	.000300	.000732
.000001	.000004	.000101	.000247	.000201	.000492	.000301	.000736
.000002	.000007	.000102	.000251	.000202	.000495	.000302	.000740
.000003	.000011	.000103	.000255	.000203	.000499	.000303	.000743
.000004	.000015	.000104	.000259	.000204	.000503	.000304	.000747
.000005	000019	.000105	.000263	.000205	.000507	.000305	.000751
.000006	000022	.000106	.000267	.000206	.000511	.000306	.000755
.000007	.000026	.000107	.000270	.000207	.000514	.000307	.000759
.000010	000030	.000110	.000274	.000210	.000518	.000310	.000762
.000011	.000034	.000111	.000278	.000211	.000522	.000311	.000766
.000012	.000038	.000112	.000282	.000212	.000526	.000312	.000770
.000013	.000041	.000113	.000286	.000213	.000530	.000313	.000774
.000014	.000045	.000114	.000289	.000214	.000534	.000314	.000778
.000015	.000049	.000115	.000293	.000215	.000537	.000315	.000782
.000016	000053	.000116	.000297	.000216	.000541	.000316	.000785
.000017	000057	.000117	.000301	.000217	.000545	.000317	.000789
.000020	000061	.000120	.000305	.000220	.000549	.000320	.000793
.000021	.000064	.000121	.000308	.000221	.000553	.000321	.000797
.000022	000068	.000122	.000312	.000222	.000556	.000322	.000801
.000023	.000072	.000123	.000316	.000223	.000560	.000323	.000805
.000024	.000076	.000124	.000320	.000224	.000564	.000324	.000808
.000025	.000080	.000125	.000324	.000225	.000568	.000325	.000812
.000026	.000083	.000126	.000328	.000226	.000572	.000326	.000816
.000027	.000087	.000127	.000331	.000227	.000576	.000327	.000820
.000030	.000091	.000130	000335	.000230	.000570	000330	.000823
.000031	.000095	.000131	.000339	.000231	.000583	.000331	.000827
.000032	.000099	.000132	.000343	.000232	.000587	.000332	.000831
.000033	.000102	.000133	.000347	.000233	.000591	.000333	.000835
.000034	.000106	.000134	.000350	.000234	.000595	.000334	.000839
.000035	.000110	.000135	000354	.C00235	.000598	.000335	.000843
.000036	.000114	.000136	.000358	.000236	.000602	.000336	.000846
.000037	.000118	.000137	.000362	.000237	.000606	.000337	.000850
.000040	.000122	.000110	.000366	.000240	.000610	.000340	.000854
.000041	.000125	.000141	.000070	.000241	000614	.000341	.000858
.000042	.000129	.000142	.000373	.000242	.000617	.000342	.000862
.000043	.000133	.000143	.000377	.000243	.000621	.000343	.000865
.000044	.000137	.000144	.000381	.000244	.000625	.000344	.000869
.000045	.000141	.000145	.000385	.000245	.000629	.000345	.000873
.000046	.000144	.000146	.000389	.000246	.000633	.000346	.000877
.000047	.000148	.000147	.000392	.000247	.000637	.000347	.000881
.000050	.000152	.000150	.000396	.000250	.000640	.000350	.000885
.000051	.000156	.000151	.000400	.000251	.000644	.000351	.000888
.000052	.000160	.000152	.000404	.000252	.000648	.000352	.000892
.000053	.000164	.000153	.000408	.000253	.000652	.000353	.000896
.000054	.000167	.000154	.000411	.000254	.000656	.000354	.000900
.000055	.000171	.000155	.000415	.000255	.000659	.000355	.000904
.000056	.000175	.000156	.000419	.000256	.000663	.000356	.000907
.000057	.000179	.000157	.000423	.000257	.000667	.000357	.000911
.000060	.000183	.000160	.000427	.000260	.000671	.000360	.000915
.000061	.000186	.000161	.000431	.000261	.000675	.000361	.000919
.000062	.000190	.000162	.000434	.000262	.000679	.000362	.000923
.000063	.000194	.000163	.000438	.000263	.000682	.000363	.000926
.000064	.000198	.000164	.000442	.000264	.000686	.000364	.000930
.000065	.000202	.000165	.000446	.000265	.000690	.000365	.000934
.000066	.000205	.000166	.000450	.000266	.000694	.000366	.000938
.000067	.000209	.000167	.000453	.000267	.000698	.000367	.000942
.000070	000213	.000170	.000457	.000270	.000701	.000370	.000946
.000071	.000217	.000171	.000461	.000271	.000705	.000371	.000949
.000072	.000221	.000172	.000465	.000272	.000709	.000372	.000953
.000073	.000225	.000173	.000469	.000273	.000713	.000373	.000957
.000074	.000228	.000174	.000473	.000274	.000717	.000374	.000961
.000075	.000232	.000175	.000476	.000275	.000720	.000375	.000965
.000076	.000326	.000176	.000480	.000276	.000724	.000376	.000968
.000077	.000240	.000177	.000484	.000277	.000728	.000377	.000972

COMBINATORIAL ANALYSIS

OCTAL-DECIMAL FRACTION CONVERSION TABLE (Continued)

OCTAL	DEC.	OCTAL	DEC.	OCTAL	DEC.	OCTAL	DEC.
.000400	.000976	.000500	.001220	.000600	.001464	.000700	.001708
.000401	.000980	.000501	.001224	.000601	.001468	.000701	.001712
.000402	.000984	.000502	.001228	.000602	.001472	.000702	.001716
.000403	.000988	.000503	.001232	.000603	.001476	.000703	.001720
.000404	.000991	.000504	.001235	.000604	.001480	.000704	.001724
.000405	.000995	.000505	.001239	.000605	.001483	.000705	.001728
.000406	.000999	.000506	.001243	.000606	.001487	.000706	.001731
.000407	.001003	.000507	.001247	.000607	.001491	.000707	.001735
.000410	.001007	.000510	.001251	.000610	.001495	.000710	.001739
.000411	.001010	.000511	.001255	.000611	.001499	.000711	.001743
.000412	.001014	.000512	.001258	.000612	.001502	.000712	.001747
.000413	.001018	.000513	.001262	.000613	.001506	.000713	.001750
.000414	.001022	.000514	.001266	.000614	.001510	.000714	.001754
.000415	.001026	.000515	.001270	.000615	.001514	.000715	.001758
.000416	.001029	.000516	.001274	.000616	.001518	.000716	.001762
.000417	.001033	.000517	.001277	.000617	.001522	.000717	.001766
.000420	.001037	.000520	.001281	.000620	.001525	.000720	.001770
.000421	.001041	.000521	.001285	.000621	.001529	.000721	.001733
.000422	.001045	.000522	.001289	.000622	.001533	.000722	.001777
.000423	.001049	.000523	.001293	.000623	.001537	.000723	.991781
.000424	.001052	.000524	.001296	.000624	.001541	.000724	.001785
.000425	.001056	.000525	.001300	.000625	.001544	.000725	.001789
.000426	.001060	.000526	.001304	.000626	.001548	.000726	.001792
.000427	.001064	.000527	.001308	.000627	.001552	.000727	.001796
.000430	.001068	.000530	.001312	.000630	.001556	.000730	.001800
.000431	.001071	.000531	.001316	.000631	.001560	.000731	.001804
.000432	.001075	.000532	.001319	.000632	.001564	.000732	.001808
.000433	.001079	.000533	.001323	.000633	.001567	.000733	.001811
.000434	.001083	.000534	.001327	.000634	.001571	.000734	.001815
.000435	.001087	.000535	.001331	.000635	.001575	.000735	.001819
.000436	.001091	.000536	.001335	.000636	.001579	.000736	.001823
.000437	.001094	.000537	.001338	.000637	.001583	.000737	.001827
.000440	.001098	.000540	.001342	.000640	.001586	.000740	.001831
.000441	.001102	.000541	.001346	.000641	.001590	.000741	.001834
.000442	.001106	.000542	.001350	.000642	.001594	.000742	.001838
.000443	.001110	.000543	.001354	.000643	.001598	.000743	.001842
.000444	.001113	.000544	.001358	.000644	.001602	.000744	.001846
.000445	.001117	.000545	.001361	.000645	.001605	.000745	.001850
.000446	.001121	.000546	.001365	.000646	.001609	.000746	.001853
.000447	.001125	.000547	.001369	.000647	.001613	.000747	.001857
.000450	.001129	.000550	.001373	.000650	.001617	.000750	.001861
.000451	.001132	.000551	.001377	.000651	.001621	.000751	.001865
.000452	.001136	.000552	.001380	.000652	.001625	.000752	.001869
.000453	.001140	.000553	.001384	.000653	.001628	.000753	.001873
.000454	.001144	.000554	.001388	.000654	.001632	.000754	.001876
.000455	.001148	.000555	.001392	.000655	.001636	.000755	.001880
.000456	.001152	.000556	.001396	.000656	.001640	.000756	.001884
.000457	.001155	.000557	.001399	.000657	.001644	.000757	.001888
.000460	.001159	.000560	.001403	.000660	.001647	.000760	.001892
.000461	.001163	.000561	.001407	.000661	.001651	.000761	.001895
.000462	.001167	.000562	.001411	.000662	.001655	.000762	.001899
.000463	.001171	.000563	.001415	.000663	.001659	.000763	.001903
.000464	.001174	.000564	.001419	.000664	.001663	.000764	.001907
.000465	.001178	.000565	.001422	.000665	.001667	.000765	.001911
.000466	.001182	.000566	.001426	.000666	.001670	.000766	.001914
.000467	.001186	.000567	.001430	.000667	.001674	.000767	.001918
.000470	.001190	.000570	.001434	.000670	.001678	.000770	.001922
.000471	.001194	.000571	.001438	.000671	.001682	.000771	.001926
.000472	.001197	.000572	.001441	.000672	.001686	.000772	.001930
.000473	.001201	.000573	.001445	.000673	.001689	.000773	.001934
.000474	.001205	.000574	.001449	.000674	.001693	.000774	.001937
.000475	.001209	.000575	.001453	.000675	.001697	.000775	.001941
.000476	.001213	.000576	.001457	.000676	.001701	.000776	.001945
.000477	.001216	.000577	.001461	.000677	.001705	.000777	.001949

I. HEXADECIMAL AND DECIMAL DIRECT CONVERSION TABLE

The following tables aid in converting hexadecimal (base 16) numbers to decimal, and the reverse. Note that the base 16 digits for the decimal values 10–15 are represented by the letters A–F, respectively.

This table provides direct conversion of decimal and hexadecimal numbers in these ranges:

HEXADECIMAL	DECIMAL
000 to FFF	0000 to 4095

For numbers outside the range of the table, add the following values to the table figures:

HEXADECIMAL	DECIMAL
1000	4096
2000	8192
3000	12288
4000	16384
5000	20480
6000	24576
7000	28672
8000	32768
9000	36864
A000	40960
B000	45056
C000	49152
D000	53248
E000	57344
F000	61440

	0	1	2	3	4	5	6	7	8	9	A	B	C	D	E	F
00_	0000	0001	0002	0003	0004	0005	0006	0007	0008	0009	0010	0011	0012	0013	0014	0015
01_	0016	0017	0018	0019	0020	0021	0022	0023	0024	0025	0026	0027	0028	0029	0030	0031
02_	0032	0033	0034	0035	0036	0037	0038	0039	0040	0041	0042	0043	0044	0045	0046	0047
03_	0048	0049	0050	0051	0052	0053	0054	0055	0056	0057	0058	0059	0060	0061	0062	0063
04_	0064	0065	0066	0067	0068	0069	0070	0071	0072	0073	0074	0075	0076	0077	0078	0079
05_	0080	0081	0082	0083	0084	0085	0086	0087	0088	0089	0090	0091	0092	0093	0094	0095
06_	0096	0097	0098	0099	0100	0101	0102	0103	0104	0105	0106	0107	0108	0109	0110	0111
07_	0112	0113	0114	0115	0116	0117	0118	0119	0120	0121	0122	0123	0124	0125	0126	0127
08_	0128	0129	0130	0131	0132	0133	0134	0135	0136	0137	0138	0139	0140	0141	0142	0143
09_	0144	0145	0146	0147	0148	0149	0150	0151	0152	0153	0154	0155	0156	0157	0158	0159
0A_	0160	0161	0162	0163	0164	0165	0166	0167	0168	0169	0170	0171	0172	0173	0174	0175
0B_	0176	0177	0178	0179	0180	0181	0182	0183	0184	0185	0186	0187	0188	0189	0190	0191
0C_	0192	0193	0194	0195	0196	0197	0198	0199	0200	0201	0202	0203	0204	0205	0206	0207
0D_	0208	0209	0210	0211	0212	0213	0214	0215	0216	0217	0218	0219	0220	0221	0222	0223
0E_	0224	0225	0226	0227	0228	0229	0230	0231	0232	0233	0234	0235	0236	0237	0238	0239
0F_	0240	0241	0242	0243	0244	0245	0246	0247	0248	0249	0250	0251	0252	0253	0254	0255
10_	0256	0257	0258	0259	0260	0261	0262	0263	0264	0265	0266	0267	0268	0269	0270	0271
11_	0272	0273	0274	0275	0276	0277	0278	0279	0280	0281	0282	0283	0284	0285	0286	0287
12_	0288	0289	0290	0291	0292	0293	0294	0295	0296	0297	0298	0299	0300	0301	0302	0303
13_	0304	0305	0306	0307	0308	0309	0310	0311	0312	0313	0314	0315	0316	0317	0318	0319
14_	0320	0321	0322	0323	0324	0325	0326	0327	0328	0329	0330	0331	0332	0333	0334	0335
15_	0336	0337	0338	0339	0340	0341	0342	0343	0344	0345	0346	0347	0348	0349	0350	0351
16_	0352	0353	0354	0355	0356	0357	0358	0359	0360	0361	0362	0363	0364	0365	0366	0367
17_	0368	0369	0370	0371	0372	0373	0374	0375	0376	0377	0378	0379	0380	0381	0382	0383
18_	0384	0385	0386	0387	0388	0389	0390	0391	0392	0393	0394	0395	0396	0397	0398	0399
19_	0400	0401	0402	0403	0404	0405	0406	0407	0408	0409	0410	0411	0412	0413	0414	0415
1A_	0416	0417	0418	0419	0420	0421	0422	0423	0424	0425	0426	0427	0428	0429	0430	0431
1B_	0432	0433	0434	0435	0436	0437	0438	0439	0440	0441	0442	0443	0444	0445	0446	0447
1C_	0448	0449	0450	0451	0452	0453	0454	0455	0456	0457	0458	0459	0460	0461	0462	0463
1D_	0464	0465	0466	0467	0468	0469	0470	0471	0472	0473	0474	0475	0476	0477	0478	0479
1E_	0480	0481	0482	0483	0484	0485	0486	0487	0488	0489	0490	0491	0492	0493	0494	0495
1F_	0496	0497	0498	0499	0500	0501	0502	0503	0504	0505	0506	0507	0508	0509	0510	0511

DIRECT CONVERSION TABLE (Continued)

	0	1	2	3	4	5	6	7	8	9	A	B	C	D	E	F
20__	0512	0513	0514	0515	0516	0517	0518	0519	0520	0521	0522	0523	0524	0525	0526	0527
21__	0528	0529	0530	0531	0532	0533	0534	0535	0536	0537	0538	0539	0540	0541	0542	0543
22__	0544	0545	0546	0547	0548	0549	0550	0551	0552	0553	0554	0555	0556	0557	0558	0559
23__	0560	0561	0562	0563	0564	0565	0566	0567	0568	0569	0570	0571	0572	0573	0574	0575
24__	0576	0577	0578	0579	0580	0581	0582	0583	0584	0585	0586	0587	0588	0589	0590	0591
25__	0592	0593	0594	0595	0596	0597	0598	0599	0600	0601	0602	0603	0604	0605	0606	0607
26__	0608	0609	0610	0611	0612	0613	0614	0615	0616	0617	0618	0619	0620	0621	0622	0623
27__	0624	0625	0626	0627	0628	0629	0630	0631	0632	0633	0634	0635	0636	0637	0638	0639
28__	0640	0641	0642	0643	0644	0645	0646	0647	0648	0649	0650	0651	0652	0653	0654	0655
29__	0656	0657	0658	0659	0660	0661	0662	0663	0664	0665	0666	0667	0668	0669	0670	0671
2A__	0672	0673	0674	0675	0676	0677	0678	0679	0680	0681	0682	0683	0684	0685	0686	0687
2B__	0688	0689	0690	0691	0692	0693	0694	0695	0696	0697	0698	0699	0700	0701	0702	0703
2C__	0704	0705	0706	0707	0708	0709	0710	0711	0712	0713	0714	0715	0716	0717	0718	0719
2D__	0720	0721	0722	0723	0724	0725	0726	0727	0728	0729	0730	0731	0732	0733	0734	0735
2E__	0736	0737	0738	0739	0740	0741	0742	0743	0744	0745	0746	0747	0748	0749	0750	0751
2F__	0752	0753	0754	0755	0756	0757	0758	0759	0760	0761	0762	0763	0764	0765	0766	0767
30__	0768	0769	0770	0771	0772	0773	0774	0775	0776	0777	0778	0779	0780	0781	0782	0783
31__	0784	0785	0786	0787	0788	0789	0790	0791	0792	0793	0794	0795	0796	0797	0798	0799
32__	0800	0801	0802	0803	0804	0805	0806	0807	0808	0809	0810	0811	0812	0813	0814	0815
33__	0816	0817	0818	0819	0820	0821	0822	0823	0824	0825	0826	0827	0828	0829	0830	0831
34__	0832	0833	0834	0835	0836	0837	0838	0839	0840	0841	0842	0843	0844	0845	0846	0847
35__	0848	0849	0850	0851	0852	0853	0854	0855	0856	0857	0858	0859	0860	0861	0862	0863
36__	0864	0865	0866	0867	0868	0869	0870	0871	0872	0873	0874	0875	0876	0877	0878	0879
37__	0880	0881	0882	0883	0884	0885	0886	0887	0888	0889	0890	0891	0892	0893	0894	0895
38__	0896	0897	0898	0899	0900	0901	0902	0903	0904	0905	0906	0907	0908	0909	0910	0911
39__	0912	0913	0914	0915	0916	0917	0918	0919	0920	0921	0922	0923	0924	0925	0926	0927
3A__	0928	0929	0930	0931	0932	0933	0934	0935	0936	0937	0938	0939	0940	0941	0942	0943
3B__	0944	0945	0946	0947	0948	0949	0950	0951	0952	0953	0954	0955	0956	0957	0958	0959
3C__	0960	0961	0962	0963	0964	0965	0966	0967	0968	0969	0970	0971	0972	0973	0974	0975
3D__	0976	0977	0978	0979	0980	0981	0982	0983	0984	0985	0986	0987	0988	0989	0990	0991
3E__	0992	0993	0994	0995	0996	0997	0998	0999	1000	1001	1002	1003	1004	1005	1006	1007
3F__	1008	1009	1010	1011	1012	1013	1014	1015	1016	1017	1018	1019	1020	1021	1022	1023
40__	1024	1025	1026	1027	1028	1029	1030	1031	1032	1033	1034	1035	1036	1037	1038	1039
41__	1040	1041	1042	1043	1044	1045	1046	1047	1048	1049	1050	1051	1052	1053	1054	1055
42__	1056	1057	1058	1059	1060	1061	1062	1063	1064	1065	1066	1067	1068	1069	1070	1071
43__	1072	1073	1074	1075	1076	1077	1078	1079	1080	1081	1082	1083	1084	1085	1086	1087
44__	1088	1089	1090	1091	1092	1093	1094	1095	1096	1097	1098	1099	1100	1101	1102	1103
45__	1104	1105	1106	1107	1108	1109	1110	1111	1112	1113	1114	1115	1116	1117	1118	1119
46__	1120	1121	1122	1123	1124	1125	1126	1127	1128	1129	1130	1131	1132	1133	1134	1135
47__	1136	1137	1138	1139	1140	1141	1142	1143	1144	1145	1146	1147	1148	1149	1150	1151
48__	1152	1153	1154	1155	1156	1157	1158	1159	1160	1161	1162	1163	1164	1165	1166	1167
49__	1168	1169	1170	1171	1172	1173	1174	1175	1176	1177	1178	1179	1180	1181	1182	1183
4A__	1184	1185	1186	1187	1188	1189	1190	1191	1192	1193	1194	1195	1196	1197	1198	1199
4B__	1200	1201	1202	1203	1204	1205	1206	1207	1208	1209	1210	1211	1212	1213	1214	1215
4C__	1216	1217	1218	1219	1220	1221	1222	1223	1224	1225	1226	1227	1228	1229	1230	1231
4D__	1232	1233	1234	1235	1236	1237	1238	1239	1240	1241	1242	1243	1244	1245	1246	1247
4E__	1248	1249	1250	1251	1252	1253	1254	1255	1256	1257	1258	1259	1260	1261	1262	1263
4F__	1264	1265	1266	1267	1268	1269	1270	1271	1272	1273	1274	1275	1276	1277	1278	1279
50__	1280	1281	1282	1283	1284	1285	1286	1287	1288	1289	1290	1291	1292	1293	1294	1295
51__	1296	1297	1298	1299	1300	1301	1302	1303	1304	1305	1306	1307	1308	1309	1310	1311
52__	1312	1313	1314	1315	1316	1317	1318	1319	1320	1321	1322	1323	1324	1325	1326	1327
53__	1328	1329	1330	1331	1332	1333	1334	1335	1336	1337	1338	1339	1340	1341	1342	1343
54__	1344	1345	1346	1347	1348	1349	1350	1351	1352	1353	1354	1355	1356	1357	1358	1359
55__	1360	1361	1362	1363	1364	1365	1366	1367	1368	1369	1370	1371	1372	1373	1374	1375
56__	1376	1377	1378	1379	1380	1381	1382	1383	1384	1385	1386	1387	1388	1389	1390	1391
57__	1392	1393	1394	1395	1396	1397	1398	1399	1400	1401	1402	1403	1404	1405	1406	1407
58__	1408	1409	1410	1411	1412	1413	1414	1415	1416	1417	1418	1419	1420	1421	1422	1423
59__	1424	1425	1426	1427	1428	1429	1430	1431	1432	1433	1434	1435	1436	1437	1438	1439
5A__	1440	1441	1442	1443	1444	1445	1446	1447	1448	1449	1450	1451	1452	1453	1454	1455
5B__	1456	1457	1458	1459	1460	1461	1462	1463	1464	1465	1466	1467	1468	1469	1470	1471
5C__	1472	1473	1474	1475	1476	1477	1478	1479	1480	1481	1482	1483	1484	1485	1486	1487
5D__	1488	1489	1490	1491	1492	1493	1494	1495	1496	1497	1498	1499	1500	1501	1502	1503
5E__	1504	1505	1506	1507	1508	1509	1510	1511	1512	1513	1514	1515	1516	1517	1518	1519
5F__	1520	1521	1522	1523	1524	1525	1526	1527	1528	1529	1530	1531	1532	1533	1534	1535

DIRECT CONVERSION TABLE (Continued)

	0	1	2	3	4	5	6	7	8	9	A	B	C	D	E	F
60_	1536	1537	1538	1539	1540	1541	1542	1543	1544	1545	1546	1547	1548	1549	1550	1551
61_	1552	1553	1554	1555	1556	1557	1558	1559	1560	1561	1562	1563	1564	1565	1566	1567
62_	1568	1569	1570	1571	1572	1573	1574	1575	1576	1577	1578	1579	1580	1581	1582	1583
63_	1584	1585	1586	1587	1588	1589	1590	1591	1592	1593	1594	1595	1596	1597	1598	1599
64_	1600	1601	1602	1603	1604	1605	1606	1607	1608	1609	1610	1611	1612	1613	1614	1615
65_	1616	1617	1618	1619	1620	1621	1622	1623	1624	1625	1626	1627	1628	1629	1630	1631
66_	1632	1633	1634	1635	1636	1637	1638	1639	1640	1641	1642	1643	1644	1645	1646	1647
67_	1648	1649	1650	1651	1652	1653	1654	1655	1656	1657	1658	1659	1660	1661	1662	1663
68_	1664	1665	1666	1667	1668	1669	1670	1671	1672	1673	1674	1675	1676	1677	1678	1679
69_	1680	1681	1682	1683	1684	1685	1686	1687	1688	1689	1690	1691	1692	1693	1694	1695
6A_	1696	1697	1698	1699	1700	1701	1702	1703	1704	1705	1706	1707	1708	1709	1710	1711
6B_	1712	1713	1714	1715	1716	1717	1718	1719	1720	1721	1722	1723	1724	1725	1726	1727
6C_	1728	1729	1730	1731	1732	1733	1734	1735	1736	1737	1738	1739	1740	1741	1742	1743
6D_	1744	1745	1746	1747	1748	1749	1750	1751	1752	1753	1754	1755	1756	1757	1758	1759
6E_	1760	1761	1762	1763	1764	1765	1766	1767	1768	1769	1770	1771	1772	1773	1774	1775
6F_	1776	1777	1778	1779	1780	1781	1782	1783	1784	1785	1786	1787	1788	1789	1790	1791
70_	1792	1793	1794	1795	1796	1797	1798	1799	1800	1801	1802	1803	1804	1805	1806	1807
71_	1808	1809	1810	1811	1812	1813	1814	1815	1816	1817	1818	1819	1820	1821	1822	1823
72_	1824	1825	1826	1827	1828	1829	1830	1831	1832	1833	1834	1835	1836	1837	1838	1839
73_	1840	1841	1842	1843	1844	1845	1846	1847	1848	1849	1850	1851	1852	1853	1854	1855
74_	1856	1857	1858	1859	1860	1861	1862	1863	1864	1865	1866	1867	1868	1869	1870	1871
75_	1872	1873	1874	1875	1876	1877	1878	1879	1880	1881	1882	1883	1884	1885	1886	1887
76_	1888	1889	1890	1891	1892	1893	1894	1895	1896	1897	1898	1899	1900	1901	1902	1903
77_	1904	1905	1906	1907	1908	1909	1910	1911	1912	1913	1914	1915	1916	1917	1918	1919
78_	1920	1921	1922	1923	1924	1925	1926	1927	1928	1929	1930	1931	1932	1933	1934	1935
79_	1936	1937	1938	1939	1940	1941	1942	1943	1944	1945	1946	1947	1948	1949	1950	1951
7A_	1952	1953	1954	1955	1956	1957	1958	1959	1960	1961	1962	1963	1964	1965	1966	1967
7B_	1968	1969	1970	1971	1972	1973	1974	1975	1976	1977	1978	1979	1980	1981	1982	1983
7C_	1984	1985	1986	1987	1988	1989	1990	1991	1992	1993	1994	1995	1996	1997	1998	1999
7D_	2000	2001	2002	2003	2004	2005	2006	2007	2008	2009	2010	2011	2012	2013	2014	2015
7E_	2016	2017	2018	2019	2020	2021	2022	2023	2024	2025	2026	2027	2028	2029	2030	2031
7F_	2032	2033	2034	2035	2036	2037	2038	2039	2040	2041	2042	2043	2044	2045	2046	2047
80_	2048	2049	2050	2051	2052	2053	2054	2055	2056	2057	2058	2059	2060	2061	2062	2063
81_	2064	2065	2066	2067	2068	2069	2070	2071	2072	2073	2074	2075	2076	2077	2078	2079
82_	2080	2081	2082	2083	2084	2085	2086	2087	2088	2089	2090	2091	2092	2093	2094	2095
83_	2096	2097	2098	2099	2100	2101	2102	2103	2104	2105	2106	2107	2108	2109	2110	2111
84_	2112	2113	2114	2115	2116	2117	2118	2119	2120	2121	2122	2123	2124	2125	2126	2127
85_	2128	2129	2130	2131	2132	2133	2134	2135	2136	2137	2138	2139	2140	2141	2142	2143
86_	2144	2145	2146	2147	2148	2149	2150	2151	2152	2153	2154	2155	2156	2157	2158	2159
87_	2160	2161	2162	2163	2164	2165	2166	2167	2168	2169	2170	2171	2172	2173	2174	2175
88_	2176	2177	2178	2179	2180	2181	2182	2183	2184	2185	2186	2187	2188	2189	2190	2191
89_	2192	2193	2194	2195	2196	2197	2198	2199	2200	2201	2202	2203	2204	2205	2206	2207
8A_	2208	2209	2210	2211	2212	2213	2214	2215	2216	2217	2218	2219	2220	2221	2222	2223
8B_	2224	2225	2226	2227	2228	2229	2230	2231	2232	2233	2234	2235	2236	2237	2238	2239
8C_	2240	2241	2242	2243	2244	2245	2246	2247	2248	2249	2250	2251	2252	2253	2254	2255
8D_	2256	2257	2258	2259	2260	2261	2262	2263	2264	2265	2266	2267	2268	2269	2270	2271
8E_	2272	2273	2274	2275	2276	2277	2278	2279	2280	2281	2282	2283	2284	2285	2286	2287
8F_	2288	2289	2290	2291	2292	2293	2294	2295	2296	2297	2298	2299	2300	2301	2302	2303
90_	2304	2305	2306	2307	2308	2309	2310	2311	2312	2313	2314	2315	2316	2317	2318	2319
91_	2320	2321	2322	2323	2324	2325	2326	2327	2328	2329	2330	2331	2332	2333	2334	2335
92_	2336	2337	2338	2339	2340	2341	2342	2343	2344	2345	2346	2347	2348	2349	2350	2351
93_	2352	2353	2354	2355	2356	2357	2358	2359	2360	2361	2362	2363	2364	2365	2366	2367
94_	2368	2369	2370	2371	2372	2373	2374	2375	2376	2377	2378	2379	2380	2381	2382	2383
95_	2384	2385	2386	2387	2388	2389	2390	2391	2392	2393	2394	2395	2396	2397	2398	2399
96_	2400	2401	2402	2403	2404	2405	2406	2407	2408	2409	2410	2411	2412	2413	2414	2415
97_	2416	2417	2418	2419	2420	2421	2422	2423	2424	2425	2426	2427	2428	2429	2430	2431
98_	2432	2433	2434	2435	2436	2437	2438	2439	2440	2441	2442	2443	2444	2445	2446	2447
99_	2448	2449	2450	2451	2452	2453	2454	2455	2456	2457	2458	2459	2460	2461	2462	2463
9A_	2464	2465	2466	2467	2468	2469	2470	2471	2472	2473	2474	2475	2476	2477	2478	2479
9B_	2480	2481	2482	2483	2484	2485	2486	2487	2488	2489	2490	2491	2492	2493	2494	2495
9C_	2496	2497	2498	2499	2500	2501	2502	2503	2504	2505	2506	2507	2508	2509	2510	2511
9D_	2512	2513	2514	2515	2516	2517	2518	2519	2520	2521	2522	2523	2524	2525	2526	2527
9E_	2528	2529	2530	2531	2532	2533	2534	2535	2536	2537	2538	2539	2540	2541	2542	2543
9F_	2544	2545	2546	2547	2548	2549	2550	2551	2552	2553	2554	2555	2556	2557	2558	2559

DIRECT CONVERSION TABLE (Continued)

	0	1	2	3	4	5	6	7	8	9	A	B	C	D	E	F
A0_	2560	2561	2562	2563	2564	2565	2566	2567	2568	2569	2570	2571	2572	2573	2574	2575
A1_	2576	2577	2578	2579	2580	2581	2582	2583	2584	2585	2586	2587	2588	2589	2590	2591
A2_	2592	2593	2594	2595	2596	2597	2598	2599	2600	2601	2602	2603	2604	2605	2606	2607
A3_	2608	2609	2610	2611	2612	2613	2614	2615	2616	2617	2618	2619	2620	2621	2622	2623
A4_	2624	2625	2626	2627	2628	2629	2630	2631	2632	2633	2634	2635	2636	2637	2638	2639
A5_	2640	2641	2642	2643	2644	2645	2646	2647	2648	2649	2650	2651	2652	2653	2654	2655
A6_	2656	2657	2658	2659	2660	2661	2662	2663	2664	2665	2666	2667	2668	2669	2670	2671
A7_	2672	2673	2674	2675	2676	2677	2678	2679	2680	2681	2682	2683	2684	2685	2686	2687
A8_	2688	2689	2690	2691	2692	2693	2694	2695	2696	2697	2698	2699	2700	2701	2702	2703
A9_	2704	2705	2706	2707	2708	2709	2710	2711	2712	2713	2714	2715	2716	2717	2718	2719
AA_	2720	2721	2722	2723	2724	2725	2726	2727	2728	2729	2730	2731	2732	2733	2734	2735
AB_	2736	2737	2738	2739	2740	2741	2742	2743	2744	2745	2746	2747	2748	2749	2750	2751
AC_	2752	2753	2754	2755	2756	2757	2758	2759	2760	2761	2762	2763	2764	2765	2766	2767
AD_	2768	2769	2770	2771	2772	2773	2774	2775	2776	2777	2778	2779	2780	2781	2782	2783
AE_	2784	2785	2786	2787	2788	2789	2790	2791	2792	2793	2794	2795	2796	2797	2798	2799
AF_	2800	2801	2802	2803	2804	2805	2806	2807	2808	2809	2810	2811	2812	2813	2814	2815
B0_	2816	2817	2818	2819	2820	2821	2822	2823	2824	2825	2826	2827	2828	2829	2830	2831
B1_	2832	2833	2834	2835	2836	2837	2838	2839	2840	2841	2842	2843	2844	2845	2846	2847
B2_	2848	2849	2850	2851	2852	2853	2854	2855	2856	2857	2858	2859	2860	2861	2862	2863
B3_	2864	2865	2866	2867	2868	2869	2870	2871	2872	2873	2874	2875	2876	2877	2878	2879
B4_	2880	2881	2882	2883	2884	2885	2886	2887	2888	2889	2890	2891	2892	2893	2894	2895
B5_	2896	2897	2898	2899	2900	2901	2902	2903	2904	2905	2906	2907	2908	2909	2910	2911
B6_	2912	2913	2914	2915	2916	2917	2918	2919	2920	2921	2922	2923	2924	2925	2926	2927
B7_	2928	2929	2930	2931	2932	2933	2934	2935	2936	2937	2938	2939	2940	2941	2942	2943
B8_	2944	2945	2946	2947	2948	2949	2950	2951	2952	2953	2954	2955	2956	2957	2958	2959
B9_	2960	2961	2962	2963	2964	2965	2966	2967	2968	2969	2970	2971	2972	2973	2974	2975
BA_	2976	2977	2978	2979	2980	2981	2982	2983	2984	2985	2986	2987	2988	2989	2990	2991
BB_	2992	2993	2994	2995	2996	2997	2998	2999	3000	3001	3002	3003	3004	3005	3006	3007
BC_	3008	3009	3010	3011	3012	3013	3014	3015	3016	3017	3018	3019	3020	3021	3022	3023
BD_	3024	3025	3026	3027	3028	3029	3030	3031	3032	3033	3034	3035	3036	3037	3038	3039
BE_	3040	3041	3042	3043	3044	3045	3046	3047	3048	3049	3050	3051	3052	3053	3054	3055
BF_	3056	3057	3058	3059	3060	3061	3062	3063	3064	3065	3066	3067	3068	3069	3070	3071
C0_	3072	3073	3074	3075	3076	3077	3078	3079	3080	3081	3082	3083	3084	3085	3086	3087
C1_	3088	3089	3090	3091	3092	3093	3094	3095	3096	3097	3098	3099	3100	3101	3102	3103
C2_	3104	3105	3106	3107	3108	3109	3110	3111	3112	3113	3114	3115	3116	3117	3118	3119
C3_	3120	3121	3122	3123	3124	3125	3126	3127	3128	3129	3130	3131	3132	3133	3134	3135
C4_	3136	3137	3138	3139	3140	3141	3142	3143	3144	3145	3146	3147	3148	3149	3150	3151
C5_	3152	3153	3154	3155	3156	3157	3158	3159	3160	3161	3162	3613	3164	3165	3166	3167
C6_	3168	3169	3170	3171	3172	3173	3174	3175	3176	3177	3178	3179	3180	3181	3182	3183
C7_	3184	3185	3186	3187	3188	3189	3190	3191	3192	3193	3194	3195	3196	3197	3198	3199
C8_	3200	3201	3202	3203	3204	3205	3206	3207	3208	3209	3210	3211	3212	3213	3214	3215
C9_	3216	3217	3218	3219	3220	3221	3222	3223	3224	3225	3226	3227	3228	3229	3230	3231
CA_	3232	3233	3234	3235	3236	3237	3238	3239	3240	3241	3242	3243	3244	3245	3246	3247
CB_	3248	3249	3250	3251	3252	3253	3254	3255	3256	3257	3258	3259	3260	3261	3262	3263
CC_	3264	3265	3266	3267	3268	3269	3270	3271	3272	3273	3274	3275	3276	3277	3278	3279
CD_	3280	3281	3282	3283	3284	3285	3286	3287	3288	3289	3290	3291	3292	3293	3294	3295
CE_	3296	3297	3298	3299	3300	3301	3302	3303	3304	3305	3306	3307	3308	3309	3310	3311
CF_	3312	3313	3314	3315	3316	3317	3318	3319	3320	3321	3322	3323	3324	3325	3326	3327
D0_	3328	3329	3330	3331	3332	3333	3334	3335	3336	3337	3338	3339	3340	3341	3342	3343
D1_	3344	3345	3346	3347	3348	3349	3350	3351	3352	3353	3354	3355	3356	3357	3358	3359
D2_	3360	3361	3362	3363	3364	3365	3366	3367	3368	3369	3370	3371	3372	3373	3374	3375
D3_	3376	3377	3378	3379	3380	3381	3382	3383	3384	3385	3386	3387	3388	3389	3390	3391
D4_	3392	3393	3394	3395	3396	3397	3398	3399	3400	3401	3402	3403	3404	3405	3406	3407
D5_	3408	3409	3410	3411	3412	3413	3414	3415	3416	3417	3418	3419	3420	3421	3422	3423
D6_	3424	3425	3426	3427	3428	3429	3430	3431	3432	3433	3434	3435	3436	3437	3438	3439
D7_	3440	3441	3442	3443	3444	3445	3446	3447	3448	3449	3450	3451	3452	3453	3454	3455
D8_	3456	3457	3458	3459	3460	3461	3462	3463	3464	3465	3466	3467	3468	3469	3470	3471
D9_	3472	3473	3474	3475	3476	3477	3478	3479	3480	3481	3482	3483	3484	3485	3486	3487
DA_	3488	3489	3490	3491	3492	3493	3494	3495	3496	3497	3498	3499	3500	3501	3502	3503
DB_	3504	3505	3506	3507	3508	3509	3510	3511	3512	3513	3514	3515	3516	3517	3518	3519
DC_	3520	3521	3522	3523	3524	3525	3526	3527	3528	3529	3530	3531	3532	3533	3534	3535
DD_	3536	3537	3538	3539	3540	3541	3542	3543	3544	3545	3546	3547	3548	3549	3550	3551
DE_	3552	3553	3554	3555	3556	3557	3558	3559	3560	3561	3562	3563	3564	3565	3566	3567
DF_	3568	3569	3570	3571	3572	3573	3574	3575	3576	3577	3578	3579	3580	3581	3582	3583

DIRECT CONVERSION TABLE (Continued)

	0	1	2	3	4	5	6	7	8	9	A	B	C	D	E	F
E0__	3584	3585	3586	3587	3588	3589	3590	3591	3592	3593	3594	3595	3596	3597	3598	3599
E1__	3600	3601	3602	3603	3604	3605	3606	3607	3608	3609	3610	3611	3612	3613	3614	3615
E2__	3616	3617	3618	3619	3620	3621	3622	3623	3624	3625	3626	3627	3628	3629	3630	3631
E3__	3632	3633	3634	3635	3636	3637	3638	3639	3640	3641	3642	3643	3644	3645	3646	3647
E4__	3648	3649	3650	3651	3652	3653	3654	3655	3656	3657	3658	3659	3660	3661	3662	3663
E5__	3664	3665	3666	3667	3668	3669	3670	3671	3672	3673	3674	3675	3676	3677	3678	3679
E6__	3680	3681	3682	3683	3684	3685	3686	3687	3688	3689	3690	3691	3692	3693	3694	3695
E7__	3696	3697	3698	3699	3700	3701	3702	3703	3704	3705	3706	3707	3708	3709	3710	3711
E8__	3712	3713	3714	3715	3716	3717	3718	3719	3720	3721	3722	3723	3724	3725	3726	3727
E9__	3728	3729	3730	3731	3732	3733	3734	3735	3736	3737	3738	3739	3740	3741	3742	3743
EA__	3744	3745	3746	3747	3748	3749	3750	3751	3752	3753	3754	3755	3756	3757	3758	3759
EB__	3760	3761	3762	3763	3764	3765	3766	3767	3768	3769	3770	3771	3772	3773	3774	3775
EC__	3776	3777	3778	3779	3780	3781	3782	3783	3784	3785	3786	3787	3788	3789	3790	3791
ED__	3792	3793	3794	3795	3796	3797	3798	3799	3800	3801	3802	3803	3804	3805	3806	3807
EE__	3808	3809	3810	3811	3812	3813	3814	3815	3816	3817	3818	3819	3820	3821	3822	3823
EF__	3824	3825	3826	3827	3828	3829	3830	3831	3832	3833	3834	3835	3836	3837	3838	3839
F0__	3840	3841	3842	3843	3844	3845	3846	3847	3848	3849	3850	3851	3852	3853	3854	3855
F1__	3856	3857	3858	3859	3860	3861	3862	3863	3864	3865	3866	3867	3868	3869	3870	3871
F2__	3872	3873	3874	3875	3876	3877	3878	3879	3880	3881	3882	3883	3884	3885	3886	3887
F3__	3888	3889	3890	3891	3892	3893	3894	3895	3896	3897	3898	3899	3900	3901	3902	3903
F4__	3904	3905	3906	3907	3908	3909	3910	3911	3912	3913	3914	3915	3916	3917	3918	3919
F5__	3920	3921	3922	3923	3924	3925	3926	3927	3928	3929	3930	3931	3932	3933	3934	3935
F6__	3936	3937	3938	3939	3940	3941	3942	3943	3944	3945	3946	3947	3948	3949	3950	3951
F7__	3952	3953	3954	3955	3956	3957	3958	3959	3960	3961	3962	3963	3964	3965	3966	3967
F8__	3968	3969	3970	3971	3972	3973	3974	3975	3976	3977	3978	3979	3980	3981	3982	3983
F9__	3984	3985	3986	3987	3988	3989	3990	3991	3992	3993	3994	3995	3996	3997	3998	3999
FA__	4000	4001	4002	4003	4004	4005	4006	4007	4008	4009	4010	4011	4012	4013	4014	4015
FB__	4016	4017	4018	4019	4020	4021	4022	4023	4024	4025	4026	4027	4028	4029	4030	4031
FC__	4032	4033	4034	4035	4036	4037	4038	4039	4040	4041	4042	4043	4044	4045	4046	4047
FD__	4048	4049	4050	4051	4052	4053	4054	4055	4056	4057	4058	4059	4060	4061	4062	4063
FE__	4064	4065	4066	4067	4068	4069	4070	4071	4072	4073	4074	4075	4076	4077	4078	4079
FF__	4080	4081	4082	4083	4084	4085	4086	4087	4088	4089	4090	4091	4092	4093	4094	4095

II. HEXADECIMAL AND DECIMAL INTEGER CONVERSION TABLE

	8		7		6		5		4		3		2		1	
Hex	Decimal	Hex	Decimal	Hex	Decimal	Hex	Decimal	Hex	Decimal	Hex	Decimal	Hex	Decimal	Hex	Decimal
0	0	0	0	0	0	0	0	0	0	0	0	0	0	0	0
1	268,435,456	1	16,777,216	1	1,048,576	1	65,536	1	4,096	1	256	1	16	1	1
2	536,870,912	2	33,554,432	2	2,097,152	2	131,072	2	8,192	2	512	2	32	2	2
3	805,306,368	3	50,331,648	3	3,145,728	3	196,608	3	12,288	3	768	3	48	3	3
4	1,073,741,824	4	67,108,864	4	4,194,304	4	262,144	4	16,384	4	1,024	4	64	4	4
5	1,342,177,280	5	83,886,080	5	5,242,880	5	327,680	5	20,480	5	1,280	5	80	5	5
6	1,610,612,736	6	100,663,296	6	6,291,456	6	393,216	6	24,576	6	1,536	6	96	6	6
7	1,879,048,192	7	117,440,512	7	7,340,032	7	458,752	7	28,672	7	1,792	7	112	7	7
8	2,147,483,648	8	134,217,728	8	8,388,608	8	524,288	8	32,768	8	2,048	8	128	8	8
9	2,415,919,104	9	150,994,944	9	9,437,184	9	589,824	9	36,864	9	2,304	9	144	9	9
A	2,684,354,560	A	167,772,160	A	10,485,760	A	655,360	A	40,960	A	2,560	A	160	A	10
B	2,952,790,016	B	184,549,376	B	11,534,336	B	720,896	B	45,056	B	2,816	B	176	B	11
C	3,221,225,472	C	201,326,592	C	12,582,912	C	786,432	C	49,152	C	3,072	C	192	C	12
D	3,489,660,928	D	218,103,808	D	13,631,488	D	851,968	D	53,248	D	3,328	D	208	D	13
E	3,758,096,384	E	234,881,024	E	14,680,064	F	917,504	E	57,344	E	3,584	E	224	E	14
F	4,026,531,840	F	251,658,240	F	15,728,640	F	983,040	F	61,440	F	3,840	F	240	F	15
	8		7		6		5		4		3		2		1

INTEGER CONVERSION TABLE (Continued)

TO CONVERT HEXADECIMAL TO DECIMAL

1. Locate the column of decimal numbers corresponding to the left-most digit or letter of the hexadecimal; select from this column and record the number that corresponds to the position of the hexadecimal digit or letter.
2. Repeat step 1 for the next (second from the left) position.
3. Repeat step 1 for the units (third from the left) position.
4. Add the numbers selected from the table to form the decimal number.

To convert integer numbers greater than the capacity of table, use the techniques below:

HEXADECIMAL TO DECIMAL

Successive cumulative multiplication from left to right, adding units position.

Example: $D34_{16} = 3380_{10}$

$$
\begin{array}{rr}
D = & 13 \\
& \times 16 \\
\hline
& 208 \\
3 = & +3 \\
\hline
& 211 \\
& \times 16 \\
\hline
& 3376 \\
4 = & +4 \\
\hline
& 3380 \\
\end{array}
$$

TO CONVERT DECIMAL TO HEXADECIMAL

1. (a) Select from the table the highest decimal number that is equal to or less than the number to be converted.
   (b) Record the hexadecimal of the column containing the selected number.
   (c) Subtract the selected decimal from the number to be converted.
2. Using the remainder from step 1(c) repeat all of step 1 to develop the second position of the hexadecimal (and a remainder).
3. Using the remainder from step 2 repeat all of step 1 to develop the units position of the hexadecimal.
4. Combine terms to form the hexadecimal number.

DECIMAL TO HEXADECIMAL

Divide and collect the remainder in reverse order.

Example: $3380_{10} = X_{16}$

$$
\begin{array}{ll}
16\ \underline{|\ 3380} & \text{remainder} \\
16\ \underline{|\ 211} & 4 \\
16\ \underline{|\ 13} & 3 \\
 D & \\
\end{array}
$$

$3380_{10} = D34_{16}$

EXAMPLE

Conversion of
Hexadecimal
Value　　　　D34

1. D		3328
2. 3		48
3. 4		4
4. Decimal		3380

EXAMPLE

Conversion of
Decimal
Value　　　　3380

1. D		−3328
		52
2. 3		−48
		4
3. 4		−4
4. Hexa-decimal		D34

INTEGER CONVERSION TABLE (Continued)

POWERS OF 16 TABLE

Example: $268,435,456_{10} = (2.68435456 \times 10^8)_{10} = 1000\ 0000_{16} = (10^7)_{16}$

16^n	n
1	0
16	1
256	2
4 096	3
65 536	4
1 048 576	5
16 777 216	6
268 435 456	7
4 294 967 296	8
68 719 476 736	9
1 099 511 627 776	10 = A
17 592 186 044 416	11 = B
281 474 976 710 656	12 = C
4 503 599 627 370 496	13 = D
72 057 594 037 927 936	14 = E
1 152 921 504 606 846 976	15 = F

Decimal Values

III. HEXADECIMAL AND DECIMAL FRACTION CONVERSION TABLE

	1		2		3			4				
Hex	Decimal	Hex	Decimal		Hex	Decimal			Hex	Decimal Equivalent			
.0	.0000	.00	.0000	0000	.000	.0000	0000	0000	.0000	.0000	0000	0000	0000
.1	.0625	.01	.0039	0625	.001	.0002	4414	0625	.0001	.0000	1525	8789	0625
.2	.1250	.02	.0078	1250	.002	.0004	8828	1250	.0002	.0000	3051	7578	1250
.3	.1875	.03	.0117	1875	.003	.0007	3242	1875	.0003	.0000	4577	6367	1875
4	2500	.04	.0156	2500	.004	.0009	7656	2500	.0004	.0000	6103	5156	2500
.5	.3125	.05	.0195	3125	.005	.0012	2070	3125	.0005	.0000	7629	3945	3125
.6	.3750	.06	.0234	3750	.006	.0014	6484	3750	.0006	.0000	9155	2734	3750
.7	.4375	.07	.0273	4375	.007	.0017	0898	4375	.0007	.0001	0681	1523	4375
.8	.5000	.08	.0312	5000	.008	.0019	5312	5000	.0008	.0001	2207	0312	5000
.9	.5625	.09	.0351	5625	.009	.0021	9726	5625	.0009	.0001	3732	9101	5625
.A	.6250	.0A	.0390	6250	.00A	.0024	4140	6250	.000A	.0001	5258	7890	6250
.B	.6875	.0B	.0429	6875	.00B	.0026	8554	6875	.000B	.0001	6784	6679	6875
.C	.7500	.0C	.0468	7500	.00C	.0029	2968	7500	.000C	.0001	8310	5468	7500
.D	.8125	.0D	.0507	8125	.00D	.0031	7382	8125	.000D	.0001	9836	4257	8125
.E	.8750	.0E	.0546	8750	.00E	.0034	1796	8750	.000E	.0002	1362	3046	8750
.F	.9375	.0F	.0585	9375	.00F	.0036	6210	9375	.000F	.0002	2888	1835	9375
	1		2		3				4				

TO CONVERT .ABC HEXADECIMAL TO DECIMAL

Find .A	in position 1	.6250
Find .0B	in position 2	.0429 6875
Find .00C	in position 3	.0029 2968 7500
.ABC Hex is equal to		.6708 9843 7500

COMBINATORIAL ANALYSIS

FRACTION CONVERSION TABLE (Continued)

TO CONVERT .13 DECIMAL TO HEXADECIMAL

1. Find .1250 next lowest to   .1300
        subtract   − .1250   = .2 Hex
2. Find .0039 0625 next lowest to   .0050 0000
         − .0039 0625   = .01
3. Find .0009 7656 2500   .0010 9375 0000
         − .0009 7656 2500   = .004
4. Find .0001 0681 1523 4375   .0001 1718 7500 0000
         − .0001 0681 1523 4375 = .0007
         .0000 1037 5976 5625 = .2147 Hex

5. .13 Decimal is approximately equal to ⟶

To convert fractions beyond the capacity of table, use techniques below:

HEXADECIMAL FRACTION TO DECIMAL

Convert the hexadecimal fraction to its decimal equivalent using the same technique as for integer numbers. Divide the results by 16^n (n is the number of fraction positions).

Example: $.8A7_{16} = .540771_{10}$

$$8A7_{16} = 2215_{10}$$
$$16^3 = 4096$$

$$4096\overline{\smash{\big)}\,2215.000000} \quad .540771$$

DECIMAL FRACTION TO HEXADECIMAL

Collect integer parts of product in the order of calculation.

Example: $.5408_{10} = .8A7_{16}$

$$
\begin{array}{r}
.5408 \\
\times 16 \\
\hline
8 \leftarrow \boxed{8}\ .6528 \\
\times 16 \\
\hline
A \leftarrow \boxed{10}\ .4448 \\
\times 16 \\
\hline
7 \leftarrow \boxed{7}\ .1168
\end{array}
$$

HEXADECIMAL ADDITION AND SUBTRACTION TABLE

Example: 6 + 2 = 8, 8 − 2 = 6, and 8 − 6 = 2

	1	2	3	4	5	6	7	8	9	A	B	C	D	E	F
1	02	03	04	05	06	07	08	09	0A	0B	0C	0D	0E	0F	10
2	03	04	05	06	07	08	09	0A	0B	0C	0D	0E	0F	10	11
3	04	05	06	07	08	09	0A	0B	0C	0D	0E	0F	10	11	12
4	05	06	07	08	09	0A	0B	0C	0D	0E	0F	10	11	12	13
5	06	07	08	09	0A	0B	0C	0D	0E	0F	10	11	12	13	14
6	07	08	09	0A	0B	0C	0D	0E	0F	10	11	12	13	14	15
7	08	09	0A	0B	0C	0D	0E	0F	10	11	12	13	14	15	16
8	09	0A	0B	0C	0D	0E	0F	10	11	12	13	14	15	16	17
9	0A	0B	0C	0D	0E	0F	10	11	12	13	14	15	16	17	18
A	0B	0C	0D	0E	0F	10	11	12	13	14	15	16	17	18	19
B	0C	0D	0E	0F	10	11	12	13	14	15	16	17	18	19	1A
C	0D	0E	0F	10	11	12	13	14	15	16	17	18	19	1A	1B
D	0E	0F	10	11	12	13	14	15	16	17	18	19	1A	1B	1C
E	0F	10	11	12	13	14	15	16	17	18	19	1A	1B	1C	1D
F	10	11	12	13	14	15	16	17	18	19	1A	1B	1C	1D	1E

HEXADECIMAL MULTIPLICATION TABLE

Example: 2 × 4 = 08, F × 2 = 1E

	1	2	3	4	5	6	7	8	9	A	B	C	D	E	F
1	01	02	03	04	05	06	07	08	09	0A	0B	0C	0D	0E	0F
2	02	04	06	08	0A	0C	0E	10	12	14	16	18	1A	1C	1E
3	03	06	09	0C	0F	12	15	18	1B	1E	21	24	27	2A	2D
4	04	08	0C	10	14	18	1C	20	24	28	2C	30	34	38	3C
5	05	0A	0F	14	19	1E	23	28	2D	32	37	3C	41	46	4B
6	06	0C	12	18	1E	24	2A	30	36	3C	42	48	4E	54	5A
7	07	0E	15	1C	23	2A	31	38	3F	46	4D	54	5B	62	69
08	08	10	18	20	28	30	38	40	48	50	58	60	68	70	78
9	09	12	1B	24	2D	36	3F	48	51	5A	63	6C	75	7E	87
A	0A	14	1E	28	32	3C	46	50	5A	64	6E	78	82	8C	96
B	0B	16	21	2C	37	42	4D	58	63	6E	79	84	8F	9A	A5
C	0C	18	24	30	3C	48	54	60	6C	78	84	90	9C	A8	B4
D	0D	1A	27	34	41	4E	5B	68	75	82	8F	9C	A9	B6	C3
E	0E	1C	2A	38	46	54	62	70	7E	8C	9A	A8	B6	C4	D2
F	0F	1E	2D	3C	4B	5A	69	78	87	96	A5	B4	C3	D2	E1

COMBINATORIAL ANALYSIS

PRIMES

The following table contains all primes from 1 to 100,000. Taken from the Handbook of Mathematical Functions, Applied Mathematics Series 55, Permission received from National Bureau of Standards, Washington, D. C.

PRIMES

Row	0	1	2	3	4	5	6	7	8	9	10	11	12	13	14	15	16	17	18	19	20	21	22	23	24
1	2	547	1229	1993	2749	3581	4421	5281	6143	7001	7927	8837	9739	10663	11677	12569	13513	14533	15413	16411	17393	18329	19427	20359	21391
2	3	557	1231	1997	2753	3583	4423	5297	6151	7013	7933	8839	9743	10667	11681	12577	13523	14537	15427	16417	17401	18341	19429	20369	21397
3	5	563	1237	1999	2767	3593	4441	5303	6163	7019	7937	8849	9749	10687	11689	12583	13537	14543	15439	16421	17417	18353	19433	20389	21401
4	7	569	1249	2003	2777	3607	4447	5309	6173	7027	7949	8861	9767	10691	11699	12589	13553	14549	15443	16427	17419	18367	19441	20393	21407
5	11	571	1259	2011	2789	3613	4451	5323	6197	7039	7951	8863	9769	10709	11701	12601	13567	14551	15451	16433	17431	18371	19447	20399	21419
6	13	577	1277	2017	2791	3617	4457	5333	6199	7043	7963	8867	9781	10711	11717	12611	13577	14557	15461	16447	17443	18379	19457	20407	21433
7	17	587	1279	2027	2797	3623	4463	5347	6203	7057	7993	8887	9787	10723	11719	12613	13591	14561	15467	16451	17449	18397	19463	20411	21467
8	19	593	1283	2029	2801	3631	4481	5351	6211	7069	8009	8893	9791	10729	11731	12619	13597	14563	15473	16453	17467	18401	19469	20431	21481
9	23	599	1289	2039	2803	3637	4483	5381	6217	7079	8011	8923	9803	10733	11743	12637	13613	14591	15493	16477	17471	18413	19471	20441	21487
10	29	601	1291	2053	2819	3643	4493	5387	6221	7103	8017	8929	9811	10739	11777	12641	13619	14593	15497	16481	17477	18427	19477	20443	21491
11	31	607	1297	2063	2833	3659	4507	5393	6229	7109	8039	8933	9817	10753	11779	12647	13627	14621	15511	16487	17483	18433	19483	20477	21493
12	37	613	1301	2069	2837	3671	4513	5399	6247	7121	8053	8941	9829	10771	11783	12653	13633	14627	15527	16493	17489	18439	19489	20479	21499
13	41	617	1303	2081	2843	3673	4517	5407	6257	7127	8059	8951	9833	10781	11789	12659	13649	14629	15541	16519	17491	18443	19501	20483	21503
14	43	619	1307	2083	2851	3677	4519	5413	6263	7129	8069	8963	9839	10789	11801	12671	13669	14633	15551	16529	17497	18451	19507	20507	21517
15	47	631	1319	2087	2857	3691	4523	5417	6269	7151	8081	8969	9851	10799	11807	12689	13679	14639	15559	16547	17509	18457	19531	20509	21521
16	53	641	1321	2089	2861	3697	4547	5419	6271	7159	8087	8971	9857	10831	11813	12697	13681	14653	15569	16553	17519	18461	19541	20521	21523
17	59	643	1327	2099	2879	3701	4549	5431	6277	7177	8089	8999	9859	10837	11821	12703	13687	14657	15581	16561	17539	18481	19543	20533	21529
18	61	647	1361	2111	2887	3709	4561	5437	6287	7187	8093	9001	9871	10847	11827	12713	13691	14669	15583	16567	17551	18493	19553	20543	21557
19	67	653	1367	2113	2897	3719	4567	5441	6299	7193	8101	9007	9883	10853	11831	12721	13693	14683	15601	16573	17569	18503	19559	20549	21559
20	71	659	1373	2129	2903	3727	4583	5443	6301	7207	8111	9011	9887	10859	11833	12739	13697	14699	15607	16603	17573	18517	19571	20551	21563
21	73	661	1381	2131	2909	3733	4591	5449	6311	7211	8117	9013	9901	10861	11839	12743	13709	14713	15619	16607	17579	18521	19577	20563	21569
22	79	673	1399	2137	2917	3739	4597	5471	6317	7213	8123	9029	9907	10867	11863	12757	13711	14717	15629	16619	17581	18523	19583	20593	21577
23	83	677	1409	2141	2927	3761	4603	5477	6323	7219	8147	9041	9923	10883	11867	12763	13721	14723	15641	16631	17597	18539	19597	20599	21587
24	89	683	1423	2143	2939	3767	4621	5479	6329	7229	8161	9043	9929	10889	11887	12781	13723	14731	15643	16633	17599	18541	19603	20611	21589
25	97	691	1427	2153	2953	3769	4637	5483	6337	7237	8167	9049	9931	10891	11897	12791	13729	14737	15647	16649	17609	18553	19609	20627	21599
26	101	701	1429	2161	2957	3779	4639	5501	6343	7243	8171	9059	9941	10903	11903	12799	13751	14741	15649	16651	17623	18583	19661	20639	21601
27	103	709	1433	2179	2963	3793	4643	5503	6353	7247	8179	9067	9949	10909	11909	12809	13757	14747	15661	16657	17627	18587	19681	20641	21611
28	107	719	1439	2203	2969	3797	4649	5507	6359	7253	8191	9091	9967	10937	11923	12821	13759	14753	15667	16661	17657	18593	19687	20663	21613
29	109	727	1447	2207	2971	3803	4651	5519	6361	7283	8209	9103	9973	10939	11927	12823	13763	14759	15671	16673	17659	18617	19697	20681	21617
30	113	733	1451	2213	2999	3821	4657	5521	6367	7297	8219	9109	10007	10949	11933	12829	13781	14767	15679	16691	17669	18637	19699	20693	21647
31	127	739	1453	2221	3001	3823	4663	5527	6373	7307	8221	9127	10009	10957	11939	12841	13789	14771	15683	16693	17681	18661	19709	20707	21649
32	131	743	1459	2237	3011	3847	4673	5531	6379	7309	8231	9133	10037	10973	11941	12853	13799	14779	15727	16699	17683	18671	19717	20717	21661
33	137	751	1471	2239	3019	3851	4679	5557	6389	7321	8233	9137	10039	10979	11953	12889	13807	14783	15731	16703	17707	18679	19727	20719	21673
34	139	757	1481	2243	3023	3853	4691	5563	6397	7331	8237	9151	10061	10987	11959	12893	13829	14797	15733	16729	17713	18691	19739	20731	21683
35	149	761	1483	2251	3037	3863	4703	5569	6421	7333	8243	9157	10067	10993	11969	12899	13831	14813	15737	16741	17729	18701	19751	20743	21701
36	151	769	1487	2267	3041	3877	4721	5573	6427	7349	8263	9161	10069	11003	11971	12907	13841	14821	15739	16747	17737	18713	19753	20747	21713
37	157	773	1489	2269	3049	3881	4723	5581	6449	7351	8269	9173	10079	11027	11981	12911	13859	14827	15749	16759	17747	18719	19759	20749	21727
38	163	787	1493	2273	3061	3889	4729	5591	6451	7369	8273	9181	10091	11047	11987	12917	13873	14831	15761	16763	17749	18731	19763	20753	21737
39	167	797	1499	2281	3067	3907	4733	5623	6469	7393	8287	9187	10093	11057	12007	12919	13877	14843	15767	16787	17761	18743	19777	20759	21739
40	173	809	1511	2287	3079	3911	4751	5639	6473	7411	8291	9199	10099	11059	12011	12923	13879	14851	15773	16811	17783	18749	19793	20771	21751
41	179	811	1523	2293	3083	3917	4759	5641	6481	7417	8293	9203	10103	11069	12037	12941	13883	14867	15787	16823	17789	18757	19801	20773	21757
42	181	821	1531	2297	3089	3919	4783	5647	6491	7433	8297	9209	10111	11071	12041	12953	13901	14869	15791	16829	17791	18773	19813	20789	21767
43	191	823	1543	2309	3109	3923	4787	5651	6521	7451	8311	9221	10133	11083	12043	12959	13903	14879	15797	16831	17807	18787	19819	20807	21773
44	193	827	1549	2311	3119	3929	4789	5653	6529	7457	8317	9227	10139	11087	12049	12967	13907	14887	15803	16843	17827	18793	19841	20809	21787
45	197	829	1553	2333	3121	3931	4793	5657	6547	7459	8329	9239	10141	11093	12071	12973	13913	14891	15809	16871	17837	18797	19843	20849	21799
46	199	839	1559	2339	3137	3943	4799	5659	6551	7477	8353	9241	10151	11113	12073	12979	13921	14897	15817	16879	17839	18803	19853	20857	21803
47	211	853	1567	2341	3163	3947	4801	5669	6553	7481	8363	9257	10159	11117	12097	12983	13931	14923	15823	16883	17851	18839	19861	20873	21817
48	223	857	1571	2347	3167	3967	4813	5683	6563	7487	8369	9277	10163	11119	12101	13001	13933	14929	15859	16889	17863	18859	19867	20879	21821
49	227	859	1579	2351	3169	3989	4817	5689	6569	7489	8377	9281	10169	11131	12107	13003	13963	14939	15877	16901	17881	18869	19889	20887	21839
50	229	863	1583	2357	3181	4001	4831	5693	6571	7499	8387	9283	10177	11149	12109	13007	13967	14947	15881	16903	17891	18899	19891	20897	21841

PRIMES (continued)

	0	1	2	3	4	5	6	7	8	9	10	11	12	13	14	15	16	17	18	19	20	21	22	23	24
51	233	877	1597	2371	3187	4001	4861	5701	6571	7507	8389	9293	10181	11159	12113	13009	13997	14951	15887	16921	17903	18911	19913	20899	21851
52	239	881	1601	2377	3191	4003	4871	5711	6577	7517	8419	9311	10193	11161	12119	13033	13999	14957	15889	16927	17909	18913	19919	20903	21859
53	241	883	1607	2381	3203	4007	4877	5717	6581	7523	8423	9319	10211	11171	12143	13037	14009	14969	15901	16931	17911	18917	19927	20921	21863
54	251	887	1609	2383	3209	4013	4889	5737	6599	7529	8429	9323	10223	11173	12149	13043	14011	14983	15907	16937	17921	18919	19937	20929	21871
55	257	907	1613	2389	3217	4019	4903	5741	6607	7537	8431	9337	10243	11177	12157	13049	14029	15013	15913	16943	17923	18947	19949	20939	21881
56	263	911	1619	2393	3221	4021	4909	5743	6619	7541	8443	9341	10247	11197	12161	13063	14033	15017	15919	16963	17929	18959	19961	20947	21893
57	269	919	1621	2399	3229	4027	4919	5749	6637	7547	8447	9343	10253	11213	12163	13093	14051	15031	15923	16979	17939	18973	19963	20959	21911
58	271	929	1627	2411	3251	4049	4931	5779	6653	7549	8461	9349	10259	11239	12197	13099	14057	15053	15937	16981	17957	18979	19973	20963	21929
59	277	937	1637	2417	3253	4051	4933	5783	6659	7559	8467	9371	10267	11243	12203	13103	14071	15061	15959	16987	17959	19001	19979	20981	21937
60	281	941	1657	2423	3257	4057	4937	5791	6661	7561	8501	9377	10271	11251	12211	13109	14081	15073	15971	16993	17971	19009	19991	20983	21943
61	283	947	1663	2437	3259	4073	4943	5801	6673	7573	8513	9391	10273	11257	12227	13121	14083	15077	15973	17011	17977	19013	19993	21001	21961
62	293	953	1667	2441	3271	4079	4951	5807	6679	7577	8521	9397	10289	11261	12239	13127	14087	15083	15991	17021	17981	19031	19997	21011	21977
63	307	967	1669	2447	3299	4091	4957	5813	6689	7583	8527	9403	10301	11273	12241	13147	14107	15091	16001	17027	17987	19037	20011	21013	21991
64	311	971	1693	2459	3301	4093	4967	5821	6691	7589	8537	9413	10303	11279	12251	13151	14143	15101	16007	17029	17989	19051	20021	21017	21997
65	313	977	1697	2467	3307	4099	4969	5827	6701	7591	8539	9419	10313	11287	12253	13159	14149	15107	16033	17033	18013	19069	20023	21019	22003
66	317	983	1699	2473	3313	4111	4973	5839	6703	7603	8543	9421	10321	11299	12263	13163	14153	15121	16057	17041	18041	19073	20029	21023	22013
67	331	991	1709	2477	3319	4127	4987	5843	6709	7607	8563	9431	10331	11311	12269	13171	14159	15131	16061	17047	18043	19079	20047	21031	22027
68	337	997	1721	2503	3323	4129	4993	5849	6719	7621	8573	9433	10333	11317	12277	13177	14173	15137	16063	17053	18047	19081	20051	21059	22031
69	347	1009	1723	2521	3329	4133	4999	5851	6733	7639	8581	9437	10337	11321	12281	13183	14177	15139	16067	17077	18049	19087	20063	21061	22037
70	349	1013	1733	2531	3331	4139	5003	5857	6737	7643	8597	9439	10343	11329	12289	13187	14197	15149	16069	17093	18059	19121	20071	21067	22039
71	353	1019	1741	2539	3343	4153	5009	5861	6761	7649	8599	9461	10357	11351	12301	13217	14207	15161	16073	17099	18061	19139	20089	21089	22051
72	359	1021	1747	2543	3347	4157	5011	5867	6763	7669	8609	9463	10369	11353	12323	13219	14221	15173	16087	17107	18077	19141	20101	21101	22063
73	367	1031	1753	2549	3359	4159	5021	5869	6779	7673	8623	9467	10391	11369	12329	13229	14243	15187	16091	17117	18089	19157	20107	21107	22067
74	373	1033	1759	2551	3361	4177	5023	5879	6781	7681	8627	9473	10399	11383	12343	13241	14249	15193	16097	17123	18097	19163	20113	21121	22073
75	379	1039	1777	2557	3371	4201	5039	5881	6791	7687	8629	9479	10427	11393	12347	13249	14251	15199	16103	17137	18119	19181	20117	21139	22079
76	383	1049	1783	2579	3373	4211	5051	5897	6793	7691	8641	9491	10429	11399	12373	13259	14281	15217	16111	17159	18121	19183	20123	21143	22091
77	389	1051	1787	2591	3389	4217	5059	5903	6803	7699	8647	9497	10433	11411	12377	13267	14293	15227	16127	17167	18127	19207	20129	21149	22093
78	397	1061	1789	2593	3391	4219	5077	5923	6823	7703	8663	9511	10453	11423	12379	13291	14303	15233	16139	17183	18131	19211	20143	21157	22109
79	401	1063	1801	2609	3407	4229	5081	5927	6827	7717	8669	9521	10457	11437	12391	13297	14321	15241	16141	17189	18133	19213	20147	21163	22111
80	409	1069	1811	2617	3413	4231	5087	5939	6829	7723	8677	9533	10459	11443	12401	13309	14323	15259	16183	17191	18143	19219	20149	21169	22123
81	419	1087	1823	2621	3433	4241	5099	5953	6833	7727	8681	9539	10463	11447	12409	13313	14327	15263	16187	17203	18149	19231	20161	21179	22129
82	421	1091	1831	2633	3449	4243	5101	5981	6841	7741	8689	9547	10477	11467	12413	13327	14341	15269	16189	17207	18169	19237	20173	21187	22133
83	431	1093	1847	2647	3457	4253	5107	5987	6857	7753	8693	9551	10487	11471	12421	13331	14347	15271	16193	17209	18181	19249	20177	21191	22147
84	433	1097	1861	2657	3461	4259	5113	6007	6863	7757	8699	9587	10499	11483	12433	13337	14369	15277	16217	17231	18191	19259	20183	21193	22153
85	439	1103	1867	2659	3463	4261	5119	6011	6869	7759	8707	9601	10501	11489	12437	13339	14387	15287	16223	17239	18199	19267	20201	21211	22157
86	443	1109	1871	2663	3467	4271	5147	6029	6871	7789	8713	9613	10513	11491	12451	13367	14389	15289	16229	17257	18211	19273	20219	21221	22159
87	449	1117	1873	2671	3469	4273	5153	6037	6883	7793	8719	9619	10529	11497	12457	13381	14401	15299	16231	17291	18217	19289	20231	21227	22171
88	457	1123	1877	2677	3491	4283	5167	6043	6899	7817	8731	9623	10531	11503	12473	13397	14407	15307	16249	17293	18223	19301	20233	21247	22189
89	461	1129	1879	2683	3499	4289	5171	6047	6907	7823	8737	9629	10559	11519	12479	13399	14411	15313	16253	17299	18229	19309	20249	21269	22193
90	463	1151	1889	2687	3511	4297	5179	6053	6911	7829	8741	9631	10567	11527	12487	13411	14419	15319	16267	17317	18233	19319	20261	21277	22229
91	467	1153	1901	2689	3517	4327	5189	6067	6917	7841	8747	9643	10589	11549	12491	13417	14423	15329	16273	17321	18251	19333	20269	21283	22247
92	479	1163	1907	2693	3527	4337	5197	6073	6947	7853	8753	9649	10597	11551	12497	13421	14431	15343	16301	17327	18253	19373	20287	21313	22259
93	487	1171	1913	2699	3529	4339	5209	6079	6959	7867	8761	9661	10601	11579	12503	13441	14437	15349	16319	17333	18257	19379	20297	21317	22271
94	491	1181	1931	2707	3533	4349	5227	6089	6961	7873	8779	9677	10607	11587	12511	13451	14447	15359	16333	17341	18269	19381	20323	21319	22273
95	499	1187	1933	2711	3539	4357	5231	6091	6967	7877	8783	9679	10613	11593	12517	13457	14449	15361	16339	17351	18287	19387	20327	21323	22277
96	503	1193	1949	2713	3541	4363	5233	6101	6971	7879	8803	9689	10627	11597	12527	13463	14461	15373	16349	17359	18289	19391	20333	21341	22279
97	509	1201	1951	2719	3547	4373	5237	6113	6977	7883	8807	9697	10631	11617	12539	13469	14479	15377	16361	17377	18301	19403	20341	21347	22283
98	521	1213	1973	2729	3557	4391	5261	6121	6983	7901	8819	9719	10639	11621	12541	13477	14489	15383	16363	17383	18307	19417	20347	21377	22291
99	523	1217	1979	2731	3559	4397	5273	6131	6991	7907	8821	9721	10651	11633	12547	13487	14503	15391	16369	17387	18311	19421	20353	21379	22303
100	541	1223	1987	2741	3571	4409	5279	6133	6997	7919	8831	9733	10657	11657	12553	13499	14519	15401	16381	17389	18313	19423	20357	21383	22307

PRIMES (continued)

COMBINATORIAL ANALYSIS

PRIMES (continued)

PRIMES (continued)

	25	26	27	28	29	30	31	32	33	34	35	36	37	38	39	40	41	42	43	44	45	46	47	48	49
1	22343	23327	24317	25409	26407	27457	28513	29453	30577	31607	32603	33617	34651	35771	36787	37831	38923	39979	41113	42083	43063	44203	45317	46451	47543
2	22349	23333	24329	25411	26417	27479	28517	29473	30593	31627	32621	33619	34667	35797	36791	37847	38933	39983	41117	42089	43067	44207	45319	46457	47563
3	22367	23339	24337	25423	26423	27481	28537	29483	30631	31643	32633	33623	34673	35801	36793	37853	38953	39989	41131	42101	43093	44221	45329	46471	47569
4	22369	23357	24359	25439	26431	27487	28541	29501	30637	31649	32647	33629	34679	35803	36809	37861	38959	40009	41141	42131	43103	44249	45337	46477	47581
5	22381	23369	24371	25447	26437	27509	28547	29527	30643	31657	32653	33637	34687	35809	36821	37871	38971	40013	41143	42139	43117	44257	45341	46489	47591
6	22391	23371	24373	25453	26449	27527	28549	29531	30649	31663	32687	33641	34693	35831	36833	37879	38977	40031	41149	42157	43133	44263	45343	46499	47599
7	22397	23399	24379	25457	26459	27539	28559	29537	30661	31667	32693	33647	34703	35837	36847	37889	38993	40037	41161	42169	43151	44267	45361	46507	47609
8	22409	23417	24391	25463	26479	27541	28571	29567	30671	31687	32707	33679	34721	35839	36857	37897	39019	40039	41177	42181	43159	44269	45377	46511	47623
9	22433	23431	24407	25469	26489	27551	28573	29569	30677	31699	32713	33703	34729	35851	36871	37907	39023	40063	41179	42187	43177	44273	45389	46523	47629
10	22441	23447	24413	25471	26497	27581	28579	29573	30689	31721	32717	33713	34739	35863	36877	37951	39041	40087	41183	42193	43189	44279	45403	46549	47639
11	22447	23459	24419	25523	26501	27583	28591	29581	30697	31723	32719	33721	34757	35869	36887	37957	39043	40093	41189	42197	43201	44281	45413	46559	47653
12	22453	23473	24421	25537	26513	27611	28597	29587	30703	31727	32749	33739	34759	35879	36899	37963	39047	40099	41201	42209	43207	44293	45427	46567	47657
13	22469	23497	24439	25541	26539	27617	28603	29599	30707	31729	32771	33749	34763	35897	36901	37967	39079	40111	41203	42221	43223	44351	45433	46573	47659
14	22481	23509	24443	25561	26557	27631	28607	29611	30713	31741	32779	33751	34781	35899	36913	37987	39089	40123	41213	42223	43237	44357	45439	46589	47681
15	22483	23531	24469	25577	26561	27647	28619	29629	30727	31751	32783	33757	34807	35911	36919	37991	39097	40127	41221	42227	43261	44371	45481	46591	47699
16	22501	23537	24473	25579	26573	27653	28621	29633	30757	31769	32789	33769	34819	35923	36923	37993	39103	40129	41227	42239	43271	44381	45491	46601	47701
17	22511	23539	24481	25583	26591	27673	28627	29641	30763	31771	32797	33773	34841	35933	36929	37997	39107	40151	41231	42257	43283	44383	45497	46619	47711
18	22531	23549	24499	25589	26597	27689	28631	29663	30773	31793	32801	33791	34843	35951	36931	38011	39113	40153	41233	42281	43291	44389	45503	46633	47713
19	22541	23557	24509	25601	26627	27691	28643	29669	30781	31799	32803	33797	34847	35963	36943	38039	39119	40163	41243	42283	43313	44417	45523	46639	47717
20	22543	23561	24517	25603	26633	27697	28649	29671	30803	31817	32831	33809	34849	35969	36947	38047	39133	40169	41257	42293	43319	44449	45533	46643	47737
21	22549	23563	24527	25609	26641	27701	28657	29683	30809	31847	32833	33811	34871	35977	36973	38053	39139	40177	41263	42299	43321	44453	45541	46649	47741
22	22567	23567	24533	25621	26647	27733	28661	29717	30817	31849	32839	33827	34877	35983	36979	38069	39157	40189	41269	42307	43331	44483	45557	46663	47743
23	22571	23581	24547	25633	26669	27737	28663	29723	30829	31859	32869	33829	34883	35993	36997	38083	39161	40193	41281	42323	43391	44491	45569	46679	47777
24	22573	23593	24551	25639	26681	27739	28669	29741	30839	31873	32887	33851	34897	35999	37003	38113	39163	40213	41299	42331	43397	44497	45587	46681	47779
25	22613	23599	24571	25643	26683	27743	28687	29753	30841	31883	32909	33857	34913	36007	37013	38119	39181	40231	41333	42337	43399	44501	45589	46687	47791
26	22619	23603	24593	25657	26687	27749	28697	29759	30851	31891	32911	33863	34919	36011	37019	38149	39191	40237	41341	42349	43403	44507	45599	46691	47797
27	22621	23609	24611	25667	26693	27751	28703	29761	30853	31907	32917	33871	34939	36013	37021	38153	39199	40241	41351	42359	43411	44519	45613	46703	47807
28	22637	23623	24623	25673	26699	27763	28711	29789	30859	31957	32933	33889	34949	36017	37039	38167	39209	40253	41357	42373	43427	44531	45631	46723	47809
29	22639	23627	24631	25679	26701	27767	28723	29803	30869	31963	32939	33893	34961	36037	37049	38177	39217	40277	41381	42379	43441	44537	45641	46727	47819
30	22643	23629	24659	25693	26711	27773	28729	29819	30871	31973	32941	33911	34963	36061	37057	38189	39227	40283	41387	42391	43451	44543	45659	46747	47837
31	22651	23633	24671	25703	26713	27779	28751	29833	30881	31981	32957	33923	34981	36067	37061	38197	39229	40289	41389	42397	43457	44549	45667	46751	47843
32	22669	23663	24677	25717	26717	27791	28753	29837	30893	31991	32969	33931	35023	36073	37087	38201	39233	40343	41399	42403	43481	44563	45673	46757	47857
33	22679	23669	24683	25733	26723	27793	28759	29851	30911	32003	32971	33937	35027	36083	37097	38219	39239	40351	41411	42407	43487	44579	45677	46769	47869
34	22691	23671	24691	25741	26729	27799	28771	29863	30931	32009	32983	33941	35051	36097	37117	38231	39241	40357	41413	42409	43499	44587	45691	46771	47881
35	22697	23677	24697	25747	26731	27803	28789	29867	30937	32027	32987	33961	35053	36107	37123	38237	39251	40361	41443	42433	43517	44617	45697	46807	47903
36	22699	23687	24709	25759	26737	27809	28793	29873	30941	32029	32993	33967	35059	36109	37139	38239	39293	40387	41453	42437	43541	44621	45707	46811	47911
37	22709	23689	24733	25763	26759	27817	28807	29879	30949	32051	32999	33997	35069	36137	37159	38261	39301	40423	41467	42443	43543	44623	45737	46817	47917
38	22717	23719	24749	25771	26777	27823	28813	29881	30971	32057	33013	34019	35081	36151	37171	38273	39313	40427	41491	42451	43573	44633	45751	46819	47933
39	22721	23741	24763	25793	26783	27827	28817	29917	30977	32059	33023	34031	35083	36161	37181	38281	39317	40429	41507	42457	43577	44641	45757	46829	47939
40	22727	23743	24767	25799	26801	27847	28837	29921	30983	32063	33029	34033	35089	36187	37189	38287	39323	40433	41513	42461	43579	44647	45763	46831	47947
41	22739	23747	24781	25801	26813	27851	28843	29927	31013	32069	33037	34039	35099	36191	37199	38299	39341	40459	41519	42463	43591	44651	45767	46853	47951
42	22741	23753	24793	25819	26821	27883	28859	29947	31019	32077	33049	34057	35107	36209	37201	38303	39343	40471	41521	42467	43597	44657	45779	46861	47963
43	22751	23761	24799	25841	26833	27893	28867	29959	31033	32083	33053	34061	35111	36217	37217	38317	39359	40483	41539	42473	43607	44683	45817	46867	47969
44	22769	23767	24809	25847	26839	27901	28871	29983	31039	32089	33071	34123	35117	36229	37223	38321	39367	40487	41543	42487	43609	44687	45821	46877	47977
45	22777	23773	24821	25849	26849	27917	28879	29989	31051	32099	33073	34127	35129	36241	37243	38327	39371	40493	41549	42491	43613	44699	45823	46889	47981
46	22783	23789	24841	25867	26861	27919	28901	30011	31063	32117	33083	34129	35141	36251	37253	38329	39373	40499	41579	42499	43627	44701	45827	46901	48017
47	22787	23801	24847	25873	26863	27941	28909	30013	31069	32119	33091	34141	35149	36263	37273	38333	39383	40507	41593	42509	43633	44711	45833	46919	48023
48	22807	23813	24851	25889	26879	27943	28921	30029	31079	32141	33107	34147	35153	36269	37277	38351	39397	40519	41597	42533	43649	44729	45841	46933	48029
49	22811	23819	24859	25903	26881	27947	28927	30047	31081	32143	33113	34157	35159	36277	37307	38371	39409	40529	41603	42557	43651	44741	45853	46957	48049
50	22817	23827	24877	25913	26891	27953	28933	30059	31091	32159	33119	34159	35171	36293	37309	38377	39419	40531	41609	42569	43661	44753	45863	46993	48073

PRIMES (continued)

	25	26	27	28	29	30	31	32	33	34	35	36	37	38	39	40	41	42	43	44	45	46	47	48	49
51	22853	23831	24889	25919	26693	27953	28949	30071	31121	32173	33119	34159	35171	36293	37313	38377	39439	40543	41609	42569	43669	44753	45863	46997	48073
52	22859	23833	24907	25931	26903	27961	28961	30089	31123	32183	33149	34171	35201	36299	37321	38393	39443	40559	41611	42577	43691	44771	45869	47017	48079
53	22861	23857	24917	25933	26921	27967	28979	30091	31139	32189	33151	34183	35221	36307	37337	38431	39451	40577	41617	42589	43711	44773	45887	47041	48091
54	22871	23869	24919	25939	26927	27983	29009	30097	31147	32191	33161	34211	35227	36313	37339	38447	39461	40583	41621	42589	43717	44777	45893	47051	48109
55	22877	23873	24923	25943	26947	27997	29017	30103	31151	32203	33179	34213	35251	36319	37357	38449	39499	40591	41627	42611	43721	44789	45943	47057	48119
56	22901	23879	24943	25951	26951	28001	29021	30109	31153	32213	33181	34217	35257	36341	37361	38453	39503	40597	41641	42641	43753	44797	45949	47059	48121
57	22907	23887	24953	25969	26959	28019	29023	30113	31159	32233	33191	34231	35267	36343	37363	38459	39509	40609	41647	42643	43759	44809	45953	47087	48131
58	22921	23893	24967	25981	26981	28027	29027	30119	31177	32237	33199	34253	35279	36353	37369	38461	39511	40627	41651	42649	43777	44819	45959	47093	48157
59	22937	23899	24971	25997	26987	28031	29033	30133	31181	32251	33203	34259	35281	36373	37379	38501	39521	40637	41659	42667	43781	44839	45971	47111	48163
60	22943	23909	24977	25999	26993	28051	29059	30137	31183	32257	33211	34261	35291	36383	37397	38543	39541	40639	41669	42677	43783	44843	45979	47119	48179
61	22961	23911	24979	26003	27011	28057	29063	30139	31189	32261	33223	34267	35311	36389	37409	38557	39551	40693	41681	42683	43787	44851	45989	47123	48187
62	22963	23917	24989	26017	27017	28069	29077	30161	31193	32297	33247	34273	35317	36433	37423	38561	39563	40697	41687	42689	43789	44867	46021	47129	48193
63	22973	23929	25013	26021	27031	28081	29101	30169	31219	32299	33287	34283	35323	36451	37441	38567	39569	40699	41719	42697	43793	44879	46027	47137	48197
64	22993	23957	25031	26029	27043	28087	29123	30181	31223	32303	33289	34297	35327	36457	37447	38569	39581	40709	41729	42701	43801	44887	46049	47143	48221
65	23003	23971	25033	26041	27059	28097	29129	30187	31231	32309	33301	34301	35339	36467	37463	38593	39607	40739	41737	42703	43853	44893	46051	47147	48239
66	23011	23977	25037	26053	27061	28099	29131	30197	31237	32321	33317	34303	35353	36469	37483	38603	39619	40751	41759	42709	43867	44909	46061	47149	48247
67	23017	23981	25057	26083	27067	28109	29137	30203	31247	32323	33313	34313	35363	36473	37489	38609	39623	40759	41761	42719	43889	44917	46073	47161	48259
68	23021	23993	25073	26099	27073	28111	29147	30211	31249	32327	33329	34319	35381	36479	37493	38611	39631	40763	41771	42727	43891	44927	46091	47189	48271
69	23027	24001	25087	26107	27077	28123	29153	30223	31253	32341	33331	34327	35393	36493	37501	38629	39659	40771	41777	42737	43913	44939	46093	47207	48281
70	23029	24007	25097	26111	27091	28151	29167	30241	31259	32353	33343	34337	35401	36497	37507	38639	39667	40787	41801	42743	43933	44953	46099	47221	48299
71	23039	24019	25111	26113	27103	28163	29173	30253	31267	32359	33347	34351	35407	36523	37511	38651	39671	40801	41809	42751	43943	44959	46103	47237	48311
72	23041	24023	25117	26119	27107	28181	29179	30259	31271	32363	33349	34361	35419	36527	37517	38653	39679	40813	41813	42767	43951	44963	46133	47251	48313
73	23053	24029	25121	26141	27109	28183	29191	30269	31277	32369	33353	34367	35423	36529	37529	38669	39703	40819	41843	42773	43961	44971	46141	47269	48337
74	23057	24043	25127	26153	27127	28201	29201	30271	31307	32371	33359	34369	35437	36541	37537	38671	39709	40823	41849	42787	43963	44983	46147	47279	48341
75	23059	24049	25147	26161	27143	28211	29207	30293	31319	32377	33377	34381	35447	36551	37547	38677	39719	40829	41851	42793	43969	44987	46153	47287	48353
76	23063	24061	25153	26171	27179	28219	29209	30307	31321	32381	33391	34403	35449	36559	37549	38699	39727	40841	41863	42797	43973	45007	46171	47293	48371
77	23071	24071	25163	26177	27191	28229	29221	30313	31327	32401	33403	34421	35461	36563	37561	38707	39733	40847	41887	42821	43987	45013	46181	47297	48383
78	23081	24077	25169	26183	27197	28277	29231	30319	31333	32411	33409	34429	35491	36571	37567	38711	39749	40849	41893	42829	43991	45053	46183	47303	48397
79	23087	24083	25171	26189	27211	28279	29243	30323	31337	32413	33413	34439	35507	36583	37571	38713	39761	40853	41897	42839	43997	45061	46187	47309	48407
80	23099	24091	25183	26203	27239	28283	29251	30341	31357	32423	33427	34457	35509	36587	37573	38723	39769	40867	41903	42841	44017	45077	46199	47317	48409
81	23117	24097	25189	26209	27241	28289	29269	30347	31379	32429	33457	34469	35521	36599	37579	38729	39779	40879	41911	42853	44021	45083	46219	47339	48413
82	23131	24103	25219	26227	27253	28297	29287	30367	31387	32441	33461	34471	35527	36607	37589	38737	39791	40883	41927	42859	44027	45119	46229	47351	48437
83	23143	24107	25229	26237	27259	28307	29297	30389	31391	32443	33469	34483	35531	36629	37591	38747	39799	40897	41941	42863	44029	45121	46237	47353	48449
84	23159	24109	25237	26249	27271	28309	29303	30391	31393	32467	33479	34487	35533	36637	37607	38749	39821	40903	41947	42899	44041	45127	46261	47363	48463
85	23167	24113	25243	26251	27277	28319	29311	30403	31397	32479	33487	34499	35537	36643	37619	38767	39827	40927	41957	42901	44053	45131	46271	47381	48473
86	23173	24121	25247	26261	27281	28349	29327	30427	31469	32491	33493	34501	35543	36653	37633	38783	39829	40933	41959	42923	44059	45137	46273	47387	48479
87	23189	24133	25253	26263	27283	28351	29333	30431	31477	32497	33503	34511	35569	36671	37643	38791	39839	40939	41969	42929	44071	45139	46279	47389	48481
88	23197	24137	25261	26267	27299	28387	29339	30449	31481	32503	33521	34513	35573	36677	37649	38821	39841	40949	41981	42937	44087	45161	46301	47407	48487
89	23201	24151	25301	26293	27329	28393	29347	30467	31489	32507	33529	34519	35591	36683	37657	38833	39847	40961	41983	42953	44089	45179	46307	47417	48491
90	23203	24169	25303	26297	27337	28403	29363	30469	31511	32531	33533	34537	35593	36691	37663	38839	39857	40973	41999	42953	44101	45181	46309	47419	48497
91	23209	24179	25307	26309	27361	28409	29383	30491	31513	32533	33547	34543	35597	36697	37691	38851	39863	40993	42013	42961	44111	45191	46327	47431	48523
92	23227	24181	25309	26317	27367	28411	29387	30493	31517	32537	33563	34549	35603	36709	37693	38861	39869	41011	42017	42967	44119	45197	46337	47441	48527
93	23251	24197	25321	26321	27397	28429	29389	30497	31531	32561	33569	34583	35617	36713	37699	38867	39877	41017	42019	42979	44123	45233	46349	47459	48533
94	23269	24203	25339	26339	27407	28433	29399	30509	31541	32563	33577	34589	35671	36721	37717	38873	39883	41023	42023	42989	44129	45247	46381	47491	48539
95	23279	24223	25343	26347	27409	28439	29401	30517	31543	32569	33581	34591	35677	36739	37747	38891	39887	41039	42043	43003	44131	45259	46399	47497	48541
96	23291	24229	25349	26357	27427	28447	29411	30529	31547	32573	33587	34603	35729	36749	37781	38903	39901	41047	42061	43013	44159	45263	46411	47501	48563
97	23293	24239	25357	26371	27431	28463	29423	30539	31567	32579	33589	34607	35731	36761	37783	38917	39929	41051	42071	43037	44171	45281	46439	47507	48571
98	23297	24247	25367	26387	27437	28477	29429	30553	31573	32587	33599	34613	35747	36767	37799	38921	39937	41057	42073	43049	44179	45289	46441	47513	48589
99	23311	24251	25373	26393	27443	28493	29437	30557	31583	32603	33601	34631	35753	36779	37811	38933	39953	41077	42083	43051	44189	45293	46447	47521	48593
100	23321	24281	25391	26399	27449	28499	29443	30559	31601	32609	33613	34649	35759	36781	37813	38953	39971	41081	42089	43063	44201	45307	46471	47527	48611

COMBINATORIAL ANALYSIS

PRIMES (continued)

PRIMES (continued)

	50	51	52	53	54	55	56	57	58	59	60	61	62	63	64	65	66	67	68	69	70	71	72	73	74
1	48619	49667	50767	51817	52937	54001	55109	56197	57193	58243	59369	60509	61637	62791	63839	65071	66109	67247	68389	69497	70663	71719	72859	73999	75083
2	48623	49669	50773	51827	52951	54011	55117	56207	57203	58271	59377	60521	61643	62801	63841	65089	66137	67261	68399	69499	70667	71741	72869	74017	75109
3	48647	49681	50777	51829	52957	54013	55127	56209	57221	58309	59387	60527	61651	62819	63853	65099	66161	67271	68437	69539	70687	71761	72883	74021	75133
4	48649	49697	50789	51839	52963	54037	55147	56237	57223	58313	59393	60539	61657	62827	63857	65101	66169	67273	68443	69557	70709	71777	72889	74027	75149
5	48661	49711	50821	51853	52967	54049	55163	56239	57241	58321	59399	60589	61667	62851	63863	65119	66173	67289	68447	69593	70717	71789	72893	74047	75161
6	48673	49727	50833	51859	52973	54059	55171	56249	57251	58337	59407	60601	61673	62861	63901	65123	66179	67307	68449	69623	70729	71807	72901	74051	75167
7	48677	49739	50839	51869	52981	54083	55201	56263	57259	58363	59417	60607	61681	62869	63907	65129	66191	67339	68473	69653	70753	71809	72907	74071	75169
8	48679	49741	50849	51871	52999	54091	55207	56267	57269	58367	59419	60611	61687	62873	63913	65141	66221	67343	68477	69661	70769	71821	72911	74077	75181
9	48731	49747	50857	51893	53003	54101	55213	56269	57271	58369	59441	60617	61703	62897	63929	65167	66239	67349	68483	69677	70783	71837	72923	74093	75193
10	48733	49757	50867	51899	53017	54121	55217	56299	57283	58379	59443	60623	61717	62903	63949	65171	66271	67369	68489	69691	70793	71843	72931	74099	75209
11	48751	49783	50873	51907	53047	54133	55219	56311	57287	58391	59447	60631	61723	62921	63977	65179	66293	67391	68491	69697	70823	71849	72937	74101	75211
12	48757	49787	50891	51913	53051	54139	55229	56333	57301	58393	59453	60637	61729	62927	63997	65183	66301	67399	68501	69709	70841	71861	72949	74131	75217
13	48761	49789	50893	51929	53069	54151	55243	56359	57329	58403	59467	60647	61751	62929	64007	65203	66337	67409	68507	69737	70843	71867	72953	74143	75223
14	48767	49801	50909	51941	53077	54163	55249	56369	57331	58411	59471	60649	61757	62939	64013	65213	66343	67411	68513	69739	70849	71879	72959	74149	75227
15	48779	49807	50923	51949	53087	54167	55259	56377	57347	58417	59473	60659	61781	62969	64019	65239	66347	67421	68521	69761	70853	71881	72973	74159	75239
16	48781	49811	50929	51971	53089	54181	55291	56383	57349	58427	59497	60661	61813	62971	64033	65257	66359	67427	68531	69763	70867	71887	72977	74161	75253
17	48787	49823	50951	51973	53093	54193	55313	56393	57367	58439	59509	60679	61819	62981	64037	65267	66361	67429	68539	69779	70877	71899	72997	74167	75269
18	48799	49831	50957	51977	53101	54217	55331	56401	57373	58441	59513	60689	61837	62983	64063	65269	66373	67433	68543	69809	70879	71909	73009	74177	75277
19	48809	49843	50969	51991	53113	54251	55333	56417	57383	58451	59539	60703	61843	62987	64067	65287	66377	67447	68567	69821	70891	71917	73013	74189	75289
20	48817	49853	50971	52009	53117	54269	55337	56431	57389	58453	59557	60719	61861	62989	64081	65293	66383	67453	68597	69827	70901	71933	73019	74197	75307
21	48821	49871	50989	52021	53129	54277	55339	56437	57397	58477	59561	60727	61871	63029	64091	65309	66403	67477	68611	69829	70913	71947	73037	74201	75323
22	48823	49877	50993	52027	53147	54287	55343	56443	57413	58481	59567	60733	61879	63031	64109	65323	66413	67481	68633	69833	70919	71963	73039	74203	75329
23	48847	49891	51001	52051	53149	54293	55351	56453	57427	58511	59581	60737	61909	63059	64123	65327	66431	67489	68639	69847	70921	71971	73043	74209	75337
24	48857	49919	51031	52057	53161	54311	55373	56467	57457	58537	59611	60757	61927	63067	64151	65353	66449	67493	68659	69857	70937	71983	73061	74219	75347
25	48859	49921	51043	52067	53171	54319	55381	56473	57467	58543	59617	60761	61933	63073	64153	65357	66457	67499	68669	69859	70949	71987	73063	74231	75353
26	48869	49927	51047	52069	53173	54323	55399	56477	57487	58549	59621	60763	61949	63079	64157	65371	66463	67511	68683	69877	70951	71993	73079	74257	75367
27	48871	49937	51059	52081	53189	54331	55411	56479	57493	58567	59627	60773	61961	63097	64171	65381	66467	67523	68687	69899	70957	71999	73091	74279	75377
28	48883	49943	51061	52103	53197	54347	55439	56489	57503	58573	59629	60779	61967	63103	64187	65393	66491	67531	68699	69911	70969	72019	73121	74287	75389
29	48889	49957	51071	52121	53201	54361	55441	56501	57527	58579	59651	60793	61979	63113	64189	65407	66499	67537	68711	69929	70979	72031	73127	74293	75391
30	48907	49991	51109	52127	53231	54367	55457	56503	57529	58601	59659	60811	61981	63127	64217	65413	66509	67547	68713	69931	70981	72043	73133	74297	75401
31	48947	49993	51131	52147	53233	54371	55469	56509	57557	58603	59663	60821	61987	63131	64223	65419	66523	67559	68729	69941	70991	72047	73141	74311	75407
32	48953	49999	51137	52153	53239	54377	55487	56519	57559	58613	59669	60859	61991	63149	64231	65423	66529	67567	68737	69959	70997	72053	73181	74317	75431
33	48973	50021	51151	52163	53267	54401	55501	56527	57571	58631	59671	60869	62003	63179	64237	65437	66533	67577	68743	69991	70999	72073	73189	74323	75437
34	48989	50023	51157	52177	53269	54403	55511	56531	57587	58657	59693	60887	62011	63197	64271	65447	66541	67579	68749	69997	71011	72077	73237	74353	75479
35	48991	—	—	52181	—	54409	55529	56533	57593	58661	59699	60889	62017	63199	64279	65449	66553	67589	68767	70001	71023	72089	73243	74357	75497
36	49003	50033	51169	52183	53279	54413	55541	56543	57601	58679	59707	60899	62039	63211	64283	65479	66569	67601	68771	70003	71039	72091	73259	74363	75503
37	49009	50047	51193	52189	53281	54419	55547	56569	57637	58687	59723	60901	62047	63241	64301	65497	66571	67607	68777	70009	71059	72101	73277	74377	75511
38	49019	50051	51197	52201	53299	54421	55579	56591	57641	58693	59729	60913	62053	63247	64303	65519	66587	67619	68791	70019	71069	72103	73291	74381	75521
39	49031	50053	51199	52223	53309	54437	55589	56597	57649	58699	59743	60917	62057	63277	64319	65521	66593	67631	68813	70039	71081	72109	73303	74383	75527
40	49033	50069	51203	52237	53327	54443	55603	56599	57653	58711	59747	60919	62071	63281	64327	65537	66601	67651	68819	70051	71089	72139	73309	74411	75533
41	49037	50077	51217	52249	53353	54449	55609	56611	57667	58727	59753	60923	62081	63299	64333	65539	66617	67679	68821	70061	71119	72161	73327	74413	75539
42	49043	50087	51229	52253	53359	54469	55619	56629	57679	58733	59771	60937	62099	63311	64373	65543	66629	67699	68863	70067	71129	72167	73331	74419	75541
43	49057	50093	51239	52259	53377	54493	55621	56633	57689	58741	59779	60943	62119	63313	64381	65551	66643	67709	68879	70079	71143	72169	73351	74441	75553
44	49069	50101	51241	52267	53381	54497	55631	56659	57697	58757	59791	60953	62129	63317	64399	65557	66653	67723	68881	70099	71147	72173	73361	74449	75557
45	49081	50111	51257	52289	53401	54499	55633	56663	57709	58763	59797	60961	62131	63331	64403	65563	66683	67733	68891	70111	71153	72211	73363	74453	75571
46	49103	50119	51263	52291	53407	54503	55639	56671	57713	58771	59809	61001	62137	63337	64433	65579	66697	67741	68897	70117	71161	72221	73369	74471	75577
47	49109	50123	51283	52301	53411	54517	55661	56681	57719	58787	59833	61007	62141	63347	64439	65581	66701	67751	68899	70121	71167	72223	73379	74489	75583
48	49117	50129	51287	52313	53419	54521	55663	56687	57727	58789	59863	61027	62171	63353	64451	65587	66713	67757	68903	70123	71171	72227	73387	74507	75611
49	49121	50131	51307	52321	53437	54539	55667	56701	57731	58831	59879	61031	62189	63361	64453	65599	66721	67759	68909	70139	71191	72229	73417	74509	75617
50	49123	50147	51329	52361	53441	54541	55673	56711	57737	58889	59887	61043	62191	63367	64483	65609	66733	67763	68917	70141	71209	72251	73421	74521	75619

PRIMES (continued)

	50	51	52	53	54	55	56	57	58	59	60	61	62	63	64	65	66	67	68	69	70	71	72	73	74
51	49139	50153	51341	52363	53453	54547	55681	56713	57751	58897	59921	61051	62191	63377	64483	65687	66733	67777	68927	70141	71233	72251	73421	74527	75629
52	49157	50159	51343	52369	53479	54563	55697	56731	57773	58901	59929	61057	62201	63389	64489	65599	66739	67783	68947	70157	71237	72253	73433	74531	75641
53	49169	50177	51347	52379	53503	54583	55711	56737	57781	58907	59951	61091	62207	63391	64499	65609	66749	67789	68963	70163	71249	72269	73453	74551	75653
54	49171	50207	51349	52387	53507	54577	55717	56747	57787	58909	59957	61099	62213	63397	64513	65617	66751	67801	68993	70177	71257	72271	73459	74561	75659
55	49177	50221	51361	52391	53527	54581	55721	56767	57791	58913	59971	61121	62219	63409	64553	65629	66763	67807	69001	70181	71261	72277	73471	74567	75679
56	49193	50227	51383	52433	53549	54583	55733	56773	57793	58921	59981	61129	62233	63419	64567	65633	66791	67819	69011	70183	71263	72287	73477	74573	75683
57	49199	50231	51407	52453	53551	54601	55763	56779	57803	58937	59999	61141	62273	63421	64577	65647	66797	67829	69019	70199	71287	72307	73483	74587	75689
58	49201	50261	51413	52457	53569	54617	55787	56783	57809	58943	60013	61151	62297	63439	64579	65651	66809	67843	69029	70201	71293	72313	73517	74597	75703
59	49207	50263	51419	52489	53591	54623	55793	56807	57829	58963	60017	61153	62299	63443	64591	65657	66821	67853	69031	70207	71317	72337	73523	74609	75707
60	49211	50273	51421	52501	53593	54629	55799	56809	57839	58967	60029	61169	62303	63459	64601	65677	66841	67867	69061	70223	71327	72341	73529	74611	75709
61	49223	50287	51427	52511	53597	54631	55813	56813	57847	58979	60037	61211	62311	63467	64609	65687	66851	67883	69067	70229	71329	72353	73547	74623	75721
62	49253	50291	51431	52517	53609	54647	55817	56827	57853	58991	60041	61223	62323	63487	64613	65699	66853	67891	69073	70237	71333	72367	73553	74653	75731
63	49261	50311	51437	52529	53611	54667	55819	56843	57859	58997	60077	61231	62327	63493	64621	65701	66863	67901	69109	70241	71339	72379	73561	74687	75743
64	49277	50321	51439	52541	53617	54673	55823	56857	57881	59009	60083	61253	62347	63499	64627	65707	66877	67927	69119	70249	71341	72383	73571	74699	75767
65	49279	50329	51449	52543	53623	54679	55829	56873	57899	59011	60089	61261	62351	63559	64633	65713	66883	67931	69127	70271	71347	72421	73583	74707	75773
66	49297	50333	51461	52553	53629	54709	55837	56891	57901	59021	60091	61283	62383	63521	64661	65717	66889	67933	69143	70289	71353	72431	73589	74713	75781
67	49307	50341	51473	52561	53633	54713	55843	56893	57917	59023	60101	61291	62401	63527	64663	65719	66919	67939	69149	70297	71359	72461	73597	74717	75787
68	49331	50359	51479	52567	53639	54721	55849	56897	57923	59051	60103	61297	62417	63533	64667	65729	66923	67943	69151	70309	71363	72467	73607	74719	75793
69	49333	50363	51481	52571	53653	54727	55871	56909	57943	59053	60107	61331	62423	63541	64679	65731	66931	67957	69163	70313	71387	72469	73609	74729	75797
70	49339	50377	51487	52579	53657	54751	55889	56911	57947	59063	60127	61333	62459	63559	64693	65761	66943	67961	69191	70321	71389	72481	73613	74731	75821
71	49363	50383	51503	52583	53681	54767	55897	56921	57973	59063	60133	61339	62467	63577	64709	65777	66947	67967	69193	70327	71399	72497	73637	74747	75833
72	49367	50387	51511	52609	53693	54779	55901	56923	57977	59069	60139	61343	62473	63587	64717	65789	66949	67979	69197	70351	71411	72503	73643	74759	75853
73	49369	50411	51517	52627	53699	54787	55903	56929	57991	59077	60149	61357	62477	63589	64747	65809	66959	67987	69203	70373	71413	72533	73651	74761	75869
74	49391	50417	51521	52631	53717	54799	55921	56941	58013	59083	60161	61363	62483	63599	64763	65827	66973	67993	69221	70379	71419	72547	73673	74771	75883
75	49393	50423	51539	52639	53719	54829	55927	56951	58027	59093	60167	61379	62497	63601	64781	65831	66977	68023	69233	70381	71429	72551	73679	74779	75913
76	49409	50441	51551	52667	53731	54829	55931	56957	58031	59107	60169	61381	62501	63607	64783	65837	67003	68041	69239	70393	71437	72559	73681	74797	75931
77	49411	50459	51563	52673	53759	54833	55933	56963	58043	59113	60209	61403	62507	63611	64793	65839	67021	68053	69247	70423	71443	72577	73693	74821	75937
78	49417	50461	51577	52691	53773	54851	55949	56983	58049	59119	60217	61409	62533	63641	64811	65843	67033	68059	69257	70429	71453	72613	73699	74827	75941
79	49429	50497	51581	52697	53777	54869	55967	56989	58057	59123	60223	61417	62539	63649	64817	65851	67043	68071	69259	70439	71471	72617	73709	74831	75967
80	49433	50503	51593	52709	53783	54877	55987	56993	58061	59141	60251	61441	62549	63659	64849	65867	67049	68087	69263	70451	71473	72623	73721	74843	75979
81	49451	50513	51599	52711	53791	54881	55997	56993	58067	59149	60257	61463	62563	63659	64853	65881	67057	68099	69313	70459	71479	72643	73727	74857	75983
82	49459	50527	51607	52721	53813	54907	56003	57037	58073	59159	60259	61469	62581	63655	64871	65899	67061	68111	69317	70481	71483	72647	73751	74861	75989
83	49463	50539	51613	52727	53819	54917	56009	57047	58099	59167	60271	61471	62591	63661	64877	65921	67073	68113	69337	70481	71503	72649	73757	74869	75991
84	49477	50543	51631	52733	53831	54919	56039	57059	58109	59183	60289	61483	62597	63671	64879	65927	67079	68141	69341	70487	71527	72661	73771	74873	75997
85	49481	50549	51637	52747	53849	54941	56041	57073	58111	59197	60293	61487	62603	63689	64891	65929	67103	68147	69371	70489	71537	72671	73783	74887	76001
86	49499	50551	51647	52757	53857	54949	56053	57077	58129	59207	60317	61493	62617	63691	64901	65951	67121	68161	69379	70501	71549	72673	73819	74891	76003
87	49523	50587	51659	52769	53861	54959	56081	57089	58147	59209	60331	61507	62627	63667	64919	65957	67129	68171	69383	70507	71551	72679	73823	74897	76031
88	49529	50591	51673	52783	53881	54973	56087	57097	58151	59219	60337	61511	62633	63733	64921	65963	67139	68207	69389	70529	71563	72689	73847	74903	76039
89	49531	50593	51679	52807	53887	54979	56093	57107	58153	59221	60343	61519	62639	63779	64927	65981	67141	68209	69401	70537	71569	72689	73849	74923	76079
90	49537	50599	51683	52813	53891	54983	56099	57119	58169	59233	60353	61543	62653	63709	64937	65983	67153	68213	69403	70549	71593	72701	73859	74929	76081
91	49547	50627	51691	52817	53897	55001	56101	57131	58171	59239	60373	61547	62659	63727	64951	65993	67157	68219	69427	70571	71633	72707	73867	74933	76091
92	49559	50647	51713	52837	53899	55009	56093	57133	58189	59243	60383	61553	62683	63733	64969	66029	67169	68227	69431	70573	71647	72719	73883	74959	76099
93	49597	50651	51719	52859	53917	55021	56101	57143	58193	59263	60397	61559	62687	63743	64997	66037	67181	68239	69439	70583	71663	72727	73897	74961	76103
94	49603	50671	51721	52861	53923	55049	56113	57163	58199	59281	60413	61561	62701	63761	65003	66041	67187	68261	69457	70589	71671	72733	73907	75013	76123
95	49613	50683	51749	52879	53927	55051	56123	57173	58207	59333	60427	61583	62723	63773	65011	66047	67189	68279	69463	70607	71693	72739	73939	75017	76129
96	49627	50707	51767	52883	53939	55057	56149	57179	58211	59341	60443	61603	62731	63781	65027	66067	67211	68281	69467	70619	71699	72763	73943	75037	76147
97	49633	50723	51769	52889	53951	55061	56167	57191	58217	59351	60449	61609	62743	63793	65029	66071	67213	68311	69473	70621	71707	72767	73951	75041	76157
98	49639	50741	51787	52901	53959	55073	56171	57173	58229	59357	60457	61613	62753	63799	65033	66083	67217	68329	69481	70627	71711	72797	73961	75079	76159
99	49663	50753	51803	52903	53987	55079	56179	57191	58231	59359	60493	61627	62761	63803	65053	66089	67231	68351	69491	70639	71713	72817	73973	75011	76079
100	49667	—	—	52919	53993	55103	56207	—	58237	—	60497	61631	62773	63839	65063	66103	67231	68371	69493	70657	—	72823	—	—	76207

COMBINATORIAL ANALYSIS

PRIMES (continued)

	75	76	77	78	79	80	81	82	83	84	85	86	87	88	89	90	91	92	93	94	95
1	76213	77359	78487	79627	80737	81817	82903	84131	85243	86381	87557	88807	89867	90989	92177	93187	94351	95443	96587	97829	98953
2	76231	77369	78497	79631	80747	81839	82913	84137	85247	86389	87559	88811	89891	91003	92179	93199	94379	95461	96589	97841	98963
3	76243	77377	78509	79633	80749	81847	82939	84143	85259	86399	87583	88813	89897	91009	92189	93229	94397	95467	96601	97843	98981
4	76249	77383	78511	79657	80761	81853	82963	84163	85297	86413	87587	88817	89899	91019	92203	93239	94399	95471	96643	97847	98993
5	76253	77417	78517	79669	80777	81869	82981	84179	85303	86423	87589	88819	89909	91033	92219	93241	94421	95479	96661	97849	98999
6	76259	77419	78539	79687	80779	81883	82997	84181	85313	86441	87613	88843	89917	91079	92221	93251	94427	95483	96667	97859	99013
7	76261	77431	78541	79691	80783	81899	83003	84191	85331	86453	87623	88853	89923	91081	92227	93253	94433	95507	96671	97861	99017
8	76283	77447	78553	79693	80789	81901	83009	84199	85333	86461	87629	88861	89939	91097	92233	93257	94439	95527	96697	97871	99023
9	76289	77471	78569	79697	80803	81919	83023	84211	85361	86467	87631	88867	89959	91099	92237	93263	94441	95531	96703	97879	99041
10	76303	77477	78571	79699	80809	81929	83047	84221	85363	86477	87641	88873	89963	91121	92243	93281	94447	95539	96731	97883	99053
11	76333	77479	78577	79757	80819	81931	83059	84223	85369	86491	87643	88883	89977	91127	92251	93283	94463	95549	96737	97919	99079
12	76343	77489	78583	79769	80831	81937	83063	84229	85381	86501	87649	88897	89983	91129	92269	93287	94477	95561	96739	97927	99083
13	76367	77491	78593	79777	80833	81943	83071	84239	85411	86509	87671	88903	89989	91139	92297	93307	94483	95569	96749	97931	99089
14	76369	77509	78607	79801	80849	81953	83077	84247	85427	86531	87679	88919	90001	91141	92311	93319	94513	95581	96757	97943	99103
15	76379	77513	78623	79811	80863	81967	83089	84263	85429	86533	87683	88937	90007	91151	92317	93323	94529	95597	96763	97961	99109
16	76387	77521	78643	79813	80897	81971	83093	84299	85439	86539	87691	88951	90011	91153	92333	93329	94531	95603	96769	97967	99119
17	76403	77527	78649	79817	80909	81973	83101	84307	85447	86561	87697	88969	90017	91159	92347	93337	94541	95617	96779	97973	99131
18	76421	77543	78653	79823	80911	82003	83117	84313	85451	86573	87701	88993	90019	91163	92353	93371	94543	95621	96787	97987	99133
19	76423	77549	78691	79829	80923	82007	83137	84317	85453	86579	87719	88997	90023	91183	92357	93377	94547	95629	96797	98009	99137
20	76441	77551	78697	79841	80929	82009	83177	84319	85469	86587	87721	89003	90031	91193	92363	93383	94559	95633	96799	98011	99139
21	76463	77557	78707	79843	80933	82013	83203	84347	85487	86599	87739	89009	90053	91199	92369	93407	94561	95651	96821	98017	99149
22	76471	77563	78713	79847	80953	82021	83207	84349	85513	86627	87743	89017	90059	91229	92377	93419	94573	95701	96823	98041	99173
23	76481	77569	78721	79861	80963	82031	83219	84377	85517	86629	87751	89041	90067	91237	92381	93427	94583	95707	96827	98047	99181
24	76487	77573	78737	79867	80989	82037	83221	84389	85523	86677	87767	89051	90071	91243	92387	93463	94597	95713	96847	98057	99191
25	76493	77587	78779	79873	81001	82039	83227	84391	85531	86689	87793	89057	90073	91249	92399	93479	94603	95717	96851	98081	99223
26	76507	77591	78781	79889	81013	82051	83231	84401	85549	86693	87797	89069	90089	91253	92401	93481	94613	95723	96857	98101	99233
27	76511	77611	78787	79901	81017	82067	83233	84407	85571	86711	87803	89071	90107	91283	92413	93487	94621	95731	96893	98123	99241
28	76519	77617	78791	79903	81019	82073	83243	84421	85577	86719	87811	89083	90121	91291	92419	93491	94649	95737	96907	98129	99251
29	76537	77621	78797	79907	81023	82129	83257	84431	85597	86729	87833	89087	90127	91297	92431	93493	94651	95747	96911	98143	99257
30	76541	77641	78803	79939	81031	82139	83267	84437	85601	86743	87853	89101	90149	91303	92459	93497	94687	95773	96931	98179	99259
31	76543	77647	78809	79943	81041	82141	83269	84443	85607	86753	87869	89107	90163	91309	92461	93503	94693	95783	96953	98207	99277
32	76561	77659	78823	79967	81043	82153	83273	84449	85619	86767	87877	89113	90173	91331	92467	93523	94709	95789	96959	98213	99289
33	76579	77681	78839	79973	81047	82163	83299	84457	85621	86771	87881	89119	90187	91367	92479	93529	94723	95791	96973	98221	99317
34	76597	77687	78853	79979	81049	82171	83311	84463	85627	86783	87887	89123	90191	91369	92489	93553	94727	95801	96979	98227	99347
35	76603	77689	78857	79987	81071	82183	83339	84467	85639	86813	87911	89137	90197	91373	92503	93557	94747	95803	96989	98251	99349
36	76607	77699	78877	79997	81077	82189	83341	84481	85643	86837	87917	89153	90199	91381	92507	93559	94771	95813	96997	98257	99367
37	76631	77711	78887	79999	81083	82193	83357	84499	85661	86843	87931	89189	90203	91387	92551	93563	94777	95819	97001	98269	99371
38	76649	77713	78889	80021	81097	82207	83383	84503	85667	86851	87943	89203	90217	91393	92557	93581	94781	95857	97003	98297	99377
39	76651	77719	78893	80039	81101	82217	83389	84509	85669	86857	87959	89209	90227	91397	92567	93601	94789	95869	97007	98299	99391
40	76667	77723	78901	80051	81119	82219	83399	84521	85691	86861	87961	89213	90239	91411	92569	93607	94793	95873	97021	98317	99397
41	76673	77731	78919	80071	81131	82223	83401	84523	85703	86869	87973	89227	90247	91423	92581	93629	94811	95881	97039	98321	99401
42	76679	77743	78929	80077	81157	82231	83407	84533	85711	86923	87977	89231	90263	91433	92593	93637	94819	95891	97073	98323	99409
43	76697	77747	78941	80107	81163	82237	83417	84551	85717	86927	87991	89237	90271	91453	92623	93683	94823	95911	97081	98327	99431
44	76717	77761	78977	80111	81173	82241	83423	84559	85733	86939	88001	89261	90281	91457	92627	93701	94837	95917	97103	98347	99439
45	76733	77773	78979	80141	81181	82261	83431	84589	85751	86951	88003	89269	90289	91459	92639	93703	94841	95923	97117	98369	99469
46	76753	77783	78989	80147	81199	82267	83437	84629	85781	86959	88007	89273	90313	91463	92641	93719	94847	95929	97127	98377	99487
47	76757	77797	79031	80149	81203	82279	83443	84631	85793	86969	88019	89293	90353	91493	92647	93739	94849	95947	97151	98389	99497
48	76771	77801	79039	80153	81223	82301	83449	84649	85817	86981	88037	89303	90359	91499	92657	93761	94873	95957	97157	98407	99523
49	76777	77813	79043	80167	81233	82307	83459	84653	85819	86993	88069	89317	90371	91513	92669	93763	94889	95959	97159	98411	99527
50	76781	77839	79063	80173	81239	82339	83471	84659	85829	87011	88079	89329	90373	91529	92671	93787	94903	95971	97169	98419	99529

PRIMES (continued)

	75	76	77	78	79	80	81	82	83	84	85	86	87	88	89	90	91	92	93	94	95
51	76801	77849	79087	80177	81233	82349	83477	84673	85831	87011	88093	89329	90379	91153	92671	93803	94907	95957	97171	98467	99551
52	76819	77863	79103	80191	81239	82351	83497	84691	85837	87013	88117	89363	90397	91159	92681	93811	94933	95959	97177	98473	99559
53	76829	77867	79111	80207	81281	82361	83537	84697	85843	87037	88129	89371	90401	91163	92683	93827	94949	95971	97187	98479	99563
54	76831	77893	79133	80209	81283	82373	83557	84701	85847	87041	88169	89381	90403	91183	92693	93851	94951	95987	97213	98491	99571
55	76837	77899	79139	80221	81293	82387	83561	84713	85853	87049	88177	89387	90407	91193	92699	93871	94961	95989	97231	98507	99577
56	76847	77929	79147	80231	81299	82393	83563	84719	85889	87071	88211	89399	90437	91199	92707	93887	94993	96001	97241	98519	99581
57	76871	77933	79151	80233	81307	82421	83579	84731	85903	87083	88223	89413	90439	91229	92717	93889	94999	96013	97259	98533	99607
58	76873	77951	79153	80239	81331	82457	83591	84737	85907	87103	88237	89417	90469	91237	92723	93893	95003	96017	97283	98543	99611
59	76883	77969	79159	80251	81343	82463	83597	84751	85931	87107	88241	89431	90473	91243	92737	93901	95009	96043	97301	98561	99623
60	76907	77977	79181	80263	81349	82469	83609	84761	85933	87119	88259	89443	90481	91249	92753	93911	95021	96053	97303	98563	99643
61	76913	77983	79187	80273	81359	82471	83617	84787	85991	87121	88261	89449	90499	91253	92761	93913	95027	96059	97327	98573	99661
62	76919	77999	79193	80279	81371	82483	83621	84793	85999	87133	88289	89459	90511	91283	92767	93923	95063	96079	97367	98597	99667
63	76943	78007	79201	80287	81373	82487	83639	84809	86011	87149	88301	89477	90523	91291	92779	93937	95083	96097	97369	98621	99679
64	76949	78017	79229	80309	81401	82493	83641	84811	86017	87151	88321	89491	90527	91297	92789	93941	95087	96137	97373	98627	99689
65	76961	78031	79231	80317	81409	82499	83653	84827	86027	87179	88327	89501	90529	91303	92791	93949	95089	96149	97379	98639	99707
66	76963	78041	79241	80329	81421	82507	83663	84857	86029	87181	88337	89513	90533	91309	92801	93967	95093	96157	97381	98641	99709
67	76991	78049	79259	80341	81439	82529	83689	84859	86069	87187	88339	89519	90547	91331	92809	93971	95101	96167	97387	98663	99713
68	77003	78059	79273	80347	81457	82531	83701	84869	86077	87211	88379	89521	90583	91367	92821	93979	95107	96179	97397	98669	99719
69	77017	78079	79279	80363	81463	82549	83717	84871	86083	87221	88397	89527	90599	91369	92831	93983	95111	96181	97423	98689	99721
70	77023	78101	79283	80369	81509	82559	83719	84913	86111	87223	88411	89533	90617	91373	92849	93997	95131	96199	97429	98711	99733
71	77029	78121	79301	80387	81517	82561	83737	84919	86113	87251	88423	89561	90619	91381	92857	94007	95143	96211	97453	98713	99761
72	77041	78137	79309	80407	81527	82567	83761	84947	86117	87253	88427	89563	90631	91387	92861	94009	95153	96221	97459	98717	99767
73	77047	78139	79319	80429	81533	82571	83773	84961	86131	87257	88463	89567	90641	91393	92863	94033	95177	96223	97463	98729	99787
74	77069	78157	79333	80447	81547	82591	83777	84967	86137	87277	88469	89591	90647	91397	92867	94049	95189	96233	97499	98731	99793
75	77081	78163	79337	80461	81551	82601	83791	84977	86143	87281	88471	89597	90659	91411	92893	94057	95191	96259	97501	98737	99809
76	77093	78167	79349	80471	81553	82609	83813	84979	86161	87293	88493	89599	90677	91423	92899	94063	95203	96263	97511	98773	99817
77	77101	78173	79357	80473	81559	82613	83833	84991	86171	87299	88499	89603	90679	91433	92921	94079	95213	96269	97523	98779	99823
78	77137	78179	79367	80489	81563	82619	83843	85009	86179	87313	88513	89611	90697	91453	92927	94099	95219	96281	97547	98801	99829
79	77141	78191	79379	80491	81569	82633	83857	85021	86197	87317	88523	89627	90703	91457	92941	94109	95231	96289	97549	98807	99833
80	77153	78193	79393	80513	81611	82651	83869	85027	86201	87323	88547	89633	90709	91459	92951	94111	95233	96293	97553	98809	99839
81	77167	78203	79397	80527	81619	82657	83873	85037	86209	87337	88589	89653	90731	91463	92957	94117	95239	96323	97561	98837	99859
82	77171	78229	79399	80537	81629	82699	83891	85049	86239	87359	88591	89657	90749	91493	92959	94121	95257	96329	97571	98849	99871
83	77191	78233	79411	80557	81637	82721	83903	85061	86243	87383	88607	89659	90787	91499	92987	94151	95261	96331	97577	98867	99877
84	77201	78241	79423	80567	81647	82723	83911	85081	86249	87403	88609	89669	90793	91513	92993	94153	95267	96337	97579	98869	99881
85	77213	78259	79427	80599	81649	82727	83921	85087	86257	87407	88643	89671	90803	91529	93001	94169	95273	96353	97583	98873	99901
86	77237	78277	79433	80603	81667	82729	83933	85091	86263	87421	88651	89681	90821	91541	93047	94201	95279	96377	97607	98887	99907
87	77239	78283	79451	80611	81671	82757	83939	85093	86269	87427	88657	89689	90823	91571	93053	94207	95287	96401	97609	98893	99923
88	77243	78301	79481	80621	81677	82759	83969	85103	86287	87433	88661	89753	90833	91573	93059	94219	95311	96419	97613	98897	99929
89	77249	78307	79493	80627	81689	82763	83983	85109	86291	87443	88663	89759	90841	91577	93077	94229	95317	96431	97649	98899	99961
90	77261	78311	79531	80629	81701	82781	83987	85121	86293	87473	88667	89767	90847	91583	93083	94253	95327	96443	97651	98909	99971
91	77263	78317	79537	80651	81703	82787	84011	85133	86297	87481	88681	89779	90863	91591	93089	94261	95339	96451	97673	98911	99989
92	77267	78341	79549	80657	81707	82793	84017	85147	86311	87491	88721	89783	90887	91621	93097	94273	95369	96457	97687	98927	99991
93	77269	78347	79559	80669	81727	82799	84047	85159	86323	87509	88729	89797	90901	91631	93103	94291	95383	96461	97711	98929	
94	77279	78367	79561	80671	81737	82811	84053	85193	86341	87511	88741	89809	90907	91639	93113	94307	95393	96469	97729	98939	
95	77291	78401	79579	80677	81749	82813	84059	85199	86351	87517	88747	89819	90911	91673	93131	94309	95401	96479	97771	98947	
96	77317	78427	79589	80681	81761	82837	84061	85201	86353	87523	88771	89821	90917	91691	93133	94321	95413	96517	97777		
97	77323	78437	79601	80683	81769	82847	84067	85213	86369	87539	88789	89833	90931	91703	93139	94327	95419	96527	97787		
98	77339	78439	79609	80687	81773	82883	84089	85223	86371	87541	88793	89839	90947	91711	93151	94331	95429	96557	97789		
99	77347	78467	79613	80701	81799	82889	84121	85229	86381	87547	88799	89849	90971	91733	93169	94343	95441	96581	97813		
100	77351	78479	79621	80713	81817	82891	84127	85237	86389	87553	88801	89867	90977	91757	93179	94349	95443	96589	97829		

FACTORS AND PRIMES

This table presents the prime factors of all factorable numbers and the mantissas of the common logarithms of all prime numbers from 1 to 2,000. The table runs across two facing pages. Thus, the factors of 258 are found on a line with 25 and under vertical column 8 to be $2\cdot3\cdot43$. If n is prime, the mantissa of its common logarithm is given. If n is not prime its prime factors are given.

n	0	1	2	3	4
0	....	0000000	3010300	4771213	2^2
1	$2\cdot5$	0413927	$2^2\cdot3$	1139434	$2\cdot7$
2	$2^2\cdot5$	$3\cdot7$	$2\cdot11$	3617278	$2^3\cdot3$
3	$2\cdot3\cdot5$	3913617	2^5	$3\cdot11$	$2\cdot17$
4	$2^3\cdot5$	6127839	$2\cdot3\cdot7$	6334685	$2^2\cdot11$
5	$2\cdot5^3$	$3\cdot17$	$2^2\cdot13$	7242759	$2\cdot3^3$
6	$2^2\cdot3\cdot5$	7853298	$2\cdot31$	$3^2\cdot7$	2^6
7	$2\cdot5\cdot7$	8512583	$2^3\cdot3^2$	8633229	$2\cdot37$
8	$2^4\cdot5$	3^4	$2\cdot41$	9190781	$2^2\cdot3\cdot7$
9	$2\cdot3^2\cdot5$	$7\cdot13$	$2^2\cdot23$	$3\cdot31$	$2\cdot47$
10	$2^2\cdot5^2$	0043214	$2\cdot3\cdot17$	0128372	$2^3\cdot13$
11	$2\cdot5\cdot11$	$3\cdot37$	$2^4\cdot7$	0530784	$2\cdot3\cdot19$
12	$2^3\cdot3\cdot5$	11^2	$2\cdot61$	$3\cdot41$	$2^2\cdot31$
13	$2\cdot5\cdot13$	1172713	$2^2\cdot3\cdot11$	$7\cdot19$	$2\cdot67$
14	$2^2\cdot5\cdot7$	$3\cdot47$	$2\cdot71$	$11\cdot13$	$2^4\cdot3^2$
15	$2\cdot3\cdot5^2$	1789769	$2^3\cdot19$	$3^2\cdot17$	$2\cdot7\cdot11$
16	$2^5\cdot5$	$7\cdot23$	$2\cdot3^4$	2121876	$2^2\cdot41$
17	$2\cdot5\cdot17$	$3^2\cdot19$	$2^2\cdot43$	2380461	$2\cdot3\cdot29$
18	$2^2\cdot3^2\cdot5$	2576786	$2\cdot7\cdot13$	$3\cdot61$	$2^3\cdot23$
19	$2\cdot5\cdot19$	2810334	$2^6\cdot3$	2855573	$2\cdot97$
20	$2^3\cdot5^2$	$3\cdot67$	$2\cdot101$	$7\cdot29$	$2^2\cdot3\cdot17$
21	$2\cdot3\cdot5\cdot7$	3242825	$2^2\cdot53$	$3\cdot71$	$2\cdot107$
22	$2^2\cdot5\cdot11$	$13\cdot17$	$2\cdot3\cdot37$	3483049	$2^5\cdot7$
23	$2\cdot5\cdot23$	$3\cdot7\cdot11$	$2^3\cdot29$	3673559	$2\cdot3^2\cdot13$
24	$2^4\cdot3\cdot5$	3820170	$2\cdot11^2$	3^5	$2^2\cdot61$
25	$2\cdot5^3$	3996737	$2^2\cdot3^2\cdot7$	$11\cdot23$	$2\cdot127$
26	$2^2\cdot5\cdot13$	$3^2\cdot29$	$2\cdot131$	4199557	$2^3\cdot3\cdot11$
27	$2\cdot3^3\cdot5$	4329693	$2^4\cdot17$	$3\cdot7\cdot13$	$2\cdot137$
28	$2^3\cdot5\cdot7$	4487063	$2\cdot3\cdot47$	4517864	$2^2\cdot71$
29	$2\cdot5\cdot29$	$3\cdot97$	$2^2\cdot73$	4668676	$2\cdot3\cdot7^2$
30	$2^2\cdot3\cdot5^2$	$7\cdot43$	$2\cdot151$	$3\cdot101$	$2^4\cdot19$
31	$2\cdot5\cdot\cdot31$	4927604	$2^3\cdot3\cdot13$	4955443	$2\cdot157$
32	$2^6\cdot5$	$3\cdot107$	$2\cdot7\cdot23$	$17\cdot19$	$2^2\cdot3^4$
33	$2\cdot3\cdot5\cdot11$	5198280	$2^2\cdot83$	$3^2\cdot37$	$2\cdot167$
34	$2^2\cdot5\cdot17$	$11\cdot31$	$2\cdot3^2\cdot19$	7^3	$2^3\cdot43$
35	$2\cdot5^2\cdot7$	$3^3\cdot13$	$2^5\cdot11$	5477747	$2\cdot3\cdot59$
36	$2^3\cdot3^2\cdot5$	19^3	$2\cdot181$	$3\cdot11^2$	$2^2\cdot7\cdot13$
37	$2\cdot5\cdot37$	$7\cdot53$	$2^2\cdot3\cdot31$	5717088	$2\cdot11\cdot17$
38	$2^2\cdot5\cdot19$	$3\cdot127$	$2\cdot191$	5831988	$2^7\cdot3$
39	$2\cdot3\cdot5\cdot13$	$17\cdot23$	$2^3\cdot7^2$	$3\cdot131$	$2\cdot197$
40	$2^4\cdot5^2$	6031444	$2\cdot3\cdot67$	$13\cdot31$	$2^2\cdot101$
41	$2\cdot5\cdot41$	$3\cdot137$	$2^2\cdot103$	$7\cdot59$	$2\cdot3^2\cdot23$
42	$2^2\cdot3\cdot5\cdot7$	6242821	$2\cdot211$	$3^2\cdot47$	$2^3\cdot53$
43	$2\cdot5\cdot43$	634477733^2	$2^4\cdot3^3$	6364879	$2\cdot7\cdot31$
44	$2^3\cdot5\cdot11$	7^2	$2\cdot13\cdot17$	6464037	$2^2\cdot3\cdot37$
45	$2\cdot3^2\cdot5^2$	$11\cdot41$	$2^2\cdot113$	$3\cdot151$	$2\cdot227$
46	$2^2\cdot5\cdot23$	6637009	$2\cdot3\cdot7\cdot11$	6655810	$2^4\cdot29$
47	$2\cdot5\cdot47$	$3\cdot157$	$2^3\cdot59$	$11\cdot43$	$2\cdot3\cdot79$
48	$2^5\cdot3\cdot5$	$13\cdot37$	$2\cdot241$	$3\cdot7\cdot23$	$2^2\cdot11^2$
49	$2\cdot5\cdot7^2$	6910815	$2^2\cdot3\cdot41$	$17\cdot29$	$2\cdot13\cdot19$

FACTORS AND PRIMES (Continued)

n	5	6	7	8	9
0	6989700	$2\cdot3$	8450980	2^3	3^2
1	$3\cdot5$	2^4	2304489	$2\cdot3^2$	2787536
2	5^2	$2\cdot13$	3^3	$2^2\cdot7$	4623980
3	$5\cdot7$	$2^2\cdot3^2$	5682017	$2\cdot19$	$3\cdot13$
4	$3^2\cdot5$	$2\cdot23$	6720979	$2^4\cdot3$	7^2
5	$5\cdot11$	$2^3\cdot7$	$3\cdot19$	$2\cdot29$	7708520
6	$5\cdot13$	$2\cdot3\cdot11$	8260748	$2^2\cdot17$	$3\cdot23$
7	$3\cdot5^2$	$2^3\cdot19$	$7\cdot11$	$2\cdot3\cdot13$	8976271
8	$5\cdot17$	$2\cdot43$	$3\cdot29$	$2^3\cdot11$	9493900
9	$5\cdot19$	$2^5\cdot3$	9867717	$2\cdot7^2$	$3^2\cdot11$
10	$3\cdot5\cdot7$	$2\cdot53$	0293838	$2^2\cdot3^3$	0374265
11	$5\cdot23$	$2^2\cdot29$	$3^2\cdot13$	$2\cdot59$	$7\cdot17$
12	5^3	$2\cdot3^2\cdot7$	1038037	2^7	$3\cdot43$
13	$3^3\cdot5$	$2^3\cdot17$	1367206	$2\cdot3\cdot23$	1430148
14	$5\cdot29$	$2\cdot73$	$4\cdot7^2$	$2^2\cdot37$	1731863
15	$5\cdot31$	$2^2\cdot3\cdot13$	1958997	$2\cdot79$	$3\cdot53$
16	$3\cdot5\cdot11$	$2\cdot83$	2227165	$2^3\cdot3\cdot7$	13^2
17	$5^2\cdot7$	$2^4\cdot11$	$3\cdot50$	$2\cdot89$	2528530
18	$5\cdot37$	$2\cdot3\cdot31$	$11\cdot17$	$2^2\cdot47$	$3^3\cdot7$
19	$3\cdot5\cdot13$	$2^2\cdot7^2$	2944662	$2\cdot3^2\cdot11$	2988531
20	$5\cdot41$	$2\cdot103$	$3^2\cdot23$	$2^4\cdot13$	$11\cdot19$
21	$5\cdot43$	$2^3\cdot3^3$	$7\cdot31$	$2\cdot109$	$3\cdot73$
22	$3^2\cdot5^2$	$2\cdot113$	3560259	$2^2\cdot3\cdot19$	3598355
23	$5\cdot47$	$2^2\cdot59$	$3\cdot79$	$2\cdot7\cdot17$	3783979
24	$5\cdot7^2$	$2\cdot3\cdot41$	$13\cdot19$	$2^3\cdot31$	$3\cdot83$
25	$3\cdot5\cdot17$	2^3	4099331	$2\cdot3\cdot43$	$7\cdot37$
26	$5\cdot53$	$2\cdot7\cdot19$	$3\cdot80$	$2^2\cdot67$	4297523
27	$5^2\cdot11$	$2^3\cdot3\cdot23$	4424798	$2\cdot139$	$3^2\cdot31$
28	$3\cdot5\cdot19$	$2\cdot11\cdot13$	$7\cdot41$	$2^5\cdot3^2$	17^2
29	$5\cdot59$	$2^3\cdot37$	$3^3\cdot11$	$2\cdot149$	$13\cdot23$
30	$5\cdot61$	$2\cdot3^2\cdot17$	4871384	$2^2\cdot7\cdot11$	$3\cdot103$
31	$3^2\cdot5\cdot7$	$2^2\cdot79$	5010593	$2\cdot3\cdot53$	$11\cdot29$
32	$5^2\cdot13$	$2\cdot163$	$3\cdot109$	$2^3\cdot41$	$7\cdot47$
33	$5\cdot67$	$2^4\cdot3\cdot7$	5276299	$2\cdot13^2$	$3\cdot113$
34	$3\cdot5\cdot23$	$2\cdot173$	5403295	$2^2\cdot3\cdot29$	5428254
35	$5\cdot71$	$2^3\cdot89$	$3\cdot7\cdot17$	$2\cdot179$	5550944
36	$5\cdot73$	$2\cdot3\cdot61$	5646661	$2^4\cdot23$	$3^2\cdot41$
37	$3\cdot5^3$	$2^3\cdot47$	$13\cdot29$	$2\cdot3^3\cdot7$	5786392
38	$5\cdot7\cdot11$	$2\cdot193$	$3^2\cdot43$	$2^2\cdot97$	5899496
39	$5\cdot79$	$2^2\cdot3^2\cdot11$	5987905	$2\cdot199$	$3\cdot7\cdot19$
40	$3^4\cdot5$	$2\cdot7\cdot29$	$11\cdot37$	$2^3\cdot3\cdot17$	6117233
41	$5\cdot83$	$2^5\cdot13$	$3\cdot139$	$2\cdot11\cdot19$	6222140
42	$5^2\cdot17$	$2\cdot3\cdot71$	$7\cdot61$	$2^2\cdot107$	$3\cdot11\cdot13$
43	$3\cdot5\cdot29$	$2^2\cdot109$	$19\cdot23$	$2\cdot3\cdot73$	6424645
44	$5\cdot89$	$2\cdot223$	$3\cdot149$	$2^6\cdot7$	6522463
45	$5\cdot7\cdot13$	$2^3\cdot3\cdot19$	659162	$2\cdot229$	$3^3\cdot17$
46	$3\cdot5\cdot31$	$2\cdot233$	6693169	$2^2\cdot3^5\cdot13$	$7\cdot67$
47	$5^2\cdot19$	$2^2\cdot7\cdot17$	$3^2\cdot53$	$2\cdot239$	6803355
48	$5\cdot97$	$2\cdot3^5$	6875290	$2^3\cdot61$	$3\cdot163$
49	$3^2\cdot5\cdot11$	$2^4\cdot31$	$7\cdot71$	$2\cdot3\cdot83$	6981005

FACTORS AND PRIMES (Continued)

n	0	1	2	3	4
50	$2^2 \cdot 5^3$	$3 \cdot 167$	$2 \cdot 251$	7015680	$2^3 \cdot 3^2 \cdot 7$
51	$2 \cdot 3 \cdot 5 \cdot 17$	$7 \cdot 73$	2^9	$3^3 \cdot 19$	$2 \cdot 257$
52	$2^3 \cdot 5 \cdot 13$	7168377	$2 \cdot 3^2 \cdot 29$	7185017	$2^2 \cdot 131$
53	$2 \cdot 5 \cdot 53$	$3^2 \cdot 59$	$2^2 \cdot 7 \cdot 19$	$13 \cdot 41$	$2 \cdot 3 \cdot 89$
54	$2^2 \cdot 3^3 \cdot 5$	7331973	$2 \cdot 271$	$3 \cdot 181$	$2^5 \cdot 17$
55	$2 \cdot 5^2 \cdot 11$	$19 \cdot 29$	$2^3 \cdot 3 \cdot 23$	$7 \cdot 79$	$2 \cdot 277$
56	$2^4 \cdot 5 \cdot 7$	$3 \cdot 11 \cdot 17$	$2 \cdot 281$	7505084	$2^2 \cdot 3 \cdot 47$
57	$2 \cdot 3 \cdot 5 \cdot 19$	7566361	$2^2 \cdot 11 \cdot 13$	$3 \cdot 191$	$2 \cdot 7 \cdot 41$
58	$2^2 \cdot 5 \cdot 29$	$7 \cdot 83$	$2 \cdot 3 \cdot 97$	$11 \cdot 53$	$2^3 \cdot 73$
59	$2 \cdot 5 \cdot 59$	$3 \cdot 197$	$2^4 \cdot 37$	7730547	$2 \cdot 3^3 \cdot 11$
60	$2^3 \cdot 3 \cdot 5^2$	7788745	$2 \cdot 7 \cdot 43$	$3^2 \cdot 67$	$2^2 \cdot 151$
61	$2 \cdot 5 \cdot 61$	$13 \cdot 47$	$2^2 \cdot 3^2 \cdot 17$	7874605	$2 \cdot 307$
62	$2^2 \cdot 5 \cdot 31$	$3^3 \cdot 23$	$2 \cdot 311$	$7 \cdot 89$	$2^4 \cdot 3 \cdot 13$
63	$2 \cdot 3^2 \cdot 5 \cdot 7$	8000294	$2^3 \cdot 79$	$3 \cdot 211$	$2 \cdot 317$
64	$2^7 \cdot 5$	8068580	$2 \cdot 3 \cdot 107$	8082110	$2^2 \cdot 7 \cdot 23$
65	$2 \cdot 5^2 \cdot 13$	$3 \cdot 7 \cdot 31$	$2^2 \cdot 163$	8149132	$2 \cdot 3 \cdot 109$
66	$2^2 \cdot 3 \cdot 5 \cdot 11$	8202015	$2 \cdot 331$	$3 \cdot 13 \cdot 17$	$2^3 \cdot 83$
67	$2 \cdot 5 \cdot 67$	$11 \cdot 61$	$2^5 \cdot 3 \cdot 7$	8280151	$2 \cdot 337$
68	$2^2 \cdot 5 \cdot 17$	$3 \cdot 227$	$2 \cdot 11 \cdot 31$	8344207	$2^2 \cdot 3^2 \cdot 19$
69	$2 \cdot 3 \cdot 5 \cdot 23$	8394780	$2^2 \cdot 173$	$3^2 \cdot 7 \cdot 11$	$2 \cdot 347$
70	$2^2 \cdot 5^2 \cdot 7$	8457180	$2 \cdot 3^3 \cdot 13$	$19 \cdot 37$	$2^6 \cdot 11$
71	$2 \cdot 5 \cdot 71$	$3^2 \cdot 79$	$2^3 \cdot 89$	$23 \cdot 3$	$2 \cdot 3 \cdot 7 \cdot 17$
72	$2^4 \cdot 3^2 \cdot 5$	$7 \cdot 103$	$2 \cdot 19^2$	$3 \cdot 241$	$2^2 \cdot 181$
73	$2 \cdot 5 \cdot 73$	$17 \cdot 43$	$2^2 \cdot 3 \cdot 61$	8651040	$2 \cdot 367$
74	$2^2 \cdot 5 \cdot 37$	$3 \cdot 13 \cdot 19$	$2 \cdot 7 \cdot 53$	8709888	$2^3 \cdot 3 \cdot 31$
75	$2 \cdot 3 \cdot 5^3$	8756399	$2^4 \cdot 47$	$3 \cdot 251$	$2 \cdot 13 \cdot 29$
76	$2^3 \cdot 5 \cdot 19$	8813847	$2 \cdot 3 \cdot 127$	$7 \cdot 109$	$2^2 \cdot 191$
77	$2 \cdot 5 \cdot 7 \cdot 11$	$3 \cdot 257$	$2^2 \cdot 193$	8881795	$2 \cdot 3^2 \cdot 43$
78	$2^2 \cdot 3 \cdot 5 \cdot 13$	$11 \cdot 71$	$2 \cdot 17 \cdot 23$	$3^3 \cdot 29$	$2^4 \cdot 7^2$
79	$2 \cdot 5 \cdot 79$	$7 \cdot 113$	$2^3 \cdot 3^2 \cdot 11$	$13 \cdot 61$	$2 \cdot 397$
80	$2^5 \cdot 5^2$	$3^2 \cdot 89$	$2 \cdot 401$	$11 \cdot 73$	$2^2 \cdot 3 \cdot 67$
81	$2 \cdot 3^4 \cdot 5$	9090209	$2^2 \cdot 7 \cdot 29$	$3 \cdot 271$	$2 \cdot 11 \cdot 37$
82	$2^2 \cdot 5 \cdot 41$	9143432	$2 \cdot 3 \cdot 137$	9153998	$2^3 \cdot 103$
83	$2 \cdot 5 \cdot 83$	$3 \cdot 277$	$2^6 \cdot 13$	$7^2 \cdot 17$	$2 \cdot 3 \cdot 139$
84	$2^3 \cdot 3 \cdot 5 \cdot 7$	29^2	$2 \cdot 421$	$3 \cdot 281$	$2^2 \cdot 211$
85	$2 \cdot 5^2 \cdot 17$	$23 \cdot 37$	$2^2 \cdot 3 \cdot 71$	9309490	$2 \cdot 7 \cdot 61$
86	$2^2 \cdot 5 \cdot 43$	$3 \cdot 7 \cdot 41$	$2 \cdot 431$	9360108	$2^5 \cdot 3^3$
87	$2 \cdot 3 \cdot 5 \cdot 29$	$13 \cdot 67$	$2^3 \cdot 109$	$3^2 \cdot 97$	$2 \cdot 19 \cdot 23$
88	$2^4 \cdot 5 \cdot 11$	9449759	$2 \cdot 3^2 \cdot 7^2$	9459607	$2^2 \cdot 13 \cdot 17$
89	$2 \cdot 5 \cdot 89$	$3^4 \cdot 11$	$2^2 \cdot 223$	$19 \cdot 47$	$2 \cdot 3 \cdot 149$
90	$2^2 \cdot 3^2 \cdot 5^2$	$17 \cdot 53$	$2 \cdot 11 \cdot 41$	$3 \cdot 7 \cdot 43$	$2^3 \cdot 113$
91	$2 \cdot 5 \cdot 7 \cdot 13$	9595184	$2^4 \cdot 3 \cdot 19$	$11 \cdot 83$	$2 \cdot 457$
92	$2^3 \cdot 5 \cdot 23$	$3 \cdot 307$	$2 \cdot 461$	$13 \cdot 71$	$2^2 \cdot 3 \cdot 7 \cdot 11$
93	$2 \cdot 3 \cdot 5 \cdot 31$	$7^2 \cdot 19$	$2^2 \cdot 233$	$3 \cdot 311$	$2 \cdot 467$
94	$2^2 \cdot 5 \cdot 47$	9735896	$2 \cdot 3 \cdot 157$	$23 \cdot 41$	$2^4 \cdot 59$
95	$2 \cdot 5^2 \cdot 19$	$3 \cdot 317$	$2^3 \cdot 7 \cdot 17$	9790929	$2 \cdot 3^2 \cdot 53$
96	$2^6 \cdot 3 \cdot 5$	31^2	$2 \cdot 13 \cdot 37$	$3^2 \cdot 107$	$2^2 \cdot 241$
97	$2 \cdot 5 \cdot 97$	9872192	$2^2 \cdot 3^5$	$7 \cdot 139$	$2 \cdot 487$
98	$2^2 \cdot 5 \cdot 7^2$	$3^2 \cdot 100$	$2 \cdot 491$	9925535	$2^3 \cdot 3 \cdot 41$
99	$2 \cdot 3^2 \cdot 5 \cdot 11$	9960737	$2^5 \cdot 31$	$3 \cdot 331$	$2 \cdot 7 \cdot 71$

FACTORS AND PRIMES (Continued)

n	5	6	7	8	9
50	$5 \cdot 101$	$2 \cdot 11 \cdot 23$	$3 \cdot 13^2$	$2^2 \cdot 127$	7067178
51	$5 \cdot 103$	$2^2 \cdot 3 \cdot 43$	$11 \cdot 47$	$2 \cdot 7 \cdot 37$	$3 \cdot 173$
52	$3 \cdot 5^2 \cdot 7$	$2 \cdot 263$	$17 \cdot 31$	$2^4 \cdot 3 \cdot 11$	23^2
53	$5 \cdot 107$	$2^3 \cdot 67$	$3 \cdot 179$	$2 \cdot 269$	$7^2 \cdot 11$
54	$5 \cdot 109$	$2 \cdot 3 \cdot 7 \cdot 13$	7379873	$2^2 \cdot 137$	$3^2 \cdot 61$
55	$3 \cdot 5 \cdot 37$	$2^2 \cdot 139$	7458552	$2 \cdot 3^2 \cdot 31$	$13 \cdot 43$
56	$5 \cdot 113$	$2 \cdot 283$	$3^4 \cdot 7$	$2^3 \cdot 71$	7551123
57	$5^2 \cdot 23$	$2^6 \cdot 3^2$	7611758	$2 \cdot 17^2$	$3 \cdot 193$
58	$3^2 \cdot 5 \cdot 13$	$2 \cdot 293$	7686381	$2^2 \cdot 3 \cdot 7^2$	$19 \cdot 31$
59	$5 \cdot 7 \cdot 17$	$2^2 \cdot 149$	$3 \cdot 199$	$2 \cdot 13 \cdot 23$	7774268
60	$5 \cdot 11^2$	$2 \cdot 3 \cdot 101$	7831887	$2^5 \cdot 19$	$3 \cdot 7 \cdot 29$
61	$3 \cdot 5 \cdot 41$	$2^3 \cdot 7 \cdot 11$	7902852	$2 \cdot 3 \cdot 103$	7916906
62	5^4	$2 \cdot 313$	$3 \cdot 11 \cdot 19$	$2^2 \cdot 157$	$17 \cdot 37$
63	$5 \cdot 127$	$2^2 \cdot 3 \cdot 53$	$7^2 \cdot 13$	$2 \cdot 11 \cdot 29$	$3^2 \cdot 71$
64	$3 \cdot 5 \cdot 43$	$2 \cdot 17 \cdot 19$	8109043	$2^3 \cdot 3^4$	$11 \cdot 59$
65	$5 \cdot 131$	$2^4 \cdot 41$	$3^2 \cdot 73$	$2 \cdot 7 \cdot 47$	8188854
66	$5 \cdot 7 \cdot 19$	$2 \cdot 3^2 \cdot 37$	$23 \cdot 29$	$2^2 \cdot 167$	$3 \cdot 223$
67	$3^3 \cdot 5^2$	$2^2 \cdot 13^2$	8305887	$2 \cdot 3 \cdot 113$	$7 \cdot 97$
68	$5''137$	$2 \cdot 7^3$	$3 \cdot 229$	$2^4 \cdot 43$	$13 \cdot 53$
69	$5 \cdot 139$	$2^3 \cdot 3 \cdot 29$	$17 \cdot 41$	$2 \cdot 349$	$3 \cdot 233$
70	$3 \cdot 5 \cdot 47$	$2 \cdot 353$	$7 \cdot 101$	$2^3 \cdot 3 \cdot 59$	8506462
71	$5 \cdot 11 \cdot 13$	$2^2 \cdot 179$	$3 \cdot 239$	$2 \cdot 359$	8567289
72	$5^2 \cdot 29$	$2 \cdot 3 \cdot 11^2$	8615344	$2^3 \cdot 7 \cdot 13$	3^6
73	$3 \cdot 5 \cdot 7^2$	$2^5 \cdot 23$	$11 \cdot 67$	$2 \cdot 3^2 \cdot 41$	8686444
74	$5 \cdot 149$	$2 \cdot 373$	$3^2 \cdot 83$	$2^2 \cdot 11 \cdot 17$	$7 \cdot 107$
75	$5 \cdot 151$	$2^2 \cdot 3^3 \cdot 7$	8790959	$2 \cdot 379$	$3 \cdot 11 \cdot 23$
76	$3^2 \cdot 5 \cdot 17$	$2 \cdot 383$	$13 \cdot 59$	$2^8 \cdot 3$	8859263
77	$5^2 \cdot 31$	$2^5 \cdot 97$	$3 \cdot 7 \cdot 37$	$2 \cdot 389$	$19 \cdot 41$
78	$5 \cdot 157$	$2 \cdot 3 \cdot 131$	8959747	$2^2 \cdot 197$	$3 \cdot 263$
79	$3 \cdot 5 \cdot 53$	$2^2 \cdot 199$	9014583	$2 \cdot 3 \cdot 7 \cdot 19$	$17 \cdot 47$
80	$5 \cdot 7 \cdot 23$	$2 \cdot 13 \cdot 31$	$3 \cdot 269$	$2^3 \cdot 101$	9079485
81	$5 \cdot 163$	$2^4 \cdot 3 \cdot 17$	$19 \cdot 43$	$2 \cdot 409$	$3^2 \cdot 7 \cdot 13$
82	$3 \cdot 5^2 \cdot 11$	$2 \cdot 7 \cdot 59$	9175055	$2^2 \cdot 3^2 \cdot 23$	9185545
83	$5 \cdot 167$	$2^2 \cdot 11 \cdot 19$	$3^3 \cdot 31$	$2 \cdot 419$	9237620
84	$5 \cdot 13^2$	$2 \cdot 3^2 \cdot 47$	$7 \cdot 11^2$	$2^4 \cdot 53$	$3 \cdot 283$
85	$3^2 \cdot 5 \cdot 19$	$2^3 \cdot 107$	9329808	$2 \cdot 3 \cdot 11 \cdot 13$	9339932
86	$5 \cdot 173$	$2 \cdot 433$	$3 \cdot 17^2$	$2^2 \cdot 7 \cdot 31$	$11 \cdot 79$
87	$5^3 \cdot 7$	$2^2 \cdot 3 \cdot 73$	9429996	$2 \cdot 439$	$3 \cdot 293$
88	$3 \cdot 5 \cdot 59$	$2 \cdot 443$	9479236	$2^3 \cdot 3 \cdot 37$	$7 \cdot 127$
89	$5 \cdot 179$	$2^7 \cdot 7$	$3 \cdot 13 \cdot 23$	$2 \cdot 449$	$29 \cdot 31$
90	$5 \cdot 181$	$2 \cdot 3 \cdot 151$	9576073	$2^2 \cdot 227$	$3^2 \cdot 101$
91	$3 \cdot 5 \cdot 61$	$2^2 \cdot 229$	$7 \cdot 131$	$2 \cdot 3^3 \cdot 17$	9633155
92	$5^2 \cdot 37$	$2 \cdot 463$	$3^2 \cdot 103$	$2^5 \cdot 29$	9680157
93	$5 \cdot 11 \cdot 17$	$2^3 \cdot 3^2 \cdot 13$	9717396	$2 \cdot 7 \cdot 67$	$3 \cdot 313$
94	$3^3 \cdot 5 \cdot 7$	$2 \cdot 11 \cdot 43$	9763500	$2^2 \cdot 3 \cdot 79$	$13 \cdot 73$
95	$5 \cdot 191$	$2^2 \cdot 239$	$3 \cdot 11 \cdot 29$	$2 \cdot 479$	$7 \cdot 137$
96	$5 \cdot 193$	$2 \cdot 3 \cdot 7 \cdot 23$	9854265	$2^3 \cdot 11^2$	$3 \cdot 17 \cdot 19$
97	$3 \cdot 5^2 \cdot 13$	$2^4 \cdot 61$	9898946	$2 \cdot 3 \cdot 163$	$11 \cdot 89$
98	$5 \cdot 197$	$2 \cdot 17 \cdot 29$	$3 \cdot 7 \cdot 47$	$2^2 \cdot 13 \cdot 19$	$23 \cdot 43$
99	$5 \cdot 199$	$2^2 \cdot 3 \cdot 83$	9986952	$2 \cdot 499$	$3^3 \cdot 37$

FACTORS AND PRIMES (Continued)

n	0	1	2	3	4
100	$2^3 \cdot 5^2$	$7 \cdot 11 \cdot 13$	$2 \cdot 3 \cdot 167$	$17 \cdot 59$	$2^2 \cdot 251$
101	$2 \cdot 5 \cdot 101$	$3 \cdot 337$	$2^2 \cdot 11 \cdot 23$	0056094	$2 \cdot 3 \cdot 13^2$
102	$2^2 \cdot 3 \cdot 5 \cdot 17$	0090257	$2 \cdot 7 \cdot 73$	$3 \cdot 11 \cdot 31$	2^{10}
103	$2 \cdot 5 \cdot 103$	0132587	$2^3 \cdot 3 \cdot 43$	0141003	$2 \cdot 11 \cdot 47$
104	$2^4 \cdot 5 \cdot 13$	$3 \cdot 347$	$2 \cdot 521$	$7 \cdot 149$	$2^2 \cdot 3^2 \cdot 29$
105	$2 \cdot 3 \cdot 5^2 \cdot 7$	0216027	$2^2 \cdot 263$	$3^4 \cdot 13$	$2 \cdot 17 \cdot 31$
106	$2^2 \cdot 5 \cdot 53$	0257154	$2 \cdot 3^2 \cdot 59$	0265333	$2^3 \cdot 7 \cdot 19$
107	$2 \cdot 5 \cdot 107$	$3^2 \cdot 7 \cdot 17$	$2^4 \cdot 67$	$29 \cdot 37$	$2 \cdot 3 \cdot 179$
108	$2^3 \cdot 3^3 \cdot 5$	$23 \cdot 47$	$2 \cdot 541$	$3 \cdot 19^2$	$2^2 \cdot 271$
109	$2 \cdot 5 \cdot 109$	0378248	$2^2 \cdot 3 \cdot 7 \cdot 13$	0386202	$2 \cdot 547$
110	$2^2 \cdot 5^2 \cdot 11$	$3 \cdot 367$	$2 \cdot 19 \cdot 29$	0425755	$2^4 \cdot 3 \cdot 23$
111	$2 \cdot 3 \cdot 5 \cdot 37$	$11 \cdot 101$	$2^3 \cdot 139$	$3 \cdot 7 \cdot 53$	$2 \cdot 557$
112	$2^5 \cdot 5 \cdot 7$	$19 \cdot 59$	$2 \cdot 3 \cdot 11 \cdot 17$	0503798	$2^2 \cdot 281$
113	$2 \cdot 5 \cdot 113$	$3 \cdot 13 \cdot 29$	$2^2 \cdot 283$	$11 \cdot 103$	$2 \cdot 3^4 \cdot 7$
114	$2^2 \cdot 3 \cdot 5 \cdot 19$	$7 \cdot 163$	$2 \cdot 571$	$3^2 \cdot 127$	$2^3 \cdot 11 \cdot 13$
115	$2 \cdot 5^2 \cdot 23$	0610753	$2^7 \cdot 3^2$	0618293	$2 \cdot 577$
116	$2^3 \cdot 5 \cdot 29$	$3^3 \cdot 43$	$2 \cdot 7 \cdot 83$	0655797	$2^2 \cdot 3 \cdot 97$
117	$2 \cdot 3^2 \cdot 5 \cdot 13$	0685569	$2^2 \cdot 293$	$3 \cdot 17 \cdot 23$	$2 \cdot 587$
118	$2^2 \cdot 5 \cdot 59$	0722499	$2 \cdot 3 \cdot 197$	$7 \cdot 13^2$	$2^5 \cdot 37$
119	$2 \cdot 5 \cdot 7 \cdot 17$	$3 \cdot 397$	$2^3 \cdot 149$	0766404	$2 \cdot 3 \cdot 199$
120	$2^4 \cdot 3 \cdot 5^2$	0795430	$2 \cdot 601$	$3 \cdot 401$	$2^2 \cdot 7 \cdot 43$
121	$2 \cdot 5 \cdot 11^2$	$7 \cdot 173$	$2^2 \cdot 3 \cdot 101$	0838608	$2 \cdot 607$
122	$2^2 \cdot 5 \cdot 61$	$3 \cdot 11 \cdot 37$	$2 \cdot 13 \cdot 47$	0874265	$2^3 \cdot 3^2 \cdot 17$
123	$2 \cdot 3 \cdot 5 \cdot 41$	0902581	$2^4 \cdot 7 \cdot 11$	$3^2 \cdot 137$	$2 \cdot 617$
124	$2^3 \cdot 5 \cdot 31$	$17 \cdot 73$	$2 \cdot 3^3 \cdot 23$	$11 \cdot 113$	$2^2 \cdot 311$
125	$2 \cdot 5^4$	$3^2 \cdot 139$	$2^2 \cdot 313$	$7 \cdot 179$	$2 \cdot 3 \cdot 11 \cdot 19$
126	$2^2 \cdot 3^2 \cdot 5 \cdot 7$	$13 \cdot 97$	$2 \cdot 631$	$3 \cdot 421$	$2^4 \cdot 79$
127	$2 \cdot 5 \cdot 127$	$31 \cdot 41$	$2^3 \cdot 3 \cdot 53$	$19 \cdot 67$	$2 \cdot 7^2 \cdot 13$
128	$2^8 \cdot 5$	$3 \cdot 7 \cdot 61$	$2 \cdot 641$	1082267	$2^2 \cdot 3 \cdot 107$
129	$2 \cdot 3 \cdot 5 \cdot 43$	1109262	$2^2 \cdot 17 \cdot 19$	$3 \cdot 431$	$2 \cdot 647$
130	$2^2 \cdot 5^2 \cdot 13$	1142773	$2 \cdot 3 \cdot 7 \cdot 31$	1149444	$2^3 \cdot 163$
131	$2 \cdot 5 \cdot 131$	$3 \cdot 19 \cdot 23$	$2^5 \cdot 41$	$13 \cdot 101$	$2 \cdot 3^2 \cdot 73$
132	$2^3 \cdot 3 \cdot 5 \cdot 11$	1209028	$2 \cdot 661$	$3^3 \cdot 7^2$	$2^3 \cdot 331$
133	$2 \cdot 5 \cdot 7 \cdot 19$	11^3	$2^2 \cdot 3^2 \cdot 37$	$31 \cdot 43$	$2 \cdot 23 \cdot 29$
134	$2^2 \cdot 5 \cdot 67$	$3^2 \cdot 149$	$2 \cdot 11 \cdot 61$	$17 \cdot 79$	$2^6 \cdot 3 \cdot 7$
135	$2 \cdot 3^3 \cdot 5^2$	$7 \cdot 193$	$2^3 \cdot 13^2$	$3 \cdot 11 \cdot 41$	$2 \cdot 677$
136	$2^4 \cdot 5 \cdot 17$	1338581	$2 \cdot 3 \cdot 227$	$29 \cdot 47$	$2^2 \cdot 11 \cdot 31$
137	$2 \cdot 5 \cdot 137$	$3 \cdot 457$	$2^2 \cdot 7^3$	1376705	$2 \cdot 3 \cdot 229$
138	$2^2 \cdot 3 \cdot 5 \cdot 23$	1401937	$2 \cdot 691$	$3 \cdot 461$	$2^3 \cdot 173$
139	$2 \cdot 5 \cdot 139$	$13 \cdot 107$	$2^4 \cdot 3 \cdot 29$	$7 \cdot 199$	$2 \cdot 17 \cdot 41$
140	$2^3 \cdot 5^2 \cdot 7$	$3 \cdot 467$	$2 \cdot 701$	$23 \cdot 61$	$2^2 \cdot 3^3 \cdot 13$
141	$2 \cdot 3 \cdot 5 \cdot 47$	$17 \cdot 83$	$2^2 \cdot 353$	$3^2 \cdot 157$	$2 \cdot 7 \cdot 101$
142	$2^2 \cdot 5 \cdot 71$	$7^2 \cdot 29$	$2 \cdot 3^2 \cdot 79$	1532049	$2^4 \cdot 89$
143	$2 \cdot 5 \cdot 11 \cdot 13$	$3^3 \cdot 53$	$2^3 \cdot 179$	1562462	$2 \cdot 3 \cdot 239$
144	$2^5 \cdot 3^2 \cdot 5$	$11 \cdot 131$	$2 \cdot 7 \cdot 103$	$3 \cdot 13 \cdot 37$	$2^2 \cdot 19^2$
145	$2 \cdot 5^2 \cdot 29$	1616674	$2^2 \cdot 3 \cdot 11^2$	1622656	$2 \cdot 727$
146	$2^2 \cdot 5 \cdot 73$	$3 \cdot 487$	$2 \cdot 17 \cdot 43$	$7 \cdot 11 \cdot 19$	$2^3 \cdot 3 \cdot 61$
147	$2 \cdot 3 \cdot 5 \cdot 7^2$	1676127	$2^6 \cdot 23$	$3 \cdot 491$	$2 \cdot 11 \cdot 67$
148	$2^3 \cdot 5 \cdot 37$	1705551	$2 \cdot 3 \cdot 13 \cdot 19$	1711412	$2^2 \cdot 7 \cdot 53$
149	$2 \cdot 5 \cdot 149$	$3 \cdot 7 \cdot 71$	$2^2 \cdot 373$	1740598	$2 \cdot 3^2 \cdot 83$

FACTORS AND PRIMES (Continued)

n	5	6	7	8	9
100	$3 \cdot 5 \cdot 67$	$2 \cdot 503$	$19 \cdot 53$	$2^4 \cdot 3^2 \cdot 7$	0038912
101	$5 \cdot 7 \cdot 29$	$2^3 \cdot 127$	$3^2 \cdot 113$	$2 \cdot 509$	0081742
102	$5^2 \cdot 41$	$2 \cdot 3^3 \cdot 19$	$13 \cdot 79$	$2^2 \cdot 257$	$3 \cdot 7^3$
103	$3^2 \cdot 5 \cdot 23$	$2^2 \cdot 7 \cdot 37$	$17 \cdot 61$	$2 \cdot 3 \cdot 173$	0166155
104	$5 \cdot 11 \cdot 19$	$2 \cdot 523$	$3 \cdot 349$	$2^3 \cdot 131$	0207755
105	$5 \cdot 211$	$2^5 \cdot 3 \cdot 11$	$7 \cdot 151$	$2 \cdot 23^2$	$3 \cdot 353$
106	$3 \cdot 5 \cdot 71$	$2 \cdot 13 \cdot 41$	$11 \cdot 97$	$2^2 \cdot 3 \cdot 89$	0289777
107	$5^2 \cdot 43$	$2^2 \cdot 269$	$3 \cdot 359$	$2 \cdot 7^2 \cdot 11$	$13 \cdot 83$
108	$5 \cdot 7 \cdot 31$	$2 \cdot 3 \cdot 181$	0362295	$2^6 \cdot 17$	$3^2 \cdot 11^2$
109	$3 \cdot 5 \cdot 73$	$2^3 \cdot 137$	0402066	$2 \cdot 3^2 \cdot 61$	$7 \cdot 157$
110	$5 \cdot 13 \cdot 17$	$2 \cdot 7 \cdot 79$	$3^3 \cdot 41$	$2^2 \cdot 277$	0449315
111	$5 \cdot 223$	$2^2 \cdot 3^2 \cdot 31$	0480532	$2 \cdot 13 \cdot 43$	$3 \cdot 373$
112	$3^2 \cdot 5^3$	$2 \cdot 563$	$7^2 \cdot 23$	$2^3 \cdot 3 \cdot 47$	0526939
113	$5 \cdot 227$	$2^4 \cdot 71$	$3 \cdot 379$	$2 \cdot 569$	$17 \cdot 67$
114	$5 \cdot 229$	$2 \cdot 3 \cdot 191$	$31 \cdot 37$	$2^2 \cdot 7 \cdot 41$	$3 \cdot 383$
115	$3 \cdot 5 \cdot 7 \cdot 11$	$2^2 \cdot 17^2$	$13 \cdot 89$	$2 \cdot 3 \cdot 193$	$19 \cdot 61$
116	$5 \cdot 233$	$2 \cdot 11 \cdot 53$	$3 \cdot 389$	$2^4 \cdot 73$	$7 \cdot 167$
117	$5^2 \cdot 47$	$2^3 \cdot 3 \cdot 7^2$	$11 \cdot 107$	$2 \cdot 19 \cdot 31$	$3^2 \cdot 131$
118	$3 \cdot 5 \cdot 79$	$2 \cdot 593$	0744507	$2^2 \cdot 3^3 \cdot 11$	$29 \cdot 41$
119	$5 \cdot 239$	$2^3 \cdot 13 \cdot 23$	$3^2 \cdot 7 \cdot 19$	$2 \cdot 599$	$11 \cdot 109$
120	$5 \cdot 241$	$2 \cdot 3^2 \cdot 67$	$17 \cdot 71$	$2^3 \cdot 151$	$3 \cdot 13 \cdot 31$
121	$3^5 \cdot 5$	$2^6 \cdot 19$	0852906	$2 \cdot 3 \cdot 7 \cdot 29$	$23 \cdot 53$
122	$5^2 \cdot 7^2$	$2 \cdot 613$	$3 \cdot 409$	$2^2 \cdot 307$	0895519
123	$5 \cdot 13 \cdot 19$	$2^2 \cdot 3 \cdot 103$	0923697	$2 \cdot 619$	$3 \cdot 7 \cdot 59$
124	$3 \cdot 5 \cdot 83$	$2 \cdot 7 \cdot 89$	$29 \cdot 43$	$2^5 \cdot 3 \cdot 13$	0965624
125	$5 \cdot 251$	$2^3 \cdot 157$	$3 \cdot 419$	$2 \cdot 17 \cdot 37$	1000257
126	$5 \cdot 11 \cdot 23$	$2 \cdot 3 \cdot 211$	$7 \cdot 181$	$2^2 \cdot 317$	$3^3 \cdot 47$
127	$3 \cdot 5^2 \cdot 17$	$2^2 \cdot 11 \cdot 29$	1061909	$2 \cdot 3^2 \cdot 71$	1068705
128	$5 \cdot 257$	$2 \cdot 643$	$3^2 \cdot 11 \cdot 13$	$2^3 \cdot 7 \cdot 23$	1102529
129	$5 \cdot 7 \cdot 37$	$2^4 \cdot 3^4$	1129400	$2 \cdot 11 \cdot 59$	$3 \cdot 433$
130	$3^2 \cdot 5 \cdot 29$	$2 \cdot 653$	1162756	$2^2 \cdot 3 \cdot 109$	$7 \cdot 11 \cdot 17$
131	$5 \cdot 263$	$2^3 \cdot 7 \cdot 47$	$3 \cdot 439$	$2 \cdot 659$	1202448
132	$5^2 \cdot 53$	$2 \cdot 3 \cdot 13 \cdot 17$	1228709	$2^4 \cdot 83$	$3 \cdot 443$
133	$3 \cdot 5 \cdot 89$	$2^3 \cdot 167$	$7 \cdot 191$	$2 \cdot 3 \cdot 223$	$13 \cdot 103$
134	$5 \cdot 269$	$2 \cdot 673$	$3 \cdot 449$	$2^2 \cdot 337$	$19 \cdot 71$
135	$5 \cdot 271$	$2^2 \cdot 3 \cdot 113$	$23 \cdot 59$	$2 \cdot 7 \cdot 97$	$3^2 \cdot 151$
136	$3 \cdot 5 \cdot 7 \cdot 13$	$2 \cdot 683$	1357685	$2^3 \cdot 3^2 \cdot 19$	37^2
137	$5^3 \cdot 11$	$2^5 \cdot 43$	$3^4 \cdot 17$	$2 \cdot 13 \cdot 53$	$7 \cdot 197$
138	$5 \cdot 277$	$2 \cdot 3^2 \cdot 7 \cdot 11$	$19 \cdot 73$	$2^2 \cdot 347$	$3 \cdot 463$
139	$3^2 \cdot 5 \cdot 31$	$2^2 \cdot 349$	$11 \cdot 127$	$2 \cdot 3 \cdot 233$	1458177
140	$5 \cdot 281$	$2 \cdot 19 \cdot 37$	$3 \cdot 7 \cdot 67$	$2^7 \cdot 11$	1489110
141	$5 \cdot 283$	$2^3 \cdot 3 \cdot 59$	$13 \cdot 109$	$2 \cdot 709$	$3 \cdot 11 \cdot 43$
142	$3 \cdot 5^2 \cdot 19$	$2 \cdot 23 \cdot 31$	1544240	$2^2 \cdot 3 \cdot 7 \cdot 17$	1550322
143	$5 \cdot 7 \cdot 41$	$2^2 \cdot 359$	$3 \cdot 479$	$2 \cdot 719$	1580608
144	$5 \cdot 17^2$	$2 \cdot 3 \cdot 241$	1604685	$2^3 \cdot 181$	$3^2 \cdot 7 \cdot 23$
145	$3 \cdot 5 \cdot 97$	$2^4 \cdot 7 \cdot 13$	$31 \cdot 47$	$2 \cdot 3^6$	1640553
146	$5 \cdot 293$	$2 \cdot 733$	$3^2 \cdot 163$	$2^2 \cdot 367$	$13 \cdot 113$
147	$5^2 \cdot 59$	$2^2 \cdot 3^2 \cdot 41$	$7 \cdot 211$	$2 \cdot 739$	$3 \cdot 17 \cdot 29$
148	$3^3 \cdot 5 \cdot 11$	$2 \cdot 743$	1723110	$2^4 \cdot 3 \cdot 31$	1728947
149	$5 \cdot 13 \cdot 23$	$2^3 \cdot 11 \cdot 17$	$3 \cdot 499$	$2 \cdot 7 \cdot 107$	1758016

FACTORS AND PRIMES (Continued)

n	0	1	2	3	4
150	$2^3 \cdot 3 \cdot 5^2$	$19 \cdot 79$	$2 \cdot 751$	$3^2 \cdot 167$	$2^5 \cdot 47$
151	$2 \cdot 5 \cdot 151$	1792645	$2^3 \cdot 3^3 \cdot 7$	$17 \cdot 89$	$2 \cdot 757$
152	$2^4 \cdot 5 \cdot 19$	$3^2 \cdot 13^2$	$2 \cdot 761$	1826999	$2^2 \cdot 3 \cdot 127$
153	$2 \cdot 3^2 \cdot 5 \cdot 17$	1849752	$2^2 \cdot 383$	$3 \cdot 7 \cdot 73$	$2 \cdot 13 \cdot 59$
154	$2^2 \cdot 5 \cdot 7 \cdot 11$	$23 \cdot 67$	$2 \cdot 3 \cdot 257$	1883659	$2^3 \cdot 193$
155	$2 \cdot 5^2 \cdot 31$	$3 \cdot 11 \cdot 47$	$2^4 \cdot 97$	1911715	$2 \cdot 3 \cdot 7 \cdot 37$
156	$2^3 \cdot 3 \cdot 5 \cdot 13$	$7 \cdot 223$	$2 \cdot 11 \cdot 71$	$3 \cdot 521$	$2^2 \cdot 17 \cdot 23$
157	$2 \cdot 5 \cdot 157$	1961762	$2^2 \cdot 3 \cdot 131$	$11^2 \cdot 13$	$2 \cdot 787$
158	$2^2 \cdot 5 \cdot 79$	$3 \cdot 17 \cdot 31$	$2 \cdot 7 \cdot 113$	1994809	$2^4 \cdot 3^2 \cdot 11$
159	$2 \cdot 3 \cdot 5 \cdot 53$	$37 \cdot 43$	$2^3 \cdot 199$	$3^3 \cdot 59$	$2 \cdot 797$
160	$2^6 \cdot 5^2$	2043913	$2 \cdot 3 \cdot 89$	$7 \cdot 229$	$2^2 \cdot 401$
161	$2 \cdot 5 \cdot 7 \cdot 23$	$3^2 \cdot 179$	$2^2 \cdot 13 \cdot 31$	2076344	$2 \cdot 3 \cdot 269$
162	$2^2 \cdot 3^4 \cdot 5$	2097830	$2 \cdot 811$	$3 \cdot 541$	$2^3 \cdot 7 \cdot 29$
163	$2 \cdot 5 \cdot 163$	$7 \cdot 233$	$2^5 \cdot 3 \cdot 17$	$23 \cdot 71$	$2 \cdot 19 \cdot 43$
164	$2^2 \cdot 5 \cdot 41$	$3 \cdot 547$	$2 \cdot 821$	$31 \cdot 53$	$2^2 \cdot 3 \cdot 137$
165	$2 \cdot 3 \cdot 5^2 \cdot 11$	$13 \cdot 127$	$2^2 \cdot 7 \cdot 59$	$3 \cdot 19 \cdot 29$	$2 \cdot 827$
166	$2^2 \cdot 5 \cdot 83$	$11 \cdot 151$	$2 \cdot 3 \cdot 277$	2208922	$2^7 \cdot 13$
167	$2 \cdot 5 \cdot 167$	$3 \cdot 557$	$2^3 \cdot 11 \cdot 19$	$7 \cdot 239$	$2 \cdot 3^3 \cdot 31$
168	$2^4 \cdot 3 \cdot 5 \cdot 7$	41^2	$2 \cdot 29^2$	$3^2 \cdot 11 \cdot 17$	$2^2 \cdot 421$
169	$2 \cdot 5 \cdot 13^2$	$19 \cdot 89$	$2^2 \cdot 3^3 \cdot 47$	2286570	$2 \cdot 7 \cdot 11^2$
170	$2^2 \cdot 5^2 \cdot 17$	$3^5 \cdot 7$	$2 \cdot 23 \cdot 37$	$13 \cdot 131$	$2^3 \cdot 3 \cdot 71$
171	$2 \cdot 3^2 \cdot 5 \cdot 19$	$29 \cdot 59$	$2^4 \cdot 107$	$3 \cdot 571$	$2 \cdot 857$
172	$2^3 \cdot 5 \cdot 43$	2357809	$2 \cdot 3 \cdot 7 \cdot 41$	2362853	$2^2 \cdot 431$
173	$2 \cdot 5 \cdot 173$	$3 \cdot 577$	$2^2 \cdot 433$	2387986	$2 \cdot 3 \cdot 17^2$
174	$2^2 \cdot 3 \cdot 5 \cdot 29$	2407988	$2 \cdot 13 \cdot 67$	$3 \cdot 7 \cdot 83$	$2^4 \cdot 109$
175	$2 \cdot 5^3 \cdot 7$	$17 \cdot 103$	$2^3 \cdot 3 \cdot 73$	2437819	$2 \cdot 877$
176	$2^5 \cdot 5 \cdot 11$	$3 \cdot 587$	$2 \cdot 881$	$41 \cdot 43$	$2^2 \cdot 3^2 \cdot 7^2$
177	$2 \cdot 3 \cdot 5 \cdot 59$	$7 \cdot 11 \cdot 23$	$2^2 \cdot 443$	$3^2 \cdot 197$	$2 \cdot 887$
178	$2^2 \cdot 5 \cdot 89$	$13 \cdot 137$	$2 \cdot 3^4 \cdot 11$	2511513	$2^3 \cdot 223$
179	$2 \cdot 5 \cdot 179$	$3^2 \cdot 199$	$2^8 \cdot 7$	$11 \cdot 163$	$2 \cdot 3 \cdot 13 \cdot 23$
180	$2^3 \cdot 3^2 \cdot 5^2$	2555137	$2 \cdot 17 \cdot 53$	$3 \cdot 601$	$2^2 \cdot 11 \cdot 41$
181	$2 \cdot 5 \cdot 181$	2579185	$2^2 \cdot 3 \cdot 151$	$7^2 \cdot 37$	$2 \cdot 907$
182	$2^2 \cdot 5 \cdot 7 \cdot 13$	$3 \cdot 607$	$2 \cdot 911$	2607867	$2^5 \cdot 3 \cdot 19$
183	$2 \cdot 3 \cdot 5 \cdot 61$	2626883	$2^3 \cdot 229$	$3 \cdot 13 \cdot 47$	$2 \cdot 7 \cdot 131$
184	$2^4 \cdot 5 \cdot 23$	$7 \cdot 263$	$2 \cdot 3 \cdot 307$	$19 \cdot 97$	$2^2 \cdot 461$
185	$2 \cdot 5^2 \cdot 37$	$3 \cdot 617$	$2^2 \cdot 463$	$17 \cdot 109$	$2 \cdot 3^2 \cdot 103$
186	$2^2 \cdot 3 \cdot 5 \cdot 31$	2697464	$2 \cdot 7^2 \cdot 19$	$3^4 \cdot 23$	$2^3 \cdot 233$
187	$2 \cdot 5 \cdot 11 \cdot 17$	2720738	$2^4 \cdot 3^2 \cdot 13$	2725378	$2 \cdot 937$
188	$2^3 \cdot 5 \cdot 47$	$3^2 \cdot 11 \cdot 19$	$2 \cdot 941$	$7 \cdot 269$	$2^2 \cdot 3 \cdot 157$
189	$2 \cdot 3^3 \cdot 5 \cdot 7$	$31 \cdot 61$	$2^2 \cdot 11 \cdot 43$	$3 \cdot 631$	$2 \cdot 947$
190	$2^2 \cdot 5 \cdot 2 \cdot 19$	2789821	$2 \cdot 3 \cdot 317$	$11 \cdot 173$	$2^4 \cdot 7 \cdot 17$
191	$2 \cdot 5 \cdot 191$	$3 \cdot 7^2 \cdot 13$	$2^3 \cdot 239$	2817150	$2 \cdot 3 \cdot 11 \cdot 29$
192	$2^7 \cdot 3 \cdot 5$	$17 \cdot 113$	$2 \cdot 31^2$	$3 \cdot 641$	$2^2 \cdot 13 \cdot 37$
193	$2 \cdot 5 \cdot 193$	2857823	$2^2 \cdot 3 \cdot 7 \cdot 23$	2862319	$2 \cdot 967$
194	$2^2 \cdot 5 \cdot 97$	$3 \cdot 647$	$2 \cdot 971$	$29 \cdot 67$	$2^3 \cdot 3^5$
195	$2 \cdot 3 \cdot 5^2 \cdot 13$	2902573	$2^5 \cdot 61$	$3^2 \cdot 7 \cdot 31$	$2 \cdot 977$
196	$2^3 \cdot 5 \cdot 7^2$	$37 \cdot 53$	$2 \cdot 3^2 \cdot 109$	$13 \cdot 151$	$2^2 \cdot 491$
197	$2 \cdot 5 \cdot 197$	$3^3 \cdot 73$	$2^2 \cdot 17 \cdot 29$	2951271	$2 \cdot 3 \cdot 7 \cdot 47$
198	$2^2 \cdot 3^2 \cdot 5 \cdot 11$	$7 \cdot 283$	$2 \cdot 991$	$3 \cdot 661$	$2^6 \cdot 31$
199	$2 \cdot 5 \cdot 199$	$11 \cdot 181$	$2^3 \cdot 3 \cdot 83$	2995073	$2 \cdot 997$

FACTORS AND PRIMES (Continued)

n	5	6	7	8	9
150	$5 \cdot 7 \cdot 43$	$2 \cdot 3 \cdot 251$	$11 \cdot 137$	$2^2 \cdot 13 \cdot 29$	$3 \cdot 503$
151	$3 \cdot 5 \cdot 101$	$2^2 \cdot 379$	$37 \cdot 41$	$2 \cdot 3 \cdot 11 \cdot 23$	$7^2 \cdot 31$
152	$5^2 \cdot 61$	$2 \cdot 7 \cdot 109$	$3 \cdot 509$	$2^3 \cdot 191$	$11 \cdot 139$
153	$5 \cdot 307$	$2^9 \cdot 3$	$29 \cdot 53$	$2 \cdot 769$	$3^4 \cdot 19$
154	$3 \cdot 5 \cdot 103$	$2 \cdot 773$	$7 \cdot 13 \cdot 17$	$2^2 \cdot 3^2 \cdot 43$	1900514
155	$5 \cdot 311$	$2^2 \cdot 389$	$3^2 \cdot 173$	$2 \cdot 19 \cdot 41$	1928461
156	$5 \cdot 313$	$2 \cdot 3^3 \cdot 29$	1950690	$2^5 \cdot 7^2$	$3 \cdot 523$
157	$3^2 \cdot 5^2 \cdot 7$	$2^3 \cdot 197$	$19 \cdot 83$	$2 \cdot 3 \cdot 263$	1983821
158	$5 \cdot 317$	$2 \cdot 13 \cdot 61$	$3 \cdot 23^2$	$2^2 \cdot 397$	$7 \cdot 227$
159	$5 \cdot 11 \cdot 29$	$2^2 \cdot 3 \cdot 7 \cdot 19$	2033049	$2 \cdot 17 \cdot 47$	$3 \cdot 13 \cdot 41$
160	$3 \cdot 5 \cdot 107$	$2 \cdot 11 \cdot 73$	2060159	$2^3 \cdot 3 \cdot 67$	2065560
161	$5 \cdot 17 \cdot 19$	$2^4 \cdot 101$	$3 \cdot 7^2 \cdot 11$	$2 \cdot 809$	2092468
162	$5^3 \cdot 13$	$2 \cdot 3 \cdot 271$	2113876	$2^2 \cdot 11 \cdot 37$	$3^2 \cdot 181$
163	$3 \cdot 5 \cdot 109$	$2^2 \cdot 409$	2140487	$2 \cdot 3^2 \cdot 7 \cdot 13$	$11 \cdot 149$
164	$5 \cdot 7 \cdot 47$	$2 \cdot 823$	$3^3 \cdot 61$	$2^4 \cdot 103$	$17 \cdot 97$
165	$5 \cdot 331$	$2^3 \cdot 3^2 \cdot 23$	2193225	$2 \cdot 829$	$3 \cdot 7 \cdot 79$
166	$3^2 \cdot 5 \cdot 37$	$2 \cdot 7^2 \cdot 17$	2219356	$2^2 \cdot 3 \cdot 139$	2224563
167	$5^2 \cdot 67$	$2^2 \cdot 419$	$3 \cdot 13 \cdot 43$	$2 \cdot 839$	$23 \cdot 73$
168	$5 \cdot 337$	$2 \cdot 3 \cdot 281$	$7 \cdot 241$	$2^3 \cdot 211$	$3 \cdot 563$
169	$3 \cdot 5 \cdot 113$	$2^5 \cdot 53$	2296818	$2 \cdot 3 \cdot 283$	2301934
170	$5 \cdot 11 \cdot 31$	$2 \cdot 853$	$3 \cdot 569$	$2^2 \cdot 7 \cdot 61$	2327421
171	$5 \cdot 7^3$	$2^2 \cdot 3 \cdot 11 \cdot 13$	$17 \cdot 101$	$2 \cdot 859$	$3^2 \cdot 191$
172	$3 \cdot 5^2 \cdot 23$	$2 \cdot 863$	$11 \cdot 157$	$2^6 \cdot 3^3$	$7 \cdot 13 \cdot 19$
173	$5 \cdot 347$	$2^3 \cdot 7 \cdot 31$	$3^2 \cdot 193$	$2 \cdot 11 \cdot 79$	$37 \cdot 47$
174	$5 \cdot 349$	$2 \cdot 3^2 \cdot 97$	2422929	$2^2 \cdot 19 \cdot 23$	$3 \cdot 11 \cdot 53$
175	$3^3 \cdot 5 \cdot 13$	$2^2 \cdot 439$	$7 \cdot 251$	$2 \cdot 3 \cdot 293$	2452658
176	$5 \cdot 353$	$2 \cdot 883$	$3 \cdot 19 \cdot 31$	$2^3 \cdot 13 \cdot 17$	$29 \cdot 61$
177	$5^2 \cdot 71$	$2^4 \cdot 3 \cdot 37$	2496874	$2 \cdot 7 \cdot 127$	$3 \cdot 593$
178	$3 \cdot 5 \cdot 7 \cdot 17$	$2 \cdot 19 \cdot 47$	2521246	$2^2 \cdot 3 \cdot 149$	2526103
179	$5 \cdot 359$	$2^2 \cdot 449$	$3 \cdot 599$	$2 \cdot 29 \cdot 31$	$7 \cdot 257$
180	$5 \cdot 19^2$	$2 \cdot 3 \cdot 7 \cdot 43$	$13 \cdot 139$	$2^4 \cdot 113$	$3^3 \cdot 67$
181	$3 \cdot 5 \cdot 11^2$	$2^3 \cdot 227$	$23 \cdot 79$	$2 \cdot 3^2 \cdot 101$	$17 \cdot 107$
182	$5^2 \cdot 73$	$2 \cdot 11 \cdot 83$	$3^2 \cdot 7 \cdot 29$	$2^2 \cdot 457$	$31 \cdot 59$
183	$5 \cdot 367$	$2^2 \cdot 3^3 \cdot 17$	$11 \cdot 167$	$2 \cdot 919$	$3 \cdot 613$
184	$3^2 \cdot 5 \cdot 41$	$2 \cdot 13 \cdot 71$	2664669	$2^3 \cdot 3 \cdot 7 \cdot 11$	43^2
185	$5 \cdot 7 \cdot 53$	$2^6 \cdot 29$	$3 \cdot 619$	$2 \cdot 929$	$11 \cdot 13^2$
186	$5 \cdot 373$	$2 \cdot 3 \cdot 311$	2711443	$2^2 \cdot 467$	$3 \cdot 7 \cdot 89$
187	$3 \cdot 5^4$	$2^2 \cdot 7 \cdot 67$	2734643	$2 \cdot 3 \cdot 313$	2739268
188	$5 \cdot 13 \cdot 29$	$2 \cdot 23 \cdot 41$	$3 \cdot 17 \cdot 37$	$2^5 \cdot 59$	2762320
189	$5 \cdot 379$	$2^3 \cdot 3 \cdot 79$	$7 \cdot 271$	$2 \cdot 13 \cdot 73$	$3^2 \cdot 211$
190	$3 \cdot 5 \cdot 127$	$2 \cdot 953$	2803507	$2^2 \cdot 3^2 \cdot 53$	$23 \cdot 83$
191	$5 \cdot 383$	$2^2 \cdot 479$	$3^3 \cdot 71$	$2 \cdot 7 \cdot 137$	$19 \cdot 101$
192	$5^2 \cdot 7 \cdot 11$	$2 \cdot 3^2 \cdot 107$	$41 \cdot 47$	$2^3 \cdot 241$	$3 \cdot 643$
193	$3^2 \cdot 5 \cdot 43$	$2^4 \cdot 11^2$	$13 \cdot 149$	$2 \cdot 3 \cdot 17 \cdot 19$	$7 \cdot 277$
194	$5 \cdot 389$	$2 \cdot 7 \cdot 139$	$3 \cdot 11 \cdot 59$	$2^2 \cdot 487$	2898118
195	$5 \cdot 17 \cdot 23$	$2^2 \cdot 3 \cdot 163$	$19 \cdot 103$	$2 \cdot 11 \cdot 89$	$3 \cdot 653$
196	$3 \cdot 5 \cdot 131$	$2 \cdot 983$	$7 \cdot 281$	$2^4 \cdot 3 \cdot 41$	$11 \cdot 179$
197	$5^2 \cdot 79$	$2^3 \cdot 13 \cdot 19$	$3 \cdot 659$	$2 \cdot 23 \cdot 43$	2964458
198	$5 \cdot 397$	$2 \cdot 3 \cdot 331$	2981979	$2^2 \cdot 7 \cdot 71$	$3^2 \cdot 13 \cdot 17$
199	$3 \cdot 5 \cdot 7 \cdot 19$	$2^2 \cdot 499$	3003781	$2 \cdot 3^3 \cdot 37$	3008128

EXTENDED TABLES OF FACTORS AND PRIMES

The following procedure makes possible the determination of a number between 2009 and 19,949 as being either prime, or if not, what its factors will be. It is herewith included with the kind permission of its author, Professor Leonard Caners.

The following symbols will be used:

N is the number whose factors, if any, are to be determined.

A is N with the last digit dropped. Thus if $N = 17{,}873$, $A = 1787$.

R is the range and is the integral part of the square root of N with its last two digits dropped. Thus R of $17{,}873 = 13$.

K is the key number and is found in the table below.

P_1 is a possible factor corresponding to K.

The twin series are given in the table for the sake of completeness but are not written down in actual practice.

There are two steps as outlined below.

Step I

N	K	P_1	Series	Procedure
Any number ending in 1, 3, 7, or 9, within limits stated above	$A - R$	$10R +$ last digit	$K + n$; $P_1 - 10n$ $n = 0, 1, \ldots, 2R + 1$	Beginning with K read table of Factors and Primes from left to right and look for corresponding possible factors

Step II

N ending in	K	P_1	Series	Procedure
1	$(A - 2) - 3R$	$10R + 7$	$K + 3n$; $P_1 - 10n$ $n = 0, 1, \ldots, 2R + 1$	Same as in Step I except that only every *third* entry in table of Factors and Primes is examined for corresponding possible factors
3	$A - 3R$	$10R + 1$	$n = 0, 1, \ldots, 2R + 1$	
7	$(A - 2) - 3R$	$10R + 9$	$n = 0, 1, \ldots, 2R + 1$	
9	$A - 3R$	$10R + 3$	$n = 0, 1, \ldots, 2R + 1$	

As an example consider 7519. Using tables on preceding pages proceed as follows:

Step I

From Table $N = 7519$; $A = 751$; $R = 8$, $K = 743$; $P_1 = 89$ and the series obtained:

743	89
744	79
745	69
746	59
—	—
751	9
752	1
753	11
—	—
760	81

EXTENDED TABLES OF FACTORS AND PRIMES (Continued)

Examining the Table of Factors and Primes beginning with 743 note whether 743 has 89 as a factor; whether 744 has 79 as a factor, etc. Then continue to read from left to right keeping in mind the possible factors 69, 59, . . . 81.

Since none are found one concludes that 7519 has no factors within the given range ending in 9 or 1.

Step II

From Table $K = 727$ and $P_1 = 83$; the series obtained is:

727	83
730	73
733	63
—	—
751	3
754	7
—	—
778	87

Proceed exactly as in Step I. Begin with 727 and note whether it has 83 as a factor. Thereafter examine every *third* entry for the remaining corresponding possible factors 73, 63 . . . 3, 7, . . . 87. Since 730 yields the factor 73 one concludes that 73 is the factor of 7519. By division $7519 = 73 \times 103$. Had no factor been found in either Step I or Step II the conclusion would be that the number under consideration was prime.

Proof for Step I

Let N end in 1. Then, if N has a factor ending in 1, write

$$(10a + 1)(10b + 1) = N = 10A + 1$$
$$10ab + a + b = A$$

or

$$b(10a + 1) = (A - a)$$

This proves that any factor $10a + 1$ of N is also a factor of $A - a$. Let $a = (R - n)$, then $(10a + 1)$ becomes $(10R + 1) - 10n$ and $(A - a)$ becomes $(A - R) + n$, i.e., $P_1 - 10n$ and $K + n$ respectively as given in the table. Since $P_1 - 10n$ becomes a number ending in 9 when $P_1 < 10n$, all possible factors ending in 9 are also provided for.

The proof for Step II is similar. For possible factors ending in 3 or 7 write

$$(10a + 7)(10b + 3) = N = 10A + 1.$$

By identical reasoning this results in $P_1 - 10n$ and $K + 3n$ respectively given in the table. This completes the proof for numbers ending in 1.

Identical reasoning establishes the key numbers and series for ending in 3, 7, or 9.

A separate table giving primes from 1 to 100,000 is included on pages 84–91.

TOTIENT FUNCTION $\phi(n)$

Introductory facts

$\phi(n)$ is the number of integers not exceeding and relatively prime to n.

σ_k is the sum of the k'th powers of divisors of n.

σ_0 is usually denoted by $d(n)$ and specifies the number of divisors of n.

For example if $n = 10$, then the divisors are 1, 2, 5, 10 and the sum of the first powers is 18, i.e. $\sigma_0 = d(10) = 4$; $\sigma_1 = 18$; $\phi(10) = 4$.

n	$\phi(n)$	σ_0	σ_1	n	$\phi(n)$	σ_0	σ_1	n	$\phi(n)$	σ_0	σ_1	n	$\phi(n)$	σ_0	σ_1
1	1	1	1	41	40	2	42	81	54	5	121	121	110	3	133
2	1	2	3	42	12	8	96	82	40	4	126	122	60	4	186
3	2	2	4	43	42	2	44	83	82	2	84	123	80	4	168
4	2	3	7	44	20	6	84	84	24	12	224	124	60	6	224
5	4	2	6	45	24	6	78	85	64	4	108	125	100	4	156
6	2	4	12	46	22	4	72	86	42	4	132	126	36	12	312
7	6	2	8	47	46	2	48	87	56	4	120	127	126	2	128
8	4	4	15	48	16	10	124	88	40	8	180	128	64	8	255
9	6	3	13	49	42	3	57	89	88	2	90	129	84	4	176
10	4	4	18	50	20	6	93	90	24	12	234	130	48	8	252
11	10	2	12	51	32	4	72	91	72	4	112	131	130	2	132
12	4	6	28	52	24	6	98	92	44	6	168	132	40	12	336
13	12	2	14	53	52	2	54	93	60	4	128	133	108	4	160
14	6	4	24	54	18	8	120	94	46	4	144	134	66	4	204
15	8	4	24	55	40	4	72	95	72	4	120	135	72	8	240
16	8	5	31	56	24	8	120	96	32	12	252	136	64	8	270
17	16	2	18	57	36	4	80	97	96	2	98	137	136	2	138
18	6	6	39	58	28	4	90	98	42	6	171	138	44	8	288
19	18	2	20	59	58	2	60	99	60	6	156	139	138	2	140
20	8	6	42	60	16	12	168	100	40	9	217	140	48	12	336
21	12	4	32	61	60	2	62	101	100	2	102	141	92	4	192
22	10	4	36	62	30	4	96	102	32	8	216	142	70	4	216
23	22	2	24	63	36	6	104	103	102	2	104	143	120	4	168
24	8	8	60	64	32	7	127	104	48	8	210	144	48	15	403
25	20	3	31	65	48	4	84	105	48	8	192	145	112	4	180
26	12	4	42	66	20	8	144	106	52	4	162	146	72	4	222
27	18	4	40	67	66	2	68	107	106	2	108	147	84	6	228
28	12	6	56	68	32	6	126	108	36	12	280	148	72	6	266
29	28	2	30	69	44	4	96	109	108	2	110	149	148	2	150
30	8	8	72	70	24	8	144	110	40	8	216	150	40	12	372
31	30	2	32	71	70	2	72	111	72	4	152	151	150	2	152
32	16	6	63	72	24	12	195	112	48	10	248	152	72	8	300
33	20	4	48	73	72	2	74	113	112	2	114	153	96	6	234
34	16	4	54	74	36	4	114	114	36	8	240	154	60	8	288
35	24	4	48	75	40	6	124	115	88	4	144	155	120	4	192
36	12	9	91	76	36	6	140	116	56	6	210	156	48	12	392
37	36	2	38	77	60	4	96	117	72	6	182	157	156	2	158
38	18	4	60	78	24	8	168	118	58	4	180	158	78	4	240
39	24	4	56	79	78	2	80	119	96	4	144	159	104	4	216
40	16	8	90	80	32	10	186	120	32	16	360	160	64	12	378

TOTIENT FUNCTION $\phi(n)$ (Continued)

n	$\phi(n)$	σ_0	σ_1	n	$\phi(n)$	σ_0	σ_1	n	$\phi(n)$	σ_0	σ_1	n	$\phi(n)$	σ_0	σ_1
161	132	4	192	211	210	2	212	261	168	6	390	311	310	2	312
162	54	10	363	212	104	6	378	262	130	4	396	312	96	16	840
163	162	2	164	213	140	4	288	263	262	2	264	313	312	2	314
164	80	6	294	214	106	4	324	264	80	16	720	314	156	4	474
165	80	8	288	215	168	4	264	265	208	4	324	315	144	12	624
166	82	4	252	216	72	16	600	266	108	8	480	316	156	6	560
167	166	2	168	217	180	4	256	267	176	4	360	317	316	2	318
168	48	16	480	218	108	4	330	268	132	6	476	318	104	8	648
169	156	3	183	219	144	4	296	269	268	2	270	319	280	4	360
170	64	8	324	220	80	12	504	270	72	16	720	320	128	14	762
171	108	6	260	221	192	4	252	271	270	2	272	321	212	4	432
172	84	6	308	222	72	8	456	272	128	10	558	322	132	8	576
173	172	2	174	223	222	2	224	273	144	8	448	323	288	4	360
174	56	8	360	224	96	12	504	274	136	4	414	324	108	15	847
175	120	6	248	225	120	9	403	275	200	6	372	325	240	6	434
176	80	10	372	226	112	4	342	276	88	12	672	326	162	4	492
177	116	4	240	227	226	2	228	277	276	2	278	327	216	4	440
178	88	4	270	228	72	12	560	278	138	4	420	328	160	8	630
179	178	2	180	229	228	2	230	279	180	6	416	329	276	4	384
180	48	18	546	230	88	8	432	280	96	16	720	330	80	16	864
181	180	2	182	231	120	8	384	281	280	2	282	331	330	2	332
182	72	8	336	232	112	8	450	282	92	8	576	332	164	6	588
183	120	4	248	233	232	2	234	283	282	2	284	333	216	6	494
184	88	8	360	234	72	12	546	284	140	6	504	334	166	4	504
185	144	4	228	235	184	4	288	285	144	8	480	335	264	4	408
186	60	8	384	236	116	6	420	286	120	8	504	336	96	20	992
187	160	4	216	237	156	4	320	287	240	4	336	337	336	2	338
188	92	6	336	238	96	8	432	288	96	18	819	338	156	6	549
189	108	8	320	239	238	2	240	289	272	3	307	339	224	4	456
190	72	8	360	240	64	20	744	290	112	8	540	340	128	12	756
191	190	2	192	241	240	2	242	291	192	4	392	341	300	4	384
192	64	14	508	242	110	6	399	292	144	6	518	342	108	12	780
193	192	2	194	243	162	6	364	293	292	2	294	343	294	4	400
194	96	4	294	244	120	6	434	294	84	12	684	344	168	8	660
195	96	8	336	245	168	6	342	295	232	4	360	345	176	8	576
196	84	9	399	246	80	8	504	296	144	8	570	346	172	4	522
197	196	2	198	247	216	4	280	297	180	8	480	347	346	2	348
198	60	12	468	248	120	8	480	298	148	4	450	348	112	12	840
199	198	2	200	249	164	4	336	299	264	4	336	349	348	2	350
200	80	12	465	250	100	8	468	300	80	18	868	350	120	12	744
201	132	4	272	251	250	2	252	301	252	4	352	351	216	8	560
202	100	4	306	252	72	18	728	302	150	4	456	352	160	12	756
203	168	4	240	253	220	4	288	303	200	4	408	353	352	2	354
204	64	12	504	254	126	4	384	304	144	10	620	354	116	8	720
205	160	4	252	255	128	8	432	305	240	4	372	355	280	4	432
206	102	4	312	256	128	9	511	306	96	12	702	356	176	6	630
207	132	6	312	257	256	2	258	307	306	2	308	357	192	8	576
208	96	10	434	258	84	8	528	308	120	12	672	358	178	4	540
209	180	4	240	259	216	4	304	309	204	4	416	359	358	2	360
210	48	16	576	260	96	12	588	310	120	8	576	360	96	24	1170

COMBINATORIAL ANALYSIS

TOTIENT FUNCTION $\phi(n)$ (Continued)

n	$\phi(n)$	σ_0	σ_1	n	$\phi(n)$	σ_0	σ_1	n	$\phi(n)$	σ_0	σ_1	n	$\phi(n)$	σ_0	σ_1
361	342	3	381	411	272	4	552	461	460	2	462	511	432	4	592
362	180	4	546	412	204	6	728	462	120	16	1152	512	256	10	1023
363	220	6	532	413	348	4	480	463	462	2	464	513	324	8	800
364	144	12	784	414	132	12	936	464	224	10	930	514	256	4	774
365	288	4	444	415	328	4	504	465	240	8	768	515	408	4	624
366	120	8	744	416	192	12	882	466	232	4	702	516	168	12	1232
367	366	2	368	417	276	4	560	467	466	2	468	517	460	4	576
368	176	10	744	418	180	8	720	468	144	18	1274	518	216	8	912
369	240	6	546	419	418	2	420	469	396	4	544	519	344	4	696
370	144	8	684	420	96	24	1344	470	184	8	864	520	192	16	1260
371	312	4	432	421	420	2	422	471	312	4	632	521	520	2	522
372	120	12	896	422	210	4	636	472	232	8	900	522	168	12	1170
373	372	2	374	423	276	6	624	473	420	4	528	523	522	2	524
374	160	8	648	424	208	8	810	474	156	8	960	524	260	6	924
375	200	8	624	425	320	6	558	475	360	6	620	525	240	12	992
376	184	8	720	426	140	8	864	476	192	12	1008	526	262	4	792
377	336	4	420	427	360	4	496	477	312	6	702	527	480	4	576
378	108	16	960	428	212	6	756	478	238	4	720	528	160	20	1488
379	378	2	380	429	240	8	672	479	478	2	480	529	506	3	553
380	144	12	840	430	168	8	792	480	128	24	1512	530	208	8	972
381	252	4	512	431	430	2	432	481	432	4	532	531	348	6	780
382	190	4	576	432	144	20	1240	482	240	4	726	532	216	12	1120
383	382	2	384	433	432	2	434	483	264	8	768	533	480	4	588
384	128	16	1020	434	180	8	768	484	220	9	931	534	176	8	1080
385	240	8	576	435	224	8	720	485	384	4	588	535	424	4	648
386	192	4	582	436	216	6	770	486	162	12	1092	536	264	8	1020
387	252	6	572	437	396	4	480	487	486	2	488	537	356	4	720
388	192	6	686	438	144	8	888	488	240	8	930	538	268	4	810
389	388	2	390	439	438	2	440	489	324	4	656	539	420	6	684
390	96	16	1008	440	160	16	1080	490	168	12	1026	540	144	24	1680
391	352	4	432	441	252	9	741	491	490	2	492	541	540	2	542
392	168	12	855	442	192	8	756	492	160	12	1176	542	270	4	816
393	260	4	528	443	442	2	444	493	448	4	540	543	360	4	728
394	196	4	594	444	144	12	1064	494	216	8	840	544	256	12	1134
395	312	4	480	445	352	4	540	495	240	12	936	545	432	4	660
396	120	18	1092	446	222	4	672	496	240	10	992	546	144	16	1344
397	396	2	398	447	296	4	600	497	420	4	576	547	546	2	548
398	198	4	600	448	192	14	1016	498	164	8	1008	548	272	6	966
399	216	8	640	449	448	2	450	499	498	2	500	549	360	6	806
400	160	15	961	450	120	18	1209	500	200	12	1092	550	200	12	1116
401	400	2	402	451	400	4	504	501	332	4	672	551	504	4	600
402	132	8	816	452	224	6	798	502	250	4	756	552	176	16	1440
403	360	4	448	453	300	4	608	503	502	2	504	553	468	4	640
404	200	6	714	454	226	4	684	504	144	24	1560	554	276	4	834
405	216	10	726	455	288	8	672	505	400	4	612	555	288	8	912
406	168	8	720	456	144	16	1200	506	220	8	864	556	276	6	980
407	360	4	456	457	456	2	458	507	312	6	732	557	556	2	558
408	128	16	1080	458	228	4	690	508	252	6	896	558	180	12	1248
409	408	2	410	459	288	8	720	509	508	2	510	559	504	4	616
410	160	8	756	460	176	12	1008	510	128	16	1296	560	192	20	1488

TOTIENT FUNCTION $\phi(n)$ (Continued)

n	$\phi(n)$	σ_0	σ_1	n	$\phi(n)$	σ_0	σ_1	n	$\phi(n)$	σ_0	σ_1	n	$\phi(n)$	σ_0	σ_1
561	320	8	864	611	552	4	672	661	660	2	662	711	468	6	1040
562	280	4	846	612	192	18	1638	662	330	4	996	712	352	8	1350
563	562	2	564	613	612	2	614	663	384	8	1008	713	660	4	768
564	184	12	1344	614	306	4	924	664	328	8	1260	714	192	16	1728
565	448	4	684	615	320	8	1008	665	432	8	960	715	480	8	1008
566	282	4	852	616	240	16	1440	666	216	12	1482	716	356	6	1260
567	324	10	968	617	616	2	618	667	616	4	720	717	476	4	960
568	280	8	1080	618	204	8	1248	668	332	6	1176	718	358	4	1080
569	568	2	570	619	618	2	620	669	444	4	896	719	718	2	720
570	144	16	1440	620	240	12	1344	670	264	8	1224	720	192	30	2418
571	570	2	572	621	396	8	960	671	600	4	744	721	612	4	832
572	240	12	1176	622	310	4	936	672	192	24	2016	722	342	6	1143
573	380	4	768	623	528	4	720	673	672	2	674	723	480	4	968
574	240	8	1008	624	192	20	1736	674	336	4	1014	724	360	6	1274
575	440	6	744	625	500	5	781	675	360	12	1240	725	560	6	930
576	192	21	1651	626	312	4	942	676	312	9	1281	726	220	12	1596
577	576	2	578	627	360	8	960	677	676	2	678	727	726	2	728
578	272	6	921	628	312	6	1106	678	224	8	1368	728	288	16	1680
579	384	4	776	629	576	4	684	679	576	4	784	729	486	7	1093
580	224	12	1260	630	144	24	1872	680	256	16	1620	730	288	8	1332
581	492	4	672	631	630	2	632	681	452	4	912	731	672	4	792
582	192	8	1176	632	312	8	1200	682	300	8	1152	732	240	12	1736
583	520	4	648	633	420	4	848	683	682	2	684	733	732	2	734
584	288	8	1110	634	316	4	954	684	216	18	1820	734	366	4	1104
585	288	12	1092	635	504	4	768	685	544	4	828	735	336	12	1368
586	292	4	882	636	208	12	1512	686	294	8	1200	736	352	12	1512
587	586	2	588	637	504	6	798	687	456	4	920	737	660	4	816
588	168	18	1596	638	280	8	1080	688	336	10	1364	738	240	12	1638
589	540	4	640	639	420	6	936	689	624	4	756	739	738	2	740
590	232	8	1080	640	256	16	1530	690	176	16	1728	740	288	12	1596
591	392	4	792	641	640	2	642	691	690	2	692	741	432	8	1120
592	288	10	1178	642	212	8	1296	692	344	6	1218	742	312	8	1296
593	592	2	594	643	642	2	644	693	360	12	1248	743	742	2	744
594	180	16	1440	644	264	12	1344	694	346	4	1044	744	240	16	1920
595	384	8	864	645	336	8	1056	695	552	4	840	745	592	4	900
596	296	6	1050	646	288	8	1080	696	224	16	1800	746	372	4	1122
597	396	4	800	647	646	2	648	697	640	4	756	747	492	6	1092
598	264	8	1008	648	216	20	1815	698	348	4	1050	748	320	12	1512
599	598	2	600	649	580	4	720	699	464	4	936	749	636	4	864
600	160	24	1860	650	240	12	1302	700	240	18	1736	750	200	16	1872
601	600	2	602	651	360	8	1024	701	700	2	702	751	750	2	752
602	252	8	1056	652	324	6	1148	702	216	16	1680	752	368	10	1488
603	396	6	884	653	652	2	654	703	648	4	760	753	500	4	1008
604	300	6	1064	654	216	8	1320	704	320	14	1524	754	336	8	1260
605	440	6	798	655	520	4	792	705	368	8	1152	755	600	4	912
606	200	8	1224	656	320	10	1302	706	352	4	1062	756	216	24	2240
607	606	2	608	657	432	6	962	707	600	4	816	757	756	2	758
608	288	12	1260	658	276	8	1152	708	232	12	1680	758	378	4	1140
609	336	8	960	659	658	2	660	709	708	2	710	759	440	8	1152
610	240	8	1116	660	160	24	2016	710	280	8	1296	760	288	16	1800

COMBINATORIAL ANALYSIS
TOTIENT FUNCTION $\phi(n)$ (Continued)

n	$\phi(n)$	σ_0	σ_1	n	$\phi(n)$	σ_0	σ_1	n	$\phi(n)$	σ_0	σ_1	n	$\phi(n)$	σ_0	σ_1
761	760	2	762	811	810	2	812	861	480	8	1344	911	910	2	912
762	252	8	1536	812	336	12	1680	862	430	4	1296	912	288	20	2480
763	648	4	880	813	540	4	1088	863	862	2	864	913	820	4	1008
764	380	6	1344	814	360	8	1368	864	288	24	2520	914	456	4	1374
765	384	12	1404	815	648	4	984	865	688	4	1044	915	480	8	1488
766	382	4	1152	816	256	20	2232	866	432	4	1302	916	456	6	1610
767	696	4	840	817	756	4	880	867	544	6	1228	917	780	4	1056
768	256	18	2044	818	408	4	1230	868	360	12	1792	918	288	16	2160
769	768	2	770	819	432	12	1456	869	780	4	960	919	918	2	920
770	240	16	1728	820	320	12	1764	870	224	16	2160	920	352	16	2160
771	512	4	1032	821	820	2	822	871	792	4	952	921	612	4	1232
772	384	6	1358	822	272	8	1656	872	432	8	1650	922	460	4	1386
773	772	2	774	823	822	2	824	873	576	6	1274	923	840	4	1008
774	252	12	1716	824	408	8	1560	874	396	8	1440	924	240	24	2688
775	600	6	992	825	400	12	1488	875	600	8	1248	925	720	6	1178
776	384	8	1470	826	348	8	1440	876	288	12	2072	926	462	4	1392
777	432	8	1216	827	826	2	828	877	876	2	878	927	612	6	1352
778	388	4	1170	828	264	18	2184	878	438	4	1320	928	448	12	1890
779	720	4	840	829	828	2	830	879	584	4	1176	929	928	2	930
780	192	24	2352	830	328	8	1512	880	320	20	2232	930	240	16	2304
781	700	4	864	831	552	4	1112	881	880	2	882	931	756	6	1140
782	352	8	1296	832	384	14	1778	882	252	18	2223	932	464	6	1638
783	504	8	1200	833	672	6	1026	883	882	2	884	933	620	4	1248
784	336	15	1767	834	276	8	1680	884	384	12	1764	924	466	4	1404
785	624	4	948	835	664	4	1008	885	464	8	1440	935	640	8	1296
786	260	8	1584	836	360	12	1680	886	442	4	1332	936	288	24	2730
787	786	2	788	837	540	8	1280	887	886	2	888	937	936	2	938
788	392	6	1386	838	418	4	1260	888	288	16	2280	938	396	8	1632
789	524	4	1056	839	838	2	840	889	756	4	1024	939	624	4	1256
790	312	8	1440	840	192	32	2880	890	352	8	1620	940	368	12	2016
791	672	4	912	841	812	3	871	891	540	10	1452	941	940	2	942
792	240	24	2340	842	420	4	1266	892	444	6	1568	942	312	8	1896
793	720	4	868	843	560	4	1128	893	828	4	960	943	880	4	1008
794	396	4	1194	844	420	6	1484	894	296	8	1800	944	464	10	1860
795	416	8	1296	845	624	6	1098	895	712	4	1080	945	432	16	1920
796	396	6	1400	846	276	12	1872	896	384	16	2040	946	420	8	1584
797	796	2	798	847	660	6	1064	897	528	8	1344	947	946	2	948
798	216	16	1920	848	416	10	1674	898	448	4	1350	948	312	12	2240
799	736	4	864	849	564	4	1136	899	840	4	960	949	864	4	1036
800	320	18	1953	850	320	12	1674	900	240	27	2821	950	360	12	1860
801	528	6	1170	851	792	4	912	901	832	4	972	951	632	4	1272
802	400	4	1206	852	280	12	2016	902	400	8	1512	952	384	16	2160
803	720	4	888	853	852	2	854	903	504	8	1408	953	952	2	954
804	264	12	1904	854	360	8	1488	904	448	8	1710	954	312	12	2106
805	528	8	1152	855	432	12	1560	905	720	4	1092	955	760	4	1152
806	360	8	1344	856	424	8	1620	906	300	8	1824	956	476	6	1680
807	536	4	1080	857	856	2	858	907	906	2	908	957	560	8	1440
808	400	8	1530	858	240	16	2016	908	452	6	1596	958	478	4	1440
809	808	2	810	859	858	2	860	909	600	6	1326	959	816	4	1104
810	216	20	2178	860	336	12	1848	910	288	16	2016	960	256	28	3048

TOTIENT FUNCTION $\phi(n)$ (Continued)

n	$\phi(n)$	σ_0	σ_1	n	$\phi(n)$	σ_0	σ_1	n	$\phi(n)$	σ_0	σ_1	n	$\phi(n)$	σ_0	σ_1
961	930	3	993	971	970	2	972	981	648	6	1430	991	990	2	992
962	432	8	1596	972	324	18	2548	982	490	4	1476	992	480	12	2016
963	636	6	1404	973	828	4	1120	983	982	2	984	993	660	4	1328
964	480	6	1694	974	486	4	1464	984	320	16	2520	994	420	8	1728
965	768	4	1164	975	480	12	1736	985	784	4	1188	995	792	4	1200
966	264	16	2304	976	480	10	1922	986	448	8	1620	996	328	12	2352
967	966	2	968	977	976	2	978	987	552	8	1536	997	996	2	998
968	440	12	1995	978	324	8	1968	988	432	12	1960	998	498	4	1500
969	576	8	1440	979	880	4	1080	989	924	4	1056	999	648	8	1520
970	384	8	1764	980	336	18	2394	990	240	24	2808	1000	400	16	2340

From British Association for the Advancement of Science, Mathematical Tables, vol. VIII, Number-divisor tables, by permission of Cambridge University Press.

108

TABLES OF INDICES AND POWER RESIDUES

H. C. Williams, Ph.D.

1. Introduction

Let a and b be any two positive integers. If the largest integer that divides both a and b is unity, a and b are said to be *relatively prime*. If m is any positive integer, the *totient of m*, denoted by $\phi(m)$, is defined to be the number of positive integers less than m and relatively prime to m. If one writes

$$m = p_1^{\alpha_1} p_2^{\alpha_2} \ldots p_n^{\alpha_n},$$

where $p_1, p_2, \ldots, p_n$ are distinct primes, it can be shown that

$$\phi(m) = p_1^{\alpha_1-1} p_2^{\alpha_2-1} \ldots p_n^{\alpha_n-1}(p_1-1)(p_2-1) \ldots (p_n-1).$$

Two integers a and b are said to be *congruent modulo* an integer m if m is an integer divisor of $a-b$. This relation between a and b is denoted by

$$a \equiv b \pmod{m}.$$

If a is congruent to b modulo m and $0 \le b < m$, b is called the residue of a modulo m. a and b are said to be distinct modulo m if m is not an integer divisor of $a-b$. A well-known result is the

Theorem. If a and m ($\geqslant 0$) are relatively prime integers, then

$$a^{\phi(m)} \equiv 1 \pmod{m}.$$

Corollary. If p is a prime which is not a divisor of an integer a, then

$$a^{p-1} \equiv 1 \pmod{p}.$$

If a and m (>0) are relatively prime integers, and μ is the least positive integer for which

$$a^\mu \equiv 1 \pmod{m},$$

μ is called the *exponent* to which a belongs modulo m. If the exponent to which a belongs modulo m is $\phi(m)$, a is defined to be a *primitive root* modulo m (or a primitive root of m).

Theorem. The only integers which possess primitive roots are 2, 4, p^n, $2p^n$, where p is an odd prime.

If m does have a primitive root, it has $\phi(\phi(m))$ distinct primitive roots modulo m.

Let p be any prime, and let g be any primitive root of p. To each integer a relatively prime to p there corresponds a unique integer i such that

$$a \equiv g^i \pmod{p} \qquad (0 < i < p-1).$$

i is called the *index* to base g of a modulo p. This is written

$$i = \text{ind}_g a.$$

For example, consider the prime 11 which has a primitive root $g = 2$. Construct the table below, where

$$r_i \equiv g^i \pmod{p} \qquad (0 < r_i < p).$$

i	0	1	2	3	4	5	6	7	8	9	10
r_i	1	2	4	8	5	10	9	7	3	6	1

Note that the bottom row runs through all possible residues modulo 11, except zero. The index, then, of any number in this row is the corresponding entry in the top row. That is, $\text{ind}_2 1 = 0$, $\text{ind}_2 2 = 1$, $\text{ind}_2 3 = 8$, $\text{ind}_2 4 = 2$, $\text{ind}_2 5 = 4$, $\text{ind}_2 6 = 9$, etc. Indices possess the following important properties (analogous to those of logarithms).

TABLES OF INDICES AND POWER RESIDUES (Continued)

(1) $\text{ind}_g 1 = 0$

(2) $\text{ind}_g(-1) = (p-1)/2$

(3) $\text{ind}_g(ab) \equiv \text{ind}_g a + \text{ind}_g b \pmod{p-1}$

(4) $\text{ind}_g a^n \equiv n\, \text{ind}_g a \pmod{p-1}$

(5) $\text{ind}_g a \equiv \text{ind}_g g' \cdot \text{ind}_{g'} a \pmod{p-1}$,

where g' is any other primitive root of p.

The *power residue* of an integer b to base g is simply the integer r such that

$$r \equiv g^b \pmod{p} \qquad (0 < r < p).$$

Since the power residue of $\text{ind}_g a$ is congruent to a modulo p, it is clear that the concept of indices and power residues is analogous to that of logarithms and antilogarithms.

2. Use of the tables

In the tables that follow, the indices and power residues are given for all primes less than 100. (A table of indices and power residues for all primes and prime powers less than 2000 is given in [1].) In the tables which follow, the base g for the indices and power residues is chosen as the smallest primitive root of p.

To determine the index or power residue of an integer a modulo p, look at the appropriate table for prime p and select the entry in the column under j and in the row under i, where

$$a \equiv 10i+j \pmod{p} \qquad (0 \le i,j \le 9).$$

For example, the index modulo 97 of 134 ($\equiv 37 \pmod{97}$) is the entry under the number 7 and across from the number 3. This is found to be 91.

Indices and power residues can be used to simplify number-theoretic calculations. The following two examples illustrate the use of indices and power residues.

Example. Find an integer congruent to

$$(134)^{92}(54)^{67} \quad \text{modulo } (97).$$

Let

$$a \equiv (134)^{92}(54)^{67} \pmod{97};$$

then

$$\text{ind}_g a \equiv 92\,\text{ind}_g 37 + 67\,\text{ind}_g 54 \pmod{96}$$

$$\equiv 92\,(91) + 67\,(52) \pmod{96}$$

$$\equiv 48 \pmod{96}.$$

Hence

$$a \equiv g^{48} \equiv 96 \pmod{97}.$$

This is obtained by looking in the body of the table of indices under $p = 97$ and locating the index 48, and then reading the appropriate quadratic residue from the first column and top row.

Example. Find an integer x such that

$$44x^{21} \equiv 53 \pmod{73}.$$

If we take indices of both sides, we get

$$\text{ind}_g 44 + 21\,\text{ind}_g x \equiv \text{ind}_g 53 \pmod{72}$$

$$71 + 21\,\text{ind}_g x \equiv 53 \pmod{72}$$

$$21\,\text{ind}_g x \equiv 54 \pmod{72}$$

$$7\,\text{ind}_g x \equiv 18 \pmod{24}$$

TABLES OF INDICES AND POWER RESIDUES (Continued)

Multiplying by 7, we get

$$\text{ind}_g x \equiv 7(18) \equiv 6 \pmod{24}.$$

Hence

$$x \equiv g^6, g^{30}, g^{54} \pmod{73}$$

and

$$x \equiv 3, 24, 46 \pmod{73}.$$

REFERENCE

1. *A Table of Indices and Power Residues,* University of Oklahoma Mathematical Tables Project, W. W. Norton and Co., New York, N.Y., 1962.

INDICES FOR PRIMES 3 to 97

Prime 3
$g = 2$ Indices

	0	1	2	3	4	5	6	7	8	9
0		0	1							

Prime 5
$g = 2$ Indices

	0	1	2	3	4	5	6	7	8	9
0		0	1	3	2					

Prime 7
$g = 3$ Indices

	0	1	2	3	4	5	6	7	8	9
0		0	2	1	4	5	3			

Prime 11
$g = 2$ Indices

	0	1	2	3	4	5	6	7	8	9
0		0	1	8	2	4	9	7	3	6
1	5									

Prime 13
$g = 2$ Indices

	0	1	2	3	4	5	6	7	8	9
0		0	1	4	2	9	5	11	3	8
1	10	7	6							

Prime 17
$g = 3$ Indices

	0	1	2	3	4	5	6	7	8	9
0		0	14	1	12	5	15	11	10	2
1	3	7	13	4	9	6	8			

Prime 19
$g = 2$ Indices

	0	1	2	3	4	5	6	7	8	9
0		0	1	13	2	16	14	6	3	8
1	17	12	15	5	7	11	4	10	9	

Prime 23
$g = 5$ Indices

	0	1	2	3	4	5	6	7	8	9
0		0	2	16	4	1	18	19	6	10
1	3	9	20	14	21	17	8	7	12	15
2	5	13	11							

Prime 29
$g = 2$ Indices

	0	1	2	3	4	5	6	7	8	9
0		0	1	5	2	22	6	12	3	10
1	23	25	7	18	13	27	4	21	11	9
2	24	17	26	20	8	16	19	15	14	

Prime 31
$g = 3$ Indices

	0	1	2	3	4	5	6	7	8	9
0		0	24	1	18	20	25	28	12	2
1	14	23	19	11	22	21	6	7	26	4
2	8	29	17	27	13	10	5	3	16	9
3	14									

Prime 37
$g = 2$ Indices

	0	1	2	3	4	5	6	7	8	9
0		0	1	26	2	23	27	32	3	16
1	24	30	28	11	33	13	4	7	17	35
2	25	22	31	15	29	10	12	6	34	21
3	14	9	5	20	8	19	18			

Prime 41
$g = 6$ Indices

	0	1	2	3	4	5	6	7	8	9
0		0	26	15	12	22	1	39	38	30
1	8	3	27	31	25	37	24	33	16	9
2	34	14	29	36	13	4	17	5	11	7
3	23	28	10	18	19	21	2	32	35	6
4	20									

Prime 43
$g = 3$ Indices

	0	1	2	3	4	5	6	7	8	9
0		0	27	1	12	25	28	35	39	2
1	10	30	13	32	20	26	24	38	29	19
2	37	36	15	16	40	8	17	3	5	41
3	11	34	9	31	23	18	14	7	4	33
4	22	6	21							

Prime 47
$g = 5$ Indices

	0	1	2	3	4	5	6	7	8	9
0		0	18	20	36	1	38	32	8	40
1	19	7	10	11	4	21	26	16	12	45
2	37	6	25	5	28	2	29	14	22	35
3	39	3	44	27	34	33	30	42	17	31
4	9	15	24	13	43	41	23			

Prime 53
$g = 2$ Indices

	0	1	2	3	4	5	6	7	8	9
0		0	1	17	2	47	18	14	3	34
1	48	6	19	24	15	12	4	10	35	37
2	49	31	7	39	20	42	25	51	16	46
3	13	33	5	23	11	9	36	30	38	41
4	50	45	32	22	8	29	40	44	21	28
5	43	27	26							

Prime 59
$g = 2$ Indices

	0	1	2	3	4	5	6	7	8	9
0		0	1	50	2	6	51	18	3	42
1	7	25	52	45	19	56	4	40	43	38
2	8	10	26	15	53	12	46	34	20	28
3	57	49	5	17	41	24	44	55	39	37
4	9	14	11	33	27	48	16	23	54	36
5	13	32	47	22	35	31	21	30	29	

Prime 61
$g = 2$ Indices

	0	1	2	3	4	5	6	7	8	9
0		0	1	6	2	22	7	49	3	12
1	23	15	8	40	50	28	4	47	13	26
2	24	55	16	57	9	44	41	18	51	35
3	29	59	5	21	48	11	14	39	27	46
4	25	54	56	43	17	34	58	20	10	38
5	45	53	42	33	19	37	52	32	36	31
6	30									

INDICES FOR PRIMES 3 to 97 (Continued)

Prime 67
$g = 2$

Indices

	0	1	2	3	4	5	6	7	8	9
0		0	1	39	2	15	40	23	3	12
1	16	59	41	19	24	54	4	64	13	10
2	17	62	60	28	42	30	20	51	25	44
3	55	47	5	32	65	38	14	22	11	58
4	18	53	63	9	61	27	29	50	43	46
5	31	37	21	57	52	8	26	49	45	36
6	56	7	48	35	6	34	33			

Prime 71
$g = 7$

Indices

	0	1	2	3	4	5	6	7	8	9
0		0	6	26	12	28	32	1	18	52
1	34	31	38	39	7	54	24	49	58	16
2	40	27	37	15	44	56	45	8	13	68
3	60	11	30	57	55	29	64	20	22	65
4	46	25	33	48	43	10	21	9	50	2
5	62	5	51	23	14	59	19	42	4	3
6	66	69	17	53	36	67	63	47	61	41
7	35									

Prime 73
$g = 5$

Indices

	0	1	2	3	4	5	6	7	8	9
0		0	8	6	16	1	14	33	24	12
1	9	55	22	59	41	7	32	21	20	62
2	17	39	63	46	30	2	67	18	49	35
3	15	11	40	61	29	34	28	64	70	65
4	25	4	47	51	71	13	54	31	38	66
5	10	27	3	53	26	56	57	68	43	5
6	23	58	19	45	48	60	69	50	37	52
7	42	44	36							

Prime 79
$g = 3$

Indices

	0	1	2	3	4	5	6	7	8	9
0		0	4	1	8	62	5	43	12	2
1	66	68	9	34	57	63	16	21	6	32
2	70	54	72	26	13	46	38	3	61	11
3	67	56	20	69	25	37	10	19	36	35
4	74	75	58	49	76	64	30	59	17	28
5	50	22	42	77	7	52	65	33	15	31
6	71	45	60	55	24	18	73	48	29	27
7	41	51	14	44	23	47	40	43	39	

Prime 83
$g = 2$

Indices

	0	1	2	3	4	5	6	7	8	9
0		0	1	72	2	27	73	8	3	62
1	28	24	74	77	9	17	4	56	63	47
2	29	80	25	60	75	54	78	52	10	12
3	18	38	5	14	57	35	64	20	48	67
4	30	40	81	71	26	7	61	23	76	16
5	55	46	79	59	53	51	11	37	3	34
6	19	66	39	70	6	22	15	45	58	50
7	36	33	65	69	21	44	49	32	68	43
8	31	42	41							

Prime 89
$g = 3$

Indices

	0	1	2	3	4	5	6	7	8	9
0		0	16	1	32	70	17	81	48	2
1	86	84	33	23	9	71	64	6	18	35
2	14	82	12	57	49	52	39	3	25	59
3	87	31	80	85	22	63	34	11	51	24
4	30	21	10	29	28	72	73	54	65	74
5	68	7	55	78	19	66	41	36	75	43
6	15	69	47	83	8	5	13	56	38	58
7	79	62	50	20	27	53	67	77	40	42
8	46	4	37	61	26	76	45	60	44	

Prime 97
$g = 5$

Indices

	0	1	2	3	4	5	6	7	8	9
0		0	34	70	68	1	8	31	6	44
1	35	86	42	25	65	71	40	89	78	81
2	69	5	24	77	76	2	59	18	3	13
3	9	46	74	60	27	32	16	91	19	95
4	7	85	39	4	58	45	15	84	14	62
5	36	63	93	10	52	87	37	55	47	67
6	43	64	80	75	12	26	94	57	61	51
7	66	11	50	28	29	72	53	21	33	30
8	41	88	23	17	73	90	38	83	92	54

PRIMITIVE ROOTS FOR PRIMES 3 to 5003

In this table

g denotes the least primitive root of p

G denotes the least negative primitive root of p

ϵ denotes whether 10, -10 both or neither are primitive roots of p

Introductory Facts

As noted in the preceding Totient and Indices Tables, the number of integers not exceeding and relatively prime to a fixed integer n is represented by $\phi(n)$. These integers form a group; the group is cyclic if and only if $n = 2, 4$, or n is of the form p^k or $2p^k$ where p is an odd prime. We refer to g as a primitive root of n if it generates that group i.e. if $g, g^2, \ldots,$ $g^{\phi(n)}$ are distinct modulo n. There are $\phi(\phi(n))$ primitive roots of n. If g is a primitive root of p and $g^{p-1} \not\equiv 1 \pmod{p^2}$, then g is a primitive root of p^k for all k. If $g^{p-1} \equiv 1 \pmod{p^2}$ then $g + p$ is a primitive root of p^k for all k.

If g is a primitive root of p^k then either g or $g + p^k$, whichever is odd, is a primitive root of $2p^k$.

If g is a primitive root of n, then g^k is a primitive root of n if and only if k and $\phi(n)$ are relatively prime, and each primitive root of n is of this form, i.e. $(k, \phi(n)) = 1$.

p	$p-1$	g	$-G$	ϵ	p	$p-1$	g	$-G$	ϵ
3	2	2	1	-10	167	$2 \cdot 83$	5	2	10
5	2^2	2	2	—	173	$2^2 \cdot 43$	2	2	—
7	$2 \cdot 3$	3	2	10	179	$2 \cdot 89$	2	3	10
11	$2 \cdot 5$	2	3	—	181	$2^2 \cdot 3^2 \cdot 5$	2	2	± 10
13	$2^2 \cdot 3$	2	2	—	191	$2 \cdot 5 \cdot 19$	19	2	-10
17	2^4	3	3	$+10$	193	$2^6 \cdot 3$	5	5	± 10
19	$2 \cdot 3^2$	2	4	10	197	$2^2 \cdot 7^2$	2	2	—
23	$2 \cdot 11$	5	2	10	199	$2 \cdot 3^2 \cdot 11$	3	2	-10
29	$2^2 \cdot 7$	2	2	± 10	211	$2 \cdot 3 \cdot 5 \cdot 7$	2	4	—
31	$2 \cdot 3 \cdot 5$	3	7	-10	223	$2 \cdot 3 \cdot 37$	3	9	10
37	$2^2 \cdot 3^2$	2	2	—	227	$2 \cdot 113$	2	3	-10
41	$2^3 \cdot 5$	6	6	—	229	$2^2 \cdot 3 \cdot 19$	6	6	± 10
43	$2 \cdot 3 \cdot 7$	3	9	-10	233	$2^3 \cdot 29$	3	3	$+10$
47	$2 \cdot 23$	5	2	10	239	$2 \cdot 7 \cdot 17$	7	2	—
53	$2^2 \cdot 13$	2	2	—	241	$2^4 \cdot 3 \cdot 5$	7	7	—
59	$2 \cdot 29$	2	3	10	251	$2 \cdot 5^3$	6	3	—
61	$2^2 \cdot 3 \cdot 5$	2	2	± 10	257	2^8	3	3	± 10
67	$2 \cdot 3 \cdot 11$	2	4	-10	263	$2 \cdot 131$	5	2	10
71	$2 \cdot 5 \cdot 7$	7	2	-10	269	$2^2 \cdot 67$	2	2	± 10
73	$2^3 \cdot 3^2$	5	5	—	271	$2 \cdot 3^3 \cdot 5$	6	2	—
79	$2 \cdot 3 \cdot 13$	3	2	—	277	$2^2 \cdot 3 \cdot 23$	5	5	—
83	$2 \cdot 41$	2	3	-10	281	$2^3 \cdot 5 \cdot 7$	3	3	—
89	$2^3 \cdot 11$	3	3	—	283	$2 \cdot 3 \cdot 47$	3	6	-10
97	$2^5 \cdot 3$	5	5	± 10	293	$2^2 \cdot 73$	2	2	—
101	$2^2 \cdot 5^2$	2	2	—	307	$2 \cdot 3^2 \cdot 17$	5	7	-10
103	$2 \cdot 3 \cdot 17$	5	2	—	311	$2 \cdot 5 \cdot 31$	17	2	-10
107	$2 \cdot 53$	2	3	-10	313	$2^3 \cdot 3 \cdot 13$	10	10	± 10
109	$2^2 \cdot 3^3$	6	6	± 10	317	$2^2 \cdot 79$	2	2	—
113	$2^4 \cdot 7$	3	3	± 10	331	$2 \cdot 3 \cdot 5 \cdot 11$	3	5	—
127	$2 \cdot 3^2 \cdot 7$	3	9	—	337	$2^4 \cdot 3 \cdot 7$	10	10	± 10
131	$2 \cdot 5 \cdot 13$	2	3	10	347	$2 \cdot 173$	2	3	-10
137	$2^3 \cdot 17$	3	3	—	349	$2^2 \cdot 3 \cdot 29$	2	2	—
139	$2 \cdot 3 \cdot 23$	2	4	—	353	$2^5 \cdot 11$	3	3	—
149	$2^2 \cdot 37$	2	2	± 10	359	$2 \cdot 179$	7	2	-10
151	$2 \cdot 3 \cdot 5^2$	6	5	-10	367	$2 \cdot 3 \cdot 61$	6	2	10
157	$2^2 \cdot 3 \cdot 13$	5	5	—	373	$2^2 \cdot 3 \cdot 31$	2	2	—
163	$2 \cdot 3^4$	2	4	-10	379	$2 \cdot 3^3 \cdot 7$	2	4	10

PRIMITIVE ROOTS FOR PRIMES 3 to 5003 (Continued)

p	$p-1$	g	$-G$	ϵ	p	$p-1$	g	$-G$	ϵ
383	$2\cdot191$	5	2	10	769	$2^8\cdot3$	11	11	—
389	$2^3\cdot97$	2	2	±10	773	$2^2\cdot193$	2	2	—
397	$2^2\cdot3^2\cdot11$	5	5	—	787	$2\cdot3\cdot131$	2	4	−10
401	$2^4\cdot5^2$	3	3	—	797	$2^2\cdot199$	2	2	—
409	$2^3\cdot3\cdot17$	21	21	—	809	$2^3\cdot101$	3	3	—
419	$2\cdot11\cdot19$	2	3	10	811	$2\cdot3^4\cdot5$	3	5	10
421	$2^2\cdot3\cdot5\cdot7$	2	2	—	821	$2^2\cdot5\cdot41$	2	2	±10
431	$2\cdot5\cdot43$	7	5	−10	823	$2\cdot3\cdot137$	3	2	10
433	$2^4\cdot3^3$	5	5	±10	827	$2\cdot7\cdot59$	2	3	−10
439	$2\cdot3\cdot73$	15	5	−10	829	$2^2\cdot3^2\cdot23$	2	2	—
443	$2\cdot13\cdot17$	2	3	−10	839	$2\cdot419$	11	2	−10
449	$2^6\cdot7$	3	3	—	853	$2^2\cdot3\cdot71$	2	2	—
457	$2^3\cdot3\cdot19$	13	13	—	857	$2^3\cdot107$	3	3	±10
461	$2^2\cdot5\cdot23$	2	2	±10	859	$2\cdot3\cdot11\cdot13$	2	4	—
463	$2\cdot3\cdot7\cdot11$	3	2	—	863	$2\cdot431$	5	2	10
467	$2\cdot233$	2	3	−10	877	$2^2\cdot3\cdot73$	2	2	—
479	$2\cdot239$	13	2	−10	881	$2^4\cdot5\cdot11$	3	3	—
487	$2\cdot3^5$	3	3	10	883	$2\cdot3^2\cdot7^2$	2	4	−10
491	$2\cdot5\cdot7^2$	2	4	10	887	$2\cdot443$	5	2	10
499	$2\cdot3\cdot83$	7	5	10	907	$2\cdot3\cdot151$	2	4	—
503	$2\cdot251$	5	2	10	911	$2\cdot5\cdot7\cdot13$	17	3	−10
509	$2^2\cdot127$	2	2	±10	919	$2\cdot3^3\cdot17$	7	5	−10
521	$2^3\cdot5\cdot13$	3	3	—	929	$2^5\cdot29$	3	3	—
523	$2\cdot3^2\cdot29$	2	4	−10	937	$2^3\cdot3^2\cdot13$	5	5	±10
541	$2^2\cdot3^3\cdot5$	2	2	±10	941	$2^2\cdot5\cdot47$	2	2	±10
547	$2\cdot3\cdot7\cdot13$	2	4	—	947	$2\cdot11\cdot43$	2	3	−10
557	$2^2\cdot139$	2	2	—	953	$2^3\cdot7\cdot17$	3	3	±10
563	$2\cdot281$	2	3	−10	967	$2\cdot3\cdot7\cdot23$	5	2	—
569	$2^3\cdot71$	3	3	—	971	$2\cdot5\cdot97$	6	3	10
571	$2\cdot3\cdot5\cdot19$	3	5	10	977	$2^4\cdot61$	3	3	±10
577	$2^6\cdot3^2$	5	5	±10	983	$2\cdot491$	5	2	10
587	$2\cdot293$	2	3	−10	991	$2\cdot3^2\cdot5\cdot11$	6	2	−10
593	$2^4\cdot37$	3	3	±10	997	$2^2\cdot3\cdot83$	7	7	—
599	$2\cdot13\cdot23$	7	2	−10	1009	$2^4\cdot3^2\cdot7$	11	11	—
601	$2^3\cdot3\cdot5^2$	7	7	—	1013	$2^2\cdot11\cdot23$	3	3	—
607	$2\cdot3\cdot101$	3	2	—	1019	$2\cdot509$	2	3	10
613	$2^2\cdot3^2\cdot17$	2	2	—	1021	$2^2\cdot3\cdot5\cdot17$	10	10	±10
617	$2^3\cdot7\cdot11$	3	3	—	1031	$2\cdot5\cdot103$	14	2	—
619	$2\cdot3\cdot103$	2	4	10	1033	$2^3\cdot3\cdot43$	5	5	±10
631	$2\cdot3^2\cdot5\cdot7$	3	9	−10	1039	$2\cdot3\cdot173$	3	2	−10
641	$2^7\cdot5$	3	3	—	1049	$2^3\cdot131$	3	3	—
643	$2\cdot3\cdot107$	11	7	—	1051	$2\cdot3\cdot5^2\cdot7$	7	5	10
647	$2\cdot17\cdot19$	5	2	10	1061	$2^2\cdot5\cdot53$	2	2	—
653	$2^2\cdot163$	2	2	—	1063	$2\cdot3^2\cdot59$	3	2	10
659	$2\cdot7\cdot47$	2	3	10	1069	$2^2\cdot3\cdot89$	6	6	±10
661	$2^2\cdot3\cdot5\cdot11$	2	2	—	1087	$2\cdot3\cdot181$	3	2	10
673	$2^5\cdot3\cdot7$	5	5	—	1091	$2\cdot5\cdot109$	2	4	10
677	$2^2\cdot13^2$	2	2	—	1093	$2^2\cdot3\cdot7\cdot13$	5	5	—
683	$2\cdot11\cdot31$	5	10	−10	1097	$2^3\cdot137$	3	3	±10
691	$2\cdot3\cdot5\cdot23$	3	6	—	1103	$2\cdot19\cdot29$	5	3	10
701	$2^2\cdot5^2\cdot7$	2	2	±10	1109	$2^2\cdot277$	2	2	±10
709	$2^2\cdot3\cdot59$	2	2	±10	1117	$2^2\cdot3^2\cdot31$	2	2	—
719	$2\cdot359$	11	2	−10	1123	$2\cdot3\cdot11\cdot17$	2	4	−10
727	$2\cdot3\cdot11^2$	5	7	10	1129	$2^3\cdot3\cdot47$	11	11	—
733	$2^2\cdot3\cdot61$	6	6	—	1151	$2\cdot5^2\cdot23$	17	2	−10
739	$2\cdot3^2\cdot41$	3	6	—	1153	$2^7\cdot3^2$	5	5	±10
743	$2\cdot7\cdot53$	5	2	10	1163	$2\cdot7\cdot83$	5	3	−10
751	$2\cdot3\cdot5^3$	3	2	—	1171	$2\cdot3^2\cdot5\cdot13$	2	4	10
757	$2^2\cdot3^3\cdot7$	2	2	—	1181	$2^2\cdot5\cdot59$	7	7	±10
761	$2^3\cdot5\cdot19$	6	6	—	1187	$2\cdot593$	2	3	−10

PRIMITIVE ROOTS FOR PRIMES 3 to 5003 (Continued)

p	$p-1$	g	$-G$	ϵ	p	$p-1$	g	$-G$	ϵ
1193	$2^3 \cdot 149$	3	3	± 10	1619	$2 \cdot 809$	2	3	10
1201	$2^4 \cdot 3 \cdot 5^2$	11	11	—	1621	$2^2 \cdot 3^4 \cdot 5$	2	2	± 10
1213	$2^2 \cdot 3 \cdot 101$	2	2	—	1627	$2 \cdot 3 \cdot 271$	3	6	—
1217	$2^6 \cdot 19$	3	3	± 10	1637	$2^2 \cdot 409$	2	2	—
1223	$2 \cdot 13 \cdot 47$	5	2	10	1657	$2^3 \cdot 3^2 \cdot 23$	11	11	—
1229	$2^2 \cdot 307$	2	2	± 10	1663	$2 \cdot 3 \cdot 277$	3	2	10
1231	$2 \cdot 3 \cdot 5 \cdot 41$	3	2	—	1667	$2 \cdot 7^2 \cdot 17$	2	3	-10
1237	$2^2 \cdot 3 \cdot 103$	2	2	—	1669	$2^2 \cdot 3 \cdot 139$	2	2	—
1249	$2^5 \cdot 3 \cdot 13$	7	7	—	1693	$2^2 \cdot 3^2 \cdot 47$	2	2	—
1259	$2 \cdot 17 \cdot 37$	2	3	10	1697	$2^5 \cdot 53$	3	3	± 10
1277	$2^2 \cdot 11 \cdot 29$	2	2	—	1699	$2 \cdot 3 \cdot 283$	3	6	—
1279	$2 \cdot 3^2 \cdot 71$	3	2	-10	1709	$2^2 \cdot 7 \cdot 61$	3	3	± 10
1283	$2 \cdot 641$	2	3	-10	1721	$2^3 \cdot 5 \cdot 43$	3	3	—
1289	$2^3 \cdot 7 \cdot 23$	6	6	—	1723	$2 \cdot 3 \cdot 7 \cdot 41$	3	6	—
1291	$2 \cdot 3 \cdot 5 \cdot 43$	2	4	10	1733	$2^2 \cdot 433$	2	2	—
1297	$2^4 \cdot 3^4$	10	10	± 10	1741	$2^2 \cdot 3 \cdot 5 \cdot 29$	2	2	± 10
1301	$2^2 \cdot 5^2 \cdot 13$	2	2	± 10	1747	$2 \cdot 3^2 \cdot 97$	2	4	—
1303	$2 \cdot 3 \cdot 7 \cdot 31$	6	2	10	1753	$2^3 \cdot 3 \cdot 73$	7	7	—
1307	$2 \cdot 653$	2	3	-10	1759	$2 \cdot 3 \cdot 293$	6	2	-10
1319	$2 \cdot 659$	13	2	-10	1777	$2^4 \cdot 3 \cdot 37$	5	5	± 10
1321	$2^3 \cdot 3 \cdot 5 \cdot 11$	13	13	—	1783	$2 \cdot 3^4 \cdot 11$	10	2	10
1327	$2 \cdot 3 \cdot 13 \cdot 17$	3	9	10	1787	$2 \cdot 19 \cdot 47$	2	3	-10
1361	$2^4 \cdot 5 \cdot 17$	3	3	—	1789	$2^2 \cdot 3 \cdot 149$	6	6	± 10
1367	$2 \cdot 683$	5	2	10	1801	$2^3 \cdot 3^2 \cdot 5^2$	11	11	—
1373	$2^2 \cdot 7^3$	2	2	—	1811	$2 \cdot 5 \cdot 181$	6	3	10
1381	$2^2 \cdot 3 \cdot 5 \cdot 23$	2	2	± 10	1823	$2 \cdot 911$	5	2	10
1399	$2 \cdot 3 \cdot 233$	13	5	-10	1831	$2 \cdot 3 \cdot 5 \cdot 61$	3	9	—
1409	$2^7 \cdot 11$	3	3	—	1847	$2 \cdot 13 \cdot 71$	5	2	10
1423	$2 \cdot 3^2 \cdot 79$	3	9	—	1861	$2^2 \cdot 3 \cdot 5 \cdot 31$	2	2	± 10
1427	$2 \cdot 23 \cdot 31$	2	3	-10	1867	$2 \cdot 3 \cdot 311$	2	4	-10
1429	$2^2 \cdot 3 \cdot 7 \cdot 17$	6	6	± 10	1871	$2 \cdot 5 \cdot 11 \cdot 17$	14	2	-10
1433	$2^3 \cdot 179$	3	3	± 10	1873	$2^4 \cdot 3^2 \cdot 13$	10	10	± 10
1439	$2 \cdot 719$	7	2	-10	1877	$2^2 \cdot 7 \cdot 67$	2	2	—
1447	$2 \cdot 3 \cdot 241$	3	2	10	1879	$2 \cdot 3 \cdot 313$	6	2	—
1451	$2 \cdot 5^2 \cdot 29$	2	3	—	1889	$2^5 \cdot 59$	3	3	—
1453	$2^2 \cdot 3 \cdot 11^2$	2	2	—	1901	$2^2 \cdot 5^2 \cdot 19$	2	2	—
1459	$2 \cdot 3^6$	3	6	—	1907	$2 \cdot 953$	2	3	-10
1471	$2 \cdot 3 \cdot 5 \cdot 7^2$	6	5	-10	1913	$2^3 \cdot 239$	3	3	± 10
1481	$2^3 \cdot 5 \cdot 37$	3	3	—	1931	$2 \cdot 5 \cdot 193$	2	3	—
1483	$2 \cdot 3 \cdot 13 \cdot 19$	2	4	—	1933	$2^2 \cdot 3 \cdot 7 \cdot 23$	5	5	—
1487	$2 \cdot 743$	5	2	10	1949	$2^2 \cdot 487$	2	2	± 10
1489	$2^4 \cdot 3 \cdot 31$	14	14	—	1951	$2 \cdot 3 \cdot 5^2 \cdot 13$	3	2	—
1493	$2^2 \cdot 373$	2	2	—	1973	$2^2 \cdot 17 \cdot 29$	2	2	—
1499	$2 \cdot 7 \cdot 107$	2	3	—	1979	$2 \cdot 23 \cdot 43$	2	3	10
1511	$2 \cdot 5 \cdot 151$	11	2	-10	1987	$2 \cdot 3 \cdot 331$	2	4	—
1523	$2 \cdot 761$	2	3	-10	1993	$2^3 \cdot 3 \cdot 83$	5	5	—
1531	$2 \cdot 3^2 \cdot 5 \cdot 17$	2	4	10	1997	$2^2 \cdot 499$	2	2	—
1543	$2 \cdot 3 \cdot 257$	5	2	10	1999	$2 \cdot 3^3 \cdot 37$	3	5	-10
1549	$2^2 \cdot 3^2 \cdot 43$	2	2	± 10	2003	$2 \cdot 7 \cdot 11 \cdot 13$	5	3	-10
1553	$2^4 \cdot 97$	3	3	± 10	2011	$2 \cdot 3 \cdot 5 \cdot 67$	3	5	—
1559	$2 \cdot 19 \cdot 41$	19	2	-10	2017	$2^5 \cdot 3^2 \cdot 7$	5	5	± 10
1567	$2 \cdot 3^3 \cdot 29$	3	2	10	2027	$2 \cdot 1013$	2	3	-10
1571	$2 \cdot 5 \cdot 157$	2	3	10	2029	$2^2 \cdot 3 \cdot 13^2$	2	2	± 10
1579	$2 \cdot 3 \cdot 263$	3	5	10	2039	$2 \cdot 1019$	7	2	-10
1583	$2 \cdot 7 \cdot 113$	5	2	10	2053	$2^2 \cdot 3^3 \cdot 19$	2	2	—
1597	$2^2 \cdot 3 \cdot 7 \cdot 19$	11	11	—	2063	$2 \cdot 1031$	5	2	10
1601	$2^6 \cdot 5^2$	3	3	—	2069	$2^3 \cdot 11 \cdot 47$	2	2	± 10
1607	$2 \cdot 11 \cdot 73$	5	2	10	2081	$2^5 \cdot 5 \cdot 13$	3	3	—
1609	$2^3 \cdot 3 \cdot 67$	7	7	—	2083	$2 \cdot 3 \cdot 347$	2	4	-10
1613	$2^2 \cdot 13 \cdot 31$	3	3	—	2087	$2 \cdot 7 \cdot 149$	5	2	—

PRIMITIVE ROOTS FOR PRIMES 3 to 5003 (Continued)

p	$p-1$	g	$-G$	ϵ	p	$p-1$	g	$-G$	ϵ
2089	$2^3 \cdot 3^2 \cdot 29$	7	7	—	2579	$2 \cdot 1289$	2	3	10
2099	$2 \cdot 1049$	2	3	10	2591	$2 \cdot 5 \cdot 7 \cdot 37$	7	2	—
2111	$2 \cdot 5 \cdot 211$	7	2	-10	2593	$2^5 \cdot 3^4$	7	7	± 10
2113	$2^6 \cdot 3 \cdot 11$	5	5	± 10	2609	$2^4 \cdot 163$	3	3	—
2129	$2^4 \cdot 7 \cdot 19$	3	3	—	2617	$2^3 \cdot 3 \cdot 109$	5	5	± 10
2131	$2 \cdot 3 \cdot 5 \cdot 71$	2	4	—	2621	$2^2 \cdot 5 \cdot 131$	2	2	± 10
2137	$2^3 \cdot 3 \cdot 89$	10	10	± 10	2633	$2^3 \cdot 7 \cdot 47$	3	3	± 10
2141	$2^2 \cdot 5 \cdot 107$	2	2	± 10	2647	$2 \cdot 3^3 \cdot 7^2$	3	2	—
2143	$2 \cdot 3^2 \cdot 7 \cdot 17$	3	9	10	2657	$2^5 \cdot 83$	3	3	± 10
2153	$2^3 \cdot 269$	3	3	± 10	2659	$2 \cdot 3 \cdot 443$	2	4	—
2161	$2^4 \cdot 3^3 \cdot 5$	23	23	—	2663	$2 \cdot 11^3$	5	2	10
2179	$2 \cdot 3^2 \cdot 11^2$	7	5	10	2671	$2 \cdot 3 \cdot 5 \cdot 89$	7	5	-10
2203	$2 \cdot 3 \cdot 367$	5	7	-10	2677	$2^2 \cdot 3 \cdot 223$	2	2	—
2207	$2 \cdot 1103$	5	2	10	2683	$2 \cdot 3^2 \cdot 149$	2	4	—
2213	$2^2 \cdot 7 \cdot 79$	2	2	—	2687	$2 \cdot 17 \cdot 79$	5	3	10
2221	$2^2 \cdot 3 \cdot 5 \cdot 37$	2	2	± 10	2689	$2^7 \cdot 3 \cdot 7$	19	19	—
2237	$2^2 \cdot 13 \cdot 43$	2	2	—	2693	$2^2 \cdot 673$	2	2	—
2239	$2 \cdot 3 \cdot 373$	3	2	-10	2699	$2 \cdot 19 \cdot 71$	2	3	10
2243	$2 \cdot 19 \cdot 59$	2	3	-10	2707	$2 \cdot 3 \cdot 11 \cdot 41$	2	4	-10
2251	$2 \cdot 3^2 \cdot 5^3$	7	5	10	2711	$2 \cdot 5 \cdot 271$	7	2	-10
2267	$2 \cdot 11 \cdot 103$	2	3	-10	2713	$2^3 \cdot 3 \cdot 113$	5	5	± 10
2269	$2^2 \cdot 3^4 \cdot 7$	2	2	± 10	2719	$2 \cdot 3^2 \cdot 151$	3	2	-10
2273	$2^5 \cdot 71$	3	3	± 10	2729	$2^3 \cdot 11 \cdot 31$	3	3	—
2281	$2^3 \cdot 3 \cdot 5 \cdot 19$	7	7	—	2731	$2 \cdot 3 \cdot 5 \cdot 7 \cdot 13$	3	5	10
2287	$2 \cdot 3^2 \cdot 127$	19	7	—	2741	$2^2 \cdot 5 \cdot 137$	2	2	± 10
2293	$2^2 \cdot 3 \cdot 191$	2	2	—	2749	$2^2 \cdot 3 \cdot 229$	6	6	—
2297	$2^3 \cdot 7 \cdot 41$	5	5	± 10	2753	$2^6 \cdot 43$	3	3	± 10
2309	$2^2 \cdot 577$	2	2	± 10	2767	$2 \cdot 3 \cdot 461$	3	9	10
2311	$2 \cdot 3 \cdot 5 \cdot 7 \cdot 11$	3	2	—	2777	$2^3 \cdot 347$	3	3	± 10
2333	$2^2 \cdot 11 \cdot 53$	2	2	—	2789	$2^2 \cdot 17 \cdot 41$	2	2	± 10
2339	$2 \cdot 7 \cdot 167$	2	3	10	2791	$2 \cdot 3^2 \cdot 5 \cdot 31$	6	7	—
2341	$2^2 \cdot 3^2 \cdot 5 \cdot 13$	7	7	± 10	2797	$2^2 \cdot 3 \cdot 233$	2	2	—
2347	$2 \cdot 3 \cdot 17 \cdot 23$	3	6	-10	2801	$2^4 \cdot 5^2 \cdot 7$	3	3	—
2351	$2 \cdot 5^2 \cdot 47$	13	3	-10	2803	$2 \cdot 3 \cdot 467$	2	4	-10
2357	$2^2 \cdot 19 \cdot 31$	2	2	—	2819	$2 \cdot 1409$	2	3	10
2371	$2 \cdot 3 \cdot 5 \cdot 79$	2	4	10	2833	$2^4 \cdot 3 \cdot 59$	5	5	± 10
2377	$2^3 \cdot 3^3 \cdot 11$	5	5	—	2837	$2^2 \cdot 709$	2	2	—
2381	$2^2 \cdot 5 \cdot 7 \cdot 17$	3	3	—	2843	$2 \cdot 7^2 \cdot 29$	2	4	-10
2383	$2 \cdot 3 \cdot 397$	5	13	10	2851	$2 \cdot 3 \cdot 5^2 \cdot 19$	2	4	10
2389	$2^2 \cdot 3 \cdot 199$	2	2	± 10	2857	$2^3 \cdot 3 \cdot 7 \cdot 17$	11	11	—
2393	$2^3 \cdot 13 \cdot 23$	3	3	—	2861	$2^2 \cdot 5 \cdot 11 \cdot 13$	2	2	± 10
2399	$2 \cdot 11 \cdot 109$	11	2	-10	2879	$2 \cdot 1439$	7	2	-10
2411	$2 \cdot 5 \cdot 241$	6	3	10	2887	$2 \cdot 3 \cdot 13 \cdot 37$	5	2	10
2417	$2^4 \cdot 151$	3	3	± 10	2897	$2^4 \cdot 181$	3	3	10
2423	$2 \cdot 7 \cdot 173$	5	2	10	2903	$2 \cdot 1451$	5	2	10
2437	$2^2 \cdot 3 \cdot 7 \cdot 29$	2	2	—	2909	$2^2 \cdot 727$	2	2	± 10
2441	$2^3 \cdot 5 \cdot 61$	6	6	—	2917	$2^2 \cdot 3^6$	5	5	—
2447	$2 \cdot 1223$	5	2	10	2927	$2 \cdot 7 \cdot 11 \cdot 19$	5	2	10
2459	$2 \cdot 1229$	2	3	10	2939	$2 \cdot 13 \cdot 113$	2	3	10
2467	$2 \cdot 3^2 \cdot 137$	2	4	—	2953	$2^3 \cdot 3^2 \cdot 41$	13	13	—
2473	$2^3 \cdot 3 \cdot 103$	5	5	± 10	2957	$2^2 \cdot 739$	2	2	—
2477	$2^2 \cdot 619$	2	2	—	2963	$2 \cdot 1481$	2	3	-10
2503	$2 \cdot 3^2 \cdot 139$	3	2	—	2969	$2^3 \cdot 7 \cdot 53$	3	3	—
2521	$2^3 \cdot 3^2 \cdot 5 \cdot 7$	17	17	—	2971	$2 \cdot 3^3 \cdot 5 \cdot 11$	10	5	10
2531	$2 \cdot 5 \cdot 11 \cdot 23$	2	3	—	2999	$2 \cdot 1499$	17	2	-10
2539	$2 \cdot 3^3 \cdot 47$	2	4	10	3001	$2^3 \cdot 3 \cdot 5^3$	14	14	—
2543	$2 \cdot 31 \cdot 41$	5	2	10	3011	$2 \cdot 5 \cdot 7 \cdot 43$	2	3	10
2549	$2^2 \cdot 7^2 \cdot 13$	2	2	± 10	3019	$2 \cdot 3 \cdot 503$	2	4	10
2551	$2 \cdot 3 \cdot 5^2 \cdot 17$	6	2	—	3023	$2 \cdot 1511$	5	2	10
2557	$2^2 \cdot 3^2 \cdot 71$	2	2	—	3037	$2^2 \cdot 3 \cdot 11 \cdot 23$	2	2	—

PRIMITIVE ROOTS FOR PRIMES 3 to 5003 (Continued)

p	$p-1$	g	$-G$	ϵ	p	$p-1$	g	$-G$	ϵ
3041	$2^5 \cdot 5 \cdot 19$	3	3	—	3541	$2^2 \cdot 3 \cdot 5 \cdot 59$	7	7	—
3049	$2^3 \cdot 3 \cdot 127$	11	11	—	3547	$2 \cdot 3^2 \cdot 197$	2	4	-10
3061	$2^2 \cdot 3^2 \cdot 5 \cdot 17$	6	6	—	3557	$2^2 \cdot 7 \cdot 127$	2	2	—
3067	$2 \cdot 3 \cdot 7 \cdot 73$	2	4	-10	3559	$2 \cdot 3 \cdot 593$	3	2	-10
3079	$2 \cdot 3^4 \cdot 19$	6	2	-10	3571	$2 \cdot 3 \cdot 5 \cdot 7 \cdot 17$	2	4	10
3083	$2 \cdot 23 \cdot 67$	2	3	-10	3581	$2^2 \cdot 5 \cdot 179$	2	2	± 10
3089	$2^4 \cdot 193$	3	3	—	3583	$2 \cdot 3^2 \cdot 199$	3	2	—
3109	$2^2 \cdot 3 \cdot 7 \cdot 37$	6	6	—	3593	$2^3 \cdot 449$	3	3	± 10
3119	$2 \cdot 1559$	7	2	-10	3607	$2 \cdot 3 \cdot 601$	5	11	10
3121	$2^4 \cdot 3 \cdot 5 \cdot 13$	7	7	—	3613	$2^2 \cdot 3 \cdot 7 \cdot 43$	2	2	—
3137	$2^6 \cdot 7^2$	3	3	± 10	3617	$2^5 \cdot 113$	3	3	± 10
3163	$2 \cdot 3 \cdot 17 \cdot 31$	3	6	-10	3623	$2 \cdot 1811$	5	2	10
3167	$2 \cdot 1583$	5	2	10	3631	$2 \cdot 3 \cdot 5 \cdot 11^2$	15	10	-10
3169	$2^5 \cdot 3^2 \cdot 11$	7	7	—	3637	$2^2 \cdot 3^2 \cdot 101$	2	2	—
3181	$2^2 \cdot 3 \cdot 5 \cdot 53$	7	7	—	3643	$2 \cdot 3 \cdot 607$	2	4	-10
3187	$2 \cdot 3^3 \cdot 59$	2	4	—	3659	$2 \cdot 31 \cdot 59$	2	3	10
3191	$2 \cdot 5 \cdot 11 \cdot 29$	11	5	—	3671	$2 \cdot 5 \cdot 367$	13	2	—
3203	$2 \cdot 1601$	2	3	-10	3673	$2^3 \cdot 3^3 \cdot 17$	5	5	± 10
3209	$2^3 \cdot 401$	3	3	—	3677	$2^2 \cdot 919$	2	2	—
3217	$2^4 \cdot 3 \cdot 67$	5	5	—	3691	$2 \cdot 3^2 \cdot 5 \cdot 41$	2	4	—
3221	$2^2 \cdot 5 \cdot 7 \cdot 23$	10	10	± 10	3697	$2 \cdot 43^2$	5	5	—
3229	$2^2 \cdot 3 \cdot 269$	6	6	—	3701	$2^2 \cdot 5^2 \cdot 37$	2	2	± 10
3251	$2 \cdot 5^3 \cdot 13$	6	3	10	3709	$2^2 \cdot 3^2 \cdot 103$	2	2	± 10
3253	$2^2 \cdot 3 \cdot 271$	2	2	—	3719	$2 \cdot 11 \cdot 13^2$	7	2	-10
3257	$2^3 \cdot 11 \cdot 37$	3	3	± 10	3727	$2 \cdot 3^4 \cdot 23$	3	2	10
3259	$2 \cdot 3^2 \cdot 181$	3	5	10	3733	$2^2 \cdot 3 \cdot 311$	2	2	—
3271	$2 \cdot 3 \cdot 5 \cdot 109$	3	5	-10	3739	$2 \cdot 3 \cdot 7 \cdot 89$	7	5	—
3299	$2 \cdot 17 \cdot 97$	2	3	10	3761	$2^4 \cdot 5 \cdot 47$	3	3	—
3301	$2^2 \cdot 3 \cdot 5^2 \cdot 11$	6	6	± 10	3767	$2 \cdot 7 \cdot 269$	5	2	10
3307	$2 \cdot 3 \cdot 19 \cdot 29$	2	4	-10	3769	$2^3 \cdot 3 \cdot 157$	7	7	—
3313	$2^4 \cdot 3^2 \cdot 23$	10	10	± 10	3779	$2 \cdot 1889$	2	3	10
3319	$2 \cdot 3 \cdot 7 \cdot 79$	6	2	—	3793	$2^4 \cdot 3 \cdot 79$	5	5	—
3323	$2 \cdot 11 \cdot 151$	2	3	-10	3797	$2^2 \cdot 13 \cdot 73$	2	2	—
3329	$2^8 \cdot 13$	3	3	—	3803	$2 \cdot 1901$	2	3	-10
3331	$2 \cdot 3^2 \cdot 5 \cdot 37$	3	5	10	3821	$2^2 \cdot 5 \cdot 191$	3	3	± 10
3343	$2 \cdot 3 \cdot 557$	5	11	10	3823	$2 \cdot 3 \cdot 7^2 \cdot 13$	3	9	—
3347	$2 \cdot 7 \cdot 239$	2	3	-10	3833	$2^3 \cdot 479$	3	3	± 10
3359	$2 \cdot 23 \cdot 73$	11	2	-10	3847	$2 \cdot 3 \cdot 641$	5	2	10
3361	$2^5 \cdot 3 \cdot 5 \cdot 7$	22	22	—	3851	$2 \cdot 5^2 \cdot 7 \cdot 11$	2	4	—
3371	$2 \cdot 5 \cdot 337$	2	3	10	3853	$2^2 \cdot 3^2 \cdot 107$	2	2	—
3373	$2^2 \cdot 3 \cdot 281$	5	5	—	3863	$2 \cdot 1931$	5	2	10
3389	$2^2 \cdot 7 \cdot 11^2$	3	3	± 10	3877	$2^2 \cdot 3 \cdot 17 \cdot 19$	2	2	—
3391	$2 \cdot 3 \cdot 5 \cdot 113$	3	5	-10	3881	$2^3 \cdot 5 \cdot 97$	13	13	—
3407	$2 \cdot 13 \cdot 131$	5	2	10	3889	$2^4 \cdot 3^5$	11	11	—
3413	$2^2 \cdot 853$	2	2	—	3907	$2 \cdot 3^2 \cdot 7 \cdot 31$	2	4	-10
3433	$2^3 \cdot 3 \cdot 11 \cdot 13$	5	5	± 10	3911	$2 \cdot 5 \cdot 17 \cdot 23$	13	2	-10
3449	$2^3 \cdot 431$	3	3	—	3917	$2^2 \cdot 11 \cdot 89$	2	2	—
3457	$2^7 \cdot 3^3$	7	7	—	3919	$2 \cdot 3 \cdot 653$	3	2	—
3461	$2^2 \cdot 5 \cdot 173$	2	2	± 10	3923	$2 \cdot 37 \cdot 53$	2	3	-10
3463	$2 \cdot 3 \cdot 577$	3	9	10	3929	$2^3 \cdot 491$	3	3	—
3467	$2 \cdot 1733$	2	3	-10	3931	$2 \cdot 3 \cdot 5 \cdot 131$	2	4	—
3469	$2^2 \cdot 3 \cdot 17^2$	2	2	± 10	3943	$2 \cdot 3^3 \cdot 73$	3	9	10
3491	$2 \cdot 5 \cdot 349$	2	3	—	3947	$2 \cdot 1973$	2	3	-10
3499	$2 \cdot 3 \cdot 11 \cdot 53$	2	4	—	3967	$2 \cdot 3 \cdot 661$	6	2	10
3511	$2 \cdot 3^3 \cdot 5 \cdot 13$	7	2	-10	3989	$2^2 \cdot 997$	2	2	± 10
3517	$2^2 \cdot 3 \cdot 293$	2	2	—	4001	$2^5 \cdot 5^3$	3	3	—
3527	$2 \cdot 41 \cdot 43$	5	2	10	4003	$2 \cdot 3 \cdot 23 \cdot 29$	2	4	—
3529	$2^3 \cdot 3^2 \cdot 7^2$	17	17	—	4007	$2 \cdot 2003$	5	2	10
3533	$2^2 \cdot 883$	2	2	—	4013	$2^2 \cdot 17 \cdot 59$	2	2	—
3539	$2 \cdot 29 \cdot 61$	2	3	10	4019	$2 \cdot 7^2 \cdot 41$	2	4	10

PRIMITIVE ROOTS FOR PRIMES 3 to 5003 (Continued)

p	$p-1$	g	$-G$	ϵ	p	$p-1$	g	$-G$	ϵ
4021	$2^2 \cdot 3 \cdot 5 \cdot 67$	2	2	—	4519	$2 \cdot 3^2 \cdot 251$	3	9	—
4027	$2 \cdot 3 \cdot 11 \cdot 61$	3	6	-10	4523	$2 \cdot 7 \cdot 17 \cdot 19$	5	3	-10
4049	$2^4 \cdot 11 \cdot 23$	3	3	—	4547	$2 \cdot 2273$	2	3	-10
4051	$2 \cdot 3^4 \cdot 5^2$	10	5	10	4549	$2^2 \cdot 3 \cdot 379$	6	6	—
4057	$2^3 \cdot 3 \cdot 13^2$	5	5	±10	4561	$2^4 \cdot 3 \cdot 5 \cdot 19$	11	11	—
4073	$2^3 \cdot 509$	3	3	±10	4567	$2 \cdot 3 \cdot 761$	3	7	10
4079	$2 \cdot 2039$	11	2	-10	4583	$2 \cdot 29 \cdot 79$	5	2	10
4091	$2 \cdot 5 \cdot 409$	2	3	10	4591	$2 \cdot 3^3 \cdot 5 \cdot 17$	11	2	-10
4093	$2^2 \cdot 3 \cdot 11 \cdot 31$	2	2	—	4597	$2^2 \cdot 3 \cdot 383$	5	5	—
4099	$2 \cdot 3 \cdot 683$	2	4	10	4603	$2 \cdot 3 \cdot 13 \cdot 59$	2	4	-10
4111	$2 \cdot 3 \cdot 5 \cdot 137$	12	2	-10	4621	$2^2 \cdot 3 \cdot 5 \cdot 7 \cdot 11$	2	2	—
4127	$2 \cdot 2063$	5	2	10	4637	$2^2 \cdot 19 \cdot 61$	2	2	—
4129	$2^5 \cdot 3 \cdot 43$	13	13	—	4639	$2 \cdot 3 \cdot 773$	3	2	-10
4133	$2^2 \cdot 1033$	2	2	—	4643	$2 \cdot 11 \cdot 211$	5	3	-10
4139	$2 \cdot 2069$	2	3	10	4649	$2^3 \cdot 7 \cdot 83$	3	3	—
4153	$2^3 \cdot 3 \cdot 173$	5	5	±10	4651	$2 \cdot 3 \cdot 5^2 \cdot 31$	3	5	10
4157	$2^2 \cdot 1039$	2	2	—	4657	$2^4 \cdot 3 \cdot 97$	15	15	—
4159	$2 \cdot 3^3 \cdot 7 \cdot 11$	3	2	—	4663	$2 \cdot 3^2 \cdot 7 \cdot 37$	3	9	—
4177	$2^4 \cdot 3^2 \cdot 29$	5	5	±10	4673	$2^6 \cdot 73$	3	3	±10
4201	$2^3 \cdot 3 \cdot 5^2 \cdot 7$	11	11	—	4679	$2 \cdot 2339$	11	2	-10
4211	$2 \cdot 5 \cdot 421$	6	3	10	4691	$2 \cdot 5 \cdot 7 \cdot 67$	2	3	10
4217	$2^3 \cdot 17 \cdot 31$	3	3	±10	4703	$2 \cdot 2351$	5	2	10
4219	$2 \cdot 3 \cdot 19 \cdot 37$	2	4	10	4721	$2^4 \cdot 5 \cdot 59$	6	6	—
4229	$2^2 \cdot 7 \cdot 151$	2	2	±10	4723	$2 \cdot 3 \cdot 787$	2	4	-10
4231	$2 \cdot 3^2 \cdot 5 \cdot 47$	3	2	-10	4729	$2^3 \cdot 3 \cdot 197$	17	17	—
4241	$2^4 \cdot 5 \cdot 53$	3	3	—	4733	$2^2 \cdot 7 \cdot 13^2$	5	5	—
4243	$2 \cdot 3 \cdot 7 \cdot 101$	2	4	-10	4751	$2 \cdot 5^3 \cdot 19$	19	3	-10
4253	$2^2 \cdot 1063$	2	2	—	4759	$2 \cdot 3 \cdot 13 \cdot 61$	3	5	-10
4259	$2 \cdot 2129$	2	3	10	4783	$2 \cdot 3 \cdot 797$	6	2	10
4261	$2^2 \cdot 3 \cdot 5 \cdot 71$	2	2	±10	4787	$2 \cdot 2393$	2	3	-10
4271	$2 \cdot 5 \cdot 7 \cdot 61$	7	3	-10	4789	$2^2 \cdot 3^2 \cdot 7 \cdot$	2	2	—
4273	$2^4 \cdot 3 \cdot 89$	5	5	—	4793	$2^3 \cdot 599$	3	3	±10
4283	$2 \cdot 2141$	2	3	-10	4799	$2 \cdot 2399$	7	2	-10
4289	$2^6 \cdot 67$	3	3	—	4801	$2^6 \cdot 3 \cdot 5^2$	7	7	—
4297	$2^3 \cdot 3 \cdot 179$	5	5	—	4813	$2^2 \cdot 3 \cdot 401$	2	2	—
4327	$2 \cdot 3 \cdot 7 \cdot 103$	3	2	10	4817	$2^4 \cdot 7 \cdot 43$	3	3	±10
4337	$2^4 \cdot 271$	3	3	±10	4831	$2 \cdot 3 \cdot 5 \cdot 7 \cdot 23$	3	2	—
4339	$2 \cdot 3^2 \cdot 241$	10	5	—	4861	$2^2 \cdot 3^5 \cdot 5$	11	11	—
4349	$2^2 \cdot 1087$	2	2	±10	4871	$2 \cdot 5 \cdot 487$	11	3	-10
4357	$2^2 \cdot 3^2 \cdot 11^2$	2	2	—	4877	$2^2 \cdot 23 \cdot 53$	2	2	—
4363	$2 \cdot 3 \cdot 727$	2	4	-10	4889	$2^3 \cdot 13 \cdot 47$	3	3	—
4373	$2^2 \cdot 1093$	2	2	—	4903	$2 \cdot 3 \cdot 19 \cdot 43$	3	2	—
4391	$2 \cdot 5 \cdot 439$	14	2	-10	4909	$2^2 \cdot 3 \cdot 409$	6	6	—
4397	$2^2 \cdot 7 \cdot 157$	2	2	—	4919	$2 \cdot 2459$	13	2	-10
4409	$2^3 \cdot 19 \cdot 29$	3	3	—	4931	$2 \cdot 5 \cdot 17 \cdot 29$	6	3	10
4421	$2^2 \cdot 5 \cdot 13 \cdot 17$	3	3	±10	4933	$2^2 \cdot 3^2 \cdot 137$	2	2	—
4423	$2 \cdot 3 \cdot 11 \cdot 67$	3	7	10	4937	$2^3 \cdot 617$	3	3	±10
4441	$2^3 \cdot 3 \cdot 5 \cdot 37$	21	21	—	4943	$2 \cdot 7 \cdot 353$	7	2	10
4447	$2 \cdot 3^2 \cdot 13 \cdot 19$	3	2	10	4951	$2 \cdot 3^2 \cdot 5^2 \cdot 11$	6	2	-10
4451	$2 \cdot 5^2 \cdot 89$	2	3	10	4957	$2^2 \cdot 3 \cdot 7 \cdot 59$	2	2	—
4457	$2^3 \cdot 557$	3	3	±10	4967	$2 \cdot 13 \cdot 191$	5	2	10
4463	$2 \cdot 23 \cdot 97$	5	2	10	4969	$2^3 \cdot 3^3 \cdot 23$	11	11	—
4481	$2^7 \cdot 5 \cdot 7$	3	3	—	4973	$2^2 \cdot 11 \cdot 113$	2	2	—
4483	$2 \cdot 3^3 \cdot 83$	2	4	—	4987	$2 \cdot 3^2 \cdot 277$	2	4	-10
4493	$2^2 \cdot 1123$	2	2	—	4993	$2^7 \cdot 3 \cdot 13$	5	5	—
4507	$2 \cdot 3 \cdot 751$	2	4	—	4999	$2 \cdot 3 \cdot 7^2 \cdot 17$	3	9	—
4513	$2^5 \cdot 3 \cdot 47$	7	7	—	5003	$2 \cdot 41 \cdot 61$	2	3	-10
4517	$2^2 \cdot 1129$	2	2	—					

From Applied Mathematics Series—55, Primitive Roots, Factorization of $p-1$, pages 864, 866, 867. By permission of United States Department of Commerce, National Bureau of Standards, Washington, D.C.

MENSURATION FORMULAS

DR. HOWARD EVES

TRIANGLES

In the following: K = area, r = radius of the inscribed circle, R = radius of the circumscribed circle.

Right Triangle

$A + B = C = 90°$

$c^2 = a^2 + b^2$ (*Pythagorean relation*)

$a = \sqrt{(c + b)(c - b)}$

$K = \frac{1}{2}ab$

$r = \dfrac{ab}{a + b + c}$, $\quad R = \frac{1}{2}c$

$h = \dfrac{ab}{c}$, $\quad m = \dfrac{b^2}{c}$, $\quad n = \dfrac{a^2}{c}$

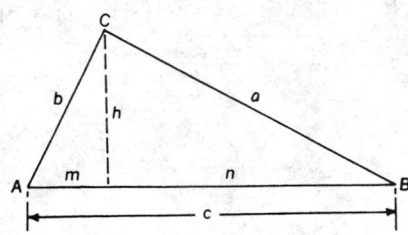

Equilateral Triangle

$A = B = C = 60°$

$K = \frac{1}{4}a^2\sqrt{3}$

$r = \frac{1}{6}a\sqrt{3}$, $\quad R = \frac{1}{3}a\sqrt{3}$

$h = \frac{1}{2}a\sqrt{3}$

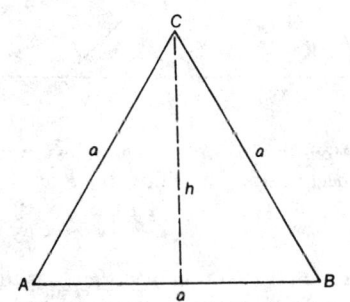

General Triangle

Let $s = \frac{1}{2}(a + b + c)$, h_c = length of altitude on side c, t_c = length of bisector of angle C, m_c = length of median to side c.

$A + B + C = 180°$

$c^2 = a^2 + b^2 - 2ab \cos C$

$\qquad$ (*law of cosines*)

$K = \frac{1}{2}h_c c = \frac{1}{2}ab \sin C$

$\quad = \dfrac{c^2 \sin A \sin B}{2 \sin C}$

$\quad = rs = \dfrac{abc}{4R}$

$\quad = \sqrt{s(s - a)(s - b)(s - c)}$ (*Heron's formula*)

$$r = c \sin \frac{A}{2} \sin \frac{B}{2} \sec \frac{C}{2} = \frac{ab \sin C}{2s} = (s - c) \tan \frac{C}{2}$$

$$= \sqrt{\frac{(s - a)(s - b)(s - c)}{s}} = \frac{K}{s} = 4R \sin \frac{A}{2} \sin \frac{B}{2} \sin \frac{C}{2}$$

$$R = \frac{c}{2 \sin C} = \frac{abc}{4\sqrt{s(s - a)(s - b)(s - c)}} = \frac{abc}{4K}$$

$$h_c = a \sin B = b \sin A = \frac{2K}{c}$$

$$t_c = \frac{2ab}{a + b} \cos \frac{C}{2} = \sqrt{ab \left\{ 1 - \frac{c^2}{(a + b)^2} \right\}}$$

$$m_c = \sqrt{\frac{a^2}{2} + \frac{b^2}{2} - \frac{c^2}{4}}$$

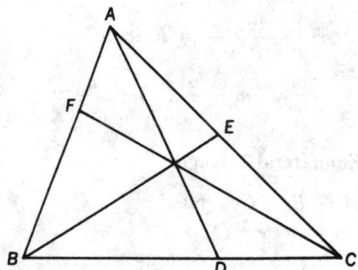

Menelaus' Theorem. A necessary and sufficient condition for points D, E, F on the respective side lines BC, CA, AB of a triangle ABC to be collinear is that

$$BD \cdot CE \cdot AF = - DC \cdot EA \cdot FB,$$

where all segments in the formula are directed segments.

Ceva's Theorem. A necessary and sufficient condition for AD, BE, CF, where D, E, F are points on the respective side lines BC, CA, AB of a triangle ABC, to be concurrent is that

$$BD \cdot CE \cdot AF = + DC \cdot EA \cdot FB,$$

where all segments in the formula are directed segments.

QUADRILATERALS

In the following: K = area, p and q are diagonals.

Rectangle

$$A = B = C = D = 90°$$
$$K = ab, \quad p = \sqrt{a^2 + b^2}$$

Parallelogram

$A = C$, $B = D$, $A + B = 180°$
$K = bh = ab \sin A = ab \sin B$
$h = a \sin A = a \sin B$
$p = \sqrt{a^2 + b^2 - 2ab \cos A}$
$q = \sqrt{a^2 + b^2 - 2ab \cos B} = \sqrt{a^2 + b^2 + 2ab \cos A}$

Rhombus

$p^2 + q^2 = 4a^2$
$K = \frac{1}{2}pq$

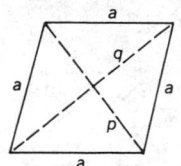

Trapezoid

$m = \frac{1}{2}(a + b)$
$K = \frac{1}{2}(a + b)h = mh$

General quadrilateral

Let $s = \frac{1}{2}(a + b + c + d)$.
$$K = \frac{1}{2}pq \sin \theta$$
$$= \frac{1}{4}(b^2 + d^2 - a^2 - c^2) \tan \theta$$
$$= \frac{1}{4}\sqrt{4p^2q^2 - (b^2 + d^2 - a^2 - c^2)^2}$$

(Bretschneider's formula)

$$= \sqrt{(s - a)(s - b)(s - c)(s - d) - abcd \cos^2\left(\frac{A + B}{2}\right)}$$

Theorem. The diagonals of a quadrilateral with consecutive sides a, b, c, d are perpendicular if and only if $a^2 + c^2 = b^2 + d^2$.

Cyclic Quadrilateral

Let R = radius of the circumscribed circle.
$A + C = B + D = 180°$
$$K = \sqrt{(s - a)(s - b)(s - c)(s - d)}$$

(Brahmagupta's formula)

$$= \frac{\sqrt{(ac + bd)(ad + bc)(ab + cd)}}{4R}$$

$$p = \sqrt{\frac{(ac + bd)(ab + cd)}{ad + bc}}, \quad q = \sqrt{\frac{(ac + bd)(ad + bc)}{ab + cd}}$$

$$R = \frac{1}{2}\sqrt{\frac{(ac + bd)(ad + bc)(ab + cd)}{(s - a)(s - b)(s - c)(s - d)}}, \quad \sin \theta = \frac{2K}{ac + bd}$$

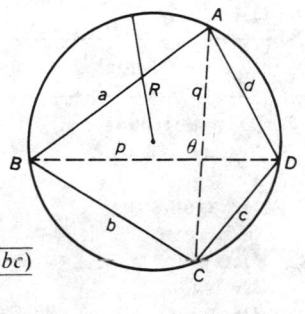

Ptolemy's Theorem. A convex quadrilateral with consecutive sides a, b, c, d and diagonals p and q is cyclic if and only if $ac + bd = pq$.

Cyclic-inscriptable Quadrilateral

Let r = radius of the inscribed circle, R = radius of the circumscribed circle, m = distance between the centers of the inscribed and the circumscribed circles.

$$A + C = B + D = 180°$$

$$a + c = b + d$$

$$K = \sqrt{abcd}$$

$$\frac{1}{(R - m)^2} + \frac{1}{(R + m)^2} = \frac{1}{r^2}$$

$$r = \frac{\sqrt{abcd}}{s}$$

$$R = \tfrac{1}{2} \sqrt{\frac{(ac + bd)(ad + bc)(ab + cd)}{abcd}}$$

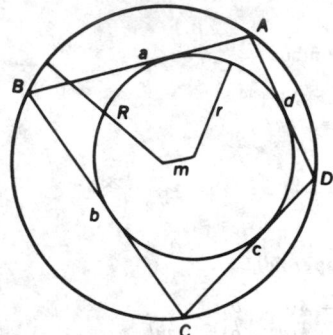

REGULAR POLYGONS

In the following: n = number of sides, s = length of each side, p = perimeter, θ = one of the vertex angles, r = radius of the inscribed circle, R = radius of the circumscribed circle, K = area.

$$\theta = \left(\frac{n - 2}{n}\right) 180°$$

$$s = 2r \tan \frac{180°}{n} = 2R \sin \frac{180°}{n}$$

$$p = ns$$

$$K = \tfrac{1}{4} ns^2 \cot \frac{180°}{n}$$

$$= nr^2 \tan \frac{180°}{n}$$

$$= \tfrac{1}{2} nR^2 \sin \frac{360°}{n}$$

$$r = \tfrac{1}{2} s \cot \frac{180°}{n}, \quad R = \tfrac{1}{2} s \csc \frac{180°}{n}$$

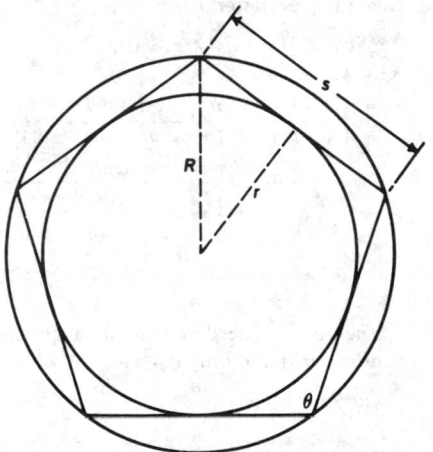

Polygon	n	K	r	R
Triangle (equilateral)	3	$0.43301s^2$	$0.28868s$	$0.57735s$
Square	4	$1.00000s^2$	$0.50000s$	$0.70711s$
Pentagon	5	$1.72048s^2$	$0.68819s$	$0.85065s$
Hexagon	6	$2.59808s^2$	$0.86603s$	$1.00000s$
Heptagon	7	$3.63391s^2$	$1.0383s$	$1.1524s$
Octagon	8	$4.82843s^2$	$1.2071s$	$1.3066s$
Nonagon	9	$6.18182s^2$	$1.3737s$	$1.4619s$
Decagon	10	$7.69421s^2$	$1.5388s$	$1.6180s$
Undecagon	11	$9.36564s^2$	$1.7028s$	$1.7747s$
Dodecagon	12	$11.19615s^2$	$1.8660s$	$1.9319s$

Mensuration Formulas

If s_k denotes the side of a regular polygon of k sides inscribed in a circle of radius R, then

$$s_{2n} = \sqrt{2R^2 - R\sqrt{4R^2 - s_n^2}}$$

If S_k denotes the side of a regular polygon of k sides circumscribed about a circle of radius r, then

$$S_{2n} = \frac{2rS_n}{2r + \sqrt{4r^2 + S_n^2}}$$

If p_k and P_k denote, respectively, the perimeters of regular polygons of k sides inscribed in and circumscribed about the same circle, then

$$P_{2n} = \frac{2p_n P_n}{p_n + P_n} \quad \text{and} \quad p_{2n} = \sqrt{p_n P_{2n}}.$$

If a_k and A_k denote, respectively, the areas of regular polygons of k sides inscribed in and circumscribed about the same circle, then

$$a_{2n} = \sqrt{a_n A_n} \quad \text{and} \quad A_{2n} = \frac{2a_{2n} A_n}{a_{2n} + A_n}.$$

CIRCLES

In the following: R = radius, D = diameter, C = circumference, K = area.

Circumference and Area of a Circle

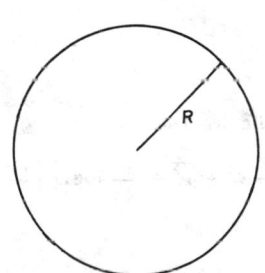

$$C = 2\pi R = \pi D \quad (\pi = 3.14159\cdots)$$
$$K = \pi R^2 = \tfrac{1}{4}\pi D^2 = 0.7854 D^2$$
$$C = 2\sqrt{\pi K} = \frac{2K}{R}$$

$$K = \frac{C^2}{4\pi} = \tfrac{1}{2}CR$$

Sector and Segment of a Circle

Let the central angle θ be measured in radians ($\theta < \pi$).

$$h = R - d, \quad d = R - h$$
$$s = R\theta$$

$$d = R\cos\frac{\theta}{2} = \tfrac{1}{2}c\cot\frac{\theta}{2}$$
$$= \tfrac{1}{2}\sqrt{4R^2 - c^2}$$

$$c = 2R\sin\frac{\theta}{2} = 2d\tan\frac{\theta}{2}$$
$$= 2\sqrt{R^2 - d^2} = \sqrt{4h(2R - h)}$$

$$\theta = \frac{s}{R} = 2\,\mathrm{Cos}^{-1}\frac{d}{R} = 2\,\mathrm{Tan}^{-1}\frac{c}{2d} = 2\,\mathrm{Sin}^{-1}\frac{c}{2R}$$

$$K\,(\text{sector}) = \tfrac{1}{2}Rs = \tfrac{1}{2}R^2\theta$$

$$K\,(\text{segment}) = \tfrac{1}{2}R^2(\theta - \sin\theta) = \tfrac{1}{2}(Rs - cd) = R^2\,\mathrm{Cos}^{-1}\frac{d}{R} - d\sqrt{R^2 - d^2}$$

$$= R^2\,\mathrm{Cos}^{-1}\frac{R - h}{R} - (R - h)\sqrt{2Rh - h^2}$$

Sector of an Annulus

$h = R_1 - R_2$

$K = \frac{1}{2}\theta(R_1 + R_2)(R_1 - R_2)$

$\quad = \frac{1}{2}\theta h(R_1 + R_2)$

$\quad = \frac{1}{2}h(s_1 + s_2)$

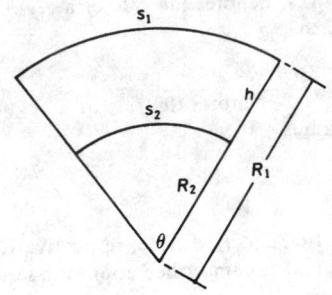

CONIC SECTIONS

Ellipse

Let p = circumference, K = area

$p = 2\pi \sqrt{\dfrac{a^2 + b^2}{2}}$ (approximately)

$\quad = 4aE$ (exactly)   See table of elliptic integral for E, using $k = \sqrt{a^2 - b^2}/a$.

$K = \pi ab$

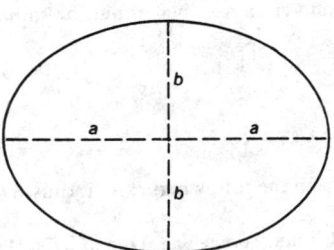

Parabolic Segment

$s = \sqrt{4x^2 + y^2} + \dfrac{y^2}{2x} \log_e \left[\dfrac{2x + \sqrt{4x^2 + y^2}}{y} \right]$

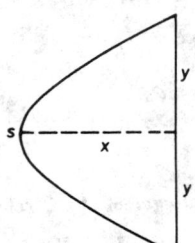

K (right segment) $= \frac{4}{3}xy$

K (oblique segment) $= \frac{4}{3}T$, where T is the area of the triangle with base along the chord of the segment and with opposite vertex at the point on the parabola at which the tangent to the parabola is parallel to the chord of the segment.

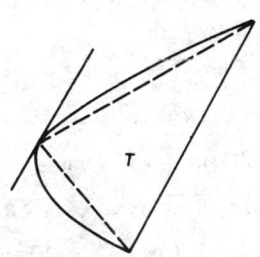

CAVALIERI'S THEOREM FOR THE PLANE

If two planar areas are included between a pair of parallel lines, and if the two segments cut off by the areas on any line parallel to the including lines are equal in length, then the two planar areas are equal.

Mensuration Formulas

PLANAR AREAS BY APPROXIMATION

Divide the planar area K into n strips by equidistant parallel chords of lengths $y_0, y_1, y_2, \ldots, y_n$ (where y_0 and/or y_n may be zero), and let h denote the common distance between the chords. Then, approximately:

Trapezoidal Rule

$$K = h(\tfrac{1}{2}y_0 + y_1 + y_2 + \cdots + y_{n-1} + \tfrac{1}{2}y_n)$$

Durand's Rule

$$K = h(\tfrac{4}{10}y_0 + \tfrac{11}{10}y_1 + y_2 + y_3 + \cdots + y_{n-2} + \tfrac{11}{10}y_{n-1} + \tfrac{4}{10}y_n)$$

Simpson's rule (n even)

$$K = \tfrac{1}{3}h(y_0 + 4y_1 + 2y_2 + 4y_3 + 2y_4 + \cdots + 2y_{n-2} + 4y_{n-1} + y_n)$$

Weddle's Rule ($n = 6$)

$$K = \tfrac{3}{10}h(y_0 + 5y_1 + y_2 + 6y_3 + y_4 + 5y_5 + y_6)$$

SOLIDS BOUNDED BY PLANES

In the following: S = lateral surface, T = total surface, V = volume.

Cube

Let a = length of each edge.

$$T = 6a^2, \quad \text{diagonal of face} = a\sqrt{2}$$
$$V = a^3, \quad \text{diagonal of cube} = a\sqrt{3}$$

Rectangular Parallelepiped (or box)

Let a, b, c be the lengths of its edges.

$$T = 2(ab + bc + ca), \quad V = abc$$
$$\text{diagonal} = \sqrt{a^2 + b^2 + c^2}$$

Prism

S = (perimeter of right section) × (lateral edge)
V = (area of right section) × (lateral edge)
   = (area of base) × (altitude)

Truncated Triangular Prism

V = (area of right section) × $\tfrac{1}{3}$ (sum of the three lateral edges)

Mensuration Formulas

Pyramid

S of regular pyramid = $\frac{1}{2}$(perimeter of base) × (slant height)

$V = \frac{1}{3}$(area of base) × (altitude)

Frustum of Pyramid

Let B_1 = area of lower base, B_2 = area of upper base, h = altitude.

S of regular figure = $\frac{1}{2}$(sum of perimeters of bases) × (slant height)

$V = \frac{1}{3}h(B_1 + B_2 + \sqrt{B_1 B_2})$

Prismatoid

A *prismatoid* is a polyhedron having for bases two polygons in parallel planes, and for lateral faces triangles or trapezoids with one side lying in one base, and the opposite vertex or side lying in the other base, of the polyhedron. Let B_1 = area of lower base, M = area of midsection, B_2 = area of upper base, h = altitude.

$$V = \frac{1}{6}h(B_1 + 4M + B_2) \quad \text{(the } prismoidal \; formula\text{)}$$

Note: Since cubes, rectangular parallelepipeds, prisms, pyramids, and frustums of pyramids are all examples of prismatoids, the formula for the volume of a prismatoid subsumes most of the above volume formulae.

Regular Polyhedra

Let v = number of vertices, e = number of edges, f = number of faces, α = each dihedral angle, a = length of each edge, r = radius of the inscribed sphere, R = radius of the circumscribed sphere, A = area of each face, T = total area, V = volume.

$v - e + f = 2$ (the *Euler-Descartes formula*—actually holds for *any* convex polyhedron)

$$T = fA$$
$$V = \frac{1}{3}rfA = \frac{1}{3}rT$$

Name	Nature of Surface	T	V
Tetrahedron	4 equilateral triangles	$1.73205a^2$	$0.11785a^3$
Hexahedron (cube)	6 squares	$6.00000a^2$	$1.00000a^3$
Octahedron	8 equilateral triangles	$3.46410a^2$	$0.47140a^3$
Dodecahedron	12 regular pentagons	$20.64573a^2$	$7.66312a^3$
Icosahedron	20 equilateral triangles	$8.66025a^2$	$2.18169a^3$

Name	v	e	f	α	a	r
Tetrahedron	4	6	4	70° 32′	$1.633R$	$0.333R$
Hexahedron	8	12	6	90°	$1.155R$	$0.577R$
Octahedron	6	12	8	109° 28′	$1.414R$	$0.577R$
Dodecahedron	20	30	12	116° 34′	$0.714R$	$0.795R$
Icosahedron	12	30	20	138° 11′	$1.051R$	$0.795R$

Mensuration Formulas

Name	A	r	R	V
Tetrahedron	$\frac{1}{4}a^2\sqrt{3}$	$\frac{1}{12}a\sqrt{6}$	$\frac{1}{4}a\sqrt{6}$	$\frac{1}{12}a^3\sqrt{2}$
Hexahedron	a^2	$\frac{1}{2}a$	$\frac{1}{2}a\sqrt{3}$	a^3
Octahedron	$\frac{1}{4}a^2\sqrt{3}$	$\frac{1}{6}a\sqrt{6}$	$\frac{1}{2}a\sqrt{2}$	$\frac{1}{3}a^3\sqrt{2}$
Dodecahedron	$\frac{1}{4}a^2\sqrt{25+10\sqrt{5}}$	$\frac{1}{20}a\sqrt{250+110\sqrt{5}}$	$\frac{1}{4}a(\sqrt{15}+\sqrt{3})$	$\frac{1}{4}a^3(15+7\sqrt{5})$
Icosahedron	$\frac{1}{4}a^2\sqrt{3}$	$\frac{1}{12}a\sqrt{42+18\sqrt{5}}$	$\frac{1}{4}a\sqrt{10+2\sqrt{5}}$	$\frac{5}{12}a^3(3+\sqrt{5})$

CYLINDERS AND CONES

In the following: B_1 = area of lower base, B_2 = area of upper base, h = altitude, S = lateral surface, T = total surface, V = volume.

Cylinder

S = (perimeter of right section) × (lateral edge)
V = (area of right section) × (lateral edge)

Right Circular Cylinder

Let R = radius of base.

$$S = 2\pi Rh, \quad T = 2\pi R(R + h), \quad V = \pi R^2 h$$

Cone

$$V = \tfrac{1}{3}B_1 h$$

Right Circular Cone

Let R = radius of base, s = slant height.

$$s = \sqrt{R^2 + h^2}$$
$$S = \pi Rs = \pi R\sqrt{R^2 + h^2}$$
$$T = \pi R(R + s) = \pi R(R + \sqrt{R^2 + h^2})$$
$$V = \tfrac{1}{3}\pi R^2 h$$

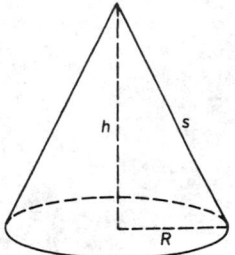

Frustum of Cone

$V = \tfrac{1}{3}h(B_1 + B_2 + \sqrt{B_1 B_2})$, where B_1 and B_2 are the areas of the bases.

Frustum of Right Circular Cone

Let R_1 = radius of lower base, R_2 = radius of upper base, s = slant height.

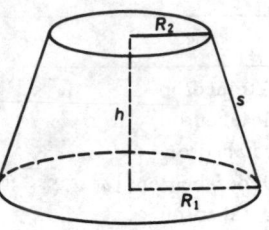

$$s = \sqrt{(R_1 - R_2)^2 + h^2}$$
$$S = \pi(R_1 + R_2)s$$
$$\quad = \pi(R_1 + R_2)\sqrt{(R_1 - R_2)^2 + h^2}$$
$$T = \pi[R_1^2 + R_2^2 + (R_1 + R_2)s]$$
$$\quad = \pi[R_1^2 + R_2^2 + (R_1 + R_2)\sqrt{(R_1 - R_2)^2 + h^2}]$$
$$V = \tfrac{1}{3}\pi h(R_1^2 + R_2^2 + R_1 R_2)$$

SPHERICAL FIGURES

In the following: R = radius of sphere, D = diameter of sphere, S = surface area, V = volume.

Sphere

$$D = 2R$$
$$S = 4\pi R^2 = \pi D^2 = 12.57 R^2$$
$$V = \tfrac{4}{3}\pi R^3 = \tfrac{1}{6}\pi D^3 = 4.189 R^3$$

Zone and Segment of One Base

$$S = 2\pi R h = \pi D h = \pi p^2$$
$$V = \tfrac{1}{3}\pi h^2(3R - h) = \tfrac{1}{6}\pi h(3a^2 + h^2)$$

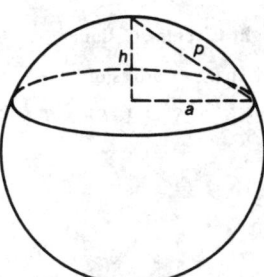

Zone and Segment of Two Bases

$$S = 2\pi R h = \pi D h$$
$$V = \tfrac{1}{6}\pi h(3a^2 + 3b^2 + h^2)$$

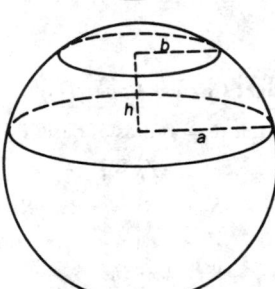

Lune

$$S = 2R^2\theta, \quad \theta \text{ in radians}$$

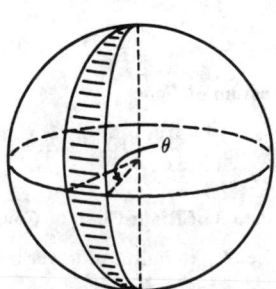

Spherical Sector

$$V = \tfrac{2}{3}\pi R^2 h = \tfrac{1}{6}\pi D^2 h$$

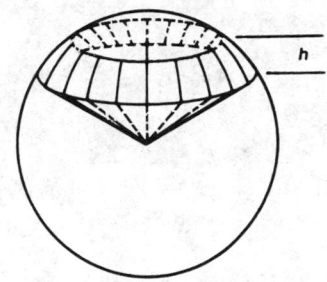

Spherical Triangle and Polygon

Let A, B, C be the angles, in radians, of the triangle; let θ = sum of angles, in radians, of a spherical polygon of n sides.

$$S = (A + B + C - \pi) R^2$$
$$S = [\theta - (n - 2)\pi] R^2$$

SPHEROIDS

Ellipsoid

Let a, b, c be the lengths of the semiaxes.

$$V = \tfrac{4}{3}\pi abc$$

Oblate Spheroid

An *oblate spheroid* is formed by the rotation of an ellipse about its minor axis. Let a and b be the major and minor semiaxes, respectively, and ϵ the eccentricity, of the revolving ellipse.

$$S = 2\pi a^2 + \pi \frac{b^2}{\epsilon} \log_e \frac{1 + \epsilon}{1 - \epsilon}$$

$$V = \tfrac{4}{3}\pi a^2 b$$

Prolate Spheroid

A *prolate spheroid* is formed by the rotation of an ellipse about its major axis. Let a and b be the major and minor semiaxes, respectively, and ϵ the eccentricity, of the revolving ellipse.

$$S = 2\pi b^2 + 2\pi \frac{ab}{\epsilon} \sin^{-1} \epsilon$$

$$V = \tfrac{4}{3}\pi ab^2$$

CIRCULAR TORUS

A *circular torus* is formed by the rotation of a circle about an axis in the plane of the circle and not cutting the circle. Let r be the radius of the revolving circle and let R be the distance of its center from the axis of rotation.

$S = 4\pi^2 R r$
$V = 2\pi^2 R r^2$

PAPPUS-GULDINUS THEOREMS

1. If a planar arc be revolved about an axis in its plane, but not cutting the arc, the area of the surface of revolution so formed is equal to the product of the length of the arc and the length of the path traced by the centroid of the arc.
2. If a planar area be revolved about an axis in its plane, but not intersecting the area, the volume of the solid of revolution so formed is equal to the product of the area and the length of the path traced by the centroid of the area.

CAVALIERI'S THEOREM FOR SPACE

If two solids are included between a pair of parallel planes, and if the two sections cut by them on any plane parallel to the including planes are equal in area, then the volumes of the solids are equal.

GENERAL PRISMATOID

A *general prismatoid* is a solid such that the area A_y of any section parallel to and distant y from a fixed plane can be expressed as a polynomial in y of degreee not higher than the third. That is,

$$A_y = ay^3 + by^2 + cy + d,$$

where a, b, c, d are constants which may be positive, zero, or negative. Let B_1 = area of lower base; M = area of midsection, B_2 = area of upper base, h = altitude.

$$V = \tfrac{1}{6}h(B_1 + 4M + B_2)$$

Note. All prismatoids, cylinders, cones, spheres, spheroids, and many other solids are general prismatoids.

CENTROIDS

If a geometrical figure possesses a center of symmetry, that point is the centroid of the figure.

If a geometrical figure possesses an axis of symmetry, the centroid of the figure lies on that axis.

Geometrical Figure	Location of Centroid
Perimeter of triangle	Center of the inscribed circle of the triangle whose vertices are the midpoints of the sides of the given triangle.
Arc of semicircle of radius R	Distance from diameter $= \dfrac{2R}{\pi}$
Arc of 2α radians of a circle of radius R	Distance from center of circle $= \dfrac{R \sin \alpha}{\alpha}$
Area of triangle	Intersection of the medians
Area of quadrilateral	Intersection of the diagonals of the parallelogram whose sides pass through adjacent trisection points of pairs of consecutive sides of the quadrilateral
Area of semicircle of radius R	Distance from diameter $= \dfrac{4R}{3\pi}$
Area of circular sector of radius R and central angle 2α radians	Distance from center of circle $= \dfrac{2R \sin \alpha}{3\alpha}$
Area of semiellipse of altitude h	Distance from base $= \dfrac{4h}{3\pi}$
Area of a quadrant of an ellipse of major and minor semiaxes a and b	Distance from minor axis $= \dfrac{4a}{3\pi}$, distance from major axis $= \dfrac{4b}{3\pi}$
Area of right parabolic segment of altitude h	Distance from base $= \frac{2}{5}h$
Lateral area of regular pyramid or right circular cone	Distance from base $= \frac{1}{3}h$
Area of hemisphere of radius R	Distance from base $= \frac{1}{2}R$
Volume of pyramid or cone	One fourth the way from the centroid of the base to the vertex of the pyramid or cone
Volume of frustum of pyramid or cone with d as the distance between the centroids of its bases and k as the ratio of similarity of the upper base to the lower base	On the line joining the centroids of the two bases at a distance from the centroid of the lower base $= \frac{1}{4}d\left(\dfrac{1 + 2k + 3k^2}{1 + k + k^2}\right)$
Volume of hemisphere of radius R	Distance from base $= \frac{3}{8}R$
Volume of revolution of altitude h obtained by revolving a semiellipse about its axis of symmetry	Distance from base $= \frac{3}{8}h$
Volume of paraboloid of revolution of altitude h	Distance from base $= \frac{1}{3}h$

V. TRIGONOMETRY

TRIGONOMETRY
Dr. Howard Eves

PLANE TRIGONOMETRY

Angle

That part of a straight line lying entirely to one side of a point O on the line is called a *ray* (or a *half-line*); the point O is called the *origin* of the ray. A ray of origin O is identified by the notation OA, where A is any point of the ray.

If a ray OA is rotated, in a plane, about its origin O onto ray OB, an *angle AOB* is said to be generated. Ray OA is called the *initial side*, ray OB the *terminal side*, and point O the *vertex* of the angle. The angle is said to be *positive* or *negative* according as the generating rotation is counterclockwise or clockwise.

An angle is said to be in *standard position* if its vertex is at the origin O and its initial side is on the positive x-axis of a rectangular Cartesian coordinate system (see p. 542). If the terminal side of the angle falls on a coordinate axis, the angle is called a *quadrantal angle*; otherwise the angle is called a *first, second, third,* or *fourth quadrant angle* according as the terminal side falls in the first, second, third, or fourth quadrant of the coordinate system.

An angle of one *degree* is an angle in which the rotation is 1/360 of one complete rotation.

A *straight angle* is an angle of 180° (180 degrees).

A *right angle* is an angle of 90°.

An *acute angle* is an angle between 0° and 90°.

An *obtuse angle* is an angle between 90° and 180°.

A *radian* is an angle subtended at the center of a circle by an arc whose length is equal to that of the radius.

$$180° = \pi \text{ radians}; \quad 1° = \frac{\pi}{180} \text{ radians}; \quad 1 \text{ radian} = \frac{180}{\pi} \text{ degrees}.$$

The Trigonometric Functions of an Acute Angle

In the right triangle ABC,
sine A = sin A = a/c
cosine A = cos A = b/c
tangent A = tan A = a/b

cosecant A = csc A = c/a
secant A = sec A = c/b
cotangent A = cot A = ctn A = b/a
exsecant A = exsec A = sec A − 1
versine A = vers A = 1 − cos A
coversine A = covers A = 1 − sin A
haversine A = hav A = $\frac{1}{2}$ vers A

Formulas for Use in Trigonometry

The Trigonometric Functions of an Arbitrary Angle

Let α be any angle in standard position and let $P(x,y)$ be any point on the terminal side of the angle. Denote the positive distance OP by r. Then

$$\sin \alpha = y/r \qquad \csc \alpha = r/y$$
$$\cos \alpha = x/r \qquad \sec \alpha = r/x$$
$$\tan \alpha = y/x \qquad \cot \alpha = \operatorname{ctn} \alpha = x/y$$
$$\operatorname{exsec} \alpha = \sec \alpha - 1 \qquad \operatorname{covers} \alpha = 1 - \sin \alpha$$
$$\operatorname{vers} \alpha = 1 - \cos \alpha \qquad \operatorname{hav} \alpha = \tfrac{1}{2}\operatorname{vers} \alpha$$
$$\operatorname{cis} \alpha = \cos \alpha + i \sin \alpha = e^{i\alpha}, \alpha \text{ in radians}, i = \sqrt{-1}$$

RELATIONS BETWEEN CIRCULAR (OR INVERSE CIRCULAR) FUNCTIONS

$$\left(0 \le x \le \frac{\pi}{2}\right)$$

	$\sin x = a$	$\cos x = a$	$\tan x = a$
$\sin x$	a	$(1 - a^2)^{1/2}$	$a(1 + a^2)^{-1/2}$
$\cos x$	$(1 - a^2)^{1/2}$	a	$(1 + a^2)^{-1/2}$
$\tan x$	$a(1 - a^2)^{-1/2}$	$a^{-1}(1 - a^2)^{1/2}$	a
$\csc x$	a^{-1}	$(1 - a^2)^{-1/2}$	$a^{-1}(1 + a^2)^{1/2}$
$\sec x$	$(1 - a^2)^{-1/2}$	a^{-1}	$(1 + a^2)^{1/2}$
$\cot x$	$a^{-1}(1 - a^2)^{1/2}$	$a(1 - a^2)^{-1/2}$	a^{-1}

	$\csc x = a$	$\sec x = a$	$\cot x = a$
$\sin x$	a^{-1}	$a^{-1}(a^2 - 1)^{1/2}$	$(1 + a^2)^{-1/2}$
$\cos x$	$a^{-1}(a^2 - 1)^{1/2}$	a^{-1}	$a(1 + a^2)^{-1/2}$
$\tan x$	$(a^2 - 1)^{-1/2}$	$(a^2 - 1)^{1/2}$	a^{-1}
$\csc x$	a	$a(a^2 - 1)^{-1/2}$	$(1 + a^2)^{1/2}$
$\sec x$	$a(a^2 - 1)^{-1/2}$	a	$a^{-1}(1 + a^2)^{1/2}$
$\cot x$	$(a^2 - 1)^{1/2}$	$(a^2 - 1)^{-1/2}$	a

Examples:

If $\sec x = a$, then $\tan x = (a^2 - 1)^{1/2}$
$\quad \arctan a = \arccos (1 + a^2)^{-1/2}$

TRIGONOMETRY
Formulas for Use in Trigonometry

SIGNS OF THE TRIGONOMETRIC FUNCTIONS

Quadrant	sin	cos	tan	cot	sec	csc
I	+	+	+	+	+	+
II	+	−	−	−	−	+
III	−	−	+	+	−	−
IV	−	+	−	−	+	−

VARIATIONS OF THE TRIGONOMETRIC FUNCTIONS

Quadrant	sin	cos	tan	cot	sec	csc
I	$0 \rightarrow +1$	$+1 \rightarrow 0$	$0 \rightarrow +\infty$	$+\infty \rightarrow 0$	$+1 \rightarrow +\infty$	$+\infty \rightarrow +1$
II	$+1 \rightarrow 0$	$0 \rightarrow -1$	$-\infty \rightarrow 0$	$0 \rightarrow -\infty$	$-\infty \rightarrow -1$	$+1 \rightarrow +\infty$
III	$0 \rightarrow -1$	$-1 \rightarrow 0$	$0 \rightarrow +\infty$	$+\infty \rightarrow 0$	$-1 \rightarrow -\infty$	$-\infty \rightarrow -1$
IV	$-1 \rightarrow 0$	$0 \rightarrow +1$	$-\infty \rightarrow 0$	$0 \rightarrow -\infty$	$+\infty \rightarrow +1$	$-1 \rightarrow -\infty$

TRIGONOMETRIC FUNCTIONS OF SOME SPECIAL ANGLES

Angle	sin	cos	tan	cot	sec	csc
$0° = 0$	0	1	0	$\cdots$	1	$\cdots$
$15° = \dfrac{\pi}{12}$	$\dfrac{\sqrt{2}}{4}(\sqrt{3}-1)$	$\dfrac{\sqrt{2}}{4}(\sqrt{3}+1)$	$2-\sqrt{3}$	$2+\sqrt{3}$	$\sqrt{2}(\sqrt{3}-1)$	$\sqrt{2}(\sqrt{3}+1)$
$30° = \dfrac{\pi}{6}$	$1/2$	$\sqrt{3}/2$	$\sqrt{3}/3$	$\sqrt{3}$	$2\sqrt{3}/3$	2
$45° = \dfrac{\pi}{4}$	$\sqrt{2}/2$	$\sqrt{2}/2$	1	1	$\sqrt{2}$	$\sqrt{2}$
$60° = \dfrac{\pi}{3}$	$\sqrt{3}/2$	$1/2$	$\sqrt{3}$	$\sqrt{3}/3$	2	$2\sqrt{3}/3$
$75° = \dfrac{5\pi}{12}$	$\dfrac{\sqrt{2}}{4}(\sqrt{3}+1)$	$\dfrac{\sqrt{2}}{4}(\sqrt{3}-1)$	$2+\sqrt{3}$	$2-\sqrt{3}$	$\sqrt{2}(\sqrt{3}+1)$	$\sqrt{2}(\sqrt{3}-1)$
$90° = \dfrac{\pi}{2}$	1	0	$\cdots$	0	$\cdots$	1
$105° = \dfrac{7\pi}{12}$	$\dfrac{\sqrt{2}}{4}(\sqrt{3}+1)$	$-\dfrac{\sqrt{2}}{4}(\sqrt{3}-1)$	$-(2+\sqrt{3})$	$-(2-\sqrt{3})$	$-\sqrt{2}(\sqrt{3}+1)$	$\sqrt{2}(\sqrt{3}-1)$
$120° = \dfrac{2\pi}{3}$	$\sqrt{3}/2$	$-1/2$	$-\sqrt{3}$	$-\sqrt{3}/3$	-2	$2\sqrt{3}/3$
$135° = \dfrac{3\pi}{4}$	$\sqrt{2}/2$	$-\sqrt{2}/2$	-1	-1	$-\sqrt{2}$	$\sqrt{2}$
$150° = \dfrac{5\pi}{6}$	$1/2$	$-\sqrt{3}/2$	$-\sqrt{3}/3$	$-\sqrt{3}$	$-2\sqrt{3}/3$	2
$165° = \dfrac{11\pi}{12}$	$\dfrac{\sqrt{2}}{4}(\sqrt{3}-1)$	$-\dfrac{\sqrt{2}}{4}(\sqrt{3}+1)$	$-(2-\sqrt{3})$	$-(2+\sqrt{3})$	$-\sqrt{2}(\sqrt{3}-1)$	$\sqrt{2}(\sqrt{3}+1)$
$180° = \pi$	0	-1	0	$\cdots$	-1	$\cdots$
$270° = \dfrac{3\pi}{2}$	-1	0	$\cdots$	0	$\cdots$	-1

Formulas for Use in Trigonometry

Relations among the Functions

$$\sin x = \frac{1}{\csc x}$$

$$\csc x = \frac{1}{\sin x}$$

$$\cos x = \frac{1}{\sec x}$$

$$\sec x = \frac{1}{\cos x}$$

$$\tan x = \frac{1}{\cot x} = \frac{\sin x}{\cos x}$$

$$\sin^2 x + \cos^2 x = 1$$
$$1 + \tan^2 x = \sec^2 x$$

$$\cot x = \frac{1}{\tan x} = \frac{\cos x}{\sin x}$$

$$1 + \cot^2 x = \csc^2 x$$

* $\sin x = \pm \sqrt{1 - \cos^2 x}$
* $\tan x = \pm \sqrt{\sec^2 x - 1}$
* $\cot x = \pm \sqrt{\csc^2 x - 1}$

* $\cos x = \pm \sqrt{1 - \sin^2 x}$
* $\sec x = \pm \sqrt{\tan^2 x + 1}$
* $\csc x = \pm \sqrt{\cot^2 x + 1}$

$$\sin x = \cos(90° - x) = \sin(180° - x)$$
$$\cos x = \sin(90° - x) = -\cos(180° - x)$$
$$\tan x = \cot(90° - x) = -\tan(180° - x)$$
$$\cot x = \tan(90° - x) = -\cot(180° - x)$$

$$\csc x = \cot \frac{x}{2} - \cot x$$

*The sign in front of radical depends on quadrant in which x falls.

Reduction Formulas

$$\sin \alpha = + \cos(\alpha - 90°) = -\sin(\alpha - 180°) = -\cos(\alpha - 270°)$$
$$\cos \alpha = -\sin(\alpha - 90°) = -\cos(\alpha - 180°) = +\sin(\alpha - 270°)$$
$$\tan \alpha = -\cot(\alpha - 90°) = +\tan(\alpha - 180°) = -\cot(\alpha - 270°)$$
$$\cot \alpha = -\tan(\alpha - 90°) = +\cot(\alpha - 180°) = -\tan(\alpha - 270°)$$
$$\sec \alpha = -\csc(\alpha - 90°) = -\sec(\alpha - 180°) = +\csc(\alpha - 270°)$$
$$\csc \alpha = +\sec(\alpha - 90°) = -\csc(\alpha - 180°) = -\sec(\alpha - 270°)$$

FURTHER REDUCTION FORMULAS

	sin	cos	tan	cot	sec	csc
$-\alpha$	$-\sin\alpha$	$+\cos\alpha$	$-\tan\alpha$	$-\cot\alpha$	$+\sec\alpha$	$-\csc\alpha$
$90° + \alpha$	$+\cos\alpha$	$-\sin\alpha$	$-\cot\alpha$	$-\tan\alpha$	$-\csc\alpha$	$+\sec\alpha$
$90° - \alpha$	$+\cos\alpha$	$+\sin\alpha$	$+\cot\alpha$	$+\tan\alpha$	$+\csc\alpha$	$+\sec\alpha$
$180° + \alpha$	$-\sin\alpha$	$-\cos\alpha$	$+\tan\alpha$	$+\cot\alpha$	$-\sec\alpha$	$-\csc\alpha$
$180° - \alpha$	$+\sin\alpha$	$-\cos\alpha$	$-\tan\alpha$	$-\cot\alpha$	$-\sec\alpha$	$+\csc\alpha$
$270° + \alpha$	$-\cos\alpha$	$+\sin\alpha$	$-\cot\alpha$	$-\tan\alpha$	$+\csc\alpha$	$-\sec\alpha$
$270° - \alpha$	$-\cos\alpha$	$-\sin\alpha$	$+\cot\alpha$	$+\tan\alpha$	$-\csc\alpha$	$-\sec\alpha$
$360° + \alpha$	$+\sin\alpha$	$+\cos\alpha$	$+\tan\alpha$	$+\cot\alpha$	$+\sec\alpha$	$+\csc\alpha$
$360° - \alpha$	$-\sin\alpha$	$+\cos\alpha$	$-\tan\alpha$	$-\cot\alpha$	$+\sec\alpha$	$-\csc\alpha$

Formulas for Use in Trigonometry

The previous table may be summarized and extended by the following easily remembered rule:

$$f(\pm\alpha + n90°) = \pm g(\alpha)$$

where n may be any integer, positive, negative, or zero

f is any one of the six trigonometric functions: sin, cos, tan, cot, sec, or csc

α may be any real angle measure

If n is even, then g is the same function as f. If n is odd, then g is the *cofunction* of f.

(Sine and cosine, tangent and cotangent, secant and cosecant, are cofunctions of each other.)

The second $\pm$ sign is not necessarily the same as the first one, but is determined as follows: For a given function f, a given value of n, and a given choice of the first $\pm$ sign, the second $\pm$ sign will be the same for all values of α. Thus it is only necessary to check the sign for any one value of α, and the formula will be complete.

EXAMPLES:

tan $(\alpha + 270°) = \pm$ cot α. Since $n(=3)$ is odd, we use the cofunction. To determine the sign, assume a value of α in the first quadrant. Then $\alpha + 270°$ is in the fourth quadrant, where the tangent is negative, so a minus sign is required. Thus the formula becomes

$$\tan(\alpha + 270°) = -\cot\alpha, \quad \text{valid for all values of } \alpha.$$

cos $(\alpha - 450°) = \pm$ sin α. Again assuming a value of α in the first quadrant, we find that $\alpha - 450°$ is in the fourth quadrant. Thus cos $(\alpha - 450°)$ would be positive, and no minus sign is needed. Hence the formula becomes

$$\cos(\alpha - 450°) = \sin\alpha.$$

sec $(180° - \alpha) = \pm$ sec α. Here $n(=2)$ is even, so we use the same function. To determine the sign, again assume a value of α in the first quadrant. Then $180° - \alpha$ is in the second quadrant, where the secant is negative. Thus the formula. valid for all values of α, is

$$\sec(180° - \alpha) = -\sec\alpha.$$

Fundamental Identities

Where a double sign appears in the following, the choice of sign depends upon the quadrant in which the angle terminates.

Reciprocal relations

$$\sin\alpha = \frac{1}{\csc\alpha}, \quad \cos\alpha = \frac{1}{\sec\alpha}, \quad \tan\alpha = \frac{1}{\cot\alpha}$$

$$\csc\alpha = \frac{1}{\sin\alpha}, \quad \sec\alpha = \frac{1}{\cos\alpha}, \quad \cot\alpha = \frac{1}{\tan\alpha}$$

Product relations

$$\sin\alpha = \tan\alpha\cos\alpha, \quad \cos\alpha = \cot\alpha\sin\alpha$$
$$\tan\alpha = \sin\alpha\sec\alpha, \quad \cot\alpha = \cos\alpha\csc\alpha$$
$$\sec\alpha = \csc\alpha\tan\alpha, \quad \csc\alpha = \sec\alpha\cot\alpha$$

Quotient relations

$$\sin\alpha = \frac{\tan\alpha}{\sec\alpha}, \quad \cos\alpha = \frac{\cot\alpha}{\csc\alpha}, \quad \tan\alpha = \frac{\sin\alpha}{\cos\alpha}$$

$$\csc\alpha = \frac{\sec\alpha}{\tan\alpha}, \quad \sec\alpha = \frac{\csc\alpha}{\cot\alpha}, \quad \cot\alpha = \frac{\cos\alpha}{\sin\alpha}$$

Formulas for Use in Trigonometry

Pythagorean relations

$$\sin^2\alpha + \cos^2\alpha = 1, \qquad 1 + \tan^2\alpha = \sec^2\alpha, \qquad 1 + \cot^2\alpha = \csc^2\alpha$$

Angle-sum and angle-difference relations

$$\sin(\alpha + \beta) = \sin\alpha\cos\beta + \cos\alpha\sin\beta$$
$$\sin(\alpha - \beta) = \sin\alpha\cos\beta - \cos\alpha\sin\beta$$
$$\cos(\alpha + \beta) = \cos\alpha\cos\beta - \sin\alpha\sin\beta$$
$$\cos(\alpha - \beta) = \cos\alpha\cos\beta + \sin\alpha\sin\beta$$
$$\tan(\alpha + \beta) = \frac{\tan\alpha + \tan\beta}{1 - \tan\alpha\tan\beta}$$
$$\tan(\alpha - \beta) = \frac{\tan\alpha - \tan\beta}{1 + \tan\alpha\tan\beta}$$
$$\cot(\alpha + \beta) = \frac{\cot\beta\cot\alpha - 1}{\cot\beta + \cot\alpha}$$
$$\cot(\alpha - \beta) = \frac{\cot\beta\cot\alpha + 1}{\cot\beta - \cot\alpha}$$
$$\sin(\alpha + \beta)\sin(\alpha - \beta) = \sin^2\alpha - \sin^2\beta = \cos^2\beta - \cos^2\alpha$$
$$\cos(\alpha + \beta)\cos(\alpha - \beta) = \cos^2\alpha - \sin^2\beta = \cos^2\beta - \sin^2\alpha$$

Double-angle relations

$$\sin 2\alpha = 2\sin\alpha\cos\alpha = \frac{2\tan\alpha}{1 + \tan^2\alpha}$$
$$\cos 2\alpha = \cos^2\alpha - \sin^2\alpha = 2\cos^2\alpha - 1 = 1 - 2\sin^2\alpha = \frac{1 - \tan^2\alpha}{1 + \tan^2\alpha}$$
$$\tan 2\alpha = \frac{2\tan\alpha}{1 - \tan^2\alpha}, \qquad \cot 2\alpha = \frac{\cot^2\alpha - 1}{2\cot\alpha}$$

Multiple-angle relations

$$\sin 3\alpha = 3\sin\alpha - 4\sin^3\alpha$$
$$\cos 3\alpha = 4\cos^3\alpha - 3\cos\alpha$$
$$\sin 4\alpha = 4\sin\alpha\cos\alpha - 8\sin^3\alpha\cos\alpha$$
$$\cos 4\alpha = 8\cos^4\alpha - 8\cos^2\alpha + 1$$
$$\sin 5\alpha = 5\sin\alpha - 20\sin^3\alpha + 16\sin^5\alpha$$
$$\cos 5\alpha = 16\cos^5\alpha - 20\cos^3\alpha + 5\cos\alpha$$
$$\sin 6\alpha = 32\cos^5\alpha\sin\alpha - 32\cos^3\alpha\sin\alpha + 6\cos\alpha\sin\alpha$$
$$\cos 6\alpha = 32\cos^6\alpha - 48\cos^4\alpha + 18\cos^2\alpha - 1$$
$$\sin n\alpha = 2\sin(n - 1)\alpha\cos\alpha - \sin(n - 2)\alpha$$
$$\cos n\alpha = 2\cos(n - 1)\alpha\cos\alpha - \cos(n - 2)\alpha$$
$$\tan 3\alpha = \frac{3\tan\alpha - \tan^3\alpha}{1 - 3\tan^2\alpha}$$
$$\tan 4\alpha = \frac{4\tan\alpha - 4\tan^3\alpha}{1 - 6\tan^2\alpha + \tan^4\alpha}$$
$$\tan n\alpha = \frac{\tan(n - 1)\alpha + \tan\alpha}{1 - \tan(n - 1)\alpha\tan\alpha}$$

Formulas for Use in Trigonometry

Function-product relations

$$\sin \alpha \sin \beta = \tfrac{1}{2}\cos(\alpha - \beta) - \tfrac{1}{2}\cos(\alpha + \beta)$$
$$\cos \alpha \cos \beta = \tfrac{1}{2}\cos(\alpha - \beta) + \tfrac{1}{2}\cos(\alpha + \beta)$$
$$\sin \alpha \cos \beta = \tfrac{1}{2}\sin(\alpha + \beta) + \tfrac{1}{2}\sin(\alpha - \beta)$$
$$\cos \alpha \sin \beta = \tfrac{1}{2}\sin(\alpha + \beta) - \tfrac{1}{2}\sin(\alpha - \beta)$$

Function-sum and function-difference relations

$$\sin \alpha + \sin \beta = 2 \sin \tfrac{1}{2}(\alpha + \beta) \cos \tfrac{1}{2}(\alpha - \beta)$$
$$\sin \alpha - \sin \beta = 2 \cos \tfrac{1}{2}(\alpha + \beta) \sin \tfrac{1}{2}(\alpha - \beta)$$
$$\cos \alpha + \cos \beta = 2 \cos \tfrac{1}{2}(\alpha + \beta) \cos \tfrac{1}{2}(\alpha - \beta)$$
$$\cos \alpha - \cos \beta = -2 \sin \tfrac{1}{2}(\alpha + \beta) \sin \tfrac{1}{2}(\alpha - \beta)$$

$$\tan \alpha + \tan \beta = \frac{\sin(\alpha + \beta)}{\cos \alpha \cos \beta}, \qquad \tan \alpha - \tan \beta = \frac{\sin(\alpha - \beta)}{\cos \alpha \cos \beta}$$

$$\cot \alpha + \cot \beta = \frac{\sin(\alpha + \beta)}{\sin \alpha \sin \beta}, \qquad \cot \alpha - \cot \beta = \frac{\sin(\beta - \alpha)}{\sin \alpha \sin \beta}$$

$$\frac{\sin \alpha + \sin \beta}{\sin \alpha - \sin \beta} = \frac{\tan \tfrac{1}{2}(\alpha + \beta)}{\tan \tfrac{1}{2}(\alpha - \beta)}, \qquad \frac{\sin \alpha + \sin \beta}{\cos \alpha - \cos \beta} = \cot \tfrac{1}{2}(\beta - \alpha)$$

$$\frac{\sin \alpha + \sin \beta}{\cos \alpha + \cos \beta} = \tan \tfrac{1}{2}(\alpha + \beta), \qquad \frac{\sin \alpha - \sin \beta}{\cos \alpha + \cos \beta} = \tan \tfrac{1}{2}(\alpha - \beta)$$

Half-angle relations

$$\sin \frac{\alpha}{2} = \pm \sqrt{\frac{1 - \cos \alpha}{2}}, \qquad \cos \frac{\alpha}{2} = \pm \sqrt{\frac{1 + \cos \alpha}{2}}$$

$$\tan \frac{\alpha}{2} = \pm \sqrt{\frac{1 - \cos \alpha}{1 + \cos \alpha}} = \frac{1 - \cos \alpha}{\sin \alpha} = \frac{\sin \alpha}{1 + \cos \alpha}$$

$$\cot \frac{\alpha}{2} = \pm \sqrt{\frac{1 + \cos \alpha}{1 - \cos \alpha}} = \frac{1 + \cos \alpha}{\sin \alpha} = \frac{\sin \alpha}{1 - \cos \alpha}$$

Power relations

$$\sin^2 \alpha = \tfrac{1}{2}(1 - \cos 2\alpha), \qquad \sin^3 \alpha = \tfrac{1}{4}(3 \sin \alpha - \sin 3\alpha)$$
$$\sin^4 \alpha = \tfrac{1}{8}(3 - 4 \cos 2\alpha + \cos 4\alpha)$$
$$\cos^2 \alpha = \tfrac{1}{2}(1 + \cos 2\alpha), \qquad \cos^3 \alpha = \tfrac{1}{4}(3 \cos \alpha + \cos 3\alpha)$$
$$\cos^4 \alpha = \tfrac{1}{8}(3 + 4 \cos 2\alpha + \cos 4\alpha)$$

$$\tan^2 \alpha = \frac{1 - \cos 2\alpha}{1 + \cos 2\alpha}, \qquad \cot^2 \alpha = \frac{1 + \cos 2\alpha}{1 - \cos 2\alpha}$$

Exponential relations (α in radians), Euler's equation

$$e^{i\alpha} = \cos \alpha + i \sin \alpha, \qquad i = \sqrt{-1}$$

$$\sin \alpha = \frac{e^{i\alpha} - e^{-i\alpha}}{2i}, \qquad \cos \alpha = \frac{e^{i\alpha} + e^{-i\alpha}}{2}$$

$$\tan \alpha = -i \left(\frac{e^{i\alpha} - e^{-i\alpha}}{e^{i\alpha} + e^{-i\alpha}} \right) = -i \left(\frac{e^{2i\alpha} - 1}{e^{2i\alpha} + 1} \right)$$

THE TRIGONOMETRIC FUNCTIONS IN TERMS OF ONE ANOTHER

Function	$\sin \alpha$	$\cos \alpha$	$\tan \alpha$	$\cot \alpha$	$\sec \alpha$	$\csc \alpha$
$\sin \alpha$	$\sin \alpha$	$\pm \sqrt{1 - \cos^2 \alpha}$	$\dfrac{\tan \alpha}{\pm \sqrt{1 + \tan^2 \alpha}}$	$\dfrac{1}{\pm \sqrt{1 + \cot^2 \alpha}}$	$\dfrac{\pm \sqrt{\sec^2 \alpha - 1}}{\sec \alpha}$	$\dfrac{1}{\csc \alpha}$
$\cos \alpha$	$\pm \sqrt{1 - \sin^2 \alpha}$	$\cos \alpha$	$\dfrac{1}{\pm \sqrt{1 + \tan^2 \alpha}}$	$\dfrac{\cot \alpha}{\pm \sqrt{1 + \cot^2 \alpha}}$	$\dfrac{1}{\sec \alpha}$	$\dfrac{\pm \sqrt{\csc^2 \alpha - 1}}{\csc \alpha}$
$\tan \alpha$	$\dfrac{\sin \alpha}{\pm \sqrt{1 - \sin^2 \alpha}}$	$\dfrac{\pm \sqrt{1 - \cos^2 \alpha}}{\cos \alpha}$	$\tan \alpha$	$\dfrac{1}{\cot \alpha}$	$\pm \sqrt{\sec^2 \alpha - 1}$	$\dfrac{1}{\pm \sqrt{\csc^2 \alpha - 1}}$
$\cot \alpha$	$\dfrac{\pm \sqrt{1 - \sin^2 \alpha}}{\sin \alpha}$	$\dfrac{\cos \alpha}{\pm \sqrt{1 - \cos^2 \alpha}}$	$\dfrac{1}{\tan \alpha}$	$\cot \alpha$	$\dfrac{1}{\pm \sqrt{\sec^2 \alpha - 1}}$	$\pm \sqrt{\csc^2 \alpha - 1}$
$\sec \alpha$	$\dfrac{1}{\pm \sqrt{1 - \sin^2 \alpha}}$	$\dfrac{1}{\cos \alpha}$	$\pm \sqrt{1 + \tan^2 \alpha}$	$\dfrac{\pm \sqrt{1 + \cot^2 \alpha}}{\cot \alpha}$	$\sec \alpha$	$\dfrac{\csc \alpha}{\pm \sqrt{\csc^2 \alpha - 1}}$
$\csc \alpha$	$\dfrac{1}{\sin \alpha}$	$\dfrac{1}{\pm \sqrt{1 - \cos^2 \alpha}}$	$\dfrac{\pm \sqrt{1 + \tan^2 \alpha}}{\tan \alpha}$	$\pm \sqrt{1 + \cot^2 \alpha}$	$\dfrac{\sec \alpha}{\pm \sqrt{\sec^2 \alpha - 1}}$	$\csc \alpha$

Note. The choice of sign depends upon the quadrant in which the angle terminates.

Principal Values of the Inverse Trigonometric Functions

The notation arcsin x (or $\sin^{-1} x$) is used to denote any angle whose sine is x; Arcsin x (or $\text{Sin}^{-1} x$) is usually used to denote the *principal value*. Similar notation is used for the other inverse trigonometric functions. The principal values of the inverse trigonometric functions are defined as follows:

$$-\pi/2 \leq \text{Arcsin } x \leq \pi/2, \qquad -1 \leq x \leq 1$$

$$0 \leq \text{Arccos } x \leq \pi, \qquad -1 \leq x \leq 1$$

$$-\pi/2 < \text{Arctan } x < \pi/2, \qquad -\infty < x < \infty$$

$$0 < \text{Arccsc } x \leq \pi/2, \qquad x \geq 1$$
$$-\pi < \text{Arccsc } x \leq -\pi/2, \qquad x \leq -1$$

$$0 \leq \text{Arcsec } x < \pi/2, \qquad x \geq 1$$
$$-\pi \leq \text{Arcsec } x < -\pi/2, \qquad x \leq -1$$

$$0 < \text{Arccot } x < \pi, \qquad -\infty < x < \infty$$

Note. There is no uniform agreement on the definitions of Arccsc x, Arcsec x, Arccot x for negative values of x.

Fundamental Identities Involving Principal Values

$$\text{Arcsin } x + \text{Arccos } x = \pi/2$$
$$\text{Arctan } x + \text{Arccot } x = \pi/2$$

If $\alpha = \text{Arcsin } x$, then

$$\sin \alpha = x, \qquad \cos \alpha = \sqrt{1 - x^2}, \qquad \tan \alpha = \frac{x}{\sqrt{1 - x^2}}$$

$$\csc \alpha = \frac{1}{x}, \qquad \sec \alpha = \frac{1}{\sqrt{1 - x^2}}, \qquad \cot \alpha = \frac{\sqrt{1 - x^2}}{x}$$

If $\alpha = \text{Arccos } x$, then

$$\sin \alpha = \sqrt{1 - x^2}, \qquad \cos \alpha = x, \qquad \tan \alpha = \frac{\sqrt{1 - x^2}}{x}$$

$$\csc \alpha = \frac{1}{\sqrt{1 - x^2}}, \qquad \sec \alpha = \frac{1}{x}, \qquad \cot \alpha = \frac{x}{\sqrt{1 - x^2}}$$

If $\alpha = \text{Arctan } x$, then

$$\sin \alpha = \frac{x}{\sqrt{1 + x^2}}, \qquad \cos \alpha = \frac{1}{\sqrt{1 + x^2}}, \qquad \tan \alpha = x$$

$$\csc \alpha = \frac{\sqrt{1 + x^2}}{x}, \qquad \sec \alpha = \sqrt{1 + x^2}, \qquad \cot \alpha = \frac{1}{x}$$

Relations Between Principal Values of Inverse Trigonometric Functions

The following additional relations between principal values of the inverse trigonometric functions are useful:

$$\text{Arcsin } x = \text{Arccos}(-x) - \frac{\pi}{2} = \frac{\pi}{2} - \text{Arccos } x$$

$$= -\text{Arcsin}(-x) = \text{Arctan } \frac{x}{\sqrt{1 - x^2}}$$

$$= \frac{\pi}{2} - \text{Arccot } \frac{x}{\sqrt{1 - x^2}}$$

$$\text{Arctan } x = \text{Arccot}(-x) - \frac{\pi}{2} = \frac{\pi}{2} - \text{Arccot } x$$

$$= -\text{Arctan}(-x) = \text{Arcsin } \frac{x}{\sqrt{x^2 + 1}}$$

$$= \frac{\pi}{2} - \text{Arccos } \frac{x}{\sqrt{x^2 + 1}}$$

$$\text{Arccos } x = \frac{\pi}{2} + \text{Arcsin}(-x) = \frac{\pi}{2} - \text{Arcsin } x$$

$$= \pi - \text{Arccos}(-x) = \text{Arccot } \frac{x}{\sqrt{1 - x^2}}$$

$$= \frac{\pi}{2} - \text{Arctan } \frac{x}{\sqrt{1 - x^2}}$$

$$\text{Arccot } x = \frac{\pi}{2} + \text{Arctan}(-x) = \frac{\pi}{2} - \text{Arctan } x$$

$$= \pi - \text{Arccot}(-x) = \text{Arccos } \frac{x}{\sqrt{x^2 + 1}}$$

$$= \frac{\pi}{2} - \text{Arcsin } \frac{x}{\sqrt{x^2 + 1}}$$

$$\text{Arccsc } x = \text{Arcsec } \frac{x}{\sqrt{x^2 - 1}}$$

$$\text{Arcsec } x = \text{Arccsc } \frac{x}{\sqrt{x^2 - 1}}$$

$$\cos(\text{Arcsin } x) = \sin(\text{Arccos } x) = \sqrt{1 - x^2}$$

$$\sec(\text{Arctan } x) = \csc(\text{Arccot } x) = \sqrt{x^2 + 1}$$

$$\tan(\text{Arcsec } x) = \cot(\text{Arccsc } x) = \sqrt{x^2 - 1}$$

The above are valid for x positive or negative. The following are valid only for $x \geq 0$.

$$\text{Arcsin } x = \text{Arccos } \sqrt{1 - x^2} = \text{Arccot } \frac{\sqrt{1 - x^2}}{x}$$

$$= \text{Arcsec } \frac{1}{\sqrt{1 - x^2}} = \text{Arccsc } \frac{1}{x}$$

$$= \frac{\pi}{2} - \text{Arcsin } \sqrt{1 - x^2}$$

$$\text{Arccos } x = \text{Arcsin } \sqrt{1 - x^2} = \text{Arctan } \frac{\sqrt{1 - x^2}}{x}$$

$$= \text{Arcsec } \frac{1}{x} = \text{Arccsc } \frac{1}{\sqrt{1 - x^2}}$$

$$= \frac{\pi}{2} - \text{Arccos } \sqrt{1 - x^2}$$

$$\text{Arctan } x = \text{Arccot } \frac{1}{x} = \text{Arccos } \frac{1}{\sqrt{x^2 + 1}}$$

$$= \text{Arcsec } \sqrt{x^2 + 1} = \text{Arccsc } \frac{\sqrt{x^2 + 1}}{x}$$

$$= \frac{\pi}{2} - \text{Arctan } \frac{1}{x}$$

$$\text{Arccot } x = \text{Arctan } \frac{1}{x} = \text{Arcsin } \frac{1}{\sqrt{x^2 + 1}}$$

$$= \text{Arcsec } \frac{\sqrt{x^2 + 1}}{x} = \text{Arccsc } \sqrt{x^2 + 1}$$

$$= \frac{\pi}{2} - \text{Arccot } \frac{1}{x}$$

$$\text{Arcsec } x = \text{Arctan } \sqrt{x^2 - 1} = \text{Arccot } \frac{1}{\sqrt{x^2 - 1}}$$

$$= \text{Arcsin } \frac{\sqrt{x^2 - 1}}{x} = \text{Arccos } \frac{1}{x} = \pi + \text{Arcsec}(-x)$$

$$= \frac{\pi}{2} - \text{Arccsc } x = -\frac{\pi}{2} - \text{Arccsc}(-x)$$

Formulas for Use in Trigonometry

$$\text{Arccsc } x = \text{Arctan } \frac{1}{\sqrt{x^2 - 1}} = \text{Arccot } \sqrt{x^2 - 1}$$

$$= \text{Arcsin } \frac{1}{x} = \text{Arccos } \frac{\sqrt{x^2 - 1}}{x} = \pi + \text{Arccsc}(-x)$$

$$= \frac{\pi}{2} - \text{Arcsec } x = -\frac{\pi}{2} - \text{Arcsec}(-x)$$

The following are valid if $x < 0$.

$$\text{Arcsin } x = -\text{Arccos } \sqrt{1 - x^2} = \text{Arccot } \frac{\sqrt{1 - x^2}}{x} - \pi$$

$$= -\pi - \text{Arccsc } \frac{1}{x} = -\text{Arcsec } \frac{1}{\sqrt{1 - x^2}}$$

$$= \text{Arcsin } \sqrt{1 - x^2} - \frac{\pi}{2}$$

$$\text{Arccos } x = \pi - \text{Arcsin } \sqrt{1 - x^2} = \pi + \text{Arctan } \frac{\sqrt{1 - x^2}}{x}$$

$$= \pi - \text{Arccsc } \frac{1}{\sqrt{1 - x^2}} = -\text{Arcsec } \frac{1}{x}$$

$$= \text{Arccos } \sqrt{1 - x^2} + \frac{\pi}{2}$$

$$\text{Arctan } x = \text{Arccot } \frac{1}{x} - \pi = -\text{Arccos } \frac{1}{\sqrt{x^2 + 1}}$$

$$= -\text{Arcsec } \sqrt{x^2 + 1} = -\pi - \text{Arccsc } \frac{\sqrt{x^2 + 1}}{x}$$

$$- \frac{\pi}{2} - \text{Arctan } \frac{1}{x}$$

$$\text{Arccot } x = \pi + \text{Arctan } \frac{1}{x} = \pi - \text{Arcsin } \frac{1}{\sqrt{x^2 + 1}}$$

$$= -\text{Arcsec } \frac{\sqrt{x^2 + 1}}{x} = \pi - \text{Arccsc } \sqrt{x^2 + 1}$$

$$= \frac{3\pi}{2} - \text{Arccot } \frac{1}{x}$$

$$\text{Arcsec } x = \text{Arctan } \sqrt{x^2 - 1} - \pi = \text{Arccot } \frac{1}{\sqrt{x^2 - 1}} - \pi$$

$$= -\pi - \text{Arcsin } \frac{\sqrt{x^2 - 1}}{x} = -\text{Arccos } \frac{1}{x} = \text{Arcsec}(-x) - \pi$$

$$= -\frac{3\pi}{2} - \text{Arccsc } x = -\frac{\pi}{2} - \text{Arccsc}(-x)$$

$$\text{Arccsc } x = \text{Arctan } \frac{1}{\sqrt{x^2 - 1}} - \pi = \text{Arccot } \sqrt{x^2 - 1} - \pi$$

$$= -\pi - \text{Arcsin } \frac{1}{x} = -\text{Arccos } \frac{\sqrt{x^2 - 1}}{x} = \text{Arccsc}(-x) - \pi$$

$$= -\frac{3\pi}{2} - \text{Arcsec } x = -\frac{\pi}{2} - \text{Arcsec}(-x)$$

Plane Triangle Formulas

In the following, A, B, C denote the angles of any plane triangle, a, b, c the corresponding opposite sides, and $s = \frac{1}{2}(a + b + c)$.

Radius of inscribed circle:

$$r = \sqrt{\frac{(s - a)(s - b)(s - c)}{s}}$$

Radius of circumscribed circle:

$$R = \frac{a}{2 \sin A} = \frac{b}{2 \sin B} = \frac{c}{2 \sin C}$$

Law of sines:

$$\frac{a}{\sin A} = \frac{b}{\sin B} = \frac{c}{\sin C}$$

Law of cosines:

$$a^2 = b^2 + c^2 - 2bc \cos A, \qquad \cos A = \frac{b^2 + c^2 - a^2}{2bc}$$

$$b^2 = c^2 + a^2 - 2ca \cos B, \qquad \cos B = \frac{c^2 + a^2 - b^2}{2ca}$$

$$c^2 = a^2 + b^2 - 2ab \cos C, \qquad \cos C = \frac{a^2 + b^2 - c^2}{2ab}$$

Law of tangents:

$$\frac{b - c}{b + c} = \frac{\tan \frac{1}{2}(B - C)}{\tan \frac{1}{2}(B + C)}, \qquad \frac{c - a}{c + a} = \frac{\tan \frac{1}{2}(C - A)}{\tan \frac{1}{2}(C + A)}$$

$$\frac{a - b}{a + b} = \frac{\tan \frac{1}{2}(A - B)}{\tan \frac{1}{2}(A + B)}$$

Half-angle formulas:

$$\tan \tfrac{1}{2}A = \frac{r}{s - a}, \qquad \tan \tfrac{1}{2}B = \frac{r}{s - b}, \qquad \tan \tfrac{1}{2}C = \frac{r}{s - c}$$

$$\sin \tfrac{1}{2}A = \sqrt{\frac{(s - b)(s - c)}{bc}}, \qquad \cos \tfrac{1}{2}A = \sqrt{\frac{s(s - a)}{bc}}$$

$$\sin \tfrac{1}{2}B = \sqrt{\frac{(s - c)(s - a)}{ca}}, \qquad \cos \tfrac{1}{2}B = \sqrt{\frac{s(s - b)}{ca}}$$

$$\sin \tfrac{1}{2}C = \sqrt{\frac{(s - a)(s - b)}{ab}}, \qquad \cos \tfrac{1}{2}C = \sqrt{\frac{s(s - c)}{ab}}$$

Formulas for Use in Trigonometry

Area:

$$K = \tfrac{1}{2}bc \sin A = \tfrac{1}{2}ca \sin B = \tfrac{1}{2}ab \sin C$$

$$K = \frac{a^2 \sin B \sin C}{2 \sin A} = \frac{b^2 \sin C \sin A}{2 \sin B} = \frac{c^2 \sin A \sin B}{2 \sin C}$$

$$K = \sqrt{s(s-a)(s-b)(s-c)} = rs = \frac{abc}{4R}$$

Mollweide's formulas:

$$\frac{b-c}{a} = \frac{\sin \tfrac{1}{2}(B-C)}{\cos \tfrac{1}{2}A}, \quad \frac{c-a}{b} = \frac{\sin \tfrac{1}{2}(C-A)}{\cos \tfrac{1}{2}B}$$

$$\frac{a-b}{c} = \frac{\sin \tfrac{1}{2}(A-B)}{\cos \tfrac{1}{2}C}$$

Newton's formulas:

$$\frac{b+c}{a} = \frac{\cos \tfrac{1}{2}(B-C)}{\sin \tfrac{1}{2}A}, \quad \frac{c+a}{b} = \frac{\cos \tfrac{1}{2}(C-A)}{\sin \tfrac{1}{2}B}$$

$$\frac{a+b}{c} = \frac{\cos \tfrac{1}{2}(A-B)}{\sin \tfrac{1}{2}C}$$

Solution of Right Triangles

(a) Given acute angle A and opposite leg a.

$$B = 90° - A, \quad b = a/\tan A = a \cot A, \quad c = a/\sin A = a \csc A$$

(b) Given acute angle A and adjacent leg b.

$$B = 90° - A, \quad a = b \tan A, \quad c = b/\cos A = b \sec A$$

(c) Given acute angle A and hypotenuse c.

$$B = 90° - A, \quad a = c \sin A, \quad b = c \cos A$$

(d) Given legs a and b.

$$c = \sqrt{a^2 + b^2}, \quad \tan A = a/b, \quad B = 90° - A$$

(e) Given hypotenuse c and leg a.

$$b = \sqrt{(c+a)(c-a)}, \quad \sin A = a/c, \quad B = 90° - A$$

Solution of Oblique Triangles

(a) Given sides b and c and included angle A.

Nonlogarithmic solution

$$a^2 = b^2 + c^2 - 2bc \cos A, \quad \cos B = (c^2 + a^2 - b^2)/2ca,$$
$$\cos C = (a^2 + b^2 - c^2)/2ab$$

Logarithmic solution

$$\tfrac{1}{2}(B+C) = 90° - \tfrac{1}{2}A, \quad \tan \tfrac{1}{2}(B-C) = \frac{b-c}{b+c} \tan \tfrac{1}{2}(B+C),$$
$$B = \tfrac{1}{2}(B+C) + \tfrac{1}{2}(B-C), \quad C = \tfrac{1}{2}(B+C) - \tfrac{1}{2}(B-C),$$
$$a = (b \sin A)/\sin B, \quad K = \tfrac{1}{2}bc \sin A$$

Formulas for Use in Spherical Trigonometry

Check. $A + B + C = 180°$, or use Newton's formula or law of sines.

(b) Given angles B and C and included side a.

$$A = 180° - (B + C), \quad b = (a \sin B)/\sin A,$$
$$c = (a \sin C)/\sin A, \quad K = \frac{a^2 \sin B \sin C}{2 \sin A}$$

Check. $a = b \cos C + c \cos B$, or use Newton's formula or law of tangents.

(c) Given sides a and c and opposite angle A.

$$\sin C = (c \sin A)/a, \quad B = 180° - (A + C),$$
$$b = (a \sin B)/\sin A, \quad K = \tfrac{1}{2}ac \sin B$$

Check. $a = b \cos C + c \cos B$, or use Newton's formula or law of tangents.

Note. In this case there may be two solutions, for C may have two values:

$C_1 < 90°$ and $C_2 = 180° - C_1 > 90°$. If $A + C_2 > 180°$, use only C_1.

(d) Given the three sides a, b, c.

Nonlogarithmic solution

$$\cos A = (b^2 + c^2 - a^2)/2bc, \quad \cos B = (c^2 + a^2 - b^2)/2ca,$$
$$\cos C = (a^2 + b^2 - c^2)/2ab$$

Logarithmic solution

$$s = \tfrac{1}{2}(a + b + c), \quad r = \sqrt{\frac{(s - a)(s - b)(s - c)}{s}},$$
$$\tan \tfrac{1}{2}A - \frac{r}{s - a}, \quad \tan \tfrac{1}{2}B - \frac{r}{s - b}, \quad \tan \tfrac{1}{2}C = \frac{r}{s - c},$$
$$K = \sqrt{s(s - a)(s - b)(s - c)}$$

Check. $A + B + C = 180°$.

Relations Between Accuracy of Computed Lengths and Angles

When solving a triangle for any of its parts, the following should be observed:

Significant figures for sides	Angles to the nearest
2	degree
3	ten minutes
4	minute
5	tenth of a minute

SPHERICAL TRIGONOMETRY

Right Spherical Triangles

Let a, b, c be the sides of a right spherical triangle with opposite angles $A, B, C = 90°$, respectively, where each side is measured by the angle subtended at the center of the sphere.

$\sin a = \tan b \cot B$,	$\sin a = \sin A \sin c$
$\sin b = \tan a \cot A$,	$\sin b = \sin B \sin c$
$\cos A = \tan b \cot c$,	$\cos A = \cos a \sin B$
$\cos B = \tan a \cot c$,	$\cos B = \cos b \sin A$
$\cos c = \cot A \cot B$,	$\cos c = \cos a \cos b$

Formulas for Use in Spherical Trigonometry

Napier's Rules of Circular Parts

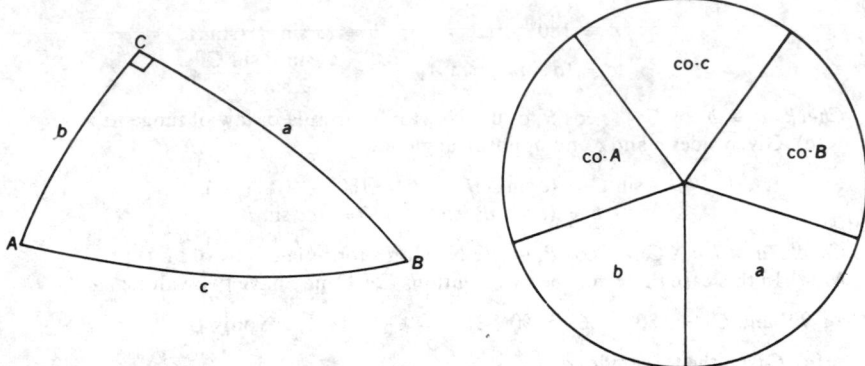

Arrange the five quantities a, b, co-A (complement of A), co-c, co-B of a right spherical triangle right-angled at C in cyclic order as pictured. If any one of these quantities is designated a *middle* part, then two of the other parts are *adjacent* to it, and the remaining two parts are *opposite* to it. The above formulas for a right spherical triangle may be recalled by the following two rules:

(a) The sine of any middle part is equal to the product of the *tangents* of the two *adjacent* parts.

(b) The sine of any middle part is equal to the product of the *cosines* of the two *opposite* parts.

Rules for Determining the Quadrant of a Calculated Part of a Right Spherical Triangle

(a) A leg and the angle opposite it are always of the same quadrant.

(b) If the hypotenuse if less than 90° the legs are of the same quadrant.

(c) If the hypotenuse is greater than 90°, the legs are of unlike quadrants.

Oblique Spherical Triangles

In the following, a, b, c represent the sides of any spherical triangle, A, B, C the corresponding opposite angles, $s = \frac{1}{2}(a + b + c)$, $S = \frac{1}{2}(A + B + C)$, Δ = area of triangle, E = spherical excess of triangle, R = radius of the sphere upon which the triangle lies, and a', b', c', A', B', C' are the corresponding parts of the polar triangle.

$$0° < a + b + c < 360°, \qquad 180° < A + B + C < 540°$$
$$E = A + B + C - 180°, \qquad \Delta = \pi R^2 E / 180$$
$$\tan \tfrac{1}{4}E = \sqrt{\tan \tfrac{s}{2} \tan \tfrac{1}{2}(s - a) \tan \tfrac{1}{2}(s - b) \tan \tfrac{1}{2}(s - c)}$$
$$A = 180° - a', \qquad B = 180° - b', \qquad C = 180° - c'$$
$$a = 180° - A', \qquad b = 180° - B', \qquad c = 180° - C'$$

Law of sines:

$$\frac{\sin a}{\sin A} = \frac{\sin b}{\sin B} = \frac{\sin c}{\sin C}$$

Law of cosines for sides:

$$\cos a = \cos b \cos c + \sin b \sin c \cos A$$
$$\cos b = \cos c \cos a + \sin c \sin a \cos B$$
$$\cos c = \cos a \cos b + \sin a \sin b \cos C$$

Formulas for Use in Spherical Trigonometry

Law of cosines for angles:

$$\cos A = -\cos B \cos C + \sin B \sin C \cos a$$
$$\cos B = -\cos C \cos A + \sin C \sin A \cos b$$
$$\cos C = -\cos A \cos B + \sin A \sin B \cos c$$

Law of tangents:

$$\frac{\tan \frac{1}{2}(B - C)}{\tan \frac{1}{2}(B + C)} = \frac{\tan \frac{1}{2}(b - c)}{\tan \frac{1}{2}(b + c)}, \quad \frac{\tan \frac{1}{2}(C - A)}{\tan \frac{1}{2}(C + A)} = \frac{\tan \frac{1}{2}(c - a)}{\tan \frac{1}{2}(c + a)}$$

$$\frac{\tan \frac{1}{2}(A - B)}{\tan \frac{1}{2}(A + B)} = \frac{\tan \frac{1}{2}(a - b)}{\tan \frac{1}{2}(a + b)}$$

Half-angle formulas:

$$\tan \frac{1}{2} A = \frac{k}{\sin(s - a)}, \quad \tan \frac{1}{2} B = \frac{k}{\sin(s - b)}, \quad \tan \frac{1}{2} C = \frac{k}{\sin(s - c)},$$

where

$$k^2 = \frac{\sin(s - a)\sin(s - b)\sin(s - c)}{\sin s} = (\tan r)^2$$

Half-side formulas:

$$\tan \frac{1}{2}\alpha = K \cos(S - A), \quad \tan \frac{1}{2}b = K \cos(S - B),$$
$$\tan \frac{1}{2}c = K \cos(S - C),$$

where

$$K^2 = \frac{-\cos S}{\cos(S - A)\cos(S - B)\cos(S - C)} = (\tan R)^2$$

Gauss's formulas:

$$\frac{\sin \frac{1}{2}(a - b)}{\sin \frac{1}{2}c} = \frac{\sin \frac{1}{2}(A - B)}{\cos \frac{1}{2}C}, \quad \frac{\cos \frac{1}{2}(a - b)}{\cos \frac{1}{2}c} = \frac{\sin \frac{1}{2}(A + B)}{\cos \frac{1}{2}C}$$

$$\frac{\sin \frac{1}{2}(a + b)}{\sin \frac{1}{2}c} = \frac{\cos \frac{1}{2}(A - B)}{\sin \frac{1}{2}C}, \quad \frac{\cos \frac{1}{2}(a + b)}{\cos \frac{1}{2}c} = \frac{\cos \frac{1}{2}(A + B)}{\sin \frac{1}{2}C}$$

Napier's analogies:

$$\frac{\sin \frac{1}{2}(A - B)}{\sin \frac{1}{2}(A + B)} = \frac{\tan \frac{1}{2}(a - b)}{\tan \frac{1}{2}c}, \quad \frac{\sin \frac{1}{2}(a - b)}{\sin \frac{1}{2}(a + b)} = \frac{\tan \frac{1}{2}(A - B)}{\cot \frac{1}{2}C}$$

$$\frac{\cos \frac{1}{2}(A - B)}{\cos \frac{1}{2}(A + B)} = \frac{\tan \frac{1}{2}(a + b)}{\tan \frac{1}{2}c}, \quad \frac{\cos \frac{1}{2}(a - b)}{\cos \frac{1}{2}(a + b)} = \frac{\tan \frac{1}{2}(A + B)}{\cot \frac{1}{2}C}$$

Haversine formulas:

$$\text{hav } a = \text{hav}(b - c) + \sin b \sin c \text{ hav } A$$

$$\text{hav } A = \frac{\sin(s - b)\sin(s - c)}{\sin b \sin c}$$

$$= \frac{\text{hav } a - \text{hav}(b - c)}{\sin b \sin c}$$

$$= \text{hav}[180° - (B + C)] + \sin B \sin C \text{ hav } a$$

148   *TRIGONOMETRY*

Formulas for Use in Spherical Trigonometry

Rules for Determining the Quadrant of a Calculated Part of an Oblique Spherical Triangle

(a) If $A > B > C$, then $a > b > c$.

(b) A side (angle) which differs more from 90° than does another side (angle) is in the same quadrant as its opposite angle (side).

(c) Half the sum of any two sides and half the sum of the opposite angles are in the same quadrant.

SUMMARY OF SOLUTION OF OBLIQUE SPHERICAL TRIANGLES

Given	Solution	Check
Three sides	Half-angle formulas.	Law of sines
Three angles	Half-side formulas.	Law of sines
Two sides and included angle	Napier's analogies (to find sum and difference of unknown angles), then law of sines (to find remaining side).	Gauss's formula
Two angles and included side	Napier's analogies (to find sum and difference of unknown sides), then law of sines (to find remaining angle).	Gauss's formula
Two sides and an opposite angle	Law of sines (to find an angle), then Napier's analogies (to find remaining angle and side). Note number of solutions.	Gauss's formula
Two angles and an opposite side	Law of sines (to find a side), then Napier's analogies (to find remaining side and angle). Note number of solutions.	Gauss's formula

DEGREES, MINUTES, AND SECONDS TO RADIANS

Units in degrees, minutes or seconds	Degrees to Radians	Minutes to Radians	Seconds to Radians
10	0.174 5329	0.002 9089	0.000 0485
20	0.349 0659	0.005 8178	0.000 0970
30	0.523 5988	0.008 7266	0.000 1454
40	0.698 1317	0.011 6355	0.000 1939
50	0.872 6646	0.014 5444	0.000 2424
60	1.047 1976	0.017 4533	0.000 2909
70	1.221 7305	(0.020 3622)	(0.000 3394)
80	1.396 2634	(0.023 2711)	(0.000 3879)
90	1.570 7963	(0.026 1799)	(0.000 4363)
100	1.745 3293	....	....
200	3.490 6585	....	....
300	5.235 9878	....	....

where n = 1, 2, 3, 4, etc. n (100°) = n (1.745 3293)

RADIANS TO DEGREES, MINUTES, AND SECONDS

Radians	1.0	0.1	0.01	0.001	0.0001
1	57° 17′ 44.8″	5° 43′ 46.5″	0° 34′ 22.6″	0° 03′ 26.3″	0° 00′ 20.6″
2	114° 35′ 29.6″	11° 27′ 33.0″	1° 08′ 45.3″	0° 06′ 52.5″	0° 00′ 41.3″
3	171° 53′ 14.4″	17° 11′ 19.4″	1° 43′ 07.9′	0° 10′ 18.8″	0° 01′ 01.9″
4	229° 10′ 59.2″	22° 55′ 05.9″	2° 17′ 30.6″	0° 13′ 45.1′	0° 01′ 22.5″
5	286° 28′ 44.0″	28° 38′ 52.4″	2° 51′ 53.2″	0° 17′ 11.3″	0° 01′ 43.1″
6	343° 46′ 28.8″	34° 22′ 38.9″	3° 26′ 15.9″	0° 20′ 37.6″	0° 02′ 03.8″
7	401° 04′ 13.6″	40° 06′ 25.4″	4° 00′ 38.5″	0° 24′ 03.9″	0° 02′ 24.4″
8	458° 21′ 58.4″	45° 50′ 11.8″	4° 35′ 01.2″	0° 27′ 30.1″	0° 02′ 45.0″
9	515° 39′ 43.3″	51° 33′ 58.3″	5° 09′ 23.8″	0° 30′ 56.4″	0° 03′ 05.6″

Example: If 3.214 is desired in degrees, minutes and seconds it is obtained as follows:

$$3 = 171° \ 53′ \ 14.4″$$
$$.2 = \ \ 11° \ 27′ \ 33.0″$$
$$.01 = \ \ \ \ 0° \ 34′ \ 22.6″$$
$$.004 = \ \ \ \ 0° \ 13′ \ 45.1″$$
$$\overline{3.214 = 184° \ \ 8′ \ 55.1″}$$

MILS—RADIANS—DEGREES

1 mil = 0.00098175 radians = 0.05625° = 3.375′ = 202.5″
1000 mils = 0.98175 radians = 56.25°
6400 mils = 360° = 2π radians
1 radian = 1018.6 mils
1° = 17.777778 mils
1′ = 0.296296 mils
1″ = 0.0049383 mils

DEGREES—RADIANS

1 radian = 57° 17′ 44″ .80625

		log
1 radian =	57.29577 95131 degrees	1.75812 26324
1 radian =	3437.74677 07849 minutes	3.53627 38828
1 radian =	206264.80625 seconds	5.31442 51332
1 degree =	0.01745 32925 19943 radians	8.24187 73676—10
1 minute =	0.00029 08882 08666 radians	6.46372 61172—10
1 second =	0.00000 48481 36811 radians	4.68557 48668—10

DEGREES AND DECIMAL FRACTIONS TO RADIANS

The table below facilitates conversion of an angle expressed in degrees and decimal fractions into radians. To convert 25.78 into radians, find the equivalents, successively, of 20°, 5°, 0°.7, 0°.08 and add.

Deg.	Radians	Deg.	Radians	Deg.	Radians	Deg.	Radians	Deg.	Radians
10	0.174533	1	0.017453	0.1	0.001745	0.01	0.000175	0.001	0.000017
20	0.349066	2	.034907	.2	.003491	.02	.000349	.002	.000035
30	0.523599	3	.052360	.3	.005236	.03	.000524	.003	.000052
40	0.698132	4	.069813	.4	.006981	.04	.000698	.004	.000070
50	0.872665	5	.087266	.5	.008727	.05	.000873	.005	.000087
60	1.047198	6	.104720	.6	.010472	.06	.001047	.006	.000105
70	1.221730	7	.122173	.7	.012217	.07	.001222	.007	.000122
80	1.396263	8	.139626	.8	.013963	.08	.001396	.008	.000140
90	1.570796	9	.157080	.9	.015708	.09	.001571	.009	.000157

RADIANS TO DEGREES AND DECIMALS

Radians	Degrees	Radians	Degrees	Radians	Degrees	Radians	Degrees
1	57.2958	0.1	5.7296	0.01	0.5730	0.001	0.0573
2	114.5916	.2	11.4592	.02	1.1459	.002	.1146
3	171.8873	.3	17.1887	.03	1.7189	.003	.1719
4	229.1831	.4	22.9183	.04	2.2918	.004	.2292
5	286.4789	.5	28.6479	.05	2.8648	.005	.2865
6	343.7747	.6	34.3775	.06	3.4377	.006	.3438
7	401.0705	.7	40.1070	.07	4.0107	.007	.4011
8	458.3662	.8	45.8366	.08	4.5837	.008	.4584
9	515.6620	.9	51.5662	.09	5.1566	.009	.5157
10	572.9578	1.0	57.2958	.10	5.7296	.010	.5730

RADIANS—DEGREES
Multiples and Fractions of π Radians in Degrees

Radians	Radians	Deg.	Radians	Radians	Deg.	Radians	Radians	Deg.
π	3.1416	180	π/2	1.5708	90	2π/3	2.0944	120
2π	6.2832	360	π/3	1.0472	60	3π/4	2.3562	135
3π	9.4248	540	π/4	0.7854	45	5π/6	2.6180	150
4π	12.5664	720	π/5	0.6283	36	7π/6	3.6652	210
5π	15.7080	900	π/6	0.5236	30	5π/4	3.9270	225
6π	18.8496	1080	π/7	0.4488	25.714	4π/3	4.1888	240
7π	21.9911	1260	π/8	0.3927	22.5	3π/2	4.7124	270
8π	25.1327	1440	π/9	0.3491	20	5π/3	5.2360	300
9π	28.2743	1620	π/10	0.3142	18	7π/4	5.4978	315
10π	31.4159	1800	π/12	0.2618	15	11π/6	5.7596	330

CONVERSION OF ANGLES FROM ARC TO TIME

Arc	Time	Arc	Time	Arc	Time	Arc	Time
°	h m	°	h m	″	s	″	s
′	m s	′	m s				
0	0 00	20	1 20	0	0.00	8	0.53
1	0 04	30	2 00	1	0.07	9	0.60
2	0 08	40	2 40	2	0.13	10	0.67
3	0 12	50	3 20	3	0.20	20	1.33
4	0 16	60	4 00	4	0.27	30	2.00
5	0 20	70	4 40	5	0.33	40	2.67
6	0 24	80	5 20	6	0.40	50	3.33
7	0 28	90	6 00	7	0.47	60	4.00
8	0 32	100	6 40				
9	0 36	200	13 20				
10	0 40	300	20 00				

NATURAL TRIGONOMETRIC FUNCTIONS TO FIVE PLACES

For degrees indicated at the top of the page use the top column headings. For degrees indicated at the bottom use the column indications at the bottom. With degrees at the left of each block (top or bottom), use the minute column at the left, with degrees at the right of each block, use the minute column at the right. Appropriate signs for the functions must be supplied in accordance with the quadrant in which the angle measure belongs. Linear interpolation may be used except in regions where the functions are rapidly changing. See "Use of Logarithm Tables" for further discussion in interpolation practice.

0° (180°) (359)° 179°

′	Sin	Tan	Cot	Cos	Sec	Csc	′
0	.00000	.00000	——	1.0000	1.0000	——	60
1	.00029	.00029	3437.7	1.0000	1.0000	3437.7	59
2	.00058	.00058	1718.9	1.0000	1.0000	1718.9	58
3	.00087	.00087	1145.9	1.0000	1.0000	1145.9	57
4	.00116	.00116	859.44	1.0000	1.0000	859.44	56
5	.00145	.00145	687.55	1.0000	1.0000	687.55	55
6	.00175	.00175	572.96	1.0000	1.0000	572.96	54
7	.00204	.00204	491.11	1.0000	1.0000	491.11	53
8	.00233	.00233	429.72	1.0000	1.0000	429.72	52
9	.00262	.00262	381.97	1.0000	1.0000	381.97	51
10	.00291	.00291	343.77	1.0000	1.0000	343.78	50
11	.00320	.00320	312.52	.99999	1.0000	312.52	49
12	.00349	.00349	286.48	.99999	1.0000	286.48	48
13	.00378	.00378	264.44	.99999	1.0000	264.44	47
14	.00407	.00407	245.55	.99999	1.0000	245.55	46
15	.00436	.00436	229.18	.99999	1.0000	229.18	45
16	.00465	.00465	214.86	.99999	1.0000	214.86	44
17	.00495	.00495	202.22	.99999	1.0000	202.22	43
18	.00524	.00524	190.98	.99999	1.0000	190.99	42
19	.00553	.00553	180.93	.99998	1.0000	180.93	41
20	.00582	.00582	171.89	.99998	1.0000	171.89	40
21	.00611	.00611	163.70	.99998	1.0000	163.70	39
22	.00640	.00640	156.26	.99998	1.0000	156.26	38
23	.00669	.00669	149.47	.99998	1.0000	149.47	37
24	.00698	.00698	143.24	.99998	1.0000	143.24	36
25	.00727	.00727	137.51	.99997	1.0000	137.51	35
26	.00756	.00756	132.22	.99997	1.0000	132.22	34
27	.00785	.00785	127.32	.99997	1.0000	127.33	33
28	.00814	.00815	122.77	.99997	1.0000	122.78	32
29	.00844	.00844	118.54	.99996	1.0000	118.54	31
30	.00873	.00873	114.59	.99996	1.0000	114.59	30
31	.00902	.00902	110.89	.99996	1.0000	110.90	29
32	.00931	.00931	107.43	.99996	1.0000	107.43	28
33	.00960	.00960	104.17	.99995	1.0000	104.18	27
34	.00989	.00989	101.11	.99995	1.0000	101.11	26
35	.01018	.01018	98.218	.99995	1.0001	98.223	25
36	.01047	.01047	95.489	.99995	1.0001	95.495	24
37	.01076	.01076	92.908	.99994	1.0001	92.914	23
38	.01105	.01105	90.463	.99994	1.0001	90.469	22
39	.01134	.01135	88.144	.99994	1.0001	88.149	21
40	.01164	.01164	85.940	.99993	1.0001	85.946	20
41	.01193	.01193	83.844	.99993	1.0001	83.849	19
42	.01222	.01222	81.847	.99993	1.0001	81.853	18
43	.01251	.01251	79.943	.99992	1.0001	79.950	17
44	.01280	.01280	78.126	.99992	1.0001	78.133	16
45	.01309	.01309	76.390	.99991	1.0001	76.397	15
46	.01338	.01338	74.729	.99991	1.0001	74.736	14
47	.01367	.01367	73.139	.99991	1.0001	73.146	13
48	.01396	.01396	71.615	.99990	1.0001	71.622	12
49	.01425	.01425	70.153	.99990	1.0001	70.160	11
50	.01454	.01455	68.750	.99989	1.0001	68.757	10
51	.01483	.01484	67.402	.99989	1.0001	67.409	9
52	.01513	.01513	66.105	.99989	1.0001	66.113	8
53	.01542	.01542	64.858	.99988	1.0001	64.866	7
54	.01571	.01571	63.657	.99988	1.0001	63.665	6
55	.01600	.01600	62.499	.99987	1.0001	62.507	5
56	.01629	.01629	61.383	.99987	1.0001	61.391	4
57	.01658	.01658	60.306	.99986	1.0001	60.314	3
58	.01687	.01687	59.266	.99986	1.0001	59.274	2
59	.01716	.01716	58.261	.99985	1.0001	58.270	1
60	.01745	.01746	57.290	.99985	1.0002	57.299	0
′	Cos	Cot	Tan	Sin	Csc	Sec	′

1° (181°) (358)° 178°

′	Sin	Tan	Cot	Cos	Sec	Csc	′
0	.01745	.01746	57.290	.99985	1.0002	57.299	60
1	.01774	.01775	56.351	.99984	1.0002	56.359	59
2	.01803	.01804	55.442	.99984	1.0002	55.451	58
3	.01832	.01833	54.561	.99983	1.0002	54.570	57
4	.01862	.01862	53.709	.99983	1.0002	53.718	56
5	.01891	.01891	52.882	.99982	1.0002	52.892	55
6	.01920	.01920	52.081	.99982	1.0002	52.090	54
7	.01949	.01949	51.303	.99981	1.0002	51.313	53
8	.01978	.01978	50.549	.99980	1.0002	50.558	52
9	.02007	.02007	49.816	.99980	1.0002	49.826	51
10	.02036	.02036	49.104	.99979	1.0002	49.114	50
11	.02065	.02066	48.412	.99979	1.0002	48.422	49
12	.02094	.02095	47.740	.99978	1.0002	47.750	48
13	.02123	.02124	47.085	.99977	1.0002	47.096	47
14	.02152	.02153	46.449	.99977	1.0002	46.460	46
15	.02181	.02182	45.829	.99976	1.0002	45.840	45
16	.02211	.02211	45.226	.99976	1.0002	45.237	44
17	.02240	.02240	44.639	.99975	1.0003	44.650	43
18	.02269	.02269	44.066	.99974	1.0003	44.077	42
19	.02298	.02298	43.508	.99974	1.0003	43.520	41
20	.02327	.02328	42.964	.99973	1.0003	42.976	40
21	.02356	.02357	42.433	.99972	1.0003	42.445	39
22	.02385	.02386	41.916	.99972	1.0003	41.928	38
23	.02414	.02415	41.411	.99971	1.0003	41.423	37
24	.02443	.02444	40.917	.99970	1.0003	40.930	36
25	.02472	.02473	40.436	.99969	1.0003	40.448	35
26	.02501	.02502	39.965	.99969	1.0003	39.978	34
27	.02530	.02531	39.506	.99968	1.0003	39.519	33
28	.02560	.02560	39.057	.99967	1.0003	39.070	32
29	.02589	.02589	38.618	.99966	1.0003	38.631	31
30	.02618	.02619	38.188	.99966	1.0003	38.202	30
31	.02647	.02648	37.769	.99965	1.0004	37.782	29
32	.02676	.02677	37.358	.99964	1.0004	37.371	28
33	.02705	.02706	36.956	.99963	1.0004	36.970	27
34	.02734	.02735	36.563	.99963	1.0004	36.576	26
35	.02763	.02764	36.178	.99962	1.0004	36.191	25
36	.02792	.02793	35.801	.99961	1.0004	35.815	24
37	.02821	.02822	35.431	.99960	1.0004	35.445	23
38	.02850	.02851	35.070	.99959	1.0004	35.084	22
39	.02879	.02881	34.715	.99959	1.0004	34.730	21
40	.02908	.02910	34.368	.99958	1.0004	34.382	20
41	.02938	.02939	34.027	.99957	1.0004	34.042	19
42	.02967	.02968	33.694	.99956	1.0004	33.708	18
43	.02996	.02997	33.366	.99955	1.0004	33.381	17
44	.03025	.03026	33.045	.99954	1.0005	33.060	16
45	.03054	.03055	32.730	.99953	1.0005	32.746	15
46	.03083	.03084	32.421	.99952	1.0005	32.437	14
47	.03112	.03114	32.118	.99952	1.0005	32.134	13
48	.03141	.03143	31.821	.99951	1.0005	31.836	12
49	.03170	.03172	31.528	.99950	1.0005	31.544	11
50	.03199	.03201	31.242	.99949	1.0005	31.258	10
51	.03228	.03230	30.960	.99948	1.0005	30.976	9
52	.03257	.03259	30.683	.99947	1.0005	30.700	8
53	.03286	.03288	30.412	.99946	1.0005	30.428	7
54	.03316	.03317	30.145	.99945	1.0006	30.161	6
55	.03345	.03346	29.882	.99944	1.0006	29.899	5
56	.03374	.03376	29.624	.99943	1.0006	29.641	4
57	.03403	.03405	29.371	.99942	1.0006	29.388	3
58	.03432	.03434	29.122	.99941	1.0006	29.139	2
59	.03461	.03463	28.877	.99940	1.0006	28.894	1
60	.03490	.03492	28.636	.99939	1.0006	28.654	0
′	Cos	Cot	Tan	Sin	Csc	Sec	′

TRIGONOMETRY

NATURAL TRIGONOMETRIC FUNCTIONS TO FIVE PLACES (continued)

2° (182°)   (357°) 177°   3° (183°)   (356°) 176°

'	Sin	Tan	Cot	Cos	Sec	Csc	'		'	Sin	Tan	Cot	Cos	Sec	Csc	'
0	.03490	.03492	28.636	.99939	1.0006	28.654	60		0	.05234	.05241	19.081	.99863	1.0014	19.107	60
1	.03519	.03521	28.399	.99938	1.0006	28.417	59		1	.05263	.05270	18.976	.99861	1.0014	19.002	59
2	.03548	.03550	28.166	.99937	1.0006	28.184	58		2	.05292	.05299	18.871	.99860	1.0014	18.898	58
3	.03577	.03579	27.937	.99936	1.0006	27.955	57		3	.05321	.05328	18.768	.99858	1.0014	18.794	57
4	.03606	.03609	27.712	.99935	1.0007	27.730	56		4	.05350	.05357	18.666	.99857	1.0014	18.692	56
5	.03635	.03638	27.490	.99934	1.0007	27.508	55		5	.05379	.05387	18.564	.99855	1.0014	18.591	55
6	.03664	.03667	27.271	.99933	1.0007	27.290	54		6	.05408	.05416	18.464	.99854	1.0015	18.492	54
7	.03693	.03696	27.057	.99932	1.0007	27.075	53		7	.05437	.05445	18.366	.99852	1.0015	18.393	53
8	.03723	.03725	26.845	.99931	1.0007	26.864	52		8	.05466	.05474	18.268	.99851	1.0015	18.295	52
9	.03752	.03754	26.637	.99930	1.0007	26.655	51		9	.05495	.05503	18.171	.99849	1.0015	18.198	51
10	.03781	.03783	26.432	.99929	1.0007	26.451	50		10	.05524	.05533	18.075	.99847	1.0015	18.103	50
11	.03810	.03812	26.230	.99927	1.0007	26.249	49		11	.05553	.05562	17.980	.99846	1.0015	18.008	49
12	.03839	.03842	26.031	.99926	1.0007	26.050	48		12	.05582	.05591	17.886	.99844	1.0016	17.914	48
13	.03868	.03871	25.835	.99925	1.0007	25.854	47		13	.05611	.05620	17.793	.99842	1.0016	17.822	47
14	.03897	.03900	25.642	.99924	1.0008	25.661	46		14	.05640	.05649	17.702	.99841	1.0016	17.730	46
15	.03926	.03929	25.452	.99923	1.0008	25.471	45		15	.05669	.05678	17.611	.99839	1.0016	17.639	45
16	.03955	.03958	25.264	.99922	1.0008	25.284	44		16	.05698	.05708	17.521	.99838	1.0016	17.549	44
17	.03984	.03987	25.080	.99921	1.0008	25.100	43		17	.05727	.05737	17.431	.99836	1.0016	17.460	43
18	.04013	.04016	24.898	.99919	1.0008	24.918	42		18	.05756	.05766	17.343	.99833	1.0017	17.372	42
19	.04042	.04046	24.719	.99918	1.0008	24.739	41		19	.05785	.05795	17.256	.99833	1.0017	17.285	41
20	.04071	.04075	24.542	.99917	1.0008	24.562	40		20	.05814	.05824	17.169	.99831	1.0017	17.198	40
21	.04100	.04104	24.368	.99916	1.0008	24.388	39		21	.05844	.05854	17.084	.99829	1.0017	17.113	39
22	.04129	.04133	24.196	.99915	1.0009	24.216	38		22	.05873	.05883	16.999	.99827	1.0017	17.028	38
23	.04159	.04162	24.026	.99913	1.0009	24.047	37		23	.05902	.05912	16.915	.99826	1.0017	16.945	37
24	.04188	.04191	23.859	.99912	1.0009	23.880	36		24	.05931	.05941	16.832	.99824	1.0018	16.862	36
25	.04217	.04220	23.695	.99911	1.0009	23.716	35		25	.05960	.05970	16.750	.99822	1.0018	16.779	35
26	.04246	.04250	23.532	.99910	1.0009	23.553	34		26	.05989	.05999	16.668	.99821	1.0018	16.698	34
27	.04275	.04279	23.372	.99909	1.0009	23.393	33		27	.06018	.06029	16.587	.99819	1.0018	16.618	33
28	.04304	.04308	23.214	.99907	1.0009	23.235	32		28	.06047	.06058	16.507	.99817	1.0018	16.538	32
29	.04333	.04337	23.058	.99906	1.0009	23.079	31		29	.06076	.06087	16.428	.99815	1.0019	16.459	31
30	.04362	.04366	22.904	.99905	1.0010	22.926	30		30	.06105	.06116	16.350	.99813	1.0019	16.380	30
31	.04391	.04395	22.752	.99904	1.0010	22.774	29		31	.06134	.06145	16.272	.99812	1.0019	16.303	29
32	.04420	.04424	22.602	.99902	1.0010	22.624	28		32	.06163	.06175	16.195	.99810	1.0019	16.226	28
33	.04449	.04454	22.454	.99901	1.0010	22.476	27		33	.06192	.06204	16.119	.99808	1.0019	16.150	27
34	.04478	.04483	22.308	.99900	1.0010	22.330	26		34	.06221	.06233	16.043	.99806	1.0019	16.075	26
35	.04507	.04512	22.164	.99898	1.0010	22.187	25		35	.06250	.06262	15.969	.99804	1.0020	16.000	25
36	.04536	.04541	22.022	.99897	1.0010	22.044	24		36	.06279	.06291	15.895	.99803	1.0020	15.926	24
37	.04565	.04570	21.881	.99896	1.0010	21.904	23		37	.06308	.06321	15.821	.99801	1.0020	15.853	23
38	.04594	.04599	21.743	.99894	1.0011	21.766	22		38	.06337	.06350	15.748	.99799	1.0020	15.780	22
39	.04623	.04628	21.606	.99893	1.0011	21.629	21		39	.06366	.06379	15.676	.99797	1.0020	15.708	21
40	.04653	.04658	21.470	.99892	1.0011	21.494	20		40	.06395	.06408	15.605	.99795	1.0021	15.637	20
41	.04682	.04687	21.337	.99891	1.0011	21.360	19		41	.06424	.06438	15.534	.99793	1.0021	15.566	19
42	.04711	.04716	21.205	.99889	1.0011	21.229	18		42	.06453	.06467	15.464	.99792	1.0021	15.496	18
43	.04740	.04745	21.075	.99888	1.0011	21.098	17		43	.06482	.06496	15.394	.99790	1.0021	15.427	17
44	.04769	.04774	20.946	.99886	1.0011	20.970	16		44	.06511	.06525	15.325	.99788	1.0021	15.358	16
45	.04798	.04803	20.819	.99885	1.0012	20.843	15		45	.06540	.06554	15.257	.99786	1.0021	15.290	15
46	.04827	.04833	20.693	.99883	1.0012	20.717	14		46	.06569	.06584	15.189	.99784	1.0022	15.222	14
47	.04856	.04862	20.569	.99882	1.0012	20.593	13		47	.06598	.06613	15.122	.99782	1.0022	15.155	13
48	.04885	.04891	20.446	.99881	1.0012	20.471	12		48	.06627	.06642	15.056	.99780	1.0022	15.089	12
49	.04914	.04920	20.325	.99879	1.0012	20.350	11		49	.06656	.06671	14.990	.99778	1.0022	15.023	11
50	.04943	.04949	20.206	.99878	1.0012	20.230	10		50	.06685	.06700	14.924	.99776	1.0022	14.958	10
51	.04972	.04978	20.087	.99876	1.0012	20.112	9		51	.06714	.06730	14.860	.99774	1.0023	14.893	9
52	.05001	.05007	19.970	.99875	1.0013	19.995	8		52	.06743	.06759	14.795	.99772	1.0023	14.829	8
53	.05030	.05037	19.855	.99873	1.0013	19.880	7		53	.06773	.06788	14.732	.99770	1.0023	14.766	7
54	.05059	.05066	19.740	.99872	1.0013	19.766	6		54	.06802	.06817	14.669	.99768	1.0023	14.703	6
55	.05088	.05095	19.627	.99870	1.0013	19.653	5		55	.06831	.06847	14.606	.99766	1.0023	14.640	5
56	.05117	.05124	19.516	.99869	1.0013	19.541	4		56	.06860	.06876	14.544	.99764	1.0024	14.578	4
57	.05146	.05153	19.405	.99867	1.0013	19.431	3		57	.06889	.06905	14.482	.99762	1.0024	14.517	3
58	.05175	.05182	19.296	.99866	1.0013	19.322	2		58	.06918	.06934	14.421	.99760	1.0024	14.456	2
59	.05205	.05212	19.188	.99864	1.0014	19.214	1		59	.06947	.06963	14.361	.99758	1.0024	14.395	1
60	.05234	.05241	19.081	.99863	1.0014	19.107	0		60	.06976	.06993	14.301	.99756	1.0024	14.336	0
'	Cos	Cot	Tan	Sin	Csc	Sec	'		'	Cos	Cot	Tan	Sin	Csc	Sec	'

92° (272°)   (267°) 87°   93° (273°)   (266°) 86°

NATURAL TRIGONOMETRIC FUNCTIONS TO FIVE PLACES (continued)

4° (184°)   (355°) 175°

′	Sin	Tan	Cot	Cos	Sec	Csc	′
0	.06976	.06993	14.301	.99756	1.0024	14.336	60
1	.07005	.07022	14.241	.99754	1.0025	14.276	59
2	.07034	.07051	14.182	.99752	1.0025	14.217	58
3	.07063	.07080	14.124	.99750	1.0025	14.159	57
4	.07092	.07110	14.065	.99748	1.0025	14.101	56
5	.07121	.07139	14.008	.99746	1.0025	14.044	55
6	.07150	.07168	13.951	.99744	1.0026	13.987	54
7	.07179	.07197	13.894	.99742	1.0026	13.930	53
8	.07208	.07227	13.838	.99740	1.0026	13.874	52
9	.07237	.07256	13.782	.99738	1.0026	13.818	51
10	.07266	.07285	13.727	.99736	1.0027	13.763	50
11	.07295	.07314	13.672	.99734	1.0027	13.708	49
12	.07324	.07344	13.617	.99731	1.0027	13.654	48
13	.07353	.07373	13.563	.99729	1.0027	13.600	47
14	.07382	.07402	13.510	.99727	1.0027	13.547	46
15	.07411	.07431	13.457	.99725	1.0028	13.494	45
16	.07440	.07461	13.404	.99723	1.0028	13.441	44
17	.07469	.07490	13.352	.99721	1.0028	13.389	43
18	.07498	.07519	13.300	.99719	1.0028	13.337	42
19	.07527	.07548	13.248	.99716	1.0028	13.286	41
20	.07556	.07578	13.197	.99714	1.0029	13.235	40
21	.07585	.07607	13.146	.99712	1.0029	13.184	39
22	.07614	.07636	13.096	.99710	1.0029	13.134	38
23	.07643	.07665	13.046	.99708	1.0029	13.084	37
24	.07672	.07695	12.996	.99705	1.0030	13.035	36
25	.07701	.07724	12.947	.99703	1.0030	12.985	35
26	.07730	.07753	12.898	.99701	1.0030	12.937	34
27	.07759	.07782	12.850	.99699	1.0030	12.888	33
28	.07788	.07812	12.801	.99696	1.0030	12.840	32
29	.07817	.07841	12.754	.99694	1.0031	12.793	31
30	.07846	.07870	12.706	.99692	1.0031	12.745	30
31	.07875	.07899	12.659	.99689	1.0031	12.699	29
32	.07904	.07929	12.612	.99687	1.0031	12.652	28
33	.07933	.07958	12.566	.99685	1.0032	12.606	27
34	.07962	.07987	12.520	.99683	1.0032	12.560	26
35	.07991	.08017	12.474	.99680	1.0032	12.514	25
36	.08020	.08046	12.429	.99678	1.0032	12.469	24
37	.08049	.08075	12.384	.99676	1.0033	12.424	23
38	.08078	.08104	12.339	.99673	1.0033	12.379	22
39	.08107	.08134	12.295	.99671	1.0033	12.335	21
40	.08136	.08163	12.251	.99668	1.0033	12.291	20
41	.08165	.08192	12.207	.99666	1.0034	12.248	19
42	.08194	.08221	12.163	.99664	1.0034	12.204	18
43	.08223	.08251	12.120	.99661	1.0034	12.161	17
44	.08252	.08280	12.077	.99659	1.0034	12.119	16
45	.08281	.08309	12.035	.99657	1.0034	12.076	15
46	.08310	.08339	11.992	.99654	1.0035	12.034	14
47	.08339	.08368	11.950	.99652	1.0035	11.992	13
48	.08368	.08397	11.909	.99649	1.0035	11.951	12
49	.08397	.08427	11.867	.99647	1.0035	11.909	11
50	.08426	.08456	11.826	.99644	1.0036	11.868	10
51	.08455	.08485	11.785	.99642	1.0036	11.828	9
52	.08484	.08514	11.745	.99639	1.0036	11.787	8
53	.08513	.08544	11.705	.99637	1.0036	11.747	7
54	.08542	.08573	11.664	.99635	1.0037	11.707	6
55	.08571	.08602	11.625	.99632	1.0037	11.668	5
56	.08600	.08632	11.585	.99630	1.0037	11.628	4
57	.08629	.08661	11.546	.99627	1.0037	11.589	3
58	.08658	.08690	11.507	.99625	1.0038	11.551	2
59	.08687	.08720	11.468	.99622	1.0038	11.512	1
60	.08716	.08749	11.430	.99619	1.0038	11.474	0
′	Cos	Cot	Tan	Sin	Csc	Sec	′

4° (274°)   (265°) 85°

5° (185°)   (354°) 174°

′	Sin	Tan	Cot	Cos	Sec	Csc	′
0	.08716	.08749	11.430	.99619	1.0038	11.474	60
1	.08745	.08778	11.392	.99617	1.0038	11.436	59
2	.08774	.08807	11.354	.99614	1.0039	11.398	58
3	.08803	.08837	11.316	.99612	1.0039	11.360	57
4	.08831	.08866	11.279	.99609	1.0039	11.323	56
5	.08860	.08895	11.242	.99607	1.0039	11.286	55
6	.08889	.08925	11.205	.99604	1.0040	11.249	54
7	.08918	.08954	11.168	.99602	1.0040	11.213	53
8	.08947	.08983	11.132	.99599	1.0040	11.176	52
9	.08976	.09013	11.095	.99596	1.0041	11.140	51
10	.09005	.09042	11.059	.99594	1.0041	11.105	50
11	.09034	.09071	11.024	.99591	1.0041	11.069	49
12	.09063	.09101	10.988	.99588	1.0041	11.034	48
13	.09092	.09130	10.953	.99586	1.0042	10.998	47
14	.09121	.09159	10.918	.99583	1.0042	10.963	46
15	.09150	.09189	10.883	.99580	1.0042	10.929	45
16	.09179	.09218	10.848	.99578	1.0042	10.894	44
17	.09208	.09247	10.814	.99575	1.0043	10.860	43
18	.09237	.09277	10.780	.99572	1.0043	10.826	42
19	.09266	.09306	10.746	.99570	1.0043	10.792	41
20	.09295	.09335	10.712	.99567	1.0043	10.758	40
21	.09324	.09365	10.678	.99564	1.0044	10.725	39
22	.09353	.09394	10.645	.99562	1.0044	10.692	38
23	.09382	.09423	10.612	.99559	1.0044	10.659	37
24	.09411	.09453	10.579	.99556	1.0045	10.626	36
25	.09440	.09482	10.546	.99553	1.0045	10.593	35
26	.09469	.09511	10.514	.99551	1.0045	10.561	34
27	.09498	.09541	10.481	.99548	1.0045	10.529	33
28	.09527	.09570	10.449	.99545	1.0046	10.497	32
29	.09556	.09600	10.417	.99542	1.0046	10.465	31
30	.09585	.09629	10.385	.99540	1.0046	10.433	30
31	.09614	.09658	10.354	.99537	1.0047	10.402	29
32	.09642	.09688	10.322	.99534	1.0047	10.371	28
33	.09671	.09717	10.291	.99531	1.0047	10.340	27
34	.09700	.09746	10.260	.99528	1.0047	10.309	26
35	.09729	.09776	10.229	.99526	1.0048	10.278	25
36	.09758	.09805	10.199	.99523	1.0048	10.248	24
37	.09787	.09834	10.168	.99520	1.0048	10.217	23
38	.09816	.09864	10.138	.99517	1.0049	10.187	22
39	.09845	.09893	10.108	.99514	1.0049	10.157	21
40	.09874	.09923	10.078	.99511	1.0049	10.128	20
41	.09903	.09952	10.048	.99508	1.0049	10.098	19
42	.09932	.09981	10.019	.99506	1.0050	10.068	18
43	.09961	.10011	9.9893	.99503	1.0050	10.039	17
44	.09990	.10040	9.9601	.99500	1.0050	10.010	16
45	.10019	.10069	9.9310	.99497	1.0051	9.9812	15
46	.10048	.10099	9.9021	.99494	1.0051	9.9525	14
47	.10077	.10128	9.8734	.99491	1.0051	9.9239	13
48	.10106	.10158	9.8448	.99488	1.0051	9.8955	12
49	.10135	.10187	9.8164	.99485	1.0052	9.8672	11
50	.10164	.10216	9.7882	.99482	1.0052	9.8391	10
51	.10192	.10246	9.7601	.99479	1.0052	9.8112	9
52	.10221	.10275	9.7322	.99476	1.0053	9.7834	8
53	.10250	.10305	9.7044	.99473	1.0053	9.7558	7
54	.10279	.10334	9.6768	.99470	1.0053	9.7283	6
55	.10308	.10363	9.6493	.99467	1.0054	9.7010	5
56	.10337	.10393	9.6220	.99464	1.0054	9.6739	4
57	.10366	.10422	9.5949	.99461	1.0054	9.6469	3
58	.10395	.10452	9.5679	.99458	1.0054	9.6200	2
59	.10424	.10481	9.5411	.99455	1.0055	9.5933	1
60	.10453	.10510	9.5144	.99452	1.0055	9.5668	0
′	Cos	Cot	Tan	Sin	Csc	Sec	′

95° (275°)   (264°) 84°

TRIGONOMETRY

NATURAL TRIGONOMETRIC FUNCTIONS TO FIVE PLACES (continued)

6° (186°)   (353°) **173°**

'	Sin	Tan	Cot	Cos	Sec	Csc	'
0	.10453	.10510	9.5144	.99452	1.0055	9.5668	60
1	.10482	.10540	9.4878	.99449	1.0055	9.5404	59
2	.10511	.10569	9.4614	.99446	1.0056	9.5141	58
3	.10540	.10599	9.4352	.99443	1.0056	9.4880	57
4	.10569	.10628	9.4090	.99440	1.0056	9.4620	56
5	.10597	.10657	9.3831	.99437	1.0057	9.4362	55
6	.10626	.10687	9.3572	.99434	1.0057	9.4105	54
7	.10655	.10716	9.3315	.99431	1.0057	9.3850	53
8	.10684	.10746	9.3060	.99428	1.0058	9.3596	52
9	.10713	.10775	9.2806	.99424	1.0058	9.3343	51
10	.10742	.10805	9.2553	.99421	1.0058	9.3092	50
11	.10771	.10834	9.2302	.99418	1.0059	9.2842	49
12	.10800	.10863	9.2052	.99415	1.0059	9.2593	48
13	.10829	.10893	9.1803	.99412	1.0059	9.2346	47
14	.10858	.10922	9.1555	.99409	1.0059	9.2100	46
15	.10887	.10952	9.1309	.99406	1.0060	9.1855	45
16	.10916	.10981	9.1065	.99402	1.0060	9.1612	44
17	.10945	.11011	9.0821	.99399	1.0060	9.1370	43
18	.10973	.11040	9.0579	.99396	1.0061	9.1129	42
19	.11002	.11070	9.0338	.99393	1.0061	9.0890	41
20	.11031	.11099	9.0098	.99390	1.0061	9.0652	40
21	.11060	.11128	8.9860	.99386	1.0062	9.0415	39
22	.11089	.11158	8.9623	.99383	1.0062	9.0179	38
23	.11118	.11187	8.9387	.99380	1.0062	8.9944	37
24	.11147	.11217	8.9152	.99377	1.0063	8.9711	36
25	.11176	.11246	8.8919	.99374	1.0063	8.9479	35
26	.11205	.11276	8.8686	.99370	1.0063	8.9248	34
27	.11234	.11305	8.8455	.99367	1.0064	8.9019	33
28	.11263	.11335	8.8225	.99364	1.0064	8.8790	32
29	.11291	.11364	8.7996	.99360	1.0064	8.8563	31
30	.11320	.11394	8.7769	.99357	1.0065	8.8337	30
31	.11349	.11423	8.7542	.99354	1.0065	8.8112	29
32	.11378	.11452	8.7317	.99351	1.0065	8.7888	28
33	.11407	.11482	8.7093	.99347	1.0066	8.7665	27
34	.11436	.11511	8.6870	.99344	1.0066	8.7444	26
35	.11465	.11541	8.6648	.99341	1.0066	8.7223	25
36	.11494	.11570	8.6427	.99337	1.0067	8.7004	24
37	.11523	.11600	8.6208	.99334	1.0067	8.6786	23
38	.11552	.11629	8.5989	.99331	1.0067	8.6569	22
39	.11580	.11659	8.5772	.99327	1.0068	8.6353	21
40	.11609	.11688	8.5555	.99324	1.0068	8.6138	20
41	.11638	.11718	8.5340	.99320	1.0068	8.5924	19
42	.11667	.11747	8.5126	.99317	1.0069	8.5711	18
43	.11696	.11777	8.4913	.99314	1.0069	8.5500	17
44	.11725	.11806	8.4701	.99310	1.0069	8.5289	16
45	.11754	.11836	8.4490	.99307	1.0070	8.5079	15
46	.11783	.11865	8.4280	.99303	1.0070	8.4871	14
47	.11812	.11895	8.4071	.99300	1.0070	8.4663	13
48	.11840	.11924	8.3863	.99297	1.0071	8.4457	12
49	.11869	.11954	8.3656	.99293	1.0071	8.4251	11
50	.11898	.11983	8.3450	.99290	1.0072	8.4047	10
51	.11927	.12013	8.3245	.99286	1.0072	8.3843	9
52	.11956	.12042	8.3041	.99283	1.0072	8.3641	8
53	.11985	.12072	8.2838	.99279	1.0073	8.3439	7
54	.12014	.12101	8.2636	.99276	1.0073	8.3238	6
55	.12043	.12131	8.2434	.99272	1.0073	8.3039	5
56	.12071	.12160	8.2234	.99269	1.0074	8.2840	4
57	.12100	.12190	8.2035	.99265	1.0074	8.2642	3
58	.12129	.12219	8.1837	.99262	1.0074	8.2446	2
59	.12158	.12249	8.1640	.99258	1.0075	8.2250	1
60	.12187	.12278	8.1443	.99255	1.0075	8.2055	0
'	Cos	Cot	Tan	Sin	Csc	Sec	'

96° (276°)   (263°) **83°**

7° (187°)   (352°) **172°**

'	Sin	Tan	Cot	Cos	Sec	Csc	'
0	.12187	.12278	8.1443	.99255	1.0075	8.2055	60
1	.12216	.12308	8.1248	.99251	1.0075	8.1861	59
2	.12245	.12338	8.1054	.99248	1.0076	8.1668	58
3	.12274	.12367	8.0860	.99244	1.0076	8.1476	57
4	.12302	.12397	8.0667	.99240	1.0077	8.1285	56
5	.12331	.12426	8.0476	.99237	1.0077	8.1095	55
6	.12360	.12456	8.0285	.99233	1.0077	8.0905	54
7	.12389	.12485	8.0095	.99230	1.0078	8.0717	53
8	.12418	.12515	7.9906	.99226	1.0078	8.0529	52
9	.12447	.12544	7.9718	.99222	1.0078	8.0342	51
10	.12476	.12574	7.9530	.99219	1.0079	8.0156	50
11	.12504	.12603	7.9344	.99215	1.0079	7.9971	49
12	.12533	.12633	7.9158	.99211	1.0079	7.9787	48
13	.12562	.12662	7.8973	.99208	1.0080	7.9604	47
14	.12591	.12692	7.8789	.99204	1.0080	7.9422	46
15	.12620	.12722	7.8606	.99200	1.0081	7.9240	45
16	.12649	.12751	7.8424	.99197	1.0081	7.9059	44
17	.12678	.12781	7.8243	.99193	1.0081	7.8879	43
18	.12706	.12810	7.8062	.99189	1.0082	7.8700	42
19	.12735	.12840	7.7882	.99186	1.0082	7.8522	41
20	.12764	.12869	7.7704	.99182	1.0082	7.8344	40
21	.12793	.12899	7.7525	.99178	1.0083	7.8168	39
22	.12822	.12929	7.7348	.99175	1.0083	7.7992	38
23	.12851	.12958	7.7171	.99171	1.0084	7.7817	37
24	.12880	.12988	7.6996	.99167	1.0084	7.7642	36
25	.12908	.13017	7.6821	.99163	1.0084	7.7469	35
26	.12937	.13047	7.6647	.99160	1.0085	7.7296	34
27	.12966	.13076	7.6473	.99156	1.0085	7.7124	33
28	.12995	.13106	7.6301	.99152	1.0086	7.6953	32
29	.13024	.13136	7.6129	.99148	1.0086	7.6783	31
30	.13053	.13165	7.5958	.99144	1.0086	7.6613	30
31	.13081	.13195	7.5787	.99141	1.0087	7.6444	29
32	.13110	.13224	7.5618	.99137	1.0087	7.6276	28
33	.13139	.13254	7.5449	.99133	1.0087	7.6109	27
34	.13168	.13284	7.5281	.99129	1.0088	7.5942	26
35	.13197	.13313	7.5113	.99125	1.0088	7.5776	25
36	.13226	.13343	7.4947	.99122	1.0089	7.5611	24
37	.13254	.13372	7.4781	.99118	1.0089	7.5446	23
38	.13283	.13402	7.4615	.99114	1.0089	7.5282	22
39	.13312	.13432	7.4451	.99110	1.0090	7.5119	21
40	.13341	.13461	7.4287	.99106	1.0090	7.4957	20
41	.13370	.13491	7.4124	.99102	1.0091	7.4795	19
42	.13399	.13521	7.3962	.99098	1.0091	7.4635	18
43	.13427	.13550	7.3800	.99094	1.0091	7.4474	17
44	.13456	.13580	7.3639	.99091	1.0092	7.4315	16
45	.13485	.13609	7.3479	.99087	1.0092	7.4156	15
46	.13514	.13639	7.3319	.99083	1.0093	7.3998	14
47	.13543	.13669	7.3160	.99079	1.0093	7.3840	13
48	.13572	.13698	7.3002	.99075	1.0093	7.3684	12
49	.13600	.13728	7.2844	.99071	1.0094	7.3527	11
50	.13629	.13758	7.2687	.99067	1.0094	7.3372	10
51	.13658	.13787	7.2531	.99063	1.0095	7.3217	9
52	.13687	.13817	7.2375	.99059	1.0095	7.3063	8
53	.13716	.13846	7.2220	.99055	1.0095	7.2909	7
54	.13744	.13876	7.2066	.99051	1.0096	7.2757	6
55	.13773	.13906	7.1912	.99047	1.0096	7.2604	5
56	.13802	.13935	7.1759	.99043	1.0097	7.2453	4
57	.13831	.13965	7.1607	.99039	1.0097	7.2302	3
58	.13860	.13995	7.1455	.99035	1.0097	7.2152	2
59	.13889	.14024	7.1304	.99031	1.0098	7.2002	1
60	.13917	.14054	7.1154	.99027	1.0098	7.1853	0
'	Cos	Cot	Tan	Sin	Csc	Sec	'

97° (277°)   (262°) **82°**

NATURAL TRIGONOMETRIC FUNCTIONS TO FIVE PLACES (continued)

8° (188°)   (351°) 171°

'	Sin	Tan	Cot	Cos	Sec	Csc	'
0	.13917	.14054	7.1154	.99027	1.0098	7.1853	60
1	.13946	.14084	7.1004	.99023	1.0099	7.1705	59
2	.13975	.14113	7.0855	.99019	1.0099	7.1557	58
3	.14004	.14143	7.0706	.99015	1.0100	7.1410	57
4	.14033	.14173	7.0558	.99011	1.0100	7.1263	56
5	.14061	.14202	7.0410	.99006	1.0100	7.1117	55
6	.14090	.14232	7.0264	.99002	1.0101	7.0972	54
7	.14119	.14262	7.0117	.98998	1.0101	7.0827	53
8	.14148	.14291	6.9972	.98994	1.0102	7.0683	52
9	.14177	.14321	6.9827	.98990	1.0102	7.0539	51
10	.14205	.14351	6.9682	.98986	1.0102	7.0396	50
11	.14234	.14381	6.9538	.98982	1.0103	7.0254	49
12	.14263	.14410	6.9395	.98978	1.0103	7.0112	48
13	.14292	.14440	6.9252	.98973	1.0104	6.9971	47
14	.14320	.14470	6.9110	.98969	1.0104	6.9830	46
15	.14349	.14499	6.8969	.98965	1.0105	6.9690	45
16	.14378	.14529	6.8828	.98961	1.0105	6.9550	44
17	.14407	.14559	6.8687	.98957	1.0105	6.9411	43
18	.14436	.14588	6.8548	.98953	1.0106	6.9273	42
19	.14464	.14618	6.8408	.98948	1.0106	6.9135	41
20	.14493	.14648	6.8269	.98944	1.0107	6.8998	40
21	.14522	.14678	6.8131	.98940	1.0107	6.8861	39
22	.14551	.14707	6.7994	.98936	1.0108	6.8725	38
23	.14580	.14737	6.7856	.98931	1.0108	6.8589	37
24	.14608	.14767	6.7720	.98927	1.0108	6.8454	36
25	.14637	.14796	6.7584	.98923	1.0109	6.8320	35
26	.14666	.14826	6.7448	.98919	1.0109	6.8186	34
27	.14695	.14856	6.7313	.98914	1.0110	6.8052	33
28	.14723	.14886	6.7179	.98910	1.0110	6.7919	32
20	.14752	.14915	6.7045	.98906	1.0111	6.7787	31
30	.14781	.14945	6.6912	.98902	1.0111	6.7655	30
31	.14810	.14975	6.6779	.98897	1.0112	6.7523	29
32	.14838	.15005	6.6646	.98893	1.0112	6.7392	28
33	.14867	.15034	6.6514	.98889	1.0112	6.7262	27
34	.14896	.15064	6.6383	.98884	1.0113	6.7132	26
35	.14925	.15094	6.6252	.98880	1.0113	6.7003	25
36	.14954	.15124	6.6122	.98876	1.0114	6.6874	24
37	.14982	.15153	6.5992	.98871	1.0114	6.6745	23
38	.15011	.15183	6.5863	.98867	1.0115	6.6618	22
39	.15040	.15213	6.5734	.98863	1.0115	6.6490	21
40	.15069	.15243	6.5606	.98858	1.0116	6.6363	20
41	.15097	.15272	6.5478	.98854	1.0116	6.6237	19
42	.15126	.15302	6.5350	.98849	1.0116	6.6111	18
43	.15155	.15332	6.5223	.98845	1.0117	6.5986	17
44	.15184	.15362	6.5097	.98841	1.0117	6.5861	16
45	.15212	.15391	6.4971	.98836	1.0118	6.5736	15
46	.15241	.15421	6.4846	.98832	1.0118	6.5612	14
47	.15270	.15451	6.4721	.98827	1.0119	6.5489	13
48	.15299	.15481	6.4596	.98823	1.0119	6.5366	12
49	.15327	.15511	6.4472	.98818	1.0120	6.5243	11
50	.15356	.15540	6.4348	.98814	1.0120	6.5121	10
51	.15385	.15570	6.4225	.98809	1.0120	6.4999	9
52	.15414	.15600	6.4103	.98805	1.0121	6.4878	8
53	.15442	.15630	6.3980	.98800	1.0121	6.4757	7
54	.15471	.15660	6.3859	.98796	1.0122	6.4637	6
55	.15500	.15689	6.3737	.98791	1.0122	6.4517	5
56	.15529	.15719	6.3617	.98787	1.0123	6.4398	4
57	.15557	.15749	6.3496	.98782	1.0123	6.4279	3
58	.15586	.15779	6.3376	.98778	1.0124	6.4160	2
59	.15615	.15809	6.3257	.98773	1.0124	6.4042	1
60	.15643	.15838	6.3138	.98769	1.0125	6.3925	0
'	Cos	Cot	Tan	Sin	Csc	Sec	'

98° (278°)   (261°) 81°

9° (189°)   (350°) 170°

'	Sin	Tan	Cot	Cos	Sec	Csc	'
0	.15643	.15838	6.3138	.98769	1.0125	6.3925	60
1	.15672	.15868	6.3019	.98764	1.0125	6.3807	59
2	.15701	.15898	6.2901	.98760	1.0126	6.3691	58
3	.15730	.15928	6.2783	.98755	1.0126	6.3574	57
4	.15758	.15958	6.2666	.98751	1.0127	6.3458	56
5	.15787	.15988	6.2549	.98746	1.0127	6.3343	55
6	.15816	.16017	6.2432	.98741	1.0127	6.3228	54
7	.15845	.16047	6.2316	.98737	1.0128	6.3113	53
8	.15873	.16077	6.2200	.98732	1.0128	6.2999	52
9	.15902	.16107	6.2085	.98728	1.0129	6.2885	51
10	.15931	.16137	6.1970	.98723	1.0129	6.2772	50
11	.15959	.16167	6.1856	.98718	1.0130	6.2659	49
12	.15988	.16196	6.1742	.98714	1.0130	6.2546	48
13	.16017	.16226	6.1628	.98709	1.0131	6.2434	47
14	.16046	.16256	6.1515	.98704	1.0131	6.2323	46
15	.16074	.16286	6.1402	.98700	1.0132	6.2211	45
16	.16103	.16316	6.1290	.98695	1.0132	6.2100	44
17	.16132	.16346	6.1178	.98690	1.0133	6.1990	43
18	.16160	.16376	6.1066	.98686	1.0133	6.1880	42
19	.16189	.16405	6.0955	.98681	1.0134	6.1770	41
20	.16218	.16435	6.0844	.98676	1.0134	6.1661	40
21	.16246	.16465	6.0734	.98671	1.0135	6.1552	39
22	.16275	.16495	6.0624	.98667	1.0135	6.1443	38
23	.16304	.16525	6.0514	.98662	1.0136	6.1335	37
24	.16333	.16555	6.0405	.98657	1.0136	6.1227	36
25	.16361	.16585	6.0296	.98652	1.0137	6.1120	35
26	.16390	.16615	6.0188	.98648	1.0137	6.1013	34
27	.16419	.16645	6.0080	.98643	1.0138	6.0906	33
28	.16447	.16674	5.9972	.98638	1.0138	6.0800	32
29	.16476	.16704	5.9865	.98633	1.0139	6.0694	31
30	.16505	.16734	5.9758	.98629	1.0139	6.0589	30
31	.16533	.16764	5.9651	.98624	1.0140	6.0483	29
32	.16562	.16794	5.9545	.98619	1.0140	6.0379	28
33	.16591	.16824	5.9439	.98614	1.0141	6.0274	27
34	.16620	.16854	5.9333	.98609	1.0141	6.0170	26
35	.16648	.16884	5.9228	.98604	1.0142	6.0067	25
36	.16677	.16914	5.9124	.98600	1.0142	5.9963	24
37	.16706	.16944	5.9019	.98595	1.0143	5.9860	23
38	.16734	.16974	5.8915	.98590	1.0143	5.9758	22
39	.16763	.17004	5.8811	.98585	1.0144	5.9656	21
40	.16792	.17033	5.8708	.98580	1.0144	5.9554	20
41	.16820	.17063	5.8605	.98575	1.0145	5.9452	19
42	.16849	.17093	5.8502	.98570	1.0145	5.9351	18
43	.16878	.17123	5.8400	.98565	1.0146	5.9250	17
44	.16906	.17153	5.8298	.98561	1.0146	5.9150	16
45	.16935	.17183	5.8197	.98556	1.0147	5.9049	15
46	.16964	.17213	5.8095	.98551	1.0147	5.8950	14
47	.16992	.17243	5.7994	.98546	1.0148	5.8850	13
48	.17021	.17273	5.7894	.98541	1.0148	5.8751	12
49	.17050	.17303	5.7794	.98536	1.0149	5.8652	11
50	.17078	.17333	5.7694	.98531	1.0149	5.8554	10
51	.17107	.17363	5.7594	.98526	1.0150	5.8456	9
52	.17136	.17393	5.7495	.98521	1.0150	5.8358	8
53	.17164	.17423	5.7396	.98516	1.0151	5.8261	7
54	.17193	.17453	5.7297	.98511	1.0151	5.8164	6
55	.17222	.17483	5.7199	.98506	1.0152	5.8067	5
56	.17250	.17513	5.7101	.98501	1.0152	5.7970	4
57	.17279	.17543	5.7004	.98496	1.0153	5.7874	3
58	.17308	.17573	5.6906	.98491	1.0153	5.7778	2
59	.17336	.17603	5.6809	.98486	1.0154	5.7683	1
60	.17365	.17633	5.6713	.98481	1.0154	5.7588	0
'	Cos	Cot	Tan	Sin	Csc	Sec	'

99° (279°)   (260°) 80°

TRIGONOMETRY

NATURAL TRIGONOMETRIC FUNCTIONS TO FIVE PLACES (continued)

10° (190°)   (349°) 169°   11° (191°)   (348°) 168°

'	Sin	Tan	Cot	Cos	Sec	Csc	'
0	.17365	.17633	5.6713	.98481	1.0154	5.7588	60
1	.17393	.17663	5.6617	.98476	1.0155	5.7493	59
2	.17422	.17693	5.6521	.98471	1.0155	5.7398	58
3	.17451	.17723	5.6425	.98466	1.0156	5.7304	57
4	.17479	.17753	5.6329	.98461	1.0156	5.7210	56
5	.17508	.17783	5.6234	.98455	1.0157	5.7117	55
6	.17537	.17813	5.6140	.98450	1.0157	5.7023	54
7	.17565	.17843	5.6045	.98445	1.0158	5.6930	53
8	.17594	.17873	5.5951	.98440	1.0158	5.6838	52
9	.17623	.17903	5.5857	.98435	1.0159	5.6745	51
10	.17651	.17933	5.5764	.98430	1.0160	5.6653	50
11	.17680	.17963	5.5671	.98425	1.0160	5.6562	49
12	.17708	.17993	5.5578	.98420	1.0161	5.6470	48
13	.17737	.18023	5.5485	.98414	1.0161	5.6379	47
14	.17766	.18053	5.5393	.98409	1.0162	5.6288	46
15	.17794	.18083	5.5301	.98404	1.0162	5.6198	45
16	.17823	.18113	5.5209	.98399	1.0163	5.6107	44
17	.17852	.18143	5.5118	.98394	1.0163	5.6017	43
18	.17880	.18173	5.5026	.98389	1.0164	5.5928	42
19	.17909	.18203	5.4936	.98383	1.0164	5.5838	41
20	.17937	.18233	5.4845	.98378	1.0165	5.5749	40
21	.17966	.18263	5.4755	.98373	1.0165	5.5660	39
22	.17995	.18293	5.4665	.98368	1.0166	5.5572	38
23	.18023	.18323	5.4575	.98362	1.0166	5.5484	37
24	.18052	.18353	5.4486	.98357	1.0167	5.5396	36
25	.18081	.18384	5.4397	.98352	1.0168	5.5308	35
26	.18109	.18414	5.4308	.98347	1.0168	5.5221	34
27	.18138	.18444	5.4219	.98341	1.0169	5.5134	33
28	.18166	.18474	5.4131	.98336	1.0169	5.5047	32
29	.18195	.18504	5.4043	.98331	1.0170	5.4960	31
30	.18224	.18534	5.3955	.98325	1.0170	5.4874	30
31	.18252	.18564	5.3868	.98320	1.0171	5.4788	29
32	.18281	.18594	5.3781	.98315	1.0171	5.4702	28
33	.18309	.18624	5.3694	.98310	1.0172	5.4617	27
34	.18338	.18654	5.3607	.98304	1.0173	5.4532	26
35	.18367	.18684	5.3521	.98299	1.0173	5.4447	25
36	.18395	.18714	5.3435	.98294	1.0174	5.4362	24
37	.18424	.18745	5.3349	.98288	1.0174	5.4278	23
38	.18452	.18775	5.3263	.98283	1.0175	5.4194	22
39	.18481	.18805	5.3178	.98277	1.0175	5.4110	21
40	.18509	.18835	5.3093	.98272	1.0176	5.4026	20
41	.18538	.18865	5.3008	.98267	1.0176	5.3943	19
42	.18567	.18895	5.2924	.98261	1.0177	5.3860	18
43	.18595	.18925	5.2839	.98256	1.0178	5.3777	17
44	.18624	.18955	5.2755	.98250	1.0178	5.3695	16
45	.18652	.18986	5.2672	.98245	1.0179	5.3612	15
46	.18681	.19016	5.2588	.98240	1.0179	5.3530	14
47	.18710	.19046	5.2505	.98234	1.0180	5.3449	13
48	.18738	.19076	5.2422	.98229	1.0180	5.3367	12
49	.18767	.19106	5.2339	.98223	1.0181	5.3286	11
50	.18795	.19136	5.2257	.98218	1.0181	5.3205	10
51	.18824	.19166	5.2174	.98212	1.0182	5.3124	9
52	.18852	.19197	5.2092	.98207	1.0183	5.3044	8
53	.18881	.19227	5.2011	.98201	1.0183	5.2963	7
54	.18910	.19257	5.1929	.98196	1.0184	5.2883	6
55	.18938	.19287	5.1848	.98190	1.0184	5.2804	5
56	.18967	.19317	5.1767	.98185	1.0185	5.2724	4
57	.18995	.19347	5.1686	.98179	1.0185	5.2645	3
58	.19024	.19378	5.1606	.98174	1.0186	5.2566	2
59	.19052	.19408	5.1526	.98168	1.0187	5.2487	1
60	.19081	.19438	5.1446	.98163	1.0187	5.2408	0
'	Cos	Cot	Tan	Sin	Csc	Sec	'

100° (280°)   (259°) 79°

'	Sin	Tan	Cot	Cos	Sec	Csc	'
0	.19081	.19438	5.1446	.98163	1.0187	5.2408	60
1	.19109	.19468	5.1366	.98157	1.0188	5.2330	59
2	.19138	.19498	5.1286	.98152	1.0188	5.2252	58
3	.19167	.19529	5.1207	.98146	1.0189	5.2174	57
4	.19195	.19559	5.1128	.98140	1.0189	5.2097	56
5	.19224	.19589	5.1049	.98135	1.0190	5.2019	55
6	.19252	.19619	5.0970	.98129	1.0191	5.1942	54
7	.19281	.19649	5.0892	.98124	1.0191	5.1865	53
8	.19309	.19680	5.0814	.98118	1.0192	5.1789	52
9	.19338	.19710	5.0736	.98112	1.0192	5.1712	51
10	.19366	.19740	5.0658	.98107	1.0193	5.1636	50
11	.19395	.19770	5.0581	.98101	1.0194	5.1560	49
12	.19423	.19801	5.0504	.98096	1.0194	5.1484	48
13	.19452	.19831	5.0427	.98090	1.0195	5.1409	47
14	.19481	.19861	5.0350	.98084	1.0195	5.1333	46
15	.19509	.19891	5.0273	.98079	1.0196	5.1258	45
16	.19538	.19921	5.0197	.98073	1.0197	5.1183	44
17	.19566	.19952	5.0121	.98067	1.0197	5.1109	43
18	.19595	.19982	5.0045	.98061	1.0198	5.1034	42
19	.19623	.20012	4.9969	.98056	1.0198	5.0960	41
20	.19652	.20042	4.9894	.98050	1.0199	5.0886	40
21	.19680	.20073	4.9819	.98044	1.0199	5.0813	39
22	.19709	.20103	4.9744	.98039	1.0200	5.0739	38
23	.19737	.20133	4.9669	.98033	1.0201	5.0666	37
24	.19766	.20164	4.9594	.98027	1.0201	5.0593	36
25	.19794	.20194	4.9520	.98021	1.0202	5.0520	35
26	.19823	.20224	4.9446	.98016	1.0202	5.0447	34
27	.19851	.20254	4.9372	.98010	1.0203	5.0375	33
28	.19880	.20285	4.9298	.98004	1.0204	5.0302	32
29	.19908	.20315	4.9225	.97998	1.0204	5.0230	31
30	.19937	.20345	4.9152	.97992	1.0205	5.0159	30
31	.19965	.20376	4.9078	.97987	1.0205	5.0087	29
32	.19994	.20406	4.9006	.97981	1.0206	5.0016	28
33	.20022	.20436	4.8933	.97975	1.0207	4.9944	27
34	.20051	.20466	4.8860	.97969	1.0207	4.9873	26
35	.20079	.20497	4.8788	.97963	1.0208	4.9803	25
36	.20108	.20527	4.8716	.97958	1.0209	4.9732	24
37	.20136	.20557	4.8644	.97952	1.0209	4.9662	23
38	.20165	.20588	4.8573	.97946	1.0210	4.9591	22
39	.20193	.20618	4.8501	.97940	1.0210	4.9521	21
40	.20222	.20648	4.8430	.97934	1.0211	4.9452	20
41	.20250	.20679	4.8359	.97928	1.0212	4.9382	19
42	.20279	.20709	4.8288	.97922	1.0212	4.9313	18
43	.20307	.20739	4.8218	.97916	1.0213	4.9244	17
44	.20336	.20770	4.8147	.97910	1.0213	4.9175	16
45	.20364	.20800	4.8077	.97905	1.0214	4.9106	15
46	.20393	.20830	4.8007	.97899	1.0215	4.9037	14
47	.20421	.20861	4.7937	.97893	1.0215	4.8969	13
48	.20450	.20891	4.7867	.97887	1.0216	4.8901	12
49	.20478	.20921	4.7798	.97881	1.0217	4.8833	11
50	.20507	.20952	4.7729	.97875	1.0217	4.8765	10
51	.20535	.20982	4.7659	.97869	1.0218	4.8697	9
52	.20563	.21013	4.7591	.97863	1.0218	4.8630	8
53	.20592	.21043	4.7522	.97857	1.0219	4.8563	7
54	.20620	.21073	4.7453	.97851	1.0220	4.8496	6
55	.20649	.21104	4.7385	.97845	1.0220	4.8429	5
56	.20677	.21134	4.7317	.97839	1.0221	4.8362	4
57	.20706	.21164	4.7249	.97833	1.0222	4.8296	3
58	.20734	.21195	4.7181	.97827	1.0222	4.8229	2
59	.20763	.21225	4.7114	.97821	1.0223	4.8163	1
60	.20791	.21256	4.7046	.97815	1.0223	4.8097	0
'	Cos	Cot	Tan	Sin	Csc	Sec	'

101° (281°)   (258°) 78°

NATURAL TRIGONOMETRIC FUNCTIONS TO FIVE PLACES (continued)

12° (192°)   (347°) 167°

′	Sin	Tan	Cot	Cos	Sec	Csc	′
0	.20791	.21256	4.7046	.97815	1.0223	4.8097	60
1	.20820	.21286	4.6979	.97809	1.0224	4.8032	59
2	.20848	.21316	4.6912	.97803	1.0225	4.7966	58
3	.20877	.21347	4.6845	.97797	1.0225	4.7901	57
4	.20905	.21377	4.6779	.97791	1.0226	4.7836	56
5	.20933	.21408	4.6712	.97784	1.0227	4.7771	55
6	.20962	.21438	4.6646	.97778	1.0227	4.7706	54
7	.20990	.21469	4.6580	.97772	1.0228	4.7641	53
8	.21019	.21499	4.6514	.97766	1.0228	4.7577	52
9	.21047	.21529	4.6448	.97760	1.0229	4.7512	51
10	.21076	.21560	4.6382	.97754	1.0230	4.7448	50
11	.21104	.21590	4.6317	.97748	1.0230	4.7384	49
12	.21132	.21621	4.6252	.97742	1.0231	4.7321	48
13	.21161	.21651	4.6187	.97735	1.0232	4.7257	47
14	.21189	.21682	4.6122	.97729	1.0232	4.7194	46
15	.21218	.21712	4.6057	.97723	1.0233	4.7130	45
16	.21246	.21743	4.5993	.97717	1.0234	4.7067	44
17	.21275	.21773	4.5928	.97711	1.0234	4.7004	43
18	.21303	.21804	4.5864	.97705	1.0235	4.6942	42
19	.21331	.21834	4.5800	.97698	1.0236	4.6879	41
20	.21360	.21864	4.5736	.97692	1.0236	4.6817	40
21	.21388	.21895	4.5673	.97686	1.0237	4.6755	39
22	.21417	.21925	4.5609	.97680	1.0238	4.6693	38
23	.21445	.21956	4.5546	.97673	1.0238	4.6631	37
24	.21474	.21986	4.5483	.97667	1.0239	4.6569	36
25	.21502	.22017	4.5420	.97661	1.0240	4.6507	35
26	.21530	.22047	4.5357	.97655	1.0240	4.6446	34
27	.21559	.22078	4.5294	.97648	1.0241	4.6385	33
28	.21587	.22108	4.5232	.97642	1.0241	4.6324	32
29	.21616	.22139	4.5169	.97636	1.0242	4.6263	31
30	.21644	.22169	4.5107	.97630	1.0243	4.6202	30
31	.21672	.22200	4.5045	.97623	1.0243	4.6142	29
32	.21701	.22231	4.4983	.97617	1.0244	4.6081	28
33	.21729	.22261	4.4922	.97611	1.0245	4.6021	27
34	.21758	.22292	4.4860	.97604	1.0245	4.5961	26
35	.21786	.22322	4.4799	.97598	1.0246	4.5901	25
36	.21814	.22353	4.4737	.97592	1.0247	4.5841	24
37	.21843	.22383	4.4676	.97585	1.0247	4.5782	23
38	.21871	.22414	4.4615	.97579	1.0248	4.5722	22
39	.21899	.22444	4.4555	.97573	1.0249	4.5663	21
40	.21928	.22475	4.4494	.97566	1.0249	4.5604	20
41	.21956	.22505	4.4434	.97560	1.0250	4.5545	19
42	.21985	.22536	4.4373	.97553	1.0251	4.5486	18
43	.22013	.22567	4.4313	.97547	1.0251	4.5428	17
44	.22041	.22597	4.4253	.97541	1.0252	4.5369	16
45	.22070	.22628	4.4194	.97534	1.0253	4.5311	15
46	.22098	.22658	4.4134	.97528	1.0253	4.5253	14
47	.22126	.22689	4.4075	.97521	1.0254	4.5195	13
48	.22155	.22719	4.4015	.97515	1.0255	4.5137	12
49	.22183	.22750	4.3956	.97508	1.0256	4.5079	11
50	.22212	.22781	4.3897	.97502	1.0256	4.5022	10
51	.22240	.22811	4.3838	.97496	1.0257	4.4964	9
52	.22268	.22842	4.3779	.97489	1.0258	4.4907	8
53	.22297	.22872	4.3721	.97483	1.0258	4.4850	7
54	.22325	.22903	4.3662	.97476	1.0259	4.4793	6
55	.22353	.22934	4.3604	.97470	1.0260	4.4736	5
56	.22382	.22964	4.3546	.97463	1.0260	4.4679	4
57	.22410	.22995	4.3488	.97457	1.0261	4.4623	3
58	.22438	.23026	4.3430	.97450	1.0262	4.4566	2
59	.22467	.23056	4.3372	.97444	1.0262	4.4510	1
60	.22495	.23087	4.3315	.97437	1.0263	4.4454	0
′	Cos	Cot	Tan	Sin	Csc	Sec	′

102° (282°)   (257°) 77°

13° (193°)   (346°) 166°

′	Sin	Tan	Cot	Cos	Sec	Csc	′
0	.22495	.23087	4.3315	.97437	1.0263	4.4454	60
1	.22523	.23117	4.3257	.97430	1.0264	4.4398	59
2	.22552	.23148	4.3200	.97424	1.0264	4.4342	58
3	.22580	.23179	4.3143	.97417	1.0265	4.4287	57
4	.22608	.23209	4.3086	.97411	1.0266	4.4231	56
5	.22637	.23240	4.3029	.97404	1.0266	4.4176	55
6	.22665	.23271	4.2972	.97398	1.0267	4.4121	54
7	.22693	.23301	4.2916	.97391	1.0268	4.4066	53
8	.22722	.23332	4.2859	.97384	1.0269	4.4011	52
9	.22750	.23363	4.2803	.97378	1.0269	4.3956	51
10	.22778	.23393	4.2747	.97371	1.0270	4.3901	50
11	.22807	.23424	4.2691	.97365	1.0271	4.3847	49
12	.22835	.23455	4.2635	.97358	1.0271	4.3792	48
13	.22863	.23485	4.2580	.97351	1.0272	4.3738	47
14	.22892	.23516	4.2524	.97345	1.0273	4.3684	46
15	.22920	.23547	4.2468	.97338	1.0273	4.3630	45
16	.22948	.23578	4.2413	.97331	1.0274	4.3576	44
17	.22977	.23608	4.2358	.97325	1.0275	4.3522	43
18	.23005	.23639	4.2303	.97318	1.0276	4.3469	42
19	.23033	.23670	4.2248	.97311	1.0276	4.3415	41
20	.23062	.23700	4.2193	.97304	1.0277	4.3362	40
21	.23090	.23731	4.2139	.97298	1.0278	4.3309	39
22	.23118	.23762	4.2084	.97291	1.0278	4.3256	38
23	.23146	.23793	4.2030	.97284	1.0279	4.3203	37
24	.23175	.23823	4.1976	.97278	1.0280	4.3150	36
25	.23203	.23854	4.1922	.97271	1.0281	4.3098	35
26	.23231	.23885	4.1868	.97264	1.0281	4.3045	34
27	.23260	.23916	4.1814	.97257	1.0282	4.2993	33
28	.23288	.23946	4.1760	.97251	1.0283	4.2941	32
29	.23316	.23977	4.1706	.97244	1.0283	4.2889	31
30	.23345	.24008	4.1653	.97237	1.0284	4.2837	30
31	.23373	.24039	4.1600	.97230	1.0285	4.2785	29
32	.23401	.24069	4.1547	.97223	1.0286	4.2733	28
33	.23429	.24100	4.1493	.97217	1.0286	4.2681	27
34	.23458	.24131	4.1441	.97210	1.0287	4.2630	26
35	.23486	.24162	4.1388	.97203	1.0288	4.2579	25
36	.23514	.24193	4.1335	.97196	1.0288	4.2527	24
37	.23542	.24223	4.1282	.97189	1.0289	4.2476	23
38	.23571	.24254	4.1230	.97182	1.0290	4.2425	22
39	.23599	.24285	4.1178	.97176	1.0291	4.2375	21
40	.23627	.24316	4.1126	.97169	1.0291	4.2324	20
41	.23656	.24347	4.1074	.97162	1.0292	4.2273	19
42	.23684	.24377	4.1022	.97155	1.0293	4.2223	18
43	.23712	.24408	4.0970	.97148	1.0294	4.2173	17
44	.23740	.24439	4.0918	.97141	1.0294	4.2122	16
45	.23769	.24470	4.0867	.97134	1.0295	4.2072	15
46	.23797	.24501	4.0815	.97127	1.0296	4.2022	14
47	.23825	.24532	4.0764	.97120	1.0297	4.1973	13
48	.23853	.24562	4.0713	.97113	1.0297	4.1923	12
49	.23882	.24593	4.0662	.97106	1.0298	4.1873	11
50	.23910	.24624	4.0611	.97100	1.0299	4.1824	10
51	.23938	.24655	4.0560	.97093	1.0299	4.1774	9
52	.23966	.24686	4.0509	.97086	1.0300	4.1725	8
53	.23995	.24717	4.0459	.97079	1.0301	4.1676	7
54	.24023	.24747	4.0408	.97072	1.0302	4.1627	6
55	.24051	.24778	4.0358	.97065	1.0302	4.1578	5
56	.24079	.24809	4.0308	.97058	1.0303	4.1529	4
57	.24108	.24840	4.0257	.97051	1.0304	4.1481	3
58	.24136	.24871	4.0207	.97044	1.0305	4.1432	2
59	.24164	.24902	4.0158	.97037	1.0305	4.1384	1
60	.24192	.24933	4.0108	.97030	1.0306	4.1336	0
′	Cos	Cot	Tan	Sin	Csc	Sec	′

103° (283°)   (256°) 76°

158

NATURAL TRIGONOMETRIC FUNCTIONS TO FIVE PLACES (continued)

14° (194°)   (345°) 165°   15° (195°)   (344°) 164°

'	Sin	Tan	Cot	Cos	Sec	Csc	'
0	.24192	.24933	4.0108	.97030	1.0306	4.1336	60
1	.24220	.24964	4.0058	.97023	1.0307	4.1287	59
2	.24249	.24995	4.0009	.97015	1.0308	4.1239	58
3	.24277	.25026	3.9959	.97008	1.0308	4.1191	57
4	.24305	.25056	3.9910	.97001	1.0309	4.1144	56
5	.24333	.25087	3.9861	.96994	1.0310	4.1096	55
6	.24362	.25118	3.9812	.96987	1.0311	4.1048	54
7	.24390	.25149	3.9763	.96980	1.0311	4.1001	53
8	.24418	.25180	3.9714	.96973	1.0312	4.0954	52
9	.24446	.25211	3.9665	.96966	1.0313	4.0906	51
10	.24474	.25242	3.9617	.96959	1.0314	4.0859	50
11	.24503	.25273	3.9568	.96952	1.0314	4.0812	49
12	.24531	.25304	3.9520	.96945	1.0315	4.0765	48
13	.24559	.25335	3.9471	.96937	1.0316	4.0718	47
14	.24587	.25366	3.9423	.96930	1.0317	4.0672	46
15	.24615	.25397	3.9375	.96923	1.0317	4.0625	45
16	.24644	.25428	3.9327	.96916	1.0318	4.0579	44
17	.24672	.25459	3.9279	.96909	1.0319	4.0532	43
18	.24700	.25490	3.9232	.96902	1.0320	4.0486	42
19	.24728	.25521	3.9184	.96894	1.0321	4.0440	41
20	.24756	.25552	3.9136	.96887	1.0321	4.0394	40
21	.24784	.25583	3.9089	.96880	1.0322	4.0348	39
22	.24813	.25614	3.9042	.96873	1.0323	4.0302	38
23	.24841	.25645	3.8995	.96866	1.0324	4.0256	37
24	.24869	.25676	3.8947	.96858	1.0324	4.0211	36
25	.24897	.25707	3.8900	.96851	1.0325	4.0165	35
26	.24925	.25738	3.8854	.96844	1.0326	4.0120	34
27	.24954	.25769	3.8807	.96837	1.0327	4.0075	33
28	.24982	.25800	3.8760	.96829	1.0327	4.0029	32
29	.25010	.25831	3.8714	.96822	1.0328	3.9984	31
30	.25038	.25862	3.8667	.96815	1.0329	3.9939	30
31	.25066	.25893	3.8621	.96807	1.0330	3.9894	29
32	.25094	.25924	3.8575	.96800	1.0331	3.9850	28
33	.25122	.25955	3.8528	.96793	1.0331	3.9805	27
34	.25151	.25986	3.8482	.96786	1.0332	3.9760	26
35	.25179	.26017	3.8436	.96778	1.0333	3.9716	25
36	.25207	.26048	3.8391	.96771	1.0334	3.9672	24
37	.25235	.26079	3.8345	.96764	1.0334	3.9627	23
38	.25263	.26110	3.8299	.96756	1.0335	3.9583	22
39	.25291	.26141	3.8254	.96749	1.0336	3.9539	21
40	.25320	.26172	3.8208	.96742	1.0337	3.9495	20
41	.25348	.26203	3.8163	.96734	1.0338	3.9451	19
42	.25376	.26235	3.8118	.96727	1.0338	3.9408	18
43	.25404	.26266	3.8073	.96719	1.0339	3.9364	17
44	.25432	.26297	3.8028	.96712	1.0340	3.9320	16
45	.25460	.26328	3.7983	.96705	1.0341	3.9277	15
46	.25488	.26359	3.7938	.96697	1.0342	3.9234	14
47	.25516	.26390	3.7893	.96690	1.0342	3.9190	13
48	.25545	.26421	3.7848	.96682	1.0343	3.9147	12
49	.25573	.26452	3.7804	.96675	1.0344	3.9104	11
50	.25601	.26483	3.7760	.96667	1.0345	3.9061	10
51	.25629	.26515	3.7715	.96660	1.0346	3.9018	9
52	.25657	.26546	3.7671	.96653	1.0346	3.8976	8
53	.25685	.26577	3.7627	.96645	1.0347	3.8933	7
54	.25713	.26608	3.7583	.96638	1.0348	3.8890	6
55	.25741	.26639	3.7539	.96630	1.0349	3.8848	5
56	.25769	.26670	3.7495	.96623	1.0350	3.8806	4
57	.25798	.26701	3.7451	.96615	1.0350	3.8763	3
58	.25826	.26733	3.7408	.96608	1.0351	3.8721	2
59	.25854	.26764	3.7364	.96600	1.0352	3.8679	1
60	.25882	.26795	3.7321	.96593	1.0353	3.8637	0
'	Cos	Cot	Tan	Sin	Csc	Sec	'

104° (284°)   (255°) 75°

'	Sin	Tan	Cot	Cos	Sec	Csc	'
0	.25882	.26795	3.7321	.96593	1.0353	3.8637	60
1	.25910	.26826	3.7277	.96585	1.0354	3.8595	59
2	.25938	.26857	3.7234	.96578	1.0354	3.8553	58
3	.25966	.26888	3.7191	.96570	1.0355	3.8512	57
4	.25994	.26920	3.7148	.96562	1.0356	3.8470	56
5	.26022	.26951	3.7105	.96555	1.0357	3.8428	55
6	.26050	.26982	3.7062	.96547	1.0358	3.8387	54
7	.26079	.27013	3.7019	.96540	1.0368	3.8346	53
8	.26107	.27044	3.6976	.96532	1.0359	3.8304	52
9	.26135	.27076	3.6933	.96524	1.0360	3.8263	51
10	.26163	.27107	3.6891	.96517	1.0361	3.8222	50
11	.26191	.27138	3.6848	.96509	1.0362	3.8181	49
12	.26219	.27169	3.6806	.96502	1.0363	3.8140	48
13	.26247	.27201	3.6764	.96494	1.0363	3.8100	47
14	.26275	.27232	3.6722	.96486	1.0364	3.8059	46
15	.26303	.27263	3.6680	.96479	1.0365	3.8018	45
16	.26331	.27294	3.6638	.96471	1.0366	3.7978	44
17	.26359	.27326	3.6596	.96463	1.0367	3.7937	43
18	.26387	.27357	3.6554	.96456	1.0367	3.7897	42
19	.26415	.27388	3.6512	.96448	1.0368	3.7857	41
20	.26443	.27419	3.6470	.96440	1.0369	3.7817	40
21	.26471	.27451	3.6429	.96433	1.0370	3.7777	39
22	.26500	.27482	3.6387	.96425	1.0371	3.7737	38
23	.26528	.27513	3.6346	.96417	1.0372	3.7697	37
24	.26556	.27545	3.6305	.96410	1.0372	3.7657	36
25	.26584	.27576	3.6264	.96402	1.0373	3.7617	35
26	.26612	.27607	3.6222	.96394	1.0374	3.7577	34
27	.26640	.27638	3.6181	.96386	1.0375	3.7538	33
28	.26668	.27670	3.6140	.96379	1.0376	3.7498	32
29	.26696	.27701	3.6100	.96371	1.0377	3.7459	31
30	.26724	.27732	3.6059	.96363	1.0377	3.7420	30
31	.26752	.27764	3.6018	.96355	1.0378	3.7381	29
32	.26780	.27795	3.5978	.96347	1.0379	3.7341	28
33	.26808	.27826	3.5937	.96340	1.0380	3.7302	27
34	.26836	.27858	3.5897	.96332	1.0381	3.7263	26
35	.26864	.27889	3.5856	.96324	1.0382	3.7225	25
36	.26892	.27921	3.5816	.96316	1.0382	3.7186	24
37	.26920	.27952	3.5776	.96308	1.0383	3.7147	23
38	.26948	.27983	3.5736	.96301	1.0384	3.7108	22
39	.26976	.28015	3.5696	.96293	1.0385	3.7070	21
40	.27004	.28046	3.5656	.96285	1.0386	3.7032	20
41	.27032	.28077	3.5616	.96277	1.0387	3.6993	19
42	.27060	.28109	3.5576	.96269	1.0388	3.6955	18
43	.27088	.28140	3.5536	.96261	1.0388	3.6917	17
44	.27116	.28172	3.5497	.96253	1.0389	3.6879	16
45	.27144	.28203	3.5457	.96246	1.0390	3.6840	15
46	.27172	.28234	3.5418	.96238	1.0391	3.6803	14
47	.27200	.28266	3.5379	.96230	1.0392	3.6765	13
48	.27228	.28297	3.5339	.96222	1.0393	3.6727	12
49	.27256	.28329	3.5300	.96214	1.0394	3.6689	11
50	.27284	.28360	3.5261	.96206	1.0394	3.6652	10
51	.27312	.28391	3.5222	.96198	1.0395	3.6614	9
52	.27340	.28423	3.5183	.96190	1.0396	3.6575	8
53	.27368	.28454	3.5144	.96182	1.0397	3.6539	7
54	.27396	.28486	3.5105	.96174	1.0398	3.6502	6
55	.27424	.28517	3.5067	.96166	1.0399	3.6465	5
56	.27452	.28549	3.5028	.96158	1.0400	3.6427	4
57	.27480	.28580	3.4989	.96150	1.0400	3.6390	3
58	.27508	.28612	3.4951	.96142	1.0401	3.6353	2
59	.27536	.28643	3.4912	.96134	1.0402	3.6316	1
60	.27564	.28675	3.4874	.96126	1.0403	3.6280	0
'	Cos	Cot	Tan	Sin	Csc	Sec	'

105° (285°)   (254°) 74°

NATURAL TRIGONOMETRIC FUNCTIONS TO FIVE PLACES (continued)

16° (196°)   (343°) 163°

'	Sin	Tan	Cot	Cos	Sec	Csc	'
0	.27564	.28675	3.4874	.96126	1.0403	3.6280	60
1	.27592	.28706	3.4836	.96118	1.0404	3.6243	59
2	.27620	.28738	3.4798	.96110	1.0405	3.6206	58
3	.27648	.28769	3.4760	.96102	1.0406	3.6169	57
4	.27676	.28801	3.4722	.96094	1.0406	3.6133	56
5	.27704	.28832	3.4684	.96086	1.0407	3.6097	55
6	.27731	.28864	3.4646	.96078	1.0408	3.6060	54
7	.27759	.28895	3.4608	.96070	1.0409	3.6024	53
8	.27787	.28927	3.4570	.96062	1.0410	3.5988	52
9	.27815	.28958	3.4533	.96054	1.0411	3.5951	51
10	.27843	.28990	3.4495	.96046	1.0412	3.5915	50
11	.27871	.29021	3.4458	.96037	1.0413	3.5879	49
12	.27899	.29053	3.4420	.96029	1.0413	3.5843	48
13	.27927	.29084	3.4383	.96021	1.0414	3.5808	47
14	.27955	.29116	3.4346	.96013	1.0415	3.5772	46
15	.27983	.29147	3.4308	.96005	1.0416	3.5736	45
16	.28011	.29179	3.4271	.95997	1.0417	3.5700	44
17	.28039	.29210	3.4234	.95989	1.0418	3.5665	43
18	.28067	.29242	3.4197	.95981	1.0419	3.5629	42
19	.28095	.29274	3.4160	.95972	1.0420	3.5594	41
20	.28123	.29305	3.4124	.95964	1.0421	3.5559	40
21	.28150	.29337	3.4087	.95956	1.0421	3.5523	39
22	.28178	.29368	3.4050	.95948	1.0422	3.5488	38
23	.28206	.29400	3.4014	.95940	1.0423	3.5453	37
24	.28234	.29432	3.3977	.95931	1.0424	3.5418	36
25	.28262	.29463	3.3941	.95923	1.0425	3.5383	35
26	.28290	.29495	3.3904	.95915	1.0426	3.5348	34
27	.28318	.29526	3.3868	.95907	1.0427	3.5313	33
28	.28346	.29558	3.3832	.95898	1.0428	3.5279	32
29	.28374	.29590	3.3796	.95890	1.0429	3.5244	31
30	.28402	.29621	3.3759	.95882	1.0429	3.5209	30
31	.28429	.29653	3.3723	.95874	1.0430	3.5175	29
32	.28457	.29685	3.3687	.95865	1.0431	3.5140	28
33	.28485	.29716	3.3652	.95857	1.0432	3.5106	27
34	.28513	.29748	3.3616	.95849	1.0433	3.5072	26
35	.28541	.29780	3.3580	.95841	1.0434	3.5037	25
36	.28569	.29811	3.3544	.95832	1.0435	3.5003	24
37	.28597	.29843	3.3509	.95824	1.0436	3.4969	23
38	.28625	.29875	3.3473	.95816	1.0437	3.4935	22
39	.28652	.29906	3.3438	.95807	1.0438	3.4901	21
40	.28680	.29938	3.3402	.95799	1.0439	3.4867	20
41	.28708	.29970	3.3367	.95791	1.0439	3.4833	19
42	.28736	.30001	3.3332	.95782	1.0440	3.4799	18
43	.28764	.30033	3.3297	.95774	1.0441	3.4766	17
44	.28792	.30065	3.3261	.95766	1.0442	3.4732	16
45	.28820	.30097	3.3226	.95757	1.0443	3.4699	15
46	.28847	.30128	3.3191	.95749	1.0444	3.4665	14
47	.28875	.30160	3.3156	.95740	1.0445	3.4632	13
48	.28903	.30192	3.3122	.95732	1.0446	3.4598	12
49	.28931	.30224	3.3087	.95724	1.0447	3.4565	11
50	.28959	.30255	3.3052	.95715	1.0448	3.4532	10
51	.28987	.30287	3.3017	.95707	1.0449	3.4499	9
52	.29015	.30319	3.2983	.95698	1.0450	3.4465	8
53	.29042	.30351	3.2948	.95690	1.0450	3.4432	7
54	.29070	.30382	3.2914	.95681	1.0451	3.4399	6
55	.29098	.30414	3.2879	.95673	1.0452	3.4367	5
56	.29126	.30446	3.2845	.95664	1.0453	3.4334	4
57	.29154	.30478	3.2811	.95656	1.0454	3.4301	3
58	.29182	.30509	3.2777	.95647	1.0455	3.4268	2
59	.29209	.30541	3.2743	.95639	1.0456	3.4236	1
60	.29237	.30573	3.2709	.95630	1.0457	3.4203	0
'	Cos	Cot	Tan	Sin	Csc	Sec	'

17° (197°)   (342°) 162°

'	Sin	Tan	Cot	Cos	Sec	Csc	'
0	.29237	.30573	3.2709	.95630	1.0457	3.4203	60
1	.29265	.30605	3.2675	.95622	1.0458	3.4171	59
2	.29293	.30637	3.2641	.95613	1.0459	3.4138	58
3	.29321	.30669	3.2607	.95605	1.0460	3.4106	57
4	.29348	.30700	3.2573	.95596	1.0461	3.4073	56
5	.29376	.30732	3.2539	.95588	1.0462	3.4041	55
6	.29404	.30764	3.2506	.95579	1.0463	3.4009	54
7	.29432	.30796	3.2472	.95571	1.0463	3.3977	53
8	.29460	.30828	3.2438	.95562	1.0464	3.3945	52
9	.29487	.30860	3.2405	.95554	1.0465	3.3913	51
10	.29515	.30891	3.2371	.95545	1.0466	3.3881	50
11	.29543	.30923	3.2338	.95536	1.0467	3.3849	49
12	.29571	.30955	3.2305	.95528	1.0468	3.3817	48
13	.29599	.30987	3.2272	.95519	1.0469	3.3785	47
14	.29626	.31019	3.2238	.95511	1.0470	3.3754	46
15	.29654	.31051	3.2205	.95502	1.0471	3.3722	45
16	.29682	.31083	3.2172	.95493	1.0472	3.3691	44
17	.29710	.31115	3.2139	.95485	1.0473	3.3659	43
18	.29737	.31147	3.2106	.95476	1.0474	3.3628	42
19	.29765	.31178	3.2073	.95467	1.0475	3.3596	41
20	.29793	.31210	3.2041	.95459	1.0476	3.3565	40
21	.29821	.31242	3.2008	.95450	1.0477	3.3534	39
22	.29849	.31274	3.1975	.95441	1.0478	3.3502	38
23	.29876	.31306	3.1943	.95433	1.0479	3.3471	37
24	.29904	.31338	3.1910	.95424	1.0480	3.3440	36
25	.29932	.31370	3.1878	.95415	1.0480	3.3409	35
26	.29960	.31402	3.1845	.95407	1.0481	3.3378	34
27	.29987	.31434	3.1813	.95398	1.0482	3.3347	33
28	.30015	.31466	3.1780	.95389	1.0483	3.3317	32
29	.30043	.31498	3.1748	.95380	1.0484	3.3286	31
30	.30071	.31530	3.1716	.95372	1.0485	3.3255	30
31	.30098	.31562	3.1684	.95363	1.0486	3.3224	29
32	.30126	.31594	3.1652	.95354	1.0487	3.3194	28
33	.30154	.31626	3.1620	.95345	1.0488	3.3163	27
34	.30182	.31658	3.1588	.95337	1.0489	3.3133	26
35	.30209	.31690	3.1556	.95328	1.0490	3.3102	25
36	.30237	.31722	3.1524	.95319	1.0491	3.3072	24
37	.30265	.31754	3.1492	.95310	1.0492	3.3042	23
38	.30292	.31786	3.1460	.95301	1.0493	3.3012	22
39	.30320	.31818	3.1429	.95293	1.0494	3.2981	21
40	.30348	.31850	3.1397	.95284	1.0495	3.2951	20
41	.30376	.31882	3.1366	.95275	1.0496	3.2921	19
42	.30403	.31914	3.1334	.95266	1.0497	3.2891	18
43	.30431	.31946	3.1303	.95257	1.0498	3.2861	17
44	.30459	.31978	3.1271	.95248	1.0499	3.2831	16
45	.30486	.32010	3.1240	.95240	1.0500	3.2801	15
46	.30514	.32042	3.1209	.95231	1.0501	3.2772	14
47	.30542	.32074	3.1178	.95222	1.0502	3.2742	13
48	.30570	.32106	3.1146	.95213	1.0503	3.2712	12
49	.30597	.32139	3.1115	.95204	1.0504	3.2683	11
50	.30625	.32171	3.1084	.95195	1.0505	3.2653	10
51	.30653	.32203	3.1053	.95186	1.0506	3.2624	9
52	.30680	.32235	3.1022	.95177	1.0507	3.2594	8
53	.30708	.32267	3.0991	.95168	1.0508	3.2565	7
54	.30736	.32299	3.0961	.95159	1.0509	3.2535	6
55	.30763	.32331	3.0930	.95150	1.0510	3.2506	5
56	.30791	.32363	3.0899	.95142	1.0511	3.2477	4
57	.30819	.32396	3.0868	.95133	1.0512	3.2448	3
58	.30846	.32428	3.0838	.95124	1.0513	3.2419	2
59	.30874	.32460	3.0807	.95115	1.0514	3.2390	1
60	.30902	.32492	3.0777	.95106	1.0515	3.2361	0
'	Cos	Cot	Tan	Sin	Csc	Sec	'

TRIGONOMETRY

NATURAL TRIGONOMETRIC FUNCTIONS TO FIVE PLACES (continued)

18° (198°)   (341°) 161°

′	Sin	Tan	Cot	Cos	Sec	Csc	′
0	.30902	.32492	3.0777	.95106	1.0515	3.2361	60
1	.30929	.32524	3.0746	.95097	1.0516	3.2332	59
2	.30957	.32556	3.0716	.95088	1.0517	3.2303	58
3	.30985	.32588	3.0686	.95079	1.0518	3.2274	57
4	.31012	.32621	3.0655	.95070	1.0519	3.2245	56
5	.31040	.32653	3.0625	.95061	1.0520	3.2217	55
6	.31068	.32685	3.0595	.95052	1.0521	3.2188	54
7	.31095	.32717	3.0565	.95043	1.0522	3.2159	53
8	.31123	.32749	3.0535	.95033	1.0523	3.2131	52
9	.31151	.32782	3.0505	.95024	1.0524	3.2102	51
10	.31178	.32814	3.0475	.95015	1.0525	3.2074	50
11	.31206	.32846	3.0445	.95006	1.0526	3.2045	49
12	.31233	.32878	3.0415	.94997	1.0527	3.2017	48
13	.31261	.32911	3.0385	.94988	1.0528	3.1989	47
14	.31289	.32943	3.0356	.94979	1.0529	3.1960	46
15	.31316	.32975	3.0326	.94970	1.0530	3.1932	45
16	.31344	.33007	3.0296	.94961	1.0531	3.1904	44
17	.31372	.33040	3.0267	.94952	1.0532	3.1876	43
18	.31399	.33072	3.0237	.94943	1.0533	3.1848	42
19	.31427	.33104	3.0208	.94933	1.0534	3.1820	41
20	.31454	.33136	3.0178	.94924	1.0535	3.1792	40
21	.31482	.33169	3.0149	.94915	1.0536	3.1764	39
22	.31510	.33201	3.0120	.94906	1.0537	3.1736	38
23	.31537	.33233	3.0090	.94897	1.0538	3.1708	37
24	.31565	.33266	3.0061	.94888	1.0539	3.1681	36
25	.31593	.33298	3.0032	.94878	1.0540	3.1653	35
26	.31620	.33330	3.0003	.94869	1.0541	3.1625	34
27	.31648	.33363	2.9974	.94860	1.0542	3.1598	33
28	.31675	.33395	2.9945	.94851	1.0543	3.1570	32
29	.31703	.33427	2.9916	.94842	1.0544	3.1543	31
30	.31730	.33460	2.9887	.94832	1.0545	3.1515	30
31	.31758	.33492	2.9858	.94823	1.0546	3.1488	29
32	.31786	.33524	2.9829	.94814	1.0547	3.1461	28
33	.31813	.33557	2.9800	.94805	1.0548	3.1433	27
34	.31841	.33589	2.9772	.94795	1.0549	3.1406	26
35	.31868	.33621	2.9743	.94786	1.0550	3.1379	25
36	.31896	.33654	2.9714	.94777	1.0551	3.1352	24
37	.31923	.33686	2.9686	.94768	1.0552	3.1325	23
38	.31951	.33718	2.9657	.94758	1.0553	3.1298	22
39	.31979	.33751	2.9629	.94749	1.0554	3.1271	21
40	.32006	.33783	2.9600	.94740	1.0555	3.1244	20
41	.32034	.33816	2.9572	.94730	1.0556	3.1217	19
42	.32061	.33848	2.9544	.94721	1.0557	3.1190	18
43	.32089	.33881	2.9515	.94712	1.0558	3.1163	17
44	.32116	.33913	2.9487	.94702	1.0559	3.1137	16
45	.32144	.33945	2.9459	.94693	1.0560	3.1110	15
46	.32171	.33978	2.9431	.94684	1.0561	3.1083	14
47	.32199	.34010	2.9403	.94674	1.0563	3.1057	13
48	.32227	.34043	2.9375	.94665	1.0564	3.1030	12
49	.32254	.34075	2.9347	.94656	1.0565	3.1004	11
50	.32282	.34108	2.9319	.94646	1.0566	3.0977	10
51	.32309	.34140	2.9291	.94637	1.0567	3.0951	9
52	.32337	.34173	2.9263	.94627	1.0568	3.0925	8
53	.32364	.34205	2.9235	.94618	1.0569	3.0898	7
54	.32392	.34238	2.9208	.94609	1.0570	3.0872	6
55	.32419	.34270	2.9180	.94599	1.0571	3.0846	5
56	.32447	.34303	2.9152	.94590	1.0572	3.0820	4
57	.32474	.34335	2.9125	.94580	1.0573	3.0794	3
58	.32502	.34368	2.9097	.94571	1.0574	3.0768	2
59	.32529	.34400	2.9070	.94561	1.0575	3.0742	1
60	.32557	.34433	2.9042	.94552	1.0576	3.0716	0
′	Cos	Cot	Tan	Sin	Csc	Sec	′

19° (199°)   (340°) 160°

′	Sin	Tan	Cot	Cos	Sec	Csc	′
0	.32557	.34433	2.9042	.94552	1.0576	3.0716	60
1	.32584	.34465	2.9015	.94542	1.0577	3.0690	59
2	.32612	.34498	2.8987	.94533	1.0578	3.0664	58
3	.32639	.34530	2.8960	.94523	1.0579	3.0638	57
4	.32667	.34563	2.8933	.94514	1.0580	3.0612	56
5	.32694	.34596	2.8905	.94504	1.0582	3.0586	55
6	.32722	.34628	2.8878	.94495	1.0583	3.0561	54
7	.32749	.34661	2.8851	.94485	1.0584	3.0535	53
8	.32777	.34693	2.8824	.94476	1.0585	3.0509	52
9	.32804	.34726	2.8797	.94466	1.0586	3.0484	51
10	.32832	.34758	2.8770	.94457	1.0587	3.0458	50
11	.32859	.34791	2.8743	.94447	1.0588	3.0433	49
12	.32887	.34824	2.8716	.94438	1.0589	3.0407	48
13	.32914	.34856	2.8689	.94428	1.0590	3.0382	47
14	.32942	.34889	2.8662	.94418	1.0591	3.0357	46
15	.32969	.34922	2.8636	.94409	1.0592	3.0331	45
16	.32997	.34954	2.8609	.94399	1.0593	3.0306	44
17	.33024	.34987	2.8582	.94390	1.0594	3.0281	43
18	.33051	.35020	2.8556	.94380	1.0595	3.0256	42
19	.33079	.35052	2.8529	.94370	1.0597	3.0231	41
20	.33106	.35085	2.8502	.94361	1.0598	3.0206	40
21	.33134	.35118	2.8476	.94351	1.0599	3.0181	39
22	.33161	.35150	2.8449	.94342	1.0600	3.0156	38
23	.33189	.35183	2.8423	.94332	1.0601	3.0131	37
24	.33216	.35216	2.8397	.94322	1.0602	3.0106	36
25	.33244	.35248	2.8370	.94313	1.0603	3.0081	35
26	.33271	.35281	2.8344	.94303	1.0604	3.0056	34
27	.33298	.35314	2.8318	.94293	1.0605	3.0031	33
28	.33326	.35346	2.8291	.94284	1.0606	3.0007	32
29	.33353	.35379	2.8265	.94274	1.0607	2.9982	31
30	.33381	.35412	2.8239	.94264	1.0608	2.9957	30
31	.33408	.35445	2.8213	.94254	1.0610	2.9933	29
32	.33436	.35477	2.8187	.94245	1.0611	2.9908	28
33	.33463	.35510	2.8161	.94235	1.0612	2.9884	27
34	.33490	.35543	2.8135	.94225	1.0613	2.9859	26
35	.33518	.35576	2.8109	.94215	1.0614	2.9835	25
36	.33545	.35608	2.8083	.94206	1.0615	2.9811	24
37	.33573	.35641	2.8057	.94196	1.0616	2.9786	23
38	.33600	.35674	2.8032	.94186	1.0617	2.9762	22
39	.33627	.35707	2.8006	.94176	1.0618	2.9738	21
40	.33655	.35740	2.7980	.94167	1.0619	2.9713	20
41	.33682	.35772	2.7955	.94157	1.0621	2.9689	19
42	.33710	.35805	2.7929	.94147	1.0622	2.9665	18
43	.33737	.35838	2.7903	.94137	1.0623	2.9641	17
44	.33764	.35871	2.7878	.94127	1.0624	2.9617	16
45	.33792	.35904	2.7852	.94118	1.0625	2.9593	15
46	.33819	.35937	2.7827	.94108	1.0626	2.9569	14
47	.33846	.35969	2.7801	.94098	1.0627	2.9545	13
48	.33874	.36002	2.7776	.94088	1.0628	2.9521	12
49	.33901	.36035	2.7751	.94078	1.0629	2.9498	11
50	.33929	.36068	2.7725	.94068	1.0631	2.9474	10
51	.33956	.36101	2.7700	.94058	1.0632	2.9450	9
52	.33983	.36134	2.7675	.94049	1.0633	2.9426	8
53	.34011	.36167	2.7650	.94039	1.0634	2.9403	7
54	.34038	.36199	2.7625	.94029	1.0635	2.9379	6
55	.34065	.36232	2.7600	.94019	1.0636	2.9355	5
56	.34093	.36265	2.7575	.94009	1.0637	2.9332	4
57	.34120	.36298	2.7550	.93999	1.0638	2.9308	3
58	.34147	.36331	2.7525	.93989	1.0640	2.9285	2
59	.34175	.36364	2.7500	.93979	1.0641	2.9261	1
60	.34202	.36397	2.7475	.93969	1.0642	2.9238	0
′	Cos	Cot	Tan	Sin	Csc	Sec	′

NATURAL TRIGONOMETRIC FUNCTIONS TO FIVE PLACES (continued)

20° (200°)   (339°) 159°   **21° (201°)**   (338°) 158°

′	Sin	Tan	Cot	Cos	Sec	Csc	′	′	Sin	Tan	Cot	Cos	Sec	Csc	′
0	.34202	.36397	2.7475	.93969	1.0642	2.9238	60	0	.35837	.38386	2.6051	.93358	1.0711	2.7904	60
1	.34229	.36430	2.7450	.93959	1.0643	2.9215	59	1	.35864	.38420	2.6028	.93348	1.0713	2.7883	59
2	.34257	.36463	2.7425	.93949	1.0644	2.9191	58	2	.35891	.38453	2.6006	.93337	1.0714	2.7862	58
3	.34284	.36496	2.7400	.93939	1.0645	2.9168	57	3	.35918	.38487	2.5983	.93327	1.0715	2.7841	57
4	.34311	.36529	2.7376	.93929	1.0646	2.9145	56	4	.35945	.38520	2.5961	.93316	1.0716	2.7820	56
5	.34339	.36562	2.7351	.93919	1.0647	2.9122	55	5	.35973	.38553	2.5938	.93306	1.0717	2.7799	55
6	.34366	.36595	2.7326	.93909	1.0649	2.9099	54	6	.36000	.38587	2.5916	.93295	1.0719	2.7778	54
7	.34393	.36628	2.7302	.93899	1.0650	2.9075	53	7	.36027	.38620	2.5893	.93285	1.0720	2.7757	53
8	.34421	.36661	2.7277	.93889	1.0651	2.9052	52	8	.36054	.38654	2.5871	.93274	1.0721	2.7736	52
9	.34448	.36694	2.7253	.93879	1.0652	2.9029	51	9	.36081	.38687	2.5848	.93264	1.0722	2.7715	51
10	.34475	.36727	2.7228	.93869	1.0653	2.9006	50	10	.36108	.38721	2.5826	.93253	1.0723	2.7695	50
11	.34503	.36760	2.7204	.93859	1.0654	2.8983	49	11	.36135	.38754	2.5804	.93243	1.0725	2.7674	49
12	.34530	.36793	2.7179	.93849	1.0655	2.8960	48	12	.36162	.38787	2.5782	.93232	1.0726	2.7653	48
13	.34557	.36826	2.7155	.93839	1.0657	2.8938	47	13	.36190	.38821	2.5759	.93222	1.0727	2.7632	47
14	.34584	.36859	2.7130	.93829	1.0658	2.8915	46	14	.36217	.38854	2.5737	.93211	1.0728	2.7612	46
15	.34612	.36892	2.7106	.93819	1.0659	2.8892	45	15	.36244	.38888	2.5715	.93201	1.0730	2.7591	45
16	.34639	.36925	2.7082	.93809	1.0660	2.8869	44	16	.36271	.38921	2.5693	.93190	1.0731	2.7570	44
17	.34666	.36958	2.7058	.93799	1.0661	2.8846	43	17	.36298	.38955	2.5671	.93180	1.0732	2.7550	43
18	.34694	.36991	2.7034	.93789	1.0662	2.8824	42	18	.36325	.38988	2.5649	.93169	1.0733	2.7529	42
19	.34721	.37024	2.7009	.93779	1.0663	2.8801	41	19	.36352	.39022	2.5627	.93159	1.0734	2.7509	41
20	.34748	.37057	2.6985	.93769	1.0665	2.8779	40	20	.36379	.39055	2.5605	.93148	1.0736	2.7488	40
21	.34775	.37090	2.6961	.93759	1.0666	2.8756	39	21	.36406	.39089	2.5583	.93137	1.0737	2.7468	39
22	.34803	.37123	2.6937	.93748	1.0667	2.8733	38	22	.36434	.39122	2.5561	.93127	1.0738	2.7447	38
23	.34830	.37157	2.6913	.93738	1.0668	2.8711	37	23	.36461	.39156	2.5539	.93116	1.0739	2.7427	37
24	.34857	.37190	2.6889	.93728	1.0669	2.8688	36	24	.36488	.39190	2.5517	.93106	1.0740	2.7407	36
25	.34884	.37223	2.6865	.93718	1.0670	2.8666	35	25	.36515	.39223	2.5495	.93095	1.0742	2.7386	35
26	.34912	.37256	2.6841	.93708	1.0671	2.8644	34	26	.36542	.39257	2.5473	.93084	1.0743	2.7366	34
27	.34939	.37289	2.6818	.93698	1.0673	2.8621	33	27	.36569	.39290	2.5452	.93074	1.0744	2.7346	33
28	.34966	.37322	2.6794	.93688	1.0674	2.8599	32	28	.36596	.39324	2.5430	.93063	1.0745	2.7325	32
29	.34993	.37355	2.0770	.93077	1.0075	2.8577	31	29	.36623	.39357	2.5408	.93052	1.0747	2.7305	31
30	.35021	.37388	2.6746	.93667	1.0676	2.8555	30	30	.36650	.39391	2.5386	.93042	1.0748	2.7285	30
31	.35048	.37422	2.6723	.93657	1.0677	2.8532	29	31	.36677	.39425	2.5365	.93031	1.0749	2.7265	29
32	.35075	.37455	2.6699	.93647	1.0678	2.8510	28	32	.36704	.39458	2.5343	.93020	1.0750	2.7245	28
33	.35102	.37488	2.6675	.93636	1.0680	2.8488	27	33	.36731	.39492	2.5322	.93010	1.0752	2.7225	27
34	.35130	.37521	2.6652	.93626	1.0681	2.8466	26	34	.36758	.39526	2.5300	.92999	1.0753	2.7205	26
35	.35157	.37554	2.6628	.93616	1.0682	2.8444	25	35	.36785	.39559	2.5279	.92988	1.0754	2.7185	25
36	.35184	.37588	2.6605	.93606	1.0683	2.8422	24	36	.36812	.39593	2.5257	.92978	1.0755	2.7165	24
37	.35211	.37621	2.6581	.93596	1.0684	2.8400	23	37	.36839	.39626	2.5236	.92967	1.0757	2.7145	23
38	.35239	.37654	2.6558	.93585	1.0685	2.8378	22	38	.36867	.39660	2.5214	.92956	1.0758	2.7125	22
39	.35266	.37687	2.6534	.93575	1.0687	2.8356	21	39	.36894	.39694	2.5193	.92945	1.0759	2.7105	21
40	.35293	.37720	2.6511	.93565	1.0688	2.8334	20	40	.36921	.39727	2.5172	.92935	1.0760	2.7085	20
41	.35320	.37754	2.6488	.93555	1.0689	2.8312	19	41	.36948	.39761	2.5150	.92924	1.0761	2.7065	19
42	.35347	.37787	2.6464	.93544	1.0690	2.8291	18	42	.36975	.39795	2.5129	.92913	1.0763	2.7046	18
43	.35375	.37820	2.6441	.93534	1.0691	2.8269	17	43	.37002	.39829	2.5108	.92902	1.0764	2.7026	17
44	.35402	.37853	2.6418	.93524	1.0692	2.8247	16	44	.37029	.39862	2.5086	.92892	1.0765	2.7006	16
45	.35429	.37887	2.6395	.93514	1.0694	2.8225	15	45	.37056	.39896	2.5065	.92881	1.0766	2.6986	15
46	.35456	.37920	2.6371	.93503	1.0695	2.8204	14	46	.37083	.39930	2.5044	.92870	1.0768	2.6967	14
47	.35484	.37953	2.6348	.93493	1.0696	2.8182	13	47	.37110	.39963	2.5023	.92859	1.0769	2.6947	13
48	.35511	.37986	2.6325	.93483	1.0697	2.8161	12	48	.37137	.39997	2.5002	.92849	1.0770	2.6927	12
49	.35538	.38020	2.6302	.93472	1.0698	2.8139	11	49	.37164	.40031	2.4981	.92838	1.0771	2.6908	11
50	.35565	.38053	2.6279	.93462	1.0700	2.8117	10	50	.37191	.40065	2.4960	.92827	1.0773	2.6888	10
51	.35592	.38086	2.6256	.93452	1.0701	2.8096	9	51	.37218	.40098	2.4939	.92816	1.0774	2.6869	9
52	.35619	.38120	2.6233	.93441	1.0702	2.8075	8	52	.37245	.40132	2.4918	.92805	1.0775	2.6849	8
53	.35647	.38153	2.6210	.93431	1.0703	2.8053	7	53	.37272	.40166	2.4897	.92794	1.0777	2.6830	7
54	.35674	.38186	2.6187	.93420	1.0704	2.8032	6	54	.37299	.40200	2.4876	.92784	1.0778	2.6811	6
55	.35701	.38220	2.6165	.93410	1.0705	2.8010	5	55	.37326	.40234	2.4855	.92773	1.0779	2.6791	5
56	.35728	.38253	2.6142	.93400	1.0707	2.7989	4	56	.37353	.40267	2.4834	.92762	1.0780	2.6772	4
57	.35755	.38286	2.6119	.93389	1.0708	2.7968	3	57	.37380	.40301	2.4813	.92751	1.0782	2.6752	3
58	.35782	.38320	2.6096	.93379	1.0709	2.7947	2	58	.37407	.40335	2.4792	.92740	1.0783	2.6733	2
59	.35810	.38353	2.6074	.93368	1.0710	2.7925	1	59	.37434	.40369	2.4772	.92729	1.0784	2.6714	1
60	.35837	.38386	2.6051	.93358	1.0711	2.7904	0	60	.37461	.40403	2.4751	.92718	1.0785	2.6695	0
′	Cos	Cot	Tan	Sin	Csc	Sec	′	′	Cos	Cot	Tan	Sin	Csc	Sec	′

110° (290°)   (249°) 69°   **111° (291°)**   (248°) 68°

NATURAL TRIGONOMETRIC FUNCTIONS TO FIVE PLACES (continued)

22° (202°)   (337°) 157°

'	Sin	Tan	Cot	Cos	Sec	Csc	'
0	.37461	.40403	2.4751	.92718	1.0785	2.6695	60
1	.37488	.40436	2.4730	.92707	1.0787	2.6675	59
2	.37515	.40470	2.4709	.92697	1.0788	2.6656	58
3	.37542	.40504	2.4689	.92686	1.0789	2.6637	57
4	.37569	.40538	2.4668	.92675	1.0790	2.6618	56
5	.37595	.40572	2.4648	.92664	1.0792	2.6599	55
6	.37622	.40606	2.4627	.92653	1.0793	2.6580	54
7	.37649	.40640	2.4606	.92642	1.0794	2.6561	53
8	.37676	.40674	2.4586	.92631	1.0796	2.6542	52
9	.37703	.40707	2.4566	.92620	1.0797	2.6523	51
10	.37730	.40741	2.4545	.92609	1.0798	2.6504	50
11	.37757	.40775	2.4525	.92598	1.0799	2.6485	49
12	.37784	.40809	2.4504	.92587	1.0801	2.6466	48
13	.37811	.40843	2.4484	.92576	1.0802	2.6447	47
14	.37838	.40877	2.4464	.92565	1.0803	2.6429	46
15	.37865	.40911	2.4443	.92554	1.0804	2.6410	45
16	.37892	.40945	2.4423	.92543	1.0806	2.6391	44
17	.37919	.40979	2.4403	.92532	1.0807	2.6372	43
18	.37946	.41013	2.4383	.92521	1.0808	2.6354	42
19	.37973	.41047	2.4362	.92510	1.0810	2.6335	41
20	.37999	.41081	2.4342	.92499	1.0811	2.6316	40
21	.38026	.41115	2.4322	.92488	1.0812	2.6298	39
22	.38053	.41149	2.4302	.92477	1.0814	2.6279	38
23	.38080	.41183	2.4282	.92466	1.0815	2.6260	37
24	.38107	.41217	2.4262	.92455	1.0816	2.6242	36
25	.38134	.41251	2.4242	.92444	1.0817	2.6223	35
26	.38161	.41285	2.4222	.92432	1.0819	2.6205	34
27	.38188	.41319	2.4202	.92421	1.0820	2.6186	33
28	.38215	.41353	2.4182	.92410	1.0821	2.6168	32
29	.38241	.41387	2.4162	.92399	1.0823	2.6150	31
30	.38268	.41421	2.4142	.92388	1.0824	2.6131	30
31	.38295	.41455	2.4122	.92377	1.0825	2.6113	29
32	.38322	.41490	2.4102	.92366	1.0827	2.6095	28
33	.38349	.41524	2.4083	.92355	1.0828	2.6076	27
34	.38376	.41558	2.4063	.92343	1.0829	2.6058	26
35	.38403	.41592	2.4043	.92332	1.0830	2.6040	25
36	.38430	.41626	2.4023	.92321	1.0832	2.6022	24
37	.38456	.41660	2.4004	.92310	1.0833	2.6003	23
38	.38483	.41694	2.3984	.92299	1.0834	2.5985	22
39	.38510	.41728	2.3964	.92287	1.0836	2.5967	21
40	.38537	.41763	2.3945	.92276	1.0837	2.5949	20
41	.38564	.41797	2.3925	.92265	1.0838	2.5931	19
42	.38591	.41831	2.3906	.92254	1.0840	2.5913	18
43	.38617	.41865	2.3886	.92243	1.0841	2.5895	17
44	.38644	.41899	2.3867	.92231	1.0842	2.5877	16
45	.38671	.41933	2.3847	.92220	1.0844	2.5859	15
46	.38698	.41968	2.3828	.92209	1.0845	2.5841	14
47	.38725	.42002	2.3808	.92198	1.0846	2.5823	13
48	.38752	.42036	2.3789	.92186	1.0848	2.5805	12
49	.38778	.42070	2.3770	.92175	1.0849	2.5788	11
50	.38805	.42105	2.3750	.92164	1.0850	2.5770	10
51	.38832	.42139	2.3731	.92152	1.0852	2.5752	9
52	.38859	.42173	2.3712	.92141	1.0853	2.5734	8
53	.38886	.42207	2.3693	.92130	1.0854	2.5716	7
54	.38912	.42242	2.3673	.92119	1.0856	2.5699	6
55	.38939	.42276	2.3654	92107	1.0857	2.5681	5
56	.38966	.42310	2.3635	92096	1.0858	2.5663	4
57	.38993	.42345	2.3616	.92085	1.0860	2.5646	3
58	.39020	.42379	2.3597	.92073	1.0861	2.5628	2
59	.39046	.42413	2.3578	.92062	1.0862	2.5611	1
60	.39073	.42447	2.3559	.92050	1.0864	2.5593	0
'	Cos	Cot	Tan	Sin	Csc	Sec	'

23° (203°)   (336°) 156°

'	Sin	Tan	Cot	Cos	Sec	Csc	'
0	.39073	.42447	2.3559	.92050	1.0864	2.5593	60
1	.39100	.42482	2.3539	.92039	1.0865	2.5576	59
2	.39127	.42516	2.3520	.92028	1.0866	2.5558	58
3	.39153	.42551	2.3501	.92016	1.0868	2.5541	57
4	.39180	.42585	2 3483	92005	1.0869	2.5523	56
5	.39207	.42619	2.3464	.91994	1.0870	2.5506	55
6	.39234	.42654	2.3445	.91982	1.0872	2.5488	54
7	.39260	.42688	2.3426	.91971	1.0873	2.5471	53
8	.39287	.42722	2.3407	.91959	1.0874	2.5454	52
9	.39314	.42757	2.3388	.91948	1.0876	2.5436	51
10	.39341	.42791	2.3369	.91936	1.0877	2.5419	50
11	.39367	.42826	2.3351	.91925	1.0878	2.5402	49
12	.39394	.42860	2.3332	.91914	1.0880	2.5384	48
13	.39421	.42894	2.3313	.91902	1.0881	2.5367	47
14	.39448	.42929	2.3294	.91891	1.0883	2.5350	46
15	.39474	.42963	2.3276	.91879	1.0884	2.5333	45
16	.39501	.42998	2.3257	.91868	1.0885	2.5316	44
17	.39528	.43032	2.3238	.91856	1.0887	2.5299	43
18	.39555	.43067	2.3220	.91845	1.0888	2.5282	42
19	.39581	.43101	2.3201	.91833	1.0889	2.5264	41
20	.39608	.43136	2.3183	.91822	1.0891	2.5247	40
21	.39635	.43170	2.3164	.91810	1.0892	2.5230	39
22	.39661	.43205	2.3146	.91799	1.0893	2.5213	38
23	.39688	.43239	2.3127	.91787	1.0895	2.5196	37
24	.39715	.43274	2.3109	.91775	1.0896	2.5180	36
25	.39741	.43308	2.3090	.91764	1.0898	2.5163	35
26	.39768	.43343	2.3072	.91752	1.0899	2.5146	34
27	.39795	.43378	2.3053	.91741	1.0900	2.5129	33
28	.39822	.43412	2.3035	.91729	1.0902	2.5112	32
29	.39848	.43447	2.3017	.91718	1.0903	2.5095	31
30	.39875	.43481	2.2998	.91706	1.0904	2.5078	30
31	.39902	.43516	2.2980	.91694	1.0906	2.5062	29
32	.39928	.43550	2.2962	.91683	1.0907	2.5045	28
33	.39955	.43585	2.2944	.91671	1.0909	2.5028	27
34	.39982	.43620	2.2925	.91659	1.0910	2.5012	26
35	.40008	.43654	2.2907	.91648	1.0911	2.4995	25
36	.40035	.43689	2.2889	.91636	1.0913	2.4978	24
37	.40062	.43724	2.2871	.91625	1.0914	2.4962	23
38	.40088	.43758	2.2853	.91613	1.0915	2.4945	22
39	.40115	.43793	2.2835	.91601	1.0917	2.4928	21
40	.40141	.43828	2.2817	.91590	1.0918	2.4912	20
41	.40168	.43862	2.2799	.91578	1.0920	2.4895	19
42	.40195	.43897	2.2781	.91566	1.0921	2.4879	18
43	.40221	.43932	2.2763	.91555	1.0922	2.4862	17
44	.40248	.43966	2.2745	.91543	1.0924	2.4846	16
45	.40275	.44001	2.2727	.91531	1.0925	2.4830	15
46	.40301	.44036	2.2709	.91519	1.0927	2.4813	14
47	.40328	.44071	2.2691	.91508	1.0928	2.4797	13
48	.40355	.44105	2.2673	.91496	1.0929	2.4780	12
49	.40381	.44140	2.2655	.91484	1.0931	2.4764	11
50	.40408	.44175	2.2637	.91472	1.0932	2.4748	10
51	.40434	.44210	2.2620	.91461	1.0934	2.4731	9
52	.40461	.44244	2.2602	.91449	1.0935	2.4715	8
53	.40488	.44279	2.2584	.91437	1.0936	2.4699	7
54	.40514	.44314	2.2566	.91425	1.0938	2.4683	6
55	.40541	.44349	2.2549	.91414	1.0939	2.4667	5
56	.40567	.44384	2.2531	.91402	1.0941	2.4650	4
57	.40594	.44418	2.2513	.91390	1.0942	2.4634	3
58	.40621	.44453	2.2496	.91378	1.0944	2.4618	2
59	.40647	.44488	2.2478	.91366	1.0945	2.4602	1
60	.40674	.44523	2.2460	.91355	1.0946	2.4586	0
'	Cos	Cot	Tan	Sin	Csc	Sec	'

NATURAL TRIGONOMETRIC FUNCTIONS TO FIVE PLACES (continued)

24° (204°)   (335°) 155°   **25° (205°)**   (334°) 154°

′	Sin	Tan	Cot	Cos	Sec	Csc	′		′	Sin	Tan	Cot	Cos	Sec	Csc	′
0	.40674	.44523	2.2460	.91355	1.0946	2.4586	60		0	.42262	.46631	2.1445	.90631	1.1034	2.3662	60
1	.40700	.44558	2.2443	.91343	1.0948	2.4570	59		1	.42288	.46666	2.1429	.90618	1.1035	2.3647	59
2	.40727	.44593	2.2425	.91331	1.0949	2.4554	58		2	.42315	.46702	2.1413	.90606	1.1037	2.3633	58
3	.40753	.44627	2.2408	.91319	1.0951	2.4538	57		3	.42341	.46737	2.1396	.90594	1.1038	2.3618	57
4	.40780	.44662	2.2390	.91307	1.0952	2.4522	56		4	.42367	.46772	2.1380	.90582	1.1040	2.3603	56
5	.40806	.44697	2.2373	.91295	1.0953	2.4506	55		5	.42394	.46808	2.1364	.90569	1.1041	2.3588	55
6	.40833	.44732	2.2355	.91283	1.0955	2.4490	54		6	.42420	.46843	2.1348	.90557	1.1043	2.3574	54
7	.40860	.44767	2.2338	.91272	1.0956	2.4474	53		7	.42446	.46879	2.1332	.90545	1.1044	2.3559	53
8	.40886	.44802	2.2320	.91260	1.0958	2.4458	52		8	.42473	.46914	2.1315	.90532	1.1046	2.3545	52
9	.40913	.44837	2.2303	.91248	1.0959	2.4442	51		9	.42499	.46950	2.1299	.90520	1.1047	2.3530	51
10	.40939	.44872	2.2286	.91236	1.0961	2.4426	50		10	.42525	.46985	2.1283	.90507	1.1049	2.3515	50
11	.40966	.44907	2.2268	.91224	1.0962	2.4411	49		11	.42552	.47021	2.1267	.90495	1.1050	2.3501	49
12	.40992	.44942	2.2251	.91212	1.0963	2.4395	48		12	.42578	.47056	2.1251	.90483	1.1052	2.3486	48
13	.41019	.44977	2.2234	.91200	1.0965	2.4379	47		13	.42604	.47092	2.1235	.90470	1.1053	2.3472	47
14	.41045	.45012	2.2216	.91188	1.0966	2.4363	46		14	.42631	.47128	2.1219	.90458	1.1055	2.3457	46
15	.41072	.45047	2.2199	.91176	1.0968	2.4348	45		15	.42657	.47163	2.1203	.90446	1.1056	2.3443	45
16	.41098	.45082	2.2182	.91164	1.0969	2.4332	44		16	.42683	.47199	2.1187	.90433	1.1058	2.3428	44
17	.41125	.45117	2.2165	.91152	1.0971	2.4316	43		17	.42709	.47234	2.1171	.90421	1.1059	2.3414	43
18	.41151	.45152	2.2148	.91140	1.0972	2.4300	42		18	.42736	.47270	2.1155	.90408	1.1061	2.3400	42
19	.41178	.45187	2.2130	.91128	1.0974	2.4285	41		19	.42762	.47305	2.1139	.90396	1.1062	2.3385	41
20	.41204	.45222	2.2113	.91116	1.0975	2.4269	40		20	.42788	.47341	2.1123	.90383	1.1064	2.3371	40
21	.41231	.45257	2.2096	.91104	1.0976	2.4254	39		21	.42815	.47377	2.1107	.90371	1.1066	2.3356	39
22	.41257	.45292	2.2079	.91092	1.0978	2.4238	38		22	.42841	.47412	2.1092	.90358	1.1067	2.3342	38
23	.41284	.45327	2.2062	.91080	1.0979	2.4222	37		23	.42867	.47448	2.1076	.90346	1.1069	2.3328	37
24	.41310	.45362	2.2045	.91068	1.0981	2.4207	36		24	.42894	.47483	2.1060	.90334	1.1070	2.3314	36
25	.41337	.45397	2.2028	.91056	1.0982	2.4191	35		25	.42920	.47519	2.1044	.90321	1.1072	2.3299	35
26	.41363	.45432	2.2011	.91044	1.0984	2.4176	34		26	.42946	.47555	2.1028	.90309	1.1073	2.3285	34
27	.41390	.45467	2.1994	.91032	1.0985	2.4160	33		27	.42972	.47590	2.1013	.90296	1.1075	2.3271	33
28	.41416	.45502	2.1977	.91020	1.0987	2.4145	32		28	.42999	.47626	2.0997	.90284	1.1076	2.3257	32
29	.41443	.45538	2.1960	.91008	1.0988	2.4130	31		29	.43025	.47662	2.0981	.90271	1.1078	2.3242	31
30	.41469	.45573	2.1943	.90996	1.0989	2.4114	30		30	.43051	.47698	2.0965	.90259	1.1079	2.3228	30
31	.41496	.45608	2.1926	.90984	1.0991	2.4099	29		31	.43077	.47733	2.0950	.90246	1.1081	2.3214	29
32	.41522	.45643	2.1909	.90972	1.0992	2.4083	28		32	.43104	.47769	2.0934	.90233	1.1082	2.3200	28
33	.41549	.45678	2.1892	.90960	1.0994	2.4068	27		33	.43130	.47805	2.0918	.90221	1.1084	2.3186	27
34	.41575	.45713	2.1876	.90948	1.0995	2.4053	26		34	.43156	.47840	2.0903	.90208	1.1085	2.3172	26
35	.41602	.45748	2.1859	.90936	1.0997	2.4038	25		35	.43182	.47876	2.0887	.90196	1.1087	2.3158	25
36	.41628	.45784	2.1842	.90924	1.0998	2.4022	24		36	.43209	.47912	2.0872	.90183	1.1089	2.3144	24
37	.41655	.45819	2.1825	.90911	1.1000	2.4007	23		37	.43235	.47948	2.0856	.90171	1.1090	2.3130	23
38	.41681	.45854	2.1808	.90899	1.1001	2.3992	22		38	.43261	.47984	2.0840	.90158	1.1092	2.3115	22
39	.41707	.45889	2.1792	.90887	1.1003	2.3977	21		39	.43287	.48019	2.0825	.90146	1.1093	2.3101	21
40	.41734	.45924	2.1775	.90875	1.1004	2.3961	20		40	.43313	.48055	2.0809	.90133	1.1095	2.3088	20
41	.41760	.45960	2.1758	.90863	1.1006	2.3946	19		41	.43340	.48091	2.0794	.90120	1.1096	2.3074	19
42	.41787	.45995	2.1742	.90851	1.1007	2.3931	18		42	.43366	.48127	2.0778	.90108	1.1098	2.3060	18
43	.41813	.46030	2.1725	.90839	1.1009	2.3916	17		43	.43392	.48163	2.0763	.90095	1.1099	2.3046	17
44	.41840	.46065	2.1708	.90826	1.1010	2.3901	16		44	.43418	.48198	2.0748	.90082	1.1101	2.3032	16
45	.41866	.46101	2.1692	.90814	1.1011	2.3886	15		45	.43445	.48234	2.0732	.90070	1.1102	2.3018	15
46	.41892	.46136	2.1675	.90802	1.1013	2.3871	14		46	.43471	.48270	2.0717	.90057	1.1104	2.3004	14
47	.41919	.46171	2.1659	.90790	1.1014	2.3856	13		47	.43497	.48306	2.0701	.90045	1.1106	2.2990	13
48	.41945	.46206	2.1642	.90778	1.1016	2.3841	12		48	.43523	.48342	2.0686	.90032	1.1107	2.2976	12
49	.41972	.46242	2.1625	.90766	1.1017	2.3826	11		49	.43549	.48378	2.0671	.90019	1.1109	2.2962	11
50	.41998	.46277	2.1609	.90753	1.1019	2.3811	10		50	.43575	.48414	2.0655	.90007	1.1110	2.2949	10
51	.42024	.46312	2.1592	.90741	1.1020	2.3796	9		51	.43602	.48450	2.0640	.89994	1.1112	2.2935	9
52	.42051	.46348	2.1576	.90729	1.1022	2.3781	8		52	.43628	.48486	2.0625	.89981	1.1113	2.2921	8
53	.42077	.46383	2.1560	.90717	1.1023	2.3766	7		53	.43654	.48521	2.0609	.89968	1.1115	2.2907	7
54	.42104	.46418	2.1543	.90704	1.1025	2.3751	6		54	.43680	.48557	2.0594	.89956	1.1117	2.2894	6
55	.42130	.46454	2.1527	.90692	1.1026	2.3736	5		55	.43706	.48593	2.0579	.89943	1.1118	2.2880	5
56	.42156	.46489	2.1510	.90680	1.1028	2.3721	4		56	.43733	.48629	2.0564	.89930	1.1120	2.2866	4
57	.42183	.46525	2.1494	.90668	1.1029	2.3706	3		57	.43759	.48665	2.0549	.89918	1.1121	2.2853	3
58	.42209	.46560	2.1478	.90655	1.1031	2.3692	2		58	.43785	.48701	2.0533	.89905	1.1123	2.2839	2
59	.42235	.46595	2.1461	.90643	1.1032	2.3677	1		59	.43811	.48737	2.0518	.89892	1.1124	2.2825	1
60	.42262	.46631	2.1445	.90631	1.1034	2.3662	0		60	.43837	.48773	2.0503	.89879	1.1126	2.2812	0
′	Cos	Cot	Tan	Sin	Csc	Sec	′		′	Cos	Cot	Tan	Sin	Csc	Sce	′

114° (294°)   (245°) 65°   **115° (295°)**   (244°) 64°

NATURAL TRIGONOMETRIC FUNCTIONS TO FIVE PLACES (continued)

26° (206°)   (333°) 153°   **27° (207°)**   (332°) 152°

′	Sin	Tan	Cot	Cos	Sec	Csc	′		′	Sin	Tan	Cot	Cos	Sec	Csc	′
0	.43837	.48773	2.0503	.89879	1.1126	2.2812	60		0	.45399	.50953	1.9626	.89101	1.1223	2.2027	60
1	.43863	.48809	2.0488	.89867	1.1128	2.2798	59		1	.45425	.50989	1.9612	.89087	1.1225	2.2014	59
2	.43889	.48845	2.0473	.89854	1.1129	2.2785	58		2	.45451	.51026	1.9598	.89074	1.1227	2.2002	58
3	.43916	.48881	2.0458	.89841	1.1131	2.2771	57		3	.45477	.51063	1.9584	.89061	1.1228	2.1989	57
4	.43942	.48917	2.0443	.89828	1.1132	2.2757	56		4	.45503	.51099	1.9570	.89048	1.1230	2.1977	56
5	.43968	.48953	2.0428	.89816	1.1134	2.2744	55		5	.45529	.51136	1.9556	.89035	1.1232	2.1964	55
6	.43994	.48989	2.0413	.89803	1.1136	2.2730	54		6	.45554	.51173	1.9542	.89021	1.1233	2.1952	54
7	.44020	.49026	2.0398	.89790	1.1137	2.2717	53		7	.45580	.51209	1.9528	.89008	1.1235	2.1939	53
8	.44046	.49062	2.0383	.89777	1.1139	2.2703	52		8	.45606	.51246	1.9514	.88995	1.1237	2.1927	52
9	.44072	.49098	2.0368	.89764	1.1140	2.2690	51		9	.45632	.51283	1.9500	.88981	1.1238	2.1914	51
10	.44098	.49134	2.0353	.89752	1.1142	2.2677	50		10	.45658	.51319	1.9486	.88968	1.1240	2.1902	50
11	.44124	.49170	2.0338	.89739	1.1143	2.2663	49		11	.45684	.51356	1.9472	.88955	1.1242	2.1890	49
12	.44151	.49206	2.0323	.89726	1.1145	2.2650	48		12	.45710	.51393	1.9458	.88942	1.1243	2.1877	48
13	.44177	.49242	2.0308	.89713	1.1147	2.2636	47		13	.45736	.51430	1.9444	.88928	1.1245	2.1865	47
14	.44203	.49278	2.0293	.89700	1.1148	2.2623	46		14	.45762	.51467	1.9430	.88915	1.1247	2.1852	46
15	.44229	.49315	2.0278	.89687	1.1150	2.2610	45		15	.45787	.51503	1.9416	.88902	1.1248	2.1840	45
16	.44255	.49351	2.0263	.89674	1.1151	2.2596	44		16	.45813	.51540	1.9402	.88888	1.1250	2.1828	44
17	.44281	.49387	2.0248	.89662	1.1153	2.2583	43		17	.45839	.51577	1.9388	.88875	1.1252	2.1815	43
18	.44307	.49423	2.0233	.89649	1.1155	2.2570	42		18	.45865	.51614	1.9375	.88862	1.1253	2.1803	42
19	.44333	.49459	2.0219	.89636	1.1156	2.2556	41		19	.45891	.51651	1.9361	.88848	1.1255	2.1791	41
20	.44359	.49495	2.0204	.89623	1.1158	2.2543	40		20	.45917	.51688	1.9347	.88835	1.1257	2.1779	40
21	.44385	.49532	2.0189	.89610	1.1159	2.2530	39		21	.45942	.51724	1.9333	.88822	1.1259	2.1766	39
22	.44411	.49568	2.0174	.89597	1.1161	2.2517	38		22	.45968	.51761	1.9319	.88808	1.1260	2.1754	38
23	.44437	.49604	2.0160	.89584	1.1163	2.2504	37		23	.45994	.51798	1.9306	.88795	1.1262	2.1742	37
24	.44464	.49640	2.0145	.89571	1.1164	2.2490	36		24	.46020	.51835	1.9292	.88782	1.1264	2.1730	36
25	.44490	.49677	2.0130	.89558	1.1166	2.2477	35		25	.46046	.51872	1.9278	.88768	1.1265	2.1718	35
26	.44516	.49713	2.0115	.89545	1.1168	2.2464	34		26	.46072	.51909	1.9265	.88755	1.1267	2.1705	34
27	.44542	.49749	2.0101	.89532	1.1169	2.2451	33		27	.46097	.51946	1.9251	.88741	1.1269	2.1693	33
28	.44568	.49786	2.0086	.89519	1.1171	2.2438	32		28	.46123	.51983	1.9237	.88728	1.1270	2.1681	32
29	.44594	.49822	2.0072	.89506	1.1172	2.2425	31		29	.46149	.52020	1.9223	.88715	1.1272	2.1669	31
30	.44620	.49858	2.0057	.89493	1.1174	2.2412	30		30	.46175	.52057	1.9210	.88701	1.1274	2.1657	30
31	.44646	.49894	2.0042	.89480	1.1176	2.2399	29		31	.46201	.52094	1.9196	.88688	1.1276	2.1645	29
32	.44672	.49931	2.0028	.89467	1.1177	2.2385	28		32	.46226	.52131	1.9183	.88674	1.1277	2.1633	28
33	.44698	.49967	2.0013	.89454	1.1179	2.2372	27		33	.46252	.52168	1.9169	.88661	1.1279	2.1621	27
34	.44724	.50004	1.9999	.89441	1.1180	2.2359	26		34	.46278	.52205	1.9155	.88647	1.1281	2.1609	26
35	.44750	.50040	1.9984	.89428	1.1182	2.2346	25		35	.46304	.52242	1.9142	.88634	1.1282	2.1596	25
36	.44776	.50076	1.9970	.89415	1.1184	2.2333	24		36	.46330	.52279	1.9128	.88620	1.1284	2.1584	24
37	.44802	.50113	1.9955	.89402	1.1185	2.2320	23		37	.46355	.52316	1.9115	.88607	1.1286	2.1572	23
38	.44828	.50149	1.9941	.89389	1.1187	2.2308	22		38	.46381	.52353	1.9101	.88593	1.1288	2.1560	22
39	.44854	.50185	1.9926	.89376	1.1189	2.2295	21		39	.46407	.52390	1.9088	.88580	1.1289	2.1549	21
40	.44880	.50222	1.9912	.89363	1.1190	2.2282	20		40	.46433	.52427	1.9074	.88566	1.1291	2.1537	20
41	.44906	.50258	1.9897	.89350	1.1192	2.2269	19		41	.46458	.52464	1.9061	.88553	1.1293	2.1525	19
42	.44932	.50295	1.9883	.89337	1.1194	2.2256	18		42	.46484	.52501	1.9047	.88539	1.1294	2.1513	18
43	.44958	.50331	1.9868	.89324	1.1195	2.2243	17		43	.46510	.52538	1.9034	.88526	1.1296	2.1501	17
44	.44984	.50368	1.9854	.89311	1.1197	2.2230	16		44	.46536	.52575	1.9020	.88512	1.1298	2.1489	16
45	.45010	.50404	1.9840	.89298	1.1198	2.2217	15		45	.46561	.52613	1.9007	.88499	1.1300	2.1477	15
46	.45036	.50441	1.9825	.89285	1.1200	2.2205	14		46	.46587	.52650	1.8993	.88485	1.1301	2.1465	14
47	.45062	.50477	1.9811	.89272	1.1202	2.2192	13		47	.46613	.52687	1.8980	.88472	1.1303	2.1453	13
48	.45088	.50514	1.9797	.89259	1.1203	2.2179	12		48	.46639	.52724	1.8967	.88458	1.1305	2.1441	12
49	.45114	.50550	1.9782	.89245	1.1205	2.2166	11		49	.46664	.52761	1.8953	.88445	1.1307	2.1430	11
50	.45140	.50587	1.9768	.89232	1.1207	2.2153	10		50	.46690	.52798	1.8940	.88431	1.1308	2.1418	10
51	.45166	.50623	1.9754	.89219	1.1208	2.2141	9		51	.46716	.52836	1.8927	.88417	1.1310	2.1406	9
52	.45192	.50660	1.9740	.89206	1.1210	2.2128	8		52	.46742	.52873	1.8913	.88404	1.1312	2.1394	8
53	.45218	.50696	1.9725	.89193	1.1212	2.2115	7		53	.46767	.52910	1.8900	.88390	1.1313	2.1382	7
54	.45243	.50733	1.9711	.89180	1.1213	2.2103	6		54	.46793	.52947	1.8887	.88377	1.1315	2.1371	6
55	.45269	.50769	1.9697	.89167	1.1215	2.2090	5		55	.46819	.52985	1.8873	.88363	1.1317	2.1359	5
56	.45295	.50806	1.9683	.89153	1.1217	2.2077	4		56	.46844	.53022	1.8860	.88349	1.1319	2.1347	4
57	.45321	.50843	1.9669	.89140	1.1218	2.2065	3		57	.46870	.53059	1.8847	.88336	1.1320	2.1336	3
58	.45347	.50879	1.9654	.89127	1.1220	2.2052	2		58	.46896	.53096	1.8834	.88322	1.1322	2.1324	2
59	.45373	.50916	1.9640	.89114	1.1222	2.2039	1		59	.46921	.53134	1.8820	.88308	1.1324	2.1312	1
60	.45399	.50953	1.9626	.89101	1.1223	2.2027	0		60	.46947	.53171	1.8807	.88295	1.1326	2.1301	0
′	Cos	Cot	Tan	Sin	Csc	Sec	′		′	Cos	Cot	Tan	Sin	Csc	Sec	′

116° (296°)   (243°) 63°   **117° (297°)**   (242°) 62°

NATURAL TRIGONOMETRIC FUNCTIONS TO FIVE PLACES (continued)

28° (208°)   (331°) 151°

′	Sin	Tan	Cot	Cos	Sec	Csc	′
0	.46947	.53171	1.8807	.88295	1.1326	2.1301	60
1	.46973	.53208	1.8794	.88281	1.1327	2.1289	59
2	.46999	.53246	1.8781	.88267	1.1329	2.1277	58
3	.47024	.53283	1.8768	.88254	1.1331	2.1266	57
4	.47050	.53320	1.8755	.88240	1.1333	2.1254	56
5	.47076	.53358	1.8741	.88226	1.1334	2.1242	55
6	.47101	.53395	1.8728	.88213	1.1336	2.1231	54
7	.47127	.53432	1.8715	.88199	1.1338	2.1219	53
8	.47153	.53470	1.8702	.88185	1.1340	2.1208	52
9	.47178	.53507	1.8689	.88172	1.1342	2.1196	51
10	.47204	.53545	1.8676	.88158	1.1343	2.1185	50
11	.47229	.53582	1.8663	.88144	1.1345	2.1173	49
12	.47255	.53620	1.8650	.88130	1.1347	2.1162	48
13	.47281	.53657	1.8637	.88117	1.1349	2.1150	47
14	.47306	.53694	1.8624	.88103	1.1350	2.1139	46
15	.47332	.53732	1.8611	.88089	1.1352	2.1127	45
16	.47358	.53769	1.8598	.88075	1.1354	2.1116	44
17	.47383	.53807	1.8585	.88062	1.1356	2.1105	43
18	.47409	.53844	1.8572	.88048	1.1357	2.1093	42
19	.47434	.53882	1.8559	.88034	1.1359	2.1082	41
20	.47460	.53920	1.8546	.88020	1.1361	2.1070	40
21	.47486	.53957	1.8533	.88006	1.1363	2.1059	39
22	.47511	.53995	1.8520	.87993	1.1365	2.1048	38
23	.47537	.54032	1.8507	.87979	1.1366	2.1036	37
24	.47562	.54070	1.8495	.87965	1.1368	2.1025	36
25	.47588	.54107	1.8482	.87951	1.1370	2.1014	35
26	.47614	.54145	1.8469	.87937	1.1372	2.1002	34
27	.47639	.54183	1.8456	.87923	1.1374	2.0991	33
28	.47665	.54220	1.8443	.87909	1.1375	2.0980	32
29	.47690	.54258	1.8430	.87896	1.1377	2.0969	31
30	.47716	.54296	1.8418	.87882	1.1379	2.0957	30
31	.47741	.54333	1.8405	.87868	1.1381	2.0946	29
32	.47767	.54371	1.8392	.87854	1.1383	2.0935	28
33	.47793	.54409	1.8379	.87840	1.1384	2.0924	27
34	.47818	.54446	1.8367	.87826	1.1386	2.0913	26
35	.47844	.54484	1.8354	.87812	1.1388	2.0901	25
36	.47869	.54522	1.8341	.87798	1.1390	2.0890	24
37	.47895	.54560	1.8329	.87784	1.1392	2.0879	23
38	.47920	.54597	1.8316	.87770	1.1393	2.0868	22
39	.47946	.54635	1.8303	.87756	1.1395	2.0857	21
40	.47971	.54673	1.8291	.87743	1.1397	2.0846	20
41	.47997	.54711	1.8278	.87729	1.1399	2.0835	19
42	.48022	.54748	1.8265	.87715	1.1401	2.0824	18
43	.48048	.54786	1.8253	.87701	1.1402	2.0813	17
44	.48073	.54824	1.8240	.87687	1,1404	2.0802	16
45	.48099	.54862	1.8228	.87673	1.1406	2.0791	15
46	.48124	.54900	1.8215	.87659	1.1408	2.0779	14
47	.48150	.54938	1.8202	.87645	1.1410	2.0768	13
48	.48175	.54975	1.8190	.87631	1.1412	2.0757	12
49	.48201	.55013	1.8177	.87617	1.1413	2.0747	11
50	.48226	.55051	1.8165	.87603	1.1415	2.0736	10
51	.48252	.55089	1.8152	.87589	1.1417	2.0725	9
52	.48277	.55127	1.8140	.87575	1.1419	2.0714	8
53	.48303	.55165	1.8127	.87561	1.1421	2.0703	7
54	.48328	.55203	1.8115	.87546	1.1423	2.0692	6
55	.48354	.55241	1.8103	.87532	1.1424	2.0681	5
56	.48379	.55279	1.8090	.87518	1.1426	2.0670	4
57	.48405	.55317	1.8078	.87504	1.1428	2.0659	3
58	.48430	.55355	1.8065	.87490	1.1430	2.0648	2
59	.48456	.55393	1.8053	.87476	1.1432	2.0637	1
60	.48481	.55431	1.8040	.87462	1.1434	2.0627	0
′	Cos	Cot	Tan	Sin	Csc	Sec	′

118° (298°)   (241°) 61°

29° (209°)   (330)° 150°

′	Sin	Tan	Cot	Cos	Sec	Csc	′
0	.48481	.55431	1.8040	.87462	1.1434	2.0627	60
1	.48506	.55469	1.8028	.87448	1.1435	2.0616	59
2	.48532	.55507	1.8016	.87434	1.1437	2.0605	58
3	.48557	.55545	1.8003	.87420	1.1439	2.0594	57
4	.48583	.55583	1.7991	.87406	1.1441	2.0583	56
5	.48608	.55621	1.7979	.87391	1.1443	2.0573	55
6	.48634	.55659	1.7966	.87377	1.1445	2.0562	54
7	.48659	.55697	1.7954	.87363	1.1446	2.0551	53
8	.48684	.55736	1.7942	.87349	1.1448	2.0540	52
9	.48710	.55774	1.7930	.87335	1.1450	2.0530	51
10	.48735	.55812	1.7917	.87321	1.1452	2.0519	50
11	.48761	.55850	1.7905	.87306	1.1454	2.0508	49
12	.48786	.55888	1.7893	.87292	1.1456	2.0498	48
13	.48811	.55926	1.7881	.87278	1.1458	2.0487	47
14	.48837	.55964	1.7868	.87264	1.1460	2.0476	46
15	.48862	.56003	1.7856	.87250	1.1461	2.0466	45
16	.48888	.56041	1.7844	.87235	1.1463	2.0455	44
17	.48913	.56079	1.7832	.87221	1.1465	2.0445	43
18	.48938	.56117	1.7820	.87207	1.1467	2.0434	42
19	.48964	.56156	1.7808	.87193	1.1469	2.0423	41
20	.48989	.56194	1.7796	.87178	1.1471	2.0413	40
21	.49014	.56232	1.7783	.87164	1.1473	2.0402	39
22	.49040	.56270	1.7771	.87150	1.1474	2.0392	38
23	.49065	.56309	1.7759	.87136	1.1476	2.0381	37
24	.49090	.56347	1.7747	.87121	1.1478	2.0371	36
25	.49116	.56385	1.7735	.87107	1.1480	2.0360	35
26	.49141	.56424	1.7723	.87093	1.1482	2.0350	34
27	.49166	.56462	1.7711	.87079	1.1484	2.0339	33
28	.49192	.56501	1.7699	.87064	1.1486	2.0329	32
29	.49217	.56539	1.7687	.87050	1.1488	2.0318	31
30	.49242	.56577	1.7675	.87036	1.1490	2.0308	30
31	.49268	.56616	1.7663	.87021	1.1491	2.0297	29
32	.49293	.56654	1.7651	.87007	1.1493	2.0287	28
33	.49318	.56693	1.7639	.86993	1.1495	2.0276	27
34	.49344	.56731	1.7627	.86978	1.1497	2.0266	26
35	.49369	.56760	1.7615	.86964	1.1499	2.0256	25
36	.49394	.56808	1.7603	.86949	1.1501	2.0245	24
37	.49419	.56846	1.7591	.86935	1.1503	2.0235	23
38	.49445	.56885	1.7579	.86921	1.1505	2.0225	22
39	.49470	.56923	1.7567	.86906	1.1507	2.0214	21
40	.49495	.56962	1.7556	.86892	1.1509	2.0204	20
41	.49521	.57000	1.7544	.86878	1.1510	2.0194	19
42	.49546	.57039	1.7532	.86863	1.1512	2.0183	18
43	.49571	.57078	1.7520	.86849	1.1514	2.0173	17
44	.49596	.57116	1.7508	.86834	1.1516	2.0163	16
45	.49622	.57155	1.7496	.86820	1.1518	2.0152	15
46	.49647	.57193	1.7485	.86805	1.1520	2.0142	14
47	.49672	.57232	1.7473	.86791	1.1522	2.0132	13
48	.49697	.57271	1.7461	.86777	1.1524	2.0122	12
49	.49723	.57309	1.7449	.86762	1.1526	2.0112	11
50	.49748	.57348	1.7437	.86748	1.1528	2.0101	10
51	.49773	.57386	1.7426	.86733	1.1530	2.0091	9
52	.49798	.57425	1.7414	.86719	1.1532	2.0081	8
53	.49824	.57464	1.7402	.86704	1.1533	2.0071	7
54	.49849	.57503	1.7391	.86690	1.1535	2.0061	6
55	.49874	.57541	1.7379	.86675	1.1537	2.0051	5
56	.49899	.57580	1.7367	.86661	1.1539	2.0040	4
57	.49924	.57619	1.7355	.86646	1.1541	2.0030	3
58	.49950	.57657	1.7344	.86632	1.1543	2.0020	2
59	.49975	.57696	1.7332	.86617	1.1545	2.0010	1
60	.50000	.57735	1.7321	.86603	1.1547	2.0000	0
′	Cos	Cot	Tan	Sin	Csc	Sec	′

119° (299°)   (240°) 60°

TRIGONOMETRY

NATURAL TRIGONOMETRIC FUNCTIONS TO FIVE PLACES (continued)

30° (210°)   (329°) 149°

'	Sin	Tan	Cot	Cos	Sec	Csc	'
0	.50000	.57735	1.7321	.86603	1.1547	2.0000	60
1	.50025	.57774	1.7309	.86588	1.1549	1.9990	59
2	.50050	.57813	1.7297	.86573	1.1551	1.9980	58
3	.50076	.57851	1.7286	.86559	1.1553	1.9970	57
4	.50101	.57890	1.7274	.86544	1.1555	1.9960	56
5	.50126	.57929	1.7262	.86530	1.1557	1.9950	55
6	.50151	.57968	1.7251	.86515	1.1559	1.9940	54
7	.50176	.58007	1.7239	.86501	1.1561	1.9930	53
8	.50201	.58046	1.7228	.86486	1.1563	1.9920	52
9	.50227	.58085	1.7216	.86471	1.1565	1.9910	51
10	.50252	.58124	1.7205	.86457	1.1566	1.9900	50
11	.50277	.58162	1.7193	.86442	1.1568	1.9890	49
12	.50302	.58201	1.7182	.86427	1.1570	1.9880	48
13	.50327	.58240	1.7170	.86413	1.1572	1.9870	47
14	.50352	.58279	1.7159	.86398	1.1574	1.9860	46
15	.50377	.58318	1.7147	.86384	1.1576	1.9850	45
16	.50403	.58357	1.7136	.86369	1.1578	1.9840	44
17	.50428	.58396	1.7124	.86354	1.1580	1.9830	43
18	.50453	.58435	1.7113	.86340	1.1582	1.9821	42
19	.50478	.58474	1.7102	.86325	1.1584	1.9811	41
20	.50503	.58513	1.7090	.86310	1.1586	1.9801	40
21	.50528	.58552	1.7079	.86296	1.1588	1.9791	39
22	.50553	.58591	1.7067	.86281	1.1590	1.9781	38
23	.50578	.58631	1.7056	.86266	1.1592	1.9771	37
24	.50603	.58670	1.7045	.86251	1.1594	1.9762	36
25	.50628	.58709	1.7033	.86237	1.1596	1.9752	35
26	.50654	.58748	1.7022	.86222	1.1598	1.9742	34
27	.50679	.58787	1.7011	.86207	1.1600	1.9732	33
28	.50704	.58826	1.6999	.86192	1.1602	1.9722	32
29	.50729	.58865	1.6988	.86178	1.1604	1.9713	31
30	.50754	.58905	1.6977	.86163	1.1606	1.9703	30
31	.50779	.58944	1.6965	.86148	1.1608	1.9693	29
32	.50804	.58983	1.6954	.86133	1.1610	1.9684	28
33	.50829	.59022	1.6943	.86119	1.1612	1.9674	27
34	.50854	.59061	1.6932	.86104	1.1614	1.9664	26
35	.50879	.59101	1.6920	.86089	1.1616	1.9654	25
36	.50904	.59140	1.6909	.86074	1.1618	1.9645	24
37	.50929	.59179	1.6898	.86059	1.1620	1.9635	23
38	.50954	.59218	1.6887	.86045	1.1622	1.9625	22
39	.50979	.59258	1.6875	.86030	1.1624	1.9616	21
40	.51004	.59297	1.6864	.86015	1.1626	1.9606	20
41	.51029	.59336	1.6853	.86000	1.1628	1.9597	19
42	.51054	.59376	1.6842	.85985	1.1630	1.9587	18
43	.51079	.59415	1.6831	.85970	1.1632	1.9577	17
44	.51104	.59454	1.6820	.85956	1.1634	1.9568	16
45	.51129	.59494	1.6808	.85941	1.1636	1.9558	15
46	.51154	.59533	1.6797	.85926	1.1638	1.9549	14
47	.51179	.59573	1.6786	.85911	1.1640	1.9539	13
48	.51204	.59612	1.6775	.85896	1.1642	1.9530	12
49	.51229	.59651	1.6764	.85881	1.1644	1.9520	11
50	.51254	.59691	1.6753	.85866	1.1646	1.9511	10
51	.51279	.59730	1.6742	.85851	1.1648	1.9501	9
52	.51304	.59770	1.6731	.85836	1.1650	1.9492	8
53	.51329	.59809	1.6720	.85821	1.1652	1.9482	7
54	.51354	.59849	1.6709	.85806	1.1654	1.9473	6
55	.51379	.59888	1.6698	.85792	1.1656	1.9463	5
56	.51404	.59928	1.6687	.85777	1.1658	1.9454	4
57	.51429	.59967	1.6676	.85762	1.1660	1.9444	3
58	.51454	.60007	1.6665	.85747	1.1662	1.9435	2
59	.51479	.60046	1.6654	.85732	1.1664	1.9425	1
60	.51504	.60086	1.6643	.85717	1.1666	1.9416	0
'	Cos	Cot	Tan	Sin	Csc	Sec	'

120° (300°)   (239°) 59°

31° (211°)   (328°) 148°

'	Sin	Tan	Cot	Cos	Sec	Csc	'
0	.51504	.60086	1.6643	.85717	1.1666	1.9416	60
1	.51529	.60126	1.6632	.85702	1.1668	1.9407	59
2	.51554	.60165	1.6621	.85687	1.1670	1.9397	58
3	.51579	.60205	1.6610	.85672	1.1672	1.9388	57
4	.51604	.60245	1.6599	.85657	1.1675	1.9379	56
5	.51628	.60284	1.6588	.85642	1.1677	1.9369	55
6	.51653	.60324	1.6577	.85627	1.1679	1.9360	54
7	.51678	.60364	1.6566	.85612	1.1681	1.9351	53
8	.51703	.60403	1.6555	.85597	1.1683	1.9341	52
9	.51728	.60443	1.6545	.85582	1.1685	1.9332	51
10	.51753	.60483	1.6534	.85567	1.1687	1.9323	50
11	.51778	.60522	1.6523	.85551	1.1689	1.9313	49
12	.51803	.60562	1.6512	.85536	1.1691	1.9304	48
13	.51828	.60602	1.6501	.85521	1.1693	1.9295	47
14	.51852	.60642	1.6490	.85506	1.1695	1.9285	46
15	.51877	.60681	1.6479	.85491	1.1697	1.9276	45
16	.51902	.60721	1.6469	.85476	1.1699	1.9267	44
17	.51927	.60761	1.6458	.85461	1.1701	1.9258	43
18	.51952	.60801	1.6447	.85446	1.1703	1.9249	42
19	.51977	.60841	1.6436	.85431	1.1705	1.9239	41
20	.52002	.60881	1.6426	.85416	1.1707	1.9230	40
21	.52026	.60921	1.6415	.85401	1.1710	1.9221	39
22	.52051	.60960	1.6404	.85385	1.1712	1.9212	38
23	.52076	.61000	1.6393	.85370	1.1714	1.9203	37
24	.52101	.61040	1.6383	.85355	1.1716	1.9194	36
25	.52126	.61080	1.6372	.85340	1.1718	1.9184	35
26	.52151	.61120	1.6361	.85325	1.1720	1.9175	34
27	.52175	.61160	1.6351	.85310	1.1722	1.9166	33
28	.52200	.61200	1.6340	.85294	1.1724	1.9157	32
29	.52225	.61240	1.6329	.85279	1.1726	1.9148	31
30	.52250	.61280	1.6319	.85264	1.1728	1.9139	30
31	.52275	.61320	1.6308	.85249	1.1730	1.9130	29
32	.52299	.61360	1.6297	.85234	1.1732	1.9121	28
33	.52324	.61400	1.6287	.85218	1.1735	1.9112	27
34	.52349	.61440	1.6276	.85203	1.1737	1.9103	26
35	.52374	.61480	1.6265	.85188	1.1739	1.9094	25
36	.52399	.61520	1.6255	.85173	1.1741	1.9084	24
37	.52423	.61561	1.6244	.85157	1.1743	1.9075	23
38	.52448	.61601	1.6234	.85142	1.1745	1.9066	22
39	.52473	.61641	1.6223	.85127	1.1747	1.9057	21
40	.52498	.61681	1.6212	.85112	1.1749	1.9048	20
41	.52522	.61721	1.6202	.85096	1.1751	1.9039	19
42	.52547	.61761	1.6191	.85081	1.1753	1.9031	18
43	.52572	.61801	1.6181	.85066	1.1756	1.9022	17
44	.52597	.61842	1.6170	.85051	1.1758	1.9013	16
45	.52621	.61882	1.6160	.85035	1.1760	1.9004	15
46	.52646	.61922	1.6149	.85020	1.1762	1.8995	14
47	.52671	.61962	1.6139	.85005	1.1764	1.8986	13
48	.52696	.62003	1.6128	.84989	1.1766	1.8977	12
49	.52720	.62043	1.6118	.84974	1.1768	1.8968	11
50	.52745	.62083	1.6107	.84959	1.1770	1.8959	10
51	.52770	.62124	1.6097	.84943	1.1773	1.8950	9
52	.52794	.62164	1.6087	.84928	1.1775	1.8941	8
53	.52819	.62204	1.6076	.84913	1.1777	1.8933	7
54	.52844	.62245	1.6066	.84897	1.1779	1.8924	6
55	.52869	.62285	1.6055	.84882	1.1781	1.8915	5
56	.52893	.62325	1.6045	.84866	1.1783	1.8906	4
57	.52918	.62366	1.6034	.84851	1.1785	1.8897	3
58	.52943	.62406	1.6024	.84836	1.1788	1.8888	2
59	.52967	.62446	1.6014	.84820	1.1790	1.8880	1
60	.52992	.62487	1.6003	.84805	1.1792	1.8871	0
'	Cos	Cot	Tan	Sin	Csc	Sec	'

121° (301°)   (238°) 58°

NATURAL TRIGONOMETRIC FUNCTIONS TO FIVE PLACES (continued)

32° (212°)   (327°) 147°

'	Sin	Tan	Cot	Cos	Sec	Csc	'
0	.52992	.62487	1.6003	.84805	1.1792	1.8871	60
1	.53017	.62527	1.5993	.84789	1.1794	1.8862	59
2	.53041	.62568	1.5983	.84774	1.1796	1.8853	58
3	.53066	.62608	1.5972	.84759	1.1798	1.8844	57
4	.53091	.62649	1.5962	.84743	1.1800	1.8836	56
5	.53115	.62689	1.5952	.84728	1.1803	1.8827	55
6	.53140	.62730	1.5941	.84712	1.1805	1.8818	54
7	.53164	.62770	1.5931	.84697	1.1807	1.8810	53
8	.53189	.62811	1.5921	.84681	1.1809	1.8801	52
9	.53214	.62852	1.5911	.84666	1.1811	1.8792	51
10	.53238	.62892	1.5900	.84650	1.1813	1.8783	50
11	.53263	.62933	1.5890	.84635	1.1815	1.8775	49
12	.53288	.62973	1.5880	.84619	1.1818	1.8766	48
13	.53312	.63014	1.5869	.84604	1.1820	1.8757	47
14	.53337	.63055	1.5859	.84588	1.1822	1.8749	46
15	.53361	.63095	1.5849	.84573	1.1824	1.8740	45
16	.53386	.63136	1.5839	.84557	1.1826	1.8731	44
17	.53411	.63177	1.5829	.84542	1.1828	1.8723	43
18	.53435	.63217	1.5818	.84526	1.1831	1.8714	42
19	.53460	.63258	1.5808	.84511	1.1833	1.8706	41
20	.53484	.63299	1.5798	.84495	1.1835	1.8697	40
21	.53509	.63340	1.5788	.84480	1.1837	1.8688	39
22	.53534	.63380	1.5778	.84464	1.1839	1.8680	38
23	.53558	.63421	1.5768	.84448	1.1842	1.8671	37
24	.53583	.63462	1.5757	.84433	1.1844	1.8663	36
25	.53607	.63503	1.5747	.84417	1.1846	1.8654	35
26	.53632	.63544	1.5737	.84402	1.1848	1.8646	34
27	.53656	.63584	1.5727	.84386	1.1850	1.8637	33
28	.53681	.63625	1.5717	.84370	1.1852	1.8629	32
29	.53705	.63666	1.5707	.84355	1.1855	1.8620	31
30	.53730	.63707	1.5697	.84339	1.1857	1.8612	30
31	.53754	.63748	1.5687	.84324	1.1859	1.8603	29
32	.53779	.63789	1.5677	.84308	1.1861	1.8595	28
33	.53804	.63830	1.5667	.84292	1.1863	1.8586	27
34	.53828	.63871	1.5657	.84277	1.1866	1.8578	26
35	.53853	.63912	1.5647	.84261	1.1868	1.8569	25
36	.53877	.63953	1.5637	.84245	1.1870	1.8561	24
37	.53902	.63994	1.5627	.84230	1.1872	1.8552	23
38	.53926	.64035	1.5617	.84214	1.1875	1.8544	22
39	.53951	.64076	1.5607	.84198	1.1877	1.8535	21
40	.53975	.64117	1.5597	.84182	1.1879	1.8527	20
41	.54000	.64158	1.5587	.84167	1.1881	1.8519	19
42	.54024	.64199	1.5577	.84151	1.1883	1.8510	18
43	.54049	.64240	1.5567	.84135	1.1886	1.8502	17
44	.54073	.64281	1.5557	.84120	1.1888	1.8494	16
45	.54097	.64322	1.5547	.84104	1.1890	1.8485	15
46	.54122	.64363	1.5537	.84088	1.1892	1.8477	14
47	.54146	.64404	1.5527	.84072	1.1895	1.8468	13
48	.54171	.64446	1.5517	.84057	1.1897	1.8460	12
49	.54195	.64487	1.5507	.84041	1.1899	1.8452	11
50	.54220	.64528	1.5497	.84025	1.1901	1.8443	10
51	.54244	.64569	1.5487	.84009	1.1903	1.8435	9
52	.54269	.64610	1.5477	.83994	1.1906	1.8427	8
53	.54293	.64652	1.5468	.83978	1.1908	1.8419	7
54	.54317	.64693	1.5458	.83962	1.1910	1.8410	6
55	.54342	.64734	1.5448	.83946	1.1912	1.8402	5
56	.54366	.64775	1.5438	.83930	1.1915	1.8394	4
57	.54391	.64817	1.5428	.83915	1.1917	1.8385	3
58	.54415	.64858	1.5418	.83899	1.1919	1.8377	2
59	.54440	.64899	1.5408	.83883	1.1921	1.8369	1
60	.54464	.64941	1.5399	.83867	1.1924	1.8361	0
'	Cos	Cot	Tan	Sin	Csc	Sec	'

122° (302°)   (237°) 57°

33° (213°)   (326°) 146°

'	Sin	Tan	Cot	Cos	Sec	Csc	'
0	.54464	.64941	1.5399	.83867	1.1924	1.8361	60
1	.54488	.64982	1.5389	.83851	1.1926	1.8353	59
2	.54513	.65024	1.5379	.83835	1.1928	1.8344	58
3	.54537	.65065	1.5369	.83819	1.1930	1.8336	57
4	.54561	.65106	1.5359	.83804	1.1933	1.8328	56
5	.54586	.65148	1.5350	.83788	1.1935	1.8320	55
6	.54610	.65189	1.5340	.83772	1.9137	1.8312	54
7	.54635	.65231	1.5330	.83756	1.1939	1.8303	53
8	.54659	.65272	1.5320	.83740	1.1942	1.8295	52
9	.54683	.65314	1.5311	.83724	1.1944	1.8287	51
10	.54708	.65355	1.5301	.83708	1.1946	1.8279	50
11	.54732	.65397	1.5291	.83692	1.1949	1.8271	49
12	.54756	.65438	1.5282	.83676	1.1951	1.8263	48
13	.54781	.65480	1.5272	.83660	1.1953	1.8255	47
14	.54805	.65521	1.5262	.83645	1.1955	1.8247	46
15	.54829	.65563	1.5253	.83629	1.1958	1.8238	45
16	.54854	.65604	1.5243	.83613	1.1960	1.8230	44
17	.54878	.65646	1.5233	.83597	1.1962	1.8222	43
18	.54902	.65688	1.5224	.83581	1.1964	1.8214	42
19	.54927	.65729	1.5214	.83565	1.1967	1.8206	41
20	.54951	.65771	1.5204	.83549	1.1969	1.8198	40
21	.54975	.65813	1.5195	.83533	1.1971	1.8190	39
22	.54999	.65854	1.5185	.83517	1.1974	1.8182	38
23	.55024	.65896	1.5175	.83501	1.1976	1.8174	37
24	.55048	.65938	1.5166	.83485	1.1978	1.8166	36
25	.55072	.65980	1.5156	.83469	1.1981	1.8158	35
26	.55097	.66021	1.5147	.83453	1.1983	1.8150	34
27	.55121	.66063	1.5137	.83437	1.1985	1.8142	33
28	.55145	.66105	1.5127	.83421	1.1987	1.8134	32
29	.55169	.66147	1.5118	.83405	1.1990	1.8126	31
30	.55194	.66189	1.5108	.83389	1.1992	1.8118	30
31	.55218	.66230	1.5099	.83373	1.1994	1.8110	29
32	.55242	.66272	1.5089	.83356	1.1997	1.8102	28
33	.55266	.66314	1.5080	.83340	1.1999	1.8094	27
34	.55291	.66356	1.5070	.83324	1.2001	1.8086	26
35	.55315	.66398	1.5061	.83308	1.2004	1.8078	25
36	.55339	.66440	1.5051	.83292	1.2006	1.8070	24
37	.55363	.66482	1.5042	.83276	1.2008	1.8062	23
38	.55388	.66524	1.5032	.83260	1.2011	1.8055	22
39	.55412	.66566	1.5023	.83244	1.2013	1.8047	21
40	.55436	.66608	1.5013	.83228	1.2015	1.8039	20
41	.55460	.66650	1.5004	.83212	1.2018	1.8031	19
42	.55484	.66692	1.4994	.83195	1.2020	1.8023	18
43	.55509	.66734	1.4985	.83179	1.2022	1.8015	17
44	.55533	.66776	1.4975	.83163	1.2025	1.8007	16
45	.55557	.66818	1.4966	.83147	1.2027	1.8000	15
46	.55581	.66860	1.4957	.83131	1.2029	1.7992	14
47	.55605	.66902	1.4947	.83115	1.2032	1.7984	13
48	.55630	.66944	1.4938	.83098	1.2034	1.7976	12
49	.55654	.66986	1.4928	.83082	1.2036	1.7968	11
50	.55678	.67028	1.4919	.83066	1.2039	1.7960	10
51	.55702	.67071	1.4910	.83050	1.2041	1.7953	9
52	.55726	.67113	1.4900	.83034	1.2043	1.7945	8
53	.55750	.67155	1.4891	.83017	1.2046	1.7937	7
54	.55775	.67197	1.4882	.83001	1.2048	1.7929	6
55	.55799	.67239	1.4872	.82985	1.2050	1.7922	5
56	.55823	.67282	1.4863	.82969	1.2053	1.7914	4
57	.55847	.67324	1.4854	.82953	1.2055	1.7906	3
58	.55871	.67366	1.4844	.82936	1.2057	1.7898	2
59	.55895	.67409	1.4835	.82920	1.2060	1.7891	1
60	.55919	.67451	1.4826	.82904	1.2062	1.7883	0
'	Cos	Cot	Tan	Sin	Csc	Sec	'

123° (303°)   (236°) 56°

168

TRIGONOMETRY

NATURAL TRIGONOMETRIC FUNCTIONS TO FIVE PLACES (continued)

34° (214°)   **(325°) 145°**   **35° (215°)**   **(324°) 144°**

′	Sin	Tan	Cot	Cos	Sec	Csc	′		′	Sin	Tan	Cot	Cos	Sec	Csc	′
0	.55919	.67451	1.4826	.82904	1.2062	1.7883	60		0	.57358	.70021	1.4281	.81915	1.2208	1.7434	60
1	.55943	.67493	1.4816	.82887	1.2065	1.7875	59		1	.57381	.70064	1.4273	.81899	1.2210	1.7427	59
2	.55968	.67536	1.4807	.82871	1.2067	1.7868	58		2	.57405	.70107	1.4264	.81882	1.2213	1.7420	58
3	.55992	.67578	1.4798	.82855	1.2069	1.7860	57		3	.57429	.70151	1.4255	.81865	1.2215	1.7413	57
4	.56016	.67620	1.4788	.82839	1.2072	1.7852	56		4	.57453	.70194	1.4246	.81848	1.2218	1.7406	56
5	.56040	.67663	1.4779	.82822	1.2074	1.7844	55		5	.57477	.70238	1.4237	.81832	1.2220	1.7398	55
6	.56064	.67705	1.4770	.82806	1.2076	1.7837	54		6	.57501	.70281	1.4229	.81815	1.2223	1.7391	54
7	.56088	.67748	1.4761	.82790	1.2079	1.7829	53		7	.57524	.70325	1.4220	.81798	1.2225	1.7384	53
8	.56112	.67790	1.4751	.82773	1.2081	1.7821	52		8	.57548	.70368	1.4211	.81782	1.2228	1.7377	52
9	.56136	.67832	1.4742	.82757	1.2084	1.7814	51		9	.57572	.70412	1.4202	.81765	1.2230	1.7370	51
10	.56160	.67875	1.4733	.82741	1.2086	1.7806	50		10	.57596	.70455	1.4193	.81748	1.2233	1.7362	50
11	.56184	.67917	1.4724	.82724	1.2088	1.7799	49		11	.57619	.70499	1.4185	.81731	1.2235	1.7355	49
12	.56208	.67960	1.4715	.82708	1.2091	1.7791	48		12	.57643	.70542	1.4176	.81714	1.2238	1.7348	48
13	.56232	.68002	1.4705	.82692	1.2093	1.7783	47		13	.57667	.70586	1.4167	.81698	1.2240	1.7341	47
14	.56256	.68045	1.4696	.82675	1.2096	1.7776	46		14	.57691	.70629	1.4158	.81681	1.2243	1.7334	46
15	.56280	.68088	1.4687	.82659	1.2098	1.7768	45		15	.57715	.70673	1.4150	.81664	1.2245	1.7327	45
16	.56305	.68130	1.4678	.82643	1.2100	1.7761	44		16	.57738	.70717	1.4141	.81647	1.2248	1.7320	44
17	.56329	.68173	1.4669	.82626	1.2103	1.7753	43		17	.57762	.70760	1.4132	.81631	1.2250	1.7312	43
18	.56353	.68215	1.4659	.82610	1.2105	1.7745	42		18	.57786	.70804	1.4124	.81614	1.2253	1.7305	42
19	.56377	.68258	1.4650	.82593	1.2108	1.7738	41		19	.57810	.70848	1.4115	.81597	1.2255	1.7298	41
20	.56401	.68301	1.4641	.82577	1.2110	1.7730	40		20	.57833	.70891	1.4106	.81580	1.2258	1.7291	40
21	.56425	.68343	1.4632	.82561	1.2112	1.7723	39		21	.57857	.70935	1.4097	.81563	1.2260	1.7284	39
22	.56449	.68386	1.4623	.82544	1.2115	1.7715	38		22	.57881	.70979	1.4089	.81546	1.2263	1.7277	38
23	.56473	.68429	1.4614	.82528	1.2117	1.7708	37		23	.57904	.71023	1.4080	.81530	1.2265	1.7270	37
24	.56497	.68471	1.4605	.82511	1.2120	1.7700	36		24	.57928	.71066	1.4071	.81513	1.2268	1.7263	36
25	.56521	.68514	1.4596	.82495	1.2122	1.7693	35		25	.57952	.71110	1.4063	.81496	1.2271	1.7256	35
26	.56545	.68557	1.4586	.82478	1.2124	1.7685	34		26	.57976	.71154	1.4054	.81479	1.2273	1.7249	34
27	.56569	.68600	1.4577	.82462	1.2127	1.7678	33		27	.57999	.71198	1.4045	.81462	1.2276	1.7242	33
28	.56593	.68642	1.4568	.82446	1.2129	1.7670	32		28	.58023	.71242	1.4037	.81445	1.2278	1.7235	32
29	.56617	.68685	1.4559	.82429	1.2132	1.7663	31		29	.58047	.71285	1.4028	.81428	1.2281	1.7228	31
30	.56641	.68728	1.4550	.82413	1.2134	1.7655	30		30	.58070	.71329	1.4019	.81412	1.2283	1.7221	30
31	.56665	.68771	1.4541	.82396	1.2136	1.7648	29		31	.58094	.71373	1.4011	.81395	1.2286	1.7213	29
32	.56689	.68814	1.4532	.82380	1.2139	1.7640	28		32	.58118	.71417	1.4002	.81378	1.2288	1.7206	28
33	.56713	.68857	1.4523	.82363	1.2141	1.7633	27		33	.58141	.71461	1.3994	.81361	1.2291	1.7199	27
34	.56736	.68900	1.4514	.82347	1.2144	1.7625	26		34	.58165	.71505	1.3985	.81344	1.2293	1.7192	26
35	.56760	.68942	1.4505	.82330	1.2146	1.7618	25		35	.58189	.71549	1.3976	.81327	1.2296	1.7185	25
36	.56784	.68985	1.4496	.82314	1.2149	1.7610	24		36	.58212	.71593	1.3968	.81310	1.2299	1.7179	24
37	.56808	.69028	1.4487	.82297	1.2151	1.7603	23		37	.58236	.71637	1.3959	.81293	1.2301	1.7172	23
38	.56832	.69071	1.4478	.82281	1.2154	1.7596	22		38	.58260	.71681	1.3951	.81276	1.2304	1.7165	22
39	.56856	.69114	1.4469	.82264	1.2156	1.7588	21		39	.58283	.71725	1.3942	.81259	1.2306	1.7158	21
40	.56880	.69157	1.4460	.82248	1.2158	1.7581	20		40	.58307	.71769	1.3934	.81242	1.2309	1.7151	20
41	.56904	.69200	1.4451	.82231	1.2161	1.7573	19		41	.58330	.71813	1.3925	.81225	1.2311	1.7144	19
42	.56928	.69243	1.4442	.82214	1.2163	1.7566	18		42	.58354	.71857	1.3916	.81208	1.2314	1.7137	18
43	.56952	.69286	1.4433	.82198	1.2166	1.7559	17		43	.58378	.71901	1.3908	.81191	1.2317	1.7130	17
44	.56976	.69329	1.4424	.82181	1.2168	1.7551	16		44	.58401	.71946	1.3899	.81174	1.2319	1.7123	16
45	.57000	.69372	1.4415	.82165	1.2171	1.7544	15		45	.58425	.71990	1.3891	.81157	1.2322	1.7116	15
46	.57024	.69416	1.4406	.82148	1.2173	1.7537	14		46	.58449	.72034	1.3882	.81140	1.2324	1.7109	14
47	.57047	.69459	1.4397	.82132	1.2176	1.7529	13		47	.58472	.72078	1.3874	.81123	1.2327	1.7102	13
48	.57071	.69502	1.4388	.82115	1.2178	1.7522	12		48	.58496	.72122	1.3865	.81106	1.2329	1.7095	12
49	.57095	.69545	1.4379	.82098	1.2181	1.7515	11		49	.58519	.72167	1.3857	.81089	1.2332	1.7088	11
50	.57119	.69588	1.4370	.82082	1.2183	1.7507	10		50	.58543	.72211	1.3848	.81072	1.2335	1.7081	10
51	.57143	.69631	1.4361	.82065	1.2185	1.7500	9		51	.58567	.72255	1.3840	.81055	1.2337	1.7075	9
52	.57167	.69675	1.4352	.82048	1.2188	1.7493	8		52	.58590	.72299	1.3831	.81038	1.2340	1.7068	8
53	.57191	.69718	1.4344	.82032	1.2190	1.7485	7		53	.58614	.72344	1.3823	.81021	1.2342	1.7061	7
54	.57215	.69761	1.4335	.82015	1.2193	1.7478	6		54	.58637	.72388	1.3814	.81004	1.2345	1.7054	6
55	.57238	.69804	1.4326	.81999	1.2195	1.7471	5		55	.58661	.72432	1.3806	.80987	1.2348	1.7047	5
56	.57262	.69847	1.4317	.81982	1.2198	1.7463	4		56	.58684	.72477	1.3798	.80970	1.2350	1.7040	4
57	.57286	.69891	1.4308	.81965	1.2200	1.7456	3		57	.58708	.72521	1.3789	.80953	1.2353	1.7033	3
58	.57310	.69934	1.4299	.81949	1.2203	1.7449	2		58	.58731	.72565	1.3781	.80936	1.2355	1.7027	2
59	.57334	.69977	1.4290	.81932	1.2205	1.7442	1		59	.58755	.72610	1.3772	.80919	1.2358	1.7020	1
60	.57358	.70021	1.4281	.81915	1.2208	1.7434	0		60	.58779	.72654	1.3764	.80902	1.2361	1.7013	0
′	Cos	Cot	Tan	Sin	Csc	Sec	′		′	Cos	Cot	Tan	Sin	Csc	Sec	′

124° (304°)   **(235°) 55°**   **125° (305°)**   **(234°) 54°**

NATURAL TRIGONOMETRIC FUNCTIONS TO FIVE PLACES (continued)

36° (216°)   (323°) **143°**   **37° (217°)**   (322°) **142°**

′	Sin	Tan	Cot	Cos	Sec	Csc	′		′	Sin	Tan	Cot	Cos	Sec	Csc	′
0	.58779	.72654	1.3764	.80902	1.2361	1.7013	60		0	.60182	.75355	1.3270	.79864	1.2521	1.6616	60
1	.58802	.72699	1.3755	.80885	1.2363	1.7006	59		1	.60205	.75401	1.3262	.79846	1.2524	1.6610	59
2	.58826	.72743	1.3747	.80867	1.2366	1.6999	58		2	.60228	.75447	1.3254	.79829	1.2527	1.6604	58
3	.58849	.72788	1.3739	.80850	1.2369	1.6993	57		3	.60251	.75492	1.3246	.79811	1.2530	1.6597	57
4	.58873	.72832	1.3730	.80833	1.2371	1.6986	56		4	.60274	.75538	1.3238	.79793	1.2532	1.6591	56
5	.58896	.72877	1.3722	.80816	1.2374	1.6979	55		5	.60298	.75584	1.3230	.79776	1.2535	1.6584	55
6	.58920	.72921	1.3713	.80799	1.2376	1.6972	54		6	.60321	.75629	1.3222	.79758	1.2538	1.6578	54
7	.58943	.72966	1.3705	.80782	1.2379	1.6966	53		7	.60344	.75675	1.3214	.79741	1.2541	1.6572	53
8	.58967	.73010	1.3697	.80765	1.2382	1.6959	52		8	.60367	.75721	1.3206	.79723	1.2543	1.6565	52
9	.58990	.73055	1.3688	.80748	1.2384	1.6952	51		9	.60390	.75767	1.3198	.79706	1.2546	1.6559	51
10	.59014	.73100	1.3680	.80730	1.2387	1.6945	50		10	.60414	.75812	1.3190	.79688	1.2549	1.6553	50
11	.59037	.73144	1.3672	.80713	1.2390	1.6939	49		11	.60437	.75858	1.3182	.79671	1.2552	1.6546	49
12	.59061	.73189	1.3663	.80696	1.2392	1.6932	48		12	.60460	.75904	1.3175	.79653	1.2554	1.6540	48
13	.59084	.73234	1.3655	.80679	1.2395	1.6925	47		13	.60483	.75950	1.3167	.79635	1.2557	1.6534	47
14	.59108	.73278	1.3647	.80662	1.2397	1.6918	46		14	.60506	.75996	1.3159	.79618	1.2560	1.6527	46
15	.59131	.73323	1.3638	.80644	1.2400	1.6912	45		15	.60529	.76042	1.3151	.79600	1.2563	1.6521	45
16	.59154	.73368	1.3630	.80627	1.2403	1.6905	44		16	.60553	.76088	1.3143	.79583	1.2566	1.6515	44
17	.59178	.73413	1.3622	.80610	1.2405	1.6898	43		17	.60576	.76134	1.3135	.79565	1.2568	1.6508	43
18	.59201	.73457	1.3613	.80593	1.2408	1.6892	42		18	.60599	.76180	1.3127	.79547	1.2571	1.6502	42
19	.59225	.73502	1.3605	.80576	1.2411	1.6885	41		19	.60622	.76226	1.3119	.79530	1.2574	1.6496	41
20	.59248	.73547	1.3597	.80558	1.2413	1.6878	40		20	.60645	.76272	1.3111	.79512	1.2577	1.6489	40
21	.59272	.73592	1.3588	.80541	1.2416	1.6871	39		21	.60668	.76318	1.3103	.79494	1.2579	1.6483	39
22	.59295	.73637	1.3580	.80524	1.2419	1.6865	38		22	.60691	.76364	1.3095	.79477	1.2582	1.6477	38
23	.59318	.73681	1.3572	.80507	1.2421	1.6858	37		23	.60714	.76410	1.3087	.79459	1.2585	1.6471	37
24	.59342	.73726	1.3564	.80489	1.2424	1.6852	36		24	.60738	.76456	1.3079	.79441	1.2588	1.6464	36
25	.59365	.73771	1.3555	.80472	1.2427	1.6845	35		25	.60761	.76502	1.3072	.79424	1.2591	1.6458	35
26	.59389	.73816	1.3547	.80455	1.2429	1.6838	34		26	.60784	.76548	1.3064	.79406	1.2593	1.6452	34
27	.59412	.73861	1.3539	.80438	1.2432	1.6832	33		27	.60807	.76594	1.3056	.79388	1.2596	1.6446	33
28	.59436	.73906	1.3531	.80420	1.2435	1.6825	32		28	.60830	.76640	1.3048	.79371	1.2599	1.6439	32
29	.59459	.73951	1.3522	.80403	1.2437	1.6818	31		29	.60853	.76686	1.3040	.79353	1.2602	1.6433	31
30	.59482	.73996	1.3514	.80386	1.2440	1.6812	30		30	.60876	.76733	1.3032	.79335	1.2605	1.6427	30
31	.59506	.74041	1.3506	.80368	1.2443	1.6805	29		31	.60899	.76779	1.3024	.79318	1.2608	1.6421	29
32	.59529	.74086	1.3498	.80351	1.2445	1.6799	28		32	.60922	.76825	1.3017	.79300	1.2610	1.6414	28
33	.59552	.74131	1.3490	.80334	1.2448	1.6792	27		33	.60945	.76871	1.3009	.79282	1.2613	1.6408	27
34	.59576	.74176	1.3481	.80316	1.2451	1.6785	26		34	.60968	.76918	1.3001	.79264	1.2616	1.6402	26
35	.59599	.74221	1.3473	.80299	1.2453	1.6779	25		35	.60991	.76964	1.2993	.79247	1.2619	1.6396	25
36	.59622	.74267	1.3465	.80282	1.2456	1.6772	24		36	.61015	.77010	1.2985	.79229	1.2622	1.6390	24
37	.59646	.74312	1.3457	.80264	1.2459	1.6766	23		37	.61038	.77057	1.2977	.79211	1.2624	1.6383	23
38	.59669	.74357	1.3449	.80247	1.2462	1.6759	22		38	.61061	.77103	1.2970	.79193	1.2627	1.6377	22
39	.59693	.74402	1.3440	.80230	1.2464	1.6753	21		39	.61084	.77149	1.2962	.79176	1.2630	1.6371	21
40	.59716	.74447	1.3432	.80212	1.2467	1.6746	20		40	.61107	.77196	1.2954	.79158	1.2633	1.6365	20
41	.59739	.74492	1.3424	.80195	1.2470	1.6739	19		41	.61130	.77242	1.2946	.79140	1.2636	1.6359	19
42	.59763	.74538	1.3416	.80178	1.2472	1.6733	18		42	.61153	.77289	1.2938	.79122	1.2639	1.6353	18
43	.59786	.74583	1.3408	.80160	1.2475	1.6726	17		43	.61176	.77335	1.2931	.79105	1.2641	1.6346	17
44	.59809	.74628	1.3400	.80143	1.2478	1.6720	16		44	.61199	.77382	1.2923	.79087	1.2644	1.6340	16
45	.59832	.74674	1.3392	.80125	1.2480	1.6713	15		45	.61222	.77428	1.2915	.79069	1.2647	1.6334	15
46	.59856	.74719	1.3384	.80108	1.2483	1.6707	14		46	.61245	.77475	1.2907	.79051	1.2650	1.6328	14
47	.59879	.74764	1.3375	.80091	1.2486	1.6700	13		47	.61268	.77521	1.2900	.79033	1.2653	1.6322	13
48	.59902	.74810	1.3367	.80073	1.2489	1.6694	12		48	.61291	.77568	1.2892	.79016	1.2656	1.6316	12
49	.59926	.74855	1.3359	.80056	1.2491	1.6687	11		49	.61314	.77615	1.2884	.78998	1.2659	1.6310	11
50	.59949	.74900	1.3351	.80038	1.2494	1.6681	10		50	.61337	.77661	1.2876	.78980	1.2661	1.6303	10
51	.59972	.74946	1.3343	.80021	1.2497	1.6674	9		51	.61360	.77708	1.2869	.78962	1.2664	1.6297	9
52	.59995	.74991	1.3335	.80003	1.2499	1.6668	8		52	.61383	.77754	1.2861	.78944	1.2667	1.6291	8
53	.60019	.75037	1.3327	.79986	1.2502	1.6661	7		53	.61406	.77801	1.2853	.78926	1.2670	1.6285	7
54	.60042	.75082	1.3319	.79968	1.2505	1.6655	6		54	.61429	.77848	1.2846	.78908	1.2673	1.6279	6
55	.60065	.75128	1.3311	.79951	1.2508	1.6649	5		55	.61451	.77895	1.2838	.78891	1.2676	1.6273	5
56	.60089	.75173	1.3303	.79934	1.2510	1.6642	4		56	.61474	.77941	1.2830	.78873	1.2679	1.6267	4
57	.60112	.75219	1.3295	.79916	1.2513	1.6636	3		57	.61497	.77988	1.2822	.78855	1.2682	1.6261	3
58	.60135	.75264	1.3287	.79899	1.2516	1.6629	2		58	.61520	.78035	1.2815	.78837	1.2684	1.6255	2
59	.60158	.75310	1.3278	.79881	1.2519	1.6623	1		59	.61543	.78082	1.2807	.78819	1.2687	1.6249	1
60	.60182	.75355	1.3270	.79864	1.2521	1.6616	0		60	.61566	.78129	1.2799	.78801	1.2690	1.6243	0
′	Cos	Cot	Tan	Sin	Csc	Sec	′		′	Cos	Cot	Tan	Sin	Csc	Sec	′

126° (306°)   (233°) **53°**   **127° (307°)**   (232°) **52°**

NATURAL TRIGONOMETRIC FUNCTIONS TO FIVE PLACES (continued)

38° (218°)   (321°) 141°   **39° (219°)**   (320°) 140°

′	Sin.	Tan	Cot	Cos	Sec	Csc	′		′	Sin	Tan	Cot	Cos	Sec	Csc	′
0	.61566	.78129	1.2799	.78801	1.2690	1.6243	60		0	.62932	.80978	1.2349	.77715	1.2868	1.5890	60
1	.61589	.78175	1.2792	.78783	1.2693	1.6237	59		1	.62955	.81027	1.2342	.77696	1.2871	1.5884	59
2	.61612	.78222	1.2784	.78765	1.2696	1.6231	58		2	.62977	.81075	1.2334	.77678	1.2874	1.5879	58
3	.61635	.78269	1.2776	.78747	1.2699	1.6225	57		3	.63000	.81123	1.2327	.77660	1.2877	1.5873	57
4	.61658	.78316	1.2769	.78729	1.2702	1.6219	56		4	.63022	.81171	1.2320	.77641	1.2880	1.5867	56
5	.61681	.78363	1.2761	.78711	1.2705	1.6213	55		5	.63045	.81220	1.2312	.77623	1.2883	1.5862	55
6	.61704	.78410	1.2753	.78694	1.2708	1.6207	54		6	.63068	.81268	1.2305	.77605	1.2886	1.5856	54
7	.61726	.78457	1.2746	.78676	1.2710	1.6201	53		7	.63090	.81316	1.2298	.77586	1.2889	1.5850	53
8	.61749	.78504	1.2738	.78658	1.2713	1.6195	52		8	.63113	.81364	1.2290	.77568	1.2892	1.5845	52
9	.61772	.78551	1.2731	.78640	1.2716	1.6189	51		9	.63135	.81413	1.2283	.77550	1.2895	1.5839	51
10	.61795	.78598	1.2723	.78622	1.1719	1.6183	50		10	.63158	.81461	1.2276	.77531	1.2898	1.5833	50
11	.61818	.78645	1.2715	.78604	1.2722	1.6177	49		11	.63180	.81510	1.2268	.77513	1.2901	1.5828	49
12	.61841	.78692	1.2708	.78586	1.2725	1.6171	48		12	.63203	.81558	1.2261	.77494	1.2904	1.5822	48
13	.61864	.78739	1.2700	.78568	1.2728	1.6165	47		13	.63225	.81606	1.2254	.77476	1.2907	1.5816	47
14	.61887	.78786	1.2693	.78550	1.2731	1.6159	46		14	.63248	.81655	1.2247	.77458	1.2910	1.5811	46
15	.61909	.78834	1.2685	.78532	1.2734	1.6153	45		15	.63271	.81703	1.2239	.77439	1.2913	1.5805	45
16	.61932	.78881	1.2677	.78514	1.2737	1.6147	44		16	.63293	.81752	1.2232	.77421	1.2916	1.5800	44
17	.61955	.78928	1.2670	.78496	1.2740	1.6141	43		17	.63316	.81800	1.2225	.77402	1.2919	1.5794	43
18	.61978	.78975	1.2662	.78478	1.2742	1.6135	42		18	.63338	.81849	1.2218	.77384	1.2923	1.5788	42
19	.62001	.79022	1.2655	.78460	1.2745	1.6129	41		19	.63361	.81898	1.2210	.77366	1.2926	1.5783	41
20	.62024	.79070	1.2647	.78442	1.2748	1.6123	40		20	.63383	.81946	1.2203	.77347	1.2929	1.5777	40
21	.62046	.79117	1.2640	.78424	1.2751	1.6117	39		21	.63406	.81995	1.2196	.77329	1.2932	1.5771	39
22	.62069	.79164	1.2632	.78405	1.2754	1.6111	38		22	.63428	.82044	1.2189	.77310	1.2935	1.5766	38
23	.62092	.79212	1.2624	.78387	1.2757	1.6105	37		23	.63451	.82092	1.2181	.77292	1.2938	1.5760	37
24	.62115	.79259	1.2617	.78369	1.2760	1.6099	36		24	.63473	.82141	1.2174	.77273	1.2941	1.5755	36
25	.62138	.79306	1.2609	.78351	1.2763	1.6093	35		25	.63496	.82190	1.2167	.77255	1.2944	1.5749	35
26	.62160	.79354	1.2602	.78333	1.2766	1.6087	34		26	.63518	.82238	1.2160	.77236	1.2947	1.5744	34
27	.62183	.79401	1.2594	.78315	1.2769	1.6082	33		27	.63540	.82287	1.2153	.77218	1.2950	1.5738	33
28	.62206	.79449	1.2587	.78297	1.2772	1.6076	32		28	.63563	.82336	1.2145	.77199	1.2953	1.5732	32
29	.62229	.79496	1.2579	.78279	1.2775	1.6070	31		29	.63585	.82385	1.2138	.77181	1.2957	1.5727	31
30	.62251	.79544	1.2572	.78261	1.2778	1.6064	30		30	.63608	.82434	1.2131	.77162	1.2960	1.5721	30
31	.62274	.79591	1.2564	.78243	1.2781	1.6058	29		31	.63630	.82483	1.2124	.77144	1.2963	1.5716	29
32	.62297	.79639	1.2557	.78225	1.2784	1.6052	28		32	.63653	.82531	1.2117	.77125	1.2966	1.5710	28
33	.62320	.79686	1.2549	.78206	1.2787	1.6046	27		33	.63675	.82580	1.2109	.77107	1.2969	1.5705	27
34	.62342	.79734	1.2542	.78188	1.2790	1.6040	26		34	.63698	.82629	1.2102	.77088	1.2972	1.5699	26
35	.62365	.79781	1.2534	.78170	1.2793	1.6035	25		35	.63720	.82678	1.2095	.77070	1.2975	1.5694	25
36	.62388	.79829	1.2527	.78152	1.2796	1.6029	24		36	.63742	.82727	1.2088	.77051	1.2978	1.5688	24
37	.62411	.79877	1.2519	.78134	1.2799	1.6023	23		37	.63765	.82776	1.2081	.77033	1.2981	1.5683	23
38	.62433	.79924	1.2512	.78116	1.2802	1.6017	22		38	.63787	.82825	1.2074	.77014	1.2985	1.5677	22
39	.62456	.79972	1.2504	.78098	1.2804	1.6011	21		39	.63810	.82874	1.2066	.76996	1.2988	1.5672	21
40	.62479	.80020	1.2497	.78079	1.2807	1.6005	20		40	.63832	.82923	1.2059	.76977	1.2991	1.5666	20
41	.62502	.80067	1.2489	.78061	1.2810	1.6000	19		41	.63854	.82972	1.2052	.76959	1.2994	1.5661	19
42	.62524	.80115	1.2482	.78043	1.2813	1.5994	18		42	.63877	.83022	1.2045	.76940	1.2997	1.5655	18
43	.62547	.80163	1.2475	.78025	1.2816	1.5988	17		43	.63899	.83071	1.2038	.76921	1.3000	1.5650	17
44	.62570	.80211	1.2467	.78007	1.2819	1.5982	16		44	.63922	.83120	1.2031	.76903	1.3003	1.5644	16
45	.62592	.80258	1.2460	.77988	1.2822	1.5976	15		45	.63944	.83169	1.2024	.76884	1.3007	1.5639	15
46	.62615	.80306	1.2452	.77970	1.2825	1.5971	14		46	.63966	.83218	1.2017	.76866	1.3010	1.5633	14
47	.62638	.80354	1.2445	.77952	1.2828	1.5965	13		47	.63989	.83268	1.2009	.76847	1.3013	1.5628	13
48	.62660	.80402	1.2437	.77934	1.2831	1.5959	12		48	.64011	.83317	1.2002	.76828	1.3016	1.5622	12
49	.62683	.80450	1.2430	.77916	1.2834	1.5953	11		49	.64033	.83366	1.1995	.76810	1.3019	1.5617	11
50	.62706	.80498	1.2423	.77897	1.2837	1.5948	10		50	.64056	.83415	1.1988	.76791	1.3022	1.5611	10
51	.62728	.80546	1.2415	.77879	1.2840	1.5942	9		51	.64078	.83465	1.1981	.76772	1.3026	1.5606	9
52	.62751	.80594	1.2408	.77861	1.2843	1.5936	8		52	.64100	.83514	1.1974	.76754	1.3029	1.5601	8
53	.62774	.80642	1.2401	.77843	1.2846	1.5930	7		53	.64123	.83564	1.1967	.76735	1.3032	1.5595	7
54	.62796	.80690	1.2393	.77824	1.2849	1.5925	6		54	.64145	.83613	1.1960	.76717	1.3035	1.5590	6
55	.62819	.80738	1.2386	.77806	1.2852	1.5919	5		55	.64167	.83662	1.1953	.76698	1.3038	1.5584	5
56	.62842	.80786	1.2378	.77788	1.2855	1.5913	4		56	.64190	.83712	1.1946	.76679	1.3041	1.5579	4
57	.62864	.80834	1.2371	.77769	1.2859	1.5907	3		57	.64212	.83761	1.1939	.76661	1.3045	1.5573	3
58	.62887	.80882	1.2364	.77751	1.2862	1.5902	2		58	.64234	.83811	1.1932	.76642	1.3048	1.5568	2
59	.62909	.80930	1.2356	.77733	1.2865	1.5896	1		59	.64256	.83860	1.1925	.76623	1.3051	1.5563	1
60	.62932	.80978	1.2349	.77715	1.2868	1.5890	0		60	.64279	.83910	1 1918	.76604	1.3054	1.5557	0
′	Cos	Cot	Tan	Sin	Csc	Sec	′		′	Cos	Cot	Tan	Sin	Csc	Sec	′

NATURAL TRIGONOMETRIC FUNCTIONS TO FIVE PLACES (continued)

40° (220°)   (319°) **139°**   **41° (221°)**   (318°) **138°**

′	Sin	Tan	Cot	Cos	Sec	Csc	′	′	Sin	Tan	Cot	Cos	Sec	Csc	′
0	.64279	.83910	1.1918	.76604	1.3054	1.5557	60	0	.65606	.86929	1.1504	.75471	1.3250	1.5243	60
1	.64301	.83960	1.1910	.76586	1.3057	1.5552	59	1	.65628	.86980	1.1497	.75452	1.3253	1.5237	59
2	.64323	.84009	1.1903	.76567	1.3060	1.5546	58	2	.65650	.87031	1.1490	.75433	1.3257	1.5232	58
3	.64346	.84059	1.1896	.76548	1.3064	1.5541	57	3	.65672	.87082	1.1483	.75414	1.3260	1.5227	57
4	.64368	.84108	1.1889	.76530	1.3067	1.5536	56	4	.65694	.87133	1.1477	.75395	1.3264	1.5222	56
5	.64390	.84158	1.1882	.76511	1.3070	1.5530	55	5	.65716	.87184	1.1470	.75375	1.3267	1.5217	55
6	.64412	.84208	1.1875	.76492	1.3073	1.5525	54	6	.65738	.87236	1.1463	.75356	1.3270	1.5212	54
7	.64435	.84258	1.1868	.76473	1.3076	1.5520	53	7	.65759	.87287	1.1456	.75337	1.3274	1.5207	53
8	.64457	.84307	1.1861	.76455	1.3080	1.5514	52	8	.65781	.87338	1.1450	.75318	1.3277	1.5202	52
9	.64479	.84357	1.1854	.76436	1.3083	1.5509	51	9	.65803	.87389	1.1443	.75299	1.3280	1.5197	51
10	.64501	.84407	1.1847	.76417	1.3086	1.5504	50	10	.65825	.87441	1.1436	.75280	1.3284	1.5192	50
11	.64524	.84457	1.1840	.76398	1.3089	1.5498	49	11	.65847	.87492	1.1430	.75261	1.3287	1.5187	49
12	.64546	.84507	1.1833	.76380	1.3093	1.5493	48	12	.65869	.87543	1.1423	.75241	1.3291	1.5182	48
13	.64568	.84556	1.1826	.76361	1.3096	1.5488	47	13	.65891	.87595	1.1416	.75222	1.3294	1.5177	47
14	.64590	.84606	1.1819	.76342	1.3099	1.5482	46	14	.65913	.87646	1.1410	.75203	1.3297	1.5172	46
15	.64612	.84656	1.1812	.76323	1.3102	1.5477	45	15	.65935	.87698	1.1403	.75184	1.3301	1.5167	45
16	.64635	.84706	1.1806	.76304	1.3105	1.5472	44	16	.65956	.87749	1.1396	.75165	1.3304	1.5162	44
17	.64657	.84756	1.1799	.76286	1.3109	1.5466	43	17	.65978	.87801	1.1389	.75146	1.3307	1.5156	43
18	.64679	.84806	1.1792	.76267	1.3112	1.5461	42	18	.66000	.87852	1.1383	.75126	1.3311	1.5151	42
19	.64701	.84856	1.1785	.76248	1.3115	1.5456	41	19	.66022	.87904	1.1376	.75107	1.3314	1.5146	41
20	.64723	.84906	1.1778	.76229	1.3118	1.5450	40	20	.66044	.87955	1.1369	.75088	1.3318	1.5141	40
21	.64746	.84956	1.1771	.76210	1.3122	1.5445	39	21	.66066	.88007	1.1363	.75069	1.3321	1.5136	39
22	.04768	.85006	1.1764	.76192	1.3125	1.5440	38	22	.66088	.88059	1.1356	.75050	1.3325	1.5131	38
23	.64790	.85057	1.1757	.76173	1.3128	1.5435	37	23	.66109	.88110	1.1349	.75030	1.3328	1.5126	37
24	.64812	.85107	1.1750	.76154	1.3131	1.5429	36	24	.66131	.88162	1.1343	.75011	1.3331	1.5121	36
25	.64834	.85157	1.1743	.76135	1.3135	1.5424	35	25	.66153	.88214	1.1336	.74992	1.3335	1.5116	35
26	.64856	.85207	1.1736	.76116	1.3138	1.5419	34	26	.66175	.88265	1.1329	.74973	1.3338	1.5111	34
27	.64878	.85257	1.1729	.76097	1.3141	1.5413	33	27	.66197	.88317	1.1323	.74953	1.3342	1.5107	33
28	.64901	.85308	1.1722	.76078	1.3144	1.5408	32	28	.66218	.88369	1.1316	.74934	1.3345	1.5102	32
29	.64923	.85358	1.1715	.76059	1.3148	1.5403	31	29	.66240	.88421	1.1310	.74915	1.3348	1.5097	31
30	.64945	.85408	1.1708	.76041	1.3151	1.5398	30	30	.66262	.88473	1.1303	.74896	1.3352	1.5092	30
31	.64967	.85458	1.1702	.76022	1.3154	1.5392	29	31	.66284	.88524	1.1296	.74876	1.3355	1.5087	29
32	.64989	.85509	1.1695	.76003	1.3157	1.5387	28	32	.66306	.88576	1.1290	.74857	1.3359	1.5082	28
33	.65011	.85559	1.1688	.75984	1.3161	1.5382	27	33	.66327	.88628	1.1283	.74838	1.3362	1.5077	27
34	.65033	.85609	1.1681	.75965	1.3164	1.5377	26	34	.66349	.88680	1.1276	.74818	1.3366	1.5072	26
35	.65055	.85660	1.1674	.75946	1.3167	1.5372	25	35	.66371	.88732	1.1270	.74799	1.3369	1.5067	25
36	.65077	.85710	1.1667	.75927	1.3171	1.5366	24	36	.66393	.88784	1.1263	.74780	1.3373	1.5062	24
37	.65100	.85761	1.1660	.75908	1.3174	1.5361	23	37	.66414	.88836	1.1257	.74760	1.3376	1.5057	23
38	.65122	.85811	1.1653	.75889	1.3177	1.5356	22	38	.66436	.88888	1.1250	.74741	1.3380	1.5052	22
39	.65144	.85862	1.1647	.75870	1.3180	1.5351	21	39	.66458	.88940	1.1243	.74722	1.3383	1.5047	21
40	.65166	.85912	1.1640	.75851	1.3184	1.5345	20	40	.66480	.88992	1.1237	.74703	1.3386	1.5042	20
41	.65188	.85963	1.1633	.75832	1.3187	1.5340	19	41	.66501	.89045	1.1230	.74683	1.3390	1.5037	19
42	.65210	.86014	1.1626	.75813	1.3190	1.5335	18	42	.66523	.89097	1.1224	.74664	1.3393	1.5032	18
43	.65232	.86064	1.1619	.75794	1.3194	1.5330	17	43	.66545	.89149	1.1217	.74644	1.3397	1.5027	17
44	.65254	.86115	1.1612	.75775	1.3197	1.5325	16	44	.66566	.89201	1.1211	.74625	1.3400	1.5023	16
45	.65276	.86166	1.1606	.75756	1.3200	1.5320	15	45	.66588	.89253	1.1204	.74606	1.3404	1.5018	15
46	.65298	.86216	1.1599	.75738	1.3203	1.5314	14	46	.66610	.89306	1.1197	.74586	1.3407	1.5013	14
47	.65320	.86267	1.1592	.75719	1.3207	1.5309	13	47	.66632	.89358	1.1191	.74567	1.3411	1.5008	13
48	.65342	.86318	1.1585	.75700	1.3210	1.5304	12	48	.66653	.89410	1.1184	.74548	1.3414	1.5003	12
49	.65364	.86368	1.1578	.75680	1.3213	1.5299	11	49	.66675	.89463	1.1178	.74528	1.3418	1.4998	11
50	.65386	.86419	1.1571	.75661	1.3217	1.5294	10	50	.66697	.89515	1.1171	.74509	1.3421	1.4993	10
51	.65408	.86470	1.1565	.75642	1.3220	1.5289	9	51	.66718	.89567	1.1165	.74489	1.3425	1.4988	9
52	.65430	.86521	1.1558	.75623	1.3223	1.5283	8	52	.66740	.89620	1.1158	.74470	1.3428	1.4984	8
53	.65452	.86572	1.1551	.75604	1.3227	1.5278	7	53	.66762	.89672	1.1152	.74451	1.3432	1.4979	7
54	.65474	.86623	1.1544	.75585	1.3230	1.5273	6	54	.66783	.89725	1.1145	.74431	1.3435	1.4974	6
55	.65496	.86674	1.1538	.75566	1.3233	1.5268	5	55	.66805	.89777	1.1139	.74412	1.3439	1.4969	5
56	.65518	.86725	1.1531	.75547	1.3237	1.5263	4	56	.66827	.89830	1.1132	.74392	1.3442	1.4964	4
57	.65540	.86776	1.1524	.75528	1.3240	1.5258	3	57	.66848	.89883	1.1126	.74373	1.3446	1.4959	3
58	.65562	.86827	1.1517	.75509	1.3243	1.5253	2	58	.66870	.89935	1.1119	.74353	1.3449	1.4954	2
59	.65584	.86878	1.1510	.75490	1.3247	1.5248	1	59	.66891	.89988	1.1113	.74334	1.3453	1.4950	1
60	.65606	.86929	1.1504	.75471	1.3250	1.5243	0	60	.66913	.90040	1.1106	.74314	1.3456	1.4945	0
′	Cos	Cot	Tan	Sin	Csc	Sec	′	′	Cos	Cot	Tan	Sin	Csc	Sec	′

NATURAL TRIGONOMETRIC FUNCTIONS TO FIVE PLACES (continued)

42° (222°)   (317°) 137°

'	Sin	Tan	Cot	Cos	Sec	Csc	'
0	.66913	.90040	1.1106	.74314	1.3456	1.4945	60
1	.66935	.90093	1.1100	.74295	1.3460	1.4940	59
2	.66956	.90146	1.1093	.74276	1.3463	1.4935	58
3	.66978	.90199	1.1087	.74256	1.3467	1.4930	57
4	.66999	.90251	1.1080	.74237	1.3470	1.4925	56
5	.67021	.90304	1.1074	.74217	1.3474	1.4921	55
6	.67043	.90357	1.1067	.74198	1.3478	1.4916	54
7	.67064	.90410	1.1061	.74178	1.3481	1.4911	53
8	.67086	.90463	1.1054	.74159	1.3485	1.4906	52
9	.67107	.90516	1.1048	.74139	1.3488	1.4901	51
10	.67129	.90569	1.1041	.74120	1.3492	1.4897	50
11	.67151	.90621	1.1035	.74100	1.3495	1.4892	49
12	.67172	.90674	1.1028	.74080	1.3499	1.4887	48
13	.67194	.90727	1.1022	.74061	1.3502	1.4882	47
14	.67215	.90781	1.1016	.74041	1.3506	1.4878	46
15	.67237	.90834	1.1009	.74022	1.3510	1.4873	45
16	.67258	.90887	1.1003	.74002	1.3513	1.4868	44
17	.67280	.90940	1.0996	.73983	1.3517	1.4863	43
18	.67301	.90993	1.0990	.73963	1.3520	1.4859	42
19	.67323	.91046	1.0983	.73944	1.3524	1.4854	41
20	.67344	.91099	1.0977	.73924	1.3527	1.4849	40
21	.67366	.91153	1.0971	.73904	1.3531	1.4844	39
22	.67387	.91206	1.0964	.73885	1.3535	1.4840	38
23	.67409	.91259	1.0958	.73865	1.3538	1.4835	37
24	.67430	.91313	1.0951	.73846	1.3542	1.4830	36
25	.67452	.91366	1.0945	.73826	1.3545	1.4825	35
26	.67473	.91419	1.0939	.73806	1.3549	1.4821	34
27	.67495	.91473	1.0932	.73787	1.3553	1.4816	33
28	.67516	.91526	1.0926	.73767	1.3556	1.4811	32
29	.67538	.91580	1.0919	.73747	1.3560	1.4807	31
30	.67559	.91633	1.0913	.73728	1.3563	1.4802	30
31	.67580	.91687	1.0907	.73708	1.3567	1.4797	29
32	.67602	.91740	1.0900	.73688	1.3571	1.4792	28
33	.67623	.91794	1.0894	.73669	1.3574	1.4788	27
34	.67645	.91847	1.0888	.73649	1.3578	1.4783	26
35	.67666	.91901	1.0881	.73629	1.3582	1.4778	25
36	.67688	.91955	1.0875	.73610	1.3585	1.4774	24
37	.67709	.92008	1.0869	.73590	1.3589	1.4769	23
38	.67730	.92062	1.0862	.73570	1.3592	1.4764	22
39	.67752	.92116	1.0856	.73551	1.3596	1.4760	21
40	.67773	.92170	1.0850	.73531	1.3600	1.4755	20
41	.67795	.92224	1.0843	.73511	1.3603	1.4750	19
42	.67816	.92277	1.0837	.73491	1.3607	1.4746	18
43	.67837	.92331	1.0831	.73472	1.3611	1.4741	17
44	.67859	.92385	1.0824	.73452	1.3614	1.4737	16
45	.67880	.92439	1.0818	.73432	1.3618	1.4732	15
46	.67901	.92493	1.0812	.73413	1.3622	1.4727	14
47	.67923	.92547	1.0805	.73393	1.3625	1.4723	13
48	.67944	.92601	1.0799	.73373	1.3629	1.4718	12
49	.67965	.92655	1.0793	.73353	1.3633	1.4713	11
50	.67987	.92709	1.0786	.73333	1.3636	1.4709	10
51	.68008	.92763	1.0780	.73314	1.3640	1.4704	9
52	.68029	.92817	1.0774	.73294	1.3644	1.4700	8
53	.68051	.92872	1.0768	.73274	1.3647	1.4695	7
54	.68072	.92926	1.0761	.73254	1.3651	1.4690	6
55	.68093	.92980	1.0755	.73234	1.3655	1.4686	5
56	.68115	.93034	1.0749	.73215	1.3658	1.4681	4
57	.68136	.93088	1.0742	.73195	1.3662	1.4677	3
58	.68157	.93143	1.0736	.73175	1.3666	1.4672	2
59	.68179	.93197	1.0730	.73155	1.3670	1.4667	1
60	.68200	.93252	1.0724	.73135	1.3673	1.4663	0
'	Cos	Cot	Tan	Sin	Csc	Sec	'

43° (223°)   (316°) 136°

'	Sin	Tan	Cot	Cos	Sec	Csc	'
0	.68200	.93252	1.0724	.73135	1.3673	1.4663	60
1	.68221	.93306	1.0717	.73116	1.3677	1.4658	59
2	.68242	.93360	1.0711	.73096	1.3681	1.4654	58
3	.68264	.93415	1.0705	.73076	1.3684	1.4649	57
4	.68285	.93469	1.0699	.73056	1.3688	1.4645	56
5	.68306	.93524	1.0692	.73036	1.3692	1.4640	55
6	.68327	.93578	1.0686	.73016	1.3696	1.4635	54
7	.68349	.93633	1.0680	.72996	1.3699	1.4631	53
8	.68370	.93688	1.0674	.72976	1.3703	1.4626	52
9	.68391	.93742	1.0668	.72957	1.3707	1.4622	51
10	.68412	.93797	1.0661	.72937	1.3711	1.4617	50
11	.68434	.93852	1.0655	.72917	1.3714	1.4613	49
12	.68455	.93906	1.0649	.72897	1.3718	1.4608	48
13	.68476	.93961	1.0643	.72877	1.3722	1.4604	47
14	.68497	.94016	1.0637	.72857	1.3726	1.4599	46
15	.68518	.94071	1.0630	.72837	1.3729	1.4595	45
16	.68539	.94125	1.0624	.72817	1.3733	1.4590	44
17	.68561	.94180	1.0618	.72797	1.3737	1.4586	43
18	.68582	.94235	1.0612	.72777	1.3741	1.4581	42
19	.68603	.94290	1.0606	.72757	1.3744	1.4577	41
20	.68624	.94345	1.0599	.72737	1.3748	1.4572	40
21	.68645	.94400	1.0593	.72717	1.3752	1.4568	39
22	.68666	.94455	1.0587	.72697	1.3756	1.4563	38
23	.68688	.94510	1.0581	.72677	1.3759	1.4559	37
24	.68709	.94565	1.0575	.72657	1.3763	1.4554	36
25	.68730	.94620	1.0569	.72637	1.3767	1.4550	35
26	.68751	.94676	1.0562	.72617	1.3771	1.4545	34
27	.68772	.94731	1.0556	.72597	1.3775	1.4541	33
28	.68793	.94786	1.0550	.72577	1.3778	1.4536	32
29	.68814	.94841	1.0544	.72557	1.3782	1.4532	31
30	.68835	.94896	1.0538	.72537	1.3786	1.4527	30
31	.68857	.94952	1.0532	.72517	1.3790	1.4523	29
32	.68878	.95007	1.0526	.72497	1.3794	1.4518	28
33	.68899	.95062	1.0519	.72477	1.3797	1.4514	27
34	.68920	.95118	1.0513	.72457	1.3801	1.4510	26
35	.68941	.95173	1.0507	.72437	1.3805	1.4505	25
36	.68962	.95229	1.0501	.72417	1.3809	1.4501	24
37	.68983	.95284	1.0495	.72397	1.3813	1.4496	23
38	.69004	.95340	1.0489	.72377	1.3817	1.4492	22
39	.69025	.95395	1.0483	.72357	1.3820	1.4487	21
40	.69046	.95451	1.0477	.72337	1.3824	1.4483	20
41	.69067	.95506	1.0470	.72317	1.3828	1.4479	19
42	.69088	.95562	1.0464	.72297	1.3832	1.4474	18
43	.69109	.95618	1.0458	.72277	1.3836	1.4470	17
44	.69130	.95673	1.0452	.72257	1.3840	1.4465	16
45	.69151	.95729	1.0446	.72236	1.3843	1.4461	15
46	.69172	.95785	1.0440	.72216	1.3847	1.4457	14
47	.69193	.95841	1.0434	.72196	1.3851	1.4452	13
48	.69214	.95897	1.0428	.72176	1.3855	1.4448	12
49	.69235	.95952	1.0422	.72156	1.3859	1.4443	11
50	.69256	.96008	1.0416	.72136	1.3863	1.4439	10
51	.69277	.96064	1.0410	.72116	1.3867	1.4435	9
52	.69298	.96120	1.0404	.72095	1.3871	1.4430	8
53	.69319	.96176	1.0398	.72075	1.3874	1.4426	7
54	.69340	.96232	1.0392	.72055	1.3878	1.4422	6
55	.69361	.96288	1.0385	.72035	1.3882	1.4417	5
56	.69382	.96344	1.0379	.72015	1.3886	1.4413	4
57	.69403	.96400	1.0373	.71995	1.3890	1.4409	3
58	.69424	.96457	1.0367	.71974	1.3894	1.4404	2
59	.69445	.96513	1.0361	.71954	1.3898	1.4400	1
60	.69466	.96569	1.0355	.71934	1.3902	1.4396	0
'	Cos	Cot	Tan	Sin	Csc	Sec	'

NATURAL TRIGONOMETRIC FUNCTIONS TO FIVE PLACES (continued)

44° (224°)   (315°) 135°

′	Sin	Tan	Cot	Cos	Sec	Csc	′
0	.69466	.96569	1.0355	.71934	1.3902	1.4396	60
1	.69487	.96625	1.0349	.71914	1.3906	1.4391	59
2	.69508	.96681	1.0343	.71894	1.3909	1.4387	58
3	.69529	.96738	1.0337	.71873	1.3913	1.4383	57
4	.69549	.96794	1.0331	.71853	1.3917	1.4378	56
5	.69570	.96850	1.0325	.71833	1.3921	1.4374	55
6	.69591	.96907	1.0319	.71813	1.3925	1.4370	54
7	.69612	.96963	1.0313	.71792	1.3929	1.4365	53
8	.69633	.97020	1.0307	.71772	1.3933	1.4361	52
9	.69654	.97076	1.0301	.71752	1.3937	1.4357	51
10	.69675	.97133	1.0295	.71732	1.3941	1.4352	50
11	.69696	.97189	1.0289	.71711	1.3945	1.4348	49
12	.69717	.97246	1.0283	.71691	1.3949	1.4344	48
13	.69737	.97302	1.0277	.71671	1.3953	1.4340	47
14	.69758	.97359	1.0271	.71650	1.3957	1.4335	46
15	.69779	.97416	1.0265	.71630	1.3961	1.4331	45
16	.69800	.97472	1.0259	.71610	1.3965	1.4327	44
17	.69821	.97529	1.0253	.71590	1.3969	1.4322	43
18	.69842	.97586	1.0247	.71569	1.3972	1.4318	42
19	.69862	.97643	1.0241	.71549	1.3976	1.4314	41
20	.69883	.97700	1.0235	.71529	1.3980	1.4310	40
21	.69904	.97756	1.0230	.71508	1.3984	1.4305	39
22	.69925	.97813	1.0224	.71488	1.3988	1.4301	38
23	.69946	.97870	1.0218	.71468	1.3992	1.4297	37
24	.69966	.97927	1.0212	.71447	1.3996	1.4293	36
25	.69987	.97984	1.0206	.71427	1.4000	1.4288	35
26	.70008	.98041	1.0200	.71407	1.4004	1.4284	34
27	.70029	.98098	1.0194	.71386	1.4008	1.4280	33
28	.70049	.98155	1.0188	.71366	1.4012	1.4276	32
29	.70070	.98213	1.0182	.71345	1.4016	1.4271	31
30	.70091	.98270	1.0176	.71325	1.4020	1.4267	30
31	.70112	.98327	1.0170	.71305	1.4024	1.4263	29
32	.70132	.98384	1.0164	.71284	1.4028	1.4259	28
33	.70153	.98441	1.0158	.71264	1.4032	1.4255	27
34	.70174	.98499	1.0152	.71243	1.4036	1.4250	26
35	.70195	.98556	1.0147	.71223	1.4040	1.4246	25
36	.70215	.98613	1.0141	.71203	1.4044	1.4242	24
37	.70236	.98671	1.0135	.71182	1.4048	1.4238	23
38	.70257	.98728	1.0129	.71162	1.4052	1.4234	22
39	.70277	.98786	1.0123	.71141	1.4057	1.4229	21
40	.70298	.98843	1.0117	.71121	1.4061	1.4225	20
41	.70319	.98901	1.0111	.71100	1.4065	1.4221	19
42	.70339	.98958	1.0105	.71080	1.4069	1.4217	18
43	.70360	.99016	1.0099	.71059	1.4073	1.4213	17
44	.70381	.99073	1.0094	.71039	1.4077	1.4208	16
45	.70401	.99131	1.0088	.71019	1.4081	1.4204	15
46	.70422	.99189	1.0082	.70998	1.4085	1.4200	14
47	.70443	.99247	1.0076	.70978	1.4089	1.4196	13
48	.70463	.99304	1.0070	.70957	1.4093	1.4192	12
49	.70484	.99362	1.0064	.70937	1.4097	1.4188	11
50	.70505	.99420	1.0058	.70916	1.4101	1.4183	10
51	.70525	.99478	1.0052	.70896	1.4105	1.4179	9
52	.70546	.99536	1.0047	.70875	1.4109	1.4175	8
53	.70567	.99594	1.0041	.70855	1.4113	1.4171	7
54	.70587	.99652	1.0035	.70834	1.4118	1.4167	6
55	.70608	.99710	1.0029	.70813	1.4122	1.4163	5
56	.70628	.99768	1.0023	.70793	1.4126	1.4159	4
57	.70649	.99826	1.0017	.70772	1.4130	1.4154	3
58	.70670	.99884	1.0012	.70752	1.4134	1.4150	2
59	.70690	.99942	1.0006	.70731	1.4138	1.4146	1
60	.70711	1.0000	1.0000	.70711	1.4142	1.4142	0
′	Cos	Cot	Tan	Sin	Csc	Sec	′

134° (314°)   (225°) 45°

TRIGONOMETRY

NATURAL TRIGONOMETRIC FUNCTIONS
SINE, TANGENT, COTANGENT, COSINE
FOR ANGLES IN πx RADIANS

x		Sin (πx)	Tan (πx)	Cot (πx)	Cos (πx)
.00 or 1.00		.00000	.00000	inf	1.00000
.01	.99	.03141	.03143	31.821	.99951
.02	.98	.06279	.06291	15.895	.99803
.03	.97	.09411	.09453	10.579	.99556
.04	.96	.12533	.12633	7.9158	.99211
.05	.95	.15643	.15838	6.3138	.98769
.06	.94	.18738	.19076	5.2422	.98229
.07	.93	.21814	.22353	4.4737	.97592
.08	.92	.24869	.25676	3.8947	.96858
.09	.91	.27899	.29053	3.4420	.96029
.10	.90	.30902	.32492	3.0777	.95106
.11	.89	.33874	.36002	2.7776	.94088
.12	.88	.36812	.39593	2.5257	.92978
.13	.87	.39715	.43274	2.3109	.91775
.14	.86	.42578	.47056	2.1251	.90483
.15	.85	.45399	.50953	1.9626	.89101
.16	.84	.48175	.54975	1.8190	.87631
.17	.83	.50904	.59140	1.6909	.86074
.18	.82	.53583	.63462	1.5757	.84433
.19	.81	.56208	.67960	1.4715	.82708
.20	.80	.58779	.72654	1.3764	.80902
.21	.79	.61291	.77568	1.2892	.79016
.22	.78	.63742	.82727	1.2088	.77051
.23	.77	.66131	.88162	1.1343	.75011
.24	.76	.68455	.93906	1.0649	.72897
.25	.75	.70711	1.0000	1.0000	.70711
.26	.74	.72897	1.0649	.93906	.68455
.27	.73	.75011	1.1343	.88162	.66131
.28	.72	.77051	1.2088	.82727	.63742
.29	.71	.79016	1.2892	.77568	.61291
.30	.70	.80902	1.3764	.72654	.58779
.31	.69	.82708	1.4715	.67960	.56208
.32	.68	.84433	1.5757	.63462	.53583
.33	.67	.86074	1.6909	.59140	.50904
.34	.66	.87631	1.8190	.54975	.48175
.35	.65	.89101	1.9626	.50953	.45399
.36	.64	.90483	2.1251	.47056	.42578
.37	.63	.91775	2.3109	.43274	.39715
.38	.62	.92978	2.5257	.39593	.36812
.39	.61	.94088	2.7776	.36002	.33874
.40	.60	.95106	3.0777	.32492	.30902
.41	.59	.96029	3.4420	.29053	.27899
.42	.58	.96858	3.8947	.25676	.24869
.43	.57	.97592	4.4737	.22353	.21814
.44	.56	.98229	5.2422	.19076	.18738
.45	.55	.98769	6.3138	.15838	.15643
.46	.54	.99211	7.9158	.12633	.12533
.47	.53	.99556	10.579	.09453	.09411
.48	.52	.99803	15.895	.06291	.06279
.49	.51	.99951	31.821	.03143	.03141
.50	.50	1.0000	inf	.00000	.00000

These functions are useful in the solution of wave equations such as the displacement equation of a sound wave in the form:

$$y = A \sin 2\pi n x$$

without the necessity of reducing the angular rotation either to radians or to degrees in order to find the value of the function.

The algebraic sign of the function follows the familiar Quadrant Law for the particular function desired. Thus a numerical value of 9.13π radians becomes (by the subtraction of the greatest multiple of 2π radians) 1.13π radians. This is the same as $.13\pi$ radians in the 3rd Quadrant, which would give, from the tables above, a value of the sine function of $-.39715$.

Submitted by J. A. Blythe Jr.

LAWS OF EXPONENTS

For a any real number and m a positive integer, the exponential a^m is defined as

$$\underbrace{a \cdot a \cdot a \cdots \cdot a}_{m \text{ terms}}$$

Using this definition, it is easy to show that the following three *laws of exponents* hold:

I. $a^m \cdot a^n = a^{m+n}$

II. $\dfrac{a^m}{a^n} = \begin{cases} a^{m-n} & \text{if } m > n \\ 1 & \text{if } m = n \\ \dfrac{1}{a^{n-m}} & \text{if } m < n \end{cases}$

III. $(a^m)^n = a^{mn}$

The n-th root function is defined as the inverse of the n-th power function; that is, if

$$b^n = a, \quad \text{then} \quad b = \sqrt[n]{a}.$$

If n is odd, there will be a unique real number satisfying the above definition for $\sqrt[n]{a}$, for any real value of a. If n is even, for positive values of a there will be two real values for $\sqrt[n]{a}$, one positive and one negative. By convention, the symbol $\sqrt[n]{a}$ is understood to mean the positive value. If n is even and a is negative, there are no real values for $\sqrt[n]{a}$.

If we now attempt to extend the definition of the exponential a^t to all rational values of the exponent t, in such a way that the three laws of exponents continue to hold, it is easily shown that the required definitions are:

$$a^0 = 1$$
$$a^{p/q} = \sqrt[q]{a^p}$$
$$a^{-t} = \frac{1}{a^t}$$

In order to avoid difficulties with imaginary numbers and division by zero, a must now be restricted to be positive, if p is odd, and q is even.

With this extended definition, it is possible to restate the second law of exponents in a simpler form:

II'. $\dfrac{a^m}{a^n} = a^{m-n}$

It is shown in advanced calculus that this definition may be further extended so that the exponent may be any real number, and the laws of exponents continue to hold. When the quantity a^x thus defined is viewed as a function of the exponent x, with the base a held constant, it is a continuous function. Also, if $a > 1$, the exponential function is monotone increasing, and if $0 < a < 1$, it is monotone decreasing.

LOGARITHMS

Any monotone function has a single-valued inverse function, which is also monotone. Furthermore, if the original function is continuous, so is the inverse. Therefore, the inverse function to the exponential function a^x exists for all positive values of a, except $a = 1$. This function is given the name *logarithm to the base a*, abbreviated $\log_a$. That is, if

$$x = a^y, \quad \text{then} \quad y = \log_a x.$$

This function is defined and continuous for all positive values of x. It is monotone increasing if $a > 1$, and monotone decreasing if $0 < a < 1$.

If the laws of exponents are rewritten in terms of logarithms, they become the *laws of logarithms*:

I. $\log_a(xy) = \log_a x + \log_a y$

II. $\log_a\left(\dfrac{x}{y}\right) = \log_a x - \log_a y$

III. $\log_a(x^n) = n \log_a x$

Logarithms derive their main usefulness in computation from the above laws, since they allow multiplication, division, and exponentiation to be replaced by the simpler operations of addition, subtraction, and multiplication, respectively. See the examples which follow.

Further recourse to the definition of logarithm leads to the following formula for change of base

$$\log_a x = \log_b x / \log_b a = (\log_b x) \cdot (\log_a b)$$

Two numbers are commonly used as bases for logarithms. Logarithms to the base 10 are most convenient for use in computation. These logarithms are called common or Briggsian logarithms.

The other usual base for logarithms is an irrational number denoted by e, whose value is approximately 2.71828 .... These logarithms are called natural, Naperian, or hyperbolic logarithms, and occur in many formulas of higher mathematics. The abbreviation ln is frequently used for the natural logarithm function.

Other bases for logarithms, such as 2 and 3, occur in certain applications. These applications are rather specialized and separate tables for these bases are not given. Instead, the formulas for change of base are applied to common or natural logarithms.

If the formulas for change of base are applied to the two usual bases, the following formulas result:

$$\log_{10} x = \log_e x / \log_e 10 = (\log_{10} e)(\log_e x) = M \log_e x$$
$$= 0.43429\ 44819 \log_e x$$
$$\log_e x = \log_{10} x / \log_{10} e = (\log_e 10)(\log_{10} x) = \frac{1}{M} \log_{10} x$$
$$= 2.30258\ 50930 \log_{10} x$$

The following remarks apply to common logarithms.

Since most numbers are irrational powers of ten, a common logarithm, in general, consists of an integer, which is called the characteristic, and an endless decimal, the mantissa.

It is to be observed that the common logarithms of all numbers expressed by the same figures in the same order with the decimal point in different positions have different characteristics but the same mantissa. To illustrate:—if the decimal point stand after the first figure of a number, counting from the left, the characteristic is 0; if after two figures, it is 1; if after three figures, it is 2; and so forth. If the decimal point stands before the first significant figure the characteristic is –1, usually written $\bar{1}$; if there is one zero between the decimal point and the first significant figure it is $\bar{2}$, and so on. For example: log 256 = 2.40824, log 2.56 = 0.40824, log 0.256 = $\bar{1}$.40824, log 0.00256 = $\bar{3}$.40824. The two latter are often written log 0.256 = 9.40824 – 10, log 0.00256 = 7.40824 – 10.

Notice that, although the common logarithm of a number less than one is a negative number, it is customarily written as a negative characteristic and a *positive* mantissa, since the mantissas are usually given in tables as positive numbers. This is the reason that the negative sign is written above the characteristic, since it does not apply to the mantissa. Thus log 0.00256 = $\bar{3}$.40824 = 7.40824 – 10 = –2.59176.

A method of determining characteristics of logarithms is to write the number with one figure to the left of the decimal point multiplied by the appropriate power of 10. The characteristic is then the exponent used. For example:

$$256\ 000\ 000 = 2.56 \times 10^8 \qquad \log = 8.40824$$
$$0.000\ 000\ 256 = 2.56 \times 10^{-7} \qquad \log = \bar{7}.40824 \text{ or } 3.40824 - 10$$

Inasmuch as the characteristic may be determined by inspection, the mantissas only are given in tables of common logarithms.

USE OF LOGARITHM TABLES

To find the common logarithm of a number:

The description and examples 1–8 refer specifically to the table entitled "Five-Place Mantissas for Common Logarithms." This table is also referred to as a five-place table for common logarithms.

In the five-place table, the first two digits of each mantissa are given only once for each line. The remaining three digits are given in the correct place in the table. When the leading two digits are not given on a line, they should be taken from the last line above it on which they do appear. When a mantissa is marked with an asterisk, it indicates that the value for the leading digits is to be taken from the *next* line instead of the present line.

For a number of more than four figures, interpolation must be used. There are several precautions that must be observed when interpolating:

1. Linear interpolation, as described below, may only be used to add one extra digit to the argument (i.e., in the present case, for a four-digit argument).

2. Even though the mantissas given in the table are accurate to five decimal places, interpolated values are accurate only to the same number of places as in the argument, i.e., four places.

If the above precautions cannot be observed, then higher order interpolation should be used. Where applicable, linear interpolation is carried out as follows:

Take the tabular value of the mantissa for the first four figures; find the difference between the mantissa and the next greater tabular mantissa and multiply the difference so found by the remaining figures of the number as a decimal and add the product to the mantissa of the first four figures. For example, to find log 46.762:

$$\log 46.76 = 1.66987$$

Tabular difference between this mantissa and that for 4677 is .00010

$$\begin{aligned} \log 46.762 &= 1.66987 + .2 \times (.00010) \\ &= 1.66987 + .00002 \\ &= 1.66989 \end{aligned}$$

The last digit is not accurate, so must be rounded out. Thus

$$\log 46.762 = 1.66989$$

The accuracy will not ordinarily be affected by more than 1 in the last place if the mantissas are rounded to five decimal places before interpolation, and this makes the computation somewhat easier.

In both the four and five-place logarithmic tables, a column of proportional parts is given at the end of each line. If we use the four-place table the number in the column under the fourth digit of the argument is the amount that must be added to any mantissa in that line to interpolate for the fourth digit. This number is to be added to the last place of the mantissa. These numbers are averages for the entire line, so may be off by 1 in the last place.

For example, to find log 33.74

$$\begin{aligned} \log 33.7 &= 1.5276 \\ \text{proportional part for } 4 &= \qquad 5 \\ \therefore \log 33.74 &= 1.5281 \end{aligned}$$

To *find the number corresponding to a given logarithm*:

(*Note*: This number is called the antilogarithm, and is denoted by $\log^{-1}$. Since the logarithm function is the inverse of the exponential function, $\log_a^{-1} x = a^x$. Therefore, any procedure or table which calculates antilogarithms may also be used to calculate exponentials, and vice-versa. In particular, tables of e^x may be used to compute antilogarithms to the base e.)

The procedure given below refers to a five-place logarithm table. As before, any significant deviation for other tables will be noted.

If the mantissa is found exactly in the table, join the figure at the top which is directly above the given mantissa to the three figures on the line at the left and place the decimal point according to the characteristic of the logarithm. For example,

$$\log^{-1} 3.39967 = \text{antilog } 3.39967 = 2510.$$

If the mantissa is not found exactly in the table it is necessary to interpolate. For example, to find antilog 3.40028, we find in the table

$$\begin{aligned} \text{antilog } 3.40019 &= 2513. \\ \text{antilog } 3.40037 &= 2514. \\ \text{tabular difference} \qquad 18 \end{aligned}$$

The required difference is 9, so we must add $\frac{9}{18}$ = .5.

$$\therefore \text{antilog } 3.40028 = 2513.5$$

The same precautions must be observed for interpolation in finding antilogarithms as in finding logarithms.

A four-place antilogarithm table is also provided. When using this table, the mantissa of the logarithm is looked up on the margins of the table, and the significant digits of the antilogarithm are read from the body of the table. Lookup and interpolation are done in the same manner as when looking up logarithms in a table of logarithms.

Tables of natural logarithms are used in the same way as tables of common logarithms, except that they contain both the characteristics and the mantissas of the logarithms.

Examples of the use of logarithms in computation follow. Almost all computation with logarithms is done with common logarithms, since the computation of the characteristic is simpler, and since only the significant digits of the argument need be given in the table, without regard for the decimal point location. These examples all use the table of five-place common logarithms.

1. $52600 \times 0.00381 \times 2.74 = 549.11$

log 52600	=	4.72099
log 0.00381	=	$\overline{3}.58092$
log 2.74	=	0.43775
adding	=	2.73966
antilog	=	549.11

The sum is the logarithm of the product, the mantissa of which is 73966. On looking up this mantissa in the logarithm tables we see that it corresponds to the digits 54911. The characteristic is 2, hence there are three figures before the decimal point. The number corresponding to the logarithm, called the antilogarithm, is 549.11.

2. $0.00123 \div 52.7 = 0.000\ 023\ 34$

log 0.00123	=	$\overline{3}.08991$
log 52.7	=	1.72181
subtracting	=	$\overline{5}.36810$
antilog	=	0.000 023 34

The characteristic 5(5. − 10) shows four zeros after the decimal point before the first significant figure.

3. $\dfrac{273 \times 780}{292 \times 760} \times 15 \times 0.09 = 1.2954$

log 273	= 2.43616		log 292	= 2.46538	
log 780	= 2.89209		log 760	= 2.88081	
log 15	= 1.17609		log denominator	= 5.34619	
log 0.09	= $\overline{2}.95424$				
log numerator	= 5.45858				

log numerator	= 5.45858
log denominator	= 5.34619
subtracting	= 0.11239
antilog	= 1.2954

As division may be accomplished by multiplying by the reciprocal of a number, the above may be considerably simplified. The logarithm of the reciprocal of a number, called the cologarithm, is readily obtained from the table by subtracting the logarithm of the number from zero. This may be readily read off from the table of mantissas. Change the sign of the characteristic algebraically adding to it −1, then mentally subtract each figure of the mantissa from 9 proceeding from left to right, the last figure being subtracted from 10. The example then is:

$$
\begin{array}{ll}
\log 273 & = 2.43616 \\
\log 780 & = 2.89209 \\
\log\ 15 & = 1.17609 \\
\log\ \ 0.09 & = \bar{2}.95424 \\
\text{colog } 292 & = \bar{3}.53462 \\
\text{colog } 760 & = \underline{\bar{3}.11919} \\
& \ \ \ \ 0.11239
\end{array}
$$

4. $(0.00098)^4 = 9.224 \times 10^{-13}$ An alternative method:

$$
\begin{array}{llll}
\log 0.00098 & = \bar{4}.99123 & & \log 0.00098 = 6.99123 - 10 \\
& \underline{\quad 4} & & \underline{\qquad\qquad 4} \\
& 3.96492 & \text{(a)} & 27.96492 - 40 \\
\bar{4} \times 4 & = \bar{16}. & \text{(b)} & \text{or}\ \ 7.96492 - 20 \\
\log (0.00098)^4 & = \bar{13}.96492 & \text{(c)} & \text{or}\ \bar{13}.96492 \\
\text{antilog} & = 9.224 \times 10^{-13} & & \text{antilog}\quad = 9.224 \times 10^{-13}
\end{array}
$$

In the above it will be noted that the mantissa is always positive, hence the multiplication of the mantissa shown at (a), while (b) shows the multiplication of the characteristic. (c) is the algebraic sum.

5. $\sqrt[5]{492} = 3.4546$
$\log 492 = 2.69197$

Dividing the logarithm by 5 gives as the logarithm of the root 0.53839, the antilogarithm of which is 3.4546, both characteristic and mantissa being positive. When the characteristic is negative and not evenly divisible by the root to be taken, a modification of the logarithm is necessary, as the following example shows:

6. $\sqrt[3]{0.000372} = 0.07192$
$\log 3.72 \times 10^{-4} = \bar{4}.57054$ (a)
$\qquad\qquad\quad\ \ = 26.57054 - 30$ (b)

Dividing (b) by 3 gives 8.85685 − 10, which may be written $\bar{2}.85685$, and is the logarithm of the root sought, which is 0.07192.

7. $(0.000\ 372)^{1.2} = 0.000\ 076\ 674$
$$
\begin{array}{ll}
\log 0.000\ 372 & = \bar{4}.57054 \\
& \text{or } 6.57054 - 10 \\
& \underline{\qquad\qquad 1.2} \\
& 7.88465 - 12 \\
\text{antilog} & = 0.000\ 076\ 674
\end{array}
$$

8. $(0.000372)^{-1.32} = 33642$
$$
\begin{array}{ll}
\text{colog } 0.000372 & = 3.42946 \\
& \underline{\qquad\ \ 1.32} \\
& 4.52689 \\
\text{antilog} & = 33642
\end{array}
$$

FIVE-PLACE MANTISSAS FOR COMMON LOGARITHMS

N.	0	1	2	3	4	5	6	7	8	9	Proportional parts		
100	00 000	043	087	130	173	217	260	303	346	389	**44**	**43**	**42**
101	432	475	518	561	604	647	689	732	775	817	1 4.4	4.3	4.2
102	860	903	945	988	*030	*072	*115	*157	*199	*242	2 8.8	8.6	8.4
103	01 284	326	368	410	452	494	536	578	620	662	3 13.2	12.9	12.6
104	703	745	787	828	870	912	953	995	*036	*078	4 17.6	17.2	16.8
105	02 119	160	202	243	284	325	366	407	449	490	5 22.0	21.5	21.0
106	531	572	612	653	694	735	776	816	857	898	6 26.4	25.8	25.2
107	938	979	*019	*060	*100	*141	*181	*222	*262	*302	7 30.8	30.1	29.4
108	03 342	383	423	463	503	543	583	623	663	703	8 35.2	34.4	33.6
109	743	782	822	862	902	941	981	*021	*060	*100	9 39.6	38.7	37.8
110	04 139	179	218	258	297	336	376	415	454	493	**41**	**40**	**39**
111	532	571	610	650	689	727	766	805	844	883	1 4.1	4.0	3.9
112	922	961	999	*038	*077	*115	*154	*192	*231	*269	2 8.2	8.0	7.8
113	05 308	346	385	423	461	500	538	576	614	652	3 12.3	12.0	11.7
114	690	729	767	805	843	881	918	956	994	*032	4 16.4	16.0	15.6
115	06 070	108	145	183	221	258	296	333	371	408	5 20.5	20.0	19.5
116	446	483	521	558	595	633	670	707	744	781	6 24.6	24.0	23.4
117	819	856	893	930	967	*004	*041	*078	*115	*151	7 28.7	28.0	27.3
118	07 188	225	262	298	335	372	408	445	482	518	8 32.8	32.0	31.2
119	555	591	628	664	700	737	773	809	846	882	9 36.9	36.0	35.1
120	918	954	990	*027	*063	*099	*135	*171	*207	*243	**38**	**37**	**36**
121	08 279	314	350	386	422	458	493	529	565	600	1 3.8	3.7	3.6
122	636	672	707	743	778	814	849	884	920	955	2 7.6	7.4	7.2
123	991	*026	*061	*096	*132	*167	*202	*237	*272	*307	3 11.4	11.1	10.8
124	09 342	377	412	447	482	517	552	587	621	656	4 15.2	14.8	14.4
125	091	726	760	795	830	864	899	934	968	*003	5 19.0	18.5	18.0
126	10 037	072	106	140	175	209	243	278	312	346	6 22.8	22.2	21.6
127	380	415	449	483	517	551	585	619	653	687	7 26.6	25.9	25.2
128	721	755	789	823	857	890	924	958	992	*025	8 30.4	29.6	28.8
129	11 059	093	126	160	193	227	261	294	327	361	9 34.2	33.3	32.4
130	394	428	461	494	528	561	594	628	661	694	**35**	**34**	**33**
131	727	760	793	826	860	893	926	959	992	*024	1 3.5	3.4	3.3
132	12 057	090	123	156	189	222	254	287	320	352	2 7.0	6.8	6.6
133	385	418	450	483	516	548	581	613	646	678	3 10.5	10.2	9.9
134	710	743	775	808	840	872	905	937	969	*001	4 14.0	13.6	13.2
135	13 033	066	098	130	162	194	226	258	290	322	5 17.5	17.0	16.5
136	354	386	418	450	481	513	545	577	609	640	6 21.0	20.4	19.8
137	672	704	735	767	799	830	862	893	925	956	7 24.5	23.8	23.1
138	988	*019	*051	*082	*114	*145	*176	*208	*239	*270	8 28.0	27.2	26.4
139	14 301	333	364	395	426	457	489	520	551	582	9 31.5	30.6	29.7
140	613	644	675	706	737	768	799	829	860	891	**32**	**31**	**30**
141	922	953	983	*014	*045	*076	*106	*137	*168	*198	1 3.2	3.1	3.0
142	15 229	259	290	320	351	381	412	442	473	503	2 6.4	6.2	6.0
143	534	564	594	625	655	685	715	746	776	806	3 9.6	9.3	9.0
144	836	866	897	927	957	987	*017	*047	*077	*107	4 12.8	12.4	12.0
145	16 137	167	197	227	256	286	316	346	376	406	5 16.0	15.5	15.0
146	435	465	495	524	554	584	613	643	673	702	6 19.2	18.6	18.0
147	732	761	791	820	850	879	909	938	967	997	7 22.4	21.7	21.0
148	17 026	056	085	114	143	173	202	231	260	289	8 25.6	24.8	24.0
149	319	348	377	406	435	464	493	522	551	580	9 28.8	27.9	27.0
150	609	638	667	696	725	754	782	811	840	869			
N.	0	1	2	3	4	5	6	7	8	9	Proportional parts		

FIVE-PLACE MANTISSAS FOR COMMON LOGARITHMS (continued)

N.	0	1	2	3	4	5	6	7	8	9
150	17 609	638	667	696	725	754	782	811	840	869
151	898	926	955	984	*013	*041	*070	*099	*127	*156
152	18 184	213	241	270	298	327	355	384	412	441
153	469	498	526	554	583	611	639	667	696	724
154	752	780	808	837	865	893	921	949	977	*005
155	19 033	061	089	117	145	173	201	229	257	285
156	312	340	368	396	424	451	479	507	535	562
157	590	618	645	673	700	728	756	783	811	838
158	866	893	921	948	976	*003	*030	*058	*085	*112
159	20 140	167	194	222	249	276	303	330	358	385
160	412	439	466	493	520	548	575	602	629	656
161	683	710	737	763	790	817	844	871	898	925
162	952	978	*005	*032	*059	*085	*112	*139	*165	*192
163	21 219	245	272	299	325	352	378	405	431	458
164	484	511	537	564	590	617	643	669	696	722
165	748	775	801	827	854	880	906	932	958	985
166	22 011	037	063	089	115	141	167	194	220	246
167	272	298	324	350	376	401	427	453	479	505
168	531	557	583	608	634	660	686	712	737	763
169	789	814	840	866	891	917	943	968	994	*019
170	23 045	070	096	121	147	172	198	223	249	274
171	300	325	350	376	401	426	452	477	502	528
172	553	578	603	629	654	679	704	729	754	779
173	805	830	855	880	905	930	955	980	*005	*030
174	24 055	080	105	130	155	180	204	229	254	279
175	304	329	353	378	403	428	452	477	502	527
176	551	576	601	625	650	674	699	724	748	773
177	797	822	846	871	895	920	944	969	993	*018
178	25 042	066	091	115	139	164	188	212	237	261
179	285	310	334	358	382	406	431	455	479	503
180	527	551	575	600	624	648	672	696	720	744
181	768	792	816	840	864	888	912	935	959	983
182	26 007	031	055	079	102	126	150	174	198	221
183	245	269	293	316	340	364	387	411	435	458
184	482	505	529	553	576	600	623	647	670	694
185	717	741	764	788	811	834	858	881	905	928
186	951	975	998	*021	*045	*068	*091	*114	*138	*161
187	27 184	207	231	254	277	300	323	346	370	393
188	416	439	462	485	508	531	554	577	600	623
189	646	669	692	715	738	761	784	807	830	852
190	875	898	921	944	967	989	*012	*035	*058	*081
191	28 103	126	149	171	194	217	240	262	285	307
192	330	353	375	398	421	443	466	488	511	533
193	556	578	601	623	646	668	691	713	735	758
194	780	803	825	847	870	892	914	937	959	981
195	29 003	026	048	070	092	115	137	159	181	203
196	226	248	270	292	314	336	358	380	403	425
197	447	469	491	513	535	557	579	601	623	645
198	667	688	710	732	754	776	798	820	842	863
199	885	907	929	951	973	994	*016	*038	*060	*081
200	30 103	125	146	168	190	211	233	255	276	298

Proportional parts

	29	28		27	26		25		24	23		22	21
1	2.9	2.8	1	2.7	2.6	1	2.5	1	2.4	2.3	1	2.2	2.1
2	5.8	5.6	2	5.4	5.2	2	5.0	2	4.8	4.6	2	4.4	4.2
3	8.7	8.4	3	8.1	7.8	3	7.5	3	7.2	6.9	3	6.6	6.3
4	11.6	11.2	4	10.8	10.4	4	10.0	4	9.6	9.2	4	8.8	8.4
5	14.5	14.0	5	13.5	13.0	5	12.5	5	12.0	11.5	5	11.0	10.5
6	17.4	16.8	6	16.2	15.6	6	15.0	6	14.4	13.8	6	13.2	12.6
7	20.3	19.6	7	18.9	18.2	7	17.5	7	16.8	16.1	7	15.4	14.7
8	23.2	22.4	8	21.6	20.8	8	20.0	8	19.2	18.4	8	17.6	16.8
9	26.1	25.2	9	24.3	23.4	9	22.5	9	21.6	20.7	9	19.8	18.9

N.	0	1	2	3	4	5	6	7	8	9	Proportional parts

FIVE-PLACE MANTISSAS FOR COMMON LOGARITHMS (continued)

N.	0	1	2	3	4	5	6	7	8	9
200	30 103	125	146	168	190	211	233	255	276	298
201	320	341	363	384	406	428	449	471	492	514
202	535	557	578	600	621	643	664	685	707	728
203	750	771	792	814	835	856	878	899	920	942
204	963	984	*006	*027	*048	*069	*091	*112	*133	*154
205	31 175	197	218	239	260	281	302	323	345	366
206	387	408	429	450	471	492	513	534	555	576
207	597	618	639	660	681	702	723	744	765	785
208	806	827	848	869	890	911	931	952	973	994
209	32 015	035	056	077	098	118	139	160	181	201
210	222	243	263	284	305	325	346	366	387	408
211	428	449	469	490	510	531	552	572	593	613
212	634	654	675	695	715	736	756	777	797	818
213	838	858	879	899	919	940	960	980	*001	*021
214	33 041	062	082	102	122	143	163	183	203	224
215	244	264	284	304	325	345	365	385	405	425
216	445	465	486	506	526	546	566	586	606	626
217	646	666	686	706	726	746	766	786	806	826
218	846	866	885	905	925	945	965	985	*005	*025
219	34 044	064	084	104	124	143	163	183	203	223
220	242	262	282	301	321	341	361	380	400	420
221	439	459	479	498	518	537	557	577	596	616
222	635	655	674	694	713	733	753	772	792	811
223	830	850	869	889	908	928	947	967	986	*005
224	35 025	044	064	083	102	122	141	160	180	199
225	218	238	257	276	295	315	334	353	372	392
226	411	430	449	468	488	507	526	545	564	583
227	603	622	641	660	679	698	717	736	755	774
228	793	813	832	851	870	889	908	927	946	965
229	984	*003	*021	*040	*059	*078	*097	*116	*135	*154
230	36 173	192	211	229	248	267	286	305	324	342
231	361	380	399	418	436	455	474	493	511	530
232	549	568	586	605	624	642	661	680	698	717
233	736	754	773	791	810	829	847	866	884	903
234	922	940	959	977	996	*014	*033	*051	*070	*088
235	37 107	125	144	162	181	199	218	236	254	273
236	291	310	328	346	365	383	401	420	438	457
237	475	493	511	530	548	566	585	603	621	639
238	658	676	694	712	731	749	767	785	803	822
239	840	858	876	894	912	931	949	967	985	*003
240	38 021	039	057	075	093	112	130	148	166	184
241	202	220	238	256	274	292	310	328	346	364
242	382	399	417	435	453	471	489	507	525	543
243	561	578	596	614	632	650	668	686	703	721
244	739	757	775	792	810	828	846	863	881	899
245	917	934	952	970	987	*005	*023	*041	*058	*076
246	39 094	111	129	146	164	182	199	217	235	252
247	270	287	305	322	340	358	375	393	410	428
248	445	463	480	498	515	533	550	568	585	602
249	620	637	655	672	690	707	724	742	759	777
250	794	811	829	846	863	881	898	915	933	950

Proportional parts

	22	21
1	2.2	2.1
2	4.4	4.2
3	6.6	6.3
4	8.8	8.4
5	11.0	10.5
6	13.2	12.6
7	15.4	14.7
8	17.6	16.8
9	19.8	18.9

	20
1	2.0
2	4.0
3	6.0
4	8.0
5	10.0
6	12.0
7	14.0
8	16.0
0	18.0

	19
1	1.9
2	3.8
3	5.7
4	7.6
5	9.5
6	11.4
7	13.3
8	15.2
9	17.1

	18
1	1.8
2	3.6
3	5.4
4	7.2
5	9.0
6	10.8
7	12.6
8	14.4
9	16.2

	17
1	1.7
2	3.4
3	5.1
4	6.8
5	8.5
6	10.2
7	11.9
8	13.6
9	15.3

| N. | 0 | 1 | 2 | 3 | 4 | 5 | 6 | 7 | 8 | 9 | Proportional parts |

FIVE-PLACE MANTISSAS FOR COMMON LOGARITHMS (continued)

N.	0	1	2	3	4	5	6	7	8	9	Proportional parts
250	39 794	811	829	846	863	881	898	915	933	950	**18**
251	967	985	*002	*019	*037	*054	*071	*088	*106	*123	1 1.8
252	40 140	157	175	192	209	226	243	261	278	295	2 3.6
253	312	329	346	364	381	398	415	432	449	466	3 5.4
254	483	500	518	535	552	569	586	603	620	637	4 7.2
255	654	671	688	705	722	739	756	773	790	807	5 9.0
256	824	841	858	875	892	909	926	943	960	976	6 10.8
257	993	*010	*027	*044	*061	*078	*095	*111	*128	*145	7 12.6
258	41 162	179	196	212	229	246	263	280	296	313	8 14.4
259	330	347	363	380	397	414	430	447	464	481	9 16.2
260	497	514	531	547	564	581	597	614	631	647	**17**
261	664	681	697	714	731	747	764	780	797	814	1 1.7
262	830	847	863	880	896	913	929	946	963	979	2 3.4
263	996	*012	*029	*045	*062	*078	*095	*111	*127	*144	3 5.1
264	42 160	177	193	210	226	243	259	275	292	308	4 6.8
265	325	341	357	374	390	406	423	439	455	472	5 8.5
266	488	504	521	537	553	570	586	602	619	635	6 10.2
267	651	667	684	700	716	732	749	765	781	797	7 11.9
268	813	830	846	862	878	894	911	927	943	959	8 13.6
269	975	991	*008	*024	*040	*056	*072	*088	*104	*120	9 15.3
270	43 136	152	169	185	201	217	233	249	265	281	**16**
271	297	313	329	345	361	377	393	409	425	441	1 1.6
272	457	473	489	505	521	537	553	569	584	600	2 3.2
273	616	632	648	664	680	696	712	727	743	759	3 4.8
274	775	791	807	823	838	854	870	886	902	917	4 6.4
275	933	949	965	981	996	*012	*028	*044	*059	*075	5 8.0
276	44 091	107	122	138	154	170	185	201	217	232	6 9.6
277	248	264	279	295	311	326	342	358	373	389	7 11.2
278	404	420	436	451	467	483	498	514	529	545	8 12.8
279	560	576	592	607	623	638	654	669	685	700	9 14.4
280	716	731	747	762	778	793	809	824	840	855	**15**
281	871	886	902	917	932	948	963	979	994	*010	1 1.5
282	45 025	040	056	071	086	102	117	133	148	163	2 3.0
283	179	194	209	225	240	255	271	286	301	317	3 4.5
284	332	347	362	378	393	408	423	439	454	469	4 6.0
285	484	500	515	530	545	561	576	591	606	621	5 7.5
286	637	652	667	682	697	712	728	743	758	773	6 9.0
287	788	803	818	834	849	864	879	894	909	924	7 10.5
288	939	954	969	984	*000	*015	*030	*045	*060	*075	8 12.0
289	46 090	105	120	135	150	165	180	195	210	225	9 13.5
290	240	255	270	285	300	315	330	345	359	374	**14**
291	389	404	419	434	449	464	479	494	509	523	1 1.4
292	538	553	568	583	598	613	627	642	657	672	2 2.8
293	687	702	716	731	746	761	776	790	805	820	3 4.2
294	835	850	864	879	894	909	923	938	953	967	4 5.6
295	982	997	*012	*026	*041	*056	*070	*085	*100	*114	5 7.0
296	47 129	144	159	173	188	202	217	232	246	261	6 8.4
297	276	290	305	319	334	349	363	378	392	407	7 9.8
298	422	436	451	465	480	494	509	524	538	553	8 11.2
299	567	582	596	611	625	640	654	669	683	698	9 12.6
300	712	727	741	756	770	784	799	813	828	842	

| N. | 0 | 1 | 2 | 3 | 4 | 5 | 6 | 7 | 8 | 9 | Proportional parts |

FIVE-PLACE MANTISSAS FOR COMMON LOGARITHMS (continued)

N.	0	1	2	3	4	5	6	7	8	9	Proportional parts	
300	47 712	727	741	756	770	784	799	813	828	842		**15**
301	857	871	885	900	914	929	943	958	972	986	1	1.5
302	48 001	015	029	044	058	073	087	101	116	130	2	3.0
303	144	159	173	187	202	216	230	244	259	273	3	4.5
304	287	302	316	330	344	359	373	387	401	416	4	6.0
305	430	444	458	473	487	501	515	530	544	558	5	7.5
306	572	586	601	615	629	643	657	671	686	700	6	9.0
307	714	728	742	756	770	785	799	813	827	841	7	10.5
308	855	869	883	897	911	926	940	954	968	982	8	12.0
309	996	*010	*024	*038	*052	*066	*080	*094	*108	*122	9	13.5
310	49 136	150	164	178	192	206	220	234	248	262		
311	276	290	304	318	332	346	360	374	388	402		
312	415	429	443	457	471	485	499	513	527	541		
313	554	568	582	596	610	624	638	651	665	679		**14**
314	693	707	721	734	748	762	776	790	803	817	1	1.4
315	831	845	859	872	886	900	914	927	941	955	2	2.8
316	969	982	996	*010	*024	*037	*051	*065	*079	*092	3	4.2
317	50 106	120	133	147	161	174	188	202	215	229	4	5.6
318	243	256	270	284	297	311	325	338	352	365	5	7.0
319	379	393	406	420	433	447	461	474	488	501	6	8.4
											7	9.8
320	515	529	542	556	569	583	596	610	623	637	8	11.2
321	651	664	678	691	705	718	732	745	759	772	9	12.6
322	786	799	813	826	840	853	866	880	893	907		
323	920	934	947	961	974	987	*001	*014	*028	*041		
324	51 055	068	081	095	108	121	135	148	162	175		
325	188	202	215	228	242	255	268	282	295	308		
326	322	335	348	362	375	388	402	415	428	441		**13**
327	455	468	481	495	508	521	534	548	561	574	1	1.3
328	587	601	614	627	640	654	667	680	693	706	2	2.6
329	720	733	746	759	772	786	799	812	825	838	3	3.9
											4	5.2
330	851	865	878	891	904	917	930	943	957	970	5	6.5
331	983	996	*009	*022	*035	*048	*061	*075	*088	*101	6	7.8
332	52 114	127	140	153	166	179	192	205	218	231	7	9.1
333	244	257	270	284	297	310	323	336	349	362	8	10.4
334	375	388	401	414	427	440	453	466	479	492	9	11.7
335	504	517	530	543	556	569	582	595	608	621		
336	634	647	660	673	686	699	711	724	737	750		
337	763	776	789	802	815	827	840	853	866	879		
338	892	905	917	930	943	956	969	982	994	*007		**12**
339	53 020	033	046	058	071	084	097	110	122	135	1	1.2
											2	2.4
340	148	161	173	186	199	212	224	237	250	263	3	3.6
341	275	288	301	314	326	339	352	364	377	390	4	4.8
342	403	415	428	441	453	466	479	491	504	517	5	6.0
343	529	542	555	567	580	593	605	618	631	643	6	7.2
344	656	668	681	694	706	719	732	744	757	769	7	8.4
345	782	794	807	820	832	845	857	870	882	895	8	9.6
346	908	920	933	945	958	970	983	995	*008	*020	9	10.8
347	54 033	045	058	070	083	095	108	120	133	145		
348	158	170	183	195	208	220	233	245	258	270		
349	283	295	307	320	332	345	357	370	382	394		
350	407	419	432	444	456	469	481	494	506	518		
N.	0	1	2	3	4	5	6	7	8	9	Proportional parts	

FIVE-PLACE MANTISSAS FOR COMMON LOGARITHMS (continued)

N.	0	1	2	3	4	5	6	7	8	9
350	54 407	419	432	444	456	469	481	494	506	518
351	531	543	555	568	580	593	605	617	630	642
352	654	667	679	691	704	716	728	741	753	765
353	777	790	802	814	827	839	851	864	876	888
354	900	913	925	937	949	962	974	986	998	*011
355	55 023	035	047	060	072	084	096	108	121	133
356	145	157	169	182	194	206	218	230	242	255
357	267	279	291	303	315	328	340	352	364	376
358	388	400	413	425	437	449	461	473	485	497
359	509	522	534	546	558	570	582	594	606	618
360	630	642	654	666	678	691	703	715	727	739
361	751	763	775	787	799	811	823	835	847	859
362	871	883	895	907	919	931	943	955	967	979
363	991	*003	*015	*027	*038	*050	*062	*074	*086	*098
364	56 110	122	134	146	158	170	182	194	205	217
365	229	241	253	265	277	289	301	312	324	336
366	348	360	372	384	396	407	419	431	443	455
367	467	478	490	502	514	526	538	549	561	573
368	585	597	608	620	632	644	656	667	679	691
369	703	714	726	738	750	761	773	785	797	808
370	820	832	844	855	867	879	891	902	914	926
371	937	949	961	972	984	996	*008	*019	*031	*043
372	57 054	066	078	089	101	113	124	136	148	159
373	171	183	194	206	217	229	241	252	264	276
374	287	299	310	322	334	345	357	368	380	392
375	403	415	426	438	449	461	473	484	496	507
376	519	530	542	553	565	576	588	600	611	623
377	634	646	657	669	680	692	703	715	726	738
378	749	761	772	784	795	807	818	830	841	852
379	864	875	887	898	910	921	933	944	955	967
380	978	990	*001	*013	*024	*035	*047	*058	*070	*081
381	58 092	104	115	127	138	149	161	172	184	195
382	206	218	229	240	252	263	274	286	297	309
383	320	331	343	354	365	377	388	399	410	422
384	433	444	456	467	478	490	501	512	524	535
385	546	557	569	580	591	602	614	625	636	647
386	659	670	681	692	704	715	726	737	749	760
387	771	782	794	805	816	827	838	850	861	872
388	883	894	906	917	928	939	950	961	973	984
389	995	*006	*017	*028	*040	*051	*062	*073	*084	*095
390	59 106	118	129	140	151	162	173	184	195	207
391	218	229	240	251	262	273	284	295	306	318
392	329	340	351	362	373	384	395	406	417	428
393	439	450	461	472	483	494	506	517	528	539
394	550	561	572	583	594	605	616	627	638	649
395	660	671	682	693	704	715	726	737	748	759
396	770	780	791	802	813	824	835	846	857	868
397	879	890	901	912	923	934	945	956	966	977
398	988	999	*010	*021	*032	*043	*054	*065	*076	*086
399	60 097	108	119	130	141	152	163	173	184	195
400	206	217	228	239	249	260	271	282	293	304
N.	0	1	2	3	4	5	6	7	8	9

Proportional parts

13
1	1.3
2	2.6
3	3.9
4	5.2
5	6.5
6	7.8
7	9.1
8	10.4
9	11.7

12
1	1.2
2	2.4
3	3.6
4	4.8
5	6.0
6	7.2
7	8.4
8	9.6
9	10.8

11
1	1.1
2	2.2
3	3.3
4	4.4
5	5.5
6	6.6
7	7.7
8	8.8
9	9.9

10
1	1.0
2	2.0
3	3.0
4	4.0
5	5.0
6	6.0
7	7.0
8	8.0
9	9.0

FIVE-PLACE MANTISSAS FOR COMMON LOGARITHMS (continued)

N.	0	1	2	3	4	5	6	7	8	9	Proportional parts	
400	60 206	217	228	239	249	260	271	282	293	304		
401	314	325	336	347	358	369	379	390	401	412		
402	423	433	444	455	466	477	487	498	509	520		
403	531	541	552	563	574	584	595	606	617	627		
404	638	649	660	670	681	692	703	713	724	735		
405	746	756	767	778	788	799	810	821	831	842		
406	853	863	874	885	895	906	917	927	938	949		
407	959	970	981	991	*002	*013	*023	*034	*045	*055		**11**
408	61 066	077	087	098	109	119	130	140	151	162	1	1.1
409	172	183	194	204	215	225	236	247	257	268	2	2.2
											3	3.3
410	278	289	300	310	321	331	342	352	363	374	4	4.4
411	384	395	405	416	426	437	448	458	469	479	5	5.5
412	490	500	511	521	532	542	553	563	574	584	6	6.6
413	595	606	616	627	637	648	658	669	679	690	7	7.7
414	700	711	721	731	742	752	763	773	784	794	8	8.8
415	805	815	826	836	847	857	868	878	888	899	9	9.9
416	909	920	930	941	951	962	972	982	993	*003		
417	62 014	024	034	045	055	066	076	086	097	107		
418	118	128	138	149	159	170	180	190	201	211		
419	221	232	242	252	263	273	284	294	304	315		
420	325	335	346	356	366	377	387	397	408	418		**10**
421	428	439	449	459	469	480	490	500	511	521	1	1.0
422	531	542	552	562	572	583	593	603	613	624	2	2.0
423	634	644	655	665	675	685	696	706	716	726	3	3.0
424	737	747	757	767	778	788	798	808	818	829	4	4.0
425	839	849	859	870	880	890	900	910	921	931	5	5.0
426	941	951	961	972	982	992	*002	*012	*022	*033	6	6.0
427	63 043	053	063	073	083	094	104	114	124	134	7	7.0
428	144	155	165	175	185	195	205	215	225	236	8	8.0
429	246	256	266	276	286	296	306	317	327	337	9	9.0
430	347	357	367	377	387	397	407	417	428	438		
431	448	458	468	478	488	498	508	518	528	538		
432	548	558	568	579	589	599	609	619	629	639		
433	649	659	669	679	689	699	709	719	729	739		
434	749	759	769	779	789	799	809	819	829	839		
435	849	859	869	879	889	899	909	919	929	939		
436	949	959	969	979	988	998	*008	*018	*028	*038		
437	64 048	058	068	078	088	098	108	118	128	137		
438	147	157	167	177	187	197	207	217	227	237		**9**
439	246	256	266	276	286	296	306	316	326	335	1	0.9
											2	1.8
440	345	355	365	375	385	395	404	414	424	434	3	2.7
441	444	454	464	473	483	493	503	513	523	532	4	3.6
442	542	552	562	572	582	591	601	611	621	631	5	4.5
443	640	650	660	670	680	689	699	709	719	729	6	5.4
444	738	748	758	768	777	787	797	807	816	826	7	6.3
445	836	846	856	865	875	885	895	904	914	924	8	7.2
446	933	943	953	963	972	982	992	*002	*011	*021	9	8.1
447	65 031	040	050	060	070	079	089	099	108	118		
448	128	137	147	157	167	176	186	196	205	215		
449	225	234	244	254	263	273	283	292	302	312		
450	321	331	341	350	360	369	379	389	398	408		
N.	0	1	2	3	4	5	6	7	8	9	Proportional parts	

FIVE-PLACE MANTISSAS FOR COMMON LOGARITHMS (continued)

N.	0	1	2	3	4	5	6	7	8	9
450	65 321	331	341	350	360	369	379	389	398	408
451	418	427	437	447	456	466	475	485	495	504
452	514	523	533	543	552	562	571	581	591	600
453	610	619	629	639	648	658	667	677	686	696
454	706	715	725	734	744	753	763	772	782	792
455	801	811	820	830	839	849	858	868	877	887
456	896	906	916	925	935	944	954	963	973	982
457	992	*001	*011	*020	*030	*039	*049	*058	*068	*077
458	66 087	096	106	115	124	134	143	153	162	172
459	181	191	200	210	219	229	238	247	257	266
460	276	285	295	304	314	323	332	342	351	361
461	370	380	389	398	408	417	427	436	445	455
462	464	474	483	492	502	511	521	530	539	549
463	558	567	577	586	596	605	614	624	633	642
464	652	661	671	680	689	699	708	717	727	736
465	745	755	764	773	783	792	801	811	820	829
466	839	848	857	867	876	885	894	904	913	922
467	932	941	950	960	969	978	987	997	*006	*015
468	67 025	034	043	052	062	071	080	089	099	108
469	117	127	136	145	154	164	173	182	191	201
470	210	219	228	237	247	256	265	274	284	293
471	302	311	321	330	339	348	357	367	376	385
472	394	403	413	422	431	440	449	459	468	477
473	486	495	504	514	523	532	541	550	560	569
474	578	587	596	605	614	624	633	642	651	660
475	669	679	688	697	706	715	724	733	742	752
476	761	770	779	788	797	806	815	825	834	843
477	852	861	870	879	888	897	906	916	925	934
478	943	952	961	970	979	988	997	*006	*015	*024
479	68 034	043	052	061	070	079	088	097	106	115
480	124	133	142	151	160	169	178	187	196	205
481	215	224	233	242	251	260	269	278	287	296
482	305	314	323	332	341	350	359	368	377	386
483	395	404	413	422	431	440	449	458	467	476
484	485	494	502	511	520	529	538	547	556	565
485	574	583	592	601	610	619	628	637	646	655
486	664	673	681	690	699	708	717	726	735	744
487	753	762	771	780	789	797	806	815	824	833
488	842	851	860	869	878	886	895	904	913	922
489	931	940	949	958	966	975	984	993	*002	*011
490	69 020	028	037	046	055	064	073	082	090	099
491	108	117	126	135	144	152	161	170	179	188
492	197	205	214	223	232	241	249	258	267	276
493	285	294	302	311	320	329	338	346	355	364
494	373	381	390	399	408	417	425	434	443	452
495	461	469	478	487	496	504	513	522	531	539
496	548	557	566	574	583	592	601	609	618	627
497	636	644	653	662	671	679	688	697	705	714
498	723	732	740	749	758	767	775	784	793	801
499	810	819	827	836	845	854	862	871	880	888
500	897	906	914	923	932	940	949	958	966	975

Proportional parts

	10
1	1.0
2	2.0
3	3.0
4	4.0
5	5.0
6	6.0
7	7.0
8	8.0
9	9.0

	9
1	0.9
2	1.8
3	2.7
4	3.6
5	4.5
6	5.4
7	6.3
8	7.2
9	8.1

	8
1	0.8
2	1.6
3	2.4
4	3.2
5	4.0
6	4.8
7	5.6
8	6.4
9	7.2

N.	0	1	2	3	4	5	6	7	8	9	Proportional parts

FIVE-PLACE MANTISSAS FOR COMMON LOGARITHMS (continued)

N.	0	1	2	3	4	5	6	7	8	9	Proportional parts	
500	69 897	906	914	923	932	940	949	958	966	975		
501	984	992	*001	*010	*018	*027	*036	*044	*053	*062		
502	70 070	079	088	096	105	114	122	131	140	148		
503	157	165	174	183	191	200	209	217	226	234		
504	243	252	260	269	278	286	295	303	312	321		
505	329	338	346	355	364	372	381	389	398	406		
506	415	424	432	441	449	458	467	475	484	492		
507	501	509	518	526	535	544	552	561	569	578		**9**
508	586	595	603	612	621	629	638	646	655	663	1	0.9
509	672	680	689	697	706	714	723	731	740	749	2	1.8
											3	2.7
510	757	766	774	783	791	800	808	817	825	834	4	3.6
511	842	851	859	868	876	885	893	902	910	919	5	4.5
512	927	935	944	952	961	969	978	986	995	*003	6	5.4
513	71 012	020	029	037	046	054	063	071	079	088	7	6.3
514	096	105	113	122	130	139	147	155	164	172	8	7.2
515	181	189	198	206	214	223	231	240	248	257	9	8.1
516	265	273	282	290	299	307	315	324	332	341		
517	349	357	366	374	383	391	399	408	416	425		
518	433	441	450	458	466	475	483	492	500	508		
519	517	525	533	542	550	559	567	575	584	592		
520	600	609	617	625	634	642	650	659	667	675		
521	684	692	700	709	717	725	734	742	750	759		**8**
522	767	775	784	792	800	809	817	825	834	842	1	0.8
523	850	858	867	875	883	892	900	908	917	925	2	1.6
524	933	941	950	958	966	975	983	991	999	*008	3	2.4
525	72 016	024	032	041	049	057	066	074	082	090	4	3.2
526	099	107	115	123	132	140	148	156	165	173	5	4.0
527	181	189	198	206	214	222	230	239	247	255	6	4.8
528	263	272	280	288	296	304	313	321	329	337	7	5.6
529	346	354	362	370	378	387	395	403	411	419	8	6.4
											9	7.2
530	428	436	444	452	460	469	477	485	493	501		
531	509	518	526	534	542	550	558	567	575	583		
532	591	599	607	616	624	632	640	648	656	665		
533	673	681	689	697	705	713	722	730	738	746		
534	754	762	770	779	787	795	803	811	819	827		
535	835	843	852	860	868	876	884	892	900	908		
536	916	925	933	941	949	957	965	973	981	989		
537	997	*006	*014	*022	*030	*038	*046	*054	*062	*070		
538	73 078	086	094	102	111	119	127	135	143	151		
539	159	167	175	183	191	199	207	215	223	231		**7**
											1	0.7
540	239	247	255	263	272	280	288	296	304	312	2	1.4
541	320	328	336	344	352	360	368	376	384	392	3	2.1
542	400	408	416	424	432	440	448	456	464	472	4	2.8
543	480	488	496	504	512	520	528	536	544	552	5	3.5
544	560	568	576	584	592	600	608	616	624	632	6	4.2
545	640	648	656	664	672	679	687	695	703	711	7	4.9
546	719	727	735	743	751	759	767	775	783	791	8	5.6
547	799	807	815	823	830	838	846	854	862	870	9	6.3
548	878	886	894	902	910	918	926	933	941	949		
549	957	965	973	981	989	997	*005	*013	*020	*028		
550	74 036	044	052	060	068	076	084	092	099	107		

N.	0	1	2	3	4	5	6	7	8	9	Proportional parts

FIVE-PLACE MANTISSAS FOR COMMON LOGARITHMS (continued)

N.	0	1	2	3	4	5	6	7	8	9	Proportional parts	
550	74 036	044	052	060	068	076	084	092	099	107		
551	115	123	131	139	147	155	162	170	178	186		
552	194	202	210	218	225	233	241	249	257	265		
553	273	280	288	296	304	312	320	327	335	343		
554	351	359	367	374	382	390	398	406	414	421		
555	429	437	445	453	461	468	476	484	492	500		
556	507	515	523	531	539	547	554	562	570	578		
557	586	593	601	609	617	624	632	640	648	656		
558	663	671	679	687	695	702	710	718	726	733		
559	741	749	757	764	772	780	788	796	803	811		
560	819	827	834	842	850	858	865	873	881	889		**8**
561	896	904	912	920	927	935	943	950	958	966	1	0.8
562	974	981	989	997	*005	*012	*020	*028	*035	*043	2	1.6
563	75 051	059	066	074	082	089	097	105	113	120	3	2.4
564	128	136	143	151	159	166	174	182	189	197	4	3.2
565	205	213	220	228	236	243	251	259	266	274	5	4.0
566	282	289	297	305	312	320	328	335	343	351	6	4.8
567	358	366	374	381	389	397	404	412	420	427	7	5.6
568	435	442	450	458	465	473	481	488	496	504	8	6.4
569	511	519	526	534	542	549	557	565	572	580	9	7.2
570	587	595	603	610	618	626	633	641	648	656		
571	664	671	679	686	694	702	709	717	724	732		
572	740	747	755	762	770	778	785	793	800	808		
573	815	823	831	838	846	853	861	868	876	884		
574	891	899	906	914	921	929	937	944	952	959		
575	967	974	982	989	997	*005	*012	*020	*027	*035		
576	76 042	050	057	065	072	080	087	095	103	110		
577	118	125	133	140	148	155	163	170	178	185		
578	193	200	208	215	223	230	238	245	253	260		
579	268	275	283	290	298	305	313	320	328	335		
580	343	350	358	365	373	380	388	395	403	410		**7**
581	418	425	433	440	448	455	462	470	477	485	1	0.7
582	492	500	507	515	522	530	537	545	552	559	2	1.4
583	567	574	582	589	597	604	612	619	626	634	3	2.1
584	641	649	656	664	671	678	686	693	701	708	4	2.8
585	716	723	730	738	745	753	760	768	775	782	5	3.5
586	790	797	805	812	819	827	834	842	849	856	6	4.2
587	864	871	879	886	893	901	908	916	923	930	7	4.9
588	938	945	953	960	967	975	982	989	997	*004	8	5.6
589	77 012	019	026	034	041	048	056	063	070	078	9	6.3
590	085	093	100	107	115	122	129	137	144	151		
591	159	166	173	181	188	195	203	210	217	225		
592	232	240	247	254	262	269	276	283	291	298		
593	305	313	320	327	335	342	349	357	364	371		
594	379	386	393	401	408	415	422	430	437	444		
595	452	459	466	474	481	488	495	503	510	517		
596	525	532	539	546	554	561	568	576	583	590		
597	597	605	612	619	627	634	641	648	656	663		
598	670	677	685	692	699	706	714	721	728	735		
599	743	750	757	764	772	779	786	793	801	808		
600	815	822	830	837	844	851	859	866	873	880		
N.	0	1	2	3	4	5	6	7	8	9	Proportional parts	

FIVE-PLACE MANTISSAS FOR COMMON LOGARITHMS (continued)

N.	0	1	2	3	4	5	6	7	8	9	Proportional parts	
600	77 815	822	830	837	844	851	859	866	873	880		
601	887	895	902	909	916	924	931	938	945	952		
602	960	967	974	981	988	996	*003	*010	*017	*025		
603	78 032	039	046	053	061	068	075	082	089	097		
604	104	111	118	125	132	140	147	154	161	168		
605	176	183	190	197	204	211	219	226	233	240		
606	247	254	262	269	276	283	290	297	305	312		
607	319	326	333	340	347	355	362	369	376	383		
608	390	398	405	412	419	426	433	440	447	455		
609	462	469	476	483	490	497	504	512	519	526		**8**
											1	0.8
610	533	540	547	554	561	569	576	583	590	597	2	1.6
611	604	611	618	625	633	640	647	654	661	668	3	2.4
612	675	682	689	696	704	711	718	725	732	739	4	3.2
613	746	753	760	767	774	781	789	796	803	810	5	4.0
614	817	824	831	838	845	852	859	866	873	880	6	4.8
615	888	895	902	909	916	923	930	937	944	951	7	5.6
616	958	965	972	979	986	993	*000	*007	*014	*021	8	6.4
617	79 029	036	043	050	057	064	071	078	085	092	9	7.2
618	099	106	113	120	127	134	141	148	155	162		
619	169	176	183	190	197	204	211	218	225	232		
620	239	246	253	260	267	274	281	288	295	302		**7**
621	309	316	323	330	337	344	351	358	365	372	1	0.7
622	379	386	393	400	407	414	421	428	435	442	2	1.4
623	449	456	463	470	477	484	491	498	505	511	3	2.1
624	518	525	532	539	546	553	560	567	574	581	4	2.8
625	588	595	602	609	616	623	630	637	644	650	5	3.5
626	657	664	671	678	685	692	699	706	713	720	6	4.2
627	727	734	741	748	754	761	768	775	782	789	7	4.9
628	796	803	810	817	824	831	837	844	851	858	8	5.6
629	865	872	879	886	893	900	906	913	920	927	9	6.3
630	934	941	948	955	962	969	975	982	989	996		
631	80 003	010	017	024	030	037	044	051	058	065		
632	072	079	085	092	099	106	113	120	127	134		
633	140	147	154	161	168	175	182	188	195	202		**6**
634	209	216	223	229	236	243	250	257	264	271	1	0.6
635	277	284	291	298	305	312	318	325	332	339	2	1.2
636	346	353	359	366	373	380	387	393	400	407	3	1.8
637	414	421	428	434	441	448	455	462	468	475	4	2.4
638	482	489	496	502	509	516	523	530	536	543	5	3.0
639	550	557	564	570	577	584	591	598	604	611	6	3.6
											7	4.2
640	618	625	632	638	645	652	659	665	672	679	8	4.8
641	686	693	699	706	713	720	726	733	740	747	9	5.4
642	754	760	767	774	781	787	794	801	808	814		
643	821	828	835	841	848	855	862	868	875	882		
644	889	895	902	909	916	922	929	936	943	949		
645	956	963	969	976	983	990	996	*003	*010	*017		
646	81 023	030	037	043	050	057	064	070	077	084		
647	090	097	104	111	117	124	131	137	144	151		
648	158	164	171	178	184	191	198	204	211	218		
649	224	231	238	245	251	258	265	271	278	285		
650	291	298	305	311	318	325	331	338	345	351		
N.	0	1	2	3	4	5	6	7	8	9	Proportional parts	

FIVE-PLACE MANTISSAS FOR COMMON LOGARITHMS (continued)

N.	0	1	2	3	4	5	6	7	8	9	Proportional parts	
650	81 291	298	305	311	318	325	331	338	345	351		
651	358	365	371	378	385	391	398	405	411	418		
652	425	431	438	445	451	458	465	471	478	485		
653	491	498	505	511	518	525	531	538	544	551		
654	558	564	571	578	584	591	598	604	611	617		
655	624	631	637	644	651	657	664	671	677	684		
656	690	697	704	710	717	723	730	737	743	750		
657	757	763	770	776	783	790	796	803	809	816		
658	823	829	836	842	849	856	862	869	875	882		
659	889	895	902	908	915	921	928	935	941	948		
660	954	961	968	974	981	987	994	*000	*007	*014		**7**
661	82 020	027	033	040	046	053	060	066	073	079	1	0.7
662	086	092	099	105	112	119	125	132	138	145	2	1.4
663	151	158	164	171	178	184	191	197	204	210	3	2.1
664	217	223	230	236	243	249	256	263	269	276	4	2.8
665	282	289	295	302	308	315	321	328	334	341	5	3.5
666	347	354	360	367	373	380	387	393	400	406	6	4.2
667	413	419	426	432	439	445	452	458	465	471	7	4.9
668	478	484	491	497	504	510	517	523	530	536	8	5.6
669	543	549	556	562	569	575	582	588	595	601	9	6.3
670	607	614	620	627	633	640	646	653	659	666		
671	672	679	685	692	698	705	711	718	724	730		
672	737	743	750	756	763	769	776	782	789	795		
673	802	808	814	821	827	834	840	847	853	860		
674	866	872	879	885	892	898	905	911	918	924		
675	930	937	943	950	956	963	969	975	982	988		
676	995	*001	*008	*014	*020	*027	*033	*040	*046	*052		
677	83 059	065	072	078	085	091	097	104	110	117		
678	123	129	136	142	149	155	161	168	174	181		
679	187	193	200	206	213	219	225	232	238	245		
680	251	257	264	270	276	283	289	296	302	308		**6**
681	315	321	327	334	340	347	353	359	366	372	1	0.6
682	378	385	391	398	404	410	417	423	429	436	2	1.2
683	442	448	455	461	467	474	480	487	493	499	3	1.8
684	506	512	518	525	531	537	544	550	556	563	4	2.4
685	569	575	582	588	594	601	607	613	620	626	5	3.0
686	632	639	645	651	658	664	670	677	683	689	6	3.6
687	696	702	708	715	721	727	734	740	746	753	7	4.2
688	759	765	771	778	784	790	797	803	809	816	8	4.8
689	822	828	835	841	847	853	860	866	872	879	9	5.4
690	885	891	897	904	910	916	923	929	935	942		
691	948	954	960	967	973	979	985	992	998	*004		
692	84 011	017	023	029	036	042	048	055	061	067		
693	073	080	086	092	098	105	111	117	123	130		
694	136	142	148	155	161	167	173	180	186	192		
695	198	205	211	217	223	230	236	242	248	255		
696	261	267	273	280	286	292	298	305	311	317		
697	323	330	336	342	348	354	361	367	373	379		
698	386	392	398	404	410	417	423	429	435	442		
699	448	454	460	466	473	479	485	491	497	504		
700	510	516	522	528	535	541	547	553	559	566		
N.	0	1	2	3	4	5	6	7	8	9	Proportional parts	

FIVE-PLACE MANTISSAS FOR COMMON LOGARITHMS (continued)

N.	0	1	2	3	4	5	6	7	8	9	Proportional parts
700	84 510	516	522	528	535	541	547	553	559	566	
701	572	578	584	590	597	603	609	615	621	628	
702	634	640	646	652	658	665	671	677	683	689	
703	696	702	708	714	720	726	733	739	745	751	
704	757	763	770	776	782	788	794	800	807	813	
705	819	825	831	837	844	850	856	862	868	874	
706	880	887	893	899	905	911	917	924	930	936	
707	942	948	954	960	967	973	979	985	991	997	**7**
708	85 003	009	016	022	028	034	040	046	052	058	1| 0.7
709	065	071	077	083	089	095	101	107	114	120	2| 1.4
											3| 2.1
710	126	132	138	144	150	156	163	169	175	181	4| 2.8
711	187	193	199	205	211	217	224	230	236	242	5| 3.5
712	248	254	260	266	272	278	285	291	297	303	6| 4.2
713	309	315	321	327	333	339	345	352	358	364	7| 4.9
714	370	376	382	388	394	400	406	412	418	425	8| 5.6
715	431	437	443	449	455	461	467	473	479	485	9| 6.3
716	491	497	503	509	516	522	528	534	540	546	
717	552	558	564	570	576	582	588	594	600	606	
718	612	618	625	631	637	643	649	655	661	667	
719	673	679	685	691	697	703	709	715	721	727	**6**
											1| 0.6
720	733	739	745	751	757	763	769	775	781	788	2| 1.2
721	794	800	806	812	818	824	830	836	842	848	3| 1.8
722	854	860	866	872	878	884	890	896	902	908	4| 2.4
723	914	920	926	932	938	944	950	956	962	968	5| 3.0
724	974	980	986	992	998	*004	*010	*016	*022	*028	6| 3.6
725	86 034	040	046	052	058	064	070	076	082	088	7| 4.2
726	094	100	106	112	118	124	130	136	141	147	8| 4.8
727	153	159	165	171	177	183	189	195	201	207	9| 5.4
728	213	219	225	231	237	243	249	255	261	267	
729	273	279	285	291	297	303	308	314	320	326	
											5
730	332	338	344	350	356	362	368	374	380	386	
731	392	398	404	410	415	421	427	433	439	445	1| 0.5
732	451	457	463	469	475	481	487	493	499	504	2| 1.0
733	510	516	522	528	534	540	546	552	558	564	3| 1.5
734	570	576	581	587	593	599	605	611	617	623	4| 2.0
735	629	635	641	646	652	658	664	670	676	682	5| 2.5
736	688	694	700	705	711	717	723	729	735	741	6| 3.0
737	747	753	759	764	770	776	782	788	794	800	7| 3.5
738	806	812	817	823	829	835	841	847	853	859	8| 4.0
739	864	870	876	882	888	894	900	906	911	917	9| 4.5
740	923	929	935	941	947	953	958	964	970	976	
741	982	988	994	999	*005	*011	*017	*023	*029	*035	
742	87 040	046	052	058	064	070	075	081	087	093	
743	099	105	111	116	122	128	134	140	146	151	
744	157	163	169	175	181	186	192	198	204	210	
745	216	221	227	233	239	245	251	256	262	268	
746	274	280	286	291	297	303	309	315	320	326	
747	332	338	344	349	355	361	367	373	379	384	
748	390	396	402	408	413	419	425	431	437	442	
749	448	454	460	466	471	477	483	489	495	500	
750	506	512	518	523	529	535	541	547	552	558	

N.	0	1	2	3	4	5	6	7	8	9	Proportional parts

LOGARITHMIC, EXPONENTIAL, AND HYPERBOLIC FUNCTIONS

FIVE-PLACE MANTISSAS FOR COMMON LOGARITHMS (continued)

N.	0	1	2	3	4	5	6	7	8	9	Proportional parts	
750	87 506	512	518	523	529	535	541	547	552	558		
751	564	570	576	581	587	593	599	604	610	616		
752	622	628	633	639	645	651	656	662	668	674		
753	679	685	691	697	703	708	714	720	726	731		
754	737	743	749	754	760	766	772	777	783	789		
755	795	800	806	812	818	823	829	835	841	846		
756	852	858	864	869	875	881	887	892	898	904		
757	910	915	921	927	933	938	944	950	955	961		
758	967	973	978	984	990	996	*001	*007	*013	*018		
759	88 024	030	036	041	047	053	058	064	070	076		
760	081	087	093	098	104	110	116	121	127	133		**6**
761	138	144	150	156	161	167	173	178	184	190	1	0.6
762	195	201	207	213	218	224	230	235	241	247	2	1.2
763	252	258	264	270	275	281	287	292	298	304	3	1.8
764	309	315	321	326	332	338	343	349	355	360	4	2.4
765	366	372	377	383	389	395	400	406	412	417	5	3.0
766	423	429	434	440	446	451	457	463	468	474	6	3.6
767	480	485	491	497	502	508	513	519	525	530	7	4.2
768	536	542	547	553	559	564	570	576	581	587	8	4.8
769	593	598	604	610	615	621	627	632	638	643	9	5.4
770	649	655	660	666	672	677	683	689	694	700		
771	705	711	717	722	728	734	739	745	750	756		
772	762	767	773	779	784	790	795	801	807	812		
773	818	824	829	835	840	846	852	857	863	868		
774	874	880	885	891	897	902	908	913	919	925		
775	930	936	941	947	953	958	964	969	975	981		
776	986	992	997	*003	*009	*014	*020	*025	*031	*037		
777	89 042	048	053	059	064	070	076	081	087	092		
778	098	104	109	115	120	126	131	137	143	148		
779	154	159	165	170	176	182	187	193	198	204		
780	209	215	221	226	232	237	243	248	254	260		**5**
781	265	271	276	282	287	293	298	304	310	315	1	0.5
782	321	326	332	337	343	348	354	360	365	371	2	1.0
783	376	382	387	393	398	404	409	415	421	426	3	1.5
784	432	437	443	448	454	459	465	470	476	481	4	2.0
785	487	492	498	504	509	515	520	526	531	537	5	2.5
786	542	548	553	559	564	570	575	581	586	592	6	3.0
787	597	603	609	614	620	625	631	636	642	647	7	3.5
788	653	658	664	669	675	680	686	691	697	702	8	4.0
789	708	713	719	724	730	735	741	746	752	757	9	4.5
790	763	768	774	779	785	790	796	801	807	812		
791	818	823	829	834	840	845	851	856	862	867		
792	873	878	883	889	894	900	905	911	916	922		
793	927	933	938	944	949	955	960	966	971	977		
794	982	988	993	998	*004	*009	*015	*020	*026	*031		
795	90 037	042	048	053	059	064	069	075	080	086		
796	091	097	102	108	113	119	124	129	135	140		
797	146	151	157	162	168	173	179	184	189	195		
798	200	206	211	217	222	227	233	238	244	249		
799	255	260	266	271	276	282	287	293	298	304		
800	309	314	320	325	331	336	342	347	352	358		
N.	0	1	2	3	4	5	6	7	8	9	Proportional parts	

FIVE-PLACE MANTISSAS FOR COMMON LOGARITHMS (continued)

N.	0	1	2	3	4	5	6	7	8	9	Proportional parts	
800	90 309	314	320	325	331	336	342	347	352	358		
801	363	369	374	380	385	390	396	401	407	412		
802	417	423	428	434	439	445	450	455	461	466		
803	472	477	482	488	493	499	504	509	515	520		
804	526	531	536	542	547	553	558	563	569	574		
805	580	585	590	596	601	607	612	617	623	628		
806	634	639	644	650	655	660	666	671	677	682		
807	687	693	698	703	709	714	720	725	730	736		
808	741	747	752	757	763	768	773	779	784	789		
809	795	800	806	811	816	822	827	832	838	843		
810	849	854	859	865	870	875	881	886	891	897		**6**
811	902	907	913	918	924	929	934	940	945	950	1	0.6
812	956	961	966	972	977	982	988	993	998	*004	2	1.2
813	91 009	014	020	025	030	036	041	046	052	057	3	1.8
814	062	068	073	078	084	089	094	100	105	110	4	2.4
815	116	121	126	132	137	142	148	153	158	164	5	3.0
816	169	174	180	185	190	196	201	206	212	217	6	3.6
817	222	228	233	238	243	249	254	259	265	270	7	4.2
818	275	281	286	291	297	302	307	312	318	323	8	4.8
819	328	334	339	344	350	355	360	365	371	376	9	5.4
820	381	387	392	397	403	408	413	418	424	429		
821	434	440	445	450	455	461	466	471	477	482		
822	487	492	498	503	508	514	519	524	529	535		
823	540	545	551	556	561	566	572	577	582	587		
824	593	598	603	609	614	619	624	630	635	640		
825	645	651	656	661	666	672	677	682	687	693		
826	698	703	709	714	719	724	730	735	740	745		
827	751	756	761	766	772	777	782	787	793	798		
828	803	808	814	819	824	829	834	840	845	850		
829	855	861	866	871	876	882	887	892	897	903		
830	908	913	918	924	929	934	939	944	950	955		**5**
831	960	965	971	976	981	986	991	997	*002	*007	1	0.5
832	92 012	018	023	028	033	038	044	049	054	059	2	1.0
833	065	070	075	080	085	091	096	101	106	111	3	1.5
834	117	122	127	132	137	143	148	153	158	163	4	2.0
835	169	174	179	184	189	195	200	205	210	215	5	2.5
836	221	226	231	236	241	247	252	257	262	267	6	3.0
837	273	278	283	288	293	298	304	309	314	319	7	3.5
838	324	330	335	340	345	350	355	361	366	371	8	4.0
839	376	381	387	392	397	402	407	412	418	423	9	4.5
840	428	433	438	443	449	454	459	464	469	474		
841	480	485	490	495	500	505	511	516	521	526		
842	531	536	542	547	552	557	562	567	572	578		
843	583	588	593	598	603	609	614	619	624	629		
844	634	639	645	650	655	660	665	670	675	681		
845	686	691	696	701	706	711	716	722	727	732		
846	737	742	747	752	758	763	768	773	778	783		
847	788	793	799	804	809	814	819	824	829	834		
848	840	845	850	855	860	865	870	875	881	886		
849	891	896	901	906	911	916	921	927	932	937		
850	942	947	952	957	962	967	973	978	983	988		
N.	0	1	2	3	4	5	6	7	8	9	Proportional parts	

FIVE-PLACE MANTISSAS FOR COMMON LOGARITHMS (continued)

N.	0	1	2	3	4	5	6	7	8	9	Proportional parts	
850	92 942	947	952	957	962	967	973	978	983	988		
851	993	998	*003	*008	*013	*018	*024	*029	*034	*039		
852	93 044	049	054	059	064	069	075	080	085	090		
853	095	100	105	110	115	120	125	131	136	141		
854	146	151	156	161	166	171	176	181	186	192		
855	197	202	207	212	217	222	227	232	237	242		
856	247	252	258	263	268	273	278	283	288	293		
857	298	303	308	313	318	323	328	334	339	344		
858	349	354	359	364	369	374	379	384	389	394		**6**
859	399	404	409	414	420	425	430	435	440	445	1	0.6
											2	1.2
860	450	455	460	465	470	475	480	485	490	495	3	1.8
861	500	505	510	515	520	526	531	536	541	546	4	2.4
862	551	556	561	566	571	576	581	586	591	596	5	3.0
863	601	606	611	616	621	626	631	636	641	646	6	3.6
864	651	656	661	666	671	676	682	687	692	697	7	4.2
865	702	707	712	717	722	727	732	737	742	747	8	4.8
866	752	757	762	767	772	777	782	787	792	797	9	5.4
867	802	807	812	817	822	827	832	837	842	847		
868	852	857	862	867	872	877	882	887	892	897		
869	902	907	912	917	922	927	932	937	942	947		
870	952	957	962	967	972	977	982	987	992	997		**5**
871	94 002	007	012	017	022	027	032	037	042	047	1	0.5
872	052	057	062	067	072	077	082	086	091	096	2	1.0
873	101	106	111	116	121	126	131	136	141	146	3	1.5
874	151	156	161	166	171	176	181	186	191	196	4	2.0
875	201	206	211	216	221	226	231	236	240	245	5	2.5
876	250	255	260	265	270	275	280	285	290	295	6	3.0
877	300	305	310	315	320	325	330	335	340	345	7	3.5
878	349	354	359	364	369	374	379	384	389	394	8	4.0
879	399	404	409	414	419	424	429	433	438	443	9	4.5
880	448	453	458	463	468	473	478	483	488	493		
881	498	503	507	512	517	522	527	532	537	542		
882	547	552	557	562	567	571	576	581	586	591		
883	596	601	606	611	616	621	626	630	635	640		
884	645	650	655	660	665	670	675	680	685	689		
885	694	699	704	709	714	719	724	729	734	738		
886	743	748	753	758	763	768	773	778	783	787		**4**
887	792	797	802	807	812	817	822	827	832	836	1	0.4
888	841	846	851	856	861	866	871	876	880	885	2	0.8
889	890	895	900	905	910	915	919	924	929	934	3	1.2
											4	1.6
890	939	944	949	954	959	963	968	973	978	983	5	2.0
891	988	993	998	*002	*007	*012	*017	*022	*027	*032	6	2.4
892	95 036	041	046	051	056	061	066	071	075	080	7	2.8
893	085	090	095	100	105	109	114	119	124	129	8	3.2
894	134	139	143	148	153	158	163	168	173	177	9	3.6
895	182	187	192	197	202	207	211	216	221	226		
896	231	236	240	245	250	255	260	265	270	274		
897	279	284	289	294	299	303	308	313	318	323		
898	328	332	337	342	347	352	357	361	366	371		
899	376	381	386	390	395	400	405	410	415	419		
900	424	429	434	439	444	448	453	458	463	468		
N.	0	1	2	3	4	5	6	7	8	9	Proportional parts	

FIVE-PLACE MANTISSAS FOR COMMON LOGARITHMS (continued)

N.	0	1	2	3	4	5	6	7	8	9	Proportional parts	
900	95 424	429	434	439	444	448	453	458	463	468		
901	472	477	482	487	492	497	501	506	511	516		
902	521	525	530	535	540	545	550	554	559	564		
903	569	574	578	583	588	593	598	602	607	612		
904	617	622	626	631	636	641	646	650	655	660		
905	665	670	674	679	684	689	694	698	703	708		
906	713	718	722	727	732	737	742	746	751	756		
907	761	766	770	775	780	785	789	794	799	804		
908	809	813	818	823	828	832	837	842	847	852		
909	856	861	866	871	875	880	885	890	895	899		
910	904	909	914	918	923	928	933	938	942	947		**5**
911	952	957	961	966	971	976	980	985	990	995	1	0.5
912	999	*004	*009	*014	*019	*023	*028	*033	*038	*042	2	1.0
913	96 047	052	057	061	066	071	076	080	085	090	3	1.5
914	095	099	104	109	114	118	123	128	133	137	4	2.0
915	142	147	152	156	161	166	171	175	180	185	5	2.5
916	190	194	199	204	209	213	218	223	227	232	6	3.0
917	237	242	246	251	256	261	265	270	275	280	7	3.5
918	284	289	294	298	303	308	313	317	322	327	8	4.0
919	332	336	341	346	350	355	360	365	369	374	9	4.5
920	379	384	388	393	398	402	407	412	417	421		
921	426	431	435	440	445	450	454	459	464	468		
922	473	478	483	487	492	497	501	506	511	515		
923	520	525	530	534	539	544	548	553	558	562		
924	567	572	577	581	586	591	595	600	605	609		
925	614	619	624	628	633	638	642	647	652	656		
926	661	666	670	675	680	685	689	694	699	703		
927	708	713	717	722	727	731	736	741	745	750		
928	755	759	764	769	774	778	783	788	792	797		
929	802	806	811	816	820	825	830	834	839	844		
930	848	853	858	862	867	872	876	881	886	890		**4**
931	895	900	904	909	914	918	923	928	932	937	1	0.4
932	942	946	951	956	960	965	970	974	979	984	2	0.8
933	988	993	997	*002	*007	*011	*016	*021	*025	*030	3	1.2
934	97 035	039	044	049	053	058	063	067	072	077	4	1.6
935	081	086	090	095	100	104	109	114	118	123	5	2.0
936	128	132	137	142	146	151	155	160	165	169	6	2.4
937	174	179	183	188	192	197	202	206	211	216	7	2.8
938	220	225	230	234	239	243	248	253	257	262	8	3.2
939	267	271	276	280	285	290	294	299	304	308	9	3.6
940	313	317	322	327	331	336	340	345	350	354		
941	359	364	368	373	377	382	387	391	396	400		
942	405	410	414	419	424	428	433	437	442	447		
943	451	456	460	465	470	474	479	483	488	493		
944	497	502	506	511	516	520	525	529	534	539		
945	543	548	552	557	562	566	571	575	580	585		
946	589	594	598	603	607	612	617	621	626	630		
947	635	640	644	649	653	658	663	667	672	676		
948	681	685	690	695	699	704	708	713	717	722		
949	727	731	736	740	745	749	754	759	763	768		
950	772	777	782	786	791	795	800	804	809	813		
N.	0	1	2	3	4	5	6	7	8	9	Proportional parts	

FIVE-PLACE MANTISSAS FOR COMMON LOGARITHMS (continued)

N.	0	1	2	3	4	5	6	7	8	9	Proportional parts	
950	97 772	777	782	786	791	795	800	804	809	813		
951	818	823	827	832	836	841	845	850	855	859		
952	864	868	873	877	882	886	891	896	900	905		
953	909	914	918	923	928	932	937	941	946	950		
954	955	959	964	968	973	978	982	987	991	996		
955	98 000	005	009	014	019	023	028	032	037	041		
956	046	050	055	059	064	068	073	078	082	087		
957	091	096	100	105	109	114	118	123	127	132		
958	137	141	146	150	155	159	164	168	173	177		
959	182	186	191	195	200	204	209	214	218	223		
960	227	232	236	241	245	250	254	259	263	268		**5**
961	272	277	281	286	290	295	299	304	308	313	1	0.5
962	318	322	327	331	336	340	345	349	354	358	2	1.0
963	363	367	372	376	381	385	390	394	399	403	3	1.5
964	408	412	417	421	426	430	435	439	444	448	4	2.0
965	453	457	462	466	471	475	480	484	489	493	5	2.5
966	498	502	507	511	516	520	525	529	534	538	6	3.0
967	543	547	552	556	561	565	570	574	579	583	7	3.5
968	588	592	597	601	605	610	614	619	623	628	8	4.0
969	632	637	641	646	650	655	659	664	668	673	9	4.5
970	677	682	686	691	695	700	704	709	713	717		
971	722	726	731	735	740	744	749	753	758	762		
972	767	771	776	780	784	789	793	798	802	807		
973	811	816	820	825	829	834	838	843	847	851		
974	856	860	865	869	874	878	883	887	892	896		
975	900	905	909	914	918	923	927	932	936	941		
976	945	949	954	958	963	967	972	976	981	985		
977	989	994	998	*003	*007	*012	*016	*021	*025	*029		
978	99 034	038	043	047	052	056	061	065	069	074		
979	078	083	087	092	096	100	105	109	114	118		
980	123	127	131	136	140	145	149	154	158	162		**4**
981	167	171	176	180	185	189	193	198	202	207	1	0.4
982	211	216	220	224	229	233	238	242	247	251	2	0.8
983	255	260	264	269	273	277	282	286	291	295	3	1.2
984	300	304	308	313	317	322	326	330	335	339	4	1.6
985	344	348	352	357	361	366	370	374	379	383	5	2.0
986	388	392	396	401	405	410	414	419	423	427	6	2.4
987	432	436	441	445	449	454	458	463	467	471	7	2.8
988	476	480	484	489	493	498	502	506	511	515	8	3.2
989	520	524	528	533	537	542	546	550	555	559	9	3.6
990	564	568	572	577	581	585	590	594	599	603		
991	607	612	616	621	625	629	634	638	642	647		
992	651	656	660	664	669	673	677	682	686	691		
993	695	699	704	708	712	717	721	726	730	734		
994	739	743	747	752	756	760	765	769	774	778		
995	782	787	791	795	800	804	808	813	817	822		
996	826	830	835	839	843	848	852	856	861	865		
997	870	874	878	883	887	891	896	900	904	909		
998	913	917	922	926	930	935	939	944	948	952		
999	957	961	965	970	974	978	983	987	991	996		
1000	00 000	004	009	013	017	022	026	030	035	039		
N.	0	1	2	3	4	5	6	7	8	9	Proportional parts	

NATURAL OR NAPERIAN LOGARITHMS
0.000–0.499

N	0	1	2	3	4	5	6	7	8	9
0.00	− ∞	−6** .90776	−6 .21461	−5 .80914	−5 .52146	−5 .29832	−5 .11600	−4 .96185	−4 .82831	−4 .71053
.01	−4.60517	.50986	.42285	.34281	.26870	.19971	.13517	.07454	.01738	* .96332
.02	−3.91202	.86323	.81671	.77226	.72970	.68888	.64966	.61192	.57555	.54046
.03	.50656	.47377	.44202	.41125	.38139	.35241	.32424	.29684	.27017	.24419
.04	.21888	.19418	.17009	.14656	.12357	.10109	.07911	.05761	.03655	.01593
.05	−2.99573	.97593	.95651	.93746	.91877	.90042	.88240	.86470	.84731	.83022
.06	.81341	.79688	.78062	.76462	.74887	.73337	.71810	.70306	.68825	.67365
.07	.65926	.64508	.63109	.61730	.60369	.59027	.57702	.56395	.55105	.53831
.08	.52573	.51331	.50104	.48891	.47694	.46510	.45341	.44185	.43042	.41912
.09	.40795	.39690	.38597	.37516	.36446	.35388	.34341	.33304	.32279	.31264
0.10	−2.30259	.29263	.28278	.27303	.26336	.25379	.24432	.23493	.22562	.21641
.11	.20727	.19823	.18926	.18037	.17156	.16282	.15417	.14558	.13707	.12863
.12	.12026	.11196	.10373	.09557	.08747	.07944	.07147	.06357	.05573	.04794
.13	.04022	.03256	.02495	.01741	.00992	.00248	* .99510	* .98777	* .98050	* .97328
.14	−1.96611	.95900	.95193	.94491	.93794	.93102	.92415	.91732	.91054	.90381
.15	.89712	.89048	.88387	.87732	.87080	.86433	.85790	.85151	.84516	.83885
.16	.83258	.82635	.82016	.81401	.80789	.80181	.79577	.78976	.78379	.77786
.17	.77196	.76609	.76026	.75446	.74870	.74297	.73727	.73161	.72597	.72037
.18	.71480	.70926	.70375	.69827	.69282	.68740	.68201	.67665	.67131	.66601
.19	.66073	.65548	.65026	.64507	.63990	.63476	.62964	.62455	.61949	.61445
0.20	−1.60944	.60445	.59949	.59455	.58964	.58475	.57988	.57504	.57022	.56542
.21	.56065	.55590	.55117	.54646	.54178	.53712	.53248	.52786	.52326	.51868
.22	.51413	.50959	.50508	.50058	.49611	.49165	.48722	.48281	.47841	.47403
.23	.46968	.46534	.46102	.45672	.45243	.44817	.44392	.43970	.43548	.43129
.24	.42712	.42296	.41882	.41469	.41059	.40650	.40242	.39837	.39433	.39030
.25	.38629	.38230	.37833	.37437	.37042	.36649	.36258	.35868	.35480	.35093
.26	.34707	.34323	.33941	.33560	.33181	.32803	.32426	.32051	.31677	.31304
.27	.30933	.30564	.30195	.29828	.29463	.29098	.28735	.28374	.28013	.27654
.28	.27297	.26940	.26585	.26231	.25878	.25527	.25176	.24827	.24479	.24133
.29	.23787	.23443	.23100	.22758	.22418	.22078	.21740	.21402	.21066	.20731
0.30	−1.20397	.20065	.19733	.19402	.19073	.18744	.18417	.18091	.17766	.17441
.31	.17118	.16796	.16475	.16155	.15836	.15518	.15201	.14885	.14570	.14256
.32	.13943	.13631	.13320	.13010	.12701	.12393	.12086	.11780	.11474	.11170
.33	.10866	.10564	.10262	.09961	.09661	.09362	.09064	.08767	.08471	.08176
.34	.07881	.07587	.07294	.07002	.06711	.06421	.06132	.05843	.05555	.05268
.35	−1.04982	.04697	.04412	.04129	.03846	.03564	.03282	.03002	.02722	.02443
.36	.02165	.01888	.01611	.01335	.01060	.00786	.00512	.00239	* .99967	* .99696
.37	−0.99425	.99155	.98886	.98618	.98350	.98083	.97817	.97551	.97286	.97022
.38	.96758	.96496	.96233	.95972	.95711	.95451	.95192	.94933	.94675	.94418
.39	.94161	.93905	.93649	.93395	.93140	.92887	.92634	.92382	.92130	.91879
0.40	−0.91629	.91379	.91130	.90882	.90634	.90387	.90140	.89894	.89649	.89404
.41	.89160	.88916	.88673	.88431	.88189	.87948	.87707	.87467	.87227	.86988
.42	.86750	.86512	.86275	.86038	.85802	.85567	.85332	.85097	.84863	.84630
.43	.84397	.84165	.83933	.83702	.83471	.83241	.83011	.82782	.82554	.82326
.44	.82098	.81871	.81645	.81419	.81193	.80968	.80744	.80520	.80296	.80073
.45	.79851	.79629	.79407	.79186	.78966	.78746	.78526	.78307	.78089	.77871
.46	.77653	.77436	.77219	.77003	.76787	.76572	.76357	.76143	.75929	.75715
.47	.75502	.75290	.75078	.74866	.74655	.74444	.74234	.74024	.73814	.73605
.48	.73397	.73189	.72981	.72774	.72567	.72361	.72155	.71949	.71744	.71539
.49	.71335	.71131	.70928	.70725	.70522	.70320	.70118	.69917	.69716	.69515

**Note that the whole number values are given above the decimal values for the first line. In the second and following lines they are given at the left.

NATURAL OR NAPERIAN LOGARITHMS (Continued)
0.500–0.999

N	0	1	2	3	4	5	6	7	8	9
0.50	−0.69315	.69115	.68916	.68717	.68518	.68320	.68122	.67924	.67727	.67531
.51	.67334	.67139	.66943	.66748	.66553	.66359	.66165	.65971	.65778	.65585
.52	.65393	.65201	.65009	.64817	.64626	.64436	.64245	.64055	.63866	.63677
.53	.63488	.63299	.63111	.62923	.62736	.62549	.62362	.62176	.61990	.61804
.54	.61619	.61434	.61249	.61065	.60881	.60697	.60514	.60331	.60148	.59966
.55	.59784	.59602	.59421	.59240	.59059	.58879	.58699	.58519	.58340	.58161
.56	.57982	.57803	.57625	.57448	.57270	.57093	.56916	.56740	.56563	.56387
.57	.56212	.56037	.55862	.55687	.55513	.55339	.55165	.54991	.54818	.54645
.58	.54473	.54300	.54128	.53957	.53785	.53614	.53444	.53273	.53103	.52933
.59	.52763	.52594	.52425	.52256	.52088	.51919	.51751	.51584	.51416	.51249
0.60	−0.51083	.50916	.50750	.50584	.50418	.50253	.50088	.49923	.49758	.49594
.61	.49430	.49266	.49102	.48939	.48776	.48613	.48451	.48289	.48127	.47965
.62	.47804	.47642	.47482	.47321	.47160	.47000	.46840	.46681	.46522	.46362
.63	.46204	.46045	.45887	.45728	.45571	.45413	.45256	.45099	.44942	.44785
.64	.44629	.44473	.44317	.44161	.44006	.43850	.43696	.43541	.43386	.43232
.65	.43078	.42925	.42771	.42618	.42465	.42312	.42159	.42007	.41855	.41703
.66	.41552	.41400	.41249	.41098	.40947	.40797	.40647	.40497	.40347	.40197
.67	.40048	.39899	.39750	.39601	.39453	.39304	.39156	.39008	.38861	.38713
.68	.38566	.38419	.38273	.38126	.37980	.37834	.37688	.37542	.37397	.37251
.69	.37106	.36962	.36817	.36673	.36528	.36384	.36241	.36097	.35954	.35810
0.70	−0.35667	.35525	.35382	.35240	.35098	.34956	.34814	.34672	.34531	.34390
.71	.34249	.34108	.33968	.33827	.33687	.33547	.33408	.33268	.33129	.32989
.72	.32850	.32712	.32573	.32435	.32296	.32158	.32021	.31883	.31745	.31608
.73	.31471	.31334	.31197	.31061	.30925	.30788	.30653	.30517	.30381	.30246
.74	.30111	.29975	.29841	.29706	.29571	.29437	.29303	.29169	.29035	.28902
.75	.28768	.28635	.28502	.28369	.28236	.28104	.27971	.27839	.27707	.27575
.76	.27444	.27312	.27181	.27050	.26919	.26788	.26657	.26527	.26397	.26266
.77	.26136	.26007	.25877	.25748	.25618	.25489	.25360	.25231	.25103	.24974
.78	.24846	.24718	.24590	.24462	.24335	.24207	.24080	.23953	.23826	.23699
.79	.23572	.23446	.23319	.23193	.23067	.22941	.22816	.22690	.22565	.22439
0.80	−0.22314	.22189	.22065	.21940	.21816	.21691	.21567	.21433	.21319	.21196
.81	.21072	.20949	.20825	.20702	.20579	.20457	.20334	.20212	.20089	.19967
.82	.19845	.19723	.19601	.19480	.19358	.19237	.19116	.18995	.18874	.18754
.83	.18633	.18513	.18392	.18272	.18152	.18032	.17913	.17793	.17674	.17554
.84	.17435	.17316	.17198	.17079	.16960	.16842	.16724	.16605	.16487	.16370
.85	−0.16252	.16134	.16017	.15900	.15782	.15665	.15548	.15432	.15315	.15199
.86	.15082	.14966	.14850	.14734	.14618	.14503	.14387	.14272	.14156	.14041
.87	.13926	.13811	.13697	.13582	.13467	.13353	.13239	.13125	.13011	.12897
.88	.12783	.12670	.12556	.12443	.12330	.12217	.12104	.11991	.11878	.11766
.89	.11653	.11541	.11429	.11317	.11205	.11093	.10981	.10870	.10759	.10647
0.90	−0.10536	.10425	.10314	.10203	.10093	.09982	.09872	.09761	.09651	.09541
.91	.09431	.09321	.09212	.09102	.08992	.08883	.08774	.08665	.08556	.08447
.92	.08338	.08230	.08121	.08013	.07904	.07796	.07688	.07580	.07472	.07365
.93	.07257	.07150	.07042	.06935	.06828	.06721	.06614	.06507	.06401	.06294
.94	.06188	.06081	.05975	.05869	.05763	.05657	.05551	.05446	.05340	.05235
.95	.05129	.05024	.04919	.04814	.04709	.04604	.04500	.04395	.04291	.04186
.96	.04082	.03978	.03874	.03770	.03666	.03563	.03459	.03356	.03252	.03149
.97	.03046	.02943	.02840	.02737	.02634	.02532	.02429	.02327	.02225	.02122
.98	.02020	.01918	.01816	.01715	.01613	.01511	.01410	.01309	.01207	.01106
.99	.01005	.00904	.00803	.00702	.00602	.00501	.00401	.00300	.00200	.00100

NATURAL OR NAPERIAN LOGARITHMS (Continued)

To find the natural logarithm of a number which is $\frac{1}{10}, \frac{1}{100}, \frac{1}{1000}$, etc. of a number whose logarithm is given, subtract from the given logarithm $\log_e 10$, $2 \log_e 10$, $3 \log_e 10$, etc.

To find the natural logarithm of a number which is 10, 100, 1000, etc. times a number whose logarithm is given, add to the given logarithm $\log_e 10$, $2 \log_e 10$, $3 \log_e 10$, etc.

$\log_e 10 = 2.30258\ 50930$	$6 \log_e 10 = 13.81551\ 05580$
$2 \log_e 10 = 4.60517\ 01860$	$7 \log_e 10 = 16.11809\ 56510$
$3 \log_e 10 = 6.90775\ 52790$	$8 \log_e 10 = 18.42068\ 07440$
$4 \log_e 10 = 9.21034\ 03720$	$9 \log_e 10 = 20.72326\ 58369$
$5 \log_e 10 = 11.51292\ 54650$	$10 \log_e 10 = 23.02585\ 09299$

See pages 240–241 for logarithms for numbers between 0.000 and 0.999.

1.00–4.99

N	0	1	2	3	4	5	6	7	8	9
1.0	0.00000	.00995	.01980	.02956	.03922	.04879	.05827	.06766	.07696	.08618
.1	.09531	.10436	.11333	.12222	.13103	.13976	.14842	.15700	.16551	.17395
.2	.18232	.19062	.19885	.20701	.21511	.22314	.23111	.23902	.24686	.25464
.3	.26236	.27003	.27763	.28518	.29267	.30010	.30748	.31481	.32208	.32930
.4	.33647	.34359	.35066	.35767	.36464	.37156	.37844	.38526	.39204	.39878
.5	.40547	.41211	.41871	.42527	.43178	.43825	.44469	.45108	.45742	.46373
.6	.47000	.47623	.48243	.48858	.49470	.50078	.50682	.51282	.51879	.52473
.7	.53063	.53649	.54232	.54812	.55389	.55962	.56531	.57098	.57661	.58222
.8	.58779	.59333	.59884	.60432	.60977	.61519	.62058	.62594	.63127	.63658
.9	.64185	.64710	.65233	.65752	.66269	.66783	.67294	.67803	.68310	.68813
2.0	0.69315	.69813	.70310	.70804	.71295	.71784	.72271	.72755	.73237	.73716
.1	.74194	.74669	.75142	.75612	.76081	.76547	.77011	.77473	.77932	.78390
.2	.78846	.79299	.79751	.80200	.80648	.81093	.81536	.81978	.82418	.82855
.3	.83291	.83725	.84157	.84587	.85015	.85442	.85866	.86289	.86710	.87129
.4	.87547	.87963	.88377	.88789	.89200	.89609	.90016	.90422	.90826	.91228
.5	.91629	.92028	.92426	.92822	.93216	.93609	.94001	.94391	.94779	.95166
.6	.95551	.95935	.96317	.96698	.97078	.97456	.97833	.98208	.98582	.98954
.7	.99325	.99695	*.00063	*.00430	*.00796	*.01160	*.01523	*.01885	*.02245	*.02604
.8	1.02962	.03318	.03674	.04028	.04380	.04732	.05082	.05431	.05779	.06126
.9	.06471	.06815	.07158	.07500	.07841	.08181	.08519	.08856	.09192	.09527
3.0	1.09861	.10194	.10526	.10856	.11186	.11514	.11841	.12168	.12493	.12817
.1	.13140	.13462	.13783	.14103	.14422	.14740	.15057	.15373	.15688	.16002
.2	.16315	.16627	.16938	.17248	.17557	.17865	.18173	.18479	.18784	.19089
.3	.19392	.19695	.19996	.20297	.20597	.20896	.21194	.21491	.21788	.22083
.4	.22378	.22671	.22964	.23256	.23547	.23837	.24127	.24415	.24703	.24990
.5	.25276	.25562	.25846	.26130	.26413	.26695	.26976	.27257	.27536	.27815
.6	.28093	.28371	.28647	.28923	.29198	.29473	.29746	.30019	.30291	.30563
.7	.30833	.31103	.31372	.31641	.31909	.32176	.32442	.32708	.32972	.33237
.8	.33500	.33763	.34025	.34286	.34547	.34807	.35067	.35325	.35584	.35841
.9	.36098	.36354	.36609	.36864	.37118	.37372	.37624	.37877	.38128	.38379
4.0	1.38629	.38879	.39128	.39377	.39624	.39872	.40118	.40364	.40610	.40854
.1	.41099	.41342	.41585	.41828	.42070	.42311	.42552	.42792	.43031	.43270
.2	.43508	.43746	.43984	.44220	.44456	.44692	.44927	.45161	.45395	.45629
.3	.45862	.46094	.46326	.46557	.46787	.47018	.47247	.47476	.47705	.47933
.4	.48160	.48387	.48614	.48840	.49065	.49290	.49515	.49739	.49962	.50185
.5	.50408	.50630	.50851	.51072	.51293	.51513	.51732	.51951	.52170	.52388
.6	.52606	.52823	.53039	.53256	.53471	.53687	.53902	.54116	.54330	.54543
.7	.54756	.54969	.55181	.55393	.55604	.55814	.56025	.56235	.56444	.56653
.8	.56862	.57070	.57277	.57485	.57691	.57898	.58104	.58309	.58515	.58719
.9	.58924	.59127	.59331	.59534	.59737	.59939	.60141	.60342	.60543	.60744

NATURAL OR NAPERIAN LOGARITHMS (Continued)
5.00–9.99

N	0	1	2	3	4	5	6	7	8	9
5.0	1.60944	.61144	.61343	.61542	.61741	.61939	.62137	.62334	.62531	.62728
.1	.62924	.63120	.63315	.63511	.63705	.63900	.64094	.64287	.64481	.64673
.2	.64866	.65058	.65250	.65441	.65632	.65823	.66013	.66203	.66393	.66582
.3	.66771	.66959	.67147	.67335	.67523	.67710	.67896	.68083	.68269	.68455
.4	.68640	.68825	.69010	.69194	.69378	.69562	.69745	.69928	.70111	.70293
.5	.70475	.70656	.70838	.71019	.71199	.71380	.71560	.71740	.71919	.72098
.6	.72277	.72455	.72633	.72811	.72988	.73166	.73342	.73519	.73695	.73871
.7	.74047	.74222	.74397	.74572	.74746	.74920	.75094	.75267	.75440	.75613
.8	.75786	.75958	.76130	.76302	.76473	.76644	.76815	.76985	.77156	.77326
.9	.77495	.77665	.77834	.78002	.78171	.78339	.78507	.78675	.78842	.79009
6.0	1.79176	.79342	.79509	.79675	.79840	.80006	.80171	.80336	.80500	.80665
.1	.80829	.80993	.81156	.81319	.81482	.81645	.81808	.81970	.82132	.82294
.2	.82455	.82616	.82777	.82938	.83098	.83258	.83418	.83578	.83737	.83896
.3	.84055	.84214	.84372	.84530	.84688	.84845	.85003	.85160	.85317	.85473
.4	.85630	.85786	.85942	.86097	.86253	.86408	.86563	.86718	.86872	.87026
.5	.87180	.87334	.87487	.87641	.87794	.87947	.88099	.88251	.88403	.88555
.6	.88707	.88858	.89010	.89160	.89311	.89462	.89612	.89762	.89912	.90061
.7	.90211	.90360	.90509	.90658	.90806	.90954	.91102	.91250	.91398	.91545
.8	.91692	.91839	.91986	.92132	.92279	.92425	.92571	.92716	.92862	.93007
.9	.93152	.93297	.93442	.93586	.93730	.93874	.94018	.94162	.94305	.94448
7.0	1.94591	.94734	.94876	.95019	.95161	.95303	.95445	.95586	.95727	.95869
.1	.96009	.96150	.96291	.96431	.96571	.96711	.96851	.96991	.97130	.97269
.2	.97408	.97547	.97685	.97824	.97962	.98100	.98238	.98376	.98513	.98650
.3	.98787	.98924	.99061	.99198	.99334	.99470	.99606	.99742	.99877	*.00013
.4	2.00148	00283	00418	00553	00687	.00821	.00956	.01089	.01223	.01357
.5	.01490	.01624	.01757	.01890	.02022	.02155	.02287	.02419	.02551	.02683
.6	.02815	.02946	.03078	.03209	.03340	.03471	.03601	.03732	.03862	.03992
.7	.04122	04252	.04381	.04511	.04640	.04769	.04898	.05027	.05156	.05284
.8	.05412	.05540	.05668	.05796	.05924	.06051	.06179	.06306	.06433	.06560
.9	.06686	.06813	.06939	.07065	.07191	.07317	.07443	.07568	.07694	.07819
8.0	2.07944	.08069	.08194	.08318	.08443	.08567	.08691	.08815	.08939	.09063
.1	.09186	.09310	.09433	.09556	.09679	.09802	.09924	.10047	.10169	.10291
.2	.10413	.10535	.10657	.10779	.10900	.11021	.11142	.11263	.11384	.11505
.3	.11626	.11746	.11866	.11986	.12106	.12226	.12346	.12465	.12585	.12704
.4	.12823	.12942	.13061	.13180	.13298	.13417	.13535	.13653	.13771	.13889
.5	.14007	.14124	.14242	.14359	.14476	.14593	.14710	.14827	.14943	.15060
.6	.15176	.15292	.15409	.15524	.15640	.15756	.15871	.15987	.16102	.16217
.7	.16332	.16447	.16562	.16677	.16791	.16905	.17020	.17134	.17248	.17361
.8	.17475	.17589	.17702	.17816	.17929	.18042	.18155	.18267	.18380	.18493
.9	.18605	.18717	.18830	.18942	.19054	.19165	.19277	.19389	.19500	.19611
9.0	2.19722	.19834	.19944	.20055	.20166	.20276	.20387	.20497	.20607	.20717
.1	.20827	.20937	.21047	.21157	.21266	.21375	.21485	.21594	.21703	.21812
.2	.21920	.22029	.22138	.22246	.22354	.22462	.22570	.22678	.22786	.22894
.3	.23001	.23109	.23216	.23324	.23431	.23538	.23645	.23751	.23858	.23965
.4	.24071	.24177	.24284	.24390	.24496	.24601	.24707	.24813	.24918	.25024
.5	.25129	.25234	.25339	.25444	.25549	.25654	.25759	.25863	.25968	.26072
.6	.26176	.26280	.26384	.26488	.26592	.26696	.26799	.26903	.27006	.27109
.7	.27213	.27316	.27419	.27521	.27624	.27727	.27829	.27932	.28034	.28136
.8	.28238	.28340	.28442	.28544	.28646	.28747	.28849	.28950	.29051	.29152
.9	.29253	.29354	.29455	.29556	.29657	.29757	.29858	.29958	.30058	.30158

NATURAL OR NAPERIAN LOGARITHMS (Continued)

Constants

$\log_e 10$ = 2.30258 50930	6 $\log_e 10$ = 13.81551 05580
2 $\log_e 10$ = 4.60517 01860	7 $\log_e 10$ = 16.11809 56510
3 $\log_e 10$ = 6.90775 52790	8 $\log_e 10$ = 18.42068 07440
4 $\log_e 10$ = 9.21034 03720	9 $\log_e 10$ = 20.72326 58369
5 $\log_e 10$ = 11.51292 54650	10 $\log_e 10$ = 23.02585 09299

10.0–49.9

N	0	1	2	3	4	5	6	7	8	9
10.	2.30259	.31254	.32239	.33214	.34181	.35138	.36085	.37024	.37955	.38876
11.	.39790	.40695	.41591	.42480	.43361	.44235	.45101	.45959	.46810	.47654
12.	.48491	.49321	.50144	.50960	.51770	.52573	.53370	.54160	.54945	.55723
13.	.56495	.57261	.58022	.58776	.59525	.60269	.61007	.61740	.62467	.63189
14.	.63906	.64617	.65324	.66026	.66723	.67415	.68102	.68785	.69463	.70136
15.	.70805	.71469	.72130	.72785	.73437	.74084	.74727	.75366	.76001	.76632
16.	.77259	.77882	.78501	.79117	.79728	.80336	.80940	.81541	.82138	.82731
17.	.83321	.83908	.84491	.85071	.85647	.86220	.86790	.87356	.87920	.88480
18.	.89037	.89591	.90142	.90690	.91235	.91777	.92316	.92852	.93386	.93916
19.	.94444	.94969	.95491	.96011	.96527	.97041	.97553	.98062	.98568	.99072
20.	2.99573	*.00072	*.00568	*.01062	*.01553	*.02042	*.02529	*.03013	*.03495	*.03975
21.	3.04452	.04927	.05400	.05871	.06339	.06805	.07269	.07731	.08191	.08649
22.	.00104	.09558	.10009	.10459	.10906	.11352	.11795	.12236	.12676	.13114
23.	.13549	.13983	.14415	.14845	.15274	.15700	.10125	.16548	.16060	.17388
24.	.17805	.18221	.18635	.19048	.19458	.19867	.20275	.20680	.21084	.21487
25.	.21888	.22287	.22684	.23080	.23475	.23868	.24259	.24649	.25037	.25424
26.	.25810	.26194	.26576	.26957	.27336	.27714	.28091	.28466	.28840	.29213
27.	.29584	.29953	.30322	.30689	.31054	.31419	31782	.32143	.32504	.32863
28.	.33220	.33577	.33932	.34286	.34639	.34990	.35341	.35690	.36038	.36384
29.	.36730	.37074	.37417	.37759	.38099	.38439	.38777	.39115	.39451	.39786
30.	3.40120	.40453	.40784	.41115	.41444	.41773	.42100	.42426	.42751	.43076
31.	.43399	.43721	.44042	.44362	.44681	.44999	.45316	.45632	.45947	.46261
32.	.46574	.46886	.47197	.47507	.47816	.48124	.48431	.48738	.49043	.49347
33.	.49651	.49953	.50255	.50556	.50856	.51155	.51453	.51750	.52046	.52342
34.	.52636	.52930	.53223	.53515	.53806	.54096	.54385	.54674	.54962	.55249
35.	.55535	.55820	.56105	.56388	.56671	.56953	.57235	.57515	.57795	.58074
36.	.58352	.58629	.58906	.59182	.59457	.59731	.60005	.60278	.60550	.60821
37.	.61092	.61362	.61631	.61899	.62167	.62434	.62700	.62966	.63231	.63495
38.	.63759	.64021	.64284	.64545	.64806	.65066	.65325	.65584	.65842	.66099
39.	.66356	.66612	.66868	.67122	.67377	.67630	.67883	.68135	.68387	.68638
40.	3.68888	.69138	.69387	.69635	.69883	.70130	.70377	.70623	.70868	.71113
41.	.71357	.71601	.71844	.72086	.72328	.72569	.72810	.73050	.73290	.73529
42.	.73767	.74005	.74242	.74479	.74715	.74950	.75185	.75420	.75654	.75887
43.	.76120	.76352	.76584	.76815	.77046	.77276	.77506	.77735	.77963	.78191
44.	.78419	.78646	.78872	.79098	.79324	.79549	.79773	.79997	.80221	.80444
45.	.80666	.80888	.81110	.81331	.81551	.81771	.81991	.82210	.82428	.82647
46.	.82864	.83081	.83298	.83514	.83730	.83945	.84160	.84374	.84588	.84802
47.	.85015	.85227	.85439	.85651	.85862	.86073	.86283	.86493	.86703	.86912
48.	.87120	.87328	.87536	.87743	.87950	.88156	.88362	.88568	.88773	.88978
49.	.89182	.89386	.89589	.89792	.89995	.90197	.90399	.90600	.90801	.91002

NATURAL OR NAPERIAN LOGARITHMS (Continued)
50.0–99.9

N	0	1	2	3	4	5	6	7	8	9
50.	3.91202	.91402	.91602	.91801	.91999	.92197	.92395	.92593	.92790	.92986
51.	.93183	.93378	.93574	.93769	.93964	.94158	.94352	.94546	.94739	.94932
52.	.95124	.95316	.95508	.95700	.95891	.96081	.96272	.96462	.96651	.96840
53.	.97029	.97218	.97406	.97594	.97781	.97968	.98155	.98341	.98527	.98713
54.	.98898	.99083	.99268	.99452	.99636	.99820	*.00003	*.00186	*.00369	*.00551
55.	4.00733	.00915	.01096	.01277	.01458	.01638	.01818	.01998	.02177	.02356
56.	.02535	.02714	.02892	.03069	.03247	.03424	.03601	.03777	.03954	.04130
57.	.04305	.04480	.04655	.04830	.05004	.05178	.05352	.05526	.05699	.05872
58.	.06044	.06217	.06389	.06560	.06732	.06903	.07073	.07244	.07414	.07584
59.	.07754	.07923	.08092	.08261	.08429	.08598	.08766	.08933	.09101	.09268
60.	4.09434	.09601	.09767	.09933	.10099	.10264	.10429	.10594	.10759	.10923
61.	.11087	.11251	.11415	.11578	.11741	.11904	.12066	.12228	.12390	.12552
62.	.12713	.12875	.13036	.13196	.13357	.13517	.13677	.13836	.13996	.14155
63.	.14313	.14472	.14630	.14789	.14946	.15104	.15261	.15418	.15575	.15732
64.	.15888	.16044	.16200	.16356	.16511	.16667	.16821	.16976	.17131	.17285
65.	.17439	.17592	.17746	.17899	.18052	.18205	.18358	.18510	.18662	.18814
66.	.18965	.19117	.19268	.19419	.19570	.19720	.19870	.20020	.20170	.20320
67.	.20469	.20618	.20767	.20916	.21065	.21213	.21361	.21509	.21656	.21804
68.	.21951	.22098	.22244	.22391	.22537	.22683	.22829	.22975	.23120	.23266
69.	.23411	.23555	.23700	.23844	.23989	.24133	.24276	.24420	.24563	.24707
70.	4.24850	.24992	.25135	.25277	.25419	.25561	.25703	.25845	.25986	.26127
71.	.26268	.26409	.26549	.26690	.26830	.26970	.27110	.27249	.27388	.27528
72.	.27667	.27805	.27944	.28082	.28221	.28359	.28496	.28634	.28772	.28909
73.	.29046	.29183	.29320	.29456	.29592	.29729	.29865	.30000	.30136	.30271
74.	.30407	.30542	.30676	.30811	.30946	.31080	.31214	.31348	.31482	.31615
75.	.31749	.31882	.32015	.32149	.32281	.32413	.32546	.32678	.32810	.32942
76.	.33073	.33205	.33336	.33467	.33598	.33729	.33860	.33990	.34120	.34251
77.	.34381	.34510	.34640	.34769	.34899	.35028	.35157	.35286	.35414	.35543
78.	.35671	.35800	.35927	.36055	.36182	.36310	.36437	.36564	.36691	.36818
79.	.36945	.37071	.37198	.37324	.37450	.37576	.37701	.37827	.37952	.38078
80.	4.38203	38328	.38452	.38577	.38701	.38826	.38950	.39074	.39198	.39321
81.	.39445	.39568	.39692	.39815	.39938	.40060	.40183	.40305	.40428	.40550
82.	.40672	.40794	.40916	.41037	.41159	.41280	.41401	.41522	.41643	.41764
83.	.41884	.42004	.42125	.42245	.42365	.42485	.42604	.42724	.42843	.42963
84.	.43082	.43201	.43319	.43438	.43557	.43675	.43793	.43912	.44030	.44147
85.	.44265	.44383	.44500	.44617	.44735	.44852	.44969	.45085	.45202	.45318
86.	.45435	.45551	.45667	.45783	.45899	.46014	.46130	.46245	.46361	.46476
87.	.46591	.46706	.46820	.46935	.47050	.47164	.47278	.47392	.47506	.47620
88.	.47734	.47847	.47961	.48074	.48187	.48300	.48413	.48526	.48639	.48751
89.	.48864	.48976	.49088	.49200	.49312	.49424	.49536	.49647	.49758	.49870
90.	4.49981	.50092	.50203	.50314	.50424	.50535	.50645	.50756	.50866	.50976
91.	.51086	.51196	.51305	.51415	.51525	.51634	.51743	.51852	.51961	.52070
92.	.52179	.52287	.52396	.52504	.52613	.52721	.52829	.52937	.53045	.53152
93.	.53260	.53367	.53475	.53582	.53689	.53796	.53903	.54010	.54116	.54223
94.	.54329	.54436	.54542	.54648	.54754	.54860	.54966	.55071	.55177	.55282
95.	.55388	.55493	.55598	.55703	.55808	.55913	.56017	.56122	.56226	.56331
96.	.56435	.56539	.56643	.56747	.56851	.56954	.57058	.57161	.57265	.57368
97.	.57471	.57574	.57677	.57780	.57883	.57985	.58088	.58190	.58292	.58395
98.	.58497	.58599	.58701	.58802	.58904	.59006	.59107	.59208	.59310	.59411
99.	.59512	.59613	.59714	.59815	.59915	.60016	.60116	.60217	.60317	.60417

NATURAL OR NAPERIAN LOGARITHMS (Continued)
0–499

N	0	1	2	3	4	5	6	7	8	9
0	− ∞	0.00000	0.69315	1.09861	.38629	.60944	.79176	.94591	*.07944	*.19722
1	2.30259	.39790	.48491	.56495	.63906	.70805	.77259	.83321	.89037	.94444
2	.99573	*.04452	*.09104	*.13549	*.17805	*.21888	*.25810	*.29584	*.33220	*.36730
3	3.40120	.43399	.46574	.49651	.52636	.55535	.58352	.61092	.63759	.66356
4	.68888	.71357	.73767	.76120	.78419	.80666	.82864	.85015	.87120	.89182
5	.91202	.93183	.95124	.97029	.98898	*.00733	*02535	*.04305	*.06044	*.07754
6	4.09434	.11087	.12713	.14313	.15888	.17439	.18965	.20469	.21951	.23411
7	.24850	.26268	.27667	.29046	.30407	.31749	.33073	.34381	.35671	.36945
8	.38203	.39445	.40672	.41884	.43082	.44265	.45435	.46591	.47734	.48864
9	.49981	.51086	.52179	.53260	.54329	.55388	.56435	.57471	.58497	.59512
10	4.60517	.61512	.62497	.63473	.64439	.65396	.66344	.67283	.68213	.69135
11	.70048	.70953	.71850	.72739	.73620	.74493	.75359	.76217	.77068	.77912
12	.78749	.79579	.80402	.81218	.82028	.82831	.83628	.84419	.85203	.85981
13	.86753	.87520	.88280	.89035	.89784	.90527	.91265	.91998	.92725	.93447
14	.94164	.94876	.95583	.96284	.96981	.97673	.98361	.99043	.99721	*.00395
15	5.01064	.01728	.02388	.03044	.03695	.04343	.04986	.05625	.06260	.06890
16	.07517	.08140	.08760	.09375	.09987	.10595	.11199	.11799	.12396	.12990
17	.13580	.14166	.14749	.15329	.15906	.16479	.17048	.17615	.18178	.18739
18	.19296	.19850	.20401	.20949	.21494	.22036	.22575	.23111	.23644	.24175
19	.24702	.25227	.25750	.26269	.26786	.27300	.27811	.28320	.28827	.29330
20	5.29832	.30330	.30827	.31321	.31812	.32301	.32788	.33272	.33754	.34233
21	.34711	.35186	.35659	.36129	.36598	.37064	.37528	.37990	.38450	.38907
22	.39363	.39816	.40268	.40717	.41165	.41610	.42053	.42495	.42935	.43372
23	.43808	.44242	.44674	.45104	.45532	.45959	.46383	.46806	.47227	.47646
24	.48064	.48480	.48894	.49306	.49717	.50126	.50533	.50939	.51343	.51745
25	.52146	.52545	.52943	.53339	.53733	.54126	.54518	.54908	.55296	.55683
26	.56068	.56452	.56834	.57215	.57595	.57973	.58350	.58725	.59099	.59471
27	.59842	.60212	.60580	.60947	.61313	.61677	.62040	.62402	.62762	.63121
28	.63479	.63835	.64191	.64545	.64897	.65249	.65599	.65948	.66296	.66643
29	.66988	.67332	.67675	.68017	.68358	.68698	.69036	.69373	.69709	.70044
30	5.70378	.70711	.71043	.71373	.71703	.72031	.72359	.72685	.73010	.73334
31	.73657	.73979	.74300	.74620	.74939	.75257	.75574	.75890	.76205	.76519
32	.76832	.77144	.77455	.77765	.78074	.78383	.78690	.78996	.79301	.79606
33	.79909	.80212	.80513	.80814	.81114	.81413	.81711	.82008	.82305	.82600
34	.82895	.83188	.83481	.83773	.84064	.84354	.84644	.84932	.85220	.85507
35	.85793	.86079	.86363	.86647	.86930	.87212	.87493	.87774	.88053	.88332
36	.88610	.88888	.89164	.89440	.89715	.89990	.90263	.90536	.90808	.91080
37	.91350	.91620	.91889	.92158	.92426	.92693	.92959	.93225	.93489	.93754
38	.94017	.94280	.94542	.94803	.95064	.95324	.95584	.95842	.96101	.96358
39	.96615	.96871	.97126	.97381	.97635	.97889	.98141	.98394	.98645	.98896
40	5.99146	.99396	.99645	.99894	*.00141	*.00389	*.00635	*.00881	*.01127	*.01372
41	6.01616	.01859	.02102	.02345	.02587	.02828	.03069	.03309	.03548	.03787
42	.04025	.04263	.04501	.04737	.04973	.05209	.05444	.05678	.05912	.06146
43	.06379	.06611	.06843	.07074	.07304	.07535	.07764	.07993	.08222	.08450
44	.08677	.08904	.09131	.09357	.09582	.09807	.10032	.10256	.10479	.10702
45	.10925	.11147	.11368	.11589	.11810	.12030	.12249	.12468	.12687	.12905
46	.13123	.13340	.13556	.13773	.13988	.14204	.14419	.14633	.14847	.15060
47	.15273	.15486	.15698	.15910	.16121	.16331	.16542	.16752	.16961	.17170
48	.17379	.17587	.17794	.18002	.18208	.18415	.18621	.18826	.19032	.19236
49	.19441	.19644	.19848	.20051	.20254	.20456	.20658	.20859	.21060	.21261

NATURAL OR NAPERIAN LOGARITHMS (Continued)

500–999

N	0	1	2	3	4	5	6	7	8	9
50	6.21461	.21661	.21860	.22059	.22258	.22456	.22654	.22851	.23048	.23245
51	.23441	.23637	.23832	.24028	.24222	.24417	.24611	.24804	.24998	.25190
52	.25383	.25575	.25767	.25958	.26149	.26340	.26530	.26720	.26910	.27099
53	.27288	.27476	.27664	.27852	.28040	.28227	.28413	.28600	.28786	.28972
54	.29157	.29342	.29527	.29711	.29895	.30079	.30262	.30445	.30628	.30810
55	.30992	.31173	.31355	.31536	.31716	.31897	.32077	.32257	.32436	.32615
56	.32794	.32972	.33150	.33328	.33505	.33683	.33859	.34036	.34212	.34388
57	.34564	.34739	.34914	.35089	.35263	.35437	.35611	.35784	.35957	.36130
58	.36303	.36475	.36647	.36819	.36990	.37161	.37332	.37502	.37673	.37843
59	.38012	.38182	.38351	.38519	.38688	.38856	.39024	.39192	.39359	.39526
60	6.39693	.39859	.40026	.40192	.40357	.40523	.40688	.40853	.41017	.41182
61	.41346	.41510	.41673	.41836	.41999	.42162	.42325	.42487	.42649	.42811
62	.42972	.43133	.43294	.43455	.43615	.43775	.43935	.44095	.44254	.44413
63	.44572	.44731	.44889	.45047	.45205	.45362	.45520	.45677	.45834	.45990
64	.46147	.46303	.46459	.46614	.46770	.46925	.47080	.47235	.47389	.47543
65	.47697	.47851	.48004	.48158	.48311	.48464	.48616	.48768	.48920	.49072
66	.49224	.49375	.49527	.49677	.49828	.49979	.50129	.50279	.50429	.50578
67	.50728	.50877	.51026	.51175	.51323	.51471	.51619	.51767	.51915	.52062
68	.52209	.52356	.52503	.52649	.52796	.52942	.53088	.53233	.53379	.53524
69	.53669	.53814	.53959	.54103	.54247	.54391	.54535	.54679	.54822	.54965
70	6.55108	.55251	.55393	.55536	.55678	.55820	.55962	.56103	.56244	.56386
71	.56526	.56667	.56808	.56948	.57088	.57228	.57368	.57508	.57647	.57786
72	.57925	.58064	.58203	.58341	.58479	.58617	.58755	.58893	.59030	.59167
73	.59304	.59441	.59578	.59715	.59851	.59987	.60123	.60259	.60394	.60530
74	.60665	.60800	.60935	.61070	.61204	.61338	.61473	.61607	.61740	.61874
75	.62007	.62141	.62274	.62407	.62539	.62672	.62804	.62936	.63068	.63200
76	.63332	.63463	.63595	.63726	.63857	.63988	.64118	.64249	.64379	.64509
77	.64639	.64769	.64898	.65028	.65157	.65286	.65415	.65544	.65673	.65801
78	.65929	.66058	.66185	.66313	.66441	.66568	.66696	.66823	.66950	.67077
79	.67203	.67330	.67456	.67582	.67708	.67834	.67960	.68085	.68211	.68336
80	6.68461	.68586	.68711	.68835	.68960	.69084	.69208	.69332	.69456	.69580
81	.69703	.69827	.69950	.70073	.70196	.70319	.70441	.70564	.70686	.70808
82	.70930	.71052	.71174	.71296	.71417	.71538	.71659	.71780	.71901	.72022
83	.72143	.72263	.72383	.72503	.72623	.72743	.72863	.72982	.73102	.73221
84	.73340	.73459	.73578	.73697	.73815	.73934	.74052	.74170	.74288	.74406
85	.74524	.74641	.74759	.74876	.74993	.75110	.75227	.75344	.75460	.75577
86	.75693	.75809	.75926	.76041	.76157	.76273	.76388	.76504	.76619	.76734
87	.76849	.76964	.77079	.77194	.77308	.77422	.77537	.77651	.77765	.77878
88	.77992	.78106	.78219	.78333	.78446	.78559	.78672	.78784	.78897	.79010
89	.79122	.79234	.79347	.79459	.79571	.79682	.79794	.79906	.80017	.80128
90	6.80239	.80351	.80461	.80572	.80683	.80793	.80904	.81014	.81124	.81235
91	.81344	.81454	.81564	.81674	.81783	.81892	.82002	.82111	.82220	.82329
92	.82437	.82546	.82655	.82763	.82871	.82979	.83087	.83195	.83303	.83411
93	.83518	.83626	.83733	.83841	.83948	.84055	.84162	.84268	.84375	.84482
94	.84588	.84694	.84801	.84907	.85013	.85118	.85224	.85330	.85435	.85541
95	.85646	.85751	.85857	.85961	.86066	.86171	.86276	.86380	.86485	.86589
96	.86693	.86797	.86901	.87005	.87109	.87213	.87316	.87420	.87523	.87626
97	.87730	.87833	.87936	.88038	.88141	.88244	.88346	.88449	.88551	.88653
98	.88755	.88857	.88959	.89061	.89163	.89264	.89366	.89467	.89568	.89669
99	.89770	.89871	.89972	.90073	.90174	.90274	.90375	.90475	.90575	.90675

EXPONENTIAL FUNCTIONS

Values of e^x, $\log e^x$ and e^{-x} where e is the base of the natural system of logarithms 2.71828... and x has values from 0 to 10. Facilitating the solution of exponential equations, these tables also serve as a table of natural or Naperian antilogarithms. For instance, if the logarithm or exponent x = 3.26, the corresponding number or value of e^x is 26.050. Its reciprocal e^{-x} is .038388.

x	e^x	$\text{Log}_{10}(e^x)$	e^{-x}	x	e^x	$\text{Log}_{10}(e^x)$	e^{-x}
0.00	1.0000	0.00000	1.000000	**0.50**	1.6487	0.21715	0.606531
0.01	1.0101	.00434	0.990050	0.51	1.6653	.22149	.600496
0.02	1.0202	.00869	.980199	0.52	1.6820	.22583	.594521
0.03	1.0305	.01303	.970446	0.53	1.6989	.23018	.588605
0.04	1.0408	.01737	.960789	0.54	1.7160	.23452	.582748
0.05	1.0513	0.02171	0.951229	**0.55**	1.7333	0.23886	0.576950
0.06	1.0618	.02606	.941765	0.56	1.7507	.24320	.571209
0.07	1.0725	.03040	.932394	0.57	1.7683	.24755	.565525
0.08	1.0833	.03474	.923116	0.58	1.7860	.25189	.559898
0.09	1.0942	.03909	.913931	0.59	1.8040	.25623	.554327
0.10	1.1052	0.04343	0.904837	**0.60**	1.8221	0.26058	0.548812
0.11	1.1163	.04777	.895834	0.61	1.8404	.26492	.543351
0.12	1.1275	.05212	.886920	0.62	1.8589	.26926	.537944
0.13	1.1388	.05646	.878095	0.63	1.8776	.27361	.532592
0.14	1.1503	.06080	.869358	0.64	1.8965	.27795	.527292
0.15	1.1618	0.06514	0.860708	**0.65**	1.9155	0.28229	0.522046
0.16	1.1735	.06949	.852144	0.66	1.9348	.28663	.516851
0.17	1.1853	.07383	.843665	0.67	1.9542	.29098	.511709
0.18	1.1972	.07817	.835270	0.68	1.9739	.29532	.506617
0.19	1.2092	.08252	.826959	0.69	1.9937	.29966	.501576
0.20	1.2214	0.08080	0.818731	**0.70**	2.0138	0.30401	0.496585
0.21	1.2337	.09120	.810584	0.71	2.0340	.30835	.491644
0.22	1.2461	.09554	.802519	0.72	2.0544	.31269	.486752
0.23	1.2586	.09989	.794534	0.73	2.0751	.31703	.481909
0.24	1.2712	.10423	.786628	0.74	2.0959	.32138	.477114
0.25	1.2840	0.10857	0.778801	**0.75**	2.1170	0.32572	0.472367
0.26	1.2969	.11292	.771052	0.76	2.1383	.33006	.167666
0.27	1.3100	.11726	.763379	0.77	2.1598	.33441	.463013
0.28	1.3231	.12160	.755784	0.78	2.1815	.33875	.458406
0.29	1.3364	.12595	.748264	0.79	2.2034	.34309	.453845
0.30	1.3499	0.13029	0.740818	**0.80**	2.2255	0.34744	0.449329
0.31	1.3634	.13463	.733447	0.81	2.2479	.35178	.444858
0.32	1.3771	.13897	.726149	0.82	2.2705	.35612	.440432
0.33	1.3910	.14332	.718924	0.83	2.2933	.36046	.436049
0.34	1.4049	.14766	.711770	0.84	2.3164	.36481	.431711
0.35	1.4191	0.15200	0.704688	**0.85**	2.3396	0.36915	0.427415
0.36	1.4333	.15635	.697676	0.86	2.3632	.37349	.423162
0.37	1.4477	.16069	.690734	0.87	2.3869	.37784	.418952
0.38	1.4623	.16503	.683861	0.88	2.4109	.38218	.414783
0.39	1.4770	.16937	.677057	0.89	2.4351	.38652	.410656
0.40	1.4918	0.17372	0.670320	**0.90**	2.4596	0.39087	0.406570
0.41	1.5068	.17806	.663650	0.91	2.4843	.39521	.402524
0.42	1.5220	.18240	.657047	0.92	2.5093	.39955	.398519
0.43	1.5373	.18675	.650509	0.93	2.5345	.40389	.394554
0.44	1.5527	.19109	.644036	0.94	2.5600	.40824	.390628
0.45	1.5683	0.19543	0.637628	**0.95**	2.5857	0.41258	0.386741
0.46	1.5841	.19978	.631284	0.96	2.6117	.41692	.382893
0.47	1.6000	.20412	.625002	0.97	2.6379	.42127	.379083
0.48	1.6161	.20846	.618783	0.98	2.6645	.42561	.375311
0.49	1.6323	.21280	.612626	0.99	2.6912	.42995	.371577
0.50	1.6487	0.21715	0.606531	**1.00**	2.7183	0.43429	0.367870

EXPONENTIAL FUNCTIONS (Continued)

x	e^x	$\text{Log}_{10}(e^x)$	e^{-x}	x	e^x	$\text{Log}_{10}(e^x)$	e^{-x}
1.00	2.7183	0.43429	0.367879	**1.50**	4.4817	0.65144	0.223130
1.01	2.7456	.43864	.364219	1.51	4.5267	.65578	.220910
1.02	2.7732	.44298	.360595	1.52	4.5722	.66013	.218712
1.03	2.8011	.44732	.357007	1.53	4.6182	.66447	.216536
1.04	2.8292	.45167	.353455	1.54	4.6646	.66881	.214381
1.05	2.8577	0.45601	0.349938	**1.55**	4.7115	0.67316	0.212248
1.06	2.8864	.46035	.346456	1.56	4.7588	.67750	.210136
1.07	2.9154	.46470	.343009	1.57	4.8066	.68184	.208045
1.08	2.9447	.46904	.339596	1.58	4.8550	.68619	.205975
1.09	2.9743	.47338	.336216	1.59	4.9037	.69053	.203926
1.10	3.0042	0.47772	0.332871	**1.60**	4.9530	0.69487	0.201897
1.11	3.0344	.48207	.329559	1.61	5.0028	.69921	.199888
1.12	3.0649	.48641	.326280	1.62	5.0531	.70356	.197899
1.13	3.0957	.49075	.323033	1.63	5.1039	.70790	.195930
1.14	3.1268	.49510	.319819	1.64	5.1552	.71224	.193980
1.15	3.1582	0.49944	0.316637	**1.65**	5.2070	0.71659	0.192050
1.16	3.1899	.50378	.313486	1.66	5.2593	.72093	.190139
1.17	3.2220	.50812	.310367	1.67	5.3122	.72527	.188247
1.18	3.2544	.51247	.307279	1.68	5.3656	.72961	.186374
1.19	3.2871	.51681	.304221	1.69	5.4195	.73396	.184520
1.20	3.3201	0.52115	0.301194	**1.70**	5.4739	0.73830	0.182684
1.21	3.3535	.52550	.298197	1.71	5.5290	.74264	.180866
1.22	3.3872	.52984	.295230	1.72	5.5845	.74699	.179066
1.23	3.4212	.53418	.292293	1.73	5.6407	.75133	.177284
1.24	3.4556	.53853	.289384	1.74	5.6973	.75567	.175520
1.25	3.4903	0.54287	0.286505	**1.75**	5.7546	0.76002	0.173774
1.26	3.5254	.54721	.283654	1.76	5.8124	.76436	.172045
1.27	3.5609	.55155	.280832	1.77	5.8709	.76870	.170333
1.28	3.5966	.55590	.278037	1.78	5.9299	.77304	.168638
1.29	3.6328	.56024	.275271	1.79	5.9895	.77739	.166960
1.30	3.6693	0.56458	0.272532	**1.80**	6.0496	0.78173	0.165299
1.31	3.7062	.56893	.269820	1.81	6.1104	.78607	.163654
1.32	3.7434	.57327	.267135	1.82	6.1719	.79042	.162026
1.33	3.7810	.57761	.264477	1.83	6.2339	.79476	.160414
1.34	3.8190	.58195	.261846	1.84	6.2965	.79910	.158817
1.35	3.8574	0.58630	0.259240	**1.85**	6.3598	0.80344	0.157237
1.36	3.8962	.59064	.256661	1.86	6.4237	.80779	.155673
1.37	3.9354	.59498	.254107	1.87	6.4883	.81213	.154124
1.38	3.9749	.59933	.251579	1.88	6.5535	.81647	.152590
1.39	4.0149	.60367	.249075	1.89	6.6194	.82082	.151072
1.40	4.0552	0.60801	0.246597	**1.90**	6.6859	0.82516	0.149569
1.41	4.0960	.61236	.244143	1.91	6.7531	.82950	.148080
1.42	4.1371	.61670	.241714	1.92	6.8210	.83385	.146607
1.43	4.1787	.62104	.239309	1.93	6.8895	.83819	.145148
1.44	4.2207	.62538	.236928	1.94	6.9588	.84253	.143704
1.45	4.2631	0.62973	0.234570	**1.95**	7.0287	0.84687	0.142274
1.46	4.3060	.63407	.232236	1.96	7.0993	.85122	.140858
1.47	4.3492	.63841	.229925	1.97	7.1707	.85556	.139457
1.48	4.3929	.64276	.227638	1.98	7.2427	.85990	.138069
1.49	4.4371	.64710	.225373	1.99	7.3155	.86425	.136695
1.50	4.4817	0.65144	0.223130	**2.00**	7.3891	0.86859	0.135335

EXPONENTIAL FUNCTIONS (Continued)

x	e^x	$Log_{10}(e^x)$	e^{-x}	x	e^x	$Log_{10}(e^x)$	e^{-x}
2.00	7.3891	0.86859	0.135335	2.50	12.182	1.08574	0.082085
2.01	7.4633	.87293	.133989	2.51	12.305	1.09008	.081268
2.02	7.5383	.87727	.132655	2.52	12.429	1.09442	.080460
2.03	7.6141	.88162	.131336	2.53	12.554	1.09877	.079659
2.04	7.6906	.88596	.130029	2.54	12.680	1.10311	.078866
2.05	7.7679	0.89030	0.128735	2.55	12.807	1.10745	0.078082
2.06	7.8460	.89465	.127454	2.56	12.936	1.11179	.077305
2.07	7.9248	.89899	.126186	2.57	13.066	1.11614	.076536
2.08	8.0045	.90333	.124930	2.58	13.197	1.12048	.075774
2.09	8.0849	.90768	.123687	2.59	13.330	1.12482	.075020
2.10	8.1662	0.91202	0.122456	2.60	13.464	1.12917	0.074274
2.11	8.2482	.91636	.121238	2.61	13.599	1.13351	.073535
2.12	8.3311	.92070	.120032	2.62	13.736	1.13785	.072803
2.13	8.4149	.92505	.118837	2.63	13.874	1.14219	.072078
2.14	8.4994	.92939	.117655	2.64	14.013	1.14654	.071361
2.15	8.5849	0.93373	0.116484	2.65	14.154	1.15088	0.070651
2.16	8.6711	.93808	.115325	2.66	14.296	1.15522	.069948
2.17	8.7583	.94242	.114178	2.67	14.440	1.15957	.069252
2.18	8.8463	.94676	.113042	2.68	14.585	1.16391	.068563
2.19	8.9352	.95110	.111917	2.69	14.732	1.16825	.067881
2.20	9.0250	0.95545	0.110803	2.70	14.880	1.17260	0.067206
2.21	9.1157	.95979	.109701	2.71	15.029	1.17694	.066537
2.22	9.2073	.96413	.108609	2.72	15.180	1.18128	.065875
2.23	9.2999	.96848	.107528	2.73	15.333	1.18562	.065219
2.24	9.3933	.97282	.106459	2.74	15.487	1.18997	.064570
2.25	9.4877	0.97716	0.105399	2.75	15.643	1.19431	0.063928
2.26	9.5831	.98151	.104350	2.76	15.800	1.19865	.063292
2.27	9.6794	.98585	.103312	2.77	15.959	1.20300	.062662
2.28	9.7767	.99019	.102284	2.78	16.119	1.20734	.062039
2.29	9.8749	.99453	.101266	2.79	16.281	1.21168	.061421
2.30	9.9742	0.99888	0.100259	2.80	16.445	1.21602	0.060810
2.31	10.074	1.00322	.099261	2.81	16.610	1.22037	.060205
2.32	10.176	1.00756	.098274	2.82	16.777	1.22471	.059606
2.33	10.278	1.01191	.097296	2.83	16.945	1.22905	.059013
2.34	10.381	1.01625	.096328	2.84	17.116	1.23340	.058426
2.35	10.486	1.02059	0.095369	2.85	17.288	1.23774	0.057844
2.36	10.591	1.02493	.094420	2.86	17.462	1.24208	.057269
2.37	10.697	1.02928	.093481	2.87	17.637	1.24643	.056699
2.38	10.805	1.03362	.092551	2.88	17.814	1.25077	.056135
2.39	10.913	1.03796	.091630	2.89	17.993	1.25511	.055576
2.40	11.023	1.04231	0.090718	2.90	18.174	1.25945	0.055023
2.41	11.134	1.04665	.089815	2.91	18.357	1.26380	.054476
2.42	11.246	1.05099	.088922	2.92	18.541	1.26814	.053934
2.43	11.359	1.05534	.088037	2.93	18.728	1.27248	.053397
2.44	11.473	1.05968	.087161	2.94	18.916	1.27683	.052866
2.45	11.588	1.06402	0.086294	2.95	19.106	1.28117	0.052340
2.46	11.705	1.06836	.085435	2.96	19.298	1.28551	.051819
2.47	11.822	1.07271	.084585	2.97	19.492	1.28985	.051303
2.48	11.941	1.07705	.083743	2.98	19.688	1.29420	.050793
2.49	12.061	1.08139	.082910	2.99	19.886	1.29854	.050287
2.50	12.182	1.08574	0.082085	3.00	20.086	1.30288	0.049787

EXPONENTIAL FUNCTIONS (Continued)

x	e^x	$Log_{10}(e^x)$	e^{-x}	x	e^x	$Log_{10}(e^x)$	e^{-x}
3.00	20.086	1.30288	0.049787	**3.50**	33.115	1.52003	0.030197
3.01	20.287	1.30723	.049292	3.51	33.448	1.52437	.029897
3.02	20.491	1.31157	.048801	3.52	33.784	1.52872	.029599
3.03	20.697	1.31591	.048316	3.53	34.124	1.53306	.029305
3.04	20.905	1.32026	.047835	3.54	34.467	1.53740	.029013
3.05	21.115	1.32460	0.047359	**3.55**	34.813	1.54175	0.028725
3.06	21.328	1.32894	.046888	3.56	35.163	1.54609	.028439
3.07	21.542	1.33328	.046421	3.57	35.517	1.55043	.028156
3.08	21.758	1.33763	.045959	3.58	35.874	1.55477	.027876
3.09	21.977	1.34197	.045502	3.59	36.234	1.55912	.027598
3.10	22.198	1.34631	0.045049	**3.60**	36.598	1.56346	0.027324
3.11	22.421	1.35066	.044601	3.61	36.966	1.56780	.027052
3.12	22.646	1.35500	.044157	3.62	37.338	1.57215	.026783
3.13	22.874	1.35934	.043718	3.63	37.713	1.57649	.026516
3.14	23.104	1.36368	.043283	3.64	38.092	1.58083	.026252
3.15	23.336	1.36803	0.042852	**3.65**	38.475	1.58517	0.025991
3.16	23.571	1.37237	.042426	3.66	38.861	1.58952	.025733
3.17	23.807	1.37671	.042004	3.67	39.252	1.59386	.025476
3.18	24.047	1.38106	.041586	3.68	39.646	1.59820	.025223
3.19	24.288	1.38540	.041172	3.69	40.045	1.60255	.024972
3.20	24.533	1.38974	0.040762	**3.70**	40.447	1.60689	0.024724
3.21	24.779	1.39409	.040357	3.71	40.854	1.61123	.024478
3.22	25.028	1.39843	.039955	3.72	41.264	1.61558	.024234
3.23	25.280	1.40277	.039557	3.73	41.679	1.61992	.023993
3.24	25.534	1.40711	.039164	3.74	42.098	1.62426	.023754
3.25	25.790	1.41146	0.038774	**3.75**	42.521	1.62860	0.023518
3.26	26.050	1.41580	.038388	3.76	42.948	1.63295	.023284
3.27	26.311	1.42014	.038006	3.77	43.380	1.63729	.023052
3.28	26.576	1.42449	.037628	3.78	43.816	1.64163	.022823
3.29	26.843	1.42883	.037254	3.79	44.256	1.64598	.022596
3.30	27.113	1.43317	0.036883	**3.80**	44.701	1.65032	0.022371
3.31	27.385	1.43751	.036516	3.81	45.150	1.65466	.022148
3.32	27.660	1.44186	.036153	3.82	45.604	1.65900	.021928
3.33	27.938	1.44620	.035793	3.83	46.063	1.66335	.021710
3.34	28.219	1.45054	.035437	3.84	46.525	1.66769	.021494
3.35	28.503	1.45489	0.035084	**3.85**	46.993	1.67203	0.021280
3.36	28.789	1.45923	.034735	3.86	47.465	1.67638	.021068
3.37	29.079	1.46357	.034390	3.87	47.942	1.68072	.020858
3.38	29.371	1.46792	.034047	3.88	48.424	1.68506	.020651
3.39	29.666	1.47226	.033709	3.89	48.911	1.68941	.020445
3.40	29.964	1.47660	0.033373	**3.90**	49.402	1.69375	0.020242
3.41	30.265	1.48094	.033041	3.91	49.899	1.69809	.020041
3.42	30.569	1.48529	.032712	3.92	50.400	1.70243	.019841
3.43	30.877	1.48963	.032387	3.93	50.907	1.70678	.019644
3.44	31.187	1.49397	.032065	3.94	51.419	1.71112	.019448
3.45	31.500	1.49832	0.031746	**3.95**	51.935	1.71546	0.019255
3.46	31.817	1.50266	.031430	3.96	52.457	1.71981	.019063
3.47	32.137	1.50700	.031117	3.97	52.985	1.72415	.018873
3.48	32.460	1.51134	.030807	3.98	53.517	1.72849	.018686
3.49	32.786	1.51569	.030501	3.99	54.055	1.73283	.018500
3.50	33.115	1.52003	0.030197	**4.00**	54.598	1.73718	0.018316

EXPONENTIAL FUNCTIONS (Continued)

x	e^x	$Log_{10}(e^x)$	e^{-x}	x	e^x	$Log_{10}(e^x)$	e^{-x}
4.00	54.598	1.73718	0.018316	**4.50**	90.017	1.95433	0.011109
4.01	55.147	1.74152	.018133	4.51	90.922	1.95867	.010998
4.02	55.701	1.74586	.017953	4.52	91.836	1.96301	.010889
4.03	56.261	1.75021	.017774	4.53	92.759	1.96735	.010781
4.04	56.826	1.75455	.017597	4.54	93.691	1.97170	.010673
4.05	57.397	1.75889	0.017422	**4.55**	94.632	1.97604	0.010567
4.06	57.974	1.76324	.017249	4.56	95.583	1.98038	.010462
4.07	58.557	1.76758	.017077	4.57	96.544	1.98473	.010358
4.08	59.145	1.77192	.016907	4.58	97.514	1.98907	.010255
4.09	59.740	1.77626	.016739	4.59	98.494	1.99341	.010153
4.10	60.340	1.78061	0.016573	**4.60**	99.484	1.99775	0.010052
4.11	60.947	1.78495	.016408	4.61	100.48	2.00210	.009952
4.12	61.559	1.78929	.016245	4.62	101.49	2.00644	.009853
4.13	62.178	1.79364	.016083	4.63	102.51	2.01078	.009755
4.14	62.803	1.79798	.015923	4.64	103.54	2.01513	.009658
4.15	63.434	1.80232	0.015764	**4.65**	104.58	2.01947	0.009562
4.16	64.072	1.80667	.015608	4.66	105.64	2.02381	.009466
4.17	64.715	1.81101	.015452	4.67	106.70	2.02816	.009372
4.18	65.366	1.81535	.015299	4.68	107.77	2.03250	.009279
4.19	66.023	1.81969	.015146	4.69	108.85	2.03684	.009187
4.20	66.686	1.82404	0.014996	**4.70**	109.95	2.04118	0.009095
4.21	67.357	1.82838	.014846	4.71	111.05	2.04553	.009005
4.22	68.033	1.83272	.014699	4.72	112.17	2.04987	.008915
4.23	68.717	1.83707	.014552	4.73	113.30	2.05421	.008826
4.24	69.408	1.84141	.014408	4.74	114.43	2.05856	.008739
4.25	70.105	1.84575	0.014264	**4.75**	115.58	2.06290	0.008652
4.26	70.810	1.85009	.014122	4.76	116.75	2.06724	.008566
4.27	71.522	1.85444	.013982	4.77	117.92	2.07158	.008480
4.28	72.240	1.85878	.013843	4.78	119.10	2.07593	.008396
4.29	72.966	1.86312	.013705	4.79	120.30	2.08027	.008312
4.30	73.700	1.86747	0.013569	**4.80**	121.51	2.08461	0.008230
4.31	74.440	1.87181	.013434	4.81	122.73	2.08896	.008148
4.32	75.189	1.87615	.013300	4.82	123.97	2.09330	.008067
4.33	75.944	1.88050	.013168	4.83	125.21	2.09764	.007987
4.34	76.708	1.88484	.013037	4.84	126.47	2.10199	.007907
4.35	77.478	1.88918	0.012907	**4.85**	127.74	2.10633	0.007828
4.36	78.257	1.89352	.012778	4.86	129.02	2.11067	.007750
4.37	79.044	1.89787	.012651	4.87	130.32	2.11501	.007673
4.38	79.838	1.90221	.012525	4.88	131.63	2.11936	.007597
4.39	80.640	1.90655	.012401	4.89	132.95	2.12370	.007521
4.40	81.451	1.91090	0.012277	**4.90**	134.29	2.12804	0.007447
4.41	82.269	1.91524	.012155	4.91	135.64	2.13239	.007372
4.42	83.096	1.91958	.012034	4.92	137.00	2.13673	.007299
4.43	83.931	1.92392	.011914	4.93	138.38	2.14107	.007227
4.44	84.775	1.92827	.011796	4.94	139.77	2.14541	.007155
4.45	85.627	1.93261	0.011679	**4.95**	141.17	2.14976	0.007083
4.46	86.488	1.93695	.011562	4.96	142.59	2.15410	.007013
4.47	87.357	1.94130	.011447	4.97	144.03	2.15844	.006943
4.48	88.235	1.94564	.011333	4.98	145.47	2.16279	.006874
4.49	89.121	1.94998	.011221	4.99	146.94	2.16713	.006806
4.50	90.017	1.95433	0.011109	**5.00**	148.41	2.17147	0.006738

EXPONENTIAL FUNCTIONS (Continued)

x	e^x	$\text{Log}_{10}(e^x)$	e^{-x}	x	e^x	$\text{Log}_{10}(e^x)$	e^{-x}
5.00	148.41	2.17147	0.006738	**5.50**	244.69	2.38862	0.0040868
5.01	149.90	2.17582	.006671	5.55	257.24	2.41033	.0038875
5.02	151.41	2.18016	.006605	5.60	270.43	2.43205	.0036979
5.03	152.93	2.18450	.006539	5.65	284.29	2.45376	.0035175
5.04	154.47	2.18884	.006474	5.70	298.87	2.47548	.0033460
5.05	156.02	2.19319	0.006409	**5.75**	314.19	2.49719	0.0031828
5.06	157.59	2.19753	.006346	5.80	330.30	2.51891	.0030276
5.07	159.17	2.20187	.006282	5.85	347.23	2.54062	.0028799
5.08	160.77	2.20622	.006220	5.90	365.04	2.56234	.0027394
5.09	162.39	2.21056	.006158	5.95	383.75	2.58405	.0026058
5.10	164.02	2.21490	0.006097	**6.00**	403.43	2.60577	0.0024788
5.11	165.67	2.21924	.006036	6.05	424.11	2.62748	.0023579
5.12	167.34	2.22359	.005976	6.10	445.86	2.64920	.0022429
5.13	169.02	2.22793	.005917	6.15	468.72	2.67091	.0021335
5.14	170.72	2.23227	.005858	6.20	492.75	2.69263	.0020294
5.15	172.43	2.23662	0.005799	**6.25**	518.01	2.71434	0.0019305
5.16	174.16	2.24096	.005742	6.30	544.57	2.73606	.0018363
5.17	175.91	2.24530	.005685	6.35	572.49	2.75777	.0017467
5.18	177.68	2.24965	.005628	6.40	601.85	2.77948	.0016616
5.19	179.47	2.25399	.005572	6.45	632.70	2.80120	.0015805
5.20	181.27	2.25833	0.005517	**6.50**	665.14	2.82291	0.0015034
5.21	183.09	2.26267	.005462	6.55	699.24	2.84463	.0014301
5.22	184.93	2.26702	.005407	6.60	735.10	2.86634	.0013604
5.23	186.79	2.27136	.905354	6.65	772.78	2.88806	.0012940
5.24	188.67	2.27570	.005300	6.70	812.41	2.90977	.0012309
5.25	190.57	2.28005	0.005248	**6.75**	854.06	2.93149	0.0011709
5.26	192.48	2.28439	.005195	6.80	897.85	2.95320	.9011138
5.27	194.42	2.28873	.005144	6.85	943.88	2.97492	.0010595
5.28	196.37	2.29307	.005092	6.90	992.27	2.99663	.0010078
5.29	198.34	2.29742	.005042	6.95	1043.1	3.01835	.0009586
5.30	200.34	2.30176	0.004992	**7.00**	1096.6	3.04006	0.0009119
5.31	202.35	2.30610	.004942	7.05	1152.9	3.06178	.0008674
5.32	204.38	2.31045	.004893	7.10	1212.0	3.08349	.0008251
5.33	206.44	2.31479	.004844	7.15	1274.1	3.10521	.0007849
5.34	208.51	2.31913	.004796	7.20	1339.4	3.12692	.0007466
5.35	210.61	2.32348	0.004748	**7.25**	1408.1	3.14863	0.0007102
5.36	212.72	2.32782	.004701	7.30	1480.3	3.17035	.0006755
5.37	214.86	2.33216	.004654	7.35	1556.2	3.19206	.0006426
5.38	217.02	2.33650	.004608	7.40	1636.0	3.21378	.0006113
5.39	219.20	2.34085	.004562	7.45	1719.9	3.23549	.0005814
5.40	221.41	2.34519	0.004517	**7.50**	1808.0	3.25721	0.0005531
5.41	223.63	2.34953	.004472	7.55	1900.7	3.27892	.0005261
5.42	225.88	2.35388	.004427	7.60	1998.2	3.30064	.0005005
5.43	228.15	2.35822	.004383	7.65	2100.6	3.32235	.0004760
5.44	230.44	2.36256	.004339	7.70	2208.3	3.34407	.0004528
5.45	232.76	2.36690	0.004296	**7.75**	2321.6	3.36578	0.0004307
5.46	235.10	2.37125	.004254	7.80	2440.6	3.38750	.0004097
5.47	237.46	2.37559	.004211	7.85	2565.7	3.40921	.0003898
5.48	239.85	2.37993	.004169	7.90	2697.3	3.43093	.0003707
5.49	242.26	2.38428	.004128	7.95	2835.6	3.45264	.0003527
5.50	244.69	2.38862	0.004087	**8.00**	2981.0	3.47436	0.0003355

EXPONENTIAL FUNCTIONS (Continued)

x	e^x	$\text{Log}_{10}(e^x)$	e^{-x}
8.00	2981.0	3.47436	0.0003355
8.05	3133.8	3.49607	.0003191
8.10	3294.5	3.51779	.0003035
8.15	3463.4	3.53950	.0002887
8.20	3641.0	3.56121	.0002747
8.25	3827.6	3.58293	0.0002613
8.30	4023.9	3.60464	.0002485
8.35	4230.2	3.62636	.0002364
8.40	4447.1	3.64807	.0002249
8.45	4675.1	3.66979	.0002139
8.50	4914.8	3.69150	0.0002035
8.55	5166.8	3.71322	.0001935
8.60	5431.7	3.73493	.0001841
8.65	5710.1	3.75665	.0001751
8.70	6002.9	3.77836	.0001666
8.75	6310.7	3.80008	0.0001585
8.80	6634.2	3.82179	.0001507
8.85	6974.4	3.84351	.0001434
8.90	7332.0	3.86522	.0001364
8.95	7707.9	3.88694	.0001297
9.00	8103.1	3.90865	0.0001234
9.05	8518.5	3.93037	.0001174
9.10	8955.3	3.95208	.0001117
9.15	9414.4	3.97379	.0001062
9.20	9897.1	3.99551	.0001010
9.25	10405	4.01722	0.0000961
9.30	10938	4.03894	.0000914
9.35	11499	4.06065	.0000870
9.40	12088	4.08237	.0000827
9.45	12708	4.10408	.0000787
9.50	13360	4.12580	0.0000749
9.55	14045	4.14751	.0000712
9.60	14765	4.16923	.0000677
9.65	15522	4.19094	.0000644
9.70	16318	4.21266	.0000613
9.75	17154	4.23437	0.0000583
9.80	18034	4.25609	.0000555
9.85	18958	4.27780	.0000527
9.90	19930	4.29952	.0000502
9.95	20952	4.32123	0.0000477
10.00	22026	4.34294	0.0000454

HYPERBOLIC AND RELATED FUNCTIONS

Dr. Madhu S. Gupta

HYPERBOLIC FUNCTIONS

Geometrical Definition

Let O be the center, A the vertex, and P any point with coordinates (x,y) on the branch B′AB of the rectangular hyperbola $X^2 - Y^2 = a^2$. Set OM = x, MP = y and OA = a. The shaded area shown in the figure is given by

$$\text{Area OPAP}' = a^2 \log_e \frac{(x + y)}{a}$$

If the angle POP′ in hyperbolic radians is denoted by u,

$$u = \frac{\text{area OPAP}'}{a^2} \text{ hyperbolic radians.}$$

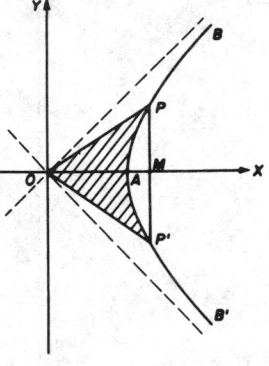

The hyperbolic functions are defined by

hyperbolic sine of u = sinh u = y/a
hyperbolic cosine of u = cosh u - x/a

The approximate length of the hyperbolic curve is given by

$$\text{arc AP} = \frac{3}{2} y - \frac{1}{2} \tan^{-1} y$$

while the straight line distance is

$$\text{line AP} = \sqrt{\sinh^2 u + (\cosh u - 1)^2}$$

Exponential Definition

hyperbolic sine of u = sinh u = $1/2(e^u - e^{-u})$

hyperbolic cosine of u = cosh u = $1/2(e^u + e^{-u})$

hyperbolic tangent of u = tanh u = $\dfrac{\sinh u}{\cosh u}$ = $\dfrac{e^u - e^{-u}}{e^u + e^{-u}}$

$$\text{csch } u = \frac{1}{\sinh u},$$

$$\text{sech } u = \frac{1}{\cosh u},$$

$$\text{coth } u = \frac{1}{\tanh u},$$

Domain and Range for Real Argument

Function	Domain (interval of u)	Range (interval of function)	Remarks
sinh u	$(-\infty, +\infty)$	$(-\infty, +\infty)$	
cosh u	$(-\infty, +\infty)$	$[1, +\infty)$	
tanh u	$(-\infty, +\infty)$	$(-1, +1)$	
cosech u	$(-\infty, 0)$	$(0, -\infty)$	Two branches, pole
	$(0, +\infty)$	$(+\infty, 0)$	at $u = 0$.
sech u	$(-\infty, +\infty)$	$(0, 1]$	
coth u	$(-\infty, 0)$	$(-1, -\infty)$	Two branches, pole
	$(0, +\infty)$	$(+\infty, 1)$	at $u = 0$

Hyperbolic Functions In Terms of One Another

Function	$\sinh x$	$\cosh x$	$\tanh x$
$\sinh x =$	$\sinh x$	$\pm \sqrt{\cosh^2 x - 1}$	$\dfrac{\tanh x}{\sqrt{1 - \tanh^2 x}}$
$\cosh x =$	$\sqrt{1 + \sinh^2 x}$	$\cosh x$	$\dfrac{1}{\sqrt{1 - \tanh^2 x}}$
$\tanh x =$	$\dfrac{\sinh x}{\sqrt{1 + \sinh^2 x}}$	$\pm \dfrac{\sqrt{\cosh^2 x - 1}}{\cosh x}$	$\tanh x$
$\operatorname{cosech} x =$	$\dfrac{1}{\sinh x}$	$\pm \dfrac{1}{\sqrt{\cosh^2 x - 1}}$	$\dfrac{\sqrt{1 - \tanh^2 x}}{\tanh x}$
$\operatorname{sech} x =$	$\dfrac{1}{\sqrt{1 + \sinh^2 x}}$	$\dfrac{1}{\cosh x}$	$\sqrt{1 - \tanh^2 x}$
$\coth x =$	$\dfrac{\sqrt{1 + \sinh^2 x}}{\sinh x}$	$\pm \dfrac{\cosh x}{\sqrt{\cosh^2 x - 1}}$	$\dfrac{1}{\tanh x}$

Function	$\operatorname{cosech} x$	$\operatorname{sech} x$	$\coth x$
$\sinh x =$	$\dfrac{1}{\operatorname{cosech} x}$	$\pm \dfrac{\sqrt{1 - \operatorname{sech}^2 x}}{\operatorname{sech} x}$	$\dfrac{\pm 1}{\sqrt{\coth^2 x - 1}}$
$\cosh x =$	$\pm \dfrac{\sqrt{\operatorname{cosech}^2 x + 1}}{\operatorname{cosech} x}$	$\dfrac{1}{\operatorname{sech} x}$	$\pm \dfrac{\coth x}{\sqrt{\coth^2 x - 1}}$
$\tanh x =$	$\dfrac{1}{\sqrt{\operatorname{cosech}^2 x + 1}}$	$\pm \sqrt{1 - \operatorname{sech}^2 x}$	$\dfrac{1}{\coth x}$
$\operatorname{cosech} x =$	$\operatorname{cosech} x$	$\pm \dfrac{\operatorname{sech} x}{\sqrt{1 - \operatorname{sech}^2 x}}$	$\pm \dfrac{\sqrt{\coth^2 x - 1}}{1}$
$\operatorname{sech} x =$	$\pm \dfrac{\operatorname{cosec} x}{\sqrt{\operatorname{cosech}^2 x + 1}}$	$\operatorname{sech} x$	$\pm \dfrac{\sqrt{\coth^2 x - 1}}{\coth x}$
$\coth x =$	$\sqrt{\operatorname{cosech}^2 x + 1}$	$\pm \dfrac{1}{\sqrt{1 - \operatorname{sech}^2 x}}$	$\coth x$

Whenever two signs are shown, choose $+$ sign if x is positive, $-$ sign if x is negative.

Special Values of Hyperbolic Functions

x	0	$\dfrac{\pi}{2} i$	πi	$\dfrac{3\pi}{2} i$	∞
$\sinh x$	0	i	0	$-i$	∞
$\cosh x$	1	0	-1	0	∞
$\tanh x$	0	∞i	0	$-\infty i$	1
$\operatorname{csch} x$	∞	$-i$	∞	i	0
$\operatorname{sech} x$	1	∞	-1	∞	0
$\coth x$	∞	0	∞	0	1

Symmetry and Periodicity

$$\sinh(-u) = -\sinh u, \qquad \operatorname{csch}(-u) = -\operatorname{csch} u$$
$$\cosh(-u) = \cosh u, \qquad \operatorname{sech}(-u) = \operatorname{sech} u$$
$$\tanh(-u) = -\tanh u, \qquad \coth(-u) = -\coth u$$

$\sinh u$, $\cosh u$, $\operatorname{cosech} u$ and $\operatorname{sech} u$ are periodic with a period $2\pi i$; $\tanh u$ and $\coth u$ are periodic with a period πi.

Fundamental Identities

Reciprocal Relations

$$\operatorname{csch} u = \frac{1}{\sinh u}, \qquad \operatorname{sech} u = \frac{1}{\cosh u}, \qquad \coth u = \frac{1}{\tanh u}$$

Product Relations

$$\sinh u = \tanh u \cosh u \qquad\qquad \cosh u = \coth u \sinh u$$
$$\tanh u = \sinh u \operatorname{sech} u \qquad\qquad \coth u = \cosh u \operatorname{cosech} u$$
$$\operatorname{sech} u = \operatorname{cosech} u \tanh u \qquad\qquad \operatorname{cosech} u = \operatorname{sech} u \coth u$$

Quotient Relations

$$\sinh u = \frac{\tanh u}{\operatorname{sech} u} \qquad \cosh u = \frac{\coth u}{\operatorname{cosech} u} \qquad \tanh u = \frac{\sinh u}{\cosh u}$$

$$\operatorname{cosech} u = \frac{\operatorname{sech} u}{\tanh u} \qquad \operatorname{sech} u = \frac{\operatorname{cosech} u}{\coth u} \qquad \coth u = \frac{\cosh u}{\sinh u}$$

Relations Between Squares of Functions

$$\cosh^2 u - \sinh^2 u = 1, \qquad \tanh^2 u + \operatorname{sech}^2 u = 1$$
$$\coth^2 u - \operatorname{csch}^2 u = 1, \qquad \operatorname{csch}^2 u - \operatorname{sech}^2 u = \operatorname{csch}^2 u \operatorname{sech}^2 u$$

Angle-Sum and Angle-Difference Relations

$$\sinh (u + v) = \sinh u \cosh v + \cosh u \sinh v$$
$$\sinh (u - v) = \sinh u \cosh v - \cosh u \sinh v$$
$$\cosh (u + v) = \cosh u \cosh v + \sinh u \sinh v$$
$$\cosh (u - v) = \cosh u \cosh v - \sinh u \sinh v$$

$$\tanh (u + v) = \frac{\tanh u + \tanh v}{1 + \tanh u \tanh v} = \frac{\sinh 2u + \sinh 2v}{\cosh 2u + \cosh 2v}$$

$$\tanh (u - v) = \frac{\tanh u - \tanh v}{1 - \tanh u \tanh v} = \frac{\sinh 2u - \sinh 2v}{\cosh 2u - \cosh 2v}$$

$$\coth (u + v) = \frac{1 + \coth u \coth v}{\coth u + \coth v} = \frac{\sinh 2u - \sinh 2v}{\cosh 2u - \cosh 2v}$$

$$\coth (u - v) = \frac{1 - \coth u \coth v}{\coth u - \coth v} = \frac{\sinh 2u + \sinh 2v}{\cosh 2u - \cosh 2v}$$

Multiple Angle Relations

$$\sinh 2u = 2 \sinh u \cosh u = \frac{2 \tanh u}{1 - \tanh^2 u}$$

$$\cosh 2u = \cosh^2 u + \sinh^2 u = 2 \cosh^2 u - 1 = 1 + 2 \sinh^2 u = \frac{1 + \tanh^2 u}{1 - \tanh^2 u}$$

$$\tanh 2u = \frac{2 \tanh u}{1 + \tanh^2 u}$$

$$\coth 2u = \frac{\coth^2 u + 1}{2 \coth u}$$

$$\sinh 3u = 3 \sinh u + 4 \sinh^3 u = \sinh u \, (4 \cosh^2 u - 1)$$
$$\cosh 3u = 4 \cosh^3 u - 3 \cosh u = \cosh u \, (1 + 4 \sinh^2 u)$$

$$\tanh 3u = \frac{3 \tanh u + \tanh^3 u}{1 + 3 \tanh^2 u}$$

$$\coth 3u = \frac{3 \coth u + \coth^3 u}{1 + 3 \coth^2 u}$$

$$\sinh 4u = 4 \sinh u \cosh u (2 \cosh^2 u - 1)$$
$$= 4 \sinh u \cosh u (1 + 2 \sinh^2 u)$$
$$= 4 \sinh u \cosh u (\cosh^2 u + \sinh^2 u)$$

$$\cosh 4u = 1 + 8 \cosh^2 u (\cosh^2 u - 1)$$
$$= 1 + 8 \sinh^2 u (\sinh^2 u + 1)$$
$$= \cosh^4 u + 6 \sinh^2 u \cosh^2 u + \sinh^4 u$$

$$\tanh 4u = \frac{4 \tanh u (1 + \tanh^2 u)}{1 + 6 \tanh^2 u + \tanh^4 u}$$

$$\coth 4u = \frac{\coth^4 u + 6 \coth^2 u + 1}{4 \coth u (\coth^2 u + 1)}$$

$$\sinh 5u = \sinh u (16 \sinh^4 u + 20 \sinh^2 u + 5)$$
$$= \sinh u (16 \cosh^4 u - 12 \cosh^2 u + 1)$$

$$\cosh 5u = \cosh u (16 \cosh^4 u - 20 \cosh^2 u + 5)$$
$$= \cosh u (16 \sinh^4 u + 12 \sinh^2 u + 1)$$

$$\sinh 6u = 2 \sinh u \cosh u (16 \cosh^4 u - 16 \cosh^2 u + 3)$$
$$= 2 \sinh u \cosh u (16 \sinh^4 u + 16 \sinh^2 u + 3)$$

$$\cosh 6u = 32 \cosh^6 u - 48 \cosh^4 u + 18 \cosh^2 u - 1$$
$$= 32 \sinh^6 u + 48 \sinh^4 u + 18 \sinh^2 u + 1$$

$$\sinh nu = \sinh u \left[(2 \cosh u)^{n-1} - \frac{(n-2)}{1!} \cdot (2 \cosh u)^{n-3} + \frac{(n-3)(n-4)}{2!} \right.$$
$$\left. \cdot (2 \cosh u)^{n-5} - \frac{(n-4)(n-5)(n-6)}{3!} (2 \cosh u)^{n-7} + \cdots \right].$$

$$\cosh nu = \frac{1}{2} \left[(2 \cosh u)^n - \frac{n}{1!} (2 \cosh u)^{n-2} + \frac{n(n-3)}{2!} (2 \cosh u)^{n-4} \right.$$
$$\left. - \frac{n(n-4)(n-5)}{3!} (2 \cosh u)^{n-6} + \cdots \right].$$

Half Angle Relations

$$\sinh \tfrac{1}{2}u = \pm \sqrt{\tfrac{1}{2}(\cosh u - 1)}$$
$$\cosh \tfrac{1}{2}u = \sqrt{\tfrac{1}{2}(\cosh u + 1)}$$

$$\tanh \frac{u}{2} = \frac{\sinh u}{1 + \cosh u} = \frac{\cosh u - 1}{\sinh u} = \pm \sqrt{\frac{\cosh u - 1}{\cosh u + 1}}$$

$$\coth \frac{u}{2} = \frac{1 + \cosh u}{\sinh u} = \frac{\sinh u}{\cosh u - 1} = \pm \sqrt{\frac{\cosh u + 1}{\cosh u - 1}}$$

Choose $+$ sign if u is positive, otherwise choose the $-$ sign.

Function Sum and Function Difference Relations

$$\sinh u + \sinh v = 2 \sinh \tfrac{1}{2}(u + v) \cosh \tfrac{1}{2}(u - v)$$
$$\sinh u - \sinh v = 2 \cosh \tfrac{1}{2}(u + v) \sinh \tfrac{1}{2}(u - v)$$
$$\cosh u + \cosh v = 2 \cosh \tfrac{1}{2}(u + v) \cosh \tfrac{1}{2}(u - v)$$
$$\cosh u - \cosh v = 2 \sinh \tfrac{1}{2}(u + v) \sinh \tfrac{1}{2}(u - v)$$

$$\tanh u + \tanh v = (1 + \tanh u \tanh v) \tanh (u + v) = \frac{\sinh (u + v)}{\cosh u \cosh v}$$

$$\tanh u - \tanh v = (1 - \tanh u \tanh v) \tanh (u - v) = \frac{\sinh (u - v)}{\cosh u \cosh v}$$

$$\coth u + \coth v = \frac{1 + \coth u \coth v}{\coth (u + v)} = \frac{\sinh (u + v)}{\sinh u \sinh v}$$

$$\coth u - \coth v = \frac{1 - \coth u \coth v}{\coth (u - v)} = \frac{\sinh (u - v)}{\sinh u \sinh v}$$

$$\sinh u + \cosh u = \frac{1 + \tanh \tfrac{1}{2}u}{1 - \tanh \tfrac{1}{2}u} = e^{u} \qquad \cosh u - \sinh u = \frac{1 - \tanh \tfrac{1}{2}u}{1 + \tanh \tfrac{1}{2}u} = e^{-u}$$

Function Product Relations

$$\sinh u \cosh v = \tfrac{1}{2} \sinh (u + v) + \tfrac{1}{2} \sinh (u - v)$$
$$\cosh u \sinh v = \tfrac{1}{2} \sinh (u + v) - \tfrac{1}{2} \sinh (u - v)$$
$$\cosh u \cosh v = \tfrac{1}{2} \cosh (u + v) + \tfrac{1}{2} \cosh (u - v)$$
$$\sinh u \sinh v = \tfrac{1}{2} \cosh (u + v) - \tfrac{1}{2} \cosh (u - v)$$
$$\sinh (u + v) \sinh (u - v) = \sinh^2 u - \sinh^2 v = \cosh^2 u - \cosh^2 v$$
$$\cosh (u + v) \cosh (u - v) = \sinh^2 u + \cosh^2 v = \cosh^2 u + \sinh^2 v$$

Power Relations

$$\sinh^2 u = \tfrac{1}{2}(\cosh 2u - 1)$$
$$\cosh^2 u = \tfrac{1}{2}(\cosh 2u + 1)$$
$$\sinh^3 u = \tfrac{1}{4}(-3 \sinh u + \sinh 3u)$$
$$\cosh^3 u = \tfrac{1}{4}(3 \cosh u + \cosh 3u)$$
$$\sinh^4 u = \tfrac{1}{8}(3 - 4 \cosh 2u + \cosh 4u)$$
$$\cosh^4 u = \tfrac{1}{8}(3 + 4 \cosh 2u + \cosh 4u)$$
$$\sinh^5 u = \tfrac{1}{16}(10 \sinh u - 5 \sinh 3u + \sinh 5u)$$
$$\cosh^5 u = \tfrac{1}{16}(10 \cosh u + 5 \cosh 3u + \cosh 5u)$$
$$\sinh^6 u = \tfrac{1}{32}(-10 + 15 \cosh 2u - 6 \cosh 4u + \cosh 6u)$$
$$\cosh^6 u = \tfrac{1}{32}(10 + 15 \cosh 2u + 6 \cosh 4u + \cosh 6u)$$
$$(\cosh u \pm \sinh u)^n = \cosh nu \pm \sinh nu$$

Relations with Circular Functions

$$\begin{array}{ll}
\sinh iu = i \sin u, & \sinh u = -i \sin iu \\
\cosh iu = \cos u, & \cosh u = \cos iu \\
\tanh iu = i \tan u, & \tanh u = -i \tan iu \\
\operatorname{cosech} iu = -i \operatorname{cosec} u & \operatorname{cosech} u = i \operatorname{cosec} iu \\
\operatorname{sech} iu = \sec u & \operatorname{sech} u = \sec iu \\
\coth iu = -i \cot u & \coth u = i \coth iu
\end{array}$$

Hyperbolic Functions of Complex Argument

$$\sinh (u + iv) = \sinh u \cos v + i \cosh u \sin v$$
$$\sinh (u - iv) = \sinh u \cos v - i \cosh u \sin v$$
$$\cosh (u + iv) = \cosh u \cos v + i \sinh u \sin v$$
$$\cosh (u - iv) = \cosh u \cos v - i \sinh u \sin v$$

$$\tanh (u + iv) = \frac{\sinh 2u + i \sin 2v}{\cosh 2u + \cos 2v}$$

$$\tanh (u - iv) = \frac{\sinh 2u - i \sin 2v}{\cosh 2u + \cos 2v}$$

$$\coth (u + iv) = \frac{\sinh 2u - i \sin 2v}{\cosh 2u - \cos 2v}$$

$$\coth (u - iv) = \frac{\sinh 2u + i \sin 2v}{\cosh 2u - \cos 2v}$$

$$\sinh (u + \tfrac{1}{2}\pi i) = i \cosh u, \qquad \cosh (u + \tfrac{1}{2}\pi i) = i \sinh u$$
$$\sinh (u + \pi i) = -\sinh u, \qquad \cosh (u + \pi i) = -\cosh u$$
$$\sinh (u + 2\pi i) = \sinh u, \qquad \cosh (u + 2\pi i) = \cosh u$$

Series for Hyperbolic Functions

(see series expansions for $\sinh nu$ and $\cosh nu$ under multiple angle relations).

$$\sinh x = x + \frac{x^3}{3!} + \frac{x^5}{5!} + \frac{x^7}{7!} + \cdots + \frac{x^{2n+1}}{(2n+1)!} + \cdots \qquad |x| < \infty$$

$$\sinh ax = \frac{2}{\pi} \sinh \pi a \left[\frac{\sin x}{a^2 + 1^2} - \frac{2 \sin 2x}{a^2 + 2^2} + \frac{3 \sin 3x}{a^2 + 3^2} - + \cdots \right]$$

$$= \frac{2}{\pi} \sinh \pi a \sum_{n=1}^{\infty} (-1)^{n+1} \frac{n \sin nx}{n^2 + a^2}, \qquad |x| < \pi$$

$$\cosh x = 1 + \frac{x^2}{2!} + \frac{x^4}{4!} + \frac{x^6}{6!} + \cdots + \frac{x^{2n}}{(2n)!} + \cdots \qquad |x| < \infty$$

$$\cosh ax = \frac{2a}{\pi} \sinh \pi a \left[\frac{1}{2a^2} - \frac{\cos x}{a^2 + 1^2} + \frac{\cos 2x}{a^2 + 2^2} - \frac{\cos 3x}{a^2 + 3^2} + - \cdots \right]$$

$$= \frac{\sinh \pi a}{a\pi} + \frac{2a}{\pi} \sinh \pi a \sum_{n=1}^{\infty} (-1)^n \frac{\cos nx}{n^2 + a^2}, \qquad |x| < \pi$$

$$\tanh x = x - \frac{1}{3} x^3 + \frac{2}{15} x^5 - \frac{17}{315} x^7 + \frac{62}{2835} x^9 - \cdots$$
$$+ \frac{2^{2n}(2^{2n} - 1) B_{2n} x^{2n-1}}{(2n)!} \pm \cdots \quad \text{(1)} \qquad |x| < \frac{\pi}{2}$$

$$\tanh x = 1 - 2e^{-2x} + 2e^{-4x} - 2e^{-6x} + - \cdots$$

$$= 1 + 2 \sum_{n=1}^{\infty} (-1)^n e^{-2nx}, \qquad \text{Re }(x) > 0$$

$$\tanh x = 2x \left[\frac{1}{\left(\frac{\pi}{2}\right)^2 + x^2} + \frac{1}{\left(\frac{3\pi}{2}\right)^2 + x^2} + \frac{1}{\left(\frac{5\pi}{2}\right)^2 + x^2} + \cdots \right]$$

$$= 2x \sum_{n=0}^{\infty} \frac{1}{\left(n + \frac{1}{2}\right)^2 \pi^2 + x^2}$$

$$\coth x = \frac{1}{x} + \frac{x}{3} - \frac{x^3}{45} + \frac{2x^5}{945} - \frac{x^7}{4725} + \cdots + \frac{2^{2n}}{(2n)!} B_{2n} x^{2n-1} \pm \cdots \quad \text{(1)}$$
$$0 < |x| < \pi$$

$$\coth x = 1 + 2e^{-2x} + 2e^{-4x} + 2e^{-6x} + \cdots$$

$$= 1 + 2 \sum_{n=1}^{\infty} e^{-2nx} \qquad \text{Re }(x) > 0$$

$$\coth x = \frac{1}{x} + 2x \left[\frac{1}{\pi^2 + x^2} + \frac{1}{(2\pi)^2 + x^2} + \frac{1}{(3\pi)^2 + x^2} + \cdots \right]$$

$$= \frac{1}{x} + 2x \sum_{n=1}^{\infty} \frac{1}{(n\pi)^2 + x^2}$$

(1) B_{2n} denotes Bernoulli numbers.

$$\text{sech } x = 1 - \frac{1}{2!} x^2 + \frac{5}{4!} x^4 - \frac{61}{6!} x^6 + \frac{1385}{8!} x^8 - \cdots + \frac{E_{2n} x^{2n}}{(2n)!} \pm \cdots \text{ (2)}$$

$$|x| < \frac{\pi}{2}$$

$$\text{sech } x = 2e^{-x} - 2e^{-3x} + 2e^{-5x} - 2e^{-7x} + - \cdots$$

$$= 2 \sum_{n=0}^{\infty} (-1)^n e^{-(2n+1)x}, \text{ Re } (x) > 0$$

$$\text{sech } x = 4\pi \left[\frac{1}{\pi^2 + 4x^2} - \frac{3}{(3\pi)^2 + 4x^2} + \frac{5}{(5\pi)^2 + 4x^2} - + \cdots \right]$$

$$= 4\pi \sum_{n=0}^{\infty} \frac{(-1)^n (2n + 1)}{(2n + 1)^2 \pi^2 + 4x^2}$$

$$\text{cosech } x = \frac{1}{x} - \frac{x}{6} + \frac{7x^3}{360} - \frac{31x^5}{15,120} + \cdots - \frac{2(2^{2n-1} - 1)}{(2n)!} B_{2n} x^{2n-1} + \cdots \text{ (1)}$$

$$0 < |x| < \pi$$

$$\text{cosech } x = 2e^{-x} + 2e^{-3x} + 2e^{-5x} + \cdots$$

$$= 2 \sum_{n=0}^{\infty} e^{-(2n+1)x}, \text{ Re } (x) > 0$$

$$\text{cosech } x = \frac{1}{x} - \frac{2x}{\pi^2 + x^2} + \frac{2x}{(2\pi)^2 + x^2} - \frac{2x}{(3\pi)^2 + x^2} + - \cdots$$

$$= \frac{1}{x} + 2x \sum_{n=1}^{\infty} \frac{(-1)^n}{(n\pi)^2 + x^2}$$

[2] E_{2n} denotes Euler numbers.

INVERSE HYPERBOLIC FUNCTIONS

Definitions

If $z = \sinh y$, then $y = \sinh^{-1} z$. Other inverse functions are denned similarly.

Domain and Range

Function	Domain	Range	Remarks
$\sinh^{-1} z$	$(-\infty, +\infty)$	$(-\infty, +\infty)$	odd function
$\cosh^{-1} z$	$[1, +\infty)$	$(-\infty, +\infty)$	even function double valued
$\tanh^{-1} z$	$(-1, 1)$	$(-\infty, +\infty)$	odd function
$\operatorname{cosech}^{-1} z$	$(-\infty, 0), (0, \infty)$	$(0, -\infty), (\infty, 0)$	odd function two branches, pole at $z = 0$
$\operatorname{sech}^{-1} z$	$(0, 1]$	$(-\infty, +\infty)$	double valued
$\coth^{-1} z$	$(-\infty, -1), (1, \infty)$	$(0, -\infty), (\infty, 0)$	odd function two branches

Graph of the Inverse Functions

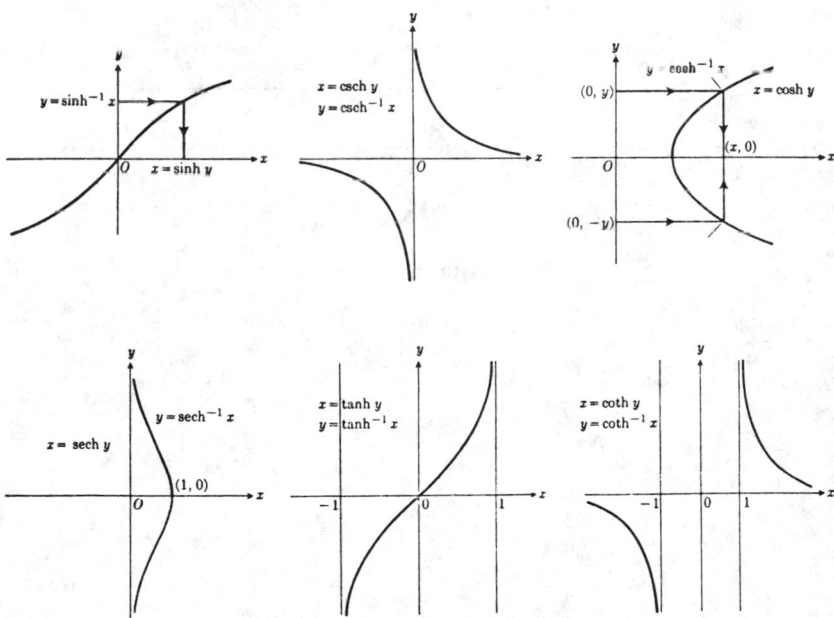

Inverse Hyperbolic Functions In Terms of One Another

Function	$\sinh^{-1} x$	$\cosh^{-1} x$ $+$ if $x > 0$ $-$ if $x < 0$	$\tanh^{-1} x$
$\sinh^{-1} x =$	$\sinh^{-1} x$	$\pm \cosh^{-1} \sqrt{x^2 + 1}$	$\tanh^{-1} \dfrac{x}{\sqrt{1 + x^2}}$
$\cosh^{-1} x =$	$\pm \sinh^{-1} \sqrt{x^2 - 1}$	$\cosh^{-1} x$	$\pm \tanh^{-1} \dfrac{\sqrt{x^2 - 1}}{x}$
$\tanh^{-1} x =$	$\sinh^{-1} \dfrac{x}{\sqrt{1 - x^2}}$	$\pm \cosh^{-1} \dfrac{1}{\sqrt{1 - x^2}}$	$\tanh^{-1} x$
$\operatorname{cosech}^{-1} x =$	$\sinh^{-1} \dfrac{1}{x}$	$\pm \cosh^{-1} \dfrac{\sqrt{x^2 + 1}}{x}$	$\tanh^{-1} \dfrac{1}{\sqrt{1 + x^2}}$
$\operatorname{sech}^{-1} x =$	$\pm \sinh^{-1} \dfrac{\sqrt{1 - x^2}}{x}$	$\cosh^{-1} \dfrac{1}{x}$	$\pm \tanh^{-1} \sqrt{1 - x^2}$
$\coth^{-1} x =$	$\sinh^{-1} \dfrac{1}{\sqrt{x^2 - 1}}$	$\pm \cosh^{-1} \dfrac{x}{\sqrt{x^2 - 1}}$	$\tanh^{-1} \dfrac{1}{x}$

Function	$\operatorname{cosech}^{-1} x$	$\operatorname{sech}^{-1} x$ $+$ if $x > 0$ $-$ if $x < 0$	$\coth^{-1} x$
$\sinh^{-1} x =$	$\operatorname{cosech}^{-1} \dfrac{1}{x}$	$\pm \operatorname{sech}^{-1} \dfrac{1}{\sqrt{1 + x^2}}$	$\coth^{-1} \dfrac{\sqrt{1 + x^2}}{x}$
$\cosh^{-1} x =$	$\pm \operatorname{cosech}^{-1} \dfrac{1}{\sqrt{x^2 - 1}}$	$\operatorname{sech}^{-1} \dfrac{1}{x}$	$\pm \coth^{-1} \dfrac{x}{\sqrt{x^2 - 1}}$
$\tanh^{-1} x =$	$\operatorname{cosech}^{-1} \dfrac{\sqrt{1 - x^2}}{x}$	$\pm \operatorname{sech}^{-1} \sqrt{1 - x^2}$	$\coth^{-1} \dfrac{1}{x}$
$\operatorname{cosech}^{-1} x =$	$\operatorname{cosech}^{-1} x$	$\pm \operatorname{sech}^{-1} \dfrac{x}{\sqrt{x^2 + 1}}$	$\coth^{-1} \sqrt{1 + x^2}$
$\operatorname{sech}^{-1} x =$	$\pm \operatorname{cosech}^{-1} \dfrac{x}{\sqrt{1 - x^2}}$	$\operatorname{sech}^{-1} x$	$\pm \coth^{-1} \dfrac{1}{\sqrt{1 - x^2}}$
$\coth^{-1} x =$	$\operatorname{cosech}^{-1} \sqrt{x^2 - 1}$	$\operatorname{sech}^{-1} \dfrac{\sqrt{x^2 - 1}}{x}$	$\coth^{-1} x$

Fundamental Identities

Relations with Logarithmic Functions

$$\sinh^{-1} x = \log_e (x + \sqrt{x^2 + 1})$$

$$\cosh^{-1} x = \log_e (x \pm \sqrt{x^2 - 1}), \quad x \geq 1. \quad \text{The plus sign is used for the principal value.}$$

$$\tanh^{-1} x = \tfrac{1}{2} \log_e \left(\frac{1 + x}{1 - x}\right), \quad x^2 < 1$$

$$\operatorname{csch}^{-1} x = \log_e \left(\frac{1 \pm \sqrt{1 + x^2}}{x}\right). \quad \text{The plus sign is used if } x > 0, \text{ the minus sign if } x < 0.$$

$$\operatorname{sech}^{-1} x = \log_e \left(\frac{1 \pm \sqrt{1 - x^2}}{x}\right), \quad 0 < x \leq 1. \quad \text{The plus sign is used for the principal values.}$$

$$\coth^{-1} x = \tfrac{1}{2} \log_e \left(\frac{x + 1}{x - 1}\right), \quad x^2 > 1$$

Relations with Circular Functions

$$\sinh^{-1} x = -i \sin^{-1} (ix) \qquad\qquad \sinh^{-1} (ix) = i \sin^{-1} x$$
$$\cosh^{-1} x = \pm i \cos^{-1} x \qquad\qquad \cosh^{-1} (ix) = \pm i \cos^{-1} (ix)$$
$$\tanh^{-1} x = -i \tan^{-1} (ix) \qquad\qquad \tanh^{-1} (ix) = i \tan^{-1} x$$
$$\operatorname{cosech}^{-1} x = i \operatorname{cosec}^{-1} (ix) \qquad \operatorname{cosech}^{-1} (ix) = -i \operatorname{cosec}^{-1} x$$
$$\operatorname{sech}^{-1} x = \pm i \sec^{-1} x \qquad\qquad \operatorname{sech}^{-1} (ix) = \pm i \sec^{-1} (ix)$$
$$\coth^{-1} x = i \coth^{-1} (ix) \qquad\qquad \coth^{-1} (ix) = -i \cot^{-1} (x)$$

Function Sum and Function Difference Relations

$$\sinh^{-1} x + \sinh^{-1} y = \sinh^{-1} (x \sqrt{1 + y^2} + y \sqrt{1 + x^2})$$
$$\sinh^{-1} x - \sinh^{-1} y = \sinh^{-1} (x \sqrt{1 + y^2} - y \sqrt{1 + x^2})$$
$$\cosh^{-1} x + \cosh^{-1} y = \cosh^{-1} (xy + \sqrt{(x^2 - 1)(y^2 - 1)})$$
$$\cosh^{-1} x - \cosh^{-1} y = \cosh^{-1} (xy - \sqrt{(x^2 - 1)(y^2 - 1)})$$

$$\tanh^{-1} x + \tanh^{-1} y = \tanh^{-1} \left(\frac{x + y}{1 + xy}\right)$$

$$\tanh^{-1} x - \tanh^{-1} y = \tanh^{-1} \left(\frac{x - y}{1 - xy}\right)$$

$$\sinh^{-1} x + \cosh^{-1} y = \sinh^{-1} (xy + \sqrt{(1 + x^2)(y^2 - 1)})$$
$$= \cosh^{-1} (y \sqrt{1 + x^2} + x \sqrt{y^2 - 1})$$

$$\sinh^{-1} x - \cosh^{-1} y = \sinh^{-1} (xy - \sqrt{(1 + x^2)(y^2 - 1)})$$
$$= \cosh^{-1} (y \sqrt{1 + x^2} - x \sqrt{y^2 - 1})$$

$$\tanh^{-1} x + \coth^{-1} y = \tanh^{-1} \left(\frac{xy + 1}{y + x}\right)$$

$$= \coth^{-1} \left(\frac{y + x}{xy + 1}\right)$$

$$\tanh^{-1} x - \coth^{-1} y = \tanh^{-1} \left(\frac{xy - 1}{y - x}\right)$$

$$= \coth^{-1} \left(\frac{y - x}{xy - 1}\right)$$

Series Expansions

$$\sinh^{-1} x = x - \frac{1}{2 \cdot 3} x^3 + \frac{1 \cdot 3}{2 \cdot 4 \cdot 5} x^5 - \frac{1 \cdot 3 \cdot 5}{2 \cdot 4 \cdot 6 \cdot 7} x^7 + \cdots$$

$$+ (-1)^n \cdot \frac{1 \cdot 3 \cdot 5 \ldots (2n-1)}{2 \cdot 4 \cdot 6 \ldots 2n(2n+1)} x^{2n+1} \pm \cdots \qquad |x| < 1$$

$$\sinh^{-1} x = \ln (2x) + \frac{1}{2} \cdot \frac{1}{2x^2} - \frac{1 \cdot 3}{2 \cdot 4} \cdot \frac{1}{4x^4} + \frac{1 \cdot 3 \cdot 5}{2 \cdot 4 \cdot 6} \cdot \frac{1}{6x^6} - \cdots$$

$$= \ln (2x) + \sum_{n=1}^{\infty} (-1)^{n+1} \frac{(2n)! x^{-2n}}{2^{2n} (n!)^2 2n}, \qquad |x| > 1$$

$$\cosh^{-1} x = \pm \left[\ln (2x) - \frac{1}{2 \cdot 2x^2} - \frac{1 \cdot 3}{2 \cdot 4 \cdot 4x^4} - \frac{1 \cdot 3 \cdot 5}{2 \cdot 4 \cdot 6 \cdot 6x^6} - \cdots \right] \qquad x > 1$$

$$\operatorname{cosech}^{-1} x = \frac{1}{x} - \frac{1}{2} \cdot \frac{1}{3x^3} + \frac{1 \cdot 3}{2 \cdot 4} \cdot \frac{1}{5x^5} - \frac{1 \cdot 3 \cdot 5}{2 \cdot 4 \cdot 6} \cdot \frac{1}{7x^7} + - \cdots$$

$$= \sum_{n=0}^{\infty} (-1)^n \frac{(2n)! x^{-2n-1}}{2^{2n} (n!)^2 (2n+1)}, \qquad |x| > 1$$

$$\operatorname{cosech}^{-1} x = \ln \frac{2}{x} + \frac{1}{2} \cdot \frac{x^2}{2} - \frac{1 \cdot 3}{2 \cdot 4} \cdot \frac{x^4}{4} + \frac{1 \cdot 3 \cdot 5}{2 \cdot 4 \cdot 6} \cdot \frac{x^6}{6} - + \cdots$$

$$= \ln \frac{2}{x} + \sum_{n=1}^{\infty} \frac{(-1)^{n+1} (2n)! x^{2n}}{2^{2n} (n!)^2 2n}, \qquad 0 < x < 1$$

$$\operatorname{sech}^{-1} x = \ln \frac{2}{x} - \frac{1}{2} \frac{x^2}{2} - \frac{1 \cdot 3}{2 \cdot 4} \cdot \frac{x^4}{4} - \frac{1 \cdot 3 \cdot 5}{2 \cdot 4 \cdot 6} \cdot \frac{x^6}{6} - - \cdots$$

$$= \ln \frac{2}{x} - \sum_{n=1}^{\infty} \frac{(2n)! x^{2n}}{2^{2n} (n!)^2 2n}, \qquad 0 < x < 1$$

$$* \tanh^{-1} x = x + \frac{x^3}{3} + \frac{x^5}{5} + \frac{x^7}{7} + \cdots + \frac{x^{2n+1}}{2n+1} + \cdots \qquad |x| < 1$$

$$* \coth^{-1} x = \frac{1}{x} + \frac{1}{3x^3} + \frac{1}{5x^5} + \frac{1}{7x^7} + \cdots + \frac{1}{(2n+1)x^{2n+1}} + \cdots \qquad |x| > 1$$

GUDERMANNIAN FUNCTION

This function is useful because it relates circular and hyperbolic functions without the use of functions of imaginary argument.

Definition

$$\gamma = \text{the gudermannian of } x, \text{ written as } \mathbf{gd} \, x$$

$$= 2 \tan^{-1} e^x - \frac{\pi}{2}$$

$$x = \mathbf{gd}^{-1} \gamma$$

$$= \ln \tan \left(\frac{\pi}{4} + \frac{\gamma}{2} \right)$$

$$= \ln (\sec \gamma + \tan \gamma)$$

* Can also be written as arg tanh x and arg coth x respectively.

Special Values of Gudermannian Function

x	0	∞	$-\infty$
gd x	0	$\frac{1}{2}\pi$	$-\frac{1}{2}\pi$

Relations with Hyperbolic and Circular Functions

$$\sinh x = \tan (\text{gd } x) \qquad \text{cosech } x = \cot (\text{gd } x)$$
$$\cosh x = \sec (\text{gd } x) \qquad \text{sech } (x) = \cos (\text{gd } x)$$
$$\tanh x = \sin (\text{gd } x) \qquad \coth (x) = \text{cosec } (\text{gd } x)$$

Derivatives

$$\frac{d}{dx} (\text{gd } x) = \text{sech } x$$

$$\frac{d}{dx} (\text{gd}^{-1} \gamma) = \sec \gamma$$

Fundamental Identities

$$\tanh (\tfrac{1}{2}x) = \tan (\tfrac{1}{2}\text{gd } x)$$
$$e^x = \cosh x + \sinh x = \sec (\text{gd } x) + \tan (\text{gd } x)$$
$$= \tan \left(\frac{\pi}{4} + \frac{1}{2} \text{gd } x\right) = \frac{1 + \sin (\text{gd } x)}{\cos (\text{gd } x)} = \frac{1 + \tan (\tfrac{1}{2}\text{gd } x)}{1 - \tan (\tfrac{1}{2}\text{gd } x)}$$
$$\text{gd } x = 2 \tan^{-1}\left(\tanh \frac{1}{2} x\right) = \int_0^x \frac{dt}{\cosh t}$$
$$\text{gd}^{-1} \gamma = \int_0^\gamma \frac{dt}{\cos t}$$
$$i \, \text{gd}^{-1} \gamma = \text{gd } i\gamma, \text{ where } i = \sqrt{-1}.$$

If

$$\alpha + i\beta = \text{gd}(x + iy)$$

then

$$\tan \alpha = \frac{\sinh x}{\cos y} \qquad \tanh \beta = \frac{\sin y}{\cosh x}$$

$$\tanh x = \frac{\sin \alpha}{\cosh \beta} \qquad \tan y = \frac{\sinh \beta}{\cosh \alpha}$$

Series Expansions

$$\text{gd } x = x - \frac{1}{6} x^3 + \frac{1}{24} x^5 - \frac{61}{5040} x^7 + - \cdots$$

$$= \sum_{n=0}^{\infty} \frac{E_{2n}}{(2n+1)!} x^{2n+1}, \quad |x| < 1 \text{ *}$$

$$\text{gd } x = \frac{\pi}{2} - \text{sech } x - \frac{1}{2} \frac{\text{sech}^3 x}{3} - \frac{1 \cdot 3}{2 \cdot 4} \frac{\text{sech}^5 x}{5} + - \cdots$$

$$= \frac{\pi}{2} - \sum_{n=0}^{\infty} \frac{(2n)!}{2^{2n}(n!)^2} \frac{\text{sech}^{2n+1} x}{(2n+1)} \quad x \text{ large}$$

*E_n denotes Euler numbers.

$$\text{gd } x = \frac{2}{1} \tanh \frac{x}{2} - \frac{2}{3} \tanh^3 \frac{x}{2} + \frac{2}{5} \tanh^5 \frac{x}{2} - \frac{2}{7} \tanh^7 \frac{x}{2} + - \cdots$$

$$= 2 \sum_{n=0}^{\infty} \frac{(-1)^n}{2n+1} \tanh^{2n+1} \frac{x}{2}$$

$$\text{gd}^{-1} \gamma = \gamma + \frac{1}{6} \gamma^3 + \frac{1}{24} \gamma^5 + \frac{61}{5040} \gamma^7 + \cdots$$

$$= \sum_{n=0}^{\infty} \frac{(-1)^n E_{2n} \gamma^{2n+1}}{(2n+1)!}, \qquad |\gamma| < \frac{\pi}{2} \ *$$

$$\text{gd}^{-1} \gamma = \frac{2}{1} \tan \frac{\gamma}{2} + \frac{2}{3} \tan^3 \frac{\gamma}{2} + \frac{2}{5} \tan^5 \frac{\gamma}{2} + \cdots$$

$$= 2 \sum_{n=0}^{\infty} \frac{1}{2n+1} \tan^{2n+1} \frac{\gamma}{2}$$

*E_n are Euler's numbers.

References

1. For extensive tables of inverse hyperbolic functions: Tables of Inverse Hyperbolic Functions, Harvard University Computation Laboratory (Cambridge, Harvard University Press, 1949).
2. For tables of hyperbolic functions with complex argument: Tables of Complex Hyperbolic and Circular Functions, A. E. Kennelly (Cambridge, Harvard University Press, 1914).

HYPERBOLIC FUNCTIONS AND THEIR COMMON LOGARITHMS

The logarithms given below show the mantissa only. The proper characteristic must be added.

x	Sinh x Value	$\log_{10}$	Cosh x Value	$\log_{10}$	Tanh x Value	$\log_{10}$	Coth x Value	$\log_{10}$
0.00	0.00000	$-\infty$	1.00000	.00000	0.00000	$-\infty$	∞	∞
0.01	.01000	.00001	1.00005	.00002	.01000	.99999	100.003	.00001
0.02	.02000	.30106	1.00020	.00009	.02000	.30097	50.007	.69903
0.03	.03000	.47719	1.00045	.00020	.02999	.47699	33.343	.52301
0.04	.04001	.60218	1.00080	.00035	.03998	.60183	25.013	.39817
0.05	0.05002	.69915	1.00125	.00054	0.04996	.69861	20.017	.30139
0.06	.06004	.77841	1.00180	.00078	.05993	.77763	16.687	.22237
0.07	.07006	.84545	1.00245	.00106	.06989	.84439	14.309	.15561
0.08	.08009	.90355	1.00320	.00139	.07983	.90216	12.527	.09784
0.09	.09012	.95483	1.00405	.00176	.08976	.95307	11.141	.04693
0.10	0.10017	.00072	1.00500	.00217	0.09967	.99856	10.0333	.00144
0.11	.11022	.04227	1.00606	.00262	.10956	.03965	9.1275	.96035
0.12	.12029	.08022	1.00721	.00312	.11943	.07710	8.3733	.92290
0.13	.13037	.11517	1.00846	.00366	.12927	.11151	7.7356	.88849
0.14	.14046	.14755	1.00982	.00424	.13909	.14330	7.1895	.85670
0.15	0.15056	.17772	1.01127	.00487	0.14889	.17285	6.7166	.82715
0.16	.16068	.20597	1.01283	.00554	.15865	.20044	6.3032	.79956
0.17	.17082	.23254	1.01448	.00625	.16838	.22629	5.9389	.77371
0.18	.18097	.25762	1.01624	.00700	.17808	.25062	5.6154	.74938
0.19	.19115	.28136	1.01810	.00779	.18775	.27357	5.3263	.72643
0.20	0.20134	.30392	1.02007	.00863	0.19738	.29529	5.0665	.70471
0.21	.21155	.32541	1.02213	.00951	.20697	.31590	4.8317	.68410
0.22	.22178	.34592	1.02430	.01043	.21652	.33549	4.6186	.66451
0.23	.23203	.36555	1.02657	.01139	.22603	.35416	4.4242	.64584
0.24	.24231	.38437	1.02894	.01239	.23550	.37198	4.2464	.62802
0.25	0.25261	.40245	1.03141	.01343	0.24492	.38902	4.0830	.61098
0.26	.26294	.41986	1.03399	.01452	.25430	.40534	3.9324	.59466
0.27	.27329	.43663	1.03667	.01564	.26362	.42099	3.7933	.57901
0.28	.28367	.45282	1.03946	.01681	.27291	.43601	3.6643	.56399
0.29	.29408	.46847	1.04235	.01801	.28213	.45046	3.5444	.54954
0.30	0.30452	.48362	1.04534	.01926	0.29131	.46436	3.4327	.53564
0.31	.31499	.49830	1.04844	.02054	.30044	.47775	3.3285	.52225
0.32	.32549	.51254	1.05164	.02187	.30951	.49067	3.2309	.50933
0.33	.33602	.52637	1.05495	.02323	.31852	.50314	3.1395	.49686
0.34	.34659	.53981	1.05836	.02463	.32748	.51518	3.0536	.48482
0.35	0.35719	.55290	1.06188	.02607	0.33638	.52682	2.9729	.47318
0.36	.36783	.56564	1.06550	.02755	.34521	.53809	2.8968	.46191
0.37	.37850	.57807	1.06923	.02907	.35399	.54899	2.8249	.45101
0.38	.38921	.59019	1.07307	.03063	.36271	.55956	2.7570	.44044
0.39	.39996	.60202	1.07702	.03222	.37136	.56980	2.6928	.43020
0.40	0.41075	.61358	1.08107	.03385	0.37995	.57973	2.6319	.42027
0.41	.42158	.62488	1.08523	.03552	.38847	.58936	2.5742	.41064
0.42	.43246	.63594	1.08950	.03723	.39693	.59871	2.5193	.40129
0.43	.44337	.64677	1.09388	.03897	.40532	.60780	2.4672	.39220
0.44	.45434	.65738	1.09837	.04075	.41364	.61663	2.4175	.38337
0.45	0.46534	.66777	1.10297	.04256	0.42190	.62521	2.3702	.37479
0.46	.47640	.67797	1.10768	.04441	.43008	.63355	2.3251	.36645
0.47	.48750	.68797	1.11250	.04630	.43820	.64167	2.2821	.35833
0.48	.49865	.69779	1.11743	.04822	.44624	.64957	2.2409	.35043
0.49	.50984	.70744	1.12247	.05018	.45422	.65726	2.2016	.34274

HYPERBOLIC FUNCTIONS AND THEIR COMMON LOGARITHMS (Continued)

x	Sinh x Value	Sinh x $\log_{10}$	Cosh x Value	Cosh x $\log_{10}$	Tanh x Value	Tanh x $\log_{10}$	Coth x Value	Coth x $\log_{10}$
0.50	0.52110	.71692	1.12763	.05217	0.46212	.66475	2.1640	.33525
0.51	.53240	.72624	1.13289	.05419	.46995	.67205	2.1279	.32795
0.52	.54375	.73540	1.13827	.05625	.47770	.67916	2.0934	.32084
0.53	.55516	.74442	1.14377	.05834	.48538	.68608	2.0602	.31392
0.54	.56663	.75330	1.14938	.06046	.49299	.69284	2.0284	.30716
0.55	0.57815	.76204	1.15510	.06262	0.50052	.69942	1.9979	.30058
0.56	.58973	.77065	1.16094	.06481	.50798	.70584	1.9686	.29416
0.57	.60137	.77914	1.16690	.06703	.51536	.71211	1.9404	.28789
0.58	.61307	.78751	1.17297	.06929	.52267	.71822	1.9133	.28178
0.59	.62483	.79576	1.17916	.07157	.52990	.72419	1.8872	.27581
0.60	0.63665	.80390	1.18547	.07389	0.53705	.73001	1.8620	.26999
0.61	.64854	.81194	1.19189	.07624	.54413	.73570	1.8378	.26430
0.62	.66049	.81987	1.19844	.07861	.55113	.74125	1.8145	.25875
0.63	.67251	.82770	1.20510	.08102	.55805	.74667	1.7919	.25333
0.64	.68459	.83543	1.21189	.08346	.56490	.75197	1.7702	.24803
0.65	0.69675	.84308	1.21879	.08593	0.57167	.75715	1.7493	.24285
0.66	.70897	.85063	1.22582	.08843	.57836	.76220	1.7290	.23780
0.67	.72126	.85809	1.23297	.09095	.58498	.76714	1.7095	.23286
0.68	.73363	.86548	1.24025	.09351	.59152	.77197	1.6906	.22803
0.69	.74607	.87278	1.24765	.09609	.59798	.77669	1.6723	.22331
0.70	0.75858	.88000	1.25517	.09870	0.60437	.78130	1.6546	.21870
0.71	.77117	.88715	1.26282	.10134	.61068	.78581	1.6375	.21419
0.72	.78384	.89423	1.27059	.10401	.61691	.79022	1.6210	.20978
0.73	.79659	.90123	1.27849	.10670	.62307	.79453	1.6050	.20547
0.74	.80941	.90817	1.28652	.10942	.62915	.79875	1.5895	.20125
0.75	0.82232	.91504	1.29468	.11216	0.63515	.80288	1.5744	.19712
0.76	.83530	.92185	1.30297	.11493	.64108	.80691	1.5599	.19309
0.77	.84838	.92859	1.31139	.11773	.64693	.81086	1.5458	.18914
0.78	.86153	.93527	1.31994	.12055	.65271	.81472	1.5321	.18528
0.79	.87478	.94190	1.32862	.12340	.65841	.81850	1.5188	.18150
0.80	0.88811	.94846	1.33743	.12627	0.66404	.82219	1.5059	.17781
0.81	.90152	.95498	1.34638	.12917	.66959	.82581	1.4935	.17419
0.82	.91503	.96144	1.35547	.13209	.67507	.82935	1.4813	.17065
0.83	.92863	.96784	1.36468	.13503	.68048	.83281	1.4696	.16719
0.84	.94233	.97420	1.37404	.13800	.68581	.83620	1.4581	.16380
0.85	0.95612	.98051	1.38353	.14099	0.69107	.83952	1.4470	.16048
0.86	.97000	.98677	1.39316	.14400	.69626	.84277	1.4362	.15723
0.87	.98398	.99299	1.40293	.14704	.70137	.84595	1.4258	.15405
0.88	.99806	.99916	1.41284	.15009	.70642	.84906	1.4156	.15094
0.89	1.01224	.00528	1.42289	.15317	.71139	.85211	1.4057	.14789
0.90	1.02652	.01137	1.43309	.15627	0.71630	.85509	1.3961	.14491
0.91	1.04090	.01741	1.44342	.15939	.72113	.85801	1.3867	.14199
0.92	1.05539	.02341	1.45390	.16254	.72590	.86088	1.3776	.13912
0.93	1.06998	.02937	1.46453	.16570	.73059	.86368	1.3687	.13632
0.94	1.08468	.03530	1.47530	.16888	.73522	.86642	1.3601	.13358
0.95	1.09948	.04119	1.48623	.17208	0.73978	.86910	1.3517	.13090
0.96	1.11440	.04704	1.49729	.17531	.74428	.87173	1.3436	.12827
0.97	1.12943	.05286	1.50851	.17855	.74870	.87431	1.3356	.12569
0.98	1.14457	.05864	1.51988	.18181	.75307	.87683	1.3279	.12317
0.99	1.15983	.06439	1.53141	.18509	.75736	.87930	1.3204	.12070

HYPERBOLIC FUNCTIONS AND THEIR COMMON LOGARITHMS (Continued)

x	Sinh x Value	Sinh x $\log_{10}$	Cosh x Value	Cosh x $\log_{10}$	Tanh x Value	Tanh x $\log_{10}$	Coth x Value	Coth x $\log_{10}$
1.00	1.17520	.07011	1.54308	.18839	0.76159	.88172	1.3130	.11828
1.01	1.19069	.07580	1.55491	.19171	.76576	.88409	1.3059	.11591
1.02	1.20630	.08146	1.56689	.19504	.76987	.88642	1.2989	.11358
1.03	1.22203	.08708	1.57904	.19839	.77391	.88869	1.2921	.11131
1.04	1.23788	.09268	1.59134	.20176	.77789	.89092	1.2855	.10908
1.05	1.25386	.09825	1.60379	.20515	0.78181	.89310	1.2791	.10690
1.06	1.26996	.10379	1.61641	.20855	.78566	.89524	1.2728	.10476
1.07	1.28619	.10930	1.62919	.21197	.78946	.89733	1.2667	.10267
1.08	1.30254	.11479	1.64214	.21541	.79320	.89938	1.2607	.10062
1.09	1.31903	.12025	1.65525	.21886	.79688	.90139	1.2549	.09861
1.10	1.33565	.12569	1.66852	.22233	0.80050	.90336	1.2492	.09664
1.11	1.35240	.13111	1.68196	.22582	.80406	.90529	1.2437	.09471
1.12	1.36929	.13649	1.69557	.22931	.80757	.90718	1.2383	.09282
1.13	1.38631	.14186	1.70934	.23283	.81102	.90903	1.2330	.09097
1.14	1.40347	.14720	1.72329	.23636	.81441	.91085	1.2279	.08915
1.15	1.42078	.15253	1.73741	.23990	0.81775	.91262	1.2229	.08738
1.16	1.43822	.15783	1.75171	.24346	.82104	.91436	1.2180	.08564
1.17	1.45581	.16311	1.76618	.24703	.82427	.91607	1.2132	.08393
1.18	1.47355	.16836	1.78083	.25062	.82745	.91774	1.2085	.08226
1.19	1.49143	.17360	1.79565	.25422	.83058	.91938	1.2040	.08062
1.20	1.50946	.17882	1.81066	.25784	0.83365	.92099	1.1995	.07901
1.21	1.52764	.18402	1.82584	.26146	.83668	.92256	1.1952	.07744
1.22	1.54598	.18920	1.84121	.26510	.83965	.92410	1.1910	.07590
1.23	1.56447	.19437	1.85676	.26876	.84258	.92561	1.1868	.07439
1.24	1.58311	.19951	1.87250	.27242	.84546	.92709	1.1828	.07291
1.25	1.60192	.20464	1.88842	.27610	0.84828	.92854	1.1789	.07146
1.26	1.62088	.20975	1.90454	.27979	.85106	.92996	1.1750	.07004
1.27	1.64001	.21485	1.92084	.28349	.85380	.93135	1.1712	.06865
1.28	1.65930	.21993	1.93734	.28721	.85648	.93272	1.1676	.06728
1.29	1.67876	.22499	1.95403	.29093	.85913	.93406	1.1640	.06594
1.30	1.69838	.23004	1.97091	.29467	0.86172	.93537	1.1605	.06463
1.31	1.71818	.23507	1.98800	.29842	.86428	.93665	1.1570	.06335
1.32	1.73814	.24009	2.00528	.30217	.86678	.93791	1.1537	.06209
1.33	1.75828	.24509	2.02276	.30594	.86925	.93914	1.1504	.06086
1.34	1.77860	.25008	2.04044	.30972	.87167	.94035	1.1472	.05965
1.35	1.79909	.25505	2.05833	.31352	0.87405	.94154	1.1441	.05846
1.36	1.81977	.26002	2.07643	.31732	.87639	.94270	1.1410	.05730
1.37	1.84062	.26496	2.09473	.32113	.87869	.94384	1.1381	.05616
1.38	1.86166	.26990	2.11324	.32495	.88095	.94495	1.1351	.05505
1.39	1.88289	.27482	2.13196	.32878	.88317	.94604	1.1323	.05396
1.40	1.90430	.27974	2.15090	.33262	0.88535	.94712	1.1295	.05288
1.41	1.92591	.28464	2.17005	.33647	.88749	.94817	1.1268	.05183
1.42	1.94770	.28952	2.18942	.34033	.88960	.94919	1.1241	.05081
1.43	1.96970	.29440	2.20900	.34420	.89167	.95020	1.1215	.04980
1.44	1.99188	.29926	2.22881	.34807	.89370	.95119	1.1189	.04881
1.45	2.01427	.30412	2.24884	.35196	0.89569	.95216	1.1165	.04784
1.46	2.03686	.30896	2.26910	.35585	.89765	.95311	1.1140	.04689
1.47	2.05965	.31379	2.28958	.35976	.89958	.95404	1.1116	.04596
1.48	2.08265	.31862	2.31029	.36367	.90147	.95495	1.1093	.04505
1.49	2.10586	.32343	2.33123	.36759	.90332	.95584	1.1070	.04416

HYPERBOLIC FUNCTIONS AND THEIR COMMON LOGARITHMS (Continued)

x	Sinh x Value	$\log_{10}$	Cosh x Value	$\log_{10}$	Tanh x Value	$\log_{10}$	Coth x Value	$\log_{10}$
1.50	2.12928	.32823	2.35241	.37151	0.90515	.95672	1.1048	.04328
1.51	2.15291	.33303	2.37382	.37545	.90694	.95758	1.1026	.04242
1.52	2.17676	.33781	2.39547	.37939	.90870	.95842	1.1005	.04158
1.53	2.20082	.34258	2.41736	.38334	.91042	.95924	1.0984	.04076
1.54	2.22510	.34735	2.43949	.38730	.91212	.96005	1.0963	.03995
1.55	2.24961	.35211	2.46186	.39126	0.91379	.96084	1.0943	.03916
1.56	2.27434	.35686	2.48448	.39524	.91542	.96162	1.0924	.03838
1.57	2.29930	.36160	2.50735	.39921	.91703	.96238	1.0905	.03762
1.58	2.32449	.36633	2.53047	.40320	.91860	.96313	1.0886	.03687
1.59	2.34991	.37105	2.55384	.40719	.92015	.96386	1.0868	.03614
1.60	2.37557	.37577	2.57746	.41119	0.92167	.96457	1.0850	.03543
1.61	2.40146	.38048	2.60135	.41520	.92316	.96528	1.0832	.03472
1.62	2.42760	.38518	2.62549	.41921	.92462	.96597	1.0815	.03403
1.63	2.45397	.38987	2.64990	.42323	.92606	.96664	1.0798	.03336
1.64	2.48059	.39456	2.67457	.42725	.92747	.96730	1.0782	.03270
1.65	2.50746	.39923	2.69951	.43129	0.92886	.96795	1.0766	.03205
1.66	2.53459	.40391	2.72472	.43532	.93022	.96858	1.0750	.03142
1.67	2.56196	.40857	2.75021	.43937	.93155	.96921	1.0735	.03079
1.68	2.58959	.41323	2.77596	.44341	.93286	.96982	1.0720	.03018
1.69	2.61748	.41788	2.80200	.44747	.93415	.97042	1.0705	.02958
1.70	2.64563	.42253	2.82832	.45153	.93541	.97100	1.0691	.02900
1.71	2.67405	.42717	2.85491	.45559	.93665	.97158	1.0676	.02842
1.72	2.70273	.43180	2.88180	.45966	.93786	.97214	1.0663	.02786
1.73	2.73168	.43643	2.90897	.46374	.93906	.97269	1.0649	.02731
1.74	2.76091	.44105	2.93643	.46782	.94023	.97323	1.0636	.02677
1.75	2.79041	.44567	2.96419	.47191	0.94138	.97376	1.0623	.02624
1.76	2.82020	.45028	2.99224	.47600	.94250	.97428	1.0610	.02572
1.77	2.85026	.45488	3.02059	.48009	.94361	.97479	1.0598	.02521
1.78	2.88061	.45948	3.04925	.48419	.94470	.97529	1.0585	.02471
1.79	2.91125	.46408	3.07821	.48830	.94576	.97578	1.0574	.02422
1.80	2.94217	.46867	3.10747	.49241	0.94681	.97626	1.0562	.02374
1.81	2.97340	.47325	3.13705	.49652	.94783	.97673	1.0550	.02327
1.82	3.00492	.47783	3.16694	.50064	.94884	.97719	1.0539	.02281
1.83	3.03674	.48241	3.19715	.50476	.94983	.97764	1.0528	.02236
1.84	3.06886	.48698	3.22768	.50889	.95080	.97809	1.0518	.02191
1.85	3.10129	.49154	3.25853	.51302	0.95175	.97852	1.0507	.02148
1.86	3.13403	.49610	3.28970	.51716	.95268	.97895	1.0497	.02105
1.87	3.16709	.50066	3.32121	.52130	.95359	.97936	1.0487	.02064
1.88	3.20046	.50521	3.35305	.52544	.95449	.97977	1.0477	.02023
1.89	3.23415	.50976	3.38522	.52959	.95537	.98017	1.0467	.01983
1.90	3.26816	.51430	3.41773	.53374	0.95624	.98057	1.0458	.01943
1.91	3.30250	.51884	3.45058	.53789	.95709	.98095	1.0448	.01905
1.92	3.33718	.52338	3.48378	.54205	.95792	.98133	1.0439	.01867
1.93	3.37218	.52791	3.51733	.54621	.95873	.98170	1.0430	.01830
1.94	3.40752	.53244	3.55123	.55038	.95953	.98206	1.0422	.01794
1.95	3.44321	.53696	3.58548	.55455	0.96032	.98242	1.0413	.01758
1.96	3.47923	.54148	3.62009	.55872	.96109	.98276	1.0405	.01724
1.97	3.51561	.54600	3.65507	.56290	.96185	.98311	1.0397	.01689
1.98	3.55234	.55051	3.69041	.56707	.96259	.98344	1.0389	.01656
1.99	3.58942	.55502	3.72611	.57126	.96331	.98377	1.0381	.01623

HYPERBOLIC FUNCTIONS AND THEIR COMMON LOGARITHMS (Continued)

x	Sinh x Value	Sinh x $\log_{10}$	Cosh x Value	Cosh x $\log_{10}$	Tanh x Value	Tanh x $\log_{10}$	Coth x Value	Coth x $\log_{10}$
2.00	3.62686	.55953	3.76220	.57544	0.96403	.98409	1.0373	.01591
2.01	3.66466	.56403	3.79865	.57963	.96473	.98440	1.0366	.01560
2.02	3.70283	.56853	3.83549	.58382	.96541	.98471	1.0358	.01529
2.03	3.74138	.57303	3.87271	.58802	.96609	.98502	1.0351	.01498
2.04	3.78029	.57753	3.91032	.59221	.96675	.98531	1.0344	.01469
2.05	3.81958	.58202	3.94832	.59641	0.96740	.98560	1.0337	.01440
2.06	3.85926	.58650	3.98671	.60061	.96803	.98589	1.0330	.01411
2.07	3.89932	.59099	4.02550	.60482	.96865	.98617	1.0324	.01383
2.08	3.93977	.59547	4.06470	.60903	.96926	.98644	1.0317	.01356
2.09	3.98061	.59995	4.10430	.61324	.96986	.98671	1.0311	.01329
2.10	4.02186	.60443	4.14431	.61745	0.97045	.98697	1.0304	.01303
2.11	4.06350	.60890	4.18474	.62167	.97103	.98723	1.0298	.01277
2.12	4.10555	.61337	4.22558	.62589	.97159	.98748	1.0292	.01252
2.13	4.14801	.61784	4.26685	.63011	.97215	.98773	1.0286	.01227
2.14	4.19089	.62231	4.30855	.63433	.97269	.98798	1.0281	.01202
2.15	4.23419	.62677	4.35067	.63856	0.97323	.98821	1.0275	.01179
2.16	4.27791	.63123	4.39323	.64278	.97375	.98845	1.0270	.01155
2.17	4.32205	.63569	4.43623	.64701	.97426	.98868	1.0264	.01132
2.18	4.36663	.64015	4.47967	.65125	.97477	.98890	1.0259	.01110
2.19	4.41165	.64460	4.52356	.65548	.97526	.98912	1.0254	.01088
2.20	4.45711	.64905	4.56791	.65972	0.97574	.98934	1.0249	.01066
2.21	4.50301	.65350	4.61271	.66396	.97622	.98955	1.0244	.01045
2.22	4.54936	.65795	4.65797	.66820	.97668	.98975	1.0239	.01025
2.23	4.59617	.66240	4.70370	.67244	.97714	.98996	1.0234	.01004
2.24	4.64344	.66684	4.74989	.67668	.97759	.99016	1.0229	.00984
2.25	4.69117	.67128	4.79657	.68093	0.97803	.99035	1.0225	.00965
2.26	4.73937	.67572	4.84372	.68518	.97846	.99054	1.0220	.00946
2.27	4.78804	.08010	4.89136	.68043	.97888	.99073	1.0216	.00927
2.28	4.83720	.68459	4.93948	.69368	.97929	.99091	1.0211	.00909
2.29	4.88684	.68903	4.98810	.69794	.97970	.99109	1.0207	.00891
2.30	4.93696	.69346	5.03722	.70219	0.98010	.99127	1.0203	.00873
2.31	4.98758	.69789	5.08684	.70645	.98049	.99144	1.0199	.00856
2.32	5.03870	.70232	5.13697	.71071	.98087	.99161	1.0195	.00839
2.33	5.09032	.70675	5.18762	.71497	.98124	.99178	1.0191	.00822
2.34	5.14245	.71117	5.23878	.71923	.98161	.99194	1.0187	.00806
2.35	5.19510	.71559	5.29047	.72349	0.98197	.99210	1.0184	.00790
2.36	5.24827	.72002	5.34269	.72776	.98233	.99226	1.0180	.00774
2.37	5.30196	.72444	5.39544	.73203	.98267	.99241	1.0176	.00759
2.38	5.35618	.72885	5.44873	.73630	.98301	.99256	1.0173	.00744
2.39	5.41093	.73327	5.50256	.74056	.98335	.99271	1.0169	.00729
2.40	5.46623	.73769	5.55695	.74484	0.98367	.99285	1.0166	.00715
2.41	5.52207	.74210	5.61189	.74911	.98400	.99299	1.0163	.00701
2.42	5.57847	.74652	5.66739	.75338	.98431	.99313	1.0159	.00687
2.43	5.63542	.75093	5.72346	.75766	.98462	.99327	1.0156	.00673
2.44	5.69294	.75534	5.78010	.76194	.98492	.99340	1.0153	.00660
2.45	5.75103	.75975	5.83732	.76621	0.98522	.99353	1.0150	.00647
2.46	5.80969	.75415	5.89512	.77049	.98551	.99366	1.0147	.00634
2.47	5.86893	.76856	5.95352	.77477	.98579	.99379	1.0144	.00621
2.48	5.92876	.77296	6.01250	.77906	.98607	.99391	1.0141	.00609
2.49	5.98918	.77737	6.07209	.78334	.98635	.99403	1.0138	.00597

HYPERBOLIC FUNCTIONS AND THEIR COMMON LOGARITHMS (Continued)

x	Sinh x Value	Sinh x log$_{10}$	Cosh x Value	Cosh x log$_{10}$	Tanh x Value	Tanh x log$_{10}$	Coth x Value	Coth x log$_{10}$
2.50	6.05020	.78177	6.13229	.78762	0.98661	.99415	1.0136	.00585
2.51	6.11183	.78617	6.19310	.79191	.98688	.99426	1.0133	.00574
2.52	6.17407	.79057	6.25453	.79619	.98714	.99438	1.0130	.00562
2.53	6.23692	.79497	6.31658	.80048	.98739	.99449	1.0128	.00551
2.54	6.30040	.79937	6.37927	.80477	.98764	.99460	1.0125	.00540
2.55	6.36451	.80377	6.44259	.80906	0.98788	.99470	1.0123	.00530
2.56	6.42926	.80816	6.50656	.81335	.98812	.99481	1.0120	.00519
2.57	6.49464	.81256	6.57118	.81764	.98835	.99491	1.0118	.00509
2.58	6.56068	.81695	6.63646	.82194	.98858	.99501	1.0115	.00499
2.59	6.62738	.82134	6.70240	.82623	.98881	.99511	1.0113	.00489
2.60	6.69473	.82573	6.76901	.83052	0.98903	.99521	1.0111	.00479
2.61	6.76276	.83012	6.83629	.83482	.98924	.99530	1.0109	.00470
2.62	6.83146	.83451	6.90426	.83912	.98946	.99540	1.0107	.00460
2.63	6.90085	.83890	6.97292	.84341	.98966	.99549	1.0104	.00451
2.64	6.97092	.84329	7.04228	.84771	.98987	.99558	1.0102	.00442
2.65	7.04169	.84768	7.11234	.85201	0.99007	.99566	1.0100	.00434
2.66	7.11317	.85206	7.18312	.85631	.99026	.99575	1.0098	.00425
2.67	7.18536	.85645	7.25461	.86061	.99045	.99583	1.0096	.00417
2.68	7.25827	.86083	7.32683	.86492	.99064	.99592	1.0094	.00408
2.69	7.33190	.86522	7.39978	.86922	.99083	.99600	1.0093	.00400
2.70	7.40626	.86960	7.47347	.87352	0.99101	.99608	1.0091	.00392
2.71	7.48137	.87398	7.54791	.87783	.99118	.99615	1.0089	.00385
2.72	7.55722	.87836	7.62310	.88213	.99136	.99623	1.0087	.00377
2.73	7.63383	.88274	7.69905	.88644	.99153	.99631	1.0085	.00369
2.74	7.71121	.88712	7.77578	.89074	.99170	.99638	1.0084	.00362
2.75	7.78935	.89150	7.85328	.89505	0.99186	.99645	1.0082	.00355
2.76	7.86828	.89588	7.93157	.89936	.99202	.99652	1.0080	.00348
2.77	7.94799	.90026	8.01065	.90367	.99218	.99659	1.0079	.00341
2.78	8.02849	.90463	8.09053	.90798	.99233	.99666	1.0077	.00334
2.79	8.10980	.90901	8.17122	.91229	.99248	.99672	1.0076	.00328
2.80	8.19192	.91339	8.25273	.91660	0.99263	.99679	1.0074	.00321
2.81	8.27486	.91776	8.33506	.92091	.99278	.99685	1.0073	.00315
2.82	8.35862	.92213	8.41823	.92522	.99292	.99691	1.0071	.00309
2.83	8.44322	.92651	8.50224	.92953	.99306	.99698	1.0070	.00302
2.84	8.52867	.93088	8.58710	.93385	.99320	.99704	1.0069	.00296
2.85	8.61497	.93525	8.67281	.93816	0.99333	.99709	1.0067	.00291
2.86	8.70213	.93963	8.75940	.94247	.99346	.99715	1.0066	.00285
2.87	8.79016	.94400	8.84686	.94679	.99359	.99721	1.0065	.00279
2.88	8.87907	.94837	8.93520	.95110	.99372	.99726	1.0063	.00274
2.89	8.96887	.95274	9.02444	.95542	.99384	.99732	1.0062	.00268
2.90	9.05956	.95711	9.11458	.95974	0.99396	.99737	1.0061	.00263
2.91	9.15116	.96148	9.20564	.96405	.99408	.99742	1.0060	.00258
2.92	9.24368	.96584	9.29761	.96837	.99420	.99747	1.0058	.00253
2.93	9.33712	.97021	9.39051	.97269	.99431	.99752	1.0057	.00248
2.94	9.43149	.97458	9.48436	.97701	.99443	.99757	1.0056	.00243
2.95	9.52681	.97895	9.57915	.98133	0.99454	.99762	1.0055	.00238
2.96	9.62308	.98331	9.67490	.98565	.99464	.99767	1.0054	.00233
2.97	9.72031	.98768	9.77161	.98997	.99475	.99771	1.0053	.00229
2.98	9.81851	.99205	9.86930	.99429	.99485	.99776	1.0052	.00224
2.99	9.91770	.99641	9.96798	.99861	.99496	.99780	1.0051	.00220

HYPERBOLIC FUNCTIONS AND THEIR COMMON LOGARITHMS (Continued)

z	Sinh z Value	Sinh z $\log_{10}$	Cosh z Value	Cosh z $\log_{10}$	Tanh z Value	Tanh z $\log_{10}$	Coth z Value	Coth z $\log_{10}$
3.0	10.0179	.00078	10.0677	.00293	0.99505	.99785	1.0050	.00215
3.1	11.0765	.04440	11.1215	.04616	.99595	.99824	1.0041	.00176
3.2	12.2459	.08799	12.2866	.08943	.99668	.99856	1.0033	.00144
3.3	13.5379	.13155	13.5748	.13273	.99728	.99882	1.0027	.00118
3.4	14.9654	.17509	14.9987	.17605	.99777	.99903	1.0022	.00097
3.5	16.5426	.21860	16.5728	.21940	0.99818	.99921	1.0018	.00079
3.6	18.2855	.26211	18.3128	.26275	.99851	.99935	1.0015	.00065
3.7	20.2113	.30559	20.2360	.30612	.99878	.99947	1.0012	.00053
3.8	22.3394	.34907	22.3618	.34951	.99900	.99957	1.0010	.00043
3.9	24.6911	.39254	24.7113	.39290	.99918	.99964	1.0008	.00036
4.0	27.2899	.43600	27.3082	.43629	0.99933	.99971	1.0007	.00029
4.1	30.1619	.47946	30.1784	.47970	.99945	.99976	1.0005	.00024
4.2	33.3357	.52291	33.3507	.52310	.99955	.99980	1.0004	.00020
4.3	36.8431	.56636	36.8567	.56652	.99963	.99984	1.0004	.00016
4.4	40.7193	.60980	40.7316	.60993	.99970	.99987	1.0003	.00013
4.5	45.0030	.65324	45.0141	.65335	0.99975	.99989	1.0002	.00011
4.6	49.7371	.69668	49.7472	.69677	.99980	.99991	1.0002	.00009
4.7	54.9690	.74012	54.9781	.74019	.99983	.99993	1.0002	.00007
4.8	60.7511	.78355	60.7593	.78361	.99986	.99994	1.0001	.00006
4.9	67.1412	.82699	67.1486	.82704	.99989	.99995	1.0001	.00005
5.0	74.2032	.87042	74.2099	.87046	0.99991	.99996	1.0001	.00004
5.1	82.008	.91386	82.014	.91389	.99993	.99997	1.0001	.00003
5.2	90.633	.95730	90.630	.95731	.99994	.99997	1.0001	.00003
5.3	100.17	.00072	100.17	.00074	.99995	.99998	1.0000	.00002
5.4	110.70	.04415	110.71	.04417	.99996	.99998	1.0000	.00002
5.5	122.34	.08758	122.35	.08760	0.99997	.99999	1.0000	.00001
5.6	135.21	.13101	135.22	.13103	.99997	.99999	1.0000	.00001
5.7	149.43	.17444	149.44	.17445	.99998	.99999	1.0000	.00001
5.8	165.15	.21787	165.15	.21788	.99998	.99999	1.0000	.00001
5.9	182.52	.26130	182.52	.26131	.99998	.99999	1.0000	.00001
6.0	201.71	.30473	201.72	.30474	0.99999	.00000	1.0000	.00000
6.1	222.93	.34817	222.93	.34817	.99999	.00000	1.0000	.00000
6.2	246.37	.39159	246.38	.39161	.99999	.00000	1.0000	.00000
6.3	272.29	.43503	272.29	.43503	.99999	.00000	1.0000	.00000
6.4	300.92	.47845	300.92	.47845	.99999	.00000	1.0000	.00000
6.5	332.57	.52188	332.57	.52188	1.0000	.00000	1.0000	.00000
6.6	367.55	.56532	367.55	.56532	1.0000	.00000	1.0000	.00000
6.7	406.20	.60874	406.20	.60874	1.0000	.00000	1.0000	.00000
6.8	448.92	.65217	448.92	.65217	1.0000	.00000	1.0000	.00000
6.9	496.14	.69560	496.14	.69560	1.0000	.00000	1.0000	.00000
7.0	548.32	.73903	548.32	.73903	1.0000	.00000	1.0000	.00000
7.1	605.98	.78246	605.98	.78246	1.0000	.00000	1.0000	.00000
7.2	669.72	.82589	669.72	.82589	1.0000	.00000	1.0000	.00000
7.3	740.15	.86932	740.15	.86932	1.0000	.00000	1.0000	.00000
7.4	817.99	.91275	817.99	.91275	1.0000	.00000	1.0000	.00000
7.5	904.02	.95618	904.02	.95618	1.0000	.00000	1.0000	.00000
7.6	999.10	.99961	999.10	.99961	1.0000	.00000	1.0000	.00000
7.7	1104.2	.04305	1104.2	.04305	1.0000	.00000	1.0000	.00000
7.8	1220.3	.08647	1220.3	.08647	1.0000	.00000	1.0000	.00000
7.9	1348.6	.12988	1348.6	.12988	1.0000	.00000	1.0000	.00000

HYPERBOLIC FUNCTIONS AND THEIR COMMON LOGARITHMS (Continued)

x	Sinh x		Cosh x		Tanh x		Coth x	
	Value	$\log_{10}$	Value	$\log_{10}$	Value	$\log_{10}$	Value	$\log_{10}$
8.0	1490.5	.17333	1490.5	.17333	1.0000	.00000	1.0000	.00000
8.1	1647.2	.21675	1647.2	.21675	1.0000	.00000	1.0000	.00000
8.2	1820.5	.26019	1820.5	.26019	1.0000	.00000	1.0000	.00000
8.3	2011.9	.30360	2011.9	.30360	1.0000	.00000	1.0000	.00000
8.4	2223.5	.34704	2223.5	.34704	1.0000	.00000	1.0000	.00000
8.5	2457.4	.39048	2457.4	.39048	1.0000	.00000	1.0000	.00000
8.6	2715.8	.43390	2715.8	.43390	1.0000	.00000	1.0000	.00000
8.7	3001.5	.47734	3001.5	.47734	1.0000	.00000	1.0000	.00000
8.8	3317.1	.52076	3317.1	.52076	1.0000	.00000	1.0000	.00000
8.9	3666.0	.56419	3666.0	.56419	1.0000	.00000	1.0000	.00000
9.0	4051.5	.60762	4051.5	.60762	1.0000	.00000	1.0000	.00000
9.1	4477.6	.65105	4477.6	.65105	1.0000	.00000	1.0000	.00000
9.2	4948.6	.69448	4948.6	.69448	1.0000	.00000	1.0000	.00000
9.3	5469.0	.73791	5469.0	.73791	1.0000	.00000	1.0000	.00000
9.4	6044.2	.78134	6044.2	.78134	1.0000	.00000	1.0000	.00000
9.5	6679.9	.82477	6679.9	.82477	1.0000	.00000	1.0000	.00000
9.6	7382.4	.86820	7382.4	.86820	1.0000	.00000	1.0000	.00000
9.7	8158.8	.91163	8158.8	.91163	1.0000	.00000	1.0000	.00000
9.8	9016.9	.95506	9016.9	.95506	1.0000	.00000	1.0000	.00000
9.9	9965.2	.99849	9965.2	.99849	1.0000	.00000	1.0000	.00000
10.0	11013.2	.04191	11013.2	.04191	1.0000	.00000	1.0000	.00000

EXPONENTIAL AND HYPERBOLIC FUNCTIONS FOR πx

x	$e^{\pi x}$	$e^{-\pi x}$	$\sinh \pi x$	$\cosh \pi x$	$\tanh \pi x$
0.00	1.00000	1.00000	0.00000	1.00000	0.00000
0.01	1.03191	0.96907	0.03142	1.00049	0.03141
0.02	1.06485	0.93910	0.06287	1.00197	0.06275
0.03	1.09883	0.91006	0.09439	1.00444	0.09397
0.04	1.13390	0.88191	0.12599	1.00791	0.12501
0.05	1.17009	0.85464	0.15773	1.01236	0.15580
0.06	1.20743	0.82820	0.18961	1.01782	0.18629
0.07	1.24597	0.80259	0.22169	1.02428	0.21643
0.08	1.28573	0.77777	0.25398	1.03175	0.24617
0.09	1.32676	0.75371	0.28653	1.04024	0.27544
0.10	1.36911	0.73040	0.31935	1.04976	0.30422
0.11	1.41280	0.70781	0.35249	1.06031	0.33245
0.12	1.45789	0.68592	0.38598	1.07191	0.36009
0.13	1.50442	0.66471	0.41986	1.08456	0.38712
0.14	1.55243	0.64415	0.45414	1.09829	0.41350
0.15	1.60198	0.62423	0.48887	1.11310	0.43920
0.16	1.65310	0.60492	0.52409	1.12901	0.46420
0.17	1.70586	0.58621	0.55982	1.14604	0.48849
0.18	1.76030	0.56808	0.59611	1.16419	0.51204
0.19	1.81648	0.55051	0.63298	1.18350	0.53484
0.20	1.87446	0.53349	0.67048	1.20397	0.55689
0.21	1.93428	0.51699	0.70865	1.22563	0.57819
0.22	1.99601	0.50100	0.74751	1.24850	0.59872
0.23	2.05971	0.48550	0.78710	1.27261	0.61850
0.24	2.12545	0.47049	0.82748	1.29797	0.63752
0.25	2.19328	0.45594	0.86867	1.32461	0.65579
0.26	2.26328	0.44184	0.91072	1.35256	0.67333
0.27	2.33551	0.42817	0.95367	1.38184	0.69014
0.28	2.41005	0.41493	0.99756	1.41249	0.70624
0.29	2.48696	0.40210	1.04243	1.44453	0.72164
0.30	2.56633	0.38966	1.08834	1.47800	0.73636
0.31	2.64824	0.37761	1.13531	1.51292	0.75041
0.32	2.73275	0.36593	1.18341	1.54934	0.76382
0.33	2.81997	0.35461	1.23268	1.58729	0.77659
0.34	2.90997	0.34365	1.28316	1.62681	0.78876
0.35	3.00284	0.33302	1.33491	1.66793	0.80034
0.36	3.09867	0.32272	1.38798	1.71070	0.81135
0.37	3.19756	0.31274	1.44241	1.75515	0.82182
0.38	3.29961	0.30307	1.49827	1.80134	0.83176
0.39	3.40492	0.29369	1.55561	1.84931	0.84119
0.40	3.51359	0.28461	1.61449	1.89910	0.85013
0.41	3.62572	0.27581	1.67496	1.95076	0.85862
0.42	3.74143	0.26728	1.73708	2.00436	0.86665
0.43	3.86084	0.25901	1.80091	2.05993	0.87426
0.44	3.98406	0.25100	1.86653	2.11753	0.88147
0.45	4.11121	0.24324	1.93398	2.17722	0.88828
0.46	4.24241	0.23571	2.00335	2.23906	0.89473
0.47	4.37781	0.22842	2.07469	2.30312	0.90082
0.48	4.51753	0.22136	2.14808	2.36944	0.90658
0.49	4.66170	0.21451	2.22359	2.43811	0.91202
0.50	4.81048	0.20788	2.30130	2.50918	0.91715
0.51	4.96400	0.20145	2.38128	2.58273	0.92200
0.52	5.12243	0.19522	2.46360	2.65882	0.92658
0.53	5.28591	0.18918	2.54836	2.73754	0.93089
0.54	5.45460	0.18333	2.63564	2.81897	0.93497

EXPONENTIAL AND HYPERBOLIC FUNCTIONS FOR πx (Continued)

x	$e^{\pi x}$	$e^{-\pi x}$	sinh πx	cosh πx	tanh πx
0.55	5.62869	0.17766	2.72551	2.90317	0.93880
0.56	5.80832	0.17217	2.81808	2.99024	0.94242
0.57	5.99369	0.16684	2.91343	3.08027	0.94584
0.58	6.18498	0.16168	3.01165	3.17333	0.94905
0.59	6.38237	0.15668	3.11284	3.26953	0.95208
0.60	6.58606	0.15184	3.21711	3.36895	0.95493
0.61	6.79625	0.14714	3.32456	3.47170	0.95762
0.62	7.01315	0.14259	3.43528	3.57787	0.96015
0.63	7.23698	0.13818	3.54940	3.68758	0.96253
0.64	7.46794	0.13391	3.66702	3.80092	0.96477
0.65	7.70628	0.12976	3.78826	3.91802	0.96688
0.66	7.95222	0.12575	3.91323	4.03899	0.96887
0.67	8.20601	0.12186	4.04208	4.16394	0.97073
0.68	8.46790	0.11809	4.17491	4.29300	0.97249
0.69	8.73815	0.11444	4.31186	4.42630	0.97415
0.70	9.01703	0.11090	4.45306	4.56396	0.97570
0.71	9.30480	0.10747	4.59867	4.70614	0.97716
0.72	9.60176	0.10415	4.74881	4.85296	0.97854
0.73	9.90820	0.10093	4.90364	5.00456	0.97983
0.74	10.22442	0.09781	5.06331	5.16111	0.98105
0.75	10.55072	0.09478	5.22797	5.32275	0.98219
0.76	10.88745	0.09185	5.39780	5.48965	0.98327
0.77	11.23492	0.08901	5.57295	5.66196	0.98428
0.78	11.59347	0.08626	5.75361	5.83986	0.98523
0.79	11.96347	0.08359	5.93994	6.02353	0.98612
0.80	12.34528	0.08100	6.13214	6.21314	0.98696
0.81	12.73928	0.07850	6.33039	6.40889	0.98775
0.82	13.14585	0.07607	6.53489	6.61096	0.98849
0.83	13.56539	0.07372	6.74584	6.81955	0.98919
0.84	13.99833	0.07144	6.96344	7.03488	0.98985
0.85	14.44508	0.06923	7.18793	7.25715	0.99046
0.86	14.90609	0.06709	7.41950	7.48659	0.99104
0.87	15.38181	0.06501	7.65840	7.72341	0.99158
0.88	15.87271	0.06300	7.90486	7.96786	0.99209
0.89	16.37929	0.06105	8.15912	8.22017	0.99257
0.90	16.90202	0.05916	8.41243	8.48059	0.99302
0.91	17.44145	0.05733	8.69206	8.74939	0.99345
0.92	17.99808	0.05556	8.97126	9.02682	0.99384
0.93	18.57248	0.05384	9.25932	9.31316	0.99422
0.94	19.16522	0.05218	9.55652	9.60870	0.99457
0.95	19.77687	0.05056	9.86315	9.91372	0.99490
0.96	20.40804	0.04900	10.17952	10.22852	0.99521
0.97	21.05935	0.04748	10.50594	10.55342	0.99550
0.98	21.73146	0.04602	10.84272	10.88874	0.99577
0.99	22.42501	0.04459	11.19021	11.23480	0.99603
1.00	23.14069	0.04321	11.54874	11.59195	0.99627

INVERSE HYPERBOLIC FUNCTIONS

x	$\sinh^{-1} x$	$\tanh^{-1} x$	$\operatorname{cosech}^{-1} x$	$\operatorname{sech}^{-1} x$
0.0	0.00000	0.00000	∞	∞
0.01	0.01000	0.01000	5.29834	5.29829
0.02	0.02000	0.02000	4.60527	4.60507
0.03	0.03000	0.03001	4.19993	4.19948
0.04	0.03999	0.04002	3.91242	3.91162
0.05	0.04998	0.05004	3.68950	3.68825
0.06	0.05996	0.06007	3.50746	3.50566
0.07	0.06994	0.07011	3.35363	3.35118
0.08	0.07991	0.08017	3.22047	3.21727
0.09	0.08988	0.09024	3.10311	3.09906
0.10	0.09983	0.10034	2.99822	2.99322
0.11	0.10978	0.11045	2.90343	2.89738
0.12	0.11971	0.12058	2.81699	2.80979
0.13	0.12964	0.13074	2.73757	2.72912
0.14	0.13955	0.14093	2.66412	2.65432
0.15	0.14944	0.15114	2.59585	2.58459
0.16	0.15933	0.16139	2.53207	2.51927
0.17	0.16919	0.17167	2.47225	2.45780
0.18	0.17904	0.18198	2.41595	2.39975
0.19	0.18888	0.19234	2.36278	2.34473
0.20	0.19869	0.20273	2.31244	2.29243
0.21	0.20849	0.21317	2.26464	2.24258
0.22	0.21826	0.22366	2.21916	2.19495
0.23	0.22802	0.23419	2.17579	2.14933
0.24	0.23775	0.24477	2.13436	2.10554
0.25	0.24747	0.25541	2.09471	2.06344
0.26	0.25716	0.26611	2.05671	2.02288
0.27	0.26682	0.27686	2.02023	1.98374
0.28	0.27646	0.28768	1.98516	1.94591
0.29	0.28608	0.29857	1.95141	1.90930
0.30	0.29567	0.30952	1.91890	1.87382
0.31	0.30524	0.32055	1.88753	1.83939
0.32	0.31478	0.33165	1.85725	1.80594
0.33	0.32429	0.34283	1.82799	1.77340
0.34	0.33377	0.35409	1.79968	1.74172
0.35	0.34322	0.36544	1.77228	1.71083
0.36	0.35265	0.37689	1.74573	1.68070
0.37	0.36204	0.38842	1.71999	1.65127
0.38	0.37140	0.40006	1.69502	1.62250
0.39	0.38073	0.41180	1.67078	1.59436
0.40	0.39004	0.42365	1.64723	1.56680
0.41	0.39930	0.43561	1.62434	1.53979
0.42	0.40854	0.44769	1.60209	1.51331
0.43	0.41774	0.45990	1.58043	1.48731
0.44	0.42691	0.47223	1.55935	1.46178
0.45	0.43605	0.48470	1.53882	1.43669
0.46	0.44515	0.49731	1.51881	1.41200
0.47	0.45422	0.51007	1.49931	1.38771
0.48	0.46325	0.52298	1.48029	1.36379
0.49	0.47225	0.53606	1.46174	1.34021
0.50	0.48121	0.54931	1.44364	1.31696

INVERSE HYPERBOLIC FUNCTIONS (Continued)

x	$\sinh^{-1} x$	$\tanh^{-1} x$	$\operatorname{cosech}^{-1} x$	$\operatorname{sech}^{-1} x$
0.50	0.48121	0.54931	1.44364	1.31696
0.51	0.49014	0.56273	1.42596	1.29401
0.52	0.49903	0.57634	1.40870	1.27136
0.53	0.50788	0.59015	1.39183	1.24898
0.54	0.51670	0.60416	1.37535	1.22686
0.55	0.52548	0.61838	1.35924	1.20497
0.56	0.53422	0.63283	1.34348	1.18331
0.57	0.54293	0.64752	1.32807	1.16186
0.58	0.55160	0.66246	1.31299	1.14060
0.59	0.56023	0.67767	1.29824	1.11952
0.60	0.56882	0.69315	1.28380	1.09861
0.61	0.57738	0.70892	1.26965	1.07785
0.62	0.58590	0.72501	1.25580	1.05723
0.63	0.59438	0.74142	1.24223	1.03673
0.64	0.60282	0.75817	1.22894	1.01635
0.65	0.61122	0.77530	1.21591	0.99606
0.66	0.61959	0.79281	1.20314	0.97585
0.67	0.62792	0.81074	1.19062	0.95572
0.68	0.63620	0.82911	1.17834	0.93564
0.69	0.64446	0.84796	1.16629	0.91560
0.70	0.65267	0.86730	1.15448	0.89559
0.71	0.66084	0.88718	1.14288	0.87559
0.72	0.66897	0.90764	1.13151	0.85558
0.73	0.67707	0.92873	1.12034	0.83555
0.74	0.68513	0.95048	1.10938	0.81549
0.75	0.69315	0.97296	1.09861	0.79537
0.76	0.70113	0.99622	1.08804	0.77517
0.77	0.70907	1.02033	1.07766	0.75487
0.78	0.71697	1.04537	1.06746	0.73445
0.79	0.72484	1.07143	1.05744	0.71388
0.80	0.73267	1.09861	1.04759	0.69315
0.81	0.74046	1.12703	1.03792	0.67221
0.82	0.74821	1.15682	1.02840	0.65103
0.83	0.75592	1.18814	1.01905	0.62958
0.84	0.76360	1.22117	1.00986	0.60781
0.85	0.77124	1.25615	1.00082	0.58568
0.86	0.77884	1.29334	0.99193	0.56313
0.87	0.78640	1.33308	0.98319	0.54008
0.88	0.79393	1.37577	0.97459	0.51647
0.89	0.80142	1.42193	0.96613	0.49220
0.90	0.80887	1.47222	0.95780	0.46715
0.91	0.81628	1.52752	0.94961	0.44116
0.92	0.82366	1.58903	0.94154	0.41406
0.93	0.83100	1.65839	0.93361	0.38560
0.94	0.83830	1.73805	0.92580	0.35542
0.95	0.84557	1.83178	0.91810	0.32304
0.96	0.85280	1.94591	0.91053	0.28768
0.97	0.86000	2.09230	0.90307	0.24807
0.98	0.86716	2.29756	0.89573	0.20169
0.99	0.87428	2.64665	0.88850	0.14201
1.00	0.88137	∞	0.88137	0.00000

INVERSE HYPERBOLIC FUNCTIONS (Continued)

x	$\sinh^{-1} x$	$\cosh^{-1} x$	$\operatorname{cosech}^{-1} x$	$\coth^{-1} x$
1.00	0.88137	0.00000	0.88137	∞
1.01	0.88843	0.14130	0.87436	2.65165
1.02	0.89545	0.19967	0.86744	2.30756
1.03	0.90243	0.24434	0.86063	2.10730
1.04	0.90938	0.28191	0.85391	1.96591
1.05	0.91629	0.31492	0.84730	1.85679
1.06	0.92317	0.34470	0.84078	1.76806
1.07	0.93002	0.37202	0.83435	1.69340
1.08	0.93683	0.39738	0.82801	1.62905
1.09	0.94360	0.42114	0.82177	1.57255
1.10	0.95035	0.44357	0.81561	1.52226
1.11	0.95706	0.46485	0.80954	1.47698
1.12	0.96373	0.48513	0.80355	1.43584
1.13	0.97038	0.50453	0.79764	1.39817
1.14	0.97699	0.52316	0.79182	1.36346
1.15	0.98357	0.54110	0.78607	1.33129
1.16	0.99011	0.55840	0.78041	1.30134
1.17	0.99663	0.57514	0.77482	1.27334
1.18	1.00311	0.59135	0.76930	1.24706
1.19	1.00956	0.60708	0.76386	1.22232
1.20	1.01597	0.62236	0.75849	1.19895
1.21	1.02236	0.63724	0.75319	1.17682
1.22	1.02871	0.65173	0.74796	1.15582
1.23	1.03504	0.66586	0.74279	1.13584
1.24	1.04133	0.67966	0.73770	1.11680
1.25	1.04759	0.69315	0.73267	1.09861
1.26	1.05382	0.70634	0.72770	1.08122
1.27	1.06003	0.71924	0.72280	1.06456
1.28	1.06620	0.73189	0.71796	1.04857
1.29	1.07234	0.74428	0.71318	1.03321
1.30	1.07845	0.75643	0.70846	1.01844
1.31	1.08453	0.76836	0.70380	1.00422
1.32	1.09059	0.78007	0.69920	0.99050
1.33	1.09661	0.79157	0.69465	0.97727
1.34	1.10261	0.80288	0.69016	0.96448
1.35	1.10857	0.81400	0.68572	0.95212
1.36	1.11451	0.82494	0.68134	0.94016
1.37	1.12042	0.83570	0.67701	0.92857
1.38	1.12630	0.84630	0.67273	0.91734
1.39	1.13216	0.85673	0.66851	0.90645
1.40	1.13798	0.86701	0.66433	0.89588
1.41	1.14378	0.87715	0.66020	0.88561
1.42	1.14955	0.88714	0.65612	0.87563
1.43	1.15530	0.89699	0.65209	0.86593
1.44	1.16101	0.90670	0.64811	0.85649
1.45	1.16670	0.91629	0.64417	0.84730
1.46	1.17237	0.92575	0.64028	0.83835
1.47	1.17801	0.93509	0.63643	0.82962
1.48	1.18362	0.94432	0.63263	0.82111
1.49	1.18920	0.95343	0.62886	0.81282
1.50	1.19476	0.96242	0.62515	0.80472

INVERSE HYPERBOLIC FUNCTIONS (Continued)

x	$\sinh^{-1} x$	$\cosh^{-1} x$	$\operatorname{cosech}^{-1} x$	$\coth^{-1} x$
1.50	1.19476	0.96242	0.62515	0.80472
1.51	1.20030	0.97131	0.62147	0.79681
1.52	1.20581	0.98010	0.61783	0.78909
1.53	1.21129	0.98879	0.61424	0.78155
1.54	1.21675	0.99737	0.61068	0.77418
1.55	1.22218	1.00587	0.60716	0.76697
1.56	1.22759	1.01426	0.60368	0.75991
1.57	1.23298	1.02257	0.60024	0.75301
1.58	1.23834	1.03079	0.59684	0.74626
1.59	1.24367	1.03892	0.59347	0.73965
1.60	1.24898	1.04697	0.59014	0.73317
1.61	1.25427	1.05493	0.58685	0.72682
1.62	1.25954	1.06282	0.58359	0.72061
1.63	1.26478	1.07063	0.58036	0.71451
1.64	1.26999	1.07836	0.57717	0.70853
1.65	1.27519	1.08601	0.57401	0.70267
1.66	1.28036	1.09360	0.57089	0.69692
1.67	1.28551	1.10111	0.56780	0.69128
1.68	1.29064	1.10855	0.56474	0.68574
1.69	1.29574	1.11592	0.56171	0.68030
1.70	1.30082	1.12323	0.55871	0.67496
1.71	1.30588	1.13047	0.55574	0.66972
1.72	1.31092	1.13765	0.55281	0.66457
1.73	1.31593	1.14476	0.54990	0.65951
1.74	1.32093	1.15182	0.54702	0.65453
1.75	1.32590	1.15881	0.54417	0.64964
1.76	1.33085	1.16574	0.54135	0.64483
1.77	1.33578	1.17262	0.53856	0.64011
1.78	1.34069	1.17944	0.53579	0.63546
1.79	1.34557	1.18620	0.53305	0.63088
1.80	1.35044	1.19291	0.53034	0.62638
1.81	1.35529	1.19957	0.52766	0.62195
1.82	1.36011	1.20617	0.52500	0.61759
1.83	1.36492	1.21272	0.52237	0.61330
1.84	1.36970	1.21922	0.51976	0.60908
1.85	1.37447	1.22567	0.51718	0.60492
1.86	1.37921	1.23207	0.51462	0.60082
1.87	1.38394	1.23842	0.51208	0.59679
1.88	1.38864	1.24473	0.50957	0.59281
1.89	1.39333	1.25098	0.50709	0.58890
1.90	1.39800	1.25720	0.50462	0.58504
1.91	1.40265	1.26336	0.50218	0.58123
1.92	1.40728	1.26949	0.49977	0.57748
1.93	1.41188	1.27557	0.49737	0.57379
1.94	1.41648	1.28160	0.49500	0.57014
1.95	1.42105	1.28760	0.49265	0.56655
1.96	1.42560	1.29355	0.49032	0.56301
1.97	1.43014	1.29946	0.48801	0.55951
1.98	1.43466	1.30533	0.48572	0.55606
1.99	1.43915	1.31117	0.48346	0.55266
2.00	1.44364	1.31696	0.48121	0.54931

INVERSE HYPERBOLIC FUNCTIONS (Continued)

x	$\sinh^{-1} x$	$\cosh^{-1} x$	$\operatorname{cosech}^{-1} x$	$\coth^{-1} x$
2.00	1.44364	1.31696	0.48121	0.54931
2.10	1.48748	1.37286	0.45982	0.51805
2.20	1.52966	1.42542	0.44019	0.49041
2.30	1.57028	1.47504	0.42213	0.46578
2.40	1.60944	1.52208	0.40547	0.44365
2.50	1.64723	1.56680	0.39004	0.42365
2.60	1.68374	1.60944	0.37571	0.40547
2.70	1.71905	1.65019	0.36239	0.38885
2.80	1.75323	1.68924	0.34996	0.37361
2.90	1.78634	1.72671	0.33834	0.35956
3.00	1.81845	1.76275	0.32745	0.34657
3.10	1.84960	1.79746	0.31723	0.33452
3.20	1.87986	1.83094	0.30763	0.32331
3.30	1.90927	1.86328	0.29857	0.31285
3.40	1.93788	1.89456	0.29003	0.30307
3.50	1.96572	1.92485	0.28196	0.29389
3.60	1.99284	1.95421	0.27432	0.28527
3.70	2.01926	1.98270	0.26708	0.27716
3.80	2.04503	2.01037	0.26021	0.26950
3.90	2.07017	2.03727	0.25368	0.26226
4.00	2.09471	2.06344	0.24747	0.25541
4.10	2.11869	2.08892	0.24155	0.24892
4.20	2.14211	2.11375	0.23590	0.24275
4.30	2.16502	2.13796	0.23051	0.23689
4.40	2.18742	2.16158	0.22536	0.23131
4.50	2.20935	2.18464	0.22043	0.22599
4.60	2.23081	2.20717	0.21571	0.22092
4.70	2.25184	2.22920	0.21119	0.21607
4.80	2.27244	2.25073	0.20685	0.21143
4.90	2.29264	2.27180	0.20269	0.20699
5.00	2.31244	2.29243	0.19869	0.20273
5.10	2.33186	2.31263	0.19484	0.19865
5.20	2.35093	2.33243	0.19114	0.19473
5.30	2.36964	2.35183	0.18758	0.19097
5.40	2.38801	2.37086	0.18414	0.18735
5.50	2.40606	2.38953	0.18083	0.18386
5.60	2.42379	2.40784	0.17764	0.18051
5.70	2.44122	2.42583	0.17455	0.17727
5.80	2.45836	2.44349	0.17157	0.17415
5.90	2.47521	2.46084	0.16869	0.17114
6.00	2.49178	2.47789	0.16590	0.16824
6.10	2.50809	2.49465	0.16321	0.16543
6.20	2.52414	2.51113	0.16060	0.16271
6.30	2.53994	2.52734	0.15807	0.16008
6.40	2.55549	2.54329	0.15562	0.15754
6.50	2.57081	2.55898	0.15325	0.15508
6.60	2.58591	2.57443	0.15094	0.15269
6.70	2.60078	2.58964	0.14871	0.15038
6.80	2.61543	2.60462	0.14653	0.14813
6.90	2.62988	2.61938	0.14442	0.14596
7.00	2.64412	2.63392	0.14238	0.14384

GUDERMANNIAN FUNCTION

x	0	1	2	3	4	5	6	7	8	9
0.0	0.00000	0.01000	0.02000	0.03000	0.03999	0.04998	0.05996	0.06994	0.07991	0.08988
0.1	0.09983	0.10978	0.11971	0.12964	0.13954	0.14944	0.15932	0.16919	0.17904	0.18887
0.2	0.19868	0.20847	0.21825	0.22800	0.23773	0.24744	0.25712	0.26678	0.27641	0.28602
0.3	0.29560	0.30515	0.31467	0.32417	0.33363	0.34307	0.35247	0.36184	0.37117	0.38047
0.4	0.38974	0.39897	0.40817	0.41733	0.42645	0.43554	0.44459	0.45359	0.46256	0.47149
0.5	0.48038	0.48923	0.49803	0.50680	0.51552	0.52420	0.53284	0.54143	0.54997	0.55848
0.6	0.56694	0.57535	0.58372	0.59204	0.60031	0.60854	0.61672	0.62486	0.63294	0.64098
0.7	0.64897	0.65692	0.66481	0.67266	0.68045	0.68820	0.69590	0.70355	0.71115	0.71870
0.8	0.72620	0.73366	0.74106	0.74841	0.75571	0.76297	0.77017	0.77732	0.78443	0.79148
0.9	0.79848	0.80544	0.81234	0.81919	0.82599	0.83275	0.83945	0.84611	0.85271	0.85926
1.0	0.86577	0.87223	0.87863	0.88499	0.89130	0.89756	0.90377	0.90993	0.91604	0.92211
1.1	0.92813	0.93410	0.94002	0.94589	0.95172	0.95750	0.96323	0.96892	0.97455	0.98015
1.2	0.98569	0.99119	0.99665	1.00205	1.00742	1.01274	1.01801	1.02324	1.02842	1.03356
1.3	1.03866	1.04371	1.04872	1.05368	1.05860	1.06348	1.06832	1.07312	1.07787	1.08258
1.4	1.08725	1.09188	1.09647	1.10101	1.10552	1.10999	1.11441	1.11880	1.12315	1.12746
1.5	1.13173	1.13596	1.14015	1.14431	1.14843	1.15251	1.15655	1.16056	1.16453	1.16846
1.6	1.17236	1.17622	1.18005	1.18384	1.18760	1.19132	1.19500	1.19866	1.20228	1.20586
1.7	1.20941	1.21293	1.21642	1.21987	1.22330	1.22668	1.23004	1.23337	1.23666	1.23993
1.8	1.24316	1.24636	1.24954	1.25268	1.25579	1.25888	1.26193	1.26496	1.26795	1.27092
1.9	1.27386	1.27677	1.27966	1.28251	1.28534	1.28815	1.29092	1.29367	1.29639	1.29909
2.0	1.30176	1.30441	1.30703	1.30962	1.31219	1.31473	1.31726	1.31975	1.32222	1.32467
2.1	1.32710	1.32950	1.33188	1.33423	1.33656	1.33887	1.34116	1.34343	1.34567	1.34789
2.2	1.35009	1.35227	1.35443	1.35656	1.35868	1.36077	1.36285	1.36490	1.36694	1.36895
2.3	1.37095	1.37292	1.37488	1.37682	1.37873	1.38063	1.38251	1.38438	1.38622	1.38805
2.4	1.38986	1.39165	1.39342	1.39518	1.39691	1.39864	1.40034	1.40203	1.40370	1.40535
2.5	1.40699	1.40862	1.41022	1.41181	1.41339	1.41495	1.41649	1.41802	1.41954	1.42104
2.6	1.42252	1.42399	1.42545	1.42689	1.42832	1.42973	1.43113	1.43251	1.43388	1.43524
2.7	1.43659	1.43792	1.43924	1.44054	1.44183	1.44311	1.44438	1.44564	1.44688	1.44811
2.8	1.44933	1.45053	1.45173	1.45291	1.45408	1.45524	1.45638	1.45752	1.45864	1.45976
2.9	1.46086	1.46195	1.46303	1.46410	1.46516	1.46621	1.46725	1.46828	1.46930	1.47031
3.0	1.47130	1.47229	1.47327	1.47424	1.47520	1.47615	1.47709	1.47802	1.47894	1.47986
3.1	1.48076	1.48165	1.48254	1.48342	1.48428	1.48514	1.48600	1.48684	1.48767	1.48850
3.2	1.48932	1.49013	1.49093	1.49172	1.49251	1.49329	1.49406	1.49482	1.49558	1.49632
3.3	1.49706	1.49780	1.49852	1.49924	1.49995	1.50066	1.50135	1.50204	1.50273	1.50340
3.4	1.50407	1.50474	1.50539	1.50605	1.50669	1.50733	1.50796	1.50858	1.50920	1.50981
3.5	1.51042	1.51102	1.51161	1.51220	1.51279	1.51336	1.51393	1.51450	1.51506	1.51561
3.6	1.51616	1.51671	1.51724	1.51778	1.51830	1.51883	1.51934	1.51985	1.52036	1.52086
3.7	1.52136	1.52185	1.52234	1.52282	1.52330	1.52377	1.52424	1.52470	1.52516	1.52561
3.8	1.52606	1.52651	1.52695	1.52738	1.52782	1.52824	1.52867	1.52909	1.52950	1.52991
3.9	1.53032	1.53072	1.53112	1.53151	1.53190	1.53229	1.53267	1.53305	1.53343	1.53380
4.0	1.53417	1.53453	1.53489	1.53525	1.53561	1.53596	1.53630	1.53664	1.53698	1.53732
4.1	1.53765	1.53798	1.53831	1.53863	1.53895	1.53927	1.53958	1.53989	1.54020	1.54051
4.2	1.54081	1.54111	1.54140	1.54169	1.54198	1.54227	1.54255	1.54283	1.54311	1.54339
4.3	1.54366	1.54393	1.54420	1.54446	1.54472	1.54498	1.54524	1.54550	1.54575	1.54600
4.4	1.54624	1.54649	1.54673	1.54697	1.54721	1.54744	1.54767	1.54790	1.54813	1.54836
4.5	1.54858	1.54880	1.54902	1.54924	1.54945	1.54966	1.54987	1.55008	1.55029	1.55049
4.6	1.55069	1.55089	1.55109	1.55129	1.55148	1.55167	1.55186	1.55205	1.55224	1.55242
4.7	1.55261	1.55279	1.55297	1.55314	1.55332	1.55349	1.55367	1.55384	1.55400	1.55417
4.8	1.55434	1.55450	1.55466	1.55482	1.55498	1.55514	1.55530	1.55545	1.55560	1.55575
4.9	1.55590	1.55605	1.55620	1.55634	1.55649	1.55663	1.55677	1.55691	1.55705	1.55719

GUDERMANNIAN FUNCTION (Continued)

x	0	1	2	3	4	5	6	7	8	9
5.0	1.55732	1.55745	1.55759	1.55772	1.55785	1.55798	1.55811	1.55823	1.55836	1.55848
5.1	1.55860	1.55872	1.55884	1.55896	1.55908	1.55920	1.55931	1.55943	1.55954	1.55965
5.2	1.55976	1.55987	1.55998	1.56009	1.56020	1.56030	1.56041	1.56051	1.56061	1.56071
5.3	1.56081	1.56091	1.56101	1.56111	1.56120	1.56130	1.56139	1.56149	1.56158	1.56167
5.4	1.56176	1.56185	1.56194	1.56203	1.56212	1.56220	1.56229	1.56237	1.56246	1.56254
5.5	1.56262	1.56270	1.56278	1.56286	1.56294	1.56302	1.56310	1.56318	1.56325	1.56333
5.6	1.56340	1.56347	1.56355	1.56362	1.56369	1.56376	1.56383	1.56390	1.56397	1.56404
5.7	1.56410	1.56417	1.56424	1.56430	1.56437	1.56443	1.56449	1.56456	1.56462	1.56468
5.8	1.56474	1.56480	1.56486	1.56492	1.56498	1.56504	1.56509	1.56515	1.56521	1.56526
5.9	1.56532	1.56537	1.56543	1.56548	1.56553	1.56558	1.56564	1.56569	1.56574	1.56579
6.0	1.56584	1.56589	1.56594	1.56599	1.56603	1.56608	1.56613	1.56617	1.56622	1.56627
6.1	1.56631	1.56636	1.56640	1.56644	1.56649	1.56653	1.56657	1.56661	1.56666	1.56670
6.2	1.56674	1.56678	1.56682	1.56686	1.56690	1.56694	1.56697	1.56701	1.56705	1.56709
6.3	1.56712	1.56716	1.56720	1.56723	1.56727	1.56730	1.56734	1.56737	1.56741	1.56744
6.4	1.56747	1.56751	1.56754	1.56757	1.56760	1.56764	1.56767	1.56770	1.56773	1.56776
6.5	1.56779	1.56782	1.56785	1.56788	1.56791	1.56794	1.56796	1.56799	1.56802	1.56805
6.6	1.56808	1.56810	1.56813	1.56816	1.56818	1.56821	1.56823	1.56826	1.56828	1.56831
6.7	1.56833	1.56836	1.56838	1.56841	1.56843	1.56845	1.56848	1.56850	1.56852	1.56855
6.8	1.56857	1.56859	1.56861	1.56863	1.56866	1.56868	1.56870	1.56872	1.56874	1.56876
6.9	1.56878	1.56880	1.56882	1.56884	1.56886	1.56888	1.56890	1.56892	1.56894	1.56895
7.0	1.56897	1.56899	1.56901	1.56903	1.56904	1.56906	1.56908	1.56910	1.56911	1.56913
7.1	1.56915	1.56916	1.56918	1.56919	1.56921	1.56923	1.56924	1.56926	1.56927	1.56929
7.2	1.56930	1.56932	1.56933	1.56935	1.56936	1.56938	1.56939	1.56940	1.56942	1.56943
7.3	1.56945	1.56946	1.56947	1.56949	1.56950	1.56951	1.56952	1.56954	1.56955	1.56956
7.4	1.56957	1.56959	1.56960	1.56961	1.56962	1.56963	1.56965	1.56966	1.56967	1.56968
7.5	1.56969	1.56970	1.56971	1.56972	1.56973	1.56974	1.56975	1.56976	1.56978	1.56979
7.6	1.56980	1.56981	1.56982	1.56983	1.56983	1.56984	1.56985	1.56986	1.56987	1.56988
7.7	1.56989	1.56990	1.56991	1.56992	1.56993	1.56993	1.56994	1.56995	1.56996	1.56997
7.8	1.56998	1.56999	1.56999	1.57000	1.57001	1.57002	1.57002	1.57003	1.57004	1.57005
7.9	1.57005	1.57006	1.57007	1.57008	1.57008	1.57009	1.57010	1.57010	1.57011	1.57012
8.0	1.57013	1.57013	1.57014	1.57015	1.57015	1.57016	1.57016	1.57017	1.57018	1.57018
8.1	1.57019	1.57020	1.57020	1.57021	1.57021	1.57022	1.57022	1.57023	1.57024	1.57024
8.2	1.57025	1.57025	1.57026	1.57026	1.57027	1.57027	1.57028	1.57028	1.57029	1.57029
8.3	1.57030	1.57030	1.57031	1.57031	1.57032	1.57032	1.57033	1.57033	1.57034	1.57034
8.4	1.57035	1.57035	1.57036	1.57036	1.57036	1.57037	1.57037	1.57038	1.57038	1.57039
8.5	1.57039	1.57039	1.57040	1.57040	1.57041	1.57041	1.57041	1.57042	1.57042	1.57042
8.6	1.57043	1.57043	1.57044	1.57044	1.57044	1.57045	1.57045	1.57045	1.57046	1.57046
8.7	1.57046	1.57047	1.57047	1.57047	1.57048	1.57048	1.57048	1.57049	1.57049	1.57049
8.8	1.57049	1.57050	1.57050	1.57050	1.57051	1.57051	1.57051	1.57052	1.57052	1.57052
8.9	1.57052	1.57053	1.57053	1.57053	1.57053	1.57054	1.57054	1.57054	1.57054	1.57055
9.0	1.57055	1.57055	1.57055	1.57056	1.57056	1.57056	1.57056	1.57057	1.57057	1.57057
9.1	1.57057	1.57058	1.57058	1.57058	1.57058	1.57058	1.57059	1.57059	1.57059	1.57059
9.2	1.57059	1.57060	1.57060	1.57060	1.57060	1.57060	1.57061	1.57061	1.57061	1.57061
9.3	1.57061	1.57062	1.57062	1.57062	1.57062	1.57062	1.57062	1.57063	1.57063	1.57063
9.4	1.57063	1.57063	1.57063	1.57064	1.57064	1.57064	1.57064	1.57064	1.57064	1.57065
9.5	1.57065	1.57065	1.57065	1.57065	1.57065	1.57065	1.57066	1.57066	1.57066	1.57066
9.6	1.57066	1.57066	1.57066	1.57066	1.57067	1.57067	1.57067	1.57067	1.57067	1.57067
9.7	1.57067	1.57067	1.57068	1.57068	1.57068	1.57068	1.57068	1.57068	1.57068	1.57068
9.8	1.57069	1.57069	1.57069	1.57069	1.57069	1.57069	1.57069	1.57069	1.57069	1.57069
9.9	1.57070	1.57070	1.57070	1.57070	1.57070	1.57070	1.57070	1.57070	1.57070	1.57070

INVERSE GUDERMANNIAN FUNCTION

x	0	1	2	3	4	5	6	7	8	9
0.0	0.00000	0.01000	0.02000	0.03000	0.04001	0.05002	0.06004	0.07006	0.08009	0.09012
0.1	0.10017	0.11022	0.12029	0.13037	0.14046	0.15057	0.16069	0.17082	0.18098	0.19115
0.2	0.20135	0.21156	0.22180	0.23206	0.24234	0.25265	0.26298	0.27334	0.28373	0.29415
0.3	0.30460	0.31509	0.32561	0.33616	0.34675	0.35737	0.36804	0.37874	0.38949	0.40028
0.4	0.41111	0.42199	0.43292	0.44390	0.45493	0.46600	0.47714	0.48833	0.49957	0.51087
0.5	0.52224	0.53366	0.54515	0.55671	0.56834	0.58003	0.59180	0.60364	0.61555	0.62755
0.6	0.63962	0.65178	0.66402	0.67636	0.68878	0.70129	0.71390	0.72661	0.73942	0.75233
0.7	0.76535	0.77848	0.79172	0.80508	0.81856	0.83217	0.84590	0.85976	0.87376	0.88790
0.8	0.90218	0.91660	0.93118	0.94592	0.96082	0.97589	0.99113	1.00654	1.02215	1.03794
0.9	1.05392	1.07011	1.08651	1.10313	1.11997	1.13704	1.15435	1.17192	1.18974	1.20783
1.0	1.22619	1.24485	1.26380	1.28306	1.30265	1.32258	1.34285	1.36349	1.38451	1.40593
1.1	1.42776	1.45003	1.47275	1.49594	1.51963	1.54384	1.56860	1.59394	1.61987	1.64645
1.2	1.67370	1.70166	1.73037	1.75987	1.79022	1.82147	1.85367	1.88689	1.92120	1.95667
1.3	1.99340	2.03147	2.07100	2.11210	2.15491	2.19959	2.24630	2.29524	2.34666	2.40080
1.4	2.45800	2.51861	2.58307	2.65193	2.72583	2.80558	2.89219	2.98695	3.09160	3.20843
1.5	3.34068	3.49307	3.67286	3.89217	4.17343	4.56609	5.22169	7.82865		

ANALYTIC GEOMETRY

Dr. Howard Eves

RECTANGULAR COORDINATES IN A PLANE

Rectangular (Cartesian) Coordinates

Let $X'X$ (called the *x-axis*) and $Y'Y$ (called the *y-axis*) be two perpendicular lines (here taken horizontally and vertically, respectively) intersecting in point O (called the *origin*). Then any point P in the plane of the axes is located by the distance x (called the *abscissa*) and the distance y (called the *ordinate*) from $Y'Y$ and $X'X$, respectively, to P, where x is taken as positive to the right and negative to the left of $Y'Y$, and y is taken as positive above and negative below $X'X$. The ordered pair of numbers, (x,y), are called *rectangular coordinates* of the point P.

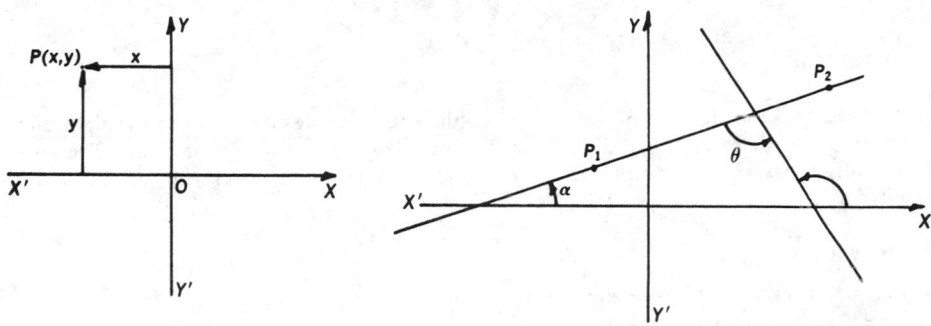

Points, Slopes, Angles

Let $P_1(x_1,y_1)$ and $P_2(x_2,y_2)$ be any two points and let α be the angle measured counterclockwise from $X'X$ to P_1P_2.

Distance between P_1 and P_2:
$$\sqrt{(x_2 - x_1)^2 + (y_2 - y_1)^2}$$

Point dividing P_1P_2 in ratio $\dfrac{r}{s}$:
$$\left(\frac{rx_2 + sx_1}{r + s} , \frac{ry_2 + sy_1}{r + s}\right)$$

Midpoint of P_1P_2:
$$\left(\frac{x_1 + x_2}{2} , \frac{y_1 + y_2}{2}\right)$$

Slope m of P_1P_2:
$$m = \tan \alpha = \frac{y_2 - y_1}{x_2 - x_1}$$

Angle θ between two lines of slopes m_1 and m_2:
$$\tan \theta = \frac{m_2 - m_1}{1 + m_1 m_2}$$

For parallel lines: $m_1 = m_2$
For perpendicular lines: $m_1 m_2 = -1$

Points P_1, P_2, P_3 are collinear if and only if
$$\begin{vmatrix} x_1 & y_1 & 1 \\ x_2 & y_2 & 1 \\ x_3 & y_3 & 1 \end{vmatrix} = 0.$$

Polygonal Areas

Area of triangle $P_1 P_2 P_3$:

$$\frac{1}{2}\begin{vmatrix} x_1 & y_1 & 1 \\ x_2 & y_2 & 1 \\ x_3 & y_3 & 1 \end{vmatrix} = \tfrac{1}{2}(x_1 y_2 + x_2 y_3 + x_3 y_1 - y_1 x_2 - y_2 x_3 - y_3 x_1)$$

Area of polygon $P_1 P_2 \cdots P_n$:

$$\tfrac{1}{2}(x_1 y_2 + x_2 y_3 + \cdots + x_{n-1} y_n + x_n y_1 - y_1 x_2 - y_2 x_3 - \cdots - y_{n-1} x_n - y_n x_1)$$

Note. The parenthesis in the last formula is remembered by the device

Here one adds the products of coordinates on the lines slanting downward to the right and subtracts the products of coordinates on the lines slanting upward to the right. The area is positive or negative according as $P_1 P_2 \cdots P_n$ is a counterclockwise or clockwise polygon.

Straight Lines

Line parallel to y-axis: $x = a$
Line parallel to x-axis: $y = b$
Slope y-intercept form: $y = mx + b$

Intercept form: $\dfrac{x}{a} + \dfrac{y}{b} = 1$

Point-slope form: $y - y_1 = m(x - x_1)$

Two-point form: $\dfrac{y - y_1}{x - x_1} = \dfrac{y_2 - y_1}{x_2 - x_1}$

or

$$\begin{vmatrix} x & y & 1 \\ x_1 & y_1 & 1 \\ x_2 & y_2 & 1 \end{vmatrix} = 0$$

Normal form: $x \cos \omega + y \sin \omega = p$
General form: $Ax + By + C = 0$

Slope: $m = -\dfrac{A}{B}$

Intercepts: $a = -\dfrac{C}{A}, \quad b = -\dfrac{C}{B}$

To reduce $Ax + By + C = 0$ to normal form, divide by $\pm \sqrt{A^2 + B^2}$, where the sign of the radical is chosen opposite to the sign of C when $C \neq 0$ and the same as the sign of B when $C = 0$.

Distance from $Ax + By + C = 0$ to P_1: $\dfrac{Ax_1 + By_1 + C}{\pm \sqrt{A^2 + B^2}}$

Formulas for Use in Analytic Geometry

Angle θ between lines $A_1 x + B_1 y + C_1 = 0$
and $A_2 x + B_2 y + C_2 = 0$

$$\tan \theta = \frac{A_1 B_2 - A_2 B_1}{A_1 A_2 + B_1 B_2}$$

Lines parallel: $\qquad A_1 B_2 = A_2 B_1$

Lines perpendicular: $\qquad A_1 A_2 = - B_1 B_2$

Lines $A_1 x + B_1 y + C_1 = 0, A_2 x + B_2 y + C_2 = 0, A_3 x + B_3 y + C_3 = 0$ are concurrent if and only if

$$\begin{vmatrix} A_1 & B_1 & C_1 \\ A_2 & B_2 & C_2 \\ A_3 & B_3 & C_3 \end{vmatrix} = 0.$$

Line of Best Fit

In seeking the straight line which best fits a given set of n points $P_1(x_1, y_1), P_2(x_2, y_2),$ $\cdots, P_n(x_n, y_n)$, calculate

$$\overline{x} = \frac{x_1 + x_2 + \cdots + x_n}{n}, \qquad \overline{y} = \frac{y_1 + y_2 + \cdots + y_n}{n},$$

$$m = \frac{(x_1 y_1 + x_2 y_2 + \cdots + x_n y_n) - n\overline{x}\overline{y}}{(x_1^2 + x_2^2 + \cdots + x_n^2) - n\overline{x}^2}.$$

Then the sought line is given by

$$y - \overline{y} = m(x - \overline{x}).$$

Circles

Center at origin, radius r: $\qquad x^2 + y^2 = r^2$

Center at (h, k), radius r: $\qquad (x - h)^2 + (y - k)^2 = r^2$

General form: $\qquad \begin{cases} Ax^2 + Ay^2 + Dx + Ey + F = 0, A \neq 0 \\ x^2 + y^2 + 2dx + 2ey + f = 0 \end{cases}$

Center: $\qquad (-d, -e)$

Radius: $\qquad r = \sqrt{d^2 + e^2 - f}$

Circle on $P_1 P_2$ as diameter: $\qquad (x - x_1)(x - x_2) + (y - y_1)(y - y_2) = 0$

Three-point form: $\qquad \begin{vmatrix} x^2 + y^2 & x & y & 1 \\ x_1^2 + y_1^2 & x_1 & y_1 & 1 \\ x_2^2 + y_2^2 & x_2 & y_2 & 1 \\ x_3^2 + y_3^2 & x_3 & y_3 & 1 \end{vmatrix} = 0$

Conic Sections

A *conic section* is the locus of a point P that moves in the plane of a fixed point F (called a *focus*) and a fixed line d (called a *directrix*), F not on d, such that the ratio of the distance of P from F to its distance from d is a constant e (called the *eccentricity*).

If $e = 1$, the conic is a *parabola*; if $e < 1$, an *ellipse*; if $e > 1$, a *hyperbola*.

Focus, $(0,0)$; directrix, $x = -a$: $\quad x^2 + y^2 = e^2(x + a)^2$

Formulas for Use in Analytic Geometry

Parabolas $(e = 1)$

Let p = distance from the vertex to the focus, e = eccentricity.

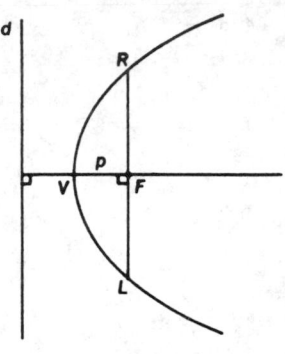

V: vertex
F: focus
d: directrix
LR: latus rectum
line VF: axis

Latus rectum:	$4p$
Distance from vertex to directrix:	p
Vertex at origin, focus at $(p,0)$:	$y^2 = 4px$
Vertex at origin, focus at $(-p,0)$:	$y^2 = -4px$
Vertex at origin, focus at $(0,p)$:	$x^2 = 4py$
Vertex at origin, focus at $(0,-p)$:	$x^2 = -4py$
Vertex, (h,k); focus, $(h + p,k)$:	$(y - k)^2 = 4p(x - h)$
Vertex, (h,k); focus, $(h - p,k)$:	$(y - k)^2 = -4p(x - h)$
Vertex, (h,k); focus, $(h,k + p)$:	$(x - h)^2 = 4p(y - k)$
Vertex, (h,k); focus, $(h,k - p)$:	$(x - h)^2 = -4p(y - k)$
General form, axis parallel to $X'X$:	$Cy^2 + Dx + Ey + F = 0$
General form, axis parallel to $Y'Y$:	$\begin{cases} Ax^2 + Dx + Ey + F = 0 \\ y = ax^2 + bx + c \end{cases}$
General form, axis oblique to coordinate axes:	$Ax^2 + Bxy + Cy^2 + Dx + Ey + F = 0,$ $B^2 - 4AC = 0$

Ellipses $(e < 1)$

Let $2a$ = major axis, $2b$ = minor axis, e = eccentricity.

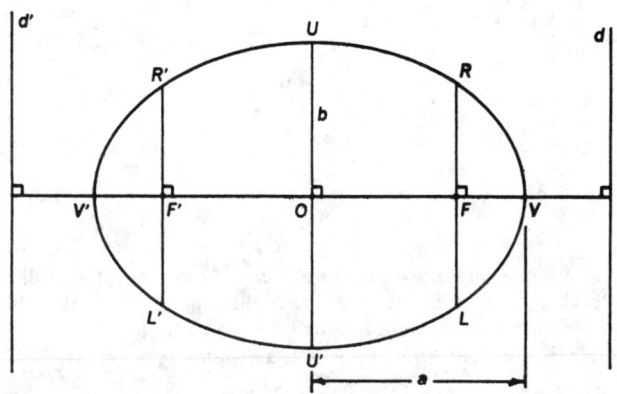

O: center
V, V': vertices
$V'V$: major axis = $2a$
$U'U$: minor axis = $2b$
F, F': foci
d, d': directrices
$LR, L'R'$: latera recta

Eccentricity: $$e = \frac{\sqrt{a^2 - b^2}}{a}$$

Latus rectum: $$\frac{2b^2}{a}$$

Distance from center to either focus: $$\sqrt{a^2 - b^2}$$

Distance from center to either directrix: $$\frac{a}{e}$$

Sum of distances from any point on ellipse to the foci: $2a$

Center at origin, foci on $X'X$: $$\frac{x^2}{a^2} + \frac{y^2}{b^2} = 1$$

Center at origin, foci on $Y'Y$: $$\frac{x^2}{b^2} + \frac{y^2}{a^2} = 1$$

Center at (h,k), major axis parallel to $X'X$: $$\frac{(x - h)^2}{a^2} + \frac{(y - k)^2}{b^2} = 1$$

Center at (h,k), major axis parallel to $Y'Y$: $$\frac{(x - h)^2}{b^2} + \frac{(y - k)^2}{a^2} = 1$$

General form, axes parallel to coordinate axes: $$Ax^2 + Cy^2 + Dx + Ey + F = 0, AC > 0$$

General form, axes oblique to coordinate axes: $$Ax^2 + Bxy + Cy^2 + Dx + Ey + F = 0,$$ $$B^2 - 4AC < 0$$

For a *circle*: $a = b$, $e = 0$, foci coincide at the center of the circle, directrices are at infinity.

Hyperbolas ($e > 1$)

Let $2a$ = transverse axis, $2b$ = conjugate axis, e = eccentricity.

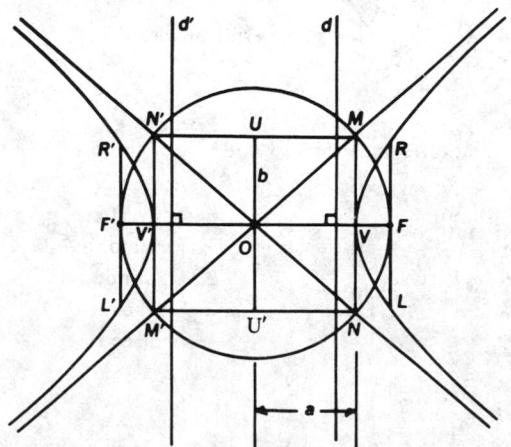

O: center
V, V': vertices
$V'V$: transverse axis = $2a$
$U'U$: conjugate axis = $2b$
F, F': foci
d, d': directrices
$LR, L'R'$: latera recta
lines $M'M$ and $N'N$: asymptotes

Eccentricity: $$e = \frac{\sqrt{a^2 + b^2}}{a}$$

Latus rectum: $$\frac{2b^2}{a}$$

Distance from center to either focus: $\sqrt{a^2 + b^2}$

Distance from center to either directrix: $\dfrac{a}{e}$

Difference of distances of any point on hyperbola from foci: $2a$

Center at origin, foci on $X'X$: $$\frac{x^2}{a^2} - \frac{y^2}{b^2} = 1$$

   Slopes of asymptotes: $\pm \dfrac{b}{a}$

Center at origin, foci on $Y'Y$: $$\frac{y^2}{a^2} - \frac{x^2}{b^2} = 1$$

   Slopes of asymptotes: $\pm \dfrac{a}{b}$

Center at (h,k), transverse axis parallel
to $X'X$: $$\frac{(x - h)^2}{a^2} - \frac{(y - k)^2}{b^2} = 1$$

   Slopes of asymptotes: $\pm \dfrac{b}{a}$

Center at (h, k), transverse axis parallel
to $Y'Y$: $$\frac{(y - k)^2}{a^2} - \frac{(x - h)^2}{b^2} = 1$$

   Slopes of asymptotes: $\pm \dfrac{a}{b}$

Center at origin, $X'X$ and $Y'Y$ for
asymptotes: $xy = c$

Center at (h,k), asymptotes parallel to $X'X$
and $Y'Y$: $(x - h)(y - k) = c$

General form, axes parallel to coordinate
axes: $Ax^2 + Cy^2 + Dx + Ey + F = 0, AC < 0$

General form, axes oblique to coordinate
axes: $Ax^2 + Bxy + Cy^2 + Dx + Ey + F = 0,$
$B^2 - 4AC > 0$

For a *rectangular hyperbola*: $a = b, e = \sqrt{2}$, asymptotes are perpendicular.

General Equation of Second Degree

The nature of the graph of the general quadratic equation in x and y,

$$ax^2 + 2hxy + by^2 + 2gx + 2fy + c = 0,$$

is described in the following table in terms of the values of

$$\Delta = \begin{vmatrix} a & h & g \\ h & b & f \\ g & f & c \end{vmatrix}, \quad J = \begin{vmatrix} a & h \\ h & b \end{vmatrix},$$

$$I = a + b, \quad K = \begin{vmatrix} a & g \\ g & c \end{vmatrix} + \begin{vmatrix} b & f \\ f & c \end{vmatrix}.$$

Formulas for Use in Analytic Geometry

Case	$\triangle$	J	$\triangle/I$	K	Conic
1	$\neq 0$	> 0	< 0		real ellipse
2	$\neq 0$	> 0	> 0		imaginary ellipse
3	$\neq 0$	< 0			hyperbola
4	$\neq 0$	0			parabola
5	0	< 0			real intersecting lines
6	0	> 0			conjugate complex intersecting lines
7	0	0		< 0	real distinct parallel lines
8	0	0		> 0	conjugate complex parallel lines
9	0	0		0	coincident lines

In cases 1, 2, and 3, the center (x_0, y_0) of the conic is given by the simultaneous solution of the equations

$$ax + hy + g = 0, \quad hx + by + f = 0.$$

The equations of the axes of the conic are

$$y - y_0 = m(x - x_0), \quad y - y_0 = -\frac{1}{m}(x - x_0),$$

where m is the positive root of

$$hm^2 + (a - b)m - h = 0.$$

Transformation of Coordinates

To transform an equation of a curve from an old system of rectangular coordinates (x, y) to a new system of rectangular coordinates (x', y'), substitute for each old variable in the equation of the curve its expression in terms of the new variables.

Translation: $\begin{cases} x = x' + h \\ y = y' + k \end{cases}$ The new axes are parallel to the old axes and the coordinates of the new origin in terms of the old system are (h, k).

Rotation: $\begin{cases} x = x' \cos \theta - y' \sin \theta \\ y = x' \sin \theta + y' \cos \theta \end{cases}$ The new origin is coincident with the old origin and the new axes make an angle θ with the old axes.

To remove the xy-term from the equation

$$ax^2 + 2hxy + by^2 + 2gx + 2fy + c = 0,$$

rotate the coordinate axes about the origin through the acute angle $\theta = \arctan m$, where m is the positive root of

$$hm^2 + (a - b)m - h = 0.$$

OBLIQUE COORDINATES IN A PLANE

Oblique (Cartesian) Coordinates

Let $X'X$ (called the *x-axis*, here taken horizontally) and $Y'Y$ (called the *y-axis*) be two lines intersecting in point O (called the *origin*), and denote by ω the counterclockwise angle from $X'X$ to $Y'Y$. Then any point P in the plane of the axes is located by the distance x (called the *abscissa*) measured parallel to the x-axis and the distance y (called

Formulas for Use in Analytic Geometry

the *ordinate*) measured parallel to the y-axis from $Y'Y$ and $X'X$, respectively, to P, **where** x is taken as positive to the right and negative to the left of $Y'Y$, and y is taken as positive above and negative below $X'X$. The ordered pair of numbers, (x,y), are called *oblique co-ordinates* of the point P. If $\omega = 90°$, this coordinate system becomes a rectangular (Cartesian) coordinate system.

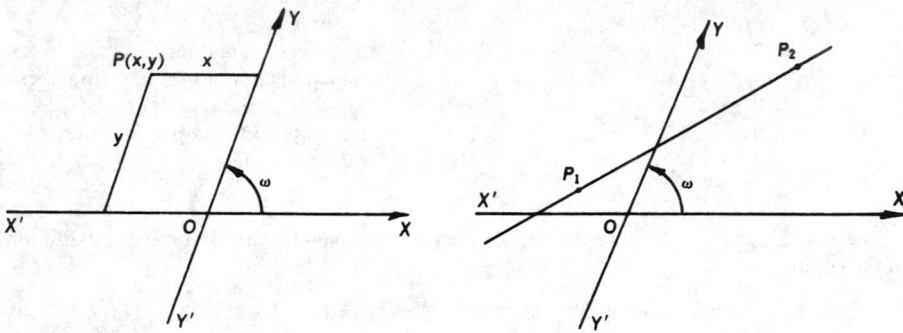

Points

Let $P_1(x_1,y_1)$ and $P_2(x_2,y_2)$ be any two points.
Distance between P_1 and P_2:

$$\sqrt{(x_2 - x_1)^2 + (y_2 - y_1)^2 + 2(x_2 - x_1)(y_2 - y_1)\cos \omega}$$

Point dividing $P_1 P_2$ in ratio $\dfrac{r}{s}$:
$$\left(\frac{rx_2 + sx_1}{r + s}, \frac{ry_2 + sy_1}{r + s}\right)$$

Midpoint of $P_1 P_2$:
$$\left(\frac{x_1 + x_2}{2}, \frac{y_1 + y_2}{2}\right)$$

Points P_1, P_2, P_3 are collinear if and only if $\begin{vmatrix} x_1 & y_1 & 1 \\ x_2 & y_2 & 1 \\ x_3 & y_3 & 1 \end{vmatrix} = 0.$

Polygonal Areas

Area of triangle $P_1 P_2 P_3$:

$$\sin \omega \begin{vmatrix} x_1 & y_1 & 1 \\ x_2 & y_3 & 1 \\ x_3 & y_3 & 1 \end{vmatrix} = \tfrac{1}{2}(\sin \omega)(x_1 y_2 + x_2 y_3 + x_3 y_1 - y_1 x_2 - y_2 x_3 - y_3 x_1)$$

Area of polygon $P_1 P_2 \cdots P_n$:

$$\tfrac{1}{2}(\sin \omega)(x_1 y_2 + x_2 y_3 + \cdots + x_{n-1} y_n + x_n y_1 - y_1 x_2 - y_2 x_3 - \cdots - y_{n-1} x_n - y_n x_1)$$

The area is positive or negative according as $P_1 P_2 \cdots P_n$ is a counterclockwise or clockwise polygon.

Straight Lines

Line parallel to y-axis: $x = a$

Line parallel to x-axis: $y = b$

Intercept form: $\dfrac{x}{a} + \dfrac{y}{b} = 1$

Two-point form: $\dfrac{y - y_1}{x - x_1} = \dfrac{y_2 - y_1}{x_2 - x_1}$, or $\begin{vmatrix} x & y & 1 \\ x_1 & y_1 & 1 \\ x_2 & y_2 & 1 \end{vmatrix} = 0$

General form: $Ax + By + C = 0$

   Intercepts: $a = -\dfrac{C}{A}$, $b = -\dfrac{C}{B}$

Distance from $Ax + By + C = 0$ to P_1: $\dfrac{(Ax_1 + By_1 + C)\sin \omega}{\pm \sqrt{A^2 + B^2 - 2AB \cos \omega}}$

Angle θ between lines $A_1 x + B_1 y + C_1 = 0$ and $A_2 x + B_2 y + C_2 = 0$:

$$\tan \theta = \frac{(A_1 B_2 - A_2 B_1)\sin \omega}{A_1 A_2 + B_1 B_2 - (A_1 B_2 + A_2 B_1)\cos \omega}$$

   Lines parallel: $A_1 B_2 = A_2 B_1$

   Lines perpendicular: $A_1 A_2 + B_1 B_2 = (A_1 B_2 + A_2 B_1)\cos \omega$

Lines $A_1 x + B_1 y + C_1 = 0$, $A_2 x + B_2 y + C_2 = 0$, $A_3 x + B_3 y + C_3 = 0$ are concurrent if and only if

$$\begin{vmatrix} A_1 & B_1 & C_1 \\ A_2 & B_2 & C_2 \\ A_3 & B_3 & C_3 \end{vmatrix} = 0.$$

Circles

Center at (h, k), radius r: $(x - h)^2 + (y - k)^2 + 2(x - h)(y - k)\cos \omega = r^2$

Transformation of Coordinates

Translation: $\begin{cases} x = x' + h \\ y = y' + k \end{cases}$ The new axes are parallel to the old axes and the coordinates of the new origin in terms of the old system are (h, k).

From one oblique system to another, origin fixed: $\begin{cases} x = \dfrac{x' \sin(\omega - \theta) + y' \sin(\omega - \omega' - \theta)}{\sin \omega} \\ y = \dfrac{x' \sin \theta + y' \sin(\omega' + \theta)}{\sin \omega} \end{cases}$

The old and new origins coincide; the old axes intersect at angle ω; the new axes intersect at angle ω'; the counterclockwise angle from x-axis to x'-axis is θ.

POLAR COORDINATES IN A PLANE

Polar Coordinates

In a plane, let OX (called the *initial line*) be a fixed ray radiating from point O (called the *pole* or *origin*). Then any point P, other than O, in the plane is located by angle θ (called the *vectorial angle*) measured from OX to the line determined by O and P and the distance r (called the *radius vector*) from O to P, where θ is taken as positive if measured counterclockwise and negative if measured clockwise, and r is taken as positive if measured along the terminal side of angle θ and negative if measured along the terminal side of θ produced through the pole. Such an ordered pair of numbers, (r, θ), are called *polar coordinates* of the point P. The polar coordinates of the pole O are taken as $(0, \theta)$, where θ is arbitrary. It follows that, for a given initial line and pole, each point of the plane has infinitely many polar coordinates, but each pair of coordinates corresponds to only one point.

Example

Some polar coordinates of P are: $(2, 60°)$, $(2, 420°)$, $(2, -300°)$, $(-2, 240°)$, $(-2, -120°)$.

Points

Distance between P_1 and P_2: $\sqrt{r_1^2 + r_2^2 - 2r_1 r_2 \cos(\theta_1 - \theta_2)}$
Points P_1, P_2, P_3 are collinear if and only if

$$r_2 r_3 \sin(\theta_3 - \theta_2) + r_3 r_1 \sin(\theta_1 - \theta_3) + r_1 r_2 \sin(\theta_2 - \theta_1) = 0.$$

Polygonal Areas

Area of triangle $P_1 P_2 P_3$:

$$\tfrac{1}{2}[r_1 r_2 \sin(\theta_2 - \theta_1) + r_2 r_3 \sin(\theta_3 - \theta_2) + r_3 r_1 \sin(\theta_1 - \theta_3)]$$

Area of polygon $P_1 P_2 \cdots P_n$:

$$\tfrac{1}{2}[r_1 r_2 \sin(\theta_2 - \theta_1) + r_2 r_3 \sin(\theta_3 - \theta_2) + \cdots + r_{n-1} r_n \sin(\theta_n - \theta_{n-1}) + r_n r_1 \sin(\theta_1 - \theta_n)]$$

The area is positive or negative according as $P_1 P_2 \cdots P_n$ is a counterclockwise or clockwise polygon.

Straight Lines

Let p = distance of line from O, ω = counterclockwise angle from OX to the perpendicular through O to the line.

Normal form: $r \cos(\theta - \omega) = p$
Two-point form: $r[r_1 \sin(\theta - \theta_1) - r_2 \sin(\theta - \theta_2)] = r_1 r_2 \sin(\theta_2 - \theta_1)$

Circles

Center at pole, radius a: $r = a$
Center at $(a, 0)$ and passing
  through the pole: $r = 2a \cos \theta$

Center at $\left(a, \dfrac{\pi}{2}\right)$ and passing

through the pole: $\qquad\qquad\qquad r = 2a \sin \theta$

Center (h, α), radius a: $\qquad\quad r^2 - 2hr \cos (\theta - \alpha) + h^2 - a^2 = 0$

Conics

Let $2p$ = distance from directrix to focus, e = eccentricity.

Focus at pole, directrix to left of pole: $\qquad r = \dfrac{2ep}{1 - e \cos \theta}$

Focus at pole, directrix to right of pole: $\qquad r = \dfrac{2ep}{1 + e \cos \theta}$

Focus at pole, directrix below pole: $\qquad r = \dfrac{2ep}{1 - e \sin \theta}$

Focus at pole, directrix above pole: $\qquad r = \dfrac{2ep}{1 + e \sin \theta}$

Parabola with vertex at pole,
directrix to left of pole: $\qquad r = \dfrac{4p \cos \theta}{\sin^2 \theta}$

Ellipse with center at pole, semiaxes a and b horizontal and vertical, respectively: $\qquad r^2 = \dfrac{a^2 b^2}{a^2 \sin^2 \theta + b^2 \cos^2 \theta}$

Hyperbola with center at pole, semiaxes a and b horizontal and vertical, respectively: $\qquad r^2 = \dfrac{a^2 b^2}{a^2 \sin^2 \theta - b^2 \cos^2 \theta}$

Relations Between Rectangular and Polar Coordinates

Let the positive x-axis coincide with the initial line and let r be nonnegative.

$$x = r \cos \theta, \quad y = r \sin \theta,$$

$$r = \sqrt{x^2 + y^2}, \quad \theta = \arctan \dfrac{y}{x},$$

$$\sin \theta = \dfrac{y}{\sqrt{x^2 + y^2}}, \quad \cos \theta = \dfrac{x}{\sqrt{x^2 + y^2}}$$

RECTANGULAR COORDINATES IN SPACE

Rectangular (Cartesian) Coordinates

Let $X'X$, $Y'Y$, $Z'Z$ (called the *x-axis*, the *y-axis*, and the *z-axis*, respectively) be three mutually perpendicular lines in space intersecting in a point O (called the *origin*), forming in this way three mutually perpendicular planes XOY, XOZ, YOZ (called the *xy-*

Formulas for Use in Analytic Geometry

plane, the *xz-plane*, and the *yz-plane*, respectively). Then any point *P* of space is located by its signed distances *x*, *y*, *z* from the *yz*-plane, the *xz*-plane, and the *xy*-plane, respectively, where *x* and *y* are the rectangular coordinates with respect to the axes $X'X$ and $Y'Y$ of the orthogonal projection P' of *P* on the *xy*-plane (here taken horizontally) and *z* is taken as positive above and negative below the *xy*-plane. The ordered triple of numbers, (x, y, z), are called *rectangular coordinates* of the point *P*.

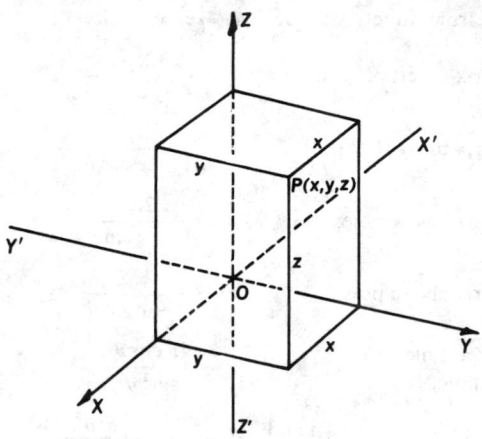

Points

Let $P_1(x_1, y_1, z_1)$ and $P_2(x_2, y_2, z_2)$ be any two points.

Distance between P_1 and P_2: $\sqrt{(x_2 - x_1)^2 + (y_2 - y_1)^2 + (z_2 - z_1)^2}$

Point dividing P_1P_2 in ratio $\dfrac{r}{s}$: $\left(\dfrac{rx_2 + sx_1}{r + s}, \ \dfrac{ry_2 + sy_1}{r + s}, \ \dfrac{rz_2 + sz_1}{r + s} \right)$

Midpoint of P_1P_2: $\left(\dfrac{x_1 + x_2}{2}, \ \dfrac{y_1 + y_2}{2} \ \dfrac{z_1 + z_2}{2} \right)$

Points P_1, P_2, P_3 are collinear if and only if

$$x_2 - x_1 : y_2 - y_1 : z_2 - z_1 = x_3 - x_1 : y_3 - y_1 : z_3 - z_1.$$

Points P_1, P_2, P_3, P_4 are coplanar if and only if $\begin{vmatrix} x_1 & y_1 & z_1 & 1 \\ x_2 & y_2 & z_2 & 1 \\ x_3 & y_3 & z_3 & 1 \\ x_4 & y_4 & z_4 & 1 \end{vmatrix} = 0.$

Area of triangle $P_1P_2P_3$:

$$\frac{1}{2} \sqrt{ \begin{vmatrix} y_1 & z_1 & 1 \\ y_2 & z_2 & 1 \\ y_3 & z_3 & 1 \end{vmatrix}^2 + \begin{vmatrix} z_1 & x_1 & 1 \\ z_2 & x_2 & 1 \\ z_3 & x_3 & 1 \end{vmatrix}^2 + \begin{vmatrix} x_1 & y_1 & 1 \\ x_2 & y_2 & 1 \\ x_3 & y_3 & 1 \end{vmatrix}^2 }$$

Volume of tetrahedron $P_1 P_2 P_3 P_4$: $\frac{1}{6}$ $\begin{vmatrix} x_1 & y_1 & z_1 & 1 \\ x_2 & y_2 & z_2 & 1 \\ x_3 & y_3 & z_3 & 1 \\ x_4 & y_4 & z_4 & 1 \end{vmatrix}$

Direction Numbers and Direction Cosines

Let α, β, γ (called *direction angles*) be the angles that $P_1 P_2$, or any line parallel to $P_1 P_2$, makes with the x-, y-, and z-axis, respectively. Let d = distance between P_1 and P_2.
Direction cosines of $P_1 P_2$:

$$\cos \alpha = \frac{x_2 - x_1}{d}, \quad \cos \beta = \frac{y_2 - y_1}{d}, \quad \cos \gamma = \frac{z_2 - z_1}{d}$$

$$\cos^2 \alpha + \cos^2 \beta + \cos^2 \gamma = 1$$

If a, b, c are direction numbers of $P_1 P_2$, then:

$$a : b : c = x_2 - x_1 : y_2 - y_1 : z_2 - z_1$$
$$= \cos \alpha : \cos \beta : \cos \gamma$$

$$\cos \alpha = \frac{a}{\pm \sqrt{a^2 + b^2 + c^2}}, \quad \cos \beta = \frac{b}{\pm \sqrt{a^2 + b^2 + c^2}},$$

$$\cos \gamma = \frac{c}{\pm \sqrt{a^2 + b^2 + c^2}}$$

Angle between two lines with direction angles $\alpha_1, \beta_1, \gamma_1$ and $\alpha_2, \beta_2, \gamma_2$:

$$\cos \theta = \cos \alpha_1 \cos \alpha_2 + \cos \beta_1 \cos \beta_2 + \cos \gamma_1 \cos \gamma_2$$

For parallel lines: $\alpha_1 = \alpha_2$, $\beta_1 = \beta_2$, $\gamma_1 = \gamma_2$
For perpendicular lines:

$$\cos \alpha_1 \cos \alpha_2 + \cos \beta_1 \cos \beta_2 + \cos \gamma_1 \cos \gamma_2 = 0$$

Angle between two lines with directions (a_1, b_1, c_1) and (a_2, b_2, c_2):

$$\cos \theta = \frac{a_1 a_2 + b_1 b_2 + c_1 c_2}{\sqrt{a_1^2 + b_1^2 + c_1^2} \, \sqrt{a_2^2 + b_2^2 + c_2^2}}$$

$$\sin \theta = \frac{\sqrt{(b_1 c_2 - c_1 b_2)^2 + (c_1 a_2 - a_1 c_2)^2 + (a_1 b_2 - b_1 a_2)^2}}{\sqrt{a_1^2 + b_1^2 + c_1^2} \, \sqrt{a_2^2 + b_2^2 + c_2^2}}$$

For parallel lines: $a_1 : b_1 : c_1 = a_2 : b_2 : c_2$
For perpendicular lines: $a_1 a_2 + b_1 b_2 + c_1 c_2 = 0$

The direction

$$(b_1 c_2 - c_1 b_2, c_1 a_2 - a_1 c_2, a_1 b_2 - b_1 a_2)$$

is perpendicular to both directions (a_1, b_1, c_1) and (a_2, b_2, c_2).
The directions (a_1, b_1, c_1), (a_2, b_2, c_2), (a_3, b_3, c_3) are parallel to a common plane if and only if

$$\begin{vmatrix} a_1 & b_1 & c_1 \\ a_2 & b_2 & c_2 \\ a_3 & b_3 & c_3 \end{vmatrix} = 0.$$

ANALYTIC GEOMETRY
Formulas for Use in Analytic Geometry

Straight Lines

Point-direction form:
$$\frac{x - x_1}{a} = \frac{y - y_1}{b} = \frac{z - z_1}{c}$$

Two-point form:
$$\frac{x - x_1}{x_2 - x_1} = \frac{y - y_1}{y_2 - y_1} = \frac{z - z_1}{z_2 - z_1}$$

Parametric form:
$$x = x_1 + ta, \quad y = y_1 + tb, \quad z = z_1 + tc$$

General form:
$$\begin{cases} A_1 x + B_1 y + C_1 z + D_1 = 0 \\ A_2 x + B_2 y + C_2 z + D_2 = 0 \end{cases}$$

Direction of line:
$$(B_1 C_2 - C_1 B_2, C_1 A_2 - A_1 C_2, A_1 B_2 - B_1 A_2)$$

Projection of segment $P_1 P_2$ on any line having the direction (a, b, c):

$$\frac{(x_2 - x_1)a + (y_2 - y_1)b + (z_2 - z_1)c}{\sqrt{a^2 + b^2 + c^2}}$$

Distance from point P_0 to line through P_1 in direction (a, b, c):

$$\sqrt{\frac{\begin{vmatrix} y_0 - y_1 & z_0 - z_1 \\ b & c \end{vmatrix}^2 + \begin{vmatrix} z_0 - z_1 & x_0 - x_1 \\ c & a \end{vmatrix}^2 + \begin{vmatrix} x_0 - x_1 & y_0 - y_1 \\ a & b \end{vmatrix}^2}{a^2 + b^2 + c^2}}$$

Distance between line through P_1 in direction (a_1, b_1, c_1) and line through P_2 in direction (a_2, b_2, c_2):

$$\pm \frac{\begin{vmatrix} x_2 - x_1 & y_2 - y_1 & z_2 - z_1 \\ a_1 & b_1 & c_1 \\ a_2 & b_2 & c_2 \end{vmatrix}}{\sqrt{\begin{vmatrix} b_1 & c_1 \\ b_2 & c_2 \end{vmatrix}^2 + \begin{vmatrix} c_1 & a_1 \\ c_2 & a_2 \end{vmatrix}^2 + \begin{vmatrix} a_1 & b_1 \\ a_2 & b_2 \end{vmatrix}^2}}$$

The line through P_1 in direction (a_1, b_1, c_1) and the line through P_2 in direction (a_2, b_2, c_2) intersect if and only if

$$\begin{vmatrix} x_2 - x_1 & y_2 - y_1 & z_2 - z_1 \\ a_1 & b_1 & c_1 \\ a_2 & b_2 & c_2 \end{vmatrix} = 0.$$

Planes

General form: $\quad Ax + By + Cz + D = 0$

Direction of normal: $\quad (A, B, C)$

Perpendicular to yz-plane: $\quad By + Cz + D = 0$

Perpendicular to xz-plane: $\quad Ax + Cz + D = 0$

Perpendicular to xy-plane: $\quad Ax + By + D = 0$

Perpendicular to x-axis: $\quad Ax + D = 0$

Perpendicular to y-axis: $\quad By + D = 0$

Perpendicular to z-axis: $\quad Cz + D = 0$

Intercept form:
$$\frac{x}{a} + \frac{y}{b} + \frac{z}{c} = 1$$

Plane through point P_1 and perpendicular to direction (a, b, c):

$$a(x - x_1) + b(y - y_1) + c(z - z_1) = 0$$

Formulas for Use in Analytic Geometry

Plane through point P_1 and parallel to directions (a_1,b_1,c_1) and (a_2,b_2,c_2):

$$\begin{vmatrix} x - x_1 & y - y_1 & z - z_1 \\ a_1 & b_1 & c_1 \\ a_2 & b_2 & c_2 \end{vmatrix} = 0$$

Plane through points P_1 and P_2 parallel to direction (a,b,c):

$$\begin{vmatrix} x - x_1 & y - y_1 & z - z_1 \\ x_2 - x_1 & y_2 - y_1 & z_2 - z_1 \\ a & b & c \end{vmatrix} = 0$$

Three-point form:

$$\begin{vmatrix} x & y & z & 1 \\ x_1 & y_1 & z_1 & 1 \\ x_2 & y_2 & z_2 & 1 \\ x_3 & y_3 & z_3 & 1 \end{vmatrix} = 0 \quad \text{or} \quad \begin{vmatrix} x - x_1 & y - y_1 & z - z_1 \\ x_2 - x_1 & y_2 - y_1 & z_2 - z_1 \\ x_3 - x_1 & y_3 - y_1 & z_3 - z_1 \end{vmatrix} = 0$$

Normal form (p = distance from origin to plane; α, β, γ are direction angles of perpendicular to plane from origin):

$$x \cos \alpha + y \cos \beta + z \cos \gamma = p$$

To reduce $Ax + By + Cz + D = 0$ to normal form, divide by $\pm \sqrt{A^2 + B^2 + C^2}$, where the sign of the radical is chosen opposite to the sign of D when $D \neq 0$, the same as the sign of C when $D = 0$ and $C \neq 0$, the same as the sign of B when $C = D = 0$.

Distance from point P_1 to plane $Ax + By + Cz + D = 0$:

$$\frac{Ax_1 + By_1 + Cz_1 + D}{\pm \sqrt{A^2 + B^2 + C^2}}$$

Angle θ between planes $A_1x + B_1y + C_1z + D_1 = 0$ and $A_2x + B_2y + C_2z + D_2 = 0$:

$$\cos \theta = \frac{A_1A_2 + B_1B_2 + C_1C_2}{\sqrt{A_1^2 + B_1^2 + C_1^2}\ \sqrt{A_2^2 + B_2^2 + C_2^2}}$$

Planes parallel: $A_1 : B_1 : C_1 = A_2 : B_2 : C_2$

Planes perpendicular: $A_1A_2 + B_1B_2 + C_1C_2 = 0$

Spheres

Center at origin, radius r: $x^2 + y^2 + z^2 = r^2$

Center at (g,h,k), radius r: $(x - g)^2 + (y - h)^2 + (z - k)^2 = r^2$

General form: $\begin{cases} Ax^2 + Ay^2 + Az^2 + Dx + Ey + Fz + M = 0, & A \neq 0 \\ x^2 + y^2 + z^2 + 2dx + 2ey + 2fz + m = 0 \end{cases}$

Center: $(-d, -e, -f)$

Radius: $r = \sqrt{d^2 + e^2 + f^2 - m}$

Formulas for Use in Analytic Geometry

Sphere on $P_1 P_2$ as diameter:

$$(x - x_1)(x - x_2) + (y - y_1)(y - y_2) + (z - z_1)(z - z_2) = 0$$

Four-point form:
$$\begin{vmatrix} x^2 + y^2 + z^2 & x & y & z & 1 \\ x_1^2 + y_1^2 + z_1^2 & x_1 & y_1 & z_1 & 1 \\ x_2^2 + y_2^2 + z_2^2 & x_2 & y_2 & z_2 & 1 \\ x_3^2 + y_3^2 + z_3^2 & x_3 & y_3 & z_3 & 1 \\ x_4^2 + y_4^2 + z_4^2 & x_4 & y_4 & z_4 & 1 \end{vmatrix} = 0$$

The Seventeen Quadric Surfaces in Standard Form

1. Real ellipsoid: $\qquad\qquad\quad x^2/a^2 + y^2/b^2 + z^2/c^2 = 1$
2. Imaginary ellipsoid: $\qquad\quad x^2/a^2 + y^2/b^2 + z^2/c^2 = -1$
3. Hyperboloid of one sheet: $\quad x^2/a^2 + y^2/b^2 - z^2/c^2 = 1$
4. Hyperboloid of two sheets: $\quad x^2/a^2 + y^2/b^2 - z^2/c^2 = -1$
5. Real quadric cone: $\qquad\quad\; x^2/a^2 + y^2/b^2 - z^2/c^2 = 0$
6. Imaginary quadric cone: $\quad\; x^2/a^2 + y^2/b^2 + z^2/c^2 = 0$
7. Elliptic paraboloid: $\qquad\quad x^2/a^2 + y^2/b^2 + 2z = 0$
8. Hyperbolic paraboloid: $\qquad x^2/a^2 - y^2/b^2 + 2z = 0$
9. Real elliptic cylinder: $\qquad\; x^2/a^2 + y^2/b^2 = 1$
10. Imaginary elliptic cylinder: $\; x^2/a^2 + y^2/b^2 = -1$
11. Hyperbolic cylinder: $\qquad\; x^2/a^2 - y^2/b^2 = -1$
12. Real intersecting planes: $\quad x^2/a^2 - y^2/b^2 = 0$
13. Imaginary intersecting planes: $\; x^2/a^2 + y^2/b^2 = 0$
14. Parabolic cylinder: $\qquad\quad x^2 + 2rz = 0$
15. Real parallel planes: $\qquad\quad x^2 = a^2$
16. Imaginary parallel planes: $\quad x^2 = -a^2$
17. Coincident planes: $\qquad\qquad x^2 = 0$

General Equation of Second Degree

The nature of the graph of the general quadratic equation in x, y, z,

$$ax^2 + by^2 + cz^2 + 2fyz + 2gzx + 2hxy + 2px + 2qy + 2rz + d = 0,$$

is described in the following table in terms of $\rho_3, \rho_4, \Delta, k_1, k_2, k_3$, where

$$e = \begin{bmatrix} a & h & g \\ h & b & f \\ g & f & c \end{bmatrix}, \quad E = \begin{bmatrix} a & h & g & p \\ h & b & f & q \\ g & f & c & r \\ p & q & r & d \end{bmatrix},$$

$$\rho_3 = \text{rank } e, \quad \rho_4 = \text{rank } E,$$
$$\Delta = \text{determinant of } E,$$

k_1, k_2, k_3 are the roots of $\begin{vmatrix} a - x & h & g \\ h & b - x & f \\ g & f & c - x \end{vmatrix} = 0.$

Formulas for Use in Analytic Geometry

Case	ρ_3	ρ_4	Sign of Δ	Nonzero k's same sign?	Quadric Surface
1	3	4	−	yes	Real ellipsoid
2	3	4	+	yes	Imaginary ellipsoid
3	3	4	+	no	Hyperboloid of one sheet
4	3	4	−	no	Hyperboloid of two sheets
5	3	3		no	Real quadric cone
6	3	3		yes	Imaginary quadric cone
7	2	4	−	yes	Elliptic paraboloid
8	2	4	+	no	Hyperbolic paraboloid
9	2	3		yes	Real elliptic cylinder
10	2	3		yes	Imaginary elliptic cylinder
11	2	3		no	Hyperbolic cylinder
12	2	2		no	Real intersecting planes
13	2	2		yes	Imaginary intersecting planes
14	1	3			Parabolic cylinder
15	1	2			Real parallel planes
16	1	2			Imaginary parallel planes
17	1	1			Coincident planes

Cylindrical and Conical Surfaces

Any equation in just two of the variables x, y, z represents a *cylindrical surface* whose elements are parallel to the axis of the missing variable.

Any equation homogeneous in the variables x, y, z represents a *conical surface* whose vertex is at the origin.

Transformation of Coordinates

To transform an equation of a surface from an old system of rectangular coordinates (x, y, z) to a new system of rectangular coordinates (x', y', z'), substitute for each old variable in the equation of the surface its expression in terms of the new variables.

Translation:
$$x = x' + h$$
$$y = y' + k$$
$$z = z' + l$$
The new axes are parallel to the old axes and the coordinates of the new origin in terms of the old system are (h, k, l).

Rotation about the origin:
$$x = \lambda_1 x' + \lambda_2 y' + \lambda_3 z'$$
$$y = \mu_1 x' + \mu_2 y' + \mu_3 z'$$
$$z = \nu_1 x' + \nu_2 y' + \nu_3 z'$$
The new origin is coincident with the old origin and the x'-axis, y'-axis, z'-axis have direction cosines $(\lambda_1, \mu_1, \nu_1), (\lambda_2, \mu_2, \nu_2), (\lambda_3, \mu_3, \nu_3)$, respectively, with respect to the old system of axes.

$$x' = \lambda_1 x + \mu_1 y + \nu_1 z$$
$$y' = \lambda_2 x + \mu_2 y + \nu_2 z$$
$$z' = \lambda_3 x + \mu_3 y + \nu_3 z$$

Formulas for Use in Analytic Geometry

Cylindrical Coordinates

If (r, θ, z) are the cylindrical coordinates and (x, y, z) the rectangular coordinates of a point P, then

$$x = r \cos \theta, \quad r = \sqrt{x^2 + y^2},$$

$$y = r \sin \theta, \quad \theta = \arctan \frac{y}{x},$$

$$z = z, \qquad z = z.$$

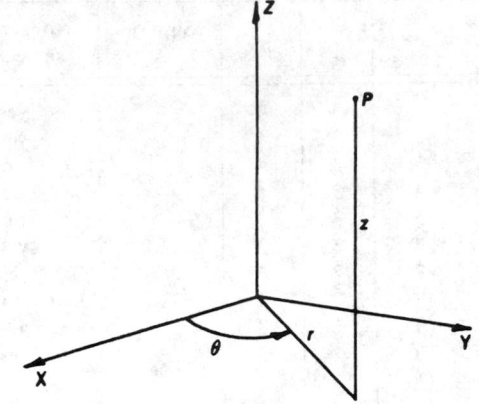

Spherical Coordinates

If (ρ, θ, ϕ) are the spherical coordinates and (x, y, z) the rectangular coordinates of a point P, then

$$x = \rho \cos \theta \sin \phi$$

$$y = \rho \sin \theta \sin \phi$$

$$z = \rho \cos \phi$$

$$\phi = \text{arc cos} \ \frac{z}{\sqrt{x^2 + y^2 + z^2}},$$

$$\theta = \arctan \frac{y}{x}$$

$$p^2 = x^2 + y^2 + z^2$$

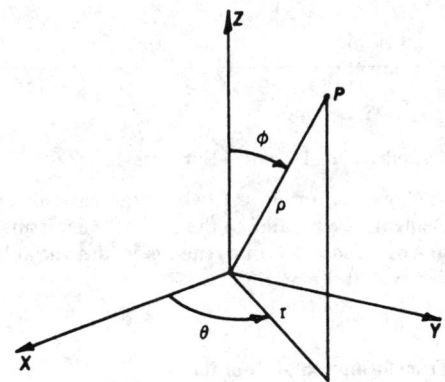

CURVES AND SURFACES

Dr. Howard Eves

The curves and surfaces collected here for reference appear frequently in mathematical literature. The equations most generally associated with each figure are given. The equation of a plane curve when placed otherwise on the coordinate frame of reference may often be found from the given equation by the following rules.

RECTANGULAR COORDINATES

1. If a given curve is reflected in the x-axis, the new equation is obtained from the old by replacing y by $-y$.

2. If a given curve is reflected in the y-axis, the new equation is obtained from the old by replacing x by $-x$.

3. If a given curve is reflected in the origin, the new equation is obtained from the old by replacing x by $-x$ and y by $-y$.

4. If a given curve is reflected in the line $y = x$, the new equation is obtained from the old by interchanging x and y.

5. If a given curve is rotated about the origin through 90°, the new equation is obtained from the old by replacing x by y and y by $-x$.

6. If a given curve is rotated about the origin through $-90°$, the new equation is obtained from the old by replacing x by $-y$ and y by x.

7. If a given curve is translated a distance h in the x-direction, the new equation is obtained from the old by replacing x by $x - h$.

8. If a given curve is translated a distance k in the y-direction, the new equation is obtained from the old by replacing y by $y - k$.

9. If a given curve is altered by multiplying all the abscissas by a, the new equation is obtained from the old by replacing x by x/a.

10. If a given curve is altered by multiplying all the ordinates by b, the new equation is obtained from the old by replacing y by y/b.

POLAR COORDINATES

1. If a given curve is reflected in the polar axis, the new equation is obtained from the old by replacing θ by $-\theta$, or by replacing r by $-r$ and θ by $\pi - \theta$.

2. If a given curve is reflected in the 90° axis, the new equation is obtained from the old by replacing θ by $\pi - \theta$, or by replacing r by $-r$ and θ by $-\theta$.

3. If a given curve is reflected in the pole, the new equation is obtained from the old by replacing θ by $\pi + \theta$, or by replacing r by $-r$.

4. If a given curve is rotated about the pole through an angle α, the new equation is obtained from the old by replacing θ by $\theta - \alpha$.

PLANE CURVES

Archimedean spiral
  See: Spiral of Archimedes

Astroid
  See: Hypocycloid of four cusps

Curves and Surfaces

Bifolium

$$(x^2 + y^2)^2 = ax^2y$$
$$r = a \sin \theta \cos^2 \theta$$

Circle

(a)

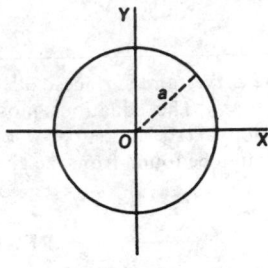

$$x^2 + y^2 = a^2$$
$$r = a$$

Cardioid

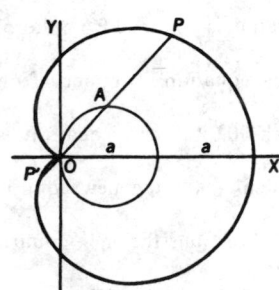

$$(x^2 + y^2 - ax)^2 = a^2(x^2 + y^2)$$
$$r = a(\cos \theta + 1)$$
or
$$r = a(\cos \theta - 1)$$
$$[P'A = AP = a]$$

(b)

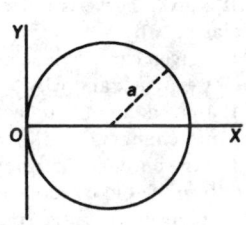

$$x^2 + y^2 = 2ax$$
$$r = 2a \cos \theta$$

(c)

$$x^2 + y^2 = ax + by$$
$$r = a \cos \theta + b \sin \theta$$

Cassinian curves
  See: Ovals of Cassini

Catenary, Hyperbolic cosine

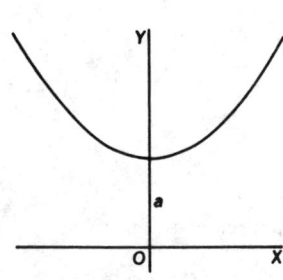

$$y = \frac{a}{2}(e^{x/a} + e^{-x/a}) = a \cosh \frac{x}{a}$$

Cissoid of Diocles

$$y^2(a - x) = x^3$$
$$r = a \sin \theta \tan \theta$$
$$[OP = AB]$$

Cochleoid, Oui-ja board curve

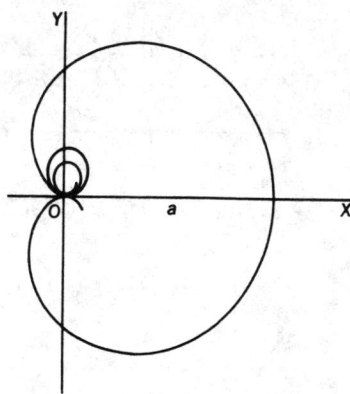

$$(x^2 + y^2)\tan^{-1}(y/x) = ay$$
$$r\theta = a\sin\theta$$

Companion to the cycloid

$$\begin{cases} x = a\phi \\ y = a(1 - \cos\phi) \end{cases}$$
$$[OB = \widehat{AB}]$$
(This is a sinusoid)

Conchoid of Nicomedes
(a) $a < b$

(b) $a > b$

$$(y - a)^2(x^2 + y^2) = b^2 y^2$$
$$r = a\csc\theta \pm b$$
$$[P'A = AP = b]$$

Conic sections
  See: Circle; Ellipse; Hyperbola; Parabola

Cosecant curve

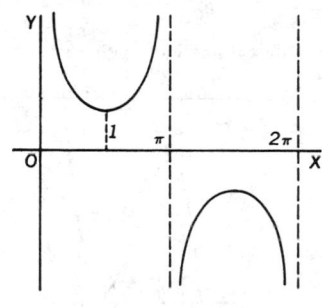

$$y = \csc x$$

Cosine curve

$$y = \cos x$$

Cotangent curve

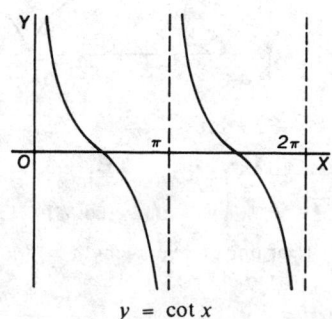

$$y = \cot x$$

Cubical parabola (special)

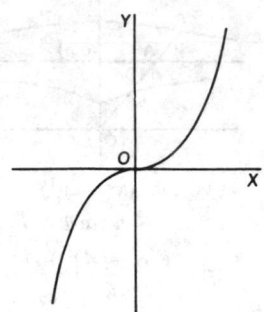

$$y = ax^3, \quad a > 0$$

$$r^2 = \frac{1}{a} \sec^2 \theta \tan \theta, \quad a > 0$$

Cubical parabola (general)

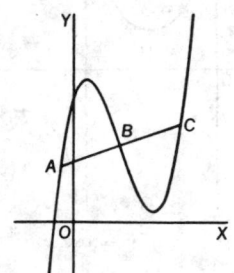

$$y = ax^3 + bx^2 + cx + d, \quad a > 0$$
$$[AB = BC]$$
$$(\text{abscissa of } B = -b/3a)$$

Curtate cycloid, Trochoids
See: Cycloid, curtate

Cycloid (cusp at origin)

$$x = a \arccos \frac{a - y}{a} \mp \sqrt{2ay - y^2}$$

$$\begin{cases} x = a(\phi - \sin \phi) \\ y = a(1 - \cos \phi) \end{cases}$$

(For one arch: arc length = $8a$;
area = $3\pi a^2$)

Cycloid (vertex at origin)

$$x = 2a \arcsin \sqrt{y/2a} + \sqrt{2ay - y^2}$$

$$\begin{cases} x = a(\phi + \sin \phi) \\ y = a(1 - \cos \phi) \end{cases}$$

Cycloid, curtate

$$\begin{cases} x = a\phi - b \sin \phi \\ y = a - b \cos \phi \end{cases}$$
$$a > b$$

Cycloid, prolate

$$\begin{cases} x = a\phi - b \sin \phi \\ y = a - b \cos \phi \end{cases}$$
$$a < b$$

Deltoid
See: Hypocycloid of three cusps

Ellipse

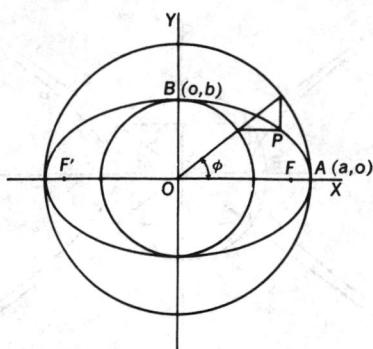

$$x^2/a^2 + y^2/b^2 = 1$$

$$\begin{cases} x = a\cos\phi \\ y = b\sin\phi \end{cases}$$

$$[BF' = BF = a, \quad PF' + PF = 2a]$$

Evolute of ellipse

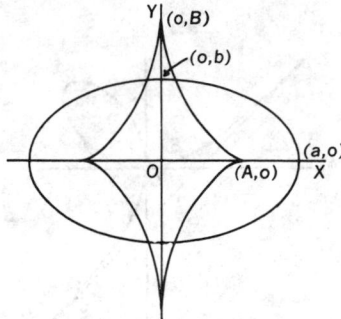

$$(ax)^{2/3} + (by)^{2/3} = (a^2 - b^2)^{2/3}$$

$$\begin{cases} x = A\cos^3\phi \\ y = B\sin^3\phi \end{cases}$$

$$[A = (a^2 - b^2)/a, \quad B = (a^2 - b^2)/b]$$

Exponential curve
(1) $a > 0$

Epicycloid

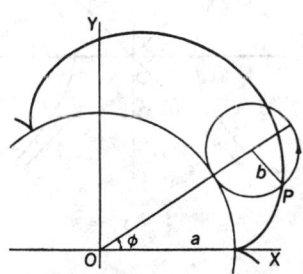

$$\begin{cases} x = (a + b)\cos\phi - b\cos\left(\dfrac{a + b}{b}\,\phi\right) \\ y = (a + b)\sin\phi - b\sin\left(\dfrac{a + b}{b}\,\phi\right) \end{cases}$$

(2) $a < 0$

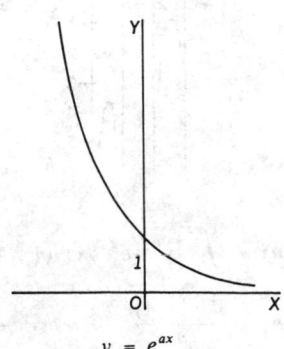

$$y = e^{ax}$$

Equiangular spiral
See: Spiral, logarithmic or equiangular

Equilateral hyperbola
See: Hyperbola, equilateral or rectangular

Folium of Descartes

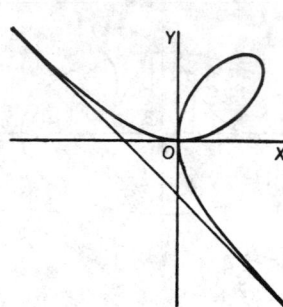

$$x^3 + y^3 - 3axy = 0$$

$$\begin{cases} x = 3a\phi/(1 + \phi^3) \\ y = 3a\phi^2/(1 + \phi^3) \end{cases}$$

$$r = \frac{3a \sin\theta \cos\theta}{\sin^3\theta + \cos^3\theta}$$

[asymptote: $x + y + a = 0$]

Gamma function

$$\Gamma(n) = \int_0^\infty x^{n-1} e^{-x} dx \quad (n > 0)$$

$$\Gamma(n) = \frac{\Gamma(n + 1)}{n} \quad (0 > n \neq -1, -2, -3, \ldots)$$

Hyperbola

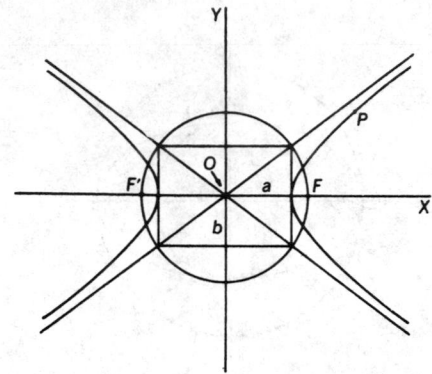

$$x^2/a^2 - y^2/b^2 = 1$$
$$[F'P - FP = 2a]$$

Hyperbola, equilateral or **rectangular**
(1)

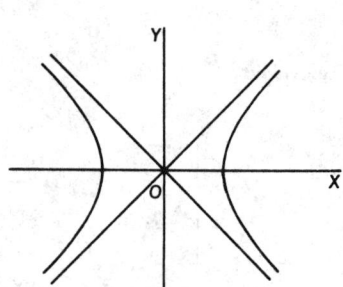

$$x^2 - y^2 = a^2$$

(2)

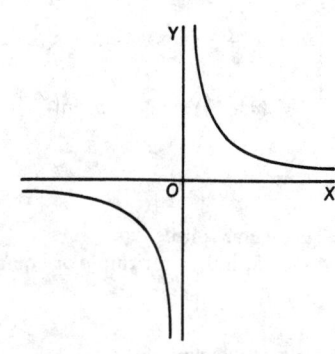

$$xy = k, \quad k > 0$$

(3)

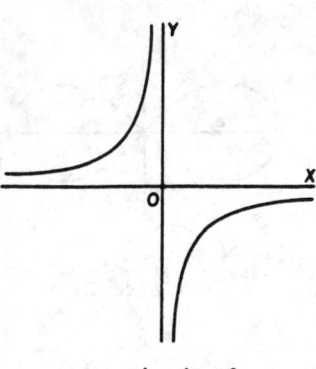

$$xy = k, \quad k < 0$$

Hyperbolic functions*

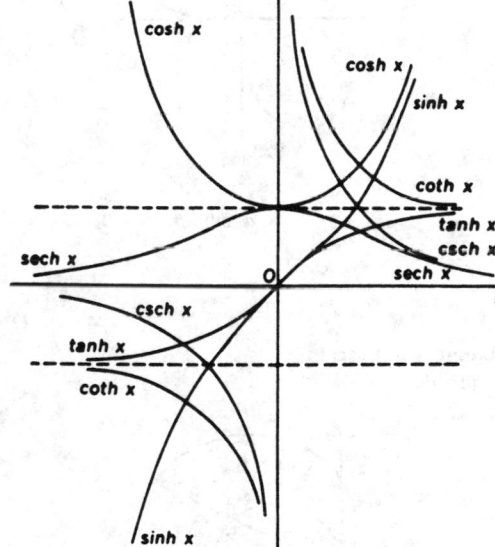

$$\sinh x = \frac{e^x - e^{-x}}{2} \qquad \operatorname{csch} x = \frac{2}{e^x - e^{-x}}$$

$$\cosh x = \frac{e^x + e^{-x}}{2} \qquad \operatorname{sech} x = \frac{2}{e^x + e^{-x}}$$

$$\tanh x = \frac{e^x - e^{-x}}{e^x + e^{-x}} \qquad \coth x = \frac{e^x + e^{-x}}{e^x - e^{-x}}$$

Hyperbolic spiral
   See: Spiral, hyperbolic or reciprocal

*See page 221 for inverse hyperbolic functions.

Hypocycloid of three cusps, Deltoid

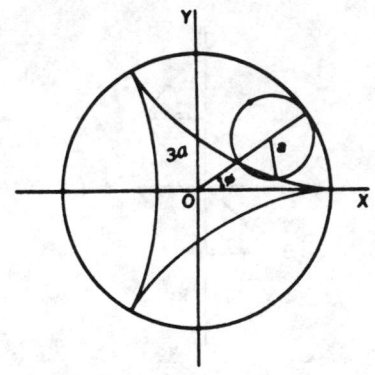

$$\begin{cases} x = 2a \cos \phi + a \cos 2\phi \\ y = 2a \sin \phi - a \sin 2\phi \end{cases}$$

Hypocycloid of four cusps, Astroid

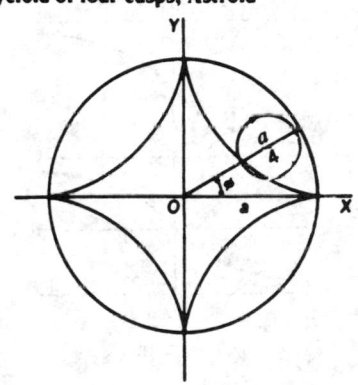

$$x^{2/3} + y^{2/3} = a^{2/3}$$
$$\begin{cases} x = a \cos^3 \phi \\ y = a \sin^3 \phi \end{cases}$$

Inverse cosine curve

$$y = \arccos x$$

Inverse sine curve

$$y = \arcsin x$$

Inverse tangent curve

$$y = \arctan x$$

Involute of circle

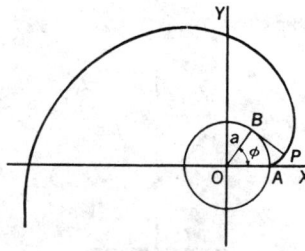

$$\begin{cases} x = a\cos\phi + a\phi\sin\phi \\ y = a\sin\phi - a\phi\cos\phi \end{cases}$$

$$[BP = \widehat{BA}]$$

Lemniscate of Bernoulli, Two-leaved rose

(a)

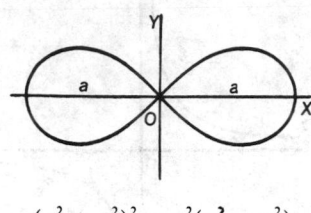

$$(x^2 + y^2)^2 = a^2(x^2 - y^2)$$
$$r^2 = a^2\cos 2\theta$$

(b)

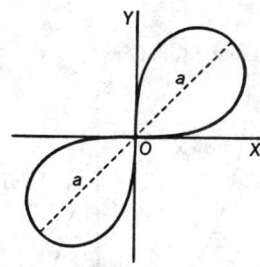

$$(x^2 + y^2)^2 = 2a^2 xy$$
$$r^2 = a^2\sin 2\theta$$

Limacon of Pascal

(1) $a > b$

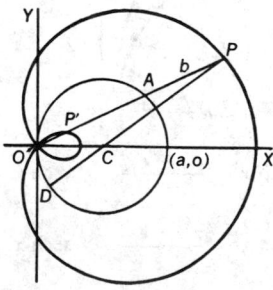

[If $a = 2b$, the curve is called the *trisectrix*, since then $\sphericalangle OPD = \frac{1}{3}\sphericalangle OCD$.]

(2) $a = b$

　　See: Cardioid

(3) $a < b$

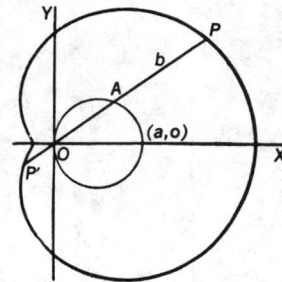

$$(x^2 + y^2 - ax)^2 = b^2(x^2 + y^2)$$
$$r = b + a \cos \theta$$
$$[P'A = AP = b]$$

Lituus

$$r^2\theta = a^2$$

Logarithmic curve

(1) $a > 1$

(2) $0 < a < 1$

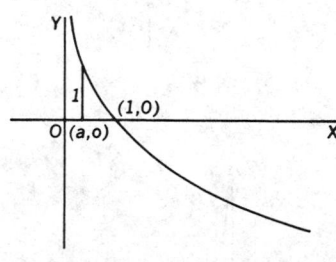

$$y = \log_a x$$

Logarithmic spiral
See: Spiral, logarithmic or equiangular

Nephroid

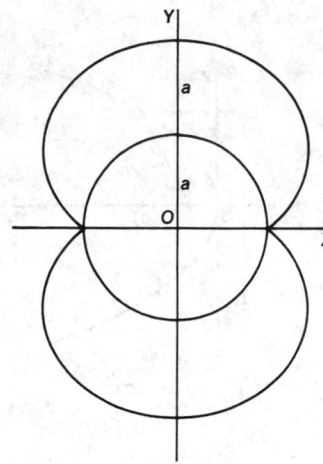

$$\begin{cases} x = \tfrac{1}{2}a(3 \cos \phi - \cos 3\phi) \\ y = \tfrac{1}{2}a(3 \sin \phi - \sin 3\phi) \end{cases}$$

[The nephroid is a 2-cusped epicycloid.]

Oui-ja board curve
See: Cochleoid

Ovals of Cassini
(1) $b > k$

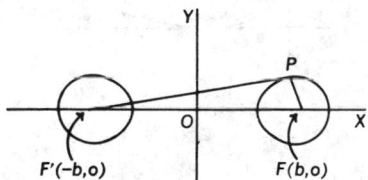

(2) $b = k$
See: Lemniscate of Bernoulli

(3) $b < k$

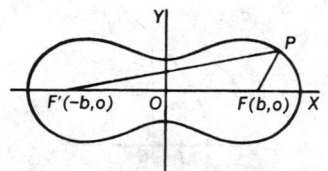

$$(x^2 + y^2 + b^2)^2 - 4b^2x^2 = k^4$$
$$r^4 + b^4 - 2r^2b^2 \cos 2\theta = k^4$$
$$[F'P \cdot FP = k^2]$$

[These curves are sections of a torus on planes parallel to the axis of the torus.]

Parabola

(1)

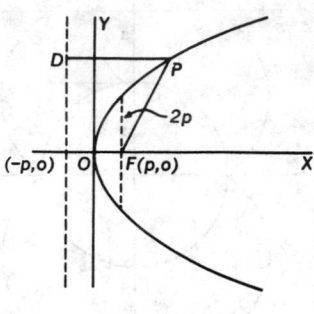

$$y^2 = 4px$$
$$[DP = FP]$$

(2)

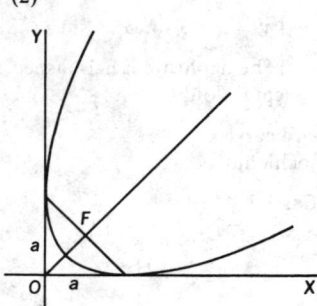

$$\pm x^{1/2} \pm y^{1/2} = a^{1/2}$$
$$(x - y)^2 - 2a(x + y) + a^2 = 0$$

(3)

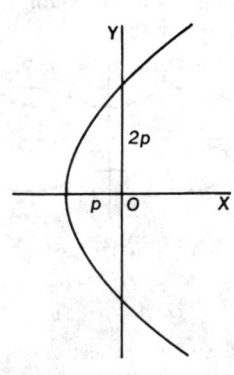

$$r = 2p/(1 - \cos \theta)$$

(4)

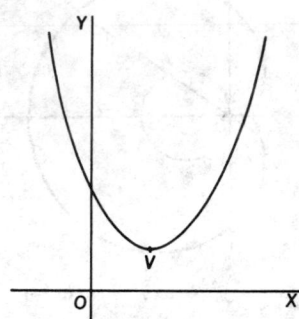

$$y = ax^2 + bx + c, \quad a > 0$$
$$[\text{abscissa of vertex} = -b/2a]$$

Parabolic spiral
See: Spiral, parabolic

Power functions

(1)

$$y = x^{-2}$$

(2) Equilateral hyperbola

$$y = x^{-1}$$

(3)

$$y = x^{-1/2}$$

(4) Cubical parabola

$$y = x^{1/3}$$

(5) Half of a parabola

$$y = x^{1/2}$$

(6) Semicubical parabola

$$y = x^{2/3}$$

(7) Half of semicubical parabola

$$y = x^{3/2}$$

(8) Parabola

$$y = x^2$$

(9) Cubical parabola

$$y = x^3$$

Probability curve

$$y = \frac{1}{\sqrt{2\pi}}\, e^{-x^2/2}$$

Prolate cycloid
See: Cycloid, prolate.

Pursuit curve
See: Tractrix

Quadratrix of Hippias

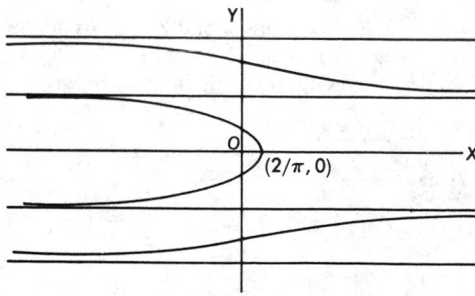

$$y = x \tan(\pi y/2)$$

Reciprocal spiral
See: Spiral, hyperbolic or reciprocal

Rectangular hyperbola
See: Hyperbola, equilateral or rectangular

Rose curves
(1) Two-leaved
See: Lemniscate of Bernoulli, Two-leaved rose

(2) Three-leaved

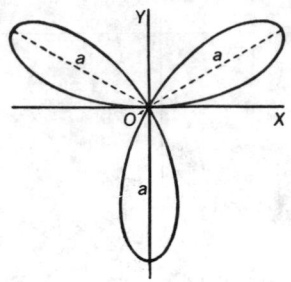

$$r = a \sin 3\theta$$

(3) Three-leaved

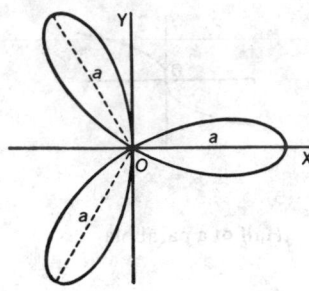

$$r = a \cos 3\theta$$

(4) Four-leaved

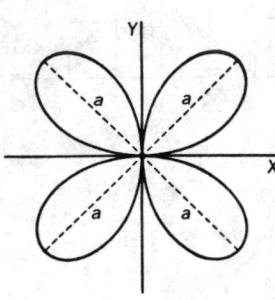

$$r = a \sin 2\theta$$

(5) Four-leaved

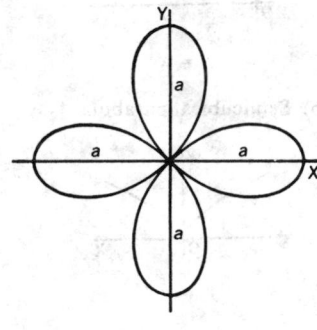

$$r = a \cos 2\theta$$

(6) *n*-leaved

The roses $r = a \sin n\theta$ and $r = a \cos n\theta$, have, for n an even integer, $2n$ leaves; for n an odd integer, n leaves. The roses $r^2 = a \sin n\theta$ and $r^2 = a \cos n\theta$, have, for n an even integer, n leaves; for n an odd integer, $2n$ leaves.

Secant curve

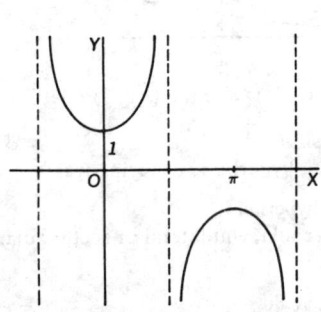

$$y = \sec x$$

Semicubical parabola

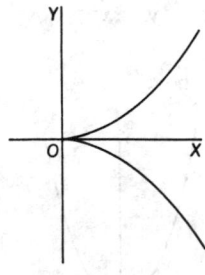

$$y^2 = ax^3$$
$$r = \frac{1}{a} \tan^2 \theta \sec \theta$$

Serpentine curve

$$(a^2 + x^2)y = abx$$

$$\begin{cases} x = a \cot \phi \\ y = b \sin \phi \cos \phi \end{cases}$$

Sine curve

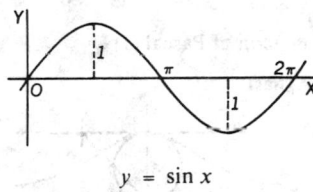

$$y = \sin x$$

Sinusoid

$$y = a \sin (bx + c)$$

Spiral of Archimedes

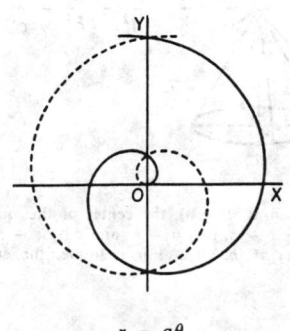

$$r = a\theta$$

Spiral, hyperbolic or reciprocal

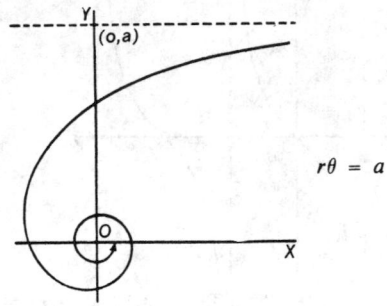

$$r\theta = a$$

Spiral, logarithmic or equiangular

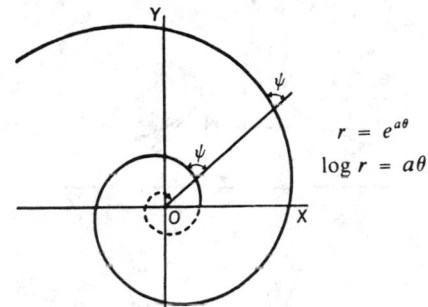

$$r = e^{a\theta}$$
$$\log r = a\theta$$

Spiral, parabolic

$$(r - a)^2 = 4ak\theta$$

Strophoid

$$y^2 = x^2 \frac{a - x}{a + x}$$
$$r = a \cos 2\theta \sec \theta$$
$$[P'A = AP = OA]$$

Tangent curve

$$y = \tan x$$

Tractrix, Pursuit curve

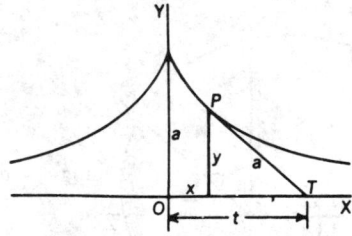

$$x = a \operatorname{sech}^{-1}(y/a) - \sqrt{a^2 - y^2}$$

$$\begin{cases} x = t - a \tanh(t/a) \\ y = a \operatorname{sech}(t/a) \end{cases}$$

$$[PT = a]$$

Trajectory (a parabola)

$$y = x \tan \alpha - gx^2/(2v_0^2 \cos^2 \alpha)$$

$$x = (v_0 \cos \alpha) t$$

$$y = (v_0 \sin \alpha) t - gt^2/2$$

Trigonometric functions

See: Cosecant curve; Cosine curve; Cotangent curve; Secant curve Sine curve; Tangent curve

Trisectrix

See: Limaçon of Pascal (1)

Witch of Agnesi

$$y = a^3/(x^2 + a^2)$$

$$\begin{cases} x = a \cot \phi \\ y = a \sin^2 \phi \end{cases}$$

QUADRIC SURFACES*

Ellipsoid

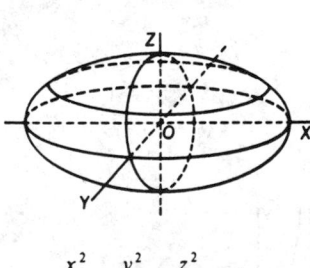

$$\frac{x^2}{a^2} + \frac{y^2}{b^2} + \frac{z^2}{c^2} = 1$$

Elliptic cone

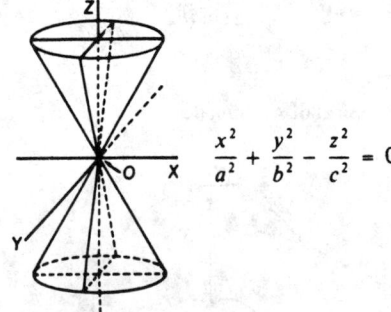

$$\frac{x^2}{a^2} + \frac{y^2}{b^2} - \frac{z^2}{c^2} = 0$$

*Each of the equations is given for the case where the origin is located at $(0, 0, 0)$, the center of the quadric surface. If, however, the center of the surface is at (h, k, l), replace x by $x - h$, y by $y - k$, and z by $z - l$, and the particular standardized form will be that of the surface with center at (h, k, l). For example, the elliptic paraboloid would be

$$\frac{(x - h)^2}{a^2} + \frac{(y - k)^2}{b^2} = c(z - l).$$

Elliptic cylinder

$$\frac{x^2}{a^2} + \frac{y^2}{b^2} = 1$$

Hyperboloid of one sheet

$$\frac{x^2}{a^2} + \frac{y^2}{b^2} - \frac{z^2}{c^2} = 1$$

Elliptic paraboloid

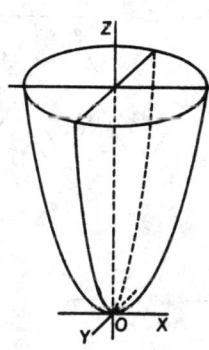

$$\frac{x^2}{a^2} + \frac{y^2}{b^2} = cz$$

Hyperboloid of two sheets

$$\frac{z^2}{c^2} - \frac{x^2}{a^2} - \frac{y^2}{b^2} = 1$$

Hyperbolic paraboloid

$$\frac{x^2}{a^2} - \frac{y^2}{b^2} = cz$$

Sphere

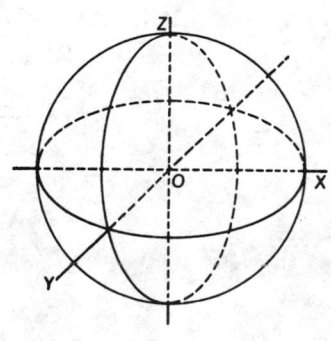

$$x^2 + y^2 + z^2 = a^2$$

ANALYTIC GEOMETRY

Curves and Surfaces

PATTERNS OF REGULAR POLYHEDRA

Tetrahedron

Octahedron

Hexahedron, or Cube

Icosahedron

Dodecahedron

VIII. CALCULUS

Derivatives *

In the following formulas u, v, w represent functions of x, while a, c, n represent fixed real numbers. All arguments in the trigonometric functions are measured in radians, and all inverse trigonometric and hyperbolic functions represent principal values.

1. $\dfrac{d}{dx}(a) = 0$

2. $\dfrac{d}{dx}(x) = 1$

3. $\dfrac{d}{dx}(au) = a\dfrac{du}{dx}$

4. $\dfrac{d}{dx}(u + v - w) = \dfrac{du}{dx} + \dfrac{dv}{dx} - \dfrac{dw}{dx}$

5. $\dfrac{d}{dx}(uv) = u\dfrac{dv}{dx} + v\dfrac{du}{dx}$

6. $\dfrac{d}{dx}(uvw) = uv\dfrac{dw}{dx} + vw\dfrac{du}{dx} + uw\dfrac{dv}{dx}$ and so on to n factors

7. $\dfrac{d}{dx}\left(\dfrac{u}{v}\right) = \dfrac{v\dfrac{du}{dx} - u\dfrac{dv}{dx}}{v^2} = \dfrac{1}{v}\dfrac{du}{dx} - \dfrac{u}{v^2}\dfrac{dv}{dx}$

8. $\dfrac{d}{dx}(u^n) = nu^{n-1}\dfrac{du}{dx}$

9. $\dfrac{d}{dx}(\sqrt{u}) = \dfrac{1}{2\sqrt{u}}\dfrac{du}{dx}$

10. $\dfrac{d}{dx}\left(\dfrac{1}{u}\right) = -\dfrac{1}{u^2}\dfrac{du}{dx}$

11. $\dfrac{d}{dx}\left(\dfrac{1}{u^n}\right) = -\dfrac{n}{u^{n+1}}\dfrac{du}{dx}$

12. $\dfrac{d}{dx}\left(\dfrac{u^n}{v^m}\right) = \dfrac{u^{n-1}}{v^{m+1}}\left(nv\dfrac{du}{dx} - mu\dfrac{dv}{dx}\right)$

13. $\dfrac{d}{dx}(u^n v^m) = u^{n-1}v^{m-1}\left(nv\dfrac{du}{dx} + mu\dfrac{dv}{dx}\right)$

14. $\dfrac{d}{dx}[f(u)] = \dfrac{d}{du}[f(u)]\cdot\dfrac{du}{dx}$

*Let $y = f(x)$ and $\dfrac{dy}{dx} = \dfrac{d[f(x)]}{dx} = f'(x)$ define respectively a function and its derivative for any value x in their common domain. The differential for the function at such a value x is accordingly defined as

$$dy = d[f(x)] = \dfrac{dy}{dx}\,dx = \dfrac{d[f(x)]}{dx}\,dx = f'(x)dx$$

Each derivative formula has an associated differential formula. For example, formula 6 above has the differential formula

$$d(uvw) = uv\,dw + vw\,du + uw\,dv$$

DERIVATIVES (Continued)

15. $\dfrac{d^2}{dx^2}[f(u)] = \dfrac{df(u)}{du} \cdot \dfrac{d^2u}{dx^2} + \dfrac{d^2f(u)}{du^2} \cdot \left(\dfrac{du}{dx}\right)^2$

16. $\dfrac{d^n}{dx^n}[uv] = \binom{n}{0}v\dfrac{d^nu}{dx^n} + \binom{n}{1}\dfrac{dv}{dx}\dfrac{d^{n-1}u}{dx^{n-1}} + \binom{n}{2}\dfrac{d^2v}{dx^2}\dfrac{d^{n-2}u}{dx^{n-2}}$

$$+ \cdots + \binom{n}{k}\dfrac{d^kv}{dx^k}\dfrac{d^{n-k}u}{dx^{n-k}} + \cdots + \binom{n}{n}u\dfrac{d^nv}{dx^n}$$

where $\binom{n}{r} = \dfrac{n!}{r!(n-r)!}$ the binomial coefficient, n non-negative integer and $\binom{n}{0} = 1$

17. $\dfrac{du}{dx} = \dfrac{1}{\dfrac{dx}{du}}$　　　if $\dfrac{dx}{du} \neq 0$

18. $\dfrac{d}{dx}(\log_a u) = (\log_a e)\dfrac{1}{u}\dfrac{du}{dx}$

19. $\dfrac{d}{dx}(\log_e u) = \dfrac{1}{u}\dfrac{du}{dx}$

20. $\dfrac{d}{dx}(a^u) = a^u(\log_e a)\dfrac{du}{dx}$

21. $\dfrac{d}{dx}(e^u) = e^u\dfrac{du}{dx}$

22. $\dfrac{d}{dx}(u^v) = vu^{v-1}\dfrac{du}{dx} + (\log_e u)u^v\dfrac{dv}{dx}$

23. $\dfrac{d}{dx}(\sin u) = \dfrac{du}{dx}(\cos u)$

24. $\dfrac{d}{dx}(\cos u) = -\dfrac{du}{dx}(\sin u)$

25. $\dfrac{d}{dx}(\tan u) = \dfrac{du}{dx}(\sec^2 u)$

26. $\dfrac{d}{dx}(\cot u) = -\dfrac{du}{dx}(\csc^2 u)$

27. $\dfrac{d}{dx}(\sec u) = \dfrac{du}{dx}\sec u \cdot \tan u$

28. $\dfrac{d}{dx}(\csc u) = -\dfrac{du}{dx}\csc u \cdot \cot u$

29. $\dfrac{d}{dx}(\text{vers } u) = \dfrac{du}{dx}\sin u$

30. $\dfrac{d}{dx}(\arcsin u) = \dfrac{1}{\sqrt{1-u^2}}\dfrac{du}{dx},$　　$\left(-\dfrac{\pi}{2} \leq \arcsin u \leq \dfrac{\pi}{2}\right)$

DERIVATIVES (Continued)

31. $\dfrac{d}{dx}(\text{arc cos } u) = -\dfrac{1}{\sqrt{1 - u^2}}\dfrac{du}{dx}$, $(0 \le \text{arc cos } u \le \pi)$

32. $\dfrac{d}{dx}(\text{arc tan } u) = \dfrac{1}{1 + u^2}\dfrac{du}{dx}$, $\left(-\dfrac{\pi}{2} < \text{arc tan } u < \dfrac{\pi}{2}\right)$

33. $\dfrac{d}{dx}(\text{arc cot } u) = -\dfrac{1}{1 + u^2}\dfrac{du}{dx}$, $(0 \le \text{arc cot } u \le \pi)$

34. $\dfrac{d}{dx}(\text{arc sec } u) = \dfrac{1}{u\sqrt{u^2 - 1}}\dfrac{du}{dx}$, $\left(0 \le \text{arc sec } u < \dfrac{\pi}{2}, -\pi \le \text{arc sec } u < -\dfrac{\pi}{2}\right)$

35. $\dfrac{d}{dx}(\text{arc csc } u) = -\dfrac{1}{u\sqrt{u^2 - 1}}\dfrac{du}{dx}$, $\left(0 < \text{arc csc } u \le \dfrac{\pi}{2}, -\pi < \text{arc csc } u \le -\dfrac{\pi}{2}\right)$

36. $\dfrac{d}{dx}(\text{arc vers } u) = \dfrac{1}{\sqrt{2u - u^2}}\dfrac{du}{dx}$, $(0 \le \text{arc vers } u \le \pi)$

37. $\dfrac{d}{dx}(\sinh u) = \dfrac{du}{dx}(\cosh u)$

38. $\dfrac{d}{dx}(\cosh u) = \dfrac{du}{dx}(\sinh u)$

39. $\dfrac{d}{dx}(\tanh u) = \dfrac{du}{dx}(\text{sech}^2 u)$

40. $\dfrac{d}{dx}(\coth u) = -\dfrac{du}{dx}(\text{csch}^2 u)$

41. $\dfrac{d}{dx}(\text{sech } u) = -\dfrac{du}{dx}(\text{sech } u \cdot \tanh u)$

42. $\dfrac{d}{dx}(\text{csch } u) = -\dfrac{du}{dx}(\text{csch } u \cdot \coth u)$

43. $\dfrac{d}{dx}(\sinh^{-1} u) = \dfrac{d}{dx}[\log (u + \sqrt{u^2 + 1})] = \dfrac{1}{\sqrt{u^2 + 1}}\dfrac{du}{dx}$

44. $\dfrac{d}{dx}(\cosh^{-1} u) = \dfrac{d}{dx}[\log (u + \sqrt{u^2 - 1})] = \dfrac{1}{\sqrt{u^2 - 1}}\dfrac{du}{dx}$, $(u > 1, \cosh^{-1} u > 0)$

45. $\dfrac{d}{dx}(\tanh^{-1} u) = \dfrac{d}{dx}\left[\dfrac{1}{2}\log\dfrac{1 + u}{1 - u}\right] = \dfrac{1}{1 - u^2}\dfrac{du}{dx}$, $(u^2 < 1)$

46. $\dfrac{d}{dx}(\coth^{-1} u) = \dfrac{d}{dx}\left[\dfrac{1}{2}\log\dfrac{u + 1}{u - 1}\right] = \dfrac{1}{1 - u^2}\dfrac{du}{dx}$, $(u^2 > 1)$

47. $\dfrac{d}{dx}(\text{sech}^{-1} u) = \dfrac{d}{dx}\left[\log\dfrac{1 + \sqrt{1 - u^2}}{u}\right] = -\dfrac{1}{u\sqrt{1 - u^2}}\dfrac{du}{dx}$, $(0 < u < 1, \text{sech}^{-1} u > 0)$

48. $\dfrac{d}{dx}(\text{csch}^{-1} u) = \dfrac{d}{dx}\left[\log\dfrac{1 + \sqrt{1 + u^2}}{u}\right] = -\dfrac{1}{|u|\sqrt{1 + u^2}}\dfrac{du}{dx}$

49. $\dfrac{d}{dq} \displaystyle\int_{p}^{q} f(x)\,dx = f(q),$ [p constant]

50. $\dfrac{d}{dp} \displaystyle\int_{p}^{q} f(x)\,dx = -f(p),$ [q constant]

51. $\dfrac{d}{da} \displaystyle\int_{p}^{q} f(x,a)\,dx = \displaystyle\int_{p}^{q} \dfrac{\partial}{\partial a}[f(x,a)]\,dx + f(q,a)\dfrac{dq}{da} - f(p,a)\dfrac{dp}{da}$

INTEGRATION

The following is a brief discussion of some integration techniques. A more complete discussion can be found in a number of good text books. However, the purpose of this introduction is simply to discuss a few of the important techniques which may be used, in conjunction with the integral table which follows, to integrate particular functions.

No matter how extensive the integral table, it is a fairly uncommon occurrence to find in the table the exact integral desired. Usually some form of transformation will have to be made. The simplest type of transformation, and yet the most general, is substitution. Simple forms of substitution, such as $y = ax$, are employed almost unconsciously by experienced users of integral tables. Other substitutions may require more thought. In some sections of the tables, appropriate substitutions are suggested for integrals which are similar to, but not exactly like, integrals in the table. Finding the right substitution is largely a matter of intuition and experience.

Several precautions must be observed when using substitutions:

1. Be sure to make the substitution in the dx term, as well as everywhere else in the integral.
2. Be sure that the function substituted is one-to-one and continuous. If this is not the case, the integral must be restricted in such a way as to make it true. See the example following.
3. With definite integrals, the limits should also be expressed in terms of the new dependent variable. With indefinite integrals, it is necessary to perform the reverse substitution to obtain the answer in terms of the original independent variable. This may also be done for definite integrals, but it is usually easier to change the limits.

Example:

$$\int \frac{x^4}{\sqrt{a^2 - x^2}}\, dx$$

Here we make the substitution $x = |a| \sin \theta$. Then $dx = |a| \cos \theta\, d\theta$, and

$$\sqrt{a^2 - x^2} = \sqrt{a^2 - a^2 \sin^2 \theta} = |a|\sqrt{1 - \sin^2 \theta} = |a \cos \theta|$$

Notice the absolute value signs. It is very important to keep in mind that a square root radical always denotes the positive square root, and to assure the sign is always kept positive. Thus $\sqrt{x^2} = |x|$. Failure to observe this is a common cause of errors in integration.

Notice also that the indicated substitution is not a one-to-one function, that is, it does not have a unique inverse. Thus we must restrict the range of θ in such a way as to make the function one-to-one. Fortunately, this is easily done by solving for θ

$$\theta = \sin^{-1} \frac{x}{|a|}$$

and restricting the inverse sine to the principal values, $-\frac{\pi}{2} \le \theta \le \frac{\pi}{2}$.

Thus the integral becomes

$$\int \frac{a^4 \sin^4 \theta |a| \cos \theta\, d\theta}{|a|\, |\cos \theta|}$$

Now, however, in the range of values chosen for θ, $\cos \theta$ is always positive. Thus we may remove the absolute value signs from $\cos \theta$ in the denominator. (This is one of the reasons that the principal values of the inverse trigonometric functions are defined as they are.)

Then the cos θ terms cancel, and the integral becomes

$$a^4 \int \sin^4 \theta \, d\theta$$

By application of integral formulas 299 and 296, we integrate this to

$$-a^4 \frac{\sin^3 \theta \cos \theta}{4} - \frac{3a^4}{8} \cos \theta \sin \theta + \frac{3a^4}{8} \theta + C$$

We now must perform the inverse substitution to get the result in terms of x. We have

$$\theta = \sin^{-1} \frac{x}{|a|}$$

$$\sin \theta = \frac{x}{|a|}$$

Then

$$\cos \theta = \pm \sqrt{1 - \sin^2 \theta} = \pm \sqrt{1 - \frac{x^2}{a^2}} = \pm \frac{\sqrt{a^2 - x^2}}{|a|}.$$

Because of the previously mentioned fact that cos θ is positive, we may omit the ± sign. The reverse substitution then produces the final answer

$$\int \frac{x^4}{\sqrt{a^2 - x^2}} dx = -\tfrac{1}{4}x^3 \sqrt{a^2 - x^2} - \tfrac{3}{8}a^2 x \sqrt{a^2 - x^2} + \frac{3a^4}{8} \sin^{-1} \frac{x}{|a|} + C.$$

Any rational function of x may be integrated, if the denominator is factored into linear and irreducible quadratic factors. The function may then be broken into partial fractions, and the individual partial fractions integrated by use of the appropriate formula from the integral table. See the section on partial fractions for further information.

Many integrals may be reduced to rational functions by proper substitutions. For example,

$$z = \tan \frac{x}{2}$$

will reduce any rational function of the six trigonometric functions of x to a rational function of z. (Frequently there are other substitutions which are simpler to use, but this one will always work. See integral formula number 484.)

Any rational function of x and $\sqrt{ax + b}$ may be reduced to a rational function of z by making the substitution

$$z = \sqrt{ax + b}.$$

Other likely substitutions will be suggested by looking at the form of the integrand.

The other main method of transforming integrals is integration by parts. This involves applying formula number 5 or 6 in the accompanying integral table. The critical factor in this method is the choice of the functions u and v. In order for the method to be successful, $v = \int dv$ and $\int v \, du$ must be easier to integrate than the original integral. Again, this choice is largely a matter of intuition and experience.

Example:

$$\int x \sin x \, dx$$

Two obvious choices are $u = x$, $dv = \sin x \, dx$, or $u = \sin x$, $dv = x \, dx$. Since a preliminary mental calculation indicates that $\int v \, du$ in the second choice would be more, rather than less, complicated than the original integral (it would contain x^2), we use the first choice.

$$u = x \qquad\qquad du = dx$$
$$dv = \sin x\, dx \qquad\qquad v = -\cos x$$
$$\int x \sin x\, dx = \int u\, dv = uv - \int v\, du = -x \cos x + \int \cos x\, dx$$
$$= \sin x - x \cos x$$

Of course, this result could have been obtained directly from the integral table, but it provides a simple example of the method. In more complicated examples the choice of u and v may not be so obvious, and several different choices may have to be tried Of course, there is no guarantee that any of them will work.

Integration by parts may be applied more than once, or combined with substitution. A fairly common case is illustrated by the following example.

Example:

$$\int e^x \sin x\, dx$$

Let

$$u = e^x \qquad \text{Then} \quad du = e^x\, dx$$
$$dv = \sin x\, dx \qquad\qquad v = -\cos x$$
$$\int e^x \sin x\, dx = \int u\, dv = uv - \int v\, du = -e^x \cos x + \int e^x \cos x\, dx$$

In this latter integral,

$$\text{let} \quad u = e^x \qquad \text{Then} \quad du = e^x\, dx$$
$$dv = \cos x\, dx \qquad\qquad v = \sin x$$
$$\int e^x \sin x\, dx = -e^x \cos x + \int e^x \cos x\, dx = -e^x \cos x + \int u\, dv$$
$$= -e^x \cos x + uv - \int v\, du$$
$$= -e^x \cos x + e^x \sin x - \int e^x \sin x\, dx$$

This looks as if a circular transformation has taken place, since we are back at the same integral we started from. However, the above equation can be solved algebraically for the required integral:

$$\int e^x \sin x\, dx = \tfrac{1}{2}(e^x \sin x - e^x \cos x)$$

In the second integration by parts, if the parts had been chosen as $u = \cos x$, $dv = e^x\, dx$, we would indeed have made a circular transformation, and returned to the starting place. In general, when doing repeated integration by parts, one should never choose the function u at any stage to be the same as the function v at the previous stage, or a constant times the previous v.

The following rule is called the extended rule for integration by parts. It is the result of $n + 1$ successive applications of integration by parts.

If

$$g_1(x) = \int g(x)\,dx, \qquad g_2(x) = \int g_1(x)\,dx,$$

$$g_3(x) = \int g_2(x)\,dx, \ldots, g_m(x) = \int g_{m-1}(x)\,dx, \ldots,$$

then

$$\int f(x) \cdot g(x)\,dx = f(x) \cdot g_1(x) - f'(x) \cdot g_2(x) + f''(x) \cdot g_3(x) - + \cdots$$

$$+ (-1)^n f^{(n)}(x) g_{n+1}(x) + (-1)^{n+1} \int f^{(n+1)}(x) g_{n+1}(x)\,dx.$$

A useful special case of the above rule is when $f(x)$ is a polynomial of degree n. Then $f^{(n+1)}(x) = 0$, and

$$\int f(x) \cdot g(x)\,dx = f(x) \cdot g_1(x) - f'(x) \cdot g_2(x) + f''(x) \cdot g_3(x) - + \cdots + (-1)^n f^{(n)}(x) g_{n+1}(x) + C$$

Example:

If $f(x) = x^2$, $g(x) = \sin x$

$$\int x^2 \sin x\,dx = -x^2 \cos x + 2x \sin x + 2 \cos x + C$$

Another application of this formula occurs if

$$f''(x) = af(x) \quad \text{and} \quad g''(x) = bg(x),$$

where a and b are unequal constants. In this case, by a process similar to that used in the above example for $\int e^x \sin x\,dx$, we get the formula

$$\int f(x)g(x)\,dx = \frac{f(x) \cdot g'(x) - f'(x) \cdot g(x)}{b - a} + C$$

This formula could have been used in the example mentioned. Here is another example.

Example:

If $f(x) = e^{2x}$, $g(x) = \sin 3x$, then $a = 4, b = -9$, and

$$\int e^{2x} \sin 3x\,dx = \frac{3\,e^{2x} \cos 3x - 2\,e^{2x} \sin 3x}{-9 - 4} + C = \frac{e^{2x}}{13}(2 \sin 3x - 3 \cos 3x) + C$$

The following additional points should be observed when using this table.

1. A constant of integration is to be supplied with the answers for indefinite integrals.
2. Logarithmic expressions are to base $e = 2.71828\ldots$, unless otherwise specified, and are to be evaluated for the absolute value of the arguments involved therein.
3. All angles are measured in radians, and inverse trigonometric and hyperbolic functions represent principal values, unless otherwise indicated.
4. If the application of a formula produces either a zero denominator or the square root of a negative number in the result, there is usually available another form of the answer which avoids this difficulty. In many of the results, the excluded values are specified, but when such are omitted it is presumed that one can tell what these should be, especially when difficulties of the type herein mentioned are obtained.
5. When inverse trigonometric functions occur in the integrals, be sure that any replacements made for them are strictly in accordance with the rules for such functions. This causes

little difficulty when the argument of the inverse trigonometric function is positive, since then all angles involved are in the first quadrant. However, if the argument is negative, special care must be used. Thus if $u > 0$,

$$\sin^{-1} u = \cos^{-1}\sqrt{1 - u^2} = \csc^{-1}\frac{1}{u}, \text{ etc.}$$

However, if $u < 0$,

$$\sin^{-1} u = -\cos^{-1}\sqrt{1 - u^2} = -\pi - \csc^{-1}\frac{1}{u}, \text{ etc.}$$

See the section on inverse trigonometric functions for a full treatment of the allowable substitutions.

6. In integrals 340–345 and some others, the right side includes expressions of the form

$$A \tan^{-1}[B + C \tan f(x)].$$

In these formulas, the $\tan^{-1}$ does not necessarily represent the principal value. Instead of always employing the principal branch of the inverse tangent function, one must instead use that branch of the inverse tangent function upon which $f(x)$ lies for any particular choice of x.

Example:

$$\int_0^{4\pi} \frac{dx}{2 + \sin x} = \frac{2}{\sqrt{3}}\tan^{-1}\frac{2\tan\frac{x}{2} + 1}{\sqrt{3}}\Bigg]_0^{4\pi}$$

$$= \frac{2}{\sqrt{3}}\left[\tan^{-1}\frac{2\tan 2\pi + 1}{\sqrt{3}} - \tan^{-1}\frac{2\tan 0 + 1}{\sqrt{3}}\right]$$

$$= \frac{2}{\sqrt{3}}\left[\frac{13\pi}{6} - \frac{\pi}{6}\right] = \frac{4\pi}{\sqrt{3}} = \frac{4\sqrt{3}\pi}{3}$$

Here

$$\tan^{-1}\frac{2\tan 2\pi + 1}{\sqrt{3}} = \tan^{-1}\frac{1}{\sqrt{3}} = \frac{13\pi}{6},$$

since $f(x) = 2\pi$; and

$$\tan^{-1}\frac{2\tan 0 + 1}{\sqrt{3}} = \tan^{-1}\frac{1}{\sqrt{3}} = \frac{\pi}{6},$$

since $f(x) = 0$.

7. B_n and E_n where used in Integrals represents the Bernoulli and Euler numbers as defined in the tables contained from pages 442 to 448.

INTEGRALS

ELEMENTARY FORMS

1. $\displaystyle\int a\,dx = ax$

2. $\displaystyle\int a \cdot f(x)\,dx = a\int f(x)\,dx$

3. $\displaystyle\int \phi(y)\,dx = \int \frac{\phi(y)}{y'}\,dy,$ where $y' = \dfrac{dy}{dx}$

4. $\displaystyle\int (u+v)\,dx = \int u\,dx + \int v\,dx,$ where u and v are any functions of x

5. $\displaystyle\int u\,dv = u\int dv - \int v\,du = uv - \int v\,du$

6. $\displaystyle\int u\frac{dv}{dx}\,dx = uv - \int v\frac{du}{dx}\,dx$

7. $\displaystyle\int x^n\,dx = \frac{x^{n+1}}{n+1},$ except $n = -1$

8. $\displaystyle\int \frac{f'(x)\,dx}{f(x)} = \log f(x),$ $(df(x) = f'(x)\,dx)$

9. $\displaystyle\int \frac{dx}{x} = \log x$

10. $\displaystyle\int \frac{f'(x)\,dx}{2\sqrt{f(x)}} = \sqrt{f(x)},$ $(df(x) = f'(x)\,dx)$

11. $\displaystyle\int e^x\,dx = e^x$

12. $\displaystyle\int e^{ax}\,dx = e^{ax}/a$

13. $\displaystyle\int b^{ax}\,dx = \frac{b^{ax}}{a\log b},$ $(b > 0)$

14. $\displaystyle\int \log x\,dx = x\log x - x$

15. $\displaystyle\int a^x \log a\,dx = a^x,$ $(a > 0)$

16. $\displaystyle\int \frac{dx}{a^2 + x^2} = \frac{1}{a}\tan^{-1}\frac{x}{a}$

INTEGRALS (Continued)

17. $\displaystyle\int \frac{dx}{a^2 - x^2} = \begin{cases} \dfrac{1}{a}\tanh^{-1}\dfrac{x}{a} \\ \text{or} \\ \dfrac{1}{2a}\log\dfrac{a + x}{a - x}, \quad (a^2 > x^2) \end{cases}$

18. $\displaystyle\int \frac{dx}{x^2 - a^2} = \begin{cases} -\dfrac{1}{a}\coth^{-1}\dfrac{x}{a} \\ \text{or} \\ \dfrac{1}{2a}\log\dfrac{x - a}{x + a}, \quad (x^2 > a^2) \end{cases}$

19. $\displaystyle\int \frac{dx}{\sqrt{a^2 - x^2}} = \begin{cases} \sin^{-1}\dfrac{x}{|a|} \\ \text{or} \\ -\cos^{-1}\dfrac{x}{|a|}, \quad (a^2 > x^2) \end{cases}$

20. $\displaystyle\int \frac{dx}{\sqrt{x^2 \pm a^2}} = \log\left(x + \sqrt{x^2 \pm a^2}\right)$

21. $\displaystyle\int \frac{dx}{x\sqrt{x^2 - a^2}} = \frac{1}{|a|}\sec^{-1}\frac{x}{a}$

22. $\displaystyle\int \frac{dx}{x\sqrt{a^2 \pm x^2}} = -\frac{1}{a}\log\left(\frac{a + \sqrt{a^2 \pm x^2}}{x}\right)$

FORMS CONTAINING $(a + bx)$

For forms containing $a + bx$, but not listed in the table, the substitution $u = \dfrac{a + bx}{x}$ may prove helpful.

23. $\displaystyle\int (a + bx)^n\, dx = \frac{(a + bx)^{n+1}}{(n + 1)b}, \quad (n \neq -1)$

24. $\displaystyle\int x(a + bx)^n\, dx$

$$= \frac{1}{b^2(n + 2)}(a + bx)^{n+2} - \frac{a}{b^2(n + 1)}(a + bx)^{n+1}, \quad (n \neq -1, -2)$$

25. $\displaystyle\int x^2(a + bx)^n\, dx = \frac{1}{b^3}\left[\frac{(a + bx)^{n+3}}{n + 3} - 2a\frac{(a + bx)^{n+2}}{n + 2} + a^2\frac{(a + bx)^{n+1}}{n + 1}\right]$

INTEGRALS (Continued)

26. $\displaystyle\int x^m(a + bx)^n\, dx =$
$$
\begin{cases}
\dfrac{x^{m+1}(a + bx)^n}{m + n + 1} + \dfrac{an}{m + n + 1}\displaystyle\int x^m(a + bx)^{n-1}\, dx \\[2mm]
\qquad\qquad\text{or} \\[2mm]
\dfrac{1}{a(n + 1)}\left[-x^{m+1}(a + bx)^{n+1}\right. \\[3mm]
\qquad\qquad\qquad \left. + (m + n + 2)\displaystyle\int x^m(a + bx)^{n+1}\, dx\right] \\[3mm]
\qquad\qquad\text{or} \\[2mm]
\dfrac{1}{b(m + n + 1)}\left[x^m(a + bx)^{n+1} - ma\displaystyle\int x^{m-1}(a + bx)^n\, dx\right]
\end{cases}
$$

27. $\displaystyle\int \frac{dx}{a + bx} = \frac{1}{b}\log(a + bx)$

28. $\displaystyle\int \frac{dx}{(a + bx)^2} = -\frac{1}{b(a + bx)}$

29. $\displaystyle\int \frac{dx}{(a + bx)^3} = -\frac{1}{2b(a + bx)^2}$

30. $\displaystyle\int \frac{x\, dx}{a + bx} =$
$$
\begin{cases}
\dfrac{1}{b^2}[a + bx - a\log(a + bx)] \\[3mm]
\qquad\qquad\text{or} \\[3mm]
\dfrac{x}{b} - \dfrac{a}{b^2}\log(a + bx)
\end{cases}
$$

31. $\displaystyle\int \frac{x\, dx}{(a + bx)^2} = \frac{1}{b^2}\left[\log(a + bx) + \frac{a}{a + bx}\right]$

32. $\displaystyle\int \frac{x\, dx}{(a + bx)^n} = \frac{1}{b^2}\left[\frac{-1}{(n - 2)(a + bx)^{n-2}} + \frac{a}{(n - 1)(a + bx)^{n-1}}\right], \quad n \neq 1, 2$

33. $\displaystyle\int \frac{x^2\, dx}{a + bx} = \frac{1}{b^3}\left[\frac{1}{2}(a + bx)^2 - 2a(a + bx) + a^2\log(a + bx)\right]$

34. $\displaystyle\int \frac{x^2\, dx}{(a + bx)^2} = \frac{1}{b^3}\left[a + bx - 2a\log(a + bx) - \frac{a^2}{a + bx}\right]$

35. $\displaystyle\int \frac{x^2\, dx}{(a + bx)^3} = \frac{1}{b^3}\left[\log(a + bx) + \frac{2a}{a + bx} - \frac{a^2}{2(a + bx)^2}\right]$

36. $\displaystyle\int \frac{x^2\, dx}{(a + bx)^n} = \frac{1}{b^3}\left[\frac{-1}{(n - 3)(a + bx)^{n-3}}\right.$
$$
\left. + \frac{2a}{(n - 2)(a + bx)^{n-2}} - \frac{a^2}{(n - 1)(a + bx)^{n-1}}\right], \quad n \neq 1, 2, 3
$$

INTEGRALS (Continued)

37. $\displaystyle\int \frac{dx}{x(a + bx)} = -\frac{1}{a}\log\frac{a + bx}{x}$

38. $\displaystyle\int \frac{dx}{x(a + bx)^2} = \frac{1}{a(a + bx)} - \frac{1}{a^2}\log\frac{a + bx}{x}$

39. $\displaystyle\int \frac{dx}{x(a + bx)^3} = \frac{1}{a^3}\left[\frac{1}{2}\left(\frac{2a + bx}{a + bx}\right)^2 + \log\frac{x}{a + bx}\right]$

40. $\displaystyle\int \frac{dx}{x^2(a + bx)} = -\frac{1}{ax} + \frac{b}{a^2}\log\frac{a + bx}{x}$

41. $\displaystyle\int \frac{dx}{x^3(a + bx)} = \frac{2bx - a}{2a^2x^2} + \frac{b^2}{a^3}\log\frac{x}{a + bx}$

42. $\displaystyle\int \frac{dx}{x^2(a + bx)^2} = -\frac{a + 2bx}{a^2x(a + bx)} + \frac{2b}{a^3}\log\frac{a + bx}{x}$

FORMS CONTAINING $c^2 \pm x^2$, $x^2 - c^2$

43. $\displaystyle\int \frac{dx}{c^2 + x^2} = \frac{1}{c}\tan^{-1}\frac{x}{c}$

44. $\displaystyle\int \frac{dx}{c^2 - x^2} = \frac{1}{2c}\log\frac{c + x}{c - x}, \qquad (c^2 > x^2)$

45. $\displaystyle\int \frac{dx}{x^2 - c^2} = \frac{1}{2c}\log\frac{x - c}{x + c}, \qquad (x^2 > c^2)$

46. $\displaystyle\int \frac{x\,dx}{c^2 \pm x^2} = \pm\frac{1}{2}\log(c^2 \pm x^2)$

47. $\displaystyle\int \frac{x\,dx}{(c^2 \pm x^2)^{n+1}} = \mp\frac{1}{2n(c^2 \pm x^2)^n}$

48. $\displaystyle\int \frac{dx}{(c^2 \pm x^2)^n} = \frac{1}{2c^2(n - 1)}\left[\frac{x}{(c^2 \pm x^2)^{n-1}} + (2n - 3)\int\frac{dx}{(c^2 \pm x^2)^{n-1}}\right]$

49. $\displaystyle\int \frac{dx}{(x^2 - c^2)^n} = \frac{1}{2c^2(n - 1)}\left[-\frac{x}{(x^2 - c^2)^{n-1}} - (2n - 3)\int\frac{dx}{(x^2 - c^2)^{n-1}}\right]$

50. $\displaystyle\int \frac{x\,dx}{x^2 - c^2} = \frac{1}{2}\log(x^2 - c^2)$

51. $\displaystyle\int \frac{x\,dx}{(x^2 - c^2)^{n+1}} = -\frac{1}{2n(x^2 - c^2)^n}$

INTEGRALS (Continued)

FORMS CONTAINING $a + bx$ and $c + dx$

$$u = a + bx, \qquad v = c + dx, \qquad k = ad - bc$$

If $k = 0$, then $v = \dfrac{c}{a} u$

52. $\displaystyle \int \frac{dx}{u \cdot v} = \frac{1}{k} \cdot \log\left(\frac{v}{u}\right)$

53. $\displaystyle \int \frac{x\, dx}{u \cdot v} = \frac{1}{k}\left[\frac{a}{b}\log(u) - \frac{c}{d}\log(v)\right]$

54. $\displaystyle \int \frac{dx}{u^2 \cdot v} = \frac{1}{k}\left(\frac{1}{u} + \frac{d}{k}\log\frac{v}{u}\right)$

55. $\displaystyle \int \frac{x\, dx}{u^2 \cdot v} = \frac{-a}{bku} - \frac{c}{k^2}\log\frac{v}{u}$

56. $\displaystyle \int \frac{x^2\, dx}{u^2 \cdot v} = \frac{a^2}{b^2 ku} + \frac{1}{k^2}\left[\frac{c^2}{d}\log(v) + \frac{a(k - bc)}{b^2}\log(u)\right]$

57. $\displaystyle \int \frac{dx}{u^n \cdot v^m} = \frac{1}{k(m-1)}\left[\frac{-1}{u^{n-1} \cdot v^{m-1}} - (m + n - 2)b\int \frac{dx}{u^n \cdot v^{m-1}}\right]$

58. $\displaystyle \int \frac{u}{v}\, dx = \frac{bx}{d} + \frac{k}{d^2}\log(v)$

59. $\displaystyle \int \frac{u^m\, dx}{v^n} = \begin{cases} \dfrac{-1}{k(n-1)}\left[\dfrac{u^{m+1}}{v^{n-1}} + b(n - m - 2)\displaystyle\int \dfrac{u^m}{v^{n-1}}\, dx\right] \\ \text{or} \\ \dfrac{-1}{d(n - m - 1)}\left[\dfrac{u^m}{v^{n-1}} + mk\displaystyle\int \dfrac{u^{m-1}}{v^n}\, dx\right] \\ \text{or} \\ \dfrac{-1}{d(n-1)}\left[\dfrac{u^m}{v^{n-1}} - mb\displaystyle\int \dfrac{u^{m-1}}{v^{n-1}}\, dx\right] \end{cases}$

FORMS CONTAINING $(a + bx^n)$

60. $\displaystyle \int \frac{dx}{a + bx^2} = \frac{1}{\sqrt{ab}}\tan^{-1}\frac{x\sqrt{ab}}{a}, \qquad (ab > 0)$

61. $\displaystyle \int \frac{dx}{a + bx^2} = \begin{cases} \dfrac{1}{2\sqrt{-ab}}\log\dfrac{a + x\sqrt{-ab}}{a - x\sqrt{-ab}}, & (ab < 0) \\ \text{or} \\ \dfrac{1}{\sqrt{-ab}}\tanh^{-1}\dfrac{x\sqrt{-ab}}{a}, & (ab < 0) \end{cases}$

INTEGRALS (Continued)

62. $\displaystyle\int \frac{dx}{a^2 + b^2 x^2} = \frac{1}{ab} \tan^{-1} \frac{bx}{a}$

63. $\displaystyle\int \frac{x\,dx}{a + bx^2} = \frac{1}{2b} \log (a + bx^2)$

64. $\displaystyle\int \frac{x^2\,dx}{a + bx^2} = \frac{x}{b} - \frac{a}{b} \int \frac{dx}{a + bx^2}$

65. $\displaystyle\int \frac{dx}{(a + bx^2)^2} = \frac{x}{2a(a + bx^2)} + \frac{1}{2a} \int \frac{dx}{a + bx^2}$

66. $\displaystyle\int \frac{dx}{a^2 - b^2 x^2} = \frac{1}{2ab} \log \frac{a + bx}{a - bx}$

67. $\displaystyle\int \frac{dx}{(a + bx^2)^{m+1}} = \begin{cases} \dfrac{1}{2ma} \dfrac{x}{(a + bx^2)^m} + \dfrac{2m - 1}{2ma} \displaystyle\int \dfrac{dx}{(a + bx^2)^m} \\[2mm] \text{or} \\[2mm] \dfrac{(2m)!}{(m!)^2} \left[\dfrac{x}{2a} \displaystyle\sum_{r=1}^{m} \dfrac{r!(r - 1)!}{(4a)^{m-r}(2r)!(a + bx^2)^r} + \dfrac{1}{(4a)^m} \displaystyle\int \dfrac{dx}{a + bx^2} \right] \end{cases}$

68. $\displaystyle\int \frac{x\,dx}{(a + bx^2)^{m+1}} = -\frac{1}{2bm(a + bx^2)^m}$

69. $\displaystyle\int \frac{x^2\,dx}{(a + bx^2)^{m+1}} = \frac{-x}{2mb(a + bx^2)^m} + \frac{1}{2mb} \int \frac{dx}{(a + bx^2)^m}$

70. $\displaystyle\int \frac{dx}{x(a + bx^2)} = \frac{1}{2a} \log \frac{x^2}{a + bx^2}$

71. $\displaystyle\int \frac{dx}{x^2(a + bx^2)} = -\frac{1}{ax} - \frac{b}{a} \int \frac{dx}{a + bx^2}$

72. $\displaystyle\int \frac{dx}{x(a + bx^2)^{m+1}} = \begin{cases} \dfrac{1}{2am(a + bx^2)^m} + \dfrac{1}{a} \displaystyle\int \dfrac{dx}{x(a + bx^2)^m} \\[2mm] \text{or} \\[2mm] \dfrac{1}{2a^{m+1}} \left[\displaystyle\sum_{r=1}^{m} \dfrac{a^r}{r(a + bx^2)^r} + \log \dfrac{x^2}{a + bx^2} \right] \end{cases}$

73. $\displaystyle\int \frac{dx}{x^2(a + bx^2)^{m+1}} = \frac{1}{a} \int \frac{dx}{x^2(a + bx^2)^m} - \frac{b}{a} \int \frac{dx}{(a + bx^2)^{m+1}}$

74. $\displaystyle\int \frac{dx}{a + bx^3} = \frac{k}{3a} \left[\frac{1}{2} \log \frac{(k + x)^3}{a + bx^3} + \sqrt{3} \tan^{-1} \frac{2x - k}{k\sqrt{3}} \right], \quad \left(k = \sqrt[3]{\frac{a}{b}} \right)$

75. $\displaystyle\int \frac{x\,dx}{a + bx^3} = \frac{1}{3bk} \left[\frac{1}{2} \log \frac{a + bx^3}{(k + x)^3} + \sqrt{3} \tan^{-1} \frac{2x - k}{k\sqrt{3}} \right], \quad \left(k = \sqrt[3]{\frac{a}{b}} \right)$

INTEGRALS (Continued)

76. $\displaystyle \int \frac{x^2\, dx}{a + bx^3} = \frac{1}{3b} \log (a + bx^3)$

77. $\displaystyle \int \frac{dx}{a + bx^4} = \frac{k}{2a} \left[\frac{1}{2} \log \frac{x^2 + 2kx + 2k^2}{x^2 - 2kx + 2k^2} + \tan^{-1} \frac{2kx}{2k^2 - x^2} \right],$

$$\left(ab > 0, k = \sqrt[4]{\frac{a}{4b}} \right)$$

78. $\displaystyle \int \frac{dx}{a + bx^4} = \frac{k}{2a} \left[\frac{1}{2} \log \frac{x + k}{x - k} + \tan^{-1} \frac{x}{k} \right], \qquad \left(ab < 0, k = \sqrt[4]{-\frac{a}{b}} \right)$

79. $\displaystyle \int \frac{x\, dx}{a + bx^4} = \frac{1}{2bk} \tan^{-1} \frac{x^2}{k}, \qquad \left(ab > 0, k = \sqrt{\frac{a}{b}} \right)$

80. $\displaystyle \int \frac{x\, dx}{a + bx^4} = \frac{1}{4bk} \log \frac{x^2 - k}{x^2 + k}, \qquad \left(ab < 0, k = \sqrt{-\frac{a}{b}} \right)$

81. $\displaystyle \int \frac{x^2\, dx}{a + bx^4} = \frac{1}{4bk} \left[\frac{1}{2} \log \frac{x^2 - 2kx + 2k^2}{x^2 + 2kx + 2k^2} + \tan^{-1} \frac{2kx}{2k^2 - x^2} \right],$

$$\left(ab > 0, k = \sqrt[4]{\frac{a}{4b}} \right)$$

82. $\displaystyle \int \frac{x^2\, dx}{a + bx^4} = \frac{1}{4bk} \left[\log \frac{x - k}{x + k} + 2 \tan^{-1} \frac{x}{k} \right], \qquad \left(ab < 0, k = \sqrt[4]{-\frac{a}{b}} \right)$

83. $\displaystyle \int \frac{x^3\, dx}{a + bx^4} = \frac{1}{4b} \log (a + bx^4)$

84. $\displaystyle \int \frac{dx}{x(a + bx^n)} = \frac{1}{an} \log \frac{x^n}{a + bx^n}$

85. $\displaystyle \int \frac{dx}{(a + bx^n)^{m+1}} = \frac{1}{a} \int \frac{dx}{(a + bx^n)^m} - \frac{b}{a} \int \frac{x^n\, dx}{(a + bx^n)^{m+1}}$

86. $\displaystyle \int \frac{x^m\, dx}{(a + bx^n)^{p+1}} = \frac{1}{b} \int \frac{x^{m-n}\, dx}{(a + bx^n)^p} - \frac{a}{b} \int \frac{x^{m-n}\, dx}{(a + bx^n)^{p+1}}$

87. $\displaystyle \int \frac{dx}{x^m (a + bx^n)^{p+1}} = \frac{1}{a} \int \frac{dx}{x^m (a + bx^n)^p} - \frac{b}{a} \int \frac{dx}{x^{m-n}(a + bx^n)^{p+1}}$

INTEGRALS (Continued)

88. $\displaystyle\int x^m(a + bx^n)^p\, dx =$

$$\frac{1}{b(np + m + 1)}\left[x^{m-n+1}(a + bx^n)^{p+1}\right.$$

$$\left. - a(m - n + 1)\int x^{m-n}(a + bx^n)^p\, dx\right]$$

or

$$\frac{1}{np + m + 1}\left[x^{m+1}(a + bx^n)^p\right.$$

$$\left. + anp\int x^m(a + bx^n)^{p-1}\, dx\right]$$

or

$$\frac{1}{a(m + 1)}\left[x^{m+1}(a + bx^n)^{p+1}\right.$$

$$\left. - (m + 1 + np + n)b\int x^{m+n}(a + bx^n)^p\, dx\right]$$

or

$$\frac{1}{an(p + 1)}\left[-x^{m+1}(a + bx^n)^{p+1}\right.$$

$$\left. + (m + 1 + np + n)\int x^m(a + bx^n)^{p+1}\, dx\right]$$

FORMS CONTAINING $c^3 \pm x^3$

89. $\displaystyle\int \frac{dx}{c^3 \pm x^3} = \pm\frac{1}{6c^2}\log\frac{(c \pm x)^3}{c^3 \pm x^3} + \frac{1}{c^2\sqrt{3}}\tan^{-1}\frac{2x \mp c}{c\sqrt{3}}$

90. $\displaystyle\int \frac{dx}{(c^3 \pm x^3)^2} = \frac{x}{3c^3(c^3 \pm x^3)} + \frac{2}{3c^3}\int \frac{dx}{c^3 \pm x^3}$

91. $\displaystyle\int \frac{dx}{(c^3 \pm x^3)^{n+1}} = \frac{1}{3nc^3}\left[\frac{x}{(c^3 \pm x^3)^n} + (3n - 1)\int \frac{dx}{(c^3 \pm x^3)^n}\right]$

92. $\displaystyle\int \frac{x\, dx}{c^3 \pm x^3} = \frac{1}{6c}\log\frac{c^3 \pm x^3}{(c \pm x)^3} \pm \frac{1}{c\sqrt{3}}\tan^{-1}\frac{2x \mp c}{c\sqrt{3}}$

93. $\displaystyle\int \frac{x\, dx}{(c^3 \pm x^3)^2} = \frac{x^2}{3c^3(c^3 \pm x^3)} + \frac{1}{3c^3}\int \frac{x\, dx}{c^3 \pm x^3}$

94. $\displaystyle\int \frac{x\, dx}{(c^3 + x^3)^{p+1}} = \frac{1}{3nc^3}\left[\frac{x^2}{(c^3 \pm x^3)^n} + (3n - 2)\int \frac{x\, dx}{(c^3 \pm x^3)^n}\right]$

95. $\displaystyle\int \frac{x^2\, dx}{c^3 \pm x^3} = \pm\frac{1}{3}\log(c^3 \pm x^3)$

INTEGRALS (Continued)

96. $\int \dfrac{x^2\,dx}{(c^3 \pm x^3)^{n+1}} = \mp \dfrac{1}{3n(c^3 \pm x^3)^n}$

97. $\int \dfrac{dx}{x(c^3 \pm x^3)} = \dfrac{1}{3c^3} \log \dfrac{x^3}{c^3 \pm x^3}$

98. $\int \dfrac{dx}{x(c^3 \pm x^3)^2} = \dfrac{1}{3c^3(c^3 \pm x^3)} + \dfrac{1}{3c^6} \log \dfrac{x^3}{c^3 \pm x^3}$

99. $\int \dfrac{dx}{x(c^3 \pm x^3)^{n+1}} = \dfrac{1}{3nc^3(c^3 \pm x^3)^n} + \dfrac{1}{c^3} \int \dfrac{dx}{x(c^3 \pm x^3)^n}$

100. $\int \dfrac{dx}{x^2(c^3 \pm x^3)} = -\dfrac{1}{c^3 x} \mp \dfrac{1}{c^3} \int \dfrac{x\,dx}{c^3 \pm x^3}$

101. $\int \dfrac{dx}{x^2(c^3 \pm x^3)^{n+1}} = \dfrac{1}{c^3} \int \dfrac{dx}{x^2(c^3 \pm x^3)^n} \mp \dfrac{1}{c^3} \int \dfrac{x\,dx}{(c^3 \pm x^3)^{n+1}}$

FORMS CONTAINING $c^4 \pm x^4$

102. $\int \dfrac{dx}{c^4 + x^4} = \dfrac{1}{2c^3\sqrt{2}} \left[\dfrac{1}{2} \log \dfrac{x^2 + cx\sqrt{2} + c^2}{x^2 - cx\sqrt{2} + c^2} + \tan^{-1} \dfrac{cx\sqrt{2}}{c^2 - x^2} \right]$

103. $\int \dfrac{dx}{c^4 - x^4} = \dfrac{1}{2c^3} \left[\dfrac{1}{2} \log \dfrac{c + x}{c - x} + \tan^{-1} \dfrac{x}{c} \right]$

104. $\int \dfrac{x\,dx}{c^4 + x^4} = \dfrac{1}{2c^2} \tan^{-1} \dfrac{x^2}{c^2}$

105. $\int \dfrac{x\,dx}{c^4 - x^4} = \dfrac{1}{4c^2} \log \dfrac{c^2 + x^2}{c^2 - x^2}$

106. $\int \dfrac{x^2\,dx}{c^4 + x^4} = \dfrac{1}{2c\sqrt{2}} \left[\dfrac{1}{2} \log \dfrac{x^2 - cx\sqrt{2} + c^2}{x^2 + cx\sqrt{2} + c^2} + \tan^{-1} \dfrac{cx\sqrt{2}}{c^2 - x^2} \right]$

107. $\int \dfrac{x^2\,dx}{c^4 - x^4} = \dfrac{1}{2c} \left[\dfrac{1}{2} \log \dfrac{c + x}{c - x} - \tan^{-1} \dfrac{x}{c} \right]$

108. $\int \dfrac{x^3\,dx}{c^4 \pm x^4} = \pm \dfrac{1}{4} \log (c^4 \pm x^4)$

FORMS CONTAINING $(a + bx + cx^2)$

$$X = a + bx + cx^2 \text{ and } q = 4ac - b^2$$

If $q = 0$, then $X = c\left(x + \dfrac{b}{2c}\right)^2$, and formulas starting with 23 should be used in place of these.

109. $\int \dfrac{dx}{X} = \dfrac{2}{\sqrt{q}} \tan^{-1} \dfrac{2cx + b}{\sqrt{q}}, \qquad (q > 0)$

INTEGRALS (Continued)

110. $\displaystyle \int \frac{dx}{X} = \begin{cases} \dfrac{-2}{\sqrt{-q}} \tanh^{-1} \dfrac{2cx + b}{\sqrt{-q}} \\[4mm] \qquad\text{or} \\[4mm] \dfrac{1}{\sqrt{-q}} \log \dfrac{2cx + b - \sqrt{-q}}{2cx + b + \sqrt{-q}}, \qquad (q < 0) \end{cases}$

111. $\displaystyle \int \frac{dx}{X^2} = \frac{2cx + b}{qX} + \frac{2c}{q} \int \frac{dx}{X}$

112. $\displaystyle \int \frac{dx}{X^3} = \frac{2cx + b}{q}\left(\frac{1}{2X^2} + \frac{3c}{qX} \right) + \frac{6c^2}{q^2} \int \frac{dx}{X}$

113. $\displaystyle \int \frac{dx}{X^{n+1}} = \begin{cases} \dfrac{2cx + b}{nqX^n} + \dfrac{2(2n - 1)c}{qn} \int \dfrac{dx}{X^n} \\[4mm] \qquad\text{or} \\[4mm] \dfrac{(2n)!}{(n!)^2}\left(\dfrac{c}{q}\right)^n\left[\dfrac{2cx + b}{q} \displaystyle\sum_{r=1}^{n} \left(\dfrac{q}{cX}\right)^r\left(\dfrac{(r - 1)!\,r!}{(2r)!}\right) + \displaystyle\int \dfrac{dx}{X} \right] \end{cases}$

114. $\displaystyle \int \frac{x\,dx}{X} - \frac{1}{2c} \log X - \frac{b}{2c} \int \frac{dx}{X}$

115. $\displaystyle \int \frac{x\,dx}{X^2} = -\frac{bx + 2a}{qX} - \frac{b}{q} \int \frac{dx}{X}$

116. $\displaystyle \int \frac{x\,dx}{X^{n+1}} = -\frac{2a + bx}{nqX^n} - \frac{b(2n - 1)}{nq} \int \frac{dx}{X^n}$

117. $\displaystyle \int \frac{x^2}{X}\,dx = \frac{x}{c} - \frac{b}{2c^2} \log X + \frac{b^2 - 2ac}{2c^2} \int \frac{dx}{X}$

118. $\displaystyle \int \frac{x^2}{X^2}\,dx = \frac{(b^2 - 2ac)x + ab}{cqX} + \frac{2a}{q} \int \frac{dx}{X}$

119. $\displaystyle \int \frac{x^m\,dx}{X^{n+1}} = -\frac{x^{m-1}}{(2n - m + 1)cX^n} - \frac{n - m + 1}{2n - m + 1}\cdot\frac{b}{c} \int \frac{x^{m-1}\,dx}{X^{n+1}}$

$$+ \frac{m - 1}{2n - m + 1}\cdot\frac{a}{c} \int \frac{x^{m-2}\,dx}{X^{n+1}}$$

120. $\displaystyle \int \frac{dx}{xX} = \frac{1}{2a} \log \frac{x^2}{X} - \frac{b}{2a} \int \frac{dx}{X}$

121. $\displaystyle \int \frac{dx}{x^2 X} = \frac{b}{2a^2} \log \frac{X}{x^2} - \frac{1}{ax} + \left(\frac{b^2}{2a^2} - \frac{c}{a} \right) \int \frac{dx}{X}$

122. $\displaystyle \int \frac{dx}{xX^n} = \frac{1}{2a(n - 1)X^{n-1}} - \frac{b}{2a} \int \frac{dx}{X^n} + \frac{1}{a} \int \frac{dx}{xX^{n-1}}$

INTEGRALS (Continued)

123. $\displaystyle\int \frac{dx}{x^m X^{n+1}} = -\frac{1}{(m-1)ax^{m-1}X^n} - \frac{n+m-1}{m-1}\cdot\frac{b}{a}\int \frac{dx}{x^{m-1}X^{n+1}}$

$$-\frac{2n+m-1}{m-1}\cdot\frac{c}{a}\int \frac{dx}{x^{m-2}X^{n+1}}$$

FORMS CONTAINING $\sqrt{a + bx}$

124. $\displaystyle\int \sqrt{a+bx}\,dx = \frac{2}{3b}\sqrt{(a+bx)^3}$

125. $\displaystyle\int x\sqrt{a+bx}\,dx = -\frac{2(2a-3bx)\sqrt{(a+bx)^3}}{15b^2}$

126. $\displaystyle\int x^2\sqrt{a+bx}\,dx = \frac{2(8a^2-12abx+15b^2x^2)\sqrt{(a+bx)^3}}{105b^3}$

127. $\displaystyle\int x^m\sqrt{a+bx}\,dx = \begin{cases} \dfrac{2}{b(2m+3)}\left[x^m\sqrt{(a+bx)^3} - ma\int x^{m-1}\sqrt{a+bx}\,dx\right] \\[2ex] \text{or} \\[2ex] \dfrac{2}{b^{m+1}}\sqrt{a+bx}\displaystyle\sum_{r=0}^{m}\dfrac{m!(-a)^{m-r}}{r!(m-r)!(2r+3)}(a+bx)^{r+1} \end{cases}$

128. $\displaystyle\int \frac{\sqrt{a+bx}}{x}\,dx = 2\sqrt{a+bx} + a\int \frac{dx}{x\sqrt{a+bx}}$

129. $\displaystyle\int \frac{\sqrt{a+bx}}{x^2}\,dx = -\frac{\sqrt{a+bx}}{x} + \frac{b}{2}\int \frac{dx}{x\sqrt{a+bx}}$

130. $\displaystyle\int \frac{\sqrt{a+bx}}{x^m}\,dx = -\frac{1}{(m-1)a}\left[\frac{\sqrt{(a+bx)^3}}{x^{m-1}} + \frac{(2m-5)b}{2}\int \frac{\sqrt{a+bx}}{x^{m-1}}\,dx\right]$

131. $\displaystyle\int \frac{dx}{\sqrt{a+bx}} = \frac{2\sqrt{a+bx}}{b}$

132. $\displaystyle\int \frac{x\,dx}{\sqrt{a+bx}} = -\frac{2(2a-bx)}{3b^2}\sqrt{a+bx}$

133. $\displaystyle\int \frac{x^2\,dx}{\sqrt{a+bx}} = \frac{2(8a^2-4abx+3b^2x^2)}{15b^3}\sqrt{a+bx}$

INTEGRALS (Continued)

134. $\displaystyle\int \frac{x^m\,dx}{\sqrt{a+bx}} = \begin{cases} \dfrac{2}{(2m+1)b}\left[x^m\sqrt{a+bx} - ma\displaystyle\int \frac{x^{m-1}\,dx}{\sqrt{a+bx}}\right] \\[3mm] \text{or} \\[3mm] \dfrac{2(-a)^m\sqrt{a+bx}}{b^{m+1}}\displaystyle\sum_{r=0}^{m}\frac{(-1)^r m!(a+bx)^r}{(2r+1)r!(m-r)!a^r} \end{cases}$

135. $\displaystyle\int \frac{dx}{x\sqrt{a+bx}} = \frac{1}{\sqrt{a}}\log\left(\frac{\sqrt{a+bx}-\sqrt{a}}{\sqrt{a+bx}+\sqrt{a}}\right), \qquad (a>0)$

136. $\displaystyle\int \frac{dx}{x\sqrt{a+bx}} = \frac{2}{\sqrt{-a}}\tan^{-1}\sqrt{\frac{a+bx}{-a}}, \qquad (a<0)$

137. $\displaystyle\int \frac{dx}{x^2\sqrt{a+bx}} = -\frac{\sqrt{a+bx}}{ax} - \frac{b}{2a}\int \frac{dx}{x\sqrt{a+bx}}$

138. $\displaystyle\int \frac{dx}{x^n\sqrt{a+bx}} = \begin{cases} -\dfrac{\sqrt{a+bx}}{(n-1)ax^{n-1}} - \dfrac{(2n-3)b}{(2n-2)a}\displaystyle\int \frac{dx}{x^{n-1}\sqrt{a+bx}} \\[3mm] \text{or} \\[3mm] \dfrac{(2n-2)!}{[(n-1)!]^2}\left[-\dfrac{\sqrt{a+bx}}{a}\displaystyle\sum_{r=1}^{n-1}\frac{r!(r-1)!}{x^r(2r)!}\left(-\frac{b}{4a}\right)^{n-r-1} \right. \\[3mm] \left. \qquad\qquad\qquad + \left(-\dfrac{b}{4a}\right)^{n-1}\displaystyle\int \frac{dx}{x\sqrt{a+bx}}\right] \end{cases}$

139. $\displaystyle\int (a+bx)^{\pm\frac{n}{2}}\,dx = \frac{2(a+bx)^{\frac{2\pm n}{2}}}{b(2\pm n)}$

140. $\displaystyle\int x(a+bx)^{\pm\frac{n}{2}}\,dx = \frac{2}{b^2}\left[\frac{(a+bx)^{\frac{4\pm n}{2}}}{4\pm n} - \frac{a(a+bx)^{\frac{2\pm n}{2}}}{2\pm n}\right]$

141. $\displaystyle\int \frac{dx}{x(a+bx)^{\frac{m}{2}}} = \frac{1}{a}\int \frac{dx}{x(a+bx)^{\frac{m-2}{2}}} - \frac{b}{a}\int \frac{dx}{(a+bx)^{\frac{m}{2}}}$

142. $\displaystyle\int \frac{(a+bx)^{\frac{n}{2}}\,dx}{x} = b\int (a+bx)^{\frac{n-2}{2}}\,dx + a\int \frac{(a+bx)^{\frac{n-2}{2}}}{x}\,dx$

143. $\displaystyle\int f(x,\sqrt{a+bx})\,dx = \frac{2}{b}\int f\left(\frac{z^2-a}{b},z\right)z\,dz, \qquad (z=\sqrt{a+bx})$

INTEGRALS (Continued)

FORMS CONTAINING $\sqrt{a + bx}$ and $\sqrt{c + dx}$

$$u = a + bx \qquad v = c + dx \qquad k = ad - bc$$

If $k = 0$, then $v = \dfrac{c}{a} u$, and formulas starting with 124 should be used in place of these.

144. $\displaystyle\int \frac{dx}{\sqrt{uv}} = \begin{cases} \dfrac{2}{\sqrt{bd}} \tanh^{-1} \dfrac{\sqrt{bduv}}{bv}, \ bd > o, k < o \\[1em] \text{or} \\[1em] \dfrac{2}{\sqrt{bd}} \tanh^{-1} \dfrac{\sqrt{bduv}}{du}, \ bd > o, k > o. \\[1em] \text{or} \\[1em] \dfrac{1}{\sqrt{bd}} \log \dfrac{(bv + \sqrt{bduv})^2}{v}, \quad (bd > 0) \end{cases}$

145. $\displaystyle\int \frac{dx}{\sqrt{uv}} = \begin{cases} \dfrac{2}{\sqrt{-bd}} \tan^{-1} \dfrac{\sqrt{-bduv}}{bv} \\[1em] \text{or} \\[1em] -\dfrac{1}{\sqrt{-bd}} \sin^{-1}\left(\dfrac{2bdx + ad + bc}{|k|}\right), \quad (bd < 0) \end{cases}$

146. $\displaystyle\int \sqrt{uv}\, dx = \frac{k + 2bv}{4bd} \sqrt{uv} - \frac{k^2}{8bd} \int \frac{dx}{\sqrt{uv}}$

147. $\displaystyle\int \frac{dx}{v\sqrt{u}} = \begin{cases} \dfrac{1}{\sqrt{kd}} \log \dfrac{d\sqrt{u} - \sqrt{kd}}{d\sqrt{u} + \sqrt{kd}} \\[1em] \text{or} \\[1em] \dfrac{1}{\sqrt{kd}} \log \dfrac{(d\sqrt{u} - \sqrt{kd})^2}{v}, \quad (kd > 0) \end{cases}$

148. $\displaystyle\int \frac{dx}{v\sqrt{u}} = \frac{2}{\sqrt{-kd}} \tan^{-1} \frac{d\sqrt{u}}{\sqrt{-kd}}, \quad (kd < 0)$

149. $\displaystyle\int \frac{x\, dx}{\sqrt{uv}} = \frac{\sqrt{uv}}{bd} - \frac{ad + bc}{2bd} \int \frac{dx}{\sqrt{uv}}$

150. $\displaystyle\int \frac{dx}{v\sqrt{uv}} = \frac{-2\sqrt{uv}}{kv}$

INTEGRALS (Continued)

151. $\displaystyle\int \frac{v\,dx}{\sqrt{uv}} = \frac{\sqrt{uv}}{b} - \frac{k}{2b}\int \frac{dx}{\sqrt{uv}}$

152. $\displaystyle\int \sqrt{\frac{v}{u}}\,dx = \frac{v}{|v|}\int \frac{v\,dx}{\sqrt{uv}}$

153. $\displaystyle\int v^m \sqrt{u}\,dx = \frac{1}{(2m+3)d}\left(2v^{m+1}\sqrt{u} + k\int \frac{v^m\,dx}{\sqrt{u}}\right)$

154. $\displaystyle\int \frac{dx}{v^m \sqrt{u}} = -\frac{1}{(m-1)k}\left(\frac{\sqrt{u}}{v^{m-1}} + \left(m - \frac{3}{2}\right)b \int \frac{dx}{v^{m-1}\sqrt{u}}\right)$

155. $\displaystyle\int \frac{v^m\,dx}{\sqrt{u}} = \begin{cases} \dfrac{2}{b(2m+1)}\left[v^m\sqrt{u} - mk\int \dfrac{v^{m-1}}{\sqrt{u}}\,dx\right] \\[2ex] \qquad\qquad \text{or} \\[1ex] \dfrac{2(m!)^2\sqrt{u}}{b(2m+1)!}\displaystyle\sum_{r=0}^{m}\left(-\frac{4k}{b}\right)^{m-r}\frac{(2r)!}{(r!)^2}v^r \end{cases}$

FORMS CONTAINING $\sqrt{x^2 \pm a^2}$

156. $\displaystyle\int \sqrt{x^2 \pm a^2}\,dx = \tfrac{1}{2}[x\sqrt{x^2 \pm a^2} \pm a^2 \log(x + \sqrt{x^2 \pm a^2})]$

157. $\displaystyle\int \frac{dx}{\sqrt{x^2 \pm a^2}} = \log(x + \sqrt{x^2 \pm a^2})$

158. $\displaystyle\int \frac{dx}{x\sqrt{x^2 - a^2}} = \frac{1}{|a|}\sec^{-1}\frac{x}{a}$

159. $\displaystyle\int \frac{dx}{x\sqrt{x^2 + a^2}} = -\frac{1}{a}\log\left(\frac{a + \sqrt{x^2 + a^2}}{x}\right)$

160. $\displaystyle\int \frac{\sqrt{x^2 + a^2}}{x}\,dx = \sqrt{x^2 + a^2} - a\log\left(\frac{a + \sqrt{x^2 + a^2}}{x}\right)$

161. $\displaystyle\int \frac{\sqrt{x^2 - a^2}}{x}\,dx = \sqrt{x^2 - a^2} - |a|\sec^{-1}\frac{x}{a}$

162. $\displaystyle\int \frac{x\,dx}{\sqrt{x^2 \pm a^2}} = \sqrt{x^2 \pm a^2}$

163. $\displaystyle\int x\sqrt{x^2 \pm a^2}\,dx = \tfrac{1}{3}\sqrt{(x^2 \pm a^2)^3}$

INTEGRALS (Continued)

164. $\displaystyle\int \sqrt{(x^2 \pm a^2)^3}\, dx = \frac{1}{4}\left[x\sqrt{(x^2 \pm a^2)^3} \pm \frac{3a^2 x}{2}\sqrt{x^2 \pm a^2} \right.$

$$\left. + \frac{3a^4}{2}\log\left(x + \sqrt{x^2 \pm a^2}\right) \right]$$

165. $\displaystyle\int \frac{dx}{\sqrt{(x^2 \pm a^2)^3}} = \frac{\pm x}{a^2\sqrt{x^2 \pm a^2}}$

166. $\displaystyle\int \frac{x\,dx}{\sqrt{(x^2 \pm a^2)^3}} = \frac{-1}{\sqrt{x^2 \pm a^2}}$

167. $\displaystyle\int x\sqrt{(x^2 \pm a^2)^3}\, dx = \tfrac{1}{5}\sqrt{(x^2 \pm a^2)^5}$

168. $\displaystyle\int x^2\sqrt{x^2 \pm a^2}\, dx = \frac{x}{4}\sqrt{(x^2 \pm a^2)^3} \mp \frac{a^2}{8}x\sqrt{x^2 \pm a^2} - \frac{a^4}{8}\log\left(x + \sqrt{x^2 \pm a^2}\right)$

169. $\displaystyle\int x^3\sqrt{x^2 + a^2}\, dx = \left(\tfrac{1}{5}x^2 - \tfrac{2}{15}a^2\right)\sqrt{(a^2 + x^2)^3}$

170. $\displaystyle\int x^3\sqrt{x^2 - a^2}\, dx = \frac{1}{5}\sqrt{(x^2 - a^2)^5} + \frac{a^2}{3}\sqrt{(x^2 - a^2)^3}$

171. $\displaystyle\int \frac{x^2\,dx}{\sqrt{x^2 \pm a^2}} = \frac{x}{2}\sqrt{x^2 \pm a^2} \mp \frac{a^2}{2}\log\left(x + \sqrt{x^2 \pm a^2}\right)$

172. $\displaystyle\int \frac{x^3\,dx}{\sqrt{x^2 \pm a^2}} = \frac{1}{3}\sqrt{(x^2 \pm a^2)^3} \mp a^2\sqrt{x^2 \pm a^2}$

173. $\displaystyle\int \frac{dx}{x^2\sqrt{x^2 \pm a^2}} = \mp\frac{\sqrt{x^2 \pm a^2}}{a^2 x}$

174. $\displaystyle\int \frac{dx}{x^3\sqrt{x^2 + a^2}} = -\frac{\sqrt{x^2 + a^2}}{2a^2 x^2} + \frac{1}{2a^3}\log\frac{a + \sqrt{x^2 + a^2}}{x}$

175. $\displaystyle\int \frac{dx}{x^3\sqrt{x^2 - a^2}} = \frac{\sqrt{x^2 - a^2}}{2a^2 x^2} + \frac{1}{2|a^3|}\sec^{-1}\frac{x}{a}$

176. $\displaystyle\int x^2\sqrt{(x^2 \pm a^2)^3}\, dx = \frac{x}{6}\sqrt{(x^2 \pm a^2)^5} \mp \frac{a^2 x}{24}\sqrt{(x^2 \pm a^2)^3} - \frac{a^4 x}{16}\sqrt{x^2 \pm a^2}$

$$\mp \frac{a^6}{16}\log\left(x + \sqrt{x^2 \pm a^2}\right)$$

177. $\displaystyle\int x^3\sqrt{(x^2 \pm a^2)^3}\, dx = \frac{1}{7}\sqrt{(x^2 \pm a^2)^7} \mp \frac{a^2}{5}\sqrt{(x^2 \pm a^2)^5}$

INTEGRALS (Continued)

178. $\displaystyle\int \frac{\sqrt{x^2 \pm a^2}\, dx}{x^2} = -\frac{\sqrt{x^2 \pm a^2}}{x} + \log\left(x + \sqrt{x^2 \pm a^2}\right)$

179. $\displaystyle\int \frac{\sqrt{x^2 + a^2}}{x^3}\, dx = -\frac{\sqrt{x^2 + a^2}}{2x^2} - \frac{1}{2a}\log\frac{a + \sqrt{x^2 + a^2}}{x}$

180. $\displaystyle\int \frac{\sqrt{x^2 - a^2}}{x^3}\, dx = -\frac{\sqrt{x^2 - a^2}}{2x^2} + \frac{1}{2|a|}\sec^{-1}\frac{x}{a}$

181. $\displaystyle\int \frac{\sqrt{x^2 \pm a^2}}{x^4}\, dx = \mp\frac{\sqrt{(x^2 \pm a^2)^3}}{3a^2 x^3}$

182. $\displaystyle\int \frac{x^2\, dx}{\sqrt{(x^2 \pm a^2)^3}} = \frac{-x}{\sqrt{x^2 \pm a^2}} + \log\left(x + \sqrt{x^2 \pm a^2}\right)$

183. $\displaystyle\int \frac{x^3\, dx}{\sqrt{(x^2 \pm a^2)^3}} = \sqrt{x^2 \pm a^2} \pm \frac{a^2}{\sqrt{x^2 \pm a^2}}$

184. $\displaystyle\int \frac{dx}{x\sqrt{(x^2 + a^2)^3}} = \frac{1}{a^2\sqrt{x^2 + a^2}} - \frac{1}{a^3}\log\frac{a + \sqrt{x^2 + a^2}}{x}$

185. $\displaystyle\int \frac{dx}{x\sqrt{(x^2 - a^2)^3}} = -\frac{1}{a^2\sqrt{x^2 - a^2}} - \frac{1}{|a^3|}\sec^{-1}\frac{x}{a}$

186. $\displaystyle\int \frac{dx}{x^2\sqrt{(x^2 \pm a^2)^3}} = -\frac{1}{a^4}\left[\frac{\sqrt{x^2 \pm a^2}}{x} + \frac{x}{\sqrt{x^2 \pm a^2}}\right]$

187. $\displaystyle\int \frac{dx}{x^3\sqrt{(x^2 + a^2)^3}} = -\frac{1}{2a^2 x^2\sqrt{x^2 + a^2}} - \frac{3}{2a^4\sqrt{x^2 + a^2}}$

$$+ \frac{3}{2a^5}\log\frac{a + \sqrt{x^2 + a^2}}{x}$$

188. $\displaystyle\int \frac{dx}{x^3\sqrt{(x^2 - a^2)^3}} = \frac{1}{2a^2 x^2\sqrt{x^2 - a^2}} - \frac{3}{2a^4\sqrt{x^2 - a^2}} - \frac{3}{2|a^5|}\sec^{-1}\frac{x}{a}$

189. $\displaystyle\int \frac{x^m}{\sqrt{x^2 \pm a^2}}\, dx = \frac{1}{m}x^{m-1}\sqrt{x^2 \pm a^2} \mp \frac{m-1}{m}a^2\int \frac{x^{m-2}}{\sqrt{x^2 \pm a^2}}\, dx$

190. $\displaystyle\int \frac{x^{2m}}{\sqrt{x^2 \pm a^2}}\, dx = \frac{(2m)!}{2^{2m}(m!)^2}\left[\sqrt{x^2 \pm a^2}\sum_{r=1}^{m}\frac{r!(r-1)!}{(2r)!}(\mp a^2)^{m-r}(2x)^{2r-1}\right.$

$$\left. + (\mp a^2)^m \log\left(x + \sqrt{x^2 \pm a^2}\right)\right]$$

191. $\displaystyle\int \frac{x^{2m+1}}{\sqrt{x^2 \pm a^2}}\, dx = \sqrt{x^2 \pm a^2}\sum_{r=0}^{m}\frac{(2r)!(m!)^2}{(2m+1)!(r!)^2}(\mp 4a^2)^{m-r}x^{2r}$

INTEGRALS (Continued)

192. $\displaystyle \int \frac{dx}{x^m \sqrt{x^2 \pm a^2}} = \mp \frac{\sqrt{x^2 \pm a^2}}{(m-1)a^2 x^{m-1}} \mp \frac{(m-2)}{(m-1)a^2} \int \frac{dx}{x^{m-2}\sqrt{x^2 \pm a^2}}$

193. $\displaystyle \int \frac{dx}{x^{2m}\sqrt{x^2 \pm a^2}} = \sqrt{x^2 \pm a^2} \sum_{r=0}^{m-1} \frac{(m-1)!\,m!\,(2r)!\,2^{2m-2r-1}}{(r!)^2 (2m)!\,(\mp a^2)^{m-r} x^{2r+1}}$

194. $\displaystyle \int \frac{dx}{x^{2m+1}\sqrt{x^2 + a^2}} = \frac{(2m)!}{(m!)^2}\left[\frac{\sqrt{x^2 + a^2}}{a^2} \sum_{r=1}^{m} (-1)^{m-r+1} \frac{r!(r-1)!}{2(2r)!(4a^2)^{m-r}x^{2r}} \right.$

$$\left. + \frac{(-1)^{m+1}}{2^{2m}a^{2m+1}} \log \frac{\sqrt{x^2 + a^2} + a}{x} \right]$$

195. $\displaystyle \int \frac{dx}{x^{2m+1}\sqrt{x^2 - a^2}} = \frac{(2m)!}{(m!)^2}\left[\frac{\sqrt{x^2 - a^2}}{a^2} \sum_{r=1}^{m} \frac{r!(r-1)!}{2(2r)!(4a^2)^{m-r}x^{2r}} \right.$

$$\left. + \frac{1}{2^{2m}|a|^{2m+1}} \sec^{-1}\frac{x}{a} \right]$$

196. $\displaystyle \int \frac{dx}{(x-a)\sqrt{x^2 - a^2}} = -\frac{\sqrt{x^2 - a^2}}{a(x-a)}$

197. $\displaystyle \int \frac{dx}{(x+a)\sqrt{x^2 - a^2}} = \frac{\sqrt{x^2 - a^2}}{a(x+a)}$

198. $\displaystyle \int f(x, \sqrt{x^2 + a^2})\,dx = a \int f(a \tan u, a \sec u) \sec^2 u\,du, \quad \left(u = \tan^{-1}\frac{x}{a}, a > 0 \right)$

199. $\displaystyle \int f(x, \sqrt{x^2 - a^2})\,dx = a \int f(a \sec u, a \tan u) \sec u \tan u\,du, \quad \left(u = \sec^{-1}\frac{x}{a}, \right.$

$$\left. a > 0 \right)$$

FORMS CONTAINING $\sqrt{a^2 - x^2}$

200. $\displaystyle \int \sqrt{a^2 - x^2}\,dx = \frac{1}{2}\left[x\sqrt{a^2 - x^2} + a^2 \sin^{-1}\frac{x}{|a|} \right]$

201. $\displaystyle \int \frac{dx}{\sqrt{a^2 - x^2}} = \begin{cases} \sin^{-1}\dfrac{x}{|a|} \\ \text{or} \\ -\cos^{-1}\dfrac{x}{|a|} \end{cases}$

202. $\displaystyle \int \frac{dx}{x\sqrt{a^2 - x^2}} = -\frac{1}{a}\log\left(\frac{a + \sqrt{a^2 - x^2}}{x} \right)$

INTEGRALS (Continued)

203. $\displaystyle\int \frac{\sqrt{a^2 - x^2}}{x}\,dx = \sqrt{a^2 - x^2} - a\log\left(\frac{a + \sqrt{a^2 - x^2}}{x}\right)$

204. $\displaystyle\int \frac{x\,dx}{\sqrt{a^2 - x^2}} = -\sqrt{a^2 - x^2}$

205. $\displaystyle\int x\sqrt{a^2 - x^2}\,dx = -\tfrac{1}{3}\sqrt{(a^2 - x^2)^3}$

206. $\displaystyle\int \sqrt{(a^2 - x^2)^3}\,dx = \frac{1}{4}\left[x\sqrt{(a^2 - x^2)^3} + \frac{3a^2 x}{2}\sqrt{a^2 - x^2} + \frac{3a^4}{2}\sin^{-1}\frac{x}{|a|}\right]$

207. $\displaystyle\int \frac{dx}{\sqrt{(a^2 - x^2)^3}} = \frac{x}{a^2\sqrt{a^2 - x^2}}$

208. $\displaystyle\int \frac{x\,dx}{\sqrt{(a^2 - x^2)^3}} = \frac{1}{\sqrt{a^2 - x^2}}$

209. $\displaystyle\int x\sqrt{(a^2 - x^2)^3}\,dx = -\tfrac{1}{5}\sqrt{(a^2 - x^2)^5}$

210. $\displaystyle\int x^2\sqrt{a^2 - x^2}\,dx = -\frac{x}{4}\sqrt{(a^2 - x^2)^3} + \frac{a^2}{8}\left(x\sqrt{a^2 - x^2} + a^2\sin^{-1}\frac{x}{|a|}\right)$

211. $\displaystyle\int x^3\sqrt{a^2 - x^2}\,dx = (-\tfrac{1}{5}x^2 - \tfrac{2}{15}a^2)\sqrt{(a^2 - x^2)^3}$

212. $\displaystyle\int x^2\sqrt{(a^2 - x^2)^3}\,dx = -\frac{1}{6}x\sqrt{(a^2 - x^2)^5} + \frac{a^2 x}{24}\sqrt{(a^2 - x^2)^3}$

$$+ \frac{a^4 x}{16}\sqrt{a^2 - x^2} + \frac{a^6}{16}\sin^{-1}\frac{x}{|a|}$$

213. $\displaystyle\int x^3\sqrt{(a^2 - x^2)^3}\,dx = \frac{1}{7}\sqrt{(a^2 - x^2)^7} - \frac{a^2}{5}\sqrt{(a^2 - x^2)^5}$

214. $\displaystyle\int \frac{x^2\,dx}{\sqrt{a^2 - x^2}} = -\frac{x}{2}\sqrt{a^2 - x^2} + \frac{a^2}{2}\sin^{-1}\frac{x}{|a|}$

215. $\displaystyle\int \frac{dx}{x^2\sqrt{a^2 - x^2}} = -\frac{\sqrt{a^2 - x^2}}{a^2 x}$

216. $\displaystyle\int \frac{\sqrt{a^2 - x^2}}{x^2}\,dx = -\frac{\sqrt{a^2 - x^2}}{x} - \sin^{-1}\frac{x}{|a|}$

217. $\displaystyle\int \frac{\sqrt{a^2 - x^2}}{x^3}\,dx = -\frac{\sqrt{a^2 - x^2}}{2x^2} + \frac{1}{2a}\log\frac{a + \sqrt{a^2 - x^2}}{x}$

218. $\displaystyle\int \frac{\sqrt{a^2 - x^2}}{x^4}\,dx = -\frac{\sqrt{(a^2 - x^2)^3}}{3a^2 x^3}$

INTEGRALS (Continued)

219. $\displaystyle\int \frac{x^2\,dx}{\sqrt{(a^2-x^2)^3}} = \frac{x}{\sqrt{a^2-x^2}} - \sin^{-1}\frac{x}{|a|}$

220. $\displaystyle\int \frac{x^3\,dx}{\sqrt{a^2-x^2}} = -\frac{2}{3}(a^2-x^2)^{\frac{1}{2}} - x^2(a^2-x^2)^{\frac{1}{2}} = -\frac{1}{3}\sqrt{a^2-x^2}(x^2+2a^2)$

221. $\displaystyle\int \frac{x^3\,dx}{\sqrt{(a^2-x^2)^3}} = 2(a^2-x^2)^{\frac{1}{2}} + \frac{x^2}{(a^2-x^2)^{\frac{1}{2}}} = -\frac{a^2}{\sqrt{a^2-x^2}} + \sqrt{a^2-x^2}$

222. $\displaystyle\int \frac{dx}{x^3\sqrt{a^2-x^2}} = -\frac{\sqrt{a^2-x^2}}{2a^2x^2} - \frac{1}{2a^3}\log\frac{a+\sqrt{a^2-x^2}}{x}$

223. $\displaystyle\int \frac{dx}{x\sqrt{(a^2-x^2)^3}} = \frac{1}{a^2\sqrt{a^2-x^2}} - \frac{1}{a^3}\log\frac{a+\sqrt{a^2-x^2}}{x}$

224. $\displaystyle\int \frac{dx}{x^2\sqrt{(a^2-x^2)^3}} = \frac{1}{a^4}\left[-\frac{\sqrt{a^2-x^2}}{x} + \frac{x}{\sqrt{a^2-x^2}}\right]$

225. $\displaystyle\int \frac{dx}{x^3\sqrt{(a^2-x^2)^3}} = -\frac{1}{2a^2x^2\sqrt{a^2-x^2}} + \frac{3}{2a^4\sqrt{a^2-x^2}}$

$$-\frac{3}{2a^5}\log\frac{a+\sqrt{a^2-x^2}}{x}$$

226. $\displaystyle\int \frac{x^m}{\sqrt{a^2-x^2}}\,dx = -\frac{x^{m-1}\sqrt{a^2-x^2}}{m} + \frac{(m-1)a^2}{m}\int \frac{x^{m-2}}{\sqrt{a^2-x^2}}\,dx$

227. $\displaystyle\int \frac{x^{2m}}{\sqrt{a^2-x^2}}\,dx = \frac{(2m)!}{(m!)^2}\left[-\sqrt{a^2-x^2}\sum_{r=1}^{m}\frac{r!(r-1)!}{2^{2m-2r+1}(2r)!}a^{2m-2r}x^{2r-1}\right.$

$$\left.+\frac{a^{2m}}{2^{2m}}\sin^{-1}\frac{x}{|a|}\right]$$

228. $\displaystyle\int \frac{x^{2m+1}}{\sqrt{a^2-x^2}}\,dx = -\sqrt{a^2-x^2}\sum_{r=0}^{m}\frac{(2r)!(m!)^2}{(2m+1)!(r!)^2}(4a^2)^{m-r}x^{2r}$

229. $\displaystyle\int \frac{dx}{x^m\sqrt{a^2-x^2}} = -\frac{\sqrt{a^2-x^2}}{(m-1)a^2x^{m-1}} + \frac{m-2}{(m-1)a^2}\int \frac{dx}{x^{m-2}\sqrt{a^2-x^2}}$

230. $\displaystyle\int \frac{ax}{x^{2m}\sqrt{a^2-x^2}} = -\sqrt{a^2-x^2}\sum_{r=0}^{m-1}\frac{(m-1)!m!(2r)!2^{2m-2r-1}}{(r!)^2(2m)!a^{2m-2r}x^{2r+1}}$

231. $\displaystyle\int \frac{dx}{x^{2m+1}\sqrt{a^2-x^2}} = \frac{(2m)!}{(m!)^2}\left[-\frac{\sqrt{a^2-x^2}}{a^2}\sum_{r=1}^{m}\frac{r!(r-1)!}{2(2r)!(4a^2)^{m-r}x^{2r}}\right.$

$$\left.+\frac{1}{2^{2m}a^{2m+1}}\log\frac{a-\sqrt{a^2-x^2}}{x}\right]$$

INTEGRALS (Continued)

232. $\displaystyle\int \frac{dx}{(b^2 - x^2)\sqrt{a^2 - x^2}} = \frac{1}{2b\sqrt{a^2 - b^2}} \log \frac{(b\sqrt{a^2 - x^2} + x\sqrt{a^2 - b^2})^2}{b^2 - x^2},$

$$(a^2 > b^2)$$

233. $\displaystyle\int \frac{dx}{(b^2 - x^2)\sqrt{a^2 - x^2}} = \frac{1}{b\sqrt{b^2 - a^2}} \tan^{-1} \frac{x\sqrt{b^2 - a^2}}{b\sqrt{a^2 - x^2}}, \qquad (b^2 > a^2)$

234. $\displaystyle\int \frac{dx}{(b^2 + x^2)\sqrt{a^2 - x^2}} = \frac{1}{b\sqrt{a^2 + b^2}} \tan^{-1} \frac{x\sqrt{a^2 + b^2}}{b\sqrt{a^2 - x^2}}$

235. $\displaystyle\int \frac{\sqrt{a^2 - x^2}}{b^2 + x^2} dx = \frac{\sqrt{a^2 + b^2}}{|b|} \sin^{-1} \frac{x\sqrt{a^2 + b^2}}{|a|\sqrt{x^2 + b^2}} - \sin^{-1} \frac{x}{|a|}$

236. $\displaystyle\int f(x, \sqrt{a^2 - x^2})\, dx = a \int f(a \sin u, a \cos u) \cos u\, du, \qquad \left(u = \sin^{-1} \frac{x}{a}, a > 0 \right)$

FORMS CONTAINING $\sqrt{a + bx + cx^2}$

$$X = a + bx + cx^2, q = 4ac - b^2, \text{ and } k = \frac{4c}{q}$$

If $q = 0$, then $\sqrt{X} = \sqrt{c} \left| x + \dfrac{b}{2c} \right|$

237. $\displaystyle\int \frac{dx}{\sqrt{X}} = \begin{cases} \dfrac{1}{\sqrt{c}} \log (2\sqrt{cX} + 2cx + b) \\[2ex] \qquad \text{or} \\[2ex] \dfrac{1}{\sqrt{c}} \sinh^{-1} \dfrac{2cx + b}{\sqrt{q}}, \qquad (c > 0) \end{cases}$

238. $\displaystyle\int \frac{dx}{\sqrt{X}} = -\frac{1}{\sqrt{-c}} \sin^{-1} \frac{2cx + b}{\sqrt{-q}}, \qquad (c < 0)$

239. $\displaystyle\int \frac{dx}{X\sqrt{X}} = \frac{2(2cx + b)}{q\sqrt{X}}$

240. $\displaystyle\int \frac{dx}{X^2\sqrt{X}} = \frac{2(2cx + b)}{3q\sqrt{X}} \left(\frac{1}{X} + 2k \right)$

241. $\displaystyle\int \frac{dx}{X^n\sqrt{X}} = \begin{cases} \dfrac{2(2cx + b)\sqrt{X}}{(2n - 1)qX^n} + \dfrac{2k(n - 1)}{2n - 1} \displaystyle\int \dfrac{dx}{X^{n-1}\sqrt{X}} \\[2ex] \qquad \text{or} \\[2ex] \dfrac{(2cx + b)(n!)(n - 1)!4^n k^{n-1}}{q[(2n)!]\sqrt{X}} \displaystyle\sum_{r=0}^{n-1} \dfrac{(2r)!}{(4kX)^r (r!)^2} \end{cases}$

INTEGRALS (Continued)

242. $\displaystyle \int \sqrt{X}\, dx = \frac{(2cx + b)\sqrt{X}}{4c} + \frac{1}{2k}\int \frac{dx}{\sqrt{X}}$

243. $\displaystyle \int X\sqrt{X}\, dx = \frac{(2cx + b)\sqrt{X}}{8c}\left(X + \frac{3}{2k}\right) + \frac{3}{8k^2}\int \frac{dx}{\sqrt{X}}$

244. $\displaystyle \int X^2\sqrt{X}\, dx = \frac{(2cx + b)\sqrt{X}}{12c}\left(X^2 + \frac{5X}{4k} + \frac{15}{8k^2}\right) + \frac{5}{16k^3}\int \frac{dx}{\sqrt{X}}$

245. $\displaystyle \int X^n\sqrt{X}\, dx = \begin{cases} \dfrac{(2cx + b)X^n\sqrt{X}}{4(n + 1)c} + \dfrac{2n + 1}{2(n + 1)k}\displaystyle\int X^{n-1}\sqrt{X}\, dx \\[2mm] \text{or} \\[2mm] \dfrac{(2n + 2)!}{[(n + 1)!]^2(4k)^{n+1}}\left[\dfrac{k(2cx + b)\sqrt{X}}{c}\displaystyle\sum_{r=0}^{n}\dfrac{r!(r + 1)!(4kX)^r}{(2r + 2)!} \right.\\ \hspace{6cm}\left. + \displaystyle\int \dfrac{dx}{\sqrt{X}}\right] \end{cases}$

246. $\displaystyle \int \frac{x\, dx}{\sqrt{X}} = \frac{\sqrt{X}}{c} - \frac{b}{2c}\int \frac{dx}{\sqrt{X}}$

247. $\displaystyle \int \frac{x\, dx}{X\sqrt{X}} = -\frac{2(bx + 2a)}{q\sqrt{X}}$

248. $\displaystyle \int \frac{x\, dx}{X^n\sqrt{X}} = -\frac{\sqrt{X}}{(2n - 1)cX^n} - \frac{b}{2c}\int \frac{dx}{X^n\sqrt{X}}$

249. $\displaystyle \int \frac{x^2\, dx}{\sqrt{X}} = \left(\frac{x}{2c} - \frac{3b}{4c^2}\right)\sqrt{X} + \frac{3b^2 - 4ac}{8c^2}\int \frac{dx}{\sqrt{X}}$

250. $\displaystyle \int \frac{x^2\, dx}{X\sqrt{X}} = \frac{(2b^2 - 4ac)x + 2ab}{cq\sqrt{X}} + \frac{1}{c}\int \frac{dx}{\sqrt{X}}$

251. $\displaystyle \int \frac{x^2\, dx}{X^n\sqrt{X}} = \frac{(2b^2 - 4ac)x + 2ab}{(2n - 1)cqX^{n-1}\sqrt{X}} + \frac{4ac + (2n - 3)b^2}{(2n - 1)cq}\int \frac{dx}{X^{n-1}\sqrt{X}}$

252. $\displaystyle \int \frac{x^3\, dx}{\sqrt{X}} = \left(\frac{x^2}{3c} - \frac{5bx}{12c^2} + \frac{5b^2}{8c^3} - \frac{2a}{3c^2}\right)\sqrt{X} + \left(\frac{3ab}{4c^2} - \frac{5b^3}{16c^3}\right)\int \frac{dx}{\sqrt{X}}$

253. $\displaystyle \int \frac{x^n\, dx}{\sqrt{X}} = \frac{1}{nc}x^{n-1}\sqrt{X} - \frac{(2n - 1)b}{2nc}\int \frac{x^{n-1}\, dx}{\sqrt{X}} - \frac{(n - 1)a}{nc}\int \frac{x^{n-2}\, dx}{\sqrt{X}}$

254. $\displaystyle\int x\sqrt{X}\,dx = \frac{X\sqrt{X}}{3c} - \frac{b(2cx + b)}{8c^2}\sqrt{X} - \frac{b}{4ck}\int\frac{dx}{\sqrt{X}}$

255. $\displaystyle\int xX\sqrt{X}\,dx = \frac{X^2\sqrt{X}}{5c} - \frac{b}{2c}\int X\sqrt{X}\,dx$

256. $\displaystyle\int xX^n\sqrt{X}\,dx = \frac{X^{n+1}\sqrt{X}}{(2n+3)c} - \frac{b}{2c}\int X^n\sqrt{X}\,dx$

257. $\displaystyle\int x^2\sqrt{X}\,dx = \left(x - \frac{5b}{6c}\right)\frac{X\sqrt{X}}{4c} + \frac{5b^2 - 4ac}{16c^2}\int\sqrt{X}\,dx$

258. $\displaystyle\int\frac{dx}{x\sqrt{X}} = -\frac{1}{\sqrt{a}}\log\frac{2\sqrt{aX} + bx + 2a}{x}, \qquad (a > 0)$

259. $\displaystyle\int\frac{dx}{x\sqrt{X}} = \frac{1}{\sqrt{-a}}\sin^{-1}\left(\frac{bx + 2a}{|x|\sqrt{-q}}\right), \qquad (a < 0)$

260. $\displaystyle\int\frac{dx}{x\sqrt{X}} = -\frac{2\sqrt{X}}{bx}, \qquad (a = 0)$

261. $\displaystyle\int\frac{dx}{x^2\sqrt{X}} = -\frac{\sqrt{X}}{ax} - \frac{b}{2a}\int\frac{dx}{x\sqrt{X}}$

262. $\displaystyle\int\frac{\sqrt{X}\,dx}{x} = \sqrt{X} + \frac{b}{2}\int\frac{dx}{\sqrt{X}} + a\int\frac{dx}{x\sqrt{X}}$

263. $\displaystyle\int\frac{\sqrt{X}\,dx}{x^2} = -\frac{\sqrt{X}}{x} + \frac{b}{2}\int\frac{dx}{x\sqrt{X}} + c\int\frac{dx}{\sqrt{X}}$

FORMS INVOLVING $\sqrt{2ax - x^2}$

264. $\displaystyle\int\sqrt{2ax - x^2}\,dx = \frac{1}{2}\left[(x - a)\sqrt{2ax - x^2} + a^2\sin^{-1}\frac{x - a}{|a|}\right]$

265. $\displaystyle\int\frac{dx}{\sqrt{2ax - x^2}} = \begin{cases} \cos^{-1}\dfrac{a - x}{|a|} \\[2mm] \text{or} \\[2mm] \sin^{-1}\dfrac{x - a}{|a|} \end{cases}$

INTEGRALS (Continued)

266. $\displaystyle\int x^n \sqrt{2ax - x^2}\, dx = \begin{cases} -\dfrac{x^{n-1}(2ax - x^2)^{\frac{3}{2}}}{n + 2} + \dfrac{(2n + 1)a}{n + 2}\displaystyle\int x^{n-1}\sqrt{2ax - x^2}\, dx \\[2mm] \text{or} \\[2mm] \sqrt{2ax - x^2}\left[\dfrac{x^{n+1}}{n + 2} - \displaystyle\sum_{r=0}^{n}\dfrac{(2n + 1)!(r!)^2 a^{n-r+1}}{2^{n-r}(2r + 1)!(n + 2)!n!}x^r\right] \\[3mm] \qquad\qquad + \dfrac{(2n + 1)!a^{n+2}}{2^n n!(n + 2)!}\sin^{-1}\dfrac{x - a}{|a|} \end{cases}$

267. $\displaystyle\int \frac{\sqrt{2ax - x^2}}{x^n}\, dx = \frac{(2ax - x^2)^{\frac{3}{2}}}{(3 - 2n)ax^n} + \frac{n - 3}{(2n - 3)a}\int \frac{\sqrt{2ax - x^2}}{x^{n-1}}\, dx$

268. $\displaystyle\int \frac{x^n\, dx}{\sqrt{2ax - x^2}} = \begin{cases} \dfrac{-x^{n-1}\sqrt{2ax - x^2}}{n} + \dfrac{a(2n - 1)}{n}\displaystyle\int \frac{x^{n-1}}{\sqrt{2ax - x^2}}\, dx \\[2mm] \text{or} \\[2mm] -\sqrt{2ax - x^2}\displaystyle\sum_{r=1}^{n}\dfrac{(2n)!r!(r - 1)!a^{n-r}}{2^{n-r}(2r)!(n!)^2}x^{r-1} \\[3mm] \qquad\qquad + \dfrac{(2n)!a^n}{2^n(n!)^2}\sin^{-1}\dfrac{x - a}{|a|} \end{cases}$

269. $\displaystyle\int \frac{dx}{x^n\sqrt{2ax - x^2}} = \begin{cases} \dfrac{\sqrt{2ax - x^2}}{a(1 - 2n)x^n} + \dfrac{n - 1}{(2n - 1)a}\displaystyle\int \frac{dx}{x^{n-1}\sqrt{2ax - x^2}} \\[2mm] \text{or} \\[2mm] -\sqrt{2ax - x^2}\displaystyle\sum_{r=0}^{n-1}\dfrac{2^{n-r}(n - 1)!n!(2r)!}{(2n)!(r!)^2 a^{n-r}x^{r+1}} \end{cases}$

270. $\displaystyle\int \frac{dx}{(2ax - x^2)^{\frac{3}{2}}} = \frac{x - a}{a^2\sqrt{2ax - x^2}}$

271. $\displaystyle\int \frac{x\, dx}{(2ax - x^2)^{\frac{3}{2}}} = \frac{x}{a\sqrt{2ax - x^2}}$

MISCELLANEOUS ALGEBRAIC FORMS

272. $\displaystyle\int \frac{dx}{\sqrt{2ax + x^2}} = \log\left(x + a + \sqrt{2ax + x^2}\right)$

273. $\displaystyle\int \sqrt{ax^2 + c}\, dx = \frac{x}{2}\sqrt{ax^2 + c} + \frac{c}{2\sqrt{a}}\log\left(x\sqrt{a} + \sqrt{ax^2 + c}\right), \qquad (a > 0)$

274. $\displaystyle\int \sqrt{ax^2 + c}\, dx = \frac{x}{2}\sqrt{ax^2 + c} + \frac{c}{2\sqrt{-a}}\sin^{-1}\left(x\sqrt{-\frac{a}{c}}\right), \qquad (a < 0)$

INTEGRALS (Continued)

275. $\int \sqrt{\dfrac{1+x}{1-x}}\, dx = \sin^{-1} x - \sqrt{1-x^2}$

276. $\displaystyle\int \dfrac{dx}{x\sqrt{ax^n + c}} = \begin{cases} \dfrac{1}{n\sqrt{c}} \log \dfrac{\sqrt{ax^n + c} - \sqrt{c}}{\sqrt{ax^n + c} + \sqrt{c}} \\[2mm] \quad\text{or} \\[2mm] \dfrac{2}{n\sqrt{c}} \log \dfrac{\sqrt{ax^n + c} - \sqrt{c}}{\sqrt{x^n}}, \quad (c > 0) \end{cases}$

277. $\displaystyle\int \dfrac{dx}{x\sqrt{ax^n + c}} = \dfrac{2}{n\sqrt{-c}} \sec^{-1}\sqrt{-\dfrac{ax^n}{c}}, \quad (c < 0)$

278. $\displaystyle\int \dfrac{dx}{\sqrt{ax^2 + c}} = \dfrac{1}{\sqrt{a}} \log(x\sqrt{a} + \sqrt{ax^2 + c}), \quad (a > 0)$

279. $\displaystyle\int \dfrac{dx}{\sqrt{ax^2 + c}} = \dfrac{1}{\sqrt{-a}} \sin^{-1}\left(x\sqrt{-\dfrac{a}{c}}\right), \quad (a < 0)$

280. $\displaystyle\int (ax^2 + c)^{m+\frac{1}{2}}\, dx = \begin{cases} \dfrac{x(ax^2 + c)^{m+\frac{1}{2}}}{2(m+1)} + \dfrac{(2m+1)c}{2(m+1)} \displaystyle\int (ax^2 + c)^{m-\frac{1}{2}}\, dx \\[2mm] \quad\text{or} \\[2mm] x\sqrt{ax^2 + c}\displaystyle\sum_{r=0}^{m} \dfrac{(2m+1)!(r!)^2 c^{m-r}}{2^{2m-2r+1} m!(m+1)!(2r+1)!}(ax^2 + c)^r \\[2mm] \quad + \dfrac{(2m+1)! c^{m+1}}{2^{2m+1} m!(m+1)!}\displaystyle\int \dfrac{dx}{\sqrt{ax^2 + c}} \end{cases}$

281. $\displaystyle\int x(ax^2 + c)^{m+\frac{1}{2}}\, dx = \dfrac{(ax^2 + c)^{m+\frac{3}{2}}}{(2m+3)a}$

282. $\displaystyle\int \dfrac{(ax^2 + c)^{m+\frac{1}{2}}}{x}\, dx = \begin{cases} \dfrac{(ax^2 + c)^{m+\frac{1}{2}}}{2m+1} + c\displaystyle\int \dfrac{(ax^2 + c)^{m-\frac{1}{2}}}{x}\, dx \\[2mm] \quad\text{or} \\[2mm] \sqrt{ax^2 + c}\displaystyle\sum_{r=0}^{m} \dfrac{c^{m-r}(ax^2 + c)^r}{2r+1} + c^{m+1}\displaystyle\int \dfrac{dx}{x\sqrt{ax^2 + c}} \end{cases}$

283. $\displaystyle\int \dfrac{dx}{(ax^2 + c)^{m+\frac{1}{2}}} = \begin{cases} \dfrac{x}{(2m-1)c(ax^2 + c)^{m-\frac{1}{2}}} + \dfrac{2m-2}{(2m-1)c}\displaystyle\int \dfrac{dx}{(ax^2 + c)^{m-\frac{1}{2}}} \\[2mm] \quad\text{or} \\[2mm] \dfrac{x}{\sqrt{ax^2 + c}}\displaystyle\sum_{r=0}^{m-1} \dfrac{2^{2m-2r-1}(m-1)! m!(2r)!}{(2m)!(r!)^2 c^{m-r}(ax^2 + c)^r} \end{cases}$

INTEGRALS (Continued)

284. $\displaystyle\int \frac{dx}{x^m\sqrt{ax^2+c}} = -\frac{\sqrt{ax^2+c}}{(m-1)cx^{m-1}} - \frac{(m-2)a}{(m-1)c}\int \frac{dx}{x^{m-2}\sqrt{ax^2+c}}$

285. $\displaystyle\int \frac{1+x^2}{(1-x^2)\sqrt{1+x^4}}\, dx = \frac{1}{\sqrt{2}}\log\frac{x\sqrt{2}+\sqrt{1+x^4}}{1-x^2}$

286. $\displaystyle\int \frac{1-x^2}{(1+x^2)\sqrt{1+x^4}}\, dx = \frac{1}{\sqrt{2}}\tan^{-1}\frac{x\sqrt{2}}{\sqrt{1+x^4}}$

287. $\displaystyle\int \frac{dx}{x\sqrt{x^n+a^2}} = -\frac{2}{na}\log\frac{a+\sqrt{x^n+a^2}}{\sqrt{x^n}}$

288. $\displaystyle\int \frac{dx}{x\sqrt{x^n-a^2}} = -\frac{2}{na}\sin^{-1}\frac{a}{\sqrt{x^n}}$

289. $\displaystyle\int \sqrt{\frac{x}{a^3-x^3}}\, dx = \frac{2}{3}\sin^{-1}\left(\frac{x}{a}\right)^{\frac{3}{2}}$

FORMS INVOLVING TRIGONOMETRIC FUNCTIONS

290. $\displaystyle\int (\sin ax)\, dx = -\frac{1}{a}\cos ax$

291. $\displaystyle\int (\cos ax)\, dx = \frac{1}{a}\sin ax$

292. $\displaystyle\int (\tan ax)\, dx = -\frac{1}{a}\log\cos ax = \frac{1}{a}\log\sec ax$

293. $\displaystyle\int (\cot ax)\, dx = \frac{1}{a}\log\sin ax = -\frac{1}{a}\log\csc ax$

294. $\displaystyle\int (\sec ax)\, dx = \frac{1}{a}\log(\sec ax+\tan ax) = \frac{1}{a}\log\tan\left(\frac{\pi}{4}+\frac{ax}{2}\right)$

295. $\displaystyle\int (\csc ax)\, dx = \frac{1}{a}\log(\csc ax-\cot ax) = \frac{1}{a}\log\tan\frac{ax}{2}$

296. $\displaystyle\int (\sin^2 ax)\, dx = -\frac{1}{2a}\cos ax\sin ax + \frac{1}{2}x = \frac{1}{2}x - \frac{1}{4a}\sin 2ax$

297. $\displaystyle\int (\sin^3 ax)\, dx = -\frac{1}{3a}(\cos ax)(\sin^2 ax+2)$

298. $\displaystyle\int (\sin^4 ax)\, dx = \frac{3x}{8} - \frac{\sin 2ax}{4a} + \frac{\sin 4ax}{32a}$

299. $\displaystyle\int (\sin^n ax)\, dx = -\frac{\sin^{n-1} ax\cos ax}{na} + \frac{n-1}{n}\int (\sin^{n-2} ax)\, dx$

INTEGRALS (Continued)

300. $\displaystyle\int (\sin^{2m} ax)\, dx = -\frac{\cos ax}{a} \sum_{r=0}^{m-1} \frac{(2m)!(r!)^2}{2^{2m-2r}(2r+1)!(m!)^2} \sin^{2r+1} ax + \frac{(2m)!}{2^{2m}(m!)^2} x$

301. $\displaystyle\int (\sin^{2m+1} ax)\, dx = -\frac{\cos ax}{a} \sum_{r=0}^{m} \frac{2^{2m-2r}(m!)^2(2r)!}{(2m+1)!(r!)^2} \sin^{2r} ax$

302. $\displaystyle\int (\cos^2 ax)\, dx = \frac{1}{2a} \sin ax \cos ax + \frac{1}{2} x = \frac{1}{2} x + \frac{1}{4a} \sin 2ax$

303. $\displaystyle\int (\cos^3 ax)\, dx = \frac{1}{3a}(\sin ax)(\cos^2 ax + 2)$

304. $\displaystyle\int (\cos^4 ax)\, dx = \frac{3x}{8} + \frac{\sin 2ax}{4a} + \frac{\sin 4ax}{32a}$

305. $\displaystyle\int (\cos^n ax)\, dx = \frac{1}{na} \cos^{n-1} ax \sin ax + \frac{n-1}{n} \int (\cos^{n-2} ax)\, dx$

306. $\displaystyle\int (\cos^{2m} ax)\, dx = \frac{\sin ax}{a} \sum_{r=0}^{m-1} \frac{(2m)!(r!)^2}{2^{2m-2r}(2r+1)!(m!)^2} \cos^{2r+1} ax + \frac{(2m)!}{2^{2m}(m!)^2} x$

307. $\displaystyle\int (\cos^{2m+1} ax)\, dx = \frac{\sin ax}{a} \sum_{r=0}^{m} \frac{2^{2m-2r}(m!)^2(2r)!}{(2m+1)!(r!)^2} \cos^{2r} ax$

308. $\displaystyle\int \frac{dx}{\sin^2 ax} = \int (\csc^2 ax)\, dx = -\frac{1}{a} \cot ax$

309. $\displaystyle\int \frac{dx}{\sin^m ax} = \int (\csc^m ax)\, dx = -\frac{1}{(m-1)a} \cdot \frac{\cos ax}{\sin^{m-1} ax} + \frac{m-2}{m-1} \int \frac{dx}{\sin^{m-2} ax}$

310. $\displaystyle\int \frac{dx}{\sin^{2m} ax} = \int (\csc^{2m} ax)\, dx = -\frac{1}{a} \cos ax \sum_{r=0}^{m-1} \frac{2^{2m-2r-1}(m-1)!m!(2r)!}{(2m)!(r!)^2 \sin^{2r+1} ax}$

311. $\displaystyle\int \frac{dx}{\sin^{2m+1} ax} = \int (\csc^{2m+1} ax)\, dx =$

$$-\frac{1}{a} \cos ax \sum_{r=0}^{m-1} \frac{(2m)!(r!)^2}{2^{2m-2r}(m!)^2(2r+1)! \sin^{2r+2} ax} + \frac{1}{a} \cdot \frac{(2m)!}{2^{2m}(m!)^2} \log \tan \frac{ax}{2}$$

312. $\displaystyle\int \frac{dx}{\cos^2 ax} = \int (\sec^2 ax)\, dx = \frac{1}{a} \tan ax$

313. $\displaystyle\int \frac{dx}{\cos^n ax} = \int (\sec^n ax)\, dx = \frac{1}{(n-1)a} \cdot \frac{\sin ax}{\cos^{n-1} ax} + \frac{n-2}{n-1} \int \frac{dx}{\cos^{n-2} ax}$

314. $\displaystyle\int \frac{dx}{\cos^{2m} ax} = \int (\sec^{2m} ax)\, dx = \frac{1}{a} \sin ax \sum_{r=0}^{m-1} \frac{2^{2m-2r-1}(m-1)!m!(2r)!}{(2m)!(r!)^2 \cos^{2r+1} ax}$

315. $\displaystyle\int \frac{dx}{\cos^{2m+1} ax} = \int (\sec^{2m+1} ax)\,dx =$

$$\frac{1}{a}\sin ax \sum_{r=0}^{m-1} \frac{(2m)!(r!)^2}{2^{2m-2r}(m!)^2(2r+1)!\cos^{2r+2} ax}$$

$$+ \frac{1}{a}\cdot\frac{(2m)!}{2^{2m}(m!)^2}\log(\sec ax + \tan ax)$$

316. $\displaystyle\int (\sin mx)(\sin nx)\,dx = \frac{\sin(m-n)x}{2(m-n)} - \frac{\sin(m+n)x}{2(m+n)}, \qquad (m^2 \neq n^2)$

317. $\displaystyle\int (\cos mx)(\cos nx)\,dx = \frac{\sin(m-n)x}{2(m-n)} + \frac{\sin(m+n)x}{2(m+n)}, \qquad (m^2 \neq n^2)$

318. $\displaystyle\int (\sin ax)(\cos ax)\,dx = \frac{1}{2a}\sin^2 ax$

319. $\displaystyle\int (\sin mx)(\cos nx)\,dx = -\frac{\cos(m-n)x}{2(m-n)} - \frac{\cos(m+n)x}{2(m+n)}, \qquad (m^2 \neq n^2)$

320. $\displaystyle\int (\sin^2 ax)(\cos^2 ax)\,dx = -\frac{1}{32a}\sin 4ax + \frac{x}{8}$

321. $\displaystyle\int (\sin ax)(\cos^m ax)\,dx = -\frac{\cos^{m+1} ax}{(m+1)a}$

322. $\displaystyle\int (\sin^m ax)(\cos ax)\,dx = \frac{\sin^{m+1} ax}{(m+1)a}$

323. $\displaystyle\int (\cos^m ax)(\sin^n ax)\,dx = \begin{cases} \dfrac{\cos^{m-1} ax \sin^{n+1} ax}{(m+n)a} \\[2mm] \qquad + \dfrac{m-1}{m+n}\displaystyle\int (\cos^{m-2} ax)(\sin^n ax)\,dx \\[2mm] \text{or} \\[2mm] -\dfrac{\sin^{n-1} ax \cos^{m+1} ax}{(m+n)a} \\[2mm] \qquad + \dfrac{n-1}{m+n}\displaystyle\int (\cos^m ax)(\sin^{n-2} ax)\,dx \end{cases}$

324. $\displaystyle\int \frac{\cos^m ax}{\sin^n ax}\,dx = \begin{cases} -\dfrac{\cos^{m+1} ax}{(n-1)a \sin^{n-1} ax} - \dfrac{m-n+2}{n-1}\displaystyle\int \dfrac{\cos^m ax}{\sin^{n-2} ax}\,dx \\[2mm] \text{or} \\[2mm] \dfrac{\cos^{m-1} ax}{a(m-n)\sin^{n-1} ax} + \dfrac{m-1}{m-n}\displaystyle\int \dfrac{\cos^{m-2} ax}{\sin^n ax}\,dx \end{cases}$

INTEGRALS (Continued)

325. $\displaystyle\int \frac{\sin^m ax}{\cos^n ax}\, dx = \begin{cases} \dfrac{\sin^{m+1} ax}{a(n-1)\cos^{n-1} ax} - \dfrac{m-n+2}{n-1}\displaystyle\int \frac{\sin^m ax}{\cos^{n-2} ax}\, dx \\[2mm] \quad\text{or} \\[2mm] -\dfrac{\sin^{m-1} ax}{a(m-n)\cos^{n-1} ax} + \dfrac{m-1}{m-n}\displaystyle\int \frac{\sin^{m-2} ax}{\cos^n ax}\, dx \end{cases}$

326. $\displaystyle\int \frac{\sin ax}{\cos^2 ax}\, dx = \frac{1}{a\cos ax} = \frac{\sec ax}{a}$

327. $\displaystyle\int \frac{\sin^2 ax}{\cos ax}\, dx = -\frac{1}{a}\sin ax + \frac{1}{a}\log\tan\left(\frac{\pi}{4} + \frac{ax}{2}\right)$

328. $\displaystyle\int \frac{\cos ax}{\sin^2 ax}\, dx = -\frac{1}{a\sin ax} = -\frac{\csc ax}{a}$

329. $\displaystyle\int \frac{dx}{(\sin ax)(\cos ax)} = \frac{1}{a}\log\tan ax$

330. $\displaystyle\int \frac{dx}{(\sin ax)(\cos^2 ax)} = \frac{1}{a}\left(\sec ax + \log\tan\frac{ax}{2}\right)$

331. $\displaystyle\int \frac{dx}{(\sin ax)(\cos^n ax)} = \frac{1}{a(n-1)\cos^{n-1} ax} + \int \frac{dx}{(\sin ax)(\cos^{n-2} ax)}$

332. $\displaystyle\int \frac{dx}{(\sin^2 ax)(\cos ax)} = -\frac{1}{a}\csc ax + \frac{1}{a}\log\tan\left(\frac{\pi}{4} + \frac{ax}{2}\right)$

333. $\displaystyle\int \frac{dx}{(\sin^2 ax)(\cos^2 ax)} = -\frac{2}{a}\cot 2ax$

334. $\displaystyle\int \frac{dx}{\sin^m ax \cos^n ax} = \begin{cases} -\dfrac{1}{a(m-1)(\sin^{m-1} ax)(\cos^{n-1} ax)} \\[3mm] \quad + \dfrac{m+n-2}{m-1}\displaystyle\int \frac{dx}{(\sin^{m-2} ax)(\cos^n ax)} \\[3mm] \quad\text{or} \\[3mm] \dfrac{1}{a(n-1)\sin^{m-1} ax \cos^{n-1} ax} \\[3mm] \quad - \dfrac{m+n-2}{n-1}\displaystyle\int \frac{dx}{\sin^m ax \cos^{n-2} ax} \end{cases}$

335. $\displaystyle\int \sin(a+bx)\, dx = -\frac{1}{b}\cos(a+bx)$

336. $\displaystyle\int \cos(a+bx)\, dx = \frac{1}{b}\sin(a+bx)$

337. $\displaystyle\int \frac{dx}{1 \pm \sin ax} = \mp\frac{1}{a}\tan\left(\frac{\pi}{4} \mp \frac{ax}{2}\right)$

INTEGRALS (Continued)

338. $\displaystyle\int \frac{dx}{1 + \cos ax} = \frac{1}{a}\tan\frac{ax}{2}$

339. $\displaystyle\int \frac{dx}{1 - \cos ax} = -\frac{1}{a}\cot\frac{ax}{2}$

*340. $\displaystyle\int \frac{dx}{a + b\sin x} = \begin{cases} \dfrac{2}{\sqrt{a^2 - b^2}}\tan^{-1}\dfrac{a\tan\frac{x}{2} + b}{\sqrt{a^2 - b^2}} \\[4mm] \text{or} \\[4mm] \dfrac{1}{\sqrt{b^2 - a^2}}\log\dfrac{a\tan\frac{x}{2} + b - \sqrt{b^2 - a^2}}{a\tan\frac{x}{2} + b + \sqrt{b^2 - a^2}} \end{cases}$

*341. $\displaystyle\int \frac{dx}{a + b\cos x} = \begin{cases} \dfrac{2}{\sqrt{a^2 - b^2}}\tan^{-1}\dfrac{\sqrt{a^2 - b^2}\tan\frac{x}{2}}{a + b} \\[4mm] \text{or} \\[4mm] \dfrac{1}{\sqrt{b^2 - a^2}}\log\left(-\dfrac{\sqrt{b^2 - a^2}\tan\frac{x}{2} + a + b}{\sqrt{b^2 - a^2}\tan\frac{x}{2} - a - b}\right) \end{cases}$

*342. $\displaystyle\int \frac{dx}{a + b\sin x + c\cos x}$

$= \begin{cases} \dfrac{1}{\sqrt{b^2 + c^2 - a^2}}\log\dfrac{b - \sqrt{b^2 + c^2 - a^2} + (a - c)\tan\frac{x}{2}}{b + \sqrt{b^2 + c^2 - a^2} + (a - c)\tan\frac{x}{2}}, & \text{if } a^2 < b^2 + c^2, a \neq c \\[6mm] \text{or} \\[4mm] \dfrac{2}{\sqrt{a^2 - b^2 - c^2}}\tan^{-1}\dfrac{b + (a - c)\tan\frac{x}{2}}{\sqrt{a^2 - b^2 - c^2}}, & \text{if } a^2 > b^2 + c^2 \\[6mm] \text{or} \\[4mm] \dfrac{1}{a}\left[\dfrac{a - (b + c)\cos x - (b - c)\sin x}{a - (b - c)\cos x + (b + c)\sin x}\right], & \text{if } a^2 = b^2 + c^2, a \neq c. \end{cases}$

*343. $\displaystyle\int \frac{\sin^2 x\, dx}{a + b\cos^2 x} = \frac{1}{b}\sqrt{\frac{a + b}{a}}\tan^{-1}\left(\sqrt{\frac{a}{a + b}}\tan x\right) - \frac{x}{b}, \qquad (ab > 0, \text{ or } |a| > |b|)$

*See note 6--page 287.

INTEGRALS (Continued)

*344. $\int \dfrac{dx}{a^2 \cos^2 x + b^2 \sin^2 x} = \dfrac{1}{ab} \tan^{-1}\left(\dfrac{b \tan x}{a}\right)$

*345. $\int \dfrac{\cos^2 cx}{a^2 + b^2 \sin^2 cx}\,dx = \dfrac{\sqrt{a^2 + b^2}}{ab^2 c} \tan^{-1}\dfrac{\sqrt{a^2 + b^2}\,\tan cx}{a} - \dfrac{x}{b^2}$

346. $\int \dfrac{\sin cx \cos cx}{a \cos^2 cx + b \sin^2 cx}\,dx = \dfrac{1}{2c(b - a)} \log(a \cos^2 cx + b \sin^2 cx)$

347. $\int \dfrac{\cos cx}{a \cos cx + b \sin cx}\,dx = \int \dfrac{dx}{a + b \tan cx} =$

$$\dfrac{1}{c(a^2 + b^2)}[acx + b \log(a \cos cx + b \sin cx)]$$

348. $\int \dfrac{\sin cx}{a \sin cx + b \cos cx}\,dx = \int \dfrac{dx}{a + b \cot cx} =$

$$\dfrac{1}{c(a^2 + b^2)}[acx - b \log(a \sin cx + b \cos cx)]$$

*349. $\int \dfrac{dx}{a \cos^2 x + 2b \cos x \sin x + c \sin^2 x} = \begin{cases} \dfrac{1}{2\sqrt{b^2 - ac}}\log\dfrac{c \tan x + b - \sqrt{b^2 - ac}}{c \tan x + b + \sqrt{b^2 - ac}}, \\ \qquad\qquad\qquad\qquad\qquad (b^2 > ac) \\[4pt] \text{or} \\[4pt] \dfrac{1}{\sqrt{ac - b^2}}\tan^{-1}\dfrac{c \tan x + b}{\sqrt{ac - b^2}}, \quad (b^2 < ac) \\[4pt] \text{or} \\[4pt] -\dfrac{1}{c \tan x + b}, \qquad (b^2 = ac) \end{cases}$

350. $\int \dfrac{\sin ax}{1 \pm \sin ax}\,dx = \pm x + \dfrac{1}{a}\tan\left(\dfrac{\pi}{4} \mp \dfrac{ax}{2}\right)$

351. $\int \dfrac{dx}{(\sin ax)(1 \pm \sin ax)} = \dfrac{1}{a}\tan\left(\dfrac{\pi}{4} \mp \dfrac{ax}{2}\right) + \dfrac{1}{a}\log \tan\dfrac{ax}{2}$

352. $\int \dfrac{dx}{(1 + \sin ax)^2} = -\dfrac{1}{2a}\tan\left(\dfrac{\pi}{4} - \dfrac{ax}{2}\right) - \dfrac{1}{6a}\tan^3\left(\dfrac{\pi}{4} - \dfrac{ax}{2}\right)$

353. $\int \dfrac{dx}{(1 - \sin ax)^2} = \dfrac{1}{2a}\cot\left(\dfrac{\pi}{4} - \dfrac{ax}{2}\right) + \dfrac{1}{6a}\cot^3\left(\dfrac{\pi}{4} - \dfrac{ax}{2}\right)$

354. $\int \dfrac{\sin ax}{(1 + \sin ax)^2}\,dx = -\dfrac{1}{2a}\tan\left(\dfrac{\pi}{4} - \dfrac{ax}{2}\right) + \dfrac{1}{6a}\tan^3\left(\dfrac{\pi}{4} - \dfrac{ax}{2}\right)$

*See note 6--page 287.

INTEGRALS (Continued)

355. $\int \dfrac{\sin ax}{(1 - \sin ax)^2} dx = -\dfrac{1}{2a} \cot\left(\dfrac{\pi}{4} - \dfrac{ax}{2}\right) + \dfrac{1}{6a} \cot^3\left(\dfrac{\pi}{4} - \dfrac{ax}{2}\right)$

356. $\int \dfrac{\sin x\, dx}{a + b \sin x} = \dfrac{x}{b} - \dfrac{a}{b} \int \dfrac{dx}{a + b \sin x}$

357. $\int \dfrac{dx}{(\sin x)(a + b \sin x)} = \dfrac{1}{a} \log \tan\dfrac{x}{2} - \dfrac{b}{a} \int \dfrac{dx}{a + b \sin x}$

358. $\int \dfrac{dx}{(a + b \sin x)^2} = \dfrac{b \cos x}{(a^2 - b^2)(a + b \sin x)} + \dfrac{a}{a^2 - b^2} \int \dfrac{dx}{a + b \sin x}$

359. $\int \dfrac{\sin x\, dx}{(a + b \sin x)^2} = \dfrac{a \cos x}{(b^2 - a^2)(a + b \sin x)} + \dfrac{b}{b^2 - a^2} \int \dfrac{dx}{a + b \sin x}$

*360. $\int \dfrac{dx}{a^2 + b^2 \sin^2 cx} = \dfrac{1}{ac\sqrt{a^2 + b^2}} \tan^{-1} \dfrac{\sqrt{a^2 + b^2}\, \tan cx}{a}$

*361. $\int \dfrac{dx}{a^2 - b^2 \sin^2 cx} = \begin{cases} \dfrac{1}{ac\sqrt{a^2 - b^2}} \tan^{-1} \dfrac{\sqrt{a^2 - b^2}\, \tan cx}{a}, & (a^2 > b^2) \\[2mm] \text{or} \\[2mm] \dfrac{1}{2ac\sqrt{b^2 - a^2}} \log \dfrac{\sqrt{b^2 - a^2}\, \tan cx + a}{\sqrt{b^2 - a^2}\, \tan cx - a}, & (a^2 < b^2) \end{cases}$

362. $\int \dfrac{\cos ax}{1 + \cos ax} dx = x - \dfrac{1}{a} \tan\dfrac{ax}{2}$

363. $\int \dfrac{\cos ax}{1 - \cos ax} dx = -x - \dfrac{1}{a} \cot\dfrac{ax}{2}$

364. $\int \dfrac{dx}{(\cos ax)(1 + \cos ax)} = \dfrac{1}{a} \log \tan\left(\dfrac{\pi}{4} + \dfrac{ax}{2}\right) - \dfrac{1}{a} \tan\dfrac{ax}{2}$

365. $\int \dfrac{dx}{(\cos ax)(1 - \cos ax)} = \dfrac{1}{a} \log \tan\left(\dfrac{\pi}{4} + \dfrac{ax}{2}\right) - \dfrac{1}{a} \cot\dfrac{ax}{2}$

366. $\int \dfrac{dx}{(1 + \cos ax)^2} = \dfrac{1}{2a} \tan\dfrac{ax}{2} + \dfrac{1}{6a} \tan^3\dfrac{ax}{2}$

367. $\int \dfrac{dx}{(1 - \cos ax)^2} = -\dfrac{1}{2a} \cot\dfrac{ax}{2} - \dfrac{1}{6a} \cot^3\dfrac{ax}{2}$

368. $\int \dfrac{\cos ax}{(1 + \cos ax)^2} dx = \dfrac{1}{2a} \tan\dfrac{ax}{2} - \dfrac{1}{6a} \tan^3\dfrac{ax}{2}$

369. $\int \dfrac{\cos ax}{(1 - \cos ax)^2} dx = \dfrac{1}{2a} \cot\dfrac{ax}{2} - \dfrac{1}{6a} \cot^3\dfrac{ax}{2}$

*See note 6–page 287.

INTEGRALS (Continued)

370. $\displaystyle\int \frac{\cos x\, dx}{a + b\cos x} = \frac{x}{b} - \frac{a}{b}\int \frac{dx}{a + b\cos x}$

371. $\displaystyle\int \frac{dx}{(\cos x)(a + b\cos x)} = \frac{1}{a}\log\tan\left(\frac{x}{2} + \frac{\pi}{4}\right) - \frac{b}{a}\int \frac{dx}{a + b\cos x}$

372. $\displaystyle\int \frac{dx}{(a + b\cos x)^2} = \frac{b\sin x}{(b^2 - a^2)(a + b\cos x)} - \frac{a}{b^2 - a^2}\int \frac{dx}{a + b\cos x}$

373. $\displaystyle\int \frac{\cos x}{(a + b\cos x)^2}\, dx = \frac{a\sin x}{(a^2 - b^2)(a + b\cos x)} - \frac{b}{a^2 - b^2}\int \frac{dx}{a + b\cos x}$

*374. $\displaystyle\int \frac{dx}{a^2 + b^2 - 2ab\cos cx} = \frac{2}{c(a^2 - b^2)}\tan^{-1}\left(\frac{a + b}{a - b}\tan\frac{cx}{2}\right)$

*375. $\displaystyle\int \frac{dx}{a^2 + b^2\cos^2 cx} = \frac{1}{ac\sqrt{a^2 + b^2}}\tan^{-1}\frac{a\tan cx}{\sqrt{a^2 + b^2}}$

*376. $\displaystyle\int \frac{dx}{a^2 - b^2\cos^2 cx} = \begin{cases} \dfrac{1}{uc\sqrt{a^2 - b^2}}\tan^{-1}\dfrac{a\tan cx}{\sqrt{a^2 - b^2}}, & (a^2 > b^2) \\[1em] \text{or} \\[1em] \dfrac{1}{2ac\sqrt{b^2 - a^2}}\log\dfrac{a\tan cx - \sqrt{b^2 - a^2}}{a\tan cx + \sqrt{b^2 - a^2}}, & (b^2 > a^2) \end{cases}$

377. $\displaystyle\int \frac{\sin ax}{1 \pm \cos ax}\, dx = \mp\frac{1}{a}\log(1 \pm \cos ax)$

378. $\displaystyle\int \frac{\cos ax}{1 \pm \sin ax}\, dx = \pm\frac{1}{a}\log(1 \pm \sin ax)$

379. $\displaystyle\int \frac{dx}{(\sin ax)(1 \pm \cos ax)} = \pm\frac{1}{2a(1 \pm \cos ax)} + \frac{1}{2a}\log\tan\frac{ax}{2}$

380. $\displaystyle\int \frac{dx}{(\cos ax)(1 \pm \sin ax)} = \mp\frac{1}{2a(1 \pm \sin ax)} + \frac{1}{2a}\log\tan\left(\frac{\pi}{4} + \frac{ax}{2}\right)$

381. $\displaystyle\int \frac{\sin ax}{(\cos ax)(1 \pm \cos ax)}\, dx = \frac{1}{a}\log(\sec ax \pm 1)$

382. $\displaystyle\int \frac{\cos ax}{(\sin ax)(1 \pm \sin ax)}\, dx = -\frac{1}{a}\log(\csc ax \pm 1)$

383. $\displaystyle\int \frac{\sin ax}{(\cos ax)(1 \pm \sin ax)}\, dx = \frac{1}{2a(1 \pm \sin ax)} \pm \frac{1}{2a}\log\tan\left(\frac{\pi}{4} + \frac{ax}{2}\right)$

384. $\displaystyle\int \frac{\cos ax}{(\sin ax)(1 \pm \cos ax)}\, dx = -\frac{1}{2a(1 \pm \cos ax)} \pm \frac{1}{2a}\log\tan\frac{ax}{2}$

*See note 6–page 287.

INTEGRALS (Continued)

385. $\displaystyle\int \frac{dx}{\sin ax \pm \cos ax} = \frac{1}{a\sqrt{2}} \log \tan\left(\frac{ax}{2} \pm \frac{\pi}{8}\right)$

386. $\displaystyle\int \frac{dx}{(\sin ax \pm \cos ax)^2} = \frac{1}{2a} \tan\left(ax \mp \frac{\pi}{4}\right)$

387. $\displaystyle\int \frac{dx}{1 + \cos ax \pm \sin ax} = \pm \frac{1}{a} \log\left(1 \pm \tan\frac{ax}{2}\right)$

388. $\displaystyle\int \frac{dx}{a^2 \cos^2 cx - b^2 \sin^2 cx} = \frac{1}{2abc} \log \frac{b \tan cx + a}{b \tan cx - a}$

389. $\displaystyle\int x(\sin ax)\, dx = \frac{1}{a^2} \sin ax - \frac{x}{a} \cos ax$

390. $\displaystyle\int x^2(\sin ax)\, dx = \frac{2x}{a^2} \sin ax - \frac{a^2 x^2 - 2}{a^3} \cos ax$

391. $\displaystyle\int x^3(\sin ax)\, dx = \frac{3a^2 x^2 - 6}{a^4} \sin ax - \frac{a^2 x^3 - 6x}{a^3} \cos ax$

392. $\displaystyle\int x^m \sin ax\, dx = \begin{cases} -\dfrac{1}{a} x^m \cos ax + \dfrac{m}{a} \displaystyle\int x^{m-1} \cos ax\, dx \\[2mm] \text{or} \\[2mm] \cos ax \displaystyle\sum_{r=0}^{\left[\frac{m}{2}\right]} (-1)^{r+1} \frac{m!}{(m-2r)!} \cdot \frac{x^{m-2r}}{a^{2r+1}} \\[4mm] + \sin ax \displaystyle\sum_{r=0}^{\left[\frac{m-1}{2}\right]} (-1)^r \frac{m!}{(m-2r-1)!} \cdot \frac{x^{m-2r-1}}{a^{2r+2}} \end{cases}$

Note: $[s]$ means greatest integer $\le s$; $[3\frac{1}{2}] = 3$, $[\frac{1}{2}] = 0$, etc.

393. $\displaystyle\int x(\cos ax)\, dx = \frac{1}{a^2} \cos ax + \frac{x}{a} \sin ax$

394. $\displaystyle\int x^2(\cos ax)\, dx = \frac{2x \cos ax}{a^2} + \frac{a^2 x^2 - 2}{a^3} \sin ax$

395. $\displaystyle\int x^3(\cos ax)\, dx = \frac{3a^2 x^2 - 6}{a^4} \cos ax + \frac{a^2 x^3 - 6x}{a^3} \sin ax$

396. $\displaystyle\int x^m(\cos ax)\, dx = \begin{cases} \dfrac{x^m \sin ax}{a} - \dfrac{m}{a} \displaystyle\int x^{m-1} \sin ax\, dx \\[2mm] \text{or} \\[2mm] \sin ax \displaystyle\sum_{r=0}^{\left[\frac{m}{2}\right]} (-1)^r \frac{m!}{(m-2r)!} \cdot \frac{x^{m-2r}}{a^{2r+1}} \\[4mm] + \cos ax \displaystyle\sum_{r=0}^{\left[\frac{m-1}{2}\right]} (-1)^r \frac{m!}{(m-2r-1)!} \cdot \frac{x^{m-2r-1}}{a^{2r+2}} \end{cases}$

See note integral 392.

INTEGRALS (Continued)

397. $\displaystyle\int \frac{\sin ax}{x}\,dx = \sum_{n=0}^{x} (-1)^n \frac{(ax)^{2n+1}}{(2n+1)(2n+1)!}$

398. $\displaystyle\int \frac{\cos ax}{x}\,dx = \log x + \sum_{n=1}^{x} (-1)^n \frac{(ax)^{2n}}{2n(2n)!}$

399. $\displaystyle\int x(\sin^2 ax)\,dx = \frac{x^2}{4} - \frac{x\sin 2ax}{4a} - \frac{\cos 2ax}{8a^2}$

400. $\displaystyle\int x^2(\sin^2 ax)\,dx = \frac{x^3}{6} - \left(\frac{x^2}{4a} - \frac{1}{8a^3}\right)\sin 2ax - \frac{x\cos 2ax}{4a^2}$

401. $\displaystyle\int x(\sin^3 ax)\,dx = \frac{x\cos 3ax}{12a} - \frac{\sin 3ax}{36a^2} - \frac{3x\cos ax}{4a} + \frac{3\sin ax}{4a^2}$

402. $\displaystyle\int x(\cos^2 ax)\,dx = \frac{x^2}{4} + \frac{x\sin 2ax}{4a} + \frac{\cos 2ax}{8a^2}$

403. $\displaystyle\int x^2(\cos^2 ax)\,dx = \frac{x^3}{6} + \left(\frac{x^2}{4a} - \frac{1}{8a^3}\right)\sin 2ax + \frac{x\cos 2ax}{4a^2}$

404. $\displaystyle\int x(\cos^3 ax)\,dx = \frac{x\sin 3ax}{12a} + \frac{\cos 3ax}{36a^2} + \frac{3x\sin ax}{4a} + \frac{3\cos ax}{4a^2}$

405. $\displaystyle\int \frac{\sin ax}{x^m}\,dx = -\frac{\sin ax}{(m-1)x^{m-1}} + \frac{a}{m-1}\int \frac{\cos ax}{x^{m-1}}\,dx$

406. $\displaystyle\int \frac{\cos ax}{x^m}\,dx = -\frac{\cos ax}{(m-1)x^{m-1}} - \frac{a}{m-1}\int \frac{\sin ax}{x^{m-1}}\,dx$

407. $\displaystyle\int \frac{x}{1 \pm \sin ax}\,dx = \mp \frac{x\cos ax}{a(1 \pm \sin ax)} + \frac{1}{a^2}\log(1 \pm \sin ax)$

408. $\displaystyle\int \frac{x}{1 + \cos ax}\,dx = \frac{x}{a}\tan\frac{ax}{2} + \frac{2}{a^2}\log\cos\frac{ax}{2}$

409. $\displaystyle\int \frac{x}{1 - \cos ax}\,dx = -\frac{x}{a}\cot\frac{ax}{2} + \frac{2}{a^2}\log\sin\frac{ax}{2}$

410. $\displaystyle\int \frac{x + \sin x}{1 + \cos x}\,dx = x\tan\frac{x}{2}$

411. $\displaystyle\int \frac{x - \sin x}{1 - \cos x}\,dx = -x\cot\frac{x}{2}$

412. $\displaystyle\int \sqrt{1 - \cos ax}\,dx = -\frac{2\sin ax}{a\sqrt{1 - \cos ax}} = -\frac{2\sqrt{2}}{a}\cos\left(\frac{ax}{2}\right)$

413. $\displaystyle\int \sqrt{1 + \cos ax}\,dx = \frac{2\sin ax}{a\sqrt{1 + \cos ax}} = \frac{2\sqrt{2}}{a}\sin\left(\frac{ax}{2}\right)$

INTEGRALS (Continued)

414. $\displaystyle\int \sqrt{1 + \sin x}\, dx = \pm 2\left(\sin\frac{x}{2} - \cos\frac{x}{2}\right),$

[use + if $(8k - 1)\dfrac{\pi}{2} < x \le (8k + 3)\dfrac{\pi}{2}$, otherwise $-$; k an integer]

415. $\displaystyle\int \sqrt{1 - \sin x}\, dx = \pm 2\left(\sin\frac{x}{2} + \cos\frac{x}{2}\right),$

[use + if $(8k - 3)\dfrac{\pi}{2} < x \le (8k + 1)\dfrac{\pi}{2}$, otherwise $-$; k an integer]

416. $\displaystyle\int \frac{dx}{\sqrt{1 - \cos x}} = \pm \sqrt{2}\,\log\tan\frac{x}{4},$

[use + if $4k\pi < x < (4k + 2)\pi$, otherwise $-$; k an integer]

417. $\displaystyle\int \frac{dx}{\sqrt{1 + \cos x}} = \pm \sqrt{2}\,\log\tan\left(\frac{x + \pi}{4}\right),$

[use + if $(4k - 1)\pi < x < (4k + 1)\pi$, otherwise $-$; k an integer]

418. $\displaystyle\int \frac{dx}{\sqrt{1 - \sin x}} = \pm \sqrt{2}\,\log\tan\left(\frac{x}{4} - \frac{\pi}{8}\right),$

[use + if $(8k + 1)\dfrac{\pi}{2} < x < (8k + 5)\dfrac{\pi}{2}$, otherwise $-$; k an integer]

419. $\displaystyle\int \frac{dx}{\sqrt{1 + \sin x}} = \pm \sqrt{2}\,\log\tan\left(\frac{x}{4} + \frac{\pi}{8}\right),$

[use + if $(8k - 1)\dfrac{\pi}{2} < x < (8k + 3)\dfrac{\pi}{2}$, otherwise $-$; k an integer]

420. $\displaystyle\int (\tan^2 ax)\, dx = \frac{1}{a}\tan ax - x$

421. $\displaystyle\int (\tan^3 ax)\, dx = \frac{1}{2a}\tan^2 ax + \frac{1}{a}\log\cos ax$

422. $\displaystyle\int (\tan^4 ax)\, dx = \frac{\tan^3 ax}{3a} - \frac{1}{a}\tan x + x$

423. $\displaystyle\int (\tan^n ax)\, dx = \frac{\tan^{n-1} ax}{a(n - 1)} - \int (\tan^{n-2} ax)\, dx$

424. $\displaystyle\int (\cot^2 ax)\, dx = -\frac{1}{a}\cot ax - x$

425. $\displaystyle\int (\cot^3 ax)\, dx = -\frac{1}{2a}\cot^2 ax - \frac{1}{a}\log\sin ax$

426. $\displaystyle\int (\cot^4 ax)\, dx = -\frac{1}{3a}\cot^3 ax + \frac{1}{a}\cot ax + x$

INTEGRALS (Continued)

427. $\int (\cot^n ax)\, dx = -\dfrac{\cot^{n-1} ax}{a(n-1)} - \int (\cot^{n-2} ax)\, dx$

428. $\int \dfrac{x}{\sin^2 ax}\, dx = \int x(\csc^2 ax)\, dx = -\dfrac{x \cot ax}{a} + \dfrac{1}{a^2} \log \sin ax$

429. $\int \dfrac{x}{\sin^n ax}\, dx = \int x(\csc^n ax)\, dx = -\dfrac{x \cos ax}{a(n-1)\sin^{n-1} ax}$

$$-\dfrac{1}{a^2(n-1)(n-2)\sin^{n-2} ax} + \dfrac{(n-2)}{(n-1)}\int \dfrac{x}{\sin^{n-2} ax}\, dx$$

430. $\int \dfrac{x}{\cos^2 ax}\, dx = \int x(\sec^2 ax)\, dx = \dfrac{1}{a} x \tan ax + \dfrac{1}{a^2} \log \cos ax$

431. $\int \dfrac{x}{\cos^n ax}\, dx = \int x(\sec^n ax)\, dx = \dfrac{x \sin ax}{a(n-1)\cos^{n-1} ax}$

$$-\dfrac{1}{a^2(n-1)(n-2)\cos^{n-2} ax} + \dfrac{n-2}{n-1}\int \dfrac{x}{\cos^{n-2} ax}\, dx$$

432. $\int \dfrac{\sin ax}{\sqrt{1 + b^2 \sin^2 ax}}\, dx = -\dfrac{1}{ab} \sin^{-1} \dfrac{b \cos ax}{\sqrt{1 + b^2}}$

433. $\int \dfrac{\sin ax}{\sqrt{1 - b^2 \sin^2 ax}}\, dx = -\dfrac{1}{ab} \log (b \cos ax + \sqrt{1 - b^2 \sin^2 ax})$

434. $\int (\sin ax)\sqrt{1 + b^2 \sin^2 ax}\, dx - -\dfrac{\cos ax}{2a}\sqrt{1 + b^2 \sin^2 ax}$

$$-\dfrac{1 + b^2}{2ab} \sin^{-1} \dfrac{b \cos ax}{\sqrt{1 + b^2}}$$

435. $\int (\sin ax)\sqrt{1 - b^2 \sin^2 ax}\, dx = -\dfrac{\cos ax}{2a}\sqrt{1 - b^2 \sin^2 ax}$

$$-\dfrac{1 - b^2}{2ab} \log (b \cos ax + \sqrt{1 - b^2 \sin^2 ax})$$

436. $\int \dfrac{\cos ax}{\sqrt{1 + b^2 \sin^2 ax}}\, dx = \dfrac{1}{ab} \log (b \sin ax + \sqrt{1 + b^2 \sin^2 ax})$

437. $\int \dfrac{\cos ax}{\sqrt{1 - b^2 \sin^2 ax}}\, dx = \dfrac{1}{ab} \sin^{-1} (b \sin ax)$

438. $\int (\cos ax) \sqrt{1 + b^2 \sin^2 ax}\, dx = \dfrac{\sin ax}{2a}\sqrt{1 + b^2 \sin^2 ax}$

$$+\dfrac{1}{2ab} \log (b \sin ax + \sqrt{1 + b^2 \sin^2 ax})$$

INTEGRALS (Continued)

439. $\displaystyle\int (\cos ax)\sqrt{1 - b^2\sin^2 ax}\,dx = \frac{\sin ax}{2a}\sqrt{1 - b^2\sin^2 ax} + \frac{1}{2ab}\sin^{-1}(b\sin ax)$

440. $\displaystyle\int \frac{dx}{\sqrt{a + b\tan^2 cx}} = \frac{\pm 1}{c\sqrt{a - b}}\sin^{-1}\left(\sqrt{\frac{a - b}{a}}\sin cx\right), \qquad (a > |b|)$

$\left[\text{use} + \text{ if } (2k - 1)\dfrac{\pi}{2} < x \le (2k + 1)\dfrac{\pi}{2}, \text{ otherwise } -\,; k \text{ an integer}\right]$

FORMS INVOLVING INVERSE TRIGONOMETRIC FUNCTIONS

441. $\displaystyle\int (\sin^{-1} ax)\,dx = x\sin^{-1} ax + \frac{\sqrt{1 - a^2 x^2}}{a}$

442. $\displaystyle\int (\cos^{-1} ax)\,dx = x\cos^{-1} ax - \frac{\sqrt{1 - a^2 x^2}}{a}$

443. $\displaystyle\int (\tan^{-1} ax)\,dx = x\tan^{-1} ax - \frac{1}{2a}\log(1 + a^2 x^2)$

444. $\displaystyle\int (\cot^{-1} ax)\,dx = x\cot^{-1} ax + \frac{1}{2a}\log(1 + a^2 x^2)$

445. $\displaystyle\int (\sec^{-1} ax)\,dx = x\sec^{-1} ax - \frac{1}{a}\log(ax + \sqrt{a^2 x^2 - 1})$

446. $\displaystyle\int (\csc^{-1} ax)\,dx = x\csc^{-1} ax + \frac{1}{a}\log(ax + \sqrt{a^2 x^2 - 1})$

447. $\displaystyle\int \left(\sin^{-1}\frac{x}{a}\right) dx = x\sin^{-1}\frac{x}{a} + \sqrt{a^2 - x^2}, \qquad (a > 0)$

448. $\displaystyle\int \left(\cos^{-1}\frac{x}{a}\right) dx = x\cos^{-1}\frac{x}{a} - \sqrt{a^2 - x^2}, \qquad (a > 0)$

449. $\displaystyle\int \left(\tan^{-1}\frac{x}{a}\right) dx = x\tan^{-1}\frac{x}{a} - \frac{a}{2}\log(a^2 + x^2)$

450. $\displaystyle\int \left(\cot^{-1}\frac{x}{a}\right) dx = x\cot^{-1}\frac{x}{a} + \frac{a}{2}\log(a^2 + x^2)$

451. $\displaystyle\int x[\sin^{-1}(ax)]\,dx = \frac{1}{4a^2}[(2a^2 x^2 - 1)\sin^{-1}(ax) + ax\sqrt{1 - a^2 x^2}]$

452. $\displaystyle\int x[\cos^{-1}(ax)]\,dx = \frac{1}{4a^2}[(2a^2 x^2 - 1)\cos^{-1}(ax) - ax\sqrt{1 - a^2 x^2}]$

INTEGRALS (Continued)

453. $\displaystyle\int x^n[\sin^{-1}(ax)]\,dx = \frac{x^{n+1}}{n+1}\sin^{-1}(ax) - \frac{a}{n+1}\int \frac{x^{n+1}\,dx}{\sqrt{1-a^2x^2}},$ $(n \neq -1)$

454. $\displaystyle\int x^n[\cos^{-1}(ax)]\,dx = \frac{x^{n+1}}{n+1}\cos^{-1}(ax) + \frac{a}{n+1}\int \frac{x^{n+1}\,dx}{\sqrt{1-a^2x^2}},$ $(n \neq -1)$

455. $\displaystyle\int x(\tan^{-1}ax)\,dx = \frac{1+a^2x^2}{2a^2}\tan^{-1}ax - \frac{x}{2a}$

456. $\displaystyle\int x^n(\tan^{-1}ax)\,dx = \frac{x^{n+1}}{n+1}\tan^{-1}ax - \frac{a}{n+1}\int \frac{x^{n+1}}{1+a^2x^2}\,dx$

457. $\displaystyle\int x(\cot^{-1}ax)\,dx = \frac{1+a^2x^2}{2a^2}\cot^{-1}ax + \frac{x}{2a}$

458. $\displaystyle\int x^n(\cot^{-1}ax)\,dx = \frac{x^{n+1}}{n+1}\cot^{-1}ax + \frac{a}{n+1}\int \frac{x^{n+1}}{1+a^2x^2}\,dx$

459. $\displaystyle\int \frac{\sin^{-1}(ax)}{x^2}\,dx = a\log\left(\frac{1-\sqrt{1-a^2x^2}}{x}\right) - \frac{\sin^{-1}(ax)}{x}$

460. $\displaystyle\int \frac{\cos^{-1}(ax)\,dx}{x^2} = -\frac{1}{x}\cos^{-1}(ax) + a\log\frac{1+\sqrt{1-a^2x^2}}{x}$

461. $\displaystyle\int \frac{\tan^{-1}(ax)\,dx}{x^2} = -\frac{1}{x}\tan^{-1}(ax) - \frac{a}{2}\log\frac{1+a^2x^2}{x^2}$

462. $\displaystyle\int \frac{\cot^{-1}ax}{x^2}\,dx = -\frac{1}{x}\cot^{-1}ax - \frac{a}{2}\log\frac{x^2}{a^2x^2+1}$

463. $\displaystyle\int (\sin^{-1}ax)^2\,dx = x(\sin^{-1}ax)^2 - 2x + \frac{2\sqrt{1-a^2x^2}}{a}\sin^{-1}ax$

464. $\displaystyle\int (\cos^{-1}ax)^2\,dx = x(\cos^{-1}ax)^2 - 2x - \frac{2\sqrt{1-a^2x^2}}{a}\cos^{-1}ax$

465. $\displaystyle\int (\sin^{-1}ax)^n\,dx = \begin{cases} x(\sin^{-1}ax)^n + \dfrac{n\sqrt{1-a^2x^2}}{a}(\sin^{-1}ax)^{n-1} \\ \qquad\qquad\qquad -n(n-1)\displaystyle\int (\sin^{-1}ax)^{n-2}\,dx \\ \qquad\qquad\text{or} \\ \displaystyle\sum_{r=0}^{\left[\frac{n}{2}\right]} (-1)^r \frac{n!}{(n-2r)!}x(\sin^{-1}ax)^{n-2r} \\ \quad + \displaystyle\sum_{r=0}^{\left[\frac{n-1}{2}\right]} (-1)^r \frac{n!\sqrt{1-a^2x^2}}{(n-2r-1)!a}(\sin^{-1}ax)^{n-2r-1} \end{cases}$

Note: $[s]$ means greatest integer $\leq s$. Thus $[3.5]$ means 3; $[5] = 5$, $[\frac{1}{2}] = 0$.

INTEGRALS (Continued)

466. $\displaystyle\int (\cos^{-1} ax)^n \, dx = \begin{cases} x(\cos^{-1} ax)^n - \dfrac{n\sqrt{1 - a^2x^2}}{a}(\cos^{-1} ax)^{n-1} \\ \qquad\qquad -n(n-1)\displaystyle\int (\cos^{-1} ax)^{n-2} \, dx \\ \qquad\qquad \text{or} \\ \displaystyle\sum_{r=0}^{\left[\frac{n}{2}\right]} (-1)^r \dfrac{n!}{(n-2r)!} x(\cos^{-1} ax)^{n-2r} \\ \qquad -\displaystyle\sum_{r=0}^{\left[\frac{n-1}{2}\right]} (-1)^r \dfrac{n!\sqrt{1 - a^2x^2}}{(n-2r-1)!a}(\cos^{-1} ax)^{n-2r-1} \end{cases}$

467. $\displaystyle\int \frac{1}{\sqrt{1 - a^2x^2}}(\sin^{-1} ax) \, dx = \frac{1}{2a}(\sin^{-1} ax)^2$

468. $\displaystyle\int \frac{x^n}{\sqrt{1 - a^2x^2}}(\sin^{-1} ax) \, dx = -\frac{x^{n-1}}{na^2}\sqrt{1 - a^2x^2}\,\sin^{-1} ax + \frac{x^n}{n^2a}$
$$+ \frac{n-1}{na^2}\int \frac{x^{n-2}}{\sqrt{1 - a^2x^2}}\sin^{-1} ax \, dx$$

469. $\displaystyle\int \frac{1}{\sqrt{1 - a^2x^2}}(\cos^{-1} ax) \, dx = -\frac{1}{2a}(\cos^{-1} ax)^2$

470. $\displaystyle\int \frac{x^n}{\sqrt{1 - a^2x^2}}(\cos^{-1} ax) \, dx = -\frac{x^{n-1}}{na^2}\sqrt{1 - a^2x^2}\,\cos^{-1} ax - \frac{x^n}{n^2a}$
$$+ \frac{n-1}{na^2}\int \frac{x^{n-2}}{\sqrt{1 - a^2x^2}}\cos^{-1} ax \, dx$$

471. $\displaystyle\int \frac{\tan^{-1} ax}{a^2x^2 + 1} \, dx = \frac{1}{2a}(\tan^{-1} ax)^2$

472. $\displaystyle\int \frac{\cot^{-1} ax}{a^2x^2 + 1} \, dx = -\frac{1}{2a}(\cot^{-1} ax)^2$

473. $\displaystyle\int x \sec^{-1} ax \, dx = \frac{x^2}{2}\sec^{-1} ax - \frac{1}{2a^2}\sqrt{a^2x^2 - 1}$

474. $\displaystyle\int x^n \sec^{-1} ax \, dx = \frac{x^{n+1}}{n+1}\sec^{-1} ax - \frac{1}{n+1}\int \frac{x^n \, dx}{\sqrt{a^2x^2 - 1}}$

475. $\displaystyle\int \frac{\sec^{-1} ax}{x^2} \, dx = -\frac{\sec^{-1} ax}{x} + \frac{\sqrt{a^2x^2 - 1}}{x}$

476. $\displaystyle\int x \csc^{-1} ax \, dx = \frac{x^2}{2}\csc^{-1} ax + \frac{1}{2a^2}\sqrt{a^2x^2 - 1}$

477. $\displaystyle\int x^n \csc^{-1} ax \, dx = \frac{x^{n+1}}{n+1}\csc^{-1} ax + \frac{1}{n+1}\int \frac{x^n \, dx}{\sqrt{a^2x^2 - 1}}$

INTEGRALS (Continued)

478. $\displaystyle\int \frac{\csc^{-1} ax}{x^2}\, dx = -\frac{\csc^{-1} ax}{x} - \frac{\sqrt{a^2 x^2 - 1}}{x}$

FORMS INVOLVING TRIGONOMETRIC SUBSTITUTIONS

479. $\displaystyle\int f(\sin x)\, dx = 2\int f\!\left(\frac{2z}{1+z^2}\right)\frac{dz}{1+z^2}, \qquad \left(z = \tan\frac{x}{2}\right)$

480. $\displaystyle\int f(\cos x)\, dx = 2\int f\!\left(\frac{1-z^2}{1+z^2}\right)\frac{dz}{1+z^2}, \qquad \left(z = \tan\frac{x}{2}\right)$

*481. $\displaystyle\int f(\sin x)\, dx = \int f(u)\frac{du}{\sqrt{1-u^2}}, \qquad (u = \sin x)$

*482. $\displaystyle\int f(\cos x)\, dx = -\int f(u)\frac{du}{\sqrt{1-u^2}}, \qquad (u = \cos x)$

*483. $\displaystyle\int f(\sin x, \cos x)\, dx = \int f(u, \sqrt{1-u^2})\frac{du}{\sqrt{1-u^2}}, \qquad (u = \sin x)$

484. $\displaystyle\int f(\sin x, \cos x)\, dx = 2\int f\!\left(\frac{2z}{1+z^2}, \frac{1-z^2}{1+z^2}\right)\frac{dz}{1+z^2}, \qquad \left(z = \tan\frac{x}{2}\right)$

LOGARITHMIC FORMS

485. $\displaystyle\int (\log x)\, dx = x\log x - x$

486. $\displaystyle\int x(\log x)\, dx = \frac{x^2}{2}\log x - \frac{x^2}{4}$

487. $\displaystyle\int x^2(\log x)\, dx = \frac{x^3}{3}\log x - \frac{x^3}{9}$

488. $\displaystyle\int x^n(\log ax)\, dx = \frac{x^{n+1}}{n+1}\log ax - \frac{x^{n+1}}{(n+1)^2}$

489. $\displaystyle\int (\log x)^2\, dx = x(\log x)^2 - 2x\log x + 2x$

490. $\displaystyle\int (\log x)^n\, dx = \begin{cases} x(\log x)^n - n\displaystyle\int (\log x)^{n-1}\, dx, \qquad (n \neq -1) \\[2mm] \text{or} \\[2mm] (-1)^n n! \, x \displaystyle\sum_{r=0}^{n} \frac{(-\log x)^r}{r!} \end{cases}$

---

*The square roots appearing in these formulas may be plus or minus, depending on the quadrant of *x*. Care must be used to give them the proper sign.

INTEGRALS (Continued)

491. $\displaystyle\int \frac{(\log x)^n}{x}\,dx = \frac{1}{n+1}(\log x)^{n+1}$

492. $\displaystyle\int \frac{dx}{\log x} = \log(\log x) + \log x + \frac{(\log x)^2}{2\cdot 2!} + \frac{(\log x)^3}{3\cdot 3!} + \cdots$

493. $\displaystyle\int \frac{dx}{x\log x} = \log(\log x)$

494. $\displaystyle\int \frac{dx}{x(\log x)^n} = -\frac{1}{(n-1)(\log x)^{n-1}}$

495. $\displaystyle\int \frac{x^m\,dx}{(\log x)^n} = -\frac{x^{m+1}}{(n-1)(\log x)^{n-1}} + \frac{m+1}{n-1}\int \frac{x^m\,dx}{(\log x)^{n-1}}$

496. $\displaystyle\int x^m(\log x)^n\,dx = \begin{cases} \dfrac{x^{m+1}(\log x)^n}{m+1} - \dfrac{n}{m+1}\displaystyle\int x^m(\log x)^{n-1}\,dx \\[2mm] \text{or} \\[2mm] (-1)^n \dfrac{n!}{m+1}x^{m+1}\displaystyle\sum_{r=0}^{n}\frac{(-\log x)^r}{r!(m+1)^{n-r}} \end{cases}$

497. $\displaystyle\int x^p\cos(b\,\ln x)\,dx = \frac{x^{p+1}}{(p+1)^2+b^2}\cdot[b\sin(b\,\ln x)+(p+1)\cos(b\,\ln x)]+c$

498. $\displaystyle\int x^p\sin(b\,\ln x)\,dx = \frac{x^{p+1}}{(p+1)^2+b^2}\cdot[(p+1)\sin(b\,\ln x)-b\cos(b\,\ln x)]+c$

499. $\displaystyle\int [\log(ax+b)]\,dx = \frac{ax+b}{a}\log(ax+b)-x$

500. $\displaystyle\int \frac{\log(ax+b)}{x^2}\,dx = \frac{a}{b}\log x - \frac{ax+b}{bx}\log(ax+b)$

501. $\displaystyle\int x^m[\log(ax+b)]\,dx = \frac{1}{m+1}\left[x^{m+1}-\left(-\frac{b}{a}\right)^{m+1}\right]\log(ax+b)$

$$-\frac{1}{m+1}\left(-\frac{b}{a}\right)^{m+1}\sum_{r=1}^{m+1}\frac{1}{r}\left(-\frac{ax}{b}\right)^r$$

502. $\displaystyle\int \frac{\log(ax+b)}{x^m}\,dx = -\frac{1}{m-1}\frac{\log(ax+b)}{x^{m-1}} + \frac{1}{m-1}\left(-\frac{a}{b}\right)^{m-1}\log\frac{ax+b}{x}$

$$+\frac{1}{m-1}\left(-\frac{a}{b}\right)^{m-1}\sum_{r=1}^{m-2}\frac{1}{r}\left(-\frac{b}{ax}\right)^r,\ (m>2)$$

503. $\displaystyle\int \left[\log\frac{x+a}{x-a}\right]\,dx = (x+a)\log(x+a)-(x-a)\log(x-a)$

504. $\displaystyle\int x^m\left[\log\frac{x+a}{x-a}\right]\,dx = \frac{x^{m+1}-(-a)^{m+1}}{m+1}\log(x+a) - \frac{x^{m+1}-a^{m+1}}{m+1}\log(x-a)$

$$+\frac{2a^{m+1}}{m+1}\sum_{r=1}^{\left[\frac{m+1}{2}\right]}\frac{1}{m-2r+2}\left(\frac{x}{a}\right)^{m-2r+2}$$

See note integral 392.

INTEGRALS (Continued)

505. $\displaystyle \int \frac{1}{x^2}\left[\log\frac{x+a}{x-a}\right]dx = \frac{1}{x}\log\frac{x-a}{x+a} - \frac{1}{a}\log\frac{x^2-a^2}{x^2}$

506. $\displaystyle \int (\log X)\,dx = \begin{cases} \left(x+\dfrac{b}{2c}\right)\log X - 2x + \dfrac{\sqrt{4ac-b^2}}{c}\tan^{-1}\dfrac{2cx+b}{\sqrt{4ac-b^2}}, \\ \hfill (b^2-4ac<0) \\[2pt] \text{or} \\[2pt] \left(x+\dfrac{b}{2c}\right)\log X - 2x + \dfrac{\sqrt{b^2-4ac}}{c}\tanh^{-1}\dfrac{2cx+b}{\sqrt{b^2-4ac}}, \\ \hfill (b^2-4ac>0) \\[2pt] \text{where} \\[2pt] X = a+bx+cx^2 \end{cases}$

507. $\displaystyle \int x^n(\log X)\,dx = \frac{x^{n+1}}{n+1}\log X - \frac{2c}{n+1}\int\frac{x^{n+2}}{X}dx - \frac{b}{n+1}\int\frac{x^{n+1}}{X}dx$

$\text{where } X = a+bx+cx^2$

508. $\displaystyle \int [\log(x^2+a^2)]\,dx = x\log(x^2+a^2) - 2x + 2a\tan^{-1}\frac{x}{a}$

509. $\displaystyle \int [\log(x^2-a^2)]\,dx = x\log(x^2-a^2) - 2x + a\log\frac{x+a}{x-a}$

510. $\displaystyle \int x[\log(x^2\pm a^2)]\,dx = \tfrac{1}{2}(x^2\pm a^2)\log(x^2\pm a^2) - \tfrac{1}{2}x^2$

511. $\displaystyle \int [\log(x+\sqrt{x^2\pm a^2})]\,dx = x\log(x+\sqrt{x^2\pm a^2}) - \sqrt{x^2\pm a^2}$

512. $\displaystyle \int x[\log(x+\sqrt{x^2\pm a^2})]\,dx = \left(\frac{x^2}{2}\pm\frac{a^2}{4}\right)\log(x+\sqrt{x^2\pm a^2}) - \frac{x\sqrt{x^2\pm a^2}}{4}$

513. $\displaystyle \int x^m[\log(x+\sqrt{x^2\pm a^2})]\,dx = \frac{x^{m+1}}{m+1}\log(x+\sqrt{x^2\pm a^2})$

$\displaystyle \hspace{5cm} -\frac{1}{m+1}\int\frac{x^{m+1}}{\sqrt{x^2\pm a^2}}dx$

514. $\displaystyle \int\frac{\log(x+\sqrt{x^2+a^2})}{x^2}dx = -\frac{\log(x+\sqrt{x^2+a^2})}{x} - \frac{1}{a}\log\frac{a+\sqrt{x^2+a^2}}{x}$

515. $\displaystyle \int\frac{\log(x+\sqrt{x^2-a^2})}{x^2}dx = -\frac{\log(x+\sqrt{x^2-a^2})}{x} + \frac{1}{|a|}\sec^{-1}\frac{x}{a}$

516. $\displaystyle\int x^n \log{(x^2 - a^2)}\,dx = \frac{1}{n+1}\bigg[x^{n+1} \log{(x^2 - a^2)} - a^{n+1} \log{(x - a)}$

See note integral 392. $\displaystyle -(-a)^{n+1} \log{(x + a)} - 2 \sum_{r=0}^{\left[\frac{n}{2}\right]} \frac{a^{2r} x^{n-2r+1}}{n - 2r + 1} \bigg]$

EXPONENTIAL FORMS

517. $\displaystyle\int e^x\,dx = e^x$

518. $\displaystyle\int e^{-x}\,dx = -e^{-x}$

519. $\displaystyle\int e^{ax}\,dx = \frac{e^{ax}}{a}$

520. $\displaystyle\int x\,e^{ax}\,dx = \frac{e^{ax}}{a^2}(ax - 1)$

521. $\displaystyle\int x^m e^{ax}\,dx = \begin{cases} \dfrac{x^m e^{ax}}{a} - \dfrac{m}{a}\displaystyle\int x^{m-1} e^{ax}\,dx \\[2mm] \qquad\qquad \text{or} \\[2mm] e^{ax} \displaystyle\sum_{r=0}^{m} (-1)^r \dfrac{m!\,x^{m-r}}{(m-r)!\,a^{r+1}} \end{cases}$

522. $\displaystyle\int \frac{e^{ax}\,dx}{x} = \log x + \frac{ax}{1!} + \frac{a^2 x^2}{2 \cdot 2!} + \frac{a^3 x^3}{3 \cdot 3!} + \cdots$

523. $\displaystyle\int \frac{e^{ax}}{x^m}\,dx = -\frac{1}{m-1}\frac{e^{ax}}{x^{m-1}} + \frac{a}{m-1}\int \frac{e^{ax}}{x^{m-1}}\,dx$

524. $\displaystyle\int e^{ax} \log x\,dx = \frac{e^{ax} \log x}{a} - \frac{1}{a}\int \frac{e^{ax}}{x}\,dx$

525. $\displaystyle\int \frac{dx}{1 + e^x} = x - \log{(1 + e^x)} = \log\frac{e^x}{1 + e^x}$

526. $\displaystyle\int \frac{dx}{a + be^{px}} = \frac{x}{a} - \frac{1}{ap}\log{(a + be^{px})}$

527. $\displaystyle\int \frac{dx}{ae^{mx} + be^{-mx}} = \frac{1}{m\sqrt{ab}}\tan^{-1}\left(e^{mx}\sqrt{\frac{a}{b}}\right), \qquad (a > 0, b > 0)$

528. $\displaystyle\int \frac{dx}{ae^{mx} - be^{-mx}} = \begin{cases} \dfrac{1}{2m\sqrt{ab}}\log\dfrac{\sqrt{a}\,e^{mx} - \sqrt{b}}{\sqrt{a}\,e^{mx} + \sqrt{b}} \\[3mm] \qquad\qquad \text{or} \\[3mm] \dfrac{-1}{m\sqrt{ab}}\tanh^{-1}\left(\sqrt{\dfrac{a}{b}}\,e^{mx}\right), \qquad (a > 0, b > 0) \end{cases}$

INTEGRALS (Continued)

529. $\displaystyle\int (a^x - a^{-x})\, dx = \frac{a^x + a^{-x}}{\log a}$

530. $\displaystyle\int \frac{e^{ax}}{b + ce^{ax}}\, dx = \frac{1}{ac}\log(b + ce^{ax})$

531. $\displaystyle\int \frac{x\, e^{ax}}{(1 + ax)^2}\, dx = \frac{e^{ax}}{a^2(1 + ax)}$

532. $\displaystyle\int x\, e^{-x^2}\, dx = -\tfrac{1}{2} e^{-x^2}$

533. $\displaystyle\int e^{ax}[\sin(bx)]\, dx = \frac{e^{ax}[a\sin(bx) - b\cos(bx)]}{a^2 + b^2}$

534. $\displaystyle\int e^{ax}[\sin(bx)][\sin(cx)]\, dx = \frac{e^{ax}[(b - c)\sin(b - c)x + a\cos(b - c)x]}{2[a^2 + (b - c)^2]}$

$$- \frac{e^{ax}[(b + c)\sin(b + c)x + a\cos(b + c)x]}{2[a^2 + (b + c)^2]}$$

535. $\displaystyle\int e^{ax}[\sin(bx)][\cos(cx)]\, dx = \left\{ \begin{array}{l} \dfrac{e^{ax}[a\sin(b - c)x - (b - c)\cos(b - c)x]}{2[a^2 + (b - c)^2]} \\[2mm] \quad + \dfrac{e^{ax}[a\sin(b + c)x - (b + c)\cos(b + c)x]}{2[a^2 + (b + c)^2]} \\[4mm] \text{or} \\[2mm] \dfrac{e^{ax}}{\rho}[(a\sin bx - b\cos bx)[\cos(cx - \alpha)] \\[2mm] \quad\quad\quad\quad\quad\quad - c(\sin bx)\sin(cx - \alpha)] \\[4mm] \text{where} \\[2mm] \rho = \sqrt{(a^2 + b^2 - c^2)^2 + 4a^2c^2}, \\[2mm] \rho\cos\alpha = a^2 + b^2 - c^2, \quad \rho\sin\alpha = 2ac \end{array} \right.$

536. $\displaystyle\int e^{ax}[\sin(bx)][\sin(bx + c)]\, dx$

$$= \frac{e^{ax}\cos c}{2a} - \frac{e^{ax}[a\cos(2bx + c) + 2b\sin(2bx + c)]}{2(a^2 + 4b^2)}$$

537. $\displaystyle\int e^{ax}[\sin(bx)][\cos(bx + c)]\, dx$

$$= \frac{-e^{ax}\sin c}{2a} + \frac{e^{ax}[a\sin(2bx + c) - 2b\cos(2bx + c)]}{2(a^2 + 4b^2)}$$

538. $\displaystyle\int e^{ax}[\cos(bx)]\, dx = \frac{e^{ax}}{a^2 + b^2}[a\cos(bx) + b\sin(bx)]$

INTEGRALS (Continued)

539. $\displaystyle\int e^{ax}[\cos{(bx)}][\cos{(cx)}]\,dx = \frac{e^{ax}[(b-c)\sin{(b-c)x} + a\cos{(b-c)x}]}{2[a^2 + (b-c)^2]}$

$$+ \frac{e^{ax}[(b+c)\sin{(b+c)x} + a\cos{(b+c)x}]}{2[a^2 + (b+c)^2]}$$

540. $\displaystyle\int e^{ax}[\cos{(bx)}][\cos{(bx+c)}]\,dx$

$$= \frac{e^{ax}\cos{c}}{2a} + \frac{e^{ax}[a\cos{(2bx+c)} + 2b\sin{(2bx+c)}]}{2(a^2 + 4b^2)}$$

541. $\displaystyle\int e^{ax}[\cos{(bx)}][\sin{(bx+c)}]\,dx$

$$= \frac{e^{ax}\sin{c}}{2a} + \frac{e^{ax}[a\sin{(2bx+c)} - 2b\cos{(2bx+c)}]}{2(a^2 + 4b^2)}$$

542. $\displaystyle\int e^{ax}[\sin^n{bx}]\,dx = \frac{1}{a^2 + n^2b^2}\left[(a\sin{bx} - nb\cos{bx})e^{ax}\sin^{n-1}{bx}\right.$

$$\left. + n(n-1)b^2\int e^{ax}[\sin^{n-2}{bx}]\,dx\right]$$

543. $\displaystyle\int e^{ax}[\cos^n{bx}]\,dx = \frac{1}{a^2 + n^2b^2}\left[(a\cos{bx} + nb\sin{bx})e^{ax}\cos^{n-1}{bx}\right.$

$$\left. + n(n-1)b^2\int e^{ax}[\cos^{n-2}{bx}]\,dx\right]$$

544. $\displaystyle\int x^m e^x \sin{x}\,dx = \frac{1}{2}x^m e^x(\sin{x} - \cos{x}) - \frac{m}{2}\int x^{m-1} e^x \sin{x}\,dx$

$$+ \frac{m}{2}\int x^{m-1} e^x \cos{x}\,dx$$

545. $\displaystyle\int x^m e^{ax}[\sin{bx}]\,dx = \begin{cases} x^m e^{ax}\dfrac{a\sin{bx} - b\cos{bx}}{a^2 + b^2} \\[2ex] \qquad - \dfrac{m}{a^2 + b^2}\displaystyle\int x^{m-1} e^{ax}(a\sin{bx} - b\cos{bx})\,dx \\[2ex] \qquad\qquad\text{or} \\[1ex] e^{ax}\displaystyle\sum_{r=0}^{m} \dfrac{(-1)^r m!\,x^{m-r}}{\rho^{r+1}(m-r)!}\sin{[bx - (r+1)\alpha]} \\[2ex] \qquad\qquad\text{where} \\[1ex] \rho = \sqrt{a^2 + b^2}, \qquad \rho\cos{\alpha} = a, \qquad \rho\sin{\alpha} = b \end{cases}$

546. $\displaystyle\int x^m e^x \cos{x}\,dx = \frac{1}{2}x^m e^x(\sin{x} + \cos{x})$

$$- \frac{m}{2}\int x^{m-1} e^x \sin{x}\,dx - \frac{m}{2}\int x^{m-1} e^x \cos{x}\,dx$$

INTEGRALS (Continued)

547. $\int x^m e^{ax} \cos bx \, dx =$

$$\begin{cases} x^m e^{ax} \dfrac{a \cos bx + b \sin bx}{a^2 + b^2} \\ \qquad - \dfrac{m}{a^2 + b^2} \int x^{m-1} e^{ax}(a \cos bx + b \sin bx) \, dx \\ \qquad \text{or} \\ e^{ax} \displaystyle\sum_{r=0}^{m} \dfrac{(-1)^r m! x^{m-r}}{\rho^{r+1}(m-r)!} \cos\left[bx - (r+1)\alpha\right] \\ \qquad \text{where} \\ \rho = \sqrt{a^2 + b^2}, \qquad \rho \cos \alpha = a, \qquad \rho \sin \alpha = b \end{cases}$$

548. $\int e^{ax}(\cos^m x)(\sin^n x) \, dx =$

$$\begin{cases} \dfrac{e^{ax} \cos^{m-1} x \sin^n x [a \cos x + (m+n)\sin x]}{(m+n)^2 + a^2} \\ \qquad - \dfrac{na}{(m+n)^2 + a^2} \int e^{ax}(\cos^{m-1} x)(\sin^{n-1} x) \, dx \\ \qquad + \dfrac{(m-1)(m+n)}{(m+n)^2 + a^2} \int e^{ax}(\cos^{m-2} x)(\sin^n x) \, dx \\ \qquad \text{or} \\ \dfrac{e^{ax} \cos^m x \sin^{n-1} x [a \sin x - (m+n)\cos x]}{(m+n)^2 + a^2} \\ \qquad + \dfrac{ma}{(m+n)^2 + a^2} \int e^{ax}(\cos^{m-1} x)(\sin^{n-1} x) \, dx \\ \qquad + \dfrac{(n-1)(m+n)}{(m+n)^2 + a^2} \int e^{ax}(\cos^m x)(\sin^{n-2} x) \, dx \\ \qquad \text{or} \\ \dfrac{e^{ax}(\cos^{m-1} x)(\sin^{n-1} x)(a \sin x \cos x + m \sin^2 x - n \cos^2 x)}{(m+n)^2 + a^2} \\ \qquad + \dfrac{m(m-1)}{(m+n)^2 + a^2} \int e^{ax}(\cos^{m-2} x)(\sin^n x) \, dx \\ \qquad + \dfrac{n(n-1)}{(m+n)^2 + a^2} \int e^{ax}(\cos^m x)(\sin^{n-2} x) \, dx \\ \qquad \text{or} \\ \dfrac{e^{ax}(\cos^{m-1} x)(\sin^{n-1} x)(a \cos x \sin x + m \sin^2 x - n \cos^2 x)}{(m+n)^2 + a^2} \\ \qquad + \dfrac{m(m-1)}{(m+n)^2 + a^2} \int e^{ax}(\cos^{m-2} x)(\sin^{n-2} x) \, dx \\ \qquad + \dfrac{(n-m)(n+m-1)}{(m+n)^2 + a^2} \int e^{ax}(\cos^m x)(\sin^{n-2} x) \, dx \end{cases}$$

INTEGRALS (Continued)

549. $\displaystyle\int x\, e^{ax}(\sin bx)\, dx = \frac{x\, e^{ax}}{a^2 + b^2}(a \sin bx - b \cos bx)$

$$- \frac{e^{ax}}{(a^2 + b^2)^2}[(a^2 - b^2)\sin bx - 2ab \cos bx]$$

550. $\displaystyle\int x\, e^{ax}(\cos bx)\, dx = \frac{x\, e^{ax}}{a^2 + b^2}(a \cos bx + b \sin bx)$

$$- \frac{e^{ax}}{(a^2 + b^2)^2}[(a^2 - b^2)\cos bx + 2ab \sin bx]$$

551. $\displaystyle\int \frac{e^{ax}}{\sin^n x}\, dx = - \frac{e^{ax}[a \sin x + (n - 2)\cos x]}{(n - 1)(n - 2)\sin^{n-1} x} + \frac{a^2 + (n - 2)^2}{(n - 1)(n - 2)}\int \frac{e^{ax}}{\sin^{n-2} x}\, dx$

552. $\displaystyle\int \frac{e^{ax}}{\cos^n x}\, dx = - \frac{e^{ax}[a \cos x - (n - 2)\sin x]}{(n - 1)(n - 2)\cos^{n-1} x} + \frac{a^2 + (n - 2)^2}{(n - 1)(n - 2)}\int \frac{e^{ax}}{\cos^{n-2} x}\, dx$

553. $\displaystyle\int e^{ax} \tan^n x\, dx = e^{ax}\frac{\tan^{n-1} x}{n - 1} - \frac{a}{n - 1}\int e^{ax} \tan^{n-1} x\, dx - \int e^{ax} \tan^{n-2} x\, dx$

HYPERBOLIC FORMS

554. $\displaystyle\int (\sinh x)\, dx = \cosh x$

555. $\displaystyle\int (\cosh x)\, dx = \sinh x$

556. $\displaystyle\int (\tanh x)\, dx = \log \cosh x$

557. $\displaystyle\int (\coth x)\, dx = \log \sinh x$

558. $\displaystyle\int (\operatorname{sech} x)\, dx = \tan^{-1}(\sinh x)$

559. $\displaystyle\int \operatorname{csch} x\, dx = \log \tanh \left(\frac{x}{2}\right)$

560. $\displaystyle\int x(\sinh x)\, dx = x \cosh x - \sinh x$

561. $\displaystyle\int x^n(\sinh x)\, dx = x^n \cosh x - n \int x^{n-1}(\cosh x)\, dx$

562. $\displaystyle\int x(\cosh x)\, dx = x \sinh x - \cosh x$

563. $\displaystyle\int x^n(\cosh x)\, dx = x^n \sinh x - n \int x^{n-1}(\sinh x)\, dx$

INTEGRALS (Continued)

564. $\displaystyle\int (\text{sech } x)(\tanh x)\, dx = -\text{sech } x$

565. $\displaystyle\int (\text{csch } x)(\coth x)\, dx = -\text{csch } x$

566. $\displaystyle\int (\sinh^2 x)\, dx = \frac{\sinh 2x}{4} - \frac{x}{2}$

567. $\displaystyle\int (\sinh^m x)(\cosh^n x)\, dx =$
$$\begin{cases} \dfrac{1}{m+n}(\sinh^{m+1} x)(\cosh^{n-1} x) \\[2mm] \qquad\qquad + \dfrac{n-1}{m+n}\displaystyle\int (\sinh^m x)(\cosh^{n-2} x)\, dx \\[2mm] \qquad\qquad\text{or} \\[2mm] \dfrac{1}{m+n}\sinh^{m-1} x \cosh^{n+1} x \\[2mm] \qquad\qquad - \dfrac{m-1}{m+n}\displaystyle\int (\sinh^{m-2} x)(\cosh^n x)\, dx, \qquad (m+n \neq 0) \end{cases}$$

568. $\displaystyle\int \frac{dx}{(\sinh^m x)(\cosh^n x)} =$
$$\begin{cases} -\dfrac{1}{(m-1)(\sinh^{m-1} x)(\cosh^{n-1} x)} \\[2mm] \qquad - \dfrac{m+n-2}{m-1}\displaystyle\int \dfrac{dx}{(\sinh^{m-2} x)(\cosh^n x)}, \qquad (m \neq 1) \\[2mm] \qquad\qquad\text{or} \\[2mm] \dfrac{1}{(n-1)\sinh^{m-1} x \cosh^{n-1} x} \\[2mm] \qquad + \dfrac{m+n-2}{n-1}\displaystyle\int \dfrac{dx}{(\sinh^m x)(\cosh^{n-2} x)}, \qquad (n \neq 1) \end{cases}$$

569. $\displaystyle\int (\tanh^2 x)\, dx = x - \tanh x$

570. $\displaystyle\int (\tanh^n x)\, dx = -\frac{\tanh^{n-1} x}{n-1} + \int (\tanh^{n-2} x)\, dx, \qquad (n \neq 1)$

571. $\displaystyle\int (\text{sech}^2 x)\, dx = \tanh x$

572. $\displaystyle\int (\cosh^2 x)\, dx = \frac{\sinh 2x}{4} + \frac{x}{2}$

573. $\displaystyle\int (\coth^2 x)\, dx = x - \coth x$

574. $\displaystyle\int (\coth^n x)\, dx = -\frac{\coth^{n-1} x}{n-1} + \int \coth^{n-2} x\, dx, \qquad (n \neq 1)$

575. $\int (\text{csch}^2\, x)\, dx = -\text{ctnh}\, x$

576. $\int (\sinh mx)(\sinh nx)\, dx = \dfrac{\sinh (m+n)x}{2(m+n)} - \dfrac{\sinh (m-n)x}{2(m-n)}, \quad (m^2 \neq n^2)$

577. $\int (\cosh mx)(\cosh nx)\, dx = \dfrac{\sinh (m+n)x}{2(m+n)} + \dfrac{\sinh (m-n)x}{2(m-n)}, \quad (m^2 \neq n^2)$

578. $\int (\sinh mx)(\cosh nx)\, dx = \dfrac{\cosh (m+n)x}{2(m+n)} + \dfrac{\cosh (m-n)x}{2(m-n)}, \quad (m^2 \neq n^2)$

579. $\int \left(\sinh^{-1} \dfrac{x}{a}\right) dx = x \sinh^{-1} \dfrac{x}{a} - \sqrt{x^2 + a^2}, \quad (a > 0)$

580. $\int x\left(\sinh^{-1} \dfrac{x}{a}\right) dx = \left(\dfrac{x^2}{2} + \dfrac{a^2}{4}\right) \sinh^{-1} \dfrac{x}{a} - \dfrac{x}{4}\sqrt{x^2 + a^2}, \quad (a > 0)$

581. $\int x^n(\sinh^{-1} x)\, dx = \dfrac{x^{n+1}}{n+1} \sinh^{-1} x - \dfrac{1}{n+1} \int \dfrac{x^{n+1}}{(1+x^2)^{\frac{1}{2}}}\, dx, \quad (n \neq -1)$

582. $\int \left(\cosh^{-1} \dfrac{x}{a}\right) dx = \begin{cases} x \cosh^{-1} \dfrac{x}{a} - \sqrt{x^2 - a^2}, & \left(\cosh^{-1} \dfrac{x}{a} > 0\right) \\[2ex] \qquad\quad \text{or} \\[2ex] x \cosh^{-1} \dfrac{x}{a} + \sqrt{x^2 - a^2}, & \left(\cosh^{-1} \dfrac{x}{a} < 0\right), \quad (a > 0) \end{cases}$

583. $\int x\left(\cosh^{-1} \dfrac{x}{a}\right) dx = \dfrac{2x^2 - a^2}{4} \cosh^{-1} \dfrac{x}{a} - \dfrac{x}{4}(x^2 - a^2)^{\frac{1}{2}}$

584. $\int x^n(\cosh^{-1} x)\, dx = \dfrac{x^{n+1}}{n+1} \cosh^{-1} x - \dfrac{1}{n+1} \int \dfrac{x^{n+1}}{(x^2 - 1)^{\frac{1}{2}}}\, dx, \quad (n \neq -1)$

585. $\int \left(\tanh^{-1} \dfrac{x}{a}\right) dx = x \tanh^{-1} \dfrac{x}{a} + \dfrac{a}{2} \log (a^2 - x^2), \quad \left(\left|\dfrac{x}{a}\right| < 1\right)$

586. $\int \left(\coth^{-1} \dfrac{x}{a}\right) dx = x \coth^{-1} \dfrac{x}{a} + \dfrac{a}{2} \log (x^2 - a^2), \quad \left(\left|\dfrac{x}{a}\right| > 1\right)$

587. $\int x\left(\tanh^{-1} \dfrac{x}{a}\right) dx = \dfrac{x^2 - a^2}{2} \tanh^{-1} \dfrac{x}{a} + \dfrac{ax}{2}, \quad \left(\left|\dfrac{x}{a}\right| < 1\right)$

588. $\int x^n\left(\tanh^{-1} x\right) dx = \dfrac{x^{n+1}}{n+1} \tanh^{-1} x - \dfrac{1}{n+1} \int \dfrac{x^{n+1}}{1 - x^2}\, dx, \quad (n \neq -1)$

589. $\int x\left(\coth^{-1} \dfrac{x}{a}\right) dx = \dfrac{x^2 - a^2}{2} \coth^{-1} \dfrac{x}{a} + \dfrac{ax}{2}, \quad \left(\left|\dfrac{x}{a}\right| > 1\right)$

590. $\int x^n(\coth^{-1} x)\, dx = \dfrac{x^{n+1}}{n+1} \coth^{-1} x + \dfrac{1}{n+1} \int \dfrac{x^{n+1}}{x^2 - 1}\, dx, \quad (n \neq -1)$

INTEGRALS (Continued)

591. $\displaystyle\int (\text{sech}^{-1}x)\,dx = x\,\text{sech}^{-1}x + \sin^{-1}x$

592. $\displaystyle\int x\,\text{sech}^{-1}x\,dx = \frac{x^2}{2}\,\text{sech}^{-1}x - \frac{1}{2}\sqrt{1-x^2}$

593. $\displaystyle\int x^n\,\text{sech}^{-1}x\,dx = \frac{x^{n+1}}{n+1}\,\text{sech}^{-1}x + \frac{1}{n+1}\int \frac{x^n}{(1-x^2)^{\frac{1}{2}}}\,dx,\qquad (n \ne -1)$

594. $\displaystyle\int \text{csch}^{-1}x\,dx = x\,\text{csch}^{-1}x + \frac{x}{|x|}\sinh^{-1}x$

595. $\displaystyle\int x\,\text{csch}^{-1}x\,dx = \frac{x^2}{2}\,\text{csch}^{-1}x + \frac{1}{2}\frac{x}{|x|}\sqrt{1+x^2}$

596. $\displaystyle\int x^n\,\text{csch}^{-1}x\,dx = \frac{x^{n+1}}{n+1}\,\text{csch}^{-1}x + \frac{1}{n+1}\frac{x}{|x|}\int \frac{x^n}{(x^2+1)^{\frac{1}{2}}}\,dx,\qquad (n \ne -1)$

DEFINITE INTEGRALS

597. $\displaystyle\int_0^\infty x^{n-1}e^{-x}\,dx = \int_0^1 \left(\log\frac{1}{x}\right)^{n-1}\,dx = \frac{1}{n}\prod_{m=1}^\infty \frac{\left(1+\dfrac{1}{m}\right)^n}{1+\dfrac{n}{m}}$

$$= \Gamma(n),\, n \ne 0,\, -1,\, -2,\, -3,\ldots \qquad \text{(Gamma Function)}$$

598. $\displaystyle\int_0^\infty t^n p^{-t}\,dt = \frac{n!}{(\log p)^{n+1}},\qquad (n = 0,1,2,3,\ldots \text{ and } p > 0)$

599. $\displaystyle\int_0^\infty t^{n-1}e^{-(a+1)t}\,dt = \frac{\Gamma(n)}{(a+1)^n},\qquad (n > 0, a > -1)$

600. $\displaystyle\int_0^1 x^m\left(\log\frac{1}{x}\right)^n\,dx = \frac{\Gamma(n+1)}{(m+1)^{n+1}},\qquad (m > -1, n > -1)$

601. $\Gamma(n)$ is finite if $n > 0$, $\Gamma(n+1) = n\Gamma(n)$

602. $\Gamma(n)\cdot\Gamma(1-n) = \dfrac{\pi}{\sin n\pi}$

603. $\Gamma(n) = (n-1)!$ if $n = $ integer > 0

604. $\Gamma(\tfrac{1}{2}) = 2\displaystyle\int_0^\infty e^{-t^2}\,dt = \sqrt{\pi} = 1.7724538509\cdots = (-\tfrac{1}{2})!$

605. $\Gamma(n+\tfrac{1}{2}) = \dfrac{1\cdot3\cdot5\ldots(2n-1)}{2^n}\sqrt{\pi}\qquad n = 1,2,3,\ldots$

606. $\Gamma(-n+\tfrac{1}{2}) = \dfrac{(-1)^n 2^n\sqrt{\pi}}{1\cdot3\cdot5\ldots(2n-1)}\qquad n = 1,2,3,\ldots$

DEFINITE INTEGRALS (Continued)

607. $\displaystyle\int_0^1 x^{m-1}(1-x)^{n-1}\,dx = \int_0^\infty \frac{x^{m-1}}{(1+x)^{m+n}}\,dx = \frac{\Gamma(m)\Gamma(n)}{\Gamma(m+n)} = B(m,n)$

(Beta function)

608. $B(m,n) = B(n,m) = \dfrac{\Gamma(m)\Gamma(n)}{\Gamma(m+n)}$, where m and n are any positive real numbers.

609. $\displaystyle\int_a^b (x-a)^m(b-x)^n\,dx = (b-a)^{m+n+1}\frac{\Gamma(m+1)\cdot\Gamma(n+1)}{\Gamma(m+n+2)}$,

$(m > -1, n > -1, b > a)$

610. $\displaystyle\int_1^\infty \frac{dx}{x^m} = \frac{1}{m-1}, \qquad [m>1]$

611. $\displaystyle\int_0^\infty \frac{dx}{(1+x)x^p} = \pi \csc p\pi, \qquad [p<1]$

612. $\displaystyle\int_0^\infty \frac{dx}{(1-x)x^p} = -\pi \cot p\pi, \qquad [p<1]$

613. $\displaystyle\int_0^\infty \frac{x^{p-1}\,dx}{1+x} = \frac{\pi}{\sin p\pi}$

$\qquad\qquad = B(p, 1-p) = \Gamma(p)\Gamma(1-p), \qquad [0<p<1]$

614. $\displaystyle\int_0^\infty \frac{x^{m-1}\,dx}{1+x^n} = \frac{\pi}{n\sin\dfrac{m\pi}{n}}, \qquad [0<m<n]$

615. $\displaystyle\int_0^\infty \frac{x^a\,dx}{(m+x^b)^c} = \frac{m^{\frac{a+1-bc}{b}}}{b}\left[\frac{\Gamma\!\left(\dfrac{a+1}{b}\right)\Gamma\!\left(c-\dfrac{a+1}{b}\right)}{\Gamma(c)}\right]$

$\left(a>-1, b>0, m>0, c>\dfrac{a+1}{b}\right)$

616. $\displaystyle\int_0^\infty \frac{dx}{(1+x)\sqrt{x}} = \pi$

617. $\displaystyle\int_0^\infty \frac{a\,dx}{a^2+x^2} = \frac{\pi}{2}$, if $a>0$; 0, if $a=0$; $-\dfrac{\pi}{2}$, if $a<0$

618. $\displaystyle\int_0^a (a^2-x^2)^{\frac{n}{2}}\,dx = \frac{1}{2}\int_{-a}^a (a^2-x^2)^{\frac{n}{2}}\,dx = \frac{1\cdot3\cdot5\ldots n}{2\cdot4\cdot6\ldots(n+1)}\cdot\frac{\pi}{2}\cdot a^{n+1}$ (n odd)

619. $\displaystyle\int_0^a x^m(a^2-x^2)^{\frac{n}{2}}\,dx = \begin{cases} \dfrac{1}{2}a^{m+n+1}B\!\left(\dfrac{m+1}{2}, \dfrac{n+2}{2}\right) \\[2mm] \text{or} \\[2mm] \dfrac{1}{2}a^{m+n+1}\dfrac{\Gamma\!\left(\dfrac{m+1}{2}\right)\Gamma\!\left(\dfrac{n+2}{2}\right)}{\Gamma\!\left(\dfrac{m+n+3}{2}\right)} \end{cases}$

DEFINITE INTEGRALS (Continued)

620. $\displaystyle\int_0^{\pi/2} (\sin^n x)\, dx = \begin{cases} \displaystyle\int_0^{\pi/2} (\cos^n x)\, dx \\ \\ \text{or} \\ \\ \dfrac{1 \cdot 3 \cdot 5 \cdot 7 \ldots (n-1)}{2 \cdot 4 \cdot 6 \cdot 8 \ldots (n)} \dfrac{\pi}{2}, \quad (n \text{ an even integer, } n \neq 0) \\ \\ \text{or} \\ \\ \dfrac{2 \cdot 4 \cdot 6 \cdot 8 \ldots (n-1)}{1 \cdot 3 \cdot 5 \cdot 7 \ldots (n)}, \quad (n \text{ an odd integer, } n \neq 1) \\ \\ \text{or} \\ \\ \dfrac{\sqrt{\pi}}{2} \dfrac{\Gamma\left(\dfrac{n+1}{2}\right)}{\Gamma\left(\dfrac{n}{2}+1\right)}, \quad (n > -1) \end{cases}$

621. $\displaystyle\int_0^\infty \frac{\sin mx\, dx}{x} = \frac{\pi}{2}, \text{ if } m > 0; 0, \text{ if } m = 0; -\frac{\pi}{2}, \text{ if } m < 0$

622. $\displaystyle\int_0^\infty \frac{\cos x\, dx}{x} = \infty$

623. $\displaystyle\int_0^\infty \frac{\tan x\, dx}{x} = \frac{\pi}{2}$

624. $\displaystyle\int_0^\pi \sin ax \cdot \sin bx\, dx = \int_0^\pi \cos ax \cdot \cos bx\, dx = 0, \quad (a \neq b; a, b \text{ integers})$

625. $\displaystyle\int_0^{\pi/a} [\sin (ax)][\cos (ax)]\, dx = \int_0^\pi [\sin (ax)][\cos (ax)]\, dx = 0$

626. $\displaystyle\int_0^\pi [\sin (ax)][\cos (bx)]\, dx = \frac{2a}{a^2 - b^2}, \text{ if } a - b \text{ is odd, or } 0 \text{ if } a - b \text{ is even}$

627. $\displaystyle\int_0^\infty \frac{\sin x \cos mx\, dx}{x}$

$$= 0, \text{ if } m < -1 \text{ or } m > 1; \frac{\pi}{4}, \text{ if } m = \pm 1; \frac{\pi}{2}, \text{ if } m^2 < 1$$

628. $\displaystyle\int_0^\infty \frac{\sin ax \sin bx}{x^2}\, dx = \frac{\pi a}{2}, \quad (a \leq b)$

629. $\displaystyle\int_0^\pi \sin^2 mx\, dx = \int_0^\pi \cos^2 mx\, dx = \frac{\pi}{2}$

630. $\displaystyle\int_0^\infty \frac{\sin^2 (px)}{x^2}\, dx = \frac{\pi p}{2}$

DEFINITE INTEGRALS (Continued)

631. $\displaystyle\int_0^\infty \frac{\sin x}{x^p}\, dx = \frac{\pi}{2\Gamma(p)\sin(p\pi/2)}, \qquad 0 < p < 1$

632. $\displaystyle\int_0^\infty \frac{\cos x}{x^p}\, dx = \frac{\pi}{2\Gamma(p)\cos(p\pi/2)}, \qquad 0 < p < 1$

633. $\displaystyle\int_0^\infty \frac{1 - \cos px}{x^2}\, dx = \frac{\pi p}{2}$

634. $\displaystyle\int_0^\infty \frac{\sin px \cos qx}{x}\, dx = \left\{0, \quad q > p > 0; \quad \frac{\pi}{2}, \; p > q > 0; \quad \frac{\pi}{4}, \; p = q > 0\right\}$

635. $\displaystyle\int_0^\infty \frac{\cos(mx)}{x^2 + a^2}\, dx = \frac{\pi}{2|a|}\, e^{-|ma|}$

636. $\displaystyle\int_0^\infty \cos(x^2)\, dx = \int_0^\infty \sin(x^2)\, dx = \frac{1}{2}\sqrt{\frac{\pi}{2}}$

637. $\displaystyle\int_0^\infty \sin ax^n\, dx = \frac{1}{na^{1/n}}\, \Gamma(1/n)\sin\frac{\pi}{2n}, \qquad n > 1$

638. $\displaystyle\int_0^\infty \cos ax^n\, dx = \frac{1}{na^{1/n}}\, \Gamma(1/n)\cos\frac{\pi}{2n}, \qquad n > 1$

639. $\displaystyle\int_0^\infty \frac{\sin x}{\sqrt{x}}\, dx = \int_0^\infty \frac{\cos x}{\sqrt{x}}\, dx = \sqrt{\frac{\pi}{2}}$

640. (a) $\displaystyle\int_0^\infty \frac{\sin^3 x}{x}\, dx = \frac{\pi}{4}$ (b) $\displaystyle\int_0^\infty \frac{\sin^3 x}{x^2}\, dx\, \frac{3}{4}\log 3$

641. $\displaystyle\int_0^\infty \frac{\sin^3 x}{x^3}\, dx = \frac{3\pi}{8}$

642. $\displaystyle\int_0^\infty \frac{\sin^4 x}{x^4}\, dx = \frac{\pi}{3}$

643. $\displaystyle\int_0^{\pi/2} \frac{dx}{1 + a\cos x} = \frac{\cos^{-1} a}{\sqrt{1 - a^2}}, \qquad (a < 1)$

644. $\displaystyle\int_0^\pi \frac{dx}{a + b\cos x} = \frac{\pi}{\sqrt{a^2 - b^2}}, \qquad (a > b \geq 0)$

645. $\displaystyle\int_0^{2\pi} \frac{dx}{1 + a\cos x} = \frac{2\pi}{\sqrt{1 - a^2}}, \qquad (a^2 < 1)$

646. $\displaystyle\int_0^\infty \frac{\cos ax - \cos bx}{x}\, dx = \log\frac{b}{a}$

647. $\displaystyle\int_0^{\pi/2} \frac{dx}{a^2 \sin^2 x + b^2 \cos^2 x} = \frac{\pi}{2ab}$

DEFINITE INTEGRALS (Continued)

648. $\displaystyle\int_0^{\pi/2} \frac{dx}{(a^2 \sin^2 x + b^2 \cos^2 x)^2} = \frac{\pi(a^2 + b^2)}{4a^3 b^3}$, $(a, b > 0)$

649. $\displaystyle\int_0^{\pi/2} \sin^{n-1} x \cos^{m-1} x\, dx = \frac{1}{2} B\left(\frac{n}{2}, \frac{m}{2}\right)$, m and n positive integers

650. $\displaystyle\int_0^{\pi/2} (\sin^{2n+1} \theta)\, d\theta = \frac{2 \cdot 4 \cdot 6 \ldots (2n)}{1 \cdot 3 \cdot 5 \ldots (2n+1)}$, $(n = 1, 2, 3 \ldots)$

651. $\displaystyle\int_0^{\pi/2} (\sin^{2n} \theta)\, d\theta = \frac{1 \cdot 3 \cdot 5 \ldots (2n-1)}{2 \cdot 4 \ldots (2n)}\left(\frac{\pi}{2}\right)$, $(n = 1, 2, 3 \ldots)$

652. $\displaystyle\int_0^{\pi/2} \frac{x}{\sin x}\, dx = 2\left\{\frac{1}{1^2} - \frac{1}{3^2} + \frac{1}{5^2} - \frac{1}{7^2} + \cdots\right\}$

653. $\displaystyle\int_0^{\pi/2} \frac{dx}{1 + \tan^m x} = \frac{\pi}{4}$

654. $\displaystyle\int_0^{\pi/2} \sqrt{\cos\theta}\, d\theta = \frac{(2\pi)^{\frac{3}{2}}}{[\Gamma(\frac{1}{4})]^2}$

655. $\displaystyle\int_0^{\pi/2} (\tan^h \theta)\, d\theta = \frac{\pi}{2\cos\left(\dfrac{h\pi}{2}\right)}$, $(0 < h < 1)$

656. $\displaystyle\int_0^{\infty} \frac{\tan^{-1}(ax) - \tan^{-1}(bx)}{x}\, dx = \frac{\pi}{2}\log\frac{a}{b}$, $(a, b > 0)$

657. The area enclosed by a curve defined through the equation $x^{\frac{b}{c}} + y^{\frac{b}{c}} = a^{\frac{b}{c}}$ where $a > 0$, c a positive odd integer and b a positive even integer is given by

$$\frac{\left[\Gamma\left(\dfrac{c}{b}\right)\right]^2}{\Gamma\left(\dfrac{2c}{b}\right)}\left(\frac{2ca^2}{b}\right)$$

658. $I = \displaystyle\iiint_R x^{h-1} y^{m-1} z^{n-1}\, dv$, where R denotes the region of space bounded by

the co-ordinate planes and that portion of the surface $\left(\dfrac{x}{a}\right)^p + \left(\dfrac{y}{b}\right)^q + \left(\dfrac{z}{c}\right)^k = 1$,

which lies in the first octant and where $h, m, n, p, q, k, a, b, c$, denote positive real numbers is given by

$$\int_0^a x^{h-1}\, dx \int_0^{b\left[1 - \left(\frac{x}{a}\right)^p\right]^{\frac{1}{q}}} y^m\, dy \int_0^{c\left[1 - \left(\frac{x}{a}\right)^p - \left(\frac{y}{b}\right)^q\right]^{\frac{1}{k}}} z^{n-1}\, dz$$

$$= \frac{a^h b^m c^n}{pqk}\, \frac{\Gamma\left(\dfrac{h}{p}\right)\Gamma\left(\dfrac{m}{q}\right)\Gamma\left(\dfrac{n}{k}\right)}{\Gamma\left(\dfrac{h}{p} + \dfrac{m}{q} + \dfrac{n}{k} + 1\right)}$$

659. $\displaystyle\int_0^\infty e^{-ax}\,dx = \frac{1}{a}$, $(a > 0)$

660. $\displaystyle\int_0^\infty \frac{e^{-ax} - e^{-bx}}{x}\,dx = \log\frac{b}{a}$, $(a, b > 0)$

661. $\displaystyle\int_0^\infty x^n e^{-ax}\,dx = \begin{cases} \dfrac{\Gamma(n+1)}{a^{n+1}}, & (n > -1, a > 0) \\ \qquad\text{or} \\ \dfrac{n!}{a^{n+1}}, & (a > 0, n \text{ positive integer}) \end{cases}$

662. $\displaystyle\int_0^\infty x^n \exp(-ax^p)\,dx = \frac{\Gamma(k)}{pa^k}$, $\left(n > -1, p > 0, a > 0, k = \dfrac{n+1}{p}\right)$

663. $\displaystyle\int_0^\infty e^{-a^2x^2}\,dx = \frac{1}{2a}\sqrt{\pi} = \frac{1}{2a}\Gamma\left(\frac{1}{2}\right)$, $(a > 0)$

663a. $\displaystyle\int_0^b e^{-ax^2}\,dx = \frac{1}{2}\sqrt{\frac{\pi}{a}}\ \text{erf}\,(b\sqrt{a})$ Error Function (see page 400)

663b. $\displaystyle\int_b^\infty e^{-ax^2}\,dx = \frac{1}{2}\sqrt{\frac{\pi}{a}}\ \text{erfc}\,(b\sqrt{a})$ Complimentary Error Function (see page 400)

664. $\displaystyle\int_0^\infty x e^{-x^2}\,dx = \tfrac{1}{2}$

665. $\displaystyle\int_0^\infty x^2 e^{-x^2}\,dx = \frac{\sqrt{\pi}}{4}$

666. $\displaystyle\int_0^\infty x^{2n} e^{-ax^2}\,dx = \frac{1 \cdot 3 \cdot 5 \ldots (2n-1)}{2^{n+1}a^n}\sqrt{\frac{\pi}{a}}$

667. $\displaystyle\int_0^\infty x^{2n+1} e^{-ax^2}\,dx = \frac{n!}{2a^{n+1}}$, $(a > 0)$

668. $\displaystyle\int_0^1 x^m e^{-ax}\,dx = \frac{m!}{a^{m+1}}\left[1 - e^{-a}\sum_{r=0}^m \frac{a^r}{r!}\right]$

669. $\displaystyle\int_0^\infty e^{\left(-x^2 - \frac{a^2}{x^2}\right)}\,dx = \frac{e^{-2a}\sqrt{\pi}}{2}$, $(a \geq 0)$

670. $\displaystyle\int_0^\infty e^{-nx}\sqrt{x}\,dx = \frac{1}{2n}\sqrt{\frac{\pi}{n}}$

DEFINITE INTEGRALS (Continued)

671. $\displaystyle\int_0^\infty \frac{e^{-nx}}{\sqrt{x}}\,dx = \sqrt{\frac{\pi}{n}}$

672. $\displaystyle\int_0^\infty e^{-ax}(\cos mx)\,dx = \frac{a}{a^2 + m^2}, \qquad (a > 0)$

673. $\displaystyle\int_0^\infty e^{-ax}(\sin mx)\,dx = \frac{m}{a^2 + m^2}, \qquad (a > 0)$

674. $\displaystyle\int_0^\infty x\,e^{-ax}[\sin (bx)]\,dx = \frac{2ab}{(a^2 + b^2)^2}, \qquad (a > 0)$

675. $\displaystyle\int_0^\infty x\,e^{-ax}[\cos (bx)]\,dx = \frac{a^2 - b^2}{(a^2 + b^2)^2}, \qquad (a > 0)$

676. $\displaystyle\int_0^\infty x^n\,e^{-ax}[\sin (bx)]\,dx = \frac{n![(a + ib)^{n+1} - (a - ib)^{n+1}]}{2i(a^2 + b^2)^{n+1}}, \qquad (i^2 = -1, a > 0)$

677. $\displaystyle\int_0^\infty x^n\,e^{-ax}[\cos (bx)]\,dx = \frac{n![(a - ib)^{n+1} + (a + ib)^{n+1}]}{2(a^2 + b^2)^{n+1}}, \qquad (i^2 = -1, a > 0)$

678. $\displaystyle\int_0^\infty \frac{e^{-ax}\sin x}{x}\,dx = \cot^{-1} a, \qquad (a > 0)$

679. $\displaystyle\int_0^\infty e^{-a^2x^2} \cos bx\,dx = \frac{\sqrt{\pi}}{2a}\exp\left(-\frac{b^2}{4a^2}\right), \qquad (ab \neq 0)$

680. $\displaystyle\int_0^\infty e^{-t\cos\phi}\,t^{b-1}\sin (t\sin\phi)\,dt = [\Gamma(b)]\sin (b\phi), \qquad \left(b > 0, -\frac{\pi}{2} < \phi < \frac{\pi}{2}\right)$

681. $\displaystyle\int_0^\infty e^{-t\cos\phi}\,t^{b-1}[\cos (t\sin\phi)]\,dt = [\Gamma(b)]\cos (b\phi), \qquad \left(b > 0, -\frac{\pi}{2} < \phi < \frac{\pi}{2}\right)$

682. $\displaystyle\int_0^\infty t^{b-1}\cos t\,dt = [\Gamma(b)]\cos\left(\frac{b\pi}{2}\right), \qquad (0 < b < 1)$

683. $\displaystyle\int_0^\infty t^{b-1}(\sin t)\,dt = [\Gamma(b)]\sin\left(\frac{b\pi}{2}\right), \qquad (0 < b < 1)$

684. $\displaystyle\int_0^1 (\log x)^n\,dx = (-1)^n \cdot n!$

685. $\displaystyle\int_0^1 \left(\log\frac{1}{x}\right)^{\frac{1}{2}}\,dx = \frac{\sqrt{\pi}}{2}$

686. $\displaystyle\int_0^1 \left(\log\frac{1}{x}\right)^{-\frac{1}{2}}\,dx = \sqrt{\pi}$

687. $\displaystyle\int_0^1 \left(\log\frac{1}{x}\right)^n\,dx = n!$

688. $\displaystyle\int_0^1 x\log (1 - x)\,dx = -\frac{3}{4}$

DEFINITE INTEGRALS (Continued)

689. $\displaystyle\int_0^1 x \log (1 + x)\, dx = \tfrac{1}{4}$

690. $\displaystyle\int_0^1 x^m (\log x)^n\, dx = \frac{(-1)^n n!}{(m + 1)^{n+1}}, \qquad m > -1, n = 0, 1, 2, \ldots$

If $n \neq 0, 1, 2, \ldots$ replace $n!$ by $\Gamma(n + 1)$.

691. $\displaystyle\int_0^1 \frac{\log x}{1 + x}\, dx = -\frac{\pi^2}{12}$

692. $\displaystyle\int_0^1 \frac{\log x}{1 - x}\, dx = -\frac{\pi^2}{6}$

693. $\displaystyle\int_0^1 \frac{\log (1 + x)}{x}\, dx = \frac{\pi^2}{12}$

694. $\displaystyle\int_0^1 \frac{\log (1 - x)}{x}\, dx = -\frac{\pi^2}{6}$

695. $\displaystyle\int_0^1 (\log x)[\log (1 + x)]\, dx = 2 - 2 \log 2 - \frac{\pi^2}{12}$

696. $\displaystyle\int_0^1 (\log x)[\log (1 - x)]\, dx = 2 - \frac{\pi^2}{6}$

697. $\displaystyle\int_0^1 \frac{\log x}{1 - x^2}\, dx = -\frac{\pi^2}{8}$

698. $\displaystyle\int_0^1 \log \left(\frac{1 + x}{1 - x}\right) \frac{dx}{x} = \frac{\pi^2}{4}$

699. $\displaystyle\int_0^1 \frac{\log x\, dx}{\sqrt{1 - x^2}} = -\frac{\pi}{2} \log 2$

700. $\displaystyle\int_0^1 x^m \left[\log \left(\frac{1}{x}\right) \right]^n dx = \frac{\Gamma(n + 1)}{(m + 1)^{n+1}}, \qquad \text{if } m + 1 > 0, n + 1 > 0$

701. $\displaystyle\int_0^1 \frac{(x^p - x^q)\, dx}{\log x} = \log \left(\frac{p + 1}{q + 1}\right), \qquad (p + 1 > 0, q + 1 > 0)$

702. $\displaystyle\int_0^1 \frac{dx}{\sqrt{\log \left(\frac{1}{x}\right)}} = \sqrt{\pi}$, (same as integral 686)

703. $\displaystyle\int_0^\infty \log \left(\frac{e^x + 1}{e^x - 1}\right) dx = \frac{\pi^2}{4}$

704. $\displaystyle\int_0^{\pi/2} (\log \sin x)\, dx = \int_0^{\pi/2} \log \cos x\, dx = -\frac{\pi}{2} \log 2$

705. $\displaystyle\int_0^{\pi/2} (\log \sec x)\, dx = \int_0^{\pi/2} \log \csc x\, dx = \frac{\pi}{2} \log 2$

DEFINITE INTEGRALS (Continued)

706. $\displaystyle\int_0^\pi x(\log \sin x)\,dx = -\frac{\pi^2}{2}\log 2$

707. $\displaystyle\int_0^{\pi/2} (\sin x)(\log \sin x)\,dx = \log 2 - 1$

708. $\displaystyle\int_0^{\pi/2} (\log \tan x)\,dx = 0$

709. $\displaystyle\int_0^\pi \log(a \pm b\cos x)\,dx = \pi\log\left(\frac{a + \sqrt{a^2 - b^2}}{2}\right), \quad (a \geq b)$

710. $\displaystyle\int_0^\pi \log(a^2 - 2ab\cos x + b^2)\,dx = \begin{cases} 2\pi\log a, & a \geq b > 0 \\ 2\pi\log b, & b \geq a > 0 \end{cases}$

711. $\displaystyle\int_0^\infty \frac{\sin ax}{\sinh bx}\,dx = \frac{\pi}{2b}\tanh\frac{a\pi}{2b}$

712. $\displaystyle\int_0^\infty \frac{\cos ax}{\cosh bx}\,dx = \frac{\pi}{2b}\operatorname{sech}\frac{a\pi}{2b}$

713. $\displaystyle\int_0^\infty \frac{dx}{\cosh ax} = \frac{\pi}{2a}$

714. $\displaystyle\int_0^\infty \frac{x\,dx}{\sinh ax} = \frac{\pi^2}{4a^2}$

715. $\displaystyle\int_0^\infty e^{-ax}(\cosh bx)\,dx = \frac{a}{a^2 - b^2}, \quad (0 \leq |b| < a)$

716. $\displaystyle\int_0^\infty e^{-ax}(\sinh bx)\,dx = \frac{b}{a^2 - b^2}, \quad (0 \leq |b| < a)$

717. $\displaystyle\int_0^\infty \frac{\sinh ax}{e^{bx} + 1}\,dx = \frac{\pi}{2b}\csc\frac{a\pi}{b} - \frac{1}{2a}$

718. $\displaystyle\int_0^\infty \frac{\sinh ax}{e^{bx} - 1}\,dx = \frac{1}{2a} - \frac{\pi}{2b}\cot\frac{a\pi}{b}$

719. $\displaystyle\int_0^{\pi/2} \frac{dx}{\sqrt{1 - k^2\sin^2 x}} = \frac{\pi}{2}\left[1 + \left(\frac{1}{2}\right)^2 k^2 + \left(\frac{1\cdot 3}{2\cdot 4}\right)^2 k^4\right.$

$\left. + \left(\frac{1\cdot 3\cdot 5}{2\cdot 4\cdot 6}\right)^2 k^6 + \cdots\right]$, if $k^2 < 1$

719a. $\displaystyle\int_0^{\frac{\pi}{2}} \frac{dx}{(1 - k^2 Sin^2 x)^{3/2}} = \frac{\pi}{2}\left[1 + \left(\frac{1}{2}\right)^2\cdot 3k^2 + \left(\frac{1\cdot 3}{2\cdot 4}\right)^2\cdot 5k^4 +\right.$

$\left.\left(\frac{1\cdot 3\cdot 5}{2\cdot 4\cdot 6}\right)^2\cdot 7k^6 + \cdots\right]$, if $k^2 < 1$

DEFINITE INTEGRALS (Continued)

720. $\displaystyle\int_0^{\pi/2} \sqrt{1 - k^2 \sin^2 x}\, dx = \frac{\pi}{2}\left[1 - \left(\frac{1}{2}\right)^2 k^2 - \left(\frac{1\cdot 3}{2\cdot 4}\right)^2 \frac{k^4}{3}\right.$

$$\left. - \left(\frac{1\cdot 3\cdot 5}{2\cdot 4\cdot 6}\right)^2 \frac{k^6}{5} - \cdots\right],\ \text{if } k^2 < 1$$

721. $\displaystyle\int_0^{\infty} e^{-x} \log x\, dx = -\gamma = -0.5772157\ldots$

722. $\displaystyle\int_0^{\infty} e^{-x^2} \log x\, dx = -\frac{\sqrt{\pi}}{4}(\gamma + 2\log 2)$

723. $\displaystyle\int_0^{\infty} \left(\frac{1}{1 - e^{-x}} - \frac{1}{x}\right) e^{-x}\, dx = \gamma = 0.5772157\ldots$ [Euler's Constant]

724. $\displaystyle\int_0^{\infty} \frac{1}{x}\left(\frac{1}{1 + x} - e^{-x}\right) dx = \gamma = 0.5772157\ldots$

For n even:

725. $\displaystyle\int\cos^n x\, dx = \frac{1}{2^{n-1}} \sum_{k=0}^{\frac{n}{2}-1} \binom{n}{k} \frac{\sin (n - 2k) x}{(n - 2k)} + \frac{1}{2^n}\binom{n}{\frac{n}{2}} x$

726. $\displaystyle\int\sin^n x\, dx = \frac{1}{2^{n-1}} \sum_{k=0}^{\frac{n}{2}-1} \binom{n}{k} \frac{\sin \left[(n - 2k)\left(\frac{\pi}{2} - x\right)\right]}{2k - n} + \frac{1}{2^n}\binom{n}{\frac{n}{2}} x$

For n odd:

727. $\displaystyle\int\cos^n x\, dx = \frac{1}{2^{n-1}} \sum_{k=0}^{\frac{n-1}{2}} \binom{n}{k} \frac{\sin (n - 2k) x}{(n - 2k)}$

728. $\displaystyle\int\sin^n x\, dx = \frac{1}{2^{n-1}} \sum_{k=0}^{\frac{n-1}{2}} \binom{n}{k} \frac{\sin \left[n - 2k)\left(\frac{\pi}{2} - x\right)\right]}{2k - n}$

SERIES EXPANSION

The expression in parentheses following certain of the series indicates the region of convergence. If not otherwise indicated it is to be understood that the series converges for all finite values of x.

BINOMIAL

$$(x + y)^n = x^n + nx^{n-1}y + \frac{n(n-1)}{2!} x^{n-2}y^2$$

$$+ \frac{n(n-1)(n-2)}{3!} x^{n-3}y^3 + \cdots \quad (y^2 < x^2)$$

$$(1 \pm x)^n = 1 \pm nx + \frac{n(n-1)x^2}{2!} \pm \frac{n(n-1)(n-2)x^3}{3!} + \cdots \text{ etc. } \quad (x^2 < 1)$$

$$(1 \pm x)^{-n} = 1 \mp nx + \frac{n(n+1)x^2}{2!} \mp \frac{n(n+1)(n+2)x^3}{3!} + \cdots \text{ etc. } \quad (x^2 < 1)$$

$$(1 \pm x)^{-1} = 1 \mp x + x^2 \mp x^3 + x^4 \mp x^5 + \cdots \qquad (x^2 < 1)$$
$$(1 \pm x)^{-2} = 1 \mp 2x + 3x^2 \mp 4x^3 + 5x^4 \mp 6x^5 + \cdots \qquad (x^2 < 1)$$

REVERSION OF SERIES

Let a series be represented by

$$y = a_1 x + a_2 x^2 + a_3 x^3 + a_4 x^4 + a_5 x^5 + a_6 x^6 + \cdots \quad (a_1 \neq 0)$$

to find the coefficients of the series

$$x = A_1 y + A_2 y^2 + A_3 y^3 + A_4 y^4 + \cdots$$

$$A_1 = \frac{1}{a_1} \qquad A_2 = -\frac{a_2}{a_1^3} \qquad A_3 = \frac{1}{a_1^5} (2a_2^2 - a_1 a_3)$$

$$A_4 = \frac{1}{a_1^7} (5a_1 a_2 a_3 - a_1^2 a_4 - 5a_2^3)$$

$$A_5 = \frac{1}{a_1^9} (6a_1^2 a_2 a_4 + 3a_1^2 a_3^2 + 14a_2^4 - a_1^3 a_5 - 21a_1 a_2^2 a_3)$$

$$A_6 = \frac{1}{a_1^{11}} (7a_1^3 a_2 a_5 + 7a_1^3 a_3 a_4 + 84a_1 a_2^3 a_3 - a_1^4 a_6 - 28a_1^2 a_2^2 a_4 - 28a_1^2 a_2 a_3^2 - 42a_2^5)$$

$$A_7 = \frac{1}{a_1^{13}} (8a_1^4 a_2 a_6 + 8a_1^4 a_3 a_5 + 4a_1^4 a_4^2 + 120a_1^2 a_2^3 a_4$$

$$+ 180a_1^2 a_2^2 a_3^2 + 132a_2^6 - a_1^5 a_7$$
$$- 36a_1^3 a_2^2 a_5 - 72a_1^3 a_2 a_3 a_4 - 12a_1^3 a_3^3 - 330a_1 a_2^4 a_3)$$

TAYLOR

1. $f(x) = f(a) + (x - a)f'(a) + \dfrac{(x - a)^2}{2!} f''(a) + \dfrac{(x - a)^3}{3!} f'''(a)$

$$+ \cdots + \frac{(x - a)^n}{n!} f^{(n)}(a) + \cdots \quad \text{(Taylor's Series)}$$

(Increment form)

2. $f(x + h) = f(x) + hf'(x) + \dfrac{h^2}{2!} f''(x) + \dfrac{h^3}{3!} f'''(x) + \cdots$

$$= f(h) + xf'(h) + \dfrac{x^2}{2!} f''(h) + \dfrac{x^3}{3!} f'''(h) + \cdots$$

3. If $f(x)$ is a function possessing derivatives of all orders throughout the interval $a \leqq x \leqq b$, then there is a value X, with $a < X < b$, such that

$$f(b) = f(a) + (b - a)f'(a) + \dfrac{(b - a)^2}{2!} f''(a) + \cdots$$

$$+ \dfrac{(b - a)^{n-1}}{(n - 1)!} f^{(n-1)}(a) + \dfrac{(b - a)^n}{n!} f^{(n)}(X)$$

$$f(a + h) = f(a) + hf'(a) + \dfrac{h^2}{2!} f''(a) + \cdots + \dfrac{h^{n-1}}{(n - 1)!} f^{(n-1)}(a)$$

$$+ \dfrac{h^n}{n!} f^{(n)}(a + \theta h), \quad b = a + h, 0 < \theta < 1.$$

or

$$f(x) = f(a) + (x - a)f'(a) + \dfrac{(x - a)^2}{2!} f''(a) + \cdots + (x - a)^{n-1}\dfrac{f^{(n-1)}(a)}{(n - 1)!} + R_n,$$

where

$$R_n = \dfrac{f^{(n)}[a + \theta \cdot (x - a)]}{n!} (x - a)^n, \quad 0 < \theta < 1.$$

The above forms are known as Taylor's series with the remainder term.

4. *Taylor's series for a function of two variables*

If $\left(h \dfrac{\partial}{\partial x} + k \dfrac{\partial}{\partial y}\right) f(x, y) = h \dfrac{\partial f(x, y)}{\partial x} + k \dfrac{\partial f(x, y)}{\partial y};$

$$\left(h \dfrac{\partial}{\partial x} + k \dfrac{\partial}{\partial y}\right)^2 f(x, y) = h^2 \dfrac{\partial^2 f(x, y)}{\partial x^2} + 2hk \dfrac{\partial^2 f(x, y)}{\partial x \partial y} + k^2 \dfrac{\partial^2 f(x, y)}{\partial y^2}$$

etc., and if $\left(h \dfrac{\partial}{\partial x} + k \dfrac{\partial}{\partial y}\right)^n f(x, y)\Big|_{\substack{x=a \\ y=b}}$ with the bar and subscripts means that after differentiation we are to replace x by a and y by b,

$$f(a + h, b + k) = f(a, b) + \left(h \dfrac{\partial}{\partial x} + k \dfrac{\partial}{\partial y}\right) f(x, y)\Big|_{\substack{x=a \\ y=b}} + \cdots$$

$$+ \dfrac{1}{n!} \left(h \dfrac{\partial}{\partial x} + k \dfrac{\partial}{\partial y}\right)^n f(x, y)\Big|_{\substack{x=a \\ y=b}} + \cdots$$

MACLAURIN

$$f(x) = f(0) + xf'(0) + \dfrac{x^2}{2!} f''(0) + \dfrac{x^3}{3!} f'''(0) + \cdots + x^{n-1}\dfrac{f^{(n-1)}(0)}{(n - 1)!} + R_n,$$

where

$$R_n = \dfrac{x^n f^{(n)}(\theta x)}{n!}, \quad 0 < \theta < 1.$$

Series

EXPONENTIAL

$$e = 1 + \frac{1}{1!} + \frac{1}{2!} + \frac{1}{3!} + \frac{1}{4!} + \cdots$$

$$e^x = 1 + x + \frac{x^2}{2!} + \frac{x^3}{3!} + \frac{x^4}{4!} + \cdots \qquad \text{(all real values of } x)$$

$$a^x = 1 + x \log_e a + \frac{(x \log_e a)^2}{2!} + \frac{(x \log_e a)^3}{3!} + \cdots$$

$$e^x = e^a \left[1 + (x - a) + \frac{(x - a)^2}{2!} + \frac{(x - a)^3}{3!} + \cdots \right]$$

LOGARITHMIC

$$\log_e x = \frac{x - 1}{x} + \frac{1}{2} \left(\frac{x - 1}{x} \right)^2 + \frac{1}{3} \left(\frac{x - 1}{x} \right)^3 + \cdots \qquad (x > \tfrac{1}{2})$$

$$\log_e x = (x - 1) - \tfrac{1}{2}(x - 1)^2 + \tfrac{1}{3}(x - 1)^3 - \cdots \qquad (2 \geq x > 0)$$

$$\log_e x = 2 \left[\frac{x - 1}{x + 1} + \frac{1}{3} \left(\frac{x - 1}{x + 1} \right)^3 + \frac{1}{5} \left(\frac{x - 1}{x + 1} \right)^5 + \cdots \right] \qquad (x > 0)$$

$$\log_e (1 + x) = x - \tfrac{1}{2}x^2 + \tfrac{1}{3}x^3 - \tfrac{1}{4}x^4 + \cdots \qquad (-1 < x \leq 1)$$

$$\log_e (n + 1) - \log_e (n - 1) = 2 \left[\frac{1}{n} + \frac{1}{3n^3} + \frac{1}{5n^5} + \cdots \right]$$

$$\log_e (a + x) = \log_e a + 2 \left[\frac{x}{2a + x} + \frac{1}{3} \left(\frac{x}{2a + x} \right)^3 + \frac{1}{5} \left(\frac{x}{2a + x} \right)^5 + \cdots \right]$$

$$(a > 0, -a < x < +\infty)$$

$$\log_e \frac{1 + x}{1 - x} = 2 \left[x + \frac{x^3}{3} + \frac{x^5}{5} + \cdots + \frac{x^{2n-1}}{2n - 1} + \cdots \right], \qquad -1 < x < 1$$

$$\log_e x = \log_e a + \frac{(x - a)}{a} - \frac{(x - a)^2}{2a^2} + \frac{(x - a)^3}{3a^3} - + \cdots, \qquad 0 < x \leq 2a$$

TRIGONOMETRIC

$$\sin x = x - \frac{x^3}{3!} + \frac{x^5}{5!} - \frac{x^7}{7!} + \cdots \qquad \text{(all real values of } x)$$

$$\cos x = 1 - \frac{x^2}{2!} + \frac{x^4}{4!} - \frac{x^6}{6!} + \cdots \qquad \text{(all real values of } x)$$

$$\tan x = x + \frac{x^3}{3} + \frac{2x^5}{15} + \frac{17x^7}{315} + \frac{62x^9}{2835} + \cdots + \frac{(-1)^{n-1} 2^{2n} (2^{2n} - 1) B_{2n}}{(2n)!} x^{2n-1} + \cdots,$$

$$\left[x^2 < \frac{\pi^2}{4}, \text{ and } B_n \text{ represents the } n\text{'th Bernoulli number.} \right]$$

$$\cot x = \frac{1}{x} - \frac{x}{3} - \frac{x^3}{45} - \frac{2x^5}{945} - \frac{x^7}{4725} - \cdots - \frac{(-1)^{n+1} 2^{2n}}{(2n)!} B_{2n} x^{2n-1} - \cdots,$$

$$[x^2 < \pi^2, \text{ and } B_n \text{ represents the } n\text{'th Bernoulli number.}]$$

$$\sec x = 1 + \frac{x^2}{2} + \frac{5}{24} x^4 + \frac{61}{720} x^6 + \frac{277}{8064} x^8 + \cdots + \frac{(-1)^n}{(2n)!} E_{2n} x^{2n} + \cdots,$$

$$\left[x^2 < \frac{\pi^2}{4}, \text{ and } E_n \text{ represents the } n\text{'th Euler number.} \right]$$

$$\csc x = \frac{1}{x} + \frac{x}{6} + \frac{7}{360} x^3 + \frac{31}{15,120} x^5 + \frac{127}{604,800} x^7 + \cdots$$

$$+ \frac{(-1)^{n+1} 2 (2^{2n-1} - 1)}{(2n)!} B_{2n} x^{2n-1} + \cdots,$$

$$[x^2 < \pi^2, \text{ and } B_n \text{ represents } n\text{'th Bernoulli number.}]$$

$$\sin x = x \left(1 - \frac{x^2}{\pi^2} \right) \left(1 - \frac{x^2}{2^2 \pi^2} \right) \left(1 - \frac{x^2}{3^2 \pi^2} \right) \cdots \qquad (x^2 < \infty)$$

$$\cos x = \left(1 - \frac{4x^2}{\pi^2} \right) \left(1 - \frac{4x^2}{3^2 \pi^2} \right) \left(1 - \frac{4x^2}{5^2 \pi^2} \right) \cdots \qquad (x^2 < \infty)$$

$$\sin^{-1} x = x + \frac{x^3}{2\cdot 3} + \frac{1\cdot 3}{2\cdot 4\cdot 5} x^5 + \frac{1\cdot 3\cdot 5}{2\cdot 4\cdot 6\cdot 7} x^7 + \cdots \qquad \left(x^2 < 1, -\frac{\pi}{2} < \sin^{-1} x < \frac{\pi}{2} \right)$$

$$\cos^{-1} x = \frac{\pi}{2} - \left(x + \frac{x^3}{2\cdot 3} + \frac{1\cdot 3}{2\cdot 4\cdot 5} x^5 + \frac{1\cdot 3\cdot 5 x^7}{2\cdot 4\cdot 6\cdot 7} + \cdots \right) \qquad (x^2 < 1, 0 < \cos^{-1} x < \pi)$$

$$\tan^{-1} x = x - \frac{x^3}{3} + \frac{x^5}{5} - \frac{x^7}{7} + \cdots \qquad (x^2 < 1)$$

$$\tan^{-1} x = \frac{\pi}{2} - \frac{1}{x} + \frac{1}{3x^3} - \frac{1}{5x^5} + \frac{1}{7x^7} - \cdots \qquad (x > 1)$$

$$\tan^{-1} x = -\frac{\pi}{2} - \frac{1}{x} + \frac{1}{3x^3} - \frac{1}{5x^5} + \frac{1}{7x^7} - \cdots \qquad (x < -1)$$

$$\cot^{-1} x = \frac{\pi}{2} - x + \frac{x^3}{3} - \frac{x^5}{5} + \frac{x^7}{7} - \cdots \qquad (x^2 < 1)$$

$$\log_e \sin x = \log_e x - \frac{x^2}{6} - \frac{x^4}{180} - \frac{x^6}{2835} - \cdots \qquad (x^2 < \pi^2)$$

$$\log_e \cos x = -\frac{x^2}{2} - \frac{x^4}{12} - \frac{x^6}{45} - \frac{17x^8}{2520} - \cdots \qquad \left(x^2 < \frac{\pi^2}{4} \right)$$

$$\log_e \tan x = \log_e x + \frac{x^2}{3} + \frac{7x^4}{90} + \frac{62x^6}{2835} + \cdots \qquad \left(x^2 < \frac{\pi^2}{4} \right)$$

$$e^{\sin x} = 1 + x + \frac{x^2}{2!} - \frac{3x^4}{4!} - \frac{8x^5}{5!} - \frac{3x^6}{6!} + \frac{56x^7}{7!} + \cdots$$

$$e^{\cos x} = e \left(1 - \frac{x^2}{2!} + \frac{4x^4}{4!} - \frac{31x^6}{6!} + \cdots \right)$$

$$e^{\tan x} = 1 + x + \frac{x^2}{2!} + \frac{3x^3}{3!} + \frac{9x^4}{4!} + \frac{37x^5}{5!} + \cdots \qquad \left(x^2 < \frac{\pi^2}{4} \right)$$

$$\sin x = \sin a + (x - a) \cos a - \frac{(x - a)^2}{2!} \sin a$$

$$- \frac{(x - a)^3}{3!} \cos a + \frac{(x - a)^4}{4!} \sin a + \cdots$$

VECTOR ANALYSIS

Definitions

Any quantity which is completely determined by its magnitude is called a *scalar*. Examples of such are mass, density, temperature, etc. Any quantity which is completely determined by its magnitude and direction is called a *vector*. Examples of such are velocity, acceleration, force, etc. A vector quantity is represented by a directed line segment, the length of which represents the magnitude of the vector. A vector quantity is usually represented by a boldfaced letter such as $\mathbf{V}$. Two vectors $\mathbf{V}_1$ and $\mathbf{V}_2$ are equal to one another if they have equal magnitudes and are acting in the same directions. A negative vector, written as $-\mathbf{V}$, is one which acts in the opposite direction to $\mathbf{V}$, but is of equal magnitude to it. If we represent the magnitude of $\mathbf{V}$ by v, we write $|\mathbf{V}| = v$. A vector parallel to $\mathbf{V}$, but equal to the reciprocal of its magnitude is written as $\mathbf{V}^{-1}$ or as $\frac{1}{\mathbf{V}}$.

The *unit vector* $\frac{\mathbf{V}}{v}$ $(v \neq 0)$ is that vector which has the same direction as $\mathbf{V}$, but has a magnitude of unity (sometimes represented as $\mathbf{V}_0$ or $\hat{v}$).

Vector Algebra

The vector sum of $\mathbf{V}_1$ and $\mathbf{V}_2$ is represented by $\mathbf{V}_1 + \mathbf{V}_2$. The vector sum of $\mathbf{V}_1$ and $-\mathbf{V}_2$, or the difference of the vector $\mathbf{V}_2$ from $\mathbf{V}_1$ is represented by $\mathbf{V}_1 - \mathbf{V}_2$.

If r is a scalar, then $r\mathbf{V} = \mathbf{V}r$, and represents a vector r times the magnitude of $\mathbf{V}$, in the same direction as $\mathbf{V}$ if r is positive, and in the opposite direction if r is negative. If r and s are scalars, $\mathbf{V}_1$, $\mathbf{V}_2$, $\mathbf{V}_3$, vectors, then the following rules of scalars and vectors hold:

$$\mathbf{V}_1 + \mathbf{V}_2 = \mathbf{V}_2 + \mathbf{V}_1$$
$$(r + s)\mathbf{V}_1 = r\mathbf{V}_1 + s\mathbf{V}_1; \qquad r(\mathbf{V}_1 + \mathbf{V}_2) = r\mathbf{V}_1 + r\mathbf{V}_2$$
$$\mathbf{V}_1 + (\mathbf{V}_2 + \mathbf{V}_3) = (\mathbf{V}_1 + \mathbf{V}_2) + \mathbf{V}_3 = \mathbf{V}_1 + \mathbf{V}_2 + \mathbf{V}_3$$

Vectors in Space

A plane is described by two distinct vectors $\mathbf{V}_1$ and $\mathbf{V}_2$. Should these vectors not intersect each other, then one is displaced parallel to itself until they do (fig. 1.) Any other vector $\mathbf{V}$ lying in this plane is given by

$$\mathbf{V} = r\mathbf{V}_1 + s\mathbf{V}_2$$

VECTOR ANALYSIS (Continued)

A *position vector* specifies the position in space of a point relative to a fixed origin. If therefore V_1 and V_2 are the position vectors of the points A and B, relative to the origin O, then any point P on the line AB has a position vector V given by

$$V = rV_1 + (1 - r)V_2$$

The scalar "r" can be taken as the parametric representation of P since $r = 0$ implies $P = B$ and $r = 1$ implies $P = A$. (fig. 2). If P divides the line AB in the ratio $r:s$ then

$$V = \left(\frac{r}{r+s}\right)V_1 + \left(\frac{s}{r+s}\right)V_2$$

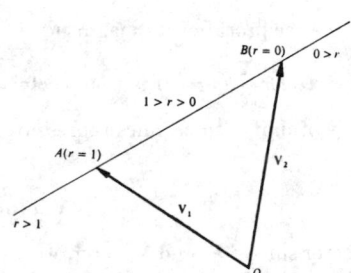

Fig. 1.   Fig. 2.

The vectors V_1, V_2, V_3, . . . , V_n are said to be *linearly dependent* if there exist scalars $r_1, r_2, r_3, \ldots, r_n$, not all zero, such that

$$r_1V_1 + r_2V_2 + \cdots + r_nV_n = 0$$

A vector V is linearly dependent upon the set of vectors V_1, V_2, V_3, . . . , V_n if

$$V = r_1V_1 + r_2V_2 + r_3V_3 + \cdots + r_nV_n$$

Three vectors are linearly dependent if and only if they are co-planar.

All points in space can be uniquely determined by linear dependence upon three *base vectors* i.e. three vectors any one of which is linearly independent of the other two. The simplest set of base vectors are the unit vectors along the coordinate Ox, Oy and Oz axes. These are usually designated by i, j and k respectively.

If V is a vector in space, and a, b and c are the respective magnitudes of the projections of the vector along the axes then

$$V = ai + bj + ck$$

and

$$v = \sqrt{a^2 + b^2 + c^2}$$

and the direction cosines of V are

$$\cos \alpha = a/v, \quad \cos \beta = b/v, \quad \cos \gamma = c/v.$$

The law of addition yields

$$V_1 + V_2 = (a_1 + a_2)i + (b_1 + b_2)j + (c_1 + c_2)k$$

The Scalar, Dot, or Inner Product of Two Vectors V_1 and V_2

This product is represented as $V_1 \cdot V_2$ and is defined to be equal to $v_1 v_2 \cos \theta$, where θ is the angle from V_1 to V_2, i.e.,

$$V_1 \cdot V_2 = v_1 v_2 \cos \theta$$

The following rules apply for this product:

$$V_1 \cdot V_2 = a_1 a_2 + b_1 b_2 + c_1 c_2 = V_2 \cdot V_1$$

It should be noted that this verifies that scalar multiplication is commutative.

$$(V_1 + V_2) \cdot V_3 = V_1 \cdot V_3 + V_2 \cdot V_3$$
$$V_1 \cdot (V_2 + V_3) = V_1 \cdot V_2 + V_1 \cdot V_3$$

If V_1 is perpendicular to V_2 then $V_1 \cdot V_2 = 0$, and if V_1 is parallel to V_2 then $V_1 \cdot V_2 = v_1 v_2 = r w_1{}^2$

In particular

$$i \cdot i = j \cdot j = k \cdot k = 1,$$
and
$$i \cdot j = j \cdot k = k \cdot i = 0$$

The Vector or Cross Product of Vectors V_1 and V_2

This product is represented as $V_1 \times V_2$ and is defined to be equal to $v_1 v_2 (\sin \theta) \mathbf{1}$, where θ is the angle from V_1 to V_2 and $\mathbf{1}$ is a unit vector perpendicular to the plane of V_1 and V_2 and so directed that a right-handed screw driven in the direction of $\mathbf{1}$ would carry V_1 into V_2, i.e.,

$$V_1 \times V_2 = v_1 v_2 (\sin \theta) \mathbf{1}$$
and
$$\tan \theta = \frac{|V_1 \times V_2|}{V_1 \cdot V_2}$$

The following rules apply for vector products:

$$V_1 \times V_2 = -V_2 \times V_1$$
$$V_1 \times (V_2 + V_3) = V_1 \times V_2 + V_1 \times V_3$$
$$(V_1 + V_2) \times V_3 = V_1 \times V_3 + V_2 \times V_3$$
$$V_1 \times (V_2 \times V_3) = V_2(V_3 \cdot V_1 - V_3(V_1 \cdot V_2)$$
$$i \times i = j \times j = k \times k = 0.1 \text{ (zero vector)}$$
$$= 0$$

$$i \times j = k, \qquad j \times k = i, \qquad k \times i = j$$

If $V_1 = a_1 i + b_1 j + c_1 k$, $\qquad V_2 = a_2 i + b_2 j + c_2 k$, $\qquad V_3 = a_3 i + b_3 j + c_3 k$,

then

$$V_1 \times V_2 = \begin{vmatrix} i & j & k \\ a_1 & b_1 & c_1 \\ a_2 & b_2 & c_2 \end{vmatrix} = (b_1 c_2 - b_2 c_1) i + (c_1 a_2 - c_2 a_1) j + (a_1 b_2 - a_2 b_1) k$$

It should be noted that, since $V_1 \times V_2 = -V_2 \times V_1$, the vector product is not commutative.

Scalar Triple Product

There is only one possible interpretation of the expression $\mathbf{V}_1 \cdot \mathbf{V}_2 \times \mathbf{V}_3$ and that is $\mathbf{V}_1 \cdot (\mathbf{V}_2 \times \mathbf{V}_3)$ which is obviously a scalar.

Further $\mathbf{V}_1 \cdot (\mathbf{V}_2 \times \mathbf{V}_3) = (\mathbf{V}_1 \times \mathbf{V}_2) \cdot \mathbf{V}_3 = \mathbf{V}_2 \cdot (\mathbf{V}_3 \times \mathbf{V}_1)$

$$= \begin{vmatrix} a_1 & b_1 & c_1 \\ a_2 & b_2 & c_2 \\ a_3 & b_3 & c_3 \end{vmatrix}$$

$$= v_1 v_2 v_3 \cos \phi \sin \theta,$$

Where θ is the angle between $\mathbf{V}_2$ and $\mathbf{V}_3$ and ϕ is the angle between $\mathbf{V}_1$ and the normal to the plane of $\mathbf{V}_2$ and $\mathbf{V}_3$.

This product is called the *scalar triple product* and is written as $[\mathbf{V}_1 \mathbf{V}_2 \mathbf{V}_3]$.

The determinant indicates that it can be considered as the volume of the parallelepiped whose three determining edges are $\mathbf{V}_1$, $\mathbf{V}_2$ and $\mathbf{V}_3$.

It also follows that cyclic permutation of the subscripts does not change the value of the scalar triple product so that

$$[\mathbf{V}_1 \mathbf{V}_2 \mathbf{V}_3] = [\mathbf{V}_2 \mathbf{V}_3 \mathbf{V}_1] = [\mathbf{V}_3 \mathbf{V}_1 \mathbf{V}_2]$$

but $\quad [\mathbf{V}_1 \mathbf{V}_2 \mathbf{V}_3] = -[\mathbf{V}_2 \mathbf{V}_1 \mathbf{V}_3]$ etc. $\qquad$ and $\quad [\mathbf{V}_1 \mathbf{V}_1 \mathbf{V}_2] \equiv 0$ etc.

Given three non-coplanar reference vectors $\mathbf{V}_1$, $\mathbf{V}_2$ and $\mathbf{V}_3$, the *reciprocal system is* given by $\mathbf{V}_1^*$, $\mathbf{V}_2^*$ and $\mathbf{V}_3^*$, where

$$1 = v_1 v_1^* = v_2 v_2^* = v_3 v_3^*$$

$$0 = v_1 v_2^* = v_1 v_3^* = v_2 v_1^* \quad \text{etc.}$$

$$\mathbf{V}_1^* = \frac{\mathbf{V}_2 \times \mathbf{V}_3}{[\mathbf{V}_1 \mathbf{V}_2 \mathbf{V}_3]}, \qquad \mathbf{V}_2^* = \frac{\mathbf{V}_3 \times \mathbf{V}_1}{[\mathbf{V}_1 \mathbf{V}_2 \mathbf{V}_3]}, \qquad \mathbf{V}_3^* = \frac{\mathbf{V}_1 \times \mathbf{V}_2}{[\mathbf{V}_1 \mathbf{V}_2 \mathbf{V}_3]}$$

The system $\mathbf{i}$, $\mathbf{j}$, $\mathbf{k}$ is its own reciprocal.

Vector Triple Product

The product $\mathbf{V}_1 \times (\mathbf{V}_2 \times \mathbf{V}_3)$ defines the *vector triple product*. Obviously, in this case, the brackets are vital to the definition.

$$\mathbf{V}_1 \times (\mathbf{V}_2 \times \mathbf{V}_3) = (\mathbf{V}_1 \cdot \mathbf{V}_3)\mathbf{V}_2 - (\mathbf{V}_1 \cdot \mathbf{V}_2)\mathbf{V}_3$$

$$= \begin{vmatrix} \mathbf{i} & \mathbf{j} & \mathbf{k} \\ a_1 & b_1 & c_1 \\ \begin{vmatrix} b_2 & c_2 \\ b_3 & c_3 \end{vmatrix} & \begin{vmatrix} c_2 & a_2 \\ c_3 & a_3 \end{vmatrix} & \begin{vmatrix} a_2 & b_2 \\ a_3 & b_3 \end{vmatrix} \end{vmatrix}$$

i.e. it is a vector, perpendicular to $\mathbf{V}_1$, lying in the plane of $\mathbf{V}_2$, $\mathbf{V}_3$.

Similarly $\qquad (\mathbf{V}_1 \times \mathbf{V}_2) \times \mathbf{V}_3 = \begin{vmatrix} \mathbf{i} & \mathbf{j} & \mathbf{k} \\ \begin{vmatrix} b_1 & c_1 \\ b_2 & c_2 \end{vmatrix} & \begin{vmatrix} c_1 & a_1 \\ c_2 & a_2 \end{vmatrix} & \begin{vmatrix} a_1 & b_1 \\ a_2 & b_2 \end{vmatrix} \\ a_3 & b_3 & c_3 \end{vmatrix}$

$$\mathbf{V}_1 \times (\mathbf{V}_2 \times \mathbf{V}_3) + \mathbf{V}_2 \times (\mathbf{V}_3 \times \mathbf{V}_1) + \mathbf{V}_3 \times (\mathbf{V}_1 + \mathbf{V}_2) \equiv 0$$

If $\mathbf{V}_1 \times (\mathbf{V}_2 \times \mathbf{V}_3) = (\mathbf{V}_1 \times \mathbf{V}_2) \times \mathbf{V}_3$ then $\mathbf{V}_1$, $\mathbf{V}_2$, $\mathbf{V}_3$ form an *orthogonal set*. Thus $\mathbf{i}$, $\mathbf{j}$, $\mathbf{k}$ form an orthogonal set.

Geometry of the Plane, Straight Line and Sphere

The position vectors of the fixed points A, B, C, D relative to O are $\mathbf{V}_1$, $\mathbf{V}_2$, $\mathbf{V}_3$, $\mathbf{V}_4$ and the position vector of the variable point P is $\mathbf{V}$.

The vector form of the equation of the straight line through A parallel to $\mathbf{V}_2$ is

$$\mathbf{V} = \mathbf{V}_1 + r\mathbf{V}_2$$

$$\text{or} \quad (\mathbf{V} - \mathbf{V}_1) = r\mathbf{V}_2$$

$$\text{or} \quad (\mathbf{V} - \mathbf{V}_1) \times \mathbf{V}_2 = 0$$

while that of the plane through A perpendicular to $\mathbf{V}_2$ is

$$(\mathbf{V} - \mathbf{V}_1) \cdot \mathbf{V}_2 = 0$$

The equation of the line AB is

$$\mathbf{V} = r\mathbf{V}_1 + (1 - r)\mathbf{V}_2$$

and those of the bisectors of the angles between $\mathbf{V}_1$ and $\mathbf{V}_2$ are

$$\mathbf{V} = r\left(\frac{\mathbf{V}_1}{v} \pm \frac{\mathbf{V}_2}{v_2}\right)$$

$$\text{or} \quad \mathbf{V} = r(\hat{\mathbf{v}}_1 \pm \hat{\mathbf{v}}_2)$$

The perpendicular from C to the line through A parallel to $\mathbf{V}_2$ has as its equation

$$\mathbf{V} = \mathbf{V}_1 - \mathbf{V}_3 - \hat{\mathbf{v}}_2 \cdot (\mathbf{V}_1 - \mathbf{V}_3)\hat{\mathbf{v}}_2.$$

The condition for the intersection of the two lines,

$$\mathbf{V} = \mathbf{V}_1 + r\mathbf{V}_3$$

$$\text{and} \quad \mathbf{V} = \mathbf{V}_2 + s\mathbf{V}_4$$

$$\text{is} \quad [(\mathbf{V}_1 - \mathbf{V}_2)\mathbf{V}_3\mathbf{V}_4] = 0.$$

The common perpendicular to the above two lines is the line of intersection of the two planes

$$[(\mathbf{V} - \mathbf{V}_1)\mathbf{V}_3(\mathbf{V}_3 \times \mathbf{V}_4)] = 0$$

$$\text{and} \quad [(\mathbf{V} - \mathbf{V}_2)\mathbf{V}_4(\mathbf{V}_3 \times \mathbf{V}_4)] = 0$$

and the length of this perpendicular is

$$\frac{[(\mathbf{V}_1 - \mathbf{V}_2)\mathbf{V}_3\mathbf{V}_4]}{|\mathbf{V}_3 \times \mathbf{V}_4|}.$$

The equation of the line perpendicular to the plane ABC is

$$\mathbf{V} = \mathbf{V}_1 \times \mathbf{V}_2 + \mathbf{V}_2 \times \mathbf{V}_3 + \mathbf{V}_3 \times \mathbf{V}_1$$

and the distance of the plane from the origin is

$$\frac{[\mathbf{V}_1\mathbf{V}_2\mathbf{V}_3]}{|(\mathbf{V}_2 - \mathbf{V}_1) \times (\mathbf{V}_3 - \mathbf{V}_1)|}.$$

In general the vector equation

$$\mathbf{V} \cdot \mathbf{V}_2 = r$$

defines the plane which is perpendicular to $\mathbf{V}_2$, and the perpendicular distance from A to this plane is

$$\frac{r - \mathbf{V}_1 \cdot \mathbf{V}_2}{v_2}.$$

The distance from A, measured along a line parallel to $\mathbf{V}_3$, is

$$\frac{r - \mathbf{V}_1 \cdot \mathbf{V}_2}{\mathbf{V}_2 \cdot \hat{\mathbf{v}}_3} \quad \text{or} \quad \frac{r - \mathbf{V}_1 \cdot \mathbf{V}_2}{v_2 \cos \theta}$$

where θ is the angle beween $\mathbf{V}_2$ and $\mathbf{V}_3$.
(If this plane contains the point C then $r = \mathbf{V}_3 \cdot \mathbf{V}_2$ and if it passes through the origin then $r = 0$.)

Given two planes

$$\mathbf{V} \cdot \mathbf{V}_1 = r$$

$$\mathbf{V} \cdot \mathbf{V}_2 = s$$

then any plane through the line of intersection of these two planes is given by

$$\mathbf{V} \cdot (\mathbf{V}_1 + \lambda \mathbf{V}_2) = r + \lambda s$$

where λ is a scalar parameter. In particular $\lambda = \pm r_1/r_2$ yields the equation of the two planes bisecting the angle between the given planes.

The plane through A parallel to the plane of $\mathbf{V}_2$, $\mathbf{V}_3$ is

$$\mathbf{V} = \mathbf{V}_1 + r\mathbf{V}_2 + s\mathbf{V}_3$$

$$\text{or} \quad (\mathbf{V} - \mathbf{V}_1) \cdot \mathbf{V}_2 \times \mathbf{V}_3 = 0$$

$$\text{or} \quad [\mathbf{V}\mathbf{V}_2\mathbf{V}_3] - [\mathbf{V}_1\mathbf{V}_2\mathbf{V}_3] = 0$$

so that the expansion in rectangular Cartesian coordinates yields

$$\begin{vmatrix} (x - a_1) & (y - b_1) & (z - c_1) \\ a_2 & b_2 & c_2 \\ a_3 & b_3 & c_3 \end{vmatrix} = 0 \qquad (\mathbf{V} \equiv x\mathbf{i} + y\mathbf{j} + z\mathbf{k})$$

which is obviously the usual linear equation in x, y and z.

The plane through AB parallel to $\mathbf{V}_3$ is given by

$$[(\mathbf{V} - \mathbf{V}_1)(\mathbf{V}_1 - \mathbf{V}_2)\mathbf{V}_3] = 0$$

$$\text{or} \quad [\mathbf{V}\mathbf{V}_2\mathbf{V}_3] - [\mathbf{V}\mathbf{V}_1\mathbf{V}_3] - [\mathbf{V}_1\mathbf{V}_2\mathbf{V}_3] = 0$$

The plane through the three points A, B and C is

$$\mathbf{V} = \mathbf{V}_1 + s(\mathbf{V}_2 - \mathbf{V}_1) + t(\mathbf{V}_3 - \mathbf{V}_1)$$

$$\text{or} \qquad \mathbf{V} = r\mathbf{V}_1 + s\mathbf{V}_2 + t\mathbf{V}_3 \qquad (r + s + t \equiv 1)$$

$$\text{or} \qquad [(\mathbf{V} - \mathbf{V}_1)(\mathbf{V}_1 - \mathbf{V}_2)(\mathbf{V}_2 - \mathbf{V}_3)] = 0$$

$$\text{or} \qquad [\mathbf{V}\mathbf{V}_1\mathbf{V}_2] + [\mathbf{V}\mathbf{V}_2\mathbf{V}_3] + [\mathbf{V}\mathbf{V}_3\mathbf{V}_1] - [\mathbf{V}_1\mathbf{V}_2\mathbf{V}_3] = 0$$

For four points A, B, C, D to be coplanar, then

$$r\mathbf{Y}_1 + s\mathbf{V}_2 + t\mathbf{V}_3 + u\mathbf{V}_4 \equiv 0 \equiv r + s + t + u$$

The following formulae relate to a sphere when the vectors are taken to lie in three dimensional space and to a circle when the space is two dimensional. For a circle in three dimensions take the intersection of the sphere with a plane.

The equation of a sphere with center O and radius OA is

$$\mathbf{V} \cdot \mathbf{V} = v_1^2 \qquad \text{(not } \mathbf{V} = \mathbf{V}_1\text{)}$$

or $\qquad (\mathbf{V} - \mathbf{V}_1) \cdot (\mathbf{V} + \mathbf{V}_1) = 0$

while that of a sphere with center B radius v_1 is

$$(\mathbf{V} - \mathbf{V}_2) \cdot (\mathbf{V} - \mathbf{V}_2) = v_1^2$$

or $\qquad \mathbf{V} \cdot (\mathbf{V} - 2\mathbf{V}_2) = v_1^2 - v_2^2$

If the above sphere passes through the origin then

$$\mathbf{V} \cdot (\mathbf{V} - 2\mathbf{V}_2) = 0$$

(note that in two dimensional polar coordinates this is simply)

$$r = 2a \cdot \cos \theta$$

while in three dimensional Cartesian coordinates it is

$$x^2 + y^2 + z^2 - 2(a_2 x + b_2 y + c_2 x) = 0.$$

The equation of a sphere having the points A and B as the extremities of a diameter is

$$(\mathbf{V} - \mathbf{V}_1) \cdot (\mathbf{V} - \mathbf{V}_2) = 0.$$

The square of the length of the tangent from C to the sphere with center B and radius v_1 is given by

$$(\mathbf{V}_3 - \mathbf{V}_2) \cdot (\mathbf{V}_3 - \mathbf{V}_2) = v_1^2$$

The condition that the plane $\mathbf{V} \cdot \mathbf{V}_3 = s$ is tangential to the sphere $(\mathbf{V} - \mathbf{V}_2) \cdot (\mathbf{V} - \mathbf{V}_2) = v_1^2$ is

$$(s - \mathbf{V}_3 \cdot \mathbf{V}_2) \cdot (s - \mathbf{V}_3 \cdot \mathbf{V}_2) = v_1^2 v_3^2.$$

The equation of the tangent plane at D, on the surface of sphere $(\mathbf{V} - \mathbf{V}_2) \cdot (\mathbf{V} - \mathbf{V}_2) = v_1^2$, is

$$(\mathbf{V} - \mathbf{V}_4) \cdot (\mathbf{V}_4 - \mathbf{V}_2) = 0$$

or $\qquad \mathbf{V} \cdot \mathbf{V}_4 - \mathbf{V}_2 \cdot (\mathbf{V} + \mathbf{V}_4) = v_1^2 - v_2^2$

The condition that the two circles $(\mathbf{V} - \mathbf{V}_2) \cdot (\mathbf{V} - \mathbf{V}_2) = v_1^2$ and $(\mathbf{V} - \mathbf{V}_4) \cdot (\mathbf{V} - \mathbf{V}_4) = v_3^2$ intersect orthogonally is clearly

$$(\mathbf{V}_2 - \mathbf{V}_4) \cdot (\mathbf{V}_2 - \mathbf{V}_4) = v_1^2 + v_3^2$$

The polar plane of D with respect to the circle

$$(\mathbf{V} - \mathbf{V}_2) \cdot (\mathbf{V} - \mathbf{V}_2) = v_1^2 \quad \text{is}$$

$$\mathbf{V} \cdot \mathbf{V}_4 - \mathbf{V}_2 \cdot (\mathbf{V} + \mathbf{V}_4) = v_1^2 - v_2^2$$

Any sphere through the intersection of the two spheres $(\mathbf{V} - \mathbf{V}_2) \cdot (\mathbf{V} - \mathbf{V}_2) = v_1^2$ and $(\mathbf{V} - \mathbf{V}_4) \cdot (\mathbf{V} - \mathbf{V}_4) = v_3^2$ is given by

$$(\mathbf{V} - \mathbf{V}_2) \cdot (\mathbf{V} - \mathbf{V}_2) + \lambda(\mathbf{V} - \mathbf{V}_4) \cdot (\mathbf{V} - \mathbf{V}_4) = v_1^2 + \lambda v_3^2$$

while the radical plane of two such spheres is

$$\mathbf{V} \cdot (\mathbf{V}_2 - \mathbf{V}_4) = -\tfrac{1}{2}(v_1^2 - r_2^2 - v_3^2 + v_4^2)$$

Differentiation of Vectors

If $\mathbf{V}_1 = a_1\mathbf{i} + b_1\mathbf{j} + c_1\mathbf{k}$, and $\mathbf{V}_2 = a_2\mathbf{i} + b_2\mathbf{j} + c_2\mathbf{k}$, and if $\mathbf{V}_1$ and $\mathbf{V}_2$ are functions of the scalar t, then

$$\frac{d}{dt}(\mathbf{V}_1 + \mathbf{V}_2 + \cdots) = \frac{d\mathbf{V}_1}{dt} + \frac{d\mathbf{V}_2}{dt} + \cdots,$$

where $\dfrac{d\mathbf{V}_1}{dt} = \dfrac{da_1}{dt}\mathbf{i} + \dfrac{db_1}{dt}\mathbf{j} + \dfrac{dc_1}{dt}\mathbf{k}$, etc.

$$\frac{d}{dt}(\mathbf{V}_1 \cdot \mathbf{V}_2) = \frac{d\mathbf{V}_1}{dt} \cdot \mathbf{V}_2 + \mathbf{V}_1 \cdot \frac{d\mathbf{V}_2}{dt}$$

$$\frac{d}{dt}(\mathbf{V}_1 \times \mathbf{V}_2) = \frac{d\mathbf{V}_1}{dt} \times \mathbf{V}_2 + \mathbf{V}_1 \times \frac{d\mathbf{V}_2}{dt}$$

$$\mathbf{V} \cdot \frac{d\mathbf{V}}{dt} = v \cdot \frac{dv}{dt}$$

In particular, if $\mathbf{V}$ is a vector of constant length then the right hand side of the last equation is identically zero showing that $\mathbf{V}$ is perpendicular to its derivative.

The derivatives of the triple products are

$$\frac{d}{dt}[\mathbf{V}_1\mathbf{V}_2\mathbf{V}_3] = \left[\left(\frac{d\mathbf{V}_1}{dt}\right)\mathbf{V}_2\mathbf{V}_3\right] + \left[\mathbf{V}_1\left(\frac{d\mathbf{V}_2}{dt}\right)\mathbf{V}_3\right] + \left[\mathbf{V}_1\mathbf{V}_2\left(\frac{d\mathbf{V}_3}{dt}\right)\right]$$

and $\dfrac{d}{dt}\left\{\mathbf{V}_1 \times (\mathbf{V}_2 \times \mathbf{V}_3)\right\} = \left(\dfrac{d\mathbf{V}_1}{dt}\right) \times (\mathbf{V}_2 \times \mathbf{V}_3) + \mathbf{V}_1$

$$\times\left(\left(\frac{d\mathbf{V}_2}{dt}\right) \times \mathbf{V}_3\right) + \mathbf{V}_1 \times \left(\mathbf{V}_2 \times \left(\frac{d\mathbf{V}_3}{dt}\right)\right)$$

Geometry of Curves in Space

s = the *length of arc*, measured from some fixed point on the curve (fig. 3).
$\mathbf{V}_1$ = the position vector of the point A on the curve
$\mathbf{V}_1 + \delta\mathbf{V}_1$ = the position vector of the point P in the neighborhood of A
$\hat{\mathbf{t}}$ = the *unit tangent* to the curve at the point A, measured in the direction of s increasing.

The *normal plane* is that plane which is perpendicular to the unit tangent. The principal normal is defined as the intersection of the normal plane with the plane defined by $\mathbf{V}_1$ and $\mathbf{V}_1 + \delta\mathbf{V}_1$ in the limit as $\delta\mathbf{V}_1 - 0$.

$\hat{\mathbf{n}}$ = the *unit normal* (principal) at the point A. The plane defined by $\hat{\mathbf{t}}$ and $\hat{\mathbf{n}}$ is called the *osculating plane* (alternatively plane of curvature or local plane).

ρ = the radius of curvature at A
$\delta\theta$ = the angle subtended at the origin by $\delta\mathbf{V}_1$.

$$\kappa = \frac{d\theta}{ds} = \frac{1}{\rho}$$

$\hat{\mathbf{b}}$ = the *unit binormal* i.e. the unit vector which is parallel to $\hat{\mathbf{t}} \times \hat{\mathbf{n}}$ at the point A:
λ = the *torsion* of the curve at A

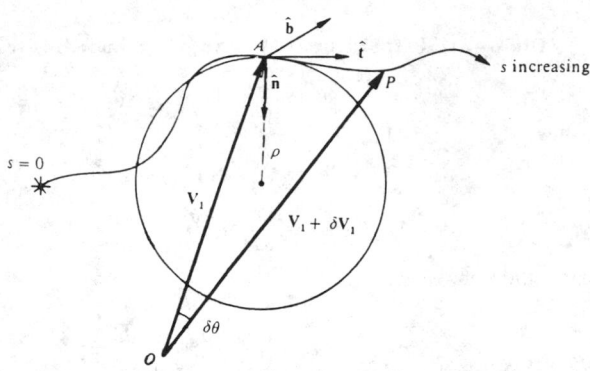

Figure 3.

Frenet's Formulae:

$$\frac{d\hat{\mathbf{t}}}{ds} = \kappa\hat{\mathbf{n}}$$

$$\frac{d\hat{\mathbf{n}}}{ds} = -\kappa\hat{\mathbf{t}} + \lambda\hat{\mathbf{b}}$$

$$\frac{d\hat{\mathbf{b}}}{ds} = -\lambda\hat{\mathbf{n}}$$

The following formulae are also applicable:

Unit tangent $\qquad\qquad \hat{\mathbf{t}} = \dfrac{d\mathbf{V}_1}{ds}$

Equation of the tangent $\qquad (\mathbf{V} - \mathbf{V}_1) \times \hat{\mathbf{t}} = 0$
$\qquad\qquad$ or $\qquad \mathbf{V} = \mathbf{V}_1 + q\hat{\mathbf{t}}$

Unit normal $\qquad\qquad \hat{\mathbf{n}} = \dfrac{1}{\kappa}\dfrac{d^2\mathbf{V}_1}{ds^2}$

Equation of the normal plane $\qquad (\mathbf{V} - \mathbf{V}_1) \cdot \hat{\mathbf{t}} = 0$

Equation of the normal $\qquad (\mathbf{V} - \mathbf{V}_1) \times \hat{\mathbf{n}} = 0$
$\qquad\qquad$ or $\qquad \mathbf{V} = \mathbf{V}_1 + r\hat{\mathbf{n}}$

Unit binormal $\qquad \hat{\mathbf{b}} = \hat{\mathbf{t}} \times \hat{\mathbf{n}}$

Equation of the binormal $\qquad (\mathbf{V} - \mathbf{V}_1) \times \hat{\mathbf{b}} = 0$
$\qquad\qquad$ or $\qquad \mathbf{V} = \mathbf{V}_1 + u\hat{\mathbf{b}}$
$\qquad\qquad$ or $\qquad \mathbf{V} = \mathbf{V}_1 + w\dfrac{d\mathbf{V}_1}{ds} \times \dfrac{d^2\mathbf{V}_1}{ds^2}$

Equation of the osculating plane:

$$[(\mathbf{V} - \mathbf{V}_1)\hat{\mathbf{t}}\hat{\mathbf{n}}] = 0$$

$$\text{or} \qquad \left[(\mathbf{V} - \mathbf{V}_1)\left(\frac{d\mathbf{V}_1}{ds}\right)\left(\frac{d^2\mathbf{V}_1}{ds^2}\right)\right] = 0$$

A *geodetic line* on a surface is a curve, the osculating plane of which is everywhere normal to the surface.

The differential equation of the geodetic is

$$[\hat{n}d\mathbf{V}_1 \, d^2\mathbf{V}_1] = \cdot 0$$

Differential Operators—Rectangular Coordinates

$$dS = \frac{\partial S}{\partial x} \cdot dx + \frac{\partial S}{\partial y} \cdot dy + \frac{\partial S}{\partial z} \cdot dz$$

By definition

$$\nabla \equiv \mathrm{del} \equiv \mathbf{i}\frac{\partial}{\partial x} + \mathbf{j}\frac{\partial}{\partial y} + \mathbf{k}\frac{\partial}{\partial z}$$

$$\nabla^2 \equiv \mathrm{Laplacian} \equiv \frac{\partial^2}{\partial x^2} + \frac{\partial^2}{\partial y^2} + \frac{\partial^2}{\partial z^2}$$

If S is a scalar function, then

$$\nabla S \equiv \mathrm{grad}\, S \equiv \frac{\partial S}{\partial x}\mathbf{i} + \frac{\partial S}{\partial y}\mathbf{j} + \frac{\partial S}{\partial z}\mathbf{k}$$

Grad S defines both the direction and magnitude of the maximum rate of increase of S at any point. Hence the name *gradient* and also its vectorial nature. ∇S is independent of the choice of rectangular coordinates.

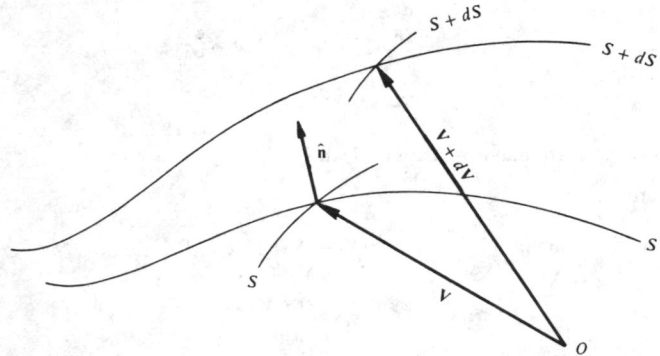

Figure 4

$$\nabla S = \frac{\partial S}{\partial n}\hat{n}$$

where $\hat{n}$ is the unit normal to the surface $S = $ constant, in the direction of S increasing. The total derivative of S at a point having the position vector $\mathbf{V}$ is given by (fig. 4)

$$dS = \frac{\partial S}{\partial n}\hat{n} \cdot d\mathbf{V}$$

$$= d\mathbf{V} \cdot \nabla S$$

and the directional derivative of S in the direction of $\mathbf{U}$ is

$$\mathbf{U} \cdot \nabla S = \mathbf{U} \cdot (\nabla S) = (\mathbf{U} \cdot \nabla)S$$

Similarly the directional derivative of the vector $\mathbf{V}$ in the direction of $\mathbf{U}$ is

$$(\mathbf{U} \cdot \nabla)\mathbf{V}$$

The *distributive* law holds for finding a gradient. Thus if S and T are scalar functions

$$\nabla(S + T) = \nabla S + \nabla T$$

The *associative* law becomes the rule for differentiating a product:

$$\nabla(ST) = S\nabla T + T\nabla S$$

If $\mathbf{V}$ is a vector function with the magnitudes of the components parallel to the three coordinate axes V_x, V_y, V_z, then

$$\nabla \cdot \mathbf{V} \equiv \text{div } \mathbf{V} \equiv \frac{\partial V_x}{\partial x} + \frac{\partial V_y}{\partial y} + \frac{\partial V_z}{\partial z}$$

The divergence obeys the distributive law. Thus, if $\mathbf{V}$ and $\mathbf{U}$ are vectors functions, then

$$\nabla \cdot (\mathbf{V} + \mathbf{U}) = \nabla \cdot \mathbf{V} + \nabla \cdot \mathbf{U}$$
$$\nabla \cdot (S\mathbf{V}) = (\nabla S) \cdot \mathbf{V} + S(\nabla \cdot \mathbf{V})$$
$$\nabla \cdot (\mathbf{U} \times \mathbf{V}) = \mathbf{V} \cdot (\nabla \times \mathbf{U}) - \mathbf{U} \cdot (\nabla \times \mathbf{V})$$

As with the gradient of a scalar, the divergence of a vector is invariant under a transformation from one set of rectangular coordinates to another.

$$\nabla \times \mathbf{V} \equiv \text{curl } \mathbf{V} \quad (\text{sometimes } \nabla \wedge \mathbf{V} \text{ or rot } \mathbf{V})$$

$$\equiv \left(\frac{\partial \mathbf{V}_z}{\partial y} - \frac{\partial \mathbf{V}_y}{\partial z}\right)\mathbf{i} + \left(\frac{\partial \mathbf{V}_x}{\partial z} - \frac{\partial \mathbf{V}_z}{\partial x}\right)\mathbf{j} + \left(\frac{\partial \mathbf{V}_y}{\partial x} - \frac{\partial \mathbf{V}_x}{\partial y}\right)\mathbf{k}$$

$$= \begin{vmatrix} \mathbf{i} & \mathbf{j} & \mathbf{k} \\ \dfrac{\partial}{\partial x} & \dfrac{\partial}{\partial y} & \dfrac{\partial}{\partial z} \\ V_x & V_y & V_z \end{vmatrix}$$

The *curl* (or *rotation*) of a vector is a vector which is invariant under a transformation from one set of rectangular coordinates to another.

$$\nabla \times (\mathbf{U} + \mathbf{V}) = \nabla \times \mathbf{U} + \nabla \times \mathbf{V}$$
$$\nabla \times (S\mathbf{V}) - (\nabla S) \times \mathbf{V} + S(\nabla \times \mathbf{V})$$
$$\nabla \times (\mathbf{U} \times \mathbf{V}) = (\mathbf{V} \cdot \nabla)\mathbf{U} - (\mathbf{U} \cdot \nabla)\mathbf{V} + \mathbf{U}(\nabla \cdot \mathbf{V}) - \mathbf{V}(\nabla \cdot \mathbf{U})$$

$$\text{grad } (\mathbf{U} \cdot \mathbf{V}) = \nabla(\mathbf{U} \cdot \mathbf{V})$$
$$= (\mathbf{V} \cdot \nabla)\mathbf{U} + (\mathbf{U} \cdot \nabla)\mathbf{V} + \mathbf{V} \times (\nabla \times \mathbf{U}) + \mathbf{U} \times (\nabla \times \mathbf{V})$$

If
$$\mathbf{V} = V_x\mathbf{i} + V_y\mathbf{j} + V_z\mathbf{k}$$
$$\nabla \cdot \mathbf{V} = \nabla V_x \cdot \mathbf{i} + \nabla V_y \cdot \mathbf{j} + \nabla V_z \cdot \mathbf{k}$$

and
$$\nabla \times \mathbf{V} = \nabla V_x \times \mathbf{i} + \nabla V_y \times \mathbf{j} + \nabla V_z \times \mathbf{k}$$

The operator ∇ can be used more than once. The number of possibilities where ∇ is used twice are

$$\nabla \cdot (\nabla\theta) \equiv \text{div grad } \theta$$
$$\nabla \times (\nabla\theta) \equiv \text{curl grad } \theta$$
$$\nabla(\nabla \cdot \mathbf{V}) \equiv \text{grad div } \mathbf{V}$$
$$\nabla \cdot (\nabla \times \mathbf{V}) \equiv \text{div curl } \mathbf{V}$$
$$\nabla \times (\nabla \times \mathbf{V}) \equiv \text{curl curl } \mathbf{V}$$

Thus: div grad $S \equiv \nabla \cdot (\nabla S) \equiv$ Laplacian $S \equiv \nabla^2 S$

$$\equiv \frac{\partial^2 S}{\partial x^2} + \frac{\partial^2 S}{\partial y^2} + \frac{\partial^2 S}{\partial z^2}$$

curl grad $S \equiv 0$;   curl curl $\mathbf{V} \equiv$ grad div $\mathbf{V} - \nabla^2 \mathbf{V}$;

div curl $\mathbf{V} \equiv \qquad 0$

Taylor's expansion in three dimensions can be written

$$f(\mathbf{V} + \boldsymbol{\varepsilon}) = e^{\boldsymbol{\varepsilon} \cdot \nabla} f(\mathbf{V})$$

where $\mathbf{V} = x\mathbf{i} + y\mathbf{j} + z\mathbf{k}$

and $\boldsymbol{\varepsilon} = h\mathbf{i} + l\mathbf{j} + m\mathbf{k}$

(note the analogy with $f_p = e^{phD} f_0$ in finite difference methods).

Orthogonal Curvilinear Coordinates

If at a point P there exist three uniform point functions u, v and w so that the surfaces $u = $ const., $v = $ const., and $w = $ const., intersect in three distinct curves through P then the surfaces are called the *coordinate surfaces* through P. The three lines of intersection are referred to as the *coordinate lines* and their tangents a, b, and c as the *coordinate axes*. When the coordinate axes form an orthogonal set the system is said to define *orthogonal curvilinear coordinates* at P.

Consider an infinitesimal volume enclosed by the surfaces u, v, w, $u + du$, $v + dv$, and $w + dw$ (fig. 5).

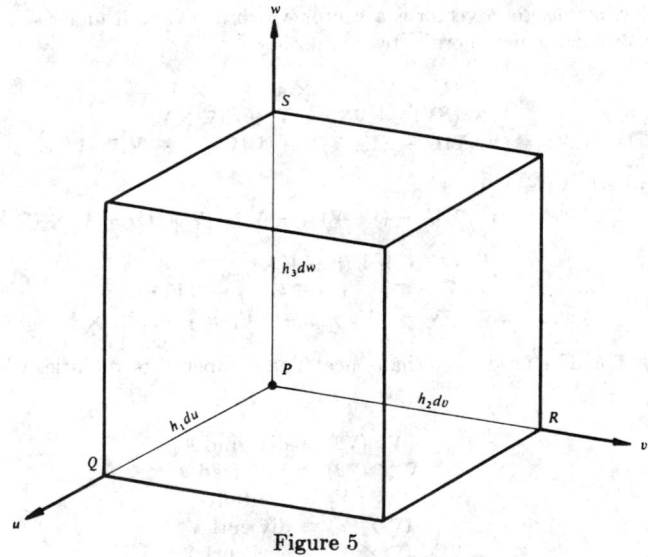

Figure 5

The surface $PRS \equiv u = \text{const.}$, and the face of the curvilinear figure immediately opposite this is $u + du = \text{const.}$ etc.

In terms of these surface constants

$$
\begin{aligned}
P &= P(u,v,w) \\
Q &= Q(u + du,v,w) \\
R &= R(u,v + dv,w) \\
S &= S(u,v,w + dw)
\end{aligned}
\qquad \text{and} \qquad
\begin{aligned}
PQ &= h_1 du \\
PR &= h_2 dv \\
PS &= h_3 dw
\end{aligned}
$$

where h_1, h_2, and h_3 are functions of u, v, and w.

In rectangular Cartesians **i, j, k**

$$h_1 = 1, \qquad h_2 = 1, \qquad h_3 = 1.$$

$$\frac{\hat{\mathbf{a}}}{h_1}\frac{\partial}{\partial u} = \mathbf{i}\,\frac{\partial}{\partial x}, \qquad \frac{\hat{\mathbf{b}}}{h_2}\frac{\partial}{\partial v} = \mathbf{j}\,\frac{\partial}{\partial y}, \qquad \frac{\hat{\mathbf{c}}}{h_3}\frac{\partial}{\partial w} = \mathbf{k}\,\frac{\partial}{\partial z}.$$

In cylindrical coordinates $\hat{\mathbf{r}}$, $\hat{\boldsymbol{\phi}}$, $\hat{\mathbf{k}}$

$$h_1 = 1, \qquad h_2 = r \qquad h_3 = 1.$$

$$\frac{\hat{\mathbf{a}}}{h_1}\frac{\partial}{\partial u} = \hat{\mathbf{r}}\,\frac{\partial}{\partial r}, \qquad \frac{\hat{\mathbf{b}}}{h_2}\frac{\partial}{\partial v} = \frac{\hat{\boldsymbol{\phi}}}{r}\frac{\partial}{\partial \phi} \qquad \frac{\hat{\mathbf{c}}}{h_3}\frac{\partial}{\partial w} = \hat{\mathbf{k}}\,\frac{\partial}{\partial z}$$

In spherical coordinates $\hat{\mathbf{r}}$, $\hat{\boldsymbol{\theta}}$, $\hat{\boldsymbol{\phi}}$

$$h_1 = 1, \qquad h_2 = r, \qquad h_3 = r \sin \theta$$

$$\frac{\hat{\mathbf{a}}}{h_1}\frac{\partial}{\partial u} = \hat{\mathbf{r}}\,\frac{\partial}{\partial r}, \qquad \frac{\mathbf{b}}{h_2}\frac{\partial}{\partial v} = \frac{\hat{\boldsymbol{\phi}}}{r}\frac{\partial}{\partial \theta}, \qquad \frac{\hat{\mathbf{c}}}{h_3}\frac{\partial}{\partial w} = \frac{\hat{\boldsymbol{\phi}}}{r \sin \theta}\frac{\partial}{\partial \phi}$$

The general expressions for grad, div and curl together with those for ∇^2 and the directional derivative are, in orthogonal curvilinear coordinates, given by

$$\nabla S = \frac{\hat{\mathbf{a}}}{h_1}\frac{\partial S}{\partial u} + \frac{\hat{\mathbf{b}}}{h_2}\frac{\partial S}{\partial v} + \frac{\hat{\mathbf{c}}}{h_3}\frac{\partial S}{\partial w}$$

$$(\mathbf{V} \cdot \nabla)S = \frac{V_1}{h_1}\frac{\partial S}{\partial u} + \frac{V_2}{h_2}\frac{\partial S}{\partial v} + \frac{V_3}{h_3}\frac{\partial S}{\partial w}$$

$$\nabla \cdot \mathbf{V} = \frac{1}{h_1 h_2 h_3}\left\{ \frac{\partial}{\partial u}(h_2 h_3 V_1) + \frac{\partial}{\partial v}(h_3 h_1 V_2) + \frac{\partial}{\partial w}(h_1 h_2 V_3) \right.$$

$$\nabla \times \mathbf{V} = \frac{\hat{\mathbf{a}}}{h_2 h_3}\left\{ \frac{\partial}{\partial v}(h_3 V_3) - \frac{\partial}{\partial w}(h_2 V_2) \right\} + \frac{\hat{\mathbf{b}}}{h_3 h_1}\left\{ \frac{\partial}{\partial w}(h_1 V_1) - \frac{\partial}{\partial u}(h_3 V_3) \right\}$$

$$+ \frac{\hat{\mathbf{c}}}{h_1 h_2}\left\{ \frac{\partial}{\partial u}(h_2 V_2) - \frac{\partial}{\partial v}(h_1 V_1) \right\}$$

$$\nabla^2 S = \frac{1}{h_1 h_2 h_3}\left\{ \frac{\partial}{\partial u}\left(\frac{h_2 h_3}{h_1}\frac{\partial S}{\partial u} \right) + \frac{\partial}{\partial v}\left(\frac{h_3 h_1}{h_2}\frac{\partial S}{\partial v} \right) + \frac{\partial}{\partial w}\left(\frac{h_1 h_2}{h_3}\frac{\partial S}{\partial w} \right) \right\}$$

FORMULAS OF VECTOR ANALYSIS

	Rectangular coordinates	Cylindrical coordinates	Spherical coordinates
Conversion to rectangular coordinates		$x = r\cos\varphi \quad y = r\sin\varphi \quad z = z$	$x = r\cos\varphi\sin\theta \quad y = r\sin\varphi\sin\theta$ $z = r\cos\theta$
Gradient	$\nabla\phi = \dfrac{\partial\phi}{\partial x}\mathbf{i} + \dfrac{\partial\phi}{\partial y}\mathbf{j} + \dfrac{\partial\phi}{\partial z}\mathbf{k}$	$\nabla\phi = \dfrac{\partial\phi}{\partial r}\mathbf{r} + \dfrac{1}{r}\dfrac{\partial\phi}{\partial\varphi}\boldsymbol{\phi} + \dfrac{\partial\phi}{\partial z}\mathbf{k}$	$\nabla\phi = \dfrac{\partial\phi}{\partial r}\mathbf{r} + \dfrac{1}{r}\dfrac{\partial\phi}{\partial\theta}\boldsymbol{\theta} + \dfrac{1}{r\sin\theta}\dfrac{\partial\phi}{\partial\varphi}\boldsymbol{\phi}$
Divergence	$\nabla\cdot\mathbf{A} = \dfrac{\partial A_x}{\partial x} + \dfrac{\partial A_y}{\partial y} + \dfrac{\partial A_z}{\partial z}$	$\nabla\cdot\mathbf{A} = \dfrac{1}{r}\dfrac{\partial(rA_r)}{\partial r} + \dfrac{1}{r}\dfrac{\partial A_\varphi}{\partial\varphi} + \dfrac{\partial A_z}{\partial z}$	$\nabla\cdot\mathbf{A} = \dfrac{1}{r^2}\dfrac{\partial(r^2 A_r)}{\partial r} + \dfrac{1}{r\sin\theta}\dfrac{\partial(A_\theta\sin\theta)}{\partial\theta} + \dfrac{1}{r\sin\theta}\dfrac{\partial A_\varphi}{\partial\varphi}$
Curl	$\nabla\times\mathbf{A} = \begin{vmatrix} \mathbf{i} & \mathbf{j} & \mathbf{k} \\ \dfrac{\partial}{\partial x} & \dfrac{\partial}{\partial y} & \dfrac{\partial}{\partial z} \\ A_x & A_y & A_z \end{vmatrix}$	$\nabla\times\mathbf{A} = \begin{vmatrix} \dfrac{1}{r}\mathbf{r} & \boldsymbol{\phi} & \dfrac{1}{r}\mathbf{k} \\ \dfrac{\partial}{\partial r} & \dfrac{\partial}{\partial\varphi} & \dfrac{\partial}{\partial z} \\ A_r & rA_\varphi & A_z \end{vmatrix}$	$\nabla\times\mathbf{A} = \begin{vmatrix} \dfrac{\mathbf{r}}{r^2\sin\theta} & \dfrac{\boldsymbol{\theta}}{r\sin\theta} & \dfrac{\boldsymbol{\phi}}{r} \\ \dfrac{\partial}{\partial r} & \dfrac{\partial}{\partial\theta} & \dfrac{\partial}{\partial\varphi} \\ A_r & rA_\theta & rA_\varphi\sin\theta \end{vmatrix}$
Laplacian	$\nabla^2\phi = \dfrac{\partial^2\phi}{\partial x^2} + \dfrac{\partial^2\phi}{\partial y^2} + \dfrac{\partial^2\phi}{\partial z^2}$	$\nabla^2\phi = \dfrac{1}{r}\dfrac{\partial}{\partial r}\left(r\dfrac{\partial\phi}{\partial r}\right) + \dfrac{1}{r^2}\dfrac{\partial^2\phi}{\partial\varphi^2} + \dfrac{\partial^2\phi}{\partial z^2}$	$\nabla^2\phi = \dfrac{1}{r^2}\dfrac{\partial}{\partial r}\left(r^2\dfrac{\partial\phi}{\partial r}\right) + \dfrac{1}{r^2\sin\theta}\dfrac{\partial}{\partial\theta}\left(\sin\theta\dfrac{\partial\phi}{\partial\theta}\right) + \dfrac{1}{r^2\sin^2\theta}\dfrac{\partial^2\phi}{\partial\varphi^2}$

Transformation of Integrals

s = the distance along some curve "C" in space and is measured from some fixed point.
S = a surface area
V = a volume contained by a specified surface
$\hat{t}$ = the unit tangent to C at the point P
$\hat{n}$ = the unit outward pointing normal
F = some vector function
ds = the vector element of curve ($= \hat{t}\, ds$)
dS = the vector element of surface ($= \hat{n}\, dS$)

Then
$$\int_{(c)} \mathbf{F} \cdot \hat{t}\, ds = \int_{(c)} \mathbf{F} \cdot d\mathbf{s}$$

and when
$$\mathbf{F} = \nabla\phi$$

$$\int_{(c)} (\nabla\phi) \cdot \hat{t}\, ds = \int_{(c)} d\phi$$

Gauss' Theorem (Green's Theorem)

When S defines a closed region having a volume V

$$\iiint_{(v)} (\nabla \cdot \mathbf{F})\, dV = \iint_{(s)} (\mathbf{F} \cdot \hat{n})\, dS = \iint_{(s)} \mathbf{F} \cdot d\mathbf{S}$$

also
$$\iiint_{(v)} (\nabla\phi)\, dV = \iint_{(s)} \phi\hat{n}\, dS$$

and
$$\iiint_{(v)} (\nabla \times \mathbf{F})\, dV = \iint_{(s)} (\hat{n} \times \mathbf{F})\, dS$$

Stokes' Theorem

When C is closed and bounds the open surface S.

$$\iint_{(s)} \hat{n} \cdot (\nabla \times \mathbf{F})\, dS = \int_{(c)} \mathbf{F} \cdot d\mathbf{s}$$

also
$$\iint_{(s)} (\hat{n} \times \nabla\phi)\, dS = \int_{(c)} \phi\, d\mathbf{s}$$

Green's Theorem

$$\iint_{(s)} (\nabla\phi \cdot \nabla\theta)\, dS = \iint_{(s)} \phi\hat{n} \cdot (\nabla\theta)\, dS = \iiint_{(v)} \phi(\nabla^2\theta)\, dV$$

$$= \iint_{(s)} \theta \cdot \hat{n}(\nabla\phi)\, dS = \iiint_{(v)} \theta(\nabla^2\phi)\, dV$$

MOMENT OF INERTIA FOR VARIOUS BODIES OF MASS

The mass of the body is indicated by m.

Body	Axis	Moment of inertia
Uniform thin rod	Normal to the length, at one end	$m\dfrac{l^2}{3}$
Uniform thin rod	Normal to the length, at the center	$m\dfrac{l^2}{12}$
Thin rectangular sheet, sides a and b	Through the center parallel to b	$m\dfrac{a^2}{12}$
Thin rectangular sheet, sides a and b	Through the center perpendicular to the sheet	$m\dfrac{a^2 + b^2}{12}$
Thin circular sheet of radius r	Normal to the plate through the center	$m\dfrac{r^2}{2}$
Thin circular sheet of radius r	Along any diameter	$m\dfrac{r^2}{4}$
Thin circular ring. Radii r_1 and r_2	Through center normal to plane of ring	$m\dfrac{r_1^2 + r_2^2}{2}$
Thin circular ring. Radii r_1 and r_2	Any diameter	$m\dfrac{r_1^2 + r_2^2}{4}$
Rectangular parallelopiped, edges a, b, and c	Through center perpendicular to face ab, (parallel to edge c)	$m\dfrac{a^2 + b^2}{12}$
Sphere, radius r	Any diameter	$m\dfrac{2}{5}r^2$
Spherical shell, external radius, r_1, internal radius r_2	Any diameter	$m\dfrac{2}{5}\dfrac{(r_1^5 - r_2^5)}{(r_1^3 - r_2^3)}$
Spherical shell, very thin, mean radius, r	Any diameter	$m\dfrac{2}{3}r^2$
Right circular cylinder of radius r, length l	The longitudinal axis of the solid	$m\dfrac{r^2}{2}$
Right circular cylinder of radius r, length l	Transverse diameter	$m\left(\dfrac{r^2}{4} + \dfrac{l^2}{12}\right)$
Hollow circular cylinder, length l, radii r_1 and r_2	The longitudinal axis of the figure	$m\dfrac{(r_1^2 + r_2^2)}{2}$
Thin cylindrical shell, length l, mean radius, r	The longitudinal axis of the figure	mr^2
Hollow circular cylinder, length l, radii r_1 and r_2	Transverse diameter	$m\left[\dfrac{r_1^2 + r_2^2}{4} + \dfrac{l^2}{12}\right]$
Hollow circular cylinder, length l, very thin, mean radius	Transverse diameter	$m\left(\dfrac{r^2}{2} + \dfrac{l^2}{12}\right)$
Elliptic cylinder, length l, transverse semiaxes a and b	Longitudinal axis	$m\left(\dfrac{a^2 + b^2}{4}\right)$
Right cone, altitude h, radius of base r	Axis of the figure	$m\dfrac{3}{10}r^2$
Spheroid of revolution, equatorial radius r	Polar axis	$m\dfrac{2r^2}{5}$
Ellipsoid, axes $2a$, $2b$, $2c$	Axis $2a$	$m\dfrac{(b^2 + c^2)}{5}$

Differential equation	Method of solution
Separation of variables $f_1(x) g_1(y) dx + f_2(x) g_2(y) dy = 0$	$$\int \frac{f_1(x)}{f_2(x)} dx + \int \frac{g_2(y)}{g_1(y)} dy = c$$
Exact equation $M(x,y)dx + N(x,y)dy = 0$ where $\partial M / \partial y = \partial N / \partial x$	$$\int M \, \partial x + \int \left(N - \frac{\partial}{\partial y} \int M \, \partial x \right) dy = c$$ where ∂x indicates that the integration is to be performed with respect to x keeping y constant.
Linear first order equation $\frac{dy}{dx} + P(x)y = Q(x)$	$$ye^{\int P dx} = \int Q e^{\int P dx} \, dx + c$$
Bernoulli's equation $\frac{dy}{dx} + P(x)y = Q(x)y^n$	$$ve^{(1-n)\int P dx} = (1-n) \int Q e^{(1-n)\int P dx} \, dx + c$$ where $v = y^{1-n}$. If $n = 1$, the solution is $$\ln y = \int (Q - P) \, dx + c$$
Homogeneous equation $\frac{dy}{dx} = F\left(\frac{y}{x}\right)$	$$\ln x = \int \frac{dv}{F(v) - v} + c$$ where $v = y/x$. If $F(v) = v$, the solution is $y = cx$
Reducible to homogeneous $(a_1 x + b_1 y + c_1) dx + (a_2 x + b_2 y + c_2) dy = 0$ $\frac{a_1}{a_2} \neq \frac{b_1}{b_2}$	Set $u = a_1 x + b_1 y + c_1$ $v = a_2 x + b_2 y + c_2$ Eliminate x and y and the equation becomes homogenous
Reducible to separable $(a_1 x + b_1 y + c_1) dx + (a_2 x + b_2 y + c_2) dy = 0$ $\frac{a_1}{a_2} = \frac{b_1}{b_2}$	Set $u = a_1 x + b_1 y$ Eliminate x or y and equation becomes separable

$y\,F(xy)\,dx + x\,G(xy)dy = 0$	$\ln x = \displaystyle\int \frac{G(v)\,dv}{v\,\{G(v) - F(v)\,\}} + c$ where $v = xy$. If $G(v) = F(v)$, the solution is $xy = c$.
Linear, homogeneous second order equation $\dfrac{d^2 y}{dx^2} + b\,\dfrac{dy}{dx} + cy = 0$ b, c are real constants	Let m_1, m_2 be the roots of $m^2 + bm + c = 0$. Then there are 3 cases: Case 1. m_1, m_2 real and distinct: $$y = c_1 e^{m_1 x} + c_2 e^{m_2 x}$$ Case 2. m_1, m_2 real and equal: $$y = c_1 e^{m_1 x} + c_2 x e^{m_1 x}$$ Case 3. $m_1 = p + qi,\ m_2 = p - qi$: $$y = e^{px}(c_1 \cos qx + c_2 \sin qx)$$ where $p = -b/2,\ q = \sqrt{4c - b^2}/2$
Linear, nonhomogeneous second order equation $\dfrac{d^2 y}{dx^2} + b\,\dfrac{dy}{dx} + cy = R(x)$ $b,\ c$ are real constants	There are 3 cases corresponding to those immediately above: Case 1. $$y = c_1 e^{m_1 x} + c_2 e^{m_2 x}$$ $$+ \frac{e^{m_1 x}}{m_1 - m_2} \int e^{-m_1 x} R(x)\,dx$$ $$+ \frac{e^{m_2 x}}{m_2 - m_1} \int e^{-m_2 x} R(x)\,dx$$ Case 2. $$y = c_1 e^{m_1 x} + c_2 x e^{m_1 x}$$ $$+ x e^{m_1 x} \int e^{-m_1 x} R(x)\,dx$$ $$- e^{m_1 x} \int x e^{-m_1 x} R(x)\,dx$$ Case 3. $$y = e^{px}(c_1 \cos qx + c_2 \sin qx)$$ $$+ \frac{e^{px} \sin qx}{q} \int e^{-px} R(x) \cos qx\,dx$$ $$- \frac{e^{px} \cos qx}{q} \int e^{-px} R(x) \sin qx$$

Euler or Cauchy equation $$x^2 \frac{d^2 y}{dx^2} + bx \frac{dy}{dx} + cy = S(x)$$	Putting $x = e^t$, the equation becomes $$\frac{d^2 y}{dt^2} + (b - 1) \frac{dy}{dt} + cy = S(e^t)$$ and can then be solved as a linear second order equation.
Bessel's equation $$x^2 \frac{d^2 y}{dx^2} + x \frac{dy}{dx} + (\lambda^2 x^2 - n^2)y = 0$$	$$y = c_1 J_n(\lambda x) + c_2 Y_n(\lambda x)$$ See pages 401, 405, 406, 435.
Transformed Bessel's equation $$x^2 \frac{d^2 y}{dx^2} + (2p + 1)x \frac{dy}{dx} + (\alpha^2 x^{2r} + \beta^2)y = 0$$	$$y = x^{-p} \left\{ c_1 J_{q/r}\left(\frac{\alpha}{r} x^r\right) + c_2 Y_{q/r}\left(\frac{\alpha}{r} x^r\right) \right\}$$ where $q = \sqrt{p^2 - \beta^2}$.
Legendre's equation $$(1 - x^2) \frac{d^2 y}{dx^2} - 2x \frac{dy}{dx} + n(n + 1)y = 0$$	$$y = c_1 P_n(x) + c_2 Q_n(x)$$ See page 424

SPECIAL FORMULAS

Certain types of differential equations occur sufficiently often to justify the use of formulas for the corresponding particular solutions. The following set of tables I to XIV covers all first, second and nth order ordinary linear differential equations with constant coefficients for which the right members are of the form $P(x) e^{rx} \sin sx$ or $P(x) e^{rx} \cos sx$, where r and s are constants and $P(x)$ is a polynomial of degree n.

When the right member of a reducible linear partial differential equation with constant coefficients is not zero, particular solutions for certain types of right members are contained in tables XV to XXI. In these tables both F and P are used to denote polynomials, and it is assumed that no denominator is zero. In any formula the roles of x and y may be reversed throughout, changing a formula in which x dominates to one in which y dominates. Tables XIX, XX, XXI are applicable whether the equations are reducible or not.

The symbol $\binom{m}{n}$ stands for $\dfrac{m!}{(m-n)!\,n!}$ and is the $n+1$ st coefficient in the expansion of $(a+b)^m$. Also $0! = 1$ by definition.

The tables as herewith given are those contained in the text *Differential Equations* by Ginn and Company (1955) and are published with their kind permission and that of the author, Professor Frederick H. Steen.

Solution of Linear Differential Equations with Constant Coefficients

Any linear differential equation with constant coefficients may be written in the form

$$p(D)y = R(x),$$

where D is the differential operator

$$Dy = \frac{dy}{dx},$$

$p(D)$ is a polynomial in D,
y is the dependent variable,
x is the independent variable,
$R(x)$ is an arbitrary function of x.

A power of D represents repeated differentiation, that is

$$D^n y = \frac{d^n y}{dx^n}.$$

For such an equation, the general solution may be written in the form

$$y = y_c + y_p,$$

where y_p is any particular solution, and y_c is called the *complementary function*. This complementary function is defined as the general solution of the *homogeneous equation*, which is the original differential equation with the right side replaced by zero, i.e.

$$p(D)y = 0$$

The complementary function y_c may be determined as follows:

1. Factor the polynomial $p(D)$ into real and complex linear factors, just as if D were a variable instead of an operator.

2. For each non-repeated linear factor of the form $(D - a)$, where a is real, write down a term of the form

$$ce^{ax},$$

where c is an arbitrary constant.

3. For each repeated real linear factor of the form $(D - a)^n$, write down n terms of the form

$$c_1e^{ax} + c_2xe^{ax} + c_3x^2e^{ax} + \cdots + c_nx^{n-1}e^{ax},$$

where the c_i's are arbitrary constants.

4. For each non-repeated conjugate complex pair of factors of the form $(D - a + ib)(D - a - ib)$, write down 2 terms of the form

$$c_1e^{ax}\cos bx + c_2e^{ax}\sin bx$$

5. For each repeated conjugate complex pair of factors of the form $(D - a + ib)^n(D - a - ib)^n$, write down $2n$ terms of the form

$$c_1e^{ax}\cos bx + c_2e^{ax}\sin bx + c_3xe^{ax}\cos bx + c_4xe^{ax}\sin bx + \cdots$$
$$+ c_{2n-1}x^{n-1}e^{ax}\cos bx + c_{2n}x^{n-1}e^{ax}\sin bx$$

6. The sum of all the terms thus written down is the complementary function y_c.

To find the particular solution y_p, use the following tables, as shown in the examples. For cases not shown in the tables, there are various methods of finding y_p. The most general method is called *variation of parameters*. The following example illustrates the method:

Find y_p for $(D^2 - 4)y = e^x$.

This example can be solved most easily by use of equation 63 in the tables following. However it is given here as an example of the method of variation of parameters.

The complementary function is

$$y_c = c_1e^{2x} + c_2e^{-2x}$$

To find y_p, replace the constants in the complementary function with unknown functions,

$$y_p = ue^{2x} + ve^{-2x}.$$

We now prepare to substitute this assumed solution into the original equation. We begin by taking all the necessary derivatives:

$$y_p = ue^{2x} + ve^{-2x}$$
$$y_p' = 2ue^{2x} - 2ve^{-2x} + u'e^{2x} + v'e^{-2x}$$

For each derivative of y_p except the highest, we set the sum of all the terms containing u' and v' to 0. Thus the above equation becomes

$$u'e^{2x} + v'e^{-2x} = 0 \quad \text{and} \quad y_p' = 2ue^{2x} - 2ve^{-2x}$$

Continuing to differentiate, we have

$$y_p'' = 4ue^{2x} + 4ve^{-2x} + 2u'e^{2x} - 2v'e^{-2x}$$

When we substitute into the original equation, all the terms not containing u' or v' cancel out. This is a consequence of the method by which y_p was set up.

Thus all that is necessary is to write down the terms containing u' or v' in the highest order derivative of y_p, multiply by the constant coefficient of the highest power of D in $p(D)$, and set it equal to $R(x)$. Together with the previous terms in u' and v' which were set equal to 0, this gives us as many linear equations in the first derivatives of the unknown functions as there are unknown functions. The first derivatives may then be solved for by algebra, and the unknown functions found by integration. **In the present example, this becomes**

$$u'e^{2x} + v'e^{-2x} = 0$$
$$2u'e^{2x} - 2v'e^{-2x} = e^x.$$

We eliminate v' and u' separately, getting

$$4u'e^{2x} = e^x$$
$$4v'e^{-2x} = -e^x.$$

Thus

$$u' = \tfrac{1}{4}e^{-x}$$
$$v' = -\tfrac{1}{4}e^{3x}.$$

Therefore, by integrating

$$u = -\tfrac{1}{4}e^{-x}$$
$$v = -\tfrac{1}{12}e^{3x}.$$

A constant of integration is not needed, since we need only one particular solution. **Thus**

$$y_p = ue^{2x} + ve^{-2x} = -\tfrac{1}{4}e^{-x}e^{2x} - \tfrac{1}{12}e^{3x}e^{-2x}$$
$$= -\tfrac{1}{4}e^x - \tfrac{1}{12}e^x = -\tfrac{1}{3}e^x,$$

and the general solution is

$$y = y_c + y_p = c_1e^{2x} + c_2e^{-2x} - \tfrac{1}{3}e^x.$$

The following examples illustrate the use of the tables.
Example 1.   Solve $(D^2 - 4)y = \sin 3x$.
Substitution of $q = -4, s = 3$ in formula 24 gives

$$y_p = \frac{\sin 3x}{-9 - 4},$$

wherefore the general solution is

$$y = c_1e^{2x} + c_2e^{-2x} - \frac{\sin 3x}{13}.$$

Example 2.   Obtain a particular solution of $(D^2 - 4D + 5)y = x^2e^{3x}\sin x$.
Applying formula 40 with $a = 2, b = 1, r = 3, s = 1, P(x) = x^2, s + b = 2, s - b = 0,$ $a - r = -1, (a - r)^2 + (s + b)^2 = 5, (a - r)^2 + (s - b)^2 = 1,$ **we have**

$$y_p = \frac{e^{3x}\sin x}{2}\left[\left(\frac{2}{5} - \frac{0}{1}\right)x^2 + \left(\frac{2(-1)2}{25} - \frac{2(-1)0}{1}\right)2x\right.$$
$$\left. + \left(\frac{3\cdot1\cdot2 - 2^3}{125} - \frac{3\cdot1\cdot0 - 0}{1}\right)2\right]$$
$$- \frac{e^{3x}\cos x}{2}\left[\left(\frac{-1}{5} - \frac{-1}{1}\right)x^2 + \left(\frac{1 - 4}{25} - \frac{1 - 0}{1}\right)2x\right.$$
$$\left. + \left(\frac{-1 - 3(-1)4}{125} - \frac{-1 - 3(-1)0}{1}\right)2\right]$$
$$= (\tfrac{1}{5}x^2 - \tfrac{4}{25}x - \tfrac{2}{125})e^{3x}\sin x + (-\tfrac{2}{5}x^2 + \tfrac{28}{25}x - \tfrac{138}{125})e^{3x}\cos x.$$

The special formulas effect a very considerable saving of time in problems of this type.

Example 3. Obtain a particular solution of $(D^2 - 4D + 5)y = x^2 e^{2x} \cos x$. (Compare with Example 2.)

Formula 40 is not applicable here since for this equation $r = a$, $s = b$, wherefore the denominator $(a - r)^2 + (s - b)^2 = 0$. We turn instead to formula 44. Substituting $a = 2, b = 1, P(x) = x^2$ and replacing sin by cos, cos by $-\sin$, we obtain

$$y_p = \frac{e^{2x} \cos x}{4} (x^2 - \tfrac{2}{4}) + \frac{e^{2x} \sin x}{2} \int (x^2 - \tfrac{1}{2}) \, dx$$

$$= \left(\frac{x^2}{4} - \frac{1}{8}\right) e^{2x} \cos x + \left(\frac{x^3}{6} - \frac{x}{4}\right) e^{2x} \sin x,$$

which is the required solution.

Example 4. Find z_p for $(D_x - 3D_y)z = \ln(y + 3x)$.

Referring to Table XV we note that formula 69 (not 68) is applicable. This gives

$$z_p = x \ln(y + 3x).$$

It is easily seen that $-\dfrac{y}{3} \ln(y + 3x)$ would serve equally well.

Example 5. Solve $(D_x + 2D_y - 4)z = y \cos(y - 2x)$.

Since R in formula 76 contains a polynomial in x, not y, we rewrite the given equation in the form

$$(D_y + \tfrac{1}{2}D_x - 2)z = \tfrac{1}{2}y \cos(y - 2x).$$

Then

$$z_c = e^{2y}F(x - \tfrac{1}{2}y) = e^{2y}f(2x - y),$$

and by the formula

$$z_p = -\tfrac{1}{2}\cos(y - 2x) \cdot \left(\frac{y}{2} + \frac{\tfrac{1}{2}}{2}\right)$$

$$= -\tfrac{1}{8}(2y + 1)\cos(y - 2x).$$

Example 6. Find z_p for $(D_x + 4D_y)^3 z = (2x - y)^2$.

Using formula 79, we obtain

$$z_p = \frac{\int\int\int u^2 du^3}{[2 + 4(-1)]^3} = \frac{u^5}{5 \cdot 4 \cdot 3 \cdot (-8)} = -\frac{(2x - y)^5}{480}.$$

Example 7. Find z_p for $(D_x^3 + 5D_x^2 D_y - 7D_x + 4)z = e^{2x+3y}$.

By formula 87

$$z_p = \frac{e^{2x+3y}}{2^3 + 5 \cdot 2^2 \cdot 3 - 7 \cdot 2 + 4} = \frac{e^{2x+3y}}{58}.$$

Example 8. Find z_p for

$$(D_x^4 + 6D_x^3 D_y + D_x D_y + D_y^2 + 9)z = \sin(3x + 4y).$$

Since every term in the left member is of *even* degree in the two operators D_x and D_y, formula 90 is applicable.

It gives

$$z_p = \frac{\sin(3x + 4y)}{(-9)^2 + 6(-9)(-12) + (-12) + (-16) + 9}$$

$$= \frac{\sin(3x + 4y)}{710}$$

DIFFERENTIAL EQUATIONS

TABLE I: $(D-a)y = R$

R	y_p
1. e^{rz}	$\dfrac{e^{rz}}{r-a}$
2. $\sin sz$*	$-\dfrac{a\sin sz + s\cos sz}{a^2+s^2} = -\dfrac{1}{\sqrt{a^2+s^2}}\sin\left(sz + \tan^{-1}\dfrac{s}{a}\right)$
3. $P(x)$	$-\dfrac{1}{a}\left[P(x) + \dfrac{P'(x)}{a} + \dfrac{P''(x)}{a^2} + \cdots + \dfrac{P^{(n)}(x)}{a^n}\right]$
4. $e^{rz}\sin sz$*	Replace a by $a-r$ in formula 2 and multiply by e^{rz}.
5. $P(x)e^{rz}$	Replace a by $a-r$ in formula 3 and multiply by e^{rz}.
6. $P(x)\sin sz$*	$-\sin sz\left[\dfrac{a}{a^2+s^2}P(x) + \dfrac{a^2-s^2}{(a^2+s^2)^2}P'(x) + \dfrac{a^3-3as^2}{(a^2+s^2)^3}P''(x) + \cdots\right.$ $\left. + \dfrac{a^k - \binom{k}{2}a^{k-2}s^2 + \binom{k}{4}a^{k-4}s^4 - \cdots}{(a^2+s^2)^k}P^{(k-1)}(x) + \cdots\right]$ $\qquad -\cos sz\left[\dfrac{s}{a^2+s^2}P(x) + \dfrac{2as}{(a^2+s^2)^2}P'(x) + \dfrac{3a^2s-s^3}{(a^2+s^2)^3}P''(x) + \cdots\right.$ $\left. + \dfrac{\binom{k}{1}a^{k-1}s - \binom{k}{3}a^{k-3}s^3 + \cdots}{(a^2+s^2)^k}P^{(k-1)}(x) + \cdots\right]$
7. $P(x)e^{rz}\sin sz$*	Replace a by $a-r$ in formula 6 and multiply by e^{rz}.
8. e^{az}	ze^{az}
9. $e^{az}\sin sz$*	$-\dfrac{e^{az}\cos sz}{s}$
10. $P(x)e^{az}$	$e^{az}\displaystyle\int P(x)\,dx$
11. $P(x)e^{az}\sin sz$*	$\dfrac{e^{az}\sin sz}{s}\left[P'(x) - \dfrac{P'''(x)}{s^3} + \dfrac{P^v(x)}{s^5} - \cdots\right] - \cdots - \dfrac{e^{az}\cos sz}{s}\left[P(x) - \dfrac{P''(x)}{s^2} + \dfrac{P^{iv}(x)}{s^4} - \cdots\right]$

* For $\cos sz$ in R replace "sin" by "cos" and "cos" by "$-$ sin" in y_p.

$$D^n = \frac{d^n}{dx^n} \qquad \binom{m}{n} = \frac{m!}{(m-n)!\,n!} \qquad 0! = 1$$

DIFFERENTIAL EQUATIONS (Continued)

TABLE II: $(D-a)^2 y = R$

R	y_p
12. e^{rx}	$\dfrac{e^{rx}}{(r-a)^2}$
13. $\sin sx^*$	$\dfrac{1}{(a^2+s^2)^2}[(a^2-s^2)\sin sx + 2as\cos sx] = \dfrac{1}{a^2+s^2}\sin\left(sx+\tan^{-1}\dfrac{2as}{a^2-s^2}\right)$
14. $P(x)$	$\dfrac{1}{a^2}\left[P(x)+\dfrac{2P'(x)}{a}+\dfrac{3P''(x)}{a^2}+\cdots+\dfrac{(n+1)P^{(n)}(x)}{a^n}\right]$
15. $P(x)e^{rx}\sin sx^*$	Replace a by $a-r$ in formula 13 and multiply by e^{rx}.
16. $P(x)e^{rx}$	Replace a by $a-r$ in formula 14 and multiply by e^{rx}.
17. $P(x)\sin sx^*$	$\sin sx\left[\dfrac{a^2-s^2}{(a^2+s^2)^2}P(x)+2\dfrac{a^3-3as^2}{(a^2+s^2)^3}P'(x)+3\dfrac{a^4-6a^2s^2+s^4}{(a^2+s^2)^4}P''(x)+\cdots\right.$ $\left. +(k-1)\dfrac{a^k-\binom{k}{2}a^{k-2}s^2+\binom{k}{4}a^{k-4}s^4-\cdots}{(a^2+s^2)^k}P^{(k-2)}(x)+\cdots\right]$ $+\cos sx\left[\dfrac{2as}{(a^2+s^2)^2}P(x)+2\dfrac{3a^2s-s^3}{(a^2+s^2)^3}P'(x)+3\dfrac{4a^3s-4as^3}{(a^2+s^2)^4}P''(x)+\cdots\right.$ $\left. +(k-1)\dfrac{\binom{k}{1}a^{k-1}s-\binom{k}{3}a^{k-3}s^3+\cdots}{(a^2+s^2)^k}P^{(k-2)}(x)+\cdots\right]$
18. $P(x)e^{rx}\sin sx^*$	Replace a by $a-r$ in formula 17 and multiply by e^{rx}.
19. e^{ax}	$\tfrac{1}{2}x^2e^{ax}$
20. $e^{ax}\sin sx^*$	$-\dfrac{e^{ax}\sin sx}{s^2}$
21. $P(x)e^{ax}$	$e^{ax}\iint P(x)\,dx\,dx$
22. $P(x)e^{ax}\sin sx^*$	$\dfrac{e^{ax}\sin sx}{s^2}\left[P(x)-\dfrac{3P''(x)}{s^2}+\dfrac{5P^{iv}(x)}{s^4}-\dfrac{7P^{vi}(x)}{s^6}+\cdots\right]$ $-\dfrac{e^{ax}\cos sx}{s^2}\left[\dfrac{2P'(\cdot)}{s}-\dfrac{4P'''(x)}{s^3}+\dfrac{6P^v(x)}{s^5}-\cdots\right]$

* For $\cos sx$ in R replace "sin" by "cos" and "cos" by "$-$ sin" in y_p.

DIFFERENTIAL EQUATIONS (Continued)

TABLE III: $(D^2 + q)y = R$

R	y_p
23. e^{rx}	$\dfrac{e^{rx}}{r^2 + q}$
24. $\sin sx*$	$\dfrac{\sin sx}{-s^2 + q}$
25. $P(x)$	$\dfrac{1}{q}\left[P(x) - \dfrac{P''(x)}{q} + \dfrac{P^{iv}(x)}{q^2} - \cdots + (-1)^k\,\dfrac{P^{(2k)}(x)}{q^k}\cdots\right]$
26. $e^{rx}\sin sx*$	$\dfrac{(r^2 - s^2 + q)e^{rx}\sin sx - 2rse^{rx}\cos sx}{(r^2 - s^2 + q)^2 + (2rs)^2}$

$$= \frac{e^{rx}}{\sqrt{(r^2 - s^2 + q)^2 + (2rs)^2}}\sin\left[sx - \tan^{-1}\frac{2rs}{r^2 - s^2 + q}\right]$$

27. $P(x)e^{rx}$

$$\frac{e^{rx}}{r^2 + q}\left[P(x) - \frac{2r}{r^2 + q}P'(x) + \frac{3r^2 - q}{(r^2 + q)^2}P''(x) - \frac{4r^3 - 4qr}{(r^2 + q)^3}P'''(x)\cdots\right.$$
$$+ \cdots + (-1)^{k-1}\frac{\binom{k}{1}r^{k-1} - \binom{k}{3}r^{k-3}q + \binom{k}{5}r^{k-5}q^2 - \cdots}{(r^2 + q)^{k-1}}P^{(k-1)}(x) + \cdots\left.\right]$$

28. $P(x)\sin sx*$

$$\frac{\sin sx}{(-s^2 + q)}\left[P(x) - \frac{3s^2 + q}{(-s^2 + q)^2}P''(x) + \frac{5s^4 + 10s^2q + q^2}{(-s^2 + q)^4}P^{iv}(x) + \cdots\right.$$
$$+ (-1)^k\frac{\binom{2k+1}{1}s^{2k} + \binom{2k+1}{3}s^{2k-2}q + \binom{2k+1}{5}s^{2k-4}q^2 + \cdots}{(-s^2 + q)^{2k}}P^{(2k)}(x) + \cdots\left.\right]$$
$$- \frac{s\cos sx}{(-s^2 + q)}\left[\frac{2P'(x)}{(-s^2 + q)} - \frac{4s^2 + 4q}{(-s^2 + q)^3}P'''(x) + \cdots\right.$$
$$+ (-1)^{k+1}\frac{\binom{2k}{1}s^{2k-2} + \binom{2k}{3}s^{2k-4}q + \cdots}{(-s^2 + q)^{2k-1}}P^{(2k-1)}(x) + \cdots\left.\right]$$

TABLE IV: $(D^2 + b^2)y = R$

R	y_p
29. $\sin bx*$	$-\dfrac{x\cos bx}{2b}$

30. $P(x)\sin bx*$

$$\frac{\sin bx}{(2b)^2}\left[P(x) - \frac{P''(x)}{(2b)^2} + \frac{P^{iv}(x)}{(2b)^4} - \cdots\right] - \frac{x\cos bx}{2b} - \frac{\cos bx}{2b}\int\left[P(x) - \frac{P''(x)}{(2b)^2} + \cdots\right]dx$$

* For cos sx in R replace "sin" by "cos" and "cos" by "− sin" in y_p.

DIFFERENTIAL EQUATIONS (Continued)

TABLE V: $(D^2 + pD + q)y = R$

R	y_p
31. e^{rx}	$\dfrac{e^{rx}}{r^2 + pr + q}$
32. $\sin sx$*	$\dfrac{(q - s^2)\sin sx - ps\cos sx}{(q - s^2)^2 + (ps)^2} = \dfrac{1}{\sqrt{(q - s^2)^2 + (ps)^2}}\sin\left(sx - \tan^{-1}\dfrac{ps}{q - s^2}\right)$
33. $P(x)$	$\dfrac{1}{q}\Big[P(x) - \dfrac{p}{q}P'(x) + \dfrac{p^2 - q}{q^2}P''(x) - \dfrac{p^3 - 2pq}{q^3}P'''(x) + \cdots$ $+ (-1)^n\,\dfrac{p^n - \binom{n-1}{1}p^{n-2}q + \binom{n-2}{2}p^{n-4}q^2 - \cdots}{q^n}P^{(n)}(x)\Big]$
34. $e^{rx}\sin sx$*	Replace p by $p + 2r$, q by $q + pr + r^2$ in formula 32 and multiply by e^{rx}.
35. $P(x)e^{rx}$	Replace p by $p + 2r$, q by $q + pr + r^2$ in formula 33 and multiply by e^{rx}.

TABLE VI: $(D - b)(D - a)y = R$

R	y_p
36. $P(x)\sin sx$*	$\dfrac{\sin sx}{b - a}\left[\left(\dfrac{a}{a^2 + s^2} - \dfrac{b}{b^2 + s^2}\right)P(x) + \left(\dfrac{a^2 - s^2}{(a^2 + s^2)^2} - \dfrac{b^2 - s^2}{(b^2 + s^2)^2}\right)P'(x)\right.$ $\left. + \left(\dfrac{a^3 - 3as^2}{(a^2 + s^2)^3} - \dfrac{b^3 - 3bs^2}{(b^2 + s^2)^3}\right)P''(x) + \cdots\right]$ $+ \dfrac{\cos sx}{b - a}\left[\left(\dfrac{s}{a^2 + s^2} - \dfrac{s}{b^2 + s^2}\right)P(x) + \left(\dfrac{2as}{(a^2 + s^2)^2} - \dfrac{2bs}{(b^2 + s^2)^2}\right)P'(x)\right.$ $\left. + \left(\dfrac{3a^2s - s^3}{(a^2 + s^2)^3} - \dfrac{3b^2s - s^3}{(b^2 + s^2)^3}\right)P''(x) + \cdots\right]$†
37. $P(x)e^{rx}\sin sx$*	Replace a by $a - r$, b by $b - r$ in formula 36 and multiply by e^{rx}.
38. $P(x)e^{ax}$	$\dfrac{e^{ax}}{a - b}\left[\int\!\!\int P(x)dx + \dfrac{P(x)}{(b - a)} + \dfrac{P'(x)}{(b - a)^2} + \dfrac{P''(x)}{(b - a)^3} + \cdots + \dfrac{P^{(n)}(x)}{(b - a)^{n+1}}\right]$

* For cos sx in R replace "sin" by "cos" and "cos" by "− sin" in y_p.

† For additional terms, compare with formula 6.

DIFFERENTIAL EQUATIONS (Continued)

TABLE VII: $(D^2 - 2aD + a^2 + b^2)y = R$

R	y_p
39. $P(x)\sin sx$*	$\dfrac{\sin sx}{2b}\left[\left(\dfrac{s+b}{a^2+(s+b)^2} - \dfrac{s-b}{a^2+(s-b)^2}\right)P(x) + \left(\dfrac{2a(s+b)}{[a^2+(s+b)^2]^2} - \dfrac{2a(s-b)}{[a^2+(s-b)^2]^2}\right)P'(x)\right.$ $\left. + \left(\dfrac{3a^2(s+b)-(s+b)^3}{[a^2+(s+b)^2]^3} - \dfrac{3a^2(s-b)-(s-b)^3}{[a^2+(s-b)^2]^3}\right)P''(x) + \cdots\right]$ $- \dfrac{\cos sx}{2b}\left[\left(\dfrac{a}{a^2+(s+b)^2} - \dfrac{a}{a^2+(s-b)^2}\right)P(x) + \left(\dfrac{a^2-(s+b)^2}{[a^2+(s+b)^2]^2} - \dfrac{a^2-(s-b)^2}{[a^2+(s-b)^2]^2}\right)P'(x)\right.$ $\left. + \left(\dfrac{a^3-3a(s+b)^2}{[a^2+(s+b)^2]^3} - \dfrac{a^3-3a(s-b)^2}{[a^2+(s-b)^2]^3}\right)P''(x) + \cdots\right]$†
40. $P(x)e^{rx}\sin sx$*	Replace a by $a-r$ in formula 39 and multiply by e^{rx}.
41. $P(x)e^{ax}$	$\dfrac{e^{ax}}{b^2}\left[P(x) - \dfrac{P''(x)}{b^2} + \dfrac{P^{iv}(x)}{b^4} - \cdots\right]$
42. $e^{ax}\sin sx$*	$\dfrac{e^{ax}\sin sx}{-s^2+b^2}$
43. $e^{ax}\sin bx$*	$-\dfrac{xe^{ax}\cos bx}{2b}$
44. $P(x)e^{ax}\sin bx$*	$\dfrac{e^{ax}\sin bx}{(2b)^2}\left[P(x) - \dfrac{P''(x)}{(2b)^2} + \dfrac{P^{iv}(x)}{(2b)^4} - \cdots\right] - \dfrac{e^{ax}\cos bx}{2b}\int\left[P(x) - \dfrac{P''(x)}{(2b)^2} + \dfrac{P^{iv}(x)}{(2b)^4} - \cdots\right]dx$

* For cos sx in R replace "sin" by "cos" and "cos" by "— sin" in y_p.
† For additional terms, compare with formula 6.

DIFFERENTIAL EQUATIONS (Continued)

TABLE VIII: $f(D)y = [D^n + a_{n-1}D^{n-1} + \cdots + a_1 D + a_0]y = R$

R	y_p
45. e^{rx}	$\dfrac{e^{rx}}{f(r)}$
46. $\sin sx$*	$\dfrac{[a_0 - a_2 s^2 + a_4 s^4 - \cdots]\sin sx - [a_1 s - a_3 s^3 + a_5 s^5 + \cdots]\cos sx}{[a_0 - a_2 s^2 + a_4 s^4 - \cdots]^2 + [a_1 s - a_3 s^3 + a_5 s^5 - \cdots]^2}$

TABLE IX: $f(D^2)y = R$

R	y_p
47. $\sin sx$*	$\dfrac{\sin sx}{f(-s^2)} = \dfrac{\sin sx}{a_0 - a_2 s^2 + \cdots \pm s^{2n}}$

TABLE X: $(D - a)^n y = R$

R	y_p
48. e^{rx}	$\dfrac{e^{rx}}{(r - a)^n}$
49. $\sin sx$*	$\dfrac{(-1)^n}{(a^2 + s^2)^n}\{[a^n - \binom{n}{2}a^{n-2}s^2 + \binom{n}{4}a^{n-4}s^4 - \cdots]\sin sx + [\binom{n}{1}a^{n-1}s - \binom{n}{3}a^{n-3}s^3 + \cdots]\cos sx\}$
50. $P(x)$	$\dfrac{(-1)^n}{a^n}\left[P(x) + \binom{n}{1}\dfrac{P'(x)}{a} + \binom{n+1}{2}\dfrac{P''(x)}{a^2} + \binom{n+2}{3}\dfrac{P'''(x)}{a^3} + \cdots\right]$
51. $e^{rx}\sin sx$*	Replace a by $a - r$ in formula 49 and multiply by e^{rx}.
52. $e^{rx}P(x)$	Replace a by $a - r$ in formula 50 and multiply by e^{rx}.

* For $\cos sx$ in R replace "sin" by "cos" and "cos" by "− sin" in y_p.

DIFFERENTIAL EQUATIONS (Continued)

53. $P(x) \sin sx$*

$(-1)^n \sin sx[A_nP(x) + \binom{n}{1}A_{n+1}P'(x) + \binom{n+1}{2}A_{n+2}P''(x) + \binom{n+1}{3}A_{n+3}P'''(x) + \cdots]$

$+ (-1)^n \cos sx[B_nP(x) + \binom{n}{1}B_{n+1}P'(x) + \binom{n+1}{2}B_{n+2}P''(x) + \binom{n+1}{3}B_{n+3}P'''(x) + \cdots]$

$A_1 = \dfrac{a}{a^2 + s^2}$, $A_2 = \dfrac{a^2 - s^2}{(a^2+s^2)^2}$, $\cdots$, $A_k = \dfrac{a^k - \binom{k}{2}a^{k-2}s^2 + \binom{k}{4}a^{k-4}s^4 - \cdots}{(a^2+s^2)^k}$

$B_1 = \dfrac{s}{a^2+s^2}$, $B_2 = \dfrac{2as}{(a^2+s^2)^2}$, $\cdots$, $B_k = \dfrac{\binom{k}{1}a^{k-1}s - \binom{k}{3}a^{k-3}s^3 + \cdots}{(a^2+s^2)^k}$

54. $P(x)e^{rx}\sin sx$* Replace a by $a - r$ in formula 53 and multiply by e^{rx}.

55. $e^{ax}P(x)$ $e^{ax}\displaystyle\int\int\cdots\int P(x)dx^n$

56. $P(x)e^{ax}\sin sx$*

$\dfrac{(-1)^{\frac{n-1}{2}}e^{ax}\sin sx}{s^n}\left[\binom{n}{n-1}\dfrac{P'(x)}{s} - \binom{n+2}{n-1}\dfrac{P'''(x)}{s^3} + \binom{n+4}{n-1}\dfrac{P^\mathrm{v}(x)}{s^5} - \cdots\right]$

$+ \dfrac{(-1)^{\frac{n+1}{2}}e^{ax}\cos sx}{s^n}\left[\binom{n-1}{n-1}P(x) - \binom{n+1}{n-1}\dfrac{P''(x)}{s^2} + \binom{n+3}{n-1}\dfrac{P^\mathrm{iv}(x)}{s^4} - \cdots\right]$ (n odd)

$\dfrac{(-1)^{\frac{n}{2}}e^{ax}\sin sx}{s^n}\left[\binom{n-1}{n-1}P(x) - \binom{n+1}{n-1}\dfrac{P''(x)}{s^2} + \binom{n+3}{n-1}\dfrac{P^\mathrm{iv}(x)}{s^4} - \cdots\right]$

$+ \dfrac{(-1)^{\frac{n}{2}}e^{ax}\cos sx}{s^n}\left[\binom{n}{n-1}\dfrac{P'(x)}{s} - \binom{n+2}{n-1}\dfrac{P'''(x)}{s^3} + \binom{n+4}{n-1}\dfrac{P^\mathrm{v}(x)}{s^5} - \cdots\right]$ (n even)

TABLE XI: $(D - a)^nf(D)y = R$

57. e^{ax} $\dfrac{x^n}{n!}\cdot\dfrac{e^{ax}}{f(a)}$

* For cos sx in R replace "sin" by "cos" and "cos" by " − sin" in y_p.

DIFFERENTIAL EQUATIONS (Continued)

TABLE XII: $(D^2 + q)^n y = R$

R	y_p
58. e^{rx}	$e^{rx}/(r^2+q)^n$
59. $\sin sx$*	$\sin sx/(q - s^2)^n$
60. $P(x)$	$\dfrac{1}{q^n}\left[P(x) - \binom{n}{1}\dfrac{P''(x)}{q} + \binom{n+1}{2}\dfrac{P^{iv}(x)}{q^2} - \binom{n+2}{3}\dfrac{P^{vi}(x)}{q^3} + \cdots\right]$
61. $e^{rx}\sin sx$*	$\dfrac{e^{rx}}{(A^2+B^2)^n}\left\{\left[A^n - \binom{n}{2}A^{n-2}B^2 + \binom{n}{4}A^{n-4}B^4 - \cdots\right]\sin sx - \left[\binom{n}{1}A^{n-1}B - \binom{n}{3}A^{n-2}B^3 + \cdots\right]\cos sx\right\}$

$$A = r^2 - s^2 + q, \quad B = 2rs$$

TABLE XIII: $(D^2 + b^2)^n y = R$

62. $\sin bx$*	$(-1)^{\frac{n+1}{2}}\dfrac{x^n\cos bx}{n!(2b)^n}$ (n odd), $\qquad (-1)^{\frac{n}{2}}\dfrac{x^n\sin bx}{n!(2b)^n}$ (n even)

TABLE XIV: $(D^n - q)y = R$

63. e^{rx}	$e^{rx}/(r^n - q)$
64. $P(x)$	$-\dfrac{1}{q}\left[P(x) + \dfrac{P^{(n)}(x)}{q} + \dfrac{P^{(2n)}(x)}{q^2} + \cdots\right]$
65. $\sin sx$*	$-\dfrac{q\sin sx + (-1)^{\frac{n-1}{2}}s^n\cos sx}{q^2 + s^{2n}}$ (n odd), $\qquad \dfrac{\sin sx}{(-s^2)^{n/2} - q}$ (n even)
66. $e^{rx}\sin sx$*	$\dfrac{Ae^{rx}\sin sx - Be^{rx}\cos sx}{A^2 + B^2} = \dfrac{e^{rx}}{\sqrt{A^2+B^2}}\sin\left(sx - \tan^{-1}\dfrac{B}{A}\right)$

$$A = \left[r^n - \binom{n}{2}r^{n-2}s^2 + \binom{n}{4}r^{n-4}s^4 - \cdots\right] - q, \quad B = \left[\binom{n}{1}r^{n-1}s - \binom{n}{3}r^{n-3}s^3 + \cdots\right]$$

* For cos sx in R replace "sin" by "cos" and "cos" by "− sin" in y_p.

DIFFERENTIAL EQUATIONS (Continued)

TABLE XV: $(D_x + mD_y)z = R$

R	z_p
67. e^{ax+by}	$\dfrac{e^{ax+by}}{a+mb}$
68. $f(ax+by)$	$\dfrac{\int f(u)du}{a+mb}$, $u = ax+by$
69. $f(y-mx)$	$xf(y-mx)$
70. $\phi(x,y)f(y-mx)$	$f(y-mx)\int \phi(x, a+mx)dx$ $(a = y - mx$ after integration)

TABLE XVI: $(D_x + mD_y - k)z = R$

71. e^{ax+by}	$\dfrac{e^{ax+by}}{a+mb-k}$
72. $\sin(ax+by)$*	$-\dfrac{(a+bm)\cos(ax+by)+k\sin(ax+by)}{(a+bm)^2+k^2}$
73. $e^{\alpha x+\beta y}\sin(ax+by)$*	Replace k in 72 by $k - \alpha - m\beta$ and multiply by $e^{\alpha x+\beta y}$
74. $e^{kx}f(ax+by)$	$\dfrac{e^{kx}\int f(u)du}{a+mb}$, $u = ax+by$
75. $f(y-mx)$	$-\dfrac{f(y-mx)}{k}$
76. $P(x)f(y-mx)$	$-\dfrac{1}{k}f(y-mx)\left[P(x) + \dfrac{P'(x)}{k} + \dfrac{P''(x)}{k^2} + \cdots + \dfrac{P^{(n)}(x)}{k^n}\right]$
77. $e^{kx}f(y-mx)$	$xe^{kx}f(y-mx)$

* For $\cos(ax+by)$ replace "sin" by "cos," and "cos" by "$-\sin$" in z_p.

$$D_x = \frac{\partial}{\partial x}; \quad D_y = \frac{\partial}{\partial y}; \quad D_x^k D_y^r = \frac{\partial^{k+r}}{\partial_x^k \partial_y^r}$$

DIFFERENTIAL EQUATIONS (Continued)

TABLE XVII: $(D_x + mD_y)^n z = R$

R	z
78. e^{ax+by}	$\dfrac{e^{ax+by}}{(a+mb)^n}$
79. $f(ax+by)$	$\dfrac{\int\int \cdots \int f(u)du^n}{(a+mb)^n}$, $\ u = ax + by$
80. $f(y - mx)$	$\dfrac{x^n}{n!} f(y - mx)$
81. $\phi(x,y)f(y+mx)$	$f(y - mx)\int\int \cdots \int \phi(x, a+mx)dx^n$ $(a = y - mx$ after integration)

TABLE XVIII: $(D_x + mD_y - k)^n z = R$

R	z
82. e^{ax+by}	$\dfrac{e^{ax+by}}{(a+mb-k)^n}$
83. $f(y - mx)$	$\dfrac{(-1)^n f(y-mx)}{k^n}$
84. $P(x)f(y - mx)$	$\dfrac{(-1)^n}{k^n} f(y-mx)\left[P(x) + \binom{n}{1}\dfrac{P'(x)}{k} + \binom{n+1}{2}\dfrac{P''(x)}{k^2} + \binom{n+2}{3}\dfrac{P'''(x)}{k^3} + \cdots \right]$
85. $e^{kx}f(ax + by)$	$\dfrac{e^{kx}\int\int \cdots \int f(u)du^n}{(a+mb)^n}$, $\ u = ax + by$
86. $e^{kx}f(y - mx)$	$\dfrac{x^n}{n!} e^{kx}f(y - mx)$

DIFFERENTIAL EQUATIONS (Continued)

TABLE XIX: $[D_x^n + a_1 D_x^{n-1} D_y + a_2 D_x^{n-2} D_y^2 + \cdots + a^n D_y^n] z = R$

87. e^{ax+by} $\dfrac{e^{ax+by}}{a + a_1 a^{n-1} b + a_2 a^{n-2} b^2 + \cdots + a_n b^n}$

88. $f(ax + by)$ $\dfrac{\int\int \cdots \int f(u) du^n}{a^n + a_1 a^{n-1} b + a_2 a^{n-2} b^2 + \cdots + a^n b^n}$ $(u = ax + by)$

TABLE XX: $F(D_x, D_y) z = R$

89. e^{ax+by} $\dfrac{e^{ax+by}}{F(a, b)}$

TABLE XXI: $F(D_x^2, D_x D_y, D_y^2) z = R$

90. $\sin (ax + by)$* $\dfrac{\sin (ax + by)}{F(-a^2, -ab, -b^2)}$

* For $\cos (ax + by)$ replace "sin" by "cos", and "cos" by "$-\sin$" in z.

COMPLEX VARIABLES

Complex Numbers

Cartesian Form

The cartesian form of a complex number is $z = x + iy$, where x and y are real numbers and i, called the imaginary unit, has the property that $i^2 = -1$. The real numbers x and y are called the real and imaginary parts of $x + iy$, respectively.

Polar Form

$$z = re^{i\theta} = r (\cos \theta + i \sin \theta)$$

Modulus

$$r = |z| = (x^2 + y^2)^{1/2}$$

Argument

$$\theta = \arg z = \arctan \frac{y}{x}$$

Complex Conjugate

$$\bar{z} = x - iy, \quad |\bar{z}| = |z|, \quad \arg \bar{z} = -\arg z$$

Addition and Subtraction

$$z_1 \pm z_2 = (x_1 + iy_1) \pm (x_2 + iy_2) = (x_1 \pm x_2) + (y_1 \pm y_2)$$

Multiplication

$$z_1 z_2 = (x_1 + iy_1)(x_2 + iy_2) = (x_1 x_2 - y_1 y_2) + i(x_1 y_2 + x_2 y_1)$$

$$|z_1 z_2| = |z_1| \, |z_2|, \quad \arg(z_1 z_2) = \arg z_1 + \arg z_2$$

Division

$$\frac{z_1}{z_2} = \frac{z_1 \bar{z}_2}{z_2 \bar{z}_2} = \frac{(x_1 x_2 + y_1 y_2) + i(x_2 y_1 - x_1 y_2)}{x_2^2 + y_2^2}$$

$$\left|\frac{z_1}{z_2}\right| = \frac{|z_1|}{|z_2|}, \quad \arg \left(\frac{z_1}{z_2}\right) = \arg z_1 - \arg z_2$$

Powers

$$z^n = r^n e^{in\theta} = r^n (\cos n\theta + i \sin n\theta) \quad \text{(DeMoivre's Theorem)}$$

Roots

$$z^{\frac{1}{n}} = r^{\frac{1}{n}} e^{i\theta/n} = r^{\frac{1}{n}} \left[\cos \frac{\theta + 2k\pi}{n} + i \sin \frac{\theta + 2k\pi}{n}, k = 1, 2, \ldots, n-1 \right]$$

(Principal root if $-\pi < \theta < \pi$)

Functions of a Complex Variable

A complex function

$$w = f(z) = u(x, y) + iv(x, y) = |w| e^{i\phi}$$

$$(z = x + iy = |z| e^{i\theta})$$

associates one or more values of the complex dependent variable w with each value of the complex independent variable z in a given domain of definition.

Cauchy-Riemann Equations

$$\frac{\partial u}{\partial x} = \frac{\partial v}{\partial y}, \quad \frac{\partial u}{\partial y} = -\frac{\partial v}{\partial x}$$

A function $w = f(z)$ differentiable at each point of a certain neighborhood of a point z_o (i.e., at each point of a circle with center z_o and an arbitrary small radius) is said to be analytic at the point z_o. A function is called analytic in a connected domain if it is analytic at each point in the domain.

A necessary and sufficient condition that the function $f(z) = u(x,y) + iv(x,y)$ is analytic is that it satisfy the Cauchy-Riemann conditions.

Cauchy-Goursat Theorem

If a function $f(z)$ is analytic at all points within and on a simple closed curve C, then

$$\int_c f(z) \, dz = 0$$

Cauchy's Integral Formula

If $f(z)$ is analytic inside and on a simple closed curve C and if z_o is interior to C, then

$$f(z) = \frac{1}{2\pi i} \int_c \frac{f(z)}{z - z_0} \, dz$$

Also, the derivatives $f'(z)$, $f''(z)$, ... of all orders exist, and

$$f^n(z) = \frac{n!}{2\pi i} \int_c \frac{f(z)}{(z - z_0)^{n+1}} \, dz$$

Taylor-Series Expansion

If $f(z)$ is analytic inside and on a circle C of radius r about $z = z_o$, then there exists a unique and uniformly convergent series expansion in powers of $z - z_o$,

$$f(z) = \sum_{n=0}^{\infty} a_n (z - z_0)^n, \quad (|z - z_0| < r, z_0 \neq \infty)$$

where

$$a_n = \frac{1}{n!} \, f^n(z_0) = \frac{1}{2\pi i} \int_C \frac{f(s)}{(s-z_0)^{n+1}} \, ds$$

If $M(r)$ is an upper bound on $|f(z)|$ on C, then

$$|a_n| = \frac{1}{n!} \, |f^{(n)}(z_0)| \leqslant \frac{M(r)}{r^n} \qquad \text{(Cauchy's Inequality)}$$

If Taylor's series is terminated with the term $a_{n-1}(z-z_0)^{n-1}$, the remainder $R_n(z)$ is given by

$$R_n(z) = \frac{(z-z_0)^n}{2\pi i} \int_C \frac{f(s) \, ds}{(s-z_0)^n (s-z)}$$

$$|R_n(z)| \leqslant \left(\frac{|z-z_0|}{r} \right)^n \frac{M(r) \, r}{r - |z-z_0|}$$

Laurent-Series Expansion

If $f(z)$ is analytic throughout the annular region between and on the concentric circles C_1 and C_2 centered at $z = z_0$ and radii r_1 and $r_2 < r_1$, respectively, there exists a unique series expansion in terms of positive and negative powers of $z - z_0$

$$f(z) = \sum_{n=0}^{\infty} a_n (z-z_0)^n + \sum_{n=1}^{\infty} b_n (z-z_0)^{-n}$$

where

$$a_n = \frac{1}{2\pi i} \int_C \frac{f(s)}{(s-z_0)^{n+1}} \, ds, \qquad bn = \frac{1}{2\pi i} \int_C \frac{f(s)}{(s-z_0)^{-n+1}} \, ds$$

$$(n = 0, 1, 2, \dots) \qquad\qquad (n = 1, 2, \dots)$$

Zeros and Isolated Singularities

Zeros

The points z for which $f(z) = 0$ are called the zeros of $f(z)$. A function $f(z)$ analytic at $z = z_0$ has a zero of order m, where m is a positive integer at $z = z_0$ if and only if the first m coefficients $a_0, a_1, \dots, a_{m-1}$ in the Taylor-series expansion about $z = z_0$ vanish.

Singularities

A singular point or singularity of the function $f(z)$ is any point where $f(z)$ is not analytic. An isolated singularity of $f(z)$ at $z = z_0$ is

1. A removable singularity if and only if all coefficients b_n in the Laurent-series expansion of $f(z)$ about $z = z_0$ vanish.
2. A pole of order m (m = 1,2,...) if and only if $(z-z_0)^m f(z)$, but not $(z-z_0)^{m-1} f(z)$ is analytic at $z = z_0$, i.e., if and only if $b_m \neq 0$, $b_{m+1} = b_{m+2} = \dots = 0$ in the Laurent-series expansion of $f(z)$ about $z = z_0$, or if $\frac{1}{f(z)}$ is analytic and has a zero of order m at $z = z_0$.
3. An isolated essential singularity if and only if the Laurent-series expansion of $f(z)$ about $z = z_0$ has an infinite number of terms involving negative powers of $z - z_0$.

Residues

Given a point $z = z_0$, where $f(z)$ is either analytic or has an isolated singularity, the residue of $f(z)$ at $z = z_0$ is the coefficient of $(z - z_0)^{-1}$ in the Laurent-series expansion of $f(z)$ about $z = z_0$, or

$$R_k = b_1 = \frac{1}{2\pi i} \int_c f(z)\, dz$$

If $f(z)$ is either analytic or has a removable singularity at $z = z_0$, then $b_1 = 0$. If $z = z_0$ is a pole of order m, then

$$b_1 = \frac{1}{(m-1)!} \; \frac{d^{m-1}}{dz^{m-1}} \left[(z - z_0)^m f(z) \right]_{z = z_0}$$

For every simple closed contour C enclosing at most a finite number of singularities $z_1, z_2, \ldots, z_n$ of a single-valued function $f(z)$ continuous on C,

$$\frac{1}{2\pi i} \int_c f(z)\, dz = \sum_{k=1}^{n} R_k$$

where the R_k are the residues of $f(z)$ at the singularities.

Mappings or Transformations

A function $w = f(z)$ maps points of the z-plane into corresponding points of the w-plane. At every point z such that $f(z)$ is analytic and $f'(z) \neq 0$, the mapping is conformal, i.e., the angle between two curves through such a point is reproduced in magnitude and sense by the angle between the corresponding curves in the w-plane.

A table giving real and imaginary parts, zeros, and singularities for frequently used functions of a complex variable and a table illustrating a number of special transformations of interest in various applications are provided.

Real and Imaginary Parts, Zeros, and Singularities for a Number of Frequently Used Functions

$$f(z) = u(x,y) + iv(x,y)$$

Note: $|f(z)| = \sqrt{u^2 + v^2}$ $\arg f(z) = \arctan \dfrac{v}{u}$ of a Complex Variable $z = x + iy$

Function $f(z)$	$u(x,y)$	$v(x,y)$	Zeros (order m)	Isolated singularities
z	x	y	$z=0$ $m=1$	Pole ($m=1$) at $z=\infty$
z^2	x^2-y^2	$2xy$	$z=0$ $m=2$	Pole ($m=2$) at $z=\infty$
$\dfrac{1}{z}$	$\dfrac{x}{x^2+y^2}$	$-\dfrac{y}{x^2+y^2}$	$z=\infty$ $m=1$	Pole ($m=1$) at $z=0$
$\dfrac{1}{z^2}$	$\dfrac{x^2-y^2}{(x^2+y^2)^2}$	$\dfrac{-2xy}{(x^2+y^2)^2}$	$z=\infty$ $m=2$	Pole ($m=2$) at $z=0$
$\dfrac{1}{z-(a+ib)}$ (a, b real)	$\dfrac{(x-a)}{(x-a)^2+(y-b)^2}$	$-\dfrac{(y-b)}{(x-a)^2+(y-b)^2}$	$z=\infty$ $m=1$	Pole ($m=1$) at $z=a+ib$
$\sqrt{z}$	$\pm\left(\dfrac{x+\sqrt{x^2+y^2}}{2}\right)^{1/2}$	$\pm\left(\dfrac{-x+\sqrt{x^2+y^2}}{2}\right)^{1/2}$	Zero of order 1 at $z=0$ (branch point)	Branch point ($m=1$) at $z=0$ Branch point ($m=1$) at $z=\infty$
e^z	$e^x\cos y$	$e^x \sin y$	. . . .	Essential singularity at $z=\infty$
$\sin z$	$\sin x \cosh y$	$\cos x \sinh y$	$z=k\pi$ ($k=0,\pm1,\pm2,\pm\ldots$) $m=1$	Essential singularity at $z=\infty$
$\cos z$	$\cos x \cosh y$	$-\sin x \sinh y$	$z=(k+\tfrac{1}{2})\pi$ ($k=0,\pm1,\pm2,\pm\ldots$) $m=1$	Essential singularity at $z=\infty$

Real and Imaginary Parts, Zeros, and Singularities for a Number of Frequently Used Functions

$$f(z) = u(x,y) + iv(x,y) \text{ of a Complex Variable } z = x + iy$$

Function $f(z)$	$u(x,y)$	$v(x,y)$	Zeros (order m)	Isolated singularities
$\sinh z$	$\sinh x \cos y$	$\cosh x \sin y$	$z = k\pi i$ $m = 1$ $(k = 0, \pm 1, \pm 2, \pm \dots)$	Essential singularity at $z = \infty$
$\cosh z$	$\cosh x \cos y$	$\sinh x \sin y$	$z = (k+\tfrac{1}{2})\,\pi i$ $m = 1$ $(k = 0, \pm 1, \pm 2, \pm \dots)$	Essential singularity at $z = \infty$
$\tan z$	$\dfrac{\sin 2x}{\cos 2x + \cosh 2y}$	$\dfrac{\sinh 2y}{\cos 2x + \cosh 2y}$	$z = k\pi$ $m = 1$ $(k = 0, \pm 1, \pm 2, \pm \dots)$	Essential singularity at $z = \infty$ Poles ($m = 1$) at $z = (k+\tfrac{1}{2})\pi$ $(k = 0, \pm 1, \pm 2, \pm \dots)$
$\tanh z$	$\dfrac{\sinh 2x}{\cosh 2x + \cos 2y}$	$\dfrac{\sin 2y}{\cosh 2x + \cos 2y}$	$z = k\pi i$ $m = 1$ $(k = 0, \pm 1, \pm 2, \pm \dots)$	Essential singularity at $z = \infty$ Poles ($m = 1$) at $z = (k+\tfrac{1}{2})\pi i$ $(k = 0, \pm 1, \pm 2, \pm \dots)$
$\log_e z$	$\tfrac{1}{2}\log_e (x^2+y^2)$	$\arctan \dfrac{y}{x} + 2k\pi$ $(k = 0, \pm 1, \pm 2, + \dots)$	$z = 1$ $m = 1$ (branch corresponding to $k=0$ only)	Branch points of infinite order at $z = 0$, $z = \infty$; both are essential singularities

From Korn, G. A. and Korn, T. M., *Mathematical Handbook for Scientists & Engineers*, 2nd ed., McGraw-Hill, New York, 1968, 189. With permission.

Table of Transformations of Regions

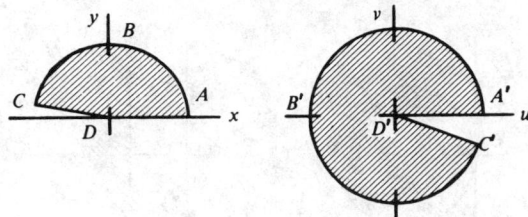

FIGURE 1. $\omega = z^2$.

FIGURE 2. $\omega = z^2$.

FIGURE 3. $\omega = z^2$; A' B' on parabola $\rho = \dfrac{2k^2}{1 + \cos\phi}$

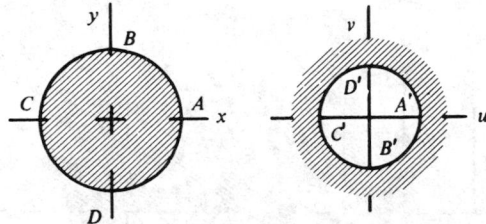

FIGURE 4. $\omega = 1/z$.

(From Churchill, R. V., Brown, J. W., and Verhey, R. F., *Complex Variables and Applications*, 34d ed., McGraw-Hill, New York, 1974, 314 – 324. With permission.)

Table of Transformations of Regions (continued)

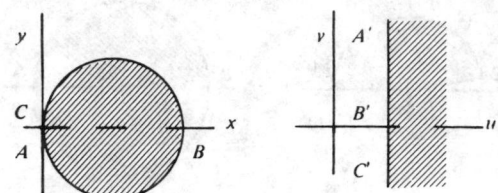

FIGURE 5. $\omega = 1/z$.

FIGURE 6. $\omega = e^z$.

FIGURE 7. $\omega = e^z$.

FIGURE 8. $\omega = e^z$.

FIGURE 9. $\omega = \sin z$.

Table of Transformations of Regions (continued)

FIGURE 10. $\omega = \sin z$.

FIGURE 11. $\omega = \sin z$; BCD: $y = k$, $B'\,C'\,D'$ is on ellipse $\left(\dfrac{u}{\cosh k}\right)^2 + \left(\dfrac{v}{\sinh k}\right)^2 = 1$.

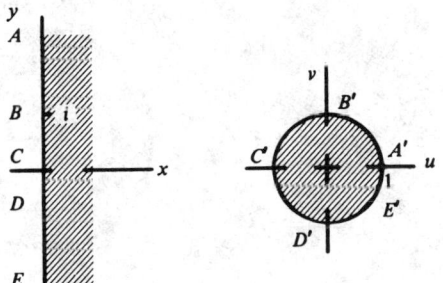

FIGURE 12. $\omega = \dfrac{z-1}{z+1}$.

FIGURE 13. $\omega = \dfrac{i-z}{i+z}$.

Table of Transformations of Regions (continued)

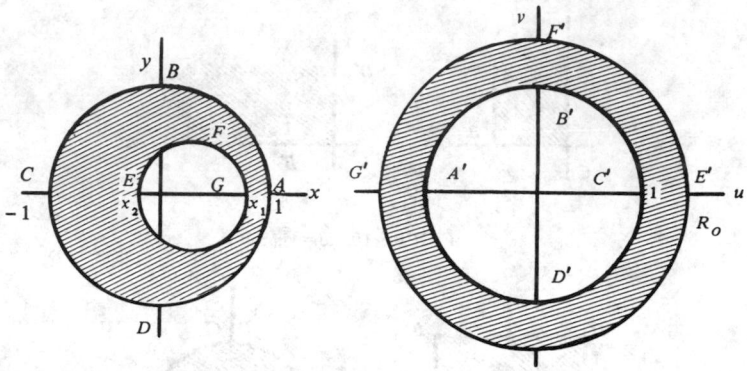

FIGURE 14. $\omega = \dfrac{z-a}{az-1}$; $a = \dfrac{1 + x_1 x_2 + \sqrt{(1-x_1{}^2)(1-x_2{}^2)}}{x_1 + x_2}$;

$$R_0 = \dfrac{1 - x_1 x_2 + \sqrt{(1-x_1{}^2)(1-x_2{}^2)}}{x_1 - x_2} \quad (a > 1 \text{ and } R_0 > 1 \text{ when } -1 < x_2 < x_1 < 1).$$

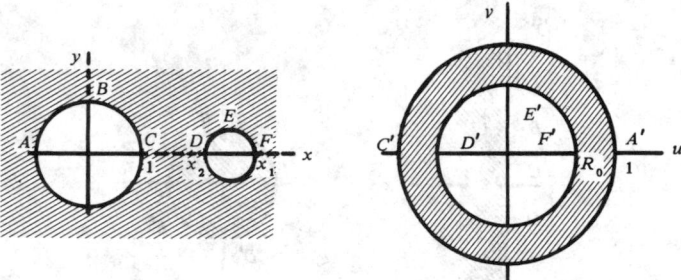

FIGURE 15. $\omega = \dfrac{z-a}{az-1}$; $a = \dfrac{1 + x_1 x_2 + \sqrt{(x_1{}^2 - 1)(x_2{}^2 - 1)}}{x_1 + x_2}$;

$$R_0 = \dfrac{x_1 x_2 - 1 - \sqrt{(x_1{}^2 - 1)(x_2{}^2 - 1)}}{x_1 - x_2} \quad (x_2 < a < x_1 \text{ and } 0 < R_0 < 1 \text{ when } 1 < x_2 < x_1).$$

FIGURE 16. $\omega = z + 1/z$.

Table of Transformations of Regions (continued)

FIGURE 17. $\omega = z + 1/z$.

FIGURE 18. $\omega = z + \dfrac{1}{z}$; B' C' D' is on ellipse $\left(\dfrac{ku}{k^2+1}\right)^2 + \left(\dfrac{kv}{k^2-1}\right)^2 = 1$.

FIGURE 19. $\omega = \log_e \dfrac{z-1}{z+1}$; $z = -\coth \dfrac{\omega}{2}$.

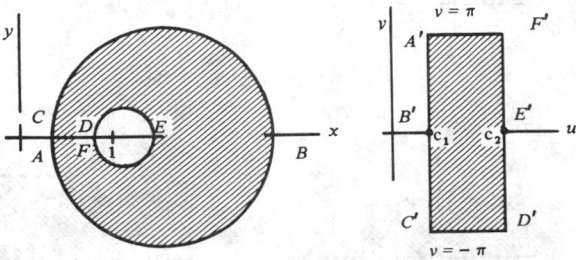

FIGURE 20. $\omega = \log_e \dfrac{z-1}{z+1}$; ABC is on circle $x^2 + y^2 - 2y \cot k = 1$.

FIGURE 21. $\omega = \log_e \dfrac{z+1}{z-1}$; centers of circles at $z = \coth c_n$, radii: csch c_n ($n = 1, 2$).

Table of Transformations of Regions (continued)

FIGURE 22. $\omega = k \log_e \dfrac{k}{1-k} + \log_e 2(1-k) + i\pi - k \log_e(z+1)$

$- (1-k)\log_e(z-1); \; x_1 = 2k - 1.$

FIGURE 23. $\omega = \tan^2 \dfrac{z}{2} = \dfrac{1 - \cos z}{1 + \cos z}$

FIGURE 24. $\omega = \coth \dfrac{z}{2} = \dfrac{e^z + 1}{e^z - 1}$.

FIGURE 25. $\omega = \log_e \coth \dfrac{z}{2}$.

FIGURE 26. $\omega = \pi i + z - \log_e z$.

Table of Transformations of Regions (continued)

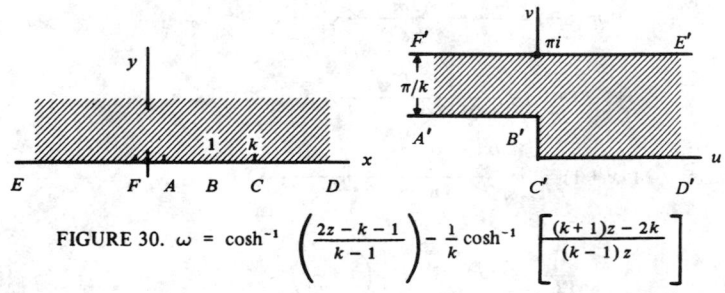

FIGURE 27. $\omega = 2(z + 1)^{1/2} + \log_e \dfrac{(z + 1)^{1/2} - 1}{(z + 1)^{1/2} + 1}$

FIGURE 28. $\omega = \dfrac{i}{k} \log_e \dfrac{1 + ikt}{1 - ikt} + \log_e \dfrac{1 + t}{1 - t} \; ; t = \left(\dfrac{z - 1}{z + k^2} \right)^{1/2}$

FIGURE 29. $\omega = \dfrac{h}{\pi} [(z^2 - 1)^{1/2} + \cosh^{-1} z]$.

FIGURE 30. $\omega = \cosh^{-1} \left(\dfrac{2z - k - 1}{k - 1} \right) - \dfrac{1}{k} \cosh^{-1} \left[\dfrac{(k + 1)z - 2k}{(k - 1) z} \right]$

The Gamma Function

Definition: $\Gamma(n) = \int_0^\infty t^{n-1} e^{-t} dt \quad n > 0$

Recursion Formula: $\Gamma(n + 1) = n\Gamma(n)$

$\Gamma(n + 1) = n!$ if $n = 0, 1, 2, \ldots$ where $0! = 1$

For $n < 0$ the gamma function can be defined by using

$$\Gamma(n) = \frac{\Gamma(n + 1)}{n}$$

Graph:

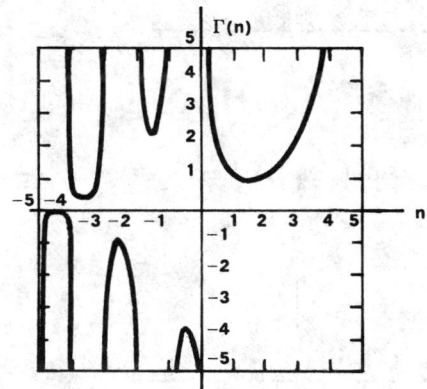

Special Values: $\quad\quad\quad\quad\quad \Gamma(\tfrac{1}{2}) = \sqrt{\pi}$

$$\Gamma(m + \tfrac{1}{2}) = \frac{1 \cdot 3 \cdot 5 \cdots (2m - 1)}{2^m} \sqrt{\pi} \quad\quad m = 1, 2, 3, \ldots$$

$$\Gamma(-m + \tfrac{1}{2}) = \frac{(-1)^m 2^m \sqrt{\pi}}{1 \cdot 3 \cdot 5 \cdots (2m - 1)} \quad\quad m = 1, 2, 3, \ldots$$

Definition: $\quad \Gamma(x + 1) = \lim_{k \to \infty} \frac{1 \cdot 2 \cdot 3 \cdots k}{(x + 1)(x + 2) \cdots (x + k)} k^x$

$$\frac{1}{\Gamma(x)} = xe^{\gamma x} \prod_{m = 1}^\infty \left\{ \left(1 + \frac{x}{m}\right) e^{-x/m} \right\}$$

This is an infinite product representation for the gamma function where γ is Euler's constant.

Properties:

$$\Gamma'(1) = \int_0^\infty e^{-x} \ln x \, dx = -\gamma$$

$$\frac{\Gamma'(x)}{\Gamma(x)} = -\gamma + \left(\frac{1}{1} - \frac{1}{x}\right) + \left(\frac{1}{2} - \frac{1}{x+1}\right) + \ldots + \left(\frac{1}{n} - \frac{1}{x+n-1}\right) + \ldots$$

$$\Gamma(x + 1) = \sqrt{2\pi x} \; x^x e^{-x} \left\{ 1 + \frac{1}{12x} + \frac{1}{288x^2} - \frac{139}{51,840x^3} + \ldots \right\}$$

This is called *Stirling's asymptotic series.*

If we let $x = n$ a positive integer, then a useful approximation for $n!$ where n is large [e.g. $n > 10$] is given by *Stirling's formula*

$$n! \approx \sqrt{2\pi n} \; n^n e^{-n}$$

GAMMA FUNCTION*

Values of $\Gamma(n) = \int_0^\infty e^{-x} x^{n-1} \, dx; \; \Gamma(n+1) = n\Gamma(n)$

n	$\Gamma(n)$	n	$\Gamma(n)$	n	$\Gamma(n)$	n	$\Gamma(n)$
1.00	1.00000	1.25	.90640	1.50	.88623	1.75	.91906
1.01	.99433	1.26	.90440	1.51	.88659	1.76	.92137
1.02	.98884	1.27	.90250	1.52	.88704	1.77	.92376
1.03	.98355	1.28	.90072	1.53	.88757	1.78	.92623
1.04	.97844	1.29	.89904	1.54	.88818	1.79	.92877
1.05	.97350	1.30	.89747	1.55	.88887	1.80	.93138
1.06	.96874	1.31	.89600	1.56	.88964	1.81	.93408
1.07	.96415	1.32	.89464	1.57	.89049	1.82	.93685
1.08	.95973	1.33	.89338	1.58	.89142	1.83	.93969
1.09	.95546	1.34	.89222	1.59	.89243	1.84	.94261
1.10	.95135	1.35	.89115	1.60	.89352	1.85	.94561
1.11	.94740	1.36	.89018	1.61	.89468	1.86	.94869
1.12	.94359	1.37	.88931	1.62	.89592	1.87	.95184
1.13	.93993	1.38	.88854	1.63	.89724	1.88	.95507
1.14	.93642	1.39	.88785	1.64	.89864	1.89	.95838
1.15	.93304	1.40	.88726	1.65	.90012	1.90	.96177
1.16	.92980	1.41	.88676	1.66	.90167	1.91	.96523
1.17	.92670	1.42	.88636	1.67	.90330	1.92	.96877
1.18	.92373	1.43	.88604	1.68	.90500	1.93	.97240
1.19	.92089	1.44	.88581	1.69	.90678	1.94	.97610
1.20	.91817	1.45	.88566	1.70	.90864	1.95	.97988
1.21	.91558	1.46	.88560	1.71	.91057	1.96	.98374
1.22	.91311	1.47	.88563	1.72	.91258	1.97	.98768
1.23	.91075	1.48	.88575	1.73	.91466	1.98	.99171
1.24	.90852	1.49	.88595	1.74	.91683	1.99	.99581
						2.00	1.00000

* For large positive values of x, $\Gamma(x)$ approximates Stirling's asymptotic series

$$x^x e^{-x} \sqrt{\frac{2\pi}{x}} \left[1 + \frac{1}{12x} + \frac{1}{288x^2} - \frac{139}{51840x^3} - \frac{571}{2488320x^4} + \cdot\right]$$

The Beta Function

Definition: $B(m,n) = \displaystyle\int_0^1 t^{n-1} (1-t)^{m-1} \, dt \quad m > 0, n > 0$

Relationship with
Gamma Function $\qquad B(m,n) = \dfrac{\Gamma(m)\Gamma(n)}{\Gamma(m+n)}$

Properties:

$$B(m,n) = B(n,m)$$

$$B(m,n) = 2\int_0^{\pi/2} \sin^{2m-1}\theta \, \cos^{2n-1}\theta \, d\theta$$

$$B(m,n) = \int_0^\infty \frac{t^{m-1}}{(1+t)^{m+n}} \, dt$$

$$B(m,n) = r^n (r+1)^m \int_0^1 \frac{t^{m-1}(1-t)^{n-1}}{(r+t)^{m+n}} \, dt$$

The Error Function

Definition: $\operatorname{erf} x = \dfrac{2}{\sqrt{\pi}} \displaystyle\int_0^x e^{-t^2} \, dt$

Series: $\operatorname{erf} x = \dfrac{2}{\sqrt{\pi}} \left(x - \dfrac{x^3}{3} + \dfrac{1}{2!}\dfrac{x^5}{5} - \dfrac{1}{3!}\dfrac{x^7}{7} + \ldots \right)$

Property: $\operatorname{erf} x = -\operatorname{erf}(-x)$

Relationship with Normal Probability Function f(t): $\displaystyle\int_0^x f(t) \, dt = \dfrac{1}{2} \operatorname{erf}\left(\dfrac{x}{\sqrt{2}}\right)$

To evaluate $\operatorname{erf}(2.3)$, one proceeds as follows: Since $\dfrac{x}{\sqrt{2}} = 2.3$, one finds $x = (2.3)(\sqrt{2}) = 3.25$. In the normal probability function table (page 527), one finds the entry 0.4994 opposite the value 3.25. Thus $\operatorname{erf}(2.3) = 2(0.4994) = 0.9988$.

$$\operatorname{erfc} z = 1 - \operatorname{erf} z = \frac{2}{\sqrt{\pi}} \int_z^\infty e^{-t^2} \, dt$$

is known as the complimentary error function.

BESSEL FUNCTIONS

1. **Bessel's differential equation for a real variable** x is

$$x^2 \frac{d^2y}{dx^2} + x \frac{dy}{dx} + (x^2 - n^2)y = 0$$

2. When n is not an integer, two independent solutions of the equation are $J_n(x)$ and $J_{-n}(x)$, where

$$J_n(x) = \sum_{k=0}^{\infty} \frac{(-1)^k}{k!\,\Gamma(n + k + 1)} \left(\frac{x}{2}\right)^{n+2k}$$

3. If n is an integer $J_{-n}(x) = (-1)^n J_n(x)$, where

$$J_n(x) = \frac{x^n}{2^n n!} \left\{ 1 - \frac{x^2}{2^2 \cdot 1!(n + 1)} + \frac{x^4}{2^4 \cdot 2!(n + 1)(n + 2)} \right.$$
$$\left. - \frac{x^6}{2^6 \cdot 3!(n + 1)(n + 2)(n + 3)} + \cdots \right\}$$

4. For $n = 0$ and $n = 1$, this formula becomes

$$J_0(x) = 1 - \frac{x^2}{2^2(1!)^2} + \frac{x^4}{2^4(2!)^2} - \frac{x^6}{2^6(3!)^2} + \frac{x^8}{2^8(4!)^2} - \cdots$$

$$J_1(x) = \frac{x}{2} - \frac{x^3}{2^3 \cdot 1!2!} + \frac{x^5}{2^5 \cdot 2!3!} - \frac{x^7}{2^7 \cdot 3!4!} + \frac{x^9}{2^9 \cdot 4!5!} - \cdots$$

5. When x is large and positive, the following asymptotic series may be used

$$J_0(x) = \left(\frac{2}{\pi x}\right)^{\frac{1}{2}} \left\{ P_0(x) \cos\left(x - \frac{\pi}{4}\right) - Q_0(x) \sin\left(x - \frac{\pi}{4}\right) \right\}$$

$$J_1(x) = \left(\frac{2}{\pi x}\right)^{\frac{1}{2}} \left\{ P_1(x) \cos\left(x - \frac{3\pi}{4}\right) - Q_1(x) \sin\left(x - \frac{3\pi}{4}\right) \right\},$$

where

$$P_0(x) \sim 1 - \frac{1^2 \cdot 3^2}{2!(8x)^2} + \frac{1^2 \cdot 3^2 \cdot 5^2 \cdot 7^2}{4!(8x)^4} - \frac{1^2 \cdot 3^2 \cdot 5^2 \cdot 7^2 \cdot 9^2 \cdot 11^2}{6!(8x)^6} + \cdots$$

$$Q_0(x) \sim -\frac{1^2}{1!8x} + \frac{1^2 \cdot 3^2 \cdot 5^2}{3!(8x)^3} - \frac{1^2 \cdot 3^2 \cdot 5^2 \cdot 7^2 \cdot 9^2}{5!(8x)^5} + - \cdots$$

$$P_1(x) \sim 1 + \frac{1^2 \cdot 3 \cdot 5}{2!(8x)^2} - \frac{1^2 \cdot 3^2 \cdot 5^2 \cdot 7 \cdot 9}{4!(8x)^4} + \frac{1^2 \cdot 3^2 \cdot 5^2 \cdot 7^2 \cdot 9^2 \cdot 11 \cdot 13}{6!(8x)^6} - + \cdots$$

$$Q_1(x) \sim \frac{1 \cdot 3}{1!8x} - \frac{1^2 \cdot 3^2 \cdot 5 \cdot 7}{3!(8x)^3} + \frac{1^2 \cdot 3^2 \cdot 5^2 \cdot 7^2 \cdot 9 \cdot 11}{5!(8x)^5} - \cdots$$

[In $P_1(x)$ the signs alternate from $+$ to $-$ after the first term]

6. If $x > 25$, it is convenient to use the formulas

$$J_0(x) = A_0(x) \sin x + B_0(x) \cos x$$
$$J_1(x) = B_T(x) \sin x - A_1(x) \cos x,$$

where

$$A_0(x) = \frac{P_0(x) - Q_0(x)}{(\pi x)^{\frac{1}{2}}} \quad \text{and} \quad A_1(x) = \frac{P_1(x) - Q_1(x)}{(\pi x)^{\frac{1}{2}}}$$

$$B_0(x) = \frac{P_0(x) + Q_0(x)}{(\pi x)^{\frac{1}{2}}} \quad \text{and} \quad B_1(x) = \frac{P_1(x) + Q_1(x)}{(\pi x)^{\frac{1}{2}}}$$

7. The zeros of $J_0(x)$ and $J_1(x)$

If $j_{0,s}$ and $j_{1,s}$ are the s'th zeros of $J_0(x)$ and $J_1(x)$ respectively, and if $a = 4s - 1$, $b = 4s + 1$

$$j_{0,s} \sim \frac{1}{4}\pi a \left\{ 1 + \frac{2}{\pi^2 a^2} - \frac{62}{3\pi^4 a^4} + \frac{15,116}{15\pi^6 a^6} - \frac{12,554,474}{105\pi^8 a^8} + \frac{8,368,654,292}{315\pi^{10} a^{10}} - + \cdots \right\}$$

$$j_{1,s} \sim \frac{1}{4}\pi b \left\{ 1 - \frac{6}{\pi^2 b^2} + \frac{6}{\pi^4 b^4} - \frac{4716}{5\pi^6 b^6} + \frac{3,902,418}{35\pi^8 b^8} - \frac{895,167,324}{35\pi^{10} b^{10}} + \cdots \right\}$$

$$J_1(j_{0,s}) \sim \frac{(-1)^{s+1} 2^{\frac{3}{2}}}{\pi a^{\frac{1}{2}}} \left\{ 1 - \frac{56}{3\pi^4 a^4} + \frac{9664}{5\pi^6 a^6} - \frac{7,381,280}{21\pi^8 a^8} + \cdots \right\}$$

$$J_0(j_{1,s}) \sim \frac{(-1)^s 2^{\frac{3}{2}}}{\pi b^{\frac{1}{2}}} \left\{ 1 + \frac{24}{\pi^4 b^4} - \frac{19,584}{10\pi^6 b^6} + \frac{2,466,720}{7\pi^8 b^8} - \cdots \right\}$$

8. Table of zeros for $J_0(x)$ and $J_1(x)$

$$J_1(\alpha_n) = 0 \qquad J_0(\beta_n) = 0$$

Roots α_n	$J_1(\alpha_n)$	Roots β_n	$J_0(\beta_n)$
2.4048	0.5191	0.0000	1.0000
5.5201	−0.3403	3.8317	−0.4028
8.6537	0.2715	7.0156	0.3001
11.7915	−0.2325	10.1735	−0.2497
14.9309	0.2065	13.3237	0.2184
18.0711	−0.1877	16.4706	−0.1965
21.2116	0.1733	19.6159	0.1801

9. Recurrence formulas

$$J_{n-1}(x) + J_{n+1}(x) = \frac{2n}{x} J_n(x) \qquad\qquad n J_n(x) + x J_n'(x) = x J_{n-1}(x)$$

$$J_{n-1}(x) - J_{n+1}(x) = 2J_n'(x) \qquad\qquad\quad n J_n(x) - x J_n'(x) = x J_{n+1}(x)$$

10. If J_n is written for $J_n(x)$ and $J_n^{(k)}$ is written for $\dfrac{d^k}{dx^k}\{J_n(x)\}$, then the following derivative relationships are important

$$J_0^{(r)} = -J_1^{(r-1)}$$

$$J_0^{(2)} = -J_0 + \frac{1}{x} J_1 = \frac{1}{2}(J_2 - J_0)$$

$$J_0^{(3)} = \frac{1}{x} J_0 + \left(1 - \frac{2}{x^2}\right) J_1 = \frac{1}{4}(-J_3 + 3J_1)$$

$$J_0^{(4)} = \left(1 - \frac{3}{x^2}\right) J_0 - \left(\frac{2}{x} - \frac{6}{x^3}\right) J_1 = \frac{1}{8}(J_4 - 4J_2 + 3J_0), \text{ etc.}$$

11. Half order Bessel functions

$$J_{\frac{1}{2}}(x) = \sqrt{\frac{2}{\pi x}} \sin x$$

$$J_{-\frac{1}{2}}(x) = \sqrt{\frac{2}{\pi x}} \cos x$$

$$J_{n+\frac{1}{2}}(x) = -x^{n+\frac{1}{2}} \frac{d}{dx} \{x^{-(n+\frac{1}{2})} J_{n+\frac{1}{2}}(x)\}$$

$$J_{n-\frac{1}{2}}(x) = x^{-(n+\frac{1}{2})} \frac{d}{dx} \{x^{n+\frac{1}{2}} J_{n+\frac{1}{2}}(x)\}$$

n	$\left(\frac{\pi x}{2}\right)^{\frac{1}{2}} J_{n+\frac{1}{2}}(x)$	$\left(\frac{\pi x}{2}\right)^{\frac{1}{2}} J_{-(n+\frac{1}{2})}(x)$
0	$\sin x$	$\cos x$
1	$\dfrac{\sin x}{x} - \cos x$	$-\dfrac{\cos x}{x} - \sin x$
2	$\left(\dfrac{3}{x^2} - 1\right)\sin x - \dfrac{3}{x}\cos x$	$\left(\dfrac{3}{x^2} - 1\right)\cos x + \dfrac{3}{x}\sin x$
3	$\left(\dfrac{15}{x^3} - \dfrac{6}{x}\right)\sin x - \left(\dfrac{15}{x^2} - 1\right)\cos x$	$-\left(\dfrac{15}{x^3} - \dfrac{6}{x}\right)\cos x - \left(\dfrac{15}{x^2} - 1\right)\sin x$

etc.

12. Additional solutions to Bessel's equation are

$Y_n(x)$ (also called Weber's function, and sometimes denoted by $N_n(x)$)

$H_n^{(1)}(x)$ and $H_n^{(2)}(x)$ (also called Hankel functions)

These solutions are defined as follows

$$Y_n(x) = \begin{cases} \dfrac{J_n(x)\cos(n\pi) - J_{-n}(x)}{\sin(n\pi)} & n \text{ not an integer} \\[2ex] \lim_{\nu \to n} \dfrac{J_\nu(x)\cos(\nu\pi) - J_{-\nu}(x)}{\sin(\nu\pi)} & n \text{ an integer} \end{cases}$$

$$H_n^{(1)}(x) = J_n(x) + iY_n(x)$$
$$H_n^{(2)}(x) = J_n(x) - iY_n(x)$$

The additional properties of these functions may all be derived from the above relations and the known properties of $J_n(x)$.

13. Complete solutions to Bessel's equation may be written as

$$c_1 J_n(x) + c_2 J_{-n}(x) \qquad \text{if } n \text{ is not an integer,}$$

or

or

$$\left.\begin{array}{l} c_1 J_n(x) + c_2 Y_n(x) \\[2mm] c_1 H_n^{(1)}(x) + c_2 H_n^{(2)}(x) \end{array}\right\} \text{ for any value of } n$$

14. The modified (or hyperbolic) Bessel's differential equation is

$$x^2 \frac{d^2 y}{dx^2} + x \frac{dy}{dx} - (x^2 + n^2)y = 0$$

15. When n is not an integer, two independent solutions of the equation are $I_n(x)$ and $I_{-n}(x)$, where

$$I_n(x) = \sum_{k=0}^{\infty} \frac{1}{k! \Gamma(n + k + 1)} \left(\frac{x}{2}\right)^{n+2k}$$

16. If n is an integer,

$$I_n(x) = I_{-n}(x) = \frac{x^n}{2^n n!} \left\{ 1 + \frac{x^2}{2^2 \cdot 1!(n + 1)} + \frac{x^4}{2^4 \cdot 2!(n + 1)(n + 2)} \right.$$

$$\left. + \frac{x^6}{2^6 \cdot 3!(n + 1)(n + 2)(n + 3)} + \cdots \right\}$$

17. For $n = 0$ and $n = 1$, this formula becomes

$$I_0(x) = 1 + \frac{x^2}{2^2(1!)^2} + \frac{x^4}{2^4(2!)^2} + \frac{x^6}{2^6(3!)^2} + \frac{x^8}{2^8(4!)^2} + \cdots$$

$$I_1(x) = \frac{x}{2} + \frac{x^3}{2^3 \cdot 1!2!} + \frac{x^5}{2^5 \cdot 2!3!} + \frac{x^7}{2^7 \cdot 3!4!} + \frac{x^9}{2^9 \cdot 4!5!} + \cdots$$

18. Another solution to the modified Bessel's equation is

$$K_n(x) = \begin{cases} \dfrac{1}{2}\pi \, \dfrac{I_{-n}(x) - I_n(x)}{\sin(n\pi)} & n \text{ not an integer} \\[4mm] \lim_{v \to n} \dfrac{1}{2}\pi \, \dfrac{I_{-v}(x) - I_v(x)}{\sin(v\pi)} & n \text{ an integer} \end{cases}$$

This function is linearly independent of $I_n(x)$ for all values of n. Thus the **complete** solution to the modified Bessel's equation may be written as

$$c_1 I_n(x) + c_2 I_{-n}(x) \qquad n \text{ not an integer}$$

or

$$c_1 I_n(x) + c_2 K_n(x) \qquad \text{any } n$$

19. The following relations hold among the various Bessel functions:

$$I_n(z) = i^{-m} J_m(iz)$$

$$Y_n(iz) = (i)^{n+1} I_n(z) - \frac{2}{\pi} i^{-n} K_n(z)$$

Most of the properties of the modified Bessel function may be deduced from the known properties of $J_n(x)$ by use of these relations and those previously given.

20. Recurrence formulas

$$I_{n-1}(x) - I_{n+1}(x) = \frac{2n}{x} I_n(x) \qquad\qquad I_{n-1}(x) + I_{n+1}(x) = 2I_n'(x)$$

$$I_{n-1}(x) - \frac{n}{x} I_n(x) = I_n'(x) \qquad\qquad I_n'(x) = I_{n+1}(x) + \frac{n}{x} I_n(z)$$

BESSEL FUNCTIONS $J_0(x)$ AND $J_1(x)$

x	$J_0(x)$	$J_1(x)$	x	$J_0(x)$	$J_1(x)$	x	$J_0(x)$	$J_1(x)$
0.0	1.0000	.0000	5.0	−.1776	−.3276	10.0	−.2459	.0435
0.1	.9975	.0499	5.1	−.1443	−.3371	10.1	−.2490	.0184
0.2	.9900	.0995	5.2	−.1103	−.3432	10.2	−.2496	−.0066
0.3	.9776	.1483	5.3	−.0758	−.3460	10.3	−.2477	−.0313
0.4	.9604	.1960	5.4	−.0412	−.3453	10.4	−.2434	−.0555
0.5	.9385	.2423	5.5	−.0068	−.3414	10.5	−.2366	−.0789
0.6	.9120	.2867	5.6	.0270	−.3343	10.6	−.2276	−.1012
0.7	.8812	.3290	5.7	.0599	−.3241	10.7	−.2164	−.1224
0.8	.8463	.3688	5.8	.0917	−.3110	10.8	−.2032	−.1422
0.9	.8075	.4059	5.9	.1220	−.2951	10.9	−.1881	−.1603
1.0	.7652	.4401	6.0	.1506	−.2767	11.0	−.1712	−.1768
1.1	.7196	.4709	6.1	.1773	−.2559	11.1	−.1528	−.1913
1.2	.6711	.4983	6.2	.2017	−.2329	11.2	−.1330	−.2039
1.3	.6201	.5220	6.3	.2238	−.2081	11.3	−.1121	−.2143
1.4	.5669	.5419	6.4	.2433	−.1816	11.4	−.0902	−.2225
1.5	.5118	.5579	6.5	.2601	−.1538	11.5	−.0677	−.2284
1.6	.4554	.5699	6.6	.2740	−.1250	11.6	−.0446	−.2320
1.7	.3980	.5778	6.7	.2851	−.0953	11.7	−.0213	−.2333
1.8	.3400	.5815	6.8	.2931	−.0652	11.8	.0020	−.2323
1.9	.2818	.5812	6.9	.2981	−.0349	11.9	.0250	−.2290
2.0	.2239	.5767	7.0	.3001	−.0047	12.0	.0477	−.2234
2.1	.1666	.5683	7.1	.2991	.0252	12.1	.0697	−.2157
2.2	.1104	.5560	7.2	.2951	.0543	12.2	.0908	−.2060
2.3	.0555	.5399	7.3	.2882	.0826	12.3	.1108	−.1943
2.4	.0025	.5202	7.4	.2786	.1096	12.4	.1296	−.1807
2.5	−.0484	.4971	7.5	.2663	.1352	12.5	.1469	−.1655
2.6	−.0968	.4708	7.6	.2516	.1592	12.6	.1626	−.1487
2.7	−.1424	.4416	7.7	.2346	.1813	12.7	.1766	−.1307
2.8	−.1850	.4097	7.8	.2154	.2014	12.8	.1887	−.1114
2.9	−.2243	.3754	7.9	.1944	.2192	12.9	.1988	−.0912
3.0	−.2601	.3391	8.0	.1717	.2346	13.0	.2069	−.0703
3.1	−.2921	.3009	8.1	.1475	.2476	13.1	.2129	−.0489
3.2	−.3202	.2613	8.2	.1222	.2580	13.2	.2167	−.0271
3.3	−.3443	.2207	8.3	.0960	.2657	13.3	.2183	−.0052
3.4	−.3643	.1792	8.4	.0692	.2708	13.4	.2177	.0166
3.5	−.3801	.1374	8.5	.0419	.2731	13.5	.2150	.0380
3.6	−.3918	.0955	8.6	.0146	.2728	13.6	.2101	.0590
3.7	−.3992	.0538	8.7	−.0125	.2697	13.7	.2032	.0791
3.8	−.4026	.0128	8.8	−.0392	.2641	13.8	.1943	.0984
3.9	−.4018	−.0272	8.9	−.0653	.2559	13.9	.1836	.1165
4.0	−.3971	−.0660	9.0	−.0903	.2453	14.0	.1711	.1334
4.1	−.3887	−.1033	9.1	−.1142	.2324	14.1	.1570	.1488
4.2	−.3766	−.1386	9.2	−.1367	.2174	14.2	.1414	.1626
4.3	−.3610	−.1719	9.3	−.1577	.2004	14.3	.1245	.1747
4.4	−.3423	−.2028	9.4	−.1768	.1816	14.4	.1065	.1850
4.5	−.3205	−.2311	9.5	−.1939	.1613	14.5	.0875	.1934
4.6	−.2961	−.2566	9.6	−.2090	.1395	14.6	.0679	.1999
4.7	−.2693	−.2791	9.7	−.2218	.1166	14.7	.0476	.2043
4.8	−.2404	−.2985	9.8	−.2323	.0928	14.8	.0271	.2066
4.9	−.2097	−.3147	9.9	−.2403	.0684	14.9	.0064	.2069

SPECIAL FUNCTIONS

BESSEL FUNCTIONS FOR SPHERICAL COORDINATES

$$j_n(x) = \sqrt{\frac{\pi}{2x}}\, J_{(n+\frac{1}{2})}(x), \quad y_n(x) = \sqrt{\frac{\pi}{2x}}\, Y_{(n+\frac{1}{2})}(x) = (-1)^{n+1}\sqrt{\frac{\pi}{2x}}\, J_{-(n+\frac{1}{2})}(x)$$

x	$j_0(x)$	$y_0(x)$	$j_1(x)$	$y_1(x)$	$j_2(x)$	$y_2(x)$
0.0	1.0000	$-\infty$	0.0000	$-\infty$	0.0000	$-\infty$
0.1	0.9983	-9.9500	0.0333	-100.50	0.0007	-3005.0
0.2	0.9933	-4.9003	0.0664	-25.495	0.0027	-377.52
0.4	0.9735	-2.3027	0.1312	-6.7302	0.0105	-48.174
0.6	0.9411	-1.3756	0.1929	-3.2337	0.0234	-14.793
0.8	0.8967	-0.8709	0.2500	-1.9853	0.0408	-6.5740
1.0	0.8415	-0.5403	0.3012	-1.3818	0.0620	-3.6050
1.2	0.7767	-0.3020	0.3453	-1.0283	0.0865	-2.2689
1.4	0.7039	-0.1214	0.3814	-0.7906	0.1133	-1.5728
1.6	0.6247	$+0.0182$	0.4087	-0.6133	0.1416	-1.1682
1.8	0.5410	0.1262	0.4268	-0.4709	0.1703	-0.9111
2.0	0.4546	0.2081	0.4354	-0.3506	0.1984	-0.7340
2.2	0.3675	0.2675	0.4345	-0.2459	0.2251	-0.6028
2.4	0.2814	0.3072	0.4245	-0.1534	0.2492	-0.4990
2.6	0.1983	0.3296	0.4058	-0.0715	0.2700	-0.4121
2.8	0.1196	0.3365	0.3792	$+0.0005$	0.2867	-0.3359
3.0	$+0.0470$	0.3300	0.3457	0.0630	0.2986	-0.2670
3.2	-0.0182	0.3120	0.3063	0.1157	0.3054	-0.2035
3.4	-0.0752	0.2844	0.2622	0.1588	0.3066	-0.1442
3.6	-0.1229	0.2491	0.2150	0.1921	0.3021	-0.0890
3.8	-0.1610	0.2081	0.1658	0.2158	0.2919	-0.0378
4.0	-0.1892	0.1634	0.1161	0.2301	0.2763	$+0.0091$
4.2	-0.2075	0.1167	0.0673	0.2353	0.2556	0.0514
4.4	-0.2163	0.0698	$+0.0207$	0.2321	0.2304	0.0884
4.6	-0.2160	$+0.0244$	-0.0226	0.2213	0.2013	0.1200
4.8	-0.2075	-0.0182	-0.0615	0.2037	0.1691	0.1456
5.0	-0.1918	-0.0567	-0.0951	0.1804	0.1347	0.1650
5.2	-0.1699	-0.0901	-0.1228	0.1526	0.0991	0.1781
5.4	-0.1431	-0.1175	-0.1440	0.1213	0.0631	0.1850
5.6	-0.1127	-0.1385	-0.1586	0.0880	$+0.0277$	0.1856
5.8	-0.0801	-0.1527	-0.1665	0.0538	-0.0060	0.1805
6.0	-0.0466	-0.1600	-0.1678	$+0.0199$	-0.0373	0.1700
6.2	-0.0134	-0.1607	-0.1629	-0.0125	-0.0654	0.1547
6.4	$+0.0182$	-0.1552	-0.1523	-0.0425	-0.0896	0.1353
6.6	0.0472	-0.1440	-0.1368	-0.0690	-0.1094	0.1126
6.8	0.0727	-0.1278	-0.1172	-0.0915	-0.1243	0.0875
7.0	0.0939	-0.1077	-0.0943	-0.1092	-0.1343	0.0609
7.2	0.1102	-0.0845	-0.0692	-0.1220	-0.1391	0.0337
7.4	0.1215	-0.0593	-0.0429	-0.1295	-0.1388	$+0.0068$
7.6	0.1274	-0.0331	-0.0163	-0.1317	-0.1338	-0.0189
7.8	0.1280	-0.0069	$+0.0095$	-0.1289	-0.1244	-0.0427
8.0	0.1237	$+0.0182$	0.0336	-0.1214	-0.1111	-0.0637

Taken from Vibration and Sound with the permission of Philip Morse, author, and McGraw-Hill Book Company, Inc., publisher.

HYPERBOLIC BESSEL FUNCTIONS

$$I_m(x) = i^{-m}J_m(ix)$$

x	$I_o(x)$	$I_1(x)$	$I_2(x)$
0.0	1.0000	0.0000	0.0000
0.1	1.0025	0.0501	0.0012
0.2	1.0100	0.1005	0.0050
0.4	1.0404	0.2040	0.0203
0.6	1.0920	0.3137	0.0464
0.8	1.1665	0.4329	0.0844
1.0	1.2661	0.5652	0.1357
1.2	1.3937	0.7147	0.2026
1.4	1.5534	0.8861	0.2875
1.6	1.7500	1.0848	0.3940
1.8	1.9896	1.3172	0.5260
2.0	2.2796	1.5906	0.6889
2.2	2.6291	1.9141	0.8891
2.4	3.0493	2.2981	1.1342
2.6	3.5533	2.7554	1.4337
2.8	4.1573	3.3011	1.7994
3.0	4.8808	3.9534	2.2452
3.2	5.7472	4.7343	2.7883
3.4	6.7848	5.6701	3.4495
3.6	8.0277	6.7927	4.2540
3.8	9.5169	8.1404	5.2325
4.0	11.302	9.7595	6.4222
4.2	13.442	11.706	7.8684
4.4	16.010	14.046	9.6258
4.6	19.093	16.863	11.761
4.8	22.794	20.253	14.355
5.0	27.240	24.336	17.506
5.2	32.584	29.254	21.332
5.4	39.009	35.182	25.978
5.6	46.738	42.328	31.620
5.8	56.038	50.946	38.470
6.0	67.234	61.342	46.787
6.2	80.718	73.886	56.884
6.4	96.962	89.026	69.141
6.6	116.54	107.30	84.021
6.8	140.14	129.38	102.08
7.0	168.59	156.04	124.01
7.2	202.92	188.25	150.63
7.4	244.34	227.17	182.94
7.6	294.33	274.22	222.17
7.8	354.68	331.10	269.79
8.0	427.56	399.87	327.60

Taken from Vibration and Sound with the permission of Philip Morse, author, and McGraw-Hill Book Company, Inc., publisher.

ELLIPTIC INTEGRALS OF THE FIRST, SECOND AND THIRD KIND

An elliptic integral has the form $\int R(x, \sqrt{f(x)})dx$, where R represents a rational function and $f(x) = a + bx + cx^2 + dx^3 + ex^4$, an algebraic function of the third or fourth degree.

1. Elliptic integrals of the *first kind* are represented by

$$F(k, \phi) = \int_0^\phi \frac{d\Phi}{\sqrt{1 - k^2 \sin^2 \Phi}}$$

$$= \int_0^x \frac{d\xi}{\sqrt{(1 - \xi^2)(1 - k^2\xi^2)}}, \quad x = \sin \phi, \ k^2 < 1.$$

2. Elliptic integrals of the second kind are represented by

$$E(k, \phi) = \int_0^\phi \sqrt{1 - k^2 \sin^2 \Phi} \, d\Phi$$

$$= \int_0^x \frac{\sqrt{1 - k^2\xi^2}}{\sqrt{1 - \xi^2}} \, d\xi, \quad x = \sin \phi, \ k^2 < 1.$$

3. Elliptic integrals of the third kind are represented as

$$\pi(k, n, \phi) = \int_0^\phi \frac{d\Phi}{(1 + n \sin^2 \Phi) \sqrt{1 - k^2 \sin^2 \Phi}},$$

$$k^2 < 1, n \text{ an integer.}$$

Elliptic integrals of the third kind are also presented as

$$\pi_1(k, n, x) = \int_0^x \frac{d\xi}{(1 + n\xi^2) \sqrt{(1 - \xi^2)(1 - k^2\xi^2)}},$$

$$x = \sin \phi, \ k^2 < 1, n \text{ an integer.}$$

4. The complete integrals are

$$K = F\left(k, \frac{\pi}{2}\right) = \frac{\pi}{2}\left[1 + \left(\frac{1}{2}\right)^2 k^2 + \left(\frac{3}{2 \cdot 4}\right)^2 k^4\right.$$

$$\left. + \left(\frac{3 \cdot 5}{2 \cdot 4 \cdot 6}\right)^2 k^6 + \cdots\right]$$

$$E = E\left(k, \frac{\pi}{2}\right) = \frac{\pi}{2}\left[1 - \left(\frac{1}{2^2}\right)k^2 - \left(\frac{3^2}{2^2 \cdot 4^2}\right)\frac{k^4}{3}\right.$$

$$\left. - \left(\frac{3^2 \cdot 5^2}{2^2 \cdot 4^2 \cdot 6^2}\right)\frac{k^6}{5} - \left(\frac{3^2 \cdot 5^2 \cdot 7^2}{2^2 \cdot 4^2 \cdot 6^2 \cdot 8^2}\right)\frac{k^8}{7} - \cdots\right].$$

$$K' = F\left(\sqrt{1 - k^2}, \frac{\pi}{2}\right), \ E' = E\left(\sqrt{1 - k^2}, \frac{\pi}{2}\right)$$

5. The following relation holds between K, K', E, E', namely

$$KE' + EK' - KK' = \frac{\pi}{2} \quad \text{Legendre's relation}$$

E: see 2 above. $\quad E' = \int_0^{\pi/2} (1 - k'^2 \sin^2 \phi)^{\frac{1}{2}}d\phi. \quad k' = \sqrt{(1 - k^2)}$

$K = \int_0^{\pi/2} (1 - k^2 \sin^2 \phi)^{-\frac{1}{2}}d\phi \qquad K' = \int_0^{\pi/2} (1 - k'^2 \sin^2 \phi)^{-\frac{1}{2}}d\phi$

6. To evaluate elliptic integrals for values outside the range contained in the following tables, these relations are useful

$$F(k, \pi) = 2K; E(k, \pi) = 2E$$

$$F(k, \phi + m\pi) = mF(k, \pi) + F(k, \phi) = 2mK + F(k, \phi)$$
$$m = 0, 1, 2, 3, \ldots$$

$$E(k, \phi + m\pi) = mE(k, \pi) + E(k, \phi) = 2mE + E(k, \phi)$$
$$m = 0, 1, 2, 3, \ldots$$

7. If $u = F(k, \phi) = \int_0^\phi \dfrac{d\Phi}{\sqrt{1 - k^2 \sin^2 \Phi}}$ $\qquad (k^2 < 1)$,

= elliptic integral of the first kind.

$$u = \int_0^x \frac{dx}{\sqrt{(1 - \xi^2)(1 - k^2\xi^2)}}, \text{ where } x = \sin \phi.$$

ϕ is called the amplitude of u or am u.
k is called the modulus.

$$k' = \sqrt{1 - k^2} = \text{the complementary modulus.}$$

$\sin \phi = \operatorname{sn} u = x \qquad \tan \phi = \operatorname{tn} u = \dfrac{x}{\sqrt{1 - x^2}}.$

$\cos \phi = \operatorname{cn} u = \sqrt{1 - x^2}.$ $\qquad \Delta\phi = \operatorname{dn} u = \sqrt{1 - k^2x^2}.$
am $0 = 0.$ $\qquad$ sn $0 = 0.$
cn $0 = 1.$ $\qquad$ dn $0 = 1.$
am $(-u) = -$am $u.$ $\qquad$ sn $(-u) = -$sn $u.$
cn $(-u) = $ cn $u.$ $\qquad$ dn $(-u) = $ dn $u.$

tn $(-u) = -$tn $u.$
$\operatorname{sn}^2 u + \operatorname{cn}^2 u = 1.$
$\operatorname{dn}^2 u + k^2 \operatorname{sn}^2 u = 1.$
$\operatorname{dn}^2 u - k^2 \operatorname{cn}^2 u = 1 - k^2 = k'^2.$

$$\operatorname{sn} u = u - (1 + k^2) \frac{u^3}{3!} + (1 + 14k^2 + k^4) \frac{u^5}{5!}$$
$$- (1 + 135k^2 + 135k^4 + k^6) \frac{u^7}{7!} + \cdots$$

Periods: $4k$ and $2ik'$

$$\operatorname{cn} u = 1 - \frac{u^2}{2!} + (1 + 4k^2) \frac{u^4}{4!} - (1 + 44k^2 + 16k^4) \frac{u^6}{6!} + \cdots$$

Periods: $4k$ and $2k + 2ik'$

$$\operatorname{dn} u = 1 - k^2 \frac{u^2}{2!} + k^2(4 + k^2) \frac{u^4}{4!} - k^2(16 + 44k^2 + k^4) \frac{u^6}{6!} + \cdots$$

Periods: $2k$ and $4ik'$

SPECIAL FUNCTIONS

ELLIPTIC INTEGRALS OF THE FIRST KIND: $F(k, \phi)$*

$$F(k, \phi) = \int_0^\phi \frac{d\Phi}{\sqrt{1 - k^2 \sin^2 \Phi}}, \qquad \theta = \sin^{-1} k$$

ϕ \ θ	5°	10°	15°	20°	25°	30°	35°	40°	45°
1°	0.0175	0.0175	0.0175	0.0175	0.0175	0.0175	0.0175	0.0175	0.0175
2°	0.0349	0.0349	0.0349	0.0349	0.0349	0.0349	0.0349	0.0349	0.0349
3°	0.0524	0.0524	0.0524	0.0524	0.0524	0.0524	0.0524	0.0524	0.0524
4°	0.0698	0.0698	0.0698	0.0698	0.0698	0.0698	0.0698	0.0698	0.0698
5°	0.0873	0.0873	0.0873	0.0873	0.0873	0.0873	0.0873	0.0873	0.0873
6°	0.1047	0.1047	0.1047	0.1047	0.1048	0.1048	0.1048	0.1048	0.1048
7°	0.1222	0.1222	0.1222	0.1222	0.1222	0.1222	0.1223	0.1223	0.1223
8°	0.1396	0.1396	0.1397	0.1397	0.1397	0.1397	0.1398	0.1398	0.1399
9°	0.1571	0.1571	0.1571	0.1572	0.1572	0.1572	0.1573	0.1573	0.1574
10°	0.1745	0.1746	0.1746	0.1746	0.1747	0.1748	0.1748	0.1749	0.1750
11°	0.1920	0.1920	0.1921	0.1921	0.1922	0.1923	0.1924	0.1925	0.1926
12°	0.2095	0.2095	0.2095	0.2096	0.2097	0.2098	0.2099	0.2101	0.2102
13°	0.2269	0.2270	0.2270	0.2271	0.2272	0.2274	0.2275	0.2277	0.2279
14°	0.2444	0.2444	0.2445	0.2446	0.2448	0.2450	0.2451	0.2453	0.2456
15°	0.2618	0.2619	0.2620	0.2621	0.2623	0.2625	0.2628	0.2630	0.2633
16°	0.2793	0.2794	0.2795	0.2797	0.2799	0.2802	0.2804	0.2808	0.2811
17°	0.2967	0.2968	0.2970	0.2972	0.2975	0.2978	0.2981	0.2985	0.2989
18°	0.3142	0.3143	0.3145	0.3148	0.3151	0.3154	0.3159	0.3163	0.3167
19°	0.3317	0.3318	0.3320	0.3323	0.3327	0.3331	0.3336	0.3341	0.3347
20°	0.3491	0.3493	0.3495	0.3499	0.3503	0.3508	0.3514	0.3520	0.3526
21°	0.3666	0.3668	0.3671	0.3675	0.3680	0.3685	0.3692	0.3699	0.3706
22°	0.3840	0.3842	0.3846	0.3851	0.3856	0.3863	0.3871	0.3879	0.3887
23°	0.4015	0.4017	0.4021	0.4027	0.4033	0.4041	0.4049	0.4059	0.4068
24°	0.4190	0.4192	0.4197	0.4203	0.4210	0.4219	0.4229	0.4239	0.4250
25°	0.4364	0.4367	0.4372	0.4379	0.4387	0.4397	0.4408	0.4420	0.4433
26°	0.4539	0.4542	0.4548	0.4556	0.4565	0.4576	0.4588	0.4602	0.4616
27°	0.4714	0.4717	0.4724	0.4732	0.4743	0.4755	0.4769	0.4784	0.4800
28°	0.4888	0.4893	0.4899	0.4909	0.4921	0.4934	0.4950	0.4967	0.4985
29°	0.5063	0.5068	0.5075	0.5086	0.5099	0.5114	0.5132	0.5150	0.5170
30°	0.5238	0.5243	0.5251	0.5263	0.5277	0.5294	0.5313	0.5334	0.5356
31°	0.5412	0.5418	0.5427	0.5440	0.5456	0.5475	0.5496	0.5519	0.5543
32°	0.5587	0.5593	0.5603	0.5617	0.5635	0.5656	0.5679	0.5704	0.5731
33°	0.5762	0.5769	0.5780	0.5795	0.5814	0.5837	0.5862	0.5890	0.5920
34°	0.5937	0.5944	0.5956	0.5973	0.5994	0.6018	0.6046	0.6077	0.6109
35°	0.6111	0.6119	0.6133	0.6151	0.6173	0.6200	0.6231	0.6264	0.6300
36°	0.6286	0.6295	0.6309	0.6329	0.6353	0.6383	0.6416	0.6452	0.6491
37°	0.6461	0.6470	0.6486	0.6507	0.6534	0.6565	0.6602	0.6641	0.6684
38°	0.6636	0.6646	0.6662	0.6685	0.6714	0.6749	0.6788	0.6831	0.6877
39°	0.6810	0.6821	0.6839	0.6864	0.6895	0.6932	0.6975	0.7021	0.7071
40°	0.6985	0.6997	0.7016	0.7043	0.7076	0.7116	0.7162	0.7213	0.7267
41°	0.7160	0.7173	0.7193	0.7222	0.7258	0.7301	0.7350	0.7405	0.7463
42°	0.7335	0.7348	0.7370	0.7401	0.7440	0.7486	0.7539	0.7598	0.7661
43°	0.7510	0.7524	0.7548	0.7580	0.7622	0.7671	0.7728	0.7791	0.7859
44°	0.7685	0.7700	0.7725	0.7760	0.7804	0.7857	0.7918	0.7986	0.8059
45°	0.7859	0.7876	0.7903	0.7940	0.7987	0.8044	0.8109	0.8181	0.8260

* For useful information about these tables see preceding page.

ELLIPTIC INTEGRALS OF THE FIRST KIND: $F(k, \phi)$ (Continued)

$$F(k, \phi) = \int_0^\phi \frac{d\Phi}{\sqrt{1 - k^2 \sin^2 \Phi}}, \qquad \theta = \sin^{-1} k$$

θ \ ϕ	50°	55°	60°	65°	70°	75°	80°	85°	90°
1°	0.0175	0.0175	0.0175	0.0175	0.0175	0.0175	0.0175	0.0175	0.0175
2°	0.0349	0.0349	0.0349	0.0349	0.0349	0.0349	0.0349	0.0349	0.0349
3°	0.0524	0.0524	0.0524	0.0524	0.0524	0.0524	0.0524	0.0524	0.0524
4°	0.0698	0.0699	0.0699	0.0699	0.0699	0.0699	0.0699	0.0699	0.0699
5°	0.0873	0.0873	0.0873	0.0874	0.0874	0.0874	0.0874	0.0874	0.0874
6°	0.1048	0.1048	0.1049	0.1049	0.1049	0.1049	0.1049	0.1049	0.1049
7°	0.1224	0.1224	0.1224	0.1224	0.1224	0.1225	0.1225	0.1225	0.1225
8°	0.1399	0.1399	0.1400	0.1400	0.1400	0.1401	0.1401	0.1401	0.1401
9°	0.1575	0.1575	0.1576	0.1576	0.1577	0.1577	0.1577	0.1577	0.1577
10°	0.1751	0.1751	0.1752	0.1753	0.1753	0.1754	0.1754	0.1754	0.1754
11°	0.1927	0.1928	0.1929	0.1930	0.1930	0.1931	0.1931	0.1932	0.1932
12°	0.2103	0.2105	0.2106	0.2107	0.2108	0.2109	0.2109	0.2110	0.2110
13°	0.2280	0.2282	0.2284	0.2285	0.2286	0.2287	0.2288	0.2288	0.2289
14°	0.2458	0.2460	0.2462	0.2464	0.2465	0.2466	0.2467	0.2468	0.2468
15°	0.2636	0.2638	0.2641	0.2643	0.2645	0.2646	0.2647	0.2648	0.2648
16°	0.2814	0.2817	0.2820	0.2823	0.2825	0.2827	0.2828	0.2829	0.2830
17°	0.2993	0.2997	0.3000	0.3003	0.3006	0.3008	0.3010	0.3011	0.3012
18°	0.3172	0.3177	0.3181	0.3185	0.3188	0.3191	0.3193	0.3194	0.3195
19°	0.3352	0.3357	0.3362	0.3367	0.3371	0.3374	0.3377	0.3378	0.3379
20°	0.3533	0.3539	0.3545	0.3550	0.3555	0.3559	0.3561	0.3563	0.3564
21°	0.3714	0.3721	0.3728	0.3734	0.3740	0.3744	0.3747	0.3749	0.3750
22°	0.3896	0.3904	0.3912	0.3919	0.3926	0.3931	0.3935	0.3937	0.3938
23°	0.4078	0.4088	0.4097	0.4105	0.4113	0.4119	0.4123	0.4126	0.4127
24°	0.4261	0.4272	0.4283	0.4292	0.4301	0.4308	0.4313	0.4316	0.4317
25°	0.4446	0.4458	0.4470	0.4481	0.4490	0.4498	0.4504	0.4508	0.4509
26°	0.4630	0.4645	0.4658	0.4670	0.4681	0.4690	0.4697	0.4701	0.4702
27°	0.4816	0.4832	0.4847	0.4861	0.4873	0.4884	0.4891	0.4896	0.4897
28°	0.5003	0.5021	0.5038	0.5053	0.5067	0.5079	0.5087	0.5092	0.5094
29°	0.5190	0.5210	0.5229	0.5247	0.5262	0.5275	0.5285	0.5291	0.5293
30°	0.5379	0.5401	0.5422	0.5442	0.5459	0.5474	0.5484	0.5491	0.5493
31°	0.5568	0.5593	0.5617	0.5639	0.5658	0.5674	0.5686	0.5693	0.5696
32°	0.5759	0.5786	0.5812	0.5837	0.5858	0.5876	0.5889	0.5898	0.5900
33°	0.5950	0.5980	0.6010	0.6037	0.6060	0.6080	0.6095	0.6104	0.6107
34°	0.6143	0.6176	0.6208	0.6238	0.6265	0.6287	0.6303	0.6313	0.6317
35°	0.6336	0.6373	0.6408	0.6441	0.6471	0.6495	0.6513	0.6525	0.6528
36°	0.6531	0.6571	0.6610	0.6647	0.6679	0.6706	0.6726	0.6739	0.6743
37°	0.6727	0.6771	0.6814	0.6854	0.6890	0.6919	0.6941	0.6955	0.6960
38°	0.6925	0.6973	0.7019	0.7063	0.7102	0.7135	0.7159	0.7175	0.7180
39°	0.7123	0.7176	0.7227	0.7275	0.7318	0.7353	0.7380	0.7397	0.7403
40°	0.7323	0.7380	0.7436	0.7488	0.7535	0.7575	0.7604	0.7623	0.7629
41°	0.7524	0.7586	0.7647	0.7704	0.7756	0.7799	0.7831	0.7852	0.7859
42°	0.7727	0.7794	0.7860	0.7922	0.7979	0.8026	0.8062	0.8084	0.8092
43°	0.7931	0.8004	0.8075	0.8143	0.8204	0.8256	0.8295	0.8320	0.8328
44°	0.8136	0.8215	0.8293	0.8367	0.8433	0.8490	0.8533	0.8560	0.8569
45°	0.8343	0.8428	0.8512	0.8592	0.8665	0.8727	0.8774	0.8804	0.8814

ELLIPTIC INTEGRALS OF THE FIRST KIND: $F(k, \phi)$ (Continued)

$$F(k, \phi) = \int_0^\phi \frac{d\Phi}{\sqrt{1 - k^2 \sin^2 \Phi}}, \qquad \theta = \sin^{-1} k$$

ϕ \ θ	5°	10°	15°	20°	25°	30°	35°	40°	45°
46°	0.8034	0.8052	0.8080	0.8120	0.8170	0.8230	0.8300	0.8378	0.8462
47°	0.8209	0.8227	0.8258	0.8300	0.8353	0.8418	0.8492	0.8575	0.8666
48°	0.8384	0.8403	0.8436	0.8480	0.8537	0.8606	0.8685	0.8773	0.8870
49°	0.8559	0.8579	0.8614	0.8661	0.8721	0.8794	0.8878	0.8972	0.9076
50°	0.8734	0.8756	0.8792	0.8842	0.8905	0.8982	0.9072	0.9173	0.9283
51°	0.8909	0.8932	0.8970	0.9023	0.9090	0.9172	0.9267	0.9374	0.9491
52°	0.9084	0.9108	0.9148	0.9204	0.9275	0.9361	0.9462	0.9575	0.9701
53°	0.9259	0.9284	0.9326	0.9385	0.9460	0.9551	0.9658	0.9778	0.9912
54°	0.9434	0.9460	0.9505	0.9567	0.9646	0.9742	0.9855	0.9982	1.0124
55°	0.9609	0.9637	0.9683	0.9748	0.9832	0.9933	1.0052	1.0187	1.0337
56°	0.9784	0.9813	0.9862	0.9930	1.0018	1.0125	1.0250	1.0393	1.0552
57°	0.9959	0.9989	1.0041	1.0112	1.0204	1.0317	1.0449	1.0600	1.0768
58°	1.0134	1.0166	1.0219	1.0295	1.0391	1.0509	1.0648	1.0807	1.0985
59°	1.0309	1.0342	1.0398	1.0477	1.0578	1.0702	1.0848	1.1016	1.1204
60°	1.0484	1.0519	1.0577	1.0660	1.0766	1.0896	1.1049	1.1226	1.1424
61°	1.0659	1.0695	1.0757	1.0843	1.0953	1.1089	1.1250	1.1436	1.1646
62°	1.0834	1.0872	1.0936	1.1026	1.1141	1.1284	1.1452	1.1648	1.1868
63°	1.1009	1.1049	1.1115	1.1209	1.1330	1.1478	1.1655	1.1860	1.2093
64°	1.1184	1.1225	1.1295	1.1392	1.1518	1.1674	1.1859	1.2073	1.2318
65°	1.1359	1.1402	1.1474	1.1575	1.1707	1.1869	1.2063	1.2288	1.2545
66°	1.1534	1.1579	1.1654	1.1759	1.1896	1.2065	1.2267	1.2503	1.2773
67°	1.1709	1.1756	1.1833	1.1943	1.2085	1.2262	1.2472	1.2719	1.3002
68°	1.1884	1.1932	1.2013	1.2127	1.2275	1.2458	1.2678	1.2936	1.3232
69°	1.2059	1.2109	1.2193	1.2311	1.2465	1.2655	1.2885	1.3154	1.3464
70°	1.2234	1.2286	1.2373	1.2495	1.2655	1.2853	1.3092	1.3372	1.3697
71°	1.2410	1.2463	1.2553	1.2680	1.2845	1.3051	1.3299	1.3592	1.3931
72°	1.2585	1.2640	1.2733	1.2864	1.3036	1.3249	1.3507	1.3812	1.4167
73°	1.2760	1.2817	1.2913	1.3049	1.3226	1.3448	1.3715	1.4033	1.4403
74°	1.2935	1.2994	1.3093	1.3234	1.3417	1.3647	1.3924	1.4254	1.4640
75°	1.3110	1.3171	1.3273	1.3418	1.3608	1.3846	1.4134	1.4477	1.4879
76°	1.3285	1.3348	1.3454	1.3603	1.3800	1.4045	1.4344	1.4700	1.5118
77°	1.3460	1.3525	1.3634	1.3788	1.3991	1.4245	1.4554	1.4923	1.5359
78°	1.3636	1.3702	1.3814	1.3974	1.4183	1.4445	1.4765	1.5147	1.5600
79°	1.3811	1.3879	1.3995	1.4159	1.4374	1.4645	1.4976	1.5372	1.5842
80°	1.3986	1.4056	1.4175	1.4344	1.4566	1.4846	1.5187	1.5597	1.6085
81°	1.4161	1.4234	1.4356	1.4530	1.4758	1.5046	1.5399	1.5823	1.6328
82°	1.4336	1.4411	1.4536	1.4715	1.4950	1.5247	1.5611	1.6049	1.6572
83°	1.4512	1.4588	1.4717	1.4901	1.5143	1.5448	1.5823	1.6276	1.6817
84°	1.4687	1.4765	1.4897	1.5086	1.5335	1.5649	1.6035	1.6502	1.7062
85°	1.4862	1.4942	1.5078	1.5272	1.5527	1.5850	1.6248	1.6730	1.7308
86°	1.5037	1.5120	1.5259	1.5457	1.5720	1.6052	1.6461	1.6957	1.7554
87°	1.5212	1.5297	1.5439	1.5643	1.5912	1.6253	1.6673	1.7184	1.7801
88°	1.5388	1.5474	1.5620	1.5829	1.6105	1.6454	1.6886	1.7412	1.8047
89°	1.5563	1.5651	1.5801	1.6015	1.6297	1.6656	1.7099	1.7640	1.8294
90°	1.5738	1.5828	1.5981	1.6200	1.6490	1.6858	1.7312	1.7863	1.8541

ELLIPTIC INTEGRALS OF THE FIRST KIND: $F(k, \phi)$ (Continued)

$$F(k, \phi) = \int_0^\phi \frac{d\Phi}{\sqrt{1 - k^2 \sin^2 \Phi}}, \qquad \theta = \sin^{-1} k$$

θ \ φ	50°	55°	60°	65°	70°	75°	80°	85°	90°
46°	0.8552	0.8643	0.8734	0.8821	0.8900	0.8968	0.9019	0.9052	0.9063
47°	0.8761	0.8860	0.8958	0.9053	0.9139	0.9212	0.9269	0.9304	0.9316
48°	0.8973	0.9079	0.9185	0.9287	0.9381	0.9461	0.9523	0.9561	0.9575
49°	0.9186	0.9300	0.9415	0.9525	0.9627	0.9714	0.9781	0.9824	0.9838
50°	0.9401	0.9523	0.9647	0.9766	0.9876	0.9971	1.0044	1.0091	1.0107
51°	0.9617	0.9748	0.9881	1.0010	1.0130	1.0233	1.0313	1.0364	1.0381
52°	0.9835	0.9976	1.0118	1.0258	1.0387	1.0499	1.0587	1.0642	1.0662
53°	1.0055	1.0205	1.0359	1.0509	1.0649	1.0771	1.0866	1.0927	1.0948
54°	1.0277	1.0437	1.0602	1.0764	1.0915	1.1048	1.1152	1.1219	1.1242
55°	1.0500	1.0672	1.0848	1.1022	1.1186	1.1331	1.1444	1.1517	1.1542
56°	1.0725	1.0908	1.1097	1.1285	1.1462	1.1619	1.1743	1.1823	1.1851
57°	1.0952	1.1147	1.1349	1.1551	1.1743	1.1914	1.2049	1.2136	1.2167
58°	1.1180	1.1389	1.1605	1.1822	1.2030	1.2215	1.2362	1.2458	1.2492
59°	1.1411	1.1632	1.1864	1.2097	1.2321	1.2522	1.2684	1.2789	1.2826
60°	1.1643	1.1879	1.2125	1.2376	1.2619	1.2837	1.3014	1.3129	1.3170
61°	1.1877	1.2128	1.2392	1.2660	1.2922	1.3159	1.3352	1.3480	1.3524
62°	1.2113	1.2379	1.2661	1.2949	1.3231	1.3490	1.3701	1.3841	1.3890
63°	1.2351	1.2633	1.2933	1.3242	1.3547	1.3828	1.4059	1.4214	1.4268
64°	1.2591	1.2890	1.3209	1.3541	1.3870	1.4175	1.4429	1.4599	1.4659
65°	1.2833	1.3149	1.3489	1.3844	1.4199	1.4532	1.4810	1.4998	1.5065
66°	1.3076	1.3411	1.3773	1.4153	1.4536	1.4898	1.5203	1.5411	1.5485
67°	1.3321	1.3675	1.4060	1.4467	1.4880	1.5274	1.5610	1.5840	1.5923
68°	1.3568	1.3942	1.4351	1.4786	1.5232	1.5661	1.6030	1.6287	1.6379
69°	1.3817	1.4212	1.4646	1.5111	1.5591	1.6059	1.6466	1.6752	1.6856
70°	1.4068	1.4484	1.4944	1.5441	1.5959	1.6468	1.6918	1.7237	1.7354
71°	1.4320	1.4759	1.5246	1.5777	1.6335	1.6891	1.7388	1.7745	1.7877
72°	1.4574	1.5036	1.5552	1.6118	1.6720	1.7326	1.7876	1.8277	1.8427
73°	1.4830	1.5315	1.5862	1.6465	1.7113	1.7774	1.8384	1.8837	1.9008
74°	1.5087	1.5597	1.6175	1.6818	1.7516	1.8237	1.8915	1.9427	1.9623
75°	1.5345	1.5882	1.6492	1.7176	1.7927	1.8715	1.9468	2.0050	2.0276
76°	1.5606	1.6168	1.6812	1.7540	1.8347	1.9207	2.0047	2.0711	2.0973
77°	1.5867	1.6457	1.7136	1.7909	1.8777	1.9716	2.0653	2.1414	2.1721
78°	1.6130	1.6748	1.7462	1.8284	1.9215	2.0240	2.1288	2.2164	2.2528
79°	1.6394	1.7040	1.7792	1.8664	1.9663	2.0781	2.1954	2.2969	2.3404
80°	1.6660	1.7335	1.8125	1.9048	2.0119	2.1339	2.2653	2.3836	2.4362
81°	1.6926	1.7631	1.8461	1.9438	2.0584	2.1913	2.3387	2.4775	2.5421
82°	1.7193	1.7929	1.8799	1.9831	2.1057	2.2504	2.4157	2.5795	2.6603
83°	1.7462	1.8228	1.9140	2.0229	2.1537	2.3110	2.4965	2.6911	2.7942
84°	1.7731	1.8528	1.9482	2.0630	2.2024	2.3731	2.5811	2.8136	2.9487
85°	1.8001	1.8830	1.9826	2.1035	2.2518	2.4366	2.6694	2.9487	3.1313
86°	1.8271	1.9132	2.0172	2.1442	2.3017	2.5013	2.7612	3.0978	3.3547
87°	1.8542	1.9435	2.0519	2.1852	2.3520	2.5670	2.8561	3.2620	3.6425
88°	1.8813	1.9739	2.0867	2.2263	2.4026	2.6336	2.9537	3.4412	4.0481
89°	1.9084	2.0043	2.1216	2.2675	2.4535	2.7007	3.0530	3.6328	4.7413
90°	1.9356	2.0347	2.1565	2.3088	2.5046	2.7681	3.1534	3.8317	———

SPECIAL FUNCTIONS

ELLIPTIC INTEGRALS OF THE SECOND KIND: $E(k, \phi)$

$$E(k, \phi) = \int_0^\phi \sqrt{(1 - k^2 \sin^2 \Phi)}\,d\Phi, \quad \theta = \sin^{-1} k$$

ϕ \ θ	5°	10°	15°	20°	25°	30°	35°	40°	45°
1°	0.0175	0.0175	0.0175	0.0175	0.0175	0.0175	0.0175	0.0175	0.0175
2°	0.0349	0.0349	0.0349	0.0349	0.0349	0.0349	0.0349	0.0349	0.0349
3°	0.0524	0.0524	0.0524	0.0524	0.0524	0.0524	0.0524	0.0523	0.0523
4°	0.0698	0.0698	0.0698	0.0698	0.0698	0.0698	0.0698	0.0698	0.0698
5°	0.0873	0.0873	0.0873	0.0873	0.0872	0.0872	0.0872	0.0872	0.0872
6°	0.1047	0.1047	0.1047	0.1047	0.1047	0.1047	0.1047	0.1046	0.1046
7°	0.1222	0.1222	0.1222	0.1221	0.1221	0.1221	0.1221	0.1220	0.1220
8°	0.1396	0.1396	0.1396	0.1396	0.1395	0.1395	0.1395	0.1394	0.1394
9°	0.1571	0.1571	0.1570	0.1570	0.1570	0.1569	0.1569	0.1568	0.1568
10°	0.1745	0.1745	0.1745	0.1744	0.1744	0.1743	0.1742	0.1742	0.1741
11°	0.1920	0.1920	0.1919	0.1918	0.1918	0.1917	0.1916	0.1915	0.1914
12°	0.2094	0.2094	0.2093	0.2093	0.2092	0.2091	0.2089	0.2088	0.2087
13°	0.2269	0.2268	0.2268	0.2267	0.2265	0.2264	0.2263	0.2261	0.2259
14°	0.2443	0.2443	0.2442	0.2441	0.2439	0.2437	0.2436	0.2433	0.2431
15°	0.2618	0.2617	0.2616	0.2615	0.2613	0.2611	0.2608	0.2606	0.2603
16°	0.2792	0.2791	0.2790	0.2788	0.2786	0.2784	0.2781	0.2778	0.2775
17°	0.2967	0.2966	0.2964	0.2962	0.2959	0.2956	0.2953	0.2949	0.2946
18°	0.3141	0.3140	0.3138	0.3136	0.3133	0.3129	0.3125	0.3121	0.3116
19°	0.3316	0.3314	0.3312	0.3309	0.3305	0.3301	0.3296	0.3291	0.3286
20°	0.3490	0.3489	0.3486	0.3483	0.3478	0.3473	0.3468	0.3462	0.3456
21°	0.3665	0.3663	0.3660	0.3656	0.3651	0.3645	0.3639	0.3632	0.3625
22°	0.3839	0.3837	0.3834	0.3829	0.3823	0.3817	0.3809	0.3802	0.3793
23°	0.4013	0.4011	0.4007	0.4002	0.3996	0.3988	0.3980	0.3971	0.3961
24°	0.4188	0.4185	0.4181	0.4175	0.4168	0.4159	0.4150	0.4139	0.4129
25°	0.4362	0.4359	0.4354	0.4348	0.4339	0.4330	0.4319	0.4308	0.4296
26°	0.4537	0.4533	0.4528	0.4520	0.4511	0.4500	0.4488	0.4475	0.4462
27°	0.4711	0.4707	0.4701	0.4693	0.4682	0.4670	0.4657	0.4643	0.4628
28°	0.4886	0.4881	0.4874	0.4865	0.4854	0.4840	0.4825	0.4809	0.4793
29°	0.5060	0.5055	0.5048	0.5037	0.5025	0.5010	0.4993	0.4975	0.4957
30°	0.5234	0.5229	0.5221	0.5209	0.5195	0.5179	0.5161	0.5141	0.5120
31°	0.5409	0.5403	0.5394	0.5381	0.5366	0.5348	0.5327	0.5306	0.5283
32°	0.5583	0.5577	0.5567	0.5553	0.5536	0.5516	0.5494	0.5470	0.5446
33°	0.5757	0.5751	0.5740	0.5725	0.5706	0.5684	0.5660	0.5634	0.5607
34°	0.5932	0.5924	0.5912	0.5896	0.5876	0.5852	0.5826	0.5797	0.5768
35°	0.6106	0.6098	0.6085	0.6067	0.6045	0.6019	0.5991	0.5960	0.5928
36°	0.6280	0.6272	0.6258	0.6238	0.6214	0.6186	0.6155	0.6122	0.6087
37°	0.6455	0.6445	0.6430	0.6409	0.6383	0.6353	0.6319	0.6283	0.6245
38°	0.6629	0.6619	0.6602	0.6580	0.6552	0.6519	0.6483	0.6444	0.6403
39°	0.6803	0.6792	0.6775	0.6750	0.6720	0.6685	0.6646	0.6604	0.6559
40°	0.6977	0.6966	0.6947	0.6921	0.6888	0.6851	0.6808	0.6763	0.6715
41°	0.7152	0.7139	0.7119	0.7091	0.7056	0.7016	0.6970	0.6921	0.6870
42°	0.7326	0.7313	0.7291	0.7261	0.7224	0.7180	0.7132	0.7079	0.7024
43°	0.7500	0.7486	0.7463	0.7431	0.7391	0.7345	0.7293	0.7237	0.7178
44°	0.7674	0.7659	0.7634	0.7600	0.7558	0.7508	0.7453	0.7393	0.7330
45°	0.7849	0.7832	0.7806	0.7770	0.7725	0.7672	0.7613	0.7549	0.7482

ELLIPTIC INTEGRALS OF THE SECOND KIND: $E(k, \phi)$ (Continued)

$$E(k, \phi) = \int_0^\phi \sqrt{(1 - k^2 \sin^2 \Phi)}\,d\Phi, \quad \theta = \sin^{-1} k$$

ϕ \ θ	50°	55°	60°	65°	70°	75°	80°	85°	90°
1°	0.0175	0.0175	0.0175	0.0175	0.0175	0.0175	0.0175	0.0175	0.0175
2°	0.0349	0.0349	0.0349	0.0349	0.0349	0.0349	0.0349	0.0349	0.0349
3°	0.0523	0.0523	0.0523	0.0523	0.0523	0.0523	0.0523	0.0523	0.0523
4°	0.0698	0.0698	0.0698	0.0698	0.0698	0.0698	0.0698	0.0698	0.0698
5°	0.0872	0.0872	0.0872	0.0872	0.0872	0.0872	0.0872	0.0872	0.0872
6°	0.1046	0.1046	0.1046	0.1046	0.1046	0.1045	0.1045	0.1045	0.1045
7°	0.1220	0.1220	0.1219	0.1219	0.1219	0.1219	0.1219	0.1219	0.1219
8°	0.1394	0.1393	0.1393	0.1393	0.1392	0.1392	0.1392	0.1392	0.1392
9°	0.1567	0.1566	0.1566	0.1566	0.1565	0.1565	0.1565	0.1564	0.1564
10°	0.1740	0.1739	0.1739	0.1738	0.1738	0.1737	0.1737	0.1737	0.1736
11°	0.1913	0.1912	0.1911	0.1910	0.1909	0.1909	0.1908	0.1908	0.1908
12°	0.2085	0.2084	0.2083	0.2082	0.2081	0.2080	0.2080	0.2079	0.2079
13°	0.2258	0.2256	0.2254	0.2253	0.2252	0.2251	0.2250	0.2250	0.2250
14°	0.2429	0.2427	0.2425	0.2424	0.2422	0.2421	0.2420	0.2419	0.2419
15°	0.2601	0.2598	0.2596	0.2594	0.2592	0.2590	0.2589	0.2588	0.2588
16°	0.2771	0.2768	0.2765	0.2763	0.2761	0.2759	0.2757	0.2757	0.2756
17°	0.2942	0.2938	0.2935	0.2932	0.2929	0.2927	0.2925	0.2924	0.2924
18°	0.3112	0.3107	0.3103	0.3099	0.3096	0.3094	0.3092	0.3091	0.3090
19°	0.3281	0.3276	0.3271	0.3267	0.3263	0.3260	0.3258	0.3256	0.3256
20°	0.3450	0.3444	0.3438	0.3433	0.3429	0.3425	0.3422	0.3421	0.3420
21°	0.3618	0.3611	0.3604	0.3598	0.3593	0.3589	0.3586	0.3584	0.3584
22°	0.3785	0.3777	0.3770	0.3763	0.3757	0.3752	0.3749	0.3747	0.3746
23°	0.3952	0.3943	0.3935	0.3927	0.3920	0.3915	0.3911	0.3908	0.3907
24°	0.4118	0.4108	0.4098	0.4090	0.4082	0.4076	0.4071	0.4068	0.4067
25°	0.4284	0.4272	0.4261	0.4251	0.4243	0.4236	0.4230	0.4227	0.4226
26°	0.4449	0.4436	0.4423	0.4412	0.4402	0.4394	0.4389	0.4385	0.4384
27°	0.4613	0.4598	0.4584	0.4572	0.4561	0.4552	0.4545	0.4541	0.4540
28°	0.4776	0.4760	0.4744	0.4730	0.4718	0.4708	0.4701	0.4696	0.4695
29°	0.4938	0.4920	0.4903	0.4887	0.4874	0.4863	0.4855	0.4850	0.4848
30°	0.5100	0.5080	0.5061	0.5044	0.5029	0.5016	0.5007	0.5002	0.5000
31°	0.5261	0.5239	0.5218	0.5199	0.5182	0.5169	0.5159	0.5152	0.5150
32°	0.5421	0.5396	0.5373	0.5352	0.5334	0.5319	0.5308	0.5301	0.5299
33°	0.5580	0.5553	0.5528	0.5505	0.5485	0.5468	0.5456	0.5449	0.5446
34°	0.5738	0.5709	0.5681	0.5656	0.5634	0.5616	0.5603	0.5595	0.5592
35°	0.5895	0.5863	0.5833	0.5806	0.5782	0.5762	0.5748	0.5739	0.5736
36°	0.6051	0.6017	0.5984	0.5954	0.5928	0.5907	0.5891	0.5881	0.5878
37°	0.6207	0.6169	0.6134	0.6101	0.6073	0.6050	0.6032	0.6022	0.6018
38°	0.6361	0.6321	0.6282	0.6247	0.6216	0.6191	0.6172	0.6160	0.6157
39°	0.6515	0.6471	0.6429	0.6391	0.6357	0.6330	0.6310	0.6297	0.6293
40°	0.6667	0.6620	0.6575	0.6533	0.6497	0.6468	0.6446	0.6432	0.6428
41°	0.6818	0.6767	0.6719	0.6674	0.6636	0.6604	0.6580	0.6566	0.6561
42°	0.6969	0.6914	0.6862	0.6814	0.6772	0.6738	0.6712	0.6697	0.6691
43°	0.7118	0.7059	0.7003	0.6952	0.6907	0.6870	0.6843	0.6826	0.6820
44°	0.7266	0.7204	0.7144	0.7088	0.7040	0.7000	0.6971	0.6953	0.6947
45°	0.7414	0.7346	0.7282	0.7223	0.7171	0.7129	0.7097	0.7078	0.7071

ELLIPTIC INTEGRALS OF THE SECOND KIND: $E(k, \phi)$ (Continued)

$$E(k, \phi) = \int_0^\phi \sqrt{(1 - k^2 \sin^2 \Phi)}\, d\Phi, \quad \theta = \sin^{-1} k$$

θ / ϕ	5°	10°	15°	20°	25°	30°	35°	40°	45°
46°	0.8023	0.8006	0.7977	0.7939	0.7891	0.7835	0.7772	0.7704	0.7633
47°	0.8197	0.8179	0.8149	0.8108	0.8057	0.7998	0.7931	0.7858	0.7782
48°	0.8371	0.8352	0.8320	0.8277	0.8223	0.8160	0.8089	0.8012	0.7931
49°	0.8545	0.8525	0.8491	0.8446	0.8389	0.8322	0.8247	0.8165	0.8079
50°	0.8719	0.8698	0.8663	0.8614	0.8554	0.8483	0.8404	0.8317	0.8227
51°	0.8894	0.8871	0.8834	0.8783	0.8719	0.8644	0.8560	0.8469	0.8373
52°	0.9068	0.9044	0.9004	0.8951	0.8884	0.8805	0.8716	0.8620	0.8518
53°	0.9242	0.9217	0.9175	0.9119	0.9048	0.8965	0.8872	0.8770	0.8663
54°	0.9416	0.9389	0.9345	0.9287	0.9212	0.9125	0.9026	0.8919	0.8806
55°	0.9590	0.9562	0.9517	0.9454	0.9376	0.9284	0.9181	0.9068	0.8949
56°	0.9764	0.9735	0.9687	0.9622	0.9540	0.9443	0.9335	0.9216	0.9091
57°	0.9938	0.9908	0.9858	0.9789	0.9703	0.9602	0.9488	0.9363	0.9232
58°	1.0112	1.0080	1.0028	0.9956	0.9866	0.9760	0.9641	0.9510	0.9372
59°	1.0286	1.0253	1.0198	1.0123	1.0029	1.9918	0.9793	0.9656	0.9511
60°	1.0460	1.0426	1.0368	1.0290	1.0191	1.0076	0.9945	0.9801	0.9650
61°	1.0634	1.0598	1.0538	1.0456	1.0354	1.0233	1.0096	0.9946	0.9787
62°	1.0808	1.0771	1.0708	1.0623	1.0516	1.0389	1.0246	1 0090	0.9924
63°	1.0982	1.0943	1.0878	1.0789	1.0678	1.0546	1.0397	1.0233	1.0060
64°	1.1156	1.1115	1.1048	1.0955	1.0839	1.0702	1.0547	1.0376	1.0195
65°	1.1330	1.1288	1.1218	1.1121	1.1001	1.0858	1.0696	1.0518	1.0329
66°	1.1504	1.1460	1.1387	1.1287	1.1162	1.1013	1.0845	1.0660	1.0463
67°	1.1678	1.1632	1.1557	1.1453	1.1323	1.1168	1.0993	1.0801	1.0596
68°	1.1852	1.1805	1.1726	1.1618	1.1483	1.1323	1.1141	1.0941	1.0728
69°	1.2026	1.1977	1.1896	1.1784	1.1644	1.1478	1.1289	1.1081	1.0859
70°	1.2200	1.2149	1.2065	1.1949	1.1804	1.1632	1.1436	1.1221	1.0990
71°	1.2374	1.2321	1.2234	1.2114	1.1964	1.1786	1.1583	1.1359	1.1120
72°	1.2548	1.2493	1.2403	1.2280	1.2124	1.1939	1.1729	1.1498	1.1250
73°	1.2722	1.2666	1.2573	1.2445	1.2284	1.2093	1.1875	1.1636	1.1379
74°	1.2896	1.2838	1.2742	1.2609	1.2443	1.2246	1.2021	1.1773	1.1507
75°	1.3070	1.3010	1.2911	1.2774	1.2603	1.2399	1.2167	1.1910	1.1635
76°	1.3244	1.3182	1.3080	1.2939	1.2762	1.2552	1.2312	1.2047	1.1762
77°	1.3418	1.3354	1.3249	1.3104	1.2921	1.2704	1.2457	1.2183	1.1889
78°	1.3592	1.3526	1.3417	1.3268	1.3080	1.2856	1.2601	1.2319	1.2015
79°	1.3765	1.3698	1.3586	1.3432	1.3239	1.3009	1.2746	1.2454	1.2141
80°	1.3939	1.3870	1.3755	1.3597	1.3398	1.3161	1.2890	1.2590	1.2266
81°	1.4113	1.4042	1.3924	1.3761	1.3556	1.3312	1.3034	1.2725	1.2391
82°	1.4287	1.4214	1.4093	1.3925	1.3715	1.3464	1.3177	1.2859	1.2516
83°	1.4461	1.4386	1.4261	1.4090	1.3873	1.3616	1.3321	1.2994	1.2640
84°	1.4635	1.4558	1.4430	1.4254	1.4032	1.3767	1.3464	1.3128	1.2765
85°	1.4809	1.4729	1.4598	1.4418	1.4190	1.3919	1.3608	1.3262	1.2889
86°	1.4983	1.4901	1.4767	1.4582	1.4348	1.4070	1.3751	1.3396	1.3012
87°	1.5156	1.5073	1.4936	1.4746	1.4507	1.4221	1.3894	1.3530	1.3136
88°	1.5330	1.5245	1.5104	1.4910	1.4665	1.4372	1.4037	1.3664	1.3260
89°	1.5504	1.5417	1.5273	1.5074	1.4823	1.4523	1.4180	1.3798	1.3383
90°	1.5678	1.5589	1.5442	1.5238	1.4981	1.4675	1.4323	1.3931	1.3506

ELLIPTIC INTEGRALS OF THE SECOND KIND: $E(k, \phi)$ (Continued)

$$E(k, \phi) = \int_0^\phi \sqrt{(1 - k^2 \sin^2 \Phi)} d\Phi, \quad \theta = \sin^{-1} k$$

θ / ϕ	50°	55°	60°	65°	70°	75°	80°	85°	90°
46°	0.7560	0.7488	0.7419	0.7356	0.7301	0.7255	0.7221	0.7200	0.7193
47°	0.7705	0.7628	0.7555	0.7488	0.7429	0.7380	0.7344	0.7321	0.7314
48°	0.7849	0.7768	0.7690	0.7618	0.7555	0.7502	0.7464	0.7440	0.7431
49°	0.7992	0.7905	0.7822	0.7746	0.7679	0.7623	0.7581	0.7556	0.7547
50°	0.8134	0.8042	0.7954	0.7872	0.7801	0.7741	0.7697	0.7670	0.7660
51°	0.8275	0.8177	0.8084	0.7997	0.7921	0.7858	0.7811	0.7781	0.7771
52°	0.8414	0.8311	0.8212	0.8120	0.8039	0.7972	0.7922	0.7891	0.7880
53°	0.8553	0.8444	0.8339	0.8241	0.8155	0.8084	0.8031	0.7998	0.7986
54°	0.8690	0.8575	0.8464	0.8361	0.8270	0.8194	0.8137	0.8102	0.8090
55°	0.8827	0.8705	0.8588	0.8479	0.8382	0.8302	0.8242	0.8204	0.8192
56°	0.8962	0.8834	0.8710	0.8595	0.8493	0.8408	0.8344	0.8304	0.8290
57°	0.9096	0.8961	0.8831	0.8709	0.8601	0.8511	0.8443	0.8401	0.8387
58°	0.9230	0.9088	0.8950	0.8822	0.8707	0.8612	0.8540	0.8496	0.8480
59°	0.9362	0.9213	0.9068	0.8932	0.8812	0.8711	0.8635	0.8588	0.8572
60°	0.9493	0.9336	0.9184	0.9042	0.8914	0.8808	0.8728	0.8677	0.8660
61°	0.9623	0.9459	0.9299	0.9149	0.9015	0.8903	0.8817	0.8764	0.8746
62°	0.9752	0.9580	0.9412	0.9254	0.9113	0.8995	0.8905	0.8849	0.8829
63°	0.9880	0.9700	0.9524	0.9358	0.9210	0.9085	0.8990	0.8930	0.8910
64°	1.0007	0.9818	0.9634	0.9460	0.9304	0.9173	0.9072	0.9009	0.8988
65°	1.0133	0.9936	0.9743	0.9561	0.9397	0.9258	0.9152	0.9086	0.9063
66°	1.0258	1.0052	0.9850	0.9659	0.9487	0.9341	0.9230	0.9159	0.9135
67°	1.0383	1.0167	0.9956	0.9756	0.9576	0.9422	0.9305	0.9230	0.9205
68°	1.0506	1.0281	1.0061	0.9852	0.9662	0.9501	0.9377	0.9299	0.9272
69°	1.0628	1.0394	1.0164	0.9946	0.9747	0.9578	0.9447	0.9364	0.9336
70°	1.0750	1.0506	1.0266	1.0038	0.9830	0.9652	0.9514	0.9427	0.9397
71°	1.0871	1.0617	1.0367	1.0129	0.9911	0.9724	0.9579	0.9487	0.9455
72°	1.0991	1.0727	1.0467	1.0218	0.9990	0.9794	0.9642	0.9544	0.9511
73°	1.1110	1.0836	1.0565	1.0306	1.0067	0.9862	0.9702	0.9599	0.9563
74°	1.1228	1.0944	1.0662	1.0392	1.0143	0.9928	0.9759	0.9650	0.9613
75°	1.1346	1.1051	1.0759	1.0477	1.0217	0.9992	0.9814	0.9699	0.9659
76°	1.1463	1.1158	1.0854	1.0561	1.0290	1.0053	0.9867	0.9745	0.9703
77°	1.1580	1.1263	1.0948	1.0643	1.0361	1.0113	0.9917	0.9789	0.9744
78°	1.1695	1.1368	1.1041	1.0724	1.0430	1.0171	0.9965	0.9829	0.9781
79°	1.1811	1.1472	1.1133	1.0805	1.0498	1.0228	1.0011	0.9867	0.9816
80°	1.1926	1.1576	1.1225	1.0884	1.0565	1.0282	1.0054	0.9902	0.9848
81°	1.2040	1.1678	1.1316	1.0962	1.0630	1.0335	1.0096	0.9935	0.9877
82°	1.2154	1.1781	1.1406	1.1040	1.0695	1.0387	1.0135	0.9965	0.9903
83°	1.2267	1.1883	1.1495	1.1116	1.0758	1.0437	1.0173	0.9992	0.9925
84°	1.2381	1.1984	1.1584	1.1192	1.0821	1.0486	1.0209	1.0017	0.9945
85°	1.2493	1.2085	1.1673	1.1267	1.0882	1.0534	1.0244	1.0039	0.9962
86°	1.2606	1.2186	1.1761	1.1342	1.0944	1.0581	1.0277	1.0060	0.9976
87°	1.2719	1.2286	1.1848	1.1417	1.1004	1.0628	1.0309	1.0078	0.9986
88°	1.2831	1.2386	1.1936	1.1491	1.1064	1.0673	1.0340	1.0095	0.9994
89°	1.2943	1.2487	1.2023	1.1565	1.1124	1.0719	1.0371	1.0111	0.9998
90°	1.3055	1.2587	1.2111	1.1638	1.1184	1.0764	1.0401	1.0127	1.0000

COMPLETE ELLIPTIC INTEGRALS

$$K = \int_0^{\pi/2} \frac{d\Phi}{\sqrt{1 - k^2 \sin^2 \Phi}} = F\left(k, \frac{\pi}{2}\right)$$

$\sin^{-1} k$	K	$\log K$	$\sin^{-1} k$	K	$\log K$
0°	1.5708	0.196120	**40°**	1.7868	0.252068
1	1.5709	0.196153	41	1.7992	0.255085
2	1.5713	0.196252	42	1.8122	0.258197
3	1.5719	0.196418	43	1.8256	0.261406
4	1.5727	0.196649	44	1.8396	0.264716
5	1.5738	0.196947	**45**	1.8541	0.268127
6	1.5751	0.197312	46	1.8691	0.271644
7	1.5767	0.197743	47	1.8848	0.275267
8	1.5785	0.198241	48	1.9011	0.279001
9	1.5805	0.198806	49	1.9180	0.282848
10	1.5828	0.199438	**50**	1.9356	0.286811
11	1.5854	0.200137	51	1.9539	0.290895
12	1.5882	0.200904	52	1.9729	0.295101
13	1.5913	0.201740	53	1.9927	0.299435
14	1.5946	0.202643	54	2.0133	0.303901
15	1.5981	0.203615	**55**	2.0347	0.308504
16	1.6020	0.204657	56	2.0571	0.313247
17	1.6061	0.205768	57	2.0804	0.318138
18	1.6105	0.206948	58	2.1047	0.323182
19	1.6151	0.208200	59	2.1300	0.328384
20	1.6200	0.209522	**60**	2.1565	0.333753
21	1.6252	0.210916	61	2.1842	0.339295
22	1.6307	0.212382	62	2.2132	0.345020
23	1.6365	0.213921	63	2.2435	0.350936
24	1.6426	0.215533	64	2.2754	0.357053
25	1.6490	0.217219	**65**	2.3088	0.363384
26	1.6557	0.218981	66	2.3439	0.369940
27	1.6627	0.220818	67	2.3809	0.376736
28	1.6701	0.222732	68	2.4198	0.383787
29	1.6777	0.224723	69	2.4610	0.391112
30	1.6858	0.226793	**70**	2.5046	0.398730
31	1.6941	0.228943	71	2.5507	0.406665
32	1.7028	0.231173	72	2.5998	0.414943
33	1.7119	0.233485	73	2.6521	0.423596
34	1.7214	0.235880	74	2.7081	0.432660
35	1.7312	0.238359	**75**	2.7681	0.442176
36	1.7415	0.240923	76	2.8327	0.452196
37	1.7522	0.243575	77	2.9026	0.462782
38	1.7633	0.246315	78	2.9786	0.474008
39	1.7748	0.249146	79	3.0617	0.485967
40	1.7868	0.252068	**80**	3.1534	0.498777

COMPLETE ELLIPTIC INTEGRALS (Continued)

$$K = \int_0^{\pi/2} \frac{d\Phi}{\sqrt{1 - k^2 \sin^2 \Phi}} = F\left(k, \frac{\pi}{2}\right)$$

$\sin^{-1} k$	K	$\log K$	$\sin^{-1} k$	K	$\log K$
80°	3.1534	0.498777	85°	3.8317	0.583396
81	3.2553	0.512591	86	4.0528	0.607751
82	3.3699	0.527613	87	4.3387	0.637355
83	3.5004	0.544120	88	4.7427	0.676027
84	3.6519	0.562514	89	5.4349	0.735192
85	3.8317	0.583396	90	∞	∞

Values of K for $\sin^{-1} k = 85°$ to $89°$ by $0.1°$ and $89°$ to $90°$ by minutes

$\sin^{-1} k$	K	$\log K$	$\sin^{-1} k$		K	$\log K$
85.0°	3.832	0.58343	89°	0′	5.435	0.73520
85.1	3.852	0.58569	89	2	5.469	0.73791
85.2	3.872	0.58794	89	4	5.504	0.74068
85.3	3.893	0.59028	89	6	5.540	0.74351
85.4	3.914	0.59262	89	8	5.578	0.74648
85.5	3.936	0.59506	89	10	5.617	0.74950
85.6	3.958	0.59748	89	12	5.658	0.75266
85.7	3.981	0.59999	89	14	5.700	0.75587
85.8	4.004	0.60249	89	16	5.745	0.75929
85.9	4.028	0.60509	89	18	5.791	0.76275
86.0	4.053	0.60778	89	20	5.840	0.76641
86.1	4.078	0.61045	89	22	5.891	0.77019
86.2	4.104	0.61321	89	24	5.946	0.77422
86.3	4.130	0.61595	89	26	6.003	0.77837
86.4	4.157	0.61878	89	28	6.063	0.78269
86.5	4.185	0.62170	89	30	6.128	0.78732
86.6	4.214	0.62469	89	32	6.197	0.79218
86.7	4.244	0.62778	89	34	6.271	0.79734
86.8	4.274	0.63083	89	36	6.351	0.80284
86.9	4.306	0.63407	89	38	6.438	0.80875
87.0	4.339	0.63739	89	40	6.533	0.81511
87.1	4.372	0.64068	89	41	6.584	0.81849
87.2	4.407	0.64414	89	42	6.639	0.82210
87.3	4.444	0.64777	89	43	6.696	0.82582
87.4	4.481	0.65137	89	44	6.756	0.82969
87.5	4.520	0.65514	89	45	6.821	0.83385
87.6	4.561	0.65916	89	46	6.890	0.83822
87.7	4.603	0.66304	89	47	6.964	0.84286
87.8	4.648	0.66727	89	48	7.044	0.84782
87.9	4.694	0.67154	89	49	7.131	0.85315
88.0	4.743	0.67605	89	50	7.226	0.85890
88.1	4.794	0.68070	89	51	7.332	0.86522
88.2	4.848	0.68556	89	52	7.449	0.87210
88.3	4.905	0.69064	89	53	7.583	0.87984
88.4	4.965	0.69592	89	54	7.737	0.88857
88.5	5.030	0.70157	89	55	7.919	0.89867
88.6	5.099	0.70749	89	56	8.143	0.91078
88.7	5.173	0.71374	89	57	8.430	0.92583
88.8	5.253	0.72041	89	58	8.836	0.94626
88.9	5.340	0.72754	89	59	9.529	0.97905
89.0	5.435	0.73520	90	0	∞	∞

420

COMPLETE ELLIPTIC INTEGRALS (Continued)

$$E = \int_0^{\pi/2} \sqrt{1 - k^2 \sin^2 \Phi}\cdot d\Phi = E\left(k, \frac{\pi}{2}\right)$$

$\sin^{-1} k$	E	$\log E$	$\sin^{-1} k$	E	$\log E$
0°	1.5708	0.196120	45°	1.3506	0.130541
1	1.5707	0.196087	46	1.3418	0.127690
2	1.5703	0.195988	47	1.3329	0.124788
3	1.5697	0.195822	48	1.3238	0.121836
4	1.5689	0.195591	49	1.3147	0.118836
5	1.5678	0.195293	50	1.3055	0.115790
6	1.5665	0.194930	51	1.2963	0.112698
7	1.5649	0.194500	52	1.2870	0.109563
8	1.5632	0.194004	53	1.2776	0.106386
9	1.5611	0.193442	54	1.2681	0.103169
10	1.5589	0.192815	55	1.2587	0.099915
11	1.5564	0.192121	56	1.2492	0.096626
12	1.5537	0.191362	57	1.2397	0.093303
13	1.5507	0.190537	58	1.2301	0.089950
14	1.5476	0.189646	59	1.2206	0.086569
15	1.5442	0.188690	60	1.2111	0.083164
16	1.5405	0.187668	61	1.2015	0.079738
17	1.5367	0.186581	62	1.1920	0.076293
18	1.5326	0.185428	63	1.1826	0.072834
19	1.5283	0.184210	64	1.1732	0.069364
20	1.5238	0.182928	65	1.1638	0.065889
21	1.5191	0.181580	66	1.1545	0.062412
22	1.5141	0.180168	67	1.1453	0.058937
23	1.5090	0.178691	68	1.1362	0.055472
24	1.5037	0.177150	69	1.1272	0.052020
25	1.4981	0.175545	70	1.1184	0.048589
26	1.4924	0.173876	71	1.1096	0.045183
27	1.4864	0.172144	72	1.1011	0.041812
28	1.4803	0.170348	73	1.0927	0.038481
29	1.4740	0.168489	74	1.0844	0.035200
30	1.4675	0.166567	75	1.0764	0.031976
31	1.4608	0.164583	76	1.0686	0.028819
32	1.4539	0.162537	77	1.0611	0.025740
33	1.4469	0.160429	78	1.0538	0.022749
34	1.4397	0.158261	79	1.0468	0.019858
35	1.4323	0.156031	80	1.0401	0.017081
36	1.4248	0.153742	81	1.0338	0.014432
37	1.4171	0.151393	82	1.0278	0.011927
38	1.4092	0.148985	83	1.0223	0.009584
39	1.4013	0.146519	84	1.0172	0.007422
40	1.3931	0.143995	85	1.0127	0.005465
41	1.3849	0.141414	86	1.0086	0.003740
42	1.3765	0.138778	87	1.0053	0.002278
43	1.3680	0.136086	88	1.0026	0.001121
44	1.3594	0.133340	89	1.0008	0.000326
45	1.3506	0.130541	90	1.0000	0.000000

SINE, COSINE, AND EXPONENTIAL INTEGRALS

$$Si(x) = \int_0^x \frac{\sin v}{v}\, dv; \qquad Ci(x) = \int_\infty^x \frac{\cos v}{v}\, dv;$$

$$Ei(x) = \int_{-\infty}^x \frac{e^v}{v}\, dv; \qquad -Ei(-x) = \int_x^\infty \frac{e^{-v}}{v}\, dv$$

x	Si(x)	Ci(x)	Ei(x)	−Ei(−x)
0.0	0.00000	−∞	−∞	+∞
0.1	0.09994	−1.72787	−1.62281	1.82292
0.2	.19956	−1.04221	− .82176	1.22265
0.3	.29850	− .64917	− .30267	.90568
0.4	.39646	− .37881	.10477	.70238
0.5	.49311	− .17778	.45422	.55977
0.6	.58813	− .02227	.76988	.45438
0.7	.68122	.10051	1.06491	.37377
0.8	.77210	.19828	1.34740	.31060
0.9	.86047	.27607	1.62281	.26018
1.0	.94608	.33740	1.89512	.21938
1.1	1.02869	.38487	2.16738	.18599
1.2	1.10805	.42046	2.44209	.15841
1.3	1.18396	.44574	2.72140	.13545
1.4	1.25623	.46201	3.00721	.11622
1.5	1.32468	.47036	3.30129	.10002
1.6	1.38918	.47173	3.60532	.08631
1.7	1.44959	.46697	3.92096	.07465
1.8	1.50582	.45681	4.24987	.06471
1.9	1.55778	.44194	4.59371	.05620
2.0	1.60541	.42298	4.95423	.04890
2.1	1.64870	.40051	5.33324	.04261
2.2	1.68762	.37507	5.73261	.03719
2.3	1.72221	.34718	6.15438	.03250
2.4	1.75249	.31729	6.60067	.02844
2.5	1.77852	.28587	7.07377	.02491
2.6	1.80039	.25334	7.57611	.02185
2.7	1.81821	.22008	8.11035	.01918
2.8	1.83210	.18649	8.67930	.01686
2.9	1.84219	.15290	9.28602	.01482
3.0	1.84865	.11963	9.93383	.01305
3.1	1.85166	.08699	10.6263	.01149
3.2	1.85140	.05526	11.3673	.01013
3.3	1.84808	.02468	12.1610	.00894
3.4	1.84191	− .00452	13.0121	.00789
3.5	1.83313	− .03213	13.9254	.00697
3.6	1.82195	− .05797	14.9063	.00616
3.7	1.80862	− .08190	15.9606	.00545
3.8	1.79339	− .10378	17.0948	.00482
3.9	1.77650	− .12350	18.3157	.00427
4.0	1.75820	− .14098	19.6309	.00378
4.1	1.73874	− .15617	21.0485	.00335
4.2	1.71837	− .16901	22.5774	.00297
4.3	1.69732	− .17951	24.2274	.00263
4.4	1.67583	− .18766	26.0090	.00234

SINE, COSINE, AND EXPONENTIAL INTEGRALS (Continued)

x	$Si(x)$	$Ci(x)$	$Ei(x)$	$-Ei(-x)$
4.5	1.65414	− .19349	27.9337	.00207
4.6	1.63246	− .19705	30.0141	.00184
4.7	1.61100	− .19839	32.2639	.00164
4.8	1.58998	− .19760	34.6979	.00145
4.9	1.56956	− .19478	37.3325	.00129
5.0	1.54993	− .19003	40.1853	.00115
5.1	1.53125	− .18348	43.2757	.00102
5.2	1.51367	− .17525	46.6249	.00091
5.3	1.49732	− .16551	50.2557	.00081
5.4	1.48230	− .15439	54.1935	.00072
5.5	1.46872	− .14205	58.4655	.00064
5.6	1.45667	− .12867	63.1018	.00057
5.7	1.44620	− .11441	68.1350	.00051
5.8	1.43736	− .09944	73.6008	.00045
5.9	1.43018	− .08393	79.5382	.00040
6.0	1.42469	− .06806	85.9898	.00036
6.1	1.42087	− .05198	93.0020	.00032
6.2	1.41871	− .03587	100.626	.00029
6.3	1.41817	− .01989	108.916	.00026
6.4	1.41922	− .00418	117.935	.00023
6.5	1.42179	+ .01110	127.747	.00020
6.6	1.42582	+ .02582	138.426	.00018
6.7	1.43121	.03986	150.050	.00016
6.8	1.43787	.05308	162.707	.00014
6.9	1.44570	.06539	176.491	.00013
7.0	1.45460	.07670	191.505	.00012
7.1	1.46443	.08691	207.863	.00010
7.2	1.47509	.09596	225.688	.00009
7.3	1.48644	.10379	245.116	.00008
7.4	1.49834	.11036	266.296	.00007
7.5	1.51068	.11563	289.388	.00007
7.6	1.52331	.11960	314.572	.00006
7.7	1.53611	.12225	342.040	.00005
7.8	1.54894	.12359	372.006	.00005
7.9	1.56167	.12364	404.701	.00004
8.0	1.57419	.12243	440.380	.00004
8.1	1.58637	.12002	479.322	.00003
8.2	1.59810	.11644	521.831	.00003
8.3	1.60928	.11177	568.242	.00003
8.4	1.61981	.10607	681.919	.00002
8.5	1.62960	.09943	674.264	.00002
8.6	1.63857	.09194	734.714	.00002
8.7	1.64665	.08368	800.749	.00002
8.8	1.65379	.07476	872.895	.00002
8.9	1.65993	.06528	951.728	.00001
9.0	1.66504	.05535	1037.88	.00001
9.1	1.66908	.04507	1132.04	.00001
9.2	1.67205	.03455	1234.96	.00001
9.3	1.67393	.02391	1347.48	.00001
9.4	1.67473	.01325	1470.51	.00001

SINE, COSINE, AND EXPONENTIAL INTEGRALS (Continued)

x	Si(x)	Ci(x)	Ei(x)	−Ei(−x)
9.5	1.67446	.00268	1605.03	.00001
9.6	1.67316	− .00771	1752.14	.00001
9.7	1.67084	− .01780	1913.05	.00001
9.8	1.66757	− .02752	2089.05	.00001
9.9	1.66338	− .03676	2281.58	.00000
10.0	1.65835	− .04546	2492.23	.00000
10.5	1.62294	− .07828	3883.74	.00000
11.0	1.57831	− .08956	6071.41	.00000
11.5	1.53572	− .07857	9518.20	.00000
12.0	1.50497	− .04978	14959.5	.00000
12.5	1.49234	− .01141	23565.1	.00000
13.0	1.49936	+ .02676	37197.7	.00000
13.5	1.52291	+ .05576	58827.0	.00000
14.0	1.55621	.06940	93193.0	.00000
14.5	1.59072	.06554	147866.	.00000
15.0	1.61819	.04628	234955.	.00000

ORTHOGONAL POLYNOMIALS

I

Name: Legendre   *Symbol*: $P_n(x)$ *Interval*: $[-1, 1]$

Differential Equation: $(1 - x^2)y'' - 2xy' + n(n + 1)y = 0$
$$y = P_n(x)$$

Explicit Expression: $P_n(x) = \dfrac{1}{2^n} \sum_{m=0}^{[n/2]} (-1)^m \binom{n}{m}\binom{2n - 2m}{n} x^{n-2m}$

Recurrence Relation: $(n + 1)P_{n+1}(x) = (2n + 1)xP_n(x) - nP_{n-1}(x)$

Weight: 1   *Standardization*: $P_n(1) = 1$

Norm: $\displaystyle\int_{-1}^{+1} [P_n(x)]^2\, dx = \dfrac{2}{2n + 1}$

Rodrigues' Formula: $P_n(x) = \dfrac{(-1)^n}{2^n n!} \dfrac{d^n}{dx^n} \{(1 - x^2)^n\}$

Generating Function: $R^{-1} = \displaystyle\sum_{n=0}^{\infty} P_n(x)z^n; \quad -1 < x < 1, \quad |z| < 1,$
$$R = \sqrt{1 - 2xz + z^2}.$$

Inequality: $|P_n(x)| \le 1, \; -1 \le x \le 1.$

II

Name: Tschebysheff, First Kind   *Symbol*: $T_n(x)$ *Interval*: $[-1, 1]$

Differential Equation: $(1 - x^2)y'' - xy' + n^2 y = 0$
$$y = T_n(x)$$

Explicit Expression: $\dfrac{n}{2} \displaystyle\sum_{m=0}^{[n/2]} (-1)^m \dfrac{(n - m - 1)!}{m!(n - 2m)!} (2x)^{n-2m} = \cos(n \arccos x) = T_n(x)$

Recurrence Relation: $T_{n+1}(x) = 2xT_n(x) - T_{n-1}(x)$

Weight: $(1 - x^2)^{-1/2}$ *Standardization*: $T_n(1) = 1$

Norm: $\displaystyle\int_{-1}^{+1} (1 - x^2)^{-1/2}[T_n(x)]^2\, dx = \begin{cases} \pi/2, & n \ne 0 \\ \pi, & n = 0 \end{cases}$

Rodrigues' Formula: $\dfrac{(-1)^n(1 - x^2)^{1/2}\sqrt{\pi}}{2^{n+1}\Gamma(n + \frac{1}{2})} \dfrac{d^n}{dx^n} \{(1 - x^2)^{n-(1/2)}\} = T_n(x)$

Generating Function: $\dfrac{1 - xz}{1 - 2xz + z^2} = \displaystyle\sum_{n=0}^{\infty} T_n(x)z^n, \quad -1 < x < 1, \quad |z| < 1.$

Inequality: $|T_n(x)| \le 1, \; -1 \le x \le 1.$

III

Name: Tschebysheff, Second Kind   *Symbol*: $U_n(x)$ *Interval*: $[-1, 1]$

Differential Equation: $(1 - x^2)y'' - 3xy' + n(n + 2)y = 0$

$$y = U_n(x)$$

Explicit Expression: $U_n(x) = \sum_{m=0}^{[n/2]} (-1)^m \frac{(m - n)!}{m!(n - 2m)!} (2x)^{n-2m}$

$$U_n(\cos \theta) = \frac{\sin[(n + 1)\theta]}{\sin \theta}$$

Recurrence Relation: $U_{n+1}(x) = 2x U_n(x) - U_{n-1}(x)$

Weight: $(1 - x^2)^{1/2}$ *Standardization*: $U_n(1) = n + 1$

Norm: $\displaystyle\int_{-1}^{+1} (1 - x^2)^{1/2}[U_n(x)]^2 dx = \frac{\pi}{2}$

Rodrigues' Formula: $U_n(x) = \dfrac{(-1)^n(n + 1)\sqrt{\pi}}{(1 - x^2)^{1/2}2^{n+1}\Gamma(n + \frac{3}{2})} \dfrac{d^n}{dx^n}\{(1 - x^2)^{n+(1/2)}\}$

Generating Function: $\dfrac{1}{1 - 2xz + z^2} = \sum_{n=0}^{\infty} U_n(x)z^n, \ -1 < x < 1, \ |z| < 1.$

Inequality: $|U_n(x)| \le n + 1, \ -1 \le x \le 1.$

IV

Name: Jacobi   *Symbol*: $P_n^{(\alpha,\beta)}(x)$ *Interval*: $[-1, 1]$

Differential Equation:

$$(1 - x^2)y'' + [\beta - \alpha - (\alpha + \beta + 2)x]y' + n(n + \alpha + \beta + 1)y = 0$$
$$y = P_n^{(\alpha,\beta)}(x)$$

Explicit Expression: $P_n^{(\alpha,\beta)}(x) = \dfrac{1}{2^n} \sum_{m=0}^{n} \binom{n + \alpha}{m}\binom{n + \beta}{n - m}(x - 1)^{n-m}(x + 1)^m$

Recurrence Relation: $2(n + 1)(n + \alpha + \beta + 1)(2n + \alpha + \beta)P_{n+1}^{(\alpha,\beta)}(x)$

$$= (2n + \alpha + \beta + 1)[(\alpha^2 - \beta^2) + (2n + \alpha + \beta + 2)$$
$$\times (2n + \alpha + \beta)x]P_n^{(\alpha,\beta)}(x)$$
$$- 2(n + \alpha)(n + \beta)(2n + \alpha + \beta + 2)P_{n-1}^{(\alpha,\beta)}(x)$$

Weight: $(1 - x)^{\alpha}(1 + x)^{\beta}; \ \alpha, \beta > 1$ *Standardization*: $P_n^{(\alpha,\beta)}(x) = \binom{n + \alpha}{n}$

Norm: $\displaystyle\int_{-1}^{+1} (1 - x)^{\alpha}(1 + x)^{\beta}[P_n^{(\alpha,\beta)}(x)]^2 dx = \dfrac{2^{\alpha+\beta+1}\Gamma(n + \alpha + 1)\Gamma(n + \beta + 1)}{(2n + \alpha + \beta + 1)n!\Gamma(n + \alpha + \beta + 1)}$

Rodrigues' Formula: $P_n^{(\alpha,\beta)}(x) = \dfrac{(-1)^n}{2^n n!(1 - x)^{\alpha}(1 + x)^{\beta}} \dfrac{d^n}{dx^n}\{(1 - x)^{n+\alpha}(1 + x)^{n+\beta}\}$

IV (Continued)

Generating Function: $R^{-1}(1 - z + R)^{-\alpha}(1 + z + R)^{-\beta} = \sum_{n=0}^{\infty} 2^{-\alpha-\beta} P_n^{(\alpha,\beta)}(x) z^n$,

$$R = \sqrt{1 - 2xz + z^2}, \ |z| < 1$$

Inequality: $\max_{-1 \le x \le 1} |P_n^{(\alpha,\beta)}(x)| = \begin{cases} \binom{n+q}{n} \sim n^q \text{ if } q = \max(\alpha, \beta) \ge -\frac{1}{2} \\ |P_n^{(\alpha,\beta)}(x')| \sim n^{-1/2} \text{ if } q < -\frac{1}{2} \\ x' \text{ is one of the two maximum points nearest} \\ \dfrac{\beta - \alpha}{\alpha + \beta + 1} \end{cases}$

V

Name: Generalized Laguerre   *Symbol*: $L_n^{(\alpha)}(x)$ *Interval*: $[0, \infty]$

Differential Equation: $xy'' + (\alpha + 1 - x)y' + ny = 0$

$$y = L_n^{(\alpha)}(x)$$

Explicit Expression: $L_n^{(\alpha)}(x) = \sum_{m=0}^{n} (-1)^m \binom{n + \alpha}{n - m} \frac{1}{m!} x^m$

Recurrence Relation: $(n + 1) L_{n+1}^{(\alpha)}(x) = [(2n + \alpha + 1) - x] L_n^{(\alpha)}(x) - (n + \alpha) L_{n-1}^{(\alpha)}(x)$

Weight: $x^\alpha e^{-x}, \ \alpha > -1$ *Standardization*: $L_n^{(\alpha)}(x) = \dfrac{(-1)^n}{n!} x^n + \cdots$

Norm: $\displaystyle\int_0^\infty x^\alpha e^{-x} [L_n^{(\alpha)}(x)]^2 \, dx = \dfrac{\Gamma(n + \alpha + 1)}{n!}$

Rodrigues' Formula: $L_n^{(\alpha)}(x) = \dfrac{1}{n! x^\alpha e^{-x}} \dfrac{d^n}{dx^n} \{x^{n+\alpha} e^{-x}\}$

Generating Function: $(1 - z)^{-\alpha-1} \exp\left(\dfrac{xz}{z - 1}\right) = \sum_{n=0}^{\infty} L_n^{(\alpha)}(x) z^n$

Inequality: $|L_n^{(\alpha)}(x)| \le \dfrac{\Gamma(n + \alpha + 1)}{n! \Gamma(\alpha + 1)} e^{x/2}; \quad \begin{array}{l} x \ge 0 \\ \alpha > 0 \end{array}$

$$|L_n^{(\alpha)}(x)| \le \left[2 - \dfrac{\Gamma(\alpha + n + 1)}{n! \Gamma(\alpha + 1)}\right] e^{x/2}; \quad \begin{array}{l} x \ge 0 \\ -1 < \alpha < 0 \end{array}$$

Orthogonal Polynomials

Name: Hermite *Symbol:* $H_n(x)$ *Interval:* $[-\infty, \infty]$

Differential Equation: $y'' - 2xy' + 2ny = 0$

Explicit Expression: $H_n(x) = \displaystyle\sum_{m=0}^{[n/2]} \frac{(-1)^m \, n! \, (2x)^{n-2m}}{m! \, (n-2m)!}$

Recurrence Relation: $H_{n+1}(x) = 2x \, H_n(x) - 2n H_{n-1}(x)$

Weight: e^{-x^2} *Standardization:* $H_n(1) = 2^n x^n + \ldots$

Norm: $\displaystyle\int_{-\infty}^{\infty} e^{-x^2} \, [H_n(x)]^2 \, dx = 2^n \, n! \, \sqrt{\pi}$

Rodriques' Formula: $H_n(x) = (-1)^n \, e^{x^2} \dfrac{d^n}{dx^n} \left(e^{-x^2} \right)$

Generating Function: $e^{-z^2 + 2zx} = \displaystyle\sum_{n=0}^{\infty} H_n(x) \, \frac{z^n}{n!}$

Inequality: $|H_n(x)| < e^{\frac{x^2}{2}} \, k \, 2^{n/2} \, \sqrt{n!}$ $k \approx 1.086435$

Coefficients for Orthogonal Polynomials, and for x^n in Terms of Orthogonal Polynomials*

I. Legendre Polynomials: $P_n(x) = a_n^{-1} \sum_{m=0}^{n} c_m x^m$ $\qquad x^n = b_n^{-1} \sum_{m=0}^{n} d_m P_m(x)$

	a_n	x^0	x^1	x^2	x^3	x^4	x^5	x^6	x^7
b_n		1	1	3	5	35	63	231	429
P_0	1	1 1		1		7		33	
P_1	1		1 1		3		27		143
P_2	2	-1		3 2		20		110	
P_3	2		-3		5 2		28		182
P_4	8	3		-30		35 8		72	
P_5	8		15		-70		63 8		88
P_6	16	-5		105		-315		231 16	
P_7	16		-35		315		-693		429 16

$P_6(x) = \dfrac{1}{16}[231x^6 - 315x^4 + 105x^2 - 5]$ $\qquad x^6 = \dfrac{1}{231}[33P_0 + 110P_2 + 72P_4 + 16P_6]$

II. Tschebysheff Polynomials: $T_n(x) = \sum_{m=0}^{n} c_m x^m$ $\qquad x^n = b_n^{-1} \sum_{m=0}^{n} d_m T_m(x)$

	x^0	x^1	x^2	x^3	x^4	x^5	x^6	x^7
b_n	1	1	2	4	8	16	32	64
T_0	1 1		1		3		10	
T_1		1 1		3		10		35
T_2	-1		2 1		4		15	
T_3		-3		4 1		5		21
T_4	1		-8		8 1		6	
T_5		5		-20		16 1		7
T_6	-1		18		-48		32 1	
T_7		-7		56		-112		64 1

$T_6(x) = 32x^6 - 48x^4 + 18x^2 - 1$ $\qquad x^6 = \dfrac{1}{32}[10T_0 + 15T_2 + 6T_4 + T_6]$

III. Tschebysheff Polynomials: $U_n(x) = \sum_{m=0}^{n} c_m x^m$ $\qquad x^n = b_n^{-1} \sum_{m=0}^{n} d_m U_m(x)$

	x^0	x^1	x^2	x^3	x^4	x^5	x^6	x^7
b_n	1	2	4	8	16	32	64	128
U_0	1 1		1		2		5	
U_1		2 1		2		5		14
U_2	-1		4 1		3		9	
U_3		-4		8 1		4		14
U_4	1		-12		16 1		5	
U_5		6		-32		32 1		6
U_6	-1		24		-80		64 1	
U_7		-8		80		-192		128 1

$U_6(x) = 64x^6 - 80x^4 + 24x^2 - 1$ $\qquad x^6 = \dfrac{1}{64}[5U_0 + 9U_2 + 5U_4 + U_6]$

Coefficients for Orthogonal Polynomials, and for x^n in Terms of Orthogonal Polynomials* (continued)

IV. Jacobi Polynomials $P_n^{(\alpha,\beta)}(x) = a_n^{-1} \sum_{m=0}^{n} c_m (x-1)^m$

	α_n	$(x-1)^0$	$(x-1)^1$	$(x-1)^2$	$(x-1)^3$	$(x-1)^4$	$(x-1)^5$	$(x-1)^6$
$P_0^{(\alpha,\beta)}$	1	1						
$P_1^{(\alpha,\beta)}$	2	$2(\alpha+1)$	$\alpha+\beta+2$					
$P_2^{(\alpha,\beta)}$	8	$4(\alpha+1)_2$	$4(\alpha+\beta+3)(\alpha+2)$	$(\alpha+\beta+3)_2$				
$P_3^{(\alpha,\beta)}$	48	$8(\alpha+1)_3$	$12(\alpha+\beta+4)(\alpha+2)_2$	$6(\alpha+\beta+4)_2(\alpha+3)$	$(\alpha+\beta+4)_3$			
$P_4^{(\alpha,\beta)}$	384	$16(\alpha+1)_4$	$32(\alpha+\beta+5)(\alpha+2)_3$	$24(\alpha+\beta+5)_2(\alpha+3)_2$	$8(\alpha+\beta+5)_3(\alpha+4)$	$(\alpha+\beta+5)_4$		
$P_5^{(\alpha,\beta)}$	3840	$32(\alpha+1)_5$	$80(\alpha+\beta+6)(\alpha+2)_4$	$80(\alpha+\beta+6)_2(\alpha+3)_3$	$40(\alpha+\beta+6)_3(\alpha+4)_2$	$10(\alpha+\beta+6)_4(\alpha+5)$	$(\alpha+\beta+6)_5$	
$P_6^{(\alpha,\beta)}$	46080	$64(\alpha+1)_6$	$192(\alpha+\beta+7)(\alpha+2)_5$	$240(\alpha+\beta+7)_2(\alpha+3)_4$	$160(\alpha+\beta+7)_3(\alpha+4)_3$	$60(\alpha+\beta+7)_4(\alpha+5)_2$	$12(\alpha+\beta+7)_5(\alpha+6)$	$(\alpha+\beta+7)_6$

$$(m)_n = m(m+1)(m+2)\cdots(m+n-1)$$

$$P_5^{(1,1)}(x) = \frac{1}{3840}\left[(8)_5(x-1)^5 + 10(8)_4(6)(x-1)^4 + 40(8)_2(5)_2(x-1)^3 + 80(8)_2(4)_3(x-1)^2 + 80(8)(3)_4(x-1) + 32(2)_5\right]$$

$$P_5^{(1,1)}(x) = \frac{1}{3840}\left[95040(x-1)^5 + 475200(x-1)^4 + 864000(x-1)^3 + 691200(x-1)^2 + 230400(x-1) + 23040\right]$$

Coefficients for Orthogonal Polynomials, and for x^n in Terms of Orthogonal Polynomials[a] (continued)

V. Laguerre Polynomials: $L_n(x) = \displaystyle\sum_{m=0}^{n} c_m x^m \qquad x^n = b_n^{-1} \sum_{m=0}^{n} d_m L_m(x)$

	x^0	x^1	x^2	x^3	x^4	x^5	x^6	x^7
b_m	1	1	2	6	24	120	720	5040
L_0	1 1	1	2	6	24	120	720	5040
L_1	1	−1 −1	−4	−18	−96	−600	−4320	−35280
L_2	2	−4	1 2	18	144	1200	10800	105840
L_3	6	−18	9	−1 −6	−96	−1200	−14400	−176400
L_4	24	−96	72	−16	1 24	600	10800	17640
L_5	120	−600	600	−200	25	−1 −120	−4320	−105840
L_6	720	−4320	5400	−2400	450	−36	1 720	35280
L_7	5040	−35280	52920	−29400	7350	−882	49	−1 −5040

$$L_6(x) = x^6 - 36x^5 + 450x^4 - 2400x^3 + 5400x^2 - 4320x + 720$$

$$x^6 = \frac{1}{720}\,[720L_0 - 4320L_1 + 10800L_2 - 14400L_3 + 10800L_4 - 4320L_5 + 720L_6]$$

VI. Hermite polynomials: $H_n(x) = \displaystyle\sum_{m=0}^{n} c_m x^m \qquad x^n = b_n^{-1} \sum_{m=0}^{n} d_m H_m(x)$

	x^0	x^1	x^2	x^3	x^4	x^5	x^6	x^7
b_n	1	2	4	8	16	32	64	128
H_0	1 1		2		12		120	
H_1		2 1		6		60		840
H_2	−2		4 1		12		180	
H_3		−12		8 1		20		420
H_4	12		−48		16 1		30	
H_5		120		−160		32 1		42
H_6	−120		720		−480		64 1	
H_7		−1680		3360		−1344		128 1

$$H_6(x) = 64x^6 - 480x^4 + 720x^2 - 120 \qquad x^6 = \frac{1}{64}\,[120\,H_0 + 180\,H_2 + 30H_4 + H_6]$$

Abridged from Abramowitz, M. and Stegun, I. A., Eds., *Handbook of Mathematical Functions,* National Bureau of Standards, Washington, D. C., 1964.

LEGENDRE FUNCTIONS

In the following, m and n are positive integers.

1. The differential equation $(1 - z^2)\dfrac{d^2w}{dz^2} - 2z\dfrac{dw}{dz} + n(n + 1)z = 0$ is known as Legendre's differential equation. The Legendre polynomials $P_0(z)$, $P_1(z)$, $P_2(z)$, . . . are solutions of this equation, and are referred to as Surface Zonal Harmonics, for which tables are herewith given.

2. The solution of Legendre's equation can be stated as

$$w = AP_n(z) + BQ_n(z) \qquad |z| < 1$$
or
$$w = AP_n(z) + B\mathbf{Q}_n(z) \qquad |z| > 1,$$

where A and B are arbitrary constants, $P_n(z)$ the Legendre polynomials, $Q_n(z)$ and $\mathbf{Q}_n(z)$ Legendre functions of the second kind.

3. The differential equation $(1 - z^2)\dfrac{d^2u}{dz^2} - 2z\dfrac{du}{dz} + \left[n(n + 1) - \dfrac{m^2}{1 - z^2} \right] u = 0$ which is obtained from the Legendre equation in 1. after it has been differentiated m times and u replaced by $(1 - z^2)^{m/2}\left(\dfrac{d^m w}{dz^m} \right)$ is referred to as the "associated Legendre differential equation."

4. The solution for the "associated Legendre differential equation" is given by

$$u = AP_n^m(z) + BQ_n^m(z),$$

where A and B are arbitrary constants and $P_n^m(z)$, $Q_n^m(z)$ are called associated Legendre functions.

5. $P_n(z) = \displaystyle\sum_{r=0}^{m} (-1)^r \frac{(2n - 2r)!}{2^n(r!)(n - r_r)!(n - 2r)!} z^{n-2r} \qquad m = \frac{1}{2}n,\ n \text{ even}$

$$m = \frac{1}{2}(n - 1),\ n \text{ odd}$$

$$Q_n(z) = \frac{1}{2}P_n(z) \log_e \frac{1 + z}{1 - z} - \sum_{r=1}^{n} \frac{1}{r} P_{r-1}(z)P_{n-r}(z)$$

$$= P_n(z)Q_0(z) - \sum_{r=1}^{n} \frac{1}{r} P_{r-1}(z)P_{n-r}(z)$$

$$Q_0(z) = \frac{1}{2} \log_e \frac{1 + z}{1 - z} = \tanh^{-1} z$$

$$\mathbf{Q}_n(z) = \frac{1}{2}P_n(z) \log_e \frac{z + 1}{z - 1} - \sum_{r=1}^{n} \frac{1}{r} P_{r-1}(z)\, P_{n-r}(z)$$

$$= P_n(z)\mathbf{Q}_0(z) - \sum_{r=1}^{n} \frac{1}{r} P_{r-1}(z)P_{n-r}(z)$$

$$\mathbf{Q}_0(z) = \frac{1}{2} \log_e \frac{z + 1}{z - 1} = \operatorname{ctnh}^{-1} z$$

LEGENDRE FUNCTIONS (Continued)

6.

n	$P_n(z)$	$Q_n(z)$
0	1	$\dfrac{\log_e}{2}\dfrac{(1+z)}{(1-z)}$
1	z	$zQ_0(z) - 1$
2	$\dfrac{1}{2}(3z^2 - 1)$	$\dfrac{1}{2}(3z^2 - 1)Q_0(z) - \dfrac{3}{2}z$
3	$\dfrac{1}{2}(5z^3 - 3z)$	$\dfrac{1}{2}(5z^3 - 3z)Q_0(z) - \dfrac{5}{2}z^2 + \dfrac{2}{3}$
4	$\dfrac{1}{8}(35z^4 - 30z^2 + 3)$	$\dfrac{1}{8}(35z^4 - 30z^2 + 3)Q_0(z) - \dfrac{35}{8}z^3 + \dfrac{55}{24}z$
5	$\dfrac{1}{8}(63z^5 - 70z^3 + 15z)$	$\dfrac{1}{8}(63z^5 - 70z^3 + 15z)Q_0(z) - \dfrac{63}{8}z^4 + \dfrac{49}{8}z^2 - \dfrac{8}{15}$

7. $\dfrac{1}{(1 - 2xZ + Z^2)^{1/2}} = \sum P_n(x)Z^n$

If $x = \cos\theta = \frac{1}{2}(e^{i\theta} + e^{-i\theta})$, $i = \sqrt{-1}$, then

$$[1 - Z(e^{i\theta} + e^{-i\theta}) + Z^2]^{-1/2} = \sum_{n=0}^{\infty} P_n(\cos\theta)Z^n$$

$P_0(\cos\theta) = 1$

$P_1(\cos\theta) = \cos\theta$

$P_2(\cos\theta) = \dfrac{1}{4}(3\cos 2\theta + 1)$

$P_3(\cos\theta) = \dfrac{1}{8}(5\cos 3\theta + 3\cos\theta)$

$P_4(\cos\theta) = \dfrac{1}{64}(35\cos 4\theta + 20\cos 2\theta + 9)$

$$P_n(\cos\theta) = \frac{(2n)!}{2^{2n}(n!)^2}\left[\cos n\theta + 1/1\,\frac{n}{2n-1}\cos(n-2)\theta\right.$$
$$+ \frac{1\cdot 3}{1\cdot 2}\frac{n(n-1)}{(2n-1)(2n-3)}\cos(n-4)\theta$$
$$\left.+ \frac{1\cdot 3\cdot 5}{1\cdot 2\cdot 3}\frac{n(n-1)(n-2)}{(2n-1)(2n-3)(2n-5)}\cos(n-6)\theta + \cdots\right]$$

LEGENDRE FUNCTIONS (Continued)

8. $P_{2n+1}(0) = 0$

$$P_{2n}(0) = (-1)^n \frac{1 \cdot 3 \cdot 5 \, \cdots \, (2n-1)}{2 \cdot 4 \cdot 6 \cdot 8 \, \cdots \, (2n)}$$

$P_n(1) = 1$

$P_n(-1) = (-1)^n$

$$P_n(z) = \frac{1}{2^n n!} \frac{d^n}{dz^n} (z^2 - 1)^n \text{ (Rodrigues' formula)}$$

$$P_n^m(z) = (1 - z^2)^{m/2} \frac{d^m[P_n(z)]}{dz^m}$$

$$Q_n^m(z) = (1 - z^2)^{m/2} \frac{d^m[Q_n(z)]}{dz^m}$$

9. Important recurrence and orthogonality relations (note that w_n stands for either P_n or Q_n in the following)

$$(n + 1)[w_{n+1}(z)] - (2n + 1)z[w_n(z)] + n[w_{n-1}(z)] = 0$$

$$z \frac{d[w_n(z)]}{dz} - z \frac{d[w_{n-1}(z)]}{dz} = n[w_n(z)]$$

$$\frac{d[w_{n+1}(z)]}{dz} - z \frac{d[w_n(z)]}{dz} = (n + 1)[w_n(z)]$$

$$\frac{d[w_{n+1}(z)]}{dz} - \frac{d}{dz}[w_{n-1}(z)] = (2n + 1)[w_n(z)]$$

$$(z^2 - 1) \frac{d}{dz}[w_n(z)] = nz[w_n(z)] - n[w_{n-1}(z)]$$

$$\int_{-1}^{1} [P_m(z)][P_n(z)]dz \begin{cases} = 0 & m \neq n \\ = \dfrac{2}{2n + 1} & m = n \end{cases}$$

$$\int_{-1}^{1} [P_n^m(z)][P_r^m(z)]dz \begin{cases} = 0 & r \neq n \\ = \dfrac{2}{2n + 1} \dfrac{(n + m)!}{(n - m)!} & r = n > m. \end{cases}$$

SURFACE ZONAL HARMONICS

$$P_n(x) = \frac{1}{2^n n!} \frac{d^n}{dx^n}(x^2 - 1)^n$$

x	$P_1(x)$	$P_2(x)$	$P_3(x)$	$P_4(x)$	$P_5(x)$
0	0	− .5	0	.375	0
.01	.01	− .49985	− .0149975	.37463	.018741
.02	.02	− .4994	− .02998	.37350	.037430
.03	.03	− .49865	− .0449325	.37163	.056014
.04	.04	− .4976	− .05984	.36901	.074441
.05	.05	− .49625	− .0746875	.36565	.092659
.06	.06	− .4946	− .08946	.36156	.11062
.07	.07	− .49265	− .10414	.35673	.12826
.08	.08	− .4904	− .11872	.35118	.14555
.09	.09	− .48785	− .13318	.34491	.16242
.10	.10	− .485	− .1475	.33794	.17883
.11	.11	− .48185	− .16167	.33027	.19473
.12	.12	− .4784	− .17568	.32191	.21008
.13	.13	− .47465	− .18951	.31287	.22482
.14	.14	− .4706	− .20314	.30318	.23891
.15	.15	− .46625	− .21656	.29284	.25232
.16	.16	− .4616	− .22976	.28187	.26499
.17	.17	− .45665	− .24272	.27028	.27688
.18	.18	− .4514	− .25542	.25809	.28796
.19	.19	− .44585	− .26785	.24533	.29818
.20	.20	− .44	− .28	.232	.30752
.21	.21	− .43385	− .29185	.21813	.31593
.22	.22	− .4274	− .30338	.20375	.32339
.23	.23	− .42065	− .31458	.18887	.32986
.24	.24	− .4136	− .32544	.17352	.33531
.25	.25	− .40625	− .33594	.15771	.33972
.26	.26	− .3986	− .34606	.14149	.34307
.27	.27	− .39065	− .35579	.12488	.34532
.28	.28	− .3824	− .36512	.10789	.34647
.29	.29	− .37385	− .37403	.09057	.34650
.30	.30	− .365	− .3825	.07294	.34539
.31	.31	− .35585	− .39052	.05503	.34312
.32	.32	− .3464	− .39808	.03688	.33970
.33	.33	− .33665	− .40516	.01851	.33512
.34	.34	− .3266	− .41174	− .00004	.32937
.35	.35	− .31625	− .41781	− .01872	.32245
.36	.36	− .3056	− .42336	− .03752	.31438
.37	.37	− .29465	− .42837	− .05638	.30514
.38	.38	− .2834	− .43282	− .07528	.29477
.39	.39	− .27185	− .43670	− .09416	.28326
.40	.40	− .26	− .44	− .113	.27064
.41	.41	− .24785	− .44270	− .13175	.25693
.42	.42	− .2354	− .44478	− .15036	.24215
.43	.43	− .22265	− .44623	− .16880	.22633
.44	.44	− .2096	− .44704	− .18702	.20951
.45	.45	− .19625	− .44719	− .20497	.19172
.46	.46	− .1826	− .44666	− .22261	.17301
.47	.47	− .16865	− .44544	− .23989	.15341
.48	.48	− .1544	− .44352	− .25676	.13298
.49	.49	− .13985	− .44088	− .27316	.11177

SURFACE ZONAL HARMONICS (Continued)

x	$P_6(x)$	$P_7(x)$	$P_8(x)$	$P_9(x)$	$P_{10}(x)$
0	$-$ .3125	$-$ 0	.2734	0	$-$ .2461
.01	$-$ .31184	$-$ .021855	.2725	.0246	$-$ .2447
.02	$-$ .30988	$-$ .043593	.2695	.0489	$-$ .2407
.03	$-$ .30661	$-$ .065094	.2646	.0729	$-$ .2340
.04	$-$ .30205	$-$ .086244	.2578	.0961	$-$ .2247
.05	$-$ .29622	$-$ .10693	.2492	.1186	$-$ .2130
.06	$-$ .28913	$-$ .12703	.2387	.1400	$-$ .1989
.07	$-$ .28081	$-$ .14644	.2265	.1601	$-$ .1826
.08	$-$ .27130	$-$ .16506	.2126	.1789	$-$ .1642
.09	$-$ .26063	$-$ .18278	.1972	.1960	$-$ .1440
.10	$-$ .24883	$-$ .19949	.1803	.2114	$-$ .1221
.11	$-$ .23595	$-$ .21511	.1621	.2249	$-$ .0989
.12	$-$ .22204	$-$ .22955	.1426	.2364	$-$ .0745
.13	$-$ .20715	$-$ .24271	.1221	.2457	$-$ .0492
.14	$-$ .19133	$-$ .25453	.1006	.2529	$-$ .0233
.15	$-$ .17465	$-$ .26492	.0783	.2577	.0020
.16	$-$ .15716	$-$ .27383	.0554	.2601	.0293
.17	$-$ .13894	$-$ .28119	.0319	.2602	.0553
.18	$-$ .12005	$-$ .28695	.0082	.2579	.0808
.19	$-$ .10057	$-$ .29107	$-$ .0157	.2531	.1055
.20	$-$ .080576	$-$ .29352	$-$ .0396	.2460	.1291
.21	$-$ .060144	$-$ .29426	$-$ .0632	.2365	.1513
.22	$-$ .039357	$-$ .29327	$-$ .0865	.2247	.1718
.23	$-$ .018300	$-$ .29055	$-$ .1093	.2108	.1905
.24	.002941	$-$ .28610	$-$ .1313	.1948	.2070
.25	.024277	$-$ .27992	$-$ .1525	.1768	.2212
.26	.045618	$-$ .27203	$-$ .1725	.1571	.2329
.27	.066872	$-$ .26246	$-$ .1914	.1357	.2419
.28	.087947	$-$ 25124	$-$ .2089	.1129	.2480
.29	.10875	$-$ .23843	$-$ .2248	.0888	.2513
.30	.12918	$-$ .22407	$-$ .2391	.0637	.2515
.31	.14915	$-$ .20824	$-$ .2515	.0378	.2487
.32	.16856	$-$ .19100	$-$ .2621	.0114	.2428
.33	.18732	$-$ .17244	$-$ .2706	$-$ .0154	.2339
.34	.20534	$-$ .15266	$-$ .2770	$-$ .0422	.2220
.35	.22251	$-$ .13176	$-$ .2812	$-$ .0688	.2073
.36	.23875	$-$ .10984	$-$ .2831	$-$ .0948	.1899
.37	.25397	$-$ .087036	$-$ .2826	$-$ .1201	.1699
.38	.26808	$-$ .063467	$-$ .2798	$-$ .1444	.1476
.39	.28100	$-$ .039271	$-$ .2746	$-$ .1674	.1231
.40	.29264	$-$ .014590	$-$ .2670	$-$ .1888	.0968
.41	.30291	.010424	$-$ .2570	$-$ .2083	.0690
.42	.31176	.035614	$-$ .2447	$-$ .2258	.0401
.43	.31909	.060820	$-$ .2302	$-$ .2410	.0103
.44	32486	.085873	$-$ .2134	$-$ .2537	$-$ .0200
.45	.32898	.11060	$-$ .1945	$-$ .2637	$-$ .0503
.46	.33141	.13483	$-$ .1737	$-$ .2708	$-$ .0803
.47	.33209	.15838	$-$ .1510	$-$ .2749	$-$ .1095
.48	.33098	.18107	$-$ .1267	$-$ .2758	$-$ .1375
.49	.32804	.20272	$-$ .1008	$-$ .2735	$-$ .1639

SURFACE ZONAL HARMONICS (Continued)

x	$P_1(x)$	$P_2(x)$	$P_3(x)$	$P_4(x)$	$P_5(x)$
0.50	0.50	−0.125	−0.4375	−0.28906	0.08984
.51	.51	−.10985	−.43337	−.30440	.06726
.52	.52	−.0944	−.42848	−.31912	.04409
.53	.53	−.07865	−.42281	−.33317	.02041
.54	.54	−.0626	−.41634	−.34649	−.00372
.55	.55	−.04625	−.40906	−.35904	−.02819
.56	.56	−.0296	−.40096	−.37074	−.05294
.57	.57	−.01265	−.39202	−.38155	−.07786
.58	.58	.0046	−.38222	−.39140	−.10285
.59	.59	.02215	−.37155	−.40024	−.12781
.60	.60	.04	−.36	−.408	−.15264
.61	.61	.05815	−.34755	−.41462	−.17721
.62	.62	.0766	−.33418	−.42004	−.20142
.63	.63	.09535	−.31988	−.42418	−.22512
.64	.64	.1144	−.30464	−.42700	−.24819
.65	.65	.13375	−.28844	−.42841	−.27049
.66	.66	.1534	−.27126	−.42836	−.29188
.67	.67	.17335	−.25309	−.42676	−.31220
.68	.68	.1936	−.23392	−.42356	−.33131
.69	.69	.21415	−.21373	−.41869	−.34903
.70	.70	.235	−.1925	−.41206	−.36520
.71	.71	.25615	−.17022	−.40361	−.37964
.72	.72	.2776	−.14688	−.39327	−.39217
.73	.73	.29935	−.12246	−.38095	−.40260
.74	.74	.3214	−.09694	−.36659	−.41074
.75	.75	.34375	−.07031	−.35010	−.41638
.76	.76	.3664	−.04256	−.33140	−.41931
.77	.77	.38935	−.01367	−.31043	−.41932
.78	.78	.4126	.01638	−.28709	−.41618
.79	.79	.43615	.04760	−.26131	−.40966
.80	.80	.46	.08	−.233	−.39952
.81	.81	.48415	.11360	−.20208	−.38552
.82	.82	.5086	.14842	−.16847	−.36739
.83	.83	.53335	.18447	−.13207	−.34489
.84	.84	.5584	.22176	−.09281	−.31774
.85	.85	.58375	.26031	−.05060	−.28566
.86	.86	.6094	.30014	−.00534	−.24838
.87	.87	.63535	.34126	.04305	−.20559
.88	.88	.6616	.38368	.09467	−.15699
.89	.89	.68815	.42742	.14960	−.10228
.90	.90	.715	.4725	.20794	−.04114
.91	.91	.74215	.51893	.26978	0.02676
.92	.92	.7696	.56672	.33522	.10175
.93	.93	.79735	.61589	.40435	.18417
.94	.94	.8254	66646	.47728	.27438
.95	.95	.85373	.71844	.55409	.37274
.96	.96	.8824	.77184	.63489	.47962
.97	.97	.91135	.82668	.71978	.59539
.98	.98	.9406	.88298	.80886	.72045
.99	.99	.97015	.94075	.90223	.85518
100	1	1	1	1	1

SURFACE ZONAL HARMONICS (Continued)

x	$P_6(x)$	$P_7(x)$	$P_8(x)$	$P_9(x)$	$P_{10}(x)$
.50	0.32324	0.22314	−0.0736	−0.2679	−0.1882
.51	.31655	.24217	−.0454	−.2590	−.2101
.52	.30796	.25961	−.0163	−.2468	−.2291
.53	.29747	.27530	.0133	−.2314	−.2450
.54	.28506	.28906	.0432	−.2128	−.2573
.55	.27077	.30074	.0732	−.1913	−.2658
.56	.25460	.31016	.1029	−.1669	−.2701
.57	.23660	.31719	.1320	−.1399	−.2702
.58	.21681	.32169	.1601	−.1105	−.2659
.59	.19528	.32353	.1870	−.0791	−.2570
.60	.17210	.32260	.2133	−.0450	−.2433
.61	.14733	.31880	.2357	−.0118	−.2258
.63	.12109	.31207	.2568	.0234	−.2036
.63	.093475	.30232	.2753	.0589	−.1773
.64	.064623	.28954	.2909	.0943	−.1471
.65	.034675	.27371	.3032	.1290	−.1136
.66	.003790	.25483	.3120	.1625	−.0771
.67	−.027853	.23295	.3170	.1941	−.0382
.68	−.060059	.20813	.3179	.2233	.0024
.69	−.092615	.18049	.3145	.2495	.0440
.70	−.12529	.15016	.3067	.2721	.0858
.71	−.15782	.11731	.2943	.2904	.1269
.72	−.18994	.082166	.2771	.3039	.1663
.73	−.22136	.044990	.2553	.3120	.2030
.74	−.25175	.006087	.2287	.3143	.2360
.75	−.28078	−.034184	.1976	.3103	.2644
.76	−.30807	−.075411	.1621	.2997	.2869
.77	−.33325	−.11713	.1225	.2823	.3027
.78	−.35589	−.15881	.0791	.2578	.3108
.79	−.37557	−.19987	.0326	.2263	.3103
.80	−.39180	−.23965	−.0167	.1879	.3005
.81	−.40409	−.27743	−.0678	.1429	.2810
.82	−.41193	−.31240	−.1199	.0920	.2513
.83	−.41475	−.34368	−.1720	.0359	.1942
.84	−.41198	−.37033	−.2228	−.0243	.1617
.85	−.40300	−.39130	−.2710	−.0873	.1029
.86	−.38716	−.40545	−.3150	−.1513	.0362
.87	−.36379	−.41156	−.3530	−.2143	−.0366
.88	−.33217	−.40829	−.3830	−.2738	−.1130
.89	−.29156	−.39423	−.4028	−.3267	−.1899
.90	−.24116	−.36782	−.4097	−.3695	−.2631
.91	−.18018	−.32743	−.4010	−.3983	−.3277
.92	−.10774	−.27129	−.3737	−.4083	−.3773
.93	−.02295	−.19749	−.3243	−.3941	−.4046
.94	.07512	−.10404	−.2491	−.3498	−.4006
.95	.18745	.01123	−.1440	−.2684	−.3549
.96	.31506	.15060	−.0046	−.1422	−.2552
.97	.45899	.31650	.1740	.0375	−.0875
.98	.62035	.51151	.3971	.2804	.1647
.99	.80029	.73838	.6704	.5973	.5201
1.00	1	1	1	1	1

SURFACE ZONAL HARMONICS (Continued)

θ deg.	$P_1(\cos\theta)$	$P_2(\cos\theta)$	$P_3(\cos\theta)$	$P_4(\cos\theta)$	$P_5(\cos\theta)$
0	1	1	1	1	1
1	0.99985	0.99954	0.99909	0.99848	0.99772
2	.99939	.99817	.99635	.99392	.99088
3	.99863	.99589	.99179	.98634	.97954
4	.99756	.99270	.98543	.97577	.96377
5	.99619	.98861	.97728	.96227	.94368
6	.99452	.98361	.96736	.94589	.91939
7	.99255	.97772	.95569	.92670	.89108
8	.99027	.97095	.94232	.90480	.85893
9	.98769	.96329	.92726	.88026	.82315
10	.98481	.95477	.91057	.85321	.78399
11	.98163	.94539	.89228	.82376	.74170
12	.97815	.93516	.87244	.79204	.69656
13	.97437	.92410	.85111	.75819	.64888
14	.97030	.91221	.82833	.72235	.59895
15	.96593	.89952	.80416	.68470	.54713
16	.96126	.88604	.77868	.64537	.49373
17	.95630	.87178	.75194	.60456	.43911
18	.95106	.85676	.72401	.56244	.38363
19	.94552	.84101	.69497	.51918	.32763
20	.93969	.82453	.66488	.47498	.27149
21	.93358	.80736	.63384	.43002	.21556
22	.92718	.78950	.60190	.38450	.16019
23	.92050	.77099	.56917	.33862	.10573
24	.91355	.75185	.53572	.29256	.05252
25	.90631	.73209	.50163	.24653	.00088
26	.89879	.71175	.46699	.20072	− .04887
27	.89101	.69084	.43190	.15531	− .09642
28	.88295	.66939	.39644	.11051	− .14151
29	.87462	.64744	.36069	.06649	− .18388
30	.86603	.62500	.32476	.02344	− .22327
31	.85717	.60210	.28873	− .01847	− .25949
32	.84805	.57878	.25269	− .05907	− .29233
33	.83867	.55505	.21673	− .09820	− .32163
34	.82904	.53095	.18094	− .13570	− .34726
35	.81915	.50652	.14542	− .17142	− .36910
36	.80902	.48176	.11025	− .20524	− .38707
37	.79864	.45673	.07551	− .23701	− .40113
38	.78801	.43144	.04129	− .26664	− .41124
39	.77715	.40593	.00769	− .29400	− .41741
40	.76604	.38024	− .02523	− .31900	− .41968
41	.75471	.35438	− .05738	− .34157	− .41811
42	.74314	.32840	− .08869	− .36163	− .41279
43	.73135	.30232	− .11907	− .37913	− .40385
44	.71934	.27617	− .14845	− .39401	− .39141
45	.70711	.25000	− .17678	− .40625	− .37565
46	.69466	.22383	− .20397	− .41582	− .35677
47	.68200	.19768	− .22997	− .42273	− .33496
48	.66913	.17160	− .25471	− .42696	− .31048
49	.65606	.14562	− .27815	− .42856	− .28357

For $P_n(\cos\theta)$, use the definition of $P_n(x)$ given on the first page of this table, and put $x = \cos\theta$.

SURFACE ZONAL HARMONICS (Continued)

$\theta°$	$P_6(\cos \theta)$	$P_7(\cos \theta)$	$P_8(\cos \theta)$	$P_9(\cos \theta)$	$P_{10}(\cos \theta)$
0	1	1	1	1	1
1	0.99680	0.99574	0.9945	0.9932	0.9916
2	.98725	.98301	.9782	.9728	.9668
3	.97142	.96198	.9513	.9393	.9260
4	.94947	.93291	.9142	.8933	.8704
5	.92160	.89616	.8675	.8358	.8012
6	.88808	.85220	.8121	.7680	.7203
7	.84922	.80158	.7487	.6911	.6296
8	.80538	.74493	.6784	.6069	.5312
9	.75698	.68296	.6024	.5168	.4277
10	.70447	.61644	.5218	.4228	.3214
11	.64833	.54619	.4380	.3266	.2150
12	.58909	.47307	.3522	.2302	.1108
13	.52729	.39798	.2657	.1353	.0113
14	.46350	.32183	.1799	.0437	− .0813
15	.39831	.24554	.0962	− .0428	− .1651
16	.33229	.17001	.0157	− .1227	− .2381
17	.26606	.09614	− .0604	− .1946	− .2992
18	.20020	.02477	− .1310	− .2573	− .3471
19	.13529	− .04327	− .1951	− .3100	− .3813
20	.07190	− .10723	− .2518	− .3517	− .4013
21	.01059	− .16640	− .3005	− .3821	− .4073
22	− .04813	− .22017	− .3407	− .4009	− .3997
23	− .10376	− .26800	− .3718	− .4082	− .3793
24	− .15585	− .30942	− .3936	− .4042	− .3473
25	− .20398	− .34408	− .4062	− .3896	− .3052
26	− .24779	− .37172	− .4096	− .3650	− .2547
27	− .28694	− .39216	− .4041	− .3515	− .1975
28	− .32117	− .40534	− .3900	− .2902	− .1358
29	− .35025	− .41130	− .3680	− .2424	− .0716
30	− .37402	− .41018	− .3388	− .1896	− .0070
31	− .39238	− .40221	− .3031	− .1332	.0559
32	− .40527	− .38771	− .2619	− .0749	.1150
33	− .41209	− .36710	− .2162	− .0161	.1689
34	− .41471	− .34086	− .1670	.0415	.2157
35	− .41145	− .30956	− .1154	.0966	.2542
36	− .40307	− .27382	− .0627	.1476	.2833
37	− .38980	− .23432	− .0098	.1935	.3024
38	− .37191	− .19178	.0421	.2331	.3111
39	− .34972	− .14695	.0919	.2655	.3093
40	− .32357	− .10060	.1386	.2900	.2973
41	− .29387	− .05351	.1814	.3062	.2758
42	− .26104	− .00645	.2194	.3137	.2455
43	− .22554	.03982	.2519	.3127	.2077
44	− .18784	.08455	.2784	.3031	.1637
45	− .14844	.12706	.2983	.2855	.1151
46	− .10783	.16668	.3115	.2605	.0635
47	− .06654	.20283	.3176	.2288	.0107
48	− .02508	.23497	.3167	.1915	− .0416
49	.01606	.26263	.3090	.1495	− .0918

SURFACE ZONAL HARMONICS (Continued)

$\theta°$	$P_1(\cos\theta)$	$P_2(\cos\theta)$	$P_3(\cos\theta)$	$P_4(\cos\theta)$	$P_5(\cos\theta)$
50	0.64279	0.11976	−0.30022	−0.42753	−0.25449
51	.62932	.09407	−.32088	−.42394	−.22353
52	.61566	.06856	−.34009	−.41784	−.19097
53	.60182	.04327	−.35781	−.40929	−.15712
54	.58779	.01824	−.37399	−.39837	−.12229
55	.57358	−.00652	−.38861	−.38519	−.08679
56	.55919	−.03095	−.40164	−.36983	−.05093
57	.54464	−.05505	−.41307	−.35241	−.01503
58	.52992	−.07878	−.42286	−.33306	.02060
59	.51504	−.10210	−.43100	−.31189	.05566
60	.50000	−.12500	−.43750	−.28906	.08984
61	.48481	−.14744	−.44234	−.26471	.12287
62	.46947	−.16939	−.44552	−.23899	.15446
63	.45399	−.19084	−.44706	−.21205	.18436
64	.43837	−.21175	−.44695	−.18407	.21232
65	.42262	−.23209	−.44522	−.15521	.23811
66	.40674	−.25185	−.44188	−.12564	.26152
67	.39073	−.27099	−.43696	−.09554	.28238
68	.37461	−.28950	−.43049	−.06508	.30051
69	.35837	−.30736	−.42249	−.03444	.31577
70	.34202	−.32453	−.41301	−.00380	.32807
71	.32557	−.34101	−.40208	.02667	.33730
72	.30902	−.35676	−.38975	.05680	.34340
73	.29237	−.37178	−.37608	.08641	.34634
74	.27564	−.38604	−.36110	.11534	.34611
75	.25882	−.39952	−.34488	.14343	.34273
76	.24192	−.41221	−.32749	.17051	.33624
77	.22495	−.42410	−.30897	.19644	.32672
78	.20791	−.43516	−.28940	.22107	.31425
79	.19081	−.44539	−.26885	.24427	.29897
80	.17365	−.45477	−.24738	.26590	.28102
81	.15643	−.46329	−.22508	.28585	.26056
82	.13917	−.47095	−.20202	.30401	.23777
83	.12187	−.47772	−.17828	.32027	.21288
84	.10453	−.48361	−.15394	.33455	.18610
85	.08716	−.48861	−.12908	.34677	.15766
86	.06976	−.49270	−.10379	.35686	.12784
87	.05234	−.49589	−.07815	.36476	.09688
88	.03490	−.49817	−.05224	.37044	.06507
89	.01745	−.49954	−.02617	.37386	.03268
90	0	−.50000	0	.37500	0

SURFACE ZONAL HARMONICS (Continued)

$\theta°$	$P_6(\cos\theta)$	$P_7(\cos\theta)$	$P_8(\cos\theta)$	$P_9(\cos\theta)$	$P_{10}(\cos\theta)$
50	0.05638	0.28543	0.2947	0.1041	−0.1381
51	.09539	.30308	.2742	.0565	−.1792
52	.13265	.31535	.2480	.0080	−.2137
53	.16772	.32213	.2167	−.0400	−.2407
54	.20020	.32336	.1812	−.0863	−.2594
55	.22972	.31910	.1422	−.1296	−.2692
56	.25597	.30949	.1005	−.1689	−.2700
57	.27866	.29475	.0572	−.2032	−.2617
58	.29756	.27518	.0131	−.2315	−.2449
59	.31246	.25117	−.0309	−.2533	−.2201
60	.32324	.22314	−.0736	−.2679	−.1882
61	.32980	.19162	−.1144	−.2751	−.1504
62	.33210	.15715	−.1523	−.2747	−.1080
63	.33016	.12034	−.1865	−.2669	−.0624
64	.32403	.08181	−.2163	−.2518	−.0151
65	.31383	.04222	−.2411	−.2300	.0323
66	.29971	.00223	−.2605	−.2022	.0783
67	.28189	−.03748	−.2741	−.1690	.1213
68	.26062	−.07627	−.2816	−.1315	.1599
69	.23617	−.11348	−.2829	−.0906	.1929
70	.20888	−.14853	−.2780	−.0476	.2193
71	.17910	−.18082	−.2671	−.0035	.2382
72	.14721	−.20986	−.2504	.0404	.2491
73	.11363	−.23516	−.2283	.0829	.2516
74	.07878	−.25634	−.2014	.1230	.2457
75	.04310	−.27305	−.1702	.1595	.2316
76	.00704	−.28504	−.1355	.1915	.2099
77	−.02896	−.29214	.0079	.2181	.1813
78	−.06444	−.29424	−.0583	.2387	.1468
79	−.09897	−.29133	−.0176	.2526	.1074
80	−.13212	−.28348	.0233	.2596	.0647
81	−.16348	−.27083	.0636	.2595	.0199
82	−.19267	−.25360	.1024	.2523	−.0254
83	−.21933	−.23211	.1389	.2383	−.0698
84	−.24313	.20671	.1722	.2177	−.1118
85	−.26378	−.17784	.2017	.1913	−.1499
86	−.28103	−.14598	2268	.1597	−.1830
87	−.29467	−.11168	.2469	.1237	−.2099
88	−.30454	−.07551	.2615	.0844	−.2298
89	−.31050	−.03807	.2704	.0428	−.2420
90	−.31250	0	.2734	0	−.2461

For larger tables, see "Report of the British Association for the Advancement of Science," 1879, pp. 54–57, Philosophical Magazine, Dec., 1891, "Tafeln der Besselschen, Theta-, Kugel- und anderer Funktionen" by K. Hayashi and "Fünfstellige Funktionentafeln" by K. Hayashi.

BERNOULLI AND EULER NUMBERS—POLYNOMIALS

There are numerous sets of defined numbers and polynomials, among which the more important ones are those classified under the names of Bernoulli and Euler.

The *Bernoulli Polynomials* are generated by the defining condition

$$\frac{te^{tx}}{e^t - 1} = B_0(x) + B_1(x)t + B_2(x)\frac{t^2}{2!} + B_3(x)\frac{t^3}{3!} + \cdots$$

$$B_0(x) = 1, \quad B_1(x) = x - \tfrac{1}{2}, \quad B_2(x) = x^2 - x + \tfrac{1}{6}$$

$$B_3(x) = x^3 - \frac{3}{2}x^2 + \frac{x}{2}, \quad B_4(x) = x^4 - 2x^3 + x^2 - \tfrac{1}{30}, \ldots$$

The following useful relations should be noted

$$B_n'(x) = nB_{n-1}(x), \quad \int_a^x B_n(x)\,dx = \frac{1}{n+1}[B_{n+1}(x) - B_{n+1}(a)]$$

The Bernoulli numbers $B_0, B_1, B_2, \ldots$, can be obtained by putting $x = 0$ in the respective polynomials. A simpler method is to use the generating function

$$\frac{x}{e^x - 1} = \sum_{n=0}^{\infty} \frac{B_n x^n}{n!}$$

and read the coefficients from the expansion. Here,

$$B_0 = 1, \quad B_1 = -\tfrac{1}{2}, \quad B_2 = \tfrac{1}{6}, \quad B_4 = -\tfrac{1}{30},$$

$$B_6 = \tfrac{1}{42}, \quad B_8 = -\tfrac{1}{30}, \ldots B_{2n+1} = 0 \qquad (n \geq 1)$$

An important application for the Bernoulli coefficients is in its use involved for the Euler-Maclaurin Sum Formula:

$$\sum_{x=1}^{n-1} f(x) = \int_1^n f(x)\,dx + \left[\sum_{i=1}^{\infty} \frac{B_i}{i!} f^{(i-1)}(x)\right]_{x=1}^{x=n}$$

$$= \left[\int f(x)\,dx - \frac{1}{2}f(x) + \frac{1}{12}f'(x) - \frac{1}{720}f'''(x)\right.$$

$$\left. + \frac{1}{30,240}f^{(V)}(x) - \frac{1}{1,209,600}f^{(VII)}(x)\right]_{x=1}^{x=n}$$

Examples:

$$\sum_{x=1}^{n-1} \sqrt{x} = \sqrt{n}\left\{\frac{2}{3}n - \frac{1}{2} + \frac{1}{24n} - \frac{1}{1,920n^3}\right.$$

$$\left. + \frac{1}{9,216n^5} - \frac{11}{163,840n^7}\right\} - 0.207,886,224,977,355$$

correct to 12 places for $n \geq 10$.

$$\log_e(x!) = \log_e \Gamma(x+1)$$

$$= \left(x + \frac{1}{2}\right)\log_e x - x + \frac{1}{12x} - \frac{1}{360x^3} + \frac{1}{1,260x^5} - \frac{1}{1,680x^7}$$

$$+ 0.918,938,533,205$$

accurate to 12 places for $x \geq 10$.

$f(x)$ (digamma function) $= \dfrac{d \log \Gamma(x)}{dx}$

$$= 1 + \frac{1}{2} + \frac{1}{3} + \cdots + \frac{1}{x-1} - \gamma$$

(Euler Constant) for x integer.

By Euler-Maclaurin

$$f(x) = \log_e x - \frac{1}{2x} - \frac{1}{12x^2} + \frac{1}{120x^4} - \frac{1}{252x^6} + \frac{1}{240x^8} - \frac{5}{660x^{10}} + \frac{691}{32,760x^{12}}$$

correct to 12 places for $x \geq 10$.

The Euler numbers together with their respective polynomials can be generated from

$$\frac{2e^{tx}}{e^t + 1} = \sum_{i=0}^{\infty} E_i(x) \frac{t^i}{i!}$$

and the relation

$$x^n = \frac{1}{2}[E_n(x + 1) + E_n(x)]$$

The Euler polynomials are

$$E_0(x) = 1, \quad E_1(x) = x - \frac{1}{2}, \quad E_2(x) = x^2 - x$$

$$E_3(x) = x^3 - \frac{3}{2}x^2 + \frac{1}{4}, \quad E_4(x) = x^4 - 2x^3 + x$$

$$E_5(x) = x^5 - \frac{5}{2}x^4 + \frac{5}{2}x^2 - \frac{1}{2}$$

Tables which follow record the first fifteen polynomials of $B_k(x)$ and $E_k(x)$. The first sixty Bernoulli and Euler numbers are given in a separate table. The value of $x^n/n!$ is also important and this is given in a succeeding table for values of x from 1 to 9 and n from 1 to 50.

COEFFICIENTS b_k OF THE BERNOULLI POLYNOMIALS $B_n(x) = \sum_{k=0}^{n} b_k x^k$

n/k	0	1	2	3	4	5	6	7	8	9	10	11	12	13	14	15
0	1															
1	$-\frac{1}{2}$	1														
2	$\frac{1}{6}$	-1	1													
3	0	$\frac{1}{2}$	$-\frac{3}{2}$	1												
4	$-\frac{1}{30}$	0	1	-2	1											
5	0	$-\frac{1}{6}$	0	$\frac{5}{3}$	$-\frac{5}{2}$	1										
6	$\frac{1}{42}$	0	$-\frac{1}{2}$	0	$\frac{5}{2}$	-3	1									
7	0	$\frac{1}{6}$	0	$-\frac{7}{6}$	0	$\frac{7}{2}$	$-\frac{7}{2}$	1								
8	$-\frac{1}{30}$	0	$\frac{2}{3}$	0	$-\frac{7}{3}$	0	$\frac{14}{3}$	-4	1							
9	0	$-\frac{3}{10}$	0	2	0	$-\frac{21}{5}$	0	6	$-\frac{9}{2}$	1						
10	$\frac{5}{66}$	0	$-\frac{3}{2}$	0	5	0	-7	0	$\frac{15}{2}$	-5	1					
11	0	$\frac{5}{6}$	0	$-\frac{11}{2}$	0	11	0	-11	0	$\frac{55}{6}$	$-\frac{11}{2}$	1				
12	$-\frac{691}{2730}$	0	5	0	$-\frac{33}{2}$	0	22	0	$-\frac{33}{2}$	0	11	-6	1			
13	0	$-\frac{691}{210}$	0	$\frac{65}{3}$	0	$-\frac{429}{10}$	0	$-\frac{286}{7}$	0	$-\frac{143}{6}$	0	13	$-\frac{13}{2}$	1		
14	$\frac{7}{6}$	0	$-\frac{691}{30}$	0	$\frac{455}{6}$	0	$-\frac{1001}{10}$	0	$\frac{143}{2}$	0	$-\frac{1001}{30}$	0	$\frac{91}{6}$	-7	1	
15	0	$\frac{35}{2}$	0	$-\frac{691}{6}$	0	$\frac{455}{2}$	0	$-\frac{429}{2}$	0	$\frac{715}{6}$	0	$-\frac{91}{2}$	0	$\frac{35}{2}$	$-\frac{15}{2}$	1

COEFFICIENTS e_k OF THE EULER POLYNOMIALS $E_n(x) = \sum_{k=0}^{n} e_k x^k$

n/k	0	1	2	3	4	5	6	7	8	9	10	11	12	13	14	15
0	1															
1	$-\frac{1}{2}$	1														
2	0	-1	1													
3	$\frac{1}{4}$	0	$-\frac{3}{2}$	1												
4	0	1	0	-2	1											
5	$-\frac{1}{2}$	0	$\frac{5}{2}$	0	$-\frac{5}{2}$	1										
6	0	-3	0	5	0	-3	1									
7	$\frac{17}{8}$	0	$-\frac{21}{2}$	0	$\frac{35}{4}$	0	$-\frac{7}{2}$	1								
8	0	17	0	-28	0	14	0	-4	1							
9	$-\frac{31}{2}$	0	$\frac{153}{2}$	0	-63	0	21	0	$-\frac{9}{2}$	1						
10	0	-155	0	255	0	-126	0	30	0	-5	1					
11	$\frac{691}{4}$	0	$-\frac{1705}{2}$	0	$\frac{2805}{4}$	0	-231	0	$\frac{165}{4}$	0	$-\frac{11}{2}$	1				
12	0	2073	0	-3410	0	1683	0	-396	0	55	0	-6	1			
13	$-\frac{5461}{2}$	0	$\frac{26949}{2}$	0	$-\frac{22165}{2}$	0	$\frac{7293}{2}$	0	$-\frac{1287}{2}$	0	$\frac{143}{2}$	0	$-\frac{13}{2}$	1		
14	0	-38227	0	62881	0	-31031	0	7293	0	-1001	0	91	0	-7	1	
15	$\frac{929569}{16}$	0	$-\frac{573405}{2}$	0	$\frac{943215}{4}$	0	$-\frac{155155}{2}$	0	$\frac{109395}{8}$	0	$-\frac{3003}{2}$	0	$\frac{455}{4}$	0	$-\frac{15}{2}$	1

BERNOULLI NUMBERS
$$B_n = N/D$$

n	N	D	B_n
0	1	1	(0) 1.0000 00000
1	−1	2	(−1) −5.0000 00000*
2	1	6	(−1) 1.6666 66667
4	−1	30	(−2) −3.3333 33333
6	1	42	(−2) 2.3809 52381
8	−1	30	(−2) −3.3333 33333
10	5	66	(−2) 7.5757 57576
12	−691	2730	(−1) −2.5311 35531
14	7	6	(0) 1.1666 66667
16	−3617	510	(0) −7.0921 56863
18	43867	798	(1) 5.4971 17794
20	−1 74611	330	(2) −5.2912 42424
22	8 54513	138	(3) 6.1921 23188
24	−2363 64091	2730	(4) −8.6580 25311
26	85 53103	6	(6) 1.4255 17167
28	−2 37494 61029	870	(7) −2.7298 23107
30	861 58412 76005	14322	(8) 6.0158 08739
32	−770 93210 41217	510	(10) −1.5116 31577
34	257 76878 58367	6	(11) 4.2961 46431
36	−26315 27155 30534 77373	19 19190	(13) −1.3711 65521
38	2 92999 39138 41559	6	(14) 4.8833 23190
40	−2 61082 71849 64491 22051	13530	(16) −1.9296 57934
42	15 20097 64391 80708 02691	1806	(17) 8.4169 30476
44	−278 33269 57930 10242 35023	690	(19) −4.0338 07185
46	5964 51111 59391 21632 77961	282	(21) 2.1150 74864
48	−560 94033 68997 81768 62491 27547	46410	(23) −1.2086 62652
50	49 50572 05241 07964 82124 77525	66	(24) 7.5008 66746
52	−80116 57181 35489 95734 79249 91853	1590	(26) −5.0387 78101
54	29 14996 36348 84862 42141 81238 12691	798	(28) 3.6528 77648
56	−2479 39292 93132 26753 68541 57396 63229	870	(30) −2.8498 76930
58	84483 61334 88800 41862 04677 59940 36021	354	(32) 2.3865 42750
60	−121 52331 40483 75557 20403 04994 07982 02460 41491	567 86730	(34) −2.1399 94926

*The floating decimal point notation is used here. For example for $n = 1$, $B_1 = -\frac{1}{2} = -.500000$ $= (-5.00000)(10^{-1}) = (-1) - 5.0000,0000$.

SPECIAL FUNCTIONS

EULER NUMBERS

n	En
0	1
2	−1
4	5
6	−61
8	1385
10	−50521
12	27 02765
14	−1993 60981
16	1 93915 12145
18	−240 48796 75441
20	37037 11882 37525
22	−69 34887 43931 37901
24	15514 53416 35570 86905
26	−40 87072 50929 31238 92361
28	12522 59641 40362 98654 68285
30	−44 15438 93249 02310 45536 82821
32	17751 93915 79539 28943 66647 89665
34	−80 72329 92358 87898 06216 82474 53281
36	41222 06033 95177 02122 34707 96712 59045
38	−234 89580 52704 31082 52017 82857 61989 47741
40	1 48511 50718 11498 00178 77156 78140 58266 84425
42	−1036 46227 33519 61211 93979 57304 74518 59763 10201
44	7 94757 94225 97592 70360 80405 10088 07061 95192 73805
46	−6667 53751 66855 44977 43502 84747 73748 19752 41076 84661
48	60 96278 64556 85421 58691 68574 28768 43153 97653 90444 35185
50	−60532 85248 18862 18963 14383 78511 16490 88103 49822 51468 15121
52	650 61624 86684 60884 77158 70634 08082 29834 83644 23676 53855 76565
54	−7 54665 99390 08739 09806 14325 65889 73674 42122 40024 71169 98586 45581
56	9420 32189 64202 41204 20228 62376 90583 22720 93888 52599 64600 93949 05945
58	−126 22019 25180 62187 19903 40923 72874 89255 48234 10611 91825 59406 99649 20041
60	181089 11496 57923 04965 45807 74165 21586 88733 48734 92363 14106 00809 54542 31325

BERNOULLI AND EULER POLYNOMIALS

$$x^n/n!$$

$n \backslash x$	2		3		4		5	
1	(0) 2.0000	00000	(0) 3.0000	00000	(0) 4.0000	00000	(0) 5.0000	00000
2	(0) 2.0000	00000	(0) 4.5000	00000	(0) 8.0000	00000	(1) 1.2500	00000
3	(0) 1.3333	33333	(0) 4.5000	00000	(1) 1.0666	66667	(1) 2.0833	33333
4	(− 1) 6.6666	66667	(0) 3.3750	00000	(1) 1.0666	66667	(1) 2.6041	66667
5	(− 1) 2.6666	66667	(0) 2.0250	00000	(0) 8.5333	33333	(1) 2.6041	66667
6	(− 2) 8.8888	88889	(0) 1.0125	00000	(0) 5.6888	88889	(1) 2.1701	38889
7	(− 2) 2.5396	82540	(− 1) 4.3392	85714	(0) 3.2507	93651	(1) 1.5500	99206
8	(− 3) 6.3492	06349	(− 1) 1.6272	32143	(0) 1.6253	96825	(0) 9.6881	20040
9	(− 3) 1.4109	34744	(− 2) 5.4241	07143	(− 1) 7.2239	85891	(0) 5.3822	88911
10	(− 4) 2.8218	69489	(− 2) 1.6272	32144	(− 1) 2.8895	94356	(0) 2.6911	44455
11	(− 5) 5.1306	71797	(− 3) 4.4379	05844	(− 1) 1.0507	61584	(0) 1.2232	47480
12	(− 6) 8.5511	19662	(− 3) 1.1094	76461	(− 2) 3.5025	38614	(− 1) 5.0968	64499
13	(− 6) 1.3155	56871	(− 4) 2.5603	30295	(− 2) 1.0777	04189	(− 1) 1.9603	32500
14	(− 7) 1.8793	66959	(− 5) 5.4864	22060	(− 3) 3.0791	54825	(− 2) 7.0011	87499
15	(− 8) 2.5058	22612	(− 5) 1.0972	84412	(− 4) 8.2110	79534	(− 2) 2.3337	29166
16	(− 9) 3.1322	78264	(− 6) 2.0574	08272	(− 4) 2.0527	69883	(− 3) 7.2929	03644
17	(−10) 3.6850	33252	(− 7) 3.6307	20481	(− 5) 4.8300	46785	(− 3) 2.1449	71660
18	(−11) 4.0944	81391	(− 8) 6.0512	00801	(− 5) 1.0733	43730	(− 4) 5.9582	54611
19	(−12) 4.3099	80412	(− 9) 9.5545	27582	(− 6) 2.2596	71011	(− 4) 1.5679	61740
20	(−13) 4.3099	80413	(− 9) 1.4331	79137	(− 7) 4.5193	42021	(− 5) 3.9199	04350
21	(−14) 4.1047	43250	(−10) 2.0473	98768	(− 8) 8.6082	70516	(− 6) 9.3331	05595
22	(−15) 3.7315	84772	(−11) 2.7919	07410	(− 8) 1.5651	40093	(− 6) 2.1211	60362
23	(−16) 3.2448	56324	(−12) 3.6416	18361	(− 9) 2.7219	82772	(− 7) 4.6112	18179
24	(−17) 2.7040	46937	(−13) 4.5520	22952	(−10) 4.5366	37953	(− 8) 9.6067	04540
25	(−18) 2.1632	37550	(−14) 5.4624	27543	(−11) 7.2586	20726	(− 8) 1.9213	40908
26	(−19) 1.6640	28884	(−15) 6.3028	01010	(−11) 1.1167	10881	(− 9) 3.6948	86362
27	(−20) 1.2326	13988	(−16) 7.0031	12233	(−12) 1.6543	86490	(−10) 6.8423	82151
28	(−22) 8.8043	85630	(−17) 7.5033	34535	(−13) 2.3634	09271	(−10) 1.2218	53956
29	(−23) 6.0719	90089	(−18) 7.7620	70209	(−14) 3.2598	74857	(−11) 2.1066	44751
30	(−24) 4.0479	93393	(−19) 7.7620	70209	(−15) 4.3464	99810	(−12) 3.5110	74585
31	(−25) 2.6116	08641	(−20) 7.5116	80847	(−16) 5.6083	86851	(−13) 5.6630	23524
32	(−26) 1.6322	55401	(−21) 7.0422	00795	(−17) 7.0104	83564	(−14) 8.8484	74257
33	(−28) 9.8924	56972	(−22) 6.4020	00722	(−18) 8.4975	55834	(−14) 1.3406	77918
34	(−29) 5.8190	92337	(−23) 5.6488	24167	(−19) 9.9971	24511	(−15) 1.9715	85173
35	(−30) 3.3251	95620	(−24) 4.8418	49286	(−19) 1.1425	28515	(−16) 2.8165	50246
36	(−31) 1.8473	30900	(−25) 4.0348	74405	(−20) 1.2694	76128	(−17) 3.9118	75343
37	(−33) 9.9855	72436	(−26) 3.2715	19788	(−21) 1.3724	06625	(−18) 5.2863	18032
38	(−34) 5.2555	64439	(−27) 2.5827	78779	(−22) 1.4446	38552	(−19) 6.9556	81619
39	(−35) 2.6951	61251	(−28) 1.9867	52908	(−23) 1.4816	80567	(−20) 8.9175	40539
40	(−36) 1.3475	80626	(−29) 1.4900	64681	(−24) 1.4816	80567	(−20) 1.1146	92567
41	(−38) 6.5735	64028	(−30) 1.0902	91230	(−25) 1.4455	42017	(−21) 1.3593	81180
42	(−39) 3.1302	68584	(−32) 7.7877	94498	(−26) 1.3767	06682	(−22) 1.6183	10928
43	(−40) 1.4559	38876	(−33) 5.4333	44999	(−27) 1.2806	57379	(−23) 1.8817	56893
44	(−42) 6.6179	03983	(−34) 3.7045	53408	(−28) 1.1642	33981	(−24) 2.1383	60106
45	(−43) 2.9412	90659	(−35) 2.4697	02271	(−29) 1.0348	74650	(−25) 2.3759	55673
46	(−44) 1.2788	22026	(−36) 1.6106	75395	(−31) 8.9989	09998	(−26) 2.5825	60514
47	(−46) 5.4417	95855	(−37) 1.0280	90677	(−32) 7.6586	46807	(−27) 2.7474	04803
48	(−47) 2.2674	14940	(−39) 6.4255	66736	(−33) 6.3822	05674	(−28) 2.8618	80003
49	(−49) 9.2547	54855	(−40) 3.9340	20450	(−34) 5.2099	63815	(−29) 2.9202	85717
50	(−50) 3.7019	10942	(−41) 2.3604	12270	(−35) 4.1679	71052	(−30) 2.9202	85717

BERNOULLI AND EULER POLYNOMIALS
$x^n/n!$

$n\backslash x$	6		7		8		9	
1	(0) 6.0000	00000	(0) 7.0000	00000	(0) 8.0000	00000	(0) 9.0000	00000
2 *	(1) 1.8000	00000	(1) 2.4500	00000	(1) 3.2000	00000	(1) 4.0500	00000
3	(1) 3.6000	00000	(1) 5.7166	66667	(1) 8.5333	33333	(2) 1.2150	00000
4	(1) 5.4000	00000	(2) 1.0004	16667	(2) 1.7066	66667	(2) 2.7337	50000
5	(1) 6.4800	00000	(2) 1.4005	83333	(2) 2.7306	66667	(2) 4.9207	50000
6	(1) 6.4800	00000	(2) 1.6340	13889	(2) 3.6408	88889	(2) 7.3811	25000
7	(1) 5.5542	85714	(2) 1.6340	13889	(2) 4.1610	15873	(2) 9.4900	17857
8	(1) 4.1657	14286	(2) 1.4297	62153	(2) 4.1610	15873	(3) 1.0676	27009
9	(1) 2.7771	42857	(2) 1.1120	37230	(2) 3.6986	80776	(3) 1.0676	27009
10	(1) 1.6662	85714	(1) 7.7842	60610	(2) 2.9589	44621	(2) 9.6086	43080
11	(0) 9.0888	31169	(1) 4.9536	20388	(2) 2.1519	59724	(2) 7.8616	17066
12	(0) 4.5444	15584	(1) 2.8896	11893	(2) 1.4346	39816	(2) 5.8962	12799
13	(0) 2.0974	22577	(1) 1.5559	44865	(1) 8.8285	52715	(2) 4.0819	93476
14	(− 1) 8.9889	53903	(0) 7.7797	24327	(1) 5.0448	87266	(2) 2.6241	38663
15	(− 1) 3.5955	81561	(0) 3.6305	38019	(1) 2.6906	06542	(2) 1.5744	83198
16	(− 1) 1.3483	43085	(0) 1.5883	60383	(1) 1.3453	03271	(1) 8.8564	67988
17	(− 2) 4.7588	57949	(− 1) 6.5403	07461	(0) 6.3308	38921	(1) 4.6887	18347
18	(− 2) 1.5862	85983	(− 1) 2.5434	52902	(0) 2.8137	06187	(1) 2.3443	59173
19	(− 3) 5.0093	24157	(− 2) 9.3706	15954	(0) 1.1847	18395	(1) 1.1104	85924
20	(− 3) 1.5027	97247	(− 2) 3.2797	15584	(− 1) 4.7388	73579	(0) 4.9971	86660
21	(− 4) 4.2937	06421	(− 2) 1.0932	38528	(− 1) 1.8052	85173	(0) 2.1416	51426
22	(− 4) 1.1710	10841	(− 3) 3.4784	86224	(− 2) 6.5646	73356	(− 1) 8.7613	01286
23	(− 5) 3.0548	10892	(− 3) 1.0586	69721	(− 2) 2.2833	64645	(− 1) 3.4283	35286
24	(− 6) 7.6370	27230	(− 4) 3.0877	86685	(− 3) 7.6112	15485	(− 1) 1.2856	25732
25	(− 6) 1.8328	86535	(− 5) 8.6458	02719	(− 3) 2.4355	88956	(− 2) 4.6282	52637
26	(− 7) 4.2297	38158	(− 5) 2.3277	16117	(− 4) 7.4941	19863	(− 2) 1.6020	87451
27	(− 8) 9.3994	18129	(− 6) 6.0348	19562	(− 4) 2.2204	79959	(− 3) 5.3402	91503
28	(− 8) 2.0141	61028	(− 6) 1.5087	04890	(− 5) 6.3442	28454	(− 3) 1.7165	22269
29	(− 9) 4.1672	29712	(− 7) 3.6417	01460	(− 5) 1.7501	31987	(− 4) 5.3271	38075
30	(−10) 8.3344	59424	(− 8) 8.4973	03406	(− 6) 4.6670	18634	(− 4) 1.5981	41423
31	(−10) 1.6131	21179	(− 8) 1.9187	45930	(− 6) 1.2043	91905	(− 5) 4.6397	65421
32	(−11) 3.0246	02211	(− 9) 4.1972	56723	(− 7) 3.0109	79764	(− 5) 1.3049	34025
33	(−12) 5.4992	76746	(−10) 8.9032	71836	(− 8) 7.2993	44881	(− 6) 3.5589	10976
34	(−13) 9.7046	06022	(−10) 1.8330	26555	(− 8) 1.7174	92913	(− 7) 9.4206	46701
35	(−13) 1.6636	46746	(−11) 3.6660	53108	(− 9) 3.9256	98086	(− 7) 2.4224	52008
36	(−14) 2.7727	44578	(−12) 7.1284	36600	(−10) 8.7237	73527	(− 8) 6.0561	30022
37	(−15) 4.4963	42559	(−12) 1.3486	23141	(−10) 1.8862	21303	(− 8) 1.4731	12708
38	(−16) 7.0994	88250	(−13) 2.4843	05785	(−11) 3.9709	92217	(− 9) 3.4889	51151
39	(−16) 1.0922	28962	(−14) 4.4590	10384	(−12) 8.1456	25061	(−10) 8.0514	25733
40	(−17) 1.6383	43443	(−15) 7.8032	68172	(−12) 1.6291	25012	(−10) 1.8115	70790
41	(−18) 2.3975	75770	(−15) 1.3322	65298	(−13) 3.1787	80512	(−11) 3.9766	18807
42	(−19) 3.4251	08241	(−16) 2.2204	42162	(−14) 6.0548	20021	(−12) 8.5213	26014
43	(−20) 4.7792	20803	(−17) 3.6146	73288	(−14) 1.1264	78144	(−12) 1.7835	33352
44	(−21) 6.5171	19276	(−18) 5.7506	16594	(−15) 2.0481	42079	(−13) 3.6481	36401
45	(−22) 8.6894	92369	(−19) 8.9454	03592	(−16) 3.6411	41473	(−14) 7.2962	72804
46	(−22) 1.1334	12048	(−19) 1.3612	57068	(−17) 6.3324	19955	(−14) 1.4275	31635
47	(−23) 1.4469	08998	(−20) 2.0274	04144	(−17) 1.0778	58716	(−15) 2.7335	71217
48	(−24) 1.8086	36247	(−21) 2.9566	31045	(−18) 1.7964	31193	(−16) 5.1254	46033
49	(−25) 2.2146	56629	(−22) 4.2237	58634	(−19) 2.9329	48887	(−17) 9.4140	84548
50	(−26) 2.6575	87955	(−23) 5.9132	62088	(−20) 4.6927	18219	(−17) 1 6945	35219

*The floating decimal point notation is used here. For example (1) 1.8000,000 = (1.8000,000) 10.

STIRLING NUMBERS

Stirling numbers

The Stirling numbers are used for reducing factorials such as $x^{(n)}$ to polynomials in x and vice versa.

Stirling numbers of the first kind

The factorial polynomial $x^{(n)}$ is defined and represented by

$$x^{(n)} = x(x - 1)(x - 2) \cdots (x - n + 1)$$

where $x^{(0)}$ is 1 by definition.

If n is a non-negative integer, then

$$x^{(n)} = s_{n1}x + s_{n2}x^2 + \cdots + s_{nn}x^n.$$

Here the numbers $s_{n1}, s_{n2}, s_{n3}, \ldots$, are Stirling numbers of the first kind. A table listing these numbers with an example using same follows.

STIRLING NUMBERS OF THE FIRST KIND

n	s_{n1}	s_{n2}	s_{n3}	s_{n4}	s_{n5}	s_{n6}	s_{n7}	s_{n8}
1	1	0	0	0	0	0	0	0
2	-1	1	0	0	0	0	0	0
3	2	-3	1	0	0	0	0	0
4	-6	11	-6	1	0	0	0	0
5	24	-50	35	-10	1	0	0	0
6	-120	274	-225	85	-15	1	0	0
7	720	-1 764	1 624	-735	175	-21	1	0
8	-5 040	13 068	-13 132	6 769	-1 960	322	-28	1

The table may be continued by using the recurrence formula

$$s_{ni} = s_{n-1,i-1} - (n - 1)s_{n-1,i}, \quad i = 1, 2, \ldots, n$$

where $s_{n0} = 0$ for all n.

Example

Express $3x^{(3)} + 2x^{(1)}$ using Stirling's numbers of the first kind

$$3x^{(3)} = 3(2x - 3x^2 + x^3)$$
$$= 6x - 9x^2 + 3x^3$$
$$2x^{(1)} = 2x$$
$$\therefore 3x^{(3)} + 2x^{(1)} = 8x - 9x^2 + 3x^3$$

This may be verified by carrying out the indicated operations as follows

$$3x^{(3)} + 2x^{(1)} = 3(x)(x - 1)(x - 2) + 2x$$
$$= 3x^3 - 9x^2 + 6x + 2x$$
$$= 8x - 9x^2 + 3x^3$$

Stirling numbers of the second kind

For every non-negative integer n the function defined by x^n can be expressed as a linear combination of factorial powers of x not higher than the n'th. In other words

$$x^n = t_{n1}x^{(1)} + t_{n2}x^{(2)} + \cdots + t_{nn}x^{(n)}.$$

The numbers $t_{n1}, t_{n2}, \ldots, t_{nn}$ are called Stirling's numbers of the second kind.

A table listing these numbers with an example using same follows.

STIRLING NUMBERS OF THE SECOND KIND

n	t_{n1}	t_{n2}	t_{n3}	t_{n4}	t_{n5}	t_{n6}	t_{n7}	t_{n8}
1	1	0	0	0	0	0	0	0
2	1	1	0	0	0	0	0	0
3	1	3	1	0	0	0	0	0
4	1	7	6	1	0	0	0	0
5	1	15	25	10	1	0	0	0
6	1	31	90	65	15	1	0	0
7	1	63	301	350	140	21	1	0
8	1	127	966	1 701	1 050	266	28	1

This table may be continued by using the recurrence formula

$$t_{ni} = it_{n-1,i} + t_{n-1,i-1}, \quad i = 1, 2, \ldots, n$$

where $t_{n0} = 0$ for all n.

Example

Express $2x^3 - 3x^2 + x + 2$ by use of factorial powers

$$2x^3 = 2[x^{(1)} + 3x^{(2)} + x^{(3)}] = 2x^{(1)} + 6x^{(2)} + 2x^{(3)}$$
$$-3x^2 = -3[x^{(1)} + x^{(2)}] \qquad = -3x^{(1)} - 3x^{(2)}$$
$$x = x^{(1)} \qquad\qquad\qquad = x^{(1)}$$
$$+2 = +2x^{(0)} \qquad\qquad\quad = 2x^{(0)}$$

$$\therefore 2x^3 - 3x^2 + x + 2 = 2 + 3x^{(2)} + 2x^{(3)}$$

Fourier Series

(Also see Index for Cosine and Sine Transforms)
1. If $f(x)$ is a bounded periodic function of period 2L (i.e. $f(x + 2L) = f(x)$), and satisfies the Dirichlet conditions:

a) In any period $f(x)$ is continuous, except possibly for a finite number of jump discontinuities

b) In any period $f(x)$ has only a finite number of maxima and minima.
Then $f(x)$ may be represented by the Fourier series

$$\frac{a_0}{2} + \sum_{n=1}^{\infty} \left(a_n \cos \frac{n\pi x}{L} + b_n \sin \frac{n\pi x}{L} \right)$$

where a_n and b_n are as determined below. This series will converge to $f(x)$ at every point where $f(x)$ is continuous, and to

$$\frac{f(x^+) + f(x^-)}{2}$$

(i.e. the average of the left-hand and right-hand limits) at every point where $f(x)$ has a jump discontinuity.

$$a_n = \frac{1}{L} \int_{-L}^{L} f(x) \cos \frac{n\pi x}{L} \, dx, \; n = 0,1,2,3, \ldots ;$$

$$b_n = \frac{1}{L} \int_{-L}^{L} f(x) \sin \frac{n\pi x}{L} \, dx, \; n = 1,2,3, \ldots$$

We may also write

$$a_n = \frac{1}{L} \int_{a}^{a+2L} f(x) \cos \frac{n\pi x}{L} \, dx \text{ and } b_n = \frac{1}{L} \int_{a}^{a+2L} f(x) \sin \frac{n\pi x}{L} \, dx,$$

where a is any real number. This if $a = 0$,

$$a_n = \frac{1}{L} \int_{0}^{2L} f(x) \cos \frac{n\pi x}{L} \, dx, \; n = 0,1,2,3, \ldots ;$$

$$b_n = \frac{1}{L} \int_{0}^{2L} f(x) \sin \frac{n\pi x}{L} \, dx, \; n = 1,2,3, \ldots$$

2. If in addition to the above restrictions, $f(x)$ is even (i.e. $f(-x) = f(x)$), the Fourier series reduces to

$$\frac{a_0}{2} + \sum_{n=1}^{\infty} a_n \cos \frac{n\pi x}{L}$$

That is, $b_n = 0$. In this case, a simple formula for a_n, is

$$a_n = \frac{2}{L} \int_{0}^{L} f(x) \cos \frac{n\pi x}{L} \, dx, \; n = 0,1,2,3, \ldots$$

3. If in addition to the restrictions in (1), $f(x)$ is an odd function (i.e. $f(-x) = f(x)$), then the Fourier series reduces to

$$\sum_{n=1}^{\infty} b_n \sin \frac{n\pi x}{L}$$

That is, $a_n = 0$. In this case, a simpler formula for the b_n is

$$b_n = \frac{2}{L} \int_0^L f(x) \sin \frac{n\pi x}{L} \, dx, \quad n = 1,2,3,\ldots$$

4. If in addition to the restrictions in (2) above, $f(x) = f(L - x)$, then a_n will be 0 for all even values of n, including n = 0. Thus in this case, the expansion reduces to

$$\sum_{n=1}^{\infty} a_{2m-1} \cos \frac{(2m - 1)\pi x}{L}$$

5. If in addition to the restrictions in (3) above, $f(x) = f(L - x)$, then b_n will be 0 for all even values of n. Thus in this case, the expansion reduces to

$$\sum_{n=1}^{\infty} b_{2m-1} \sin \frac{(2m - 1)\pi x}{L}$$

(The series in (4) and (5) are known as odd-harmonic series, since only the odd harmonics appear. Similar rules may be stated for even-harmonic series, but when a series appears in the even-harmonic form, it means that 2L has not been taken as the smallest period of $f(x)$. Since any integral multiple of a period is also a period, series obtained in this way will also work, but in general computation is simplified if 2L is taken to be the smallest period.)

6. If we write the Euler definitions for $\cos \theta$ and $\sin \theta$, we obtain the complex form of the Fourier Series known either as the "Complex Fourier Series" or the "Exponential Fourier Series" of $f(x)$. It is represented as

$$f(x) = \frac{1}{2} \sum_{n=-\infty}^{n=+\infty} c_n e^{i\omega_n x},$$

where

$$c_n = \frac{1}{L} \int_{-L}^{L} f(x) \, e^{-i\omega_n x} dx, \quad n = 0, \pm 1, \pm 2, \pm 3, \ldots$$

$$\text{with } \omega_n = \frac{n\pi}{L}, \quad n = 0, \pm 1, \pm 2, \ldots$$

The set of coefficients c_n is often referred to as the Fourier spectrum.

7. If both sine and cosine terms are present and if $f(x)$ is of period 2L and expandable by a Fourier series, it can be represented as

$$f(x) = \frac{a_0}{2} + \sum_{n=1}^{\infty} c_n \cos \left(\frac{n\pi x}{L} + \phi_n \right)$$

where $a_n = c_n \cos \phi_n$, $\quad b_n = c_n \sin \phi_n$, $\quad c_n = \sqrt{a_n^2 + b_n^2}$, $\quad \phi_n = \arctan \left(-\frac{b_n}{a_n} \right)$

It can also be represented as

$$f(x) = \frac{a_0}{2} + \sum_{n=1}^{\infty} c_n \sin\left(\frac{n\pi x}{L} + \phi_n\right)$$

where $a_n = c_n \sin\phi_n$, $\quad b_n = -c_n \cos\phi_n$, $\quad c_n = \sqrt{a_n^2 + b_n^2}$, $\quad \phi_n = \arctan\left(\frac{a_n}{b_n}\right)$

where the quadrant of ϕ_n is chosen so as to make the formulas for a_n, b_n, and c_n hold.

8. The following table of trigonometric identities should be helpful for developing Fourier Series.

	n**	n even	n odd	n/2 odd	n/2 even
$\sin n\pi$	0	0	0	0	0
$\cos n\pi$	$(-1)^n$	$+1$	-1	$+1$	$+1$
$\sin n\pi/2$*		0	$(-1)^{(n-1)/2}$	0	0
$\cos n\pi/2$*	$(-1)^{n/2}$		0	-1	$+1$
$\sin n\pi/4$		$\sqrt{2}/2\,(-1)^{(n3+4n+11)/8}$		$(-1)^{(n-2)/4}$	0

*   A useful formula for $\sin n\pi/2$ and $\cos n\pi/2$ is given by

$$\sin\frac{n\pi}{2} = \frac{(i)^{n+1}}{2}\,[(-1)^n - 1] \text{ and } \cos\frac{n\pi}{2} = \frac{(i)^n}{2}\,[(-1)^n + 1], \text{ where } i^2 = -1.$$

**   n any integer.

AUXILIARY FORMULAS FOR FOURIER SERIES

$$1 = \frac{4}{\pi}\left[\sin\frac{\pi x}{k} + \frac{1}{3}\sin\frac{3\pi x}{k} + \frac{1}{5}\sin\frac{5\pi x}{k} + \cdots\right] \qquad [0 < x < k]$$

$$x = \frac{2k}{\pi}\left[\sin\frac{\pi x}{k} - \frac{1}{2}\sin\frac{2\pi x}{k} + \frac{1}{3}\sin\frac{3\pi x}{k} - \cdots\right] \qquad [-k < x < k]$$

$$x = \frac{k}{2} - \frac{4k}{\pi^2}\left[\cos\frac{\pi x}{k} + \frac{1}{3^2}\cos\frac{3\pi x}{k} + \frac{1}{5^2}\cos\frac{5\pi x}{k} + \cdots\right] \qquad [0 < x < k]$$

$$x^2 = \frac{2k^2}{\pi^3}\left[\left(\frac{\pi^2}{1} - \frac{4}{1}\right)\sin\frac{\pi x}{k} - \frac{\pi^2}{2}\sin\frac{2\pi x}{k} + \left(\frac{\pi^2}{3} - \frac{4}{3^3}\right)\sin\frac{3\pi x}{k}\right.$$
$$\left. - \frac{\pi^2}{4}\sin\frac{4\pi x}{k} + \left(\frac{\pi^2}{5} - \frac{4}{5^3}\right)\sin\frac{5\pi x}{k} + \cdots\right] \quad [0 < x < k]$$

$$x^2 = \frac{k^2}{3} - \frac{4k^2}{\pi^2}\left[\cos\frac{\pi x}{k} - \frac{1}{2^2}\cos\frac{2\pi x}{k} + \frac{1}{3^2}\cos\frac{3\pi x}{k} - \frac{1}{4^2}\cos\frac{4\pi x}{k} + \cdots\right]$$
$$[-k < x < k]$$

$$1 - \frac{1}{3} + \frac{1}{5} - \frac{1}{7} + \cdots = \frac{\pi}{4}$$

$$1 + \frac{1}{2^2} + \frac{1}{3^2} + \frac{1}{4^2} + \cdots = \frac{\pi^2}{6}$$

$$1 - \frac{1}{2^2} + \frac{1}{3^2} - \frac{1}{4^2} + \cdots = \frac{\pi^2}{12}$$

$$1 + \frac{1}{3^2} + \frac{1}{5^2} + \frac{1}{7^2} + \cdots = \frac{\pi^2}{8}$$

$$\frac{1}{2^2} + \frac{1}{4^2} + \frac{1}{6^2} + \frac{1}{8^2} + \cdots = \frac{\pi^2}{24}$$

FOURIER EXPANSIONS FOR BASIC PERIODIC FUNCTIONS

$$f(x) = \frac{4}{\pi} \sum_{n=1,3,5\ldots} \frac{1}{n} \sin \frac{n\pi x}{L}$$

$$f(x) = \frac{2}{\pi} \sum_{n=1}^{\infty} \frac{(-1)^n}{n} \left(\cos \frac{n\pi c}{L} - 1 \right) \sin \frac{n\pi x}{L}$$

$$f(x) = \frac{c}{L} + \frac{2}{\pi} \sum_{n=1}^{\infty} \frac{(-1)^n}{n} \sin \frac{n\pi c}{L} \cos \frac{n\pi x}{L}$$

$$f(x) = \frac{2}{L} \sum_{n=1}^{\infty} \sin \frac{n\pi}{2} \frac{\sin \left(\frac{1}{2} n\pi c/L \right)}{\frac{1}{2} n\pi c/L} \sin \frac{n\pi x}{L}$$

$$f(x) = \frac{2}{\pi} \sum_{n=1}^{\infty} \frac{(-1)^{n+1}}{n} \sin \frac{n\pi x}{L}$$

$$f(x) = \frac{1}{2} - \frac{4}{\pi^2} \sum_{n=1,3,5\ldots} \frac{1}{n^2} \cos \frac{n\pi x}{L}$$

$$f(x) = \frac{8}{\pi^2} \sum_{n=1,3,5\ldots} \frac{(-1)^{(n-1)/2}}{n^2} \sin \frac{n\pi x}{L}$$

$$f(x) = \frac{1}{2} - \frac{1}{\pi} \sum_{n=1}^{\infty} \frac{1}{n} \sin \frac{n\pi x}{L}$$

456

SPECIAL FUNCTIONS

FOURIER EXPANSIONS FOR BASIC PERIODIC FUNCTIONS (Continued)

$$f(x) = \tfrac{1}{2}(1 + a) + \frac{2}{\pi^2 (1-a)} \sum_{n=1}^{\infty} \frac{1}{n^2} [(-1)^n \cos n\pi a - 1]$$
$$\cos \frac{n\pi x}{L} ; \left(a = \frac{c}{2L}\right)$$

$$f(x) = \frac{2}{\pi} \sum_{n=1}^{\infty} \frac{(-1)^{n-1}}{n} \left[1 + \frac{\sin n\pi a}{n\pi (1-a)}\right] \sin \frac{n\pi x}{L} ; \left(a = \frac{c}{2L}\right)$$

$$f(x) = \frac{1}{2} - \frac{4}{\pi^2 (1-2a)} \sum_{n=1,3,5...} \frac{1}{n^2} \cos n\pi a \cos \frac{n\pi x}{L} ; \left(a = \frac{c}{2L}\right)$$

$$f(x) = \frac{2}{\pi} \sum_{n=1}^{\infty} \frac{(-1)^n}{n} \left[1 + \frac{1+(-1)^n}{n\pi(1-2a)} \sin n\pi a\right] \sin \frac{n\pi x}{L} ; \left(a = \frac{c}{2L}\right)$$

$$f(x) = \frac{4}{\pi} \sum_{n=1}^{\infty} \frac{1}{n} \sin \frac{n\pi}{4} \sin n\pi a \sin \frac{n\pi x}{L} ; \left(a = \frac{c}{2L}\right)$$

$$f(x) = \frac{9}{\pi^2} \sum_{n=1}^{\infty} \frac{1}{n^2} \sin \frac{n\pi}{3} \sin \frac{n\pi x}{L} ; \left(a = \frac{c}{2L}\right)$$

$$f(x) = \frac{32}{3\pi^2} \sum_{n=1}^{\infty} \frac{1}{n^2} \sin \frac{n\pi}{4} \sin \frac{n\pi x}{L} ; \left(a = \frac{c}{2L}\right)$$

$$f(x) = \frac{1}{\pi} + \frac{1}{2} \sin \omega t - \frac{2}{\pi} \sum_{n=2,4,6...} \frac{1}{n^2 -1} \cos n\omega t$$

Extracted from graphs and formulas, pages 372, 373, Differential Equations in Engineering Problems, Salvadori and Schwarz, published by Prentice-Hall, Inc., 1954.

THE FOURIER TRANSFORMS

Dr. R. E. Gaskell

For a piecewise continuous function $F(x)$ over a finite interval $0 \leqq x \leqq \pi$, the *finite Fourier cosine transform* of $F(x)$ is

$$f_c(n) = \int_0^\pi F(x) \cos nx \, dx \quad (n = 0, 1, 2, \ldots). \tag{4}$$

If x ranges over the interval $0 \leqq x \leqq L$, the substitution $x' = \pi x/L$ allows the use of this definition, also. The inverse transform is written

$$\bar{F}(x) = \frac{1}{\pi} f_c(0) - \frac{2}{\pi} \sum_{n=1}^{\infty} f_c(n) \cos nx \quad (0 < x < \pi) \tag{5}$$

where $\bar{F}(x) = \dfrac{[F(x + 0) + F(x - 0)]}{2}$. We observe that $\bar{F}(x) = F(x)$ at points of continuity. The formula

$$f_c^{(2)}(n) = \int_0^\pi F''(x) \cos nx \, dx$$
$$= -n^2 f_c(n) - F'(0) + (-1)^n F'(\pi) \tag{6}$$

makes the finite Fourier cosine transform useful in certain boundary value problems.

Analogously, the *finite Fourier sine transform* of $F(x)$ is

$$f_s(n) = \int_0^\pi F(x) \sin nx \, dx \quad (n = 1, 2, 3, \ldots) \tag{7}$$

and

$$\bar{F}(x) = \frac{2}{\pi} \sum_{n=1}^{\infty} f_s(n) \sin nx \quad (0 < x < \pi) \tag{8}$$

Corresponding to (6) we have

$$f_s^{(2)}(n) = \int_0^\pi F''(x) \sin nx \, dx$$
$$= -n^2 f_s(n) - nF(0) - n(-1)^n F(\pi). \tag{9}$$

FOURIER TRANSFORMS

If $F(x)$ is defined for $x \geqq 0$ and is piecewise continuous over any finite interval, and if

$$\int_0^\infty F(x) \, dx$$

is absolutely convergent, then

$$f_c(\alpha) = \sqrt{\frac{2}{\pi}} \int_0^\infty F(x) \cos (\alpha x) \, dx \tag{10}$$

is the *Fourier cosine transform* of $F(x)$. Furthermore,

$$\bar{F}(x) = \sqrt{\frac{2}{\pi}} \int_0^\infty f_c(\alpha) \cos (\alpha x) \, d\alpha. \tag{11}$$

If $\lim\limits_{x \to \infty} \dfrac{d^n F}{dx^n} = 0$, an important property of the Fourier cosine transform

$$f_c^{(2r)}(\alpha) = \sqrt{\frac{2}{\pi}} \int_0^{\infty} \left(\frac{d^{2r} F}{dx^{2r}} \right) \cos{(\alpha x)}\, dx$$

$$= -\sqrt{\frac{2}{\pi}} \sum_{n=0}^{r-1} (-1)^n a_{2r-2n-1} \alpha^{2n} + (-1)^r \alpha^{2r} f_c(\alpha) \tag{12}$$

where $\lim\limits_{x \to 0} \dfrac{d^r F}{dx^r} = a_r$, makes it useful in the solution of many problems.

Under the same conditions,

$$f_s(\alpha) = \sqrt{\frac{2}{\pi}} \int_0^{\infty} F(x) \sin{(\alpha x)}\, dx \tag{13}$$

defines the *Fourier sine transform* of $F(x)$, and

$$\bar{F}(x) = \sqrt{\frac{2}{\pi}} \int_0^{\infty} f_s(\alpha) \sin{(\alpha x)}\, d\alpha. \tag{14}$$

Corresponding to (12) we have

$$f_s^{(2r)}(\alpha) = \sqrt{\frac{2}{\pi}} \int_0^{\infty} \frac{d^{2r} F}{dx^{2r}} \sin{(\alpha x)}\, dx$$

$$= -\sqrt{\frac{2}{\pi}} \sum_{n=1}^{r} (-1)^n \alpha^{2n-1} a_{2r-2n} + (-1)^{r-1} \alpha^{2r} f_s(\alpha). \tag{15}$$

Similarly, if $F(x)$ is defined for $-\infty < x < \infty$, and if $\displaystyle\int_{-\infty}^{\infty} F(x)\, dx$ is absolutely convergent, then

$$f(\alpha) = \frac{1}{\sqrt{2\pi}} \int_{-\infty}^{\infty} F(x) e^{i\alpha x}\, dx \tag{16}$$

is the *Fourier transform* of $F(x)$, and

$$\bar{F}(x) = \frac{1}{\sqrt{2\pi}} \int_{-\infty}^{\infty} f(\alpha) e^{-i\alpha x}\, d\alpha. \tag{17}$$

Also, if

$$\lim_{|x| \to \infty} \left| \frac{d^n F}{dx^n} \right| = 0 \quad (n = 1, 2, \ldots, r-1),$$

then

$$f^{(r)}(\alpha) = \frac{1}{\sqrt{2\pi}} \int_{-\infty}^{\infty} F^{(r)}(x) e^{i\alpha x}\, dx = (-i\alpha)^r f(\alpha). \tag{18}$$

FINITE SINE TRANSFORMS

	$f_s(n)$	$F(x)$
1	$f_s(n) = \displaystyle\int_0^\pi F(x) \sin nx\, dx \ (n = 1, 2, \cdots)$	$F(x)$
2	$(-1)^{n+1} f_s(n)$	$F(\pi - x)$
3	$\dfrac{1}{n}$	$\dfrac{\pi - x}{\pi}$
4	$\dfrac{(-1)^{n+1}}{n}$	$\dfrac{x}{\pi}$
5	$\dfrac{1 - (-1)^n}{n}$	1
6	$\dfrac{2}{n^2} \sin \dfrac{n\pi}{2}$	$\begin{cases} x & \text{when } 0 < x < \pi/2 \\ \pi - x & \text{when } \pi/2 < x < \pi \end{cases}$
7	$\dfrac{(-1)^{n+1}}{n^3}$	$\dfrac{x(\pi^2 - x^2)}{6\pi}$
8	$\dfrac{1 - (-1)^n}{n^3}$	$\dfrac{x(\pi - x)}{2}$
9	$\dfrac{\pi^2(-1)^{n-1}}{n} - \dfrac{2[1 - (-1)^n]}{n^3}$	x^2
10	$\pi(-1)^n \left(\dfrac{6}{n^3} - \dfrac{\pi^2}{n} \right)$	x^3
11	$\dfrac{n}{n^2 + c^2} [1 - (-1)^n e^{c\pi}]$	e^{cx}
12	$\dfrac{n}{n^2 + c^2}$	$\dfrac{\sinh c(\pi - x)}{\sinh c\pi}$
13	$\dfrac{n}{n^2 - k^2} \ (k \neq 0, 1, 2, \cdots)$	$\dfrac{\sin k(\pi - x)}{\sin k\pi}$
14	$\begin{cases} \dfrac{\pi}{2} & \text{when } n = m \\ 0 & \text{when } n \neq m \end{cases} \quad (m = 1, 2, \cdots)$	$\sin mx$
15	$\dfrac{n}{n^2 - k^2} [1 - (-1)^n \cos k\pi]$ $(k \neq 1, 2, \cdots)$	$\cos kx$
16	$\begin{cases} \dfrac{n}{n^2 - m^2} [1 - (-1)^{n+m}] \\ \quad \text{when } n \neq m = 1, 2, \cdots \\ 0 \quad \text{when } n = m \end{cases}$	$\cos mx$
17	$\dfrac{n}{(n^2 - k^2)^2} \ (k \neq 0, 1, 2, \cdots)$	$\dfrac{\pi \sin kx}{2k \sin^2 k\pi} - \dfrac{x \cos k(\pi - x)}{2k \sin k\pi}$
18	$\dfrac{b^n}{n} \ (|b| \leq 1)$	$\dfrac{2}{\pi} \arctan \dfrac{b \sin x}{1 - b \cos x}$
19	$\dfrac{1 - (-1)^n}{n} b^n \ (|b| \leq 1)$	$\dfrac{2}{\pi} \arctan \dfrac{2b \sin x}{1 - b^2}$

SPECIAL FUNCTIONS
The Fourier Transforms

FINITE COSINE TRANSFORMS

	$f_c(n)$	$F(x)$
1	$f_c(n) = \displaystyle\int_0^\pi F(x) \cos nx \, dx \quad (n = 0, 1, 2, \cdots)$	$F(x)$
2	$(-1)^n f_c(n)$	$F(\pi - x)$
3	0 when $n = 1, 2, \cdots$; $f_c(0) = \pi$	1
4	$\dfrac{2}{n} \sin \dfrac{n\pi}{2}$; $f_c(0) = 0$	$\begin{cases} 1 \text{ when } 0 < x < \pi/2 \\ -1 \text{ when } \pi/2 < x < \pi \end{cases}$
5	$-\dfrac{1 - (-1)^n}{n^2}$; $f_c(0) = \dfrac{\pi^2}{2}$	x
6	$\dfrac{(-1)^n}{n^2}$; $f_c(0) = \dfrac{\pi^2}{6}$	$\dfrac{x^2}{2\pi}$
7	$\dfrac{1}{n^2}$; $f_c(0) = 0$	$\dfrac{(\pi - x)^2}{2\pi} - \dfrac{\pi}{6}$
8	$3\pi^2 \dfrac{(-1)^n}{n^2} - 6 \dfrac{1 - (-1)^n}{n^4}$; $f_c(0) = \dfrac{\pi^4}{4}$	x^3
9	$\dfrac{(-1)^n e^c \pi - 1}{n^2 + c^2}$	$\dfrac{1}{c} e^{cx}$
10	$\dfrac{1}{n^2 + c^2}$	$\dfrac{\cosh c(\pi - x)}{c \sinh c\pi}$
11	$\dfrac{k}{n^2 - k^2} [(-1)^n \cos \pi k - 1]$ $(k \neq 0, 1, 2, \cdots)$	$\sin kx$
12	$\dfrac{(-1)^{n+m} - 1}{n^2 - m^2}$; $f_c(m) = 0 \quad (m = 1, 2, \cdots)$	$\dfrac{1}{m} \sin mx$
13	$\dfrac{1}{n^2 - k^2} \quad (k \neq 0, 1, 2, \cdots)$	$-\dfrac{\cos k(\pi - x)}{k \sin k\pi}$
14	0 when $n = 1, 2, \cdots$; $f_c(m) = \dfrac{\pi}{2} \quad (m = 1, 2, \cdots)$	$\cos mx$

The Fourier Transforms

FOURIER SINE TRANSFORMS**

	$F(x)$	$f_s(\alpha)$
1	$\begin{cases} 1 & (0 < x < a) \\ 0 & (x > a) \end{cases}$	$\sqrt{\dfrac{2}{\pi}}\left[\dfrac{1 - \cos \alpha}{\alpha}\right]$
2	$x^{p-1} \ (0 < p < 1)$	$\sqrt{\dfrac{2}{\pi}}\dfrac{\Gamma(p)}{\alpha^p}\sin\dfrac{p\pi}{2}$
3	$\begin{cases} \sin x & (0 < x < a) \\ 0 & (x > a) \end{cases}$	$\dfrac{1}{\sqrt{2\pi}}\left[\dfrac{\sin[a(1 - \alpha)]}{1 - \alpha} - \dfrac{\sin[a(1 + \alpha)]}{1 + \alpha}\right]$
4	e^{-x}	$\sqrt{\dfrac{2}{\pi}}\left[\dfrac{\alpha}{1 + \alpha^2}\right]$
5	$xe^{-x^2/2}$	$\alpha e^{-\alpha^2/2}$
6	$\cos\dfrac{x^2}{2}$	$\sqrt{2}\left[\sin\dfrac{\alpha^2}{2}C\left(\dfrac{\alpha^2}{2}\right) - \cos\dfrac{\alpha^2}{2}S\left(\dfrac{\alpha^2}{2}\right)\right]^{*}$
7	$\sin\dfrac{x^2}{2}$	$\sqrt{2}\left[\cos\dfrac{\alpha^2}{2}C\left(\dfrac{\alpha^2}{2}\right) + \sin\dfrac{\alpha^2}{2}S\left(\dfrac{\alpha^2}{2}\right)\right]^{*}$

*$C(y)$ and $S(y)$ are the Fresnel integrals

$$C(y) = \frac{1}{\sqrt{2\pi}}\int_0^y \frac{1}{\sqrt{t}}\cos t\,dt,$$

$$S(y) = \frac{1}{\sqrt{2\pi}}\int_0^y \frac{1}{\sqrt{t}}\sin t\,dt.$$

**More extensive tables of the Fourier sine and cosine transforms can be found in Fritz Oberhettinger, "Tabellen zur-Fourier Transformation," Springer (1957).

FOURIER COSINE TRANSFORMS

	$F(x)$	$f_c(\alpha)$
1	$\begin{cases} 1 & (0 < x < a) \\ 0 & (x > a) \end{cases}$	$\sqrt{\dfrac{2}{\pi}}\dfrac{\sin a\alpha}{\alpha}$
2	$x^{p-1} \ (0 < p < 1)$	$\sqrt{\dfrac{2}{\pi}}\dfrac{\Gamma(p)}{\alpha^p}\cos\dfrac{p\pi}{2}$
3	$\begin{cases} \cos x & (0 < x < a) \\ 0 & (x > a) \end{cases}$	$\dfrac{1}{\sqrt{2\pi}}\left[\dfrac{\sin[a(1 - \alpha)]}{1 - \alpha} + \dfrac{\sin[a(1 + \alpha)]}{1 + \alpha}\right]$
4	e^{-x}	$\sqrt{\dfrac{2}{\pi}}\left(\dfrac{1}{1 + \alpha^2}\right)$
5	$e^{-x^2/2}$	$e^{-\alpha^2/2}$
6	$\cos\dfrac{x^2}{2}$	$\cos\left(\dfrac{\alpha^2}{2} - \dfrac{\pi}{4}\right)$
7	$\sin\dfrac{x^2}{2}$	$\cos\left(\dfrac{\alpha^2}{2} + \dfrac{\pi}{4}\right)$

SPECIAL FUNCTIONS
The Fourier Transforms

FOURIER TRANSFORMS*

	$F(x)$	$f(\alpha)$
1	$\dfrac{\sin ax}{x}$	$\begin{cases} \sqrt{\dfrac{\pi}{2}} & \|\alpha\| < a \\ 0 & \|\alpha\| > a \end{cases}$
2	$\begin{cases} e^{iwx} & (p < x < q) \\ 0 & (x < p,\ x > q) \end{cases}$	$\dfrac{i}{\sqrt{2\pi}}\dfrac{e^{ip(w+\alpha)} - e^{iq(w+\alpha)}}{(w+\alpha)}$
3	$\begin{cases} e^{-cx+iwx} & (x > 0) \\ 0 & (x < 0) \end{cases}\quad (c > 0)$	$\dfrac{i}{\sqrt{2\pi}(w+\alpha+ic)}$
4	$e^{-px^2}\quad R(p) > 0$	$\dfrac{1}{\sqrt{2p}}\,e^{-\alpha^2/4p}$
5	$\cos px^2$	$\dfrac{1}{\sqrt{2p}}\cos\left[\dfrac{\alpha^2}{4p} - \dfrac{\pi}{4}\right]$
6	$\sin px^2$	$\dfrac{1}{\sqrt{2p}}\cos\left[\dfrac{\alpha^2}{4p} + \dfrac{\pi}{4}\right]$
7	$\|x\|^{-p}\quad (0 < p < 1)$	$\sqrt{\dfrac{2}{\pi}}\,\dfrac{\Gamma(1-p)\sin\dfrac{p\pi}{2}}{\|\alpha\|^{(1-p)}}$
8	$\dfrac{e^{-a\|x\|}}{\sqrt{\|x\|}}$	$\dfrac{\sqrt{\sqrt{(a^2+\alpha^2)} + a}}{\sqrt{a^2+\alpha^2}}$
9	$\dfrac{\cosh ax}{\cosh \pi x}\quad (-\pi < a < \pi)$	$\sqrt{\dfrac{2}{\pi}}\,\dfrac{\cos\dfrac{a}{2}\cosh\dfrac{\alpha}{2}}{\cosh\alpha + \cos a}$
10	$\dfrac{\sinh ax}{\sinh \pi x}\quad (-\pi < a < \pi)$	$\dfrac{1}{\sqrt{2\pi}}\,\dfrac{\sin a}{\cosh\alpha + \cos a}$
11	$\begin{cases} \dfrac{1}{\sqrt{a^2-x^2}} & (\|x\| < a) \\ 0 & (\|x\| > a) \end{cases}$	$\sqrt{\dfrac{\pi}{2}}\,J_0(a\alpha)$
12	$\dfrac{\sin[b\sqrt{a^2+x^2}]}{\sqrt{a^2+x^2}}$	$\begin{cases} 0 & (\|\alpha\| > b) \\ \sqrt{\dfrac{\pi}{2}}\,J_0(a\sqrt{b^2-\alpha^2}) & (\|\alpha\| < b) \end{cases}$
13	$\begin{cases} P_n(x) & (\|x\| < 1) \\ 0 & (\|x\| > 1) \end{cases}$	$\dfrac{i^n}{\sqrt{\alpha}}\,J_{n+\frac{1}{2}}(\alpha)$
14	$\begin{cases} \dfrac{\cos[b\sqrt{a^2-x^2}]}{\sqrt{a^2-x^2}} & (\|x\| < a) \\ 0 & (\|x\| > a) \end{cases}$	$\sqrt{\dfrac{\pi}{2}}\,J_0(a\sqrt{\alpha^2+b^2})$
15	$\begin{cases} \dfrac{\cosh[b\sqrt{a^2-x^2}]}{\sqrt{a^2-x^2}} & (\|x\| < a) \\ 0 & (\|x\| > a) \end{cases}$	$\sqrt{\dfrac{\pi}{2}}\,J_0(a\sqrt{\alpha^2-b^2})$

*More extensive tables of Fourier transforms can be found in W. Magnus and F. Oberhettinger, "Formulas and Theorems of the Special Functions of Mathematical Physics," pp. 116–120. Chelsea (1949).

The following functions appear among the entries of the tables on transforms.

Function	Definition	Name
$Ei(x)$	$\displaystyle\int_{-\infty}^{x} \frac{e^v}{v}\, dv$; or sometimes defined as $$-Ei(-x) = \int_{x}^{\infty} \frac{e^{-v}}{v}\, dv$$	Sine, Cosine, and Exponential Integral tables pages 421--423
$Si(x)$	$\displaystyle\int_{0}^{x} \frac{\sin v}{v}\, dv$	Sine, Cosine, and Exponential Integral tables pages 421--423
$Ci(x)$	$\displaystyle\int_{\infty}^{x} \frac{\cos v}{v}\, dv$; or sometimes defined as negative of this integral	Sine, Cosine, and Exponential Integral tables pages 421--423
$erf(x)$	$\displaystyle\frac{2}{\sqrt{\pi}} \int_{0}^{x} e^{-v^2}\, dv$	Error function
$erfc(x)$	$\displaystyle 1 - erf(x) = \frac{2}{\sqrt{\pi}} \int_{x}^{\infty} e^{-v^2}\, dv$	Complementary function to error function
$L_n(x)$	$\displaystyle\frac{e^x}{n!} \frac{d^n}{dx^n} (x^n e^{-x}),\quad n = 0, 1, \cdots$	Laguerre polynomial of degree n

THE LAPLACE TRANSFORM

Dr. R. E. Gaskell

If $F(t)$ is a piecewise continuous real-valued function of the real variable $t (0 \leq t < \infty)$, and if $F(t)$ is of exponential order, that is if $|F(t)| < Me^{at} (t > T; M, a, T$ positive constants) then the *Laplace transform* of $F(t)$,

$$L\{F(t)\} = f(s) = \int_0^\infty e^{-st} F(t) \, dt, \tag{1}$$

exists in the half-plane of the complex variable s for which the real part of s is greater than some fixed value s_0, i.e., $R(s) \geq s_0$. Furthermore

$$F(t) = \frac{1}{2\pi i} \int_{a-i\infty}^{a+i\infty} e^{st} f(s) \, ds, \tag{2}$$

where $a > s_0$. The important property

$$
\begin{aligned}
L\{F^{(r)}(t)\} &= \int_0^\infty e^{-st} \left(\frac{d^r F}{dt^r} \right) dt \\
&= s^r f(s) - \sum_{n=0}^{r-1} s^{r-1-n} F^{(n)}(+0)
\end{aligned}
\tag{3}
$$

makes the Laplace transform very useful for solving linear differential equations with constant coefficients, and many boundary value problems.

	$F(t)$	$f(s)$
1	$F(t)$	$\displaystyle\int_0^\infty e^{-st} F(t)\, dt$
2	$AF(t) + BG(t)$	$Af(s) + Bg(s)$
3	$F'(t)$	$sf(s) - F(+0)$
4	$F^{(n)}(t)$	$s^n f(s) - s^{n-1} F(+0) - s^{n-2} F'(+0) - \cdots$ $\qquad\qquad - F^{(n-1)}(+0)$
5	$\displaystyle\int_0^t F(\tau)\, d\tau$	$\dfrac{1}{s} f(s)$
6	$\displaystyle\int_0^t \int_0^\tau F(\lambda)\, d\lambda\, d\tau$	$\dfrac{1}{s^2} f(s)$
7	$\displaystyle\int_0^t F_1(t - \tau) F_2(\tau)\, d\tau = F_1{}^*F_2$	$f_1(s) f_2(s)$
8	$tF(t)$	$-f'(s)$
9	$t^n F(t)$	$(-1)^n f^{(n)}(s)$
10	$\dfrac{1}{t} F(t)$	$\displaystyle\int_s^\infty f(x)\, dx$
11	$e^{at} F(t)$	$f(s - a)$
12	$F(t - b)$, where $F(t) = 0$ when $t < 0$	$e^{-bs} f(s)$
13	$\dfrac{1}{c} F\left(\dfrac{t}{c}\right)$	$f(cs)$
14	$\dfrac{1}{c} e^{(bt)/c}\ F\left(\dfrac{t}{c}\right)$	$f(cs - b)$
15	$F(t + a) = F(t)$	$\dfrac{\displaystyle\int_0^a e^{-st} F(t)\, dt}{1 - e^{-as}}$
16	$F(t + a) = -F(t)$	$\dfrac{\displaystyle\int_0^a e^{-st} F(t)\, dt}{1 + e^{-as}}$
17	$F_1(t)$, the half-wave rectification of $F(t)$ in No. 16	$\dfrac{f(s)}{1 - e^{-as}}$
18	$F_2(t)$, the full-wave rectification of $F(t)$ in No. 16	$f(s) \coth \dfrac{as}{2}$
19	$\displaystyle\sum_1^m \dfrac{p(a_n)}{q'(a_n)} e^{a_n t}$	$\dfrac{p(s)}{q(s)}$, $q(s) = (s - a_1)(s - a_2)\cdots(s - a_m)$
20	$\displaystyle e^{at} \sum_{n-1}^r \dfrac{\phi^{(r-n)}(a)}{(r - n)!} \dfrac{t^{n-1}}{(n - 1)!} + \cdots$	$\dfrac{p(s)}{q(s)} = \dfrac{\phi(s)}{(s - a)^r}$

*These tables of Laplace Operations, Laplace Transforms, and Finite Fourier sine and cosine transforms were taken from "Modern Operational Mathematics in Engineering", by permission of the author, R. V. Churchill, and the publisher, McGraw-Hill Book Company, Inc.

SPECIAL FUNCTIONS
The Laplace Transforms

LAPLACE TRANSFORMS

	$f(s)$	$F(t)$
1	$\dfrac{1}{s}$	$\mu(t)$, unit step function
2	$\dfrac{1}{s^2}$	t
3	$\dfrac{1}{s^n}$ $(n = 1, 2, \ldots)$	$\dfrac{t^{n-1}}{(n-1)!}$
4	$\dfrac{1}{\sqrt{s}}$	$\dfrac{1}{\sqrt{\pi t}}$
5	$s^{-3/2}$	$2\sqrt{\dfrac{t}{\pi}}$
6	$s^{-[n+(1/2)]}$ $(n = 1, 2, \ldots)$	$\dfrac{2^n t^{n-(1/2)}}{1 \cdot 3 \cdot 5 \cdots (2n-1)\sqrt{\pi}}$
7	$\dfrac{\Gamma(k)}{s^k}$ $(k > 0)$	t^{k-1}
8	$\dfrac{1}{s-a}$	e^{at}
9	$\dfrac{1}{(s-a)^2}$	te^{at}
10	$\dfrac{1}{(s-a)^n}$ $(n = 1, 2, \ldots)$	$\dfrac{1}{(n-1)!}\,t^{n-1}e^{at}$
11	$\dfrac{\Gamma(k)}{(s-a)^k}$ $(k > 0)$	$t^{k-1}e^{at}$
12*	$\dfrac{1}{(s-a)(s-b)}$	$\dfrac{1}{a-b}(e^{at} - e^{bt})$
13*	$\dfrac{s}{(s-a)(s-b)}$	$\dfrac{1}{a-b}(ae^{at} - be^{bt})$
14*	$\dfrac{1}{(s-a)(s-b)(s-c)}$	$-\dfrac{(b-c)e^{at} + (c-a)e^{bt} + (a-b)e^{ct}}{(a-b)(b-c)(c-a)}$
15	$\dfrac{1}{s^2 + a^2}$	$\dfrac{1}{a}\sin at$
16	$\dfrac{s}{s^2 + a^2}$	$\cos at$
17	$\dfrac{1}{s^2 - a^2}$	$\dfrac{1}{a}\sinh at$
18	$\dfrac{s}{s^2 - a^2}$	$\cosh at$

*Here a, b, and (in 14) c represent distinct constants.

LAPLACE TRANSFORMS (Continued)

	$f(s)$	$F(t)$
19	$\dfrac{1}{s(s^2 + a^2)}$	$\dfrac{1}{a^2}(1 - \cos at)$
20	$\dfrac{1}{s^2(s^2 + a^2)}$	$\dfrac{1}{a^3}(at - \sin at)$
21	$\dfrac{1}{(s^2 + a^2)^2}$	$\dfrac{1}{2a^3}(\sin at - at\cos at)$
22	$\dfrac{s}{(s^2 + a^2)^2}$	$\dfrac{t}{2a}\sin at$
23	$\dfrac{s^2}{(s^2 + a^2)^2}$	$\dfrac{1}{2a}(\sin at + at\cos at)$
24	$\dfrac{s^2 - a^2}{(s^2 + a^2)^2}$	$t\cos at$
25	$\dfrac{s}{(s^2 + a^2)(s^2 + b^2)}\ (a^2 \neq b^2)$	$\dfrac{\cos at - \cos bt}{b^2 - a^2}$
26	$\dfrac{1}{(s - a)^2 + b^2}$	$\dfrac{1}{b}e^{at}\sin bt$
27	$\dfrac{s - a}{(s - a)^2 + b^2}$	$e^{at}\cos bt$
27.1	$\dfrac{1}{[(s + a)^2 + b^2]^n}$	$\dfrac{-e^{-at}}{4^{n-1}b^{2n}}\sum_{r=1}^{n}\binom{2n - r - 1}{n - 1}(-2t)^{r-1}\dfrac{d^r}{dt^r}[\cos(bt)]$
27.2	$\dfrac{s}{[(s + a)^2 + b^2]^n}$	$\dfrac{e^{-at}}{4^{n-1}b^{2n}}\left\{\sum_{r=1}^{n}\binom{2n - r - 1}{n - 1}\dfrac{1}{(r-1)!}\right.$ $(-2t)^{r-1}\dfrac{d^r}{dt^r}[a\cos(bt) + b\sin(bt)]$ $-2b\sum_{r=1}^{n-1}\dfrac{1}{(r-1)!}\binom{2n - r - 2}{n - 1}$ $\left.(-2t)^{r-1}\dfrac{d^r}{dt^r}[\sin bt]\right\}$
28	$\dfrac{3a^2}{s^3 + a^3}$	$e^{-at} - e^{(at)/2}\left(\cos\dfrac{at\sqrt{3}}{2} - \sqrt{3}\sin\dfrac{at\sqrt{3}}{2}\right)$
29	$\dfrac{4a^3}{s^4 + 4a^4}$	$\sin at\cosh at - \cos at\sinh at$
30	$\dfrac{s}{s^4 + 4a^4}$	$\dfrac{1}{2a^2}\sin at\sinh at$

LAPLACE TRANSFORMS (Continued)

	$f(s)$	$F(t)$
31	$\dfrac{1}{s^4 - a^4}$	$\dfrac{1}{2a^3}(\sinh at - \sin at)$
32	$\dfrac{s}{s^4 - a^4}$	$\dfrac{1}{2a^2}(\cosh at - \cos at)$
33	$\dfrac{8a^3 s^2}{(s^2 + a^2)^3}$	$(1 + a^2 t^2)\sin at - \cos at$
34*	$\dfrac{1}{s}\left(\dfrac{s-1}{s}\right)^n$	$L_n(t) = \dfrac{e^t}{n!}\dfrac{d^n}{dt^n}(t^n e^{-t})$
35	$\dfrac{s}{(s-a)^{3/2}}$	$\dfrac{1}{\sqrt{\pi t}}e^{at}(1 + 2at)$
36	$\sqrt{s-a} - \sqrt{s-b}$	$\dfrac{1}{2\sqrt{\pi t^3}}(e^{bt} - e^{at})$
37	$\dfrac{1}{\sqrt{s}+a}$	$\dfrac{1}{\sqrt{\pi t}} - ae^{a^2 t}\,\text{erfc}\,(a\sqrt{t})$
38	$\dfrac{\sqrt{s}}{s-a^2}$	$\dfrac{1}{\sqrt{\pi t}} + ae^{a^2 t}\,\text{erf}\,(a\sqrt{t})$
39	$\dfrac{\sqrt{s}}{s+a^2}$	$\dfrac{1}{\sqrt{\pi t}} - \dfrac{2a}{\sqrt{\pi}}e^{-a^2 t}\displaystyle\int_0^{a\sqrt{t}}e^{\lambda^2}d\lambda$
40	$\dfrac{1}{\sqrt{s}(s-a^2)}$	$\dfrac{1}{a}e^{a^2 t}\,\text{erf}\,(a\sqrt{t})$
41	$\dfrac{1}{\sqrt{s}(s+a^2)}$	$\dfrac{2}{a\sqrt{\pi}}e^{-a^2 t}\displaystyle\int_0^{a\sqrt{t}}e^{\lambda^2}d\lambda$
42	$\dfrac{b^2 - a^2}{(s-a^2)(b+\sqrt{s})}$	$e^{a^2 t}[b - a\,\text{erf}\,(a\sqrt{t})] - be^{b^2 t}\,\text{erfc}\,(b\sqrt{t})$
43	$\dfrac{1}{\sqrt{s}(\sqrt{s}+a)}$	$e^{a^2 t}\,\text{erfc}\,(a\sqrt{t})$
44	$\dfrac{1}{(s+a)\sqrt{s+b}}$	$\dfrac{1}{\sqrt{b-a}}e^{-at}\,\text{erf}\,(\sqrt{b-a}\,\sqrt{t})$
45	$\dfrac{b^2 - a^2}{\sqrt{s}(s-a^2)(\sqrt{s}+b)}$	$e^{a^2 t}\left[\dfrac{b}{a}\,\text{erf}\,(a\sqrt{t}) - 1\right] + e^{b^2 t}\,\text{erfc}\,(b\sqrt{t})$
46†	$\dfrac{(1-s)^n}{s^{n+(1/2)}}$	$\dfrac{n!}{(2n)!\,\sqrt{\pi t}}H_{2n}(\sqrt{t})$
47	$\dfrac{(1-s)^n}{s^{n+(3/2)}}$	$-\dfrac{n!}{\sqrt{\pi}(2n+1)!}H_{2n+1}(\sqrt{t})$

*$L_n(t)$ is the Laguerre polynomial of degree n.

†$H_n(x)$ is the Hermite polynomial, $H_n(x) = e^{x^2}\dfrac{d^n}{dx^n}(e^{-x^2})$.

LAPLACE TRANSFORMS (Continued)

	$f(s)$	$F(t)$
48*	$\dfrac{\sqrt{s + 2a}}{\sqrt{s}} - 1$	$ae^{-at}[I_1(at) + I_0(at)]$
49	$\dfrac{1}{\sqrt{s + a}\,\sqrt{s + b}}$	$e^{-(1/2)(a+b)t}I_0\left(\dfrac{a - b}{2}\,t\right)$
50	$\dfrac{\Gamma(k)}{(s + a)^k(s + b)^k}\ (k \geq 0)$	$\sqrt{\pi}\left(\dfrac{t}{a - b}\right)^{k-(1/2)}e^{-(1/2)(a+b)t}I_{k-(1/2)}\left(\dfrac{a - b}{2}\,t\right)$
51	$\dfrac{1}{(s + a)^{1/2}(s + b)^{3/2}}$	$te^{-(1/2)(a+b)t}\left[I_0\left(\dfrac{a - b}{2}\,t\right) + I_1\left(\dfrac{a - b}{2}\,t\right)\right]$
52	$\dfrac{\sqrt{s + 2a} - \sqrt{s}}{\sqrt{s + 2a} + \sqrt{s}}$	$\dfrac{1}{t}\,e^{-at}I_1(at)$
53	$\dfrac{(a - b)^k}{(\sqrt{s + a} + \sqrt{s + b})^{2k}}\ (k > 0)$	$\dfrac{k}{t}\,e^{-(1/2)(a+b)t}I_k\left(\dfrac{a - b}{2}\,t\right)$
54	$\dfrac{(\sqrt{s + a} + \sqrt{s})^{-2\nu}}{\sqrt{s}\sqrt{s + a}}\ (\nu > -1)$	$\dfrac{1}{a^\nu}\,e^{-(1/2)(at)}I_\nu\left(\dfrac{1}{2}\,at\right)$
55	$\dfrac{1}{\sqrt{s^2 + a^2}}$	$J_0(at)$
56	$\dfrac{(\sqrt{s^2 + a^2} - s)^\nu}{\sqrt{s^2 + a^2}}\ (\nu > -1)$	$a^\nu J_\nu(at)$
57	$\dfrac{1}{(s^2 + a^2)^k}\ (k > 0)$	$\dfrac{\sqrt{\pi}}{\Gamma(k)}\left(\dfrac{t}{2a}\right)^{k-(1/2)}J_{k-(1/2)}(at)$
58	$(\sqrt{s^2 + a^2} - s)^k\ (k > 0)$	$\dfrac{ka^k}{t}\,J_k(at)$
59	$\dfrac{(s - \sqrt{s^2 - a^2})^\nu}{\sqrt{s^2 - a^2}}\ (\nu > -1)$	$a^\nu I_\nu(at)$
60	$\dfrac{1}{(s^2 - a^2)^k}\ (k > 0)$	$\dfrac{\sqrt{\pi}}{\Gamma(k)}\left(\dfrac{t}{2a}\right)^{k-(1/2)}I_{k-(1/2)}(at)$
61	$\dfrac{e^{-ks}}{s}$	$S_k(t) = \begin{cases} 0 \text{ when } 0 < t < k \\ 1 \text{ when } t > k \end{cases}$
62	$\dfrac{e^{-ks}}{s^2}$	$\begin{cases} 0 \text{ when } 0 < t < k \\ t - k \text{ when } t > k \end{cases}$
63	$\dfrac{e^{-ks}}{s^\mu}\ (\mu > 0)$	$\begin{cases} 0 \quad\quad\quad \text{ when } 0 < t < k \\ \dfrac{(t - k)^{\mu-1}}{\Gamma(\mu)} \text{ when } t > k \end{cases}$
64	$\dfrac{1 - e^{-ks}}{s}$	$\begin{cases} 1 \text{ when } 0 < t < k \\ 0 \text{ when } t > k \end{cases}$

*$I_n(x) = i^{-n}J_n(ix)$, where J_n is Bessel's function of the first kind.

LAPLACE TRANSFORMS (Continued)

	$f(s)$	$F(t)$
65	$\dfrac{1}{s(1 - e^{-ks})} = \dfrac{1 + \coth \frac{1}{2}ks}{2s}$	$S(k,t) = n$ when $(n - 1)k < t < nk \, (n = 1, 2, \ldots)$
66	$\dfrac{1}{s(e^{ks} - a)}$	$\begin{cases} 0 \text{ when } 0 < t < k \\ 1 + a + a^2 + \cdots + a^{n-1} \\ \quad \text{when } nk < t < (n + 1)k \, (n = 1, 2, \ldots) \end{cases}$
67	$\dfrac{1}{s} \tanh ks$	$M(2k,t) = (-1)^{n-1}$ when $2k(n - 1) < t < 2kn$ $(n = 1, 2, \ldots)$
68	$\dfrac{1}{s(1 + e^{-ks})}$	$\dfrac{1}{2} M(k,t) + \dfrac{1}{2} = \dfrac{1 - (-1)^n}{2}$ when $(n - 1)k < t < nk$
69*	$\dfrac{1}{s^2} \tanh ks$	$H(2k,t)$
70	$\dfrac{1}{s \sinh ks}$	$2S(2k, t + k) - 2 = 2(n - 1)$ when $(2n - 3)k < t < (2n - 1)k \ (t > 0)$
71	$\dfrac{1}{s \cosh ks}$	$M(2k, t + 3k) + 1 = 1 + (-1)^n$ when $(2n - 3)k < t < (2n - 1)k \ (t > 0)$
72	$\dfrac{1}{s} \coth ks$	$2S(2k,t) - 1 = 2n - 1$ when $2k(n - 1) < t < 2kn$
73	$\dfrac{k}{s^2 + k^2} \coth \dfrac{\pi s}{2k}$	$\lvert \sin kt \rvert$
74	$\dfrac{1}{(s^2 + 1)(1 - e^{-\pi s})}$	$\begin{cases} \sin t \text{ when } (2n - 2)\pi < t < (2n - 1)\pi \\ 0 \quad \text{when } (2n - 1)\pi < t < 2n\pi \end{cases}$
75	$\dfrac{1}{s} e^{-k/s}$	$J_0(2\sqrt{kt})$
76	$\dfrac{1}{\sqrt{s}} e^{-k/s}$	$\dfrac{1}{\sqrt{\pi t}} \cos 2\sqrt{kt}$
77	$\dfrac{1}{\sqrt{s}} e^{k/s}$	$\dfrac{1}{\sqrt{\pi t}} \cosh 2\sqrt{kt}$
78	$\dfrac{1}{s^{3/2}} e^{-k/s}$	$\dfrac{1}{\sqrt{\pi k}} \sin 2\sqrt{kt}$
79	$\dfrac{1}{s^{3/2}} e^{k/s}$	$\dfrac{1}{\sqrt{\pi k}} \sinh 2\sqrt{kt}$
80	$\dfrac{1}{s^{\mu}} e^{-k/s} (\mu > 0)$	$\left(\dfrac{t}{k}\right)^{(\mu - 1)/2} J_{\mu - 1}(2\sqrt{kt})$

*$H(2k,t) = k + (r - k)(-1)^n$ where $t = 2kn + r; 0 \leq r < 2k; n = 0, 1, 2, \ldots$.

LAPLACE TRANSFORMS (Continued)

	$f(s)$	$F(t)$
81	$\dfrac{1}{s^\mu} e^{k/s} (\mu > 0)$	$\left(\dfrac{t}{k}\right)^{(\mu-1)/2} I_{\mu-1}(2\sqrt{kt})$
82	$e^{-k\sqrt{s}} (k > 0)$	$\dfrac{k}{2\sqrt{\pi t^3}} \exp\left(-\dfrac{k^2}{4t}\right)$
83	$\dfrac{1}{s} e^{-k\sqrt{s}} (k \geqq 0)$	$\operatorname{erfc}\left(\dfrac{k}{2\sqrt{t}}\right)$
84	$\dfrac{1}{\sqrt{s}} e^{-k\sqrt{s}} (k \geqq 0)$	$\dfrac{1}{\sqrt{\pi t}} \exp\left(-\dfrac{k^2}{4t}\right)$
85	$s^{-3/2} e^{-k\sqrt{s}} (k \geqq 0)$	$2\sqrt{\dfrac{t}{\pi}} \exp\left(-\dfrac{k^2}{4t}\right) - k \operatorname{erfc}\left(\dfrac{k}{2\sqrt{t}}\right)$
86	$\dfrac{ae^{-k\sqrt{s}}}{s(a + \sqrt{s})} (k \geqq 0)$	$-e^{ak} e^{a^2 t} \operatorname{erfc}\left(a\sqrt{t} + \dfrac{k}{2\sqrt{t}}\right) + \operatorname{erfc}\left(\dfrac{k}{2\sqrt{t}}\right)$
87	$\dfrac{e^{-k\sqrt{s}}}{\sqrt{s}(a + \sqrt{s})} (k \geqq 0)$	$e^{ak} e^{a^2 t} \operatorname{erfc}\left(a\sqrt{t} + \dfrac{k}{2\sqrt{t}}\right)$
88	$\dfrac{e^{-k\sqrt{s(s+a)}}}{\sqrt{s(s+a)}}$	$\begin{cases} 0 & \text{when } 0 < t < k \\ e^{-(1/2)(at)} I_0(\tfrac{1}{2}a\sqrt{t^2 - k^2}) & \text{when } t > k \end{cases}$
89	$\dfrac{e^{-k\sqrt{s^2+a^2}}}{\sqrt{s^2 + a^2}}$	$\begin{cases} 0 & \text{when } 0 < t < k \\ J_0(a\sqrt{t^2 - k^2}) & \text{when } t > k \end{cases}$
90	$\dfrac{e^{k\sqrt{s^2-a^2}}}{\sqrt{s^2 - a^2}}$	$\begin{cases} 0 & \text{when } 0 < t < k \\ I_0(a\sqrt{t^2 - k^2}) & \text{when } t > k \end{cases}$
91	$\dfrac{e^{-k(\sqrt{s^2+a^2}-s)}}{\sqrt{s^2 + a^2}} (k \geqq 0)$	$J_0(a\sqrt{t^2 + 2kt})$
92	$e^{-ks} - e^{-k\sqrt{s^2+a^2}}$	$\begin{cases} 0 & \text{when } 0 < t < k \\ \dfrac{ak}{\sqrt{t^2 - k^2}} J_1(a\sqrt{t^2 - k^2}) & \text{when } t > k \end{cases}$
93	$e^{-k\sqrt{s^2+a^2}} - e^{-ks}$	$\begin{cases} 0 & \text{when } 0 < t < k \\ \dfrac{ak}{\sqrt{t^2 - k^2}} I_1(a\sqrt{t^2 - k^2}) & \text{when } t > k \end{cases}$
94	$\dfrac{a^\nu e^{-k\sqrt{s^2-a^2}}}{\sqrt{s^2 + a^2}(\sqrt{s^2 + a^2} + s)^\nu}$ $(\nu > -1)$	$\begin{cases} 0 & \text{when } 0 < t < k \\ \left(\dfrac{t-k}{t+k}\right)^{(1/2)\nu} J_\nu(a\sqrt{t^2 - k^2}) & \text{when } t > k \end{cases}$
95	$\dfrac{1}{s} \log s$	$\Gamma'(1) - \log t \ [\Gamma'(1) = -0.5772]$
96	$\dfrac{1}{s^k} \log s \ (k > 0)$	$t^{k-1}\left\{\dfrac{\Gamma'(k)}{[\Gamma(k)]^2} - \dfrac{\log t}{\Gamma(k)}\right\}$
97	$\dfrac{\log s}{s - a} \ (a > 0)$	$e^{at}[\log a - \operatorname{Ei}(-at)]$

LAPLACE TRANSFORMS (Continued)

	$f(s)$	$F(t)$
98	$\dfrac{\log s}{s^2 + 1}$	$\cos t \, \mathrm{Si}(t) - \sin t \, \mathrm{Ci}(t)$
99	$\dfrac{s \log s}{s^2 + 1}$	$-\sin t \, \mathrm{Si}(t) - \cos t \, \mathrm{Ci}(t)$
100	$\dfrac{1}{s} \log (1 + ks)\,(k > 0)$	$-\mathrm{Ei}\left(-\dfrac{t}{k}\right)$
101	$\log \dfrac{s - a}{s - b}$	$\dfrac{1}{t} \left(e^{bt} - e^{at}\right)$
102	$\dfrac{1}{s} \log (1 + k^2 s^2)$	$-2\mathrm{Ci}\left(\dfrac{t}{k}\right)$
103	$\dfrac{1}{s} \log (s^2 + a^2)\ \ (a > 0)$	$2 \log a - 2\mathrm{Ci}(at)$
104	$\dfrac{1}{s^2} \log (s^2 + a^2)\ \ (a > 0)$	$\dfrac{2}{a} [at \log a + \sin at - at \, \mathrm{Ci}(at)]$
105	$\log \dfrac{s^2 + a^2}{s^2}$	$\dfrac{2}{t} (1 - \cos at)$
106	$\log \dfrac{s^2 - a^2}{s^2}$	$\dfrac{2}{t} (1 - \cosh at)$
107	$\arctan \dfrac{k}{s}$	$\dfrac{1}{t} \sin kt$
108	$\dfrac{1}{s} \arctan \dfrac{k}{s}$	$\mathrm{Si}(kt)$
109	$e^{k^2 s^2} \mathrm{erfc}\,(ks)\ \ (k > 0)$	$\dfrac{1}{k\sqrt{\pi}} \exp\left(-\dfrac{t^2}{4k^2}\right)$
110	$\dfrac{1}{s} e^{k^2 s^2} \mathrm{erfc}\,(ks)\ \ (k > 0)$	$\mathrm{erf}\left(\dfrac{t}{2k}\right)$
111	$e^{ks} \mathrm{erfc}\,(\sqrt{ks})\ \ (k > 0)$	$\dfrac{\sqrt{k}}{\pi \sqrt{t}(t + k)}$
112	$\dfrac{1}{\sqrt{s}} \mathrm{erfc}\,(\sqrt{ks})$	$\begin{cases} 0 & \text{when } 0 < t < k \\ (\pi t)^{-1/2} & \text{when } t > k \end{cases}$
113	$\dfrac{1}{\sqrt{s}} e^{ks} \mathrm{erfc}\,(\sqrt{ks})\,(k > 0)$	$\dfrac{1}{\sqrt{\pi(t + k)}}$
114	$\mathrm{erf}\left(\dfrac{k}{\sqrt{s}}\right)$	$\dfrac{1}{\pi t} \sin (2k \sqrt{t})$
115	$\dfrac{1}{\sqrt{s}} e^{k^2/s} \mathrm{erfc}\left(\dfrac{k}{\sqrt{s}}\right)$	$\dfrac{1}{\sqrt{\pi t}} e^{-2k\sqrt{t}}$

LAPLACE TRANSFORMS (Continued)

	$f(s)$	$F(t)$
115.1	$-e^{as}\,\mathrm{Ei}(-as)$	$\dfrac{1}{t+a}\,;(a>0)$
115.2	$\dfrac{1}{a}+se^{as}\,\mathrm{Ei}(-as)$	$\dfrac{1}{(t+a)^2}\,;(a>0)$
115.3	$\left[\dfrac{\pi}{2}-\mathrm{Si}(s)\right]\cos s+\mathrm{Ci}(s)\sin s$	$\dfrac{1}{t^2+1}$
116*	$K_0(ks)$	$\begin{cases}0 & \text{when } 0<t<k\\ (t^2-k^2)^{-1/2} & \text{when } t>k\end{cases}$
117	$K_0(k\sqrt{s})$	$\dfrac{1}{2t}\exp\left(-\dfrac{k^2}{4t}\right)$
118	$\dfrac{1}{s}\,e^{ks}K_1(ks)$	$\dfrac{1}{k}\sqrt{t(t+2k)}$
119	$\dfrac{1}{\sqrt{s}}\,K_1(k\sqrt{s})$	$\dfrac{1}{k}\exp\left(-\dfrac{k^2}{4t}\right)$
120	$\dfrac{1}{\sqrt{s}}\,e^{k/s}K_0\left(\dfrac{k}{s}\right)$	$\dfrac{2}{\sqrt{\pi t}}\,K_0(2\sqrt{2kt})$
121	$\pi e^{-ks}I_0(ks)$	$\begin{cases}[t(2k-t)]^{-1/2} & \text{when } 0<t<2k\\ 0 & \text{when } t>2k\end{cases}$
122**	$e^{-ks}I_1(ks)$	$\begin{cases}\dfrac{k-t}{\pi k\sqrt{t(2k-t)}} & \text{when } 0<t<2k\\ 0 & \text{when } t>2k\end{cases}$

*$K_n(x)$ is Bessel's function of the second kind for the imaginary argument.
**Several additional transforms, especially those involving other Bessel functions, can be found in the tables by G. A. Campbell and R. M. Foster, "Fourier Integrals for Practical Applications", or "Vol. 1, Bateman Manuscript Project, Transform Tables, McGraw-Hill, 1955", or N. W. McLachlan and P. Humbert, "Formulaire pour le calcul symbolique". In the tables by Campbell and Foster, only those entries containing the condition $0<g$ or $k<g$, where g is our t, are Laplace transforms.

SPECIAL FUNCTIONS

THE Z TRANSFORM

B. Girling

When $F(t)$, a continuous function of time, is sampled at regular intervals of period T the usual Laplace transform techniques are modified. The diagramatic form of a simple sampler together with its associated input-output waveforms is shown below

$$\frac{1}{T} \equiv F_s \quad \text{the sampling frequency}$$

Defining the set of impulse functions $\delta_T(t)$ by

$$\delta_T(t) \equiv \sum_{n=0}^{\infty} \delta(t - nT)$$

the input-output relationship of the sampler becomes

$$F^*(t) = F(t) \cdot \delta_T(t)$$

$$= \sum_{n=0}^{\infty} F(nT) \cdot \delta(t - nT).$$

While for a given $F(t)$ and T the $F^*(t)$ is unique, the converse is not true.

The Laplace transform can be used to define $F^*(s)$ as follows

$$L\{F^*(t)\} \equiv f^*(s)$$

$$= \sum_{n=0}^{\infty} F(nT) \cdot e^{-nTs}.$$

The variable 'z' is introduced by means of the transformation

$$z = e^{Ts}$$

and since any function of s can now be replaced by a corresponding function of z we have

$$f(z) = \sum_{n=0}^{\infty} F(nT) \cdot z^{-n}$$

where

$$f^*(s) \equiv f(z)$$

and

$$s = \frac{1}{T} \ln z$$

The Z operator can now be defined in terms of the Laplace operator by the relationship

$$Z\{F(t)\} \equiv L\{F^*(t)\}$$

THE Z TRANSFORM (Continued)

An alternative definition (quoted without proof) is

$$Z\{F(t)\} = \sum \text{ residues of } \left[\left(\frac{1}{1 - e^{Tz}z^{-1}} \right) \cdot f(z) \right]$$

The inverse z transform

$$Z^{-1}\{f(z)\} \equiv F^*(t)$$

$$= \frac{1}{2\pi j} \oint f(z) \cdot z^{n-1} \, dz$$

where the contour of integration encloses all the singularities of the integrand. In the following table Greek letters denote constants.

$F(t)$	$f(z) = Z\{F(t)\}$
$\alpha F(t)$	$\alpha f(z)$
$F(t) + G(t)$	$f(z) + g(z)$
$F(t + T)$	$z f(z) - z F(0)$
$F(t + 2T)$	$z^2 f(z) - z^2 F(0) - z F(T)$
$F(t + mT)$	$z^m f(z) - \displaystyle\sum_{r=0}^{m-1} z^{m-r} F(rT)$ $= z^m f(z)$ when $F(rT) = 0, 0 \le r \le m - 1$
$F(t - mT)$	$z^{-m} f(z)$
$e^{\alpha t} F(t)$	$f(e^{-\alpha T}z)$
$e^{-\alpha t} F(t)$	$f(e^{\alpha T}z)$
$t \cdot F(t)$	$-Tz \dfrac{d}{dz} f(z)$
$t^{-1} F(t)$	$-\dfrac{1}{T} \displaystyle\int_0^z \dfrac{f(z)}{z} \, dz$
$\displaystyle\sum_{m=0}^{t/t} F(mT)$	$\left(\dfrac{z}{z - 1} \right) f(z)$

The following limits are also valid

$$\lim_{t \to 0} F(t) = \lim_{z \to \infty} f(z)$$

$$\lim_{t \to \infty} F^*(t) = \lim_{z \to 1} \left[\left(\frac{z - 1}{z} \right) f(z) \right]$$

In the table which follows, the Heavyside unit step function is defined by

$$H(t - nT) \equiv \begin{cases} 1; t \ge nT \\ 0; t < nT. \end{cases}$$

THE Z TRANSFORM (Continued)

$F(t)$	$f(z)$
$\delta(t)$	1
$\delta(t - mT)$	$\dfrac{1}{z^m}$
$H(t)$	$\dfrac{z}{z - 1}$
$H(t - T)$	$\dfrac{1}{z - 1}$
$H(t - mT)$	$\dfrac{z}{z^m \cdot (z - 1)}$
$H(t) - H(t - T)$	1
$H(t) - H(t - 2T)$	$1 + \dfrac{1}{z}$
$H(t - mT) - H(t - \overline{m + 1}T)$	$\dfrac{1}{z}m$
$\dfrac{T}{t} H(t - T)$	$\ln\left(\dfrac{z}{z - 1}\right)$
t	$\dfrac{Tz}{(z - 1)^2}$
t^2	$\dfrac{T^2 z(z + 1)}{(z - 1)^3}$
t^3	$\dfrac{T^3 z(z^2 + 4z + 1)}{(z - 1)^4}$
t^n	$(-1)^n \lim\limits_{\chi \to 0} \dfrac{\partial^n}{\partial \chi^n}\left(\dfrac{z}{z - e^{-\chi T}}\right)$
$1 - a^{\omega t}$	$\dfrac{z(1 - a^{\omega T})}{(z - 1)(z - a^{\omega T})}$
$a^{\omega t}$	$\dfrac{z}{(z - a^{\omega T})}$
$t a^{\omega t}$	$\dfrac{T z a^{\omega T}}{(z - a^{\omega T})^2}$
$t^2 a^{\omega t}$	$\dfrac{T^2 a^{\omega T} z(z + a^{\omega T})}{(z - a^{\omega T})^3}$
$\sin \omega t$	$\dfrac{z \sin \omega T}{z^2 - 2z \cos \omega T + 1}$
$\cos \omega t$	$\dfrac{z(z - \cos \omega T)}{z^2 - 2z \cos \omega T + 1}$
$\sinh \omega t$	$\dfrac{z \sinh \omega T}{z^2 - 2z \cosh \omega T + 1}$
$\cosh \omega t$	$\dfrac{z(z - \cosh \omega T)}{z^2 - 2z \cosh \omega T + 1}$
$e^{-\alpha t} \sin \omega t$	$\dfrac{z e^{-\alpha T} \sin \omega T}{z^2 - 2z e^{-T} \cos \omega T + e^{-2\alpha T}}$
$e^{-\alpha t} \cos \omega t$	$\dfrac{z(z - e^{-\alpha T} \cos \omega T)}{z^2 - 2z e^{-T} \cos \omega T + e^{-2\alpha T}}$
$e^{-\alpha t} \sinh \omega t$	$\dfrac{z e^{-\alpha T} \sinh \omega T}{z^2 - 2z e^{-\alpha T} \cosh \omega T + e^{-2\alpha T}}$
$e^{-\alpha t} \cosh \omega t$	$\dfrac{z(z - e^{-\alpha T} \cosh \omega T)}{z^2 - 2z e^{-\alpha T} \cosh \omega T + e^{-2\alpha T}}$

THE Z TRANSFORM (Continued)

$F(t)$	$f(z)$
$-\dfrac{1}{a}\left[\delta(t) - a^{t/T}\right]$	$\dfrac{1}{z - a}$
$\dfrac{1}{(a - b)}\left[a^{\left(\frac{t}{T}-1\right)} - b^{\left(\frac{t}{T}-1\right)}\right]$	$\dfrac{1}{(z - a)(z - b)}$
$\dfrac{1}{(a - b)}\left[a^{\frac{t}{T}} - b^{\frac{t}{T}}\right]$	$\dfrac{z}{(z - a)(z - b)}$
$\dfrac{1}{(a - b)}\left[(a - c)a^{\left(\frac{t}{T}-1\right)} - (b - c)b^{\left(\frac{t}{T}-1\right)}\right]$	$\dfrac{z - c}{(z - a)(z - b)}$
$\dfrac{1}{(a - b)}\left[a^{\left(\frac{t}{T}+1\right)} - b^{\left(\frac{T}{t}+1\right)}\right]$	$\dfrac{z^2}{(z - a)(z - b)}$
$\dfrac{1}{\left(\frac{t}{T}\right)!}$	$e^{1/z}$
$\dfrac{1}{\left(\frac{2t}{T}\right)!}$	$\cosh\left(z^{-\frac{1}{2}}\right)$

Methods of evaluating inverse z transforms.

(1) Cauchy's residue theorem.

For $t = nT$,

$$G(nT) = \sum_{\text{all } z_k} \left[\text{residues of } g(z)z^{n-1} \text{ at } z_k\right]$$

where the z_k define all the poles of $g(z)z^{n-1}$.

(2) Partial fractions.

Expand $g(z)/z$ into partial fractions. The product of z with each of the partial fractions will then be recognizable from the standard forms in the table of z transforms. Note however that the continuous functions obtained are only valid at the sampling instants.

(3) Power series expansion by long division using detached coefficients.

$g(z)$ is expanded into a power series in z^{-1} and the coefficient of the term in z^{-n} is the value of $g(nT)$ i.e. the value of $G(t)$ at the nth sampling instant.

The z transform as a means of determining approximately the inverse Laplace transform.

Since
$$z \equiv e^{Ts}$$

$$s^{-1} = \frac{T}{2}\left[\frac{1}{v} - \frac{v}{3} - \frac{4v^3}{45} - \frac{44v^5}{945} - \cdots \right]$$

where

$$v \equiv \frac{1 - z^{-1}}{1 + z^{-1}},$$

the series being very rapid in its convergence. Given $g(s)$, to find its inverse Laplace transform the following operations are carried out:-

(i) Divide the numerator and denominator of $g(s)$ by the highest power of s yielding as an alternative form for $g(s)$ the quotient of two polynomials in s^{-1}.

(ii) Chose as a numerical value of T, that which makes $2\pi/T$ much larger than the imaginary part of the poles of $G(s)$.

(iii) Substitute into the alternative form for $g(s)$ obtained in (i) above the expansion for s^{-n} determined from the following short table of approximations.

THE Z TRANSFORM (Continued)

Do not at this stage insert the numerical value for T as tabulations with different intervals may be required.

(iv) Divide by T.

(v) Insert the chosen value for T and divide the numerator by the denominator.

(vi) The coefficient of z^{-n} is the required value of the function at $t = nT$.

s^{-n}	z transform (approximate)
s^{-1}	$\dfrac{T}{2}\left[\dfrac{1 + z^{-1}}{1 - z^{-1}}\right]$
s^{-2}	$\dfrac{T^2}{12}\left[\dfrac{1 + 10z^{-1} + z^{-2}}{(1 - z^{-1})^2}\right]$
s^{-3}	$\dfrac{T^3}{3}\left[\dfrac{z^{-1} + z^{-2}}{(1 - z^{-1})^3}\right]$
s^{-4}	$\dfrac{T^4}{144}\left[\dfrac{1 + 20z^{-1} + 102z^{-2} + 20z^{-3} + z^{-4}}{(1 - z^{-1})^4}\right]$
s^{-5}	$\dfrac{T^5}{24}\left[\dfrac{z^{-1} + 11z^{-2} + 11z^{-3} + z^{-4}}{(1 - z^{-1})^5}\right]$
s^{-6}	$\dfrac{T^6}{4}\left[\dfrac{z^{-2} + 2z^{-3} + z^{-4}}{(1 - z^{-1})^6}\right]$
s^{-7}	$\dfrac{T^7}{8}\left[\dfrac{z^{-2} + 3z^{-3} + 3z^{-4} + z^{-5}}{(1 - z^{-1})^7}\right]$

Additional material on Z-transforms can be found in the papers by Boxer, R. and Thaler, S., A simplified method of solving linear and nonlinear systems. *Proc. IEE*, 1956, 89–101, and Boxer, R., A note on numerical transform calculus, *Proc. IEE*, 1957, 1401–1406.

FINITE DIFFERENCES

Uniform interval h.

If a function $f(x)$ is tabulated at a uniform interval h, that is, for arguments given by $x_n = x_0 + nh$, where n is an integer, then the function $f(x)$ may be denoted by f_n. This can be generalized so that for all values of p, and in particular for $0 \leqq p \leqq 1$,

$$f(x_0 + ph) = f(x_p) = f_p ,$$

where the argument designated x_0 can be chosen quite arbitrarily.

The notation $x = a(h)b(H)C$ means that the arguments run from a to b inclusive at the uniform interval h and thereafter continue to C at the uniform interval H.

The statement $f(x)$ $4D$ means that the function $f(x)$ is to be tabulated and the answer rounded off to four places of decimals.

The statement $f(x)$ $5S$ means that the function $f(x)$ is to be tabulated with the answer being rounded off to five significant figures.

The following table lists and defines the standard operators used in numerical analysis.

Symbol	Function	Definition
E	Displacement	$Ef_p = f_{p+1}$
Δ	Forward difference	$\Delta f_p = f_{p+1} - f_p$
∇	Backward difference	$\nabla f_p = f_p - f_{p-1}$
$\wedge$	Divided difference	
δ	Central difference	$\delta f_p = f_{p+\frac{1}{2}} - f_{p-\frac{1}{2}}$
μ	Average	$\mu f_p = \frac{1}{2}(f_{p+\frac{1}{2}} + f_{p-\frac{1}{2}})$
Δ^{-1}	Backward sum	$\Delta^{-1} f_p = \Delta^{-1} f_{p-1} + f_{p-1}$
∇^{-1}	Forward sum	$\nabla^{-1} f_p = \nabla^{-1} f_{p-1} + f_p$
δ^{-1}	Central sum	$\delta^{-1} f_p = \delta^{-1} f_{p-1} + f_{p-\frac{1}{2}}$
D	Differentiation	$Df_p = \frac{d}{dx} f(x) = \frac{1}{h} \cdot \frac{d}{dp} f_p$
$I(= D^{-1})$	Integration	$If_p = \int^x f(x)dx = h \int^p f_p dp$
$J(= \Delta D^{-1})$	Definite integration	$Jf_p = h \int_p^{p+1} f_p dp$

I, Δ^{-1}, ∇^{-1} and δ^{-1} all imply the existence of an arbitrary constant which is determined by the initial conditions of the problem.

Where no confusion can arise the f can be omitted as, for example in writing Δ_p for Δf_p. Higher differences are formed by successive operations, e.g.,

$$
\begin{aligned}
\Delta^2 f_p &= \Delta_p^2 \\
&= \Delta \cdot \Delta_p \\
&= \Delta(f_{p+1} - f_p) \\
&= \Delta_{p+1} - \Delta_p \\
&= f_{p+2} - f_{p+1} - f_{p+1} + f_p .
\end{aligned}
$$

Thus $\Delta_p^2 = f_{p+2} - 2f_{p+1} + f_p$.

Note that $f_p \equiv \Delta_p^0 \equiv \nabla_p^0 \equiv \delta_p^0$.

The disposition of the differences and sums relative to the function values is as shown (the arguments are omitted in these cases in the interests of clarity). In manuscript working double spacing is recommended.

Forward difference scheme

Δ_{-1}^{-2} f_{-2} Δ_{-3}^{2}
   Δ_{-1}^{-1} Δ_{-2} Δ_{-3}^{3}
Δ_{0}^{-2} f_{-1} Δ_{-2}^{2}
   Δ_{0}^{-1} Δ_{-1} Δ_{-2}^{3}
Δ_{1}^{-2} f_{0} Δ_{-1}^{2}
   Δ_{1}^{-1} Δ_{0} Δ_{-1}^{3}
Δ_{2}^{-2} f_{1} Δ_{0}^{2}
   Δ_{2}^{-1} Δ_{1} Δ_{0}^{3}
Δ_{3}^{-2} f_{2} Δ_{1}^{2}

Backward difference scheme

∇_{-3}^{-2} f_{-2} ∇_{-1}^{2}
   ∇_{-2}^{-1} ∇_{-1} ∇_{0}^{3}
∇_{-2}^{-2} f_{-1} ∇_{0}^{2}
   ∇_{-1}^{-1} ∇_{0} ∇_{1}^{3}
∇_{-1}^{-2} f_{0} ∇_{1}^{2}
   ∇_{0}^{-1} ∇_{1} ∇_{2}^{3}
∇_{0}^{-2} f_{1} ∇_{2}^{2}
   ∇_{1}^{-1} ∇_{2} ∇_{3}^{3}
∇_{1}^{-2} f_{2} ∇_{3}^{2}

Central difference scheme

δ_{-2}^{-2} f_{-2} δ_{-2}^{2} δ_{-2}^{4}
   $\delta_{-1\frac{1}{2}}^{-1}$ $\delta_{-1\frac{1}{2}}$ $\delta_{-1\frac{1}{2}}^{3}$
δ_{-1}^{-2} f_{-1} δ_{-1}^{2} δ_{-1}^{4}
   $\delta_{-\frac{1}{2}}^{-1}$ $\delta_{-\frac{1}{2}}$ $\delta_{-\frac{1}{2}}^{3}$
δ_{0}^{-2} f_{0} δ_{0}^{2} δ_{0}^{4}
   $\delta_{\frac{1}{2}}^{-1}$ $\delta_{\frac{1}{2}}$ $\delta_{\frac{1}{2}}^{3}$
δ_{1}^{-2} f_{1} δ_{1}^{2} δ_{1}^{4}
   $\delta_{1\frac{1}{2}}^{-1}$ $\delta_{1\frac{1}{2}}$ $\delta_{1\frac{1}{2}}^{3}$
δ_{2}^{-2} f_{2} δ_{2}^{2} δ_{2}^{4}

In the forward difference scheme the subscripts are seen to move forward into the difference table and no fractional subscripts occur. In the backward difference scheme the subscripts lie on diagonals slanting backwards into the table while in the central difference scheme the subscripts maintain their position and the odd order subscripts are fractional.

All three however are merely alternative ways of labeling the same numerical quantities as any difference is the result of subtracting the number diagonally above it in the preceding column from that diagonally below it in the preceding column or, alternatively, it is the sum of the number diagonally above it in the subsequent column with that immediately above it in its own column.

In general $\Delta_{p-\frac{1}{2}n}^{n} \equiv \delta_{p}^{n} \equiv \nabla_{p+\frac{1}{2}n}^{n}$.

If a polynomial of degree r is tabulated exactly i.e., without any round-off errors, then the rth differences are constant.

Tabulate $f(x)3D$ for $x = -.4$ (.2) .8 where $f(x) = x^3 - .3x^2 + .07x - .123$.

x	f	δ	δ^2	δ^3	δ^4
$-.4$	$-.263$				
		106			
$-.2$	$-.157$		-72		
		34		48	
$+0$	$-.123$		-24		0
		10		48	
$.2$	$-.113$		$+24$		0
		34		48	
$.4$	$-.079$		72		0
		106		48	
$.6$	$+.027$		120		
		226			
$.8$	$.253$				

Calculus of Finite Differences

Had the function been rounded off to two places of decimals then the third differences would not have been constant. The differences are conventionally expressed in units of the least significant decimal as shown. This obviates the need to write down non-significant zeros. In any well tabulated function the differences decrease in magnitude.

The effect of an error term e in any function value is superimposed upon the difference table in a manner shown in the following table:

f	δ	δ^2	δ^3	δ^4	δ^5
0		0		0	
	0		0		$+1e$
0		0		$+1e$	
	0		$+1e$		$-5e$
0		$+1e$		$-4e$	
	$+1e$		$-3e$		$+10e$
$+e$		$-2e$		$+6e$	
	$-1e$		$+3e$		$-10e$
0		$+1e$		$-4e$	
	0		$-1e$		$+5e$
0		0		$+1e$	
	0		0		$-1e$
0		0		0	

In particular since round-off can contribute an error of half a unit (of the least significant decimal) in the function values, reference to the table of binomial coefficients will show that the maximum effect of this upon the twelfth differences (say) is $924(\frac{1}{2}) = 462$. This could give trouble if the problem does not require this to be multiplied by a small constant and is in fact one of the reasons why numerical differentiation is unreliable.

The following table enables the simpler operators to be expressed in terms of the others:

	E	Δ	δ,μ	∇
E	—	$1 + \Delta$	$1 + \mu\delta + \frac{1}{2}\delta^2$	$(1 - \nabla)^{-1}$
Δ	$E - 1$	—	$\mu\delta + \frac{1}{2}\delta^2$	$\nabla(1 - \nabla)^{-1}$
δ	$E^{\frac{1}{2}} - E^{-\frac{1}{2}}$	$\Delta(1 + \Delta)^{-\frac{1}{2}}$	$2(\mu^2 - 1)^{\frac{1}{2}}$	$\nabla(1 - \nabla)^{-\frac{1}{2}}$
∇	$-E^{-1}$	$\Delta(1 + \Delta)^{-1}$	$\mu\delta - \frac{1}{2}\delta^2$	—
μ	$\frac{1}{2}(E^{\frac{1}{2}} + E^{-\frac{1}{2}})$	$\frac{1}{2}(2 + \Delta)(1 + \Delta)^{-\frac{1}{2}}$	$(1 + \frac{1}{4}\delta^2)^{\frac{1}{2}}$	$\frac{1}{2}(2 - \nabla)(1 - \nabla)^{-\frac{1}{2}}$

In addition to the above there are other identities by means of which the above table can be extended, viz.,

$$E = e^{hD} = \Delta\nabla^{-1}$$

$$\mu = E^{-\frac{1}{2}} + \tfrac{1}{2}\delta = E^{\frac{1}{2}} - \tfrac{1}{2}\delta = \cosh\left(\tfrac{1}{2}hD\right)$$

$$\delta = E^{-\frac{1}{2}}\Delta = E^{\frac{1}{2}}\nabla = (\Delta\nabla)^{\frac{1}{2}} = 2\sinh\left(\tfrac{1}{2}hD\right).$$

Note the emergence of Taylor's series from

$$f_p = E^p f_0$$

$$= e^{phD} f_0$$

$$= f_0 + phDf_0 + \frac{1}{2!}p^2 h^2 D^2 f_0 + \cdots.$$

Calculus of Finite Differences

FUNCTION BUILD-UP FROM DIFFERENCES

Backward differences will be used here since the notation is simpler in this case.

If a function is tabulated as far as f_0 say, then the difference table can be built up. Assuming the differences up to and including ∇_0'' have been formed, if ∇_1'' is known it follows from the definition of backward differences that f_1 can be found from

$$f_1 = f_0 + \nabla_0 + \nabla_0^2 + \nabla_0^3 + \nabla_0^4 + \cdots + \nabla_0^{n-1} + \nabla_1^n .$$

For example, in the cubic tabulated earlier if we take $x_0 = .8$ then $f_0 = .253$ and f_1 can be built-up using the above scheme as follows:

$$f_1 = .253 + (226 + 120 + 48)(10^{-3})$$
$$= .647$$

INTERPOLATION

Finite difference interpolation entails taking a given set of points and fitting a function to them. This function is usualy a polynomial, although it is rarely found explicitly as such. Central differences formulae are in general to be preferred to those involving either forward or backward differences, although the latter may be essential at the extremities of a table where the central differences may not extend far enough.

The most useful of the central differences formulae are those of *Bessel* and *Everett*, the latter particularly so since it only utilizes even order differences. Nevertheless all central difference formulae stopping at the same difference are equivalent.

Notation used:

$$\Delta_{-m_r}^r \equiv \delta_{(r/2)-m_r}^r \equiv \nabla_{r-m_r}^r .$$
$$p = \frac{x - x_0}{h} = 1 - q \quad (\text{usually } 0 \leqq p \leqq 1) .$$

Any number of finite difference interpolation formulae can be obtained from the following schematic (shown for central differences only, but easily extensible by the above identities):

$$y_p = \sum_{r=0} \binom{p + m_{r-1}}{r} \cdot \delta_{(r/2)-m_r}^r$$
$$= f_{-m_0} + \binom{p + m_0}{1} \delta_{\frac{1}{2}-m_1} + \binom{p + m_1}{2} \delta_{1-m_2}^2 + \binom{p + m_2}{3} \delta_{1\frac{1}{2}-m_3}^3 \cdots ,$$

where the m_r are integers so chosen that the binomial coefficient term $\binom{p + m_{r-1}}{r}$ contains all the linear factors occurring in the preceding term $\binom{p + m_{r-2}}{r - 1}$. Obviously $m_{r+1} = m_r$ or $1 + m_r$.

Further formulae are obtained by taking linear combinations of the formulae obtained from the above schematic, or by using the standard identity transformations upon chosen terms of one of the formulae.

Another way of obtaining interpolation formulae is to apply the standard identity transformations to the operator E in the operational form of the interpolation formula, viz., $f_p = E^p f_0$.

Newton's forward formula

$$f_p = f_0 + p\Delta_0 + \frac{1}{2!} p(p - 1)\Delta_0^2 + \frac{1}{3!} p(p - 1)(p - 2)\Delta_0^3 \cdots \quad 0 \leqq p \leqq 1$$

Newton's backward formula

$$f_p = f_0 + p\nabla_0 + \frac{1}{2!}p(p+1)\nabla_0^2 + \frac{1}{3!}p(p+1)(p+2)\nabla_0^3 \cdots \qquad 0 \leq p \leq 1$$

Gauss' forward formula

$$f_p = f_0 + p\delta_{\frac{1}{2}} + G_2\delta_0^2 + G_3\delta_{\frac{1}{2}}^3 + G_4\delta_0^4 + G_5\delta_{\frac{1}{2}}^5 \cdots \qquad 0 \leq p \leq 1$$

Gauss' backward formula

$$f_p = f_0 + p\delta_{-\frac{1}{2}} + G_2^*\delta_0^2 + G_3\delta_{-\frac{1}{2}}^3 + G_4^*\delta_0^4 + G_5\delta_{-\frac{1}{2}}^5 \cdots \qquad 0 \leq p \leq 1$$

In the above $G_{2n} = \begin{pmatrix} p+n-1 \\ 2n \end{pmatrix}$

$$G_{2n}^* = \begin{pmatrix} p+n \\ 2n \end{pmatrix}$$

$$G_{2n+1} = \begin{pmatrix} p+n \\ 2n+1 \end{pmatrix}$$

Stirling's formula

$$f_p = f_0 + \frac{1}{2}p(\delta_{\frac{1}{2}} + \delta_{-\frac{1}{2}}) + \frac{1}{2}p^2\delta_0^2 + S_3(\delta_{\frac{1}{2}}^3 + \delta_{-\frac{1}{2}}^3) + S_4\delta_0^4 + \cdots \qquad -\frac{1}{2} \leq p \leq \frac{1}{2}$$

Steffenson's formula

$$f_p = f_0 + \frac{1}{2}p(p+1)\delta_{\frac{1}{2}} - \frac{1}{2}(p-1)p\delta_{-\frac{1}{2}} + (S_3+S_4)\delta_{\frac{1}{2}}^3 + (S_3-S_4)\delta_{-\frac{1}{2}}^3 \cdots -\frac{1}{2} \leq p \leq \frac{1}{2}.$$

In the above $S_{2n+1} = \frac{1}{2}\begin{pmatrix} p+n \\ 2n+1 \end{pmatrix}$

$$S_{2n+2} = \frac{p}{2n+2}\begin{pmatrix} p+n \\ 2n+1 \end{pmatrix}$$

$$S_{2n+1} + S_{2n+2} = \begin{pmatrix} p+n+1 \\ 2n+2 \end{pmatrix}$$

$$S_{2n+1} - S_{2n+2} = -\begin{pmatrix} p+n \\ 2n+2 \end{pmatrix}$$

Bessel's formula

$$f_p = f_0 + p\delta_{\frac{1}{2}} + B_2(\delta_0^2 + \delta_1^2) + B_3\delta_{\frac{1}{2}}^3 + B_4(\delta_0^4 + \delta_1^4) + B_5\delta_{\frac{1}{2}}^5 + \cdots \qquad 0 \leq p \leq 1$$

Everett's formula

$$f_p = (1-p)f_0 + pf_1 + E_2\delta_0^2 + F_2\delta_1^2 + E_4\delta_0^4 + F_4\delta_1^4 + E_6\delta_0^6 + F_6\delta_1^6 + \cdots \qquad 0 \leq p \leq 1$$

The coefficients in the above two formulae are related to each other and to the coefficients in the Gaussian formulae by the identities

$$B_{2n} \equiv \frac{1}{2}G_{2n} \equiv \frac{1}{2}(E_{2n} + F_{2n})$$
$$B_{2n+1} \equiv G_{2n+1} - \frac{1}{2}G_{2n} \equiv \frac{1}{2}(F_{2n} - E_{2n})$$
$$E_{2n} \equiv G_{2n} - G_{2n+1} \equiv B_{2n} - B_{2n+1}$$
$$F_{2n} \equiv G_{2n+1} \equiv B_{2n} + B_{2n+1}$$

Also for $q \equiv 1 - p$ the following symmetrical relationships hold:

$$B_{2n}(p) \equiv B_{2n}(q)$$
$$B_{2n+1}(p) \equiv -B_{2n+1}(q)$$
$$E_{2n}(p) \equiv F_{2n}(q)$$
$$F_{2n}(p) \equiv E_{2n}(q)$$

as can be seen from the tables of these coefficients.

p	B_2	B_3	B_4	B_5	B_6	B_7	$(1-p)$
0.00	−0.000000	0.0000000	0.000000	−0.000000	−0.00000	0.00000	1.00
0.01	2475	8085	415	81	8	1	0.99
0.02	4900	15680	825	158	17	2	0.98
0.03	7275	22795	1230	231	25	3	0.97
0.04	9600	29440	1631	300	33	4	0.96
0.05	−0.011875	0.0035625	0.002026	−0.000365	−0.00041	0.00005	0.95
0.06	14100	41360	2416	425	49	6	0.94
0.07	16275	46655	2801	482	57	7	0.93
0.08	18400	51520	3180	534	64	8	0.92
0.09	20475	55965	3552	583	72	8	0.91
0.10	−0.022500	0.0060000	0.003919	−0.000627	−0.00080	0.00009	0.90
0.11	24475	63635	4279	667	87	10	0.89
0.12	26400	66880	4632	704	94	10	0.88
0.13	28275	69745	4979	737	101	11	0.87
0.14	30100	72240	5319	766	109	11	0.86
0.15	−0.031875	0.0074375	0.005651	−0.000791	−0.00115	0.00012	0.85
0.16	33600	76160	5976	813	122	12	0.84
0.17	35275	77605	6294	831	129	12	0.83
0.18	36900	78720	6604	845	135	12	0.82
0.19	38475	79515	6906	856	142	13	0.81
0.20	−0.040000	0.0080000	0.007200	−0.000864	−0.00148	0.00013	0.80
0.21	41475	80185	7486	868	154	13	0.79
0.22	42900	80080	7763	870	160	13	0.78
0.23	44275	79695	8033	868	165	13	0.77
0.24	45600	79040	8293	862	171	13	0.76
0.25	−0.046875	0.0078125	0.008545	−0.000554	−0.00176	0.00013	0.75
0.26	48100	76960	8788	844	181	12	0.74
0.27	49275	75555	9022	830	186	12	0.73
0.28	50400	73920	9247	814	191	12	0.72
0.29	51475	72065	9462	795	196	12	0.71
0.30	−0.052500	0.0070000	0.009669	−0.000773	−0.00200	0.00011	0.70
0.31	53475	67735	9866	750	204	11	0.69
0.32	54400	65280	10053	724	208	11	0.68
0.33	55275	62645	10231	696	212	10	0.67
0.34	56100	59840	10399	666	216	10	0.66
0.35	−0.056875	0.0056875	0.010557	−0.000633	−0.00219	0.00009	0.65
0.36	57600	53760	10706	600	222	9	0.64
0.37	58275	50505	10844	564	225	8	0.63
0.38	58900	47120	10973	527	228	8	0.62
0.39	59475	43615	11092	488	231	7	0.61
0.40	−0.060000	0.0040000	0.011200	−0.000448	−0.00233	0.00007	0.60
0.41	60475	36285	11298	407	235	6	0.59
0.42	60900	32480	11386	364	237	5	0.58
0.43	61275	28595	11464	321	239	5	0.57
0.44	61600	24640	11532	277	240	4	0.56
0.45	−0.061875	0.0020625	0.011589	−0.000232	−0.00241	0.00003	0.55
0.46	62100	16560	11635	186	242	3	0.54
0.47	62275	12455	11672	140	243	2	0.53
0.48	62400	8320	11698	94	244	1	0.52
0.49	62475	4165	11714	47	244	1	0.51
0.50	−0.062500	0.0000000	0.011719	−0.000000	−0.00244	0.00000	0.50
$(1-p)$	B_2	$-B_3$	B_4	$-B_5$	B_6	$-B_7$	p

Bessel's formula (unmodified)

(a) $f_p = f_0 + p\delta_{\frac{1}{2}} + B_2(\delta_0^2 + \delta_1^2) + B_3\delta_{\frac{1}{2}}^3 + B_4(\delta_0^4 + \delta_1^4)$
$$+ B_5\delta_{\frac{1}{2}}^5 + B_6(\delta_0^6 + \delta_1^6) + B_7\delta_{\frac{1}{2}}^7 + \cdots.$$

The end terms can be neglected, i.e., they contribute less than half a unit in the last place of decimals, if the differences are less than the values shown in the following list:

$$\begin{array}{rll}
\text{Neglect} & B_7\delta_{\frac{1}{2}}^7 \text{ if } \delta^7 < 3500 \\
& B_6(\delta_0 + \delta_1) & \delta^6 < 100 \\
& B_5\delta_{\frac{1}{2}}^5 & \delta^5 < 500 \\
& B_4(\delta_0^4 + \delta_1^4) & \delta^4 < 20 \\
& B_3\delta_{\frac{1}{2}}^3 & \delta^3 < 60 \\
& B_2(\delta_0^2 + \delta_1^2) & \delta^2 < 4
\end{array}$$

A powerful simplification is that of throwback. If the higher differences are less than a specified value they can be combined with the lower order ones as follows:

If $\delta^6 < 10,000$ use $\delta_m^4 \equiv \delta^4 - 0.20697\delta^6$

and if $\delta^6 < 1,000^*$ use $\delta_m^4 \equiv \delta^4 - 0.2\delta^6$
in the modified form of Bessel's formula

(b) $f_p = f_0 + p\delta_{\frac{1}{2}} + B_2(\delta_0^2 + \delta_1^2) + B_3\delta_{\frac{1}{2}}^3 + B_4(\delta_{m0}^4 + \delta_{m1}^4) + B_5\delta_{\frac{1}{2}}^5$.

If $\delta^4 < 1,000$ and $\delta^6 < 1,000^*$ use $\delta_m^2 \equiv \delta^2 - 0.18393\delta^4 + 0.03808\delta^6$
and if $\delta^4 < 1,000^*$ use $\delta_m^2 \equiv \delta^2 - 0.18393\delta^4$
in the modified form of Bessel's formula

(c) $f_p = f_0 + p\delta_{\frac{1}{2}} + B_2(\delta_{m0}^2 + \delta_{m1}^2) + B_3\delta_{\frac{1}{2}}^3$.

---

* Higher even order differences are negligible. Throwback involving odd order differences is of little practical use.

p	E_2	F_2	E_4	F_4	E_6	F_6	$(1-p)$
0.00	-0.000000	-0.0000000	0.000000	0.000000	-0.00000	-0.00000	1.00
0.01	32835	16665	496	33	9	7	0.99
0.02	64680	33320	983	67	19	14	0.98
0.03	95545	49955	1461	100	28	21	0.97
0.04	125440	66560	1931	133	37	29	0.96
0.05	-0.0154375	-0.0083125	0.002391	0.001661	-0.00046	-0.00036	0.95
0.06	182360	99640	2842	199	55	43	0.94
0.07	209405	116095	3283	232	64	50	0.93
0.08	235520	132480	3714	264	72	57	0.92
0.09	260715	148785	4135	297	80	64	0.91
0.10	-0.0285000	-0.0165000	0.004546	0.003292	-0.00089	-0.00070	0.90
0.11	308385	181115	4946	361	97	77	0.89
0.12	330880	197120	5336	393	105	84	0.88
0.13	352495	213005	5716	424	112	91	0.87
0.14	373240	228760	6085	455	120	97	0.86
0.15	-0.0393125	-0.0244375	0.006442	0.004860	-0.00127	-0.00104	0.85
0.16	412160	259840	6789	516	134	110	0.84
0.17	430355	275145	7125	546	141	117	0.83
0.18	447720	290280	7449	576	148	123	0.82
0.19	464265	305235	7762	605	154	129	0.81
0.20	-0.0480000	-0.0320000	0.008064	0.006336	-0.00161	-0.00135	0.80
0.21	494935	334565	8354	662	167	141	0.79
0.22	509080	348920	8633	689	172	147	0.78
0.23	522445	363055	8900	716	178	153	0.77
0.24	535040	376960	9156	743	184	158	0.76
0.25	-0.0546875	-0.0390625	0.009399	0.007690	-0.00189	-0.00164	0.75
0.26	557960	404040	9632	794	194	169	0.74
0.27	568305	417195	9852	819	199	174	0.73
0.28	577920	430080	10060	843	203	179	0.72
0.29	586815	442685	10257	867	207	184	0.71
0.30	-0.0595000	-0.0455000	0.010442	0.008895	-0.00212	-0.00189	0.70
0.31	602485	467015	10615	912	215	193	0.69
0.32	609280	478720	10777	933	219	198	0.68
0.33	615395	490105	10927	954	222	202	0.67
0.34	620840	501160	11065	973	226	206	0.66
0.35	-0.0625625	-0.0511875	0.011191	0.009924	-0.00229	-0.00210	0.65
0.36	629760	522240	11305	1011	231	213	0.64
0.37	633255	532245	11408	1028	234	217	0.63
0.38	636120	541880	11500	1045	236	220	0.62
0.39	638365	551135	11580	1060	238	223	0.61
0.40	-0.0640000	-0.0560000	0.011648	0.010752	$-.00240$	-0.00226	0.60
0.41	641035	568465	11705	1089	241	229	0.59
0.42	641480	576520	11751	1102	242	232	0.58
0.43	641345	584155	11785	1114	243	234	0.57
0.44	640640	591360	11808	1125	244	236	0.56
0.45	-0.0639375	-0.0598125	0.011820	0.011357	-0.00245	-0.00238	0.55
0.46	637560	604440	11822	1145	245	240	0.54
0.47	635205	610295	11812	1153	245	241	0.53
0.48	632320	615680	11792	1160	245	242	0.52
0.49	628915	620585	11760	1167	245	243	0.51
0.50	-0.0625000	-0.0625000	0.011719	0.011719	-0.00244	-0.00244	0.50
$(1-p)$	F_2	E_2	F_4	E_4	F_6	E_6	p

Everett's formula (unmodified)

(a) $f_p = (1 - p)f_0 + pf_1 + E_2\delta_0^2 + F_2\delta_1^2 + E_4\delta_0^4 + F_4\delta_1^4 + E_6\delta_0^6 + F_6\delta_1^6 + \cdots$

As with Bessel's formula, end terms can be neglected as follows:

$$\text{Neglect } E_6\delta_0^6 + F_6\delta_1^6 \quad \text{if} \quad \delta^6 < 100$$
$$E_4\delta_0^4 + F_4\delta_1^4 \qquad \delta^4 < 20$$
$$E_2\delta_0^2 + F_2\delta_1^2 \qquad \delta^2 < 4$$

The corresponding throwback formulae are

(b) $\quad f_p = (1 - p)f_0 + pf_1 + E_2\delta_0^2 + F_2\delta_1^2 + E_4\delta_{m0}^4 + F_4\delta_{m1}^4,$

where $\delta_m^4 \equiv \delta^4 - 0.20697\delta^6 \quad$ providing $\delta^6 < 10{,}000^*$

or $\quad \delta_m^4 \equiv \delta^4 - 0.2\delta^6 \qquad$ providing $\delta^6 < 1{,}000^*$;

(c) $\quad f_p = (1 - p)f_0 + pf_1 + E_2\delta_{m0}^4 + F_2\delta_{m1}^2$

where $\delta_m^2 \equiv \delta^2 - 0.18393\delta^4 + 0.03808\delta^6 \quad$ if $\quad \delta^4, \delta^6 < 1{,}000^*$.

Generalized Throwback

Using 0.184 instead of the more precise value of 0.18393 we obtain a very simple but nevertheless accurate interpolation formulae

$$f \equiv (1 - p)f_0 + pf_1 + E_2\delta_{m0}^2 + F_2\delta_{m1}^2 + P_4\theta_0^4 + Q_4\theta_1^4,$$

where $\quad \delta_m^2 \equiv \delta^2 - 0.184\delta^4 + 0.03882\delta^6 - 0.0083\delta^8 + 0.0019\delta^{10}$

and $\quad 100\theta^4 \equiv \delta^4 - 0.2783\delta^6 + 0.0685\delta^8 - 0.0168\delta^{10}$

and $\quad P_4 \equiv 100(E_4 + 0.184E_2); \qquad Q_4 \equiv 100(F_4 + 0.184F_2).$

Symmetric formulae for interpolation to halves (sub-tabulation to halves)

When $p = \frac{1}{2}$ is substituted into Bessel's interpolation formula the odd coefficients B_{2n+1} all become zero and the resulting formula is

$$f_{\frac{1}{2}} = \left(1 - \frac{1}{8}\delta^2 + \frac{3}{128}\delta^4 - \frac{5}{1024}\delta^6 + \frac{35}{32768}\delta^8 - \cdots\right)\mu f_{\frac{1}{2}}.$$

If this is truncated after one, two, three terms etc. the following formulae and error terms are obtained:

Error

$f_{\frac{1}{2}} = \frac{1}{2}(f_0 + f_1)$ $\qquad\qquad -\dfrac{1}{8}\mu\delta_{\frac{1}{2}}^2$

$f_{\frac{1}{2}} = (-f_{-1} + 9f_0 + 9f_1 - f_2)/16$ $\qquad +\dfrac{3}{128}\mu\delta_{\frac{1}{2}}^4$

$f_{\frac{1}{2}} = (3f_{-2} - 25f_{-1} + 150f_0 + 150f_1 - 25f_2 + 3f_3)/256$ $\qquad -\dfrac{5}{1024}\mu\delta_{\frac{1}{2}}^6$

$f_{\frac{1}{2}} = (-5f_{-3} + 49f_{-2} - 245f_{-1} + 1225f_0 + 1225f_1 - 245f_2 + 49f_3 - 5f_4)/2048$ $\qquad +\dfrac{35}{32768}\mu\delta_{\frac{1}{2}}^8$

---

* Higher order differences are negligible.

Similarly, unsymmetric formulae for subtabulation to halves can be obtained from Newton's formula by substituting $p = \frac{1}{2}$ and truncating, e.g.,

Error

$$f_{\frac{1}{2}} = \left(1 + \frac{1}{2}\Delta - \frac{1}{8}\Delta^2 + \frac{1}{16}\Delta^3 - \frac{5}{128}\Delta^4 + \frac{7}{256}\Delta^5 \cdots\right)f_0$$

$$f_{\frac{1}{2}} = (3f_0 + 6f_1 - f_2)/8 \qquad\qquad\qquad\qquad\qquad\qquad +\frac{1}{16}\Delta_0^3$$

$$f_{\frac{1}{2}} = (5f_0 + 15f_1 - 5f_2 + f_3)/16 \qquad\qquad\qquad\qquad\quad -\frac{5}{128}\Delta_0^4$$

$$f_{\frac{1}{2}} = (35f_0 + 140f_1 - 70f_2 + 28f_3 - 5f_4)/128 \qquad\qquad +\frac{7}{256}\Delta_0^5$$

The formula obtained by considering the first two terms only of the series is identical with the first one derived from Bessel's formula.

Interpolation techniques which do not require the function to be tabulated for equal interval of the argument

a) *Lagrangian Polynomials*

The interpolated value is obtained from a set of points which bridge the required point. The nearer the argument of the required value is to the center of the range of arguments of the function values used the better.

For an odd number of points
$$f(x) = \sum_{r=-n}^{n} L_r f_r \,.$$

For an even number of points
$$f(x) = \sum_{r=-n}^{n+1} L_r f_r \,.$$

The interpolating polynomial is of degree $2n$ in the first case and $2n + 1$ in the second. Further, it must pass through the given points $f_{-n}, f_{1-n}, f_{2-n}, \cdots, f_n$ or f_{1+n} as the case may be.

The coefficient L_r associated with the function value f_r is given by

$$L_r = \frac{(x - x_{-n})(x - x_{1-n}) \cdots (x - x_{r-1})(x - x_{r+1}) \cdots (x - x_{n-1})(x - x_n)}{(x_r - x_{-n})(x_r - x_{1-n}) \cdots (x_r - x_{r-1})(x_r - x_{r+1}) \cdots (x_r - x_{n-1})(x_r - x_n)}$$

for the first case with an additional factor $(x - x_{1+n})/(x_r - x_{1+n})$ in the second.

The error involved is equal to the $(2n + 2)$th. derivative, at some point in the range, multiplied by

$$(x - x_{-n})(x - x_{1-n}) \cdots (x - x_{1-n})/(2n + 2)!$$

in the case of the even number of points, and the $(2n + 1)$th. derivative, again at some point in the range, multiplied by

$$(x - x_{-n})(x - x_{1-n}) \cdots (x - x_n)/(2n + 1)!$$

in the other case.

In both cases the multipliers are functions which oscillate with an amplitude which increases substantially as the argument of the required point departs from the center of the range. When the arguments are spaced at equal intervals h and the point x is nearer to x_0 than to the other points, then $L_r(p)$ can be tabulated for $p = (x - x_0)/h$.

If the degree of the approximating polynomial is known the method is very powerful, otherwise it is best avoided.

b) *Divided differences*

The layout of a divided difference table is similar to that of an ordinary finite difference table.

x_{-1} f_{-1}	Δ_{-1}^2	Δ_{-1}^4
$\Delta_{-\frac{1}{2}}$	$\Delta_{-\frac{1}{2}}$	
x_0 f_0	Δ_0^2	Δ_0^4
$\Delta_{\frac{1}{2}}$	$\Delta_{\frac{1}{2}}^3$	
x_1 f_1	Δ_1^2	Δ_1^4

where the Δ's are defined as follows:

$$\Delta_r^0 \equiv f_r, \qquad \Delta_{r+\frac{1}{2}} \equiv (f_{r+1} - f_r)/(x_{r+1} - x_r),$$

and in general
$$\Delta_r^{2n} \equiv (\Delta_{r+\frac{1}{2}}^{2n-1} - \Delta_{r-\frac{1}{2}}^{2n-1})/(x_{r+n} - x_{r-n})$$

and
$$\Delta_{r+\frac{1}{2}}^{2n+1} \equiv (\Delta_{r+1}^{2n} - \Delta_r^{2n})/(x_{r+1+n} - x_{r-n}).$$

Divided differences can with advantage be replaced by adjusted divided differences.

c) *Adjusted divided differences*

These differences are a modified form of the above and, when the interval of tabulation becomes constant, they reduce to ordinary central differences. Formulae analogous to all the existing formulae involving forward, central and backward differences can be obtained by means of Sheppard's rules which are stated below.

x_{-1}	p_{-1}	f_{-1}		δ_{-1}^2		δ_{-1}^4
			$\delta_{-\frac{1}{2}}$		$\delta_{-\frac{1}{2}}^3$	
x_0	p_0	f_0		δ_0^2		δ_0^4
			$\delta_{\frac{1}{2}}$		$\delta_{\frac{1}{2}}^3$	
x_1	p_1	f_1		δ_1^2		δ_1^4
			$\delta_{1\frac{1}{2}}$		$\delta_{1\frac{1}{2}}^3$	
x_2	p_2	f_2		δ_2^2		δ_2^4

where h is the total range of arguments divided by the number of intervals in that range, the result being rounded to a suitable figure and $x_k \cong p_k \cdot h$ or $(x/h)_k = p_k + w_k$ where the w's are used to shift the origin, if necessary, and to round-off the p's.

Define
$$\delta_r \equiv \frac{1}{(p_{r+\frac{1}{2}} - p_{r-\frac{1}{2}})} (f_{r+\frac{1}{2}} - f_{r-\frac{1}{2}})$$

and
$$\delta_r^2 \equiv \frac{2}{(p_{r+1} - p_{r-1})} (\delta_{r+\frac{1}{2}} - \delta_{r-\frac{1}{2}}).$$

In general
$$\delta_r^{2n+1} \equiv \frac{2n+1}{(p_{r+\frac{1}{2}+n} - p_{r-\frac{1}{2}-n})} (\delta_{r+\frac{1}{2}}^{2n} - \delta_{r-\frac{1}{2}}^{2n}),$$

$$\delta_r^{2n} \equiv \frac{2n}{(p_{r+n} - p_{r-n})} (\delta_{r+\frac{1}{2}}^{2n-1} - \delta_{r+\frac{1}{2}}^{2n-1}).$$

The formula will be in the form
$$f(x) = f_0 + \sum_{r=1}^{N} \left\{ \prod (p - p_j \cdot) \frac{\delta_s^r}{r!} \right\},$$

where the product part of each term of the formula consists of r factors and the p_j's and s are chosen as follows:

a) When $r = 1$, $j = 0$ and $s = +\frac{1}{2}$, or $-\frac{1}{2}$, thereafter:—

b) r is increased by unity and s is either increased or decreased by $\frac{1}{2}$. Thus the p's, involved in the new divided difference of the series, are all those which are necessary for the computation of the divided difference used in the *preceding term* plus one more added either to the beginning or to the end of the set.

c) The product part of the term consists of factors $(p - p_i)$ such that all the p's necessary for the evaluation of the *previous* divided difference are involved. For example, Gauss' forward difference interpolation formula in terms of adjusted divided differences is obtained by following the path shown below

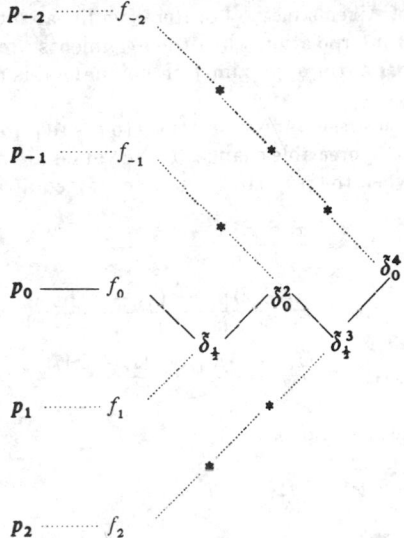

The solid line indicates the path chosen through the difference table.

$$f(x) = f_0 + (p - p_0)\delta_{\frac{1}{2}} + \frac{1}{2!} \cdot (p - p_0)(p - p_1)\delta_0^2 + \frac{1}{3!} \cdot (p - p_0)(p - p_1)(p - p_{-1})\delta_{\frac{1}{2}}^3$$
$$+ \frac{1}{4!} \cdot (p - p_0)(p - p_1)(p - p_{-1})(p - p_2)\delta_0^4 \cdots.$$

Similarly Gauss' backward difference interpolation formula is

$$f(x) = f_0 + (p - p_0)\delta_{-\frac{1}{2}} + \frac{1}{2!} \cdot (p - p_0)(p - p_{-1})\delta_0^2 + \frac{1}{3!} \cdot (p - p_0)(p - p_{-1})(p - p_1)\delta_{-\frac{1}{2}}^3$$
$$+ \frac{1}{4!} \cdot (p - p_0)(p - p_{-1})(p - p_1)(p - p_{-2})\delta_0^4 \cdots,$$

and Stirling's formula is

$$f(x) = f_0 + \frac{(p - p_0)}{2}[\delta_{\frac{1}{2}} + \delta_{-\frac{1}{2}}] + \frac{1}{2!} \cdot \frac{(p - p_0)}{2}[(p - p_{-1}) + (p - p_1)]\delta_0^2$$
$$+ \frac{1}{3!} \cdot \frac{(p - p_1)(p - p_0)(p - p_{-1})}{2}[\delta_{\frac{1}{2}}^3 + \delta_{-\frac{1}{2}}^3]$$
$$+ \frac{1}{4!} \cdot \frac{(p - p_1)(p - p_0)(p - p_{-1})}{2}[(p - p_{-2}) + (p - p_2)]\delta_0^4 \cdots$$

Calculus of Finite Differences

Replacing the odd order differences, in Gauss' forward formula, by combinations of the previous even order differences the formula analogous to Everett's is obtained, i.e.,

$$f(x) = \left[1 - \frac{(p - p_0)}{(p_1 - p_0)}\right] f_0 + \frac{1(p - p_0)}{(p_1 - p_0)} f_1 - \frac{1}{2!} \frac{(p - p_0)(p - p_1)(p - p_2)}{(p_2 - p_{-1})} \delta_0^2$$
$$+ \frac{1}{2!} \frac{(p - p_0)(p - p_1)(p - p_{-1})}{(p_2 - p_{-1})} \delta_1^2 - \cdots$$

d) Iterative linear interpolation

Neville's modification of Aiken's method of iterative linear interpolation is one of the most powerful methods of interpolation when the arguments are unevenly spaced as no prior knowledge of the order of the approximating polynomial is necessary nor is a difference table required.

The values obtained are successive approximations to the required result and the process terminates when there is no appreciable change. These values are of course useless if a new interpolation is required when the procedure must be started afresh.

Defining
$$f_{r,s} \equiv \frac{(x_s - x)f_r - (x_r - x)f_s}{(x_s - x_r)}$$

$$f_{r,s,t} \equiv \frac{(x_t - x)f_{r,s} - (x_r - x)f_{s,t}}{(x_t - x_r)}$$

$$f_{r,s,t,u} \equiv \frac{(x_u - x)f_{r,s,t} - (x_r - x)f_{s,t,u}}{(x_u - x_r)},$$

the computation is laid out as follows:

x_{-1}	$(x_{-1} - x)$	f_{-1}			
			$f_{-1,0}$		
x_0	$(x_0 - x)$	f_0		$f_{-1\,0,1}$	
			$f_{0,1}$		$f_{-1\,0\,1,2}$
x_1	$(x_1 - x)$	f_1		$f_{0,1,2}$	
			$f_{1,2}$		
x_2	$(x_2 - x)$	f_2			

As the iterates tend to their limit the common leading figures can be omitted.

e) Gauss' trigonometric interpolation formula

This is of greatest value when the function is periodic, i.e., a Fourier series expansion is possible.

$$f(x) = \sum_{r=0}^{n} C_r f_r,$$

where $C_r = N_r(x)/N_r(x_r)$ and

$$N_r(x) = \left[\sin \frac{(x - x_0)}{2}\right]\left[\sin \frac{(x - x_1)}{2}\right] \cdots \left[\sin \frac{(x - x_{r-1})}{2}\right]$$
$$\left[\sin \frac{(x - x_{r+1})}{2}\right] \cdots \left[\sin \frac{(x - x_n)}{2}\right].$$

This is similar to the Lagrangian formula.

f) Reciprocal differences

These are used when the quotient of two polynomials will give a better representation of the interpolating function than a simple polynomial expression.

A convenient layout is as shown below:

x_{-1}	f_{-1}			
		$\rho_{-\frac{1}{2}}$		
x_0	f_0		ρ_0^2	
		$\rho_{\frac{1}{2}}$		$\rho_{\frac{1}{2}}^3$
x_1	f_1		ρ_1^2	ρ_1^4
		$\rho_{1\frac{1}{2}}$		$\rho_{1\frac{1}{2}}^3$
x_2	f_2		ρ_2^2	
		$\rho_{2\frac{1}{2}}$		
x_3	f_3			

where

$$\rho_{r+\frac{1}{2}} \equiv \frac{x_{r+1} - x_r}{f_{r+1} - f_r}$$

and

$$\rho_r^2 \equiv \frac{x_{r+1} - x_{r-1}}{f_{r+\frac{1}{2}} - f_{r-\frac{1}{2}}} + f_r$$

In general

$$\rho_{r+\frac{1}{2}}^{2n+1} \equiv \frac{x_{r+n+1} - x_{r-n}}{\rho_{r+1}^{2n} - \rho_r^{2n}} + \rho_{r+\frac{1}{2}}^{2n-1}$$

$$\rho_r^{2n} \equiv \frac{x_{r+n} - x_{r-n}}{\rho_{r+\frac{1}{2}}^{2n-1} - \rho_{r-\frac{1}{2}}^{2n-1}} + \rho_r^{2n-2} \ .$$

The interpolation formula is expressed in the form of a continued fraction expansion.

The expansion corresponding to Newton's forward difference interpolation formula, in the sense of the differences involved, is

$$f(x) = f_0 + \cfrac{(x - x_0)}{\rho_{\frac{1}{2}} + \cfrac{(x_2 - x_1)}{\rho_1 - f_0 + \cfrac{(x - x_2)}{\rho_{1\frac{1}{2}}^3 - \rho_{\frac{1}{2}} + \cfrac{(x_4 - x_3)}{\rho_2^4 - \rho_1^2 + (x - x_4)}}}}$$

$$\text{etc.}$$

while that corresponding to Gauss' forward formula is

$$f(x) = f_0 + \cfrac{(x - x_0)}{\rho_{\frac{1}{2}} + \cfrac{(x_2 - x_1)}{\rho_0^2 - f_0 + \cfrac{(x_3 - x_{-1})}{\rho_{\frac{1}{2}}^3 - \rho_{\frac{1}{2}} + \cfrac{(x_4 - x_2)}{\rho_0^4 - \rho_0^2 + (x - x_{-2})}}}}$$

$$\text{etc.}$$

Inverse interpolation

Any method of interpolation which does not require the arguments to be evenly spaced will be satisfactory, by simply interchanging the roles of the arguments and the function values.

Alternative (i) Sub-tabulate the function until linear interpolation is adequate and simple ratio and proportion will yield the result.

Alternative (ii) Find an approximate value for p from

$$p \cong (f_p - f_0)/(f_1 - f_0)$$

and then iterate using *any* of the standard interpolation formulae;
e.g., Bessel

$$p = [f_p - f_0 - B_2(\delta_0^2 + \delta_1^2) - B_3\delta_{\frac{1}{2}}^3 - \cdots \text{etc.}]/\delta_{\frac{1}{2}}$$

where the Bessel's coefficients on the right hand side are evaluated for the approximate p.
Everrett

$$p = [f_p - f_0 - E_2\delta_0^2 - F_2\delta_1^2 - E_4\delta^4 - F_4\delta_0^4 \cdots]/\delta_{\frac{1}{2}}$$

Newton forward

$$p = [f_p - f_0 - \frac{1}{2!}\,p(p-1)\Delta_0^2 - \frac{1}{3!}\,p(p-1)(p-2)\Delta_0^3 \cdots]/\Delta_0$$

Newton backward

$$p = [f_p - f_0 - \frac{1}{2!}\,p(p+1)\nabla_0^2 - \frac{1}{3!}\,p(p+1)(p+2)\nabla_0^3 \cdots]/\nabla_0 \,.$$

Note that the divisors are in fact identical.

Lozenge Diagram — Interpolation

The following figure shows a scheme for writing down any of the standard interpolation formulas by associating it with a prescribed path through the diagram, called a lozenge diagram.

y_{-4} $(u+4)_1$ $\Delta^2 y_{-5}$ $(u+5)_3$ $\Delta^4 y_{-6}$ $(u+6)_5$ $\Delta^6 y_{-7}$ $(u+7)_7$

1 Δy_{-4} $(u+4)_2$ $\Delta^3 y_{-5}$ $(u+5)_4$ $\Delta^5 y_{-6}$ $(u+6)_6$ $\Delta^7 y_{-7}$

y_{-3} $(u+3)_1$ $\Delta^2 y_{-4}$ $(u+4)_3$ $\Delta^4 y_{-5}$ $(u+5)_5$ $\Delta^6 y_{-6}$ $(u+6)_7$

1 Δy_{-3} $(u+3)_2$ $\Delta^3 y_{-4}$ $(u+4)_4$ $\Delta^5 y_{-5}$ $(u+5)_6$ $\Delta^7 y_{-6}$

y_{-2} $(u+2)_1$ $\Delta^2 y_{-3}$ $(u+3)_3$ $\Delta^4 y_{-4}$ $(u+4)_5$ $\Delta^6 y_{-5}$ $(u+5)_7$

1 Δy_{-2} $(u+2)_2$ $\Delta^3 y_{-3}$ $(u+3)_4$ $\Delta^5 y_{-4}$ $(u+4)_6$ $\Delta^7 y_{-5}$

y_{-1} $(u+1)_1$ $\Delta^2 y_{-2}$ $(u+2)_3$ $\Delta^4 y_{-3}$ $(u+3)_5$ $\Delta^6 y_{-4}$ $(u+4)_7$

1 Δy_{-1} $(u+1)_2$ $\Delta^3 y_{-2}$ $(u+2)_4$ $\Delta^5 y_{-3}$ $(u+3)_6$ $\Delta^7 y_{-4}$

y_0 $(u)_1$ $\Delta^2 y_{-1}$ $(u+1)_3$ $\Delta^4 y_{-2}$ $(u+2)_5$ $\Delta^6 y_{-3}$ $(u+3)_7$

1 Δy_0 $(u)_2$ $\Delta^3 y_{-1}$ $(u+1)_4$ $\Delta^5 y_{-2}$ $(u+2)_6$ $\Delta^7 y_{-3}$

y_1 $(u-1)_1$ $\Delta^2 y_0$ $(u)_3$ $\Delta^4 y_{-1}$ $(u+1)_5$ $\Delta^6 y_{-2}$ $(u+2)_7$

1 Δy_1 $(u-1)_2$ $\Delta^3 y_0$ $(u)_4$ $\Delta^5 y_{-1}$ $(u+1)_6$ $\Delta^7 y_{-2}$

y_2 $(u-2)_1$ $\Delta^2 y_1$ $(u-1)_3$ $\Delta^4 y_0$ $(u)_5$ $\Delta^6 y_{-1}$ $(u+1)_7$

1 Δy_2 $(u-2)_2$ $\Delta^3 y_1$ $(u-1)_4$ $\Delta^5 y_0$ $(u)_6$ $\Delta^7 y_{-1}$

y_3 $(u-3)_1$ $\Delta^2 y_2$ $(u-2)_3$ $\Delta^4 y_1$ $(u-1)_5$ $\Delta^6 y_0$ $(u)_7$

1 Δy_3 $(u-3)_2$ $\Delta^3 y_2$ $(u-2)_4$ $\Delta^5 y_1$ $(u-1)_6$ $\Delta^7 y_0$

y_4 $(u-4)_1$ $\Delta^2 y_3$ $(u-3)_3$ $\Delta^4 y_2$ $(u-2)_5$ $\Delta^6 y_1$ $(u-1)_7$

——→ Gregory-Newton (Forward)   — –→ Stirling

—→→ Gregory-Newton (Backward)   ...·>.... Bessel

〜〜 Gauss (I) $(u+k)_s \equiv \dfrac{1}{s!}(u+k)^{[s]}$

Lozenge diagram showing paths for some of the standard interpolation formulas. (From Kunz, K. S., *Numerical Analysis*, McGraw-Hill, New York, 1957, 75–77, 80, 81. With permission.)

To convert a path through the lozenge to an interpolation formula the following rules are formulated:

1. Each time a difference column is crossed from left to right a term is added.
2. If a path enters a difference column (from the left) at a positive slope, the term added is the product of the difference, say $\Delta^k y_{-p}$, at which the column is crossed and the factorial $(u + p - 1)_k$ lying just below that difference.
3. If a path enters a difference column (from the left) at a positive slope, the term added is the product of the difference, say $\Delta^k y_{-p}$, at which the column is crossed and the factorial $(u + p)_k$ lying just above that difference.
4. If a path enters a difference column horizontally (from the left), the term added is the product of the difference, say $\Delta^k y_{-p}$, at which the column is crossed and the average of the two factorials $(u + p)_k$ and $(u + p-1)_k$ lying, respectively, just above and just below that difference.

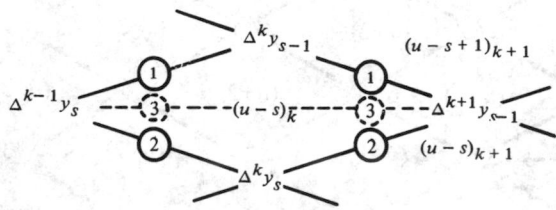

5. If a path crosses a difference column (from left to right) between two differences, say $\Delta^k y_{-(p+1)}$ and $\Delta^k y_{-p}$, then the added term is the product of the average of these two differences and the factorial $(u + p)_k$ at which the column is crossed.
6. Any portion of a path traversed from right to left gives rise to the same terms as would arise from going along this portion from left to right except that the sign of each term is changed.
7. The zero-difference column corresponding to the tabulated values may be treated by the same rules as any other difference columns provided one thinks of the lozenge as being entered just to the left of this column. Thus this column can be crossed by a path making a positive, negative, or zero slope, just as is true of the other columns.

Examples

Write down the interpolation formulas whose paths are shown on the lozenge diagram.

It is clear that, since the lozenge employs descending differences, all formulas will be in terms of descending differences. If one wishes the central-difference formulas to appear as they have been written in the earlier sections of the chapter, one should use central-difference notation for the differences.

Gregory-Newton (Forward) — Applying Rules 1, 3, and 7 to the diagonal path that slopes downward from y_0, one obtains the formula

$$y = y_0 + (u)_1 \Delta y_0 + (u)_2 \Delta^2 y_0 + (u)_3 \Delta^3 y_0 + \ldots + (u)_m \Delta^m y_0$$

Note: $(u)_r = u^{[r]}/r!$

Gregory-Newton (Backward) — Employing Rules 1, 2, and 7 to the diagonal path that slopes upward from y_0, one obtains the formula

$$y = y_0 + (u)_1 \Delta y_{-1} + (u + 1)_2 \Delta^2 y_{-2} + (u + 2)_3 \Delta^3 y_{-3} + \ldots$$

$$+ (u + m - 1)_m \Delta^m y_{-m}$$

Gauss (I) — Employing Rules 1, 2, 3, and 7 to the zigzag path shown, one has

$$y = y_0 + (u)_1 \, \Delta y_0 + (u)_2 \, \Delta^2 y_{-1} + (u+1)_3 \, \Delta^3 y_{-1} + (u+1)_4 \, \Delta^4 y_{-2} + \dots$$

$$+ (u+k-1)_{2k-1} \, \Delta^{2k-1} y_{-k+1} + (u+k-1)_{2k} \, \Delta^{2k} y_{-k} + \dots$$

where, as before, the series is terminated at any desired difference.

Stirling's — Employing Rules 1, 4, 5, and 7 to the horizontal path through y_0,

$$y = y_0 + (u)_1 \, \frac{\Delta y_{-1} + \Delta y_0}{2} + \frac{(u)_2 + (u+1)_2}{2} \, \Delta^2 y_{-1}$$

$$+ (u+1)_3 \, \frac{\Delta^2 y_{-2} + \Delta^3 y_{-1}}{2} + \dots$$

$$+ (u+k-1)_{2k-1} \, \frac{\Delta^{2k-1} y_{-k} + \Delta^{2k-1} y_{-k+1}}{2}$$

$$+ \frac{(u+k-1)_{2k} + (u+k)_{2k}}{2} \, \Delta^{2k} y_{-k} + \dots$$

Now

$$(u+k-1)_{2k-1} = \frac{1}{(2k-1)!} \, u(u^2-1)(u^2-4) \dots [u^2-(k-1)^2]$$

and

$$\frac{(u+k-1)_{2k} + (u+k)_{2k}}{2} = \frac{1}{2k!} \, u^2(u^2-1)(u^2-4) \dots [u^2-(k-1)^2]$$

Therefore, one may write

$$y = y_0 + u \, \frac{\Delta y_{-1} + \Delta y_0}{2} + \frac{1}{2!} \, u^2 \, \Delta^2 y_1 + \frac{1}{3!} \, u \, (u^2-1) \, \frac{\Delta^3 y_{-2} + \Delta^3 y_{-1}}{2}$$

$$+ \frac{1}{4!} \, u^2(u^2-1) \, \Delta^4 y_{-2} + \dots$$

$$+ \frac{1}{(2k-1)!} \, u(u^2-1)(u^2-4) \dots [u^2-(k-1)^2] \, \frac{\Delta^{2k-1} y_{-k} + \Delta^{2k-1} y_{-k+1}}{2}$$

$$+ \frac{1}{(2k)!} \, u^2(u^2-1)(u^2-4) \dots [u^2-(k-1)^2] \, \Delta^{2k} y_{-k} + \dots$$

This is the expression of Stirling's formula when descending differences are employed.

Bessel's — Employing the rules above to the horizontal path through Δy_0, one has

$$y = \frac{y_0 + y_1}{2} + \frac{(u)_1 + (u-1)_1}{2} \, \Delta y_0 + (u)_2 \, \frac{\Delta^2 y_{-1} + \Delta^2 y_0}{2}$$

$$+ \frac{(u+1)_3 + (u)_3}{2} \, \Delta^3 y_{-1} + \dots$$

$$+ \frac{(u+k-1)_{2k-1} + (u+k-2)_{2k-1}}{2} \, \Delta^{2k-1} y_{-k+1}$$

$$+ (u+k-1)_{2k} \, \frac{\Delta^{2k} y_{-k} + \Delta^{2k} y_{-k+1}}{2}$$

or

$$y = \frac{y_0 + y_1}{2} + (u - \tfrac{1}{2})\,\Delta y_0 + \frac{1}{2!}\,u^{[2]}\,\frac{\Delta^2 y_{-1} + \Delta^2 y_0}{2}$$

$$+ \frac{1}{3!}\,(u - \tfrac{1}{2})\,u^{[2]}\,\Delta^3 y_{-1} + \frac{1}{4!}\,(u+1)^{[4]}\,\frac{\Delta^4 y_{-2} + \Delta^4 y_{-1}}{2} + \ldots$$

$$+ \frac{1}{(2k)!}\,(u + k - 1)^{[2k]}\,\frac{\Delta^{2k} y_{-k} + \Delta^{2k} y_{-k+1}}{2}$$

$$+ \frac{1}{(2k+1)!}\,(u - \tfrac{1}{2})\,(u + k - 1)^{[2k]}\,\Delta^{2k+1} y_{-k} + \ldots$$

Remainder Term

The remainder term for a series terminated on a single difference, say the mth difference, is just the (m + 1)th term of the series if the series were extended, with, however, $h^{m+1} f^{m+1}(\xi)$ replacing the (m + 1)th difference (or mean difference). For a formula terminated on the mean of two differences, one must average the error terms for those formulas ending on each of the two differences.

DIFFERENTIATION FORMULAS

The notation used is defined as follows:

$$h^n \frac{d^n}{dx^n} f(x_0 + ph) \equiv h^n \frac{d^n}{dx^n} f_p \equiv \frac{d^n}{dp^n} f_p$$

Derivatives at tabular points.

$$hf_0^{(1)} = \left(\Delta - \frac{\Delta^2}{2} + \frac{\Delta^3}{3} - \frac{\Delta^4}{4} + \frac{\Delta^5}{5} - \frac{\Delta^6}{6} \cdots\right) f_0$$

$$= \left(\mu\delta - \tfrac{1}{6}\mu\delta^3 + \tfrac{1}{30}\mu\delta^5 - \tfrac{1}{140}\mu\delta^7 \cdots\right) f_0$$

$$= \left(\nabla + \frac{\nabla^2}{2} + \frac{\nabla^3}{3} + \frac{\nabla^4}{4} + \frac{\nabla^5}{5} + \frac{\nabla^6}{6} \cdots\right) f_0$$

$$h^2 f_0^{(2)} = \left(\Delta^2 - \Delta^3 + \tfrac{11}{12}\Delta^4 - \tfrac{5}{6}\Delta^5 + \tfrac{137}{180}\Delta^6 - \tfrac{7}{10}\Delta^7 \cdots\right) f_0$$

$$= \left(\delta^2 - \tfrac{1}{12}\delta^4 + \tfrac{1}{90}\delta^6 - \tfrac{1}{560}\delta^8 \cdots\right) f_0$$

$$= \left(\nabla^2 + \nabla^3 + \tfrac{11}{12}\nabla^4 + \tfrac{5}{6}\nabla^5 + \tfrac{137}{180}\nabla^6 + \tfrac{7}{10}\nabla^7 \cdots\right) f_0$$

$$h^3 f_0^{(3)} = \left(\Delta^3 - \tfrac{3}{2}\Delta^4 + \tfrac{21}{12}\Delta^5 - \tfrac{45}{24}\Delta^6 + \tfrac{348}{180}\Delta^7 - \tfrac{79}{40}\Delta^8 \cdots\right) f_0$$

$$= \left(\mu\delta^3 - \tfrac{1}{4}\mu\delta^5 + \tfrac{7}{120}\mu\delta^7 - \tfrac{41}{3024}\mu\delta^9 \cdots\right) f_0$$

$$= \left(\nabla^3 + \tfrac{3}{2}\nabla^4 + \tfrac{21}{12}\nabla^5 + \tfrac{45}{24}\nabla^6 + \tfrac{348}{180}\nabla^7 + \tfrac{79}{40}\nabla^8 \cdots\right) f_0$$

Higher derivatives can be found by direct multiplication of the above formulae e.g.

$$h^5 f_0^{(5)} = h^3 \frac{d^3}{dp^3}\left(h^2 \frac{d^2}{dp^2} f_0\right)$$

$$\therefore h^5 f_0^{(5)} = \left(\mu\delta^3 - \tfrac{1}{4}\mu\delta^5 + \tfrac{7}{120}\mu\delta^7 \cdots\right)\left(\delta^2 - \tfrac{1}{12}\delta^4 + \tfrac{1}{90}\delta^6 - \tfrac{1}{560}\delta^8 \cdots\right) f_0$$

$$= \left(\mu\delta^5 - \tfrac{1}{3}\mu\delta^7 + \tfrac{13}{141}\mu\delta^9 \cdots\right) f_0$$

The identity $\mu^2 \equiv 1 + \tfrac{1}{4}\delta^2$ is used to ensure that no powers of 'μ,' other than the first, appear in any formula.

$$h^4 f_0^{(4)} = \left(\delta^4 - \tfrac{1}{6}\delta^6 + \tfrac{7}{240}\delta^8 - \tfrac{41}{7560}\delta^{10} \cdots\right) f_0$$

$$h^6 f_0^{(6)} = \left(\delta^6 - \tfrac{1}{4}\delta^8 + \tfrac{13}{240}\delta^{10} \cdots\right) f_0$$

$$h^7 f_0^{(7)} = \left(\mu\delta^7 - \tfrac{5}{12}\mu\delta^9 + \tfrac{31}{240}\mu\delta^{11} \cdots\right) f_0$$

The preceding formulae can be truncated, as in the examples, and the differences expressed in terms of the function values. The resulting formulae are of the type

$$h^n f_0^{(n)} = \sum_{r=i}^{k} a_r \cdot f_r + \epsilon$$

where

$$\sum_{r=i}^{k} a_r = 0$$

and the principal part of the error 'ϵ' is the first neglected difference. It is still necessary for the differences to be converging.

In the following formulae only the order of the differences constituting the error term are indicated.

Numerical Differentiation

$$hf_0^{(1)} \simeq (f_1 - f_0) - \tfrac{1}{2}\Delta^2$$
$$\simeq \tfrac{1}{2}(-f_2 + 4f_1 - 3f_0) + \tfrac{1}{3}\Delta^3$$
$$\simeq \tfrac{1}{6}(2f_3 - 9f_2 + 18f_1 - 11f_0) - \tfrac{1}{4}\Delta^4$$
$$\simeq \tfrac{1}{2}(f_1 - f_{-1}) - \tfrac{1}{6}\mu\delta^3$$
$$\simeq \tfrac{1}{12}(-f_2 + 8f_1 - 8f_{-1} + f_2) + \tfrac{1}{30}\mu\delta^5$$
$$\simeq \tfrac{1}{60}(f_3 - \dot{9}f_2 + 45f_1 - 45f_{-1} + 9f_{-2} - f_{-3}) - \tfrac{1}{140}\mu\delta^7$$
$$\simeq (f_0 - f_{-1}) + \tfrac{1}{2}\nabla^2$$
$$\simeq \tfrac{1}{2}(3f_0 - 4f_{-1} + f_{-2}) + \tfrac{1}{3}\nabla^3$$
$$\simeq \tfrac{1}{6}(11f_0 - 18f_{-1} + 9f_{-2} - 2f_{-3}) + \tfrac{1}{4}\nabla^4$$

$$h^2 f_0^{(2)} \simeq (f_2 - 2f_1 + f_0) - \Delta^3$$
$$\simeq (-f_3 + 4f_2 - 5f_1 + 2f_0) + \tfrac{11}{12}\Delta^4$$
$$\simeq (11f_4 - 56f_3 + 114f_2 - 104f_1 + 35f_0) - \tfrac{5}{6}\Delta^5$$
$$\simeq (f_1 - 2f_0 + f_{-1}) - \tfrac{1}{12}\delta^4$$
$$\simeq \tfrac{1}{12}(-f_2 + 16f_1 - 30f_0 + 16f_{-1} - f_{-2}) + \tfrac{1}{90}\delta^6$$
$$\simeq \tfrac{1}{180}(2f_3 - 27f_2 + 270f_1 - 490f_0 + 270f_{-1} - 27f_{-2} + 2f_{-3}) - \tfrac{1}{560}\delta^8$$
$$\simeq (f_0 - 2f_{-1} + f_{-2}) + \nabla^3$$
$$\simeq (2f_0 - 5f_{-1} + 4f_{-2} - f_{-3}) + \tfrac{11}{12}\nabla^4$$
$$\simeq (35f_0 - 104f_{-1} + 114f_{-2} - 56f_{-3} + 11f_{-4}) + \tfrac{5}{8}\nabla^5$$

$$h^3 f_0^{(3)} \simeq (f_3 - 3f_2 + 3f_1 - f_0) - \tfrac{3}{2}\Delta^4$$
$$\simeq \tfrac{1}{2}(-3f_4 + 14f_3 - 24f_2 + 18f_1 - 5f_0) + \tfrac{21}{12}\Delta^5$$
$$\simeq \tfrac{1}{2}(f_2 - 2f_1 + 2f_{-1} - f_{-2}) - \tfrac{1}{4}\mu\delta^5$$
$$\simeq \tfrac{1}{8}(-f_3 + 8f_2 - 13f_1 + 13f_{-1} - 8f_{-2} + f_{-3}) + \tfrac{7}{120}\mu\delta^7$$
$$\simeq (f_0 - 3f_{-1} + 3f_{-2} - f_{-3}) + \tfrac{3}{2}\nabla^4$$
$$\simeq \tfrac{1}{2}(5f_0 - 18f_{-1} + 24f_{-2} - 14f_{-3} + 3f_{-4}) + \tfrac{21}{12}\nabla^5$$

For derivatives at half-way points there exist analagous formulae. The operators act upon f_0 and f_1, in the case of the forward and backward differences, and upon $f_{1/2}$ in the case of the central differences δ_0^{2n+1} being meaningless.

$$hf_{1/2}^{(1)} = (\Delta - \tfrac{1}{24}\Delta^3 + \tfrac{1}{24}\Delta^4 - \tfrac{71}{1920}\Delta^5 \cdots)f_0$$
$$= (\delta - \tfrac{1}{24}\delta^3 + \tfrac{3}{640}\delta^5 - \tfrac{5}{7168}\delta^7 \cdots)f_{1/2}$$
$$= (\nabla - \nabla^2 + \tfrac{23}{24}\nabla^3 + \tfrac{11}{12}\nabla^4 + \tfrac{563}{840}\nabla^5 \cdots)f_0$$
$$= (\nabla - \tfrac{1}{24}\nabla^3 - \tfrac{1}{24}\nabla^4 - \tfrac{71}{1920}\nabla^5 \cdots)f_1$$

$$h^2 f_{1/2}^{(2)} = (\Delta^2 - \tfrac{1}{12}\Delta^4 + \tfrac{1}{12}\Delta^5 - \tfrac{13}{180}\Delta^6 \cdots)f_0$$
$$= (\mu\delta^2 - \tfrac{5}{24}\mu\delta^4 + \tfrac{259}{5760}\mu\delta^6 - \tfrac{3229}{322560}\mu\delta^8 \cdots)f_{1/2}$$
$$= (\nabla^2 - \tfrac{1}{12}\nabla^4 - \tfrac{1}{12}\nabla^5 - \tfrac{13}{180}\nabla^6 \cdots)f_1$$

$$h^3 f_{1/2}^{(3)} = (\Delta^3 - \tfrac{1}{8}\Delta^5 + \tfrac{1}{8}\Delta^6 - \tfrac{203}{1920}\Delta^7 \cdots)f_0$$
$$= (\delta^3 - \tfrac{1}{8}\delta^5 + \tfrac{37}{1920}\delta^7 - \tfrac{2229}{967680}\delta^9 \cdots)f_{1/2}$$
$$= (\nabla^3 - \tfrac{1}{8}\nabla^5 - \tfrac{1}{8}\nabla^6 - \tfrac{203}{1920}\nabla^7 \cdots)f_1$$

Caution

Numerical differentiation is error magnifying and many guarding figures are usually necessary. This is due to two factors namely:

  a) the interval h enters each formula in the form $1/h^n$

  b) each formula starts with the nth order difference.

Both of these combine to make numerical differentiation unreliable. e.g. Suppose the formula $f^{(2)} = (\Delta^2 - \Delta^3 + \frac{11}{12}\Delta^4)/h^2$ is used with $h = 0.1$ and an error of 5×10^{-4} can be tolerated in the derivative. For an error of e in each of the tabulated values

$$5 \times 10^{-4} = \tfrac{1}{0.01}(2e + 3e + \tfrac{11}{12}6e)$$
$$= 100(\tfrac{21}{2} e)$$
$$= 1050e$$

and therefore $e \simeq 5 \times 10^{-7}$ i.e. the function must be tabulated to *six* places of decimals before the second derivative can be correct to *three*.

If, on the other hand, h had been unity in magnitude then

$$5 \times 10^{-4} = 10.5e$$

i.e. $e \simeq 5 \times 10^{-5}$ and the function would need only one guarding figure to produce the required degree of accuracy in the second derivative.

Obviously this effect is magnified for higher derivatives. In any case, the largest value of h, which permits the function to be well represented by a converging table of differences, should be used.

INTEGRATION FORMULAS

Integration is a smoothing procedure and as such errors in tabulated values tend to be cancelled out.

Notation used:

$$\int_{x_0}^{x_n} f(x) \cdot dx = h \int_0^n f_p \cdot dp$$

where

$$f_p \equiv f(x_0 + ph) \equiv f(x)$$

$$\int_0^n f_p \cdot dp = \tfrac{1}{2} f_0 + f_1 + f_2 + \cdots + f_{n-1} + \tfrac{1}{2} f_n + \tfrac{1}{12}(\Delta_0 - \nabla_n) - \tfrac{1}{24}(\Delta_0^2 + \nabla_n^2)$$
$$+ \tfrac{19}{720}(\Delta_0^3 - \nabla_n^3) - \tfrac{3}{160}(\Delta_0^4 + \nabla_n^4) \cdots$$
$$= \tfrac{1}{2} f_0 + f_1 + f_2 + \cdots + f_{n-1} + \tfrac{1}{2} f_n + \tfrac{1}{12}(\mu\delta_0 - \mu\delta_n)$$
$$- \tfrac{11}{720}(\mu\delta_0^3 - \mu\delta_n^3) + \tfrac{191}{60480}(\mu\delta_0^5 - \mu\delta_n^5) - \cdots$$

In particular

$$\int_0^1 f_p \cdot dp = (1 + \tfrac{1}{2}\Delta - \tfrac{1}{12}\Delta^2 + \tfrac{1}{24}\Delta^3 - \tfrac{19}{720}\Delta^4 + \tfrac{3}{160}\Delta^5 - \tfrac{863}{60480}\Delta^6 + \cdots) f_0$$
$$= (1 + \tfrac{1}{2}\nabla + \tfrac{5}{12}\nabla^2 + \tfrac{3}{8}\nabla^3 + \tfrac{251}{720}\nabla^4 + \tfrac{95}{288}\nabla^5 + \tfrac{19087}{60480}\nabla^6 \cdots) f_0$$
$$= (1 - \tfrac{1}{2}\nabla - \tfrac{1}{12}\nabla^2 - \tfrac{1}{24}\nabla^3 - \tfrac{19}{720}\nabla^4 - \tfrac{3}{160}\nabla^5 - \tfrac{863}{60480}\nabla^6 \cdots) f_1$$
$$= (\mu - \tfrac{1}{12}\mu\delta^2 + \tfrac{11}{720}\mu\delta^4 - \tfrac{191}{60480}\mu\delta^6 + \cdots) f_{1/2}$$

$$\int_0^{n+(1/2)} f_p \cdot dp = \tfrac{1}{2} f_0 + f_1 + f_2 \cdots + f_n + (\tfrac{1}{12}\mu\delta - \tfrac{11}{720}\mu\delta^3 + \tfrac{191}{60480}\mu\delta^5 \cdots) f_0$$
$$+ (\tfrac{1}{24}\delta - \tfrac{17}{5760}\delta^3 + \tfrac{367}{967680}\delta^5 \cdots) f_{n+(1/2)}$$

In particular

$$\int_0^{1/2} f_p \cdot dp = \tfrac{1}{2} f_0 + (\tfrac{1}{8}\delta - \tfrac{1}{24}\mu\delta^2 + \tfrac{1}{384}\delta^3 + \tfrac{11}{1440}\mu\delta^4 - \tfrac{13}{46080}\delta^5 \cdots) f_{1/2}$$

Symmetric range integrals

$$\int_{-1/2}^{1/2} f_p \cdot dp = (1 + \tfrac{1}{24}\delta^2 - \tfrac{17}{5760}\delta^4 + \tfrac{367}{967680}\delta^6 - \tfrac{27859}{464486400}\delta^8 \cdots) f_0$$

$$\int_{-1}^{1} f_p \cdot dp = 2(1 + \tfrac{1}{6}\delta^2 - \tfrac{1}{180}\delta^4 + \tfrac{1}{1512}\delta^6 - \tfrac{23}{226800}\delta^8 + \cdots) f_0$$

$$\int_{-2}^{2} f_p \cdot dp = 4(1 + \tfrac{2}{3}\delta^2 + \tfrac{7}{90}\delta^4 - \tfrac{2}{945}\delta^6 + \tfrac{13}{56700}\delta^8 - \cdots) f_0$$

$$\int_{-3}^{3} f_p \cdot dp = 6(1 + \tfrac{3}{2}\delta^2 + \tfrac{11}{20}\delta^4 + \tfrac{41}{840}\delta^6 - \tfrac{3}{2800}\delta^8 + \cdots) f_0$$

$$\int_{-4}^{4} f_p \cdot dp = 8(1 + \tfrac{8}{3}\delta^2 + \tfrac{217}{90}\delta^4 + \tfrac{92}{189}\delta^6 + \tfrac{989}{28350}\delta^8 + \cdots) f_0$$

The differences in the above formulae can be replaced by their function values, after truncation of the series, if extensive differencing is to be avoided.

When this is done, the formulae take on the forms

$$\sum_m A_m \cdot f_m$$

DESCRIPTIVE STATISTICS

a) Ungrouped Data

The formulas of this section designated as $a)$ apply to a random sample of size n, denoted by x_i, $i = 1, 2, \ldots, n$.

b) Grouped Data

The formulas of this section designated as $b)$ apply to data grouped into a frequency distribution having class marks x_i, $i = 1, 2, \ldots, k$, and corresponding class frequencies f_i, $i = 1, 2, \ldots, k$. The total number of observations given by

$$n = \sum_{i=1}^{k} f_i$$

In the formulas that follow, c denotes the width of the class interval, x_o denotes one of the class marks taken to be the computing origin, and $u_i = \dfrac{x_i - x_o}{c}$. Then coded class marks are obtained by replacing the original class marks with the integers $\ldots, -3, -2, -1, 0, 1, 2, 3, \ldots$ where 0 corresponds to class mark x_o in the original scale.

Mean (Arithmetic Mean)

$$a) \quad \bar{x} = \frac{1}{n} \sum_{i=1}^{n} x_i = \frac{x_1 + x_2 + \cdots + x_n}{n}$$

$$b.1) \quad \bar{x} = \frac{1}{n} \sum_{i=1}^{k} f_i x_i = \frac{f_1 x_1 + f_2 x_2 + \cdots + f_k x_k}{n}$$

If data is coded

$$b.2) \quad \bar{x} = x_o + c \frac{\sum_{i=1}^{k} f_i u_i}{n}$$

Weighted Mean (Weighted Arithmetic Mean)

If with each value x_i is associated a weighting factor $w_i \geq 0$, then $\sum_{i=1}^{n} w_i$ is the total weight, and

$$a) \quad \bar{x} = \frac{\sum_{i=1}^{n} w_i x_i}{\sum_{i=1}^{n} w_i} = \frac{w_1 x_1 + w_2 x_2 + \cdots + w_n x_n}{w_1 + w_2 + \cdots + w_n}$$

Geometric Mean

$$a) \quad \text{G.M.} = \sqrt[n]{x_1 \cdot x_2 \cdots x_n}$$

In logarithmic form

$$\log (\text{G.M.}) = \frac{1}{n} \sum_{i=1}^{n} \log x_i = \frac{\log x_1 + \log x_2 + \cdots + \log x_n}{n}$$

b) $\text{G.M.} = \sqrt[n]{x_1^{f_1} \cdot x_2^{f_2} \cdots x_k^{f_k}}$

In logarithmic form

$$\log (\text{G.M.}) = \frac{1}{n} \sum_{i=1}^{k} f_i \log x_i = \frac{f_1 \log x_1 + f_2 \log x_2 + \cdots + f_k \log x_k}{n}$$

Harmonic Mean

a) $\text{H.M.} = \dfrac{n}{\sum\limits_{i=1}^{n} \dfrac{1}{x_i}} = \dfrac{n}{\dfrac{1}{x_1} + \dfrac{1}{x_2} + \cdots + \dfrac{1}{x_n}}$

b) $\text{H.M.} = \dfrac{n}{\sum\limits_{i=1}^{k} \dfrac{f_i}{x_i}} = \dfrac{n}{\dfrac{f_1}{x_1} + \dfrac{f_2}{x_2} + \cdots + \dfrac{f_k}{x_k}}$

Relation Between Arithmetic, Geometric, and Harmonic Mean

$\text{H.M.} \leq \text{G.M.} \leq \bar{x}$, (Equality sign holds only if all sample values are identical.)

Mode

a) A mode M_o of a sample of size n is a value which occurs with greatest frequency, i.e., it is the most common value. A mode may not exist, and even if it does exist it may not be unique.

b) $M_o = L + c \dfrac{\Delta_1}{\Delta_1 + \Delta_2}$,

where L is the lower class boundary of the modal class (class containing the mode),
   Δ_1 is the excess of modal frequency over frequency of next lower class,
   Δ_2 is the excess of modal frequency over frequency of next higher class.

Median

a) If the sample is arranged in ascending order of magnitude, then the median M_d is given by the $\dfrac{n+1}{2}$ nd value. When n is odd, the median is the middle value of the set of ordered data; when n is even, the median is usually taken as the mean of the two middle values of the set of ordered data.

b) $M_d = L + c \dfrac{\dfrac{n}{2} - F_c}{f_m}$,

where L is lower class boundary of median class (class containing the median),
   F_c is the sum of the frequencies of all classes lower than the median class,
   f_m is the frequency of the median class.

Empirical Relation Between Mean, Median, and Mode

$$\text{Mean} - \text{Mode} = 3 \, (\text{Mean} - \text{Median})$$

Quartiles

a) If the data is arranged in ascending order of magnitude, the jth quartile Q_j, $j = 1$, 2, or 3, is given by the $\dfrac{j(n + 1)}{4}$ th value. It may be necessary to interpolate between successive values.

b) The jth quartile Q_j, $j = 1$, 2, or 3, is obtained from formula b) for the median by counting $\dfrac{jn}{4}$ cases starting at the bottom of the distribution.

Deciles

a) If the sample is arranged in ascending order of magnitude, the jth decile D_j, $j = 1, 2, \ldots$, or 9, is given by the $\dfrac{j(n + 1)}{10}$ th value. It may be necessary to interpolate between successive values.

b) The jth decile D_j, $j = 1, 2, \ldots$, or 9, is obtained from formula b) for the median by counting $\dfrac{jn}{10}$ cases starting at the bottom of the distribution.

Percentiles

a) If the sample is arranged in ascending order of magnitude, the jth percentile P_j, $j = 1, 2, \ldots$, or 99 is given by the $\dfrac{j(n + 1)}{100}$ th value. It may be necessary to interpolate between successive values.

b) The jth percentile P_j, $j = 1, 2, \ldots$, or 99, is obtained from formulas b) for the median by counting $\dfrac{jn}{100}$ cases starting at the bottom of the distribution.

Mean Deviation

a) $\text{M.D.} = \dfrac{1}{n} \sum_{i=1}^{n} |x_i - \bar{x}|$

or

$$\text{M.D.} = \frac{1}{n} \sum_{i=1}^{n} |x_i - M_d|$$

where $\bar{x}$ is the mean and M_d is the median of the sample.

b) $\text{M.D.} = \dfrac{1}{n} \sum_{i=1}^{k} f_i |x_i - \bar{x}|$

or

$$\text{M.D.} = \frac{1}{n} \sum_{i=1}^{k} f_i |x_i - M_d|$$

where $\bar{x}$ is the mean and M_d the median of the sample.

Standard Deviation

a) $s = \sqrt{\dfrac{\sum\limits_{i=1}^{n} (x_i - \bar{x})^2}{n - 1}}$, where $\bar{x}$ is the mean of the sample.

For computational purposes,

$$s = \sqrt{\dfrac{\sum\limits_{i=1}^{n} x_i^2 - n\bar{x}^2}{n - 1}}$$

$$s = \sqrt{\dfrac{n \sum\limits_{i=1}^{n} x_i^2 - \left(\sum\limits_{i=1}^{n} x_i\right)^2}{n(n - 1)}}$$

b) $s = \sqrt{\dfrac{\sum\limits_{i=1}^{k} f_i(x_i - \bar{x})^2}{n - 1}}$, where $\bar{x}$ is the mean of the sample.

For computational purposes

$$s = \sqrt{\dfrac{\sum\limits_{i=1}^{k} f_i x_i^2 - n\bar{x}^2}{n - 1}}$$

$$s = \sqrt{\dfrac{n \sum\limits_{i=1}^{k} f_i x_i^2 - \left(\sum\limits_{i=1}^{k} f_i x_i\right)^2}{n(n - 1)}}$$

If data is coded,

$$s = c \sqrt{\dfrac{n \sum\limits_{i=1}^{k} f_i u_i^2 - \left(\sum\limits_{i=1}^{k} f_i u_i\right)^2}{n(n - 1)}}$$

Variance

The variance is the square of the standard deviation.

Range

The range of a set of values is the difference between the largest and smallest values in the set.

Root Mean Square

a) $\text{R.M.S.} = \left[\dfrac{1}{n} \sum\limits_{i=1}^{n} x_i^2\right]^{\frac{1}{2}}$

b) $\text{R.M.S.} = \left[\dfrac{1}{n} \sum\limits_{i=1}^{k} f_i x_i^2\right]^{\frac{1}{2}}$

Interquartile Range

$$Q_3 - Q_1,$$

where Q_1 and Q_3 are the first and third quartiles.

Quartile Deviation (Semi-Interquartile Range)

$$\frac{Q_3 - Q_1}{2},$$

where Q_1 and Q_3 are the first and third quartiles.

Coefficient of Variation

$$V = \frac{100s}{\bar{x}},$$

where $\bar{x}$ is the mean and s the standard deviation of the sample.

Coefficient of Quartile Variation

$$V = 100 \frac{Q_3 - Q_1}{Q_3 + Q_1},$$

where Q_1 and Q_3 are the first and third quartiles.

Standardized Variable (Standard Scores)

$$z = \frac{x_i - \bar{x}}{s},$$

where $\bar{x}$ is the mean and s the standard deviation of the sample.

Moments

    a) The rth moment about the origin is given by

$$m'_r = \frac{1}{n} \sum_{i=1}^{n} x_i{}^r$$

The rth moment about the mean $\bar{x}$ is given by

$$m_r = \frac{1}{n} \sum_{i=1}^{n} (x_i - \bar{x})^r$$

If $\sum_{i=1}^{n} (x_i - \bar{x})^r$ is expanded by use of the binomial theorem, moments about the mean may be expressed in terms of moments about the origin.

    b) The rth moment about the origin is given by

$$m'_r = \frac{1}{n} \sum_{i=1}^{k} f_i x_i{}^r$$

The rth moment about the mean $\bar{x}$ is given by

$$m_r = \frac{1}{n} \sum_{i=1}^{k} f_i (x_i - \bar{x})^r$$

If $\sum_{i=1}^{k} f_i(x_i - \bar{x})^r$ is expanded by use of the binomial theorem, moments **about the**
mean may be expressed in terms of moments about the origin.

If data is coded

$$m'_r = c^r \frac{\sum_{i=1}^{k} f_i u_i^r}{n}$$

Coefficient of Skewness

$$\alpha_3 = \frac{m_3}{(m_2)^{3/2}},$$

where m_2 and m_3 are the second and third moments about the mean of the sample.

Coefficient of Momental Skewness

$$\frac{\alpha_3}{2} = \frac{m_3}{2(m_2)^{3/2}},$$

where m_2 and m_3 are the second and third moments about the mean of the sample.

Pearson's First Coefficient of Skewness

$$S_{k_1} = \frac{3(\bar{x} - M_o)}{s},$$

where $\bar{x}$ is the mean, M_o the mode, and s the standard deviation of the sample.

Pearson's Second Coefficient of Skewness

$$S_{k_2} = \frac{3(\bar{x} - M_d)}{s},$$

where $\bar{x}$ is the mean, M_d the median, and s the standard deviation of the sample.

Quartile Coefficient of Skewness

$$S_{k_Q} = \frac{Q_3 - 2Q_2 + Q_1}{Q_3 - Q_1},$$

where Q_1, Q_2, and Q_3 are the first, second, and third quartiles.

Coefficient of Kurtosis

$$\alpha_4 = \frac{m_4}{(m_2)^2},$$

where m_2 and m_4 are the second and fourth moments about the mean of the sample.

Coefficient of Excess (Kurtosis)

$$\alpha_4 - 3 = \frac{m_4}{(m_2)^2} - 3,$$

where m_2 and m_4 are the second and fourth moments about the mean of the sample.

Sheppards Corrections for Grouping

Let all class intervals be of equal length c. If the distribution of x has a high order of
contact with the x-axis at both tails, (i.e., if the distribution of x has tails which are very

nearly tangent to the x-axis), one may improve the grouped data approximation to the variance by adding Sheppard's correction $-\dfrac{c^2}{12}$. Thus

$$\text{corrected variance} = \text{grouped data variance} - \frac{c^2}{12}$$

Analogous corrections for grouped data sample moments

$$m_r' = \frac{1}{n} \sum_{i=1}^{k} f_i x_i^r \qquad \text{and} \qquad m_r = \frac{1}{n} \sum_{i=1}^{k} f_i (x_i - \bar{x})^r$$

yield improved estimates m_{r_e}' and m_{r_e} given by

$$m_{1_e}' = m_1' \qquad\qquad m_{1_e} = m_1$$

$$m_{2_e}' = m_2' - \frac{c^2}{12} \qquad\qquad m_{2_e} = m_2 - \frac{c^2}{12}$$

$$m_{3_e}' = m_3' - \frac{c^2}{4} m_1' \qquad\qquad m_{3_e} = m_3$$

$$m_{4_e}' = m_4' - \frac{c^2}{2} m_1' + \frac{7c^4}{240} \qquad m_{4_e} = m_4 - \frac{c^2}{2} m_2 + \frac{7c^4}{240}$$

Curve Fitting, Regression, and Correlation

The following formulas apply to a set of n ordered pairs $\{(x_i, y_i)\}$, $i = 1, 2, \ldots, n$. The assumptions of normal regression analysis are that the x's are fixed variables, and the y's are independent random variables having normal distributions with common variance σ^2. The assumptions of normal correlation analysis are that $\{(x_i, y_i)\}$ constitute a random sample from a bivariate normal population.

Curve Fitting

1. Polynomial Function

$$y = b_0 + b_1 x + b_2 x^2 + \cdots + b_m x^m$$

For a polynomial function fit by the method of least squares, the values of $b_0, b_1, \ldots, b_m$ are obtained by solving the system of $m + 1$ normal equations

$$nb_0 + b_1 \Sigma x_i + b_2 \Sigma x_i^2 + \cdots + b_m \Sigma x_i^m = \Sigma y_i$$
$$b_0 \Sigma x_i + b_1 \Sigma x_i^2 + b_2 \Sigma x_i^3 + \cdots + b_m \Sigma x_i^{m+1} = \Sigma x_i y_i$$
$$\cdots \cdots \cdots \cdots \cdots \cdots \cdots \cdots \cdots \cdots \cdots \cdots \cdots$$
$$b_0 \Sigma x_i^m + b_1 \Sigma x_i^{m+1} + b_2 \Sigma x_i^{m+2} + \cdots + b_m \Sigma x_i^{2m} = \Sigma x_i^m y_i$$

2. Straight Line

$$y = b_0 + b_1 x$$

For a straight line fit by the method of least squares, the values b_0 and b_1 are obtained by solving the normal equations

$$nb_0 + b_1 \Sigma x_i = \Sigma y_i$$
$$b_0 \Sigma x_i + b_1 \Sigma x_i^2 = \Sigma x_i y_i$$

The solutions of these normal equations are

$$b_1 = \frac{n \Sigma x_i y_i - (\Sigma x_i)(\Sigma y_i)}{n \Sigma x_i^2 - (\Sigma x_i)^2}$$

$$b_0 = \frac{\Sigma y_i}{n} - b_1 \frac{\Sigma x_i}{n} = \bar{y} - b_1 \bar{x}$$

3. Exponential Curve

$$y = ab^x$$

or

$$\log y = \log a + (\log b)x$$

For an exponential curve fit by the method of least squares, the values $\log a$ and $\log b$ are obtained by fitting a straight line to the set of ordered pairs $\{(x_i, \log y_i)\}$.

4. Power Function

$$y = ax^b$$

or

$$\log y = \log a + b \log x$$

For a power function fit by the method of least squares, the values $\log a$ and b are obtained by fitting a straight line to the set of ordered pairs $\{(\log x_i, \log y_i)\}$.

Regression and Correlation

1. Simple Linear Regression.

For a regression of y on x

$$E(y/x) = b_0 + b_1 x,$$

where $E(y/x)$ is the mean of the distribution of y for a given x.

Standard Error of Estimate

$$s_e = \sqrt{\frac{\Sigma[y_i - (b_0 + b_1 x_i)]^2}{n-2}},$$

where b_0 and b_1 are given by

$$b_1 = \frac{n\Sigma x_i y_i - (\Sigma x_i)(\Sigma y_i)}{n\Sigma x_i^2 - (\Sigma x_i)^2}$$

$$b_0 = \frac{\Sigma y_i}{n} - b_1 \frac{\Sigma x_i}{n} = \bar{y} - b_1 \bar{x}$$

2. Correlation.

An estimate of the population correlation coefficient ρ is given by

$$r = \frac{\Sigma(x_i - \bar{x})(y_i - \bar{y})}{\sqrt{[\Sigma(x_i - \bar{x})^2][\Sigma(y_i - \bar{y})^2]}}$$

or by the computing formula

$$r = \frac{n\Sigma x_i y_i - (\Sigma x_i)(\Sigma y_i)}{\sqrt{[n\Sigma x_i^2 - (\Sigma x_i)^2][n\Sigma y_i^2 - (\Sigma y_i)^2]}}$$

For grouped data

$$r = \frac{n\Sigma f x_i y_i - (\Sigma f_x x_i)(\Sigma f_y y_i)}{\sqrt{[n\Sigma f_x x_i^2 - (\Sigma f_x x_i)^2][n\Sigma f_y y_i^2 - (\Sigma f_y y_i)^2]}}$$

where f_x and f_y denote the frequencies corresponding to the class marks x and y, and f denotes the frequency of the corresponding cell of the correlation table.

If the data is coded

$$r = \frac{n\Sigma f uv - (\Sigma f_u u)(\Sigma f_v v)}{\sqrt{[n\Sigma f_u u^2 - (\Sigma f_u u)^2][n\Sigma f_v v^2 - (\Sigma f_v v)^2]}}$$

where the u's and v's are coded class marks. The frequencies f_u and f_v are defined analogous to f_x and f_y.

PROBABILITY

Definitions

A sample space S associated with an experiment is a set S of elements such that any outcome of the experiment corresponds to one and only one element of the set. An event E is a subset of a sample space S. An element in a sample space is called a sample point or a simple event (Unit subset of S).

Definition of Probability

If an experiment can occur in n mutually exclusive and equally likely ways, and if exactly m of these ways correspond to an event E, then the probability of E is given by

$$P(E) = \frac{m}{n}.$$

If E is a subset of S, and if to each unit subset of S, a non-negative number, called its probability, is assigned, and if E is the union of two or more different simple events, then the probability of E, denoted by $P(E)$, is the sum of the probabilities of those simple events whose union is E.

Marginal and Conditional Probability

Suppose a sample space S is partioned into rs disjoint subsets where the general subset is denoted by $E_i \cap F_j$. Then the marginal probability of E_i is defined as

$$P(E_i) = \sum_{j=1}^{s} P(E_i \cap F_j)$$

and the marginal probability of F_j is defined as

$$P(F_j) = \sum_{i=1}^{r} P(E_i \cap F_j)$$

The conditional probability of E_i, given that F_j has occurred, is defined as

$$P(E_i/F_j) = \frac{P(E_i \cap F_j)}{P(F_j)} , \qquad P(F_j) \neq 0$$

and that of F_j, given that E_i has occurred, is defined as

$$P(F_j/E_i) = \frac{P(E_i \cap F_j)}{P(E_i)} , \qquad P(E_i) \neq 0 .$$

Probability Theorems

1. If ϕ is the null set, $P(\phi) = 0$.
2. If S is the sample space, $P(S) = 1$.
3. If E and F are two events

$$P(E \cup F) = P(E) + P(F) - P(E \cap F).$$

4. If E and F are mutually exclusive events,

$$P(E \cup F) = P(E) + P(F).$$

5. If E and E' are complementary events,

$$P(E) = 1 - P(E').$$

6. The conditional probability of an event E, given an event F, is denoted by $P(E/F)$ and is defined as

$$P(E/F) = \frac{P(E \cap F)}{P(F)},$$

where $P(F) \neq 0$.

7. Two events E and F are said to be independent if and only if

$$P(E \cap F) = P(E) \cdot P(F).$$

E is said to be statistically independent of F if $P(E/F) = P(E)$ and $P(F/E) = P(F)$.

8. The events $E_1, E_2, \ldots, E_n$ are called mutually independent for all combinations if and only if every combination of these events taken any number at a time is independent.

9. *Bayes Theorem.*

   If $E_1, E_2, \ldots, E_n$ are n mutually exclusive events whose union is the sample space S, and E is any arbitrary event of S such that $P(E) \neq 0$, then

$$P(E_k/E) = \frac{P(E_k) \cdot P(E/E_k)}{\sum_{j=1}^{n} [P(E_j) \cdot P(E/E_j)]}$$

Random Variable

A function whose domain is a sample space S and whose range is some set of real numbers is called a random variable, denoted by $\mathbf{X}$. The function $\mathbf{X}$ transforms sample points of S into points on the x-axis. $\mathbf{X}$ will be called a discrete random variable if it is a random variable that assumes only a finite or denumerable number of values on the x-axis. $\mathbf{X}$ will be called a continuous random variable if it assumes a continuum of values on the x-axis.

Probability Function (Discrete Case)

The random variable $\mathbf{X}$ will be called a discrete random variable if there exists a function f such that $f(x_i) \geq 0$ and $\sum_i f(x_i) = 1$ for $i = 1, 2, 3, \ldots$ and such that for any event E,

$$P(E) = P[\mathbf{X} \text{ is in } E] = \sum_E f(x)$$

where $\sum_E$ means sum $f(x)$ over those values x_i that are in E and where $f(x) = P[\mathbf{X} = x]$. The probability that the value of $\mathbf{X}$ is some real number x, is given by $f(x) = P[\mathbf{X} = x]$, where f is called the probability function of the random variable $\mathbf{X}$.

Cumulative Distribution Function (Discrete Case)

The probability that the value of a random variable $\mathbf{X}$ is less than or equal to some real number x is defined as

$$F(x) = P(\mathbf{X} \leq x)$$
$$= \Sigma f(x_i), \qquad -\infty < x < \infty,$$

where the summation extends over those values of i such that $x_i \leq x$.

Probability Density (Continuous Case)

The random variable $\mathbf{X}$ will be called a continuous random variable if there exists a function f such that $f(x) \geq 0$ and $\int_{-\infty}^{\infty} f(x)\, dx = 1$ for all x in interval $-\infty < x < \infty$ and such that for any event E

$$P(E) = P(\mathbf{X} \text{ is in } E) = \int_E f(x)\, dx.$$

$f(x)$ is called the probability density of the random variable $\mathbf{X}$. The probability that $\mathbf{X}$ assumes any given value of x is equal to zero and the probability that it assumes a value on the interval from a to b, including or excluding either end point, is equal to

$$\int_a^b f(x)\, dx.$$

Cumulative Distribution Function (Continuous Case)

The probability that the value of a random variable $\mathbf{X}$ is less than or equal to some real number x is defined as

$$F(x) = P(\mathbf{X} \leq x), \qquad -\infty < x < \infty$$
$$= \int_{-\infty}^{x} f(x)\, dx.$$

From the cumulative distribution, the density, if it exists, can be found from

$$f(x) = \frac{dF(x)}{dx}.$$

From the cumulative distribution

$$P(a \leq \mathbf{X} \leq b) = P(\mathbf{X} \leq b) - P(\mathbf{X} \leq a)$$
$$= F(b) - F(a)$$

Mathematical Expectation

A. EXPECTED VALUE

Let $\mathbf{X}$ be a random variable with density $f(x)$. Then the expected value of $\mathbf{X}$, $E(\mathbf{X})$, is defined to be

$$E(\mathbf{X}) = \sum_x x f(x)$$

if $\mathbf{X}$ is discrete and

$$E(\mathbf{X}) = \int_{-\infty}^{\infty} x f(x)\, dx$$

if $\mathbf{X}$ is continuous. The expected value of a function g of a random variable $\mathbf{X}$ is defined as

$$E[g(\mathbf{X})] = \sum_x g(x) \cdot f(x)$$

if $\mathbf{X}$ is discrete and

$$E[g(\mathbf{X})] = \int_{-\infty}^{\infty} g(x) \cdot f(x)\, dx$$

if $\mathbf{X}$ is continuous.

Theorems

1. $E[aX + bY] = aE(X) + bE(Y)$
2. $E[X \cdot Y] = E(X) \cdot E(Y)$ if X and Y are statistically independent.

B. Moments

a. Moments About the Origin. The moments about the origin of a probability distribution are the expected values of the random variable which has the given distribution. The rth moment of X, usually denoted by μ'_r, is defined as

$$\mu'_r = E[X^r] = \sum_x x^r f(x)$$

if X is discrete and

$$\mu'_r = E[X^r] = \int_{-\infty}^{\infty} x^r f(x)\, dx$$

if X is continuous.

The first moment, μ'_1, is called the mean of the random variable X and is usually denoted by μ.

b. Moments About the Mean. The rth moment about the mean, usually denoted by μ_r, is defined as

$$\mu_r = E[(X - \mu)^r] = \sum_x (x - \mu)^r f(x)$$

if X is discrete and

$$\mu_r = E[(X - \mu)^r] = \int_{-\infty}^{\infty} (x - \mu)^r f(x)\, dx$$

if X is continuous.

The second moment about the mean, μ_2, is given by

$$\mu_2 = E[(X - \mu)^2] = \mu'_2 - \mu^2$$

and is called the variance of the random variable X, and is denoted by σ^2. The square root of the variance, σ, is called the standard deviation.

Theorems

1. $\sigma^2_{cX} = c^2 \sigma^2_X$
2. $\sigma^2_{c+X} = \sigma^2_X$
3. $\sigma^2_{aX+b} = a^2 \sigma^2_X$

c. Factorial Moments. The rth factorial moment of a probability distribution is defined as

$$\mu'_{(r)} = E[X^{[r]}] = \sum_x x^{[r]} f(x)$$

if X is discrete and

$$\mu'_{(r)} = E[X^{[r]}] = \int_{-\infty}^{\infty} x^{[r]} f(x)\, dx$$

if X is continuous, where the symbol $x^{[r]}$ denotes the factorial expression

$$x^{[r]} = x(x - 1)(x - 2) \cdots (x - r + 1), \quad r = 1, 2, 3, \ldots.$$

C. GENERATING FUNCTIONS

a. Moment Generating Functions. The moment generating function (m.g.f.) of the random variable X is defined as

$$m_x(t) = E(e^{tX}) = \sum_x e^{tx} f(x)$$

if X is discrete and

$$m_x(t) = E(e^{tX}) = \int_{-\infty}^{\infty} e^{tx} f(x)\, dx$$

if X is continuous.

$E(e^{tX})$ is the expected value of e^{tX}. If $m_x(t)$ and its derivatives exist, $|t| < h^2$, the rth moment about the origin is

$$\mu'_r = m_x^{(r)}(0), \qquad r = 0, 1, 2, \ldots$$

where $m_x^{(r)}(0)$ is the rth derivative of $m_x(t)$ with respect to t, evaluated at $t = 0$. For

$$m_x(t) = E(e^{tX})$$
$$= E\left[1 + Xt + \frac{(Xt)^2}{2!} + \cdots\right]$$
$$= 1 + \mu'_1 t + \mu'_2 \frac{t^2}{2} + \cdots$$

Thus, the moments μ'_r appear as coefficients of $\dfrac{t^r}{r!}$, and $m_x(t)$ may be regarded as generating the moments ν_r. The moments μ_r may be generated by the generating function

$$M_x(t) = E[e^{t(X-\mu)}] = e^{-\mu t} E(e^{tX}) = e^{-\mu t} m_x(t) .$$

b. Factorial Moment Generating Function. The factorial moment generating function is defined as

$$E(t^X) = \sum_x t^x f(x) \qquad \text{(probability generation function)}$$

if X is discrete and

$$E(t^X) = \int_{-\infty}^{\infty} t^x f(x)\, dx$$

if X is continuous.

The rth factorial moment is obtained from the factorial moment generating function by differentiating it r times with respect to t and then evaluating the result when $t = 1$.

Theorems

1. If c is a constant, the m.g.f. of $c + X$ is $e^{ct} m_x(t)$.
2. If c is a constant, the m.g.f. of cX is $m_x(ct)$.
3. If $Y = \sum_{i=1}^{n} X_i$, and $m_x(t)$ is the m.g.f. of X_i, where $X_1, \ldots, X_n$ is a random sample from $f(x)$, then the m.g.f. of Y is $[m_x(t)]^n$.

D. CUMULANT GENERATING FUNCTION

Let $m_x(t)$ be a m.g.f. If $\ln m_x(t)$ can be expanded in the form

$$c(t) = \ln m_x(t) = \kappa_1 t + \kappa_2 \frac{t^2}{2!} + \kappa_3 \frac{t^3}{3!} + \cdots + \kappa_r \frac{t^r}{r!} + \cdots,$$

then $c(t)$ is called the cumulant generating function (semi-invariant generating function) and κ_r are called the cumulants (semi-invariants) of a distribution.

$$\kappa_r = c^{(r)}(0)$$

where $c^{(r)}(0)$ is the rth derivative of $c(t)$ with respect to t evaluated at $t = 0$.

E. CHARACTERISTIC FUNCTIONS

The characteristic function of a distribution is defined as

$$\phi(t) = E(e^{itX}) = \sum_x e^{itx} \cdot f(x)$$

if **X** is discrete and

$$\phi(t) = E(e^{itX}) = \int_{-\infty}^{\infty} e^{itx} \cdot f(x) \, dx$$

if **X** is continuous.

Here t is a real number, $i^2 = -1$, and $e^{itX} = \cos(tX) + i\sin(tX)$. The characteristic function also generates moments, if they exist for

$$i^r \mu'_r = \phi^{(r)}(0)$$

where $\phi^{(r)}(0)$ is the rth derivative of $\phi(t)$ with respect to t evaluated at $t = 0$.

Multivariate Distributions

A. DISCRETE CASE

The k-dimensional random variable $(\mathbf{X}_1, \mathbf{X}_2, \ldots, \mathbf{X}_k)$ is a k-dimensional discrete random variable if it assumes values only at a finite or denumerable number of points $(x_1, x_2, \ldots, x_k)$. Define

$$P[\mathbf{X}_1 = x_1, \mathbf{X}_2 = x_2, \ldots, \mathbf{X}_k = x_k] = f(x_1, x_2, \ldots, x_k)$$

for every value that the random variable can assume. $f(x_1, x_2, \ldots, x_k)$ is called the joint density of the k-dimensional random variable. If E is any subset of the set of values that the random variable can assume, then

$$P(E) = P[(\mathbf{X}_1, \mathbf{X}_2, \ldots, \mathbf{X}_k) \text{ is in } E] = \sum_E f(x_1, x_2, \ldots, x_k)$$

where the sum is over all those points in E. The cumulative distribution is defined as

$$F(x_1, x_2, \ldots, x_k) = \sum_{x_1} \sum_{x_2} \cdots \sum_{x_k} f(x_1, x_2, \ldots, x_k).$$

B. CONTINUOUS CASE

The k random variables $\mathbf{X}_1, \mathbf{X}_2, \ldots, \mathbf{X}_k$ are said to be jointly distributed if there exists a function f such that $f(x_1, x_2, \ldots, x_k) \geq 0$ for all $-\infty < x_i < \infty$, $i = 1, 2, \ldots, k$ and such that for any event E

$$P(E) = P[(\mathbf{X}_1, \mathbf{X}_2, \ldots, \mathbf{X}_k) \text{ is in } E]$$
$$= \int_E f(x_1, x_2, \ldots x_k) \, dx_1 \, dx_2 \cdots dx_k.$$

$f(x_1, x_2, \ldots, x_k)$ is called the joint density of the random variables $\mathbf{X}_1, \mathbf{X}_2, \ldots, \mathbf{X}_k$. The cumulative distribution is defined as

$$F(x_1, x_2, \ldots, x_k) = \int_{-\infty}^{x_1} \int_{-\infty}^{x_2} \cdots \int_{-\infty}^{x_k} f(x_1, x_2, \ldots, x_k) \, dx_k \cdots dx_2 \, dx_1.$$

Given the cumulative distribution, the density may be found by

$$f(x_1, x_2, \ldots, x_k) = \frac{\partial}{\partial x_1} \cdot \frac{\partial}{\partial x_2} \cdots \frac{\partial}{\partial x_k} F(x_1, x_2, \ldots, x_k) .$$

Moments

The rth moment of X_i, say, is defined as

$$E(X_i^r) = \sum_{x_1} \sum_{x_2} \cdots \sum_{x_k} x_i^r f(x_1, x_2, \ldots, x_k)$$

if the X_i are discrete and

$$E(X_i^r) = \int_{-\infty}^{\infty} \int_{-\infty}^{\infty} \cdots \int_{-\infty}^{\infty} x_i^r f(x_1, x_2, \ldots, x_k) \, dx_k \cdots dx_2 \, dx_1$$

if the X_i are continuous.

Joint moments about the origin are defined as

$$E(X_1^{r_1} X_2^{r_2} \cdots X_k^{r_k})$$

where $r_1 + r_2 + \cdots + r_k$ is the order of the moment.

Joint moments about the mean are defined as

$$E[(X_1 - \mu_1)^{r_1}(X_2 - \mu_2)^{r_2} \cdots (X_k - \mu_k)^{r_k}].$$

Marginal and Conditional Distributions

If the random variables $X_1, X_2, \ldots, X_k$ have the joint density function $f(x_1, x_2, \ldots, x_k)$, then the marginal distribution of the subset of the random variables, say, $X_1, X_2, \ldots, X_p$ $(p < k)$, is given by

$$g(x_1, x_2, \ldots, x_p) = \sum_{x_{p+1}} \sum_{x_{p+2}} \cdots \sum_{x_k} f(x_1, x_2, \ldots, x_k)$$

if the X's are discrete, and

$$g(x_1, x_2, \ldots, x_p) = \int_{-\infty}^{\infty} \int_{-\infty}^{\infty} \cdots \int_{-\infty}^{\infty} f(x_1, x_2, \ldots, x_k) \, dx_{p+1} \cdots dx_{k-1} \, dx_k$$

if the X's are continuous.

The conditional distribution of a certain subset of the random variables is the joint distribution of this subset under the condition that the remaining variables are given certain values. The conditional distribution of $X_1, X_2, \ldots, X_p$ given $X_{p+1}, X_{p+2}, \ldots, X_k$ is

$$h(x_1, x_2, \ldots, x_p | x_{p+1}, x_{p+2}, \ldots, x_k) = \frac{f(x_1, x_2, \ldots, x_k)}{g(x_{p+1}, x_{p+2}, \ldots, x_k)}$$

if $g(x_{p+1}, x_{p+2}, \ldots, x_k) \neq 0$.

The variance σ_{ii} of X_i and the covariance σ_{ij} of X_i and X_j are given by

$$\sigma_{ii} = \sigma_i^2 = E[(X_i - \mu_i)^2]$$

and

$$\sigma_{ij} = \rho_{ij}\sigma_i\sigma_j = E[(X_i - \mu_i)(X_j - \mu_j)]$$

where ρ_{ij} is the correlation coefficient and σ_i and σ_j are the standard deviations of X_i and X_j.

A joint m.g.f. is defined as

$$m(t_1, t_2, \ldots, t_k) = E[e^{t_1 \mathbf{X}_1 + t_2 \mathbf{X}_2 + \cdots + t_k \mathbf{X}_k}]$$

if it exists for all values of t_i such that $|t_i| < h^2$.

The rth moment of $\mathbf{X}_i$ may be obtained by differentiating the m.g.f. r times with respect to t_i and then evaluating the result when all t's are set equal to zero. Similarly, a joint moment would be found by differentiating the m.g.f. r_1 times with respect to t_1, $\ldots$, r_k times with respect to t_k, and then evaluating the result when all t's are set equal to zero.

Probability Distributions

A. DISCRETE CASE

1. *Discrete Uniform Distribution.* If the random variable $\mathbf{X}$ has a probability function given by

$$P(\mathbf{X} = x) = f(x) = \frac{1}{n}, \qquad x = x_1, x_2, \ldots, x_n,$$

then the variable $\mathbf{X}$ is said to possess a discrete uniform distribution.

Properties

When $x_i = i$ for $i = 1, 2, \ldots$, and n

$$\text{Mean} = \mu = \frac{n + 1}{2}$$

$$\text{Variance} = \sigma^2 = \frac{n^2 - 1}{12}$$

$$\text{Standard Deviation} = \sigma = \sqrt{\frac{n^2 - 1}{12}}$$

$$\text{Moment Generating Function} = m_x(t) = \frac{e^t(1 - e^{nt})}{n(1 - e^t)}$$

2. *Binomial Distribution.* If the random variable $\mathbf{X}$ has a probability function given by

$$P(\mathbf{X} = x) = f(x) = \binom{n}{x} \theta^x (1 - \theta)^{n-x}, \qquad x = 0, 1, 2, \ldots, n$$

where

$$\binom{n}{x} = \frac{n!}{x!(n - x)!},$$

then the variable $\mathbf{X}$ is said to possess a binomial distribution. $f(x)$ is the general term of the expansion of $[\theta + (1 - \theta)]^n$.

Properties

$$\text{Mean} = \mu = n\theta$$
$$\text{Variance} = \sigma^2 = n\theta(1 - \theta)$$
$$\text{Standard Deviation} = \sigma = \sqrt{n\theta(1 - \theta)}$$
$$\text{Moment Generating Function} = m_x(t) = [\theta e^t + (1 - \theta)]^n$$

3. *Geometric Distribution.* If the random variable $\mathbf{X}$ has a probability function given by

$$P(\mathbf{X} = x) = f(x) = \theta(1 - \theta)^{x-1}, \qquad x = 1, 2, 3, \ldots,$$

then the variable $\mathbf{X}$ is said to possess a geometric distribution.

Properties

$$\text{Mean} = \mu = \frac{1}{\theta}$$

$$\text{Variance} = \sigma^2 = \frac{1 - \theta}{\theta^2}$$

$$\text{Standard Deviation} = \sigma = \sqrt{\frac{1 - \theta}{\theta^2}}$$

$$\text{Moment Generating Function} = m_x(t) = \frac{\theta e^t}{1 - e^t(1 - \theta)}$$

4. *Multinomial Distribution.* If a set of random variables $X_1, X_2, \ldots, X_n$ has a probability function given by

$$P(X_1 = x_1, X_2 = x_2, \ldots, X_n = x_n) = f(x_1, x_2, \ldots, x_n) = \frac{N!}{\prod\limits_{i=1}^{n} x_i!} \prod_{i=1}^{n} \theta_i^{x_i}$$

where x_i are positive integers and each $\theta_i > 0$ for $i = 1, 2, \ldots, n$ and

$$\sum_{i=1}^{n} \theta_i = 1, \qquad \sum_{i=1}^{n} x_i = N,$$

then the joint distribution of $X_1, X_2, \ldots, X_n$ is called the multinomial distribution. $f(x_1, x_2, \ldots, x_n)$ is the general term of the expansion of $(\theta_1 + \theta_2 + \cdots + \theta_n)^N$.

Properties

$$\text{Mean of } X_i = \mu_i = N\theta_i$$
$$\text{Variance of } X_i = \sigma_i^2 = N\theta_i(1 - \theta_i)$$
$$\text{Covariance of } X_i \text{ and } X_j = \sigma_{ij}^2 = -N\theta_i\theta_j$$
$$\text{Joint Moment Generating Function} = (\theta_1 e^{t_1} + \cdots + \theta_n e^{t_n})^N$$

5. *Poisson Distribution.* If the random variable X has a probability function given by

$$P(X = x) = f(x) = \frac{e^{-\lambda}\lambda^x}{x!}, \qquad \lambda > 0, x = 0, 1, \ldots,$$

then the variable X is said to possess a Poisson distribution.

Properties

$$\text{Mean} = \mu = \lambda$$
$$\text{Variance} = \sigma^2 = \lambda$$
$$\text{Standard Deviation} = \sigma = \sqrt{\lambda}$$
$$\text{Moment Generating Function} = m_x(t) = e^{\lambda(e^t - 1)}$$

6. *Hypergeometric Distribution.* If the random variable X has a probability function given by

$$P(X = x) = f(x) = \frac{\binom{k}{x}\binom{N - k}{n - x}}{\binom{N}{n}}, \qquad x = 0, 1, 2, \ldots, [n, k],$$

where $[n, k]$ means the smaller of the two numbers n, k, then the variable X is said to possess a hypergeometric distribution.

Properties

$$\text{Mean} = \mu = \frac{kn}{N}$$

$$\text{Variance} = \sigma^2 = \frac{k(N-k)n(N-n)}{N^2(N-1)}$$

$$\text{Standard Deviation} = \sigma = \sqrt{\frac{k(N-k)n(N-n)}{N^2(N-1)}}$$

7. *Negative Binomial Distribution.* If the random variable **X** has a probability function given by

$$\text{P}(\mathbf{X} = x) = f(x) = \binom{x+r-1}{r-1} \theta^r(1-\theta)^x, \qquad x = 0, 1, 2, \ldots;$$

then the variable **X** is said to possess a negative binomial distribution, known also as the Pascal or Pólya distribution.

Properties

$$\text{Mean} = \mu = \frac{r}{\theta}$$

$$\text{Variance} = \sigma^2 = \frac{r}{\theta}\left(\frac{1}{\theta}-1\right) = \frac{r(1-\theta)}{\theta^2}$$

$$\text{Standard Deviation} = \sigma = \sqrt{\frac{r}{\theta}\left(\frac{1}{\theta}-1\right)} = \sqrt{\frac{r(1-\theta)}{\theta^2}}$$

$$\text{Moment Generating Function} = m_x(t) = e^{tr}\theta^r[1-(1-\theta)e^t]^{-r}$$

B. CONTINUOUS CASE

1. *Uniform Distribution.* A random variable **X** is said to be distributed as the uniform distribution if the density function is given by

$$f(x) = \frac{1}{\beta-\alpha}, \qquad \alpha < x < \beta,$$

where α and β are parameters with $\alpha < \beta$.

Properties

$$\text{Mean} = \mu = \frac{\alpha+\beta}{2}$$

$$\text{Variance} = \sigma^2 = \frac{(\beta-\alpha)^2}{12}$$

$$\text{Standard Deviation} = \sigma = \sqrt{\frac{(\beta-\alpha)^2}{12}}$$

$$\text{Moment Generating Function} = m_x(t) = \frac{2}{(\beta-\alpha)t}\sin\left[\frac{(\beta-\alpha)t}{2}\right]e^{\frac{\alpha+\beta}{2}t}$$

2. *Normal Distribution.* A random variable **X** is said to be normally distributed if its density function is given by

$$f(x) = \frac{1}{\sqrt{2\pi}\,\sigma}e^{-(x-\mu)^2/2\sigma^2}, \qquad -\infty < x < \infty$$

where μ and σ are parameters, called the mean and the standard deviation or the random variable **X**, respectively.

Properties

$$\text{Mean} = \mu$$
$$\text{Variance} = \sigma^2$$
$$\text{Standard Deviation} = \sigma$$

$$\text{Moment Generating Function} = m_x(t) = e^{t\mu + \frac{\sigma^2 t^2}{2}}$$

Cumulative Distribution

$$F(x) = \int_{-\infty}^{x} \frac{1}{\sqrt{2\pi}\,\sigma} e^{-(x-\mu)^2/2\sigma^2}\,dx$$

Set $y = \dfrac{x-\mu}{\sigma}$ to obtain the cumulative standard normal.

3. *Gamma Distribution.* A random variable **X** is said to be distributed as the gamma distribution if the density function is given by

$$f(x) = \frac{1}{\Gamma(\alpha+1)\beta^{\alpha+1}}\, x^{\alpha} e^{-x/\beta}, \qquad 0 < x < \infty$$

where α and β are parameters with $\alpha > -1$ and $\beta > 0$.

Properties

$$\text{Mean} = \mu = \beta(\alpha+1)$$
$$\text{Variance} = \sigma^2 = \beta^2(\alpha+1)$$
$$\text{Standard Deviation} = \sigma = \beta\sqrt{\alpha+1}$$
$$\text{Moment Generating Function} = m_x(t) = (1-\beta t)^{-(\alpha+1)}, \qquad t < \frac{1}{\beta}.$$

4. *Exponential Distribution.* A random variable **X** is said to be distributed as the exponential distribution if the density function is given by

$$f(x) = \frac{1}{\theta} e^{-x/\theta}, \qquad x > 0$$

where θ is a parameter and $\theta > 0$.

Properties

$$\text{Mean} = \mu = \theta$$
$$\text{Variance} = \sigma^2 = \theta^2$$
$$\text{Standard Deviation} = \sigma = \sqrt{\theta^2}$$
$$\text{Moment Generating Function} = m_x(t) = (1-\theta t)^{-1}$$

5. *Beta Distribution.* A random variable **X** is said to be distributed as the beta distribution if the density function is given by

$$f(x) = \frac{\Gamma(\alpha+\beta+2)}{\Gamma(\alpha+1)\Gamma(\beta+1)}\, x^{\alpha}(1-x)^{\beta}, \qquad 0 < x < 1$$

where α and β are parameters with $\alpha > -1$ and $\beta > -1$.

Properties

$$\text{Mean} = \mu = \frac{\alpha+1}{\alpha+\beta+2}$$

$$\text{Variance} = \sigma^2 = \frac{(\alpha+1)(\beta+1)}{(\alpha+\beta+2)^2(\alpha+\beta+3)}$$

$$r\text{th moment about the origin} = \nu_r = \frac{\Gamma(\alpha+\beta+2)\Gamma(\alpha+r+1)}{\Gamma(\alpha+\beta+r+2)\Gamma(\alpha+1)}.$$

Sampling Distributions

Population—A finite or infinite set of elements of a random variable **X**.

Random Sample—If the random variables $X_1, X_2, \ldots, X_n$ have a joint density,

$$g(x_1, x_2, \ldots, x_n) = f(x_1)f(x_2) \cdots f(x_n)$$

where the density of each X_i is $f(x)$, then $X_1, X_2, \ldots, X_n$ is said to be a random sample of size n from the population with density $f(x)$.

Sampling Distributions

A random sample is selected from a population in which the form of the probability function is known, and from the joint density of the random variables a distribution, called the sampling distribution, of a function of the random variables is derived.

1. *Chi-Square Distribution.* If $Y_1, Y_2, \ldots, Y_n$ are normally and independently distributed with mean 0 and variance 1, then

$$\chi^2 = \sum_{i=1}^{n} Y_i^2$$

is distributed as Chi-Square (χ^2) with n degrees of freedom. The density function is given by

$$f(\chi^2) = \frac{(\chi^2)^{\frac{1}{2}(n-2)}}{2^{\frac{n}{2}}\Gamma\left(\frac{n}{2}\right)} e^{-\chi^2/2}, \quad 0 < \chi^2 < \infty .$$

Properties

$$\text{Mean} = \mu = n$$
$$\text{Variance} = \sigma^2 = 2n$$

Reproductive Property of χ^2 - Distribution

If $\chi_1^2, \chi_2^2, \ldots, \chi_k^2$ are independently distributed according to χ^2 - distributions with $n_1, n_2, \ldots, n_k$ degrees of freedom, respectively, then $\sum_{j=1}^{k} \chi_j^2$ is distributed according to a χ^2 - distribution with $n = \sum_{j=1}^{k} n_j$ degrees of freedom.

2. *Snedecor's F-Distribution.* If a random variable X is distributed as χ^2 with m degrees of freedom (χ_m^2) and a random variable Y is distributed as χ^2 with n degrees of freedom (χ_n^2) and if X and Y are independent, then $F = \dfrac{X/m}{Y/n}$ is distributed as Snedecor's F with m and n degrees of freedom, denoted by $F(m, n)$. The density function of the F-distribution is given by

$$f(F) = \frac{\Gamma\left(\dfrac{m+n}{2}\right)\left(\dfrac{m}{n}\right)^{m/2} F^{(m-2)/2}}{\Gamma\left(\dfrac{m}{2}\right)\Gamma\left(\dfrac{n}{2}\right)\left(1 + \dfrac{m}{n}F\right)^{(m+n)/2}}, \qquad 0 < F < \infty.$$

Properties

$$\text{Mean} = \mu = \frac{n}{n-2}, \qquad n > 2$$

$$\text{Variance} = \sigma^2 = \frac{2n^2(m+n-2)}{m(n-2)^2(n-4)}, \qquad n > 4 .$$

The transformation $w = \dfrac{mF/n}{1 + \dfrac{mF}{n}}$ transforms the F-density into a Beta density.

3. *Student's t-Distribution.* If a random variable X is normally distributed with mean 0 and variance σ^2, and if Y^2/σ^2 is distributed as χ^2 with n degrees of freedom and if X and Y are independent, then

$$t = \frac{X\sqrt{n}}{Y}$$

is distributed as Student's t with n degrees of freedom. The density function is given by

$$f(t) = \frac{\Gamma\left(\dfrac{n+1}{2}\right)}{\sqrt{n\pi}\,\Gamma\left(\dfrac{n}{2}\right)\left(1 + \dfrac{t^2}{n}\right)^{\frac{1}{2}(n+1)}}, \qquad -\infty < t < \infty .$$

Properties

$$\text{Mean} = \mu = 0$$

$$\text{Variance} = \sigma^2 = \frac{n}{n-2}, \qquad n > 2 .$$

SUMMARY OF SIGNIFICANCE TESTS: TESTING FOR THE VALUE OF A SPECIFIED PARAMETER

Hypothesis	Conditions	Test Statistic	Distribution of Test Statistic	Critical Region
1. $\mu = \mu_0$	Known σ	$z = \dfrac{(\bar{x} - \mu_0)\sqrt{n}}{\sigma}$	Normal	$z > z_\alpha$ if we wish to reject when $\mu > \mu_0$ $z < -z_\alpha$ if we wish to reject when $\mu < \mu_0$ $|z| > z_{\alpha/2}$ if we wish to reject when $\mu \neq \mu_0$
2. $\mu = \mu_0$	Unknown σ	$t = \dfrac{(\bar{x} - \mu_0)\sqrt{n}}{s}$	Student's t — with $(n-1)$ d.f.	$t > t_{\alpha,n-1}$ if we wish to reject when $\mu > \mu_0$ $t < -t_{\alpha,n-1}$ if we wish to reject when $\mu < \mu_0$ $|t| > t_{\alpha/2,n-1}$ if we wish to reject when $\mu \neq \mu_0$
3. $\sigma = \sigma_0$		$\chi^2 = \dfrac{(n-1)s^2}{\sigma_0^2}$	χ^2 with $n-1$ d.f.	$\chi^2 > \chi^2_{\alpha,n-1}$ if we wish to reject when $\sigma > \sigma_0$ $\chi^2 < \chi^2_{1-\alpha,n-1}$ if we wish to reject when $\sigma < \sigma_0$ $\chi^2 < \chi^2_{1-\alpha/2,n-1}$ or $\chi^2 > \chi^2_{\alpha/2,n-1}$ if we wish to reject when $\sigma \neq \sigma_0$
4. $\theta = \theta_0$	Large sample. (For small samples, exact tests are based on tables of binomial probabilities)	$z = \dfrac{\dfrac{x}{n} - \theta_0}{\sqrt{\dfrac{\theta_0(1 - \theta_0)}{n}}}$ Continuity correction: Replace x in numerator of formula with $x - \frac{1}{2}$ or $x + \frac{1}{2}$, whichever makes z numerically smallest.	Normal	$z > z_\alpha$ if we wish to reject when $\theta > \theta_0$ $z < -z_\alpha$ if we wish to reject when $\theta < \theta_0$ $|z| > z_{\alpha/2}$ if we wish to reject when $\theta \neq \theta_0$

SUMMARY OF SIGNIFICANCE TESTS: COMPARISON OF TWO POPULATIONS

Hypothesis	Conditions	Test Statistic	Distribution of Test Statistic	Critical Region
1. $\mu_x = \mu_y$	Known σ_x and σ_y	$z = \dfrac{\bar{x} - \bar{y}}{\sqrt{\dfrac{\sigma_x^2}{n_x} + \dfrac{\sigma_y^2}{n_y}}}$	Normal	$z > z_\alpha$ if we wish to reject when $\mu_x > \mu_y$ $z < -z_\alpha$ if we wish to reject when $\mu_x < \mu_y$ $|z| > z_{\alpha/2}$ if we wish to reject when $\mu_x \neq \mu_y$
2. $\mu_x = \mu_y$	Unknown σ_x and σ_y $\sigma_x = \sigma_y$	$t = \dfrac{\bar{x} - \bar{y}}{\sqrt{\dfrac{(n_x - 1)s_x^2 + (n_y - 1)s_y^2}{n_x + n_y - 2}}\ \sqrt{\dfrac{1}{n_x} + \dfrac{1}{n_y}}}$	Student's t with $n - 1$ d.f.	$t > t_{\alpha; n_x + n_y - 2}$ if we wish to reject when $\mu_x > \mu_y$ $t < -t_{\alpha; n_x + n_y - 2}$ if we wish to reject when $\mu_x < \mu_y$ $|t| > t_{\alpha/2; n_x + n_y - 2}$ if we wish to reject when $\mu_x \neq \mu_y$
3. $\mu_x = \mu_y$	Unknown σ_x and σ_y $\sigma_x \neq \sigma_y$	$t = \dfrac{\bar{x} - \bar{y}}{\sqrt{\dfrac{s_x^2}{n_x} + \dfrac{s_y^2}{n_y}}}$	Student's t with ν d.f.	$t > t_{\alpha}$ if we wish to reject when $\mu_x > \mu_y$ $t < -t_{\alpha}$, if we wish to reject when $\mu_x < \mu_y$ $|t| > t_{\alpha/2}$, if we wish to reject when $\mu_x \neq \mu_y$ where d.f. ν is given by closest integer to $\nu = -2 + \dfrac{\left(\dfrac{s_x^2}{n_x} + \dfrac{s_y^2}{n_y}\right)^2}{\dfrac{\left(\dfrac{s_x^2}{n_x}\right)^2}{n_x + 1} + \dfrac{\left(\dfrac{s_y^2}{n_y}\right)^2}{n_y + 1}}$
4. $\mu_x = \mu_y$	Correlated pairs	$t = \dfrac{\bar{d}\sqrt{n}}{s_d}$ where $d_i = x_i - y_i$	Student's t with $n - 1$ d.f.	$t > t_{\alpha; n-1}$ if we wish to reject when $\mu_x > \mu_y$ $t < -t_{\alpha; n-1}$ if we wish to reject when $\mu_x < \mu_y$ $|t| > t_{\alpha/2; n-1}$ if we wish to reject when $\mu_x \neq \mu_y$
5. $\sigma_x^2 = \sigma_y^2$		$F = \dfrac{s_x^2}{s_y^2}$ In a two-sided test, put larger mean square in the numerator	F with $n_x - 1$ and $n_y - 1$ d.f.	$F > F_{\alpha; n_x-1, n_y-1}$ if we wish to reject when $\sigma_x > \sigma_y$ $F > F_{\alpha/2; n_x-1, n_y-1}$ if $s_x^2 > s_y^2$ and we wish to reject when $\sigma_x \neq \sigma_y$ $F > F_{\alpha/2; n_y-1, n_x-1}$ if $s_x^2 < s_y^2$ and we wish to reject when $\sigma_x \neq \sigma_y$
6. $\theta_1 = \theta_2$	Large sample	$z = \dfrac{\dfrac{x_1}{n_1} - \dfrac{x_2}{n_2}}{\sqrt{\dfrac{\dfrac{x_1}{n_1}\left(1 - \dfrac{x_1}{n_1}\right)}{n_1} + \dfrac{\dfrac{x_2}{n_2}\left(1 - \dfrac{x_2}{n_2}\right)}{n_2}}}$ Continuity Correction: Replace x in numerator of formula with $x - \frac{1}{2}$ or $x + \frac{1}{2}$, whichever makes z numerically smallest.	Normal	$z > z_\alpha$ if we wish to reject when $\theta_1 > \theta_2$ $z < -z_\alpha$ if we wish to reject when $\theta_1 < \theta_2$ $|z| > z_{\alpha/2}$ if we wish to reject when $\theta_1 \neq \theta_2$

SUMMARY OF CONFIDENCE INTERVALS

Parameter	Conditions	Point Estimate	Confidence Interval
1. μ	Known σ	$\bar{x}$	$\bar{x} - z_{\alpha/2} \dfrac{\sigma}{\sqrt{n}} < \mu < \bar{x} + z_{\alpha/2} \dfrac{\sigma}{\sqrt{n}}$
2. μ	Unknown σ	$\bar{x}$	$\bar{x} - t_{\alpha/2} \dfrac{s}{\sqrt{n}} < \mu < \bar{x} + t_{\alpha/2} \dfrac{s}{\sqrt{n}}$
3. $\mu_x - \mu_y$	$\sigma_x = \sigma_y$ known	$\bar{x} - \bar{y}$	$\bar{x} - \bar{y} - z_{\alpha/2} \sqrt{\dfrac{\sigma_x^2}{n_x} + \dfrac{\sigma_y^2}{n_y}} < \mu_x - \mu_y$ $$< \bar{x} - \bar{y} + z_{\alpha/2} \sqrt{\dfrac{\sigma_x^2}{n_x} + \dfrac{\sigma_y^2}{n_y}}$$
4. $\mu_x - \mu_y$	$\sigma_x = \sigma_y$ unknown	$\bar{x} - \bar{y}$	$\bar{x} - \bar{y} - t_{\alpha/2} \dfrac{\sqrt{(n_x - 1)s_x^2 + (n_y - 1)s_y^2}}{\sqrt{\dfrac{n_x n_y (n_x + n_y - 2)}{n_x + n_y}}}$ $$< \mu_x - \mu_y < \bar{x} - \bar{y}$$ $$+ t_{\alpha/2} \dfrac{\sqrt{(n_x - 1)s_x^2 + (n_y - 1)s_y^2}}{\sqrt{\dfrac{n_x n_y (n_x + n_y - 2)}{n_x + n_y}}}$$
5. $\mu_d = \mu_x - \mu_y$	Correlated pairs σ_x and σ_y unknown	$\bar{d} = \bar{x} - \bar{y}$	$\bar{d} - t_{\alpha/2} \dfrac{s_d}{\sqrt{n}} < \mu_d < \bar{d} + t_{\alpha/2} \dfrac{s_d}{\sqrt{n}}$
6. σ		s	$\sqrt{\dfrac{(n-1)s^2}{\chi^2_{\alpha/2;n-1}}} < \sigma < \sqrt{\dfrac{(n-1)s^2}{\chi^2_{1-\alpha/2;n-1}}}$
7. $\dfrac{\sigma_x^2}{\sigma_y^2}$		$\dfrac{s_x^2}{s_y^2}$	$\dfrac{s_x^2}{s_y^2} \dfrac{1}{F_{\alpha/2;n_x-1,n_y-1}} < \dfrac{\sigma_x^2}{\sigma_y^2} < \dfrac{s_x^2}{s_y^2} \dfrac{1}{F_{1-\alpha/2;n_x-1,n_y-1}}$
8. θ	Large sample	$\dfrac{x}{n}$	$\dfrac{x}{n} - z_{\alpha/2} \sqrt{\dfrac{\dfrac{x}{n}\left(1 - \dfrac{x}{n}\right)}{n}} < \theta < \dfrac{x}{n}$ $$+ z_{\alpha/2} \sqrt{\dfrac{\dfrac{x}{n}\left(1 - \dfrac{x}{n}\right)}{n}}$$
9. $\theta_1 - \theta_2$	Large sample	$\dfrac{x_1}{n_1} - \dfrac{x_2}{n_2}$	$\dfrac{x_1}{n_1} - \dfrac{x_2}{n_2} - z_{\alpha/2} \sqrt{\dfrac{\dfrac{x_1}{n_1}\left(1 - \dfrac{x_1}{n_1}\right)}{n_1} + \dfrac{\dfrac{x_2}{n_2}\left(1 - \dfrac{x_2}{n_2}\right)}{n_2}}$ $$< \theta_1 - \theta_2 < \dfrac{x_1}{n_1} - \dfrac{x_2}{n_2}$$ $$+ z_{\alpha/2} \sqrt{\dfrac{\dfrac{x_1}{n_1}\left(1 - \dfrac{x_1}{n_1}\right)}{n_1} + \dfrac{\dfrac{x_2}{n_2}\left(1 - \dfrac{x_2}{n_2}\right)}{n_2}}$$

THE NORMAL PROBABILITY FUNCTION
AND RELATED FUNCTIONS

This table gives values of:

a) $f(x)$ = the probability density of a standardized random variable

$$= \frac{1}{\sqrt{2\pi}} e^{-\frac{1}{2}x^2}$$

For negative values of x, one uses the fact that $f(-x) = f(x)$.

b) $F(x)$ = the cumulative distribution function of a standardized normal random variable

$$= \int_{-\infty}^{x} \frac{1}{\sqrt{2\pi}} e^{-\frac{1}{2}t^2} dt$$

For negative values of x, one uses the relationship $F(-x) = 1 - F(x)$. Values of x corresponding to a few special values of $F(x)$ are given in a separate table following the main table. (See page 536.)

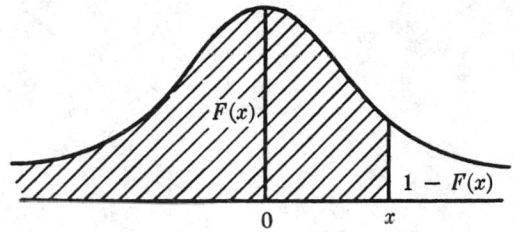

c) $f'(x)$ = the first derivative of $f(x)$ with respect to x

$$= -\frac{x}{\sqrt{2\pi}} e^{-\frac{1}{2}x^2} = -xf(x)$$

d) $f''(x)$ = the second derivative of $f(x)$ with respect to x

$$= \frac{(x^2 - 1)}{\sqrt{2\pi}} e^{-\frac{1}{2}x^2} = (x^2 - 1)f(x)$$

e) $f'''(x)$ = the third derivative of $f(x)$ with respect to x

$$= \frac{3x - x^3}{\sqrt{2\pi}} e^{-\frac{1}{2}x^2} = (3x - x^3)f(x)$$

f) $f^{\mathrm{iv}}(x)$ = the fourth derivative of $f(x)$ with respect to x

$$= \frac{x^4 - 6x^2 + 3}{\sqrt{2\pi}} e^{-\frac{1}{2}x^2} = (x^4 - 6x^2 + 3)f(x)$$

It should be noted that other probability integrals can be evaluated by the use of these tables. For example,

$$\int_0^z f(t)dt = \tfrac{1}{2}\operatorname{erf}\left(\frac{x}{\sqrt{2}}\right),$$

where $\operatorname{erf}\left(\dfrac{x}{\sqrt{2}}\right)$ represents the error function associated with the normal curve.

To evaluate erf (2.3) one proceeds as follows: Since $\dfrac{x}{\sqrt{2}} = 2.3$, one finds $x = (2.3)(\sqrt{2}) = 3.25$. In the entry opposite $x = 3.25$, the value 0.9994 is given. Subtracting 0.5000 from the tabular value, one finds the value 0.4994. Thus erf (2.3) = 2(0.4994) = 0.9988.

NORMAL DISTRIBUTION AND RELATED FUNCTIONS

x	$F(x)$	$1 - F(x)$	$f(x)$	$f'(x)$	$f''(x)$	$f'''(x)$	$f^{iv}(x)$
.00	.5000	.5000	.3989	− .0000	− .3989	.0000	1.1968
.01	.5040	.4960	.3989	− .0040	− .3989	.0120	1.1965
.02	.5080	.4920	.3989	− .0080	− .3987	.0239	1.1956
.03	.5120	.4880	.3988	− .0120	− .3984	.0359	1.1941
.04	.5160	.4840	.3986	− .0159	− .3980	.0478	1.1920
.05	.5199	.4801	.3984	− .0199	− .3975	.0597	1.1894
.06	.5239	.4761	.3982	− .0239	− .3968	.0716	1.1861
.07	.5279	.4721	.3980	− .0279	− .3960	.0834	1.1822
.08	.5319	.4681	.3977	− .0318	− .3951	.0952	1.1778
.09	.5359	.4641	.3973	− .0358	− .3941	.1070	1.1727
.10	.5398	.4602	.3970	− .0397	− .3930	.1187	1.1671
.11	.5438	.4562	.3965	− .0436	− .3917	.1303	1.1609
.12	.5478	.4522	.3961	− .0475	− .3904	.1419	1.1541
.13	.5517	.4483	.3956	− .0514	− .3889	.1534	1.1468
.14	.5557	.4443	.3951	− .0553	− .3873	.1648	1.1389
.15	.5596	.4404	.3945	− .0592	− .3856	.1762	1.1304
.16	.5636	.4364	.3939	− .0630	− .3838	.1874	1.1214
.17	.5675	.4325	.3932	− .0668	− .3819	.1986	1.1118
.18	.5714	.4286	.3925	− .0707	− .3798	.2097	1.1017
.19	.5753	.4247	.3918	− .0744	− .3777	.2206	1.0911
.20	.5793	.4207	.3910	− .0782	− .3754	.2315	1.0799
.21	.5832	.4168	.3902	− .0820	− .3730	.2422	1.0682
.22	.5871	.4129	.3894	− .0857	− .3706	.2529	1.0560
.23	.5910	.4090	.3885	− .0894	− .3680	.2634	1.0434
.24	.5948	.4052	.3876	− .0930	− .3653	.2737	1.0302
.25	.5987	.4013	.3867	− .0967	− .3625	.2840	1.0165
.26	.6026	.3974	.3857	− .1003	− .3596	.2941	1.0024
.27	.6064	.3936	.3847	− .1039	− .3566	.3040	0.9878
.28	.6103	.3897	.3836	− .1074	− .3535	.3138	0.9727
.29	.6141	.3859	.3825	− .1109	− .3504	.3235	0.9572
.30	.6179	.3821	.3814	− .1144	− .3471	.3330	0.9413
.31	.6217	.3783	.3802	− .1179	− .3437	.3423	0.9250
.32	.6255	.3745	.3790	− .1213	− .3402	.3515	0.9082
.33	.6293	.3707	.3778	− .1247	− .3367	.3605	0.8910
.34	.6331	.3669	.3765	− .1280	− .3330	.3693	0.8735
.35	.6368	.3632	.3752	− .1313	− .3293	.3779	0.8556
.36	.6406	.3594	.3739	− .1346	− .3255	.3864	0.8373
.37	.6443	.3557	.3725	− .1378	− .3216	.3947	0.8186
.38	.6480	.3520	.3712	− .1410	− .3176	.4028	0.7996
.39	.6517	.3483	.3697	− .1442	− .3135	.4107	0.7803
.40	.6554	.3446	.3683	− .1473	− .3094	.4184	0.7607
.41	.6591	.3409	.3668	− .1504	− .3051	.4259	0.7408
.42	.6628	.3372	.3653	− .1534	− .3008	.4332	0.7206
.43	.6664	.3336	.3637	− .1564	− .2965	.4403	0.7001
.44	.6700	.3300	.3621	− .1593	− .2920	.4472	0.6793
.45	.6736	.3264	.3605	− .1622	− .2875	.4539	0.6583
.46	.6772	.3228	.3589	− .1651	− .2830	.4603	0.6371
.47	.6808	.3192	.3572	− .1679	− .2783	.4666	0.6156
.48	.6844	.3156	.3555	− .1707	− .2736	.4727	0.5940
.49	.6879	.3121	.3538	− .1734	− .2689	.4785	0.5721
.50	.6915	.3085	.3521	− .1760	− .2641	.4841	0.5501

PROBABILITY AND STATISTICS

NORMAL DISTRIBUTION AND RELATED FUNCTIONS

x	$F(x)$	$1 - F(x)$	$f(x)$	$f'(x)$	$f''(x)$	$f'''(x)$	$f^{iv}(x)$
.50	.6915	.3085	.3521	− .1760	− .2641	.4841	.5501
.51	.6950	.3050	.3503	− .1787	− .2592	.4895	.5279
.52	.6985	.3015	.3485	− .1812	− .2543	.4947	.5056
.53	.7019	.2981	.3467	− .1837	− .2493	.4996	.4831
.54	.7054	.2946	.3448	− .1862	− .2443	.5043	.4605
.55	.7088	.2912	.3429	− .1886	− .2392	.5088	.4378
.56	.7123	.2877	.3410	− .1920	− .2341	.5131	.4150
.57	.7157	.2843	.3391	− .1933	− .2289	.5171	.3921
.58	.7190	.2810	.3372	− .1956	− .2238	.5209	.3691
.59	.7224	.2776	.3352	− .1978	− .2185	.5245	.3461
.60	.7257	.2743	.3332	− .1999	− .2133	.5278	.3231
.61	.7291	.2709	.3312	− .2020	− .2080	.5309	.3000
.62	.7324	.2676	.3292	− .2041	− .2027	.5338	.2770
.63	.7357	.2643	.3271	− .2061	− .1973	.5365	.2539
.64	.7389	.2611	.3251	− .2080	− .1919	.5389	.2309
.65	.7422	.2578	.3230	− .2099	− .1865	.5411	.2078
.66	.7454	.2546	.3209	− .2118	− .1811	.5431	.1849
.67	.7486	.2514	.3187	− .2136	− .1757	.5448	.1620
.68	.7517	.2483	.3166	− .2153	− .1702	.5463	.1391
.69	.7549	.2451	.3144	− .2170	− .1647	.5476	.1164
.70	.7580	.2420	.3123	− .2186	− .1593	.5486	.0937
.71	.7611	.2389	.3101	− .2201	− .1538	.5495	.0712
.72	.7642	.2358	.3079	− .2217	− .1483	.5501	.0487
.73	.7673	.2327	.3056	− .2231	− .1428	.5504	.0265
.74	.7704	.2296	.3034	− .2245	− .1373	.5506	.0043
.75	.7734	.2266	.3011	− .2259	− .1318	.5505	− .0176
.76	.7764	.2236	.2989	− .2271	− .1262	.5502	− .0394
.77	.7794	.2206	.2966	− .2284	− .1207	.5497	− .0611
.78	.7823	.2177	.2943	− .2296	− .1153	.5490	− .0825
.79	.7852	.2148	.2920	− .2307	− .1098	.5481	− .1037
.80	.7881	.2119	.2897	− .2318	− .1043	.5469	− .1247
.81	.7910	.2090	.2874	− .2328	− .0988	.5456	− .1455
.82	.7939	.2061	.2850	− .2337	− .0934	.5440	− .1660
.83	.7967	.2033	.2827	− .2346	− .0880	.5423	− .1862
.84	.7995	.2005	.2803	− .2355	− .0825	.5403	− .2063
.85	.8023	.1977	.2780	− .2363	− .0771	.5381	− .2260
.86	.8051	.1949	.2756	− .2370	− .0718	.5358	− .2455
.87	.8078	.1922	.2732	− .2377	− .0664	.5332	− .2646
.88	.8106	.1894	.2709	− .2384	− .0611	.5305	− .2835
.89	.8133	.1867	.2685	− .2389	− .0558	.5276	− .3021
.90	.8159	.1841	.2661	− .2395	− .0506	.5245	− .3203
.91	.8186	.1814	.2637	− .2400	− .0453	.5212	− .3383
.92	.8212	.1788	.2613	− .2404	− .0401	.5177	− .3559
.93	.8238	.1762	.2589	− .2408	− .0350	.5140	− .3731
.94	.8264	.1736	.2565	− .2411	− .0299	.5102	− .3901
.95	.8289	.1711	.2541	− .2414	− .0248	.5062	− .4066
.96	.8315	.1685	.2516	− .2416	− .0197	.5021	− .4228
.97	.8340	.1660	.2492	− .2417	− .0147	.4978	− .4387
.98	.8365	.1635	.2468	− .2419	− .0098	.4933	− .4541
.99	.8389	.1611	.2444	− .2420	− .0049	.4887	− .4692
1.00	.8413	.1587	.2420	− .2420	.0000	.4839	− .4839

NORMAL DISTRIBUTION AND RELATED FUNCTIONS

x	$F(x)$	$1 - F(x)$	$f(x)$	$f'(x)$	$f''(x)$	$f'''(x)$	$f^{iv}(x)$
1.00	.8413	.1587	.2420	−.2420	.0000	.4839	−.4839
1.01	.8438	.1562	.2396	−.2420	.0048	.4790	−.4983
1.02	.8461	.1539	.2371	−.2419	.0096	.4740	−.5122
1.03	.8485	.1515	.2347	−.2418	.0143	.4688	−.5257
1.04	.8508	.1492	.2323	−.2416	.0190	.4635	−.5389
1.05	.8531	.1469	.2299	−.2414	.0236	.4580	−.5516
1.06	.8554	.1446	.2275	−.2411	.0281	.4524	−.5639
1.07	.8577	.1423	.2251	−.2408	.0326	.4467	−.5758
1.08	.8599	.1401	.2227	−.2405	.0371	.4409	−.5873
1.09	.8621	.1379	.2203	−.2401	.0414	.4350	−.5984
1.10	.8643	.1357	.2179	−.2396	.0458	.4290	−.6091
1.11	.8665	.1335	.2155	−.2392	.0500	.4228	−.6193
1.12	.8686	.1314	.2131	−.2386	.0542	.4166	−.6292
1.13	.8708	.1292	.2107	−.2381	.0583	.4102	−.6386
1.14	.8729	.1271	.2083	−.2375	.0624	.4038	−.6476
1.15	.8749	.1251	.2059	−.2368	.0664	.3973	−.6561
1.16	.8770	.1230	.2036	−.2361	.0704	.3907	−.6643
1.17	.8790	.1210	.2012	−.2354	.0742	.3840	−.6720
1.18	.8810	.1190	.1989	−.2347	.0780	.3772	−.6792
1.19	.8830	.1170	.1965	−.2339	.0818	.3704	−.6861
1.20	.8849	.1151	.1942	−.2330	.0854	.3635	−.6926
1.21	.8869	.1131	.1919	−.2322	.0890	.3566	−.6986
1.22	.8888	.1112	.1895	−.2312	.0926	.3496	−.7042
1.23	.8907	.1093	.1872	−.2303	.0960	.3425	−.7094
1.24	.8925	.1075	.1849	−.2293	.0994	.3354	−.7141
1.25	.8944	.1056	.1826	−.2283	.1027	.3282	−.7185
1.26	.8962	.1038	.1804	−.2273	.1060	.3210	−.7224
1.27	.8980	.1020	.1781	−.2262	.1092	.3138	−.7259
1.28	.8997	.1003	.1758	−.2251	.1123	.3065	−.7291
1.29	.9015	.0985	.1736	−.2240	.1153	.2992	−.7318
1.30	.9032	.0968	.1714	−.2228	.1182	.2918	−.7341
1.31	.9049	.0951	.1691	−.2216	.1211	.2845	−.7361
1.32	.9066	.0934	.1669	−.2204	.1239	.2771	−.7376
1.33	.9082	.0918	.1647	−.2191	.1267	.2697	−.7388
1.34	.9099	.0901	.1626	−.2178	.1293	.2624	−.7395
1.35	.9115	.0885	.1604	−.2165	.1319	.2550	−.7399
1.36	.9131	.0869	.1582	−.2152	.1344	.2476	−.7400
1.37	.9147	.0853	.1561	−.2138	.1369	.2402	−.7396
1.38	.9162	.0838	.1539	−.2125	.1392	.2328	−.7389
1.39	.9177	.0823	.1518	−.2110	.1415	.2254	−.7378
1.40	.9192	.0808	.1497	−.2096	.1437	.2180	−.7364
1.41	.9207	.0793	.1476	−.2082	.1459	.2107	−.7347
1.42	.9222	.0778	.1456	−.2067	.1480	.2033	−.7326
1.43	.9236	.0764	.1435	−.2052	.1500	.1960	−.7301
1.44	.9251	.0749	.1415	−.2037	.1519	.1887	−.7274
1.45	.9265	.0735	.1394	−.2022	.1537	.1815	−.7243
1.46	.9279	.0721	.1374	−.2006	.1555	.1742	−.7209
1.47	.9292	.0708	.1354	−.1991	.1572	.1670	−.7172
1.48	.9306	.0694	.1334	−.1975	.1588	.1599	−.7132
1.49	.9319	.0681	.1315	−.1959	.1604	.1528	−.7089
1.50	.9332	.0668	.1295	−.1943	.1619	.1457	−.7043

NORMAL DISTRIBUTION AND RELATED FUNCTIONS

x	$F(x)$	$1 - F(x)$	$f(x)$	$f'(x)$	$f''(x)$	$f'''(x)$	$f^{iv}(x)$
1.50	.9332	.0668	.1295	−.1943	.1619	.1457	−.7043
1.51	.9345	.0655	.1276	−.1927	.1633	.1387	−.6994
1.52	.9357	.0643	.1257	−.1910	.1647	.1317	−.6942
1.53	.9370	.0630	.1238	−.1894	.1660	.1248	−.6888
1.54	.9382	.0618	.1219	−.1877	.1672	.1180	−.6831
1.55	.9394	.0606	.1200	−.1860	.1683	.1111	−.6772
1.56	.9406	.0594	.1182	−.1843	.1694	.1044	−.6710
1.57	.9418	.0582	.1163	−.1826	.1704	.0977	−.6646
1.58	.9429	.0571	.1145	−.1809	.1714	.0911	−.6580
1.59	.9441	.0559	.1127	−.1792	.1722	.0846	−.6511
1.60	.9452	.0548	.1109	−.1775	.1730	.0781	−.6441
1.61	.9463	.0537	.1092	−.1757	.1738	.0717	−.6368
1.62	.9474	.0526	.1074	−.1740	.1745	.0654	−.6293
1.63	.9484	.0516	.1057	−.1723	.1751	.0591	−.6216
1.64	.9495	.0505	.1040	−.1705	.1757	.0529	−.6138
1.65	.9505	.0495	.1023	−.1687	.1762	.0468	−.6057
1.66	.9515	.0485	.1006	−.1670	.1766	.0408	−.5975
1.67	.9525	.0475	.0989	−.1652	.1770	.0349	−.5891
1.68	.9535	.0465	.0973	−.1634	.1773	.0290	−.5806
1.69	.9545	.0455	.0957	−.1617	.1776	.0233	−.5720
1.70	.9554	.0446	.0940	−.1599	.1778	.0176	−.5632
1.71	.9564	.0436	.0925	−.1581	.1779	.0120	−.5542
1.72	.9573	.0427	.0909	−.1563	.1780	.0065	−.5452
1.73	.9582	.0418	.0893	−.1546	.1780	.0011	−.5360
1.74	.9591	.0409	.0878	−.1528	.1780	−.0042	−.5267
1.75	.9599	.0401	.0863	−.1510	.1780	−.0094	−.5173
1.76	.9608	.0392	.0848	−.1492	.1778	−.0146	−.5079
1.77	.9616	.0384	.0833	−.1474	.1777	−.0196	−.4983
1.78	.9625	.0375	.0818	−.1457	.1774	−.0245	−.4887
1.79	.9633	.0367	.0804	−.1439	.1772	−.0294	−.4789
1.80	.9641	.0359	.0790	−.1421	.1769	−.0341	−.4692
1.81	.9649	.0351	.0775	−.1403	.1765	−.0388	−.4593
1.82	.9656	.0344	.0761	−.1386	.1761	−.0433	−.4494
1.83	.9664	.0336	.0748	−.1368	.1756	−.0477	−.4395
1.84	.9671	.0329	.0734	−.1351	.1751	−.0521	−.4295
1.85	.9678	.0322	.0721	−.1333	.1746	−.0563	−.4195
1.86	.9686	.0314	.0707	−.1316	.1740	−.0605	−.4095
1.87	.9693	.0307	.0694	−.1298	.1734	−.0645	−.3995
1.88	.9699	.0301	.0681	−.1281	.1727	−.0685	−.3894
1.89	.9706	.0294	.0669	−.1264	.1720	−.0723	−.3793
1.90	.9713	.0287	.0656	−.1247	.1713	−.0761	−.3693
1.91	.9719	.0281	.0344	−.1230	.1705	−.0797	−.3592
1.92	.9726	.0274	.0632	−.1213	.1697	−.0832	−.3492
1.93	.9732	.0268	.0620	−.1196	.1688	−.0867	−.3392
1.94	.9738	.0262	.0608	−.1179	.1679	−.0900	−.3292
1.95	.9744	.0256	.0596	−.1162	.1670	−.0933	−.3192
1.96	.9750	.0250	.0584	−.1145	.1661	−.0964	−.3093
1.97	.9756	.0244	.0573	−.1129	.1651	−.0994	−.2994
1.98	.9761	.0239	.0562	−.1112	.1641	−.1024	−.2895
1.99	.9767	.0233	.0551	−.1096	.1630	−.1052	−.2797
2.00	.9772	.0228	.0540	−.1080	.1620	−.1080	−.2700

NORMAL DISTRIBUTION AND RELATED FUNCTIONS

x	$F(x)$	$1 - F(x)$	$f(x)$	$f'(x)$	$f''(x)$	$f'''(x)$	$f^{\mathrm{iv}}(x)$
2.00	.9773	.0227	.0540	−.1080	.1620	−.1080	−.2700
2.01	.9778	.0222	.0529	−.1064	.1609	−.1106	−.2603
2.02	.9783	.0217	.0519	−.1048	.1598	−.1132	−.2506
2.03	.9788	.0212	.0508	−.1032	.1586	−.1157	−.2411
2.04	.9793	.0207	.0498	−.1016	.1575	−.1180	−.2316
2.05	.9798	.0202	.0488	−.1000	.1563	−.1203	−.2222
2.06	.9803	.0197	.0478	−.0985	.1550	−.1225	−.2129
2.07	.9808	.0192	.0468	−.0969	.1538	−.1245	−.2036
2.08	.9812	.0188	.0459	−.0954	.1526	−.1265	−.1945
2.09	.9817	.0183	.0449	−.0939	.1513	−.1284	−.1854
2.10	.9821	.0179	.0440	−.0924	.1500	−.1302	−.1765
2.11	.9826	.0174	.0431	−.0909	.1487	−.1320	−.1676
2.12	.9830	.0170	.0422	−.0894	.1474	−.1336	−.1588
2.13	.9834	.0166	.0413	−.0879	.1460	−.1351	−.1502
2.14	.9838	.0162	.0404	−.0865	.1446	−.1366	−.1416
2.15	.9842	.0158	.0396	−.0850	.1433	−.1380	−.1332
2.16	.9846	.0154	.0387	−.0836	.1419	−.1393	−.1249
2.17	.9850	.0150	.0379	−.0822	.1405	−.1405	−.1167
2.18	.9854	.0146	.0371	−.0808	.1391	−.1416	−.1086
2.19	.9857	.0143	.0363	−.0794	.1377	−.1426	−.1006
2.20	.9861	.0139	.0355	−.0780	.1362	−.1436	−.0927
2.21	.9864	.0136	.0347	−.0767	.1348	−.1445	−.0850
2.22	.9868	.0132	.0339	−.0754	.1333	−.1453	−.0774
2.23	.9871	.0129	.0332	−.0740	.1319	−.1460	−.0700
2.24	.9875	.0125	.0325	−.0727	.1304	−.1467	−.0626
2.25	.9878	.0122	.0317	−.0714	.1289	−.1473	−.0554
2.26	.9881	.0119	.0310	−.0701	.1275	−.1478	−.0484
2.27	.9884	.0116	.0303	−.0689	.1260	−.1483	−.0414
2.28	.9887	.0113	.0297	−.0676	.1245	−.1486	−.0346
2.29	.9890	.0110	.0290	−.0664	.1230	−.1490	−.0279
2.30	.9893	.0107	.0283	−.0652	.1215	−.1492	−.0214
2.31	.9896	.0104	.0277	−.0639	.1200	−.1494	−.0150
2.32	.9898	.0102	.0270	−.0628	.1185	−.1495	−.0088
2.33	.9901	.0099	.0264	−.0616	.1170	−.1496	−.0027
2.34	.9904	.0096	.0258	−.0604	.1155	−.1496	.0033
2.35	.9906	.0094	.0252	−.0593	.1141	−.1495	.0092
2.36	.9909	.0091	.0246	−.0581	.1126	−.1494	.0149
2.37	.9911	.0089	.0241	−.0570	.1111	−.1492	.0204
2.38	.9913	.0087	.0235	−.0559	.1096	−.1490	.0258
2.39	.9916	.0084	.0229	−.0548	.1081	−.1487	.0311
2.40	.9918	.0082	.0224	−.0538	.1066	−.1483	.0362
2.41	.9920	.0080	.0219	−.0527	.1051	−.1480	.0412
2.42	.9922	.0078	.0213	−.0516	.1036	−.1475	.0461
2.43	.9925	.0075	.0208	−.0506	.1022	−.1470	.0508
2.44	.9927	.0073	.0203	−.0496	.1007	−.1465	.0554
2.45	.9929	.0071	.0198	−.0486	.0992	−.1459	.0598
2.46	.9931	.0069	.0194	−.0476	.0978	−.1453	.0641
2.47	.9932	.0068	.0189	−.0467	.0963	−.1446	.0683
2.48	.9934	.0066	.0184	−.0457	.0949	−.1439	.0723
2.49	.9936	.0064	.0180	−.0448	.0935	−.1432	.0762
2.50	.9938	.0062	.0175	−.0438	.0920	−.1424	.0800

NORMAL DISTRIBUTION AND RELATED FUNCTIONS

x	$F(x)$	$1 - F(x)$	$f(x)$	$f'(x)$	$f''(x)$	$f'''(x)$	$f^{iv}(x)$
2.50	.9938	.0062	.0175	−.0438	.0920	−.1424	.0800
2.51	.9940	.0060	.0171	−.0429	.0906	−.1416	.0836
2.52	.9941	.0059	.0167	−.0420	.0892	−.1408	.0871
2.53	.9943	.0057	.0163	−.0411	.0878	−.1399	.0905
2.54	.9945	.0055	.0158	−.0403	.0864	−.1389	.0937
2.55	.9946	.0054	.0155	−.0394	.0850	−.1380	.0968
2.56	.9948	.0052	.0151	−.0386	.0836	−.1370	.0998
2.57	.9949	.0051	.0147	−.0377	.0823	−.1360	.1027
2.58	.9951	.0049	.0143	−.0369	.0809	−.1350	.1054
2.59	.9952	.0048	.0139	−.0361	.0796	−.1339	.1080
2.60	.9953	.0047	.0136	−.0353	.0782	−.1328	.1105
2.61	.9955	.0045	.0132	−.0345	.0769	−.1317	.1129
2.62	.9956	.0044	.0129	−.0338	.0756	−.1305	.1152
2.63	.9957	.0043	.0126	−.0330	.0743	−.1294	.1173
2.64	.9959	.0041	.0122	−.0323	.0730	−.1282	.1194
2.65	.9960	.0040	.0119	−.0316	.0717	−.1270	.1213
2.66	.9961	.0039	.0116	−.0309	.0705	−.1258	.1231
2.67	.9962	.0038	.0113	−.0302	.0692	−.1245	.1248
2.68	.9963	.0037	.0110	−.0295	.0680	−.1233	.1264
2.69	.9964	.0036	.0107	−.0288	.0668	−.1220	.1279
2.70	.9965	.0035	.0104	−.0281	.0656	−.1207	.1293
2.71	.9966	.0034	.0101	−.0275	.0644	−.1194	.1306
2.72	.9967	.0033	.0099	−.0269	.0632	−.1181	.1317
2.73	.9968	.0032	.0096	−.0262	.0620	−.1168	.1328
2.74	.9969	.0031	.0093	−.0256	.0608	−.1154	.1338
2.75	.9970	.0030	.0091	−.0250	.0597	−.1141	.1347
2.76	.9971	.0029	.0088	−.0244	.0585	−.1127	.1356
2.77	.9972	.0028	.0086	−.0238	.0574	−.1114	.1363
2.78	.9973	.0027	.0084	−.0233	.0563	−.1100	.1369
2.79	.9974	.0026	.0081	−.0227	.0552	−.1087	.1375
2.80	.9974	.0026	.0079	−.0222	.0541	−.1073	.1379
2.81	.9975	.0025	.0077	−.0216	.0531	−.1059	.1383
2.82	.9976	.0024	.0075	−.0211	.0520	−.1045	.1386
2.83	.9977	.0023	.0073	−.0206	.0510	−.1031	.1389
2.84	.9977	.0023	.0071	−.0201	.0500	−.1017	.1390
2.85	.9978	.0022	.0069	−.0196	.0490	−.1003	.1391
2.86	.9979	.0021	.0067	−.0191	.0480	−.0990	.1391
2.87	.9979	.0021	.0065	−.0186	.0470	−.0976	.1391
2.88	.9980	.0020	.0063	−.0182	.0460	−.0962	.1389
2.89	.9981	.0019	.0061	−.0177	.0451	−.0948	.1388
2.90	.9981	.0019	.0060	−.0173	.0441	−.0934	.1385
2.91	.9982	.0018	.0058	−.0168	.0432	−.0920	.1382
2.92	.9982	.0018	.0056	−.0164	.0423	−.0906	.1378
2.93	.9983	.0017	.0055	−.0160	.0414	−.0893	.1374
2.94	.9984	.0016	.0053	−.0156	.0405	−.0879	.1369
2.95	.9984	.0016	.0051	−.0152	.0396	−.0865	.1364
2.96	.9985	.0015	.0050	−.0148	.0388	−.0852	.1358
2.97	.9985	.0015	.0048	−.0144	.0379	−.0838	.1352
2.98	.9986	.0014	.0047	−.0140	.0371	−.0825	.1345
2.99	.9986	.0014	.0046	−.0137	.0363	−.0811	.1337
3.00	.9987	.0013	.0044	−.0133	.0355	−.0798	.1330

NORMAL DISTRIBUTION AND RELATED FUNCTIONS

x	$F(x)$	$1-F(x)$	$f(x)$	$f'(x)$	$f''(x)$	$f'''(x)$	$f^{iv}(x)$
3.00	.9987	.0013	.0044	−.0133	.0355	−.0798	.1330
3.01	.9987	.0013	.0043	−.0130	.0347	−.0785	.1321
3.02	.9987	.0013	.0042	−.0126	.0339	−.0771	.1313
3.03	.9988	.0012	.0040	−.0123	.0331	−.0758	.1304
3.04	.9988	.0012	.0039	−.0119	.0324	−.0745	.1294
3.05	.9989	.0011	.0038	−.0116	.0316	−.0732	.1285
3.06	.9989	.0011	.0037	−.0113	.0309	−.0720	.1275
3.07	.9989	.0011	.0036	−.0110	.0302	−.0707	.1264
3.08	.9990	.0010	.0035	−.0107	.0295	−.0694	.1254
3.09	.9990	.0010	.0034	−.0104	.0288	−.0682	.1243
3.10	.9990	.0010	.0033	−.0101	.0281	−.0669	.1231
3.11	.9991	.0009	.0032	−.0099	.0275	−.0657	.1220
3.12	.9991	.0009	.0031	−.0096	.0268	−.0645	.1208
3.13	.9991	.0009	.0030	−.0093	.0262	−.0633	.1196
3.14	.9992	.0008	.0029	−.0091	.0256	−.0621	.1184
3.15	.9992	.0008	.0028	−.0088	.0249	−.0609	.1171
3.16	.9992	.0008	.0027	−.0086	.0243	−.0598	.1159
3.17	.9992	.0008	.0026	−.0083	.0237	−.0586	.1146
3.18	.9993	.0007	.0025	−.0081	.0232	−.0575	.1133
3.19	.9993	.0007	.0025	−.0079	.0226	−.0564	.1120
3.20	.9993	.0007	.0024	−.0076	.0220	−.0552	.1107
3.21	.9993	.0007	.0023	−.0074	.0215	−.0541	.1093
3.22	.9994	.0006	.0022	−.0072	.0210	−.0531	.1080
3.23	.9994	.0006	.0022	−.0070	.0204	−.0520	.1066
3.24	.9994	.0006	.0021	−.0068	.0199	−.0509	.1053
3.25	.9994	.0006	.0020	−.0066	.0194	−.0499	.1039
3.26	.9994	.0006	.0020	−.0064	.0189	−.0488	.1025
3.27	.9995	.0005	.0019	−.0062	.0184	−.0478	.1011
3.28	.9995	.0005	.0018	−.0060	.0180	−.0468	.0997
3.29	.9995	.0005	.0018	−.0059	.0175	−.0458	.0983
3.30	.9995	.0005	.0017	−.0057	.0170	−.0449	.0969
3.31	.9995	.0005	.0017	−.0055	.0166	−.0439	.0955
3.32	.9995	.0005	.0016	−.0054	.0162	−.0429	.0941
3.33	.9996	.0004	.0016	−.0052	.0157	−.0420	.0927
3.34	.9996	.0004	.0015	−.0050	.0153	−.0411	.0913
3.35	.9996	.0004	.0015	−.0049	.0149	−.0402	.0899
3.36	.9996	.0004	.0014	−.0047	.0145	−.0393	.0885
3.37	.9996	.0004	.0014	−.0046	.0141	−.0384	.0871
3.38	.9996	.0004	.0013	−.0045	.0138	−.0376	.0857
3.39	.9997	.0003	.0013	−.0043	.0134	−.0367	.0843
3.40	.9997	.0003	.0012	−.0042	.0130	−.0359	.0829
3.41	.9997	.0003	.0012	−.0041	.0127	−.0350	.0815
3.42	.9997	.0003	.0012	−.0039	.0123	−.0342	.0801
3.43	.9997	.0003	.0011	−.0038	.0120	−.0334	.0788
3.44	.9997	.0003	.0011	−.0037	.0116	−.0327	.0774
3.45	.9997	.0003	.0010	−.0036	.0113	−.0319	.0761
3.46	.9997	.0003	.0010	−.0035	.0110	−.0311	.0747
3.47	.9997	.0003	.0010	−.0034	.0107	−.0304	.0734
3.48	.9997	.0003	.0009	−.0033	.0104	−.0297	.0721
3.49	.9998	.0002	.0009	−.0032	.0101	−.0290	.0707
3.50	.9998	.0002	.0009	−.0031	.0098	−.0283	.0694

NORMAL DISTRIBUTION AND RELATED FUNCTIONS

x	$F(x)$	$1 - F(x)$	$f(x)$	$f'(x)$	$f''(x)$	$f'''(x)$	$f^{iv}(x)$
3.50	.9998	.0002	.0009	$-.0031$	.0098	$-.0283$	.0694
3.51	.9998	.0002	.0008	$-.0030$	.0095	$-.0276$	.0681
3.52	.9998	.0002	.0008	$-.0029$	.0093	$-.0269$	.0669
3.53	.9998	.0002	.0008	$-.0028$	.0090	$-.0262$	.0656
3.54	.9998	.0002	.0008	$-.0027$	.0087	$-.0256$	.0643
3.55	.9998	.0002	.0007	$-.0026$	.0085	$-.0249$	.0631
3.56	.9998	.0002	.0007	$-.0025$	.0082	$-.0243$	.0618
3.57	.9998	.0002	.0007	$-.0024$	.0080	$-.0237$	.0606
3.58	.9998	.0002	.0007	$-.0024$	.0078	$-.0231$	.0594
3.59	.9998	.0002	.0006	$-.0023$	.0075	$-.0225$	.0582
3.60	.9998	.0002	.0006	$-.0022$	.0073	$-.0219$	.0570
3.61	.9998	.0002	.0006	$-.0021$	.0071	$-.0214$	.0559
3.62	.9999	.0001	.0006	$-.0021$	.0069	$-.0208$	.0547
3.63	.9999	.0001	.0005	$-.0020$	.0067	$-.0203$	.0536
3.64	.9999	.0001	.0005	$-.0019$	.0065	$-.0198$	.0524
3.65	.9999	.0001	.0005	$-.0019$	.0063	$-.0192$	.0513
3.66	.9999	.0001	.0005	$-.0018$	.0061	$-.0187$	.0502
3.67	.9999	.0001	.0005	$-.0017$	.0059	$-.0182$	.0492
3.68	.9999	.0001	.0005	$-.0017$	.0057	$-.0177$	.0481
3.69	.9999	.0001	.0004	$-.0016$	.0056	$-.0173$	.0470
3.70	.9999	.0001	.0004	$-.0016$	.0054	$-.0168$	.0460
3.71	.9999	.0001	.0004	$-.0015$	.0052	$-.0164$	.0450
3.72	.9999	.0001	.0004	$-.0015$	.0051	$-.0159$	.0440
3.73	.9999	.0001	.0004	$-.0014$	.0049	$-.0155$	.0430
3.74	.9999	.0001	.0004	$-.0014$	.0048	$-.0150$	.0420
3.75	.9999	.0001	.0004	$-.0013$	.0046	$-.0146$	.0410
3.76	.9999	.0001	.0003	$-.0013$	.0045	$-.0142$	.0401
3.77	.9999	.0001	.0003	$-.0012$	.0043	$-.0138$	.0392
3.78	.9999	.0001	.0003	$-.0012$	.0042	$-.0134$	.0382
3.79	.9999	.0001	.0003	$-.0012$	.0041	$-.0131$	.0373
3.80	.9999	.0001	.0003	$-.0011$	.0039	$-.0127$	.0365
3.81	.9999	.0001	.0003	$-.0011$	.0038	$-.0123$	.0356
3.82	.9999	.0001	.0003	$-.0010$	.0037	$-.0120$	.0347
3.83	.9999	.0001	.0003	$-.0010$	.0036	$-.0116$	.0339
3.84	.9999	.0001	.0003	$-.0010$	.0034	$-.0113$	.0331
3.85	.9999	.0001	.0002	$-.0009$	.0033	$-.0110$	.0323
3.86	.9999	.0001	.0002	$-.0009$	.0032	$-.0107$	.0315
3.87	.9999	.0001	.0002	$-.0009$	.0031	$-.0104$	.0307
3.88	.9999	.0001	.0002	$-.0008$	.0030	$-.0100$	.0299
3.89	1.0000	.0000	.0002	$-.0008$	.0029	$-.0098$	.0292
3.90	1.0000	.0000	.0002	$-.0008$	.0028	$-.0095$	.0284
3.91	1.0000	.0000	.0002	$-.0008$	.0027	$-.0092$	.0277
3.92	1.0000	.0000	.0002	$-.0007$	.0026	$-.0089$	.0270
3.93	1.0000	.0000	.0002	$-.0007$	.0026	$-.0086$	.0263
3.94	1.0000	.0000	.0002	$-.0007$	.0025	$-.0084$	.0256
3.95	1.0000	.0000	.0002	$-.0006$	.0024	$-.0081$	.0250
3.96	1.0000	.0000	.0002	$-.0006$	.0023	$-.0079$	.0243
3.97	1.0000	.0000	.0002	$-.0006$	.0022	$-.0076$	.0237
3.98	1.0000	.0000	.0001	$-.0006$	.0022	$-.0074$	.0230
3.99	1.0000	.0000	.0001	$-.0006$	.0021	$-.0072$	.0224
4.00	1.0000	.0000	.0001	$-.0005$	.0020	$-.0070$	.0218

x	1.282	1.645	1.960	2.326	2.576	3.090
$F(x)$	.90	.95	.975	.99	.995	.999
$2[1 - F(x)]$	.20	.10	.05	.02	.01	.002

CUMULATIVE TERMS, BINOMIAL DISTRIBUTION

This table contains the values of $\sum_{x=x'}^{n} \binom{n}{x} \theta^x (1-\theta)^{n-x}$ for specified values of n, x', and θ. If $\theta >$ 0.5, the values for $\sum_{x=x'}^{n} \binom{n}{x} \theta^x (1-\theta)^{n-x}$ are obtained using the corresponding results obtained from

$$1 - \sum_{x=n-x'+1}^{n} \binom{n}{x} (1-\theta)^x \theta^{n-x}$$

Individual terms of the binomial distribution for specified x', n, and θ can be found from

$$\sum_{x=x}^{n} \binom{n}{x} \theta^x (1-\theta)^{n-x} - \sum_{x=x'+1}^{n} \binom{n}{x} \theta^x (1-\theta)^{n-x}$$

n	x'	.05	.10	.15	.20	.25	.30	.35	.40	.45	.50
2	1	.0975	.1900	.2775	.3600	.4375	.5100	.5775	.6400	.6975	.7500
	2	.0025	.0100	.0225	.0400	.0625	.0900	.1225	.1600	.2025	.2500
3	1	.1426	.2710	.3859	.4880	.5781	.6570	.7254	.7840	.8336	.8750
	2	.0072	.0280	.0608	.1040	.1562	.2160	.2818	.3520	.4252	.5000
	3	.0001	.0010	.0034	.0080	.0156	.0270	.0429	.0640	.0911	.1250
4	1	.1855	.3439	.4780	.5904	.6836	.7599	.8215	.8704	.9085	.9375
	2	.0140	.0523	.1095	.1808	.2617	.3483	.4370	.5248	.6090	.6875
	3	.0005	.0037	.0120	.0272	.0508	.0837	.1265	.1792	.2415	.3125
	4	.0000	.0001	.0005	.0016	.0039	.0081	.0150	.0256	.0410	.0625
5	1	.2262	.4095	.5563	.6723	.7627	.8319	.8840	.9222	.9497	.9688
	2	.0226	.0815	.1648	.2627	.3672	.4718	.5716	.6630	.7438	.8125
	3	.0012	.0086	.0266	.0579	.1035	.1631	.2352	.3174	.4069	.5000
	4	.0000	.0005	.0022	.0067	.0156	.0308	.0540	.0870	.1312	.1875
	5	.0000	.0000	.0001	.0003	.0010	.0024	.0053	.0102	.0185	.0312
6	1	.2649	.4686	.6229	.7379	.8220	.8824	.9246	.9533	.9723	.9844
	2	.0328	.1143	.2235	.3446	.4661	.5798	.6809	.7667	.8364	.8906
	3	.0022	.0158	.0473	.0989	.1694	.2557	.3529	.4557	.5585	.6562
	4	.0001	.0013	.0059	.0170	.0376	.0705	.1174	.1792	.2553	.3438
	5	.0000	.0001	.0004	.0016	.0046	.0109	.0223	.0410	.0692	.1094
	6	.0000	.0000	.0000	.0001	.0002	.0008	.0018	.0041	.0083	.0156
7	1	.3017	.5217	.6794	.7903	.8665	.9176	.9510	.9720	.9848	.9922
	2	.0444	.1497	.2834	.4233	.5551	.6706	.7662	.8414	.8976	.9375
	3	.0038	.0257	.0738	.1480	.2436	.3529	.4677	.5801	.6836	.7734
	4	.0002	.0027	.0121	.0333	.0706	.1260	.1998	.2898	.3917	.5000
	5	.0000	.0002	.0012	.0047	.0129	.0288	.0556	.0963	.1529	.2266
	6	.0000	.0000	.0001	.0004	.0013	.0038	.0090	.0188	.0357	.0625
	7	.0000	.0000	.0000	.0000	.0001	.0002	.0006	.0016	.0037	.0078
8	1	.3366	.5695	.7275	.8322	.8999	.9424	.9681	.9832	.9916	.9961
	2	.0572	.1869	.3428	.4967	.6329	.7447	.8309	.8936	.9368	.9648
	3	.0058	.0381	.1052	.2031	.3215	.4482	.5722	.6846	.7799	.8555
	4	.0004	.0050	.0214	.0563	.1138	.1941	.2936	.4059	.5230	.6367
	5	.0000	.0004	.0029	.0104	.0273	.0580	.1061	.1737	.2604	.3633
	6	.0000	.0000	.0002	.0012	.0042	.0113	.0253	.0498	.9885	.1445
	7	.0000	.0000	.0000	.0001	.0004	.0013	.0036	.0085	.0181	.0352
	8	.0000	.0000	.0000	.0000	.0000	.0001	.0002	.0007	.0017	.0039

Linear interpolation will be accurate at most to two decimal places.

CUMULATIVE TERMS, BINOMIAL DISTRIBUTION (Continued)

θ

n	x′	.05	.10	.15	.20	.25	.30	.35	.40	.45	.50
9	1	.3698	.6126	.7684	.8658	.9249	.9596	.9793	.9899	.9954	.9980
	2	.0712	.2252	.4005	.5638	.6997	.8040	.8789	.9295	.9615	.9805
	3	.0084	.0530	.1409	.2618	.3993	.5372	.6627	.7682	.8505	.9102
	4	.0006	.0083	.0339	.0856	.1657	.2703	.3911	.5174	.6386	.7461
	5	.0000	.0009	.0056	.0196	.0489	.0988	.1717	.2666	.3786	.5000
	6	.0000	.0001	.0006	.0031	.0100	.0253	.0536	.0994	.1658	.2539
	7	.0000	.0000	.0000	.0003	.0013	.0043	.0112	.0250	.0498	.0898
	8	.0000	.0000	.0000	.0000	.0001	.0004	.0014	.0038	.0091	.0195
	9	.0000	.0000	.0000	.0000	.0000	.0000	.0001	.0003	.0008	.0020
10	1	.4013	.6513	.8031	.8926	.9437	.9718	.9865	.9940	.9975	.9990
	2	.0861	.2639	.4557	.6242	.7560	.8507	.9140	.9536	.9767	.9893
	3	.0115	.0702	.1798	.3222	.4744	.6172	.7384	.8327	.9004	.9453
	4	.0010	.0128	.0500	.1209	.2241	.3504	.4862	.6177	.7340	.8281
	5	.0001	.0016	.0099	.0328	.0781	.1503	.2485	.3669	.4956	.6230
	6	.0000	.0001	.0014	.0064	.0197	.0473	.0949	.1662	.2616	.3770
	7	.0000	.0000	.0001	.0009	.0035	.0106	.0260	.0548	.1020	.1719
	8	.0000	.0000	.0000	.0001	.0004	.0016	.0048	.0123	.0274	.0547
	9	.0000	.0000	.0000	.0000	.0000	.0001	.0005	.0017	.0045	.0107
	10	.0000	.0000	.0000	.0000	.0000	.0000	.0000	.0001	.0003	.0010
11	1	.4312	.6862	.8327	.9141	.9578	.9802	.9912	.9964	.9986	.9995
	2	.1019	.3026	.5078	.6779	.8029	.8870	.9394	.9698	.9861	.9941
	3	.0152	.0896	.2212	.3826	.5448	.6873	.7999	.8811	.9348	.9673
	4	.0016	.0185	.0694	.1611	.2867	.4304	.5744	.7037	.8089	.8867
	5	.0001	.0028	.0159	.0504	.1146	.2103	.3317	.4672	.6029	.7256
	6	.0000	.0003	.0027	.0117	.0343	.0782	.1487	.2465	.3669	.5000
	7	.0000	.0000	.0003	.0020	.0076	.0216	.0501	.0994	.1738	.2744
	8	.0000	.0000	.0000	.0002	.0012	.0043	.0122	.0293	.0610	.1133
	9	.0000	.0000	.0000	.0000	.0001	.0006	.0020	.0059	.0148	.0327
	10	.0000	.0000	.0000	.0000	.0000	.0000	.0002	.0007	.0022	.0059
	11	.0000	.0000	.0000	.0000	.0000	.0000	.0000	.0000	.0002	.0005
12	1	.4596	.7176	.8578	.9313	.9683	.9862	.9943	.9978	.9992	.9998
	2	.1184	.3410	.5565	.7251	.8416	.9150	.9576	.9804	.9917	.9968
	3	.0196	.1109	.2642	.4417	.6093	.7472	.8487	.9166	.9579	.9807
	4	.0022	.0256	.0922	.2054	.3512	.5075	.6533	.7747	.8655	.9270
	5	.0002	.0043	.0239	.0726	.1576	.2763	.4167	.5618	.6956	.8062
	6	.0000	.0005	.0046	.0194	.0544	.1178	.2127	.3348	.4731	.6128
	7	.0000	.0001	.0007	.0039	.0143	.0386	.0846	.1582	.2607	.3872
	8	.0000	.0000	.0001	.0006	.0028	.0095	.0255	.0573	.1117	.1938
	9	.0000	.0000	.0000	.0001	.0004	.0017	.0056	.0153	.0356	.0730
	10	.0000	.0000	.0000	.0000	.0000	.0002	.0008	.0028	.0079	.0193
	11	.0000	.0000	.0000	.0000	.0000	.0000	.0001	.0003	.0011	.0032
	12	.0000	.0000	.0000	.0000	.0000	.0000	.0000	.0000	.0001	.0002

CUMULATIVE TERMS, BINOMIAL DISTRIBUTION (Continued)

θ

n	x′	.05	.10	.15	.20	.25	.30	.35	.40	.45	.50
13	1	.4867	.7458	.8791	.9450	.9762	.9903	.9963	.9987	.9996	.9999
	2	.1354	.3787	.6017	.7664	.8733	.9363	.9704	.9874	.9951	.9983
	3	.0245	.1339	.3080	.4983	.6674	.7975	.8868	.9421	.9731	.9888
	4	.0031	.0342	.1180	.2527	.4157	.5794	.7217	.8314	.9071	.9539
	5	.0003	.0065	.0342	.0991	.2060	.3457	.4995	.6470	.7721	.8666
	6	.0000	.0009	.0075	.0300	.0802	.1654	.2841	.4256	.5732	.7095
	7	.0000	.0001	.0013	.0070	.0243	.0624	.1295	.2288	.3563	.5000
	8	.0000	.0000	.0002	.0012	.0056	.0182	.0462	.0977	.1788	.2905
	9	.0000	.0000	.0000	.0002	.0010	.0040	.0126	.0321	.0698	.1334
	10	.0000	.0000	.0000	.0000	.0001	.0007	.0025	.0078	.0203	.0461
	11	.0000	.0000	.0000	.0000	.0000	.0001	.0003	.0013	.0041	.0112
	12	.0000	.0000	.0000	.0000	.0000	.0000	.0000	.0001	.0005	.0017
	13	.0000	.0000	.0000	.0000	.0000	.0000	.0000	.0000	.0000	.0001
14	1	.5123	.7712	.8972	.9560	.9822	.9932	.9976	.9992	.9998	.9999
	2	.1530	.4154	.6433	.8021	.8990	.9525	.9795	.9919	.9971	.9991
	3	.0301	.1584	.3521	.5519	.7189	.8392	.9161	.9602	.9830	.9935
	4	.0042	.0441	.1465	.3018	.4787	.6448	.7795	.8757	.9368	.9713
	5	.0004	.0092	.0467	.1298	.2585	.4158	.5773	.7207	.8328	.9102
	6	.0000	.0015	.0115	.0439	.1117	.2195	.3595	.5141	.6627	.7880
	7	.0000	.0002	.0022	.0116	.0383	.0933	.1836	.3075	.4539	.6047
	8	.0000	.0000	.0003	.0024	.0103	.0315	.0753	.1501	.2586	.3953
	9	.0000	.0000	.0000	.0004	.0022	.0083	.0243	.0583	.1189	.2120
	10	.0000	.0000	.0000	.0000	.0003	.0017	.0060	.0175	.0426	.0898
	11	.0000	.0000	.0000	.0000	.0000	.0002	.0011	.0039	.0114	.0287
	12	.0000	.0000	.0000	.0000	.0000	.0000	.0001	.0006	.0022	.0065
	13	.0000	.0000	.0000	.0000	.0000	.0000	.0000	.0001	.0003	.0009
	14	.0000	.0000	.0000	.0000	.0000	.0000	.0000	.0000	.0000	.0001
15	1	.5367	.7941	.9126	.9648	.9866	.9953	.9984	.9995	.9999	1.0000
	2	.1710	.4510	.6814	.8329	.9198	.9647	.9858	.9948	.9983	.9995
	3	.0362	.1841	.3958	.6020	.7639	.8732	.9383	.9729	.9893	.9963
	4	.0055	.0556	.1773	.3518	.5387	.7031	.8273	.9095	.9576	.9824
	5	.0006	.0127	.0617	.1642	.3135	.4845	.6481	.7827	.8796	.9408
	6	.0001	.0022	.0168	.0611	.1484	.2784	.4357	.5968	.7392	.8491
	7	.0000	.0003	.0036	.0181	.0566	.1311	.2452	.3902	.5478	.6964
	8	.0000	.0000	.0006	.0042	.0173	.0500	.1132	.2131	.3465	.5000
	9	.0000	.0000	.0001	.0008	.0042	.0152	.0422	.0950	.1818	.3036
	10	.0000	.0000	.0000	.0001	.0008	.0037	.0124	.0338	.0769	.1509
	11	.0000	.0000	.0000	.0000	.0001	.0007	.0028	.0093	.0255	.0592
	12	.0000	.0000	.0000	.0000	.0000	.0001	.0005	.0019	.0063	.0176
	13	.0000	.0000	.0000	.0000	.0000	.0000	.0001	.0003	.0011	.0037
	14	.0000	.0000	.0000	.0000	.0000	.0000	.0000	.0000	.0001	.0005
	15	.0000	.0000	.0000	.0000	.0000	.0000	.0000	.0000	.0000	.0000

CUMULATIVE TERMS, BINOMIAL DISTRIBUTION (Continued)

θ

n	x′	.05	.10	.15	.20	.25	.30	.35	.40	.45	.50
16	1	.5599	.8147	.9257	.9719	.9900	.9967	.9990	.9997	.9999	1.0000
	2	.1892	.4853	.7161	.8593	.9365	.9739	.9902	.9967	.9990	.9997
	3	.0429	.2108	.4386	.6482	.8029	.9006	.9549	.9817	.9934	.9979
	4	.0070	.0684	.2101	.4019	.5950	.7541	.8661	.9349	.9719	.9894
	5	.0009	.0170	.0791	.2018	.3698	.5501	.7108	.8334	.9147	.9616
	6	.0001	.0033	.0235	.0817	.1897	.3402	.5100	.6712	.8024	.8949
	7	.0000	.0005	.0056	.0267	.0796	.1753	.3119	.4728	.6340	.7228
	8	.0000	.0001	.0011	.0070	.0271	.0744	.1594	.2839	.4371	.5982
	9	.0000	.0000	.0002	.0015	.0075	.0257	.0671	.1423	.2559	.4018
	10	.0000	.0000	.0000	.0002	.0016	.0071	.0229	.0583	.1241	.2272
	11	.0000	.0000	.0000	.0000	.0003	.0016	.0062	.0191	.0486	.1051
	12	.0000	.0000	.0000	.0000	.0000	.0003	.0013	.0049	.0149	.0384
	13	.0000	.0000	.0000	.0000	.0000	.0000	.0002	.0009	.0035	.0106
	14	.0000	.0000	.0000	.0000	.0000	.0000	.0000	.0001	.0006	.0021
	15	.0000	.0000	.0000	.0000	.0000	.0000	.0000	.0000	.0001	.0003
	16	.0000	.0000	.0000	.0000	.0000	.0000	.0000	.0000	.0000	.0000
17	1	.5819	.8332	.9369	.9775	.9925	.9977	.9993	.9998	1.0000	1.0000
	2	.2078	.5182	.7475	.8818	.9499	.9807	.9933	.9979	.9994	.9999
	3	.0503	.2382	.4802	.6904	.8363	.9226	.9673	.9877	.9959	.9988
	4	.0088	.0826	.2444	.4511	.6470	.7981	.8972	.9536	.9816	.9936
	5	.0012	.0221	.0987	.2418	.4261	.6113	.7652	.8740	.9404	.9755
	6	.0001	.0047	.0319	.1057	.2347	.4032	.5803	.7361	.8529	.9283
	7	.0000	.0008	.0083	.0377	.1071	.2248	.3812	.5522	.7098	.8338
	8	.0000	.0001	.0017	.0109	.0402	.1046	.2128	.3595	.5257	.6855
	9	.0000	.0000	.0003	.0026	.0124	.0403	.0994	.1989	.3374	.5000
	10	.0000	.0000	.0000	.0005	.0031	.0127	.0383	.0919	.1834	.3145
	11	.0000	.0000	.0000	.0001	.0006	.0032	.0120	.0348	.0826	.1662
	12	.0000	.0000	.0000	.0000	.0001	.0007	.0030	.0106	.0301	.0717
	13	.0000	.0000	.0000	.0000	.0000	.0001	.0006	.0025	.0086	.0245
	14	.0000	.0000	.0000	.0000	.0000	.0000	.0001	.0005	.0019	.0064
	15	.0000	.0000	.0000	.0000	.0000	.0000	.0000	.0001	.0003	.0012
	16	.0000	.0000	.0000	.0000	.0000	.0000	.0000	.0000	.0000	.0001
	17	.0000	.0000	.0000	.0000	.0000	.0000	.0000	.0000	.0000	.0000
18	1	.6028	.8499	.9464	.9820	.9944	.9984	.9996	.9999	1.0000	1.0000
	2	.2265	.5497	.7759	.9009	.9605	.9858	.9954	.9987	.9997	.9999
	3	.0581	.2662	.5203	.7287	.8647	.9400	.9764	.9918	.9975	.9993
	4	.0109	.0982	.2798	.4990	.6943	.8354	.9217	.9672	.9880	.9962
	5	.0015	.0282	.1206	.2836	.4813	.6673	.8114	.9058	.9589	.9846
	6	.0002	.0064	.0419	.1329	.2825	.4656	.6450	.7912	.8923	.9519
	7	.0000	.0012	.0118	.0513	.1390	.2783	.4509	.6257	.7742	.8811
	8	.0000	.0002	.0027	.0163	.0569	.1407	.2717	.4366	.6085	.7597
	9	.0000	.0000	.0005	.0043	.0193	.0596	.1391	.2632	.4222	.5927
	10	.0000	.0000	.0001	.0009	.0054	.0210	.0597	.1347	.2527	.4073
	11	.0000	.0000	.0000	.0002	.0012	.0061	.0212	.0576	.1280	.2403
	12	.0000	.0000	.0000	.0000	.0002	.0014	0062	.0203	.0537	.1189
	13	.0000	.0000	.0000	.0000	.0000	.0003	.0014	.0058	.0183	.0481
	14	.0000	.0000	.0000	.0000	.0000	.0000	.0003	.0013	.0049	.0154
	15	.0000	.0000	.0000	.0000	.0000	.0000	.0000	.0002	.0010	.0038

CUMULATIVE TERMS, BINOMIAL DISTRIBUTION (Continued)

θ

n	x'	.05	.10	.15	.20	.25	.30	.35	.40	.45	.50
18	16	.0000	.0000	.0000	.0000	.0000	.0000	.0000	.0000	.0001	.0007
	17	.0000	.0000	.0000	.0000	.0000	.0000	.0000	.0000	.0000	.0001
	18	.0000	.0000	.0000	.0000	.0000	.0000	.0000	.0000	.0000	.0000
19	1	.6226	.8649	.9544	.9856	.9958	.9989	.9997	.9999	1.0000	1.0000
	2	.2453	.5797	.8015	.9171	.9690	.9896	.9969	.9992	.9998	1.0000
	3	0.665	.2946	.5587	.7631	.8887	.9538	.9830	.9945	.9985	.9996
	4	0.132	.1150	.3159	.5449	.7369	.8668	.9409	.9770	.9923	.9978
	5	.0020	.0352	.1444	.3267	.5346	.7178	.8500	.9304	.9720	.9904
	6	.0002	.0086	.0537	.1631	.3322	.5261	.7032	.8371	.9223	.9682
	7	.0000	.0017	.0163	.0676	.1749	.3345	.5188	.6919	.8273	.9165
	8	.0000	.0003	.0041	.0233	.0775	.1820	.3344	.5122	.6831	.8204
	9	.0000	.0000	.0008	.0067	.0287	.0839	.1855	.3325	.5060	.6762
	10	.0000	.0000	.0001	.0016	.0089	.0326	.0875	.1861	.3290	.5000
19	11	.0000	.0000	.0000	.0003	.0023	.0105	.0347	.0885	.1841	.3238
	12	.0000	.0000	.0000	.0000	.0005	.0028	.0114	.0352	.0871	.1796
	13	.0000	.0000	.0000	.0000	.0001	.0006	.0031	.0116	.0342	.0835
	14	.0000	.0000	.0000	.0000	.0000	.0001	.0007	.0031	.0109	.0318
	15	.0000	.0000	.0000	.0000	.0000	.0000	.0001	.0006	.0028	.0096
	16	.0000	.0000	.0000	.0000	.0000	.0000	.0000	.0001	.0005	.0022
	17	.0000	.0000	.0000	.0000	.0000	.0000	.0000	.0000	.0001	.0004
	18	.0000	.0000	.0000	.0000	.0000	.0000	.0000	.0000	.0000	.0000
	19	.0000	.0000	.0000	.0000	.0000	.0000	.0000	.0000	.0000	.0000
20	1	.6415	.8784	.9612	.9885	.9968	.9992	.9998	1.0000	1.0000	1.0000
	2	.2642	.6083	.8244	.9308	.9757	.9924	.9979	.9995	.9999	1.0000
	3	.0755	.3231	.5951	.7939	.9087	.9645	.9879	.9964	.9991	.9998
	4	.0159	.1330	.3523	.5886	.7748	.8929	.9556	.9840	.9951	.9987
	5	.0026	.0432	.1702	.3704	.5852	.7625	.8818	.9490	.9811	.9941
	6	.0003	.0113	.0673	.1958	.3828	.5836	.7546	.8744	.9447	.9793
	7	.0000	.0024	.0219	.0867	.2142	.3920	.5834	.7500	.8701	.9423
	8	.0000	.0004	.0059	.0321	.1018	.2277	.3990	.5841	.7480	.8684
	9	.0000	.0001	.0013	.0100	.0409	.1133	.2376	.4044	.5857	.7483
	10	.0000	.0000	.0002	.0026	.0139	.0480	.1218	.2447	.4086	.5881
	11	.0000	.0000	.0000	.0006	.0039	.0171	.0532	.1275	.2493	.4119
	12	.0000	.0000	.0000	.0001	.0009	.0051	.0196	.0565	.1308	.2517
	13	.0000	.0000	.0000	.0000	.0002	.0013	.0060	.0210	.0580	.1316
	14	.0000	.0000	.0000	.0000	.0000	.0003	.0015	.0065	.0214	.0577
	15	.0000	.0000	.0000	.0000	.0000	.0000	.0003	.0016	.0064	.0207
	16	.0000	.0000	.0000	.0000	.0000	.0000	.0000	.0003	.0015	.0059
	17	.0000	.0000	.0000	.0000	.0000	.0000	.0000	.0000	.0003	.0013
	18	.0000	.0000	.0000	.0000	.0000	.0000	.0000	.0000	.0000	.0002
	19	.0000	.0000	.0000	.0000	.0000	.0000	.0000	.0000	.0000	.0000
	20	.0000	.0000	.0000	.0000	.0000	.0000	.0000	.0000	.0000	.0000

CUMULATIVE TERMS, POISSON DISTRIBUTION

This table contains the values of

$$\sum_{x=x'}^{\infty} \frac{e^{-\lambda}\lambda^x}{x!}$$

for specified values of x' and λ. Individual terms of the Poisson distribution for specified x' and λ can be found from

$$\sum_{x=x'}^{\infty} \frac{e^{-\lambda}\lambda^x}{x!} - \sum_{x=x'+1}^{\infty} \frac{e^{-\lambda}\lambda^x}{x!}$$

λ

x'	0.1	0.2	0.3	0.4	0.5	0.6	0.7	0.8	0.9	1.0
0	1.0000	1.0000	1.0000	1.0000	1.0000	1.0000	1.0000	1.0000	1.0000	1.0000
1	.0952	.1813	.2592	.3297	.3935	.4512	.5034	.5507	.5934	.6321
2	.0047	.0175	.0369	.0616	.0902	.1219	.1558	.1912	.2275	.2642
3	.0002	.0011	.0036	.0079	.0144	.0231	.0341	.0474	.0629	.0803
4	.0000	.0001	.0003	.0008	.0018	.0034	.0058	.0091	.0135	.0190
5	.0000	.0000	.0000	.0001	.0002	.0004	.0008	.0014	.0023	.0037
6	.0000	.0000	.0000	.0000	.0000	.0000	.0001	.0002	.0003	.0006
7	.0000	.0000	.0000	.0000	.0000	.0000	.0000	.0000	.0000	.0001

λ

x'	1.1	1.2	1.3	1.4	1.5	1.6	1.7	1.8	1.9	2.0
0	1.0000	1.0000	1.0000	1.0000	1.0000	1.0000	1.0000	1.0000	1.0000	1.0000
1	.6671	.6988	.7275	.7534	.7769	.7981	.8173	.8347	.8504	.8647
2	.3010	.3374	.3732	.4082	.4422	.4751	.5068	.5372	.5663	.5940
3	.0996	.1205	.1429	.1665	.1912	.2166	.2428	.2694	.2963	.3233
4	.0257	.0338	.0431	.0537	.0656	.0788	.0932	.1087	.1253	.1429
5	.0054	.0077	.0107	.0143	.0186	.0237	.0296	.0364	.0441	.0527
6	.0010	.0015	.0022	.0032	.0045	.0060	.0080	.0104	.0132	.0166
7	.0001	.0003	.0004	.0006	.0009	.0013	.0019	.0026	.0034	.0045
8	.0000	.0000	.0001	.0001	.0002	.0003	.0004	.0006	.0008	.0011
9	.0000	.0000	.0000	.0000	.0000	.0000	.0001	.0001	.0002	.0002

λ

x'	2.1	2.2.	2.3	2.4	2.5	2.6	2.7	2.8	2.9	3.0
0	1.0000	1.0000	1.0000	1.0000	1.0000	1.0000	1.0000	1.0000	1.0000	1.0000
1	.8775	.8892	.8997	.9093	.9179	.9257	.9328	.9392	.9450	.9502
2	.6204	.6454	.6691	.6916	.7127	.7326	.7513	.7689	.7854	.8009
3	.3504	.3773	.4040	.4303	.4562	.4816	.5064	.5305	.5540	.5768
4	.1614	.1806	.2007	.2213	.2424	.2640	.2859	.3081	.3304	.3528
5	.0621	.0725	.0838	.0959	.1088	.1226	.1371	.1523	.1682	.1847
6	.0204	.0249	.0300	.0357	.0420	.0490	.0567	.0651	.0742	.0839
7	.0059	.0075	.0094	.0116	.0142	.0172	.0206	.0244	.0287	.0335
8	.0015	.0020	.0026	.0033	.0042	.0053	.0066	.0081	.0099	.0119
9	.0003	.0005	.0006	.0009	.0011	.0015	.0019	.0024	.0031	.0038
10	.0001	.0001	.0001	.0002	.0003	.0004	.0005	.0007	.0009	.0011
11	.0000	.0000	.0000	.0000	.0001	.0001	.0001	.0002	.0002	.0003
12	.0000	.0000	.0000	.0000	.0000	.0000	.0000	.0000	.0001	.0001

λ

x'	3.1	3.2	3.3	3.4	3.5	3.6	3.7	3.8	3.9	4.0
0	1.0000	1.0000	1.0000	1.0000	1.0000	1.0000	1.0000	1.0000	1.0000	1.0000
1	.9550	.9592	.9631	.9666	.9698	.9727	.9753	.9776	.9798	.9817
2	.8153	.8288	.8414	.8532	.8641	.8743	.8838	.8926	.9008	.9084
3	.5988	.6201	.6406	.6603	.6792	.6973	.7146	.7311	.7469	.7619
4	.3752	.3975	.4197	.4416	.4634	.4848	.5058	.5265	.5468	.5665

CUMULATIVE TERMS, POISSON DISTRIBUTION (Continued)

λ

x'	3.1	3.2	3.3	3.4	3.5	3.6	3.7	3.8	3.9	4.0
5	.2018	.2194	.2374	.2558	.2746	.2936	.3128	.3322	.3516	.3712
6	.0943	.1054	.1171	.1295	.1424	.1559	.1699	.1844	.1994	.2149
7	.0388	.0446	.0510	.0579	.0653	.0733	.0818	.0909	.1005	.1107
8	.0142	.0168	.0198	.0231	.0267	.0308	.0352	.0401	.0454	.0511
9	.0047	.0057	.0069	.0083	.0099	.0117	.0137	.0160	.0185	.0214
10	.0014	.0018	.0022	.0027	.0033	.0040	.0048	.0058	.0069	.0081
11	.0004	.0005	.0006	.0008	.0010	.0013	.0016	.0019	.0023	.0028
12	.0001	.0001	.0002	.0002	.0003	.0004	.0005	.0006	.0007	.0009
13	.0000	.0000	.0000	.0001	.0001	.0001	.0001	.0002	.0002	.0003
14	.0000	.0000	.0000	.0000	.0000	.0000	.0000	.0000	.0001	.0001

λ

x'	4.1	4.2	4.3	4.4	4.5	4.6	4.7	4.8	4.9	5.0
0	1.0000	1.0000	1.0000	1.0000	1.0000	1.0000	1.0000	1.0000	1.0000	1.0000
1	.9834	.9850	.9864	.9877	.9889	.9899	.9909	.9918	.9926	.9933
2	.9155	.9220	.9281	.9337	.9389	.9437	.9482	.9523	.9561	.9596
3	.7762	.7898	.8026	.8149	.8264	.8374	.8477	.8575	.8667	.8753
4	.5858	.6046	.6228	.6406	.6577	.6743	.6903	.7058	.7207	.7350
5	.3907	.4102	.4296	.4488	.4679	.4868	.5054	.5237	.5418	.5595
6	.2307	.2469	.2633	.2801	.2971	.3142	.3316	.3490	.3665	.3840
7	.1214	.1325	.1442	.1564	.1689	.1820	.1954	.2092	.2233	.2378
8	.0573	.0639	.0710	.0786	.0866	.0951	.1050	.1133	.1231	.1334
9	.0245	.0279	.0317	.0358	.0403	.0451	.0503	.0558	.0618	.0681
10	.0095	.0111	.0129	.0149	.0171	.0195	.0222	.0251	.0283	.0318
11	.0034	.0041	.0048	.0057	.0067	.0078	.0090	.0104	.0120	.0137
12	.0011	.0014	.0017	.0020	.0024	.0029	.0034	.0040	.0047	.0055
13	.0003	.0004	.0005	.0007	.0008	.0010	.0012	.0014	.0017	.0020
14	.0001	.0001	.0002	.0002	.0003	.0003	.0004	.0005	.0006	.0007
15	.0000	.0000	.0000	.0001	.0001	.0001	.0001	.0001	.0002	.0002
16	.0000	.0000	.0000	.0000	.0000	.0000	.0000	.0000	.0001	.0001

λ

x'	5.1	5.2	5.3	5.4	5.5	5.6	5.7	5.8	5.9	6.0
0	1.0000	1.0000	1.0000	1.0000	1.0000	1.0000	1.0000	1.0000	1.0000	1.0000
1	.9939	.9945	.9950	.9955	.9959	.9963	.9967	.9970	.9973	.9975
2	.9628	.9658	.9686	.9711	.9734	.9756	.9776	.9794	.9811	.9826
3	.8835	.8912	.8984	.9052	.9116	.9176	.9232	.9285	.9334	.9380
4	.7487	.7619	.7746	.7867	.7983	.8094	.8200	.8300	.8396	.8488
5	.5769	.5939	.6105	.6267	.6425	.6579	.6728	.6873	.7013	.7149
6	.4016	.4191	.4365	.4539	.4711	.4881	.5050	.5217	.5381	.5543
7	.2526	.2676	.2829	.2983	.3140	.3297	.3456	.3616	.3776	.3937
8	.1440	.1551	.1665	.1783	.1905	.2030	.2159	.2290	.2424	.2560
9	.0748	.0819	.0894	.0974	.1056	.1143	.1234	.1328	.1426	.1528

CUMULATIVE TERMS, POISSON DISTRIBUTION (Continued)

λ

x	5.1	5.2	5.3	5.4	5.5	5.6	5.7	5.8	5.9	6.0
10	.0356	.0397	.0441	.0488	.0538	.0591	.0648	.0708	.0772	.0839
11	.0156	.0177	.0200	.0225	.0253	.0282	.0314	.0349	.0386	.0426
12	.0063	.0073	.0084	.0096	.0110	.0125	.0141	.0160	.0179	.0201
13	.0024	.0028	.0033	.0038	.0045	.0051	.0059	.0068	.0078	.0088
14	.0008	.0010	.0012	.0014	.0017	.0020	.0023	.0027	.0031	.0036
15	.0003	.0003	.0004	.0005	.0006	.0007	.0009	.0010	.0012	.0014
16	.0001	.0001	.0001	.0002	.0002	.0002	.0003	.0004	.0004	.0005
17	.0000	.0000	.0000	.0001	.0001	.0001	.0001	.0001	.0001	.0002
18	.0000	.0000	.0000	.0000	.0000	.0000	.0000	.0000	.0000	.0001

λ

x'	6.1	6.2	6.3	6.4	6.5	6.6	6.7	6.8	6.9	7.0
0	1.0000	1.0000	1.0000	1.0000	1.0000	1.0000	1.0000	1.0000	1.0000	1.0000
1	.9978	.9980	.9982	.9983	.9985	.9986	.9988	.9989	.9990	.9991
2	.9841	.9854	.9866	.9877	.9887	.9897	.9905	.9913	.9920	.9927
3	.9423	.9464	.9502	.9537	.9570	.9600	.9629	.9656	.9680	.9704
4	.8575	.8658	.8736	.8811	.8882	.8948	.9012	.9072	.9129	.9182
5	.7281	.7408	.7531	.7649	.7763	.7873	.7978	.8080	.8177	.8270
6	.5702	.5859	.6012	.6163	.6310	.6453	.6594	.6730	.6863	.6993
7	.4098	.4258	.4418	.4577	.4735	.4892	.5047	.5201	.5353	.5503
8	.2699	.2840	.2983	.3127	.3272	.3419	.3567	.3715	.3864	.4013
9	.1633	.1741	.1852	.1967	.2084	.2204	.2327	.2452	.2580	.2709
10	.0910	.0984	.1061	.1142	.1226	.1314	.1404	.1498	.1505	.1695
11	.0469	.0514	.0563	.0614	.0668	.0726	.0786	.0849	.0916	.0985
12	.0224	.0250	.0277	.0307	.0339	.0373	.0409	.0448	.0490	.0534
13	.0100	.0113	.0127	.0143	.0160	.0179	.0199	.0221	.0245	.0270
14	.0042	.0048	.0055	.0063	.0071	.0080	.0091	.0102	.0115	.0128
15	.0016	.0019	.0022	.0026	.0030	.0034	.0039	.0044	.0050	.0057
16	.0006	.0007	.0008	.0010	.0012	.0014	.0016	.0018	.0021	.0024
17	.0002	.0003	.0003	.0004	.0004	.0005	.0006	.0007	.0008	.0010
18	.0001	.0001	.0001	.0001	.0002	.0002	.0002	.0003	.0003	.0004
19	.0000	.0000	.0000	.0000	.0001	.0001	.0001	.0001	.0001	.0001

λ

x'	7.1	7.2	7.3	7.4	7.5	7.6	7.7	7.8	7.9	8.0
0	1.0000	1.0000	1.0000	1.0000	1.0000	1.0000	1.0000	1.0000	1.0000	1.0000
1	.9992	.9993	.9993	.9994	.9994	.9995	.9995	.9996	.9996	.9997
2	.9933	.9939	.9944	.9949	.9953	.9957	.9961	.9964	.9967	.9970
3	.9725	.9745	.9764	.9781	.9797	.9812	.9826	.9839	.9851	.9862
4	.9233	.9281	.9326	.9368	.9409	.9446	.9482	.9515	.9547	.9576
5	.8359	.8445	.8527	.8605	.8679	.8751	.8819	.8883	.8945	.9004
6	.7119	.7241	.7360	.7474	.7586	.7693	.7797	.7897	.7994	.8088
7	.5651	.5796	.5940	.6080	.6218	.6354	.6486	.6616	.6743	.6866
8	.4162	.4311	.4459	.4607	.4754	.4900	.5044	.5188	.5330	.5470
9	.2840	.2973	.3108	.3243	.3380	.3518	.3657	.3796	.3935	.4075
10	.1798	.1904	.2012	.2123	.2236	.2351	.2469	.2589	.2710	.2834
11	.1058	.1133	.1212	.1293	.1378	.1465	.1555	.1648	.1743	.1841
12	.0580	.0629	.0681	.0735	.0792	.0852	.0915	.0980	.1048	.1119
13	.0297	.0327	.0358	.0391	.0427	.0464	.0504	.0546	.0591	.0638
14	.0143	.0159	.0176	.0195	.0216	.0238	.0261	.0286	.0313	.0342

CUMULATIVE TERMS, POISSON DISTRIBUTION (Continued)

λ

x'	7.1	7.2	7.3	7.4	7.5	7.6	7.7	7.8	7.9	8.0
15	.0065	.0073	.0082	.0092	.0103	.0114	.0127	.0141	.0156	.0173
16	.0028	.0031	.0036	.0041	.0046	.0052	.0059	.0066	.0074	.0082
17	.0011	.0013	.0015	.0017	.0020	.0022	.0026	.0029	.0033	.0037
18	.0004	.0005	.0006	.0007	.0008	.0009	.0011	.0012	.0014	.0016
19	.0002	.0002	.0002	.0003	.0003	.0004	.0004	.0005	.0006	.0006
20	.0001	.0001	.0001	.0001	.0001	.0001	.0002	.0002	.0002	.0003
21	.0000	.0000	.0000	.0000	.0000	.0000	.0001	.0001	.0001	.0001

λ

x'	8.1	8.2	8.3	8.4	8.5	8.6	8.7	8.8	8.9	9.0
0	1.0000	1.0000	1.0000	1.0000	1.0000	1.0000	1.0000	1.0000	1.0000	1.0000
1	.9997	.9997	.9998	.9998	.9998	.9998	.9998	.9998	.9999	.9999
2	.9972	.9975	.9977	.9979	.9981	.9982	.9984	.9985	.9987	.9988
3	.9873	.9882	.9891	.9900	.9907	.9914	.9921	.9927	.9932	.9938
4	.9604	.9630	.9654	.9677	.9699	.9719	.9738	.9756	.9772	.9788
5	.9060	.9113	.9163	.9211	.9256	.9299	.9340	.9340	.9416	.9450
6	.8178	.8264	.8347	.8427	.8504	.8578	.8648	.8716	.8781	.8843
7	.6987	.7104	.7219	.7330	.7438	.7543	.7645	.7744	.7840	.7932
8	.5609	.5746	.5881	.6013	.6144	.6272	.6398	.6522	.6643	.6761
9	.4214	.4353	.4493	.4631	.4769	.4906	.5042	.5177	.5311	.5443
10	.2959	.3085	.3212	.3341	.3470	.3600	.3731	.3863	.3994	.4126
11	.1942	.2045	.2150	.2257	.2366	.2478	.2591	.2706	.2822	.2940
12	.1193	.1269	.1348	.1429	.1513	.1600	.1689	.1780	.1874	.1970
13	.0687	.0739	.0793	.0850	.0909	.0971	.1035	.1102	.1171	.1242
14	.0372	.0405	.0439	.0476	.0514	.0555	.0597	.0642	.0689	.0739
15	.0190	.0209	.0229	.0251	.0274	.0299	.0325	.0353	.0383	.0415
16	.0092	.0102	.0113	.0125	.0138	.0152	.0168	.0184	.0202	.0220
17	.0042	.0047	.0053	.0059	.0066	.0074	.0082	.0091	.0101	.0111
18	.0018	.0021	.0023	.0027	.0030	.0034	.0038	.0043	.0048	.0053
19	.0008	.0009	.0010	.0011	.0013	.0015	.0017	.0019	.0022	.0024
20	.0003	.0003	.0004	.0005	.0005	.0006	.0007	.0008	.0009	.0011
21	.0001	.0001	.0002	.0002	.0002	.0002	.0003	.0003	.0004	.0004
22	.0000	.0000	.0001	.0001	.0001	.0001	.0001	.0001	.0002	.0002
23	.0000	.0000	.0000	.0000	.0000	.0000	.0000	.0000	.0001	.0001

λ

x'	9.1	9.2	9.3	9.4	9.5	9.6	9.7	9.8	9.9	10
0	1.0000	1.0000	1.0000	1.0000	1.0000	1.0000	1.0000	1.0000	1.0000	1.0000
1	.9999	.9999	.9999	.9999	.9999	.9999	.9999	.9999	1.0000	1.0000
2	.9989	.9990	.9991	.9991	.9992	.9993	.9993	.9994	.9995	.9995
3	.9942	.9947	.9951	.9955	.9958	.9962	.9965	.9967	.9970	.9972
4	.9802	.9816	.9828	.9840	.9851	.9862	.9871	.9880	.9889	.9897
5	.9483	.9514	.9544	.9571	.9597	.9622	.9645	.9667	.9688	.9707
6	.8902	.8959	.9014	.9065	.9115	.9162	.9207	.9250	.9290	.9329
7	.8022	.8108	.8192	.8273	.8351	.8426	.8498	.8567	.8634	.8699
8	.6877	.6990	.7101	.7208	.7313	.7416	.7515	.7612	.7706	.7798
9	.5574	.5704	.5832	.5958	.6082	.6204	.6324	.6442	.6558	.6672

CUMULATIVE TERMS, POISSON DISTRIBUTION (Continued)

λ

x'	9.1	9.2	9.3	9.4	9.5	9.6	9.7	9.8	9.9	10
10	.4258	.4389	.4521	.4651	.4782	.4911	.5040	.5168	..5295	.5421
11	.3059	.3180	.3301	.3424	.3547	.3671	.3795	.3920	.4045	.4170
12	.2068	.2168	.2270	.2374	.2480	.2588	.2697	.2807	.2919	.3032
13	.1316	.1393	.1471	.1552	.1636	.1721	.1809	.1899	.1991	.2084
14	.0790	.0844	.0900	.0958	.1019	.1081	.1147	.1214	.1284	.1355
15	.0448	.0483	.0520	.0559	.0600	.0643	.0688	.0735	.0784	.0835
16	.0240	.0262	.0285	.0309	.0335	.0362	.0391	.0421	.0454	.0487
17	.0122	.0135	.0148	.0162	.0177	.0194	.0211	.0230	.0249	.0270
18	.0059	.0066	.0073	.0081	.0089	.0098	.0108	.0119	.0130	.0143
19	.0027	.0031	.0034	.0038	.0043	.0048	.0053	.0059	.0065	.0072
20	.0012	.0014	.0015	.0017	.0020	.0022	.0025	.0028	.0031	.0035
21	.0005	.0006	.0007	.0008	.0009	.0010	.0011	.0013	.0014	.0016
22	.0002	.0002	.0003	.0003	.0004	.0004	.0005	.0005	.0006	.0007
23	.0001	.0001	.0001	.0001	.0001	.0002	.0002	.0002	.0003	.0003
24	.0000	.0000	.0000	.0000	.0001	.0001	.0001	.0001	.0001	.0001

λ

x'	11	12	13	14	15	16	17	18	19	20
0	1.0000	1.0000	1.0000	1.0000	1.0000	1.0000	1.0000	1.0000	1.0000	1.0000
1	1.0000	1.0000	1.0000	1.0000	1.0000	1.0000	1.0000	1.0000	1.0000	1.0000
2	.9998	.9999	1.0000	1.0000	1.0000	1.0000	1.0000	1.0000	1.0000	1.0000
3	.9988	.9995	.9998	.9999	1.0000	1.0000	1.0000	1.0000	1.0000	1.0000
4	.9951	.9977	.9990	.9995	.9998	.9999	1.0000	1.0000	1.0000	1.0000
5	.9849	.9924	.9963	.9982	.9991	.9996	.9998	.9999	1.0000	1.0000
6	.9625	.9797	.9893	.9945	.9972	.9986	.9993	.9997	.9998	.9999
7	.9214	.9542	.9741	.9858	.9924	.9960	.9979	.9990	.9995	.9997
8	.8568	.9105	.9460	.9684	.9820	.9900	.9946	.9971	.9985	.9992
9	.7680	.8450	.9002	.9379	.9626	.9780	.9874	.9929	.9961	.9979
10	.6595	.7576	.8342	.8906	.9301	.9567	.9739	.9846	.9911	.9950
11	.5401	.6528	.7483	.8243	.8815	.9226	.9509	.9696	.9817	.9892
12	.4207	.5384	.6468	.7400	.8152	.8730	.9153	.9451	.9653	.9786
13	.3113	.4240	.5369	.6415	.7324	.8069	.8650	.9083	.9394	.9610
14	.2187	.3185	.4270	.5356	.6368	.7255	.7991	.8574	.9016	.9339
15	.1460	.2280	.3249	.4296	.5343	.6325	.7192	.7919	.8503	.8951
16	.0926	.1556	.2364	.3306	.4319	.5333	.6285	.7133	.7852	.8435
17	.0559	.1013	.1645	.2441	.3359	.4340	.5323	.6250	.7080	.7789
18	.0322	.0630	.1095	.1728	.2511	.3407	.4360	.5314	.6216	.7030
19	.0177	.0374	.0698	.1174	.1805	.2577	.3450	.4378	.5305	.6186
20	.0093	.0213	.0427	.0765	.1248	.1878	.2637	.3491	.4394	.5297
21	.0047	.0116	.0250	.0479	.0830	.1318	.1945	.2693	.3528	.4409
22	.0023	.0061	.0141	.0288	.0531	.0892	.1385	.2009	.2745	.3563
23	.0010	.0030	.0076	.0167	.0327	.0582	.0953	.1449	.2069	.2794
24	.0005	.0015	.0040	.0093	.0195	.0367	.0633	.1011	.1510	.2125
25	.0002	.0007	.0020	.0050	.0112	.0223	.0406	.0683	.1067	.1568
26	.0001	.0003	.0010	.0026	.0062	.0131	.0252	.0446	.0731	.1122
27	.0000	.0001	.0005	.0013	.0033	.0075	.0152	.0282	.0486	.0779
28	.0000	.0001	.0002	.0006	.0017	.0041	.0088	.0173	.0313	.0525
29	.0000	.0000	.0001	.0003	.0009	.0022	.0050	.0103	.0195	.0343
30	.0000	.0000	.0000	.0001	.0004	.0011	.0027	.0059	.0118	.0218
31	.0000	.0000	.0000	.0001	.0002	.0006	.0014	.0033	.0070	.0135
32	.0000	.0000	.0000	.0000	.0001	.0003	.0007	.0018	.0040	.0081
33	.0000	.0000	.0000	.0000	.0000	.0001	.0004	.0010	.0022	.0047
34	.0000	.0000	.0000	.0000	.0000	.0001	.0002	.0005	.0012	.0027
35	.0000	.0000	.0000	.0000	.0000	.0000	.0001	.0002	.0006	.0015
36	.0000	.0000	.0000	.0000	.0000	.0000	.0000	.0001	.0003	.0008
37	.0000	.0000	.0000	.0000	.0000	.0000	.0000	.0001	.0002	.0004
38	.0000	.0000	.0000	.0000	.0000	.0000	.0000	.0000	.0001	.0002
39	.0000	.0000	.0000	.0000	.0000	.0000	.0000	.0000	.0000	.0001
40	.0000	.0000	.0000	.0000	.0000	.0000	.0000	.0000	.0000	.0001

PERCENTAGE POINTS, STUDENT'S *t*-DISTRIBUTION

This table gives values of t such that

$$F(t) = \int_{-\infty}^{t} \frac{\Gamma\left(\frac{n+1}{2}\right)}{\sqrt{n\pi}\,\Gamma\left(\frac{n}{2}\right)} \left(1 + \frac{x^2}{n}\right)^{-\frac{n+1}{2}} dx$$

for n, the number of degrees of freedom, equal to 1, 2, . . ., 30, 40, 60, 120, ∞; and for $F(t) = 0.60, 0.75, 0.90, 0.95, 0.975, 0.99, 0.995,$ and 0.9995. The *t*-distribution is symmetrical, so that $F(-t) = 1 - F(t)$

F n	.60	.75	.90	.95	.975	.99	.995	.9995
1	.325	1.000	3.078	6.314	12.706	31.821	63.657	636.619
2	.289	.816	1.886	2.920	4.303	6.965	9.925	31.598
3	.277	.765	1.638	2.353	3.182	4.541	5.841	12.924
4	.271	.741	1.533	2.132	2.776	3.747	4.604	8.610
5	.267	.727	1.476	2.015	2.571	3.365	4.032	6.869
6	.265	.718	1.440	1.943	2.447	3.143	3.707	5.959
7	.263	.711	1.415	1.895	2.365	2.998	3.499	5.408
8	.262	.706	1.397	1.860	2.306	2.896	3.355	5.041
9	.261	.703	1.383	1.833	2.262	2.821	3.250	4.781
10	.260	.700	1.372	1.812	2.228	2.764	3.169	4.587
11	.260	.697	1.363	1.796	2.201	2.718	3.106	4.437
12	.259	.695	1.356	1.782	2.179	2.681	3.055	4.318
13	.259	.694	1.350	1.771	2.160	2.650	3.012	4.221
14	.258	.692	1.345	1.761	2.145	2.624	2.977	4.140
15	.258	.691	1.341	1.753	2.131	2.602	2.947	4.073
16	.258	.690	1.337	1.746	2.120	2.583	2.921	4.015
17	.257	.689	1.333	1.740	2.110	2.567	2.898	3.965
18	.257	.688	1.330	1.734	2.101	2.552	2.878	3.922
19	.257	.688	1.328	1.729	2.093	2.539	2.861	3.883
20	.257	.687	1.325	1.725	2.086	2.528	2.845	3.850
21	.257	.686	1.323	1.721	2.080	2.518	2.831	3.819
22	.256	.686	1.321	1.717	2.074	2.508	2.819	3.792
23	.256	.685	1.319	1.714	2.069	2.500	2.807	3.767
24	.256	.685	1.318	1.711	2.064	2.492	2.797	3.745
25	.256	.684	1.316	1.708	2.060	2.485	2.787	3.725
26	.256	.684	1.315	1.706	2.056	2.479	2.779	3.707
27	.256	.684	1.314	1.703	2.052	2.473	2.771	3.690
28	.256	.683	1.313	1.701	2.048	2.467	2.763	3.674
29	.256	.683	1.311	1.699	2.045	2.462	2.756	3.659
30	.256	.683	1.310	1.697	2.042	2.457	2.750	3.646
40	.255	.681	1.303	1.684	2.021	2.423	2.704	3.551
60	.254	.679	1.296	1.671	2.000	2.390	2.660	3.460
120	.254	.677	1.289	1.658	1.980	2.358	2.617	3.373
∞	.253	.674	1.282	1.645	1.960	2.326	2.576	3.291

* This table is abridged from the "Statistical Tables" of R. A. Fisher and Frank Yates published by Oliver & Boyd. Ltd., Edinburgh and London, 1938. It is here published with the kind permission of the authors and their publishers.

PROBABILITY AND STATISTICS
CHI SQUARE DISTRIBUTION

PERCENTAGE POINTS, CHI-SQUARE DISTRIBUTION

$$F(\chi^2) = \int_0^{\chi^2} \frac{1}{2^{\frac{n}{2}} \Gamma\left(\frac{n}{2}\right)} x^{\frac{n-2}{2}} e^{-\frac{x}{2}} dx$$

n	.995	.990	.975	.950	.900	.750	.500	.250	.100	.050	.025	.010	.005
1	7.88	6.63	5.02	3.84	2.71	1.32	.455	.102	.0158	.00393	.000982	.000157	.0000393
2	10.6	9.21	7.38	5.99	4.61	2.77	1.39	.575	.211	.103	.0506	.0201	.0100
3	12.8	11.3	9.35	7.81	6.25	4.11	2.37	1.21	.584	.352	.216	.115	.0717
4	14.9	13.3	11.1	9.49	7.78	5.39	3.36	1.92	1.06	.711	.484	.297	.207
5	16.7	15.1	12.8	11.1	9.24	6.63	4.35	2.67	1.61	1.15	.831	.554	.412
6	18.5	16.8	14.4	12.6	10.6	7.84	5.35	3.45	2.20	1.64	1.24	.872	.676
7	20.3	18.5	16.0	14.1	12.0	9.04	6.35	4.25	2.83	2.17	1.69	1.24	.989
8	22.0	20.1	17.5	15.5	13.4	10.2	7.34	5.07	3.49	2.73	2.18	1.65	1.34
9	23.6	21.7	19.0	16.9	14.7	11.4	8.34	5.90	4.17	3.33	2.70	2.09	1.73
10	25.2	23.2	20.5	18.3	16.0	12.5	9.34	6.74	4.87	3.94	3.25	2.56	2.16
11	26.8	24.7	21.9	19.7	17.3	13.7	10.3	7.58	5.58	4.57	3.82	3.05	2.60
12	28.3	26.2	23.3	21.0	18.5	14.8	11.3	8.44	6.30	5.23	4.40	3.57	3.07
13	29.8	27.7	24.7	22.4	19.8	16.0	12.3	9.30	7.04	5.89	5.01	4.11	3.57
14	31.3	29.1	26.1	23.7	21.1	17.1	13.3	10.2	7.79	6.57	5.63	4.66	4.07
15	32.8	30.6	27.5	25.0	22.3	18.2	14.3	11.0	8.55	7.26	6.26	5.23	4.60
16	34.3	32.0	28.8	26.3	23.5	19.4	15.3	11.9	9.31	7.96	6.91	5.81	5.14
17	35.7	33.4	30.2	27.6	24.8	20.5	16.3	12.8	10.1	8.67	7.56	6.41	5.70
18	37.2	34.8	31.5	28.9	26.0	21.6	17.3	13.7	10.9	9.39	8.23	7.01	6.26
19	38.6	36.2	32.9	30.1	27.2	22.7	18.3	14.6	11.7	10.1	8.91	7.63	6.84
20	40.0	37.6	34.2	31.4	28.4	23.8	19.3	15.5	12.4	10.9	9.59	8.26	7.43
21	41.4	38.9	35.5	32.7	29.6	24.9	20.3	16.3	13.2	11.6	10.3	8.90	8.03
22	42.8	40.3	36.8	33.9	30.8	26.0	21.3	17.2	14.0	12.3	11.0	9.54	8.64
23	44.2	41.6	38.1	35.2	32.0	27.1	22.3	18.1	14.8	13.1	11.7	10.2	9.26
24	45.6	43.0	39.4	36.4	33.2	28.2	23.3	19.0	15.7	13.8	12.4	10.9	9.89
25	46.9	44.3	40.6	37.7	34.4	29.3	24.3	19.9	16.5	14.6	13.1	11.5	10.5
26	48.3	45.6	41.9	38.9	35.6	30.4	25.3	20.8	17.3	15.4	13.8	12.2	11.2
27	49.6	47.0	43.2	40.1	36.7	31.5	26.3	21.7	18.1	16.2	14.6	12.9	11.8
28	51.0	48.3	44.5	41.3	37.9	32.6	27.3	22.7	18.9	16.9	15.3	13.6	12.5
29	52.3	49.6	45.7	42.6	39.1	33.7	28.3	23.6	19.8	17.7	16.0	14.3	13.1
30	53.7	50.9	47.0	43.8	40.3	34.8	29.3	24.5	20.6	18.5	16.8	15.0	13.8

PERCENTAGE POINTS, F-DISTRIBUTION

$$F(F) = \int_0^F \frac{\Gamma\left(\frac{m+n}{2}\right)}{\Gamma\left(\frac{m}{2}\right)\Gamma\left(\frac{n}{2}\right)} m^{\frac{m}{2}} n^{\frac{n}{2}} x^{\frac{m}{2}-1} (n+mx)^{-\frac{m+n}{2}} dx = .90$$

n \ m	1	2	3	4	5	6	7	8	9	10	12	15	20	24	30	40	60	120	∞
1	39.86	49.50	53.59	55.83	57.24	58.20	58.91	59.44	59.86	60.19	60.71	61.22	61.74	62.00	62.26	62.53	62.79	63.06	63.33
2	8.53	9.00	9.16	9.24	9.29	9.33	9.35	9.37	9.38	9.39	9.41	9.42	9.44	9.45	9.46	9.47	9.47	9.48	9.49
3	5.54	5.46	5.39	5.34	5.31	5.28	5.27	5.25	5.24	5.23	5.22	5.20	5.18	5.18	5.17	5.16	5.15	5.14	5.13
4	4.54	4.32	4.19	4.11	4.05	4.01	3.98	3.95	3.94	3.92	3.90	3.87	3.84	3.83	3.82	3.80	3.79	3.78	3.76
5	4.06	3.78	3.62	3.52	3.45	3.40	3.37	3.34	3.32	3.30	3.27	3.24	3.21	3.19	3.17	3.16	3.14	3.12	3.10
6	3.78	3.46	3.29	3.18	3.11	3.05	3.01	2.98	2.96	2.94	2.90	2.87	2.84	2.82	2.80	2.78	2.76	2.74	2.72
7	3.59	3.26	3.07	2.96	2.88	2.83	2.78	2.75	2.72	2.70	2.67	2.63	2.59	2.58	2.56	2.54	2.51	2.49	2.47
8	3.46	3.11	2.92	2.81	2.73	2.67	2.62	2.59	2.56	2.54	2.50	2.46	2.42	2.40	2.38	2.36	2.34	2.32	2.29
9	3.36	3.01	2.81	2.69	2.61	2.55	2.51	2.47	2.44	2.42	2.38	2.34	2.30	2.28	2.25	2.23	2.21	2.18	2.16
10	3.29	2.92	2.73	2.61	2.52	2.46	2.41	2.38	2.35	2.32	2.28	2.24	2.20	2.18	2.16	2.13	2.11	2.08	2.06
11	3.23	2.86	2.66	2.54	2.45	2.39	2.34	2.30	2.27	2.25	2.21	2.17	2.12	2.10	2.08	2.05	2.03	2.00	1.97
12	3.18	2.81	2.61	2.48	2.39	2.33	2.28	2.24	2.21	2.19	2.15	2.10	2.06	2.04	2.01	1.99	1.96	1.93	1.90
13	3.14	2.76	2.56	2.43	2.35	2.28	2.23	2.20	2.16	2.14	2.10	2.05	2.01	1.98	1.96	1.93	1.90	1.88	1.85
14	3.10	2.73	2.52	2.39	2.31	2.24	2.19	2.15	2.12	2.10	2.05	2.01	1.96	1.94	1.91	1.89	1.86	1.83	1.80
15	3.07	2.70	2.49	2.36	2.27	2.21	2.16	2.12	2.09	2.06	2.02	1.97	1.92	1.90	1.87	1.85	1.82	1.79	1.76
16	3.05	2.67	2.46	2.33	2.24	2.18	2.13	2.09	2.06	2.03	1.99	1.94	1.89	1.87	1.84	1.81	1.78	1.75	1.72
17	3.03	2.64	2.44	2.31	2.22	2.15	2.10	2.06	2.03	2.00	1.96	1.91	1.86	1.84	1.81	1.78	1.75	1.72	1.69
18	3.01	2.62	2.42	2.29	2.20	2.13	2.08	2.04	2.00	1.98	1.93	1.89	1.84	1.81	1.78	1.75	1.72	1.69	1.66
19	2.99	2.61	2.40	2.27	2.18	2.11	2.06	2.02	1.98	1.96	1.91	1.86	1.81	1.79	1.76	1.73	1.70	1.67	1.63
20	2.97	2.59	2.38	2.25	2.16	2.09	2.04	2.00	1.96	1.94	1.89	1.84	1.79	1.77	1.74	1.71	1.68	1.64	1.61
21	2.96	2.57	2.36	2.23	2.14	2.08	2.02	1.98	1.95	1.92	1.87	1.83	1.78	1.75	1.72	1.69	1.66	1.62	1.59
22	2.95	2.56	2.35	2.22	2.13	2.06	2.01	1.97	1.93	1.90	1.86	1.81	1.76	1.73	1.70	1.67	1.64	1.60	1.57
23	2.94	2.55	2.34	2.21	2.11	2.05	1.99	1.95	1.92	1.89	1.84	1.80	1.74	1.72	1.69	1.66	1.62	1.59	1.55
24	2.93	2.54	2.33	2.19	2.10	2.04	1.98	1.94	1.91	1.88	1.83	1.78	1.73	1.70	1.67	1.64	1.61	1.57	1.53
25	2.92	2.53	2.32	2.18	2.09	2.02	1.97	1.93	1.89	1.87	1.82	1.77	1.72	1.69	1.66	1.63	1.59	1.56	1.52
26	2.91	2.52	2.31	2.17	2.08	2.01	1.96	1.92	1.88	1.86	1.81	1.76	1.71	1.68	1.65	1.61	1.58	1.54	1.50
27	2.90	2.51	2.30	2.17	2.07	2.00	1.95	1.91	1.87	1.85	1.80	1.75	1.70	1.67	1.64	1.60	1.57	1.53	1.49
28	2.89	2.50	2.29	2.16	2.06	2.00	1.94	1.90	1.87	1.84	1.79	1.74	1.69	1.66	1.63	1.59	1.56	1.52	1.48
29	2.89	2.50	2.28	2.15	2.06	1.99	1.93	1.89	1.86	1.83	1.78	1.73	1.68	1.65	1.62	1.58	1.55	1.51	1.47
30	2.88	2.49	2.28	2.14	2.05	1.98	1.93	1.88	1.85	1.82	1.77	1.72	1.67	1.64	1.61	1.57	1.54	1.50	1.46
40	2.84	2.44	2.23	2.09	2.00	1.93	1.87	1.83	1.79	1.76	1.71	1.66	1.61	1.57	1.54	1.51	1.47	1.42	1.38
60	2.79	2.39	2.18	2.04	1.95	1.87	1.82	1.77	1.74	1.71	1.66	1.60	1.54	1.51	1.48	1.44	1.40	1.35	1.29
120	2.75	2.35	2.13	1.99	1.90	1.82	1.77	1.72	1.68	1.65	1.60	1.55	1.48	1.45	1.41	1.37	1.32	1.26	1.19
∞	2.71	2.30	2.08	1.94	1.85	1.77	1.72	1.67	1.63	1.60	1.55	1.49	1.42	1.38	1.34	1.30	1.24	1.17	1.00

$F = \dfrac{S_1^2}{S_2^2} = \dfrac{S1/m}{S2/n}$, where $S_1^2 = S1/m$ and $S_2^2 = S2/n$ are independent mean squares estimating a common variance σ^2 and based on m and n degrees of freedom, respectively. This table also provides values corresponding to $F(F) = .10, .05, .025, .01, .005, .001$ since $F_{1-\alpha}$ for m and n degrees of freedom is the reciprocal of F_α for n and m degrees of freedom. Thus $F_{.95}(7,4) = \dfrac{1}{F_{.05}(4,7)} = \dfrac{1}{6.09} = .164$.

F-Distribution

PERCENTAGE POINTS, F-DISTRIBUTION (Continued)

$$F(F) = \int_0^F \frac{\Gamma\left(\frac{m+n}{2}\right)}{\Gamma\left(\frac{m}{2}\right)\Gamma\left(\frac{n}{2}\right)} m^{\frac{m}{2}} n^{\frac{n}{2}} x^{\frac{m}{2}-1} (n + mx)^{-\frac{m+n}{2}} dx = .95$$

n \ m	1	2	3	4	5	6	7	8	9	10	12	15	20	24	30	40	60	120	∞
1	161.4	199.5	215.7	224.6	230.2	234.0	236.8	238.9	240.5	241.9	243.9	245.9	248.0	249.1	250.1	251.1	252.2	253.3	254.3
2	18.51	19.00	19.16	19.25	19.30	19.33	19.35	19.37	19.38	19.40	19.41	19.43	19.45	19.45	19.46	19.47	19.48	19.49	19.50
3	10.13	9.55	9.28	9.12	9.01	8.94	8.89	8.85	8.81	8.79	8.74	8.70	8.66	8.64	8.62	8.59	8.57	8.55	8.53
4	7.71	6.94	6.59	6.39	6.26	6.16	6.09	6.04	6.00	5.96	5.91	5.86	5.80	5.77	5.75	5.72	5.69	5.66	5.63
5	6.61	5.79	5.41	5.19	5.05	4.95	4.88	4.82	4.77	4.74	4.68	4.62	4.56	4.53	4.50	4.46	4.43	4.40	4.36
6	5.99	5.14	4.76	4.53	4.39	4.28	4.21	4.15	4.10	4.06	4.00	3.94	3.87	3.84	3.81	3.77	3.74	3.70	3.67
7	5.59	4.74	4.35	4.12	3.97	3.87	3.79	3.73	3.68	3.64	3.57	3.51	3.44	3.41	3.38	3.34	3.30	3.27	3.23
8	5.32	4.46	4.07	3.84	3.69	3.58	3.50	3.44	3.39	3.35	3.28	3.22	3.15	3.12	3.08	3.04	3.01	2.97	2.93
9	5.12	4.26	3.86	3.63	3.48	3.37	3.29	3.23	3.18	3.14	3.07	3.01	2.94	2.90	2.86	2.83	2.79	2.75	2.71
10	4.96	4.10	3.71	3.48	3.33	3.22	3.14	3.07	3.02	2.98	2.91	2.85	2.77	2.74	2.70	2.66	2.62	2.58	2.54
11	4.84	3.98	3.59	3.36	3.20	3.09	3.01	2.95	2.90	2.85	2.79	2.72	2.65	2.61	2.57	2.53	2.49	2.45	2.40
12	4.75	3.89	3.49	3.26	3.11	3.00	2.91	2.85	2.80	2.75	2.69	2.62	2.54	2.51	2.47	2.43	2.38	2.34	2.30
13	4.67	3.81	3.41	3.18	3.03	2.92	2.83	2.77	2.71	2.67	2.60	2.53	2.46	2.42	2.38	2.34	2.30	2.25	2.21
14	4.60	3.74	3.34	3.11	2.96	2.85	2.76	2.70	2.65	2.60	2.53	2.46	2.39	2.35	2.31	2.27	2.22	2.18	2.13
15	4.54	3.68	3.29	3.06	2.90	2.79	2.71	2.64	2.59	2.54	2.48	2.40	2.33	2.29	2.25	2.20	2.16	2.11	2.07
16	4.49	3.63	3.24	3.01	2.85	2.74	2.66	2.59	2.54	2.49	2.42	2.35	2.28	2.24	2.19	2.15	2.11	2.06	2.01
17	4.45	3.59	3.20	2.96	2.81	2.70	2.61	2.55	2.49	2.45	2.38	2.31	2.23	2.19	2.15	2.10	2.06	2.01	1.96
18	4.41	3.55	3.16	2.93	2.77	2.66	2.58	2.51	2.46	2.41	2.34	2.27	2.19	2.15	2.11	2.06	2.02	1.97	1.92
19	4.38	3.52	3.13	2.90	2.74	2.63	2.54	2.48	2.42	2.38	2.31	2.23	2.16	2.11	2.07	2.03	1.98	1.93	1.88
20	4.35	3.49	3.10	2.87	2.71	2.60	2.51	2.45	2.39	2.35	2.28	2.20	2.12	2.08	2.04	1.99	1.95	1.90	1.84
21	4.32	3.47	3.07	2.84	2.68	2.57	2.49	2.42	2.37	2.32	2.25	2.18	2.10	2.05	2.01	1.96	1.92	1.87	1.81
22	4.30	3.44	3.05	2.82	2.66	2.55	2.46	2.40	2.34	2.30	2.23	2.15	2.07	2.03	1.98	1.94	1.89	1.84	1.78
23	4.28	3.42	3.03	2.80	2.64	2.53	2.44	2.37	2.32	2.27	2.20	2.13	2.05	2.01	1.96	1.91	1.86	1.81	1.76
24	4.26	3.40	3.01	2.78	2.62	2.51	2.42	2.36	2.30	2.25	2.18	2.11	2.03	1.98	1.94	1.89	1.84	1.79	1.73
25	4.24	3.39	2.99	2.76	2.60	2.49	2.40	2.34	2.28	2.24	2.16	2.09	2.01	1.96	1.92	1.87	1.82	1.77	1.71
26	4.23	3.37	2.98	2.74	2.59	2.47	2.39	2.32	2.27	2.22	2.15	2.07	1.99	1.95	1.90	1.85	1.80	1.75	1.69
27	4.21	3.35	2.96	2.73	2.57	2.46	2.37	2.31	2.25	2.20	2.13	2.06	1.97	1.93	1.88	1.84	1.79	1.73	1.67
28	4.20	3.34	2.95	2.71	2.56	2.45	2.36	2.29	2.24	2.19	2.12	2.04	1.96	1.91	1.87	1.82	1.77	1.71	1.65
29	4.18	3.33	2.93	2.70	2.55	2.43	2.35	2.28	2.22	2.18	2.10	2.03	1.94	1.90	1.85	1.81	1.75	1.70	1.64
30	4.17	3.32	2.92	2.69	2.53	2.42	2.33	2.27	2.21	2.16	2.09	2.01	1.93	1.89	1.84	1.79	1.74	1.68	1.62
40	4.08	3.23	2.84	2.61	2.45	2.34	2.25	2.18	2.12	2.08	2.00	1.92	1.84	1.79	1.74	1.69	1.64	1.58	1.51
60	4.00	3.15	2.76	2.53	2.37	2.25	2.17	2.10	2.04	1.99	1.92	1.84	1.75	1.70	1.65	1.59	1.53	1.47	1.39
120	3.92	3.07	2.68	2.45	2.29	2.17	2.09	2.02	1.96	1.91	1.83	1.75	1.66	1.61	1.55	1.50	1.43	1.35	1.25
∞	3.84	3.00	2.60	2.37	2.21	2.10	2.01	1.94	1.88	1.83	1.75	1.67	1.57	1.52	1.46	1.39	1.32	1.22	1.00

PERCENTAGE POINTS, *F*-DISTRIBUTION (Continued)

$$F(F) = \int_0^F \frac{\Gamma\left(\dfrac{m+n}{2}\right)}{\Gamma\left(\dfrac{m}{2}\right)\Gamma\left(\dfrac{n}{2}\right)} m^{\frac{m}{2}} n^{\frac{n}{2}} x^{\frac{m}{2}-1} (n+mx)^{-\frac{m+n}{2}}\,dx = .975$$

n \ m	1	2	3	4	5	6	7	8	9	10	12	15	20	24	30	40	60	120	∞
1	647.8	799.5	864.2	899.6	921.8	937.1	948.2	956.7	963.3	968.6	976.7	984.9	993.1	997.2	1001	1006	1010	1014	1018
2	38.51	39.00	39.17	39.25	39.30	39.33	39.36	39.37	39.39	39.40	39.41	39.43	39.45	39.46	39.46	39.47	39.48	39.49	39.50
3	17.44	16.04	15.44	15.10	14.88	14.73	14.62	14.54	14.47	14.42	14.34	14.25	14.17	14.12	14.08	14.04	13.99	13.95	13.90
4	12.22	10.65	9.98	9.60	9.36	9.20	9.07	8.98	8.90	8.84	8.75	8.66	8.56	8.51	8.46	8.41	8.36	8.31	8.26
5	10.01	8.43	7.76	7.39	7.15	6.98	6.85	6.76	6.68	6.62	6.52	6.43	6.33	6.28	6.23	6.18	6.12	6.07	6.02
6	8.81	7.26	6.60	6.23	5.99	5.82	5.70	5.60	5.52	5.46	5.37	5.27	5.17	5.12	5.07	5.01	4.96	4.90	4.85
7	8.07	6.54	5.89	5.52	5.29	5.12	4.99	4.90	4.82	4.76	4.67	4.57	4.47	4.42	4.36	4.31	4.25	4.20	4.14
8	7.57	6.06	5.42	5.05	4.82	4.65	4.53	4.43	4.36	4.30	4.20	4.10	4.00	3.95	3.89	3.84	3.78	3.73	3.67
9	7.21	5.71	5.08	4.72	4.48	4.32	4.20	4.10	4.03	3.96	3.87	3.77	3.67	3.61	3.56	3.51	3.45	3.39	3.33
10	6.94	5.46	4.83	4.47	4.24	4.07	3.95	3.85	3.78	3.72	3.62	3.52	3.42	3.37	3.31	3.26	3.20	3.14	3.08
11	6.72	5.26	4.63	4.28	4.04	3.88	3.76	3.66	3.59	3.53	3.43	3.33	3.23	3.17	3.12	3.06	3.00	2.94	2.88
12	6.55	5.10	4.47	4.12	3.89	3.73	3.61	3.51	3.44	3.37	3.28	3.18	3.07	3.02	2.96	2.91	2.85	2.79	2.72
13	6.41	4.97	4.35	4.00	3.77	3.60	3.48	3.39	3.31	3.25	3.15	3.05	2.95	2.89	2.84	2.78	2.72	2.66	2.60
14	6.30	4.86	4.24	3.89	3.66	3.50	3.38	3.29	3.21	3.15	3.05	2.95	2.84	2.79	2.73	2.67	2.61	2.55	2.49
15	6.20	4.77	4.15	3.80	3.58	3.41	3.29	3.20	3.12	3.06	2.96	2.86	2.76	2.70	2.64	2.59	2.52	2.46	2.40
16	6.12	4.69	4.08	3.73	3.50	3.34	3.22	3.12	3.05	2.99	2.89	2.79	2.68	2.63	2.57	2.51	2.45	2.38	2.32
17	6.04	4.62	4.01	3.66	3.44	3.28	3.16	3.06	2.98	2.92	2.82	2.72	2.62	2.56	2.50	2.44	2.38	2.32	2.25
18	5.98	4.56	3.95	3.61	3.38	3.22	3.10	3.01	2.93	2.87	2.77	2.67	2.56	2.50	2.44	2.38	2.32	2.26	2.19
19	5.92	4.51	3.90	3.56	3.33	3.17	3.05	2.96	2.88	2.82	2.72	2.62	2.51	2.45	2.39	2.33	2.27	2.20	2.13
20	5.87	4.46	3.86	3.51	3.29	3.13	3.01	2.91	2.84	2.77	2.68	2.57	2.46	2.41	2.35	2.29	2.22	2.16	2.09
21	5.83	4.42	3.82	3.48	3.25	3.09	2.97	2.87	2.80	2.73	2.64	2.53	2.42	2.37	2.31	2.25	2.18	2.11	2.04
22	5.79	4.38	3.78	3.44	3.22	3.05	2.93	2.84	2.76	2.70	2.60	2.50	2.39	2.33	2.27	2.21	2.14	2.08	2.00
23	5.75	4.35	3.75	3.41	3.18	3.02	2.90	2.81	2.73	2.67	2.57	2.47	2.36	2.30	2.24	2.18	2.11	2.04	1.97
24	5.72	4.32	3.72	3.38	3.15	2.99	2.87	2.78	2.70	2.64	2.54	2.44	2.33	2.27	2.21	2.15	2.08	2.01	1.94
25	5.69	4.29	3.69	3.35	3.13	2.97	2.85	2.75	2.68	2.61	2.51	2.41	2.30	2.24	2.18	2.12	2.05	1.98	1.91
26	5.66	4.27	3.67	3.33	3.10	2.94	2.82	2.73	2.65	2.59	2.49	2.39	2.28	2.22	2.16	2.09	2.03	1.95	1.88
27	5.63	4.24	3.65	3.31	3.08	2.92	2.80	2.71	2.63	2.57	2.47	2.36	2.25	2.19	2.13	2.07	2.00	1.93	1.85
28	5.61	4.22	3.63	3.29	3.06	2.90	2.78	2.69	2.61	2.55	2.45	2.34	2.23	2.17	2.11	2.05	1.98	1.91	1.83
29	5.59	4.20	3.61	3.27	3.04	2.88	2.76	2.67	2.59	2.53	2.43	2.32	2.21	2.15	2.09	2.03	1.96	1.89	1.81
30	5.57	4.18	3.59	3.25	3.03	2.87	2.75	2.65	2.57	2.51	2.41	2.31	2.20	2.14	2.07	2.01	1.94	1.87	1.79
40	5.42	4.05	3.46	3.13	2.90	2.74	2.62	2.53	2.45	2.39	2.29	2.18	2.07	2.01	1.94	1.88	1.80	1.72	1.64
60	5.29	3.93	3.34	3.01	2.79	2.63	2.51	2.41	2.33	2.27	2.17	2.06	1.94	1.88	1.82	1.74	1.67	1.58	1.48
120	5.15	3.80	3.23	2.89	2.67	2.52	2.39	2.30	2.22	2.16	2.05	1.94	1.82	1.76	1.69	1.61	1.53	1.43	1.31
∞	5.02	3.69	3.12	2.79	2.57	2.41	2.29	2.19	2.11	2.05	1.94	1.83	1.71	1.64	1.57	1.48	1.39	1.27	1.00

F-Distribution

PERCENTAGE POINTS, F-DISTRIBUTION (Continued)

$$F(F) = \int_0^F \frac{\Gamma\left(\frac{m+n}{2}\right)}{\Gamma\left(\frac{m}{2}\right)\Gamma\left(\frac{n}{2}\right)} m^{\frac{m}{2}} n^{\frac{n}{2}} x^{\frac{m}{2}-1} (n+mx)^{-\frac{m+n}{2}} \, dx = .99$$

m \ n	1	2	3	4	5	6	7	8	9	10	12	15	20	24	30	40	60	120	∞
1	4052	4999.5	5403	5625	5764	5859	5928	5982	6022	6056	6106	6157	6209	6235	6261	6287	6313	6339	6366
2	98.50	99.00	99.17	99.25	99.30	99.33	99.36	99.37	99.39	99.40	99.42	99.43	99.45	99.46	99.47	99.47	99.48	99.49	99.50
3	34.12	30.82	29.46	28.71	28.24	27.91	27.67	27.49	27.35	27.23	27.05	26.87	26.69	26.60	26.50	26.41	26.32	26.22	26.13
4	21.20	18.00	16.69	15.98	15.52	15.21	14.98	14.80	14.66	14.55	14.37	14.20	14.02	13.93	13.84	13.75	13.65	13.56	13.46
5	16.26	13.27	12.06	11.39	10.97	10.67	10.46	10.29	10.16	10.05	9.89	9.72	9.55	9.47	9.38	9.29	9.20	9.11	9.02
6	13.75	10.92	9.78	9.15	8.75	8.47	8.26	8.10	7.98	7.87	7.72	7.56	7.40	7.31	7.23	7.14	7.06	6.97	6.88
7	12.25	9.55	8.45	7.85	7.46	7.19	6.99	6.84	6.72	6.62	6.47	6.31	6.16	6.07	5.99	5.91	5.82	5.74	5.65
8	11.26	8.65	7.59	7.01	6.63	6.37	6.18	6.03	5.91	5.81	5.67	5.52	5.36	5.28	5.20	5.12	5.03	4.95	4.86
9	10.56	8.02	6.99	6.42	6.06	5.80	5.61	5.47	5.35	5.26	5.11	4.96	4.81	4.73	4.65	4.57	4.48	4.40	4.31
10	10.04	7.56	6.55	5.99	5.64	5.39	5.20	5.06	4.94	4.85	4.71	4.56	4.41	4.33	4.25	4.17	4.08	4.00	3.91
11	9.65	7.21	6.22	5.67	5.32	5.07	4.89	4.74	4.63	4.54	4.40	4.25	4.10	4.02	3.94	3.86	3.78	3.69	3.60
12	9.33	6.93	5.95	5.41	5.06	4.82	4.64	4.50	4.39	4.30	4.16	4.01	3.86	3.78	3.70	3.62	3.54	3.45	3.36
13	9.07	6.70	5.74	5.21	4.86	4.62	4.44	4.30	4.19	4.10	3.96	3.82	3.66	3.59	3.51	3.43	3.34	3.25	3.17
14	8.86	6.51	5.56	5.04	4.69	4.46	4.28	4.14	4.03	3.94	3.80	3.66	3.51	3.43	3.35	3.27	3.18	3.09	3.00
15	8.68	6.36	5.42	4.89	4.56	4.32	4.14	4.00	3.89	3.80	3.67	3.52	3.37	3.29	3.21	3.13	3.05	2.96	2.87
16	8.53	6.23	5.29	4.77	4.44	4.20	4.03	3.89	3.78	3.69	3.55	3.41	3.26	3.18	3.10	3.02	2.93	2.84	2.75
17	8.40	6.11	5.18	4.67	4.34	4.10	3.93	3.79	3.68	3.59	3.46	3.31	3.16	3.08	3.00	2.92	2.83	2.75	2.65
18	8.29	6.01	5.09	4.58	4.25	4.01	3.84	3.71	3.60	3.51	3.37	3.23	3.08	3.00	2.92	2.84	2.75	2.66	2.57
19	8.18	5.93	5.01	4.50	4.17	3.94	3.77	3.63	3.52	3.43	3.30	3.15	3.00	2.92	2.84	2.76	2.67	2.58	2.49
20	8.10	5.85	4.94	4.43	4.10	3.87	3.70	3.56	3.46	3.37	3.23	3.09	2.94	2.86	2.78	2.69	2.61	2.52	2.42
21	8.02	5.78	4.87	4.37	4.04	3.81	3.64	3.51	3.40	3.31	3.17	3.03	2.88	2.80	2.72	2.64	2.55	2.46	2.36
22	7.95	5.72	4.82	4.31	3.99	3.76	3.59	3.45	3.35	3.26	3.12	2.98	2.83	2.75	2.67	2.58	2.50	2.40	2.31
23	7.88	5.66	4.76	4.26	3.94	3.71	3.54	3.41	3.30	3.21	3.07	2.93	2.78	2.70	2.62	2.54	2.45	2.35	2.26
24	7.82	5.61	4.72	4.22	3.90	3.67	3.50	3.36	3.26	3.17	3.03	2.89	2.74	2.66	2.58	2.49	2.40	2.31	2.21
25	7.77	5.57	4.68	4.18	3.85	3.63	3.46	3.32	3.22	3.13	2.99	2.85	2.70	2.62	2.54	2.45	2.36	2.27	2.17
26	7.72	5.53	4.64	4.14	3.82	3.59	3.42	3.29	3.18	3.09	2.96	2.81	2.66	2.58	2.50	2.42	2.33	2.23	2.13
27	7.68	5.49	4.60	4.11	3.78	3.56	3.39	3.26	3.15	3.06	2.93	2.78	2.63	2.55	2.47	2.38	2.29	2.20	2.10
28	7.64	5.45	4.57	4.07	3.75	3.53	3.36	3.23	3.12	3.03	2.90	2.75	2.60	2.52	2.44	2.35	2.26	2.17	2.06
29	7.60	5.42	4.54	4.04	3.73	3.50	3.33	3.20	3.09	3.00	2.87	2.73	2.57	2.49	2.41	2.33	2.23	2.14	2.03
30	7.56	5.39	4.51	4.02	3.70	3.47	3.30	3.17	3.07	2.98	2.84	2.70	2.55	2.47	2.39	2.30	2.21	2.11	2.01
40	7.31	5.18	4.31	3.83	3.51	3.29	3.12	2.99	2.89	2.80	2.66	2.52	2.37	2.29	2.20	2.11	2.02	1.92	1.80
60	7.08	4.98	4.13	3.65	3.34	3.12	2.95	2.82	2.72	2.63	2.50	2.35	2.20	2.12	2.03	1.94	1.84	1.73	1.60
120	6.85	4.79	3.95	3.48	3.17	2.96	2.79	2.66	2.56	2.47	2.34	2.19	2.03	1.95	1.86	1.76	1.66	1.53	1.38
∞	6.63	4.61	3.78	3.32	3.02	2.80	2.64	2.51	2.41	2.32	2.18	2.04	1.88	1.79	1.70	1.59	1.47	1.32	1.00

PERCENTAGE POINTS, F-DISTRIBUTION (Continued)

$$F(F) = \int_0^F \frac{\Gamma\left(\frac{m+n}{2}\right)}{\Gamma\left(\frac{m}{2}\right)\Gamma\left(\frac{n}{2}\right)}\, m^{\frac{m}{2}} n^{\frac{n}{2}} x^{\frac{m}{2}-1}\,(n+mx)^{-\frac{m+n}{2}}\,dx = .995$$

n \ m	1	2	3	4	5	6	7	8	9	10	12	15	20	24	30	40	60	120	∞
1	16211	20000	21615	22500	23056	23437	23715	23925	24091	24224	24426	24630	24836	24940	25044	25148	25253	25359	25465
2	198.5	199.0	199.2	199.2	199.3	199.3	199.4	199.4	199.4	199.4	199.4	199.4	199.4	199.5	199.5	199.5	199.5	199.5	199.5
3	55.55	49.80	47.47	46.19	45.39	44.84	44.43	44.13	43.88	43.69	43.39	43.08	42.78	42.62	42.47	42.31	42.15	41.99	41.83
4	31.33	26.28	24.26	23.15	22.46	21.97	21.62	21.35	21.14	20.97	20.70	20.44	20.17	20.03	19.89	19.75	19.61	19.47	19.32
5	22.78	18.31	16.53	15.56	14.94	14.51	14.20	13.96	13.77	13.62	13.38	13.15	12.90	12.78	12.66	12.53	12.40	12.27	12.14
6	18.63	14.54	12.92	12.03	11.46	11.07	10.79	10.57	10.39	10.25	10.03	9.81	9.59	9.47	9.36	9.24	9.12	9.00	8.88
7	16.24	12.40	10.88	10.05	9.52	9.16	8.89	8.68	8.51	8.38	8.18	7.97	7.75	7.65	7.53	7.42	7.31	7.19	7.08
8	14.69	11.04	9.60	8.81	8.30	7.95	7.69	7.50	7.34	7.21	7.01	6.81	6.61	6.50	6.40	6.29	6.18	6.06	5.95
9	13.61	10.11	8.72	7.96	7.47	7.13	6.88	6.69	6.54	6.42	6.23	6.03	5.83	5.73	5.62	5.52	5.41	5.30	5.19
10	12.83	9.43	8.08	7.34	6.87	6.54	6.30	6.12	5.97	5.85	5.66	5.47	5.27	5.17	5.07	4.97	4.86	4.75	4.64
11	12.23	8.91	7.60	6.88	6.42	6.10	5.86	5.68	5.54	5.42	5.24	5.05	4.86	4.76	4.65	4.55	4.44	4.34	4.23
12	11.75	8.51	7.23	6.52	6.07	5.76	5.52	5.35	5.20	5.09	4.91	4.72	4.53	4.43	4.33	4.23	4.12	4.01	3.90
13	11.37	8.19	6.93	6.23	5.79	5.48	5.25	5.08	4.94	4.82	4.64	4.46	4.27	4.17	4.07	3.97	3.87	3.76	3.65
14	11.06	7.92	6.68	6.00	5.56	5.26	5.03	4.86	4.72	4.60	4.43	4.25	4.06	3.96	3.86	3.76	3.66	3.55	3.44
15	10.80	7.70	6.48	5.80	5.37	5.07	4.85	4.67	4.54	4.42	4.25	4.07	3.88	3.79	3.69	3.58	3.48	3.37	3.26
16	10.58	7.51	6.30	5.64	5.21	4.91	4.69	4.52	4.38	4.27	4.10	3.92	3.73	3.64	3.54	3.44	3.33	3.22	3.11
17	10.38	7.35	6.16	5.50	5.07	4.78	4.56	4.39	4.25	4.14	3.97	3.79	3.61	3.51	3.41	3.31	3.21	3.10	2.98
18	10.22	7.21	6.03	5.37	4.96	4.66	4.44	4.28	4.14	4.03	3.86	3.68	3.50	3.40	3.30	3.20	3.10	2.99	2.87
19	10.07	7.09	5.92	5.27	4.85	4.56	4.34	4.18	4.04	3.93	3.76	3.59	3.40	3.31	3.21	3.11	3.00	2.89	2.78
20	9.94	6.99	5.82	5.17	4.76	4.47	4.26	4.09	3.96	3.85	3.68	3.50	3.32	3.22	3.12	3.02	2.92	2.81	2.69
21	9.83	6.89	5.73	5.09	4.68	4.39	4.18	4.01	3.88	3.77	3.60	3.43	3.24	3.15	3.05	2.95	2.84	2.73	2.61
22	9.73	6.81	5.65	5.02	4.61	4.32	4.11	3.94	3.81	3.70	3.54	3.36	3.18	3.08	2.98	2.88	2.77	2.66	2.55
23	9.63	6.73	5.58	4.95	4.54	4.26	4.05	3.88	3.75	3.64	3.47	3.30	3.12	3.02	2.92	2.82	2.71	2.60	2.48
24	9.55	6.66	5.52	4.89	4.49	4.20	3.99	3.83	3.69	3.59	3.42	3.25	3.06	2.97	2.87	2.77	2.66	2.55	2.43
25	9.48	6.60	5.46	4.84	4.43	4.15	3.94	3.78	3.64	3.54	3.37	3.20	3.01	2.92	2.82	2.72	2.61	2.50	2.38
26	9.41	6.54	5.41	4.79	4.38	4.10	3.89	3.73	3.60	3.49	3.33	3.15	2.97	2.87	2.77	2.67	2.56	2.45	2.33
27	9.34	6.49	5.36	4.74	4.34	4.06	3.85	3.69	3.56	3.45	3.28	3.11	2.93	2.83	2.73	2.63	2.52	2.41	2.29
28	9.28	6.44	5.32	4.70	4.30	4.02	3.81	3.65	3.52	3.41	3.25	3.07	2.89	2.79	2.69	2.59	2.48	2.37	2.25
29	9.23	6.40	5.28	4.66	4.26	3.98	3.77	3.61	3.48	3.38	3.21	3.04	2.86	2.76	2.66	2.56	2.45	2.33	2.21
30	9.18	6.35	5.24	4.62	4.23	3.95	3.74	3.58	3.45	3.34	3.18	3.01	2.82	2.73	2.63	2.52	2.42	2.30	2.18
40	8.83	6.07	4.98	4.37	3.99	3.71	3.51	3.35	3.22	3.12	2.95	2.78	2.60	2.50	2.40	2.30	2.18	2.06	1.93
60	8.49	5.79	4.73	4.14	3.76	3.49	3.29	3.13	3.01	2.90	2.74	2.57	2.39	2.29	2.19	2.08	1.96	1.83	1.69
120	8.18	5.54	4.50	3.92	3.55	3.28	3.09	2.93	2.81	2.71	2.54	2.37	2.19	2.09	1.98	1.87	1.75	1.61	1.43
∞	7.88	5.30	4.28	3.72	3.35	3.09	2.90	2.74	2.62	2.52	2.36	2.19	2.00	1.90	1.79	1.67	1.53	1.36	1.00

PERCENTAGE POINTS, F-DISTRIBUTION (Continued)

$$F(F) = \int_0^F \frac{\Gamma\left(\frac{m+n}{2}\right)}{\Gamma\left(\frac{m}{2}\right)\Gamma\left(\frac{n}{2}\right)} m^{\frac{m}{2}} n^{\frac{n}{2}} x^{\frac{m}{2}-1} (n+mx)^{-\frac{m+n}{2}}\, dx = .999$$

n \ m	1	2	3	4	5	6	7	8	9	10	12	15	20	24	30	40	60	120	∞
1	4053*	5000*	5404*	5625*	5764*	5859*	5929*	5981*	6023*	6056*	6107*	6158*	6209*	6235*	6261*	6287*	6313*	6340*	6366*
2	998.5	999.0	999.2	999.2	999.3	999.3	999.4	999.4	999.4	999.4	999.4	999.4	999.4	999.5	999.5	999.5	999.5	999.5	999.5
3	167.0	148.5	141.1	137.1	134.6	132.8	131.6	130.6	129.9	129.2	128.3	127.4	126.4	125.9	125.4	125.0	124.5	124.0	123.5
4	74.14	61.25	56.18	53.44	51.71	50.53	49.66	49.00	48.47	48.05	47.41	46.76	46.10	45.77	45.43	45.09	44.75	44.40	44.05
5	47.18	37.12	33.20	31.09	29.75	28.84	28.16	27.64	27.24	26.92	26.42	25.91	25.39	25.14	24.87	24.60	24.33	24.06	23.79
6	35.51	27.00	23.70	21.92	20.81	20.03	19.46	19.03	18.69	18.41	17.99	17.56	17.12	16.89	16.67	16.44	16.21	15.99	15.75
7	29.25	21.69	18.77	17.19	16.21	15.52	15.02	14.63	14.33	14.08	13.71	13.32	12.93	12.73	12.53	12.33	12.12	11.91	11.70
8	25.42	18.49	15.83	14.39	13.49	12.86	12.40	12.04	11.77	11.54	11.19	10.84	10.48	10.30	10.11	9.92	9.73	9.53	9.33
9	22.86	16.39	13.90	12.56	11.71	11.13	10.70	10.37	10.11	9.89	9.57	9.24	8.90	8.72	8.55	8.37	8.19	8.00	7.81
10	21.04	14.91	12.55	11.28	10.48	9.92	9.52	9.20	8.96	8.75	8.45	8.13	7.80	7.64	7.47	7.30	7.12	6.94	6.76
11	19.69	13.81	11.56	10.35	9.58	9.05	8.66	8.35	8.12	7.92	7.63	7.32	7.01	6.85	6.68	6.52	6.35	6.17	6.00
12	18.64	12.97	10.80	9.63	8.89	8.38	8.00	7.71	7.48	7.29	7.00	6.71	6.40	6.25	6.09	5.93	5.76	5.59	5.42
13	17.81	12.31	10.21	9.07	8.35	7.86	7.49	7.21	6.98	6.80	6.52	6.23	5.93	5.78	5.63	5.47	5.30	5.14	4.97
14	17.14	11.78	9.73	8.62	7.92	7.43	7.08	6.80	6.58	6.40	6.13	5.85	5.56	5.41	5.25	5.10	4.94	4.77	4.60
15	16.59	11.34	9.34	8.25	7.57	7.09	6.74	6.47	6.26	6.08	5.81	5.54	5.25	5.10	4.95	4.80	4.64	4.47	4.31
16	16.12	10.97	9.00	7.94	7.27	6.81	6.46	6.19	5.98	5.81	5.55	5.27	4.99	4.85	4.70	4.54	4.39	4.23	4.06
17	15.72	10.66	8.73	7.68	7.02	6.56	6.22	5.96	5.75	5.58	5.32	5.05	4.78	4.63	4.48	4.33	4.18	4.02	3.85
18	15.38	10.39	8.49	7.46	6.81	6.35	6.02	5.76	5.56	5.39	5.13	4.87	4.59	4.45	4.30	4.15	4.00	3.84	3.67
19	15.08	10.16	8.28	7.26	6.62	6.18	5.85	5.59	5.39	5.22	4.97	4.70	4.43	4.29	4.14	3.99	3.84	3.68	3.51
20	14.82	9.95	8.10	7.10	6.46	6.02	5.69	5.44	5.24	5.08	4.82	4.56	4.29	4.15	4.00	3.86	3.70	3.54	3.38
21	14.59	9.77	7.94	6.95	6.32	5.88	5.56	5.31	5.11	4.95	4.70	4.44	4.17	4.03	3.88	3.74	3.58	3.42	3.26
22	14.38	9.61	7.80	6.81	6.19	5.76	5.44	5.19	4.99	4.83	4.58	4.33	4.06	3.92	3.78	3.63	3.48	3.32	3.15
23	14.19	9.47	7.67	6.69	6.08	5.65	5.33	5.09	4.89	4.73	4.48	4.23	3.96	3.82	3.68	3.53	3.38	3.22	3.05
24	14.03	9.34	7.55	6.59	5.98	5.55	5.23	4.99	4.80	4.64	4.39	4.14	3.87	3.74	3.59	3.45	3.29	3.14	2.97
25	13.88	9.22	7.45	6.49	5.88	5.46	5.15	4.91	4.71	4.56	4.31	4.06	3.79	3.66	3.52	3.37	3.22	3.06	2.89
26	13.74	9.12	7.36	6.41	5.80	5.38	5.07	4.83	4.64	4.48	4.24	3.99	3.72	3.59	3.44	3.30	3.15	2.99	2.82
27	13.61	9.02	7.27	6.33	5.73	5.31	5.00	4.76	4.57	4.41	4.17	3.92	3.66	3.52	3.38	3.23	3.08	2.92	2.75
28	13.50	8.93	7.19	6.25	5.66	5.24	4.93	4.69	4.50	4.35	4.11	3.86	3.60	3.46	3.32	3.18	3.02	2.86	2.69
29	13.39	8.85	7.12	6.19	5.59	5.18	4.87	4.64	4.45	4.29	4.05	3.80	3.54	3.41	3.27	3.12	2.97	2.81	2.64
30	13.29	8.77	7.05	6.12	5.53	5.12	4.82	4.58	4.39	4.24	4.00	3.75	3.49	3.36	3.22	3.07	2.92	2.76	2.59
40	12.61	8.25	6.60	5.70	5.13	4.73	4.44	4.21	4.02	3.87	3.64	3.40	3.15	3.01	2.87	2.73	2.57	2.41	2.23
60	11.97	7.76	6.17	5.31	4.76	4.37	4.09	3.87	3.69	3.54	3.31	3.08	2.83	2.69	2.55	2.41	2.25	2.08	1.89
120	11.38	7.32	5.79	4.95	4.42	4.04	3.77	3.55	3.38	3.24	3.02	2.78	2.53	2.40	2.26	2.11	1.95	1.76	1.54
∞	10.83	6.91	5.42	4.62	4.10	3.74	3.47	3.27	3.10	2.96	2.74	2.51	2.27	2.13	1.99	1.84	1.66	1.45	1.00

* Multiply these entries by 100.

RANDOM UNITS

Use of Table. If one wishes to select a random sample of N items from a universe of M items, the following procedure may be applied. ($M > N$.)

1. Decide upon some arbitrary scheme of selecting entries from the table. For example, one may decide to use the entries in the first line, second column; second line, third column; third line, fourth column; etc.

2. Assign numbers to each of the items in the universe from 1 to M. Thus, if $M = 500$, the items would be numbered from 001 to 500, and therefore, each designated item is associated with a three digit number.

3. Decide upon some arbitrary scheme of selecting positional digits from each entry chosen according to Step 1. Thus, if $M = 500$, one may decide to use the first, third, and fourth digit of each entry selected, and as a consequence a three digit number is created for each entry choice.

4. If the number formed is $\leq M$, the correspondingly designated item in the universe is chosen for the random sample of N items. If a number formed is $> M$ or is a repeated number of one already chosen, it is passed over and the next desirable number is taken. This process is continued until the random sample of N items is selected.

PROBABILITY AND STATISTICS
Table of Random Units

A TABLE OF 14,000 RANDOM UNITS

Line/Col.	(1)	(2)	(3)	(4)	(5)	(6)	(7)	(8)	(9)	(10)	(11)	(12)	(13)	(14)
1	10480	15011	01536	02011	81647	91646	69179	14194	62590	36207	20969	99570	91291	90700
2	22368	46573	25595	85393	30995	89198	27982	53402	93965	34095	52666	19174	39615	99505
3	24130	48360	22527	97265	76393	64809	15179	24830	49340	32081	30680	19655	63348	58629
4	42167	93093	06243	61680	07856	16376	39440	53537	71341	57004	00849	74917	97758	16379
5	37570	39975	81837	16656	06121	91782	60468	81305	49684	60672	14110	06927	01263	54613
6	77921	06907	11008	42751	27756	53498	18602	70659	90655	15053	21916	81825	44394	42880
7	99562	72905	56420	69994	98872	31016	71194	18738	44013	48840	63213	21069	10634	12952
8	96301	91977	05463	07972	18876	20922	94595	56869	69014	60045	18425	84903	42508	32307
9	89579	14342	63661	10281	17453	18103	57740	84378	25331	12566	58678	44947	05585	56941
10	85475	36857	43342	53988	53060	59533	38867	62300	08158	17983	16439	11458	18593	64952
11	28918	69578	88231	33276	70997	79936	56865	05859	90106	31595	01547	85590	91610	78188
12	63553	40961	48235	03427	49626	69445	18663	72695	52180	20847	12234	90511	33703	90322
13	09429	93969	52636	92737	88974	33488	36320	17617	30015	08272	84115	27156	30613	74952
14	10365	61129	87529	85689	48237	52267	67689	93394	01511	26358	85104	20285	29975	89868
15	07119	97336	71048	08178	77233	13916	47564	81056	97735	85977	29372	74461	28551	90707
16	51085	12765	51821	51259	77452	16308	60756	92144	49442	53900	70960	63990	75601	40719
17	02368	21382	52404	60268	89368	19885	55322	44819	01188	65255	64835	44919	05944	55157
18	01011	54092	33362	94904	31273	04146	18594	29852	71585	85030	51132	01915	92747	64951
19	52162	53916	46369	58586	23216	14513	83149	98736	23495	64350	94738	17752	35156	35749
20	07056	97628	33787	09998	42698	06691	76988	13602	51851	46104	88916	19509	25625	58104
21	48663	91245	85828	14346	09172	30168	90229	04734	59193	22178	30421	61666	99904	32812
22	54164	58492	22421	74103	47070	25306	76468	26384	58151	06646	21524	15227	96909	44592
23	32639	32363	05597	24200	13363	38005	94342	28728	35806	06912	17012	64161	18296	22851
24	29334	27001	87637	87308	58731	00256	45834	15398	46557	41135	10367	07684	36188	18510
25	02488	33062	28834	07351	19731	92420	60952	61280	50001	67658	32586	86679	50720	94953
26	81525	72295	04839	96423	24878	82651	66566	14778	76797	14780	13300	87074	79666	95725
27	29676	20591	68086	26432	46901	20849	89768	81536	86645	12659	92259	57102	80428	25280
28	00742	57392	39064	66432	84673	40027	32832	61362	98947	96067	64760	64584	96096	98253
29	05366	04213	25669	26422	44407	44048	37937	63904	45766	66134	75470	66520	34693	90449
30	91921	26418	64117	94305	26766	25940	39972	22209	71500	64568	91402	42416	07844	69618
31	00582	04711	87917	77341	42206	35126	74087	99547	81817	42607	43808	76655	62028	76630
32	00725	69884	62797	56170	86324	88072	76222	36086	84637	93161	76038	65855	77919	88006
33	69011	65797	95876	55293	18988	27354	26575	08625	40801	59920	29841	80150	12777	48501
34	25976	57948	29888	88604	67917	48708	18912	82271	65424	69774	33611	54262	85963	03547
35	09763	83473	73577	12908	30883	18317	28290	35797	05998	41688	34952	37888	38917	88050
36	91567	42595	27958	30134	04024	86385	29880	99730	55536	84855	29080	09250	79656	73211
37	17955	56349	90999	49127	20044	59931	06115	20542	18059	02008	73708	83517	36103	42791
38	46503	18584	18845	49618	02304	51038	20655	58727	28168	15475	56942	53389	20562	87338
39	92157	89634	94824	78171	84610	82834	09922	25417	44137	48413	25555	21246	35509	20468
40	14577	62765	35605	81263	39667	47358	56873	56307	61607	49518	89656	20103	77490	18062
41	98427	07523	33362	64270	01638	92477	66969	98420	04880	45585	46565	04102	46880	45709
42	34914	63976	88720	82765	34476	17032	87589	40836	32427	70002	70663	88863	77775	69348
43	70060	28277	39475	46473	23219	53416	94970	25832	69975	94884	19661	72828	00102	66794
44	53976	54914	06990	67245	68350	82948	11398	42878	80287	88267	47363	46634	06541	97809
45	76072	29515	40980	07391	58745	25774	22987	80059	39911	96189	41151	14222	60697	59583
46	90725	52210	83974	29992	65831	38857	50490	83765	55657	14361	31720	57375	56228	41546
47	64364	67412	33339	31926	14883	24413	59744	92351	97473	89286	35931	04110	23726	51900
48	08962	00358	31662	25388	61642	34072	81249	35648	56891	69352	48373	45578	78547	81788
49	95012	68379	93526	70765	10593	04542	76463	54328	02349	17247	28865	14777	62730	92277
50	15664	10493	20492	38391	91132	21999	59516	81652	27195	48223	46751	22923	32261	85653

Table of Random Units

A TABLE OF 14,000 RANDOM UNITS (Continued)

Line/Col.	(1)	(2)	(3)	(4)	(5)	(6)	(7)	(8)	(9)	(10)	(11)	(12)	(13)	(14)
51	16408	81899	04153	53381	79401	21438	83035	92350	36693	31238	59649	91754	72772	02338
52	18629	81953	05520	91962	04739	13092	97662	24822	94730	06496	35090	04822	86772	98289
53	73115	35101	47498	87637	99016	71060	88824	71013	18735	20286	23153	72924	35165	43040
54	57491	16703	23167	49323	45021	33132	12544	41035	80780	45393	44812	12515	98931	91202
55	30405	83946	23792	14422	15059	45799	22716	19792	09983	74353	68668	30429	70735	25499
56	16631	35006	85900	98275	32388	52390	16815	69298	82732	38480	73817	32523	41961	44437
57	96773	20206	42559	78985	05300	22164	24369	54224	35083	19687	11052	91491	60383	19746
58	38935	64202	14349	82674	66523	44133	00697	35552	35970	19124	63318	29686	03387	59846
59	31624	76384	17403	53363	44167	64486	64758	75366	76554	31601	12614	33072	60332	92325
60	78919	19474	23632	27889	47914	02584	37680	20801	72152	39339	34806	08930	85001	87820
61	03931	33309	57047	74211	63445	17361	62825	39908	05607	91284	68833	25570	38818	46920
62	74426	33278	43972	10119	89917	15665	52872	73823	73144	88662	88970	74492	51805	99378
63	09066	00903	20795	95452	92648	45454	09552	88815	16553	51125	79375	97596	16296	66092
64	42238	12426	87025	14267	20979	04508	64535	31355	86064	29472	47689	05974	52468	16834
65	16153	08002	26504	41744	81959	65642	74240	56302	00033	67107	77510	70625	28725	34191
66	21457	40742	29820	96783	29400	21840	15035	34537	33310	06116	95240	15957	16572	06004
67	21581	57802	02050	89728	17937	37621	47075	42080	97403	48626	68995	43805	33386	21597
68	55612	78095	83197	33732	05810	24813	86902	60397	16489	03264	88525	42786	05269	92532
69	44657	66999	99324	51281	84463	60563	79312	93454	68876	25471	93911	25650	12682	73572
70	91340	84979	46949	81973	37949	61023	43997	15263	80644	43942	89203	71795	99533	50501
71	91227	21199	31935	27022	84067	05462	35216	14486	29891	68607	41867	14951	91696	85065
72	50001	38140	66321	19924	72163	09538	12151	06878	91903	18749	34405	56087	82790	70925
73	65390	05224	72958	28609	81406	39147	25549	48542	42627	45233	57202	94617	23772	07896
74	27504	96131	83944	41575	10573	08619	64482	73923	36152	05184	94142	25299	84387	34925
75	37169	94851	39117	89632	00959	16487	65536	49071	39782	17095	02330	74301	00275	48280
76	11508	70225	51111	38351	19444	66499	71945	05422	13442	78675	84081	66938	93654	59894
77	37449	30362	06694	54690	04052	53115	62757	95348	78662	11163	81651	50245	34971	52924
78	46515	70331	85922	38329	57015	15765	97161	17869	45349	61796	66345	81073	49106	79860
79	30986	81223	42416	58353	21532	30502	32305	86482	05174	07901	54339	58861	74818	46942
80	63798	64995	46583	09765	44160	78128	83991	42865	92520	83531	80377	35909	81250	54238
81	82486	84846	99254	67632	43218	50076	21361	64816	51202	88124	41870	52689	51275	83556
82	21885	32906	92431	09060	64297	51674	64126	62570	26123	05155	59194	52799	28225	85762
83	60336	98782	07408	53458	13564	59089	26445	29789	85205	41001	12535	12133	14645	23541
84	43937	46891	24010	25560	86355	33941	25786	54990	71899	15475	95434	98227	21824	19585
85	97656	63175	89303	16275	07100	92063	21942	18611	47348	20203	18534	03862	78095	50136
86	03299	01221	05418	38982	55758	92237	26759	86367	21216	98442	08303	56613	91511	75928
87	79626	06486	03574	17668	07785	76020	79924	25651	83325	88428	85076	72811	22717	50585
88	85636	68335	47539	03129	65651	11977	02510	26113	99447	68645	34327	15152	55230	93448
89	18039	14367	61337	06177	12143	46609	32989	74014	64708	00533	35398	58408	13261	47908
90	08362	15656	60627	36478	65648	16764	53412	09013	07832	41574	17639	82163	60859	75567
91	79556	29068	04142	16268	15387	12856	66227	38358	22478	73373	88732	09443	82558	05250
92	92608	82674	27072	32534	17075	27698	98204	63863	11951	34648	88022	56148	34925	57031
93	23982	25835	40055	67006	12293	02753	14827	22235	35071	99704	37543	11601	35503	85171
94	09915	96306	05908	97901	28395	14186	00821	80703	70426	75647	76310	88717	37890	40129
95	50937	33300	26695	62247	69927	76123	50842	43834	86654	70959	79725	93872	28117	19233
96	42488	78077	69882	61657	34136	79180	97526	43092	04098	73571	80799	76536	71255	64239
97	46764	86273	63003	93017	31204	36692	40202	35275	57306	55543	53203	18098	47625	88684
98	03237	45430	55417	63282	90816	17349	88298	90183	36600	78406	06216	95787	42579	90730
99	86591	81482	52667	61583	14972	90053	89534	76036	49199	43716	97548	04379	46370	28672
100	38534	01715	94964	87288	65680	43772	39560	12918	86537	62738	19636	51132	25739	56947

PROBABILITY AND STATISTICS

Table of Random Units

A TABLE OF 14,000 RANDOM UNITS (Continued)

Line/Col.	(1)	(2)	(3)	(4)	(5)	(6)	(7)	(8)	(9)	(10)	(11)	(12)	(13)	(14)
101	13284	16834	74151	92027	24670	36665	00770	22878	02179	51602	07270	76517	97275	45960
102	21224	00370	30420	03883	96648	89428	41583	17564	27395	63904	41548	49197	82277	24120
103	99052	47887	81085	64933	66279	80432	65793	83287	34142	13241	30590	97760	35848	91983
104	00199	50993	98603	38452	87890	94624	69721	57484	67501	77638	44331	11257	71131	11059
105	60578	06483	28733	37867	07936	98710	98539	27186	31237	80612	44488	97819	70401	95419
106	91240	18312	17441	01929	18163	69201	31211	54288	39296	37318	65724	90401	79017	62077
107	97458	14229	12063	59611	32249	90466	33216	19358	02591	54263	88449	01912	07436	50813
108	35249	38646	34475	72417	60514	69257	12489	51924	86871	92446	36607	11458	30440	52639
109	38980	46600	11759	11900	46743	27860	77940	39298	97838	95145	32378	68038	89351	37005
110	10750	52745	38749	87365	58959	53731	89295	59062	39404	13198	59960	70408	29812	83126
111	36247	27850	73958	20673	37800	63835	71051	84724	52492	22342	78071	17456	96104	18327
112	70994	66986	99744	72438	01174	42159	11392	20724	54322	36923	70009	23233	65438	59685
113	99638	94702	11463	18148	81386	80431	90628	52506	02016	85151	88598	47821	00265	82525
114	72055	15774	43857	99805	10419	76939	25993	03544	21560	83471	43989	90770	22965	44247
115	24038	65541	85788	55835	38835	59399	13790	35112	01324	39520	76210	22467	83275	32286
116	74976	14631	35908	28221	39470	91548	12854	30166	09073	75887	36782	00268	97121	57676
117	35553	71628	70189	26436	63407	91178	90348	55359	80392	41012	36270	77786	89578	21059
118	35676	12797	51434	82976	42010	26344	92920	92155	58807	54644	58581	95331	78629	73344
119	74815	67523	72985	23183	02446	63594	98924	20633	58842	85961	07648	70164	34994	67662
120	45246	88048	65173	50989	91060	89894	36063	32819	68559	99221	49475	50558	34698	71800
121	76509	47069	86378	41797	11910	49672	88575	97966	32466	10083	54728	81972	58975	30761
122	19689	90332	04315	21358	97248	11188	39062	63312	52496	07349	79178	33692	57352	72862
123	42751	35318	97513	61537	54955	08159	00337	80778	27507	95478	21252	12746	37554	97775
124	11946	22681	45045	13964	57517	59419	58045	44067	58716	58840	45557	96345	33271	53464
125	96518	48688	20996	11090	48396	57177	83867	86464	14342	21545	46717	72364	86954	55580
126	35726	58643	76869	84622	39098	36083	72505	92265	23107	60278	05822	46760	44294	07672
127	39737	42750	48968	70536	84864	64952	38404	94317	65402	13589	01055	79044	19308	83623
128	97025	66492	56177	04049	80312	48028	26408	43591	75528	65341	49044	95495	81256	53214
129	62814	08075	09788	56350	76787	51591	54509	49295	85830	59860	30883	89660	96142	18354
130	25578	22950	15227	83291	41737	79599	96191	71845	86899	70694	24290	01551	80092	82118
131	68763	69576	88991	49662	46704	63362	56625	00481	73323	91427	15264	06969	57048	54149
132	17900	00813	64361	60725	88974	61005	99709	30666	26451	11528	44323	34778	60342	60388
133	71944	60227	63551	71109	05624	43836	58254	26160	32116	63403	35404	57146	10909	07346
134	54684	93691	85132	64399	29182	44324	14491	55226	78793	34107	30374	48429	51376	09559
135	25946	27623	11258	65204	52832	50880	22273	05554	99521	73791	85744	29276	70326	60251
136	01353	39318	44961	44972	91766	90262	56073	06606	51826	18893	83448	31915	97764	75091
137	99083	88191	27662	99113	57174	35571	99884	13951	71057	53961	61448	74909	07322	80960
138	52021	45406	37945	75234	24327	86978	22644	87779	23753	99926	63898	54886	18051	96314
139	78755	47744	43776	83098	03225	14281	83637	55984	13300	52212	58781	14905	46502	04472
140	25282	69106	59180	16257	22810	43609	12224	25643	89884	31149	85423	32581	34374	70873
141	11959	94202	02743	86847	79725	51811	12998	76844	05320	54236	53891	70226	38632	84776
142	11644	13792	98190	01424	30078	28197	55583	05197	47714	68440	22016	79204	06862	94451
143	06307	97912	68110	59812	95448	43244	31262	88880	13040	16458	43813	89416	42482	33939
144	76285	75714	89585	99296	52640	46518	55486	90754	88932	19937	57119	23251	55619	23679
145	55322	07589	39600	60866	63007	20007	66819	84164	61131	81429	60676	42807	78286	29015
146	78017	90928	90220	92503	83375	26986	74399	30885	88567	29169	72816	53357	15428	86932
147	44768	43342	20696	26331	43140	69744	82928	24988	94237	46138	77426	39039	55596	12655
148	25100	19336	14605	86603	51680	97678	24261	02464	86563	74812	60069	71674	15478	47642
149	83612	46623	62876	85197	07824	91392	58317	37726	84628	42221	10268	20692	15699	29167
150	41347	81666	82961	60413	71020	83658	02415	33322	66036	98712	46795	16308	28413	05417

Table of Random Units

A TABLE OF 14,000 RANDOM UNITS (Continued)

Line/Col.	(1)	(2)	(3)	(4)	(5)	(6)	(7)	(8)	(9)	(10)	(11)	(12)	(13)	(14)
151	38128	51178	75096	13609	16110	73533	42564	59870	29399	67834	91055	89917	51096	89011
152	60950	00455	73254	96067	50717	13878	03216	78274	65863	37011	91283	33914	91303	49326
153	90524	17320	29832	96118	75792	25326	22940	24904	80523	38928	91374	55597	97567	38914
154	49897	18278	67160	39408	97056	43517	84426	59650	20247	19293	02019	14790	02852	05819
155	18494	99209	81060	19488	65596	59787	47939	91225	98768	43688	00438	05548	09443	82897
156	65373	72984	30171	37741	70203	94094	87261	30056	58124	70133	18936	02138	59372	09075
157	40653	12843	04213	70925	95360	55774	76439	61768	52817	81151	52188	31940	54273	49032
158	51638	22238	56344	44587	83231	50317	74541	07719	25472	41602	77318	15145	57515	07633
159	69742	99303	62578	83575	30337	07488	51941	84316	42067	49692	28616	29101	03013	73449
160	58012	74072	67488	74580	47992	69482	58624	17106	47538	13452	22620	24260	40155	74716
161	18348	19855	42887	08279	43206	47077	42637	45606	00011	20662	14642	49984	94509	56380
162	59614	09193	58064	29086	44385	45740	70752	05663	49081	26960	57454	99264	24142	74648
163	75688	28630	39210	52897	62748	72658	98059	67202	72789	01869	13496	14663	87645	89713
164	13941	77802	69101	70061	35460	34576	15412	81304	58757	35498	94830	75521	00603	97701
165	96656	86420	96475	86458	54463	96419	55417	41375	76886	19008	66877	35934	59801	00497
166	03363	82042	15942	14549	38324	87094	19069	67590	11087	68570	22591	65232	85915	91499
167	70366	08390	69155	25496	13240	57407	91407	49160	07379	34444	94567	66035	38918	65708
168	47870	36605	12927	16043	53257	93796	52721	73120	48025	76074	95605	67422	41646	14557
169	79504	77606	22761	30518	28373	73898	30550	76684	77366	32276	04690	61667	64798	66276
170	46967	74841	50923	15339	37755	98995	40162	89561	69199	42257	11647	47603	48779	97907
171	14558	50769	35444	59030	87516	48193	02945	00922	48189	04724	21263	20892	92955	90251
172	12440	25057	01132	38611	28135	68089	10954	10097	54243	06460	50856	65435	79377	53890
173	32293	29938	68653	10497	98919	46587	77701	99119	93165	67788	17638	23097	21468	36992
174	10640	21875	72462	77981	56550	55999	87310	69643	45124	00349	25748	00844	96831	30651
175	47615	23169	39571	56972	20628	21788	51736	33133	72696	32605	41569	76148	91544	21121
176	16948	11128	71624	72754	49084	96303	27830	45817	67867	18062	87453	17226	72904	71474
177	21258	61092	66634	70335	92448	17354	83432	49608	66520	06442	59664	20420	39201	69549
178	15072	48853	15178	30730	47481	48490	41436	25015	49932	20474	53821	51015	79841	32405
179	99154	57412	09858	65671	70655	71479	63520	31357	56968	06729	34465	70685	04184	25250
180	08759	61089	23706	32994	35426	36666	63988	98844	37533	08269	27021	45886	22835	78451
181	67323	57839	61114	62192	47547	58023	64630	34886	98777	75442	95592	06141	45096	73117
182	09255	13986	84834	20764	72206	89393	34548	93438	88730	61805	78955	18952	46436	58740
183	36304	74712	00374	10107	85061	69228	81969	92216	03568	39630	81869	52824	50937	27954
184	15884	67429	86612	47367	10242	44880	12060	44309	46629	55105	66793	93173	00480	13311
185	18745	32031	35303	08134	33925	03044	59929	95418	04917	57596	24878	61733	92834	64454
186	72934	40086	88292	65728	38300	42323	64068	98373	48971	09049	59943	36538	05976	82118
187	17626	02944	20910	57662	80181	38579	24580	90529	52303	50436	29401	57824	86039	81062
188	27117	61399	50967	41399	81636	16663	15634	79717	94696	59240	25543	97989	63306	90946
189	93995	18678	90012	63645	85701	85269	62263	68331	00389	72571	15210	20769	44686	96176
190	67392	89421	09623	80725	62620	84162	87368	29560	00519	84545	08004	24526	41252	14521
191	04910	12261	37566	80016	21245	69377	50420	85658	55263	68667	78770	04533	14513	18099
192	81453	20283	79929	59839	23875	13245	46808	74124	74703	35769	95588	21014	37078	39170
193	19480	75790	48539	23703	15537	48885	02861	86587	74539	65227	90799	58789	96257	02708
194	21456	13162	74608	81011	55512	07481	93551	72189	76261	91206	89941	15132	37738	59284
195	89406	20912	46189	76376	25538	87212	20748	12831	57166	35026	16817	79121	18929	40628
196	09866	07414	55977	16419	01101	69343	13305	94302	80703	57910	36933	57771	42546	03003
197	86541	24681	23421	13521	28000	94917	07423	57523	97234	63951	42876	46829	09781	58160
198	10414	96941	06205	72222	57167	83902	07460	69507	10600	08858	07685	44472	64220	27040
199	49942	06683	41479	58982	56288	42853	92196	20632	62045	78812	35895	51851	83534	10689
200	23995	68882	42291	23374	24299	27024	67460	94783	40937	16961	26053	78749	46704	21983

XIII. FINANCIAL TABLES

INTEREST TABLES FROM ¼% TO 20%

SIMPLE INTEREST

If P is the principal placed at interest at a rate i (expressed as a decimal), for a period of n years

The amount,

$$A = P (1 + ni)$$

Present value,

$$P = \frac{A}{1 + ni}$$

COMPOUND INTEREST

At interest compounded annually the **amount,—**

$$A = P (1 + i)^n$$

At interest compounded q times per year,—

$$A = P \left(1 + \frac{i}{q}\right)^{nq}$$

At interest compounded annually the **present value,—**

$$P = \frac{A}{(1 + i)^n} = A (1 + i)^{-n} = Av^n. \quad v = \frac{1}{1 + i}$$

At interest compounded q times per year,—

$$P = A \left(1 + \frac{i}{q}\right)^{-nq}$$

* The **amount of an annuity of 1 per annum,—**

$$s_{\overline{n}|i} = \frac{(1 + i)^n - 1}{i}$$

* The **present value of an annuity,—**

$$a_{\overline{n}|i} = \frac{1 - (1 + i)^{-n}}{i}$$

The **annuity whose present value is 1,—**

$$\frac{1}{a_{\overline{n}|i}} = \frac{1}{s_{\overline{n}|i}} + i = \frac{i}{(1 - v^n)}$$

Compound amount of 1 for fractional periods,—$(1+i)^{1/p}$

* This information can be constructed by use of tables $(1 + i)^n$ and $(1 + i)^{-n}$ as may be observed from the formula structures. For example $s_{\overline{50}|.07} = \dfrac{(1.07)^{50} - 1}{.07}$. Here $(1.07)^{50}$ is obtained from the entries under $(1 + i)^n$, where $i = .07$ and $n = 50$ and is equal to 29.45702506. Accordingly $s_{\overline{50}|.07} = \dfrac{28.45702506}{.07} = 406.528929$ Similarly $a_{\overline{n}|i}$ may be evaluated by use of entries for $(1 + i)^{-n}$ and carrying out the appropriate arithmetic.

AMOUNT AT COMPOUND INTEREST $(1 + i)^{n}$*

The following table gives the amount after a term of n periods on unit original principal at rate of interest i.

Periods	Rate i				
n	0.0025(1/4%)	0.004167(5/12%)	0.005(1/2%)	0.005833(7/12%)	0.0075(3/4%)
1	1.00250000	1.00416667	1.00500000	1.00583333	1.00750000
2	1.00500625	1.00835069	1.01002500	1.01170069	1.01505625
3	1.00751877	1.01255216	1.01507513	1.01760228	1.02266917
4	1.01003756	1.01677112	1.02015050	1.02353830	1.03033919
5	1.01256266	1.02100767	1.02525125	1.02950894	1.03806673
6	1.01509406	1.02526187	1.03037751	1.03551440	1.04585224
7	1.01763180	1.02953379	1.03552940	1.04155490	1.05369613
8	1.02017588	1.03382352	1.04070704	1.04763064	1.06159885
9	1.02272632	1.03813111	1.04591058	1.05374182	1.06956084
10	1.02528313	1.04245666	1.05114013	1.05988865	1.07758255
11	1.02784634	1.04680023	1.05639583	1.06607133	1.08566441
12	1.03041596	1.05116190	1.06167781	1.07229008	1.09380690
13	1.03299200	1.05554174	1.06698620	1.07854511	1.10201045
14	1.03557448	1.05993983	1.07232113	1.08483662	1.11027553
15	1.03816341	1.06435625	1.07768274	1.09116483	1.11860259
16	1.04075882	1.06879106	1.08307115	1.09752996	1.12699211
17	1.04336072	1.07324436	1.08848651	1.10393222	1.13544455
18	1.04596912	1.07771621	1.09392894	1.11037182	1.14396039
19	1.04858404	1.08220670	1.09939858	1.11684899	1.15254009
20	1.05120550	1.08671589	1.10489558	1.12336395	1.16118414
21	1.05383352	1.09124387	1.11042006	1.12991690	1.16989302
22	1.05646810	1.09579072	1.11597216	1.13650808	1.17866722
23	1.05910927	1.10035652	1.12155202	1.14313771	1.18750723
24	1.06175704	1.10494134	1.12715978	1.14980602	1.19641353
25	1.06441144	1.10954526	1.13279558	1.15651322	1.20538663
26	1.06707247	1.11416836	1.13845955	1.16325955	1.21442703
27	1.06974015	1.11881073	1.14415185	1.17004523	1.22353523
28	1.07241450	1.12347244	1.14987261	1.17687049	1.23271175
29	1.07509553	1.12815358	1.15562197	1.18373557	1.24195709
30	1.07778327	1.13285422	1.16140008	1.19064069	1.25127176
31	1.08047773	1.13757444	1.16720708	1.19758610	1.26065630
32	1.08317892	1.14231434	1.17304312	1.20457202	1.27011122
33	1.08588687	1.14707398	1.17890833	1.21159869	1.27963706
34	1.08860159	1.15185346	1.18480288	1.21866634	1.28923434
35	1.09132309	1.15665284	1.19072689	1.22577523	1.29890359
36	1.09405140	1.16147223	1.19668052	1.23292559	1.30864537
37	1.09678653	1.16631170	1.20266393	1.24011765	1.31846021
38	1.09952850	1.17117133	1.20867725	1.24735167	1.32834866
39	1.10227732	1.17605121	1.21472063	1.25462789	1.33831128
40	1.10503301	1.18095142	1.22079424	1.26194655	1.34834861
41	1.10779559	1.18587206	1.22689821	1.26930791	1.35846123
42	1.11056508	1.19081319	1.23303270	1.27671220	1.36864969
43	1.11334149	1.19577491	1.23919786	1.28415969	1.37891456
44	1.11612485	1.20075731	1.24539385	1.29165062	1.38925642
45	1.11891516	1.20576046	1.25162082	1.29918525	1.39967584
46	1.12171245	1.21078446	1.25787892	1.30676383	1.41017341
47	1.12451673	1.21582940	1.26416832	1.31438662	1.42074971
48	1.12732802	1.22089536	1.27048916	1.32205388	1.43140533
49	1.13014634	1.22598242	1.27684161	1.32976586	1.44214087
50	1.13297171	1.23109068	1.28322581	1.33752283	1.45295693

*The $\textit{s}_{\overline{n}|i}$ table may be constructed from this table by use of the formula $\dfrac{(1 + i)^{n} - 1}{i}$.
See page 560.

FINANCIAL TABLES

AMOUNT AT COMPOUND INTEREST $(1 + i)^n$ (Continued)

Periods			Rate i		
n	0.0025(1/4%)	0.004167(5/12%)	0.005(1/2%)	0.005833(7/12%)	0.0075(3/4%)
51	1.13580414	1.23622022	1.28964194	1.34532504	1.46385411
52	1.23864365	1.24137114	1.29609014	1.35317277	1.47483301
53	1.14149026	1.24654352	1.30257060	1.36106628	1.48589426
54	1.14434398	1.25173745	1.30908346	1.36900583	1.49703847
55	1.14720484	1.25695302	1.31562887	1.37699170	1.50826626
56	1.15007285	1.26219033	1.32220702	1.38502415	1.51957825
57	1.15294804	1.26744946	1.32881805	1.39310346	1.53097509
58	1.15583041	1.27273050	1.33546214	1.40122990	1.54245740
59	1.15871998	1.27803354	1.34213946	1.40940374	1.55402583
60	1.16161678	1.28335868	1.34885015	1.41762526	1.56568103
61	1.16452082	1.28870601	1.35559440	1.42589474	1.57742363
62	1.16743213	1.29407561	1.36237238	1.43421246	1.58925431
63	1.17035071	1.29946760	1.36918424	1.44257870	1.60117372
64	1.17327658	1.30488204	1.37603016	1.45099374	1.61318252
65	1.17620977	1.31031905	1.38291031	1.45945787	1.62528139
66	1.17915030	1.31577872	1.38982486	1.46797138	1.63747100
67	1.18209817	1.32126113	1.39677399	1.47653454	1.64975203
68	1.18505342	1.32676638	1.40375785	1.48514766	1.66212517
69	1.18801605	1.33229458	1.41077664	1.49381102	1.67459111
70	1.19098609	1.33784580	1.41783053	1.50252492	1.68715055
71	1.19396356	1.34342016	1.42491968	1.51128965	1.69980418
72	1.19694847	1.34901774	1.43204428	1.52010550	1.71255271
73	1.19994084	1.35463865	1.43920450	1.52897279	1.72539685
74	1.20294069	1.36028298	1.44640052	1.53789179	1.73833733
75	1.20594804	1.36595082	1.45363252	1.54686283	1.75137486
76	1.20896291	1.37164229	1.46090069	1.55588620	1.76451017
77	1.21198532	1.37735746	1.46820519	1.56496220	1.77774400
78	1.21501528	1.38309645	1.47554622	1.57409115	1.79107708
79	1.21805282	1.38885935	1.48292395	1.58327334	1.80451915
80	1.22109795	1.39464627	1.49033857	1.59250910	1.81804398
81	1.22415070	1.40045729	1.49779026	1.60179874	1.83167931
82	1.22721108	1.40629253	1.50527921	1.61114257	1.84541691
83	1.23027910	1.41215209	1.51280561	1.62054090	1.85925753
84	1.23335480	1.41803605	1.52036964	1.62999405	1.87320196
85	1.23643819	1.42394454	1.52797148	1.63950235	1.88725098
86	1.23952928	1.42987764	1.53561134	1.64906612	1.90140536
87	1.24262811	1.43583546	1.54328940	1.65868567	1.91566590
88	1.24573468	1.44181811	1.55100585	1.66836134	1.93003339
89	1.24884901	1.44782568	1.55876087	1.67809344	1.94450865
90	1.25197114	1.45385829	1.56655468	1.68788232	1.95909246
91	1.25510106	1.45991603	1.57438745	1.69772830	1.97378565
92	1.25823882	1.46599902	1.58225939	1.70763172	1.98858905
93	1.26138441	1.47210735	1.59017069	1.71759290	2.00350346
94	1.26453787	1.47824113	1.59812154	1.72761219	2.01852974
95	1.26769922	1.48440047	1.60611215	1.73768993	2.03366871
96	1.27086847	1.49058547	1.61414271	1.74782646	2.04892123
97	1.27404564	1.49679624	1.62221342	1.75802211	2.06428814
98	1.27723075	1.50303289	1.63032449	1.76827724	2.07977030
99	1.28042383	1.50929553	1.63847611	1.77859219	2.09536858
100	1.28362489	1.51558426	1.64666849	1.78896731	2.11108384

AMOUNT AT COMPOUND INTEREST $(1 + i)^n$ (Continued)

Periods n	0.01(1%)	0.001125(1 1/8%)	0.0125(1 1/4%)	0.015(1 1/2%)	0.0175(1 3/4%)
1	1.01000000	1.01125000	1.01250000	1.01500000	1.01750000
2	1.02010000	1.02262656	1.02515625	1.03022500	1.03530625
3	1.03030100	1.03413111	1.03797070	1.04567838	1.05342411
4	1.04060401	1.04576509	1.05094534	1.06136355	1.07185903
5	1.05101005	1.05752994	1.06408215	1.07728400	1.09061656
6	1.06152015	1.06942716	1.07738318	1.09344326	1.10970235
7	1.07213535	1.08145821	1.09085047	1.10984491	1.12912215
8	1.08285671	1.09362462	1.10448610	1.12649259	1.14888178
9	1.09368527	1.10592789	1.11829218	1.14338998	1.16898721
10	1.10462213	1.11836958	1.13227083	1.16054083	1.18944449
11	1.11566835	1.13095124	1.14642422	1.17794894	1.21025977
12	1.12682503	1.14367444	1.16075452	1.19561817	1.23143931
13	1.13809328	1.15654078	1.17526395	1.21355244	1.25298950
14	1.14947421	1.16955186	1.18995475	1.23175573	1.27491682
15	1.16096896	1.18270932	1.20482918	1.25023207	1.29722786
16	1.17257864	1.19601480	1.21988955	1.26898555	1.31992935
17	1.18430443	1.20946997	1.23513817	1.28802033	1.34302811
18	1.19614748	1.22307650	1.25057739	1.30734064	1.36653111
19	1.20810895	1.23683611	1.26620961	1.32695075	1.39044540
20	1.22019004	1.25075052	1.28203723	1.34685501	1.41477820
21	1.23239194	1.26482146	1.29806270	1.36705783	1.43953681
22	1.24471586	1.27905071	1.31428848	1.38756370	1.46472871
23	1.25716302	1.29344003	1.33071709	1.40837715	1.49036146
24	1.26973465	1.30799123	1.34735105	1.42950281	1.51644279
25	1.28243200	1.32270613	1.36419294	1.45094535	1.54298054
26	1.29525631	1.33758657	1.38124535	1.47270953	1.56998269
27	1.30820888	1.35263442	1.39851092	1.49480018	1.59745739
28	1.32129097	1.46785156	1.41599230	1.51722218	1.62541290
29	1.33450388	1.38323989	1.43369221	1.53998051	1.65385762
30	1.34784892	1.39880134	1.45161336	1.56308022	1.68280013
31	1.36132740	1.41453785	1.46975853	1.58652642	1.71224913
32	1.37494068	1.43045140	1.48813051	1.61032432	1.74221349
33	1.38869009	1.44654398	1.50673214	1.63447918	1.77270223
34	1.40257699	1.46281760	1.52556629	1.65899637	1.80372452
35	1.41660276	1.47927430	1.54463587	1.68388132	1.83528970
36	1.43076878	1.49591613	1.56394382	1.70913954	1.86740727
37	1.44507647	1.51274519	1.58349312	1.73477663	1.90008689
38	1.45952724	1.52976357	1.60328678	1.76079828	1.93333841
39	1.47412251	1.54697341	1.62332787	1.78721025	1.96717184
40	1.48886373	1.56437687	1.64361946	1.81401841	2.00159734
41	1.50375237	1.58197611	1.66416471	1.84122868	2.03662530
42	1.51878989	1.59977334	1.68496677	1.86884712	2.07226624
43	1.53397779	1.61777079	1.70602885	1.89687982	2.10853090
44	1.54931757	1.63597071	1.72735421	1.92533302	2.14543019
45	1.56481075	1.65437538	1.74894614	1.95421301	2.18297522
46	1.58045885	1.67298710	1.77080797	1.98352621	2.22117728
47	1.59626344	1.69180821	1.79294306	2.01327910	2.26004789
48	1.61222608	1.71084105	1.81535485	2.04347829	2.29959872
49	1.62834834	1.73008801	1.83804679	2.07413046	2.33984170
50	1.64463182	1.74955150	1.86102237	2.10524242	2.38078893

AMOUNT AT COMPOUND INTEREST $(1 + i)^n$ (Continued)

Periods n	0.01(1%)	0.001125(1 1/8%)	0.0125(1 1/4%)	0.015(1 1/2%)	0.0175(1 3/4%)
			Rate i		
51	1.66107814	1.76923395	1.88428515	2.13682106	2.42245274
52	1.67768892	1.78913784	1.90783872	2.16887337	2.46484566
53	1.69446581	1.80926564	1.93168670	2.20140647	2.50798046
54	1.71141047	1.82961988	1.95583279	2.23442757	2.55187012
55	1.72852457	1.85020310	1.98028070	2.26794398	2.59652785
56	1.74580982	1.87101788	2.00503420	2.30196314	2.64196708
57	1.76326792	1.89206684	2.03009713	2.33649259	2.68820151
58	1.78090060	1.91335259	2.05547335	2.37153998	2.73524503
59	1.79870960	1.93487780	2.08116676	2.40711308	2.78311182
60	1.81669670	1.95664518	2.10718135	2.44321978	2.83181628
61	1.83486367	1.97865744	2.13352111	2.47986807	2.88137306
62	1.85321230	2.00091733	2.16019013	2.51706609	2.93179709
63	1.87174443	2.02342765	2.18719250	2.55482208	2.98310354
64	1.89046187	2.04619121	2.21453241	2.59314442	3.03530785
65	1.90936649	2.06921087	2.24221407	2.63204158	3.08842574
66	1.92846015	2.09248949	2.27024174	2.67152221	3.14247319
67	1.94774475	2.11602999	2.29861976	2.71159504	3.19746647
68	1.96722220	2.13983533	2.32735251	2.75226896	3.25342213
69	1.98689442	2.16390848	2.35644442	2.79355300	3.31035702
70	2.00676337	2.18825245	2.38589997	2.83545629	3.36828827
71	2.02683100	2.21287029	2.41572372	2.87798814	3.42723331
72	2.04709931	2.23776508	2.44592027	2.92115696	3.48720990
73	2.06757031	2.26293994	2.47649427	2.96497533	3.54823607
74	2.08824601	2.28839801	2.50745045	3.00944996	3.61033020
75	2.10912847	2.31414249	2.53879358	3.05459171	3.67351098
76	2.13021975	2.34017659	2.57052850	3.10041059	3.73779742
77	2.15152195	2.36650358	2.60266011	3.14691674	3.80320888
78	2.17303717	2.39312675	2.63519336	3.19412050	3.86976503
79	2.19476754	2.42004942	2.66813327	3.24203230	3.93748592
80	2.21671522	2.44727498	2.70148494	3.29066279	4.00639192
81	2.23888237	2.47480682	2.73525350	3.34002273	4.07650378
82	2.26127119	2.50264840	2.76944417	3.39012307	4.14784260
83	2.28388390	2.53080319	2.80406222	3.44097492	4.22042984
84	2.30672274	2.55927473	2.83911300	3.49258954	4.29428737
85	2.32978997	2.58806657	2.87460191	3.54497838	4.36943740
86	2.35308787	2.61718232	2.91053444	3.59815306	4.44590255
87	2.37661875	2.64662562	2.94691612	3.65212535	4.52370584
88	2.40038494	2.67640016	2.98375257	3.70690723	4.60287070
89	2.42438879	2.70650966	3.02104948	3.76251084	4.68342093
90	2.44863267	2.73695789	3.05881260	3.81894851	4.76538080
91	2.47311900	2.76774867	3.09704775	3.87623273	4.84877496
92	2.49785019	2.79888584	3.13576085	3.93437622	4.93362853
93	2.52282869	2.83037331	3.17495786	3.99339187	5.01996703
94	2.54805698	2.86221501	3.21464483	4.05329275	5.10781645
95	2.57353755	2.89441492	3.25482789	4.11409214	5.19720324
96	2.59927293	2.92697709	3.29551324	4.17580352	5.28815429
97	2.62526565	2.95990559	3.33670716	4.23844057	5.38069699
98	2.65151831	2.99320452	3.37841600	4.30201718	5.47485919
99	2.67803349	3.02687807	3.42064620	4.36654744	5.57066923
100	2.70481383	3.06093045	3.46340427	4.43204565	5.66815594

AMOUNT AT COMPOUND INTEREST $(1 + i)^n$ (Continued)

Periods n	Rate i			
	0.02(2%)	0.0225(2 1/4%)	0.025(2 1/2%)	0.0275(2 3/4%)
1	1.02000000	1.02250000	1.02500000	1.02750000
2	1.04040000	1.04550625	1.05062500	1.05575625
3	1.06120800	1.06903014	1.07689063	1.08478955
4	1.08243216	1.09308332	1.10381289	1.11462126
5	1.10408080	1.11767769	1.13140821	1.14527334
6	1.12616242	1.14282544	1.15969342	1.17676836
7	1.14868567	1.16853901	1.18868575	1.20912949
8	1.17165938	1.19483114	1.21840290	1.24238055
9	1.19509257	1.22171484	1.24886297	1.27654602
10	1.21899442	1.24920343	1.28008454	1.31165103
11	1.24337431	1.27731050	1.31208666	1.34772144
12	1.26824179	1.30604999	1.34488882	1.38478378
13	1.29360663	1.33543611	1.37851104	1.42286533
14	1.31947876	1.36548343	1.41297382	1.46199413
15	1.34586834	1.39620680	1.44829817	1.50219896
16	1.37278571	1.42762146	1.48450562	1.54350944
17	1.40024142	1.45974294	1.52161826	1.58595595
18	1.42824625	1.49258716	1.55965872	1.62956973
19	1.45681117	1.52617037	1.59865019	1.67438290
20	1.48594740	1.56050920	1.63861644	1.72042843
21	1.51566634	1.59562066	1.67958185	1.76774021
22	1.54597967	1.63152212	1.72157140	1.81635307
23	1.57689926	1.66823137	1.76461068	1.86630278
24	1.60843725	1.70576658	1.80872595	1.91762610
25	1.64060599	1.74414632	1.85394410	1.97036082
26	1.67341811	1.78338962	1.90029270	2.02454575
27	1.70688648	1.82351588	1.94780002	2.08022075
28	1.74102421	1.86454499	1.99649602	2.13742682
29	1.77584469	1.90649725	2.04640739	2.19620606
30	1.81136158	1.94939344	2.09756758	2.25660173
31	1.84758882	1.99325479	2.15000677	2.31865828
32	1.88454059	2.03810303	2.20375694	2.38242138
33	1.92223140	2.08396034	2.25885086	2.44793797
34	1.96067603	2.13084945	2.31532213	2.51525626
35	1.99988955	2.17879356	2.37320519	2.58442581
36	2.03988734	2.22781642	2.43253532	2.65549752
37	2.08068509	2.27794229	2.49334870	2.72852370
38	2.12229879	2.32919599	2.55568242	2.80355810
39	2.16474477	2.38160290	2.61957448	2.88065595
40	2.20803966	2.43518897	2.68506384	2.95987399
41	2.25220046	2.48998072	2.75219043	3.04127052
42	2.29724447	2.54600528	2.82099520	3.12490546
43	2.34318936	2.60329040	2.89152008	3.21084036
44	2.39005214	2.66186444	2.96380808	3.29913847
45	2.43785421	2.72175639	3.03790328	3.38986478
46	2.48661129	2.78299590	3.11385086	3.48308606
47	2.53634352	2.84561331	3.19169713	3.57887093
48	2.58707039	2.90963961	3.27148956	3.67728988
49	2.63881179	2.97510650	3.35327680	3.77841535
50	2.69158803	3.04204640	3.43710872	3.88232177

FINANCIAL TABLES

AMOUNT AT COMPOUND INTEREST $(1 + i)^n$ (Continued)

Periods n	Rate i			
	0.02(2%)	0.0225(2 1/4%)	0.025(2 1/2%)	0.0275(2 3/4%)
51	2.74541979	3.11049244	3.52303644	3.98908562
52	2.80032819	3.18047852	3.61111235	4.09878547
53	2.85633475	3.25203929	3.70139016	4.21150208
54	2.91346144	3.32521017	3.79392491	4.32721838
55	2.97173067	3.40002740	3.88877303	4.44631964
56	3.03116529	3.47652802	3.98599236	4.56859343
57	3.09178859	3.55474990	4.08564217	4.69422975
58	3.15362436	3.63473177	4.18778322	4.82332107
59	3.21669685	3.71651324	4.29247780	4.95596239
60	3.28103079	3.80013479	4.39978975	5.09225136
61	3.34665140	3.88563782	4.50978449	5.23228827
62	3.41358443	3.97306467	4.62252910	5.37617620
63	3.48185612	4.06245862	4.73809233	5.52402105
64	3.55149324	4.15386394	4.85654464	5.67593162
65	3.62252311	4.24732588	4.97795826	5.83201974
66	3.69497357	4.34289071	5.10240721	5.99240029
67	3.76887304	4.44060576	5.22996739	6.15719130
68	3.84425050	4.54051939	5.36071658	6.32651406
69	3.92113551	4.64268107	5.49473449	6.50049319
70	3.99955822	4.74714140	5.63210286	6.67925676
71	4.07954939	4.85395208	5.77290543	6.86293632
72	4.16114038	4.96316600	5.91722806	7.05166706
73	4.24436318	5.07483723	6.06515876	7.24558791
74	4.32925045	5.18902107	6.21678773	7.44484158
75	4.41583546	5.30577405	6.37220743	7.64957472
76	4.50415216	5.42515396	6.53151261	7.85993802
77	4.59423521	5.54721993	6.69480043	8.07608632
78	4.68611991	5.67203237	6.86217044	8.29817869
79	4.77984231	5.79965310	7.03372470	8.52637861
80	4.87543916	5.93014530	7.20956782	8.76085402
81	4.97294794	6.06357357	7.38980701	9.00177751
82	5.07240690	6.20000397	7.57455219	9.24932639
83	5.17385504	6.33950406	7.76391599	9.50368286
84	5.27733214	6.48214290	7.95801389	9.76503414
85	5.38287878	6.62799112	8.15695424	10.03357258
86	5.49053636	6.77712092	8.36088834	10.30949583
87	5.60034708	6.92960614	8.56991055	10.59300696
88	5.71235402	7.08552228	8.78415832	10.88421465
89	5.82660110	7.24494653	9.00376228	11.18363331
90	5.94313313	7.40795782	9.22885633	11.49118322
91	6.06199579	7.57463688	9.45957774	11.80719076
92	6.18323570	7.74506621	9.69606718	12.13188851
93	6.30690042	7.91933020	9.93846886	12.46551544
94	6.43303843	8.09751512	10.18693058	12.80831711
95	6.56169920	8.27970921	10.44160385	13.16054584
96	6.69293318	8.46600267	10.70264395	13.52246085
97	6.82679184	8.65648773	10.97021004	13.89432852
98	6.96332768	8.85125871	11.24446530	14.27642255
99	7.10359423	9.05041203	11.52557693	14.66902417
100	7.24464612	9.25404630	11.81371635	15.07242234

AMOUNT AT COMPOUND INTEREST $(1 + i)^n$ (Continued)

Periods	Rate i			
n	0.03(3%)	0.0325(3 1/4%)	0.035(3 1/2%)	0.0375(3 3/4%)
1	1.03000000	1.03250000	1.03500000	1.03750000
2	1.06090000	1.06605625	1.07122500	1.07640625
3	1.09272700	1.10070308	1.10871788	1.11677148
4	1.12550881	1.13647593	1.14752300	1.15865042
5	1.15927407	1.17341140	1.18768631	1.20209981
6	1.19405230	1.21154727	1.22925533	1.24717855
7	1.22987387	1.25092255	1.27227926	1.29394774
8	1.26677008	1.29157754	1.31680904	1.34247078
9	1.30477318	1.33355381	1.36289735	1.39281344
10	1.34391638	1.37689430	1.41059876	1.44504394
11	1.38423387	1.42164337	1.45996972	1.49923309
12	1.42576089	1.46784678	1.51106866	1.55545433
13	1.46853371	1.51555180	1.56395606	1.61378387
14	1.51258972	1.56480723	1.61869452	1.67430076
15	1.55796742	1.61566347	1.67534883	1.73708704
16	1.60470644	1.66817253	1.73398604	1.80222781
17	1.65284763	1.72238814	1.79467555	1.86981135
18	1.70243306	1.77836575	1.85748920	1.93992927
19	1.75350605	1.83616264	1.92250132	2.01267662
20	1.80611123	1.89583792	1.98978886	2.08815200
21	1.86029457	1.95745266	2.05943147	2.16645770
22	1.91610341	2.02106987	2.13151158	2.24769986
23	1.97358651	2.08675464	2.20611448	2.33198860
24	2.03279411	2.15457416	2.28332849	2.41943818
25	2.09377793	2.22459782	2.36324498	2.51016711
26	2.15659127	2.29689725	2.44595856	2.60429838
27	2.22128901	2.37154641	2.53156711	2.70195956
28	2.28792768	2.44862167	2.62017196	2.80328305
29	2.35656551	2.52820188	2.71187798	2.90840616
30	2.42726247	2.61036844	2.80679370	3.01747139
31	2.50008035	2.69520541	2.90503148	3.13062657
32	2.57508276	2.78279959	2.00670759	3.24802507
33	2.65233524	2.87324058	3.11194235	3.36982601
34	2.73190530	2.96662089	3.22086033	3.49619448
35	2.81386245	3.06303607	3.33359045	3.62730178
36	2.89827833	2.16258475	3.45026611	3.76332559
37	2.98522668	3.26536875	3.57102543	3.90445030
38	3.07478348	3.37149323	3.69601132	4.05086719
39	3.16702698	3.48106676	3.82537171	4.20277471
40	3.26203779	3.59420143	3.95925972	4.36037876
41	3.35989893	3.71101298	4.09783381	4.52389296
42	3.46069589	3.83162090	4.24125799	4.69353895
43	3.56451677	3.95614858	4.38970202	4.86954666
44	3.67145227	4.08472341	4.54334160	5.05215466
45	3.78159584	4.21747692	4.70235855	5.24161046
46	3.89504372	4.35454492	4.86694110	5.43817085
47	4.01189503	4.49606763	5.03728404	5.64210226
48	4.13225188	4.64218983	5.21358898	5.85368109
49	4.25621944	4.79306100	5.39606459	6.07319413
50	4.38390602	4.94883548	5.58492686	6.30093891

FINANCIAL TABLES

AMOUNT AT COMPOUND INTEREST $(1 + i)^n$ (Continued)

Periods		Rate i		
n	0.04(4%)	0.0425(4 1/4%)	0.045(4 1/2%)	0.0475(4 3/4%)
1	1.04000000	1.04250000	1.04500000	1.04750000
2	1.08160000	1.08680625	1.09202500	1.09725625
3	1.12486490	1.13299552	1.14116613	1.14937592
4	1.16985856	1.18114783	1.19251860	1.20397128
5	1.21665290	1.23134661	1.24618194	1.26115991
6	1.26531902	1.28367884	1.30226012	1.32106501
7	1.31593178	1.33823519	1.36086183	1.38381560
8	1.36856905	1.39511018	1.42210061	1.44954684
9	1.42331181	1.45440237	1.48609514	1.51840031
10	1.48024428	1.51621447	1.55296942	1.59052433
11	1.53945406	1.58065358	1.62285305	1.66607423
12	1.60103222	1.64783136	1.69588143	1.74521276
13	1.66507351	1.71786419	1.77219610	1.82811037
14	1.73167645	1.79087342	1.85194492	1.91494561
15	1.80094351	1.86698554	1.93528244	2.00590552
16	1.87298125	1.94633243	2.02237015	2.10118604
17	1.94790050	2.02905156	2.11337681	2.20099237
18	2.02581652	2.11528625	2.20847877	2.30553951
19	2.10684918	2.20518591	2.30786031	2.41505264
20	2.19112314	2.29890631	2.41171402	2.52976764
21	2.27876807	2.39660983	2.52024116	2.64993160
22	2.36991879	2.49846575	2.63365201	2.77580335
23	2.46471554	2.60465054	2.75216635	2.90765401
24	2.56330416	2.71534819	2.87601383	3.04576758
25	2.66583633	2.83075049	3.00543446	3.19044154
26	2.77246978	2.95105739	3.14067901	3.34198751
27	2.88336858	3.07647732	3.28200956	3.50073192
28	2.99870332	3.20722761	3.42969999	3.66701668
29	3.11865145	3.34353478	3.58403649	3.84119998
30	3.24339751	3.48563501	3.74531813	4.02365698
31	3.37313341	3.63377450	3.91385745	4.21478068
32	3.50805875	3.78820992	4.08998104	4.41498276
33	3.64838110	3.94920884	4.27403018	4.62469445
34	3.79431634	4.11705021	4.46636154	4.84436743
35	3.94608899	4.29202485	4.66734781	5.07447488
36	4.10393255	4.47443590	4.87737846	5.31551244
37	4.26808986	4.66459943	5.09686049	5.56799928
38	4.43881345	4.86284491	5.32621921	5.83247925
39	4.61636599	5.06951581	5.56589908	6.10952201
40	4.80102063	5.28497024	5.81636454	6.39972431
41	4.99306145	5.50958147	6.07810094	6.70371121
42	5.19278391	5.74373868	6.35161548	7.02213750
43	5.40049527	5.98784758	6.63743818	7.35568903
44	5.61651508	6.24233110	6.93612290	7.70508426
45	5.84117568	6.50763017	7.24824843	8.07107576
46	6.07482271	6.78420445	7.57441961	8.45445186
47	6.31781562	7.07253314	7.91526849	8.85603832
48	6.57052824	7.37311580	8.27145557	9.27670014
49	6.83334937	7.68647322	8.64367107	9.71734340
50	7.10668335	8.01314834	9.03263627	10.17891721

AMOUNT AT COMPOUND INTEREST $(1 + i)^n$ (Continued)

Periods *n*	Rate *i*			
	0.05(5%)	0.0525(5 1/4%)	0.055(5 1/2%)	0.0575(5 3/4%)
1	1.05000000	1.05250000	1.05500000	1.05750000
2	1.10250000	1.10775625	1.11302500	1.11830625
3	1.15762500	1.16591345	1.17424138	1.18260886
4	1.21550625	1.22712391	1.23882465	1.25060887
5	1.27628156	1.29154791	1.30696001	1.32251888
6	1.34009564	1.35935418	1.37884281	1.39856371
7	1.40710042	1.43072027	1.45467916	1.47898113
8	1.47745544	1.50583309	1.53468651	1.56402254
9	1.55132822	1.58488933	1.61909427	1.65395384
10	1.62889463	1.66809602	1.70814446	1.74905618
11	1.71033936	1.75567106	1.80209240	1.84962692
12	1.79585633	1.84784379	1.90120749	1.95598046
13	1.88564914	1.94485559	2.00577390	2.06844934
14	1.97993160	2.04696050	2.11609146	2.18738518
15	2.07892818	2.15442593	2.23247649	2.31315982
16	2.18287459	2.26753329	2.35526270	2.44616651
17	2.29201832	2.38657879	2.48480215	2.58682109
18	2.40661923	2.51187418	2.62146627	2.73556330
19	2.52695020	2.64374757	2.76564691	2.89285819
20	2.65329771	2.78254432	2.91775749	3.05919754
21	2.78596259	2.92862789	3.07823415	3.23510140
22	2.92526072	2.08238086	3.24753703	3.42111973
23	3.07152376	3.24420585	3.42615157	3.61783411
24	3.22509994	3.41452666	3.61458990	3.82585957
25	3.38635494	3.59378931	3.81339235	4.04584650
26	3.55567269	3.78246325	4.02312893	4.27848267
27	3.73345632	3.98104257	4.24440102	4.52449542
28	3.92012914	4.19004731	4.47784307	4.78465391
29	4.11613560	4.41002479	4.72412444	5.05977151
30	4.32194238	4.64155109	4.98395129	5.35070837
31	4.53803949	4.88523252	5.25806861	5.65837410
32	4.76494147	5.14170723	5.54726238	5.98373061
33	5.00318854	5.41164686	5.85236181	6.32779512
34	5.25334797	5.69575832	6.17424171	6.69164334
35	5.51601537	5.99478563	6.51382501	7.07641284
36	5.79181614	6.30951188	6.87208538	7.48330657
37	6.08140694	6.64076125	7.25005008	7.91359670
38	6.38547729	6.98940122	7.64880283	8.36862851
39	6.70475115	7.35634478	8.06948699	8.84982465
40	7.03998871	7.74255288	8.51330877	9.35868957
41	7.39198815	8.14903691	8.98154076	9.89681422
42	7.76158756	8.57686135	9.47552550	10.46588104
43	8.14966693	9.02714657	9.99667940	11.06766920
44	8.55715028	9.50107176	10.54649677	11.70406018
45	8.98500779	9.99987803	11.12655409	12.37704364
46	9.43425818	10.52487163	11.73851456	13.08872365
47	9.90597109	11.07742739	12.38413287	13.84132526
48	10.40126965	11.65899232	13.06526017	14.63720146
49	10.92133313	12.27108942	13.78384948	15.47884054
50	11.46739979	12.91532162	14.54196120	16.36887387

FINANCIAL TABLES

AMOUNT AT COMPOUND INTEREST $(1 + i)^n$ (Continued)

Periods	Rate i			
n	0.06(6%)	0.0625(6 1/4%)	0.065(6 1/2%)	0.0675(6 3/4%)
1	1.06000000	1.06250000	1.06500000	1.06750000
2	1.12360000	1.12890625	1.13422500	1.13955625
3	1.19101600	1.19946289	1.20794963	1.21647630
4	1.26247696	1.27442932	1.28646635	1.29858845
5	1.33822558	1.35408115	1.37008666	1.38624317
6	1.41851911	1.43871123	1.45914230	1.47981458
7	1.50363026	1.52863068	1.55398655	1.57970207
8	1.59384807	1.62417009	1.65499567	1.68633195
9	1.68947896	1.72568073	1.76257039	1.80015936
10	1.79084770	1.83353577	1.87713747	1.92167012
11	1.89829856	1.94813176	1.99915140	2.05138285
12	2.01219647	2.06988999	2.12909624	2.18985119
13	2.13292826	2.19925812	2.26748750	2.33766615
14	2.26090396	2.33671175	2.41487418	2.49545861
15	2.39655819	2.48275623	2.57184101	2.66390207
16	2.54035168	2.63792850	2.73901067	2.84371546
17	2.69277279	2.80279903	2.91704637	3.03566625
18	2.85433915	2.97797397	3.10665438	3.24057373
19	3.02559950	3.16409734	3.30858691	3.45931245
20	3.20713547	3.36185342	3.52364506	3.69281604
21	3.39956360	3.57196926	3.75268199	3.94208113
22	3.60353742	3.79521734	3.99660632	4.20817160
23	3.81974966	4.03241843	4.25638573	4.49222319
24	4.04893464	4.28444458	4.53305081	4.79544825
25	4.29187072	4.55222236	4.82769911	5.11914101
26	4.54938296	4.83673626	5.14149955	5.46468303
27	4.82234594	5.13903228	5.47569702	5.83354913
28	5.1i168670	5.46022180	5.83161733	6.22731370
29	5.41838790	5.80148566	6.21067245	6.64765737
30	5.74349117	6.16407851	6.61436616	7.90637424
31	6.08810064	6.54933342	7.04429996	7.57537950
32	6.45338668	6.95866676	7.50217946	8.08671762
33	6.84058988	7.39358343	7.98982113	8.63257106
34	7.25102528	7.85568239	8.50915950	9.21526961
35	7.68608679	8.34666254	9.06225487	9.83730031
36	8.14725200	8.86832895	9.65130143	10.50131808
37	8.63608712	9.42259951	10.27863603	11.21015705
38	9.15425235	10.01151198	10.94674737	11.96684265
39	9.70350749	10.63723148	11.65828595	12.77460453
40	10.28571794	11.30205845	12.41607453	13.63689033
41	10.90286101	12.00843710	13.22311938	14.55738043
42	11.55703267	12.75896442	14.08262214	15.54000361
43	12.25045463	13.55639970	14.99799258	16.58895385
44	12.98548191	14.40367468	15.97286209	17.70870824
45	13.76461083	15.30390434	17.01109813	18.90404604
46	14.59048748	16.26039837	18.11681951	20.18006915
47	15.46591673	17.27667326	19.29441278	21.54222382
48	16.39387173	18.35646534	20.54854961	22.99632392
49	17.37750403	19.50374443	21.88420533	24.54857579
50	18.42015427	20.72272845	23.30667868	26.20560466

AMOUNT AT COMPOUND INTEREST $(1 + i)^n$ (Continued)

Periods	Rate i			
n	0.07(7%)	0.0725(7 1/4%)	0.075(7 1/2%)	0.0775(7 3/4%)
1	1.07000000	1.07250000	1.07500000	1.07750000
2	1.14490000	1.15025625	1.15562500	1.16100625
3	1.22504300	1.23364983	1.24229687	1.25098423
4	1.31079601	1.32308944	1.33546914	1.34793551
5	1.40255173	1.41901343	1.43562933	1.45240051
6	1.50073035	1.52189190	1.54330153	1.56496155
7	1.60578148	1.63222906	1.65904914	1.68624608
8	1.71818618	1.75056567	1.78347783	1.81693015
9	1.83845921	1.87748168	1.91723866	1.95774223
10	1.96715136	2.01359910	2.06103156	2.10946726
11	2.10485195	2.15958504	2.21560893	2.27295097
12	2.25219159	2.31615495	2.38177960	2.44910467
13	2.40984500	2.48407618	2.56041307	2.63891028
14	2.57853415	2.66417171	2.75244405	2.84342583
15	2.75903154	2.85732416	2.95887735	3.06379133
16	2.95216375	3.06448016	3.18079315	3.30123516
17	3.15881521	3.28665497	3.41935264	3.55708088
18	3.37993228	3.52493745	3.67580409	3.83275465
19	3.61652754	3.78049542	3.95148940	4.12979313
20	3.86968446	4.05458134	4.24785110	4.44985210
21	4.14056237	4.34853849	4.56643993	4.79471564
22	4.43040174	4.66380753	4.90892293	5.16630610
23	4.74052986	5.00193357	5.27709215	5.56669482
24	5.07236695	5.36457375	5.67287406	5.99811367
25	5.42743264	5.75350535	6.09833961	6.46296748
26	5.80735292	6.17063449	6.55571508	6.96384746
27	6.21386763	6.61800549	7.04739371	7.50354564
28	6.64883836	7.09781089	7.57594824	8.05507043
29	7.11425705	7.61240218	8.14414436	8.71166339
30	7.61225504	8.16430134	8.75495519	9.38681730
31	8.14511290	8.75621318	9.41157683	10.11429564
32	8.71527080	9.39103864	10.11744509	10.89815355
33	9.32533975	10.07188894	10.87625347	11.74276045
34	9.97811354	10.80210089	11.69197248	12.65282439
35	10.67658148	11.58525320	12.56887042	13.63341828
36	11.42394219	12.42518406	13.51153570	14.69000819
37	12.22361814	13.32600990	14.52490088	15.82848383
38	13.07927141	14.29214562	15.61426844	17.05519132
39	13.99482041	15.32832618	16.78533858	18.37696865
40	14.97445784	16.43962983	18.04423897	19.80118372
41	16.02266989	17.63150299	19.39755689	21.33577546
42	17.14425678	18.90978696	20.85237366	22.98929806
43	18.34435475	20.28074651	22.41630168	24.77096866
44	19.62845959	21.75110063	24.09752431	26.69071873
45	21.00245176	23.32805543	25.90483863	28.75924943
46	22.47262338	25.01933945	27.84770153	30.98809126
47	24.04570702	26.83324156	29.93627915	33.38966833
48	25.72890651	28.77865157	32.18150008	35.97736763
49	27.52992997	30.86510381	34.59511259	38.76561362
50	29.45702506	33.10282384	37.18974603	41.76994868

AMOUNT AT COMPOUND INTEREST $(1 + i)^n$ (Continued)

Periods n	0.08(8%)	0.0825(8 1/4%)	0.085(8 1/2%)	0.0875(8 3/4%)
			Rate i	
1	1.08000000	1.08250000	1.08500000	1.08750000
2	1.16640000	1.17180625	1.17722500	1.18265625
3	1.25971200	1.26848027	1.27728912	1.28613867
4	1.36048896	1.37312989	1.38585870	1.39867581
5	1.46932808	1.48641310	1.50365669	1.52105994
6	1.58687432	1.60904218	1.63146751	1.65415268
7	1.71382427	1.74178816	1.77014225	1.79889104
8	1.85093021	1.88548569	1.92060434	1.95629401
9	1.99900463	2.04103826	2.08385571	2.12746974
10	2.15892500	2.20942391	2.26098344	2.31362334
11	2.33163900	2.39170139	2.45316703	2.51606538
12	2.51817012	2.58901675	2.66168623	2.73622110
13	2.71962373	2.80261063	2.88792956	2.97564045
14	2.93719362	3.03382601	3.13340357	3.23600898
15	3.17216911	3.28411666	3.39974288	3.51915977
16	3.42594264	3.55505628	3.68872102	3.82708625
17	3.70001805	3.84834842	4.00226231	4.16195630
18	3.99601950	4.16583717	4.34245461	4.52612747
19	4.31570106	4.50951873	4.71156325	4.92216363
20	4.66095714	4.88155403	5.11204612	5.35285295
21	5.03383372	5.28428224	5.54657005	5.82122758
22	5.43654041	5.72023552	6.01802850	6.33058499
23	5.87146365	6.19215495	6.52956092	6.88451118
24	6.34118074	6.70300774	7.08457360	7.48690591
25	6.84847520	7.25600587	7.68676236	8.14201017
26	7.39635321	7.85462636	8.34013716	8.85443606
27	7.98806147	8.50263303	9.04904881	9.62919922
28	8.62710639	9.20410026	9.81821796	10.47175415
29	9.31727490	9.96343853	10.65276649	11.38803264
30	10.06265689	10.78542221	11.55825164	12.38448549
31	10.86766944	11.67521954	12.54070303	13.46812797
32	11.73708300	12.63842515	13.60666279	14.64658917
33	12.67604964	13.68109523	14.76322913	15.92816573
34	13.69013361	14.80978558	16.01810360	17.32188023
35	14.78534429	16.03159290	17.37964241	18.83754475
36	15.96817184	17.35419931	18.85691201	20.48582991
37	17.24562558	18.78592075	20.45974953	22.27834003
38	18.62527563	20.33575921	22.19882824	24.22769478
39	20.11529768	22.01345935	24.08572865	26.34761807
40	21.72452150	23.82956975	26.13301558	28.65303466
41	23.46248322	25.79550925	28.35432190	31.16017519
42	25.33948187	27.92363876	30.76443927	33.88669052
43	27.36664042	30.22733896	33.37941660	36.85177594
44	29.55597166	32.72109442	36.21666702	40.07630633
45	31.92044939	35.42058471	39.29508371	43.58298314
46	34.47408534	38.34278295	42.63516583	47.39649416
47	37.23201217	41.50606255	46.25915492	51.54368740
48	40.21057314	44.93031271	50.19118309	56.05376005
49	43.42741899	48.63706351	54.45743365	60.95846405
50	46.90161251	52.64962124	59.08631551	66.29232966

AMOUNT AT COMPOUND INTEREST $(1 + i)^n$ (Continued)

Periods n	0.09(9%)	0.0925(9 1/4%)	0.095(9 1/2%)	0.0975(9 3/4%)
1	1.09000000	1.09250000	1.09500000	1.09750000
2	1.18810000	1.19355625	1.19902500	1.20450625
3	1.29502900	1.30396020	1.31293237	1.32194561
4	1.41158161	1.42457652	1.43766095	1.45083531
5	1.53862395	1.55634985	1.57423874	1.59229175
6	1.67710011	1.70031221	1.72379142	1.74754019
7	1.82803912	1.85759109	1.88755161	1.91792536
8	1.99256264	2.02941827	2.06686901	2.10492309
9	2.17189328	2.21713946	2.26322156	2.31015309
10	2.36736367	2.42222486	2.47822761	2.53539301
11	2.58042641	2.64628066	2.71365924	2.78259383
12	2.81266478	2.89106162	2.97145686	3.05389673
13	3.06580461	3.15848482	3.25374527	3.35165166
14	3.34172703	3.45064466	3.56285107	3.67843770
15	3.64248246	3.76982929	3.90132192	4.03708537
16	3.97030588	4.11853850	4.27194750	4.43070120
17	4.32763341	4.49950331	4.67778251	4.86269456
18	4.71712042	4.91570737	5.12217185	5.33680728
19	5.14166125	5.37041030	5.60877818	5.85714599
20	5.60441077	5.86717325	6.14161210	6.42821773
21	6.10880774	6.40988678	6.72506525	7.05496896
22	6.65860043	7.00280131	7.36394645	7.74282843
23	7.25787447	7.65056043	8.06352137	8.49775420
24	7.91108317	8.35823727	8.82955590	9.32628524
25	8.62308066	9.13137421	9.66836371	10.23559805
26	9.39915792	9.97602633	10.58685826	11.23356886
27	10.24508213	10.89880877	11.59260979	12.32884182
28	11.16713952	11.90694858	12.69390772	13.53090390
29	12.17218208	13.00834132	13.89982896	14.85016703
30	13.26767847	14.21161289	15.22031271	16.29805832
31	14.46176953	15.52618708	16.66624241	17.88711900
32	15.76332879	16.96235939	18.24953544	19.63111310
33	17.18202838	18.53137763	19.98324131	21.54514663
34	18.72841093	20.24553006	21.88164924	23.64579843
35	20.41396792	22.11824159	23.96040591	25.95126378
36	22.25122503	24.16417894	26.23664448	28.48151199
37	24.25383528	26.39936549	28.72912570	31.25845941
38	26.43668046	28.84130680	31.45839264	34.30615921
39	28.81598170	31.50912768	34.44693994	37.65100973
40	31.40942005	34.42372199	37.71939924	41.32198318
41	34.23626786	37.60791628	41.30274216	45.35087654
42	37.31753197	41.08664853	45.22650267	49.77258700
43	40.67610984	44.88716352	49.52302042	54.62541423
44	44.33695973	49.03922615	54.22770736	59.95139212
45	48.32728610	53.57535456	59.37933956	65.79665285
46	52.67674185	58.53107486	65.02037682	72.21182650
47	57.41764862	63.94519929	71.19731262	79.25247959
48	62.58523700	69.86013022	77.96105732	86.97959635
49	68.21790833	76.32219227	85.36735777	95.46010699
50	74.35752008	83.38199505	93.47725675	104.76746742

FINANCIAL TABLES

AMOUNT AT COMPOUND INTEREST $(1 + i)^n$ (Continued)

Periods	Rate i			
n	0.10(10%)	0.105(10 1/2%)	0.11(11%)	0.115(11 1/2%)
1	1.10000000	1.10500000	1.11000000	1.11500000
2	1.21000000	1.22102500	1.23210000	1.24322500
3	1.33100000	1.34923262	1.36763100	1.38619587
4	1.46410000	1.49090205	1.51807041	1.54560840
5	1.61051000	1.64744677	1.68505816	1.72335337
6	1.77156100	1.82042868	1.87041455	1.92153900
7	1.94871710	2.01157369	2.07616015	2.14251599
8	2.14358881	2.22278892	2.30453777	2.38890533
9	2.35794769	2.45618176	2.55803692	2.66362944
10	2.59374246	2.71408085	2.83942099	2.96994683
11	2.85311671	2.99905934	3.15175729	3.31149071
12	3.13842838	3.31396057	3.49845060	3.69231214
13	3.45227121	3.66192643	3.88328016	4.11692804
14	3.79749834	4.04642870	4.31044098	4.59037476
15	4.17724817	4.47130371	4.78458949	5.11826786
16	4.59497299	4.94079060	5.31089433	5.70686867
17	5.05447028	5.45957362	5.89509271	6.36315856
18	5.55991731	6.03282885	6.54355291	7.09492180
19	6.11590904	6.66627588	7.26334373	7.91083780
20	6.72749995	7.36623484	8.06231154	8.82058415
21	7.40024994	8.13968950	8.94916581	9.83495133
22	8.14027494	8.99435690	9.93357404	10.96597073
23	8.95430243	9.93876437	11.02626719	12.22705737
24	9.84973268	10.98233463	12.23915658	13.63316896
25	10.83470594	12.13547977	13.58546380	15.20098340
26	11.91817654	13.40970514	15.07986482	16.94909649
27	13.10999419	14.81772418	16.73864995	18.89824258
28	14.42099361	16.37358522	18.57990145	21.07154048
29	15.86309297	18.09281167	20.62369061	23.49476763
30	17.44940227	19.99255690	22.89229657	26.19666591
31	19.19434250	22.09177537	25.41044919	29.20928249
32	21.11377675	24.41141178	28.20559861	32.56834998
33	23.22515442	26.97461002	31.30821445	36.31371022
34	25.54766986	29.80694407	34.75211804	40.48978690
35	28.10243685	32.93667320	38.57485103	45.14611239
36	30.91268053	36.39502389	42.81808464	50.33791532
37	34.00394859	40.21650140	47.52807395	56.12677558
38	37.40434344	44.43923404	52.75616209	62.58135477
39	41.14477779	49.10535362	58.55933991	69.77821057
40	45.25925557	54.26141575	65.00086731	77.80270479
41	49.78518112	59.95886440	72.15096271	86.75001584
42	54.76369924	66.25454516	80.08756861	96.72626766
43	60.24006916	73.21127240	88.89720115	107.84978844
44	66.26407608	80.89845601	98.67589328	120.25251411
45	72.89048369	89.39279389	109.53024154	134.08155323
46	80.17953205	98.77903724	121.57856811	149.50093186
47	88.19748526	109.15083616	134.95221060	166.69353902
48	97.01723378	120.61167395	149.79695377	185.86329601
49	106.71895716	133.27589972	166.27461868	207.23757505
50	117.39085288	147.26986919	184.56482674	231.06989618

AMOUNT AT COMPOUND INTEREST $(1 + i)^n$ (Continued)

Periods		Rate i		
n	0.12(12%)	0.125(12 1/2%)	0.13(13%)	0.135(13 1/2%)
1	1.12000000	1.12500000	1.13000000	1.13500000
2	1.25440000	1.26562500	1.27690000	1.28822500
3	1.40492800	1.42382812	1.44289700	1.46213537
4	1.57351936	1.60180664	1.63047361	1.65952365
5	1.76234168	1.80203247	1.84243518	1.88355934
6	1.97382269	2.02728653	2.08195175	2.13783985
7	2.21068141	2.28069735	2.35260548	2.42644824
8	2.47596318	2.56578451	2.65844419	2.75401875
9	2.77307876	2.88650758	3.00404194	3.12581128
10	3.10584821	3.24732103	3.39456739	3.54779580
11	3.47854999	3.65323615	3.83586115	4.02674823
12	3.89597599	4.10989067	4.33452310	4.57035924
13	4.36349311	4.62362701	4.89801110	5.18735774
14	4.88711229	5.20158038	5.53475255	5.88765104
15	5.47356576	5.85177793	6.25427038	6.68248393
16	6.13039365	6.58325017	7.06732553	7.58461926
17	6.86604089	7.40615644	7.98607785	8.60854286
18	7.68996580	8.33192600	9.02426797	9.77069614
19	8.61276169	9.37341675	10.19742280	11.08974012
20	9.64629309	10.54509384	11.52308776	12.58685504
21	10.80384826	11.86323057	13.02108917	14.28608047
22	12.10031006	13.34613439	14.71383077	16.21470134
23	13.55234726	15.01440119	16.62662877	18.40368602
24	15.17862893	16.89120134	18.78809051	20.88818363
25	17.00006441	19.00260151	21.23054227	23.70808842
26	19.04007214	21.37792670	23.99051277	26.90868035
27	21.32488079	24.05016754	27.10927943	30.54135220
28	23.88386649	27.05643848	30.63348575	34.66443475
29	26.74993047	30.43849329	34.61583890	39.34413344
30	29.95992212	34.24330495	39.11589796	44.65559145
31	33.55511278	38.52371807	44.20096469	50.68409630
32	37.58172631	43.33918283	49.94709010	57.52644930
33	42.09153347	48.75658068	56.44021181	65.29251996
34	47.14251748	54.85115327	63.77743935	74.10701015
35	52.79961958	61.70754742	72.06850647	84.11145652
36	59.13557393	69.42099085	81.43741231	95.46650315
37	66.23184280	78.09861471	92.02427591	108.35448108
38	74.17966394	87.86094155	103.98743178	122.98233602
39	83.08122361	98.84355924	117.50579791	139.58495138
40	93.05097044	111.19900415	132.78155163	158.42891982
41	104.21708689	125.09887966	150.04315335	179.81682400
42	116.72313732	140.73623962	169.54876328	204.09209523
43	130.72991380	158.32826958	191.59010251	231.64452809
44	146.41750346	178.11930327	216.49681583	262.91653938
45	163.98760387	200.38421618	244.64140189	298.41027220
46	183.66611634	225.43224320	276.44478414	338.69565895
47	205.70605030	253.61127360	312.38260608	384.41957291
48	230.39077633	285.31268280	352.99234487	436.31621525
49	258.03766949	320.97676816	398.88134970	495.21890431
50	289.00218983	361.09886417	450.73592516	562.07345639

AMOUNT AT COMPOUND INTEREST $(1 + i)^n$ (Continued)

Periods n	Rate i			
	0.14(14%)	0.145(14 1/2%)	0.15(15%)	0.155(15 1/2%)
1	1.14000000	1.14500000	1.15000000	1.15500000
2	1.29960000	1.31102500	1.32250000	1.33402500
3	1.48154400	1.50112362	1.52087500	1.54079887
4	1.68896016	1.71878655	1.74900625	1.77962270
5	1.92541458	1.96801060	2.01135719	2.05546422
6	2.19497262	2.25337214	2.31306077	2.37406117
7	2.50226879	2.58011110	2.66001988	2.74204066
8	2.85258642	2.95422721	3.05902286	3.16705696
9	3.25194852	3.38259015	3.51787629	3.65795078
10	3.70722131	3.87306572	4.04555774	4.22493316
11	4.22623230	4.43466025	4.65239140	4.87979780
12	4.81790482	5.07768599	5.35025011	5.63616645
13	5.49241149	5.81395046	6.15278762	6.50977225
14	6.26134910	6.65697328	7.07570576	7.51878695
15	7.13793798	7.62223440	8.13706163	8.68419893
16	8.13724930	8.72745839	9.35762087	10.03024977
17	9.27646420	9.99293985	10.76126400	11.58493848
18	10.57516918	11.44191613	12.37545361	13.38060394
19	12.05569287	13.10099397	14.23177165	15.45459756
20	13.74348987	15.00063810	16.36653739	17.85006018
21	15.66757845	17.17573062	18.82151800	20.61681950
22	17.86103944	19.66621156	21.64474570	23.81242653
23	20.36158496	22.51781224	24.89145756	27.50335264
24	23.21220685	25.78289502	28.62517619	31.76637230
25	26.46191581	29.52141479	32.91895262	36.69016000
26	30.16658403	33.80201994	37.85679551	42.37713481
27	34.38990579	38.70331283	43.53531484	48.94559070
28	39.20449260	44.31529319	50.06561207	56.53215726
29	44.69312156	50.74101070	57.57545388	65.29464163
30	50.95015858	58.09845725	66.21177196	75.41531109
31	58.08318078	66.52273355	76.14353775	87.10468431
32	66.21482609	76.16852992	87.56506841	100.60591037
33	75.48490175	87.21296676	100.69982867	116.19982648
34	86.05278799	99.85884694	115.80480298	134.21079959
35	98.10017831	114.33837974	133.17552342	155.01347352
36	111.83420328	130.91744481	153.15185194	179.04056192
37	127.49099173	149.90047430	176.12462973	206.79184901
38	145.33973058	171.63604308	202.54332419	238.84458561
39	165.68729286	196.52326932	232.92482281	275.86549638
40	188.88351386	225.01914337	267.86354623	318.62464832
41	215.32720580	257.64691916	308.04307817	368.01146881
42	245.47301461	295.00572244	354.24953990	425.05324647
43	279.83923665	337.78155220	407.38697088	490.93649968
44	319.01672979	386.75987726	468.49501651	567.03165713
45	363.67907196	442.84005947	538.76926899	654.92156398
46	414.59414203	507.05186809	619.58465934	756.43440640
47	472.63732191	580.57438896	712.52235824	873.68173939
48	538.80654698	664.75767536	819.40071197	1009.10240900
49	614.23946356	761.14753829	942.31081877	1165.51328239
50	700.23298846	871.51393134	1083.65744158	1346.16784116

AMOUNT AT COMPOUND INTEREST $(1 + i)^n$ (Continued)

Periods n	Rate i			
	0.16(16%)	0.165(16 1/2%)	0.17(17%)	0.175(17 1/2%)
1	1.16000000	1.16500000	1.17000000	1.17500000
2	1.34560000	1.35722500	1.36890000	1.38062500
3	1.56089600	1.58116712	1.60161300	1.62223437
4	1.81063936	1.84205970	1.87388721	1.90612539
5	2.10034166	2.14599955	2.19244804	2.23969733
6	2.43639632	2.50008948	2.56516420	2.63164437
7	2.82621973	2.91260424	3.00124212	3.09218213
8	3.27841489	3.39318394	3.51145328	3.63331400
9	3.80296127	3.95305929	4.10840033	4.26914396
10	4.41143508	4.60531407	4.80682839	5.01624415
11	5.11726469	5.36519090	5.62398922	5.89408687
12	5.93602704	6.25044739	6.58006738	6.92555208
13	6.88579137	7.28177121	7.69867884	8.13752369
14	7.98751799	8.48326346	9.00745424	9.56159034
15	9.26552087	9.88300194	10.53872146	11.23486864
16	10.74800420	11.51369726	12.33030411	13.20097066
17	12.46768488	13.41345730	14.42645581	15.51114052
18	14.46251446	15.62667776	16.87895329	18.22559011
19	16.77651677	18.20507959	19.74837535	21.41506838
20	19.46075945	21.20891772	23.10559916	25.16270535
21	22.57448097	24.70838914	27.03355102	29.56617879
22	26.18639792	28.78527335	31.62925470	34.74026008
23	30.37622159	33.53484345	37.00622799	40.81980559
24	35.23641704	39.06809262	43.29728675	47.96327157
25	40.87424377	45.51432791	50.65782550	56.35684409
26	47.41412277	53.02419201	59.26965584	66.21929181
27	55.00038241	61.77318369	69.34549733	77.80766787
28	63.80044360	71.96575900	81.13423187	91.42400975
29	74.00851458	83.84010924	94.92705129	107.42321146
30	85.84987691	97.67372726	111.06465001	126.22227346
31	99.58585721	113.78989226	129.94564051	148.31117132
32	115.51959437	132.56522448	152.03639940	174.26562630
33	134.00272947	154.43848652	177.88258730	204.76211090
34	155.44316618	179.92083680	208.12262714	240.59548031
35	180.31407277	209.60777487	243.50347375	282.69968936
36	209.16432441	244.19305773	284.89906429	332.17213500
37	242.63061632	284.48491225	333.33190522	390.30225862
38	281.45151493	331.42492277	389.99832910	458.60515388
39	326.48375732	386.11003503	456.29804505	538.86105581
40	378.72115849	449.81819081	533.86871271	633.16174058
41	439.31654385	524.03819230	624.62639387	743.96504518
42	509.60719087	610.50449402	730.81288083	874.15892808
43	591.14434141	711.23773554	855.05107057	1027.13674050
44	685.72743603	828.59196190	1000.40975257	1206.88567009
45	795.44382580	965.30963562	1170.47941051	1418.09066235
46	922.71483793	1124.58572549	1369.46091029	1666.25652826
47	1070.34921199	1310.14237020	1602.26926504	1957.85142071
48	1241.60508591	1526.31586128	1874.65504010	2300.47541933
49	1440.26189966	1778.15797839	2193.34639691	2703.05861771
50	1670.70380360	2071.55404483	2566.21528439	3176.09387581

AMOUNT AT COMPOUND INTEREST $(1 + i)^n$ (Continued)

Periods n	Rate i				
	0.18(18%)	0.185(18 1/2%)	0.19(19%)	0.195(19 1/2%)	0.20(20%)
1	1.18000000	1.18500000	1.19000000	1.19500000	1.20000000
2	1.39240000	1.40422500	1.41610000	1.42802500	1.44000000
3	1.64303200	1.66400662	1.68515900	1.70648987	1.72800000
4	1.93877776	1.97184785	2.00533921	2.03925540	2.07360000
5	2.28775776	2.33663970	2.38635366	2.43691020	2.48832000
6	2.69955415	2.76891805	2.83976086	2.91210769	2.98598400
7	3.18547390	3.28116789	3.37931542	3.47996869	3.58318080
8	3.75885920	3.88818395	4.02138535	4.15856259	4.29981696
9	4.43545386	4.60749798	4.78544856	4.96948229	5.15978035
10	5.23383555	5.45988510	5.69468379	5.93853134	6.19173642
11	6.17592595	6.46996385	6.77667371	7.09654495	7.43008371
12	7.28759263	7.66690716	8.06424172	8.48037122	8.91610045
13	8.59935930	9.08528498	9.59644764	10.13404361	10.69932054
14	10.14724397	10.76606270	11.41977269	12.11018211	12.83918465
15	11.97374789	12.75778430	13.58952950	14.47166762	15.40702157
16	14.12902251	15.11797440	16.17154011	17.29364281	18.48842589
17	16.67224656	17.91479966	19.24413273	20.66590315	22.18611107
18	19.67325094	21.22903760	22.90051795	24.69575427	26.62333328
19	23.21443611	25.15640955	27.25161636	29.51142635	31.94799994
20	27.39303460	29.81034532	32.42942347	35.26615449	38.33759992
21	32.32378083	35.32525921	38.59101393	42.14305461	46.00511991
22	38.14206138	41.86043216	45.92330658	50.36095026	55.20614389
23	45.00763243	49.60461211	54.64873482	60.18133557	66.24737267
24	53.10900627	58.78146535	65.03199444	71.91669600	79.49684720
25	62.66862740	69.65603644	77.38807338	85.94045172	95.39621664
26	73.94898033	82.54240318	92.09180733	102.69883981	114.47545997
27	87.25979679	97.81274777	109.58925072	122.72511357	137.37055197
28	102.96656021	115.90810611	130.41120836	146.65651072	164.84466236
29	121.50054105	137.35110574	155.18933794	175.25453031	197.81359483
30	143.37063844	162.76106030	184.67531215	209.42916372	237.37631380
31	169.17735336	192.87185646	219.76362146	250.26785064	284.85157656
32	199.62927696	228.55314990	261.51870954	299.07008152	341.82189187
33	235.56254681	270.83548263	311.20726435	357.38874741	410.18627025
34	277.96380524	320.94004692	370.33664458	427.07955316	492.22352430
35	327.99729018	380.31395560	440.70060705	510.36006602	590.66822915
36	387.03680242	450.67203738	524.43372239	609.88027890	708.80187499
37	456.70342685	534.04636430	624.07612965	728.80693328	850.56224998
38	538.91004369	632.84494170	742.65059428	870.92428527	1020.67469998
39	635.91385155	749.92125591	883.75420719	1040.75452090	1224.80963997
40	750.37834483	888.65668825	1051.66750656	1243.70165248	1469.77156797
41	885.44644690	1053.05817558	1251.48433281	1486.22347471	1763.72588156
42	1044.82680734	1247.87393806	1489.26635604	1776.03705228	2116.47105788
43	1232.89563266	1478.73061660	1772.22696369	2122.36427748	2539.76526945
44	1454.81684654	1752.29578067	2108.95008679	2536.22531159	3047.71832334
45	1716.68387891	2076.47050010	2509.65060328	3030.78924734	3657.26198801
46	2025.68697712	2460.61754262	2986.48421790	3621.79315058	4388.71438561
47	2390.31063300	2915.83178800	3553.91621930	4328.04281494	5266.45726273
48	2820.56654694	3455.26066878	4229.16030097	5172.01116385	6319.74871528
49	3328.26852539	4094.48389250	5032.70075815	6180.55334080	7583.69845834
50	3927.35685996	4851.96341262	5988.91390220	7385.76124226	9100.43815000

PRESENT VALUE $1/(1 + i)^{n}$ *

The following table gives the value of unit amount due in n years at rate of interest i, compounded annually, $1/(1 + i)^n = v^n$

Periods			Rate i		
n	0.0025(1/4%)	0.004167(5/12%)	0.005(1/2%)	0.005833(7/12%)	0.0075(3/4%)
1	0.99750623	0.99585062	0.99502488	0.99420050	0.99255583
2	0.99501869	0.99171846	0.99007450	0.98843463	0.98516708
3	0.99253734	0.98760345	0.98514876	0.98270220	0.97783333
4	0.99006219	0.98350551	0.98024752	0.97700301	0.97055417
5	0.98759321	0.97942457	0.97537067	0.97133688	0.96332920
6	0.98513038	0.97536057	0.97051808	0.96570361	0.95615802
7	0.98267370	0.97131343	0.96568963	0.96010301	0.94904022
8	0.98022314	0.96728308	0.96088520	0.95453489	0.94197540
9	0.97777869	0.96326946	0.95610468	0.94899906	0.93496318
10	0.97534034	0.95927249	0.95134794	0.94349534	0.92800315
11	0.97290807	0.95529211	0.94661487	0.93802354	0.92109494
12	0.97048187	0.95132824	0.94190534	0.93258347	0.91423815
13	0.96806171	0.94738082	0.93721924	0.92717495	0.90743241
14	0.96564759	0.94344978	0.93255646	0.92179779	0.90067733
15	0.96323949	0.93953505	0.92791688	0.91645182	0.89397254
16	0.96083740	0.93563657	0.92330037	0.91113686	0.88731766
17	0.95844130	0.93175426	0.91870684	0.90585272	0.88071231
18	0.95605117	0.92788806	0.91413616	0.90059922	0.87415614
19	0.95366700	0.92403790	0.90958822	0.89537619	0.86764878
20	0.95128878	0.92020372	0.90506290	0.89018346	0.86118985
21	0.94891649	0.91638544	0.90056010	0.88502084	0.85477901
22	0.94655011	0.91258301	0.89607971	0.87988815	0.84841589
23	0.94418964	0.90879636	0.89162160	0.87478524	0.84210014
24	0.94183505	0.90502542	0.88718567	0.86971192	0.83583140
25	0.93948634	0.90127013	0.88277181	0.86466802	0.82960933
26	0.93714348	0 89753042	0.87837991	0.85965338	0.82343358
27	0.93480646	0.89380623	0.87400986	0.85466782	0.81730380
28	0.93247527	0.89009749	0.86966155	0.84971117	0.81121966
29	0.93014990	0.88640414	0.86533488	0.84478327	0.80518080
30	0.92783032	0.88272611	0.86102973	0.83988394	0.79918690
31	0.92551653	0.87906335	0.85674600	0.83501303	0.79323762
32	0.92320851	0.87541578	0.85248358	0 83017037	0.78733262
33	0.92090624	0.87178335	0.84824237	0.82535580	0.78147158
34	0.91860972	0.86816599	0.84402226	0.82056914	0.77565418
35	0.91631892	0.86456365	0.83982314	0.81581025	0.76988008
36	0.91403384	0.86097624	0.83564492	0.81107896	0.76414896
37	0.91175445	0.85740373	0.83148748	0.80637510	0.75846051
38	0.90948075	0.85384604	0.82735073	0.80169853	0.75281440
39	0.90721272	0.85030311	0.82323455	0.79704907	0.74721032
40	0.90495034	0.84677488	0.81913886	0.79242659	0.74164796
41	0.90269361	0.84326129	0.81506354	0.78783091	0.73612701
42	0.90044250	0.83976228	0.81100850	0.78326188	0.73064716
43	0.89819701	0.83627779	0.80697363	0.77871935	0.72520809
44	0.89595712	0.83280776	0.80295884	0.77420316	0.71980952
45	0.89372281	0.82935212	0.79896402	0.76971317	0.71445114
46	0.89149407	0.82591083	0.79498907	0.76524922	0.70913264
47	0.88927090	0.82248381	0.79103390	0.76081115	0.70385374
48	0.88705326	0.81907102	0.78709841	0.75639983	0.69861414
49	0.88484116	0.81567238	0.78318250	0.75201209	0.69341353
50	0.88263457	0.81228785	0.77928607	0.74765079	0.68825165

*$a_{\overline{n}|i}$ may be constructed from these tables by use of the formula $\dfrac{1 - (1 + i)^{-n}}{i}$.

See page 549 .

$a_{\overline{n}|i}$ may be constructed from these Tables by use of the formula $a_{\overline{n}|i}^{-1} = \dfrac{i}{1 - v^n} = \overline{n}|i + i.$

FINANCIAL TABLES

PRESENT VALUE $1/(1 + i)^n$ (Continued)

Periods			Rate i		
n	0.0025(1/4%)	0.004167(5/12%)	0.005(1/2%)	0.005833(7/12%)	0.0075(3/4%)
51	0.88043349	0.80891736	0.77540902	0.74331479	0.68312819
52	0.87823790	0.80556086	0.77155127	0.73900393	0.67804286
53	0.87604778	0.80221828	0.76771270	0.73471808	0.67299540
54	0.87386312	0.79888957	0.76389324	0.73045708	0.66798551
55	0.87168391	0.79557468	0.76009277	0.72622079	0.66301291
56	0.86951013	0.79227354	0.75631122	0.72200907	0.65807733
57	0.86734178	0.78898610	0.75254847	0.71782178	0.65317849
58	0.86517883	0.78571230	0.74880445	0.71365877	0.64831612
59	0.86302128	0.78245208	0.74507906	0.70951990	0.64348995
60	0.86086911	0.77920539	0.74137220	0.70540504	0.63869970
61	0.85872230	0.77597217	0.73768378	0.70131404	0.63394511
62	0.85658085	0.77275237	0.73401371	0.69724677	0.62922592
63	0.85444474	0.76954593	0.73036190	0.69320308	0.62454185
64	0.85231395	0.76635279	0.72672826	0.68918285	0.61989266
65	0.85018848	0.76317291	0.72311269	0.68518593	0.61527807
66	0.84806831	0.76000621	0.71951512	0.68121219	0.61069784
67	0.84595343	0.75685266	0.71593544	0.67726150	0.60615170
68	0.84384382	0.75371219	0.71237357	0.67333372	0.60163940
69	0.84173947	0.75058476	0.70882943	0.66942872	0.59716070
70	0.83964037	0.74747030	0.70530291	0.66554637	0.59271533
71	0.83754650	0.74436876	0.70179394	0.66168653	0.58830306
72	0.83545786	0.74128009	0.69830243	0.65784908	0.58392363
73	0.83337442	0.73820424	0.69482829	0.65403388	0.57957681
74	0.83129618	0.73514115	0.69137143	0.65024081	0.57526234
75	0.82922312	0.73209078	0.68793177	0.64646973	0.57097999
76	0.82715523	0.72905306	0.68450923	0.64272053	0.56672952
77	0.82509250	0.72602794	0.68110371	0.63899306	0.56251069
78	0.82303491	0.72301537	0.67771513	0.63528723	0.55832326
79	0.82098246	0.72001531	0.67434342	0.63160288	0.55416701
80	0.81893512	0.71702770	0.67098847	0.62793989	0.55004170
81	0.81689289	0.71405248	0.66765022	0.62429816	0.54594710
82	0.81485575	0.71108960	0.66432858	0.62067754	0.54188297
83	0.81282369	0.70813902	0.66102346	0.61707792	0.53784911
84	0.81079670	0.70520069	0.65773479	0.61349917	0.53384527
85	0.80877476	0.70227454	0.65446248	0.60994118	0.52987123
86	0.80675787	0.69936054	0.65120644	0.60640382	0.52592678
87	0.80474600	0.69645863	0.64796661	0.60288698	0.52201169
88	0.80273915	0.69356876	0.64474290	0.59939054	0.51812575
89	0.80073731	0.69069088	0.64153522	0.59591437	0.51426873
90	0.79874046	0.68782495	0.63834350	0.59245836	0.51044043
91	0.79674859	0.68497090	0.63516766	0.58902240	0.50664063
92	0.79476168	0.68212870	0.63200763	0.58560636	0.50286911
93	0.79277973	0.67929829	0.62886331	0.58221014	0.49912567
94	0.79080273	0.67647962	0.62573464	0.57883361	0.79541009
95	0.78883065	0.67367265	0.62262153	0.57547666	0.49172217
96	0.78686349	0.67087733	0.61952391	0.57213918	0.48806171
97	0.78490124	0.66809361	0.61644170	0.56882106	0.48442850
98	0.78294388	0.66532143	0.61337483	0.56552218	0.48082233
99	0.78099140	0.66256076	0.61032321	0.56224243	0.47724301
100	0.77904379	0.65981155	0.60728678	0.55898171	0.47369033

PRESENT VALUE $1/(1 + i)^n$ (Continued)

Periods n	Rate i 0.01(1%)	0.001125(1 1/8%)	0.0125(1 1/4%)	0.0150 (1½%)	0.0175(1 3/4%)
1	0.99009901	0.98887515	0.98765432	0.98522167	0.98280098
2	0.98029605	0.97787407	0.97546106	0.97066175	0.96589777
3	0.97059015	0.96699537	0.96341833	0.95631699	0.94928528
4	0.96098034	0.95623770	0.95152428	0.94218423	0.93295851
5	0.95146569	0.94559970	0.93977706	0.92826033	0.91691254
6	0.94204524	0.93508005	0.92817488	0.91454219	0.90114254
7	0.93271805	0.92467743	0.91671593	0.90102679	0.88564378
8	0.92348322	0.91439054	0.90539845	0.88771112	0.87041157
9	0.91433982	0.90421808	0.89422069	0.87459224	0.85544135
10	0.90528695	0.89415880	0.88318093	0.86166723	0.84072860
11	0.89632372	0.88421142	0.87227746	0.84893323	0.82626889
12	0.88744923	0.87437470	0.86150860	0.83638742	0.81205788
13	0.87866260	0.86464742	0.85087269	0.82402702	0.79809128
14	0.86996297	0.85502835	0.84036809	0.81184928	0.78436490
15	0.86134947	0.84551629	0.82999318	0.79985150	0.77087459
16	0.85282126	0.83611005	0.81974635	0.78803104	0.75761631
17	0.84437749	0.82680846	0.80962602	0.77638526	0.74458605
18	0.83601731	0.81761034	0.79963064	0.76491159	0.73177990
19	0.82773992	0.80851455	0.78975866	0.75360747	0.71919401
20	0.81954447	0.79951995	0.78000855	0.74247042	0.70682458
21	0.81143017	0.79062542	0.77037881	0.73149795	0.69466789
22	0.80339621	0.78182983	0.76086796	0.72068763	0.68272028
23	0.79544179	0.77313210	0.75147453	0.71003708	0.67097817
24	0.78756613	0.76453112	0.74219707	0.69954392	0.65943800
25	0.77976844	0.75602583	0.73303414	0.68920583	0.64809632
26	0.77204796	0.74761516	0.72398434	0.67092052	0.63694970
27	0.76440392	0.73929806	0.71504626	0.66898574	0.62599479
28	0.75683557	0.73107348	0.70621853	0.65909925	0.61522829
29	0.74934215	0.72294040	0.69749978	0.64935887	0.60464697
30	0.74192292	0.71489780	0.68888867	0.63976243	0.59424764
31	0.73457715	0.70694467	0.68038387	0.63030781	0.58402716
32	0.72730411	0.69908002	0.67198407	0.62099292	0.57398247
33	0.72010307	0.69130287	0.66368797	0.61181568	0.56411053
34	0.71297334	0.68361223	0.65549429	0.60277407	0.55440839
35	0.70591420	0.67600715	0.64740177	0.59386608	0.54487311
36	0.69892495	0.66848667	0.63940916	0.58508974	0.53550183
37	0.69200490	0.66104986	0.63151522	0.57644309	0.52629172
38	0.68515337	0.65369578	0.62371873	0.56792423	0.51724002
39	0.67836967	0.64642352	0.61601850	0.55953126	0.50834400
40	0.67165314	0.63923216	0.60841334	0.55126232	0.49960098
41	0.66500311	0.63212080	0.60090206	0.54311559	0.49100834
42	0.65841892	0.62508855	0.59348352	0.53508925	0.48256348
43	0.65189992	0.61813454	0.58615656	0.52718153	0.47426386
44	0.64544546	0.61125789	0.57892006	0.51939067	0.46610699
45	0.63905492	0.60445774	0.57177290	0.51171494	0.45809040
46	0.63272764	0.59773324	0.56471397	0.50415265	0.45021170
47	0.62646301	0.59108355	0.55774219	0.49670212	0.44246850
48	0.62026041	0.58450784	0.55085649	0.48936170	0.43485848
49	0.61411921	0.57800528	0.54405579	0.48212975	0.42737934
50	0.60803882	0.57157506	0.53733905	0.47500468	0.42002883

PRESENT VALUE $1/(1 + i)^n$ (Continued)

Periods n	0.01(1%)	0.001125(1 1/8%)	0.0125(1 1/4%)	0.0150 (1½%)	0.0175(1 3/4%)
			Rate i		
51	0.60201864	0.56521637	0.53070524	0.46798491	0.41280475
52	0.59605806	0.55892843	0.52415332	0.46106887	0.40570492
53	0.59015649	0.55271044	0.51768229	0.45425505	0.39872719
54	0.58431336	0.54656162	0.51129115	0.44754192	0.39186947
55	0.57852808	0.54048120	0.50497892	0.44092800	0.38512970
56	0.57280008	0.53446843	0.49874461	0.43441182	0.37850585
57	0.56712879	0.52852256	0.49258727	0.42799194	0.37199592
58	0.56151365	0.52264282	0.48650594	0.42166694	0.36559796
59	0.55595411	0.51682850	0.48049970	0.41543541	0.35931003
60	0.55044962	0.51107887	0.47456760	0.40929597	0.35313025
61	0.54499962	0.50539319	0.46870874	0.40324726	0.34705676
62	0.53960358	0.49977077	0.46292222	0.39728794	0.34108772
63	0.53426097	0.49421090	0.45720713	0.39141669	0.33522135
64	0.52897126	0.48871288	0.45156259	0.38563221	0.32945587
65	0.52373392	0.48327602	0.44598775	0.37993321	0.32378956
66	0.51854844	0.47789965	0.44048173	0.37431843	0.31822069
67	0.51341429	0.47258309	0.43504368	0.36878663	0.31274761
68	0.50833099	0.46732568	0.42967277	0.36333658	0.30736866
69	0.50329801	0.46212675	0.42436817	0.35796708	0.30208222
70	0.49831486	0.45698566	0.41912905	0.35267692	0.29688670
71	0.49338105	0.45190177	0.41395462	0.34746495	0.29178054
72	0.48849609	0.44687443	0.40884407	0.34233000	0.28676221
73	0.48365949	0.44190302	0.40379661	0.33727093	0.28183018
74	0.47887078	0.43698692	0.39881147	0.33228663	0.27698298
75	0.47412949	0.43212551	0.39388787	0.32737599	0.27221914
76	0.46943514	0.42731818	0.38902506	0.32253793	0.26753724
77	0.46478726	0.42256433	0.38422228	0.31777136	0.26293586
78	0.46018541	0.41786337	0.37947879	0.31307523	0.25841362
79	0.45562912	0.41321470	0.37479387	0.30844850	0.25396916
80	0.45111794	0.40861775	0.37016679	0.30389015	0.24960114
81	0.44665142	0.40407194	0.36559683	0.29939916	0.24530825
82	0.44222913	0.39957670	0.36108329	0.29497454	0.24108919
83	0.43785063	0.39513148	0.35662547	0.29061531	0.23694269
84	0.43351547	0.39073570	0.35222268	0.28632050	0.23286751
85	0.42922324	0.38638882	0.34787426	0.28208917	0.22886242
86	0.42497350	0.38209031	0.34357951	0.27792036	0.22492621
87	0.42076585	0.37783961	0.33933779	0.27381316	0.22105770
88	0.41659985	0.37363621	0.33514843	0.26976666	0.21725572
89	0.41247510	0.36947956	0.33101080	0.26577996	0.21351914
90	0.40839119	0.36536916	0.32692425	0.26185218	0.20984682
91	0.40434771	0.36130448	0.32288814	0.25798245	0.20623766
92	0.40034427	0.35728503	0.31890187	0.25416990	0.20269057
93	0.39638046	0.35331029	0.31496481	0.25041369	0.19920450
94	0.39245590	0.34937976	0.31107636	0.24671300	0.19577837
95	0.38857020	0.34549297	0.30723591	0.24306699	0.19241118
96	0.38472297	0.34164941	0.30344287	0.23947487	0.18910190
97	0.38091383	0.33784861	0.29969666	0.23593583	0.18584953
98	0.37714241	0.33409010	0.29599670	0.23244909	0.18265310
99	0.37340832	0.33037340	0.29234242	0.22901389	0.17951165
100	0.36971121	0.32669805	0.28873326	0.22562944	0.17642422

PRESENT VALUE $1/(1 + i)^n$ (Continued)

Periods	Rate i			
n	0.02(2%)	0.0225(2 1/4%)	0.025(2 1/2%)	0.0275(2 3/4%)
1	0.98039216	0.97799511	0.97560976	0.97323601
2	0.96116878	0.95647444	0.95181440	0.94718833
3	0.94232233	0.93542732	0.92859941	0.92183779
4	0.92384543	0.91484335	0.90595064	0.89716573
5	0.90573081	0.89471232	0.88385429	0.87315400
6	0.88797138	0.87502427	0.86229687	0.84978491
7	0.87056018	0.85576946	0.84126524	0.82704128
8	0.85349037	0.83693835	0.82074657	0.80490635
9	0.83675527	0.81852161	0.80072836	0.78336385
10	0.82034830	0.80051013	0.78119840	0.76239791
11	0.80426304	0.78289499	0.76214478	0.74199310
12	0.78849318	0.76566748	0.74355589	0.72213440
13	0.77303253	0.74881905	0.72542038	0.70280720
14	0.75787502	0.73234137	0.70772720	0.68399728
15	0.74301473	0.71622628	0.69046556	0.66569078
16	0.72844581	0.70046580	0.67362493	0.64787424
17	0.71416256	0.68505212	0.65719506	0.63053454
18	0.70015937	0.66997763	0.64115691	0.61365892
19	0.68643076	0.65523484	0.62552772	0.59723496
20	0.67297133	0.64081647	0.61027094	0.58125057
21	0.65977582	0.62671538	0.59538629	0.56569398
22	0.64683904	0.61292457	0.58086467	0.55055375
23	0.63415592	0.59943724	0.56669724	0.53581874
24	0.62172149	0.58624668	0.55287535	0.52147809
25	0.60953087	0.57334639	0.53939059	0.50752126
26	0.59757928	0.56072997	0.52623472	0.49393796
27	0.58586204	0.54839117	0.51339973	0.48071821
28	0.57437455	0.53632388	0.50087778	0.46785227
29	0.56311231	0.52452213	0.48866125	0.45533068
30	0.55207089	0.51298008	0.47674269	0.44314421
31	0.54124597	0.50169201	0.46511481	0.43128391
32	0.53063330	0.49065233	0.45377055	0.41974103
33	0.52022873	0.47985558	0.44270298	0.40850708
34	0.51002817	0.46929641	0.43190534	0.39757380
35	0.50002761	0.45896960	0.42137107	0.38693314
36	0.49022315	0.44887002	0.41109372	0.37657727
37	0.48061093	0.43899268	0.40106705	0.36649856
38	0.47118719	0.42933270	0.39128492	0.35668959
39	0.46194822	0.41988528	0.38174139	0.34714316
40	0.45289042	0.41064575	0.37243062	0.33785222
41	0.44401021	0.40160954	0.36334695	0.32880995
42	0.43530413	0.39277216	0.35448483	0.32000968
43	0.42676875	0.38412925	0.34583886	0.31144495
44	0.41840074	0.37567653	0.33740376	0.30310944
45	0.41019680	0.36740981	0.32917440	0.29499702
46	0.40215373	0.35932500	0.32114576	0.28710172
47	0.39426836	0.35141809	0.31331294	0.27941773
48	0.38653761	0.34368518	0.30567116	0.27193940
49	0.37895844	0.33612242	0.29821576	0.26466122
50	0.37152788	0.32872608	0.29094221	0.25757783

FINANCIAL TABLES

PRESENT VALUE $1/(1 + i)^n$ (Continued)

Periods		Rate i		
n	0.02(2%)	0.0225(2 1/4%)	0.025(2 1/2%)	0.0275(2 3/4%)
51	0.36424302	0.32149250	0.28384606	0.25068402
52	0.35710100	0.31441810	0.27692298	0.24397471
53	0.35009902	0.30749936	0.27016876	0.23744497
54	0.34323433	0.30073287	0.26357928	0.23109000
55	0.33650425	0.29411528	0.25715052	0.22490511
56	0.32990613	0.28764330	0.25087855	0.21888575
57	0.32343738	0.28131374	0.24475956	0.21302749
58	0.31709547	0.27512347	0.23878982	0.20732603
59	0.31087791	0.26906940	0.23296568	0.20177716
60	0.30478227	0.26314856	0.22728359	0.19637679
61	0.29880614	0.25735801	0.22174009	0.19112097
62	0.29294720	0.25169487	0.21633179	0.18600581
63	0.28720314	0.24615635	0.21105541	0.18102755
64	0.28157170	0.24073971	0.20590771	0.17618253
65	0.27605069	0.23544226	0.20088557	0.17146718
66	0.27063793	0.23026138	0.19598593	0.16687804
67	0.26533130	0.22519450	0.19120578	0.16241172
68	0.26012873	0.22023912	0.18654223	0.15806493
69	0.25502817	0.21539278	0.18199241	0.15383448
70	0.25002761	0.21065309	0.17755358	0.14971726
71	0.24512511	0.20601769	0.17322300	0.14571023
72	0.24031874	0.20148429	0.16899805	0.14181044
73	0.23560661	0.19705065	0.16487615	0.13801503
74	0.23098687	0.19271458	0.16085478	0.13432119
75	0.22645771	0.18847391	0.15693149	0.13072622
76	0.22201737	0.18432657	0.15310389	0.12722747
77	0.21766408	0.18027048	0.14936965	0.12382235
78	0.21339616	0.17630365	0.14572649	0.12050837
79	0.20921192	0.17242411	0.14217218	0.11728309
80	0.20510973	0.16862993	0.13870457	0.11414412
81	0.20108797	0.16491925	0.13532153	0.11108917
82	0.19714507	0.16129022	0.13202101	0.10811598
83	0.19327948	0.15774105	0.12880098	0.10522237
84	0.18948968	0.15426997	0.12565949	0.10240620
85	0.18577420	0.15087528	0.12259463	0.09966540
86	0.18213157	0.14755528	0.11960452	0.09699795
87	0.17856036	0.14430835	0.11668733	0.09440190
88	0.17505918	0.14113286	0.11384130	0.09187533
89	0.17162665	0.13802724	0.11106468	0.08941638
90	0.16826142	0.13498997	0.10835579	0.08702324
91	0.16496217	0.13201953	0.10571296	0.08469415
92	0.16172762	0.12911445	0.10313460	0.08242740
93	0.15855649	0.12627331	0.10061912	0.08022131
94	0.15544754	0.12349468	0.09816500	0.07807427
95	0.15239955	0.12077719	0.09577073	0.07598469
96	0.14941132	0.11811950	0.09343486	0.07395104
97	0.14648169	0.11552029	0.09115596	0.07197181
98	0.14360950	0.11297828	0.08893264	0.07004556
99	0.14079363	0.11049221	0.08676355	0.06817086
100	0.13803297	0.10806084	0.08464737	0.06634634

PRESENT VALUE $1/(1 + i)^n$ (Continued)

Periods		Rate i		
n	0.03(3%)	0.0325(3 1/4%)	0.035(3 1/2%)	0.0375(3 3/4%)
1	0.97087379	0.96852300	0.96618357	0.96385542
2	0.94259591	0.93803681	0.93351070	0.92901727
3	0.91514166	0.90851022	0.90194271	0.89543834
4	0.88848705	0.87991305	0.87144223	0.86307310
5	0.86260878	0.85221603	0.84197317	0.83187768
6	0.83748426	0.82539083	0.81350064	0.80180981
7	0.81309151	0.79941000	0.78599096	0.77282874
8	0.78940923	0.77424698	0.75941156	0.74489517
9	0.76641673	0.74987601	0.73373097	0.71797125
10	0.74409391	0.72627216	0.70891881	0.69202048
11	0.72242128	0.70341129	0.68494571	0.66700769
12	0.70137988	0.68127002	0.66178330	0.64289898
13	0.68095134	0.65982568	0.63940415	0.61966167
14	0.66111781	0.63905635	0.61778179	0.59726426
15	0.64186195	0.61894078	0.59689062	0.57567639
16	0.62316694	0.59945838	0.57670591	0.55486881
17	0.60501645	0.58058923	0.55720378	0.53481331
18	0.58739461	0.56231402	0.53836114	0.51548271
19	0.57028603	0.54461407	0.52015569	0.49685080
20	0.55367575	0.52747125	0.50256588	0.47889234
21	0.53754928	0.51086804	0.48557090	0.46158298
22	0.52189250	0.49478745	0.46915063	0.44489926
23	0.50669175	0.47921302	0.45328563	0.42881856
24	0.49193374	0.46412884	0.43795713	0.41331910
25	0.47760557	0.44951945	0.42314699	0.39837985
26	0.46369473	0.43536993	0.40883767	0.38398058
27	0.45018906	0.42166579	0.39501224	0.37010176
28	0.43707675	0.40839302	0.38165434	0.35672459
29	0.42434636	0.39553803	0.36874815	0.34389093
30	0.41198676	0.38308768	0.35627841	0.33140331
31	0.39998715	0.37102923	0.34423035	0.31942487
32	0.38833703	0.35935035	0.33258971	0.30787940
33	0.37702625	0.34803908	0.32134271	0.29675123
34	0.36604490	0.33708385	0.31047605	0.28602528
35	0.35538340	0.32647346	0.29997686	0.27568702
36	0.35503243	0.31619706	0.28983272	0.26572242
37	0.33498294	0.30624413	0.28003161	0.25611800
38	0.32522615	0.29660448	0.27056194	0.24686072
39	0.31575355	0.28726826	0.26141250	0.23793805
40	0.30655684	0.27822592	0.25257247	0.22933788
41	0.29762800	0.26946820	0.24403137	0.22104855
42	0.28895922	0.26098615	0.23577910	0.21305885
43	0.28054294	0.25277109	0.22780590	0.20535793
44	0.27237178	0.24481462	0.22010231	0.19793535
45	0.26443862	0.23710859	0.21265924	0.19078106
46	0.25673653	0.22964512	0.20546787	0.18388536
47	0.24925876	0.22241658	0.19851968	0.17723890
48	0.24199880	0.21541558	0.19180645	0.17083268
49	0.23495029	0.20863494	0.18532024	0.16465800
50	0.22810708	0.20206774	0.17905337	0.15870651

FINANCIAL TABLES

PRESENT VALUE $1/(1 + i)^n$ (Continued)

Periods n	Rate i 0.04(4%)	0.0425(4 1/4%)	0.045(4 1/2%)	0.0475(4 3/4%)
1	0.96153846	0.95923261	0.95693780	0.95465894
2	0.92455621	0.92012721	0.91572995	0.91136414
3	0.88899636	0.88261603	0.87629660	0.87003737
4	0.85480419	0.84663408	0.83856134	0.83058460
5	0.82192711	0.81211902	0.80245105	0.79292086
6	0.79031453	0.77901105	0.76789574	0.75696502
7	0.75991781	0.74725281	0.73482846	0.72263964
8	0.73069021	0.71678926	0.70318513	0.68987077
9	0.70258674	0.68756764	0.67290443	0.65858785
10	0.67556417	0.65953730	0.64392768	0.62872349
11	0.64958093	0.63264969	0.61619874	0.60021335
12	0.62459705	0.60685822	0.58966386	0.57299604
13	0.60057409	0.58211819	0.56427164	0.54701293
14	0.57747508	0.55838676	0.53997286	0.52220804
15	0.55526450	0.53562279	0.51672044	0.49852797
16	0.53390818	0.51378685	0.49446932	0.47592169
17	0.51337325	0.49284110	0.47317639	0.45434051
18	0.49362812	0.47274926	0.45280037	0.43373796
19	0.47464242	0.45347650	0.43330179	0.41406965
20	0.45638695	0.43498945	0.41464286	0.39529322
21	0.43883360	0.41725607	0.39678743	0.37736823
22	0.41295539	0.40024563	0.37970089	0.36025607
23	0.40572633	0.38392866	0.36335013	0.34391987
24	0.39012147	0.36827689	0.34770347	0.32832446
25	0.37511680	0.35326321	0.33273060	0.31343624
26	0.36068923	0.33886159	0.31840248	0.29922314
27	0.34681657	0.32504709	0.30469137	0.28565455
28	0.33347747	0.31179577	0.29157069	0.27270124
29	0.32065141	0.29908467	0.27901502	0.26033531
30	0.30831867	0.28689177	0.26700002	0.24853013
31	0.29646026	0.27519594	0.25550241	0.23726027
32	0.28505794	0.26397692	0.24449991	0.22650145
33	0.27409417	0.25321527	0.23397121	0.21623050
34	0.26355209	0.24289235	0.22389589	0.20642530
35	0.25341547	0.23299026	0.21425444	0.19706473
36	0.24366872	0.22349186	0.20502817	0.18812862
37	0.23429685	0.21438068	0.19619921	0.17959772
38	0.22528543	0.20564094	0.18775044	0.17134367
39	0.21662061	0.19725750	0.17966549	0.16367893
40	0.20828904	0.18921582	0.17192870	0.17145367
41	0.20027793	0.18150199	0.16452507	0.14917110
42	0.19257493	0.17410263	0.15744026	0.14240678
43	0.18516820	0.16700492	0.15066054	0.13594919
44	0.17804635	0.16019657	0.14417276	0.12978443
45	0.17119841	0.15366577	0.13796437	0.12389922
46	0.16461386	0.14740122	0.13202332	0.11828088
47	0.15828256	0.14139206	0.12633810	0.11291731
48	0.15219476	0.13562787	0.12089771	0.10779695
49	0.14634112	0.13009868	0.11569158	0.10290878
50	0.14071262	0.12479489	0.11070965	0.09824228

PRESENT VALUE $1/(1 + i)^n$ (Continued)

Periods n	Rate i			
	0.05(5%)	0.0525(5 1/4%)	0.055(5 1/2%)	0.0575(5 3/4%)
1	0.95238095	0.95011876	0.94786730	0.94562648
2	0.90702948	0.90272567	0.89845242	0.89420944
3	0.86383760	0.85769660	0.85161366	0.84558812
4	0.82270247	0.81491363	0.80721674	0.79961051
5	0.78352617	0.77426473	0.76513434	0.75613287
6	0.74621540	0.73564345	0.72524583	0.71501927
7	0.71068133	0.69894865	0.68743681	0.67614115
8	0.67683936	0.66408423	0.65159887	0.63937697
9	0.64460892	0.63095888	0.61762926	0.60461180
10	0.61391325	0.59948588	0.58543058	0.57173692
11	0.58467929	0.56958278	0.55491050	0.54064957
12	0.55683742	0.54117129	0.52598152	0.51125255
13	0.53032135	0.51417699	0.49856068	0.48345395
14	0.50506795	0.48852921	0.47256937	0.45716685
15	0.48101710	0.46416077	0.44793305	0.43230908
16	0.45811152	0.44100786	0.42458109	0.40880291
17	0.43629669	0.41900984	0.40244653	0.38657486
18	0.41552065	0.39810911	0.38146590	0.36555542
19	0.39573396	0.37825094	0.36157906	0.34567889
20	0.37688948	0.35938331	0.34272896	0.32688311
21	0.35894236	0.34145683	0.32486158	0.30910932
22	0.34184987	0.32442454	0.30792567	0.29230196
23	0.32557131	0.30824185	0.29187267	0.27640847
24	0.31006791	0.29286636	0.27665656	0.26137917
25	0.29530277	0.27825783	0.26223370	0.24716706
26	0.28124073	0.26437798	0.24856275	0.23372772
27	0.26784832	0.25119048	0.23560450	0.22101912
28	0.25509364	0.23866079	0.22332181	0.20900153
29	0.24294632	0.22675609	0.21167944	0.19763738
30	0.23137745	0.21544522	0.20064402	0.18689114
31	0.22035947	0.20469855	0.19018390	0.17672921
32	0.20986617	0.19448793	0.18026910	0.16711982
33	0.19987254	0.18478663	0.17087119	0.15803293
34	0.19035480	0.17556925	0.16196321	0.14944012
35	0.18129029	0.16681164	0.15341963	0.14131454
36	0.17265741	0.15849087	0.14551624	0.13363077
37	0.16443563	0.15058515	0.13793008	0.12636479
38	0.15660536	0.14307377	0.13073941	0.11949389
39	0.14914797	0.13593708	0.12392362	0.11299659
40	0.14204568	0.12915637	0.11746314	0.10685257
41	0.13528160	0.12271389	0.11133947	0.10104262
42	0.12883962	0.11659277	0.10553504	0.09554857
43	0.12270440	0.11077698	0.10003322	0.09035326
44	0.11686133	0.10525128	0.09481822	0.08544044
45	0.11129651	0.10000122	0.08987509	0.08079474
46	0.10599668	0.09501304	0.08518965	0.07640164
47	0.10094921	0.09027367	0.08074849	0.07224742
48	0.09614211	0.08577071	0.07653885	0.06831907
49	0.09156391	0.08149236	0.07254867	0.06460432
50	0.08720373	0.07742742	0.06876652	0.06109156

FINANCIAL TABLES

PRESENT VALUE $1/(1 + i)^n$ (Continued)

Periods	Rate i			
n	0.06(6%)	0.0625(6 1/4%)	0.065(6 1/2%)	0.0675(6 3/4%)
1	0.94339623	0.94117647	0.93896714	0.93676815
2	0.88999644	0.88581315	0.88165928	0.87753457
3	0.83961928	0.83370649	0.82784909	0.82204643
4	0.79209366	0.78466493	0.77732309	0.77006692
5	0.74725817	0.73850817	0.72988084	0.72137416
6	0.70496054	0.69506652	0.68533412	0.67576034
7	0.66505711	0.65418025	0.64350621	0.63303076
8	0.62741237	0.61569906	0.60423119	0.59300305
9	0.59189846	0.57948147	0.56735323	0.55550637
10	0.55839478	0.54539432	0.53272604	0.52038068
11	0.52678753	0.51331230	0.50021224	0.48747605
12	0.49696936	0.48311746	0.46968285	0.45665203
13	0.46883902	0.45469879	0.44101676	0.42777708
14	0.44230096	0.42795180	0.41410025	0.40072794
15	0.41726506	0.40277817	0.38882652	0.37538917
16	0.39364628	0.37908533	0.36509533	0.35165262
17	0.37136442	0.35678619	0.34281251	0.32941698
18	0.35034379	0.33579877	0.32188969	0.30858733
19	0.33051301	0.31604590	0.30224384	0.28907478
20	0.31180473	0.29745497	0.28379703	0.27079605
21	0.29415540	0.27995762	0.26647608	0.25367312
22	0.27750510	0.26348952	0.25021228	0.23763289
23	0.26179726	0.24799014	0.23494111	0.22260693
24	0.24697855	0.23340248	0.22060198	0.20853108
25	0.23299863	0.21967292	0.20713801	0.19534527
26	0.21981003	0.20675099	0.19449579	0.18299323
27	0.20736795	0.19458917	0.18262515	0.17142223
28	0.19563014	0.18314274	0.17147902	0.16058289
29	0.18455674	0.17236964	0.16101316	0.15042893
30	0.17411013	0.16223025	0.15118607	0.14091703
31	0.16425484	0.15268729	0.14195875	0.13200659
32	0.15495740	0.14370569	0.13329460	0.12365957
33	0.14618622	0.13525241	0.12515925	0.11584034
34	0.13791153	0.12729639	0.11752042	0.10851554
35	0.13010522	0.11980837	0.11034781	0.10165391
36	0.12274077	0.11276081	0.10361297	0.09522614
37	0.11579318	0.10612783	0.09728917	0.08920482
38	0.10923885	0.09988501	0.09135134	0.08356423
39	0.10305552	0.09400942	0.08577590	0.07828031
40	0.09722219	0.08847946	0.08054075	0.07333050
41	0.09171905	0.08327478	0.07562512	0.06869368
42	0.08652740	0.07837627	0.07100950	0.06435005
43	0.08162962	0.07376590	0.06667559	0.06028108
44	0.07700908	0.06942673	0.06260619	0.05646939
45	0.07265007	0.06534280	0.05878515	0.05289873
46	0.06853781	0.06149911	0.05519733	0.04955384
47	0.06465831	0.05788151	0.05182848	0.04642046
48	0.06099840	0.05447672	0.04866524	0.04348521
49	0.05754566	0.05127221	0.04569506	0.04073556
50	0.05428836	0.04825619	0.04290616	0.03815978

PRESENT VALUE $1/(1 + i)^n$ (Continued)

Periods n	Rate i 0.07(7%)	0.0725(7 1/4%)	0.075(7 1/2%)	0.0775(7 3/4%)
1	0.93457944	0.93240093	0.93023256	0.92807425
2	0.87343873	0.86937150	0.86533261	0.86132181
3	0.81629788	0.81060280	0.80496057	0.79937059
4	0.76289521	0.75580680	0.74880053	0.74187525
5	0.71298618	0.70471497	0.69655863	0.68851532
6	0.66634222	0.65707689	0.64796152	0.63899333
7	0.62274974	0.61265911	0.60275490	0.59303326
8	0.58200910	0.57124392	0.56070223	0.55037889
9	0.54393374	0.53262837	0.52158347	0.51079247
10	0.50834929	0.49662319	0.48519393	0.47405334
11	0.47509280	0.46305192	0.45134319	0.43995670
12	0.44401196	0.43175004	0.41985413	0.40831248
13	0.41496445	0.40256414	0.39056198	0.37894430
14	0.38781724	0.37535118	0.36331347	0.35168844
15	0.36244602	0.34997779	0.33796602	0.32639299
16	0.33873460	0.32631962	0.31438699	0.30291692
17	0.31657439	0.30426072	0.29245302	0.28112940
18	0.29586392	0.28369298	0.27204932	0.26090895
19	0.27650833	0.26451560	0.25306913	0.24214288
20	0.25841900	0.24663459	0.23541315	0.22472657
21	0.24151309	0.22996232	0.21898897	0.20856294
22	0.22571317	0.21441708	0.20371067	0.19356190
23	0.21094688	0.19992269	0.18949830	0.17963981
24	0.19714662	0.18640810	0.17627749	0.16671908
25	0.18424918	0.17380709	0.16397906	0.15472769
26	0.17219549	0.16205789	0.15253866	0.14359878
27	0.16093037	0.15110293	0.14189643	0.13327033
28	0.15040221	0.14088851	0.13199668	0.12368476
29	0.14056282	0.13136458	0.12278761	0.11478864
30	0.13136712	0.12248446	0.11422103	0.10653238
31	0.12277301	0.11420462	0.10625212	0.09886996
32	0.11474113	0.10648449	0.09883918	0.09175866
33	0.10723470	0.09928624	0.09194343	0.08515885
34	0.10021934	0.09257458	0.08552877	0.07903374
35	0.09366294	0.08631663	0.07956164	0.07334918
36	0.08753546	0.08048171	0.07401083	0.06807348
37	0.08180884	0.07504122	0.06884729	0.06317724
38	0.07645686	0.06996850	0.06404399	0.05863317
39	0.07145501	0.06523870	0.05957580	0.05441594
40	0.06678038	0.06082862	0.05541935	0.05050203
41	0.06241157	0.05671666	0.05155288	0.04686963
42	0.05832857	0.05288267	0.04795617	0.04349850
43	0.05451268	0.04930785	0.04461039	0.04036984
44	0.05094643	0.04597468	0.04149804	0.03746621
45	0.04761349	0.04286684	0.03860283	0.03477142
46	0.04449859	0.03996908	0.03590961	0.03227046
47	0.04158747	0.03726721	0.03340428	0.02994938
48	0.03886679	0.03474798	0.03107375	0.02779525
49	0.03632410	0.03239905	0.02890582	0.02579606
50	0.03394776	0.03020890	0.02688913	0.02394066

FINANCIAL TABLES

PRESENT VALUE $1/(1 + i)^n$ (Continued)

Periods n	Rate i			
	0.08(8%)	0.0825(8 1/4%)	0.085(8 1/2%)	0.0875(8 3/4%)
1	0.92592593	0.92378753	0.92165899	0.91954023
2	0.85733882	0.85338340	0.84945529	0.84555423
3	0.79383224	0.78834494	0.78290810	0.77752114
4	0.73502985	0.72826322	0.72157428	0.71496196
5	0.68058320	0.67276048	0.66504542	0.65743629
6	0.63016963	0.62148775	0.61294509	0.60453912
7	0.58349040	0.57412263	0.56492635	0.55589804
8	0.54026888	0.53036732	0.52066945	0.51117061
9	0.50024897	0.48994672	0.47987968	0.47004194
10	0.46319349	0.45260667	0.44228542	0.43222247
11	0.42888286	0.41811240	0.40763633	0.39744595
12	0.39711376	0.38624702	0.37570168	0.36546754
13	0.36769792	0.35681018	0.34626883	0.33606211
14	0.34046104	0.32961679	0.31914178	0.30902263
15	0.31524170	0.30449588	0.29413989	0.28415874
16	0.29189047	0.28128950	0.27109667	0.26129539
17	0.27026895	0.25985173	0.24985869	0.24027162
18	0.25024903	0.24004779	0.23028450	0.22093942
19	0.23171206	0.22175315	0.21224378	0.20316269
20	0.21454821	0.20485280	0.19561639	0.18681627
21	0.19865575	0.18924046	0.18029160	0.17178507
22	0.18394051	0.17481798	0.16616738	0.15796328
23	0.17031528	0.16149467	0.15314965	0.14525360
24	0.15769934	0.14918676	0.14115176	0.13356652
25	0.14601790	0.13781687	0.13009378	0.12281979
26	0.13520176	0.12731350	0.11990210	0.11293774
27	0.12518682	0.11761063	0.11050885	0.10385080
28	0.11591372	0.10864723	0.10185148	0.09549498
29	0.10732752	0.10036696	0.09387233	0.08781148
30	0.09937733	0.09271774	0.08651828	0.08074619
31	0.09201605	0.08565149	0.07974035	0.07424937
32	0.08520005	0.07912378	0.07349341	0.06827528
33	0.07888893	0.07309356	0.06773586	0.06278187
34	0.07304531	0.06752292	0.06242936	0.05773045
35	0.06763454	0.06237683	0.05753858	0.05308547
36	0.06262458	0.05762294	0.05303095	0.04881423
37	0.05798572	0.05323135	0.04887645	0.04488665
38	0.05369048	0.04917446	0.04504742	0.04127508
39	0.04971341	0.04542675	0.04151836	0.03795410
40	0.04603093	0.04196467	0.03826577	0.03490032
41	0.04262123	0.03876644	0.03526799	0.03209225
42	0.03946411	0.03581195	0.03250506	0.02951011
43	0.03654084	0.03308263	0.02995858	0.02713573
44	0.03383411	0.03056132	0.02761160	0.02495240
45	0.03132788	0.02823217	0.02544848	0.02294474
46	0.02900730	0.02608053	0.02345482	0.02109861
47	0.02685861	0.02409287	0.02161734	0.01940102
48	0.02486908	0.02225669	0.01992382	0.01784002
49	0.02302693	0.02056045	0.01836297	0.01640461
50	0.02132123	0.01899349	0.01692439	0.01508470

PRESENT VALUE $1/(1 + i)^n$ (Continued)

Periods n	Rate i 0.09(9%)	0.0925(9 1/4%)	0.095(9 1/2%)	0.0975(9 3/4%)
1	0.91743119	0.91533181	0.91324201	0.91116173
2	0.84167999	0.83783232	0.83401097	0.83021570
3	0.77218348	0.76689457	0.76165385	0.75646077
4	0.70842521	0.70196299	0.69557429	0.68925811
5	0.64993139	0.64252906	0.63522767	0.62802561
6	0.59626733	0.58812728	0.58011659	0.57223290
7	0.54703424	0.53833161	0.52978684	0.52139672
8	0.50186628	0.49275204	0.48382360	0.47507674
9	0.46042778	0.45103162	0.44184803	0.43287175
10	0.42241081	0.41284359	0.40351419	0.39441617
11	0.38753285	0.37788887	0.36850611	0.35937692
12	0.35553473	0.34589370	0.33653626	0.32745050
13	0.32617865	0.31660751	0.30733813	0.29836036
14	0.29924647	0.28980092	0.28067410	0.27185454
15	0.27453804	0.26526400	0.25632337	0.24770346
16	0.25186976	0.24280458	0.23408527	0.22569791
17	0.23107318	0.22224675	0.21377651	0.20564730
18	0.21199374	0.20342952	0.19522969	0.18737795
19	0.19448967	0.18620551	0.17829195	0.17073162
20	0.17843089	0.17043983	0.16282370	0.15556411
21	0.16369806	0.15600900	0.14869744	0.14174407
22	0.15018171	0.14280000	0.13579675	0.12915177
23	0.13778139	0.13070938	0.12401530	0.11767815
24	0.12640494	0.11964245	0.11325598	0.10722383
25	0.11596784	0.10951254	0.10343012	0.09769825
26	0.10639251	0.10024031	0.09445673	0.08901891
27	0.09760781	0.09175315	0.08626185	0.08111062
28	0.08954845	0.08398457	0.07877795	0.07390489
29	0.08215454	0.07687375	0.07194333	0.06733931
30	0.07537114	0.07036499	0.06570167	0.06135700
31	0.06914783	0.06440731	0.06000153	0.05590615
32	0.06343838	0.05895406	0.05479592	0.05093955
33	0.05820035	0.05396253	0.05004193	0.04641417
34	0.05339481	0.04939362	0.04570039	0.04229081
35	0.04898607	0.04521155	0.04173552	0.03853377
36	0.04494135	0.04138357	0.03811463	0.03511050
37	0.04123059	0.03787970	0.03480788	0.03199134
38	0.03782623	0.03467249	0.03178802	0.02914928
39	0.03470296	0.03173684	0.02903015	0.02655971
40	0.03183758	0.02904973	0.02651156	0.02420019
41	0.02920879	0.02659015	0.02421147	0.02205029
42	0.02679706	0.02433881	0.02211093	0.02009138
43	0.02458446	0.02227808	0.02019263	0.01830650
44	0.02255455	0.02039184	0.01844076	0.01668018
45	0.02069224	0.01866530	0.01684087	0.01519834
46	0.01898371	0.01708494	0.01537979	0.01384815
47	0.01741625	0.01563839	0.01404547	0.01261790
48	0.01597821	0.01431432	0.01282692	0.01149695
49	0.01465891	0.01310235	0.01171408	0.01047558
50	0.01344854	0.01199300	0.01069779	0.00954495

FINANCIAL TABLES

PRESENT VALUE $1/(1 + i)^n$ (Continued)

Periods	Rate i			
n	0.10(10%)	0.105(10 1/2%)	0.11(11%)	0.115(11 1/2%)
1	0.90909091	0.90497738	0.90090090	0.89686099
2	0.82644628	0.81898405	0.81162243	0.80435963
3	0.75131480	0.74116204	0.73119138	0.72139877
4	0.68301346	0.67073487	0.65873097	0.64699441
5	0.62092132	0.60699989	0.59345133	0.58026405
6	0.56447393	0.54932116	0.53464084	0.52041619
7	0.51315812	0.49712323	0.48165841	0.46674097
8	0.46650738	0.44988527	0.43392650	0.41860177
9	0.42409762	0.40713599	0.39092477	0.37542760
10	0.38554329	0.36844886	0.35218448	0.33670636
11	0.35049390	0.33343788	0.31728331	0.30197880
12	0.31863082	0.30175374	0.28584082	0.27083301
13	0.28966438	0.27308031	0.25751426	0.24289956
14	0.26333125	0.24713150	0.23199482	0.21784714
15	0.23939205	0.22364842	0.20900435	0.19537860
16	0.21762914	0.20239676	0.18829220	0.17522744
17	0.19784467	0.18316449	0.16963262	0.15715466
18	0.17985879	0.16575972	0.15282218	0.14094588
19	0.16350799	0.15000879	0.13767764	0.12640886
20	0.14864363	0.13575456	0.12403391	0.11337118
21	0.13513057	0.12285481	0.11174226	0.10167818
22	0.12284597	0.11118082	0.10066870	0.09119120
23	0.11167816	0.10061613	0.09069252	0.08178583
24	0.10152560	0.09105532	0.08170498	0.07335052
25	0.09229600	0.08240301	0.07360809	0.06578522
26	0.08390545	0.07457286	0.06631359	0.05900020
27	0.07627768	0.06748675	0.05974197	0.05291497
28	0.06934335	0.06107398	0.05382160	0.04745738
29	0.06303941	0.05527057	0.04848793	0.04256267
30	0.05730855	0.05001861	0.04368282	0.03817280
31	0.05209868	0.04526571	0.03935389	0.03423569
32	0.04736244	0.04096445	0.03545395	0.03070466
33	0.04305676	0.03707190	0.03194050	0.02753781
34	0.03914251	0.03354923	0.02877522	0.02469759
35	0.03558410	0.03036129	0.02592363	0.02215030
36	0.03234918	0.02747628	0.02335462	0.01986574
37	0.02940835	0.02486542	0.02104020	0.01781681
38	0.02673486	0.02250264	0.01895513	0.01597920
39	0.02430442	0.02036438	0.01707670	0.01433112
40	0.02209493	0.01842930	0.01538441	0.01285302
41	0.02008630	0.01667810	0.01385983	0.01152738
42	0.01826027	0.01509330	0.01248633	0.01033845
43	0.01660025	0.01365910	0.01124895	0.00927216
44	0.01509113	0.01236118	0.01013419	0.00831583
45	0.01371921	0.01118658	0.00912990	0.00745815
46	0.01247201	0.01012361	0.00822513	0.00668892
47	0.01133819	0.00916163	0.00741005	0.00599903
48	0.01030745	0.00829107	0.00667570	0.00538030
49	0.00937041	0.00750323	0.00601415	0.00482538
50	0.00851855	0.00679026	0.00541815	0.00432769

PRESENT VALUE $1/(1 + i)^n$ (Continued)

Periods	Rate i			
n	0.12(12%)	0.125(12 1/2%)	0.13(13%)	0.135(13 1/2%)
1	0.89285714	0.88888889	0.88495575	0.88105727
2	0.79719388	0.79012346	0.78314668	0.77626191
3	0.71178025	0.70233196	0.69305016	0.68393120
4	0.63551808	0.62429508	0.61331873	0.60258255
5	0.56742686	0.55492896	0.54275994	0.53090974
6	0.50663112	0.49327018	0.48031853	0.46776188
7	0.45234922	0.43846239	0.42506064	0.41212501
8	0.40388323	0.38974434	0.37615986	0.36310573
9	0.36061002	0.34643942	0.33288483	0.31991695
10	0.32197324	0.30794615	0.29458835	0.28186515
11	0.28747610	0.27372991	0.26069765	0.24833934
12	0.25667509	0.24331547	0.23070589	0.21880118
13	0.22917419	0.21628042	0.20416450	0.19277637
14	0.20461981	0.19224926	0.18067655	0.16984702
15	0.18269626	0.17088823	0.15989075	0.14964495
16	0.16312166	0.15190065	0.14149624	0.13184577
17	0.14564434	0.13502280	0.12521791	0.11616368
18	0.13003959	0.12002027	0.11081231	0.10234685
19	0.11610678	0.10668468	0.09806399	0.09017344
20	0.10366677	0.09483083	0.08678229	0.07944796
21	0.09255961	0.08429407	0.07679849	0.06999821
22	0.08264251	0.07492806	0.06796327	0.06167243
23	0.07378796	0.06660272	0.06014448	0.05433694
24	0.06588210	0.05920242	0.05322521	0.04787396
25	0.05882331	0.05262437	0.04710195	0.04217970
26	0.05252081	0.04677722	0.04168314	0.03716273
27	0.04689358	0.04157975	0.03688774	0.03274249
28	0.04186927	0.03695978	0.03264402	0.02884801
29	0.03738327	0.03285314	0.02888851	0.02541675
30	0.03337792	0.02920279	0.02556505	0.02239361
31	0.02980172	0.02595803	0.02262394	0.01973005
32	0.02660868	0.02307381	0.02002119	0.01738331
33	0.02375775	0.02051005	0.01771786	0.01531569
34	0.02121227	0.01823116	0.01567953	0.01349400
35	0.01893953	0.01620547	0.01387569	0.01188899
36	0.01691029	0.01440486	0.01227937	0.01047488
37	0.01509848	0.01280432	0.01086670	0.00922897
38	0.01348078	0.01138162	0.00961655	0.00813125
39	0.01203641	0.01011700	0.00851022	0.00716410
40	0.01074680	0.00899289	0.00753117	0.00631198
41	0.00959536	0.00799368	0.00666475	0.00556121
42	0.00856728	0.00710549	0.00589801	0.00489975
43	0.00764936	0.00631599	0.00521948	0.00431696
44	0.00682978	0.00561421	0.00461901	0.00380349
45	0.00609802	0.00499041	0.00408762	0.00335109
46	0.00544466	0.00443592	0.00361736	0.00295250
47	0.00486131	0.00394304	0.00320120	0.00260132
48	0.00434045	0.00350493	0.00283292	0.00229192
49	0.00387540	0.00311549	0.00250701	0.00201931
50	0.00346018	0.00276932	0.00221859	0.00177913

FINANCIAL TABLES

PRESENT VALUE $1/(1 + i)^n$ (Continued)

Periods n	Rate i			
	0.14(14%)	0.145(14 1/2%)	0.15(15%)	0.155(15 1/2%)
1	0.87719298	0.87336245	0.86956522	0.86580087
2	0.76946753	0.76276196	0.75614367	0.74961114
3	0.67497152	0.66616765	0.65751623	0.64901397
4	0.59208028	0.58180581	0.57175325	0.56191686
5	0.51936866	0.50812734	0.49717674	0.48650810
6	0.45558655	0.44377934	0.43232760	0.42121914
7	0.39963732	0.38758021	0.37593704	0.36469189
8	0.35055905	0.33849800	0.32690177	0.31575056
9	0.30750794	0.29563144	0.28426241	0.27337711
10	0.26974381	0.25819340	0.24718471	0.23669014
11	0.23661738	0.22549642	0.21494322	0.20492652
12	0.20755910	0.19694010	0.18690715	0.17742556
13	0.18206939	0.17200009	0.16252796	0.15361521
14	0.15970999	0.15021842	0.14132866	0.13300018
15	0.14009648	0.13119513	0.12289449	0.11515167
16	0.12289165	0.11458090	0.10686477	0.09969841
17	0.10779969	0.10007065	0.09292589	0.08631897
18	0.09456113	0.08739795	0.08080512	0.07473504
19	0.08294836	0.07633009	0.07026532	0.06470566
20	0.07276172	0.06666383	0.06110028	0.05602222
21	0.06382607	0.05822169	0.05313068	0.04850409
22	0.05598778	0.05084863	0.04620059	0.04199488
23	0.04911209	0.04440929	0.04017443	0.03635920
24	0.04308078	0.03878540	0.03493428	0.03147983
25	0.03779016	0.03387372	0.03037764	0.02725526
26	0.03314926	0.02958403	0.02641534	0.02359763
27	0.02907830	0.02583758	0.02296986	0.02043085
28	0.02550728	0.02256557	0.01997379	0.01768905
29	0.02237481	0.01970792	0.01736851	0.01531519
30	0.01962702	0.01721216	0.01510305	0.01325991
31	0.01721669	0.01503246	0.01313309	0.01148044
32	0.01510236	0.01312878	0.01142008	0.00993977
33	0.01324768	0.01146618	0.00993050	0.00860586
34	0.01162077	0.01001414	0.00863522	0.00745097
35	0.01019366	0.00874597	0.00750889	0.00645105
36	0.00894181	0.00763840	0.00652947	0.00558533
37	0.00784369	0.00667109	0.00567780	0.00483578
38	0.00688043	0.00582628	0.00493722	0.00418682
39	0.00603547	0.00508846	0.00429323	0.00362495
40	0.00529427	0.00444407	0.00373324	0.00313849
41	0.00464410	0.00388128	0.00324630	0.00271731
42	0.00407377	0.00338976	0.00282287	0.00235265
43	0.00357348	0.00296049	0.00245467	0.00203692
44	0.00313463	0.00258558	0.00213449	0.00176357
45	0.00274968	0.00225815	0.00185608	0.00152690
46	0.00241200	0.00197218	0.00161398	0.00132199
47	0.00211579	0.00172243	0.00140346	0.00114458
48	0.00185595	0.00150431	0.00122040	0.00099098
49	0.00162803	0.00131381	0.00106122	0.00085799
50	0.00142810	0.00114743	0.00092280	0.00074285

PRESENT VALUE $1/(1 + i)^n$ (Continued)

Periods n	Rate i 0.16(16%)	0.165(16 1/2%)	0.17(17%)	0.175(17 1/2%)
1	0.86206897	0.85836910	0.85470085	0.85106383
2	0.74316290	0.73679751	0.73051355	0.72430964
3	0.64065767	0.63244421	0.62437056	0.61643374
4	0.55229110	0.54287057	0.53365005	0.52462446
5	0.47611302	0.46598332	0.45611115	0.44648890
6	0.41044225	0.39998568	0.38983859	0.37999055
7	0.35382953	0.34333535	0.33319538	0.32339622
8	0.30502546	0.29470846	0.28478237	0.27523082
9	0.26295298	0.25296863	0.24340374	0.23423900
10	0.22668360	0.21714046	0.20803738	0.19935234
11	0.19541690	0.18638666	0.17780973	0.16966156
12	0.16846284	0.15998855	0.15197413	0.14439282
13	0.14522659	0.13732923	0.12989242	0.12288751
14	0.12519534	0.11787916	0.11101916	0.10458511
15	0.10792701	0.10118383	0.09488817	0.08900861
16	0.09304053	0.08685307	0.08110100	0.07575201
17	0.08020735	0.07455199	0.06931709	0.06446979
18	0.06914427	0.06399313	0.05924538	0.05486791
19	0.05960713	0.05492972	0.05063708	0.04669609
20	0.05138546	0.04714998	0.04327955	0.03974135
21	0.04429781	0.04047208	0.03699107	0.03382243
22	0.03818776	0.03473999	0.03161630	0.02878505
23	0.03292049	0.02981973	0.02702248	0.02449791
24	0.02837973	0.02559634	0.02309614	0.02084929
25	0.02446528	0.02197110	0.01974029	0.01774407
26	0.02109076	0.01885932	0.01687204	0.01510134
27	0.01818169	0.01618825	0.01442055	0.01285220
28	0.01567387	0.01389550	0.01232525	0.01093805
29	0.01351196	0.01192747	0.01053440	0.00930898
30	0.01164824	0.01023817	0.00900376	0.00792253
31	0.01004159	0.00878813	0.00769553	0.00674258
32	0.00865654	0.00754346	0.00657737	0.00573837
33	0.00746253	0.00647507	0.00562169	0.00488372
34	0.00643322	0.00555800	0.00480486	0.00415635
35	0.00554588	0.00477082	0.00410672	0.00353732
36	0.00478093	0.00409512	0.00351002	0.00301049
37	0.00412149	0.00351512	0.00300001	0.00256212
38	0.00355301	0.00201727	0.00256411	0.00218052
39	0.00306294	0.00258994	0.00219155	0.00185577
40	0.00264047	0.00222312	0.00187312	0.00157938
41	0.00227626	0.00190826	0.00160096	0.00134415
42	0.00196230	0.00163799	0.00136834	0.00114396
43	0.00169163	0.00140600	0.00116952	0.00097358
44	0.00145831	0.00120687	0.00099959	0.00082858
45	0.00125716	0.00103594	0.00085435	0.00070517
46	0.00109376	0.00088922	0.00073021	0.00060015
47	0.00093427	0.00076328	0.00062411	0.00051076
48	0.00080541	0.00065517	0.00053343	0.00043469
49	0.00069432	0.00056238	0.00045592	0.00036995
50	0.00059855	0.00048273	0.00038968	0.00031485

FINANCIAL TABLES

PRESENT VALUE $1/(1 + i)^n$ (Continued)

Periods			Rate i		
n	0.18(18%)	0.185(18 1/2%)	0.19(19%)	0.195(19 1/2%)	0.20(20%)
1	0.84745763	0.84388186	0.84033613	0.83682008	0.83333333
2	0.71818443	0.71213659	0.70616482	0.70026785	0.69444444
3	0.60863087	0.60095915	0.59341581	0.58599820	0.57870370
4	0.51578888	0.50713852	0.49866875	0.49037507	0.48225309
5	0.43710922	0.42796500	0.41904937	0.41035570	0.40187757
6	0.37043154	0.36115189	0.35214233	0.34339389	0.33489798
7	0.31392503	0.30476953	0.29591792	0.28735891	0.27908165
8	0.26603816	0.25718948	0.24867052	0.24046770	0.23256804
9	0.22545607	0.21703753	0.20896683	0.20122820	0.19380670
10	0.19106447	0.18315404	0.17560238	0.16839180	0.16150558
11	0.16191904	0.15456037	0.14756502	0.14091364	0.13458799
12	0.13721953	0.13043069	0.12400422	0.11791937	0.11215665
13	0.11628773	0.11006809	0.10420523	0.09867729	0.09346388
14	0.09854893	0.09288447	0.08756742	0.08257514	0.07788657
15	0.08351604	0.07838352	0.07358606	0.06910054	0.06490547
16	0.07077630	0.06614643	0.06183703	0.05782472	0.05408789
17	0.05997992	0.05581977	0.05196389	0.04838888	0.04507324
18	0.05083044	0.04710529	0.04366713	0.04049279	0.03756104
19	0.04307664	0.03975130	0.03669507	0.03388518	0.03130086
20	0.03650563	0.03354540	0.03083619	0.02835580	0.02608405
21	0.03093698	0.02830836	0.02591277	0.02372870	0.02173671
22	0.02621778	0.02388891	0.02177544	0.01985665	0.01811393
23	0.02221845	0.02015942	0.01829869	0.01661645	0.01509494
24	0.01882920	0.01701217	0.01537705	0.01390498	0.01257912
25	0.01595695	0.01435626	0.01292189	0.01163596	0.01048260
26	0.01352284	0.01211499	0.01085873	0.00973721	0.00873550
27	0.01146003	0.01022362	0.00912498	0.00814829	0.00727958
28	0.00971189	0.00862752	0.00766805	0.00681865	0.00606632
29	0.00823042	0.00728061	0.00644374	0.00570599	0.00505526
30	0.00697493	0.00614398	0.00541491	0.00477488	0.00421272
31	0.00591096	0.00518479	0.00455034	0.00399572	0.00351060
32	0.00500929	0.00437535	0.00382382	0.00334370	0.00292550
33	0.00424516	0.00369228	0.00321329	0.00279807	0.00243792
34	0.00359759	0.00311585	0.00270025	0.00234148	0.00203160
35	0.00304881	0.00262941	0.00226911	0.00195940	0.00169300
36	0.00258373	0.00221891	0.00190682	0.00163967	0.00141083
37	0.00218960	0.00187250	0.00160237	0.00137211	0.00117569
38	0.00185560	0.00158017	0.00134653	0.00114821	0.00097974
39	0.00157254	0.00133347	0.00113154	0.00096084	0.00081645
40	0.00133266	0.00112529	0.00095087	0.00080405	0.00068038
41	0.00112937	0.00094962	0.00079905	0.00067285	0.00056698
42	0.00095710	0.00080136	0.00067147	0.00056305	0.00047248
43	0.00081110	0.00067626	0.00056426	0.00047117	0.00039374
44	0.00068737	0.00057068	0.00047417	0.00039429	0.00032811
45	0.00058252	0.00048159	0.00039846	0.00032995	0.00027343
46	0.00049366	0.00040640	0.00033484	0.00027611	0.00022786
47	0.00041836	0.00034296	0.00028138	0.00023105	0.00018988
48	0.00035454	0.00028941	0.00023645	0.00019335	0.00015823
49	0.00030046	0.00024423	0.00019870	0.00016180	0.00013186
50	0.00025462	0.00020610	0.00016698	0.00013540	0.00010988

COMPOUND AMOUNT OF 1 FOR FRACTIONAL PERIODS
$$(1 + i)^{1/p}$$

p	i = ¼%	5/12%	1/2%	7/12%	3/4%
2	1.0012492	1.0020812	1.0024969	1.0029124	1.0037430
3	1.0008326	1.0013870	1.0016639	1.0019407	1.0024938
4	1.0006244	1.0010400	1.0012477	1.0014552	1.0018697
6	1.0004162	1.0006932	1.0008316	1.0009699	1.0012461
12	1.0002089	1.0003466	1.0004157	1.0004848	1.0006229
13	1.0001921	1.0003199	1.0003837	1.0004475	1.0005749
26	1.0000960	1.0001599	1.0001919	1.0002237	1.0002874
52	1.0000480	1.0000800	1.0000959	1.0001119	1.0001437
365	1.0000068	1.0000114	1.0000137	1.0000159	1.0000205

p	1%	1 1/8%	1 1/4%	1 1/2%	1 3/4%
2	1.0049876	1.0056093	1.0062306	1.0074721	1.0087121
3	1.0033223	1.0037360	1.0041494	1.0049752	1.0057996
4	1.0024907	1.0028008	1.0031105	1.0037291	1.0043466
6	1.0016598	1.0018663	1.0020726	1.0024845	1.0028956
12	1.0008295	1.0009327	1.0010357	1.0012415	1.0014468
13	1.0007657	1.0008609	1.0009560	1.0011459	1.0013354
26	1.0003828	1.0004304	1.0004779	1.0005728	1.0006675
52	1.0001914	1.0002152	1.0002389	1.0002864	1.0003337
365	1.0000273	1.0000307	1.0000340	1.0000408	1.0000475

p	2%	2 1/4%	2 1/2%	2 3/4%	3%
2	1.0099505	1.0111874	1.0124228	1.0136568	1.01488916
3	1.0066227	1.0074444	1.0082648	1.0090839	1.00990163
4	1.0049629	1.0055782	1.0061922	1.0068052	1.00741707
6	1.0033059	1.0037153	1.0041239	1.0045317	1.00493862
12	1.0016516	1.0018559	1.0020598	1.0022633	1.00246627
13	1.0015244	1.0017130	1.0019012	1.0020890	1.00227634
26	1.0007619	1.0008562	1.0009502	1.0010440	1.00113752
52	1.0003809	1.0004280	1.0004750	1.0005218	1.00056860
365	1.0000543	1.0000610	1.0000676	1.0000743	1.00008099

p	3 1/4%	3 1/2%	3 3/4%	4%	4 1/4%
2	1.01612007	1.01734950	1.01857744	1.01980390	1.02102889
3	1.01071805	1.01153314	1.01234693	1.01315940	1.01397058
4	1.00802781	1.00863745	1.00924598	1.00985341	1.01045974
6	1.00534474	1.00575004	1.00615452	1.00655820	1.00696106
12	1.00266881	1.00287090	1.00307254	1.00327374	1.00347450
13	1.00246326	1.00264977	1.00283586	1.00302153	1.00320680
26	1.00123087	1.00132401	1.00141692	1.00150963	1.00160212
52	1.00061525	1.00066178	1.00070821	1.00075453	1.00080074
365	1.00008763	1.00009425	1.00010087	1.00010746	1.00011404

p	4 1/2%	4 3/4%	5%	5 1/4%	5 1/2%
2	1.02225242	1.02347447	1.02469508	1.02591423	1.02713193
3	1.01478046	1.01558905	1.01639636	1.01720238	1.01800713
4	1.01106499	1.01166915	1.01227223	1.01287424	1.01347517
6	1.00736312	1.00776438	1.00816485	1.00856452	1.00896339
12	1.00367481	1.00387468	1.00407412	1.00427313	1.00447170
13	1.00339165	1.00357610	1.00376014	1.00394378	1.00412701
26	1.00169439	1.00178645	1.00187831	1.00196995	1.00206138
52	1.00084684	1.00089283	1.00093871	1.00098449	1.00103016
365	1.00012060	1.00012715	1.00013368	1.00014020	1.00014670

FINANCIAL TABLES

COMPOUND AMOUNT OF 1 FOR FRACTIONAL PERIODS

$$(1 + i)^{1/p} \text{ (Continued)}$$

p	5 3/4%	6%	6 1/4%	6 1/2%	6 3/4%
2	1.02834819	1.02956301	1.03077641	1.03198837	1.03319892
3	1.01881061	1.01961282	1.02041378	1.02121347	1.02201192
4	1.01407504	1.01467385	1.01527159	1.01586828	1.01646393
6	1.00936149	1.00975879	1.01015532	1.01055107	1.01094605
12	1.00466984	1.00486755	1.00506483	1.00526169	1.00545813
13	1.00430985	1.00449228	1.00467432	1.00485597	1.00503722
26	1.00215261	1.00224363	1.00233444	1.00242504	1.00251545
52	1.00107572	1.00112118	1.00116654	1.00121179	1.00125693
365	1.00015318	1.00015965	1.00016611	1.00017255	1.00017897

p	7%	7 1/4%	7 1/2%	7 3/4%	8%
2	1.03440804	1.03561576	1.03682207	1.03802697	1.03923048
3	1.02280912	1.02360508	1.02439981	1.02519330	1.02598557
4	1.01705853	1.01765208	1.01824460	1.01883609	1.01942655
6	1.01134026	1.01173370	1.01212638	1.01251830	1.01290946
12	1.00565415	1.00584974	1.00604442	1.00623968	1.00643403
13	1.00521808	1.00539855	1.00557863	1.00575833	1.00593764
26	1.00260564	1.00269564	1.00278544	1.00287503	1.00296443
52	1.00130197	1.00134691	1.00139175	1.00143648	1.00148112
365	1.00018538	1.00019178	1.00019816	1.00020452	1.00021087

p	8 1/4%	8 1/2%	8 3/4%	9%	9 1/4%
2	1.04043260	1.04163333	1.04283268	1.04403065	1.04522725
3	1.02677661	1.02756644	1.02835506	1.02914247	1.02992867
4	1.02001598	1.02060440	1.02119179	1.02177818	1.02236356
6	1.01329986	1.01368952	1.01407843	1.01446659	1.01485402
12	1.00662797	1.00682149	1.00701461	1.00720732	1.00739963
13	1.00611657	1.00629512	1.00647328	1.00665107	1.00682849
26	1.00305362	1.00314262	1.00323142	1.00332003	1.00340844
52	1.00152565	1.00157008	1.00161441	1.00165864	1.00170277
365	1.00021721	1.00022353	1.00022984	1.00023613	1.00024241

p	9 1/2%	9 3/4%	10%	10 1/4%	10 1/2%
2	1.04642248	1.04761634	1.04880885	1.05000000	1.05118980
3	1.03071368	1.03149749	1.03228012	1.03306155	1.03384181
4	1.02294793	1.02353131	1.02411369	1.02469508	1.02527548
6	1.01524070	1.01562665	1.01601187	1.01639636	1.01678012
12	1.00759153	1.00778304	1.00797414	1.00816485	1.00835516
13	1.00700553	1.00718220	1.00735849	1.00753442	1.00770998
26	1.00349665	1.00358467	1.00367250	1.00376014	1.00384759
52	1.00174680	1.00179073	1.00183457	1.00187831	1.00192195
365	1.00024867	1.00025492	1.00026116	1.00026738	1.00027359

p	10 3/4%	11%	11 1/4%	11 1/2%	11 3/4%
2	1.05237826	1.05356538	1.05475116	1.05593560	1.05711873
3	1.03462089	1.03539881	1.03617555	1.03695113	1.03772555
4	1.02585489	1.02643333	1.02701079	1.02758727	1.02816279
6	1.01716316	1.01754548	1.01792708	1.01830797	1.01868815
12	1.00854507	1.00873459	1.00892373	1.00911247	1.00930082
13	1.00788517	1.00806000	1.00823447	1.00840857	1.00858231
26	1.00393485	1.00402191	1.00410879	1.00419548	1.00428199
52	1.00196549	1.00200894	1.00205229	1.00209555	1.00213871
365	1.00027978	1.00028596	1.00029212	1.00029828	1.00030441

COMPOUND AMOUNT OF 1 FOR FRACTIONAL PERIODS

$$(1 + i)^{1/p} \text{ (Continued)}$$

p	12%	12 1/4%	12 1/2%	12 3/4%	13%
2	1.05830052	1.05948101	1.06066017	1.06183803	1.06301458
3	1.03849882	1.03927094	1.04004191	1.04081174	1.04158044
4	1.02873734	1.02931094	1.02988357	1.03045525	1.03102598
6	1.01906762	1.01944639	1.01982445	1.02020181	1.02057848
12	1.00948879	1.00967638	1.00986358	1.01005040	1.01023684
13	1.00875570	1.00892873	1.00910140	1.00927372	1.00944569
26	1.00436831	1.00445444	1.00454039	1.00462616	1.00471174
52	1.00218177	1.00222475	1.00226763	1.00231041	1.00235310
365	1.00031054	1.00031665	1.00032275	1.00032883	1.00033490

p	13 1/4%	13 1/2%	13 3/4%	14%	14 1/4%
2	1.06418983	1.06536379	1.06653645	1.06770783	1.06887792
3	1.04234800	1.04311443	1.04387974	1.04464393	1.04540700
4	1.03159577	1.03216461	1.03273252	1.03329948	1.03386552
6	1.02095445	1.02132974	1.02170433	1.02207824	1.02245146
12	1.01042291	1.01060860	1.01079391	1.01097885	1.01116342
13	1.00961730	1.00978857	1.00995949	1.01013006	1.01030029
26	1.00479715	1.00488237	1.00496741	1.00505227	1.00513695
52	1.00239570	1.00243821	1.00248063	1.00252295	1.00256519
365	1.00034096	1.00034700	1.00035303	1.00035905	1.00036505

p	14 1/2%	14 3/4%	15%	15 1/4%	15 1/2%
2	1.07004673	1.07121426	1.07238053	1.07354553	1.07470926
3	1.04616896	1.04692981	1.04768955	1.04844820	1.04920575
4	1.03443063	1.03499481	1.03555808	1.03612042	1.03668185
6	1.02282401	1.02319588	1.02356707	1.02393760	1.02430745
12	1.01134762	1.01153145	1.01171492	1.01189802	1.01208075
13	1.01047017	1.01063972	1.01080892	1.01097778	1.01114630
26	1.00522146	1.00530578	1.00538993	1.00547391	1.00555771
52	1.00260733	1.00264938	1.00269134	1.00273322	1.00277500
365	1.00037104	1.00037702	1.00038298	1.00038893	1.00039487

p	15 3/4%	16%	16 1/4%	16 1/2%	16 3/4%
2	1.07587174	1.07703296	1.07819293	1.07935166	1.08050914
3	1.04996221	1.05071757	1.05147186	1.05222506	1.05297719
4	1.03724237	1.03780199	1.03836069	1.03891850	1.03947542
6	1.02467663	1.02504516	1.02541302	1.02578022	1.02614677
12	1.01226313	1.01244514	1.01262679	1.01280809	1.01298903
13	1.01131449	1.01148235	1.01164986	1.01181705	1.01198391
26	1.00564133	1.00572479	1.00580807	1.00589117	1.00597411
52	1.00281670	1.00285831	1.00289983	1.00294126	1.00298261
365	1.00040080	1.00040671	1.00041261	1.00041850	1.00042438

p	17%	17 1/4%	17 1/2%	17 3/4%	18%
2	1.08166538	1.08282039	1.08397417	1.08512672	1.08627805
3	1.05372824	1.05447823	1.05522715	1.05597501	1.05672181
4	1.04003143	1.04058656	1.04114080	1.04169416	1.04224664
6	1.02651266	1.02687790	1.02724250	1.02760644	1.02796975
12	1.01316961	1.01334984	1.01352972	1.01370925	1.01388843
13	1.01215044	1.01231663	1.01248251	1.01264805	1.01281328
26	1.00605687	1.00613947	1.00622190	1.00630416	1.00638625
52	1.00302387	1.00306504	1.00310612	1.00314713	1.00318804
365	1.00043024	1.00043609	1.00044193	1.00044775	1.00045357

COMPOUND AMOUNT OF 1 FOR FRACTIONAL PERIODS

$(1 + i)^{1/p}$ (Continued)

p	18 1/4%	18 1/2%	18 3/4%	19%
2	1.08742816	1.08857705	1.08972474	1.09087121
3	1.05746755	1.05821225	1.05895590	1.05969850
4	1.04279823	1.04334896	1.04389881	1.04444780
6	1.02833241	1.02869444	1.02905583	1.02941658
12	1.01406726	1.01424575	1.01442389	1.01460169
13	1.01297817	1.01314275	1.01330701	1.01347095
26	1.00646817	1.00654993	1.00663152	1.00671294
52	1.00322887	1.00326962	1.00331028	1.00335086
365	1.00045937	1.00046516	1.00047093	1.00047670

p	19 1/4%	19 1/2%	19 3/4%	20%
2	1.09201648	1.09316056	1.09430343	1.09544512
3	1.06044007	1.06118060	1.06192010	1.06265857
4	1.04499593	1.04554319	1.04608959	1.04663514
6	1.02977671	1.03013620	1.03049507	1.03085332
12	1.01477914	1.01495626	1.01513303	1.01530947
13	1.01363457	1.01379788	1.01396087	1.01412354
26	1.00679421	1.00687530	1.00695624	1.00703701
52	1.00339135	1.00343176	1.00347209	1.00351234
365	1.00048245	1.00048819	1.00049392	1.00049964

INDEX

A

B

C

D

E

F

606

G

608

H

I

J

K

L

610

M

N

O

612

Q

R

S

T

U

V

618

W

XYZ